$$\int \frac{dx}{(ax^2+bx+c)^{n+1}} = \frac{2ax+b}{n(4ac-b^2)(ax^2+bx+c)^n} + \frac{2(2n-1)a}{n(4ac-b^2)} \int \frac{dx}{(ax^2+bx+c)^n}, \quad \text{if } n > 0 \text{ and } b^2 \neq 4ac$$

$$\int \frac{dx}{a^2-x^2} = \frac{1}{2a} \ln \left| \frac{x+a}{x-a} \right| + C$$

25. $$\int \frac{dx}{(a^2-x^2)^2} = \frac{x}{2a^2(a^2-x^2)} + \frac{1}{2a^2} \int \frac{dx}{a^2-x^2}$$

$$\int \sqrt{a^2+x^2}\, dx = \frac{x}{2}\sqrt{a^2+x^2} + \frac{a^2}{2} \ln (x + \sqrt{x^2+a^2}) + C$$

$$\int x^2\sqrt{a^2+x^2}\, dx = \frac{x(a^2+2x^2)\sqrt{a^2+x^2}}{8} - \frac{a^4}{8} \ln (x + \sqrt{x^2+a^2}) + C$$

$$\int \frac{\sqrt{a^2+x^2}}{x}\, dx = \sqrt{a^2+x^2} - a \ln \left(\frac{a+\sqrt{a^2+x^2}}{x} \right) + C$$

29. $$\int \frac{\sqrt{a^2+x^2}}{x^2}\, dx = -\frac{\sqrt{a^2+x^2}}{x} + \ln (x + \sqrt{a^2+x^2}) + C$$

$$\int \frac{dx}{\sqrt{a^2+x^2}} = \ln (x + \sqrt{a^2+x^2}) + C$$

31. $$\int \frac{x^2}{\sqrt{a^2+x^2}}\, dx = \frac{x}{2}\sqrt{a^2+x^2} - \frac{a^2}{2} \ln (x + \sqrt{a^2+x^2}) + C$$

$$\int \frac{dx}{x\sqrt{a^2+x^2}} = -\frac{1}{a} \ln \left| \frac{a+\sqrt{a^2+x^2}}{x} \right| + C$$

33. $$\int \frac{dx}{x^2\sqrt{a^2+x^2}} = -\frac{\sqrt{a^2+x^2}}{a^2x} + C$$

$$\int \sqrt{a^2-x^2}\, dx = \frac{x}{2}\sqrt{a^2-x^2} + \frac{a^2}{2} \sin^{-1} \frac{x}{a} + C$$

$$\int x^2\sqrt{a^2-x^2}\, dx = \frac{a^4}{8} \sin^{-1} \frac{x}{a} - \frac{1}{8} x\sqrt{a^2-x^2}(a^2-2x^2) + C$$

$$\int \frac{\sqrt{a^2-x^2}}{x}\, dx = \sqrt{a^2-x^2} - a \ln \left| \frac{a+\sqrt{a^2-x^2}}{x} \right| + C$$

37. $$\int \frac{\sqrt{a^2-x^2}}{x^2}\, dx = -\sin^{-1} \frac{x}{a} - \frac{\sqrt{a^2-x^2}}{x} + C$$

$$\int \frac{dx}{\sqrt{a^2-x^2}} = \sin^{-1} \frac{x}{a} + C$$

39. $$\int \frac{x^2}{\sqrt{a^2-x^2}}\, dx = \frac{a^2}{2} \sin^{-1} \frac{x}{a} - \frac{1}{2} x\sqrt{a^2-x^2} + C$$

$$\int \frac{dx}{x\sqrt{a^2-x^2}} = -\frac{1}{a} \ln \left| \frac{a+\sqrt{a^2-x^2}}{x} \right| + C$$

41. $$\int \frac{dx}{x^2\sqrt{a^2-x^2}} = -\frac{\sqrt{a^2-x^2}}{a^2x} + C$$

$$\int \sqrt{x^2-a^2}\, dx = \frac{x}{2}\sqrt{x^2-a^2} - \frac{a^2}{2} \ln \left| x + \sqrt{x^2-a^2} \right| + C$$

$$\int (\sqrt{x^2-a^2})^n\, dx = \frac{x(\sqrt{x^2-a^2})^n}{n+1} - \frac{na^2}{n+1} \int (\sqrt{x^2-a^2})^{n-2}\, dx, \quad n \neq -1$$

$$\int x(\sqrt{x^2-a^2})^n\, dx = \frac{(\sqrt{x^2-a^2})^{n+2}}{n+2} + C, \quad n \neq -2$$

$$\int x^2\sqrt{x^2-a^2}\, dx = \frac{x}{8}(2x^2-a^2)\sqrt{x^2-a^2} - \frac{a^4}{8} \ln \left| x + \sqrt{x^2-a^2} \right| + C$$

46. $\displaystyle\int \frac{\sqrt{x^2 - a^2}}{x}\,dx = \sqrt{x^2 - a^2} - a \sec^{-1}\left|\frac{x}{a}\right| + C$

47. $\displaystyle\int \frac{\sqrt{x^2 - a^2}}{x^2}\,dx = -\frac{\sqrt{x^2 - a^2}}{x} + \ln\left|x + \sqrt{x^2 - a^2}\right| + C$

48. $\displaystyle\int \frac{dx}{\sqrt{x^2 - a^2}} = \ln\left|x + \sqrt{x^2 - a^2}\right| + C$

49. $\displaystyle\int \frac{dx}{(\sqrt{x^2 - a^2})^n} = \frac{x(\sqrt{x^2 - a^2})^{2-n}}{(2-n)a^2} - \frac{n-3}{(n-2)a^2}\int \frac{dx}{(\sqrt{x^2 - a^2})^{n-2}}, \quad n \neq 2$

50. $\displaystyle\int \frac{x^2}{\sqrt{x^2 - a^2}}\,dx = \frac{x}{2}\sqrt{x^2 - a^2} + \frac{a^2}{2}\ln\left|x + \sqrt{x^2 - a^2}\right| + C$

51. $\displaystyle\int \frac{dx}{x\sqrt{x^2 - a^2}} = \frac{1}{a}\sec^{-1}\left|\frac{x}{a}\right| + C = \frac{1}{a}\cos^{-1}\left|\frac{a}{x}\right| + C$

52. $\displaystyle\int \frac{dx}{x^2\sqrt{x^2 - a^2}} = \frac{\sqrt{x^2 - a^2}}{a^2 x} + C$

53. $\displaystyle\int \sqrt{2ax - x^2}\,dx = \frac{x-a}{2}\sqrt{2ax - x^2} + \frac{a^2}{2}\sin^{-1}\left(\frac{x-a}{a}\right) + C$

54. $\displaystyle\int (\sqrt{2ax - x^2})^n\,dx = \frac{(x-a)(\sqrt{2ax - x^2})^n}{n+1} + \frac{na^2}{n+1}\int (\sqrt{2ax - x^2})^{n-2}\,dx$

55. $\displaystyle\int x\sqrt{2ax - x^2}\,dx = \frac{(x+a)(2x-3a)\sqrt{2ax - x^2}}{6} + \frac{a^3}{2}\sin^{-1}\frac{x-a}{a} + C$

56. $\displaystyle\int \frac{\sqrt{2ax - x^2}}{x}\,dx = \sqrt{2ax - x^2} + a\sin^{-1}\frac{x-a}{a} + C$

57. $\displaystyle\int \frac{\sqrt{2ax - x^2}}{x^2}\,dx = -2\sqrt{\frac{2a-x}{x}} - \sin^{-1}\left(\frac{x-a}{a}\right) + C$

58. $\displaystyle\int \frac{dx}{\sqrt{2ax - x^2}} = \sin^{-1}\left(\frac{x-a}{a}\right) + C$

59. $\displaystyle\int \frac{dx}{(\sqrt{2ax - x^2})^n} = \frac{(x-a)(\sqrt{2ax - x^2})^{2-n}}{(n-2)a^2} + \frac{(n-3)}{(n-2)a^2}\int \frac{dx}{(\sqrt{2ax - x^2})^{n-2}}$

60. $\displaystyle\int \frac{x\,dx}{\sqrt{2ax - x^2}} = a\sin^{-1}\frac{x-a}{a} - \sqrt{2ax - x^2} + C$

61. $\displaystyle\int \frac{dx}{x\sqrt{2ax - x^2}} = -\frac{1}{a}\sqrt{\frac{2a-x}{x}} + C$

Integrals involving trigonometric functions

62. $\displaystyle\int \sin ax\,dx = -\frac{1}{a}\cos ax + C$

63. $\displaystyle\int \cos ax\,dx = \frac{1}{a}\sin ax + C$

64. $\displaystyle\int \sin^2 ax\,dx = \frac{x}{2} - \frac{\sin 2ax}{4a} + C$

65. $\displaystyle\int \cos^2 ax\,dx = \frac{x}{2} + \frac{\sin 2ax}{4a} + C$

66. $\displaystyle\int \sin^n ax\,dx = \frac{-\sin^{n-1} ax\cos ax}{na} + \frac{n-1}{n}\int \sin^{n-2} ax\,dx$

67. $\displaystyle\int \cos^n ax\,dx = \frac{\cos^{n-1} ax\sin ax}{na} + \frac{n-1}{n}\int \cos^{n-2} ax\,dx$

calculus with analytic geometry

calculus with analytic geometry

John B. Fraleigh
University of Rhode Island

ADDISON-WESLEY PUBLISHING COMPANY
Reading, Massachusetts • Menlo Park, California
London • Amsterdam • Don Mills, Ontario • Sydney

This book is in the
ADDISON-WESLEY SERIES IN MATHEMATICS

Consulting editor:
Lynn H. Loomis

Sponsoring editor: Steve Quigley
Production editor: Mary Cafarella
Designer: Katrine Stevens
Illustrator: Carmela Ciampa
Cover design: Ann Scrimgeour

Library of Congress Cataloging in Publication Data:

Fraleigh, John B.
Calculus with analytic geometry.

Includes index.
1. Calculus. 2. Geometry, Analytic. I. Title.

QA303.F6842 15'.15 79-18693
ISBN 0-201-03041-1

Third printing, July 1981

ISBN 0-201-03041-1
CDEFGHIJKL-DO-8987654321

preface

This text is designed for a standard college calculus sequence for students in the physical or social sciences. Such a sequence typically spans three semesters or four quarters. Students are expected to have a background of high school algebra and geometry.

The college calculus sequence contains a great deal of very important mathematics, which students actually use after the course is completed. Few mathematics courses present so much new material at such a rapid pace: this poses a real challenge for the instructor. There is seldom enough time in the classroom to provide complete coverage and supervised drill. Accordingly, students must depend on the text for ideas as well as for exercises.

In this text, I have made every effort to present calculus as clearly and intuitively as possible. Subtle points and proofs of difficult theorems have been omitted. Emphasis is on development of an intuitive but accurate feeling for the subject, and on secure technical competence. The features mentioned below are typical of my efforts to meet this challenge and to provide a really useful text for students.

I introduce the derivative promptly, as opposed to some texts that spend the first hundred pages on preliminaries. So much calculus has to be learned that we must get right to work on it.

A summary at the end of each *lesson* identifies and collects the most important ideas and formulas. This makes the text exceptionally easy to use for study and review.

Reading mathematics is an art that is very different from reading a novel. Students need practice in reading mathematics, and instructors should encourage and require them to do so. All too often calculus students are tested only on material that has been thoroughly covered in the classroom. Such a practice leads students to feel that independent study from a mathematics text is impossibly difficult. Each semester, I like to assign my classes at least three text lessons for independent reading, problem solving, and testing, with no classroom coverage. The text contains several sections that may be assigned out of the sequence in which they appear, and are thus ideal for independent study. I recommend the following:

Semester 1

Section 1.5 Graphs of monomial and quadratic functions
Section 5.2 Newton's method (includes the intermediate-value theorem)
Section 6.5 Numerical methods of integration

Semester 2

Section 8.6 The hyperbolic functions
Section 9.4 Integration of rational functions of sin x and cos x
Section 12.2 Synthetic definitions of conic sections

Semester 3

Section 14.2 Quadric surfaces
Section 16.6 Differentiation of implicit functions (several variables)
Section 17.2 Lagrange multipliers

Assigning these sections for independent study allows more classroom time for basic concepts.

In place of the usual collection of miscellaneous exercises at the end of a chapter, there are two sets of review exercises, followed by a set of more challenging exercises. Each review exercise set gives students an easy way to test their mastery of basic material, and to determine areas that need more study.

Suggested step-by-step procedures are given for solving certain types of problems that cause many students difficulty, such as related rate and maximum–minimum word problems.

Calculus of the trigonometric functions appears in the first-semester portion of the text, shortly after the chain rule. Prompt introduction of this topic helps students understand and remember the chain rule. Two review lessons on the trigonometric functions are supplied for students who need them.

I feel that the use of numerical methods gives a concrete understanding and appreciation of the notions of calculus. Accordingly, the text has more emphasis on numerical methods than most. In particular, there are optional calculator exercises, designed to illustrate concepts of calculus as well as to emphasize numerical techniques.

Some instructors, myself included, begrudge the amount of time often spent on conic sections, since so much calculus must be covered in so short a time. The material in Chapter 12 is arranged so that only one lesson (on sketching) need be spent on conic sections in order to do the remaining unstarred material in the text. All the usual material on conic sections is included for those instructors who do wish to cover it.

The first semester of the calculus sequence presents powerful ideas and techniques, solving problems that students were previously unable to attack. This first semester is the most exciting part of the sequence and, indeed, is one of the most exciting semesters of undergraduate mathematics. The second semester is often a letdown, using the ideas of the first semester with more functions and different coordinate systems, and developing integration technique. I like to have at least one major, exciting topic in the second semester, so I am placing series in the middle of the text. The first series chapter (Chapter 10) is exceptional in training students to determine convergence or divergence of series at a glance, as a mathematician would, based on rigorous tests but without always writing them out. Of course, series can be left as the last topic of the sequence if the instructor prefers.

Some computer graphics are included, but each appears only as a companion to an artist's sketch. Pencil and paper are still the basic tools for studying mathematics. It is important that students develop some ability in sketching to strengthen their geometric intuition. A computer-generated picture, with its myriad precise curves, is ordinarily impossible for students to reproduce. Good pedagogy requires including a sketch by an artist, whom students may emulate. I worked out the computer graphics and programs at the URI Computer Center, where the staff was very helpful.

A Student Supplement is available. The supplement goes through the text, lesson by lesson, warning students of mistakes often made, and then giving complete solutions of every third problem. A Solutions Manual, which works out solutions to all problems, is available for the instructor. I myself prepared these manuals, as well as the answers to odd-numbered exercises at the end of the text. Consequently, I take full responsibility for mistakes; I hope their number is small.

I am indebted to the reviewers of the manuscript for their many valuable suggestions. Some read the manuscript with great care at two stages of development. Among the reviewers were James E. Arnold, Jr. (University of Wisconsin at Milwaukee), Ross A. Beaumont (University of Washington), Arthur T. Copeland (University of New Hampshire), William R. Fuller (Purdue University), Kendell Hyde (Weber State College), and Joan H. McCarter (Arizona State University).

I especially thank Steve Quigley, mathematics editor, and Lynn Loomis, consulting editor, of Addison-Wesley for their advice, encouragement, time, and patience during the entire project.

Kingston, R.I.
November, 1979

J. B. F.

contents

functions and graphs

1

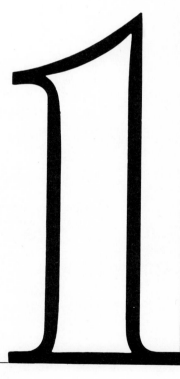

1.1 COORDINATES AND DISTANCE

1.1.1 Coordinates on the line

A **real number** is one that can be written as an unending decimal, positive or negative, or zero. For example, $3 = 3.000000\ldots$, $-\frac{2}{3} = -0.666666\ldots$, and $\pi = 3.141592\ldots$ are real numbers. It is very useful to visualize real numbers as points on the *number line.* Take a line extending infinitely in both directions. Using the real numbers, you can make this line into an infinite ruler (see Fig. 1.1). Label any point on the line with 0 and any point to the right of 0 with 1; this fixes the scale. Each positive real number r corresponds to the point a distance r units to the right of 0, while a negative number $-s$ corresponds to the point a distance s units to the left of 0. The arrow on the line indicates the positive direction. For real numbers r and s, the notation $r < s$ (read "*r is less than s*") means that r is to the left of s on the number line. For example,

$$-\tfrac{2}{3} < \tfrac{1}{2}, \qquad 2 < \pi, \qquad -3 < -\tfrac{7}{3},$$

etc. The notation $r \leq s$ is read "*r is less than or equal to s.*"

The x to the right of the arrow in Fig. 1.1 indicates that you think of x as any real number on the line. In this context, x is known as a *real variable*, and the line is called the *x-axis.*

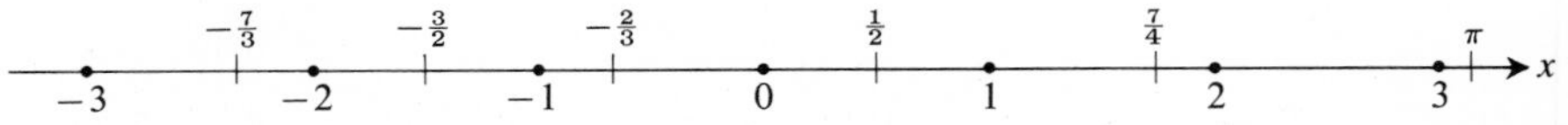

1.1

Example 1 The points x on the line that satisfy the relation $0 \leq x \leq 2$ are indicated by the heavy line and the dark points in Fig. 1.2. Both 0 and 2 satisfy this relation. ||

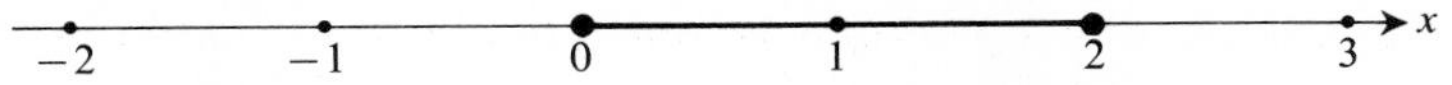

1.2

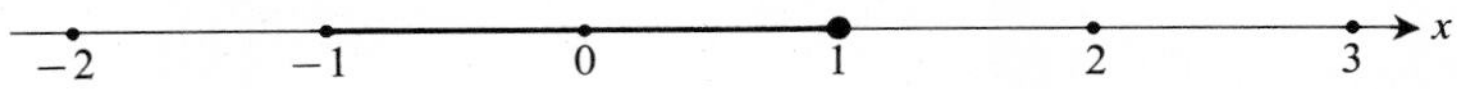

1.3

Example 2 The points satisfying $-1 < x \leq 1$ are indicated by the heavy line together with the dark point 1 in Fig. 1.3. This time -1 does not satisfy the relation, while 1 does. ||

The collection of points x satisfying a relation of the form $a \leq x \leq b$ will be important in calculus. This set of points is the **closed interval** $[a, b]$. The adjective "closed" is used to indicate that both endpoints, a and b, are considered part of the interval; that is, the doors are "closed" at both ends of the interval by these points.

The distance from a point r to the point 0 is known as the **absolute value** of the number r and is denoted by $|r|$. For example,

$$|5| = |-5| = 5,$$

for both 5 and -5 are five units from 0. Consequently,

$$|r| = r \quad \text{for any positive number } r,$$

while

$$|-s| = s \quad \text{for any negative number } -s.$$

Of course, $|0| = 0$.

Consider now the distance between any two points on the line. It is convenient to use subscripted notation x_1 and x_2 for individual numbers on the x-axis, although their values are not specifically given. The distance between the points x_1 and x_2, shown in Fig. 1.4, is surely $x_2 - x_1$. You can easily convince yourself that, for any two points x_1 and x_2, where $x_1 \leq x_2$, the distance between them is $x_2 - x_1$.

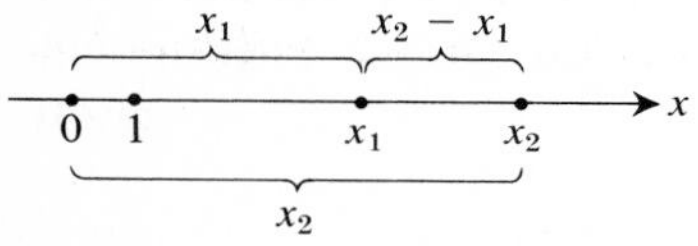

1.4

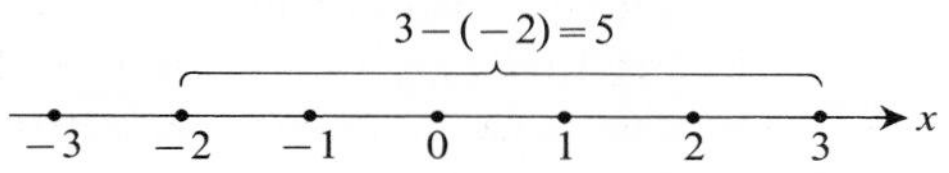

1.5

Example 3 The distance between -2 and 3 is $3 - (-2) = 5$, as indicated in Fig. 1.5. ‖

Distance on the line

Now for *any* points x_1 and x_2, the distance between them is either $x_1 - x_2$, or $x_2 - x_1$, whichever is nonnegative. This nonnegative magnitude is, of course, $|x_2 - x_1|$. Thus the distance from 3 to -2 is $|(-2) - 3| = |-5| = 5$. Another way of expressing this nonnegative difference is $\sqrt{(x_2 - x_1)^2}$, where the square root symbol $\sqrt{\ }$ always yields the *nonnegative* square root of the number. Later in this section, you will see that this square root expression extends naturally to a formula for the distance between two points in a plane.

Example 4 For the points -2 and 3,

$$\sqrt{(3 - (-2))^2} = \sqrt{5^2} = \sqrt{25} = 5;$$

and also,

$$\sqrt{((-2) - 3)^2} = \sqrt{(-5)^2} = \sqrt{25} = 5. \quad \|$$

Exercise 4 asks you to show that $(a + b)/2$ is the same distance form a as from b, so that $(a + b)/2$ is the *midpoint of* $[a, b]$.

Delta notation

Often you need to know not only the distance from x_1 to x_2 but whether x_1 is to the left or right of x_2. The change $x_2 - x_1$ in x-value going from x_1 to x_2 (in that order) is positive if $x_1 < x_2$ and negative if $x_2 < x_1$. Very soon, in calculus you will want to let Δx (read "*delta x*") be such a positive or negative change in x-value. It is a good idea to start right now so that you get used to this delta notation. Think, geometrically, of $\Delta x = x_2 - x_1$ as the *signed* length of the *directed* line segment *from* x_1 *to* x_2.

1.1.2 Coordinates in the plane

Take two copies of the number line (with equal scales) and place them perpendicular to each other in a plane, so that they intersect at the point 0 on each line (see Fig. 1.6). With each point in the plane, we associate an ordered pair (x_1, y_1) of numbers, as follows: The first number x_1 gives the left-right position of the point according to the location of x_1 on the horizontal number line. Similarly, the second number y_1 gives the up-down position of the point according to the location of y_1 on the vertical number line (see Fig. 1.6). Conversely, given any ordered pair of numbers such as $(2, -1)$, there is a unique point in our plane associated with it.

The solid lines of Fig. 1.6 are the **coordinate axes**. In particular, the horizontal axis is the **x-axis** and the vertical axis the **y-axis**, according to the labels at the arrows. For the point (x_1, y_1), the number x_1 is the **x-coordinate** of the point, while y_1 is the **y-coordinate**. The coordinate axes naturally divide the plane into four pieces or **quadrants,** according to the signs of the coordinates of the points. The quadrants are usually numbered as shown in Fig. 1.7. The point $(0, 0)$ is the origin. This introduction of coordinates allows you to use numbers and their arithmetic as a tool in studying geometry. The term *analytic geometry* is used for the study of geometry using coordinates. Of equal importance, coordinatization allows you to draw geometric pictures illustrating a great deal of numerical work.

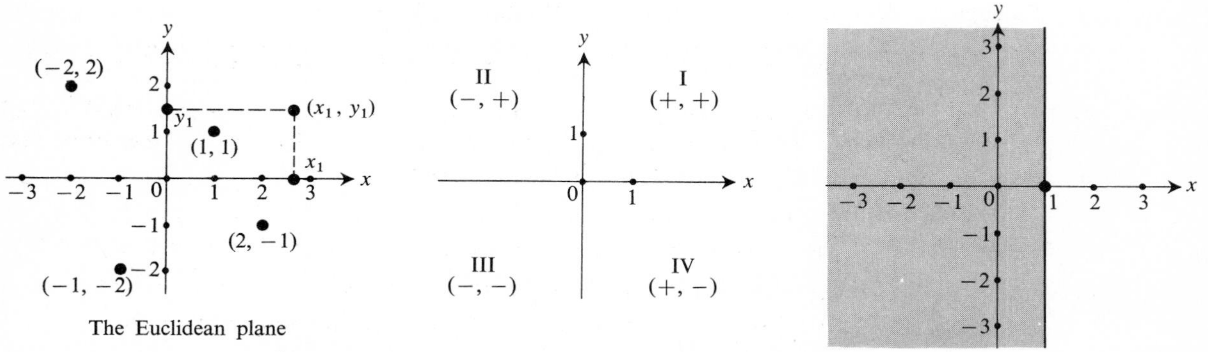

1.6 1.7 1.8

Example 5 The portion of the plane consisting of those points (x, y) satisfying the relation $x \leq 1$ is shown in Fig. 1.8. ||

Example 6 The portion of the plane consisting of the points (x, y) satisfying *both* $-2 \leq x \leq 1$ and $1 \leq y \leq 2$ is shown in Fig. 1.9. ||

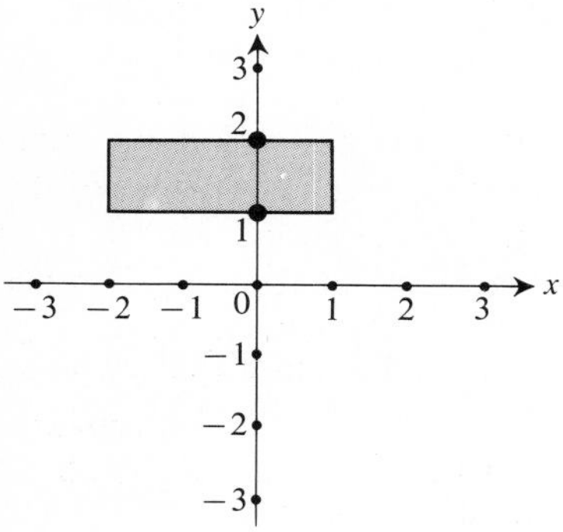

1.9

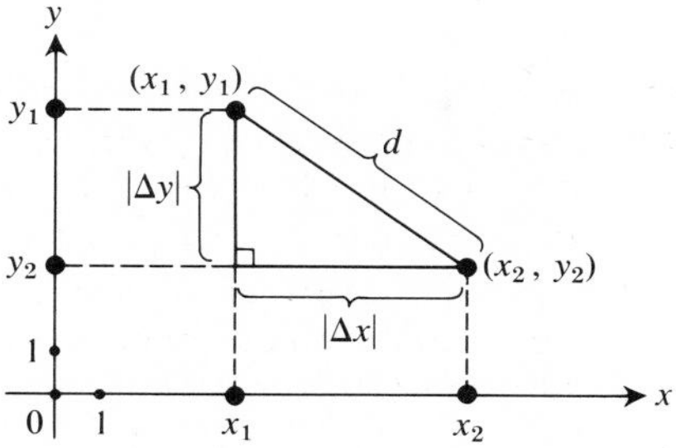

1.10

Finally, let's find the distance between two points (x_1, y_1) and (x_2, y_2) in the plane. Referring to Fig. 1.10, let $\Delta x = x_2 - x_1$ and $\Delta y = y_2 - y_1$, so that $|\Delta x|$ and $|\Delta y|$ are the lengths of the legs of the right triangle shown in the figure. The distance between (x_1, y_1) and (x_2, y_2) is the length d of the hypotenuse of this triangle; so, by the Pythagorean theorem,

$$d^2 = |\Delta x|^2 + |\Delta y|^2. \tag{1}$$

Distance in the plane

Since the terms in (1) are squared, the absolute-value symbols are not needed, so that $d^2 = (\Delta x)^2 + (\Delta y)^2$ and

$$d = \sqrt{(\Delta x)^2 + (\Delta y)^2} = \sqrt{(x_2 - x_1)^2 + (y_2 - y_1)^2}. \tag{2}$$

Example 7 The distance between $(2, -3)$ and $(-1, 1)$ is

$$\sqrt{(-1-2)^2 + (1-(-3))^2} = \sqrt{(-3)^2 + 4^2} = \sqrt{9 + 16} = \sqrt{25} = 5. \quad ||$$

The availability of inexpensive electronic calculators (the "slide-rule" models with memory) makes the computation of (2) easy. The exercises that follow conclude with a calculator portion.

SUMMARY

1. *The closed interval $[a, b]$ consists of all points x such that $a \leq x \leq b$.*
2. *The distance from x_1 to x_2 on the line is $|x_2 - x_1| = \sqrt{(x_2 - x_1)^2}$.*
3. *The signed length of the directed line segment from x_1 to x_2 is*

   $\Delta x = x_2 - x_1 =$ *(Number where you stop) − (Number where you start).*
4. *The midpoint of $[a, b]$ is $(a + b)/2$.*
5. *The distance between (x_1, y_1) and (x_2, y_2) in the plane is*

$$\sqrt{(x_2 - x_1)^2 + (y_2 - y_1)^2}.$$

EXERCISES

1. Sketch, as in Fig. 1.2 and Fig. 1.3, all points x (if there are any), that satisfy the given relation.

a) $2 \le x \le 3$ b) $x \le 0$
c) $x^2 = 4$ d) $x^2 < 4$
e) $x^2 \le 4$ f) $5 \le x \le -1$

2. Find the distance between the given points on the line.

a) 2 and 5 b) -1 and 4 c) -3 and -6

3. Find the distance between the given points on the line.

a) $-\frac{5}{2}$ and 12 b) $-\frac{8}{3}$ and $-\frac{15}{3}$
c) $\sqrt{2}$ and $-2\sqrt{2}$ d) $\sqrt{2}$ and π

4. Show that, for any a and b on the line, the distance from $(a + b)/2$ to a is the same as the distance from $(a + b)/2$ to b.

5. Find the midpoint of each of the following intervals.

a) $[-1, 1]$ b) $[-1, 4]$ c) $[-6, -3]$
d) $[-\frac{3}{2}, \frac{2}{3}]$ e) $[-2\sqrt{2}, \sqrt{2}]$ f) $[\sqrt{2}, \pi]$

6. Find the *signed* length Δx of the directed line segment

a) from 2 to 5 b) from 3 to -7
c) from -8 to -1 d) from 10 to 2

7. Sketch the points (x, y) in the plane satisfying the indicated relations, as in Examples 5 and 6.

a) $x = 1$
b) $-1 \le x \le 2$
c) $x = -1$ and $-2 \le y \le 3$
d) $x = y$ and $-1 \le x \le 1$

8. Proceed as in Exercise 7.

a) $x \le y$ b) $x = -y$
c) $y = 2x$ d) $2x \ge y$

9. Find the cordinates of the indicated point.

a) The point such that the line segment joining it to $(2, -1)$ has the x-axis as perpendicular bisector
b) The point such that the line segment joining it to $(-3, 2)$ has the y-axis as perpendicular bisector
c) The point such that the line segment joining it to $(-1, 3)$ has the origin as midpoint
d) The point such that the line segment joining it to $(2, -4)$ has $(2, 1)$ as midpoint

10. Find the distance between the given points.

a) $(-2, 5)$ and $(1, 1)$ b) $(2, -3)$ and $(-3, 5)$
c) $(2\sqrt{2}, -3)$ and $(-\sqrt{2}, 2)$
d) $(2\sqrt{3}, 5\sqrt{7})$ and $(-4\sqrt{3}, 2\sqrt{7})$

11. To reach the Edwards' home from the center of town, you drive two miles due east on Route 37 and then five miles due north on Route 101. Assuming that the surface of the earth near town is approximately flat, find the distance, as the crow flies, from the center of town to the Edwards' home.

12. Refer to Exercise 11; suppose you drive six miles due west on Route 37 and then four miles due south on Route 43 to reach the Hammonds' house from the center of town. Find the distance from the Edwards' home to the Hammonds' as the crow flies.

calculator exercises

13. Find the midpoint of $[-2\sqrt{3}, 5\sqrt{7}]$.

14. Find the signed length of the directed line segment from $22\sqrt{2}$ to π^3.

15. Find the distance between $(2, -3)$ and $(4, 1)$.

16. Find the distance between $(-3.7, 4.23)$ and $(8.61, 7.819)$.

17. Find the distance between $(\pi, -\sqrt{3})$ and $(8\sqrt{17}, -\sqrt[3]{\pi})$.

1.2 CIRCLES AND THE SLOPE OF A LINE

1.2.1 Circles

The *circle* with center (h, k) and radius r consists of all points (x, y) whose distance from (h, k) is r. Using the formula for the distance from (x, y) to (h, k), you see that this circle consists of all points (x, y) such that

$$\sqrt{(x-h)^2 + (y-k)^2} = r. \tag{1}$$

Squaring both sides of (1), you obtain the equivalent relation

$$(x-h)^2 + (y-k)^2 = r^2. \tag{2}$$

Equation (2) is known as the *equation of the circle.*

Example 1 The equation of the circle with center $(-2, 4)$ and radius 5 is $(x-(-2))^2 + (y-4)^2 = 25$, or $(x+2)^2 + (y-4)^2 = 25$. ||

Example 2 The equation $(x+3)^2 + (y+4)^2 = 18$ describes a circle with center at $(-3, -4)$ and radius $\sqrt{18} = 3\sqrt{2}$. ||

Every equation of the form $ax^2 + ay^2 + bx + cy = d$ and satisfied by at least one point (x_1, y_1) is the equation of a circle. However, the general equation may have no locus in our real plane. For example, $x^2 + y^2 = -10$ has no real locus, for a sum of squares can't be negative. You should try to put any particular such equation in the form (2) to find the center and radius of the circle.

Example 3 Let us show that $3x^2 + 3y^2 + 6x - 12y = 60$ describes a circle.

SOLUTION We start by dividing by the common coefficient 3 of x^2 and y^2, and obtain

$$x^2 + y^2 + 2x - 4y = 20.$$

Completing the square

Now we use the algebraic device of completing the square to get our equation in the form (2). The steps are as follows:

$$(x^2 + 2x) + (y^2 - 4y) = 20,$$
$$(x+1)^2 + (y-2)^2 = 20 + 1^2 + (-2)^2,$$
$$(x+1)^2 + (y-2)^2 = 25.$$

Thus our equation describes a circle with center $(-1, 2)$ and radius 5. ||

If you let $\Delta x = x - h$ and $\Delta y = y - k$, then (2) becomes

$$(\Delta x)^2 + (\Delta y)^2 = r^2. \tag{3}$$

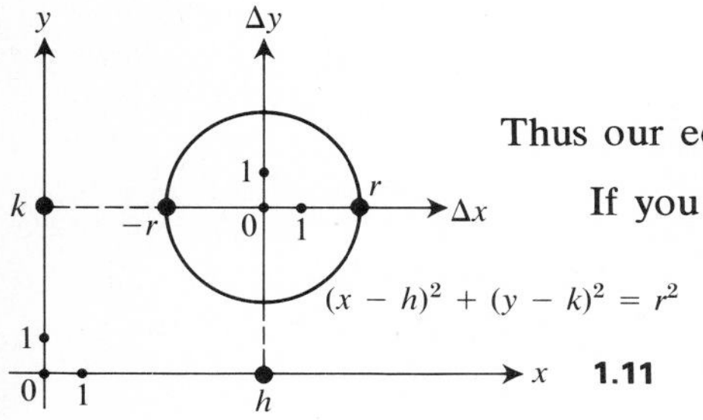

1.11

To interpret (3) geometrically, take a new Δx-axis and a new Δy-axis with the point (h, k) as new origin, as shown in Fig. 1.11. Recall that Δx is the directed distance from h to x and Δy is the directed distance from k to y.

Translating axes

Thus (3) is exactly the equation of the circle with respect to your new axes. This device is known as *translation of axes to* (h, k) and will often be useful. The equation $x^2 + y^2 = r^2$ describes a circle with center the origin of the x,y-coordinate system and radius r, while the equation $(\Delta x)^2 + (\Delta y)^2 = r^2$ describes a circle with center the origin of the $\Delta x, \Delta y$-coordinate system and radius r.

1.2.2 The slope of a line

The **slope** m of a line is the number of units the line climbs (or falls) vertically for each unit of horizontal change from left to right. To illustrate, if a line climbs upward 3 units for each unit step you go to the right, as in Fig. 1.12(a), the line has slope 3. If a line falls 2 units downward per unit step to the right, as in Fig. 1.12(b), the line has slope -2. A horizontal line neither climbs nor falls, so it has slope 0. A vertical line climbs straight up over a single point, so it is impossible to measure how much it climbs per unit horizontal change; consequently *the slope of a vertical line is undefined.*

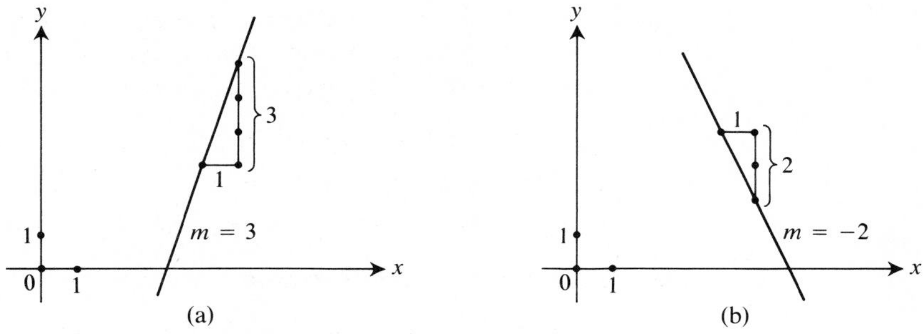

1.12

Example 4 Let's find the slope of the line through the points $(2, 4)$ and $(5, 16)$.

SOLUTION As we go from $(2, 4)$ to $(5, 16)$, we have $\Delta x = 5 - 2 = 3$ and $\Delta y = 16 - 4 = 12$. Since the line climbed $\Delta y = 12$ units while we went $\Delta x = 3$ units to the right, and since a line climbs at a uniform rate, the amount it climbs per horizontal unit to the right is $\Delta y/\Delta x = \frac{12}{3} = 4$. ‖

As illustrated in Example 4, you can find the slope m of the line through (x_1, y_1) and (x_2, y_2) if $x_1 < x_2$ by finding Δx and Δy as you go from (x_1, y_1) to (x_2, y_2), and then taking the quotient; so

$$m = \frac{\Delta y}{\Delta x} = \frac{y_2 - y_1}{x_2 - x_1}. \tag{4}$$

We assume that our line is not vertical, so $x_1 \neq x_2$. If it should happen that $x_2 < x_1$, then, to go from left to right, you should go from (x_2, y_2) to (x_1, y_1), and you obtain

$$m = \frac{\Delta y}{\Delta x} = \frac{y_1 - y_2}{x_1 - x_2} = \frac{y_2 - y_1}{x_2 - x_1}, \tag{5}$$

which is the same formula as in (4). In summary, the slope m of a nonvertical line through two points is given by

$$m = \frac{\Delta y}{\Delta x} = \frac{\text{Difference of } y\text{-coordinates}}{\text{Difference of } x\text{-coordinates in the same order}}. \tag{6}$$

Example 5 The line through (7, 5) and (−2, 8) has slope

$$m = \frac{\Delta y}{\Delta x} = \frac{8-5}{-2-7} = \frac{3}{-9} = -\frac{1}{3}. \quad \|$$

Two lines are parallel precisely when they climb (or fall) at the same rate, that is, when they have equal slopes. Now suppose that two lines are perpendicular instead. Let one line have slope m_1 and the other have slope m_2. By translating axes, we may assume that our lines intersect at the origin. Then $(1, m_1)$ and $(1, m_2)$ are points on the lines, as shown in Fig. 1.13. The lines are perpendicular if and only if the triangle with vertices (0, 0), $(1, m_1)$, and $(1, m_2)$ satisfies the Pythagorean relation $d^2 = r^2 + s^2$.

From our distance formula, we obtain

$$r^2 = (1-0)^2 + (m_1 - 0)^2 = 1 + m_1^2,$$
$$s^2 = (1-0)^2 + (m_2 - 0)^2 = 1 + m_2^2,$$
$$d^2 = (1-1)^2 + (m_2 - m_1)^2 = (m_2 - m_1)^2.$$

The Pythagorean condition becomes

$$(m_2 - m_1)^2 = (1 + m_1^2) + (1 + m_2^2)$$

or

$$m_2^2 - 2m_1m_2 + m_1^2 = 2 + m_1^2 + m_2^2.$$

Therefore

$$-2m_1m_2 = 2;$$

so

$$m_1m_2 = -1 \quad \text{or} \quad m_2 = -\frac{1}{m_1}. \tag{7}$$

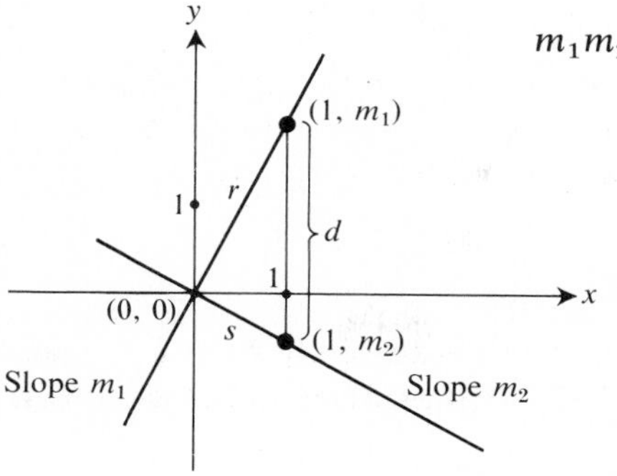

1.13

Example 6 Let's find the slope of a line perpendicular to the line through (6, −5) and (8, 3).

SOLUTION The given line has slope

$$\frac{\Delta y}{\Delta x} = \frac{3-(-5)}{8-6} = \frac{8}{2} = 4;$$

so a perpendicular line has slope $-\frac{1}{4}$. ‖

SUMMARY

1. *The circle with center (h, k) and radius r has equation*

$$(x-h)^2 + (y-k)^2 = r^2.$$

2. *To find the center (h, k) and the radius r of a circle $ax^2 + ay^2 + bx + cy = d$, complete the square on the x-terms and on the y-terms.*
3. *A vertical line has undefined slope. If $x_1 \neq x_2$, the line through (x_1, y_1) and (x_2, y_2) has slope*

$$m = \frac{\Delta y}{\Delta x} = \frac{y_2 - y_1}{x_2 - x_1}.$$

4. *Lines of slopes m_1 and m_2 are:*

parallel if and only if $m_1 = m_2$;
perpendicular if and only if $m_1 m_2 = -1$, or $m_2 = -1/m_1$.

EXERCISES

1. Find the equation of the circle with the given center and radius.

a) center (0, 0), radius 5

b) center (−1, 2), radius 3

c) center (3, −4), radius $\sqrt{30}$

2. Find the center and radius of the given circle.

a) $(x-2)^2 + (y-3)^2 = 36$

b) $(x+3)^2 + y^2 = 49$

c) $(x+1)^2 + (y+4)^2 = 50$

3. Find the center and radius of the given circle.

a) $x^2 + y^2 - 4x + 6y = 3$

b) $x^2 + y^2 + 8x = 9$

c) $4x^2 + 4y^2 - 12x - 24y = -\frac{9}{2}$

4. Find the equation of the circle with center in the second quadrant, tangent to the coordinate axes, and with radius 4.

5. Find the equation of the circle having the line segment with endpoints (−1, 2) and (5, −6) as a diameter.

6. Find the equation of the circle with center (2, −3) and passing through (5, 4).

7. Find the slope of the line through the indicated points, if the line is not vertical.

a) (−3, 4) and (2, 1)

b) (5, −2) and (−6, −3)

c) (3, 5) and (3, 8)

d) (0, 0) and (5, 4) e) (−7, 4) and (9, 4)

8. Find b so that the line through $(2, -3)$ and $(5, b)$ has slope -2.

9. Find a so that the line through $(a, -5)$ and $(3, 6)$ has slope 1.

10. Find the slope of a line perpendicular to the line through $(-3, 2)$ and $(4, 1)$.

11. Find b so that the line through $(8, 4)$ and $(4, -2)$ is parallel to the line through $(-1, 2)$ and $(2, b)$.

12. Show that the line joining the midpoints of two sides of a triangle is parallel to the third side. [*Hint.* Let the vertices of the triangle be $(0, 0)$, $(a, 0)$, and (b, c).]

13. Water freezes at 0° Celsius and 32° Fahrenheit, while it boils at 100° Celsius and 212° Fahrenheit. If points (C, F) are plotted in the plane, where F is the temperature in degrees Fahrenheit corresponding to a temperature of C degrees Celsius, then a straight line is obtained. Find the slope of the line. What does this slope represent in this situation?

14. The Easy Life Prefabricated Homes Company listed its super-deluxe ranch model for \$30,000 in 1960. The company increased the price by the same amount each year, and listed the same model for \$90,000 in 1980. Find the slope of the segment drawn through points (Y, C) in the plane, where Y could be any year from 1960 to 1980 and C is the cost of this model ranch house in that year. What does this slope represent in this situation?

calculator exercises

15. Find the center and radius of the given circle.

a) $(x - \pi)^2 + (y - \sqrt{\pi})^2 = 2.736$

b) $x^2 + y^2 + 3.1576x - 1.2354y = 3.33867$

c) $\sqrt{2}x^2 + \sqrt{2}y^2 - \pi^3 x + (\pi^2 + 3.4)y = \sqrt{17}$

16. Find the slope of the line through the indicated points.

a) $(2.367, \pi)$ and $(\sqrt{3}, 8.9)$

b) $(\pi^2, \sqrt[3]{19})$ and $(12.378, \sqrt{5.69})$

c) $(\sqrt{2} + \sqrt{3}, \pi - \sqrt{19.3})$

and

$(\sqrt{\pi} + 1.45, \sqrt{14} - \sqrt[5]{134})$

1.3 THE EQUATION OF A LINE

Let a given line have slope m and pass through the point (x_1, y_1) as shown in Fig. 1.14. Let's try to find an algebraic condition for a point (x, y) to lie on the line. If the slope of the line that joins (x_1, y_1) and (x, y) is also m, then that line is parallel to the given line, for they have the same slope. But both

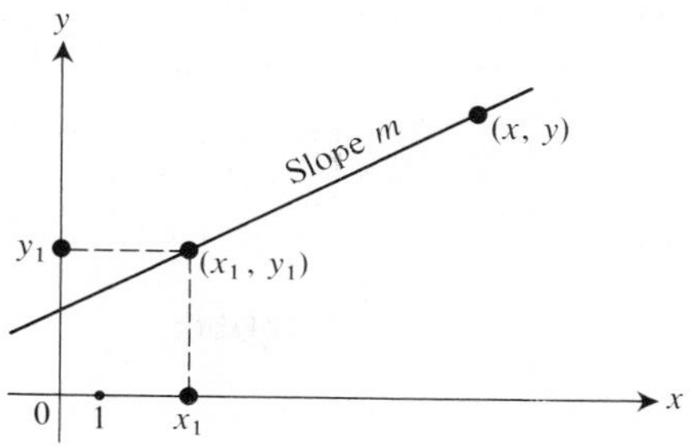

1.14

lines go through (x_1, y_1), so they must coincide. Therefore a condition for (x, y) to lie on the given line is that

$$\frac{y - y_1}{x - x_1} = m \tag{1}$$

or

$$y - y_1 = m(x - x_1). \tag{2}$$

Equation (2) is the *point–slope form* of the equation of the line.

Example 1 Let's find the equation of the line through $(2, -3)$ with slope 7.

SOLUTION The equation is $y - (-3) = 7(x - 2)$ or $y + 3 = 7(x - 2)$. This equation may be simplified to $y = 7x - 17$. The point $(3, 4)$ lies on this line, since $4 = 7 \cdot 3 - 17$. ‖

As indicated in Example 1, the point–slope equation (2) can be rewritten in the form

$$y = mx + b, \tag{3}$$

where $b = y_1 - mx_1$. The constant b in (3) has a nice interpretation. If you set $x = 0$ in (3), then $y = b$, so the point $(0, b)$ satisfies the equation and thus lies on the line. This point $(0, b)$ is on the y-axis, and b is the **y-intercept** of the line. For this reason, (3) is the *slope–intercept form* of the equation of the line. If the line crosses the x-axis at $(a, 0)$, then a is the **x-intercept** of the line.

Example 2 We find the intercepts of the line in Example 1.

SOLUTION The equation is $y = 7x - 17$, so -17 is the y-intercept. To find the x-intercept, you set $y = 0$ and obtain $7x - 17 = 0$, so $x = \frac{17}{7}$. Thus the point $(\frac{17}{7}, 0)$ lies on the line, so $\frac{17}{7}$ is the x-intercept. ‖

The vertical line through $(a, 0)$ in Fig. 1.15 has undefined slope, so it does not have an equation of the form (2) or (3). But surely a condition that (x, y) lie on the line is simply that $x = a$. Of course, $y = b$ is the horizontal line through $(0, b)$ shown in the figure. In any kind of coordinate system, it is important to know what loci are obtained by setting the coordinate variables equal to constants. We have seen that in our rectangular x,y-coordinate system, $x = a$ is a vertical line and $y = b$ is a horizontal line.

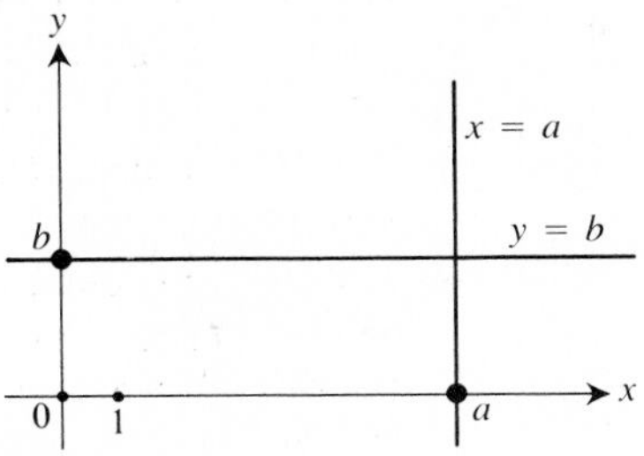

1.15

Any time you want to find the equation of a line, say to yourself, "I need to find a *point on the line* and the *slope of the line*." Then use Eq. (2).

Example 3 Let's find the equation of the line through $(-5, -3)$ and $(6, 1)$.

SOLUTION We solve the problem as follows.

POINT: $(x_1, y_1) = (-5, -3)$

SLOPE: $m = [1 - (-3)]/[6 - (-5)] = \frac{4}{11}$

EQUATION: $y + 3 = \frac{4}{11}(x + 5)$

The equation can be simplified to $11y + 33 = 4x + 20$ or $4x - 11y = 13$. ||

Finally, observe that every equation $ax + by + c = 0$, where either $a \neq 0$ or $b \neq 0$ is the equation of a line. If $b = 0$, the equation becomes $x = -c/a$, which is a vertical line. If $b \neq 0$, the equation becomes $y = -(a/b)x - c/b$, which is a line with slope $m = -a/b$ and y-intercept $-c/b$.

SUMMARY

1. *A vertical line has equation* $x = a$.
2. *A horizontal line has equation* $y = b$.
3. *To find the equation of a line, find one point* (x_1, y_1) *on the line and the slope* m *of the line. The equation is then*

$$y - y_1 = m(x - x_1).$$

4. *The line* $y = mx + b$ *has slope* m *and* y*-intercept* b.

EXERCISES

1. Find the equation of the indicated line.

a) Through $(-1, 4)$ with slope 5

b) Through $(2, 5)$ and $(-3, 5)$

c) Through $(4, -5)$ and $(-1, 1)$

d) Through $(-3, 4)$ and $(-3, -1)$

2. Find the slope, x-intercept, and y-intercept of the indicated line.

a) $x - y = 7$ b) $y = 11$

c) $x = 4$ d) $7x - 13y = 8$

3. Find the equation of the line through $(-2, 1)$ and parallel to the line $2x + 3y = 7$.

4. Find the equation of the line through $(3, -4)$ perpendicular to the line $4x - 7y = 11$.

5. Are the lines $3x + 4y = 8$ and $4x + 3y = 14$ perpendicular? Why?

6. Are the lines $7x + 8y = 10$ and $8x - 7y = -14$ perpendicular? Why?

7. Find the equation of the perpendicular bisector of the line segment joining $(-1, 5)$ and $(3, 11)$.

8. Find the point of intersection of the lines $2x + 3y = 7$ and $3x + 4y = -8$.

9. Find the distance from the point $(-2, 1)$ to the line $3x + 4y = 8$.

10. Show that the perpendicular bisectors of the sides of a triangle meet at a point. [*Hint.* Let the vertices of the triangle be $(-a, 0)$, $(a, 0)$, and (b, c).]

11. Find the equation of the circle through the points (1, 5), (2, 4), and (−2, 6).

12. Referring to Exercise 13 of the preceding section, find the linear relation giving the temperature F in degrees Fahrenheit corresponding to a temperature of C degrees Celsius.

13. A snowstorm starts at 3:00 A.M. and continues until 11:00 A.M. If there were 13 in. of old snow on the ground at the start of the storm and the new snow accumulates at a constant rate of $\frac{3}{2}$ in. per hour, find the depth d in inches at time of day t for $3 \leq t \leq 11$.

1.4 FUNCTIONS AND THEIR GRAPHS

1.4.1 Functions

The area enclosed by a circle is a *function* of the radius of the circle, meaning that the area depends on and varies with this radius. If a numerical value for the radius is given, the area enclosed by the circle is determined. For example, if the radius is 3 units, then the area is 9π square units. Similarly, the area of a rectangular region is a function of both the length and the width of the rectangle; that is, the area depends on and varies with these quantities. If the length of a rectangle is 5 units and the width is 3 units, the rectangle encloses a region that has an area of 15 square units.

The study of how one numerical quantity Q depends on and varies with other numerical quantities is one of the major concerns of science. A rule that specifies the numerical value of Q for all possible values of the other quantities is an exceedingly useful thing to have. Viewed intuitively, a *function* is such a rule.

In the next few chapters, we will be interested chiefly in the case where the value of a number y depends upon the value of some single number x, so that *y is a function of x*. This is often expressed by $y = f(x)$, and we consider *f to be* the function. We shall sometimes be sloppy and speak of the function $f(x)$, but strictly speaking, f is the function and $f(x)$ is the *value of the function f at x*. If you want to talk about several functions at once, use different letters. The letters f, g, and h are commonly used for functions.

For an example, perhaps $y = f(x) = \sqrt{x - 1}$, so that

$$f(2) = \sqrt{2 - 1} = \sqrt{1} = 1,$$

$$f(5) = \sqrt{5 - 1} = \sqrt{4} = 2,$$

and

$$f(1) = \sqrt{1 - 1} = \sqrt{0} = 0.$$

Note that $f(0)$ is not defined for this function f, for we shall allow only real numbers, and $\sqrt{0 - 1} = \sqrt{-1}$ is not a real number. The set of those x-values allowed is the **domain** of the function; and x, which may take on any number in that domain as value, is the **independent variable**. Similarly, y is the **dependent variable**; its value *depends* on the value of the variable x. The set of y-values obtained, as x goes through all values in the domain, is the **range** of the function.

If $y = f(x)$, then f should assign to each x in the domain *only one value* y. This is a very important requirement. We just worked with the function f

given by $y = f(x) = \sqrt{x - 1}$, and said that $f(5) = \sqrt{5 - 1} = \sqrt{4} = 2$. You may have wondered why we didn't say $\sqrt{4} = \pm 2$. We want $\sqrt{x - 1}$ to define a *function*, and for this reason, we shall always use the $\sqrt{}$ symbol to mean the *nonnegative* square root. If we want the negative square root, we will always use $-\sqrt{}$.

When a function f is defined by a formula and the domain of the independent variable is not specifically given, we always consider the domain to consist of all values of x for which the formula can be evaluated and yields a *real* number. In particular, *division by zero is not allowed, and square roots (or fourth roots, or any even roots) of negative numbers are not allowed.*

Example 1 Let's find the domain of the function f given by the formula $y = f(x) = \sqrt{x - 1}$.

SOLUTION The domain consists of all x such that $x - 1 \geq 0$, or such that $x \geq 1$. The range of f consists of all $y \geq 0$. ||

Example 2 At the start of this section we said that the area A of a circle is a function of its radius r. Let's describe this function.

SOLUTION If you call this function g, then r is the independent variable, A the dependent variable, and you have

$$A = g(r) = \pi r^2 \quad \text{for} \quad r \geq 0.$$

The domain constraint $r \geq 0$ must be stated because you can't have a circle of negative radius. Of course, you could compute πr^2 for negative values of r. This time, the domain restriction is due to the geometric origin of the function. ||

Example 3 We find the domain of $y = f(x) = (x^2 - 1)/(x^2 - 9)$.

SOLUTION Note that $f(2) = 3/(-5)$ and $f(5) = \frac{24}{16} = \frac{3}{2}$, but $f(3)$ is not defined since *division by zero is not allowed.* The domain of the function consists of all $x \neq \pm 3$. ||

Example 4 At the start of this section we said that the area A of a rectangular region is a function of its length ℓ and width w. If g is this function, it is customary to write

$$A = g(\ell, w) = \ell \cdot w \quad \text{for} \quad \ell \geq 0, w \geq 0.$$

This time there are *two* independent variables, ℓ and w. The domain restrictions $\ell \geq 0$ and $w \geq 0$ are again required because of the geometric origin of the function: A rectangle can't have negative length or width. ||

An intuitive picture of a function popular in elementary texts is a "black box," as in Fig. 1.16. One puts in a value of the independent variable x and a value of the dependent variable y comes out the other end. A "slide-rule" calculator is such a black (or other colored) box. For example, you punch in a value for x and then press the $\sin x$ button to "perform the function," and the y-value, where $y = f(x) = \sin x$, is shown as the display, usually to about eight-figure accuracy.

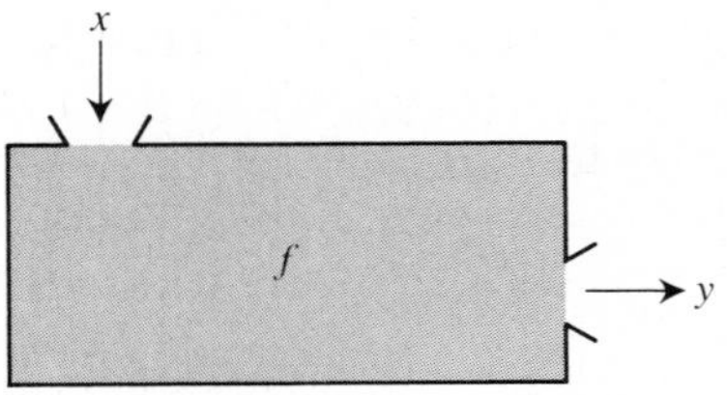

1.16

1.4.2 Graphs For a function f of one variable, we may find the points (x, y) in the plane where $y = f(x)$. These points form the **graph** of the function.

Example 5 Let $y = f(x) = 2x - 4$. The graph of this function is just the plane locus of the equation $y = 2x - 4$, which is the line with slope 2 and y-intercept -4 shown in Fig. 1.17. ||

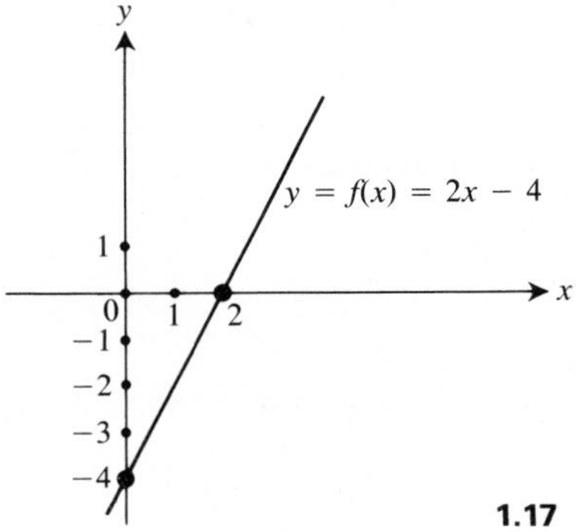

1.17

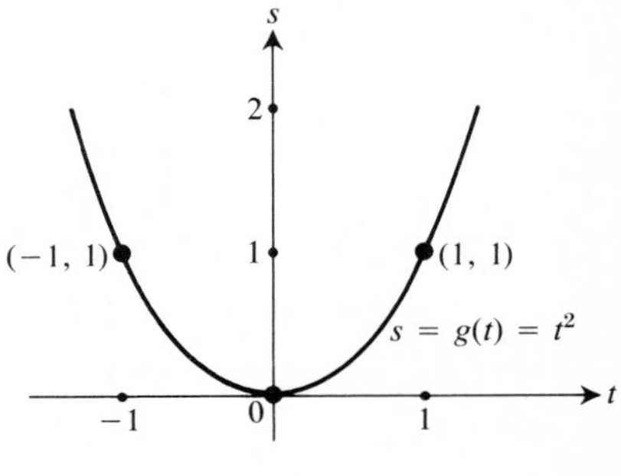

1.18

Example 6 The graph of the function $s = g(t) = t^2$ is shown in Fig. 1.18. Here we have used other letters for the variables and the function. ||

As Examples 5 and 6 illustrate, the graph of a function given by an algebraic formula as $y = f(x)$ is the plane locus of points that satisfy this equation. The second paragraph of this section states that a function is a certain type of "rule." But what *is* a "rule"? Well, perhaps a "rule" is a "law." All right—now what is a "law"? We can keep this up for pages, and will still have to leave some term undefined. Mathematicians have recognized that at least one term *must* be left undefined, and have agreed that "set" shall be taken as an undefined term. So a mathematician who says that a function is a certain type of set has the professional right to refuse to answer if you ask what a set is. Now a function, $y = f(x)$, can be evaluated at any point x_1 in its domain if you know the point (x_1, y_1) on its graph. Thus the set of all points on the graph of a function can serve as a "rule" to evaluate a function. The collection of all such points can be viewed as a set. Here is a modern definition of a real-valued function of one real variable, but please don't get carried away or confused by this definition. Keep thinking of such a function as a rule, which is frequently given by some mathematical formula.

Definition 1.1 A **real-valued function** of one real variable is a set of ordered pairs (x, y) of real numbers such that no two different pairs have the same first coordinate.

The requirement of the definition that different pairs have different x-coordinates was illustrated in the discussion of $y = f(x) = \sqrt{x - 1}$ as a function. Recall that $f(5) = 2$, not ± 2. That is, $(5, 2)$ is one of the pairs of the function, but $(5, -2)$ is not. The curve in Fig. 1.19 is *not* the graph of a function $y = g(x)$, since *three* points have the same x-coordinate a.

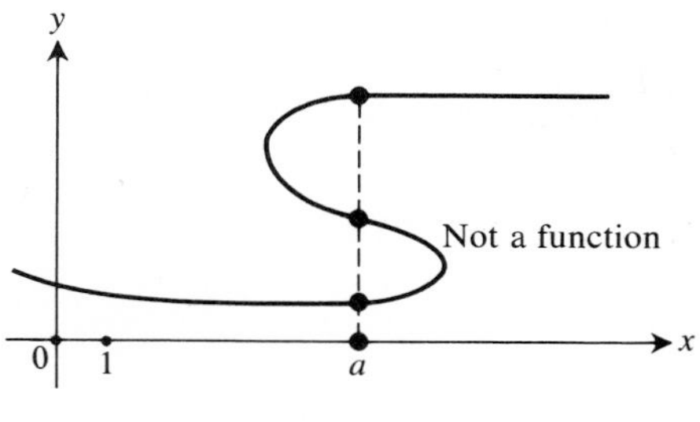

1.19

One way to sketch the graph $y = f(x)$ is to make a table of corresponding values of x and y, plot the points, and draw a curve through them. Computing y-values can be tedious, and a calculator is often helpful. The calculator exercises of this section deal with such tables and plots. A computer can easily make such a table for many important functions. Printout 1.1 shows a table of x-values and y-values for the polynomial function f given by

$$y = f(x) = x^3 + 10x^2 + 8x - 50,$$

XYVALUES

X-VALUE	Y-VALUE
-10	-130
-9.5	-80.875
-9	-41
-8.5	-9.625
-8	14
-7.5	30.625
-7	41
-6.5	45.875
-6	46
-5.5	42.125
-5	35
-4.5	25.375
-4	14
-3.5	1.625
-3	-11
-2.5	-23.125
-2	-34
-1.5	-42.875
-1	-49
-.5	-51.625
0	-50
.5	-43.375
1	-31
1.5	-12.125
2	14
2.5	48.125
3	91

Printout 1.1
$y = x^3 + 10x^2 + 8x - 50$

using 27 equally spaced x-values on the interval $[-10, 3]$. These results were obtained by using a computer program XYVALUES, written in the language BASIC and shown in Appendix 1. Tables of other functions can be found by changing just two lines (150 and 170) in the program XYVALUES.

Plotting the graph from a table is still a bit of a nuisance. The computer can also give you a pretty good plot of the graph at a terminal. Printout 1.2 shows the graph for the data in Printout 1.1 as given by a program PLOT (see Appendix 1). In PLOT, the y-axis goes across the page and the x-axis down the page, so you have to rotate the page 90° counterclockwise to bring the axes into their usual position.

Printout 1.2 $y = x^3 + 10x^2 + 8x - 50$

```
PLOT

INPUT ENDPOINTS  A,B  OF INTERVAL OF X VALUES.
?-10,3
INPUT NUMBER <= 100 OF POINTS TO BE PLOTTED.
?27


THE SMALLEST Y VALUE IS -130
THE LARGEST Y VALUE IS  91
ONE Y-AXIS MARK EQUALS 4.42     UNITS

                  -130                                             91
                                               0

Y-AXIS SCALE      +....+....+....+....+....+....+....+....+....+....+

X VALUES
-10               *
-9.5                         *
-9                                    *
-8.5                                         *
-8                                                *
-7.5                                                  *
-7                                                      *
-6.5                                                     *
-6                                                       *
-5.5                                                    *
-5                                                     *
-4.5                                                 *
-4                                                *
-3.5                                           *
-3                                          *
-2.5                                      *
-2                                     *
-1.5                                 *
-1                                  *
-.5                                *
 0                                  *
 .5                                  *
 1                                      *
 1.5                                        *
 2                                              *
 2.5                                                      *
 3                                                                  *
```

SUMMARY

1. *If* $y = f(x)$, *then* x *is the independent variable and* y *the dependent variable.*
2. *If* $y = f(x)$, *the domain of the function* f *consists of all allowable values of the variable* x. *The range of* f *consists of all values obtained for* y *as* x *goes through all values in the domain.*
3. *A function* f *assumes only* one *value* $f(x)$ *for each* x *in its domain. Thus* $\pm\sqrt{x}$ *is* not *a function.*
4. *If* $y = f(x)$ *is described by a formula, the domain of* f *consists of all* x *where* $f(x)$ *can be computed and gives a real number. For us, this usually means just excluding* x*-values that would lead to division by zero or to taking even roots of negative numbers.*
5. *The graph of* f *consists of all points* (x, y) *such that* $y = f(x)$.
6. *Graphs can be sketched by making a table of* x*- and* y*-values and plotting the points* (x, y), *although this may be hard work.*

EXERCISES

1. Express the volume V of a cube as a function of the length x of an edge of the cube.

2. Express the volume V of a cylinder as a function of the radius r of the cylinder and the length ℓ of the cylinder.

3. Express the area A enclosed by a circle as a function of the perimeter s of the circle.

4. Express the volume V of a box with square base as a function of the length x of an edge of the base and the area A of one side of the box.

5. Express the volume V of a cube as a function of the length d of a diagonal of the cube. (A diagonal of a cube joins a vertex to the *opposite vertex*, which is the vertex farthest away.)

6. Bill starts at a point A at time $t = 0$ and walks in a straight line at a constant rate of 3 mi/hr toward point B. If the distance from A to B is 21 mi, express his distance s from B as a function of the time t, measured in hours.

7. Mary and Sue start from the same point on a level plain at time $t = 0$. Mary walks north at a constant rate of 3 mi/hr, while Sue jogs west at a constant rate of 5 mi/hr. Find the distance s between them as a function of the time t, measured in hours.

8. Smith, who is 6 ft tall, starts at time $t = 0$ directly under a light 30 ft above the ground, and walks away in a straight line at a constant rate of 4 ft/sec.

a) Express the length ℓ of Smith's shadow as a function of the distance x he has walked.

b) Express the length ℓ of Smith's shadow as a function of the time t, measured in seconds.

c) Express the distance x walked as a function of the length ℓ of his shadow.

9. A portion of the graph of a function f is shown in Fig. 1.20. Estimate each of the following from the graph.

a) $f(0)$ b) $f(1)$ c) $f(-1)$

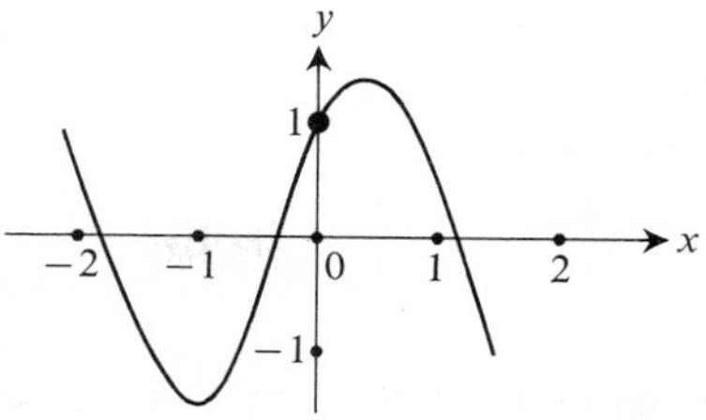

1.20

10. Let $f(x) = x^2 - 4x + 1$. Find the following.

a) $f(0)$ b) $f(-1)$ c) $f(5)$

d) A formula in terms of Δx for $f(2 + \Delta x)$

11. Let $g(t) = t/(1 - t)$. Find the following, if defined.

a) $g(0)$ b) $g(1)$ c) $g(-1)$

d) A formula in terms of Δt for

$$\frac{g(2 + \Delta t) - g(2)}{\Delta t}.$$

12. Find the domain of the function defined by the given algebraic expression.

a) $f(x) = \dfrac{1}{x}$ b) $f(x) = \dfrac{1}{x^2 - 1}$

c) $f(x) = \dfrac{x}{x^2 - 3x + 2}$ d) $g(t) = \sqrt{t + 3}$

13. Proceed as in Exercise 12.

a) $f(u) = \sqrt{u^2 - 1}$ b) $g(t) = \dfrac{\sqrt{t - 2}}{t^2 - 16}$

c) $h(x) = \sqrt{x - 4}$ d) $k(v) = \dfrac{v^2}{\sqrt{9 - v^2}}$

14. Sketch the graph of the following functions.

a) $y = f(x) = x - 1$ b) $y = g(x) = -x^2$

15. Proceed as in Exercise 14.

a) $s = g(t) = t^2 - 4$ b) $y = f(x) = \sqrt{1 - x^2}$

c) $s = f(r) = -\sqrt{1 - r^2}$

16. Proceed as in Exercise 14.

a) $y = f(x) = \dfrac{1}{x}$ b) $y = g(x) = \dfrac{1}{(x - 2)}$

c) $y = h(u) = \dfrac{-1}{u}$

17. Make a table of x-values and y-values for the function

$$y = f(x) = \frac{x + 1}{x - 1}$$

for $x = -1, -\frac{1}{2}, 0, \frac{1}{2}, \frac{3}{4}, \frac{7}{8}, \frac{9}{8}, \frac{5}{4}, \frac{3}{2}, 2, \frac{5}{2}$, and 3. Plot the points and draw the portion of the graph for all x in the domain such that $-1 \leq x \leq 3$.

calculator exercises

18. Make a table of values for the function $f(x) = (x + 1)/\sqrt{x^3 + 1}$ using 11 equally spaced x-values starting with $x = 0$ and ending with $x = 10$. Use the data to draw the graph of the function over $[0, 10]$. [*Note.* Eleven values give ten intervals.]

19. Make a table of values for the function $f(x) = \sin x^2$ using 13 equally spaced x-values starting with $x = 0$ and ending with $x = 3$. Use radian measure. Use the data to draw the graph of the function over $[0, 3]$. [*Note.* Thirteen values give twelve intervals. While we have not "defined" the function $\sin x^2$ yet, it is defined for you in your "black box" calculator.]

1.5 GRAPHS OF MONOMIAL AND QUADRATIC FUNCTIONS

The monomial functions are those given by the monomials

$$x, \quad x^2, \quad x^3, \quad x^4, \quad x^5, \quad \ldots, \quad x^n, \quad \ldots$$

or constant multiples of them. When a function is given by a formula, we often refer to the formula as the function, to save writing. Thus we may refer to the function $4x^3$ rather than the function f where $f(x) = 4x^3$.

1.5.1 Monomial functions

It is important to know the graphs of the monomial functions. You know that the graph of the function x is a straight line of slope 1 through the origin. The graph of x^2 was shown in Fig. 1.18 in the preceding section. All the monomial graphs x^n go through the origin (0, 0) and the point (1, 1). If n is even, the graph of x^n goes through (−1, 1) while, if n is odd, the graph goes through (−1, −1). In Fig. 1.21 the graphs are indicated on one set of axes for easy comparison. Note in particular that the larger the value of n, the closer the graph is to the x-axis for $-1 < x < 1$. For example, $(\frac{1}{2})^4 < (\frac{1}{2})^2$, so the graph of x^4 is closer to the x-axis where $x = \frac{1}{2}$ than the graph of x^2. If $|x| > 1$, then the larger the value of n, the farther the graph of x^n is from the x-axis. For example, $2^5 > 2^3$, so x^5 is further from the x-axis than x^3 where $x = 2$.

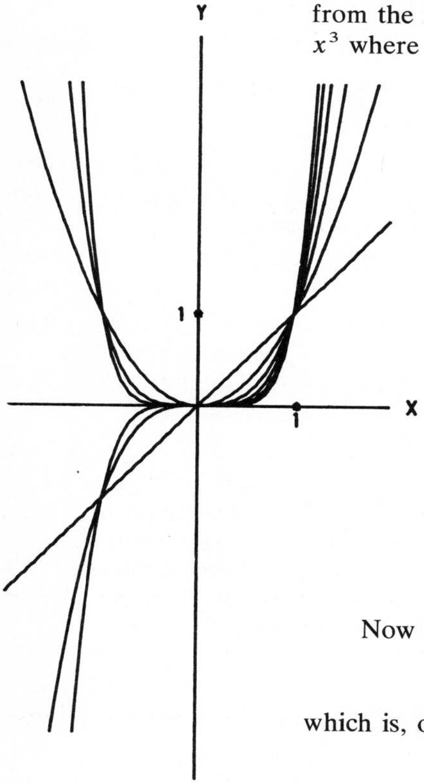

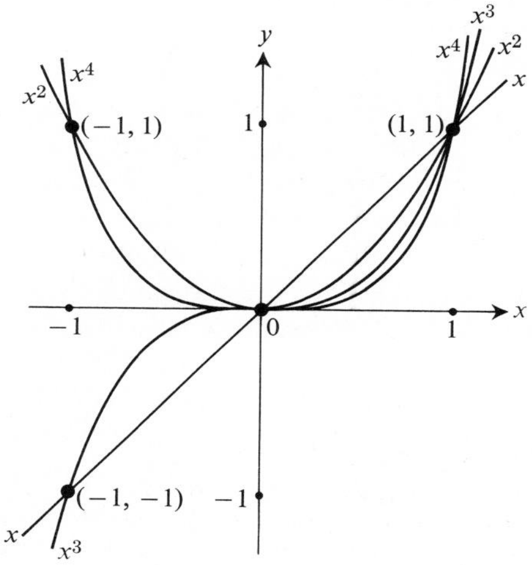

1.21. Computer-generated figure (left); artist's figure (right).

Now that you know the graph of x^n, you can easily sketch

$$y - k = (x - h)^n, \tag{1}$$

which is, of course, the graph of the function

$$y = f(x) = k + (x - h)^n.$$

You saw before that if you set $\Delta x = x - h$ and $\Delta y = y - k$, so that Eq. (1) becomes $\Delta y = (\Delta x)^n$, then you can sketch by translating to new $\Delta x, \Delta y$-axes at (h, k). In this graph-sketching context, we shall drop the Δ-notation and use more conventional notation,

Translating axes to (h, k)

$$\bar{x} = x - h, \qquad \bar{y} = y - k;$$

so Eq. (1) becomes $\bar{y} = \bar{x}^n$. The graph of $y + 2 = (x - 3)^2$ is shown in Fig. 1.22. We translate to $\bar{x}, \bar{y}$-axes at the point (3, −2), and our graph becomes $\bar{y} = \bar{x}^2$.

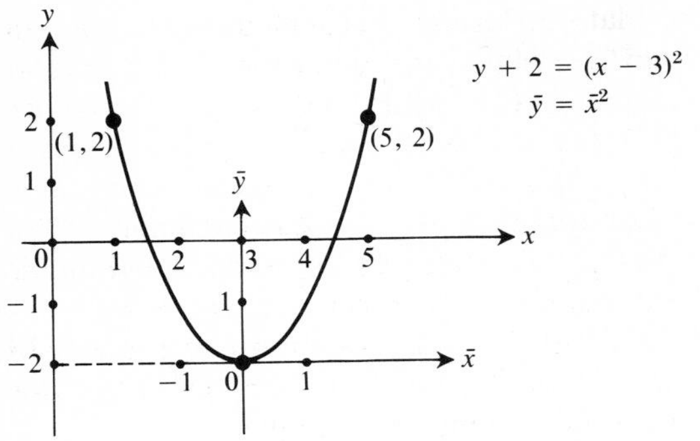

1.22

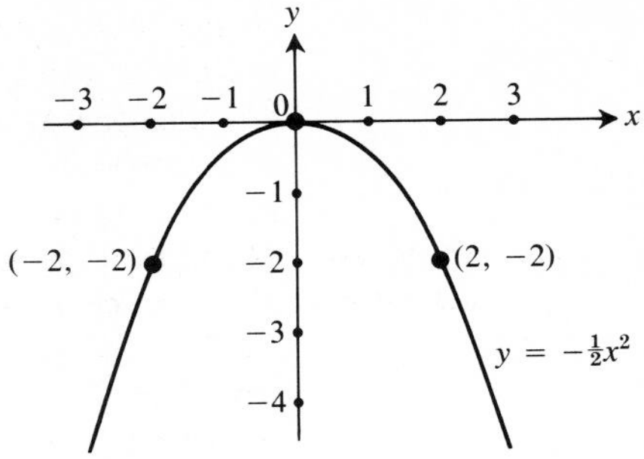

1.23

The graph of cx^n is much like the graph of x^n if $c > 0$. The points on cx^n are simply c times as far from the x-axis for each x-value as for x^n. Of course, if $c < 0$, then the graph is thrown to the other side of the x-axis. For example, the graph of $(-\frac{1}{2})x^2$ is shown in Fig. 1.23; it opens downward rather than upward.

Why are the graphs of monomial functions important? For the functions of most importance to us, we shall soon see that a very small piece of the graph near a point (h, k) on the graph looks a lot like a straight-line graph. Near that point, it can be *approximated* quite well by the line graph. We give an illustration of this in Fig. 1.24. Suppose the tangent line to the graph of f at (h, k) has slope m. Then, choosing $\bar{x},\bar{y}$-axes at (h, k), we find that the line $\bar{y} = m\bar{x}$ approximates the graph closely very near (h, k). In x,y-coordinates, the equation $\bar{y} = m\bar{x}$ becomes $y - k = m(x - h)$ or $y = k + m(x - h)$, which is the equation of the tangent line to the graph in x,y-coordinates. This problem of finding the tangent line is central to differential calculus, as we shall see in the following chapters.

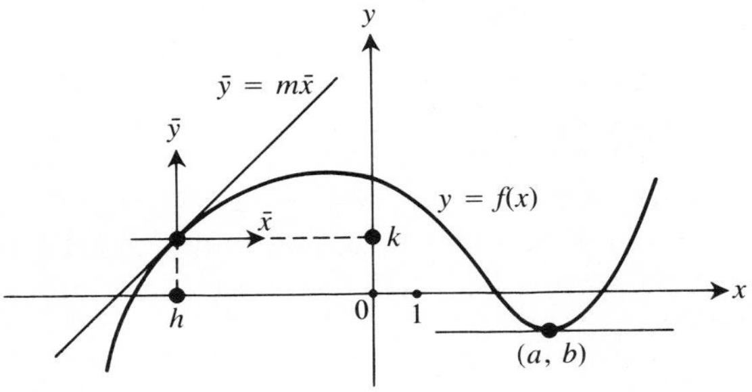

1.24

Finally, if the graph of f has a horizontal tangent of slope $m = 0$ at (a, b), as in Fig. 1.24, then you would like to know just how "flat" the graph is at (a, b). That is, is it as flat as a multiple of x^4 is at $(0, 0)$, or only as flat as a multiple of x^2? You will see much later that, for many important functions, you can measure how "flat" the graph is at such a point by finding values of c and n such that $\bar{y} = c\bar{x}^n$ gives the best monomial approximation to the graph.

1.5.2 Quadratic functions

A quadratic function f is one of the form $f(x) = ax^2 + bx + c$ where $a \neq 0$. Graphs of these functions are called *parabolas*. By completing the square and translating to $\bar{x},\bar{y}$-axes, you can put the equation $y = ax^2 + bx + c$ in the form $\bar{y} = d\bar{x}^2$ for some constant d. That is, the graph of a quadratic function is just a translation of the graph of a quadratic monomial function. The reason for this becomes clear in following an example.

Example 1 Let's sketch the graph of the function $y = f(x) = -2x^2 - 6x - 2$.

SOLUTION Dividing by the coefficient -2 of x^2 and then completing the square, you obtain

Sketching a quadratic function

$$-\frac{y}{2} = x^2 + 3x + 1,$$

$$-\frac{y}{2} + \frac{9}{4} = \left(x + \frac{3}{2}\right)^2 + 1.$$

Now move all the constant terms to the lefthand side, obtaining

$$-\frac{y}{2} + \frac{9}{4} - 1 = \left(x + \frac{3}{2}\right)^2 \quad \text{or} \quad -\frac{y}{2} + \frac{5}{4} = \left(x + \frac{3}{2}\right)^2.$$

Finally, multiply back through by the -2,

$$y - \tfrac{5}{2} = -2(x + \tfrac{3}{2})^2.$$

Now set $\bar{x} = x + \frac{3}{2}$ and $\bar{y} = y - \frac{5}{2}$, which amounts to translating axes to $(-\frac{3}{2}, \frac{5}{2})$. The equation becomes $\bar{y} = -2\bar{x}^2$. The graph is shown in Fig. 1.25. ‖

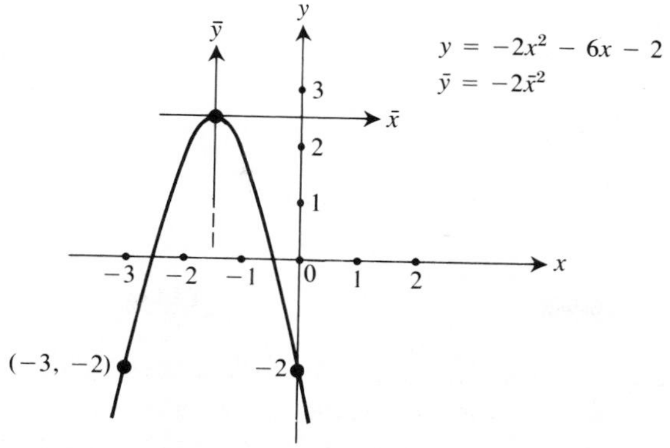

1.25

SUMMARY

1. *See Fig. 1.21 for the graphs of the monomial functions* $x, x^2, x^3, \ldots$.
2. *The graph of* $y - k = (x - h)^n$ *looks like the graph of* $y = x^n$, *but with the origin translated to* (h, k).
3. *Every quadratic equation* $y = ax^2 + bx + c$, *where* $a \neq 0$, *has a parabola as graph. The parabola can be sketched by completing the square on the x-terms and translating axes to put the equation in the form*

$$\bar{y} = d\bar{x}^2.$$

EXERCISES

In Exercises 1 *through* 14, *sketch the graph of the indicated function.*

1. $-x^3$ **2.** $x^2 + 3$ **3.** $-x^4$ **4.** $x^2/2$

5. $-x^5/3$ **6.** $4 + (x - 2)^2$ **7.** $4x^3$ **8.** $(x + 1)^3 - 3$

9. $-1 - (x + 5)^4$ **10.** $x^2 + 2x + 1$ **11.** $x^2 - 4x + 3$ **12.** $-x^2 - 6x + 5$

13. $2x^2 - 4x + 6$ **14.** $-3x^2 + 6x - 12$

15. Sketch the graph of f where $f(x) = |x|$. Can you find *one single* line graph that approximates $f(x)$ well for a short distance on *both* sides at $x = 0$?

exercise sets for chapter 1

review exercise set 1.1

1. a) Find the directed length Δx from -2 to 5.
b) Sketch all points (x, y) in the plane that satisfy $x > y + 1$.

2. a) Find the distance between $(2, -1)$ and $(-4, 7)$.
b) Find the midpoint of the line segment joining $(-1, 3)$ and $(3, 9)$.

3. a) Find the equation of the circle with center $(2, -1)$ and passing through $(4, 6)$.
b) Sketch all points (x, y) in the plane such that

$$(x - 1)^2 + (y + 2)^2 \leq 4.$$

4. a) Find the slope of the line joining $(-1, 4)$ and $(3, 7)$.
b) Find the slope of a line that is perpendicular to the line through $(4, -2)$ and $(-5, -3)$.

5. a) Find the equation of the line through $(-4, 2)$ and $(-4, 5)$.
b) Find the equation of the line through $(-1, 2)$ and parallel to the line $x - 3y = 7$.

6. a) Find the x-intercept and y-intercept of the line $3x + 4y = 12$.
b) Find the point of intersection of the lines $x - 3y = 7$ and $2x - 5y = 4$.

7. Let

$$f(x) = \frac{x^2 - 3x + 2}{x^2 - 5x}.$$

a) Find the domain of f. b) Find $f(-2)$.

8. Sketch the graph of the function $f(x) = 1/x^2$.

9. Sketch the graph of the function $3 - (x + 4)^3$.

10. Sketch the graph of the function $2x^2 + 8x - 6$.

review exercise set 1.2

1. a) Sketch on the line all x such that $|x - 1| \leq 2$.
   b) Find the midpoint of the interval $[-5.3, 2.1]$.
2. a) Find the distance from $(-6, 3)$ to $(-1, -4)$.
   b) Sketch all points (x, y) in the plane such that $x \leq y$ and also $x \geq 1$.
3. a) Find the equation of the circle with $(-2, 4)$ and $(4, 6)$ as endpoints of a diameter.
   b) Find the center and radius of the circle $x^2 + y^2 - 6x + 8y = 11$.
4. Find c such that the line through $(-1, c)$ and $(4, -6)$ is perpendicular to the line through $(-2, 3)$ and $(4, 7)$,
5. a) Find the equation of the line through $(-1, 4)$ and $(3, 5)$.
   b) Find the equation of the vertical line through $(3, -7)$.
6. a) Find the equation of the line through $(-1, 3)$ with y-intercept 5.
   b) Find the equation of the line through $(2, 4)$ parallel to the line through $(0, 5)$ and $(2, -3)$.
7. Let $f(x) = \sqrt{25 - x^2}$.
   a) Find the domain of f. b) Find $f(3)$.
   c) Sketch the graph of f.
8. Express the distance from the origin to a point (x, y) on the line $2x - 3y = 7$ as a function of x only.
9. Sketch the graph of the function $2 + (x - 1)^4$.
10. Sketch the graph of the function $4 - 2x^2$.

more challenging exercises 1

1. Show that, for all real numbers a and b,
   a) $|a + b| \leq |a| + |b|$,
   b) $|a - b| \geq |a| - |b|$.
2. Prove that, for any numbers a_1, a_2, b_1, and b_2, you have
$$(a_1a_2 + b_1b_2)^2 \leq (a_1{}^2 + b_1{}^2)(a_2{}^2 + b_2{}^2).$$
3. Prove algebraically from Exercise 2 and the formula for distance that the distance from (x_1, y_1) to (x_3, y_3) is greater than or equal to the sum of the distance from (x_1, y_1) to (x_2, y_2) and the distance from (x_2, y_2) to (x_3, y_3). This is known as the *triangle inequality* in the plane. [*Hint.* Let $a_1 = x_2 - x_1$, $a_2 = x_3 - x_2$, $b_1 = y_2 - y_1$, and $b_2 = y_3 - y_2$, so that $x_3 - x_1 = a_2 - a_1$ and $y_3 - y_1 = b_2 - b_1$.]
4. Show that if two circles
$$x^2 + y^2 + a_1x + b_1y = c_1$$
and
$$x^2 + y^2 + a_2x + b_2y = c_2$$
intersect in two points, the line through those points of intersection is
$$(a_2 - a_1)x + (b_2 - b_1)y = c_2 - c_1.$$
5. Find the distance from the point $(-3, 4)$ to the line
$$5x - 12y = 2.$$
6. Solve the inequality $x^2 + 4x < 1$ for x.
7. Find the equation of the smaller circle tangent to both coordinate axes and passing through the point $(-3, 6)$.
8. Find the distance between the lines
$$x - 2y = 15 \quad \text{and} \quad x - 2y = -3.$$
9. Find the minimum distance between the circles
$$x^2 + y^2 - 2x + 4y = 139$$
and
$$x^2 + y^2 + 4x - 6y = 3.$$
10. If $f(x) = (2x - 7)/(x + 3)$, find a function g such that $g(f(x)) = x$ for all x in the domain of f.

the derivative

2

You spent six years, in grades one through six, learning to handle real-number arithmetic, and then spent more years solving problems using that arithmetic. It is true that you learned sophistications, like carrying x along as some number to be determined; but the arithmetic operations remained really your only tools for problem solving. You will continue to use arithmetic in this course, but you will learn a powerful new tool:

computing a limit

as well. It takes much less time to master this new tool than to learn arithmetic. Four weeks from now, you will be quite expert at computing important limits known as derivatives.

The importance of calculus lies in its usefulness in the study of dynamic situations, where quantities are changing, as opposed to static situations, where quantities remain constant. To illustrate, suppose a car is driven at a constant speed of 30 mph. Then in four hours, the distance the car has traveled is $30 \cdot 4 = 120$ miles. That is just arithmetic. It was easy to find the distance, for the speed was always the same. But suppose the speed is not always the same; perhaps at time t hours the speed is $30\sqrt{t}$ mph, so it is traveling 30 mph after one hour but 60 mph after four hours, etc. Then it is not so easy to find how far the car has gone after four hours. By the end of the term, you will be able to solve this problem.

In this chapter, you will learn how to solve the converse type of problem: Suppose you know that after t hours a car has gone $15t^2$ miles, and want to find a formula for the speed of the car at time t. Surely if you know exactly where the car is at every instant, its speed should be determined.

2.1 THE SLOPE OF A GRAPH

2.1.1 The slope of a tangent line*

Suppose a car has gone a distance $s = f(t)$ at time t. Let's not worry about the units for s and t; they might be yards and seconds, for example. If the car moves with a *constant* speed of 2, then $s = 2t$, so the graph of this motion is a straight line of slope 2. Any time the graph of distance against time is a straight line of positive slope, the speed is constant and equal to that slope.

Figure 2.1 shows the graph of a function $s = f(t)$, which might give the distance s a car has gone at time t, starting from rest and accelerating gradually. The *increase in steepness* of the graph in Fig. 2.1 as t increases is due to the fact that the car is accelerating; so s is increasing faster and faster as t increases. How could you find the speed of the car at a certain time t_1,

* We do not attempt to define the *tangent line* to a curve at a point. Your work in geometry has given you an intuitive grasp of this idea, which will suffice.

when $s = s_1$? Suppose the slope of the tangent line to the graph of f at (t_1, s_1) is m_{tan}. On this tangent line, s increases m_{tan} units for each unit of increase of t. Since the line and graph have the same steepness at (t_1, s_1), the remarks above show that the distance s for the car is also increasing at the rate of m_{tan} units per unit increase of t *at that instant* t_1. That is, the speedometer reads m_{tan} at time t_1. In summary, *the slope of the tangent line gives the rate of increase of s with respect to the time t at that instant t_1.*

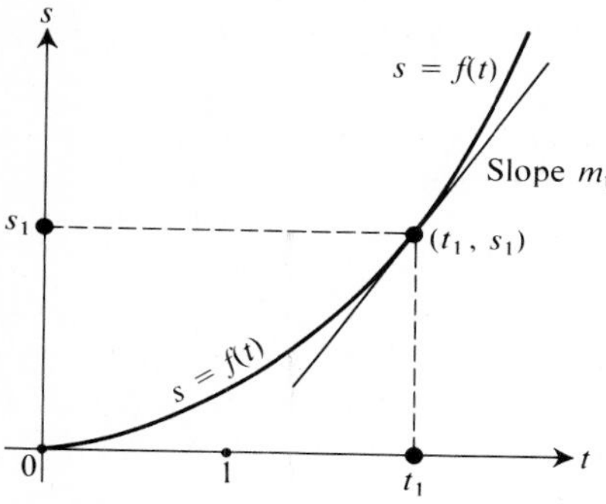

2.1

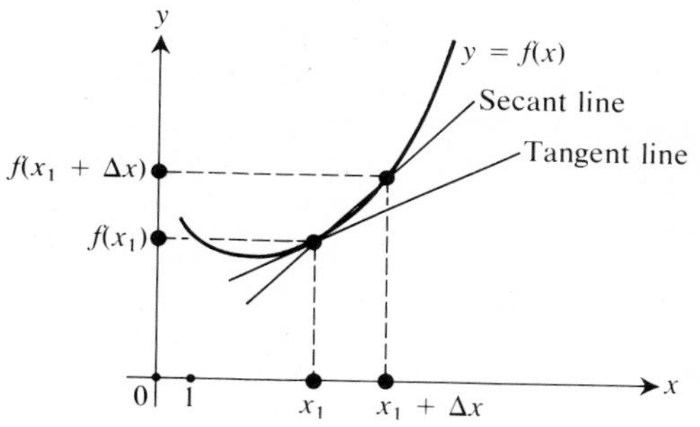

2.2

The speed of a moving body is not the only rate of change that is of interest. A manufacturer might want to know the rate at which his costs C change as the number x of items produced increases. This would be the rate of change of C with respect to x; it might be $-2¢$ per item for some value x_1 of x. You will encounter many other illustrations in the text. The mathematical formulation for all of these rates of change is as follows:

Rate of change

If $y = f(x)$ and the graph of f has a tangent line where $x = x_1$, then the slope of this tangent line gives the instantaneous rate of increase of y with respect to x at that point x_1.

Some graphs have breaks, or sharp points where there are no tangent lines. But suppose a graph does have a tangent line where $x = x_1$. How could you compute the slope m_{tan} of this tangent line? This seems to be a tough problem. Let's start by *approximating* m_{tan}, using the slope of a secant line as shown in Fig. 2.2. This secant line has slope $m_{sec} = \Delta y/\Delta x$ and passes through the points $(x_1, f(x_1))$ and $(x_1 + \Delta x, f(x_1 + \Delta x))$. The formula for the slope of a line through two points shows that

Finding m_{sec}

$$m_{sec} = \frac{\Delta y}{\Delta x} = \frac{f(x_1 + \Delta x) - f(x_1)}{\Delta x}. \tag{1}$$

Example 1 Let $f(x) = x^3 + 2x$ and $x_1 = 2$ and $\Delta x = 0.01$. The approximation (1) to the slope $m_{\tan}$ of the tangent line at (2, 12) is

$$m_{\sec} = \frac{[(2.01)^3 + 2(2.01)] - [2^3 + 2 \cdot 2]}{0.01}$$

$$= \frac{12.140601 - 12}{0.01} = \frac{0.140601}{0.01} = 14.0601.$$

You will see in Section 2.3 that the slope $m_{\tan}$ is actually 14, so 14.0601 is not a bad approximation. A calculator is handy for computing $m_{\sec}$. ||

The smaller the value of Δx (of course, $\Delta x = 0$ is not allowed), the better you would expect $m_{\sec}$ in Eq. (1) to approximate $m_{\tan}$. This is indicated by Fig. 2.3.

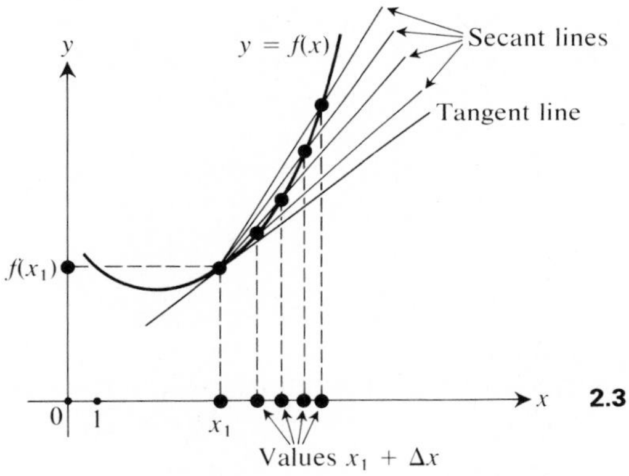

2.3

Rather than use a particular value of Δx as in Example 1, you could compute $m_{\sec}$ as a function of Δx, and then put in various values of Δx. The introduction to this chapter posed the problem of finding the speed of a car if the distance s it has traveled is given by the function $s = f(t) = 15t^2$. Let's return to t,s-notation and use (1) for this function. We have

$$m_{\sec} = \frac{f(t_1 + \Delta t) - f(t_1)}{\Delta t} = \frac{15(t_1 + \Delta t)^2 - 15t_1^2}{\Delta t}$$

$$= \frac{15(t_1^2 + 2t_1\,\Delta t + (\Delta t)^2) - 15t_1^2}{\Delta t} = \frac{15 \cdot 2t_1\,\Delta t + 15(\Delta t)^2}{\Delta t}$$

$$= \frac{\Delta t(15 \cdot 2t_1 + 15 \cdot \Delta t)}{\Delta t} = 30t_1 + 15 \cdot \Delta t. \qquad (2)$$

Here is a table showing $m_{\sec}$ for a few values of Δt.

Δt	m_{sec}	Δt	m_{sec}
-0.01	$30t_1 - 0.15$	0.001	$30t_1 + 0.015$
-0.001	$30t_1 - 0.015$	0.0001	$30t_1 + 0.0015$

As indicated by Fig. 2.3, the smaller the value of Δt, the closer m_{sec} should be to m_{tan}. As Δt gets smaller and smaller, the values $30t_1 + 15 \cdot \Delta t$ get closer and closer to $30t_1$. In symbols, this is written

$$\lim_{\Delta t \to 0} (30t_1 + 15 \cdot \Delta t) = 30t_1,$$

and is read, "*the limit as* Δt *approaches zero of* $30t_1 + 15 \cdot \Delta t$ *equals* $30t_1$." The *exact* value of m_{tan} is thus $30t_1$ and was found by computing this limit as $\Delta t \to 0$. The crucial step in computing the limit as $\Delta t \to 0$ was the legitimate cancellation of the nonzero factor Δt from the denominator in (2) with the factor in the numerator. (The symbol $\Delta t \to 0$ signifies that Δt becomes very, very small, but is *always nonzero.*) It is hard to discover what number a quotient may approach as both the numerator and denominator approach zero, and the cancellation of Δt removed this obstacle. To finish the car problem, the speed of the car at time t_1 is $30t_1$ (units distance)/(unit time). Since t_1 could be *any* time, we obtain the speed function $30t$.

Summarizing our work in x,y-notation, we have discovered that

$$m_{tan} = \lim_{\Delta x \to 0} m_{sec} = \lim_{\Delta x \to 0} \frac{f(x_1 + \Delta x) - f(x_1)}{\Delta x}. \tag{3}$$

Example 2 Let's find the slope m_{tan} of the tangent line to the graph of $f(x) = 4x - 3x^2$, where $x = x_1$.

SOLUTION We use Eq. (3) with $f(x) = 4x - 3x^2$. Then

$$\begin{aligned} m_{tan} &= \lim_{\Delta x \to 0} \frac{f(x_1 + \Delta x) - f(x_1)}{\Delta x} \\ &= \lim_{\Delta x \to 0} \frac{4(x_1 + \Delta x) - 3(x_1 + \Delta x)^2 - (4x_1 - 3x_1^2)}{\Delta x} \\ &= \lim_{\Delta x \to 0} \frac{4x_1 + 4(\Delta x) - 3x_1^2 - 6x_1(\Delta x) - 3(\Delta x)^2 - 4x_1 + 3x_1^2}{\Delta x} \\ &= \lim_{\Delta x \to 0} \frac{4(\Delta x) - 6x_1(\Delta x) - 3(\Delta x)^2}{\Delta x} \\ &= \lim_{\Delta x \to 0} \frac{\Delta x[4 - 6x_1 - 3(\Delta x)]}{\Delta x} \\ &= \lim_{\Delta x \to 0} [4 - 6x_1 - 3(\Delta x)] = 4 - 6x_1. \quad \| \end{aligned}$$

Example 3 We find $m_{\tan}$ for $f(x) = 1/(3x)$ at $x = x_1 \neq 0$.

SOLUTION This time Eq. (3) becomes

$$m_{\tan} = \lim_{\Delta x \to 0} \frac{\dfrac{1}{3(x_1 + \Delta x)} - \dfrac{1}{3x_1}}{\Delta x}$$

$$= \lim_{\Delta x \to 0} \frac{\dfrac{3x_1 - 3x_1 - 3(\Delta x)}{3(x_1 + \Delta x)(3x_1)}}{\Delta x}$$

$$= \lim_{\Delta x \to 0} \frac{-3(\Delta x)}{(\Delta x)[3(x_1 + \Delta x)(3x_1)]}$$

$$= \lim_{\Delta x \to 0} \frac{-3}{3(x_1 + \Delta x)(3x_1)} = \frac{-3}{(3x_1)(3x_1)} = \frac{-1}{3x_1^2}. \quad \|$$

2.1.2 Another numerical estimate for the slope of a graph

Referring to Fig. 2.4, we note that the chord from $x_1 - \Delta x$ to $x_1 + \Delta x$ is apt to be more nearly parallel to the tangent line than is the secant line. Let's call this slope of the chord m_{chord}. The figure indicates that m_{chord} is apt to be a better approximation to $m_{\tan}$ than $m_{\sec}$ is for the same value of Δx. Since the chord goes through the points

$$(x_1 - \Delta x, f(x_1 - \Delta x)) \quad \text{and} \quad (x_1 + \Delta x, f(x_1 + \Delta x)),$$

its slope is

$$m_{\text{chord}} = \frac{f(x_1 + \Delta x) - f(x_1 - \Delta x)}{2 \cdot \Delta x}. \tag{4}$$

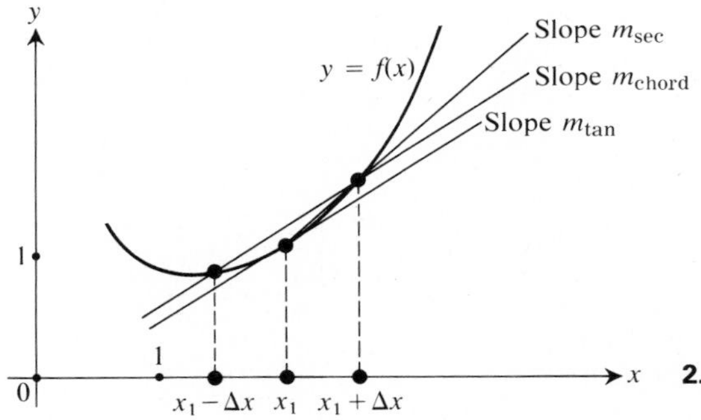

2.4

Example 4 Let's use Eq. (4) to approximate $m_{\tan}$ for the same function $f(x) = x^3 + 2x$, with $x_1 = 2$ and $x = 0.01$, that we used in Example 1.

SOLUTION The computation is

$$\frac{f(x_1+\Delta x)-f(x_1-\Delta x)}{2\cdot\Delta x}=\frac{[(2.01)^3+2(2.01)]-[(1.99)^3+2(1.99)]}{2\cdot(0.01)}$$

$$=\frac{12.140601-11.860599}{0.02}$$

$$=\frac{0.280002}{0.02}$$

$$=14.0001.$$

The correct answer is actually 14, so this approximation is significantly better than the approximation 14.0601 found in Example 1. And the computation is easy if you have a calculator handy. ‖

Printout 2.1 shows 20 values of m_{sec} and the corresponding 20 values of m_{chord} for

$$\Delta x=\frac{1}{2},\ \frac{1}{4},\ \frac{1}{8},\ \frac{1}{16},\ \ldots,\ \frac{1}{2^{20}}=\frac{1}{1{,}048{,}576}\approx 0.0000009537,$$

and the function $f(x)=\sqrt{x^2+16}$ at $x_1=2$. These results were obtained using a computer. A listing of the program SECHORD used to find the data is given in Appendix 1. See how much faster the slopes m_{chord} converge! From this data, it appears the value of $m_{\tan}$ is very close to 0.4472135955. Formula (4) is a very useful way to approximate $m_{\tan}$ using a hand calculator or a computer.

Δx value	m_{sec}	m_{chord}
.5000000000	.4897092221	.4449886934
.2500000000	.4690159307	.4466552771
.1250000000	.4582542240	.4470738849
.0625000000	.4527688296	.4471786597
.0312500000	.4499999460	.4472048610
.0156250000	.4486089543	.4472114119
.0078125000	.4479118208	.4472130496
.0039062500	.4475628446	.4472134590
.0019531250	.4473882542	.4472135614
.0009765625	.4473009334	.4472135870
.0004882813	.4472572666	.4472135934
.0002441406	.4472354316	.4472135950
.0001220703	.4472245137	.4472135954
.0000610352	.4472190546	.4472135955
.0000305176	.4472163251	.4472135955
.0000152588	.4472149603	.4472135955
.0000076294	.4472142779	.4472135955
.0000038147	.4472139367	.4472135955
.0000019073	.4472137660	.4472135955
.0000009537	.4472136807	.4472135955

Printout 2.1 $f(x)=\sqrt{x^2+16}$ at $x_1=2$

SUMMARY

1. *The slope* $m_{\tan}$ *of the tangent line to the graph* $y = f(x)$, *where* $x = x_1$ *is the instantaneous rate at which y is increasing with respect to x at the instant when* $x = x_1$.
2. *If* $s = f(t)$ *gives the distance s traveled at time t, then* $m_{\tan}$ *where* $t = t_1$ *is the velocity at time* t_1.
3. *The slope of the secant to the graph from* x_1 *to* $(x_1 + \Delta x)$ *is*

$$m_{\sec} = \frac{f(x_1 + \Delta x) - f(x_1)}{\Delta x}.$$

4. *The slope of the tangent to the graph at* x_1 *is*

$$m_{\tan} = \lim_{\Delta x \to 0} \frac{f(x_1 + \Delta x) - f(x_1)}{\Delta x}.$$

5. *The slope of the chord to the graph from* $x_1 - \Delta x$ *to* $x_1 + \Delta x$ *is*

$$m_{\text{chord}} = \frac{f(x_1 + \Delta x) - f(x_1 - \Delta x)}{2 \cdot \Delta x}.$$

6. *In general,* m_{chord} *is apt to be a better approximation to* $m_{\tan}$ *than* $m_{\sec}$ *is for the same value* Δx.

EXERCISES

In Exercises 1 through 4, find the value of $m_{\sec}$ *given by (1) and the value* m_{chord} *given by (4) for the indicated function, point, and increment.*

1. $f(x) = x^2,\ x_1 = 4,\ \Delta x = 0.01$

2. $f(x) = 1/x,\ x_1 = 2,\ \Delta x = 0.1$

3. $f(t) = t^3 - 3t,\ t_1 = 1,\ \Delta t = 0.1$

4. $f(u) = u + \dfrac{1}{u},\ u_1 = -1,\ \Delta u = -0.001$

In Exercises 5 through 9, find the exact value of $m_{\tan}$ *at the point* x_1 *by computing the limit shown in (3) as* $\Delta x \to 0$. *Compare the answers of Problems 5 through 8 with your answers for Problems 1 through 4.*

5. $f(x) = x^2,\ x_1 = 4$

6. $f(x) = 1/x,\ x_1 = 2$

7. $f(x) = x^3 - 3x,\ x_1 = 1$

8. $f(x) = x + \dfrac{1}{x},\ x_1 = -1$

9. $f(x) = \sqrt{x},\quad x_1 = 4$. [*Hint.* Multiplication of numerator and denominator of $(\sqrt{a} - \sqrt{b})/\Delta x$ by $(\sqrt{a} + \sqrt{b})$ yields $(a - b)/[\Delta x(\sqrt{a} + \sqrt{b})]$.

10. If s is distance traveled and t is time, then $\Delta s/\Delta t$ has the interpretation of the *average speed* over the time interval Δt. Suppose an object travels so that after t hours it has gone $s = f(t) = 3t^2 + 2t$ miles for $t \geq 0$.

a) Find the average speed of the object during the two-hour time interval from $t = 3$ to $t = 5$.

b) Find the average speed of the object during the one-hour time interval from $t = 3$ to $t = 4$.

c) Find the average speed of the object during the half-hour time interval from $t = 3$ to $t = 7/2$.

d) On the basis of the preceding, guess the actual speed of the object at time $t = 3$.

11. With reference to Exercise 10, find, from the expression for $m_{\tan}$ as a limit, the exact speed at time $t = 3$.

calculator exercises

Use your calculator and (4) to approximate $m_{\tan}$ for the given function at the indicated point. You choose Δx. Functions we have not discussed are defined for you by your calculator.

12. $\sin x$ at $x_1 = 0$ (Use radian measure.)

13. $\sqrt{x^2 + 2x - 3}$ at $x_1 = 2$

14. 3^x at $x_1 = 2$

15. x^x at $x_1 = 1.5$

2.2 LIMITS

2.2.1 The Notion of $\lim_{x \to x_1} f(x)$

Let the graph of f have a tangent line where $x = x_1$. From the last section, the slope $m_{\tan}$ of the tangent line is

$$m_{\tan} = \lim_{\Delta x \to 0} \frac{f(x_1 + \Delta x) - f(x_1)}{\Delta x}. \quad (1)$$

The function

$$h(\Delta x) = \frac{f(x_1 + \Delta x) - f(x_1)}{\Delta x}$$

involves just the single variable Δx, for x_1 is some constant such as -1 or 2 or π. Now $h(\Delta x)$ is not defined where $\Delta x = 0$. One of the main purposes of a limit is to describe the behavior of a function near a point where the function is not defined. As we continue the discussion of limits, we use just x rather than Δx, and talk about $\lim_{x \to x_1} f(x)$.

Example 1 Let's try to find

$$\lim_{x \to 2} \frac{x^2 - 4}{x - 2}.$$

That is, let's try to discover whether $(x^2 - 4)/(x - 2)$ gets very close to some value L as x gets very close to 2.

SOLUTION Note that $(x^2 - 4)/(x - 2)$ is not defined if $x = 2$. For all $x \neq 2$,

$$\frac{x^2 - 4}{x - 2} = \frac{(x - 2)(x + 2)}{x - 2} = x + 2;$$

so

$$\lim_{x \to 2} \frac{x^2 - 4}{x - 2} = \lim_{x \to 2} (x + 2) = 4,$$

since, if x is very close to 2, then $x + 2$ is very close to 4. ||

If $\lim_{x \to x_1} f(x) = L$, you could illustrate this by computing $f(x)$ for "random" x's closer and closer to x_1 but never equal to x_1. The values $f(x)$

then get close to L. This is illustrated in Printouts 2.2 and 2.3. The values given in the tables were obtained using a computer. It first chose a random x such that $x_1 - \frac{1}{2} < x < x_1 + \frac{1}{2}$ and printed $f(x)$. Then it chose a random x such that $x_1 - \frac{1}{4} < x < x_1 + \frac{1}{4}$ and printed $f(x)$, then again for $x_1 - \frac{1}{8} < x < x_1 + \frac{1}{8}$, etc., for twenty choices. Printout 2.2 is for $f(x) = (x^2 - 4)/(x - 2)$ as $x \to 2$, as in Example 1. The printed values get close to 4. Printout 2.3 shows data for $f(x) = (\sin x)/x$ as $x \to 0$. From Printout 2.3, it seems likely that $\lim_{x\to 0} (\sin x)/x = 1$. The program LIMIT, which gave these results, appears in Appendix 1.

x	$f(x)$
1.5177148469	3.5177148469
2.1052473784	4.1052473784
2.0572635830	4.0572635830
1.9539747313	3.9539747313
1.9979096241	3.9979096241
1.9877050277	3.9877050277
1.9948899690	3.9948899690
2.0018602884	4.0018602884
1.9988715949	3.9988715949
2.0009718261	4.0009718261
2.0000061166	4.0000061166
2.0001575557	4.0001575557
1.9998977055	3.9998977055
1.9999785303	3.9999785303
2.0000297731	4.0000297731
1.9999895183	3.9999895183
1.9999981885	3.9999981885
2.0000015930	4.0000015930
1.9999990185	3.9999990187
2.0000005198	4.0000005195

Printout 2.2 $f(x) = \dfrac{x^2 - 4}{x - 2}$

x	$f(x)$
−.4671424278	.9640244436
−.1979362558	.9934829860
−.0239035880	.9999047725
−.0416905480	.9997103415
−.0164257088	.9999550333
−.0067960371	.9999923023
.0001688687	.9999999952
.0022185121	.9999991797
−.0007549826	.9999999050
.0004569862	.9999999652
−.0000029603	1.0000000000
−.0001811096	.9999999945
.0000289490	.9999999999
.0000020952	1.0000000000
−.0000209326	.9999999999
.0000131930	1.0000000000
.0000068695	1.0000000000
−.0000007991	1.0000000000
.0000016917	1.0000000000
.0000009188	1.0000000000

Printout 2.3 $f(x) = \dfrac{\sin x}{x}$

Intuitively, $\lim_{x\to x_1} f(x) = L$ means that you can make $f(x)$ as *close* to L as you wish by making x sufficiently *close* to but different from x_1. This is a vague statement, for what does "*close*" mean? You may think that $f(x)$ is close to L if $L - 0.1 < f(x) < L + 0.1$, while a friend may say, "No, I want to have $L - 0.00001 < f(x) < L + 0.00001$." It is only proper to say $\lim_{x\to x_1} f(x) = L$ if *everyone* can be satisfied. So if someone demands to have $L - \epsilon < f(x) < L + \epsilon$ for some $\epsilon > 0$, be it 0.1 or 0.00001, you must be sure that this will be true as long as x is within a certain distance, perhaps 0.05 or 0.003, of x_1 but not equal to x_1. That is, you must be able to find a $\delta > 0$ such that $L - \epsilon < f(x) < L + \epsilon$ will be true if $x_1 - \delta < x < x_1 + \delta$ but $x \neq x_1$. The smaller the ϵ the person wants, the smaller you expect δ will have to be. Let's write this down as a definition of $\lim_{x\to x_1} f(x) = L$. [Of course, to talk about $\lim_{x \to x_1} f(x)$ at all, you want to have the domain of f contain points $x \neq x_1$ arbitrarily close to x_1. It would be absurd to talk about $\lim_{x \to -3} \sqrt{x}$. The first sentence of the definition takes care of this.]

Definition 2.1 Suppose the domain of f contains points x arbitrarily close to x_1, but different from x_1. Then $\lim_{x \to x_1} f(x) = L$ provided that, for each $\epsilon > 0$, there exists $\delta > 0$ such that $L - \epsilon < f(x) < L + \epsilon$ for any $x \neq x_1$ in the domain of f such that $x_1 - \delta < x < x_1 + \delta$.

The requirement $L - \epsilon < f(x) < L + \epsilon$ can also be expressed $|f(x) - L| < \epsilon$, while the two conditions $x_1 - \delta < x < x_1 + \delta$ but $x \neq x_1$ can be written as the single condition $0 < |x - x_1| < \delta$. This ϵ,δ-characterization is very important in theoretical work in mathematics. We told you about it here in case you should meet it in later courses. Then you would be seeing it for the second time, so it would seem easier to you. There is almost no ϵ,δ-work done in this text.

Example 2 Let's use the ϵ,δ-characterization to demonstrate that $\lim_{x \to 3} (2x - 2) = 4$. This surely should be true; if x is close to 3, then $f(x) = 2x - 2$ is close to $6 - 2 = 4$.

SOLUTION Let $\epsilon > 0$ be given; you want to be sure that

$$4 - \epsilon < f(x) < 4 + \epsilon. \tag{2}$$

You should mark $4 - \epsilon$ and $4 + \epsilon$ on the y-axis (see Fig. 2.5), for this is where $f(x)$ is plotted. Now you have to find $\delta > 0$ and mark $3 - \delta$ and $3 + \delta$ on the x-axis, so that (2) is true if $3 - \delta < x < 3 + \delta$, $x \neq 3$. Since $f(x) = 2x - 2$ has as graph a line of slope 2, each unit of change on the x-axis produces two units of change of $f(x)$ on the y-axis. Thus a change of ϵ on the y-axis is produced by a change of only $\epsilon/2$ on the x-axis, so you may let $\delta = \epsilon/2$. This was just a geometric argument.

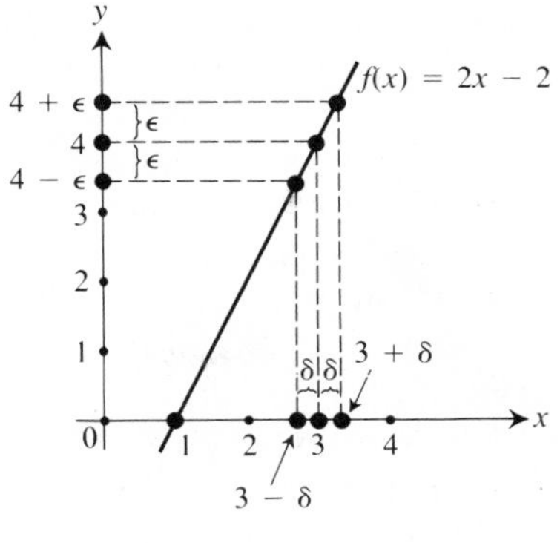

2.5

You can arrive at the same result algebraically as follows. You need

Finding $\delta > 0$, given $\epsilon > 0$

$$4 - \epsilon < 2x - 2 < 4 + \epsilon,$$

which can be written as

$$6 - \epsilon < 2x < 6 + \epsilon$$

or

$$3 - \frac{\epsilon}{2} < x < 3 + \frac{\epsilon}{2}.$$

Consequently, $\delta = \epsilon/2$ suffices. Of course, any smaller δ will work also. ||

Example 3 We give an ϵ,δ-demonstration that $\lim_{x\to 1} (5 - 7x) = -2$.

SOLUTION Let $\epsilon > 0$ be given. You need to have

$$-2 - \epsilon < (5 - 7x) < -2 + \epsilon.$$

This is true if and only if

$$-7 - \epsilon < -7x < -7 + \epsilon,$$

or if and only if

$$1 + \frac{\epsilon}{7} > x > 1 - \frac{\epsilon}{7}.$$

Thus, we can take $\delta = \epsilon/7$, for if x is within $\epsilon/7$ of 1, then $5 - 7x$ is within ϵ of -2. ||

Example 4 Let's give an ϵ,δ-demonstration that $\lim_{x\to 0} |x|/x$ *does not exist.*

SOLUTION For *every* $\delta > 0$,

$$\frac{|x|}{x} = 1 \quad \text{if} \quad 0 < x < \delta$$

and

$$\frac{|x|}{x} = -1 \quad \text{if} \quad -\delta < x < 0.$$

Now 1 and -1 are two units apart, while for any possible limit L, the numbers $L - \epsilon$ and $L + \epsilon$ are 2ϵ units apart. Consequently if $\epsilon < 1$, it is impossible to have

$$L - \epsilon < \frac{|x|}{x} < L + \epsilon \quad \text{for all} \quad -\delta < x < \delta, \quad x \neq 0,$$

for *any* choice of L and $\delta > 0$. Thus given $\epsilon = \frac{1}{2}$, there exists no $\delta > 0$ such that

$$L - \epsilon < \frac{|x|}{x} < L + \epsilon \quad \text{for} \quad -\delta < x < \delta, \quad x \neq 0,$$

no matter what L might be. This contradicts Definition 2.1, which asserts that for *each* $\epsilon > 0$, in particular for $\epsilon = \frac{1}{2}$, such $\delta > 0$ should exist. ||

2.2.2 Computing limits

Here is a theorem that you will use frequently, often without realizing it, in computing limits.

Theorem 2.1 *If* $\lim_{x \to x_1} f(x) = L$ *and* $\lim_{x \to x_1} g(x) = M$, *where the domains of f and g contain common points arbitrarily close to* x_1 *but different from* x_1, *then*

$$\lim_{x \to x_1} (f(x) + g(x)) = L + M, \tag{3}$$

$$\lim_{x \to x_1} (f(x) \cdot g(x)) = L \cdot M, \tag{4}$$

$$\lim_{x \to x_1} \left(\frac{f(x)}{g(x)}\right) = \frac{L}{M} \quad \text{if} \quad M \neq 0. \tag{5}$$

To illustrate, if $f(x)$ is very near 2 and $g(x)$ very near 5 when x is very near $x_1 = -1$, then $f(x) + g(x)$ is very near $2 + 5 = 7$, and $f(x) \cdot g(x)$ very near $2 \cdot 5 = 10$, and $f(x)/g(x)$ very near $\frac{2}{5}$ when x is near -1. Theorem 2.1 is certainly intuitively obvious. You will find an ϵ,δ-proof in any text on advanced calculus, and in some freshman-level calculus texts.

Here is a sample of what you can do using (3), (4), and (5). Surely $\lim_{x \to x_1} x = x_1$, Exercise 1 asks you to give an ϵ,δ-argument. But then, from (4),

$$\lim_{x \to x_1} x \cdot x = x_1 \cdot x_1, \qquad \lim_{x \to x_1} x^3 = \lim_{x \to x_1} x^2 \cdot x = x_1^{\,2} \cdot x_1 = x_1^{\,3}, \qquad \text{etc.}$$

From (3),

$$\lim_{x \to x_1} (x^3 + x^2) = x_1^{\,3} + x_1^{\,2}.$$

Use of the theorem on limits

Also, if $f(x) = 3$ for all x, then surely $\lim_{x \to x_1} f(x) = 3$ (see Exercise 2). Using (4),

$$\lim_{x \to x_1} 3 \cdot x^2 = 3x_1^{\,2}.$$

Similar arguments show that if $f(x)$ is any polynomial function, then $\lim_{x \to x_1} f(x) = f(x_1)$. From (5), if $g(x)$ is also a polynomial function and $g(x_1) \neq 0$, then $\lim_{x \to x_1} f(x)/g(x) = f(x_1)/g(x_1)$. (Such a quotient of polynomial functions is a **rational function**.) So the computation of the limit of a rational function as $x \to x_1$ amounts to evaluation of the function at the point x_1 provided that x_1 does not make the denominator of the function zero. The only "bad" case for a rational function $f(x)/g(x)$ is the case where $g(x_1) = 0$, as illustrated in Example 1. When a denominator approaches zero at a point, one must try some algebraic trick, such as cancelling a factor from both numerator and denominator, to try to find the limit at that point.

Example 5 We have

$$\lim_{x\to 3} \frac{x^2 - 9}{x^2 - 4x + 3} = \lim_{x\to 3} \frac{(x-3)(x+3)}{(x-3)(x-1)} = \lim_{x\to 3} \frac{x+3}{x-1} = \frac{6}{2} = 3. \quad \|$$

Example 6 The limit

$$\lim_{x\to 4} \frac{x+5}{x-4}$$

does not exist, for the numerator approaches 9 while the denominator approaches 0. Thus the quotient becomes very large in absolute value as x approaches 4 (positively large if $x > 4$ and negatively large if $x < 4$). Symbolically,

$$\lim_{x\to 4} \left| \frac{x+5}{x-4} \right| = \infty. \tag{6}$$

Using ∞ to describe a limit

This does *not* mean that ∞ (read "*infinity*") is the limit, but rather that the limit does not exist *because* the quotient becomes large in size as x approaches 4. *The purpose of finding a limit is to describe the behavior of a function near a point, and* (6) *does that for us very neatly.* ‖

There are other obvious relations similar to (3), (4), and (5), such as

$$\text{if } \lim_{x\to x_1} f(x) = L > 0,$$

then

$$\lim_{x\to x_1} \sqrt{f(x)} = \sqrt{L}.$$

The only "bad" case, where you can't just evaluate functions, that we have seen is where a denominator approaches zero. Equation (1) shows that this "bad" case *always* occurs in the attempt to find $m_{\tan}$ as a limit. Here are more illustrations.

Example 7 We have

$$\lim_{x\to 5} \frac{x^2 - 25}{x+4} = \frac{0}{9} = 0, \qquad \lim_{x\to 3} \frac{x+2}{(x-3)^2} = \infty, \qquad \lim_{x\to 2} \frac{x-7}{(x-2)^2} = -\infty.$$

The second expression becomes *positively* infinite as x approaches 3, while the last one becomes *negatively* infinite as x approaches 2. ‖

SUMMARY

1. *Limits are used to study the behavior of a function near a point x_1 where the function may not be defined.*
2. $\lim_{x \to x_1} f(x) = L$ *means that for each* $\epsilon > 0$, *there exists some* $\delta > 0$ *such that* $|f(x) - L| < \epsilon$ *provided that* $0 < |x - x_1| < \delta$.
3. *If* $\lim_{x \to x_1} f(x) = L$ *and* $\lim_{x \to x_1} g(x) = M$, *then*

$$\lim_{x \to x_1} (f(x) + g(x)) = L + M, \qquad \lim_{x \to x_1} (f(x) \cdot g(x)) = L \cdot M,$$

$$\lim_{x \to x_1} \left(\frac{f(x)}{g(x)}\right) = \frac{L}{M} \quad \text{if } M \neq 0.$$

4. *Limits as $x \to x_1$ of functions we have encountered so far can be found by evaluating at x_1 as long as a denominator does not become zero at x_1. If a denominator becomes zero at x_1, try to cancel a factor of the denominator with one in the numerator.*
5. *The symbols ∞ and $-\infty$ are used where appropriate with limit expressions to describe the behavior of a function near a point.*

EXERCISES

In Exercises 1, 2, and 3, if $\epsilon > 0$ is given, find in terms of ϵ what size δ must be taken in the ϵ,δ-characterization of a limit to establish the limit.

1. $\lim_{x \to x_1} x = x_1$

2. $\lim_{x \to x_1} c = c$, where c in $\lim_{x \to x_1} c$ is the constant function f defined by $f(x) = c$ for all x

3. $\lim_{x \to -3} (14 - 5x) = 29$

4. Does it make sense to consider $\lim_{x \to 2} \sqrt{x^2 - 9}$? Why?

In Exercises 5 through 21, find the limit of the indicated function if it exists. Use the notations ∞ or $-\infty$ to describe the behavior of the function where appropriate.

5. $\lim_{x \to 0} \dfrac{x^3 + x^2 + 2}{x}$

6. $\lim_{t \to 0} \dfrac{t^3 + t^2 + 2}{t^3 + 1}$

7. $\lim_{x \to 0} \dfrac{x^4 + 2x^2}{x^3 + x}$

8. $\lim_{s \to 0} \dfrac{s^3 - 2s^2}{s^4 + 3s^2}$

9. $\lim_{r \to 0} \dfrac{2r^2 - 3r}{r^3 + 4r^2}$

10. $\lim_{x \to 2} \dfrac{x}{x + 3}$

11. $\lim_{u \to 1} \dfrac{(u - 1)^2}{u - 1}$

12. $\lim_{s \to 1} \dfrac{2(s - 1)}{(s - 1)^2}$

13. $\lim_{x \to -1} \dfrac{x^2 + x}{x - 1}$

14. $\lim_{t \to -1} \dfrac{t^2 + t}{t + 1}$

15. $\lim_{x \to 2} \dfrac{x^2 - 4}{x^2 - x - 2}$

16. $\lim_{\Delta x \to 0} (2 + \Delta x)$

17. $\lim_{\Delta t \to 0} \dfrac{4 + \Delta t}{2}$

18. $\lim_{\Delta x \to 0} [(2 + \Delta x)^2 - 4]$

19. $\lim_{\Delta x \to 0} \dfrac{(2 + \Delta x)^2 - 4}{\Delta x}$

20. $\lim_{\Delta t \to 0} \dfrac{[1/(3 + \Delta t)] - \frac{1}{3}}{\Delta t}$

21. $\lim_{\Delta x \to 0} \dfrac{|\Delta x|}{\Delta x}$

22. Let f be defined by

$$f(x) = \begin{cases} x & \text{for } x < 0, \\ 1 & \text{for } x = 0, \\ x^2 & \text{for } x > 0. \end{cases}$$

Find each of the following limits, if it exists.
a) $\lim_{x \to -2} f(x)$ b) $\lim_{x \to 0} f(x)$ c) $\lim_{x \to 3} f(x)$

calculator exercises

Suppose $f(x_1)$ is not defined. If $f(x_1 + 0.01)$ and $f(x_1 - 0.003)$ have approximately the same value L, that is a good indication that $\lim_{x\to x_1} f(x) \approx L$. (The numbers 0.01 and −0.003 could be replaced by others of opposite sign, of different absolute value and close to zero.) Use this technique to estimate the indicated limit, if it exists. Use radian measure for all trigonometric functions.

23. $\lim_{x\to\sqrt{2}} \dfrac{x^2 + 2\sqrt{2}x - 6}{x^2 - 2}$

24. $\lim_{x\to 3} \dfrac{x - \sqrt{3x}}{27 - x^3}$

25. $\lim_{x\to 3} \dfrac{\sin (x - 3)}{x^2 - 9}$

26. $\lim_{x\to 0} (1 + x)^{1/x}$

27. $\lim_{x\to 0} \dfrac{\cos x - 1}{x^2}$

28. $\lim_{x\to 0} (1 + x)^{1/x^2}$

29. $\lim_{x\to\pi/2} (\sin x)^{1/(\pi - 2x)}$

30. $\lim_{x\to 0} \dfrac{\sin x^2}{\cos^2 x - 1}$

2.3 THE DERIVATIVE; DIFFERENTIATION OF POLYNOMIAL FUNCTIONS

2.3.1 The derivative of a function

Refer to Fig. 2.6. Let f be defined at x_1 and for at least a little way on both sides of x_1, say from $x_1 - h$ to $x_1 + h$ for some $h > 0$. The change Δy in $f(x)$ as x changes from x_1 to $x_1 + \Delta x$ is $f(x_1 + \Delta x) - f(x_1)$. In Section 2.1 you learned about $m_{\text{sec}} = \Delta y/\Delta x$ and $m_{\tan} = \lim_{\Delta x\to 0} (\Delta y/\Delta x)$. Here are two definitions that give the usual calculus terminology for m_{sec} and $m_{\tan}$.

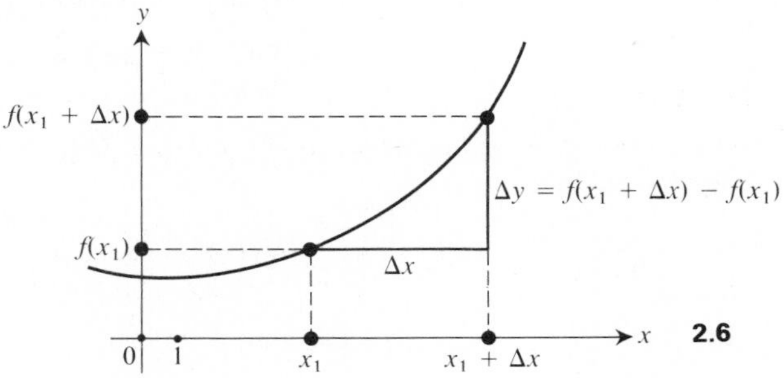

2.6

Definition 2.2 The **difference quotient** is

$$\frac{\Delta y}{\Delta x} = \frac{f(x_1 + \Delta x) - f(x_1)}{\Delta x}, \tag{1}$$

and is the **average rate of change** of $f(x)$ with respect to x from x_1 to $x_1 + \Delta x$.

Definition 2.3 The **derivative** of f at x_1 is

$$f'(x_1) = \lim_{\Delta x \to 0} \frac{f(x_1 + \Delta x) - f(x_1)}{\Delta x}, \tag{2}$$

if this limit exists, and is the **instantaneous rate of change** of $f(x)$ with respect to x at x_1. If $f'(x_1)$ exists, then f is **differentiable** at x_1. A **differentiable function** is one that is differentiable at every point x_1 in its domain.

The function f' in the notation $f'(x_1)$ is the *derived function*, and $f'(x)$ is the derivative of f at any point x where the derivative exists. The derivative $f'(x)$ is often written as dy/dx. This notation, which will turn out to be very handy for remembering some formulas, is due to Leibniz, and should be read "*the derivative of y with respect to x.*" At the moment, regard dy/dx as a single symbol, not as a quotient. (A quotient interpretation will appear in the next chapter.) Remember that $f'(x_1)$ is the slope $m_{\tan}$ of the line tangent to the graph $y = f(x)$ at $x = x_1$. It also has the interpretation of the instantaneous rate at which y is increasing compared to x at x_1. The notation dy/dx is suggestive of this rate of change of y with respect to x. The most important applications of differential calculus center around this rate-of-change interpretation of dy/dx..

The notation dy/dx

The computation of $f'(x_1)$ from (2) is exactly the same as the computation of $m_{\tan}$ given in Section 2.1. Here are two more illustrations, computing

$$\frac{dy}{dx} = f'(x) = \lim_{\Delta x \to 0} \frac{f(x + \Delta x) - f(x)}{\Delta x}$$

at any point x.

Example 1 If $y = f(x) = 1/(x - 1)$, then let us find dy/dx.

SOLUTION

$$\frac{dy}{dx} = f'(x) = \lim_{\Delta x \to 0} \frac{f(x + \Delta x) - f(x)}{\Delta x} = \lim_{\Delta x \to 0} \frac{\dfrac{1}{x + \Delta x - 1} - \dfrac{1}{x - 1}}{\Delta x}$$

$$= \lim_{\Delta x \to 0} \frac{\dfrac{x - 1 - (x + \Delta x - 1)}{(x + \Delta x - 1)(x - 1)}}{\Delta x}$$

$$= \lim_{\Delta x \to 0} \frac{-\Delta x}{\Delta x(x + \Delta x - 1)(x - 1)}$$

$$= \lim_{\Delta x \to 0} \frac{-1}{(x + \Delta x - 1)(x - 1)} = \frac{-1}{(x - 1)^2}. \quad \|$$

Example 2 If $g(x) = \sqrt{x + 3}$, then let us find $g'(x)$.

SOLUTION $$g'(x) = \lim_{\Delta x \to 0} \frac{g(x + \Delta x) - g(x)}{\Delta x} = \lim_{\Delta x \to 0} \frac{\sqrt{x + \Delta x + 3} - \sqrt{x + 3}}{\Delta x}$$

$$= \lim_{\Delta x \to 0} \frac{\sqrt{x + \Delta x + 3} - \sqrt{x + 3}}{\Delta x} \cdot \frac{\sqrt{x + \Delta x + 3} + \sqrt{x + 3}}{\sqrt{x + \Delta x + 3} + \sqrt{x + 3}}$$

$$= \lim_{\Delta x \to 0} \frac{(x + \Delta x + 3) - (x + 3)}{\Delta x(\sqrt{x + \Delta x + 3} + \sqrt{x + 3})} = \lim_{\Delta x \to 0} \frac{\Delta x}{\Delta x(\sqrt{x + \Delta x + 3} + \sqrt{x + 3})}$$

$$= \lim_{\Delta x \to 0} \frac{1}{\sqrt{x + \Delta x + 3} + \sqrt{x + 3}} = \frac{1}{2\sqrt{x + 3}}. \quad \|$$

In a moment, you will learn how to find the derivative of a polynomial function at any point. In Section 2.1, you saw that for small Δx, the approximation

$$f'(x_1) \approx m_{\text{chord}} = \frac{f(x_1 + \Delta x) - f(x_1 - \Delta x)}{2 \cdot \Delta x} \tag{3}$$

(read $\approx$ as "*approximately equals*") can be expected to be better than that given by the difference quotient (1) for the same value of Δx. The approximation (3) can be used with a computer or calculator to estimate the derivative of a function at any particular point. In Appendix 1, you will find a program DERIVE, which uses (3) with Δx set in succession equal to $\frac{1}{2}, \frac{1}{4}, \frac{1}{8}, \ldots, \frac{1}{2}^{30}$, until the approximations (3) have stabilized to six significant figures. Printout 2.4 shows the data obtained in estimating the derivative of $(x^2 - 4x)/(x + 6)$ at $x_1 = 3$. See how quickly the approximations stabilized to six significant figures; the derivative is about 0.259259. Printout 2.5 shows the data obtained for $f(x) = \sin x$ at $x_1 = 0$; the derivative seems to be 1 to six significant figures. This is an easy way to estimate the derivative of a function at a single point.

Δx	m_{chord}
.5000000000	0.2569659443
.2500000000	0.2586872587
.1250000000	0.2591163419
.0625000000	0.2592235351
.0312500000	0.2592503285
.0156250000	0.2592570266
.0078125000	0.2592587011
.0039062500	0.2592591197
.0019531250	0.2592592244

Printout 2.4 $f(x) = \dfrac{x^2 - 4x}{x + 6}$ at $x_1 = 3$

Δx(radians)	m_{chord}
.5000000000	0.9588510772
.2500000000	0.9896158370
.1250000000	0.9973978671
.0625000000	0.9993490855
.0312500000	0.9998372475
.0156250000	0.9999593104
.0078125000	0.9999898275
.0039062500	0.9999974569
.0019531250	0.9999993642
.0009765625	0.9999998411

Printout 2.5 $f(x) = \sin x$ at $x_1 = 0$

2.3.2 Differentiation of polynomial functions

This article presents the first few of many formulas that can be used to compute *exactly* the derivatives of many functions. The process of finding a derivative is *differentiation*. Mastering the technique of differentiation is as important for calculus as mastering the arithmetic operations was for all the mathematics you know now. But differentiation takes much less time to learn than arithmetic.

Let $f(x) = c$, a constant, for all x. Then for any x_1,

$$f'(x_1) = \lim_{\Delta x \to 0} \frac{f(x_1 + \Delta x) - f(x_1)}{\Delta x} = \lim_{\Delta x \to 0} \frac{c - c}{\Delta x} = \lim_{\Delta x \to 0} \frac{0}{\Delta x} = 0.$$

In d-notation,

$$\frac{d(c)}{dx} = 0. \tag{4}$$

Now suppose $f(x) = x^n$ for a positive integer n. This time the computation of $f'(x_1)$ uses the binomial theorem of algebra to expand $(x_1 + \Delta x)^n$. The binomial theorem gives an expanded formula for $(a + b)^n$ in terms of the products,

$$a^n, \quad a^{n-1}b, \quad a^{n-2}b^2, \quad a^{n-3}b^3, \quad \ldots, \quad ab^{n-1}, b^n;$$

namely,

$$(a + b)^n = a^n + na^{n-1}b + \frac{n(n-1)}{2}a^{n-2}b^2 + \frac{n(n-1)(n-2)}{3 \cdot 2}a^{n-3}b^3 + \cdots + b^n.$$

Applying this formula with $a = x_1$ and $b = \Delta x$, you obtain

$$\begin{aligned} f'(x_1) &= \lim_{\Delta x \to 0} \frac{(x_1 + \Delta x)^n - x_1{}^n}{\Delta x} \\ &= \lim_{\Delta x \to 0} \frac{[x_1{}^n + nx_1^{n-1}\Delta x + (n(n-1)/2)x_1^{n-2}(\Delta x)^2 + \cdots + (\Delta x)^n] - x_1{}^n}{\Delta x} \\ &= \lim_{\Delta x \to 0} [nx_1^{n-1} + (n(n-1)/2)x_1^{n-2}\Delta x + \cdots + (\Delta x)^{n-1}] = nx_1^{n-1}. \end{aligned}$$

Since x_1 can be any point, this shows that

$$\frac{d(x^n)}{dx} = nx^{n-1}. \tag{5}$$

Next, suppose that $u = f(x)$ and $v = g(x)$, so $u + v = f(x) + g(x)$. Suppose also that $f'(x_1)$ and $g'(x_1)$ both exist. A change Δx in x produces a change Δu in u and a change Δv in v. The total change in $u + v$ is $\Delta u + \Delta v$. Working at x_1 and using Theorem 2.1 in Section 2.2 for the limit of a sum,

$$\begin{aligned} \lim_{\Delta x \to 0} \frac{\text{Change in } (u + v)}{\Delta x} &= \lim_{\Delta x \to 0} \frac{\Delta u + \Delta v}{\Delta x} \\ &= \lim_{\Delta x \to 0} \frac{\Delta u}{\Delta x} + \lim_{\Delta x \to 0} \frac{\Delta v}{\Delta x} = \frac{du}{dx} + \frac{dv}{dx}. \end{aligned}$$

That is,

$$\frac{d(u+v)}{dx} = \frac{du}{dx} + \frac{dv}{dx} \tag{6}$$

at any point where the derivatives of both u and v exist.

For the final result before differentiating any polynomial function, let $u = f(x)$ and consider the function $c \cdot f(x)$ for a constant c. A change of Δx in x produces a change of Δu in u and therefore a change of $c \cdot \Delta u$ in $c \cdot f(x)$. This time you make use of Theorem 2.1 of Section 2.2 for limits of a product. At any point x where $f'(x)$ exists,

$$\lim_{\Delta x \to 0} \frac{\text{Change in } c \cdot f(x)}{\Delta x} = \lim_{\Delta x \to 0} \frac{c \cdot \Delta u}{\Delta x} = \left(\lim_{\Delta x \to 0} c\right)\left(\lim_{\Delta x \to 0} \frac{\Delta u}{\Delta x}\right) = c \cdot \frac{du}{dx},$$

so

$$\frac{d(c \cdot u)}{dx} = c \cdot \frac{du}{dx}. \tag{7}$$

Equations (6) and (7) are very important since they hold for any differentiable functions $u = f(x)$ and $v = g(x)$. You should "learn them in words," as they are given in the summary, independent of particular letters such as u and v. They deserve to be stated as a theorem.

Theorem 2.2 *If $u = f(x)$ and $v = g(x)$ are both differentiable at x, then so are $u + v = f(x) + g(x)$ and $c \cdot u = c \cdot f(x)$ for any constant c. Furthermore,*

$$\frac{d(u+v)}{dx} = \frac{du}{dx} + \frac{dv}{dx} \quad \text{and} \quad \frac{d(c \cdot u)}{dx} = c \cdot \frac{du}{dx}.$$

Example 3 Using (6), (7), and then (5),

$$\frac{d(4x^3 - 7x^2)}{dx} = \frac{d(4x^3)}{dx} + \frac{d(-7x^2)}{dx} = 4\frac{d(x^3)}{dx} + (-7)\frac{d(x^2)}{dx}$$
$$= 4 \cdot 3x^2 + (-7)(2x) = 12x^2 - 14x$$

for all x. ||

Example 3 can be generalized to more than two summands in the obvious way to give a very nice formula for the derivative of any polynomial function. Namely,

$$\frac{d(a_n x^n + \cdots + a_2 x^2 + a_1 x + a_0)}{dx} = n a_n x^{n-1} + \cdots + 2a_2 x + a_1. \tag{8}$$

Example 4 If $f(x) = 4x^3 - 17x^2 + 3x - 2$, then $f'(x) = 12x^2 - 34x + 3$. ||

Already you can find the derivative (compute the limit of the difference quotient) for every polynomial function! It's a lot easier than learning to add.

2.3.3 Applications These examples illustrate applications of the derivative to the slope of the tangent line and to instantaneous rate of change. Note how easily you can solve the problems in these two examples after only three lessons in calculus! The problems would have seemed formidable only one week ago.

Example 5 Let's find the equation of the line tangent to the graph of $y = f(x) = 3x^4 - 2x^2 + 3x - 7$ where $x = 1$.

SOLUTION POINT: $(1, f(1)) = (1, -3)$

SLOPE: $f'(1) = (12x^3 - 4x + 3)|_{x=1} = 12 - 4 + 3 = 11$. The notation $|_{x=1}$ means "*evaluated at* $x = 1$."

EQUATION: $y + 3 = 11(x - 1)$ or $y = 11x - 14$. ‖

Example 6 We find the velocity at time $t = 3$ if the distance s traveled by a body on a line at time t is given by $s = t^2 + 2t$.

SOLUTION The velocity of the body when $t = 3$ is

$$\text{Velocity} = \left.\frac{ds}{dt}\right|_{t=3} = (2t + 2)|_{t=3} = 6 + 2 = 8\,(\text{units distance})/(\text{unit time}). \quad \|$$

SUMMARY

1. *The derivative*

$$f'(x_1) = \lim_{\Delta x \to 0} \frac{f(x_1 + \Delta x) - f(x_1)}{\Delta x}.$$

2. *If* $f'(x_1)$ *exists, then*

$$f'(x_1) \approx \frac{f(x_1 + \Delta x) - f(x_1 - \Delta x)}{2 \cdot \Delta x}$$

for small Δx.

3. *If* $y = f(x)$, *then* $f'(x)$ *is also written as* dy/dx, *the derivative of* y *with respect to* x.

4. *The derivative of a constant function is zero; in symbols,*

$$\frac{d(c)}{dx} = 0.$$

5. *The derivative of a sum is the sum of the derivatives; in symbols,*

$$\frac{d(u + v)}{dx} = \frac{du}{dx} + \frac{dv}{dx}.$$

6. *The derivative of a constant times a function is the constant times the derivative of the function; in symbols*

$$\frac{d(c \cdot u)}{dx} = c \cdot \frac{du}{dx}.$$

7. $\dfrac{d(x^n)}{dx} = nx^{n-1}$ *for any positive integer n.*

8. *Velocity* $= v = ds/dt$ *where s is the position on a line at time t.*

EXERCISES

In Exercises 1 through 6, find $f'(x)$ using the definition in Eq. (2).

1. $f(x) = x^2 - 3x$
2. $f(x) = 4x^2 + 7$
3. $f(x) = \dfrac{1}{2x+3}$
4. $f(x) = 1/\sqrt{x}$
5. $f(x) = x/(x+1)$
6. $f(x) = \sqrt{2x-1}$

In Exercises 7 through 19, find the derivative of the given function.

7. $3x - 2$
8. $8x^3 - 7x^2 + 4$
9. $2x^7 + 4x^2 - 3$
10. $15x^3 - 4x^6 + 2x^2 + 5$
11. $\dfrac{x^2 - 3x + 4}{2}$
12. $\dfrac{x^3 - 3x^2 + 2}{4}$
13. $(3x)^4 - (2x)^5$
14. $(x^2 - 2)(x + 1)$
15. $(x^2 + 2x)^2$
16. $x(3x + 2)(3x - 2)$
17. $(2x)^2(3x + 5)$
18. $8x^3 - 3(x + 1)^2 + 2$
19. $\dfrac{x(x-1)(x+1)}{3}$

20. Find the equation of the line tangent to the curve $y = x^4 - 3x^2 - 3x$ where $x = 2$.

21. Find the equation of the line *normal* (perpendicular to the tangent line) to the curve $y = 2x^3 - 3x^2$ where $x = 2$.

22. a) Compute $d(1/x)/dx$ assuming that the formula for $d(x^n)/dx$ in (5) also holds if $n = -1$.

b) Verify your answer in (a) by computing the limit of the appropriate difference quotient.

c) Find $\dfrac{d}{dx}\left(\dfrac{3}{x} - 2x\right)$.

d) Find $\dfrac{d}{dx}\left(\dfrac{1}{4x} - \dfrac{3}{x} + \dfrac{2}{5x}\right)$.

23. a) Compute $d(\sqrt{x})/dx$ assuming that the formula for $d(x^n)/dx$ in (5) also holds if $n = \frac{1}{2}$.

b) Verify that your answer in (a) is correct by computing the limit of the appropriate difference quotient.

c) Find $\dfrac{d}{dx}(3\sqrt{x} - 2x^2)$.

d) Find $\dfrac{d}{dx}(\sqrt{5x} - \sqrt{7x})$.

24. An object travels so that after t hours, it has gone $s = f(t) = 4t^3 + 3t^2 + t$ miles for $t \geq 0$.

a) Find the velocity of the object as a function of the time t for $t \geq 0$.

b) The *acceleration* of an object is the rate of change of its velocity with respect to time. Find the acceleration of the object as a function of t for $t \geq 0$.

25. If the length of an edge of a cube increases at a rate of 1 in./sec, find the (instantaneous) rate of increase of the volume when (a) the edge is 2 in. long, (b) the edge is 5 in. long.

26. Repeat Exercise 25, assuming that the edge of the cube is increasing at a rate of 4 in./sec. (Use Exercise 25 and common sense.)

27. Suppose that, when a pebble is dropped into a large tank of fluid, a wave travels outward in a circular ring whose radius increases at a constant rate of 8 in./sec.

a) Find the area of the circular disk enclosed by the wave 2 sec after the time the pebble hits the fluid.

b) Find the (instantaneous) rate at which the area of the circular disk enclosed by the wave is increasing 2 sec after the time the pebble hits the fluid. (See Exercises 25 and 26.)

28. Let $f(x) = |x|$. Show that f is not differentiable at $x_1 = 0$. That is, show that $f'(0)$ does not exist.

29. Show that if $f'(x_1)$ exists, then

$$\lim_{\Delta x \to 0} \frac{f(x_1 + \Delta x) - f(x_1 - \Delta x)}{2 \cdot \Delta x} = f'(x_1).$$

[*Hint.* Use the fact that

$$\frac{f(x_1 + \Delta x) - f(x_1 - \Delta x)}{2 \cdot \Delta x}$$

$$= \frac{1}{2} \cdot \frac{f(x_1 + \Delta x) - f(x_1)}{\Delta x} + \frac{1}{2} \cdot \frac{f(x_1 + (-\Delta x)) - f(x_1)}{-\Delta x}$$

and Theorem 2.1 in Section 2.2.]

30. It is worth noting that $\lim_{\Delta x \to 0} [f(x_1 + \Delta x) - f(x_1 - \Delta x)]/(2 \cdot \Delta x)$ may exist while $f'(x_1)$ does not. Give an example of a function $f(x)$ and a point x_1 where this is true. [*Hint.* Consider Exercise 28.]

calculator exercises

Use the approximation (3) in the text to find the derivative of the given function at the indicated point. Use radian measure with all trigonometric functions. You choose Δx.

31. $\sin 2x$ at $x_1 = 0$

32. $[(x + 7)/(x^2 + 5)]^{1/3}$ at $x_1 = 2.374$

33. x^x at $x_1 = 2.36$

34. $\sin(\tan x)$ at $x_1 = -1.3$

35. $(\sin x)^{\cos x}$ at $x_1 = \pi/4$

36. $(x^2 - 3x)^{\sqrt{x}}$ at $x_1 = 4$

2.4 MORE LIMITS AND CONTINUITY

2.4.1 $\lim_{x \to x_1+} f(x)$ and $\lim_{x \to x_1-} f(x)$

Let $f(x)$ be defined for all x for at least a short distance to the right of x_1, say for $x_1 < x < x_1 + h$, where $h > 0$. It may happen that the values $f(x)$ approach L as x approaches x_1 from the righthand side; in symbols, $\lim_{x \to x_1+} f(x) = L$. In terms of an ϵ,δ-characterization, this means that for each $\epsilon > 0$, there is a $\delta > 0$ such that $|f(x) - L| < \epsilon$ provided that $x_1 < x < x_1 + \delta$.

Definition 2.4 Let $f(x)$ be defined for $x_1 < x < x_1 + h$, where $h > 0$. Then $\lim_{x \to x_1+} f(x) = L$ if, for each $\epsilon > 0$, there exists $\delta > 0$ such that $|f(x) - L| < \epsilon$ provided that $x_1 < x < x_1 + \delta$.

Example 1 The function $\sqrt{x}$ is not defined for x negative, but it is defined if $x \geq 0$, and

$$\lim_{x \to 0+} \frac{x - \sqrt{x}}{\sqrt{x}} = \lim_{x \to 0+} \frac{\sqrt{x}(\sqrt{x} - 1)}{\sqrt{x}} = \lim_{x \to 0+} (\sqrt{x} - 1) = -1. \quad \|$$

Of course if $f(x)$ is defined for $x_1 - h < x < x_1$ for some $h > 0$, then you may try to find $\lim_{x \to x_1-} f(x)$, the limit of $f(x)$ as x approaches x_1 from the lefthand side. The ϵ,δ-characterization of $\lim_{x \to x_1-} f(x) = L$ is just like that for $\lim_{x \to x_1+} f(x)$ except that the condition on x becomes $x_1 - \delta < x < x_1$.

Definition 2.5 Let $f(x)$ be defined for $x_1 - h < x < x_1$, where $h > 0$. Then $\lim_{x \to x_1 -} f(x) = L$ if, for each $\epsilon > 0$, there exists $\delta > 0$ such that $|f(x) - L| < \epsilon$ provided that $x_1 - \delta < x < x_1$.

Example 2 You can easily see that

$$\lim_{x \to 1+} \frac{1}{x-1} = \infty, \qquad \lim_{x \to 1-} \frac{1}{x-1} = -\infty, \qquad \text{and} \qquad \lim_{x \to 1} \left| \frac{1}{x-1} \right| = \infty. \quad \|$$

Example 3 It is possible to define a function using one formula for some values of x and a different one for other values. For example, let

$$f(x) = \begin{cases} 2 - x & \text{for} \quad x \geq 1, \\ 2x + 1 & \text{for} \quad x < 1. \end{cases}$$

The graph of $y = f(x)$ is shown in Fig. 2.7. Then

$$\lim_{x \to 1-} f(x) = \lim_{x \to 1-} (2x + 1) = 3$$

while

$$\lim_{x \to 1+} f(x) = \lim_{x \to 1+} (2 - x) = 1.$$

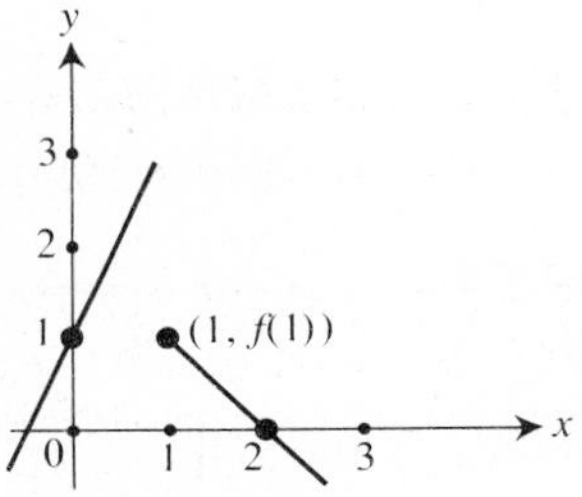

2.7

Of course the actual value $f(1) = 1$ plays no role in the computation of a limit as you approach 1. Note that $\lim_{x \to 1} f(x)$ does not exist, for $f(x)$ does not approach a *single* value as x approaches 1. ‖

Let $f(x)$ be defined for $x_1 - h < x < x_1 + h$ for some $h > 0$. As you might guess from Example 3, it is easy to see that $\lim_{x \to x_1} f(x)$ exists if and only if $\lim_{x \to x_1 +} f(x)$ and $\lim_{x \to x_1 -} f(x)$ both exist and are equal. Let's use these ideas to show that $f(x) = |x|$ is not differentiable at $x_1 = 0$. The difference quotient is

$$\frac{f(0 + \Delta x) - f(0)}{\Delta x} = \frac{|\Delta x| - |0|}{\Delta x} = \frac{|\Delta x|}{\Delta x}.$$

$|x|$ is not differentiable at 0.

Now

$$\lim_{\Delta x \to 0+} \frac{|\Delta x|}{\Delta x} = \lim_{\Delta x \to 0+} \frac{\Delta x}{\Delta x} = 1 \quad \text{while} \quad \lim_{\Delta x \to 0-} \frac{|\Delta x|}{\Delta x} = \lim_{\Delta x \to 0-} \frac{-\Delta x}{\Delta x} = -1.$$

Thus $\lim_{\Delta x \to 0} (|\Delta x|/\Delta x)$, the limit of the difference quotient of $f(x) = |x|$ at zero, does not exist, so $f'(0)$ does not exist. This function, $|x|$, is perhaps the easiest example of a nondifferentiable function.

2.4.2 Limits at infinity

Sometimes you want to know the behavior of $f(x)$ for very large values of x. A natural way to phrase this problem is to discuss the behavior of $f(x)$ "as x

approaches ∞." The assertion $\lim_{x\to\infty} f(x) = L$ means $f(x)$ will be as close to L as you wish if you take *any* sufficiently large x in the domain of f.

Definition 2.6 Let $f(x)$ be defined for all sufficiently large values of x. Then $\lim_{x\to\infty} f(x) = L$ if, for each $\epsilon > 0$, there exists a $K > 0$ such that $|f(x) - L| < \epsilon$ provided that $x > K$.

Definition 2.7 Let $f(x)$ be defined for all sufficiently small values of x. Then $\lim_{x\to-\infty} f(x) = L$ if, for each $\epsilon > 0$, there exists a $K > 0$ such that $|f(x) - L| < \epsilon$ provided that $x < -K$.

Example 4 Obviously $\lim_{x\to\infty} (1/x) = 0$. ||

Example 5 Let's find

$$\lim_{x\to\infty} \frac{2x^2 - 3x}{3x^2 + 2}.$$

SOLUTION The technique is to divide each term of both the numerator and the denominator by the highest power of x that appears, namely x^2. You obtain

$$\lim_{x\to\infty} \frac{2x^2 - 3x}{3x^2 + 2} = \lim_{x\to\infty} \frac{(2x^2/x^2) - (3x/x^2)}{(3x^2/x^2) + (2/x^2)} = \lim_{x\to\infty} \frac{2 - (3/x)}{3 + (2/x^2)} = \frac{2 - 0}{3 + 0} = \frac{2}{3}. \quad \|$$

Example 5 illustrates the standard technique for finding the limit of a rational function of one variable x as x approaches ∞. Namely, divide each term of both the numerator and denominator by the highest power of x that appears.

Example 6 Let's find

$$\lim_{x\to-\infty} \frac{2x^3 - 3x^2}{2x^2 + 4x - 7}.$$

SOLUTION Dividing both the numerator and denominator by x^3,

$$\lim_{x\to-\infty} \frac{2x^3 - 3x^2}{2x^2 + 4x - 7} = \lim_{x\to-\infty} \frac{(2x^3/x^3) - (3x^2/x^3)}{(2x^2/x^3) + (4x/x^3) - (7/x^3)}$$

$$= \lim_{x\to-\infty} \frac{2 - (3/x)}{(2/x) + (4/x^2) - (7/x^3)} = -\infty,$$

for the numerator approaches 2, while the denominator approaches 0 and is negative for negative x of large absolute value. ||

A moment of thought reveals that the trick of dividing the numerator and denominator by the largest power of x present reduces the problem to looking just at the highest-degree term of the numerator and the highest-degree term of the denominator. These monomials of highest degree *domi-*

nate the other terms near $-\infty$ and ∞. Thus the limit in Example 5 can be computed as

$$\lim_{x\to\infty}\frac{2x^2-3x}{3x^2+2}=\lim_{x\to\infty}\frac{2x^2}{3x^2}=\lim_{x\to\infty}\frac{2}{3}=\frac{2}{3},$$

and the limit in Example 6 can be computed as

$$\lim_{x\to-\infty}\frac{2x^3-3x^2}{2x^2-4x-7}=\lim_{x\to-\infty}\frac{2x^3}{2x^2}=\lim_{x\to-\infty}x=-\infty.$$

The relative degrees of the numerator and denominator determine the behavior of a rational function at ∞ and $-\infty$, as described in the summary.

2.4.3 Continuity

Roughly speaking, a function is continuous at x_1 in its domain provided that the graph has no breaks at $x = x_1$. The function in Fig. 2.7 is not continuous at 1, and the function in Fig. 2.8 is not continuous at 2. The notion of no breaks in the graph at $x = x_1$ can be best phrased in terms of a limit. If there are to be no breaks, $\lim_{x\to x_1} f(x)$ must *exist and be equal to* $f(x_1)$. For the function in Fig. 2.7, $\lim_{x\to 1} f(x)$ does not exist. For the function in Fig. 2.8, $\lim_{x\to 2} f(x) = 1.25$, but $f(2) = 2.5$. This important notion of continuity also deserves to be summarized in a definition.

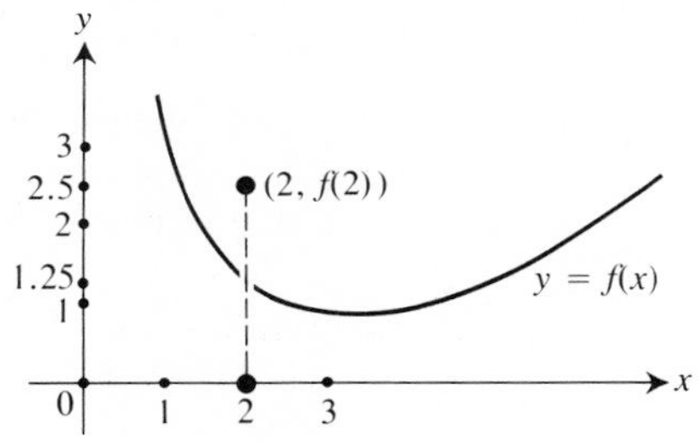

2.8

Definition 2.8 A function f is **continuous at** x_1 in its domain if $\lim_{x\to x_1} f(x)$ exists and is $f(x_1)$. If f is continuous at every point in its domain, then f is a **continuous function.***

The value $f(x_1)$ plays no role in the notion of $\lim_{x\to x_1} f(x)$, but becomes very important in the question of continuity of f at x_1. For the ϵ,δ-characterization of continuity at $x = x_1$, you may use the characterization of $\lim_{x\to x_1} f(x)$, replacing L by $f(x_1)$.

*A function is **discontinuous** at a point x_1 *in its domain* if it is not continuous there. You may find an author who says that $1/x$ is discontinuous at $x_1 = 0$ or that $\sqrt{x}$ is discontinuous at $x_1 = -3$. We prefer not even to consider the question of continuity at any point not in the domain.

Alternative Definition 2.8 A function f is **continuous at** x_1 in its domain if, for each $\epsilon > 0$, there exists $\delta > 0$ such that $|f(x) - f(x_1)| < \epsilon$ provided that $|x - x_1| < \delta$.

Example 7 Let's determine whether

$$f(x) = \begin{cases} \dfrac{x^2 - 9}{x + 3} & \text{for} \quad x \neq -3, \\ 10 & \text{for} \quad x = -3, \end{cases}$$

is continuous at $x = -3$.

SOLUTION We have

$$\lim_{x \to -3} f(x) = \lim_{x \to -3} \frac{(x - 3)(x + 3)}{x + 3} = -6.$$

But $f(-3) = 10$. Since $\lim_{x \to -3} f(x) \neq f(-3)$, this function is not continuous at -3. Hence it is not a continuous function, for -3 is a point in its domain. ‖

Example 8 Let's determine whether

$$f(x) = \begin{cases} x^2 + 2 & \text{for} \quad x > 1, \\ 5x - 1 & \text{for} \quad x \leq 1, \end{cases}$$

is continuous at $x = 1$.

SOLUTION We have

$$\lim_{x \to 1+} f(x) = \lim_{x \to 1+} (x^2 + 2) = 3 \quad \text{while} \quad \lim_{x \to 1-} f(x) = \lim_{x \to 1-} (5x - 1) = 4.$$

Therefore $\lim_{x \to 1} f(x)$ does not exist, so f is not continuous at $x = 1$. Since 1 is in the domain of f, the function is not continuous. ‖

Example 9 Let's determine whether

$$f(x) = \begin{cases} \dfrac{x^2 - x - 6}{x - 3} & \text{for} \quad x \neq 3, \\ 5 & \text{for} \quad x = 3, \end{cases}$$

is continuous at $x = 3$.

SOLUTION We have

$$\lim_{x \to 3} f(x) = \lim_{x \to 3} \frac{(x - 3)(x + 2)}{x - 3} = 5 = f(3).$$

Consequently f is continuous at $x = 3$. Theorem 2.3 will show that f is continuous at every other point also, and hence is a continuous function. ‖

Since continuity is defined in terms of limits, Theorem 2.1 (Section 2.2) on limits for sums, products, and quotients of functions has Theorem 2.3 as an immediate consequence.

Theorem 2.3 *Sums, products, and quotients of continuous functions are continuous. (Of course, quotients are not defined whenever denominators are zero.)*

The function $|x|$ is continuous at $x = 0$ but is not differentiable there. However, differentiability implies continuity.

Theorem 2.4 *If f is differentiable at $x = x_1$, then f is continuous at $x = x_1$.*

Theorem 2.4 is easy to show from the definition of the derivative. If $f'(x_1)$ exists, then

$$\lim_{\Delta x \to 0} \frac{f(x_1 + \Delta x) - f(x_1)}{\Delta x} \quad \text{exists.} \tag{1}$$

Differentiable implies continuous

Now make the substitution $x = x_1 + \Delta x$, so $\Delta x = x - x_1$. Then $\Delta x \to 0$ is equivalent to $x \to x_1$, and (1) becomes

$$\lim_{x \to x_1} \frac{f(x) - f(x_1)}{x - x_1} \quad \text{exists.} \tag{2}$$

Since the denominator in (2) approaches 0 as $x \to x_1$, the limit can exist only if $\lim_{x \to x_1} (f(x) - f(x_1)) = 0$ also, that is, only if $\lim_{x \to x_1} f(x) = f(x_1)$. Thus f is continuous at $x = x_1$.

SUMMARY

1. $\lim_{x \to x_1+}$ *means the limit as x approaches x_1 from the righthand side and* $\lim_{x \to x_1-}$ *is the limit from the lefthand side.*
2. *For a rational function (quotient of polynomial functions):*
   a) *If the degree of the numerator is less than the degree of the denominator, then the limits at ∞ and $-\infty$ are both zero.*
   b) *If the degree of the numerator is the same as the degree of the denominator, then the limits at ∞ and $-\infty$ are both the quotient of the coefficients of those terms of highest degree.*
   c) *If the degree of the numerator is greater than the degree of the denominator, then as $x \to \infty$ or $x \to -\infty$, the function approaches either ∞ or $-\infty$ according to the signs of the numerator and denominator.*
3. *A function $f(x)$ is continuous at x_1 in its domain if* $\lim_{x \to x_1} f(x)$ *exists and is $f(x_1)$. A continuous function is one that is continuous at every point in its domain.*
4. *Sums, products, and quotients of continuous functions are again continuous functions. (Quotients are not defined where denominators become zero.)*
5. *A differentiable function is continuous.*

EXERCISES

In Exercises 1 through 20, find the indicated limit if it exists. Use the symbols ∞ and −∞ to indicate the behavior near the point where appropriate.

1. $\lim_{x \to 2} \frac{1}{2-x}$

2. $\lim_{x \to 2+} \frac{1}{2-x}$

3. $\lim_{x \to 2-} \frac{1}{2-x}$

4. $\lim_{t \to 2} \frac{1}{(2-t)^2}$

5. $\lim_{u \to 5+} \frac{u+3}{u^2-25}$

6. $\lim_{x \to 2} \left|\frac{x^2+4}{x-2}\right|$

7. $\lim_{s \to 0+} \left(\frac{3}{s} - \frac{1}{s^2}\right)$

8. $\lim_{x \to 0} \left(\frac{1}{x^4} - \frac{1}{x}\right)$

9. $\lim_{x \to 0-} \left(\frac{1}{x^3} - \frac{1}{x}\right)$

10. $\lim_{x \to \infty} \frac{x+1}{x}$

11. $\lim_{x \to -\infty} \frac{3x^3-2x}{2x^3+3}$

12. $\lim_{t \to -\infty} \frac{|t|}{t}$

13. $\lim_{x \to \infty} \frac{x^3+2x}{x^2-3}$

14. $\lim_{x \to \infty} \frac{x^2-2x+1}{x^3+3x-2}$

15. $\lim_{x \to -\infty} \frac{x^2-2x+1}{x^3+3x-2}$

16. $\lim_{x \to -\infty} (x^2+3x)$

17. $\lim_{x \to -\infty} (x^3+3x^2)$

18. $\lim_{x \to \infty} (x^{1/2} - x^{1/3})$

19. $\lim_{x \to -\infty} (x^{1/5} - x^{1/3})$

20. $\lim_{x \to \infty} (x - \sqrt{x^2+1})$

21. For each condition given below, draw the graph of a function f that satisfies the condition.

a) Continuous at all points but 2, and $\lim_{x\to 2} f(x) = 3$.

b) Continuous at all points but 2, and $\lim_{x\to 2} f(x)$ undefined.

c) Not continuous at −1 with $\lim_{x\to -1} f(x) = 1$, but continuous elsewhere with $\lim_{x\to 1} f(x) = 2$.

22. Is the function f defined by

$$f(x) = \begin{cases} \frac{x^2-9}{x-3} & \text{for } x \neq 3, \\ 6 & \text{for } x = 3, \end{cases}$$

continuous? Why?

23. Is the function f defined by

$$f(x) = \begin{cases} \frac{4x^2-2x^3}{x-2} & \text{for } x \neq 2, \\ 8 & \text{for } x = 2, \end{cases}$$

continuous? Why?

24. A rubber ball has the characteristic that, when dropped on a floor from height h, it rebounds to height $h/2$. It can be shown that if the ball is dropped from a height of h feet and allowed to bounce repeatedly, then the total distance it has traveled when it hits the floor for the nth time is

$$h + 2h\left(1 - \left(\frac{1}{2}\right)^{n-1}\right) \text{ ft.}$$

Find the total distance the ball travels before it stops bouncing if it is dropped from a height of 16 ft.

25. Taking 4000 miles as radius of the earth, and 32 ft/sec^2 as gravitational acceleration at the surface of the earth, and neglecting air resistance, it can be shown that the velocity v with which a body must be fired upward from the surface of the earth to attain an altitude of s miles is given by the formula

$$v = \frac{8}{\sqrt{5280}}\sqrt{\frac{4000s}{s+(4000)(5280)}} \text{ mi/sec.}$$

Using this formula, find the velocity with which a body must be fired upward to escape the gravitational attraction of the earth.

calculator exercises

Decide whether the limit exists and estimate its value by computing the function for at least two values very near the limit point (values of very large magnitude if $x \to \infty$ or $x \to -\infty$). Use radian measure with trigonometric functions.

26. $\lim_{x\to 0+} x^x$

27. $\lim_{x\to 0+} \left(\dfrac{1}{1-x}\right)^{-1/x^2}$

28. $\lim_{x\to 0+} (\cos x)^{\cot x}$

29. $\lim_{x\to -\infty} \left(1 + \dfrac{1}{x}\right)^{2x}$

30. $\lim_{x\to \infty} \left(1 + \dfrac{1}{x}\right)^{x}$

31. $\lim_{x\to \infty} \left(1 + \dfrac{1}{x}\right)^{x^2}$

*2.5 APPLICATION TO GRAPHING RATIONAL FUNCTIONS

*2.5.1 Graphs of polynomial functions

The graphs of the first- and second-degree polynomial functions were discussed in the last chapter. Now consider a polynomial function

$$f(x) = a_n x^n + a_{n-1}x^{n-1} + \cdots + a_1 x + a_0,$$

where $n \geq 1$ and $a_n \neq 0$. We know that f is continuous, so its graph is an unbroken curve lying over (or under, or crossing) the x-axis.

Since for $x \neq 0$

$$f(x) = x^n\left(a_n + \frac{a_{n-1}}{x} + \cdots + \frac{a_1}{x^{n-1}} + \frac{a_0}{x^n}\right),$$

you see that the monomial term $a_n x^n$ dominates the other terms if x is large in absolute value. Thus $\lim_{x\to\infty} f(x)$ is either ∞ or $-\infty$, and the same is true of $\lim_{x\to-\infty} f(x)$. More precisely, if n is even, you have

$$\lim_{x\to\infty} f(x) = \lim_{x\to-\infty} f(x) = \begin{cases} \infty & \text{if } a_n > 0, \\ -\infty & \text{if } a_n < 0. \end{cases}$$

If n is odd, then

$$\lim_{x\to\infty} f(x) = \begin{cases} \infty & \text{if } a_n > 0, \\ -\infty & \text{if } a_n < 0, \end{cases} \quad \text{and} \quad \lim_{x\to-\infty} f(x) = \begin{cases} -\infty & \text{if } a_n > 0, \\ \infty & \text{if } a_n < 0. \end{cases}$$

These limits tell us whether we should "begin the graph high or low at the left" and "end the graph high or low at the right."

If $f(-x) = f(x)$, which is the case if and only if all monomial terms are of even degree, then the graph of f is symmetric about the y-axis. If $f(-x) = -f(x)$ (all monomials of odd degree), the graph is symmetric about the origin.

* This section can be omitted without loss of continuity.

Unless some linear factors of the polynomial can be found, there is not much more that you can do at the present time to sketch the graph of the function except to plot a number of points on the graph and draw an unbroken curve through them, taking into account the limits at ∞ and $-\infty$ and any possible symmetries. You will learn in Chapter 5 how to use differential calculus to obtain more information about the graph.

Suppose that the polynomial does have a linear factor of the form $x - a$. Then the graph meets the x-axis at a, and a, or $(a, 0)$, is an **x-intercept of the graph**. Suppose, furthermore, that $(x - a)^m$ is a factor of the polynomial while $(x - a)^{m+1}$ is not. Then

$$f(x) = (x - a)^m \cdot g(x),$$

where $m \geq 1$ and where g is a polynomial function and $g(a) \neq 0$. Let us study the sign (positive or negative) of $f(x)$ for x in a neighborhood of a. Since $g(a) \neq 0$ and g is continuous, there is a small neighborhood of a throughout which the sign of $g(x)$ is the same as the sign of $g(a)$. The sign of $(x - a)^m$ for $x \neq a$ is always positive if m is even, and is negative for $x < a$ and positive for $x > a$ if m is odd. Thus if m is even, the sign of $f(x)$ for $x \neq a$ in a small neighborhood of a is the same as the sign of $g(a)$, and the graph just touches the x-axis at a, but stays on the same side near a. However, if m is odd, the sign of $f(x)$ changes as you pass through a and the graph crosses the x-axis at a.

Example 1 Let's sketch the graph of the polynomial function

$$(x - 1)^2(x + 2) = x^3 - 3x + 2.$$

SOLUTION As a cubic polynomial function with dominating term x^3 for large x, its graph "starts low at the left" and "ends high at the right." The function is neither symmetric about the y-axis nor symmetric about the origin. There are x-intercepts at $(-2, 0)$ and $(1, 0)$. At $(-2, 0)$, corresponding to the factor $(x + 2)$ of *odd* degree, the graph crosses the x-axis; at $(1, 0)$ corresponding to the factor $(x - 1)^2$ of *even* degree, the graph merely touches the axis. You easily compute that $f(-1) = 4$, $f(0) = 2$, and $f(2) = 4$. Thus the graph is roughly as shown in Fig. 2.9. Later, calculus will enable us to verify that $(-1, 4)$ is actually a "high point" of the graph, as shown in Fig. 2.9. ‖

To check the accuracy of a sketch of a graph, it is sometimes useful to decide in how many points a line could cut the graph. For example, consider the graph where $y = (x - 1)^2(x + 2)$, as shown in Fig. 2.9. A nonvertical line has an equation of the form

$$y = mx + b,$$

and solving this equation simultaneously with $y = (x - 1)^2(x + 2)$ leads to a cubic equation in x that has at most three solutions, corresponding to points

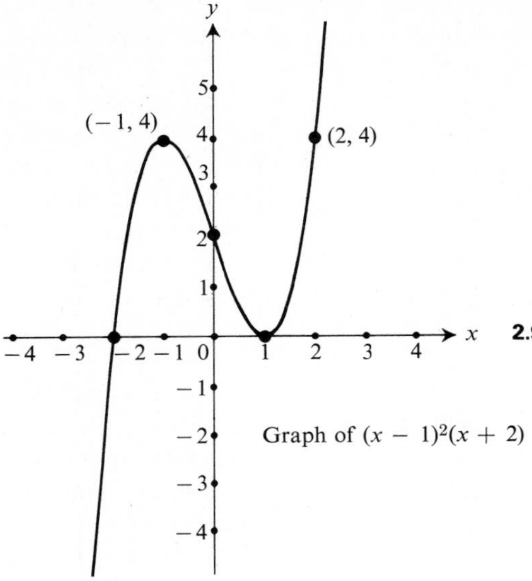

2.9

where the line cuts the graph. Thus any nonvertical line should cut the graph in at most three points. It is easy to see that this is the case for the sketch shown in Fig. 2.9. Not *every* nonvertical line need cut the graph in three points; some solutions of the cubic equation may be complex numbers. Since complex solutions of a polynomial equation with real coefficients always occur in conjugate pairs, we see that *every* nonvertical line should cut the graph in Fig. 2.9 in either one or three points. (A *tangent line* cuts the graph at "coincident points," corresponding to multiple roots of the equation in x.) Similar arguments can be made for polynomial functions of higher degree.

***2.5.2 Graphing rational functions**

We turn now to the graph of a rational function f which is not a polynomial. Such a function f is a quotient of two polynomial functions; and let's assume that the polynomial in the numerator has no factors in common with the polynomial in the denominator. (Such factors can always be cancelled; the resulting function is the same as the original except that the original function is not defined at any point where a common factor becomes zero.) Here is a technique for sketching the graph of a rational function f of one variable. We shall illustrate as we go along with the function f given by

$$y = f(x) = \frac{x-2}{x^2-1}.$$

Refer to Fig. 2.10 during the discussion.

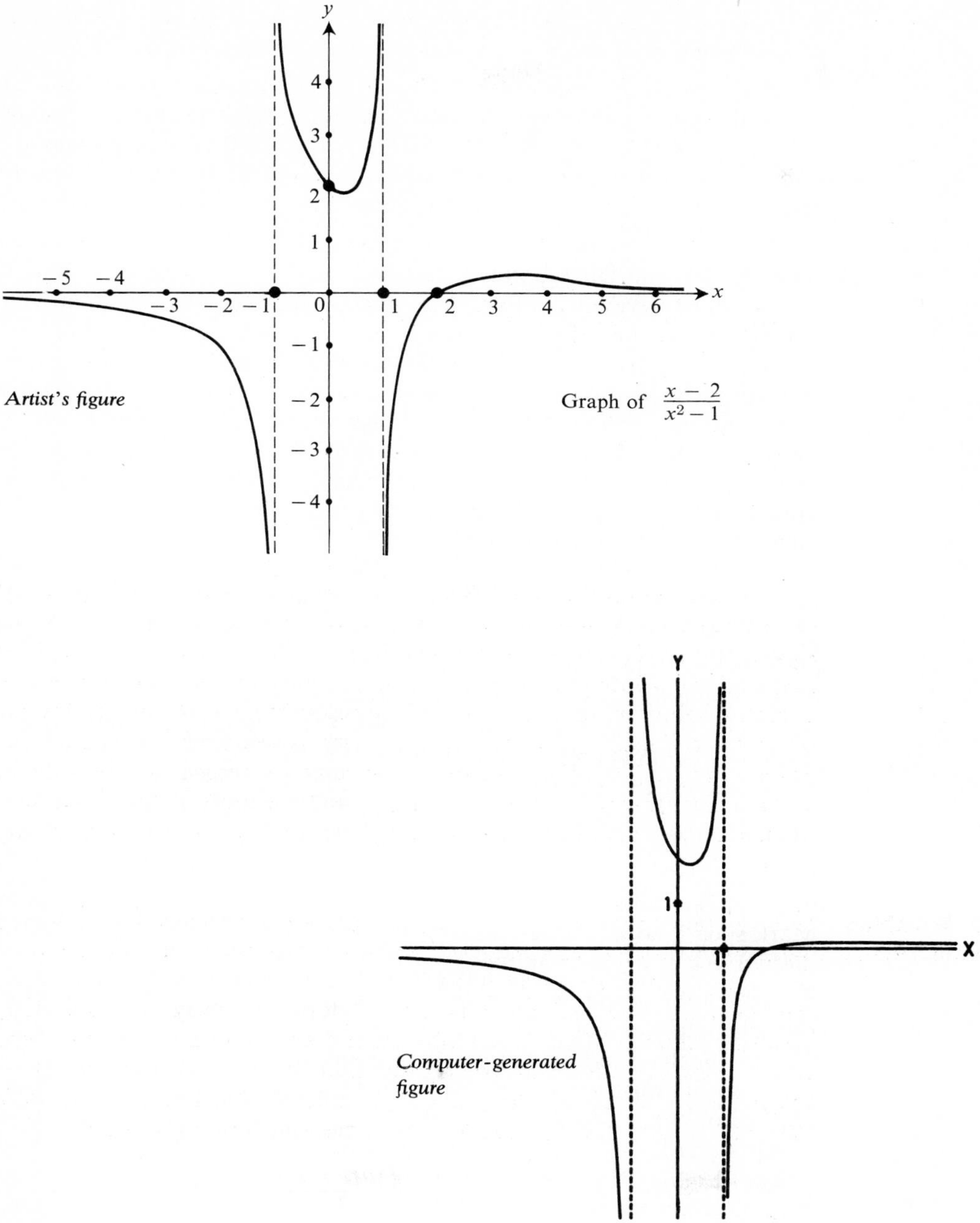

Artist's figure

Graph of $\frac{x-2}{x^2-1}$

Computer-generated figure

2.10

1. Note any symmetries that may exist. For our illustration, the graph is neither symmetric about the y-axis nor symmetric about the origin.

2. Find any x- or y-intercepts; that is, find the points where the graph meets the x-axis or the y-axis. The x-intercepts occur where the value of the numerator is 0, so that $y = 0$; this occurs at $(2, 0)$ in our example. The y-intercept occurs where $x = 0$, which gives $(0, 2)$ in our example. Mark these intercepts on your coordinate diagram.

3. A value a of x for which the denominator is zero is a number not in the domain of f and corresponds to a factor $x - a$ in the denominator. Note that we must have $\lim_{x \to a} |f(x)| = \infty$ since we are assuming the numerator is nonzero at a; otherwise a common factor $x - a$ could be cancelled. The vertical line with equation $x = a$ is a **vertical asymptote**, and the graph will run either up or down such an asymptote as it approaches it. *The graph never crosses a vertical asymptote.* Mark the vertical asymptotes on your diagram with dashed lines. In our example, the lines with equations $x = 1$ and $x = -1$ are vertical asymptotes, An analysis similar to the one made for polynomial functions in the preceding article shows that if $(x - a)^m$ is a factor of the denominator and $(x - a)^{m+1}$ is not, then the sign of $f(x)$ remains the same on the two sides of the asymptote at $x = a$ if m is even, and the sign changes if m is odd. For our example, both factors $(x - 1)$ and $(x + 1)$ are of odd degree, so we will have changes of sign as we jump across each of these asymptotes.

4. Compute $\lim_{x \to -\infty} f(x)$ and $\lim_{x \to \infty} f(x)$ to see how the graph behaves far out to the left and far out to the right. For rational functions, these two limits will be finite and equal if the degree of the numerator does not exceed the degree of the denominator. For our example, we have

$$\lim_{x \to -\infty} f(x) = \lim_{x \to \infty} f(x) = 0,$$

and the line with equation $y = 0$ (the x-axis) is a **horizontal asymptote**. Mark such a horizontal asymptote, if there is one, on your diagram. (Use a dashed line if it is not an axis.)

5. Sketch the graph *from left to right* (or vice versa, but be systematic). For our example (see Fig. 2.10), we find that as x approaches $-\infty$, the value of $f(x)$ is negative and close to 0. The graph does not cross the x-axis until $(2, 0)$ so it must go *down* the vertical asymptote at $x = -1$. The denominator changes sign at -1 (since $x + 1$ is a factor of *odd* degree), so the graph starts from *above* on the right side of the asymptote at $x = -1$. It passes through the y-intercept $(0, 2)$, and must go back *up* the vertical asymptote on the left side at $x = 1$, for it can't cross the x-axis except at $(2, 0)$. The denominator changes sign at $x = 1$, so the graph comes from

below on the right side of the asymptote at $x = 1$, goes *across* the x-axis at $(2, 0)$ since the numerator changes sign at $x = 2$, and must turn back down to run out along the x-axis, our horizontal asymptote.

In the discussion in (5) above, we said that the curve in Fig. 2.10 must go *down* the vertical asymptote at $x = -1$ since it starts below the x-axis at the left and has no x-intercept until $(2, 0)$. This argument depends on the fact that an unbroken curve can't get from one side of the x-axis to the other without intersecting it at an intercept. This is true for the graph of a continuous function.

We could check the sketch in Fig. 2.10 by counting points of intersection of the curve $y = (x - 2)/(x^2 - 1)$ with the line

$$y = mx + b.$$

A cubic equation in x is obtained if $m \neq 0$; this shows that every line that is neither vertical nor horizontal should meet the graph in either one or three points. If $m = 0$, a quadratic equation in x results, so a horizontal line other than the x-axis should meet the graph in either zero or two points.

Our outline above can be summarized briefly as follows. *To sketch the graph of a rational function, find symmetries, intercepts, and asymptotes, and then sketch the graph systematically, moving from left to right.*

Example 2 Let's sketch the graph of

$$y = f(x) = \frac{x - 2}{x + 1}.$$

SOLUTION The function is neither symmetric about the y-axis nor symmetric about the origin. We find that there is an x-intercept at $(2, 0)$ and the y-intercept is $(0, -2)$. The line with equation $x = -1$ is a vertical asymptote. Also, $\lim_{x\to\infty} f(x) = \lim_{x\to-\infty} f(x) = 1$, so the line with equation $y = 1$ is a horizontal asymptote. We put the intercepts and these asymptotes in our sketch (see Fig. 2.11). And now we start to sketch from left to right. Clearly $(x - 2)/(x + 1)$ will actually be a bit greater than 1 for negative x of large absolute value. Since we have no x-intercept until 2, the graph must go up the left side of the vertical asymptote. Since the sign of the denominator changes at $x = -1$, the graph comes from below on the right side of this asymptote, goes through the y-intercept of $(0, -2)$ and the x-intercept of $(2, 0)$, and runs out along the horizontal asymptote at $y = 1$. Any line that is neither vertical nor horizontal should meet this graph in either zero or two points. ||

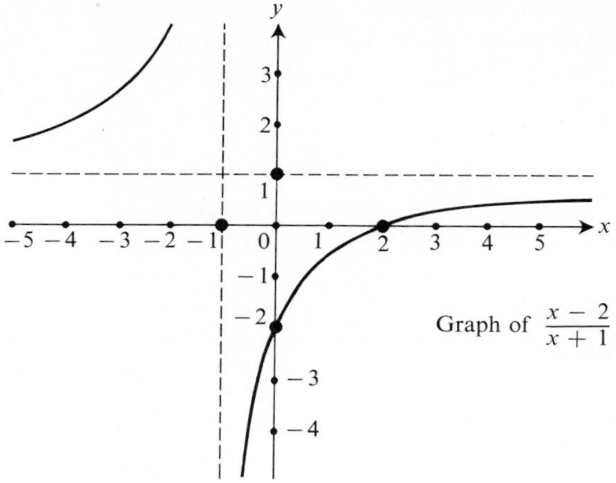

2.11

SUMMARY *To sketch the graph of a rational function $f(x)$, proceed as follows.*

1. *Check for symmetry in the y-axis $[f(-x) = f(x)]$ or in the origin $[f(-x) = -f(x)]$.*
2. *Find the x-intercepts (where the numerator is zero) and the y-intercept (where $x = 0$).*
3. *Dash in any vertical asymptotes $x = a$ where the denominator is zero and the numerator nonzero, or where $\lim_{x\to a} |f(x)| = \infty$.*
4. *Dash in any horizontal asymptote $y = b$ where $\lim_{x\to -\infty} f(x) = \lim_{x\to\infty} f(x) = b$.*
5. *Sketch the graph systematically, say from left to right.*

EXERCISES

*__1.__ Sketch the graph of each of the following polynomial functions.

a) $(x^2 - 1)(x + 2)$ b) $x^2 - 3x + 2$
c) $(x^2 - 4)(x - 1)^2$ d) $x^3 - x^2$
e) $x^2 - x^4$ f) $x - x^3$

*__2.__ Note that the graph of $x^3 - x + 1$ is the graph of $x^3 - x$ raised one unit. Sketch the graph of $x^3 - x$, and then raise it one unit to obtain the graph of $x^3 - x + 1$.

*__3.__ Using the idea developed in Exercise 2, sketch the graph of each of the following polynomial functions.

a) $x^4 + 1$ b) $x^3 - 4x + 2$
c) $-x^3 + x^2 - 1$ d) $x^4 - 4x^2 - 2$

*__4.__ For each of the following rational functions, find the x-intercepts of its graph and determine whether or not the graph *crosses* the x-axis at each intercept.

a) $\dfrac{x}{x+1}$ b) $\dfrac{x^2}{x+1}$

c) $\dfrac{x(x-1)^2}{3x+2}$ d) $\dfrac{(3x-4)(x+2)^3}{2x-1}$

e) $\dfrac{x^2(x+3)^4}{(x+1)^3}$

*5. For each of the given rational functions f, find the vertical asymptotes, and determine for each asymptote whether or not $f(x)$ changes sign as you go from one side of the asymptote to the other.

a) $\dfrac{x}{x-1}$ b) $\dfrac{x^2+2}{x(x+1)^2}$ c) $\dfrac{2x^2-3x}{(x-3)^2(x+1)^3}$

*6. Sketch the graph of each of the given rational functions.

a) $\dfrac{1}{x-1}$ b) $\dfrac{1}{x^2}$ c) $\dfrac{1}{x^2-1}$ d) $\dfrac{2x}{x}$

*7. Sketch the graph of each of the following rational functions.

a) $\dfrac{x^4}{x^2-1}$ b) $\dfrac{x^2}{(x-1)^2}$

c) $\dfrac{(x-1)^2}{(x+2)(x-3)}$ d) $\dfrac{(x-1)(x+2)}{x^2(x+1)}$

e) $\dfrac{x^3-x}{x^3+x}$

*8. Let f be any function. If there exist a and b where $a \neq 0$, such that

$$\lim_{x\to\infty} [f(x) - (ax + b)] = 0,$$

then the line with equation $y = ax + b$ is an **oblique asymptote** of the graph of f. Show that if the degree of the polynomial in the numerator of a rational function f is one greater than the degree of the denominator, then the graph of f has an oblique asymptote. [*Hint.* Argue by polynomial long division that there exist a and b, where $a \neq 0$, such that $f(x) = (ax + b) + g(x)$, where g is a rational function having denominator of degree greater than the degree of the numerator.]

*9. Use the idea developed in Exercise 8 to find the oblique asymptote of the graph of each of the following rational functions, and sketch the graph.

a) $\dfrac{x^2-1}{x}$ b) $\dfrac{(x-1)^2}{x}$ c) $\dfrac{x^3-x^2}{x^2-4}$

exercise sets for chapter 2

review exercise set 2.1

1. Let $f(x) = 1/x$. Estimate the slope m_{tan} of the graph where $x = 1$, by finding:

   a) The slope m_{sec} of the secant through the points where $x = 1$ and where $x = 1 + \frac{1}{2}$,

   b) The slope m_{chord} of the chord through the points where $x = 1 - \frac{1}{2}$ and $x = 1 + \frac{1}{2}$.

2. Find the limit, using the notations ∞, and $-\infty$ where appropriate.

   a) $\displaystyle\lim_{x\to 1}\frac{x^2-1}{x-1}$ b) $\displaystyle\lim_{x\to -2}\frac{x+1}{(x+2)^2}$

   c) $\displaystyle\lim_{x\to 4}\frac{x^2-3x-4}{x^2-16}$ d) $\displaystyle\lim_{x\to -1}\frac{x^2+2x+1}{x^2-2x}$

3. a) Define the derivative $f'(x_1)$ of f at the point where $x = x_1$.

   b) Use the definition of the derivative, not differentiation formulas, to find $f'(x)$ if $f(x) = x^2 - 3x$.

4. a) Find the equation of the line tangent to $y = x^3 - 3x^2 + 2$ where $x = 1$.

   b) If the position s of a body on a line at time t is $s = t^2 - (t/3)$, find the velocity when $t = 2$.

5. a) Find the limit, using the notations ∞ and $-\infty$ where appropriate.

   i) $\displaystyle\lim_{x\to\infty}\frac{x^4-3x^2}{2x-3x^4}$ ii) $\displaystyle\lim_{x\to-\infty}\frac{x^4+100x^2}{14-x}$

   b) Let

$$f(x) = \begin{cases} \dfrac{x^2-9}{x+3} & \text{if } x \neq -3, \\ 6 & \text{if } x = -3. \end{cases}$$

   Is f a continuous function? If so, why, and if not, why not?

*6. Graph

$$\frac{x^2-9}{x^2-4}.$$

review exercise set 2.2

1. Find the limit, using the notations ∞ and $-\infty$ where appropriate

a) $\lim_{x\to 0} \left|\frac{x^2-3}{x}\right|$ b) $\lim_{x\to 2} \frac{x^2-3x+2}{x^2-4}$

c) $\lim_{x\to 5-} \frac{|x-5|}{x^2-25}$ d) $\lim_{x\to 3} \frac{x^3-27}{x^2+9}$

2. Let $f(x)$ be differentiable and suppose that
$$f(1.99) = 7.48, \quad f(2) = 7.51,$$
$$f(2.01) = 7.534.$$
Give what you feel are the best estimates you can for

a) $f'(1.99)$, b) $f'(2)$, c) $f'(2.01)$.

3. Use the definition of the derivative, not differentiation formulas, to find $f'(x)$ if $f(x) = 1/(2x+1)$.

4. a) Find the equation of the normal (perpendicular) line to $y = 4x^2 - 3x + 2$ at the point $(-1, 9)$.

b) Let the position s of a body on a line at time t be $s = t^3 + 2t$.

i) Find the average velocity of the body from time $t = 1$ to time $t = 3$.

ii) Find the velocity of the body at time $t = 2$.

5. Find the limit, using the notations ∞ and $-\infty$ where appropriate.

a) $\lim_{x\to\infty} \frac{7-5x^2}{x^3+3x}$ b) $\lim_{x\to-\infty} \frac{14x^3-7x^2}{8x^3+4x}$

6. Let
$$f(x) = \begin{cases} \dfrac{x^2-4x-5}{x-5} & \text{if } x > 5, \\ 2x-4 & \text{if } x \le 5. \end{cases}$$

a) Find $\lim_{x\to 5-} f(x)$.

b) Find $\lim_{x\to 5+} f(x)$.

c) Is f continuous at $x = 5$? Why?

d) Is f a continuous function? Why?

*7. Graph $3x^2/(x^2 - 2x - 3)$.

more challenging exercises 2

We will see later that there is a continuous function sin x *such that*
$$\lim_{x\to 0} \frac{\sin x}{x} = 1.$$
Use this information to compute the limits in Exercises 1 *through* 10, *if they exist.*

1. $\lim_{\Delta x\to 0} \frac{\sin \Delta x}{\Delta x}$
2. $\lim_{\Delta x\to 0} \frac{\sin \Delta x}{|\Delta x|}$
3. $\lim_{x\to 0} \frac{\sin 2x}{x}$
4. $\lim_{x\to 0} \frac{\sin 2x}{\sin 3x}$
5. $\lim_{x\to\infty} \sin \frac{1}{x}$
6. $\lim_{x\to\infty} \left(x \sin \frac{1}{x}\right)$
7. $\lim_{x\to\infty} \left(x^2 \sin \frac{1}{x}\right)$
8. $\lim_{x\to\infty} \left(x \sin \frac{1}{x^2}\right)$
9. $\lim_{x\to\infty} \left(x^2 \sin \frac{1}{x^2}\right)$
10. $\lim_{x\to\infty} \left(x^3 \sin \frac{1}{x^2}\right)$

11. Students often have trouble understanding the ϵ,δ-definition of a limit. We are convinced that this difficulty is rooted in logic. The definition uses both the *universal quantifier* (for each) and the *existential quantifier* (there exists), since the definition states that *for each* $\epsilon > 0$, *there exists* $\delta > 0$. . . . This exercise deals with this logical problem.

a) Negate the statement

For each $\epsilon > 0$, there exists $\delta > 0$.

That is, write a statement synonymous with "It is not true that for each $\epsilon > 0$, there exists $\delta > 0$," without just saying, "It is not true that. . . ."

b) Negate the statement

For each apple blossom, there exists an apple.

c) Study your answer to part (a) in light of your answer to part (b), and decide whether it is correct.

d) Describe what one must do to show that $\lim_{x\to a} f(x) \neq c$.

12. Let f be a function. Classify each of the following "definitions" of the limit of f at a as either correct or incorrect. If it is incorrect, modify it to become correct.

a) The limit of f at a is c if for each $\epsilon > 0$, there exists $\delta > 0$ such that $|f(x) - c| < \epsilon$ provided that $|x - a| < \delta$.

b) The limit of f at a is c if for some $\epsilon > 0$, there exists $\delta > 0$ such that $|f(x) - c| < \epsilon$ provided that $0 < |x - a| < \delta$.

c) The limit of f at a is c if for each $\epsilon > 0$, there exists $\delta > 0$ such that $0 < |x - a| < \delta$ implies that $|f(x) - c| < \epsilon$.

d) The limit of f at a is c if there exist positive numbers ϵ and δ such that $0 < |x - a| < \delta$ implies that $|f(x) - c| < \epsilon$.

e) The limit of f at a is c if for $\delta > 0$, we have $0 < |x - a| < \delta$ implies $|f(x) - c| < \epsilon$.

f) The limit of f at a is c if $|f(x) - c|$ can be made smaller than any preassigned $\epsilon > 0$ by restricting x to elements different from a in some small interval $[a - \delta, a + \delta]$.

g) The limit of f at a is c if for each positive integer n, there exists $\delta > 0$ such that

$$|f(x) - c| < \frac{1}{n}$$

provided that $0 < |x - a| < \delta$.

13. Give an ϵ,δ-proof that if $\lim_{x\to a} f(x) = L$ and $\lim_{x\to a} g(x) = M$, then $\lim_{x\to a} (f(x) + g(x)) = L + M$.

14. Let $f(x) = 3x^{10} - 7x^8 + 5x^6 - 21x^3 + 3x^2 - 7$. Find

$$\lim_{h\to 0} \frac{f(1 - h) - f(1)}{h^3 + 3h}.$$

15. Classify each of the following "definitions" as correct or incorrect, and if it is incorrect, change it so that it becomes correct.

a) A function f is continuous at a if for each $\epsilon > 0$, there exists $\delta > 0$ such that $|x - a| < \delta$ implies that $|f(x) - a| < \epsilon$.

b) A function f is continuous at a if for each $\epsilon > 0$, there exists $\delta > 0$ such that $|f(x) - c| < \epsilon$, provided that $0 < |x - a| < \delta$.

c) A function f is continuous at a if for each $\epsilon > 0$, there exists $\delta > 0$ such that

$$|f(x) - f(a)| < \epsilon,$$

provided that $|x - a| < \delta$.

d) Let f be a function. The limit of f at a is $-\infty$ if for some γ, there exists $\delta > 0$ such that $f(x) < \gamma$, provided that $0 < |x - a| < \delta$.

e) Let f be a function. The limit of f as x approaches a from the left is $-\infty$ if for each γ, there exists $\delta > 0$ such that $f(x) < \gamma$, provided that $a - \delta < x \leq a$.

f) A function f is continuous if for each a in its domain and each positive integer n, there exists a positive integer m such that $|f(x) - f(a)| < 1/n$, provided that $|x - a| < 1/m$.

16. Give an example of two functions defined for all x, neither of which is continuous at $x = 2$, but whose sum is continuous at $x = 2$.

17. Repeat Exercise 16 but for the *product* of the functions rather than the sum.

18. Give an example of a function defined for all real numbers x but not continuous at any point.

differentiation and differentials

3

The process of finding $f'(x)$ from $f(x)$ is known as *differentiation.* One nice feature of calculus is that there are formulas that make differentiation easy, at least for the functions used most often. You have already seen how easy it is to find the derivative of a polynomial function. Section 3.1 gives easy formulas for finding the derivatives of a product $f(x)\cdot g(x)$ and a quotient $f(x)/g(x)$ in terms of the derivatives $f'(x)$ and $g'(x)$. The chain rule in Section 3.3 will complete our list of general differentiation formulas.

3.1 DIFFERENTIATION OF PRODUCTS AND QUOTIENTS

Let both f and g be differentiable at $x = x_1$, so that $f'(x_1)$ and $g'(x_1)$ exist. If we let $u = f(x)$ and $v = g(x)$, then a change Δx from x_1 to $x_1 + \Delta x$ produces changes Δu in u and Δv in v. The change in $u \cdot v$ is then $(u + \Delta u)\cdot(v + \Delta v) - u \cdot v$. Therefore, the difference quotient for $f(x)\cdot g(x)$ is

$$\frac{\text{Change in } u \cdot v}{\Delta x} = \frac{(u + \Delta u)\cdot(v + \Delta v) - u \cdot v}{\Delta x}$$

$$= \frac{uv + u\cdot\Delta v + v\cdot\Delta u + \Delta u\cdot\Delta v - uv}{\Delta x}$$

$$= \frac{u\cdot\Delta v + v\cdot\Delta u + \Delta u\cdot\Delta v}{\Delta x}$$

$$= u\cdot\frac{\Delta v}{\Delta x} + v\cdot\frac{\Delta u}{\Delta x} + \frac{\Delta u}{\Delta x}\cdot\Delta v. \tag{1}$$

We take the limit of (1) as $\Delta x \to 0$ to find the derivative of uv. Note that $\lim_{\Delta x\to 0}\Delta v = 0$, for $v = g(x)$ is a continuous function at $x = x_1$ since $g'(x_1)$ exists (Theorem 2.4). Taking the limit,

$$\lim_{\Delta x\to 0}\frac{\text{Change in } uv}{\Delta x} = \lim_{\Delta x\to 0}\left[u\cdot\frac{\Delta v}{\Delta x} + v\cdot\frac{\Delta u}{\Delta x} + \frac{\Delta u}{\Delta x}\cdot\Delta v\right]$$

$$= u\cdot\frac{dv}{dx} + v\cdot\frac{du}{dx} + \frac{du}{dx}\cdot 0 = u\cdot\frac{dv}{dx} + v\cdot\frac{du}{dx}.$$

This shows that

$$\frac{d(u\cdot v)}{dx} = u\cdot\frac{dv}{dx} + v\cdot\frac{du}{dx} \tag{2}$$

at any point where du/dx and dv/dx both exist.

Example 1 If $y = (x^2 + x)(x^3 - 7x^2 + 3x)$, then, taking $u = x^2 + x$ and $v = x^3 - 7x^2 + 3x$,

$$\frac{dy}{dx} = (x^2 + x)\frac{d(x^3 - 7x^2 + 3x)}{dx} + (x^3 - 7x^2 + 3x)\frac{d(x^2 + x)}{dx}$$

$$= (x^2 + x)(3x^2 - 14x + 3) + (x^3 - 7x^2 + 3x)(2x + 1).$$

This can be simplified algebraically, but such simplification is a waste of time if you just want the derivative at one point. For example,

$$\left.\frac{dy}{dx}\right|_{x=1} = (2)(-8) + (-3)(3) = -16 - 9 = -25. \quad \|$$

Let's turn to the differentiation of $f(x)/g(x)$ at $x = x_1$, under the assumptions that $f'(x_1)$ and $g'(x_1)$ exist and $g(x_1) \neq 0$. Again, let $u = f(x)$ and $v = g(x)$ have changes Δu and Δv produced by a change Δx in x. Then $u/v = f(x)/g(x)$ and

$$\frac{\text{Change in } (u/v)}{\Delta x} = \frac{\dfrac{u + \Delta u}{v + \Delta v} - \dfrac{u}{v}}{\Delta x}$$

$$= \frac{\dfrac{v(u + \Delta u) - u(v + \Delta v)}{v(v + \Delta v)}}{\Delta x}$$

$$= \frac{\dfrac{v \cdot \Delta u - u \cdot \Delta v}{v(v + \Delta v)}}{\Delta x}$$

$$= \frac{v \cdot \dfrac{\Delta u}{\Delta x} - u \cdot \dfrac{\Delta v}{\Delta x}}{v(v + \Delta v)}. \tag{3}$$

We take the limit as $\Delta x \to 0$ to find the derivative of u/v. Note that $\lim_{\Delta x \to 0} (v + \Delta v) = v$, for $v = g(x)$ is continuous at $x = x_1$, since $g'(x_1)$ exists (Theorem 2.4). Taking the limit,

$$\lim_{\Delta x \to 0} \frac{\text{Change in } u/v}{\Delta x} = \lim_{\Delta x \to 0} \frac{v \cdot (\Delta u/\Delta x) - u \cdot (\Delta v/\Delta x)}{v(v + \Delta v)}$$

$$= \frac{v \cdot (du/dx) - u \cdot (dv/dx)}{v^2}.$$

This shows that

$$\frac{d(u/v)}{dx} = \frac{v \cdot (du/dx) - u \cdot (dv/dx)}{v^2} \tag{4}$$

at any point where du/dx and dv/dx exist, and $v \neq 0$.

Example 2 If $y = (x^2 + 1)/(x^3 - 2x)$, then taking $u = x^2 + 1$ and $v = x^3 - 2x$,

$$\frac{dy}{dx} = \frac{(x^3 - 2x)(d(x^2 + 1)/dx) - (x^2 + 1)(d(x^3 - 2x)/dx)}{(x^3 - 2x)^2}$$

$$= \frac{(x^3 - 2x)(2x) - (x^2 + 1)(3x^2 - 2)}{(x^3 - 2x)^2} = \frac{-x^4 - 5x^2 + 2}{(x^3 - 2x)^2}. \quad \|$$

You know that the formula

$$\frac{d(x^n)}{dx} = nx^{n-1} \tag{5}$$

holds if n is a positive integer. It also holds if n is a negative integer, for then $-n$ is positive and, using (4) and (5),

$$\begin{aligned}\frac{d(x^n)}{dx} = \frac{d(1/x^{-n})}{dx} &= \frac{x^{-n}\cdot(d(1)/dx) - 1\cdot(d(x^{-n})/dx)}{(x^{-n})^2}\\ &= \frac{x^{-n}\cdot 0 - (-n)x^{-n-1}}{x^{-2n}}\\ &= \frac{nx^{-n-1}}{x^{-2n}} = nx^{-n-1+2n} = nx^{n-1}.\end{aligned}$$

Thus (5) holds for any integer n.

Example 3 If $y = 3/x^5$, then you may compute dy/dx by writing $y = 3\cdot x^{-5}$ and differentiating this expression, obtaining

$$\frac{dy}{dx} = 3(-5)x^{-6} = \frac{-15}{x^6}.$$

This is easier than using the quotient rule (4). ||

In this section, you have seen the following theorem and corollary proved.

Theorem 3.1 *If $u = f(x)$ and $v = g(x)$ are both differentiable functions, then so is $uv = f(x)g(x)$, and*

$$\frac{d(uv)}{dx} = u\frac{dv}{dx} + v\frac{du}{dx}.$$

Also, $u/v = f(x)/g(x)$ is differentiable (of course points where $g(x) = 0$ are not allowed), and

$$\frac{d(u/v)}{dx} = \frac{v(du/dx) - u(dv/dx)}{v^2}.$$

Corollary *We have $d(x^n)/dx = n\cdot x^{n-1}$ for every integer n.*

It is best to learn the differentiation formulas in Theorem 3.1 in words rather than with particular letters u and v. Such verbal renditions are given in the summary. We use "top" for "numerator" and "bottom" for "denominator" in this one instance of the quotient rule, for these short words are easier to say and more suggestive. We suggest that you learn these formulas by repeating them over and over in words, off and on, for a week or so.

SUMMARY

1. *The derivative of a product is the first times the derivative of the second, plus the second times the derivative of the first. In symbols,*

$$\frac{d(uv)}{dx} = u \cdot \frac{dv}{dx} + v \cdot \frac{du}{dx}.$$

2. *The derivative of a quotient is the bottom times the derivative of the top minus the top times the derivative of the bottom, all divided by the bottom squared. In symbols,*

$$\frac{d(u/v)}{dx} = \frac{v(du/dx) - u(dv/dx)}{v^2}.$$

3. $d(x^n)/dx = n \cdot x^{n-1}$ *for every integer n.*

EXERCISES

In Exercises 1 through 20, find the derivative of the given function. You need not simplify the answers.

1. $3x^2 + 17x - 5$ **2.** $20x^4 - \frac{3}{2}x^2 + 18$ **3.** $\frac{x^2 - 7}{3}$ **4.** $\frac{x^3 - 2x^2 + 4x}{4}$

5. $\frac{3}{x}$ **6.** $\frac{2}{x^3}$ **7.** $4x^3 - \frac{2}{x^2}$ **8.** $5x + 7 - \frac{1}{x^4}$ **9.** $(x^2 - 1)(x^2 + x + 2)$

10. $(3x^2 - 8x)(x^3 - 7x^2)$ **11.** $(x^2 + 1)[(x - 1)(x^3 + 3)]$ **12.** $[(x^2 - 5x)(2x + 3)](8 - 4x^2)$

13. $\frac{4x^2 - 3}{x}$ **14.** $\frac{8x^3 + 2x^2 + x}{x^2}$ **15.** $\frac{x^2 - 2}{x + 3}$ **16.** $\frac{4x^3 - 3x^2}{2x - 3}$

17. $\frac{(x^2 + 9)(x - 3)}{x^2 + 2}$ **18.** $\frac{(x^3 + 3x)(8x - 6)}{x^3 - 3x}$ **19.** $\frac{(2x + 3)(x^2 - 4)}{(x - 1)(4x^2 + 5)}$ **20.** $\frac{(8x - 6)(3x^2 - 2x)}{(2x + 1)(x^3 + 7)}$

The next chapter will show that there are functions $\sin x$ *and* $\cos x$ *such that*

$$\frac{d(\sin x)}{dx} = \cos x \quad \text{and} \quad \frac{d(\cos x)}{dx} = -\sin x.$$

There is also a function $\tan x$ *defined by* $\tan x = (\sin x)/(\cos x)$. *Use these facts and the formulas of this section to differentiate the functions in Exercises 21 through 30.*

21. $x(\sin x)$ **22.** $x^2(\cos x)$ **23.** $(\sin x)^2$ **24.** $\sin x \cos x$ **25.** $\tan x$

26. $\frac{\cos x}{\sin x}$ **27.** $\frac{x^3}{\sin x}$ **28.** $\frac{x^4}{\cos x}$ **29.** $\frac{\sin x}{x^2 - 4x}$ **30.** $\frac{x^3 - 3x^2}{\cos x}$

In Exercises 31 through 34, find the equations of the tangent line and the normal line to the graph of the given function at the indicated point.

31. $3x^2 - 2x$ at $(2, 8)$ **32.** $1/x$ at $(1, 1)$ **33.** $\frac{2x + 3}{x - 1}$ at $(0, -3)$ **34.** $\frac{(x^2 - 3)(x + 2)}{(x + 1)}$ at $(0, -6)$

3.2 THE DIFFERENTIAL

3.2.1 Approximation of a function near a point

One application of differential calculus is approximation of $f(x)$ at $x = x_1 + \Delta x$ if Δx is sufficiently small and the values $f(x_1)$ and $f'(x_1)$ can be computed easily. With the advent of the inexpensive pocket calculator, it is easy to find the exact (to eight figures, anyway) value of any of the most frequently used functions at *any* point, so this application is not as important now as it was a few years ago. But it still gives insight into the differential, which is the subject of this section.

Of course, $f(x_1 + \Delta x) = f(x_1) + \Delta y$, as shown in Fig. 3.1. Let's take, as approximation to $f(x_1 + \Delta x)$, the height up to the *tangent line* as shown in Fig. 3.1, rather than the height up to the actual graph. From the figure, you obtain the approximation

$$f(x_1 + \Delta x) \approx f(x_1) + \Delta y_{\tan}, \qquad (1)$$

where $\Delta y_{\tan}$ is the change in height of the tangent line from x_1 to $x_1 + \Delta x$. Since the slope of the tangent line is

$$f'(x_1) = \frac{\Delta y_{\tan}}{\Delta x},$$

you have

$$\Delta y_{\tan} = f'(x_1) \cdot \Delta x. \qquad (2)$$

From (1) and (2),

$$f(x_1 + \Delta x) \approx f(x_1) + f'(x_1) \cdot \Delta x. \qquad (3)$$

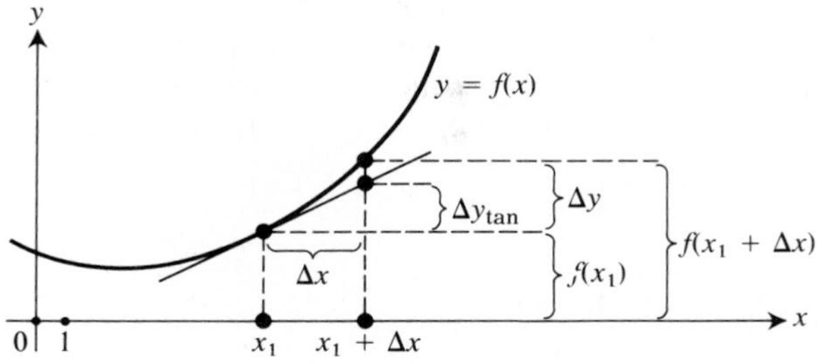

3.1

Example 1 Let us use the approximation (3) to estimate $f(2.05)$ if $f(x) = x^3/(1 + x^2)$.

SOLUTION We let $x_1 = 2$ and $\Delta x = 0.05$. It is easy to compute $f(2)$ and $f'(2)$:

$$f(2) = \frac{8}{5} = 1.6;$$

$$f'(2) = \frac{(1 + x^2)(3x^2) - x^3(2x)}{(1 + x^2)^2}\bigg|_{x=2} = \frac{5 \cdot 12 - 8 \cdot 4}{5^2} = \frac{28}{25} = 1.12.$$

From (3) we obtain

$$f(2 + 0.05) \approx f(2) + f'(2) \cdot (0.05) = 1.6 + (1.12)(0.05)$$
$$= 1.6 + 0.0560 = 1.6560.$$

A pocket calculator yields $f(2.05) = 8.615125/5.2025 = 1.6559587$, so our pencil-and-paper approximation has as error only 0.0000413. ||

3.2.2 The differential of f at x_1

In this article, we finally describe how the Leibniz notation dy/dx can be viewed as a quotient.

Definition 3.1 Let $y = f(x)$ be differentiable at x_1. The **differential of f at x_1** is the function of the single variable dx given by

$$dy = f'(x_1)\, dx. \tag{4}$$

In (4), the independent variable is dx, while dy is the dependent variable. If f is differentiable at all points in its domain, then the **differential** dy or df of $y = f(x)$ is

$$dy = f'(x)\, dx, \tag{5}$$

which associates with each point x the differential of f at that point.

Example 2 If $y = f(x) = x^3 - 3x^2$, then the differential is $dy = (3x^2 - 6x)\, dx$. ||

The differential of f at x_1 is a *linear function* of the variable dx, for $f'(x_1)$ in (4) is a *number*, perhaps 3 or -7. In a dx,dy-plane, the equation $dy = 3 \cdot dx$ gives a line through the origin, just as $y = 3x$ is a line through the origin in the x,y-plane. Comparison of (4) with (2) shows that, if you interpret dx as an increment in x, that is, if you let $dx = \Delta x$, then dy has the value $\Delta y_{\tan}$.

As indicated in Fig. 3.2, if you consider dx and dy to be new variables corresponding to translation of axes to $(x_1, f(x_1))$, then (4) is precisely the equation of the tangent line to the graph with respect to these new axes.

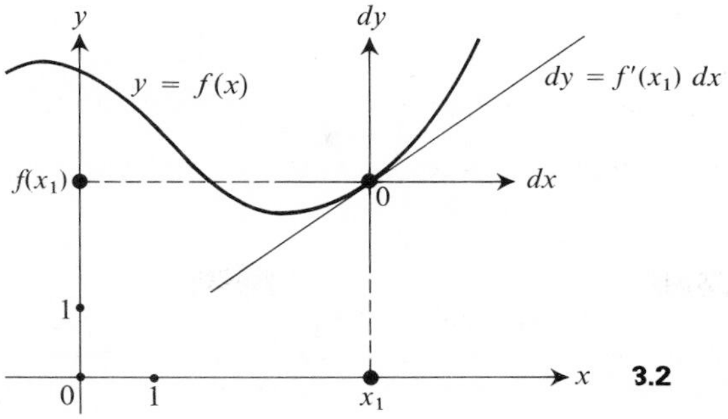

3.2

Approximation formula

In differential notation, the approximation (3) becomes

$$f(x_1 + dx) \approx f(x_1) + f'(x_1) \cdot dx. \tag{6}$$

Approximation using (6) is known as *approximation using differentials*, and should be regarded as the best *linear* approximation to $f(x)$ near x_1. Let's give another approximation example, using differential notation this time.

Example 3 Dick and Jane are lab partners in their physics course. They are supposed to measure a quantity r experimentally, and then determine the value of another quantity W using the formula

$$W = f(r) = \frac{4 + r^3}{1 + r}.$$

They know the correct value for W is 4.931. Unfortunately, one of them jiggled their apparatus, and they obtained the value $r = 2.56$, which would yield $W = 5.836$. They are taking calculus, and decide to use it to estimate the correct value of r. So they use the differential

$$dW = f'(r)\, dr,$$

with $r = 2.56$ and $dW = 4.931 - 5.836 = -0.905$, to find the approximate amount dr by which they would have to change their reading r to give a more correct value.

SOLUTION Since

$$f'(r) = \frac{(1 + r)(3r^2) - (4 + r^3)(1)}{(1 + r)^2},$$

Jane obtains, using her calculator

$$f'(2.56) \approx \frac{69.992 - 20.777}{12.674} \approx 3.883.$$

Then

$$dr = \frac{dW}{f'(2.56)} = \frac{-0.905}{3.883} = -0.233.$$

Thus they add $dr = -0.233$ to their experimental value $r = 2.56$, and they estimate the actual value of r to be $2.56 - 0.233 = 2.327$. Their calculator shows that $f(2.327) \approx 4.99$, so they feel confident they are much closer to the correct value of r. ||

3.2.3 The size of $|\Delta y - dy|$

We now show that if $f'(x_1) \neq 0$, then the approximation at $x_1 + \Delta x$ of Δy by $dy = \Delta y_{\text{tan}}$ is a good one for small Δx in the sense that if Δx is sufficiently small, the error $|\Delta y - dy|$ in the approximation will be at most 10% of Δy. By letting Δx be still smaller, the error can be made at most 1% of Δy. In fact, the error can be made as small a percentage of Δy as you please by taking Δx small enough. To see this, set

$$\epsilon = \begin{cases} \dfrac{\Delta y}{\Delta x} - f'(x_1) & \text{if} \quad \Delta x \neq 0, \\ 0 & \text{if} \quad \Delta x = 0. \end{cases} \tag{7}$$

Since $\lim_{\Delta x \to 0} (\Delta y/\Delta x) = f'(x_1)$, we discover that $\lim_{\Delta x \to 0} \epsilon = 0$. From (7), you obtain

$$\frac{\Delta y}{\Delta x} = f'(x_1) + \epsilon$$

or

$$\Delta y = f'(x_1) \cdot \Delta x + \epsilon \cdot \Delta x \qquad \text{where} \quad \lim_{\Delta x \to 0} \epsilon = 0. \tag{8}$$

If you set $dx = \Delta x$ so that $dy = f'(x_1)\, dx = f'(x_1)\, \Delta x$, then

$$\Delta y - dy = \epsilon \cdot \Delta x,$$

so

$$\frac{\Delta y - dy}{\Delta y} = \epsilon \cdot \frac{\Delta x}{\Delta y} = \frac{\epsilon}{\Delta y/\Delta x}.$$

If $f'(x_1) \neq 0$, then

$$\lim_{\Delta x \to 0} \frac{\Delta y - dy}{\Delta y} = \lim_{\Delta x \to 0} \frac{\epsilon}{\Delta y/\Delta x} = \frac{0}{f'(x_1)} = 0. \tag{9}$$

This shows that the error $|\Delta y - dy|$ will be as small a portion of Δy as you wish if you choose Δx small enough.

The relation (8) will be useful in the next section, and is so important that it deserves to be stated as a theorem.

Theorem 3.2 *Let $y = f(x)$ be differentiable at $x = x_1$, and let $\Delta y = f(x_1 + \Delta x) - f(x_1)$. Then there exists a function ϵ of Δx, defined for small Δx, such that*

$$\Delta y = f'(x_1) \cdot \Delta x + \epsilon \cdot \Delta x \qquad \textit{where} \quad \lim_{\Delta x \to 0} \epsilon = 0.$$

Example 4 Let us illustrate Theorem 3.2 for $y = f(x) = 3x^2 + 2x$ by finding the function ϵ of Δx, and showing directly that $\lim_{\Delta x \to 0} \epsilon = 0$.

SOLUTION We work at a general point x rather than x_1. From the theorem, you see that

$$\epsilon = \frac{\Delta y - f'(x)\, \Delta x}{\Delta x}.$$

Now

$$\begin{aligned}\Delta y &= f(x + \Delta x) - f(x)\\ &= [3(x + \Delta x)^2 + 2(x + \Delta x)] - [3x^2 + 2x]\\ &= 3x^2 + 6x(\Delta x) + 3(\Delta x)^2 + 2x + 2(\Delta x) - 3x^2 - 2x\\ &= 6x(\Delta x) + 3(\Delta x)^2 + 2(\Delta x)\end{aligned}$$

and

$$f'(x)\,\Delta x = (6x + 2)\,\Delta x = 6x(\Delta x) + 2(\Delta x).$$

Consequently,

$$\Delta y - f'(x)\,\Delta x = 3(\Delta x)^2.$$

Thus

$$\epsilon = \frac{\Delta y - f'(x)\,\Delta x}{\Delta x} = \frac{3(\Delta x)^2}{\Delta x} = 3(\Delta x).$$

Of course, $\lim_{\Delta x \to 0} \epsilon = \lim_{\Delta x \to 0} 3(\Delta x) = 0.$ ‖

SUMMARY

1. *If* $y = f(x)$, *then the differential* dy *is* $dy = f'(x)\,dx$.
2. *If* $f'(x_1)$ *exists and is nonzero, the approximation* $f(x_1 + dx) \approx f(x_1) + f'(x_1)\,dx$ *is a good one for small values of* dx.
3. *If* $f'(x_1)$ *exists, then* $\Delta y = f'(x_1)\,\Delta x + \epsilon \cdot \Delta x$ *where* $\lim_{\Delta x \to 0} \epsilon = 0$.

EXERCISES

In Exercises 1 *through* 5 *find the differential of the given function.*

1. $y = f(x) = \dfrac{x}{x + 1}$

2. $s = g(t) = t^3 - 2t^2 + 4t$

3. $A = f(r) = \pi r^2$

4. $y = f(x) = (x^2 + 1)(x^2 - x + 2)$

5. $x = h(t) = \dfrac{t^2 + 1}{t^2 - 1}$

In Exercises 6 *through* 8, *estimate the indicated quantity using a differential.*

6. $(0.999)^{10}$

7. $\dfrac{1}{(10.05)^5}$

8. $f(1.98)$ if $f(x) = \dfrac{x^3 + 4x}{2x - 1}$

9. Given that the derivative of $\sqrt{x}$ is $1/(2\sqrt{x})$, estimate $\sqrt{101}$. [You may think that $dx = 1$ is too large to give a good estimate, but the graph of $\sqrt{x}$ is turning so slowly at $x = 100$ that the tangent line is close to it for quite a way.]

10. If $f(x) = x^4 - 3x^2$, then $f(2) = 4$. Use differentials to find the approximate value of x such that $f(x) = 3.98$.

11. If $f(x) = x^3/(x - 2)$, then $f(4) = 32$. Use differentials to find the approximate value of x such that $f(x) = 31.8$.

12. Estimate the change in volume of a cylindrical silo 20 ft high if the radius is increased from 3 ft to 3 ft 4 in.

13. Imagine the earth to be a ball of radius 4000 mi, and imagine a string tied around the equator of the earth. The string is then cut, and six additional feet of string are inserted.

a) If the lengthened circle of string were lifted a uniform height above the equator all the way around the earth, estimate how high above the surface of the earth the string would be.

b) Discuss the accuracy of your estimate in part (a).

14. A ball has radius 4 ft, and a second ball has volume 1 ft^3 greater than the volume of the first ball. Estimate the difference in the surface area of these two balls. (The volume V and surface area A of a ball of radius r are given by $V = (\frac{4}{3})\pi r^3$ and $A = 4\pi r^2$.)

15. A rectangle is inscribed in a semicircle of radius 5 ft. Estimate the increase in the area of the rectangle if the length of the base (on the diameter) is increased from 6 ft to 6 ft 2 in. [*Hint.* Let the area be A and the length of the base be x. Express A^2 as a function of x and use $d(A^2) = 2A \cdot dA$.]

The scientist is sometimes interested in the percent of error *in the measurement of a numerical quantity Q. If the error in computing Q is h, then the* **percent of error** is

$$\left|\frac{100h}{Q}\right|.$$

Exercises 16 *through* 19 *deal with estimating percent of error using differentials.*

16. For a differentiable function f, let $Q = f(x_1) \neq 0$. Suppose Q is computed by "measuring x_1" and then computing $f(x_1) = Q$.

a) If a small error of Δx is made in the measurement of x_1, argue that the approximate resulting percent of error in the computed value for Q is $|100f'(x_1)\,\Delta x/f(x_1)|$.

b) If a small error of k percent of x_1 is made in the measurement of x_1, argue that the approximate resulting percent of error in the computed value for Q is

$$|kx_1f'(x_1)/f(x_1)|.$$

17. The radius of a sphere is found by measurement to be 2 ft plus or minus 0.04 ft. Estimate the maximum percent of error in computing the volume of the sphere from this measurement of the radius. (For a sphere, volume $V = (\frac{4}{3})\pi r^3$.)

18. One side of an equilateral triangle is found by measurement to be 8 ft with an error of at most 3%. Estimate the maximum percent of error in computing the area of the triangle from this measurement of the length of a side.

19. If it is desired to compute the area of a circle with at most 1% error by measuring its radius, estimate the allowable percent of error that may be made in measuring the radius.

20. A plane flying over the ocean at night is headed straight toward a point A on the coast where a very strong light has been placed at the water line. Visibility is excellent, and the plane is flying at low altitude with the pilot's eyes 264 ft above the ocean. Assuming the radius of the earth is 4000 miles, use differentials to estimate the distance from the light to the plane when the light first becomes visible to the pilot. [*Hint.* If x is the distance from the light to the pilot and y is the distance from the pilot's eyes to the center of the earth, then

$$x^2 = y^2 - (4000)^2. \qquad \text{(Draw a figure.)}$$

You will find that an attempt to estimate $x = \sqrt{y^2 - (4000)^2}$ by a differential for y near 4000 leads to difficulties. A successful technique is to estimate y^2 using a differential, then subtract $(4000)^2$, and finally take the square root, possibly using another differential (see Exercise 9) if the number is not a perfect square.]

21. Let $f(x) = x^2 - 2x$. Find ϵ as a function of Δx and show directly that

$$\lim_{\Delta x \to 0} \epsilon = 0.$$

(See Example 4.)

22. Repeat Exercise 21 for the function

$$f(x) = 1/x^2.$$

calculator exercises

23. If $f(x) = x^x$, then $f(2) = 4$. Use differentials to find the approximate value of x such that $f(x) = 4.15$.

24. If $f(x) = x^{\cos x}$, then, using radian measure, $f(2\pi) = 2\pi$. Use differentials to find the approximate value of x such that $f(x) = 6.3$. (Recall that $\pi \approx 3.1415927$.)

3.3 THE CHAIN RULE

Let us consider the following question.

If twelve-year-old Bill is running twice as fast as five-year-old Mary, and Mary is running three times as fast as baby Jane, how many times as fast as baby Jane is Bill running?

3.3.1 The chain rule formula

Obviously, Bill is running $2 \cdot 3 = 6$ times as fast as baby Jane. Recall that an interpretation of dy/dx is the *rate of change of y with respect to x*. If you will allow us to abuse the Leibniz notation, we can express the answer to our little problem concerning Bill, Mary, and baby Jane by

$$\frac{d(\text{Bill})}{d(\text{Jane})} = \frac{d(\text{Bill})}{d(\text{Mary})} \cdot \frac{d(\text{Mary})}{d(\text{Jane})} = 2 \cdot 3 = 6.$$

Let us take a step toward more careful mathematics. Suppose $y = f(x)$ and $x = g(t)$, so that y appears as a **composite function** of t, namely,

$$y = f(g(t)).$$

Composite functions

This composite function is defined for all t in the domain of g such that $g(t)$ is in the domain of f.*

Let $x = g(t)$ be differentiable at t_1 and let $y = f(x)$ be differentiable at $x_1 = g(t_1)$. The analogy between the following question and the one above should be obvious.

If y is increasing twice as fast as x is at x_1, and x is increasing three times as fast as t is at t_1, how many times as fast as t is y increasing when $t = t_1$?

Once again, it is really obvious that y must be increasing $2 \cdot 3 = 6$ times as fast as t is at t_1. In Leibniz notation, this becomes

Chain rule formula

$$\frac{dy}{dt} = \frac{dy}{dx} \cdot \frac{dx}{dt}. \tag{1}$$

* The notation $f \circ g$ is sometimes used for the composite function, so that $(f \circ g)(x) = f(g(x))$.

We don't claim to have proved (1), which is the *chain rule.* We shall see a more careful argument in a moment. However, good intuitive feeling for a concept is half the battle in an attempt to really understand it.

The chain rule illustrates the advantage of Leibniz notation in remembering formulas. You can remember (1) by pretending that the "dx's cancel." Again, this must not be regarded as a proof.

Example 1 Let $y = x^2 - x$ and $x = t^3$. Suppose you want to find dy/dt when $t = 2$.

SOLUTION Of course, $t = 2$ yields $x = 2^3 = 8$ and $y = 8^2 - 8 = 56$. Using the chain rule, you have

$$\frac{dy}{dt} = \frac{dy}{dx} \cdot \frac{dx}{dt} = (2x - 1)(3t^2).$$

When $t = 2$ and $x = 8$, you obtain

$$\left.\frac{dy}{dt}\right|_{t=2} = (15)(12) = 180.$$

Alternatively, you could solve the problem by expressing y directly as a function of t only and differentiating, without using the chain rule. From $y = x^2 - x$ and $x = t^3$, you obtain $y = t^6 - t^3$. Then $dy/dt = 6t^5 - 3t^2$, so $dy/dt|_{t=2} = 6 \cdot 32 - 3 \cdot 4 = 192 - 12 = 180$. ‖

Example 2 If the length of an edge of a cube increases at a rate of 4 in./sec, let's find the rate of increase of the volume per second at the instant when the edge is 2 in. long.

SOLUTION If an edge is of length x, then $V = x^3$. We are asked to find the rate of change of V *with respect to time*, that is, dV/dt. Now

$$\frac{dV}{dt} = \frac{dV}{dx} \cdot \frac{dx}{dt} = (3x^2)\frac{dx}{dt}$$

by the chain rule. We are given that $dx/dt = 4$ in./sec. Thus when $x = 2$, we have

$$\frac{dV}{dt} = 12 \cdot 4 = 48 \text{ in.}^3/\text{sec.} \quad \|$$

Example 3 Let's find the derivative of the function $(4x^3 + 7x^2)^{10}$.

SOLUTION We could raise the polynomial $4x^3 + 7x^2$ to the tenth power and obtain one polynomial expression, but this is hard work. Let $y = (4x^3 + 7x^2)^{10}$ and let $u = 4x^3 + 7x^2$. Then $y = u^{10}$ and, by the chain rule in (1)

$$\frac{dy}{dx} = \frac{dy}{du} \cdot \frac{du}{dx} = 10u^9(12x^2 + 14x) = 10(4x^3 + 7x^2)^9(12x^2 + 14x).$$

Our problem is solved. ‖

So far you have seen only an intuitive, rate-of-change explanation of the chain rule (1). A more careful argument can be made using Theorem 3.2 in the last section. Let $y = f(x)$ be differentiable at x_1 and $x = g(t)$ be differentiable at t_1 with $g(t_1) = x_1$. Then an increment Δt in t produces a corresponding increment Δx in $x = g(t)$, and Δx in turn produces an increment Δy in $y = f(x)$. From Theorem 3.2, you know that

$$\Delta y = f'(x_1) \cdot \Delta x + \epsilon \cdot \Delta x \qquad \text{where} \quad \lim_{\Delta x \to 0} \epsilon = 0. \tag{2}$$

Dividing (2) by Δt,

$$\frac{\Delta y}{\Delta t} = f'(x_1) \cdot \frac{\Delta x}{\Delta t} + \epsilon \cdot \frac{\Delta x}{\Delta t}. \tag{3}$$

Now $x = g(t)$ is continuous at t_1 since g is differentiable there, so as Δt approaches zero, Δx approaches zero also, and consequently, ϵ approaches zero. Therefore (3) yields

$$\left.\frac{dy}{dt}\right|_{t_1} = \lim_{\Delta t \to 0} \left[f'(x_1) \cdot \frac{\Delta x}{\Delta t} + \epsilon \cdot \frac{\Delta x}{\Delta t}\right] = f'(x_1) \cdot \left.\frac{dx}{dt}\right|_{t_1} + 0 \cdot \left.\frac{dx}{dt}\right|_{t_1}$$

$$= f'(x_1) \cdot \left.\frac{dx}{dt}\right|_{t_1} = \left.\frac{dy}{dx}\right|_{x_1} \cdot \left.\frac{dx}{dt}\right|_{t_1}.$$

We have proved this theorem.

Theorem 3.3 *(Chain rule) Let $y = f(x)$ be differentiable at x_1 and $x = g(t)$ be differentiable at t_1, with $g(t_1) = x_1$. Then the composite function $y = f(g(t))$ is differentiable at t_1, and*

$$\left.\frac{dy}{dt}\right|_{t_1} = \left.\frac{dy}{dx}\right|_{x_1} \cdot \left.\frac{dx}{dt}\right|_{t_1}.^{*}$$

If $y = f(x)$ and $x = g(t)$ are both differentiable functions, this shows that for $y = f(g(t))$,

$$\frac{dy}{dt} = \frac{dy}{dx} \cdot \frac{dx}{dt}. \tag{4}$$

3.3.2 Derivative of a function to a power

You know that if n is any integer, then

$$\frac{d(u^n)}{du} = nu^{n-1}. \tag{5}$$

* In the notation $f \circ g$ for the composite function, this equation becomes $(f \circ g)'(t_1) = f'(x_1) \cdot g'(t_1)$.

If u is in turn a function of x, then the chain rule (4) yields

$$\frac{d(u^n)}{dx} = \frac{d(u^n)}{du} \cdot \frac{du}{dx} = nu^{n-1} \cdot \frac{du}{dx}. \tag{6}$$

Formula (6) is used so often that it is best to memorize it.

Example 4 If $y = (x^3 - 2x)^5$, then, thinking of $x^3 - 2x$ as u in (6),

$$\frac{dy}{dx} = 5(x^3 - 2x)^4 \cdot (3x^2 - 2). \quad \|$$

Formula (6) is given in a verbal form in the summary. You should add this verbal form to the list of formulas you repeat off and on for a week or more.

We can extend formula (6) to rational exponents p/q where p and q are integers and $q \neq 0$. Using (6) for integer exponents, we have

$$\frac{d((u^{p/q})^q)}{dx} = q(u^{p/q})^{q-1} \cdot \frac{d(u^{p/q})}{dx} \tag{7}$$

and

$$\frac{d(u^p)}{dx} = pu^{p-1} \cdot \frac{du}{dx}. \tag{8}$$

Since $(u^{p/q})^q = u^p$, we obtain from (7) and (8)

$$q(u^{p/q})^{q-1} \cdot \frac{d(u^{p/q})}{dx} = pu^{p-1} \cdot \frac{du}{dx},$$

so

$$\frac{d(u^{p/q})}{dx} = \frac{pu^{p-1}}{q(u^{p/q})^{q-1}} \cdot \frac{du}{dx} = \frac{p}{q} \cdot \frac{u^{p-1}}{u^{p-(p/q)}} \cdot \frac{du}{dx}$$

$$= \frac{p}{q} \cdot u^{p-1-p+(p/q)} \cdot \frac{du}{dx}.$$

Finally we obtain

$$\frac{d(u^{p/q})}{dx} = \frac{p}{q} u^{(p/q)-1} \cdot \frac{du}{dx}. \tag{9}$$

Formula (9) is precisely what you get from (6) with $n = p/q$.

Example 5 From (9) you have

$$\frac{d(\sqrt{1 + x^2})}{dx} = \frac{d}{dx}((1 + x^2)^{1/2}) = \frac{1}{2}(1 + x^2)^{-1/2} \cdot (2x)$$

$$= \frac{x}{\sqrt{1 + x^2}}. \quad \|$$

We will see later that the formula

$$\frac{d(u^r)}{dx} = ru^{r-1} \cdot \frac{du}{dx} \qquad (10)$$

holds for any real exponent r, and we give a verbal rendition of (10) in the summary.

SUMMARY

1. *If $y = f(x)$ and $x = g(t)$ are both differentiable functions, then so is $y = f(g(t))$ and*

$$\frac{dy}{dt} = \frac{dy}{dx} \cdot \frac{dx}{dt}.$$

2. *The derivative of a function to a constant power is the power times the function to the one less power, times the derivative of the function. In symbols,*

$$\frac{d(u^n)}{dx} = nu^{n-1} \cdot \frac{du}{dx}.$$

EXERCISES

1. If $y = x^2 - 3x$ and $x = t^3 + 1$, find dy/dt when $t = 1$ by

a) using the chain rule,

b) expressing y as a function of t only and differentiating.

2. If $u = v^3 + 3v^2 - 2v$ and $v = w - 3$, find du/dw where $w = 2$ by

a) using a chain rule,

b) expressing u directly as a function of w and differentiating.

In Exercises 3 through 22, find dy/dx. You need not simplify your answers.

3. $y = \sqrt{2x + 1}$

4. $y = (3x - 2)^{4/3}$

5. $y = 4x^2 - 2x + 3x^{5/3}$

6. $y = x^3 - 2x^2 - 3\sqrt{x}$

7. $y = \dfrac{1}{\sqrt{x}}$

8. $y = \dfrac{1}{\sqrt[3]{x}}$

9. $y = x^{2/3} + x^{1/5}$

10. $y = 4x^{1/4} + 9x^{7/3}$

11. $y = \sqrt{x^2 + 1}$

12. $y = (2x^3 - 1)^{2/3}$

13. $y = \dfrac{\sqrt{x}}{x + 1}$

14. $y = \dfrac{x}{\sqrt{x + 1}}$

15. $y = (3x + 2)^4$

16. $y = (8x^2 - 17x)^3$

17. $y = (x^2 + 3x)^2(x^3 - 1)^3$

18. $y = (x^3 - 2)^3(2x^2 + 4x)^2$

19. $y = \dfrac{8x^2 - 2}{(4x^2 + 1)^2}$

20. $y = \dfrac{(3x^3 - 2)^2}{4x^3 + 2}$

21. $y = \sqrt{2x + 1}\dfrac{(4x^2 - 3x)^2}{2x + 5}$

22. $y = \dfrac{\sqrt{x^2 + 4}}{3x - 8}(2x^3 + 1)^2$

Exercises 23 through 28 depend on the fact that $d(\sin x)/dx = \cos x$, *which is shown in the next chapter. Use this fact and other differentiation formulas to find* dy/dx. *You need not simplify your answers.*

23. $y = \sin 2x$

24. $y = \sin(x^2)$

25. $y = \sin^3 x$

26. $y = \sin^3(4x + 1)$

27. $y = \sqrt{x + \sin x}$

28. $y = (x^2 + \sin x^3)^4$

29. Find the equation of the line tangent to the circle $x^2 + y^2 = 25$ at the point (3, 4).

30. Find the equation of the line normal to the graph of $1/x$ at the point $(2, \frac{1}{2})$.

31. Find the equation of the line tangent to the graph of $\sqrt{2x + 1}$ at the point (4, 3).

32. Use differentials to estimate $\sqrt[3]{26}$.

33. Use differentials to estimate $(4.01)^3 + 3(\sqrt{4.01})$.

34. Use differentials to estimate $[(1.05)^3 + 1]^4$.

35. Use differentials to estimate $\sqrt{63} + (0.95)^5$. Do the estimation in two parts.

3.4 HIGHER-ORDER DERIVATIVES AND MOTION

3.4.1 Higher-order derivatives

If $y = f(x)$ is a differentiable function, then $f'(x)$ is again a function of x, the *derived function* f'. You can attempt to find its derivative. The notations are

$$\frac{d(f'(x))}{dx} = \frac{d^2y}{dx^2} = f''(x).$$

The Leibniz notation d^2y/dx^2 is read "*the second derivative of y with respect to x.*" Of course, you can then find the derivative of $f''(x)$, which is the third derivative of $f(x)$. Table 3.1 is a summary of the various notations you may encounter for derivatives.

TABLE 3.1 Notations for derivatives of $y = f(x)$

Derivative	f'-notation	y'-notation	Leibniz notation	D-notation
1st	$f'(x)$	y'	dy/dx	Df
2nd	$f''(x)$	y''	d^2y/dx^2	D^2f
3rd	$f'''(x)$	y'''	d^3y/dx^3	D^3f
4th	$f^{iv}(x)$	y^{iv}	d^4y/dx^4	D^4f
5th	$f^{v}(x)$	y^{v}	d^5y/dx^5	D^5f
.	.	.	.	.
.	.	.	.	.
.	.	.	.	.
nth	$f^{(n)}(x)$	$y^{(n)}$	d^ny/dx^n	D^nf

Example 1 If $y = x^4 - 3x^3 + 7x^2 - 11x + 5$, then

$$\frac{dy}{dx} = 4x^3 - 9x^2 + 14x - 11, \qquad \frac{d^2y}{dx^2} = 12x^2 - 18x + 14$$

$$\frac{d^3y}{dx^3} = 24x - 18, \qquad \frac{d^4y}{dx^4} = 24, \qquad \frac{d^5y}{dx^5} = 0, \qquad \frac{d^6y}{dx^6} = 0. \quad \|$$

3.4.2 Motion on the line and in the plane

The sign of the velocity

You have seen that if s is the position at time t of a body traveling on a line, then ds/dt is the velocity v of the body at time t. If $ds/dt > 0$, then a small positive increment Δt in t produces a positive increment Δs in s, so the body is moving in the positive s-direction, while if $ds/dt < 0$, the increment Δs is negative so the body is moving in the negative s-direction.

Since $ds/dt = v$, we have

$$\frac{d^2s}{dt^2} = \frac{dv}{dt} = \text{rate of change of velocity with respect to time.}$$

The sign of the acceleration

This derivative dv/dt of the velocity is the **acceleration** of the body. If $dv/dt > 0$, then a small positive increment Δt in t produces a positive increment Δv in v, so the velocity is increasing, while if $dv/dt < 0$, the increment Δv is negative so the velocity is decreasing. The magnitude of the velocity (without regard to sign) is the **speed**, so that

$$\text{Speed} = |v|.$$

For example, if $ds/dt < 0$, then the body is moving in the negative s-direction. If also $d^2s/dt^2 > 0$, then the velocity is increasing, perhaps from -2.1 ft/sec to -1.9 ft/sec as t increases a small amount Δt, so the speed will be decreasing from 2.1 ft/sec to 1.9 ft/sec. The body is moving in the negative s-direction, but slowing down since the positive acceleration is in the positive s-direction.

Example 2 Let the position s of a body on a straight line at time $t \geq 0$ be given by

$$s = 4 - \frac{1}{t+1},$$

and let's find the velocity and acceleration when $t = 3$.

SOLUTION We have $s = 4 - (t+1)^{-1}$, so

$$v = \frac{ds}{dt} = (t+1)^{-2} \quad \text{and} \quad a = \frac{d^2s}{dt^2} = -2(t+1)^{-3}.$$

Thus

$$v\big|_{t=3} = \frac{1}{16} \quad \text{and} \quad a\big|_{t=3} = \frac{-2}{64} = -\frac{1}{32}.$$

The positive velocity indicates that the body is moving in the positive s-direction when $t = 3$, and the negative acceleration, with sign opposite to that of the velocity, indicates that the body is slowing down at $t = 3$. That is, the speed is diminishing. ‖

Now let a body be moving in the x,y-plane, perhaps in the direction of the arrows on the curve shown in Fig. 3.3. The curve need not be the graph of a function. The x-coordinate of the body's position at time t is some function $x = h(t)$, while the y-coordinate of the position is $y = k(t)$. The equations

Parametric equations

$$x = h(t),$$
$$y = k(t), \tag{1}$$

are *parametric equations* of the curve, and t is the time *parameter.* The equation $x = h(t)$ describes the motion on the x-axis of the *x-projection,* which always stays right under (or over) the main body. Similarly, $y = k(t)$ gives the motion on the y-axis of the *y-projection,* which stays opposite the main body. Then dx/dt and d^2x/dt^2 are the velocity and acceleration of the x-projection, or the *x-components of the velocity and acceleration* of the main body. Similarly, dy/dt and d^2y/dt^2 are the *y-components of the velocity and acceleration.*

If $x = h(t)$ and $y = k(t)$ are differentiable functions at t and if a piece of the curve from time $t - h$ to time $t + h$ for some $h > 0$ is the graph of a differentiable function of x, then by the chain rule

$$\frac{dy}{dt} = \frac{dy}{dx} \cdot \frac{dx}{dt}.$$

Therefore where $dx/dt \neq 0$, the slope of the curve is given by

Slope in parametric form

$$\frac{dy}{dx} = \frac{dy/dt}{dx/dt}. \tag{2}$$

Again, the handy Leibniz notation makes it easy to remember the formula.

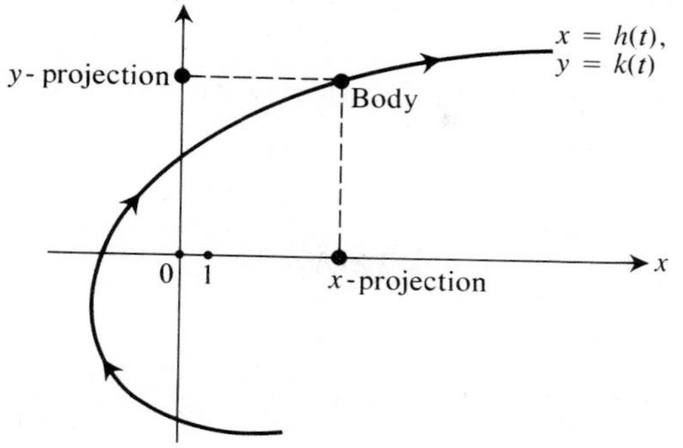

3.3

At each instant t, the body is moving in the direction tangent to the curve. That is, if one could command at any instant, "Stop curving and keep going in the same direction you are now, and at the same speed," then the body would go off along a tangent line to the curve. Now if the body were to go off on the tangent line and *keep the same speed* as it had at the instant t, then in one unit of time it would travel a distance $|dx/dt|$ in the x-direction and a distance $|dy/dt|$ in the y-direction. As shown in Fig. 3.4, the distance traveled along the hypotenuse, tangent to the curve, in one unit of time would then be $\sqrt{(dx/dt)^2 + (dy/dt)^2}$. Therefore the speed of the body at time t is

Speed in parametric form

$$\text{Speed} = \sqrt{\left(\frac{dx}{dt}\right)^2 + \left(\frac{dy}{dt}\right)^2}. \tag{3}$$

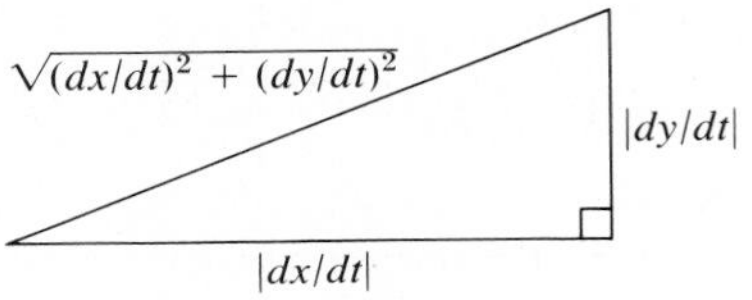

3.4

Example 3 Let the position of a body in the plane at time t be given by

$$x = t^3 - 3t,$$
$$y = 2t^2 + 7t.$$

Illustrating the discussion, you have, at any instant t,

$$x\text{-component of velocity} = \frac{dx}{dt} = 3t^2 - 3,$$

$$y\text{-component of velocity} = \frac{dy}{dt} = 4t + 7,$$

$$x\text{-component of acceleration} = \frac{d^2x}{dt^2} = 6t,$$

$$y\text{-component of acceleration} = \frac{d^2y}{dt^2} = 4,$$

$$\text{Speed} = \sqrt{(dx/dt)^2 + (dy/dt)^2} = \sqrt{(3t^2 - 3)^2 + (4t + 7)^2},$$

$$\text{Slope of curve} = \frac{dy}{dx} = \frac{dy/dt}{dx/dt} = \frac{4t + 7}{3t^2 - 3}. \quad \|$$

Example 4 Let's find the equation of the line normal to the curve given parametrically by

$$x = t^2 + 1,$$
$$y = 2t^3 - 6t,$$

at the point where $t = 2$.

SOLUTION To find the equation of a line, we need to know a point on the line and the slope of the line.

POINT: When $t = 2$, you see that $x = 5$ and $y = 4$, so the point is $(5, 4)$.

SLOPE: The tangent line has slope

$$\left.\frac{dy}{dx}\right|_{t=2} = \left.\frac{dy/dt}{dx/dt}\right|_{t=2} = \left.\frac{6t^2 - 6}{2t}\right|_{t=2} = \frac{18}{4} = \frac{9}{2}.$$

Therefore the normal line has slope $-\frac{2}{9}$.

EQUATION: $y - 4 = -\frac{2}{9}(x - 5)$ or $9y + 2x = 46$. ||

SUMMARY

1. *If $y = f(x)$, then the second derivative of y with respect to x is*

$$\frac{d^2y}{dx^2} = \frac{d(dy/dx)}{dx} = f''(x) = y'' = D^2f,$$

and the nth *derivative of y with respect to x is*

$$\frac{d^ny}{dx^n} = f^{(n)}(x) = y^{(n)} = D^nf.$$

2. *If s is the position of a body on a line (s-axis) at time t, then*

$$\text{Velocity} = v = \frac{ds}{dt}, \qquad \text{Acceleration} = a = \frac{dv}{dt} = \frac{d^2s}{dt^2},$$

$$\text{Speed} = |v|.$$

3. *Given parametric equations $x = h(t)$ and $y = k(t)$ of motion in a plane,*

$$\frac{dx}{dt} = x\text{-component of the velocity}, \qquad \frac{dy}{dt} = y\text{-component of the velocity},$$

$$\sqrt{(dx/dt)^2 + (dy/dt)^2} = \text{speed}, \qquad \frac{dy}{dx} = \frac{dy/dt}{dx/dt} = \text{slope of the curve},$$

$$\frac{d^2x}{dt^2} = x\text{-component of the acceleration},$$

$$\frac{d^2y}{dt^2} = y\text{-component of the acceleration}.$$

EXERCISES

In Exercises 1 *through* 8, *find* y', y'', *and* y'''. *You need not simplify your answers.*

1. $y = x^5 - 3x^4$

2. $y = \sqrt{x}$

3. $y = 1/\sqrt{5x}$

4. $y = x^{2/3}$

5. $y = \sqrt{x^2 + 1}$

6. $y = (3x - 2)^{3/4}$

7. $y = \dfrac{x}{x + 1}$

8. $y = x(x + 1)^4$

9. If the position s of a body on an s-axis at time t is given by $s = 3t^3 - 7t$, find the velocity and acceleration of the body when $t = 1$.

10. Let the position s on an s-axis of a body at time t be given by

$$s = 10 - \frac{20}{t^2 + 1} \quad \text{for} \quad t \geq 0.$$

a) Show that, at any time $t \geq 0$, the body has traveled less than 20 units distance since time $t = 0$.

b) Find the velocity of the body as a function of t.

c) Interpret physically the fact that the velocity is positive for all $t > 0$.

11. A body is thrown straight up from the surface of the earth at time $t = 0$. Suppose that its height s in feet after t sec is $s = -16t^2 + 48t$.

a) Find the velocity of the body at time t.

b) Find the acceleration of the body at time t.

c) Find the initial velocity v_0 with which the body was thrown upward.

d) At what time does the body reach maximum height? [*Hint.* What is the velocity at maximum height?]

e) Find the maximum height the body attains.

f) From the physics of the problem during what time interval is the equation $s = -16t^2 + 48t$ valid?

12. Let $x = h(t)$ be the position at time t of a body on the x-axis (positive direction to the right, as usual). Fill in the blanks with the proper choice from

right, left, increasing, decreasing.

a) If $dx/dt > 0$, then the body is moving to the ——— as t increases.

b) If $d^2x/dt^2 < 0$, then the velocity is ——— as t increases.

c) If $d^2x/dt^2 < 0$ and $dx/dt > 0$, then the speed of the body is ——— as t increases.

d) If $d^2x/dt^2 < 0$ and $dx/dt < 0$, then the speed of the body is ——— and the body is moving to the ——— as t increases.

e) If $d^2x/dt^2 > 0$ and $dx/dt < 0$, then the speed of the body is ——— and the body is moving to the ——— as t increases.

f) If $d^2x/dt^2 > 0$ and $dx/dt > 0$, then the speed of the body is ——— and the body is moving to the ——— as t increases.

In Exercises 13 *through* 15, *find the speed of the body and the slope of the curve at the indicated time for the given parametric motion.*

13. $x = t^3$, $y = t^2$ at time t

14. $x = 1/t^2$, $y = (t + 1)/(t - 1)$ at $t = 2$

15. $x = \sqrt{t^2 + 1}$, $y = (t^2 - 1)^3$ at $t = 1$

16. Find the equation of the tangent line to the curve $x = t^2 - 3t$, $y = \sqrt{t}$, at the point where $t = 4$.

17. Find the equation of the line normal to the curve $x = (t^2 - 3)^2$, $y = (t^2 + 3)/(t - 1)$, where $t = 2$.

18. Let $x = \sqrt{t}$ and $y = t^3$. Find dy/dx and d^2y/dx^2 in terms of t. [*Hint.*

$$\frac{d^2y}{dx^2} = \frac{d(dy/dx)}{dx} = \frac{[d(dy/dx)]/dt}{dx/dt}.\Big]$$

19. Let $x = t^2$, $y = t^3 - 2t^2 + 5$. Find dy/dx and d^2y/dx^2 when $t = 1$. (See the hint of the preceding exercise.)

Numerical approximation of $f''(x_1)$ and $f'''(x_1)$ for calculator exercises

In Section 3 of Chapter 2, we saw that $f'(x_1)$ can be approximated numerically using

$$f'(x_1) \approx \frac{f(x_1 + \Delta x) - f(x_1 - \Delta x)}{2 \cdot \Delta x} \tag{4}$$

for small Δx. Now we shall give formulas for the numerical approximation of $f''(x_1)$ and $f'''(x_1)$. First choose a small Δx and compute:

$$y_1 = f(x_1 - 2 \cdot \Delta x),$$
$$y_2 = f(x_1 - \Delta x),$$
$$y_3 = f(x_1),$$
$$y_4 = f(x_1 + \Delta x),$$
$$y_5 = f(x_1 + 2 \cdot \Delta x).$$

It can be shown that

$$f''(x_1) \approx \frac{-y_1 + 16y_2 - 30y_3 + 16y_4 - y_5}{12(\Delta x)^2} \tag{5}$$

and

$$f'''(x_1) \approx \frac{-y_1 + 2y_2 - 2y_4 + y_5}{2(\Delta x)^3}. \tag{6}$$

These approximations are derived by finding the fourth-degree polynomial through the five points on the graph where x is equal to:

$$x_1 - 2 \cdot \Delta x, \quad x_1 - \Delta x, \quad x_1, \quad x_1 + \Delta x, \quad \text{and} \quad x_1 + 2 \cdot \Delta x.$$

Formula (5) gives the second derivative of this polynomial where $x = x_1$, and (6) gives the third derivative where $x = x_1$. Note that (4) can be written as

$$f'(x_1) \approx \frac{-y_2 + y_4}{2 \cdot \Delta x}$$

in this notation. It can be shown that this is actually the derivative where $x = x_1$ of the second-degree polynomial through the points where $x = x_1 - \Delta x$, x_1, and $x_1 + \Delta x$.

Example 5 Let's use (5) with $\Delta x = 1$ to approximate $f''(3)$ if $f(x) = 1/x$. Of course, $\Delta x = 1$ is large, but it makes an easy computation for this non-calculator illustration.

SOLUTION We have $x_1 = 3$ and $\Delta x = 1$, so

$$x_1 - 2 \cdot \Delta x = 1, \quad x_1 - \Delta x = 2, \quad x_1 + 2 \cdot \Delta x = 5,$$

$$y_1 = 1, \quad y_2 = \tfrac{1}{2}, \quad y_3 = \tfrac{1}{3}, \quad y_4 = \tfrac{1}{4}, \quad y_5 = \tfrac{1}{5}.$$

The approximation (5) becomes

$$f''(3) \approx \frac{-1 + 16 \cdot \frac{1}{2} - 30 \cdot \frac{1}{3} + 16 \cdot \frac{1}{4} - \frac{1}{5}}{12(1)^2} = \frac{-1 + 8 - 10 + 4 - 0.2}{12}$$

$$= \frac{0.8}{12} \approx 0.067.$$

Of course, if $y = x^{-1}$, then $y' = -1 \cdot x^{-2}$ and $y'' = 2x^{-3}$, so $f''(3) = 2/3^3 = 2/27 \approx 0.074$. Our error in approximation is about 0.007. ‖

calculator exercises

In Exercises 20 through 23, find the second and third derivatives of the given function at the indicated point, using the approximations (5) and (6). You choose Δx.

20. $f(x) = \sqrt{x}(x^3 + 2x^2)$ at $x = 1$

21. $f(x) = \dfrac{\sin x}{x + 2}$ at $x = 3$ (radian measure)

22. $f(x) = x^x$ at $x = 1.5$

23. $f(x) = x^{\sin x}$ at $x = 2$ (radian measure)

3.5 IMPLICIT DIFFERENTIATION

The graph of a continuous function f of one variable can be viewed as a curve in the plane. The curve has a tangent line at each point where f is differentiable.

Surely it is natural to consider the circle $x^2 + y^2 = 25$ in Fig. 3.5 to be a plane curve, but this curve is not the graph of a function, since distinct points on the circle may have the same x-coordinate. For example (3, 4) and (3, −4) are both on the circle. However, if you restrict your attention to a small part of the circle containing (3, 4) and extending a short distance on both sides of (3, 4), the heavy curve in Fig. 3.5, this small piece is the graph of a function. In fact, it is part of the graph of the function given by $y = f(x) = \sqrt{25 - x^2}$. You would like to find the derivative $f'(3)$ of this function, that is, the slope of the tangent line to the circle at the point (3, 4). Of course you have

$$f'(x) = \frac{1}{2}(25 - x^2)^{-1/2}(-2x) = \frac{-x}{\sqrt{25 - x^2}}, \quad \text{so} \quad f'(3) = \frac{-3}{\sqrt{16}} = -\frac{3}{4}.$$

Implicit function

The circle in Fig. 3.5 is an example of our topic for this section. Let a plane curve be given by an x,y-equation. The curve may not be the graph of a function, but a piece of the curve extending out for a way on both sides of a point (x_1, y_1) may be the graph of a function, as shown in Fig. 3.6. For this

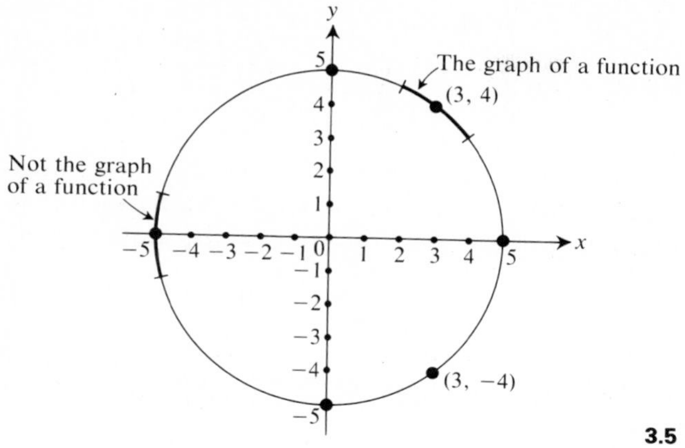

3.5

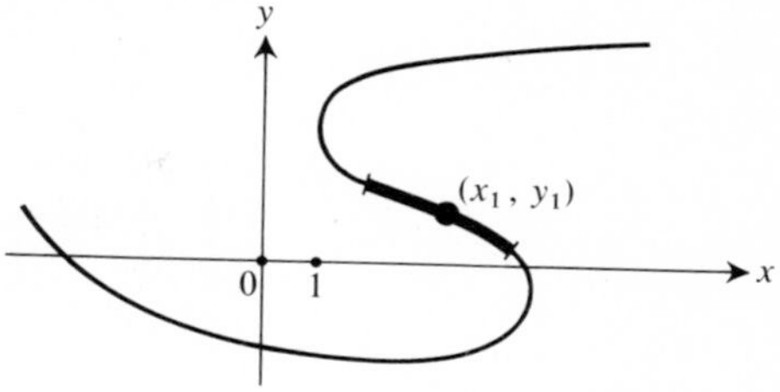

3.6

case, the x,y-equation defines y **implicitly** as a function of x near (x_1, y_1). An equation $y = f(x)$, on the other hand, defines y **explicitly** as a function of x. The terms *implicit* and *explicit* are used to distinguish between these cases. Turning back to our circle,

$$x^2 + y^2 = 25 \textit{ defines } y \textit{ implicitly as a function of } x \textit{ near } (3, 4)$$

while

$$y = \sqrt{25 - x^2} \textit{ defines } y \textit{ explicitly as a function of } x \textit{ near } (3, 4).$$

For the circle $x^2 + y^2 = 25$, you can solve explicitly for y in terms of x and then find dy/dx. But often it is very difficult to solve an equation for y. For example, it is hard to solve for y if

$$y^5 + 3y^2 - 2x^2 = -4.$$

However, assuming that y is defined implicitly as a differentiable function of x near some point (x, y) on the curve, you can use the chain rule to find dy/dx without solving for y. This technique is *implicit differentiation* and is best illustrated by an example.

Example 1 Let's find dy/dx if $y^5 + 3y^2 - 2x^2 = -4$.

SOLUTION Viewing y as a function of x and using the chain rule, we have

$$\frac{d(y^5)}{dx} = \frac{d(y^5)}{dy} \cdot \frac{dy}{dx} = 5y^4 \frac{dy}{dx}.$$

Implicit differentiation

Differentiating both sides of $y^5 + 3y^2 - 2x^2 = -4$ "with respect to x" in this fashion, we obtain

$$5y^4 \frac{dy}{dx} + 6y \frac{dy}{dx} - 4x = 0.$$

We may now solve for dy/dx, obtaining

$$\frac{dy}{dx} = \frac{4x}{5y^4 + 6y}.$$

This formula gives dy/dx at any point (x, y) on the curve where the denominator $5y^4 + 6y$ is nonzero. For example, it is easily seen that $(2, 1)$ satisfies $y^5 + 3y^2 - 2x^2 = -4$ and therefore lies on the curve. Then

$$\left.\frac{dy}{dx}\right|_{(2,1)} = \left.\frac{4x}{5y^4 + 6y}\right|_{(2,1)} = \frac{8}{11}. \quad \|$$

Example 2 If $x^3 + 2x^2y^3 + 3y^4 = 6$, then implicit differentiation yields

$$3x^2 + 2x^2 \cdot 3y^2 \frac{dy}{dx} + 4xy^3 + 12y^3 \frac{dy}{dx} = 0.$$

Therefore

$$(6x^2y^2 + 12y^3)\frac{dy}{dx} = -(3x^2 + 4xy^3)$$

and

$$\frac{dy}{dx} = -\frac{3x^2 + 4xy^3}{6x^2y^2 + 12y^3}. \quad \|$$

Example 3 Let's solve our original circle problem using implicit differentation. That is, we want to find the slope of the tangent line to $x^2 + y^2 = 25$ at $(3, 4)$.

SOLUTION Differentiating $x^2 + y^2 = 25$ implicitly,

$$2x + 2y\frac{dy}{dx} = 0,$$

so

$$\frac{dy}{dx} = \frac{-2x}{2y} = -\frac{x}{y} \quad \text{and} \quad \left.\frac{dy}{dx}\right|_{(3,4)} = -\frac{3}{4}. \quad \|$$

As illustrated by the preceding examples, one obtains a quotient when finding dy/dx by implicit differentiation of an x,y-equation. At certain points (x, y) on the curve, the denominator may become zero. It can be shown that if this calamity does not occur at (x, y) and the curve is a sufficiently "nice" one, then y is implicitly defined as a function of x near (x, y) and implicit differentiation does yield dy/dx.

Example 4 *Orthogonal curves* Let's show that the curve $y - x^2 = 0$ is orthogonal to the curve $x^2 + 2y^2 = 3$ at the point $(1, 1)$ of intersection. (Curves are **orthogonal** if their tangent lines are perpendicular.)

SOLUTION The tangent line to $y = x^2$ has slope given by

$$\left.\frac{dy}{dx}\right|_{x=1} = \left.2x\right|_{x=1} = 2.$$

Differentiating $x^2 + 2y^2 = 3$ implicitly,

$$2x + 4y\frac{dy}{dx} = 0, \qquad \text{so} \quad \frac{dy}{dx} = \frac{-2x}{4y} = \frac{-x}{2y}.$$

Then

$$\left.\frac{dy}{dx}\right|_{(1,1)} = -\frac{1}{2}.$$

Since the slopes 2 and $-\frac{1}{2}$ are negative reciprocals, the curves are orthogonal at (1, 1). ||

Higher-order derivatives of implicit functions can be found by repeated implicit differentiation, as illustrated in the next example.

Example 5 We shall find y'' if $x^2 + y^2 = 2$.

SOLUTION This time we use y' and y'' for the first and second derivative in the implicit differentiation. For y defined implicitly as a function of x by $x^2 + y^2 = 2$, you obtain

$$2x + 2yy' = 0, \qquad \text{so} \quad y' = \frac{-x}{y}.$$

Differentiating this relation again implicitly with respect to x, you obtain

$$y'' = \frac{y(-1) - (-x)y'}{y^2} = \frac{-y + xy'}{y^2}.$$

Since $y' = -x/y$,

$$y'' = \frac{-y + x(-x/y)}{y^2} = \frac{-y^2 - x^2}{y^3}.$$

Remembering that $x^2 + y^2 = 2$, you have

$$y'' = \frac{-2}{y^3}. \quad \|$$

SUMMARY 1. *You can find dy/dx from an x,y-equation by differentiating both sides with respect to x, without solving for y in terms of x. See Examples 1 and 2 for the technique.*

EXERCISES

In Exercises 1 through 3, find dy/dx when $x = x_1$ for y defined implicitly as a function of x near (x_1, y_1) by (a) finding y explicitly as a function of x and differentiating, and (b) differentiating implicitly.

1. $x^2 + y^2 = 25$ and $(x_1, y_1) = (-3, 4)$

2. $x - y^2 = 3$ and $(x_1, y_1) = (7, 2)$

3. $y^2 - 2xy + 3x^2 = 1$ and $(x_1, y_1) = (0, -1)$

In Exercises 4 through 13, find dy/dx at the given point by implicit differentiation.

4. $x^2 - y^2 = 16$; $(5, -3)$

5. $x^3 + y^3 = 7$; $(-1, 2)$

6. $xy = 12$; $(2, 6)$

7. $xy^2 = 12$; $(3, -2)$

8. $xy^2 - 3x^2y + 4 = 0$; $(-1, 1)$

9. $2x^2y^2 - 3xy + 1 = 0$; $(1, 1)$

10. $(3x + 2y)^{1/2}(x - y) = 8$; $(4, 2)$

11. $\sqrt{x + y}(x - 2y)^2 = 2$; $(3, 1)$

12. $3x^2y^3 + 4xy^2 = 6 + \sqrt{y}$; $(1, 1)$

13. $4xy^4 = 2x^2y^3 - y^{1/3} - 5$; $(-1, 1)$

14. If $x^2 - y^2 = 7$, find d^2y/dx^2 at the point $(4, 3)$.

15. Find the equations of the tangent and normal lines to the curve $y^2 = x^3 - 2x^2y + 1$ at the point $(2, 1)$.

16. Find the equation of the tangent line to the curve $y^2 - 3x^2 = 1$ at the point $(-1, 2)$.

17. Show that the curves $2x^2 + y^2 = 24$ and $y^2 = 8x$ are orthogonal at the point $(2, 4)$ of intersection.

18. Show that for any values of $c \neq 0$ and k, the curves $x^2 + 2y^2 = c$ and $y = kx^2$ are orthogonal at all points of intersection.

19. Show that for all nonzero values of c and k, the curves $y^2 - x^2 = c$ and $xy = k$ are orthogonal at all points of intersection.

20. Find all points on the curve $x^2y - xy^2 = 16$ where the curve has a vertical tangent.

exercise sets for chapter 3

review exercise set 3.1

1. Find dy/dx if $y = (x^2 - 3x)(4x^3 - 2x + 17)$. You need not simplify your answer.

2. Find dy/dx if $y = (8x^2 - 2x)/(4x^3 + 3)$.

3. If $y = 3x^2 - 6x + 7$, find dy.

4. Use differentials to estimate $(2.05)^7$.

5. Let $y = 3x^2 - 6x$ and let $x = g(t)$ be a differentiable function such that $g(14) = -2$ and $g'(14) = 8$. Find dy/dt when $t = 14$.

6. Find dy/dx if $y = \sqrt{x^2 - 17x}$.

7. Find dy/dx and d^2y/dx^2 if $y = (4x^3 - 7x)^5$. You need not simplify your answers.

8. Let the position (x, y) of a body in the plane at time t be given by the parametric equations

$$x = 3t^2 - 2t,$$
$$y = t^3 + 2.$$

a) Find the x-component of the velocity when $t = 1$.

b) Find the y-component of the acceleration when $t = 1$.

c) Find the speed of the body when $t = 1$.

9. Find the equation of the line tangent to the curve with parametric equations $x = 3t^2 - 10t$, $y = \sqrt{t + 1}$ when $t = 3$.

10. If $y^3 + x^2y - 4x^2 = 17$, find dy/dx.

review exercise set 3.2

1. Given that there is a function $\ln x$ such that $d(\ln x)/dx = 1/x$, find dy/dx if $y = (x^2 + 3)(\ln x)$.
2. Find dy/dx if
$$y = (x + 4)/(x^2 - 17).$$
3. The area A of a circle of radius r is given by $A = \pi r^2$. A circle of radius 1 has area $\pi \approx 3.14$. Use differentials to estimate the radius of a circle that has area 3.
4. If we let $dx = \Delta x$, draw the graph of a differentiable function at x_1 and label on the graph the geometric meaning at $x = x_1$ of

   a) Δy b) dy c) $|\Delta y - dy|$
5. Find dy/dx if
$$y = 1/\sqrt[3]{x^3 - 3x + 2}.$$
6. Let $y = \sqrt{x^2 - 3x}$ and $x = (t + 2)/(t - 1)$. Find dy/dt when $t = 2$.
7. Find d^3y/dx^3 if $y = \sqrt{3x + 4}$.
8. Let the position (x, y) at time t of a body in the plane be given by the parametric equations
$$x = \tfrac{1}{3}t^3 - \tfrac{3}{2}t^2 + 2t - 4,$$
$$y = t^3 - 9t^2.$$
   a) Find all times t when the direction of motion of the body is vertical.

   b) Find all times t when the direction of motion of the body is horizontal.

   c) Find the speed of the body when $t = 4$.
9. Find the equation of the line normal to the curve with parametric equations
$$x = \frac{2t}{t^2 + 1}, \qquad y = 2t - 3$$
when $t = 1$.
10. Find dy/dx and d^2y/dx^2 at the point $(1, 2)$ if $y^2 - xy = 2$.

more challenging exercises 3

1. Let f and g be differentiable at $x = a$. Show that, if $f(a) = 0$ and $g(a) = 0$, then $(fg)'(a) = 0$.
2. Find a formula for the derivative of a product $f_1(x)f_2(x)f_3(x) \cdots f_n(x)$ of n functions.
3. Let $p(x)$ be a polynomial function. Show that if $(x - a)^2$ is a factor of the polynomial expression, then both $p(a) = 0$ and $p'(a) = 0$.
4. Generalize the result in Exercise 3.
5. Find the equations of the lines through $(4, 10)$ that are tangent to the graph of $f(x) = (x^2/2) + 4$.
6. Let f be differentiable at $x = a$.

   a) Find A, B, and C such that the polynomial function (parabola) $p(x) = Ax^2 + Bx + C$ passes through the three points on the graph $y = f(x)$ whose x-coordinates are $a - \Delta x$, a, and $a + \Delta x$. (For small Δx, the quadratic function $p(x)$ approximates $f(x)$ near a.)

   b) Compute $p'(a)$ for $p(x)$ from part (a). What familiar expression do you obtain?
7. Let f and g be differentiable functions of x, and let $y = f(x)/g(x)$. Derive the formula for the derivative of the quotient $f(x)/g(x)$ using just implicit differentiation and the rule for differentiating a product.
8. You will see in the next chapter that there exist functions $\sin x$ and $\cos x$ such that
$$\frac{d(\sin x)}{dx} = \cos x \qquad \text{and} \qquad \frac{d(\cos x)}{dx} = -\sin x.$$
   Find the derivatives of the following functions.

   a) $\sin 2x$ b) $\cos 4x^3$

   c) $\sin(\cos x)$ d) $\cos(\sin 3x)$
9. The chain rule states that, under certain conditions,
$$\left.\frac{d(f(g(t)))}{dt}\right|_{t=t_2} = f'(g(t_1)) \cdot g'(t_1).$$

Find a similar formula for

$$\left.\frac{d(f(g(h(t))))}{dt}\right|_{t=t_1}$$

for suitable functions f, g, and h.

10. Let f and g be differentiable functions satisfying

$$g'(a) \neq 0, \qquad g(a) = b,$$

and

$$f(g(x)) = x.$$

Show that $f'(b) = 1/g'(a)$.

the trigonometric functions

4

Students may already be familiar with radian measure, the six elementary trigonometric functions, their graphs, and basic trigonometric identities. In that case, the first two sections of this chapter may be omitted.

4.1 TRIGONOMETRY REVIEW I: EVALUATION AND IDENTITIES

4.1.1 The six functions

The circle $u^2 + v^2 = 1$ is shown in Fig. 4.1. It is important to note that the radius of this circle is 1. The functions $\sin x$ and $\cos x$ may be evaluated as follows. If $x \geq 0$, start at the point $(1, 0)$ on the circle and go *counterclockwise* along the circle until you have traveled the distance x on the arc. As shown in Fig. 4.1, you will stop at some point (u, v) on the circle. Then

$$\sin x = v \qquad \text{and} \qquad \cos x = u. \tag{1}$$

Equations (1) are also valid if $x < 0$, except that then you travel the distance $|x|$ *clockwise* around the circle from $(1, 0)$ to arrive at a point (u, v).

Example 1 If $x = 0$, you stay at the point $(1, 0)$, so $u = 1$ and $v = 0$, $\sin 0 = 0$ and $\cos 0 = 1$. On the other hand, if $x = -\pi/2$, then you travel *clockwise* a quarter of the way around the circle (since the circle has circumference 2π) and arrive at $(u, v) = (0, -1)$. Thus $\sin(-\pi/2) = -1$ and $\cos(-\pi/2) = 0$. ||

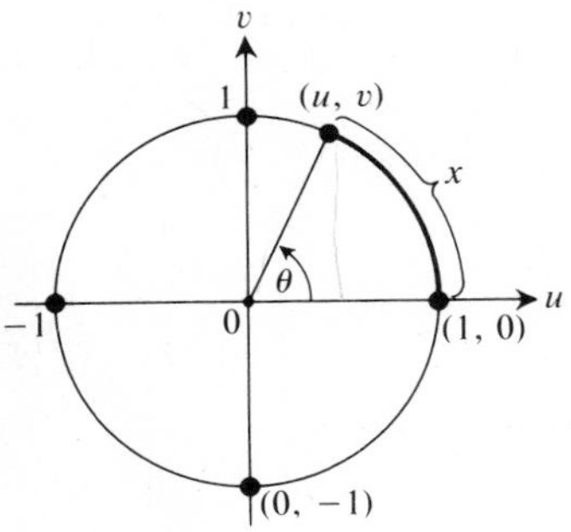

4.1

Radian measure

The arc length x shown in Fig. 4.1 is the *radian measure* of the central angle θ shown in the figure. The radian measure of a central angle θ of a circle is

$$\text{Radian measure of } \theta = \frac{\text{length of intercepted arc}}{\text{radius}}.$$

Since our radius is 1, the radian measure of θ is given by the length x of the arc. For example, a 360° angle has radian measure 2π since the length all the way around the circle is 2π. It is easy to see that

$$\text{Radian measure of } \theta = \left(\frac{\pi}{180}\right)(\text{degree measure of } \theta),$$

for 180° corresponds to π radians.

From Fig. 4.2, you see that if $0 < \theta < 90°$, then θ is an acute angle of the right triangle shown, and

$$\sin\theta = v = \frac{v}{1} = \frac{\text{opposite side length}}{\text{hypotenuse length}}$$

while

$$\cos\theta = u = \frac{u}{1} = \frac{\text{adjacent side length}}{\text{hypotenuse length}}.$$

You may have learned these definitions in high school. The advantage of the definitions (1) is that you can easily find $\sin x$ if $x > 90°$. For example, $\sin(3\pi/2) = \sin(270°) = -1$ since an arc length of $3\pi/2$ corresponds to the point $(0, -1)$ on the circle.

You should recall from geometry the lengths of the sides of the right triangles shown in Fig. 4.3.

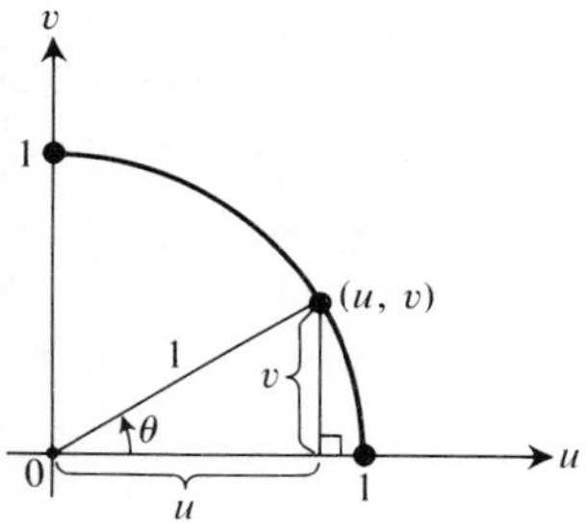

4.2

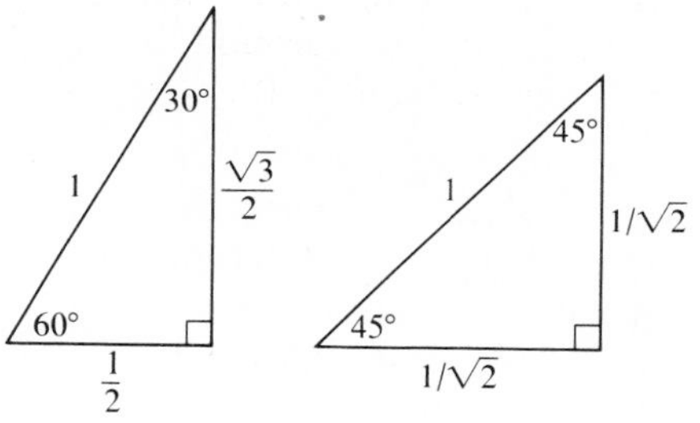

4.3

Example 2 We shall find $\sin(-2\pi/3)$ and $\cos(-2\pi/3)$.

SOLUTION Since $-2\pi/3$ radians corresponds to $-120° = -90° - 30°$, you see from Fig. 4.4 and the lefthand triangle in Fig. 4.3 that

$$\sin\left(-\frac{2\pi}{3}\right) = -\frac{\sqrt{3}}{2} \quad \text{and} \quad \cos\left(-\frac{2\pi}{3}\right) = -\frac{1}{2}. \quad \|$$

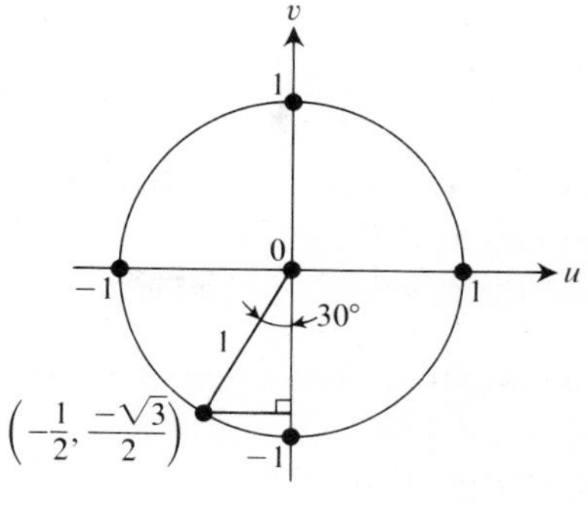

4.4

Now let's describe the remaining four basic trigonometric functions of x in terms of the point (u, v) in Fig. 4.1:*

$$\tan x = \frac{v}{u} = \frac{\sin x}{\cos x}, \qquad \cot x = \frac{u}{v} = \frac{\cos x}{\sin x} = \frac{1}{\tan x},$$
$$\sec x = \frac{1}{u} = \frac{1}{\cos x}, \qquad \csc x = \frac{1}{v} = \frac{1}{\sin x}. \tag{2}$$

Example 3 From Fig. 4.4, you see that

$$\tan\left(-\frac{2\pi}{3}\right) = \frac{-\sqrt{3}/2}{-1/2} = \sqrt{3}, \qquad \cot\left(-\frac{2\pi}{3}\right) = \frac{1}{\tan(-2\pi/3)} = \frac{1}{\sqrt{3}},$$
$$\sec\left(-\frac{2\pi}{3}\right) = \frac{1}{-1/2} = -2, \qquad \csc\left(-\frac{2\pi}{3}\right) = \frac{1}{-\sqrt{3}/2} = -\frac{2}{\sqrt{3}}. \quad \|$$

A table at the end of the text gives values of trigonometric functions of whole-numbered degree angles from 0° to 90°. The $\sin x$ and $\cos x$ columns of the table give the lengths of the legs of a right triangle of hypotenuse 1 having the given angle. You can use this data to find the functions for other angles just as you used the triangles of Fig. 4.3 in Examples 2 and 3. A quick way to evaluate trigonometric functions is to use a "scientific" pocket calculator.

4.1.2 Identities

Referring to Eqs. (1) and Fig. 4.1, you have $\sin x = v$ and $\cos x = u$, and $u^2 + v^2 = 1$. Therefore

$$\sin^2 x + \cos^2 x = 1.$$

[Note that $\sin^2 x$ means $(\sin x)^2$.] This is a fundamental trigonometric identity. There are many others, some of which are easily obtained. For example, if you divide both sides of $\sin^2 x + \cos^2 x = 1$ by $\cos^2 x$, then

$$\frac{\sin^2 x}{\cos^2 x} + \frac{\cos^2 x}{\cos^2 x} = \frac{1}{\cos^2 x}$$

or

$$\tan^2 x + 1 = \sec^2 x.$$

It is not our purpose to give a lot of drill in identities. We list in the summary the ones you might want to use in this text, and ask you to derive a few of them in the exercises.

* The trigonometric functions are sometimes called the **circular functions.**

SUMMARY 1. *Referring to Fig. 4.5, we have*

$$\sin x = v,$$

$$\cos x = u,$$

$$\tan x = \frac{v}{u} = \frac{\sin x}{\cos x},$$

$$\cot x = \frac{u}{v} = \frac{\cos x}{\sin x} = \frac{1}{\tan x},$$

$$\sec x = \frac{1}{u} = \frac{1}{\cos x},$$

$$\csc x = \frac{1}{v} = \frac{1}{\sin x}.$$

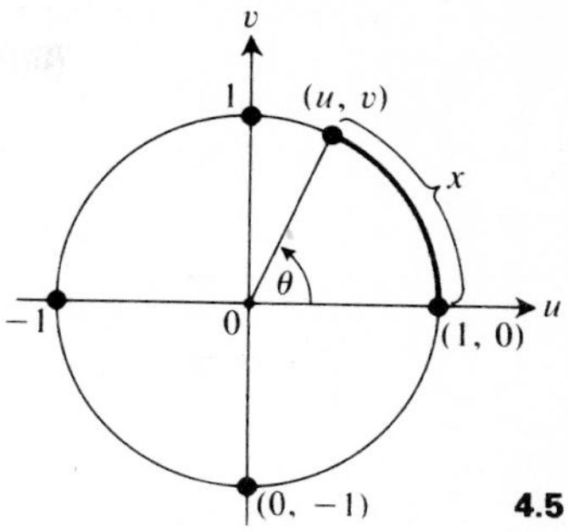

4.5

2. *Some identities:*

$$\sin^2 x + \cos^2 x = 1, \quad (3) \qquad \tan^2 x + 1 = \sec^2 x, \quad (4)$$

$$1 + \cot^2 x = \csc^2 x; \quad (5)$$

$$\sin(x + y) = \sin x \cos y + \cos x \sin y, \quad (6)$$

$$\cos(x + y) = \cos x \cos y - \sin x \sin y; \quad (7)$$

$$\sin 2x = 2 \sin x \cos x, \quad (8) \qquad \cos 2x = \cos^2 x - \sin^2 x, \quad (9)$$

$$\sin(-x) = -\sin x, \quad (10) \qquad \cos(-x) = \cos x, \quad (11)$$

$$\sin\left(x + \frac{\pi}{2}\right) = \cos x, \quad (12) \qquad \cos\left(x + \frac{\pi}{2}\right) = -\sin x, \quad (13)$$

$$\sin(x + \pi) = -\sin x, \quad (14) \qquad \cos(x + \pi) = -\cos x, \quad (15)$$

$$\sin(x + 2n\pi) = \sin x, \quad (16) \qquad \cos(x + 2n\pi) = \cos x, \quad (17)$$

$$\sin\frac{x}{2} = \pm\sqrt{\frac{1 - \cos x}{2}}, \quad (18) \qquad \cos\frac{x}{2} = \pm\sqrt{\frac{1 + \cos x}{2}}. \quad (19)$$

3. *With reference to Fig. 4.6, we have*

Law of sines: $\dfrac{\sin \theta_1}{a_1} = \dfrac{\sin \theta_2}{a_2} = \dfrac{\sin \theta_3}{a_3},$

Law of cosines: $a_3{}^2 = a_1{}^2 + a_2{}^2 - 2a_1 a_2 \cos \theta_3.$

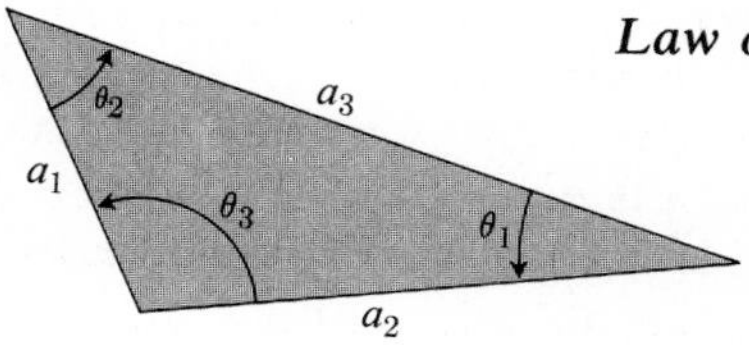

4.6

EXERCISES

In Exercises 1 *through* 20, *find the indicated value of the function.*

1. $\sin \frac{\pi}{3}$ **2.** $\cos \frac{3\pi}{2}$ **3.** $\tan \frac{5\pi}{6}$ **4.** $\sin \frac{4\pi}{3}$ **5.** $\sec \frac{5\pi}{4}$ **6.** $\tan \frac{3\pi}{4}$ **7.** $\csc \frac{7\pi}{6}$

8. $\cot \left(-\frac{2\pi}{3}\right)$ **9.** $\tan \pi$ **10.** $\cos 3\pi$ **11.** $\sec 2\pi$ **12.** $\sin 5\pi$ **13.** $\sin (-3\pi)$ **14.** $\cos (-3\pi)$

15. $\sin \left(-\frac{\pi}{2}\right)$ **16.** $\tan \frac{5\pi}{4}$ **17.** $\tan \frac{3\pi}{2}$ **18.** $\cot 5\pi$ **19.** $\sec \frac{9\pi}{4}$ **20.** $\csc \frac{23\pi}{6}$

21. If $-\pi/2 \le \theta < \pi/2$ and $\sin \theta = -1/3$, find $\cos \theta$.

22. If $\pi/2 \le \theta < 3\pi/2$ and $\tan \theta = 4$, find $\sec \theta$.

23. If $0 \le \theta < \pi$ and $\cos \theta = -1/5$, find $\cot \theta$.

24. If $\pi \le \theta < 2\pi$ and $\sec \theta = 3$, find $\tan \theta$.

25. If $\pi/2 \le \theta < 3\pi/2$ and $\sin \theta = 1/4$, find $\cot \theta$.

26. If $0 \le \theta < \pi$ and $\cos \theta = 1/3$, find $\sin 2\theta$.

27. If $-\pi/2 \le \theta < \pi/2$ and $\sin \theta = -2/3$, find $\sin 2\theta$.

28. If $0 < \theta < \pi/2$ and $\tan \theta = 3$, find $\cos 2\theta$.

29. If $0 < \theta < \pi/2$ and $\sec \theta = 4$, find $\cos 2\theta$.

30. If $0 < \theta < \pi/2$ and $\cos \theta = 1/3$, find $\sin 3\theta$.

31. As in Fig. 4.1, let the arc corresponding to x terminate at (u, v).

a) Where does the arc corresponding to $-x$ terminate?

b) Use the result in (a) to verify relations (10) and (11).

32. As in Fig. 4.1, let the arc corresponding to x terminate at (u, v).

a) Where does the arc corresponding to $x + \pi/2$ terminate?

b) Use the result in (a) to verify relations (12) and (13).

33. As in Fig. 4.1, let the arc corresponding to x terminate at (u, v).

a) Where does the arc corresponding to $x - \pi/2$ terminate?

b) Use the result in (a) to derive relations similar to (12) and (13).

In Exercises 34 *through* 42, *use the identities in the summary to verify each of the following identities.*

34. $\tan (-x) = -\tan x$ **35.** $\sec (-x) = \sec x$ **36.** $\tan \left(x + \frac{\pi}{2}\right) = -\cot x$ **37.** $\sin \left(x - \frac{\pi}{2}\right) = -\cos x$

38. $\cos \left(x - \frac{\pi}{2}\right) = \sin x$ **39.** $\sec \left(x - \frac{\pi}{2}\right) = \csc x$ **40.** $\cos (x - y) = \cos x \cos y + \sin x \sin y$

41. $\cos 2x = 2 \cos^2 x - 1 = 1 - 2 \sin^2 x$

42. $\tan (x + y) = \dfrac{\tan x + \tan y}{1 - \tan x \tan y}$

43. Let a triangle have sides of 5 and 7 units and let the angle between these two sides be $\pi/3$. Find the length of the other side of the triangle. [*Hint.* Use the law of cosines.]

4.2 TRIGONOMETRY REVIEW II: GRAPHS OF TRIGONOMETRIC FUNCTIONS

You can easily check that the graph of $y = \sin x$ is that shown in Fig. 4.7. Note that the intercepts on the x-axis are the multiples of π.

In view of the identity $\cos x = \sin (x + \pi/2)$, the graph of $\cos x$ can be obtained by translating the graph of $\sin x$ a distance $\pi/2$ to the left. The graph of $\cos x$ is shown in Fig. 4.8.

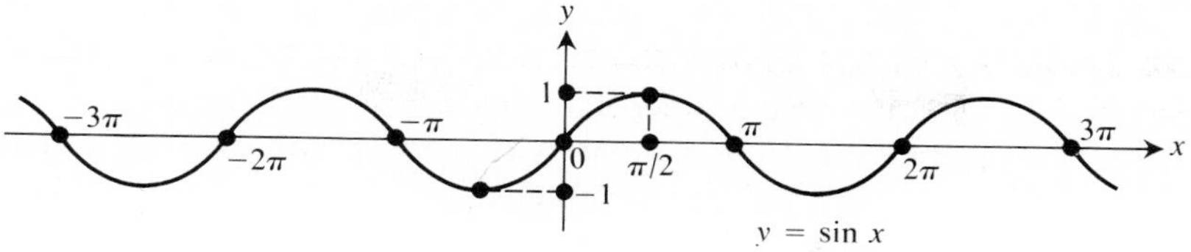

4.7

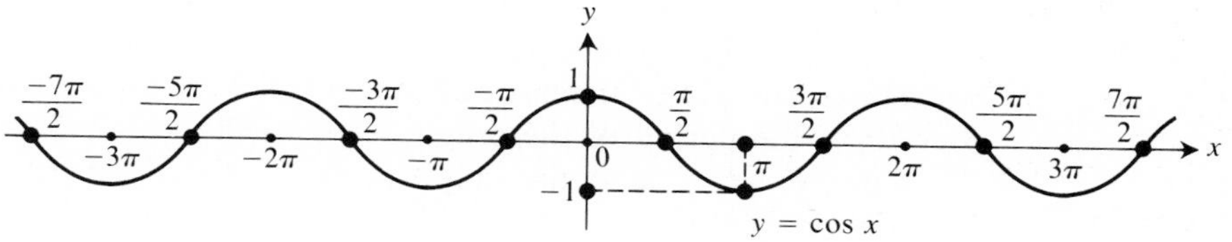

4.8

4.2.1 Graphs of the six functions

The graph of $\tan x$ has as height at each point $(\sin x)/(\cos x)$, if $\cos x \neq 0$. From the graphs of $\sin x$ and $\cos x$, you obtain the graph in Fig. 4.9 for $\tan x$. Taking the reciprocals of the heights to $\tan x$, you obtain the graph of $\cot x = 1/(\tan x)$ in Fig. 4.10. Figures 4.11 and 4.12 show the graphs of $\sec x = 1/(\cos x)$ and $\csc x = 1/(\sin x)$.

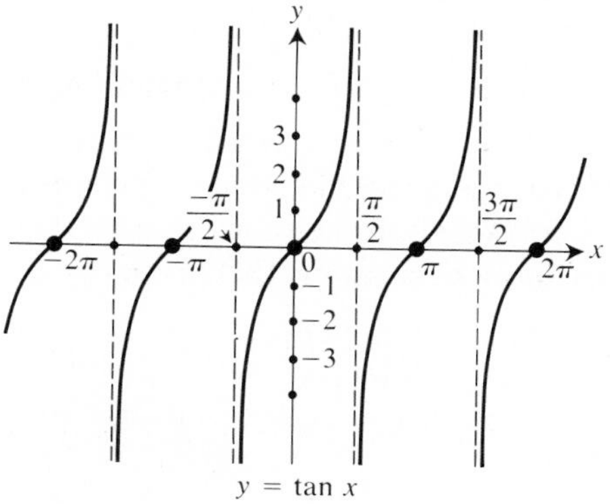

4.9

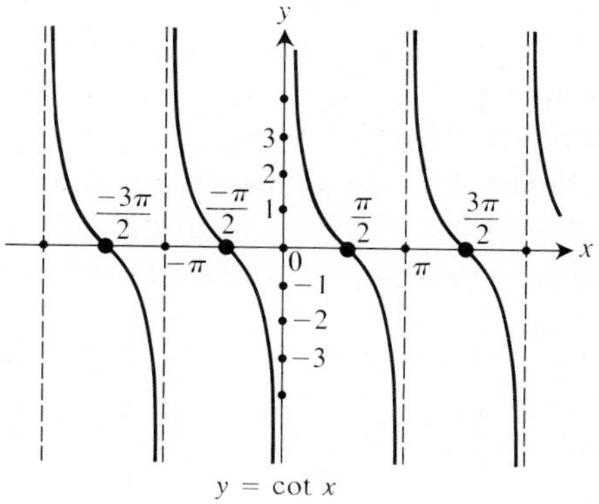

4.10

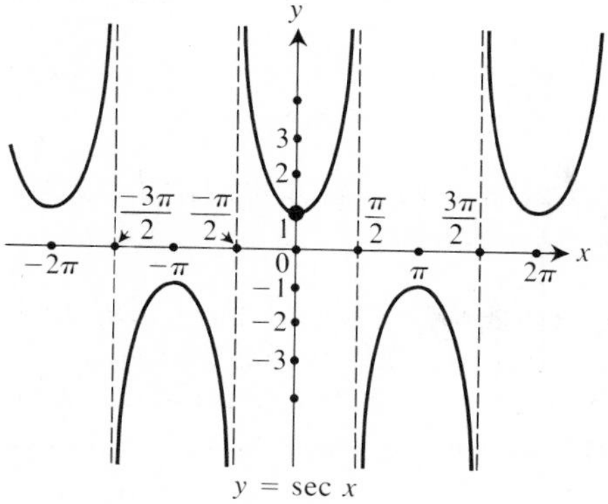

4.11

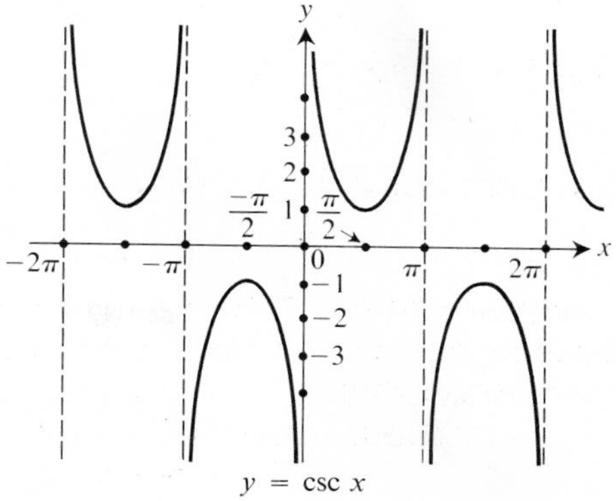

4.12

4.2.2 The graph of $y = a \cdot \sin(b(x - c))$

To find the graph of $y = a \cdot \sin(b(x - c))$, we first discover the effect of the individual constants a, b, and c on the graph.

The graph of $\sin x$ oscillates in height between -1 and 1; the graph has *amplitude* 1. It is obvious that the graph of $a(\sin x)$ oscillates between $-a$ and a, that is, has **amplitude** $|a|$. The constant a controls the amplitude of the graph.

Amplitude and period

The graph of $\sin x$ repeats itself every 2π units on the x-axis. We say it has *period* 2π. The graph of $\sin bx$ will repeat as soon as bx changes by 2π, which is as soon as x increases by $|2\pi/b|$. Thus $\sin bx$ has **period** $|2\pi/b|$, and b controls the period of the graph.

Phase

Finally, we turn to the graph of $\sin(x - c)$. You have seen that the substitution $\bar{x} = x - c$, $\bar{y} = y - 0$ amounts to translating axes to the point $(c, 0)$. Thus the graph of $\sin(x - c)$ is the graph of $\sin x$ translated c units to the right. Note that $\sin(x - c)$ is zero when $x = c$, rather than when $x = 0$. The number c is the **phase angle**.

Example 1 Let's sketch the graph of $y = 3\sin(2x + \pi)$.

SOLUTION First we rewrite:

$$y = 3\sin(2x + \pi)$$

$$= 3\sin 2\left(x - \left(-\frac{\pi}{2}\right)\right).$$

The graph has amplitude 3, period $2\pi/2 = \pi$, and phase angle $-\pi/2$. The graph is shown in Fig. 4.13. Think of moving the curve $y = \sin x$ a distance $\pi/2$ to the left, and then having it oscillate three times as high (amplitude 3 rather than 1) and twice as fast (period π rather than 2π). ||

The effect of a, b, and c on the graph of $y = a \cdot \cos(b(x - c))$ is precisely the same as for the graph of $a \cdot \sin(b(x - c))$. Indeed, multiplying any of the six functions by a and replacing x by $b(x - c)$ has an analogous significance.

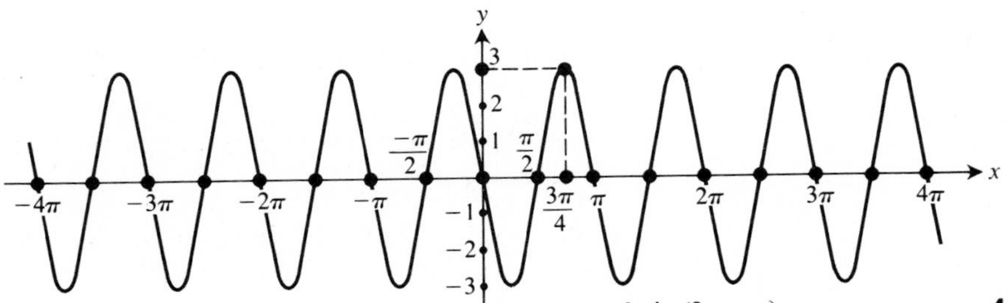

$y = 3\sin(2x + \pi)$

4.13

SUMMARY

1. *The graphs of the six trigonometric functions are shown in Figs. 4.7 through 4.12.*
2. *In the graph of $y = a \cdot \sin(b(x - c))$, the amplitude $|a|$ controls the height of the oscillation, the period (x-distance for repetition) is $|2\pi/b|$, and c is the phase angle.*

EXERCISES

In Exercises 1 through 10, find the amplitude, and period, and sketch the graph of the indicated function.

1. $\sin(-x)$ **2.** $4\cos x$ **3.** $3\sin 3x$ **4.** $2\cos(x/2)$

5. $-2\sin\left(x - \frac{\pi}{2}\right)$ **6.** $-3\cos(x + \pi)$ **7.** $3\sin(4x + \pi)$ **8.** $5\cos\left(\frac{x}{2} - \frac{\pi}{4}\right)$

9. $5\sin\left(\frac{x}{4} - \pi\right)$ **10.** $-2\cos(2x + 5\pi)$

In Exercises 11 through 18, find the period (x-distance for repetition) and sketch the graph of the indicated function.

11. $-\tan x$ **12.** $\cot 2x$ **13.** $3\sec x$ **14.** $\csc(2x - \pi)$

15. $\sin^2 x$ **16.** $4\cos^2 x$

17. $\sin x + 2\cos x$ [*Hint.* Sketch $\sin x$ and $2\cos x$ on the same axes and add their heights.]

18. $2\sin 2x - \cos(x/2)$ [*Hint.* Proceed as in Exercise 17.]

4.3 DIFFERENTIATION OF TRIGONOMETRIC FUNCTIONS

Now that you know a bit of calculus, every time you study new functions, you just can't wait to find their derivatives! When you try to compute the derivative of $\sin x$, you will run right into

4.3.1 $\lim_{x\to 0}(\sin x) = 1$

$$\lim_{x\to 0} \frac{\sin x}{x} \quad \text{and} \quad \lim_{x\to 0} \frac{\cos x - 1}{x}.$$

Note that the function $(\sin x)/x$ is not defined at 0, and that

$$\lim_{x\to 0}(\sin x) = \lim_{x\to 0} x = 0.$$

Figure 4.14 shows once more part of the circle $u^2 + v^2 = 1$. A positive value of x is indicated by the length of the dark arc. The altitude of the

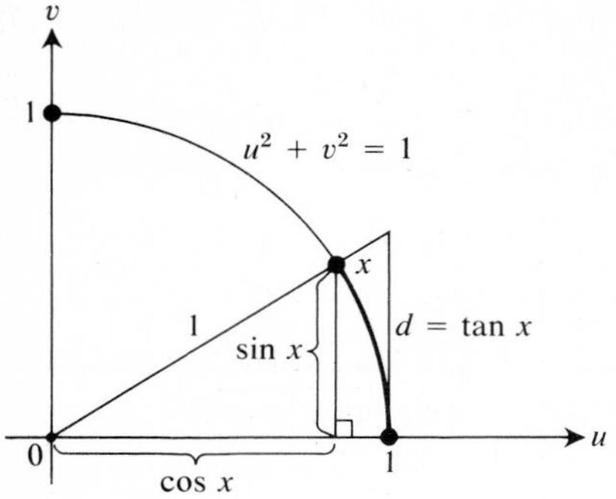

4.14

small triangle in the figure is $\sin x$. Since the large and small triangles are similar, if d is the altitude of the large triangle, you have

$$\frac{d}{\sin x} = \frac{1}{\cos x}$$

so $d = \tan x$. Clearly the area of the small triangle in Fig. 4.14 is less than the area of the sector of the circle having arc of length x, which in turn is less than the area of the large triangle. The area of the sector of the circle is the fraction $x/2\pi$ of the area $\pi \cdot 1^2 = \pi$ of the whole circle, so you have

$$\frac{\sin x \cos x}{2} < \frac{x}{2\pi} \cdot \pi < \frac{\tan x}{2}. \tag{1}$$

Multiplying (1) by $2/(\sin x)$, you obtain

$$\cos x < \frac{x}{\sin x} < \frac{1}{\cos x}. \tag{2}$$

It is easy to see that (2) is valid also for $x < 0$ but near zero; this follows at once from the relations

$$\sin(-x) = -\sin x \qquad \text{and} \qquad \cos(-x) = \cos x.$$

The definition of cosine and the graph in Fig. 4.8 show that

$$\lim_{x\to 0} (\cos x) = 1,$$

so

$$\lim_{x\to 0} \frac{1}{\cos x} = \frac{1}{1} = 1.$$

But from (2), you see that $x/(\sin x)$ is "trapped" between $\cos x$ and $1/(\cos x)$, which both approach 1 as x approaches zero, so you must have

$$\lim_{x\to 0} \frac{x}{\sin x} = 1.$$

Of course, then

$$\lim_{x\to 0}\frac{\sin x}{x} = \lim_{x\to 0}\frac{1}{[x/(\sin x)]} = \frac{1}{1} = 1.$$

Turning to $\lim_{x\to 0}(\cos x - 1)/x$, for x near zero,

$$\frac{\cos x - 1}{x} = \frac{\cos x - 1}{x}\cdot\frac{\cos x + 1}{\cos x + 1} = \frac{-\sin^2 x}{x(\cos x + 1)}$$

$$= -\frac{\sin x}{x}\cdot\frac{\sin x}{\cos x + 1}.$$

Since $\lim_{x\to 0}(\sin x)/x = 1$, you have

$$\lim_{x\to 0}\frac{\cos x - 1}{x} = \left(\lim_{x\to 0}\left(-\frac{\sin x}{x}\right)\right)\left(\lim_{x\to 0}\frac{\sin x}{\cos x + 1}\right) = (-1)\cdot\left(\frac{0}{2}\right) = 0.$$

These limits are so basic that we summarize them in a theorem.

Theorem 4.1 *For the functions* $\sin x$ *and* $\cos x$,

$$\lim_{x\to 0}\frac{\sin x}{x} = 1 \qquad \textit{and} \qquad \lim_{x\to 0}\frac{\cos x - 1}{x} = 0. \tag{3}$$

Example 1 We find $\lim_{x\to 0}(\sin 5x)/x$.

SOLUTION Using (3), we obtain

$$\lim_{x\to 0}\frac{\sin 5x}{x} = \lim_{x\to 0}\left(5\cdot\frac{\sin 5x}{5x}\right) = 5\cdot 1 = 5. \quad \|$$

More practice in using the limits in (3) is given in the exercises.

4.3.2 The derivative of sin *x*

We must go back to the definition of the derivative

$$f'(x_1) = \lim_{\Delta x\to 0}\frac{f(x_1+\Delta x) - f(x_1)}{\Delta x}$$

to find the derivative of $f(x) = \sin x$. Forming the difference quotient and using some relations for trigonometric functions, we obtain

$$\frac{f(x_1+\Delta x) - f(x_1)}{\Delta x} = \frac{\sin(x_1+\Delta x) - \sin x_1}{\Delta x}$$

$$= \frac{\sin x_1\cos\Delta x + \cos x_1\sin\Delta x - \sin x_1}{\Delta x}$$

$$= \frac{\cos x_1\sin\Delta x + (\sin x_1)(\cos\Delta x - 1)}{\Delta x}$$

$$= \cos x_1\frac{\sin\Delta x}{\Delta x} + \sin x_1\frac{\cos\Delta x - 1}{\Delta x}.$$

Therefore, using the limits (3),

$$f'(x_1) = \lim_{\Delta x \to 0} \frac{f(x_1 + \Delta x) - f(x_1)}{\Delta x}$$

$$= (\cos x_1)\left(\lim_{\Delta x \to 0} \frac{\sin \Delta x}{\Delta x}\right) + (\sin x_1)\left(\lim_{\Delta x \to 0} \frac{\cos \Delta x - 1}{\Delta x}\right)$$

$$= (\cos x_1)(1) + (\sin x_1)(0) = \cos x_1.$$

Thus

$$\frac{d(\sin x)}{dx} = \cos x.$$

By the chain rule,

$$\frac{d(\sin u)}{dx} = \frac{d(\sin u)}{du} \cdot \frac{du}{dx} = (\cos u)\frac{du}{dx} \tag{4}$$

if u is a differentiable function of x.

Example 2 The derivative of $y = \sin(x^3)$ is

$$\frac{dy}{dx} = \cos(x^3) \cdot \frac{d(x^3)}{dx} = 3x^2 \cos x^3. \quad \|$$

The reason for radian measure

Now you can see why radian measure of an angle is the convenient measure when doing calculus. Let x be radian measure and t be degree measure for an angle. Then

$$x = \frac{\pi}{180} t$$

and

$$\frac{d(\sin x)}{dt} = \frac{d(\sin x)}{dx} \cdot \frac{dx}{dt} = (\cos x)\left(\frac{\pi}{180}\right).$$

In other words, if you used degree measure, you would have the nuisance factor $\pi/180$ in your differentiation formulas for trigonometric functions.

4.3.3 Derivatives of the other trigonometric functions

Since $\cos x = \sin(x + (\pi/2))$, you have

$$\frac{d(\cos x)}{dx} = \frac{d(\sin(x + \pi/2))}{dx} = \cos\left(x + \frac{\pi}{2}\right).$$

But $\cos(x + (\pi/2)) = -\sin x$, so

$$\frac{d(\cos x)}{dx} = -\sin x.$$

Using the chain rule,

$$\frac{d(\cos u)}{du} = (-\sin u)\frac{du}{dx}. \tag{5}$$

The formula $d(\cos x)/dx = -\sin x$ can also be derived by differentiating the identity $\sin^2 x + \cos^2 x = 1$ implicitly:

$$2 \sin x \cos x + 2 \cos x \frac{d(\cos x)}{dx} = 0,$$

$$\frac{d(\cos x)}{dx} = \frac{-2 \sin x \cos x}{2 \cos x} = -\sin x.$$

The other four trigonometric functions are quotients involving only $\sin x$ and $\cos x$, so their derivatives can be found using the quotient rule for differentiation. For example,

$$\frac{d(\tan x)}{dx} = \frac{d[(\sin x)/(\cos x)]}{dx} = \frac{(\cos x)(\cos x) - (\sin x)(-\sin x)}{\cos^2 x}$$

$$= \frac{\cos^2 x + \sin^2 x}{\cos^2 x} = \frac{1}{\cos^2 x} = \sec^2 x.$$

In a similar fashion, one finds that

$$\frac{d(\cot x)}{dx} = -\csc^2 x,$$

$$\frac{d(\sec x)}{dx} = \sec x \tan x,$$

$$\frac{d(\csc x)}{dx} = -\csc x \cot x.$$

One then gets chain-rule formulas analogous to (4) and (5). The formulas are given in the summary. It is recommended that you memorize these formulas. Note that if you know the derivatives of the three functions $\sin x$, $\tan x$, $\sec x$, then the derivatives of each of the corresponding cofunctions $\cos x$, $\cot x$, $\csc x$ are found by changing the derivatives to the cofunctions and changing sign.

Example 3 We find dy/dx if $y = x^3 \tan 2x$.

SOLUTION We have

$$\frac{dy}{dx} = x^3 \frac{d(\tan 2x)}{dx} + (\tan 2x)\frac{d(x^3)}{dx} = x^3(\sec^2 2x)(2) + (\tan 2x)(3x^2)$$

$$= 2x^3(\sec^2 2x) + 3x^2(\tan 2x). \quad \|$$

.RY 1. *Limits:*

$$\lim_{x\to 0}\frac{\sin x}{x} = 1, \qquad \lim_{x\to 0}\frac{\cos x - 1}{x} = 0.$$

2. *Differentiation formulas:*

$$\frac{d(\sin u)}{dx} = (\cos u)\frac{du}{dx}, \qquad \frac{d(\cos u)}{dx} = (-\sin u)\frac{du}{dx},$$

$$\frac{d(\tan u)}{dx} = (\sec^2 u)\frac{du}{dx}, \qquad \frac{d(\cot u)}{dx} = (-\csc^2 u)\frac{du}{dx},$$

$$\frac{d(\sec u)}{dx} = (\sec u \tan u)\frac{du}{dx}, \qquad \frac{d(\csc u)}{dx} = (-\csc u \cot u)\frac{du}{dx},$$

EXERCISES

In Exercises 1 through 10, find the indicated limit, if it exists.

1. $\lim_{x\to 0}\frac{\sin x}{|x|}$ **2.** $\lim_{t\to 0}\frac{\sin 2t}{t}$ **3.** $\lim_{u\to \pi/2}\frac{\cos u}{(\pi/2) - u}$ **4.** $\lim_{x\to 0}(x \csc^2 x)$

5. $\lim_{\theta\to 0}(\theta^2 \csc^2\theta)$ **6.** $\lim_{v\to 0}\frac{\tan 3v}{v}$ **7.** $\lim_{x\to 0}\frac{\cos 2x}{\cos 3x}$ **8.** $\lim_{t\to 0}\frac{\sin 2t}{\sin 3t}$

9. $\lim_{t\to -1}\sin\left(\frac{1}{t+1}\right)$ **10.** $\lim_{x\to 0}\frac{\cos^2 x - 1}{x^2}$

In Exercises 11 through 30, find the derivative of the given function. You need not simplify your answers.

11. $x \cos x$ **12.** $x^2 \tan x$ **13.** $(x^2 + 3x)\sec x$ **14.** $\frac{\csc x}{x}$

15. $\sin^2 x$ **16.** $\sin 2x$ **17.** $\sec^2 x$ **18.** $\sin x \tan x$

19. $\frac{x}{\cot x}$ **20.** $\frac{x^2 - 2x}{\csc x}$ **21.** $y = \sin 2x$ **22.** $y = \sec(3x + 1)$

23. $y = \cos^2(2 - 3x)$ **24.** $y = \cot^2 x$ **25.** $y = \sin^2 x \cos^2 x$ **26.** $y = \tan x \sec 2x$

27. $y = \sqrt{\cot^2 x + \csc^2 x}$ **28.** $y = \sqrt{8x^2 + \cos^2 x}$ **29.** $y = \sin(\tan 3x)$ **30.** $y = \csc(x + \cos x^2)$

In Exercises 31 through 36, find dy/dx at the indicated point using implicit differentiation.

31. $x \cos y + y \sin x = \frac{\pi}{2}$; $\left(\frac{\pi}{2}, \pi\right)$ **32.** $(\sin x)(\cos y) = \frac{1}{2}$; $\left(\frac{\pi}{4}, \frac{\pi}{4}\right)$

33. $\sin(xy) + 3y = 4$; $\left(\frac{\pi}{2}, 1\right)$ **34.** $\tan xy = 1$; $\left(\frac{\pi}{4}, 1\right)$

35. $\sec x + \tan y = 1$; $(0, 0)$ **36.** $\csc\left(\frac{\pi x y^2}{2}\right) + \sin\left(\frac{\pi y}{2}\right) + y = 3$; $(1, 1)$

37. Find the equation of the tangent line to $y = \sin x$ when $x = \pi/4$.

38. Find the equation of the line normal to $y = \tan x$ when $x = 3\pi/4$.

39. Use a differential to estimate $\sin 31°$.

40. Use a differential to estimate $\tan 128°$.

exercise sets for chapter 4

review exercise set 4.1

1. Find: a) $\tan \frac{5}{6}\pi$ b) $\cos \frac{5}{4}\pi$
2. If $0 \le \theta < \pi$ and $\sec\theta = -5$, find $\sin\theta$.
3. Use the identity $\sin(x - y) = \sin x \cos y - \cos x \sin y$ to compute $\sin 15°$.
4. Find the amplitude and period, and sketch the graph of $y = 3\sin(2x - \pi)$.
5. Find the period of $|\sin 3x|$.
6. Find the indicated limit, if it exists.

   a) $\lim\limits_{x\to 0} \dfrac{\sin x}{x^2 + 4x}$ b) $\lim\limits_{x\to 3} \dfrac{\sin(x^2 - 9)}{x - 3}$
7. Find dy/dx.

   a) $y = \sin^3 2x$ b) $y = x^2 \csc x^3$
8. Find dy/dx if $x^2 \cos y + y^3 = -3$.

review exercise set 4.2

1. Find:

   a) $\sin\dfrac{11\pi}{6}$ b) $\cot\left(-\dfrac{4}{3}\pi\right)$
2. If $\pi/2 \le \theta < 3\pi/2$ and $\tan\theta = 2/3$, find $\sin\theta$.
3. If a triangle has sides of lengths 2, 4, and 5, find $\cos\theta$ if θ is the angle opposite the side of length 5.
4. Find the period and amplitude of $-\frac{1}{2}\cos(x/3)$.
5. Sketch the graph of $y = 3\sec(x/2)$.
6. Find the indicated limit, if it exists.

   a) $\lim\limits_{x\to 0} \dfrac{\sin^2 x}{x^2 + 4x}$ b) $\lim\limits_{x\to 0} (2x + 4)\cdot \sin\left(\dfrac{1}{x + 1}\right)$
7. Find dy/dx.

   a) $y = \tan(x^2 + 1)$ b) $y = \dfrac{\sin^2 x}{x - 4}$
8. Find the equation of the line tangent to $y = 2\cos(x - (\pi/2))$ at the origin.

more challenging exercises 4

In Exercises 1 *through* 6, *find the indicated limit, if it exists.*

1. $\lim\limits_{x\to\infty} x \sin\dfrac{1}{x}$
2. $\lim\limits_{x\to\infty} x^2 \sin\dfrac{1}{x}$
3. $\lim\limits_{x\to 0} \dfrac{\cos 2x - 1}{x}$
4. $\lim\limits_{x\to\pi/4} \dfrac{\tan x - 1}{x - \pi/4}$ [*Hint.* Sometimes a limit can be recognized as the definition of some derivative.]
5. $\lim\limits_{x\to\pi/6} \dfrac{2\sin x - 1}{6x - \pi}$
6. $\lim\limits_{x\to 0} \dfrac{\sin(\sin x)}{x}$

applications of the derivative

5

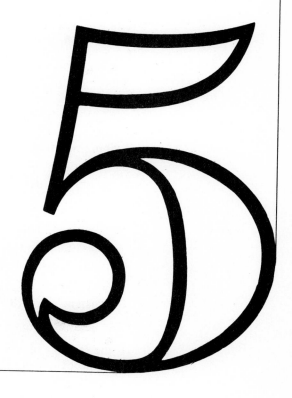

5.1 RELATED RATE PROBLEMS

Recall that the derivative dy/dx gives the instantaneous rate of change of y with respect to x. Leibniz notation is very useful in keeping track of just what rate of change you are working with. For example, imagine that a pebble is dropped into a calm pond and a circular wave spreads out from the point where the pebble was dropped. You might be interested in:

$$\frac{dr}{dt} = \text{rate of increase of } \textit{radius} \text{ per unit increase in } \textit{time},$$

$$\frac{dA}{dt} = \text{rate of increase of } \textit{area} \text{ per unit increase in } \textit{time},$$

$$\frac{dA}{dr} = \text{rate of increase of } \textit{area} \text{ per unit increase in } \textit{radius},$$

$$\frac{dC}{dt} = \text{rate of increase of } \textit{circumference} \text{ per unit increase in } \textit{time},$$

etc. The d-notation helps you remember which rate of change you want.

In many rate-of-change problems, you want to find the time rate of change of a quantity Q if you know the time rate of change of one or more related quantities, say r and s. We give one convenient step-by-step outline you might like to follow in solving such *related rate problems.* (The letters may be different from Q, r, and s in a problem, of course.)

Outline for related rate problems

STEP 1. Decide what rate of change is desired and express it in Leibniz notation:

$$\text{Find } \frac{dQ}{dt} \quad \text{when} \quad t = \underline{\qquad}.$$

STEP 2. Decide what rates of change are given and express this data in Leibniz notation:

$$\text{Given } \frac{dr}{dt} = \underline{\qquad} \quad \text{and} \quad \frac{ds}{dt} = \underline{\qquad} \quad \text{when} \quad t = \underline{\qquad}.$$

STEP 3. Find an equation relating Q, r, and s. You may have to draw a figure, or use some geometric formula.

STEP 4. Differentiate the relation in Step 3 (often implicit differentiation) to obtain a relation between dQ/dt, dr/dt, and ds/dt.

STEP 5. Put in values of r, s, and Q and of dr/dt and ds/dt corresponding to the instant when dQ/dt is desired, and solve for dQ/dt.

The following examples illustrate the use of the steps just described.

Example 1 If the radius r of a circular disk is increasing at the rate of 3 in./sec, let's find the rate of increase of its area when $r = 4$ in.

SOLUTION We let A be the area and r be the radius.

STEP 1. Find dA/dt when $r = 4$ inches.

STEP 2. Given $dr/dt = 3$ in./sec.

STEP 3. $A = \pi r^2$.

STEP 4. $dA/dt = 2\pi r \cdot dr/dt$.

STEP 5. When $r = 4$ and $dr/dt = 3$, $dA/dt = 2\pi \cdot 4 \cdot 3 = 24\pi$ in²/sec. ||

Example 2 Ship A passes a buoy at 9:00 A.M. and continues on a northward course at a rate of 12 mph. Ship B, traveling at 18 mph, passes the same buoy on its eastward course at 10:00 A.M. the same day. Let's find the rate at which the distance between the ships is increasing at 11:00 A.M. that day.

SOLUTION We draw a figure and assign letter variables as shown in Fig. 5.1.

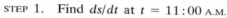

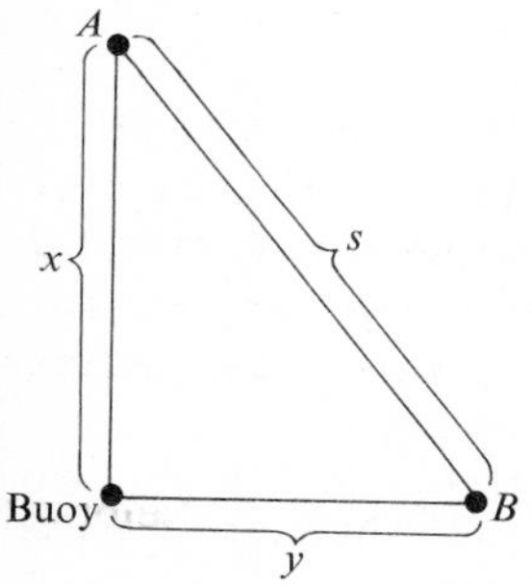

5.1

STEP 1. Find ds/dt at $t = 11{:}00$ A.M.

STEP 2. Given $dx/dt = 12$ mph and $dy/dt = 18$ mph.

STEP 3. $s^2 = x^2 + y^2$.

STEP 4. $2s(ds/dt) = 2x(dx/dt) + 2y(dy/dt)$.

STEP 5. At 11:00 A.M., $x = 24$, $y = 18$, $dx/dt = 12$, $dy/dt = 18$, and

$$s = \sqrt{x^2 + y^2} = \sqrt{(24)^2 + (18)^2} = 30.$$

Putting these values into the equation of Step 4,

$$2 \cdot 30 \frac{ds}{dt} = 2 \cdot 24 \cdot 12 + 2 \cdot 18 \cdot 18,$$

so

$$\frac{ds}{dt} = \frac{2 \cdot 24 \cdot 12 + 2 \cdot 18 \cdot 18}{2 \cdot 30} = \frac{24 \cdot 12 + 18 \cdot 18}{30}$$

$$= \frac{612}{30} = \frac{102}{5} = 20.4 \text{ mph.} \quad ||$$

SUMMARY 1. *To solve related rate problems, you could follow Step* 1 *through Step* 5 *described just before Example* 1.

EXERCISES

1. Find the rate at which the area of an equilateral triangle is increasing when it is 10 in. on a side, if the length of each side is increasing at the rate of 2 in./.min.

2. A particle starts at the origin in the plane and travels on the curve $y = \sqrt{x}$. If the x-coordinate of the particle increases at a uniform rate of 16 units/sec, find the rate of increase of the y-coordinate 9 sec after the particle started from the origin.

3. An 18-ft ladder is leaning against a vertical wall. If the bottom of the ladder is pulled away from the wall at a constant rate of 3 ft/sec, find the rate at which the top is sliding down the wall when the bottom is 8 ft from the wall.

4. A ship sails out of New York harbor at a rate of 20 ft/sec. The closest it comes to the Statue of Liberty is 1200 ft at a certain time t_0. Find the rate at which the distance from the ship to the statue is increasing 25 sec later. (Assume the ship travels in a straight line during these 25 sec.)

5. Jim is 6 ft tall and is walking at night straight toward a lighted street lamp at a rate of 5 ft/sec. If the lamp is 20 ft above the ground, find the rate at which his shadow is shortening when he is 30 ft from the lamp post.

6. A spherical balloon is being inflated so that its volume increases at a constant rate of 8 ft^3/min. Find the rate of increase of the radius when the radius is 3 ft.

7. Water is being poured into an inverted cone (vertex down) of radius 4 in. and height 10 in. at a rate of 3 in.^{3}/sec. Find the rate at which the water level is rising when the depth of water over the vertex is 5 in. (For a cone of volume V, radius r, and height h, we have $V = (\frac{1}{3})\pi r^2 h$.)

8. Sue is standing on a dock and pulling a boat in to the dock by means of a rope tied to a ring in the bow of the boat. If the ring is 2 ft above the water level and her hands are 7 ft above the water level, and if she is pulling in the rope at a uniform rate of 2 ft/sec, find the speed with which the boat is approaching the dock when it is 12 ft from the dock.

9. Sand is being poured at a rate of 2 ft^3/min to form a conical pile whose height is always three times the radius. Find the rate at which the area of the base of the cone is increasing when the cone is 4 ft high. (For a cone of volume V, height h, and radius r, we have $V = (\frac{1}{3})\pi r^2 h$.)

10. A bridge goes straight across a river at a height of 60 ft. A car on the bridge traveling at 40 ft/sec passes directly over a boat traveling up the river at 15 ft/sec at a time t_0. Find the rate at which the distance between them is increasing 3 sec later.

11. The vertex angle opposite the base of an isosceles triangle with equal sides of constant length 10 ft is increasing at a rate of $\frac{1}{2}$ radian/min.

    a) Find the rate at which the base of the triangle is increasing when the vertex angle is 60°.

    b) Find the rate at which the area of the triangle is increasing when the vertex angle is 60°.

12. A light tower is located one mile directly offshore from a point P on a straight coastline. The beam of light revolves at the rate of $\frac{1}{10}$ radian/sec. Find the rate at which the spot of light on the shore is moving along the coastline when (a) the spot is at P, and (b) the spot is two miles from P along the shore.

13. The vertex angle of a right circular cone of constant slant height l is decreasing at a constant rate of $\frac{1}{5}$ radian/sec. Find the rate at which the volume is changing when the vertex angle is 90°.

5.2 NEWTON'S METHOD

Much of high school algebra is devoted to finding a solution of an equation of the form $f(x) = 0$. Solving $f(x) = b$ is essentially the same problem, for this amounts to solving $g(x) = 0$ where $g(x) = f(x) - b$.

Suppose you wish to solve $f(x) = 0$ for a differentiable function f. This section describes *Newton's method* for finding successive approximations of a solution. First, find a number a_1, which you believe is close to a solution. You might, for example, substitute a few values of x to see where $f(x)$ is close to zero, or use a computer to plot the graph. The following fact is sometimes useful; it is proved in courses in advanced calculus.

Theorem 5.1 *(Weierstrass Intermediate-Value Theorem) If f is continuous at every point in $[a, b]$ and if $f(a)$ and $f(b)$ have opposite sign, then $f(x) = 0$ has a solution where $a < x < b$.*

Figure 5.2 shows the graph of a continuous function f where $f(a) < 0$ but $f(b) > 0$. The graph of such a continuous function can be traced with a pencil without ever removing the pencil from the page. The pencil moves from below the x-axis to above it, and must touch the axis where it crosses it. At that point x_1 where the pencil touches the axis, you have $f(x_1) = 0$. You can't get from one side of a road to the other side without crossing the road. These arguments should be viewed as an intuitive explanation of Theorem 5.1, not as a proof!

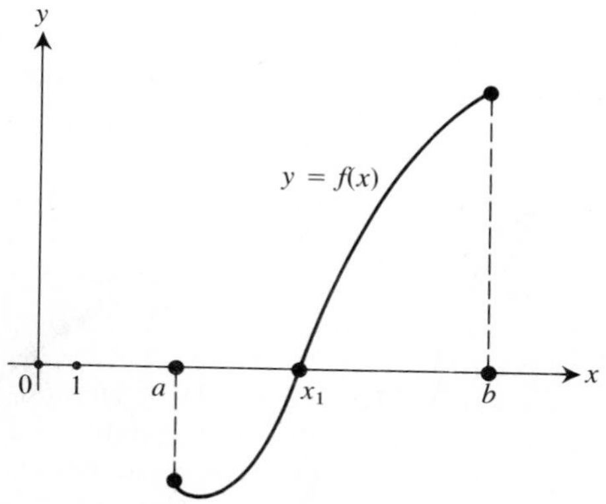

5.2

Example 1 Exercise 3 asks you to approximate a solution of $x^3 + x - 1 = 0$. Let $f(x) = x^3 + x - 1$. Then $f(0) = -1$ and $f(1) = 1$, so by the Intermediate-Value Theorem, the equation $x^3 + x - 1 = 0$ has a solution in the interval $0 < x < 1$. ‖

Suppose now you have found your approximate solution a_1 of the equation $f(x) = 0$. Look at the graph in Fig. 5.3(a). It appears that the tangent line to the graph of f at the point $(a_1, f(a_1))$ intersects the x-axis at a point a_2 which is a better approximation of a solution than a_1. Repeating this construction, starting with a_2, you expect to find a better approximation a_3, etc.

Referring to Fig. 5.3(b), we can find a formula for the next approximation a_{i+1} if we know the approximation a_i. The tangent line to $y = f(x)$ where $x = a_i$ goes through the point $(a_i, f(a_i))$ and has slope $f'(a_i)$. Its equation is therefore

$$y - f(a_i) = f'(a_i)(x - a_i). \tag{1}$$

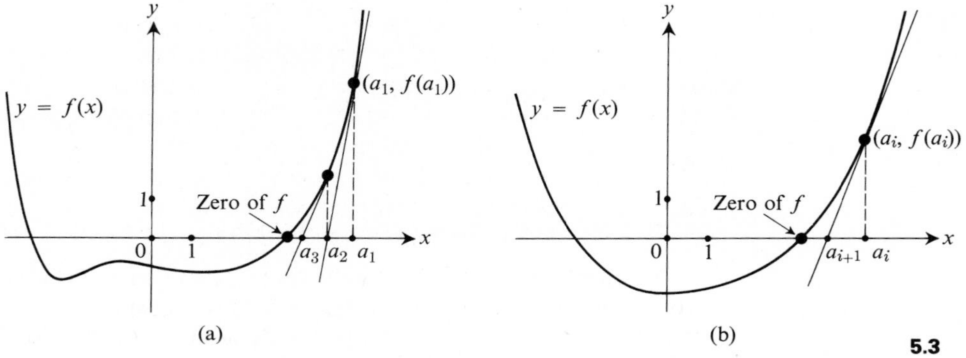

5.3

To find the point a_{i+1} where the line crosses the x-axis, set $y = 0$ in (1) and solve for x:

$$-f(a_i) = f'(a_i)(x - a_i)$$

$$x - a_i = -\frac{f(a_i)}{f'(a_i)},$$

$$x = a_i - \frac{f(a_i)}{f'(a_i)}.$$

Thus we have the recursion relation

The recursion formula

$$a_{i+1} = a_i - \frac{f(a_i)}{f'(a_i)}. \tag{2}$$

Example 2 Let's use Newton's method to approximate $\sqrt{2}$ by approximating a solution of $f(x) = x^2 - 2 = 0$.

SOLUTION We have $f'(x) = 2x$. The recursion formula (2) becomes

$$a_{i+1} = a_i - \frac{a_i^2 - 2}{2a_i}.$$

The accompanying table shows successive approximations starting with $a_1 = 2$. These were easily found using a pocket calculator.

i	a_i	a_{i+1}
1	2	$2 - \frac{2}{4} = 1.5$
2	1.5	$1.5 - \dfrac{0.25}{3} = 1.416666$
3	1.416666	$1.416666 - \dfrac{0.0069425}{2.833332} = 1.414215$
4	1.414215	$1.414215 - \dfrac{0.000004}{2.828430} = 1.414214$
5	1.414214	

Only four iterations gave us at least six significant figure accuracy. ‖

Newton's method will not always converge to a solution. For example, if you choose a_1 as shown in Fig. 5.4, the successive iterates approach ∞ rather than the solution of $f(x) = 0$. It is also possible for the iterates to oscillate back and forth without converging.

Newton's method really amounts to repeated approximation by differentials at $x = a_i$. Recall that

$$dy = f'(a_i)\,dx.$$

If you set $dy = -f(a_i)$, which is the change you desire in y to make $f(x)$ zero, then

$$dx = \frac{-f(a_i)}{f'(a_i)}$$

Thus you should change x from a_i to

$$a_i + dx = a_i - \frac{f(a_i)}{f'(a_i)},$$

which is precisely formula (2) for a_{i+1} in Newton's method.

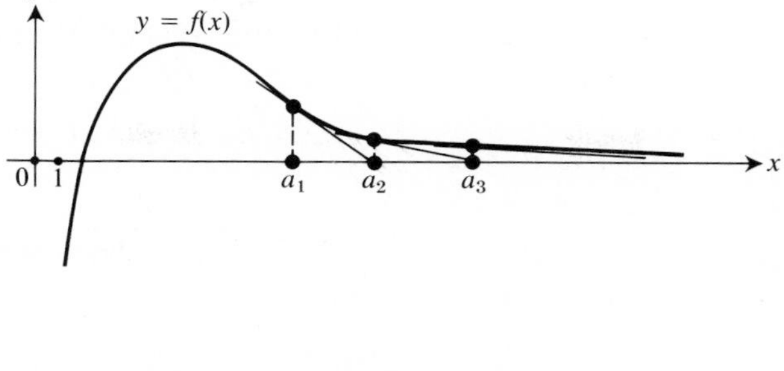

5.4

Appendix 1 contains a computer program NEWTON, which finds solutions of $f(x) = 0$ by Newton's method. Printout 5.1 show the solutions of $x^3 + 10x^2 + 8x - 50 = 0$. We started with $a_1 = -8$, which was obtained as a trial approximation from Printout 1.2 of the graph of $y = x^3 + 10x^2 + 8x - 50$, shown in Section 1.4. Two more program runs with $a_1 = -5$ and $a_1 = 2$ found the other solutions indicated by the graph in Printout 1.2. We had the computer print successive approximations, so you could see how quickly Newton's method converges and yields eight-significant-figure accuracy. It is an efficient way to find where a differentiable function is zero.

```
run

NEWTON

INITIAL APPROXIMATION?-8

APPROXIMATION X     F(X)

-8                   14
-8.350000000E+00   -1.757874998E+00
-8.314959884E+00   -1.843551389E-02
-8.314584542E+00   -2.105420137E-06
-8.314584499E+00   -3.197442311E-14
-8.314584499E+00   -2.486899575E-14

NEWTON'S METHOD CONVERGED TO A SOLUTION AT X = -8.314584499E+00


run

NEWTON

INITIAL APPROXIMATION?-5

APPROXIMATION X     F(X)

-5                   35
-2.941176470E+00   -1.246692450E+01
-3.442420374E+00    1.696476351E-01
-3.435714306E+00   -1.441581534E-05
-3.435714876E+00   -1.065814104E-13
-3.435714876E+00   -7.105427358E-15

NEWTON'S METHOD CONVERGED TO A SOLUTION AT X = -3.435714876E+00


run

NEWTON

INITIAL APPROXIMATION?2

APPROXIMATION X     F(X)

 2                   14
 1.766666667E+00    8.584074105E-01
 1.750377071E+00    4.055547129E-03
 1.750299377E+00    9.206187102E-08
 1.750299375E+00   -1.065814104E-14

NEWTON'S METHOD CONVERGED TO A SOLUTION AT X =  1.750299375E+00
```

Printout 5.1

SUMMARY

1. *(Weierstrass Intermediate-Value Theorem) If f is continuous at every point in $[a, b]$ and if $f(a)$ and $f(b)$ have opposite sign, then $f(x) = 0$ has a solution where $a < x < b$.*
2. *(Newton's method to solve $f(x) = 0$) First decide on an approximate solution a_1 of $f(x) = 0$, and then determine successive approximations a_2, a_3, a_4, ... of a solution using the recursion formula*

$$a_{i+1} = a_i - \frac{f(a_i)}{f'(a_i)}.$$

EXERCISES

1. Estimate $\sqrt{3}$ by using Newton's method for solving $f(x) = x^2 - 3 = 0$, starting with $a_1 = 2$ and finding a_3.

2. Repeat Exercise 1, but start with $a_1 = 1$.

3. Use Newton's method to approximate a solution of $x^3 + x - 1 = 0$ starting with $a_1 = 1$ and finding a_3.

4. Use Newton's method to compute $\sqrt{7}$ until the difference between successive approximations is less than 0.0005.

5. Use Newton's method to find a solution of $x^3 - x + 16 = 0$ until the difference between successive approximations is less than 0.0005.

calculator exercises

6. Use Newton's method to estimate $\sqrt{17}$, a solution of $x^2 - 17 = 0$, starting with $a_1 = 4$.

7. Use Newton's method to estimate $\sqrt[3]{25}$, a solution of $x^3 - 25 = 0$, starting with $a_1 = 3$.

8. Use Newton's method to estimate a solution of $x - 2 \sin x = 0$, starting with $a_1 = 2$.

9. Use Newton's method to estimate a solution of $x^x - 5 = 0$, starting with $a_1 = 2$. [*Hint.* You can compute the derivatives you need using the approximation

$$f'(a) \approx \frac{f(a + \Delta x) - f(a - \Delta x)}{2 \cdot \Delta x} \qquad \text{for small } \Delta x.]$$

additional exercises

Exercises 10 *through* 16 *are designed to further explore and illustrate the Intermediate-Value Theorem* (*Theorem* 5.1) *for those who are interested. At least read Exercise* 10 *before trying the other exercises.*

10. Prove the following corollary of Theorem 5.1.

If $f(x)$ is continuous for x in $[a, b]$ and if L is a number between $f(a)$ and $f(b)$, then $f(c) = L$ for some number c where $a < c < b$.

This is the way the Intermediate-Value Theorem is usually stated. A continuous function on an interval $[a, b]$ assumes every value L between $f(a)$ and $f(b)$.

11. Alice was 20 inches long when born and grew to a height of 69 inches. Use the Intermediate-Value Theorem (see Exercise 10) to argue that at some time in Alice's life, she was exactly 4 feet tall.

12. Give three more everyday applications of the Intermediate-Value Theorem, like the illustration in Exercise 11.

13. Let f be a polynomial function of odd degree, so that $f(x) = a_n x^n + \cdots + a_1 x + a_0$ where $a_n \neq 0$ and n is an odd integer. Show that $f(x) = 0$ has at least one real solution. [*Hint.* Consider $\lim_{x \to \infty} f(x)$ and $\lim_{x \to -\infty} f(x)$, and then apply the

Intermediate-Value Theorem to a sufficiently large interval $[-C, C]$.]

14. A driver on an oval racetrack passes the flag at the end of the third lap going exactly 96 mph. At the end of the fourth lap, the driver is again going exactly 96 mph when passing the flag. Use Theorem 5.1 to show that during the fourth lap, there were two diametrically opposite points of the oval where the car had equal speeds (not necessarily 96 mph of course). [*Hint.* Let S be the length of the oval track. For $0 \leq x \leq S/2$, let

$$f(x) = (\text{speed at } x) - \left(\text{speed at }\left(x + \frac{S}{2}\right)\right),$$

where x is the distance the car has traveled past the flag during that fourth lap.]

15. Use the Intermediate-Value Theorem (see Exercise 10) to show that on August 4, at some point on the 37° meridian of the earth, there must be exactly 10 hours of daylight. We define daylight to mean that some portion of the sun is above the horizon. A meridian runs from the North Pole to the South Pole and is numbered according to the degrees of longitude, measured from the prime meridian of 0° through Greenwich, England.)

16. A square table with four legs of equal length teeters on diagonally opposite legs when placed on a warped floor. Show, using Theorem 5.1, that by rotating the tables less than a quarter turn, you can make all four legs touch the floor, so that the table won't wobble any more. [*Hint.* Number the legs 1, 2, 3, 4 in a counterclockwise order. Let $f(\theta)$ equal the sum of the distances from legs one and three to the floor, minus the sum of the distances from legs two and four to the floor, when the table has been rotated counterclockwise through the angle θ for $0 \leq \theta \leq \pi/2$.]

5.3 MAXIMUM AND MINIMUM VALUES IN [*a*, *b*]

Definition 5.1 Suppose $f(x)$ is defined for all x in $[a, b]$. If there is a point x_1 in $[a, b]$ such that $f(x_1) \geq f(x)$ for all x in $[a, b]$, then $M = f(x_1)$ is the **maximum value** assumed by $f(x)$ in $[a, b]$. Similarly, if $f(x_2) \leq f(x)$ for all x in $[a, b]$, then $m = f(x_2)$ is the **minimum value** assumed by $f(x)$ in $[a, b]$.

Example 1 The maximum value assumed by $f(x) = x^2 + 1$ in $[-3, 2]$ is 10, which is assumed at $x = -3$. The minimum value is 1, which is assumed at 0. See Fig. 5.5 ‖

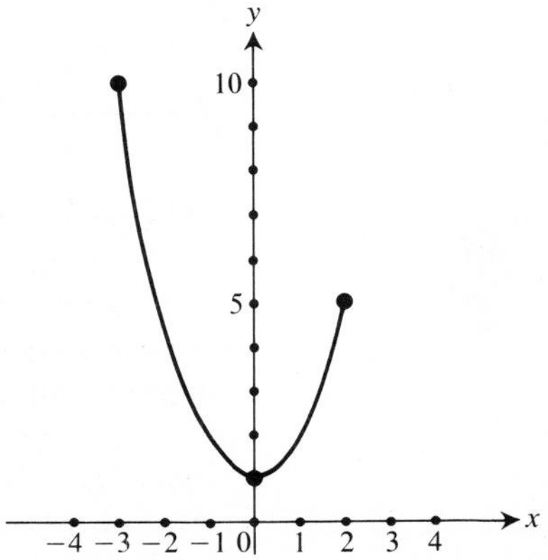

5.5

Example 2 The maximum value assumed by $\sin x$ on $[-2\pi, 2\pi]$ is 1, which is assumed at both $x = -3\pi/2$ and $x = \pi/2$. The minimum value is -1, which is assumed at both $-\pi/2$ and $3\pi/2$. ||

There is an important theorem on the *existence* of maximum and minimum values. It is proved in more advanced texts.

Theorem 5.2 *If $f(x)$ is continuous in $[a, b]$, then $f(x)$ assumes a maximum value M at some point x_1 and a minimum value m at some point x_2 in $[a, b]$.*

The fact that the interval $[a, b]$ is *closed*, so that a and b are in the interval, is an important hypothesis for this existence. For example, there is no maximum value assumed by $f(x) = x^2 + 1$ for $1 < x < 4$. The only conceivable place such a maximum could be assumed would be at "the first point x_1 to the left of 4," and it is easily seen that there can be no such "first point to the left," since $(x_1 + 4)/2$ is always halfway between x_1 and 4.

It is geometrically clear that if $f(x)$ is differentiable at x_1 and assumes a maximum value in $[a, b]$ at x_1 where $a < x_1 < b$, then $f'(x_1) = 0$. See Fig. 5.6. The tangent line where $x = x_1$ must be horizontal. You can also see this from the definition of the derivative,

$$f'(x_1) = \lim_{\Delta x \to 0} \frac{f(x_1 + \Delta x) - f(x_1)}{\Delta x}. \tag{1}$$

If $f(x_1)$ is a maximum, then for small Δx,

$$f(x_1 + \Delta x) \le f(x_1).$$

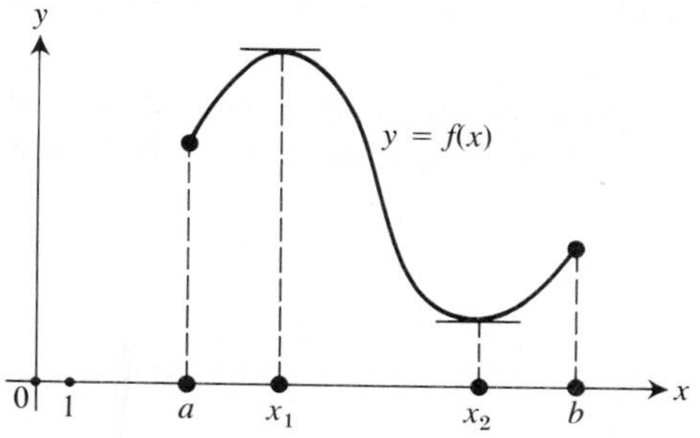

5.6

Thus the numerator of the difference quotient in (1) is always negative or zero. If $\Delta x > 0$, the whole quotient is ≤ 0, while if $\Delta x < 0$, the whole quotient is ≥ 0. The only way that $f'(x_1)$ can be simultaneously approached by numbers ≤ 0 and numbers ≥ 0 is if $f'(x_1) = 0$. An analogous argument shows that if $f(x_2)$ is minimum and $a < x_2 < b$, then $f'(x_2) = 0$.

For the function in Fig. 5.7(a), the value $f(x_3)$ is not a maximum of $f(x)$ over the whole interval $[a, b]$. The maximum occurs where $x = b$. However, $f(x_3)$ is a maximum of $f(x)$ for all x in a sufficiently small interval contained in $[a, b]$.

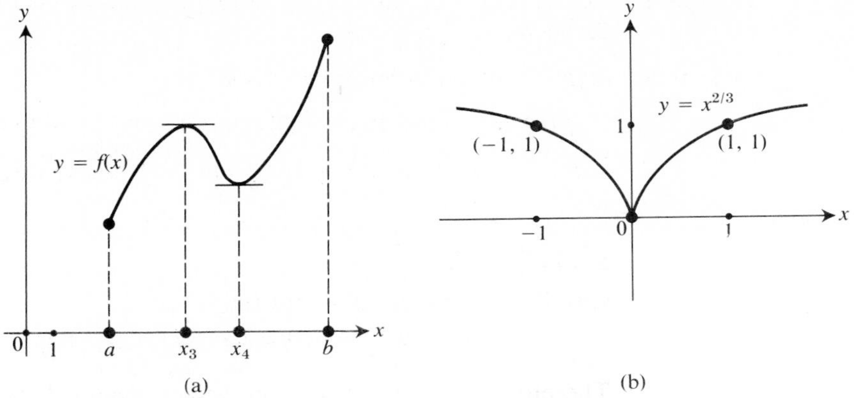

5.7

Definition 5.2 If $f(x_3)$ is a maximum of $f(x)$ for all x in $[x_3 - \Delta x, x_3 + \Delta x]$ for sufficiently small $\Delta x > 0$, then $f(x_3)$ is a **relative maximum** or **local maximum** of $f(x)$. Similarly, $f(x_4)$ in Fig. 5.7 is a **relative minimum** or **local minimum** of $f(x)$.

Intuitively, $(x_3, f(x_3))$ is a local high point and $(x_4, f(x_4))$ a local low point of the graph in Fig. 5.7(a). The argument following Eq. (1) establishes the following theorem.

Theorem 5.3 *If f is differentiable at x_1 and if $f(x_1)$ is a local maximum (or local minimum) of $f(x)$, then $f'(x_1) = 0$.*

A derivative need not exist at every point where a function has a local maximum or local minimum. Figure 5.7(b) shows the graph of $y = f(x) = x^{2/3}$, which surely has a local minimum at the origin. But

$$y' = f'(x) = \frac{2}{3}x^{-2/3} = \frac{2}{3x^{2/3}},$$

so $f'(x)$ does not exist at the origin where $x = 0$. Finding all points where $f'(x)$ exists and is zero thus does *not* guarantee that you have found all possible points where there could be a local maximum or local minimum.

Our discussion indicates that if $f(x)$ is continuous in $[a, b]$ and differentiable for $a < x < b$, then the maximum value M and minimum value m assumed by $f(x)$ in $[a, b]$ can be found as follows:

Finding extreme values over an interval

STEP 1. Find all points in $[a, b]$ where $f'(x) = 0$. Usually, there are only a finite number of such points.

STEP 2. Compute $f(x)$ for all x found in Step 1, and also compute $f(a)$ and $f(b)$. The largest value found is the maximum M and the smallest value the minimum m.

Example 3 Let's find the maximum and minimum values assumed in $[-\frac{1}{2}, 2]$ by $f(x) = 3x^4 + 4x^3 - 12x^2 + 5$.

SOLUTION STEP 1. Differentiating, we have

$$f'(x) = 12x^3 + 12x^2 - 24x = 12x(x^2 + x - 2) = 12x(x+2)(x-1).$$

Thus $f'(x) = 0$ for $x = -2$, 0, and 1. Of these three points, only

$$0 \quad \text{and} \quad 1$$

are in $[-\frac{1}{2}, 2]$.

STEP 2. Computation shows that

$$f(0) = 5, \qquad f(1) = 0, \qquad f(-\tfrac{1}{2}) = \tfrac{27}{16}, \qquad f(2) = 37.$$

Therefore 0 is the minimum value, assumed at $x = 1$, and 37 is the maximum value, assumed at $x = 2$. ||

SUMMARY

1. *If $f(x)$ is continuous at each point in $[a, b]$, then $f(x)$ assumes a maximum value M at some point x_1 in $[a, b]$ and assumes a minimum value m at some point x_2 in $[a, b]$.*
2. *If $f(x)$ is differentiable and has a local maximum at x_3, then $f'(x_3) = 0$. If $f(x)$ has a local minimum at x_4, then $f'(x_4) = 0$.*
3. *If $f(x)$ is continuous in $[a, b]$ and differentiable for $a < x < b$, then the maximum and minimum values assumed in $[a, b]$ may be found using the two-step procedure given before Example 3.*

EXERCISES

In Exercises 1 through 10, find the maximum and minimum values assumed by the function in the given interval.

1. x^4 in $[-2, 1]$

2. $\dfrac{1}{x}$ in $[2, 4]$

3. $\dfrac{1}{x^2+1}$ in $[-1, 2]$

4. $x^3 - 3x^2 + 1$ in
a) $[1, 3]$ b) $[-1, 3]$ c) $[-2, 4]$

5. $x^2 + 4x - 3$ in
a) $[-5, -3]$ b) $[-4, -1]$ c) $[-4, 0]$

6. $\dfrac{1}{x^2-1}$ in $[-3, -2]$

7. $\dfrac{x^2 - x + 1}{x^2 + 1}$ in
a) $[-3, -2]$ b) $[-2, 0]$
c) $[0, 2]$ d) $[-2, 2]$

8. $\sin x$ in the intervals:
a) $\left[0, \dfrac{\pi}{4}\right]$ b) $\left[-\dfrac{\pi}{4}, \dfrac{\pi}{4}\right]$ c) $\left[-\dfrac{3\pi}{4}, \dfrac{3\pi}{4}\right]$

9. $\sin x + \cos x$ in the intervals:
a) $\left[0, \dfrac{\pi}{2}\right]$ b) $\left[-\dfrac{\pi}{2}, \dfrac{\pi}{2}\right]$ c) $[0, \pi]$ d) $\left[\dfrac{\pi}{2}, 2\pi\right]$

10. $\sqrt{3}\sin x - \cos x$ in the intervals:
a) $\left[0, \dfrac{\pi}{2}\right]$ b) $[0, \pi]$ c) $\left[0, \dfrac{3\pi}{2}\right]$ d) $[0, 2\pi]$

11. Let f be a polynomial function of even degree n such that the coefficient a_n of x^n is positive. Show that $f(x)$ assumes a value that is minimum for all real numbers x. [*Hint.* Consider $\lim_{x\to\infty} f(x)$ and $\lim_{x\to-\infty} f(x)$, and apply Theorem 5.2 to a large interval $[-c, c]$.]

In Exercises 12 through 16, use the result of Exercise 11 and the two-step procedure outlined in the text to find the minimum value assumed by the function for all real numbers x.

12. $3x^2 + 6x - 4$ **13.** $3x^4 - 8x^3 + 2$ **14.** $x^4 - 2x^2 + 7$

15. $3x^4 - 4x^3 - 12x^2 + 5$ **16.** $x^6 - 6x^4 - 8$

calculator exercises

Find the maximum and minimum values assumed by the given function on the indicated interval.

17. $x^3 - 3x^2 + x - 2$ on $[0, 2]$ **18.** $2^x - x^3 + 100$ on $[3, 10]$ **19.** $x - \sin 2x$ on $[-1, 1]$

5.4 THE MEAN-VALUE THEOREM

The Mean-Value Theorem is one of the most useful tools, for theoretical purposes, in calculus. First we discuss a special, easy case known as Rolle's theorem, which will serve as a lemma for the main result.

Theorem 5.4 *(Rolle's Theorem) If $f(x)$ is continuous in $[a, b]$ and differentiable for $a < x < b$, and if, furthermore, $f(a) = f(b)$, then there exists c where $a < c < b$ such that $f'(c) = 0$.*

Proof. Figure 5.8 illustrates Rolle's theorem. From the last section, you know that $f(x)$ assumes a maximum value M and a minimum value m in $[a, b]$. If f is a constant function in $[a, b]$ so that $f(x) = f(a) = f(b)$ for all x in $[a, b]$, then of course $f'(c) = 0$ for any c where $a < c < b$. If f is not constant in $[a, b]$, then since $f(a) = f(b)$, either the maximum or the minimum of $f(x)$ must be assumed at a point c where $a < c < b$. From Theorem 5.3 in the last section, you then know $f'(c) = 0$. This completes the demonstration of Rolle's theorem. □

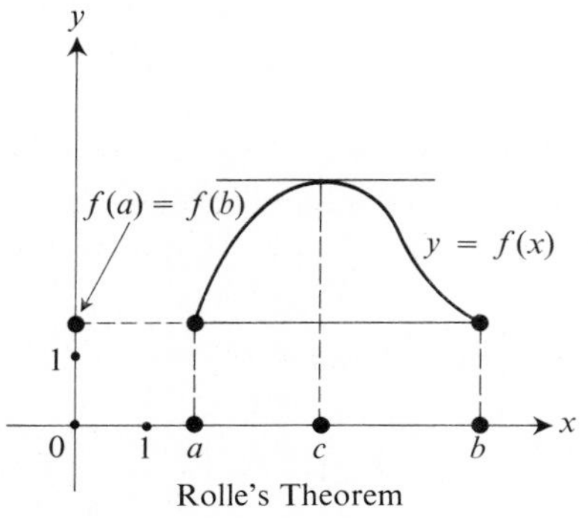

5.8 Rolle's Theorem

The Mean-Value Theorem may be regarded as a generalization of Rolle's Theorem, and is illustrated in Fig. 5.9. Both theorems assert that for f continuous in $[a, b]$ and differentiable for $a < x < b$, there exists at least one c where $a < c < b$ and where the tangent line to the graph of f is parallel to the line joining the points $(a, f(a))$ and $(b, f(b))$. The slope of this line is

$$\frac{f(b) - f(a)}{b - a}$$

so the conclusion of the Mean-Value Theorem will take the form

$$f'(c) = \frac{f(b) - f(a)}{b - a} \quad \text{or} \quad f(b) - f(a) = (b - a)f'(c)$$

for some c where $a < c < b$. There are two such points, c_1 and c_2 in Fig. 5.9.

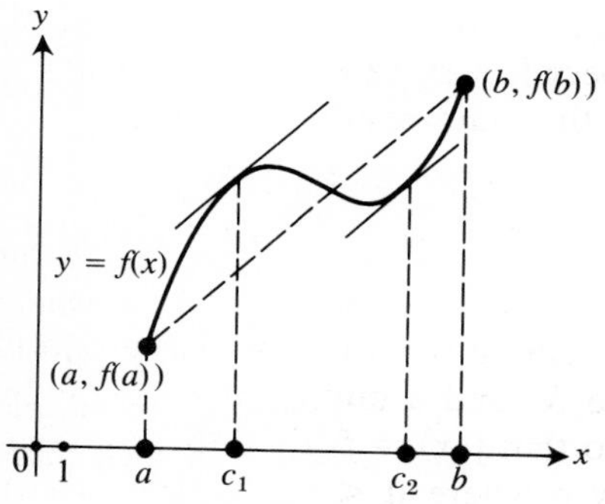

Mean-value theorem

5.9

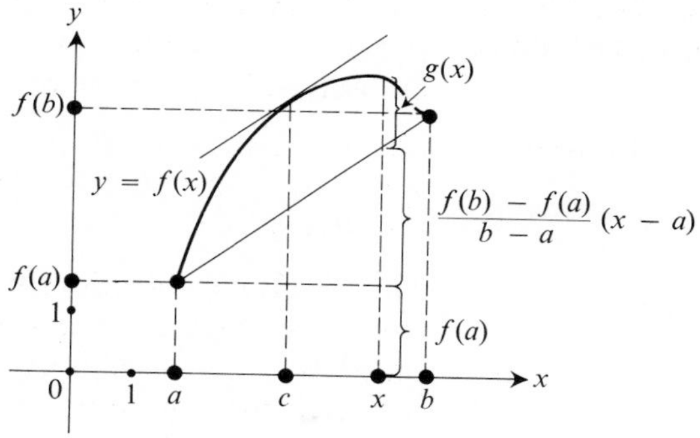

5.10

Theorem 5.5 *(Mean-Value Theorem) Let $f(x)$ be continuous in $[a, b]$ and differentiable for $a < x < b$. Then there exists c where $a < c < b$ such that*

$$f(b) - f(a) = (b - a)f'(c). \tag{1}$$

Proof. In order to obtain this result from Rolle's Theorem, we introduce the function g with domain $[a, b]$ whose value at x is indicated in Fig. 5.10. It is easy to see from Fig. 5.10 that we should define g by

$$\begin{aligned} g(x) &= f(x) - \left[f(a) + \frac{f(b) - f(a)}{b - a}(x - a)\right] \\ &= [f(x) - f(a)] - \left[\frac{f(b) - f(a)}{b - a}(x - a)\right]. \end{aligned}$$

Since f is continuous in $[a, b]$ and differentiable for $a < x < b$, we see that g is also, and $g(a) = g(b) = 0$, so Rolle's Theorem applies. Therefore for some c where $a < c < b$, we have $g'(c) = 0$. Now

$$g'(x) = f'(x) - \frac{f(b) - f(a)}{b - a},$$

so

$$0 = g'(c) = f'(c) - \frac{f(b) - f(a)}{b - a},$$

and $f(b) - f(a) = (b - a)f'(c)$. □

Example 1 Let's illustrate the Mean-Value Theorem with $f(x) = x^2$ in the interval $[0, 3]$.

SOLUTION We must find c such that

$$f(3) - f(0) = (3 - 0)f'(c),$$

or such that $9 = 3 \cdot f'(c)$. Now $f'(x) = 2x$, and $9 = 3 \cdot 2c$ when $c = 3/2$. Note that as the Mean-Value Theorem states, we found a value c satisfying $0 < c < 3$. ||

SUMMARY

1. *(Rolle's Theorem) If $f(x)$ is continuous in $[a, b]$ and differentiable for $a < x < b$, and if $f(a) = f(b)$, then there is some number c where $a < c < b$ such that $f'(c) = 0$.*
2. *(Mean-Value Theorem) If $f(x)$ is continuous in $[a, b]$ and differentiable for $a < x < b$, then there is some number c where $a < c < b$ such that*

$$\frac{f(b) - f(a)}{b - a} = f'(c).$$

EXERCISES

1. Verify that $f(x) = x^2 + x + 4$ satisfies the hypotheses for Rolle's Theorem on the interval $[-3, 2]$, and find the number c described in the theorem.
2. Repeat Exercise 1 for $f(x) = x^3 - 6x^2 + 5x + 3$ on the interval $[1, 5]$.
3. Generalize Rolle's Theorem to show that if f is a differentiable function and $f(x)$ is zero at r distinct points in an interval $[a, b]$, then $f'(x)$ is zero for at least $r - 1$ distinct points in $[a, b]$.
4. Use Rolle's Theorem to show that if f is a differentiable function and $f'(x) \neq 0$ for $a < x < b$, then $f(x)$ is zero for at most one value of x in $[a, b]$.
5. Illustrate Exercise 4 by showing that

$$f(x) = x^3 - 3x - 18 = 0$$

for at most one value of x in $[2, 4]$.

In Exercises 6 through 10, illustrate the Mean-Value Theorem for the given function over the given interval, as we did in Example 1, by finding a value of c given in the statement of the theorem.

6. $3x - 4$ on $[1, 4]$

7. x^3 on $[-1, 2]$

8. $x - \frac{1}{x}$ on $[1, 3]$

9. $x^{1/2}$ on $[9, 16]$

10. $\sqrt{1 - x}$ on $[-3, 0]$

11. Let f be a function satisfying the hypotheses of the Mean-Value Theorem on $[a, b]$.

a) Give the rate-of-change interpretation of

$$[f(b) - f(a)]/(b - a).$$

b) Give the interpretation of $f'(c)$ as a rate of change.

c) Restate the Mean-Value Theorem in terms of rates of change.

12. Two towns A and B are connected by a highway ten miles long with a 60-mph speed limit. Mr. Smith is arrested on a speeding charge, and admits having driven from A to B in eight minutes. In a speeding offense, the court is allowed to impose a fine of $15 plus $2 for each mph speed in excess of the limit. Use the Mean-Value Theorem to show that the judge is justified in imposing a fine of $45 on Mr. Smith. [*Hint.* Use the preceding exercise.]

13. For a quadratic function f given by $f(x) = ax^2 + bx + c$, show that the point between x_1 and x_2 where the tangent to the graph of f is parallel to the chord joining $(x_1, f(x_1))$ and $(x_2, f(x_2))$ is halfway between x_1 and x_2. What example in the text illustrates this?

14. Using the Mean-Value Theorem, constant bounds on $f'(x)$ can be used to give linear bounds on $f(x)$. Let f satisfy the hypotheses of the Mean-Value Theorem on $[a, b]$, and suppose that $m \leq f'(c) \leq M$ if $a < c < b$. Show that

$$f(a) + m(x - a) \leq f(x) \leq f(a) + M(x - a)$$

for all x in $[a, b]$.

5.5 SIGNS OF DERIVATIVES AND SKETCHING CURVES

It is geometrically obvious that the graph of $y = f(x)$ is rising where $f'(x) > 0$ and is falling where $f'(x) < 0$. This can be proved using the Mean-Value Theorem. First, we give a definition.

5.5.1 The sign of the first derivative

Definition 5.3 Let $f(x)$ and $g(x)$ be defined for x in $[a, b]$. If $f(x_1) < f(x_2)$ whenever $x_1 < x_2$ in $[a, b]$, then $f(x)$ is **increasing** in $[a, b]$. Similarly, if $g(x_1) > g(x_2)$ whenever $x_1 < x_2$ in $[a, b]$, then $g(x)$ is **decreasing** in $[a, b]$.

Suppose $f'(x) > 0$ for all x in $[a, b]$. Let x_1 and x_2 be in $[a, b]$ with $x_1 < x_2$. By the Mean-Value Theorem,

$$\frac{f(x_2) - f(x_1)}{x_2 - x_1} = f'(c) > 0 \tag{1}$$

where $x_1 < c < x_2$. In particular, (1) shows that $f(x_2) - f(x_1) > 0$, so $f(x_1) < f(x_2)$. Thus $f(x)$ is increasing in $[a, b]$.

Similarly, if $f'(x) < 0$ for all x in $[a, b]$, then $f(x)$ is decreasing in $[a, b]$, for if $x_1 < x_2$ in $[a, b]$, then an analogous argument shows that $f(x_1) > f(x_2)$. You already know that points where $f'(x) = 0$ are candidates for high and low points of the graph. We summarize in a theorem.

Theorem 5.6 *Let $f(x)$ be continuous on $[a, b]$ and differentiable for $a < x < b$. If $f'(x) > 0$ for $a < x < b$, then $f(x)$ is increasing on $[a, b]$. If $f'(x) < 0$ for $a < x < b$, then $f(x)$ is decreasing on $[a, b]$.*

Example 1 Let $f(x) = x^3 - 3x^2 + 2$ and find where $f(x)$ is increasing, decreasing, and has local extrema.

SOLUTION We have $f'(x) = 3x^2 - 6x = 3x(x - 2)$, so $f'(x) = 0$ where $x = 0$ or $x = 2$. The points 0 and 2 separate the x-axis into three parts, and $f'(x)$ must have the same sign throughout each individual part. In Fig. 5.11 we show the sign of $f'(x)$ on each part of the x-axis. For example, if $x < 0$, then $3x < 0$ and $x - 2 < 0$, so $f'(x) = 3x(x - 2) > 0$. Thus you see that $f(x)$ is increasing for $x < 0$ or $x > 2$, and $f(x)$ is decreasing for $0 < x < 2$.

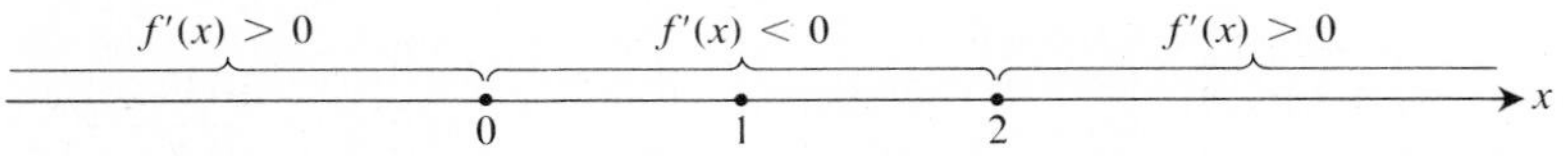

5.11

You must have a local maximum at (0, 2) where $x = 0$, for $f(x)$ is increasing from the left to attain the value $f(0)$, and starts decreasing to the right. Similarly, there is a local minimum at (2, −2) where $x = 2$ since $f(x)$ decreases coming in to $x = 2$ from the left and increases to the right. The graph $y = f(x)$ is shown in Fig. 5.12 ‖

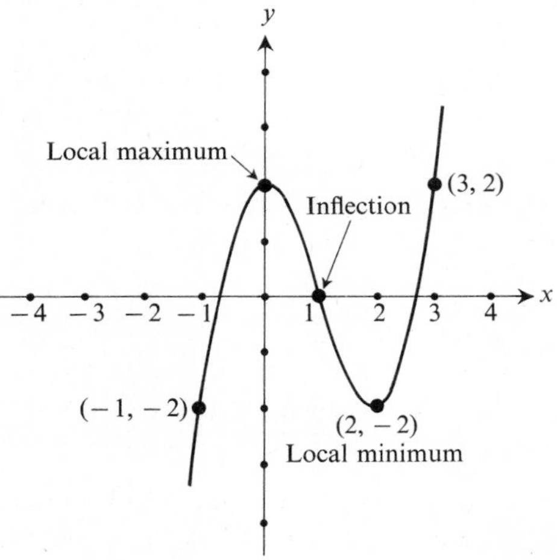

5.12

Example 2 If $f(x) = x^3$, then $f'(x) = 3x^2$ so $f'(x) \geq 0$ for all x and $f(x)$ is always increasing. Of course, $f'(x) = 0$ where $x = 0$, but $(0, 0)$ is not a high or low point since $f(x)$ increases to $(0, 0)$ from the left and continues increasing to the right of $(0, 0)$. The graph of $y = x^3$ is shown in Fig. 5.13 ‖

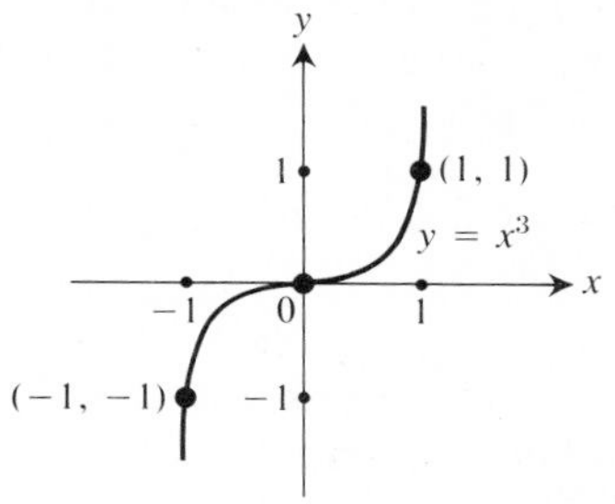

5.13

These facts concerning the first derivative are summarized as *The First Derivative Story* in the summary at the section's end.

5.5.2 The sign of the second derivative

Since

$$f''(x) = \frac{d(f'(x))}{dx},$$

you see that $f'(x)$ is increasing throughout any interval where $f''(x) > 0$. Now $f'(x)$ gives the slope of the tangent line at x. For the slope of a line to increase, the line must rotate *counterclockwise*. If the tangent line rotates counterclockwise as x increases, the graph must be bending upward as x increases. This is the case in Fig. 5.12 for $x > 1$. In this situation, we say the curve is **concave up**. If $f'(x) < 0$, the slope decreases, the tangent line turns *clockwise* as x increases, and the curve is **concave down**. This is the case in Fig. 15.12 for $x < 1$.

Concavity and inflection points

Points where $f''(x) = 0$ are candidates for places where the concavity changes from concave up to concave down, or vice versa. Such points where the concavity changes are **inflection points**.

Example 3 We discuss concavity and inflection points for the curve $y = x^3 - 3x^2 + 2$ of Example 1.

SOLUTION You have $f'(x) = 3x^2 - 6x$ and $f''(x) = 6x - 6$. Thus $f''(x) = 0$ where $6x - 6 = 0$, or where $x = 1$. Since $6x - 6 < 0$ if $x < 1$, the curve is concave down if $x < 1$. The curve is concave up if $x > 1$ since $f''(x) = 6x - 6 > 0$ there. Since the concavity does change as x increases through 1, the point $(1, 0)$ is an inflection point, as indicated in Fig. 5.12. ‖

Example 4 If $f(x) = x^4$, then $f'(x) = 4x^3$ and $f''(x) = 12x^2$. Now $f''(0) = 0$, but $(0, 0)$ is not an inflection point since $f''(x) > 0$ on both sides of $x = 0$. Actually,

(0, 0) is a local minimum since $f'(0) = 0$ and $f'(x) = 4x^3$ is <0 if $x < 0$ and >0 if $x > 0$. ‖

Finally, we mention that if $f'(x_1) = 0$ and $f''(x_1) > 0$, then $y = f(x)$ has a horizontal tangent where $x = x_1$ but is concave up there, so $y = f(x)$ has a low point where $x = x_1$. If $f'(x_2) = 0$ but $f''(x_2) < 0$, then the curve has a horizontal tangent but is concave down, so the curve has a high point where $x = x_2$. Thus if $f'(a) = 0$ but $f''(a) \neq 0$, the sign of $f''(a)$ can be used to determine whether there is a local maximum or local minimum where $x = a$. We summarize all this information about the second derivative in a theorem,

Theorem 5.7 *Let f be a twice differentiable function.*

1. *If $f''(x) > 0$ throughout an interval, the graph is concave up there.*
2. *If $f''(x) < 0$ throughout an interval, the graph is concave down there.*
3. *If $f''(a) = 0$ and $f''(x)$ changes sign at a as x increases through a, then the graph has an inflection point where $x = a$.*
4. *(Second-derivative test) If $f'(a) = 0$ and $f''(a) \neq 0$, the graph has a local maximum where $x = a$ if $f''(a) < 0$, and a local minimum there if $f''(a) > 0$.*

Example 5 Returning once more to the cubic $y = f(x) = x^3 - 3x^2 + 2$ of Fig. 5.12, you know that $f'(x) = 3x^2 - 6x = 3x(x - 2) = 0$ if $x = 0$ or $x = 2$. Also, $f''(x) = 6x - 6$. Then $f''(0) = -6 < 0$, so the curve is concave down at (0, 2), which must be a local maximum. Finally, $f''(2) = 12 - 6 = 6 > 0$, so the curve is concave up at (2, −2), which must be a local minimum. ‖

Critical points

We have shown the importance of finding places where $f'(x) = 0$. Such points, together with points where $f'(x)$ does not exist, are called **critical points** of the curve. Newton's method can be used to find where $f'(x) = 0$, using the recursion relation

$$a_{i+1} = a_i - \frac{f'(a_i)}{f''(a_i)},$$

and points where $f''(x) = 0$ can be found using the recursion relation

$$a_{i+1} = a_i - \frac{f''(a_i)}{f'''(a_i)}.$$

The derivatives $f'(a_i)$, $f''(a_i)$, and $f'''(a_i)$ can all be found numerically on a computer, so clearly one could write a computer program to find the critical points of the graph of a function.

We shall conclude with two examples illustrating the use of first and second derivatives in curve sketching. Please read the section summary before you study these examples. The summary gives in one place the information developed piece by piece during this section.

Example 6 Let's find any local maxima and minima and any inflection points of $y = f(x) = x + \sin x$.

SOLUTION Differentiating, you obtain

$$f'(x) = 1 + \cos x \qquad \text{and} \qquad f''(x) = -\sin x.$$

Now $1 + \cos x = 0$ where $\cos x = -1$, which occurs when $x = (2n+1)\pi$ for any integer n. But $f'(x) = 1 + \cos x$ cannot change sign as x increases through these points, for $1 + \cos x \geq 0$ for all x. Thus the function $x + \sin x$ increases for all x. There are no local maxima or minima.

Turning to the second derivative, $f''(x) = -\sin x$ is zero at $x = n\pi$ for all integers n, which include those points where $f'(x)$ is zero. At all these points, $f''(x) = -\sin x$ changes sign as x increases. Thus $x = n\pi$ corresponds to an inflection point for all integers n. If n is odd, then $f'(x) = 0$, and there is a horizontal tangent at the inflection point. The curve is concave up if $-\sin x > 0$, which occurs when $(2n - 1)\pi < x < 2n\pi$ for each integer n. It is concave down for $2n\pi < x < (2n + 1)\pi$. Figure 5.14 shows the curve. ‖

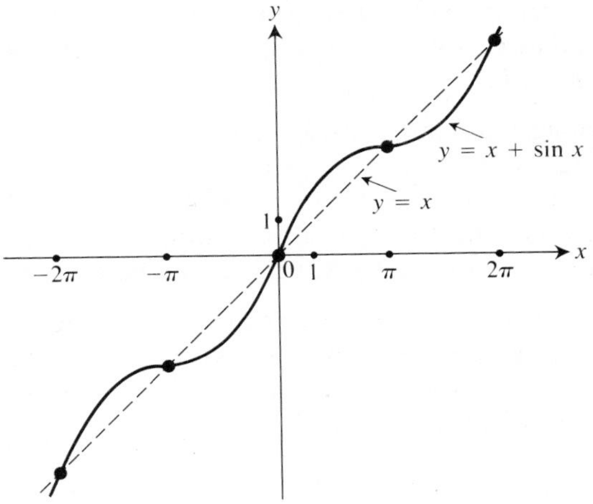

5.14

Example 7 Let's study the graph of

$$y = f(x) = \frac{x^4}{4} - \frac{4x^3}{3} + 2x^2 - 1,$$

finding all local maxima, local minima, and inflection points.

SOLUTION Differentiating,

$$\begin{aligned} f'(x) &= x^3 - 4x^2 + 4x \\ &= x(x^2 - 4x + 4) \\ &= x(x-2)^2 \end{aligned} \qquad \text{and} \qquad \begin{aligned} f''(x) &= 3x^2 - 8x + 4 \\ &= (3x - 2)(x - 2). \end{aligned}$$

Thus $f'(x) = 0$ when $x = 0$ or $x = 2$. Since the root $x = 0$ comes from the single linear factor x, you see that $f'(x)$ does change sign, from negative to positive since $(x - 2)^2 > 0$, as x increases through 0. Alternatively, $f''(0) = 4 > 0$, so the graph is concave up at $x = 0$. This shows in two ways that $f(x)$ has a local minimum when $x = 0$, corresponding to $(0, -1)$ on the graph.

Although $f'(2) = 0$, the derivative $f'(x)$ does *not* change sign as x increases through 2, since the factor $(x - 2)^2$ has an *even* exponent. Actually, $f''(2)$ is also zero, and $f''(x)$ does change sign where $x = 2$ due to the odd-powered factor $(x - 2)^1$, so $(2, f(2)) = (2, \frac{1}{3})$ is an inflection point. Another inflection point occurs when $x = \frac{2}{3}$, corresponding to the odd-powered factor $(3x - 2)^1$ in $f''(x)$.

Finally, since $(x - 2)^2 \geq 0$ for all x, you see that $f'(x) = x(x - 2)^2 > 0$ if $x > 0$ and $f'(x) < 0$ if $x < 0$. Thus $f(x)$ is increasing if $x > 0$ and decreasing if $x < 0$. A similar study of the sign of $f''(x)$ establishes the concavity described below. The graph is sketched in Fig. 5.15. Summarizing:

$f(x)$ is increasing for $x > 0$;
$f(x)$ is decreasing for $x < 0$;
$f(x)$ has a local minimum at $(0, -1)$;
$f(x)$ has inflection points at $(2, \frac{1}{3})$ and $(\frac{2}{3}, -\frac{37}{81})$;
$f(x)$ is concave up if $x > 2$ or $x < \frac{2}{3}$;
$f(x)$ is concave down for $\frac{2}{3} < x < 2$. ||

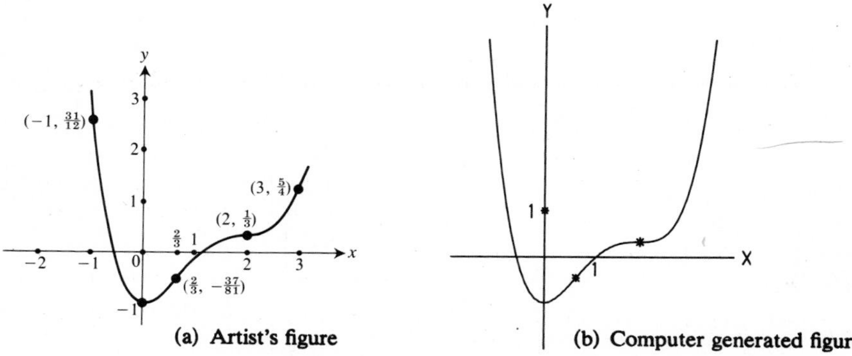

(a) Artist's figure (b) Computer generated figure

SUMMARY THE FIRST DERIVATIVE STORY

1. *If $f'(x) > 0$ throughout an interval, then $f(x)$ is increasing there.*
2. *If $f'(x) < 0$ throughout an interval, then $f(x)$ is decreasing there.*
3. *If $f'(x_1) = 0$, then x_1 is a candidate for a point where $f(x)$ has a local maximum or minimum.*

a) *If $f'(x)$ changes sign from negative to positive as x increases through x_1, then $f(x_1)$ is a local minimum.*

b) *If $f'(x)$ changes sign from positive to negative as x increases through x_1, then $f(x_1)$ is a local maximum.*

c) *If $f'(x)$ does not change sign as x increases through x_1, then $f(x_1)$ is neither a local maximum nor a local minimum.*

THE SECOND DERIVATIVE STORY

4. *If $f''(x) > 0$ throughout an interval, the graph is concave up there.*
5. *If $f''(x) < 0$ throughout an interval, the graph is concave down there.*
6. *If $f''(x_2) = 0$, then $(x_2, f(x_2))$ is a candidate for an inflection point of the graph.*

   a) *If $f''(x)$ changes sign as x increases through x_2, then $(x_2, f(x_2))$ is an inflection point.*

   b) *If $f''(x)$ does not change sign as x increases through x_2, then $(x_2, f(x_2))$ is not an inflection point.*

7. *(Second Derivative Test) If $f'(x_1) = 0$ and $f''(x_1) \neq 0$, then $f(x_1)$ is a local minimum if $f''(x_1) > 0$ and a local maximum if $f''(x_1) < 0$.*

CRITICAL POINTS

8. *Points where $f'(x) = 0$ or where $f'(x)$ does not exist are critical points of the graph.*

EXERCISES

1. Find b such that the polynomial function $x^2 + bx - 7$ has a local minimum at 4.

2. Consider the polynomial function $ax^2 + 4x + 13$.

a) Find the value of a such that the function has either a local maximum or a local minimum at 1. Which is it, a maximum or a minimum?

b) Find the value of a such that the function has either a local maximum or a local minimum at -1. Which is it, a maximum or a minimum?

3. Consider the polynomial function f given by $ax^2 + bx + 24$.

a) Find the ratio b/a if f has a local minimum at 2.

b) Find a and b if f has a local minimum at 2 and $f(2) = 12$.

c) Can you determine a and b so that f has a local maximum at 2 and $f(2) = 12$?

4. Assume that f is twice differentiable and that f'' is continuous. Mark each of the following true or false.

a) If f has a local maximum at x_0, then $f'(x_0) = 0$.

b) If f has a local maximum at x_0, then $f''(x_0) < 0$.

c) If $f''(x_0) < 0$, then f has a local maximum at x_0.

d) If $f'(x_0) = 0$ and $f''(x_0) < 0$, then f has a local maximum at x_0.

e) If $f'(x_0) > 0$, then f is increasing near x_0.

f) If $f'(x_0) = 0$, then f cannot be increasing near x_0.

g) If f has an inflection at x_0, then $f''(x_0) = 0$.

h) If $f''(x_0) = 0$, then f must have an inflection at x_0.

i) If f has a local minimum at x_0, then the tangent line to the graph of f at $(x_0, f(x_0))$ is horizontal.

j) If f has an inflection at x_0, then the tangent line to the graph of f at $(x_0, f(x_0))$ may possibly be horizontal.

5. Sketch the graph of a twice-differentiable function f such that $f(1) = 3$, $f(4) = 1$, f has a local minimum at 1, and f has a local maximum at 4.

6. Is it possible for a twice-differentiable function f to satisfy the conditions of Exercise 5 and have no local maximum or minimum for $1 < x < 4$)?

7. Sketch the graph of a twice-differentiable function f such that $f(0) = 3$, $f'(0) = 0$, $f''(0) < 0$, $f(2) = 2$, $f'(2) = -1$, $f''(2) = 0$, $f(4) = 1$, $f'(4) = 0$, and $f''(4) > 0$.

8. Sketch the graph of a twice-differentiable function f such that $f''(x) < 0$ for $x < 1$, $f(1) = -1$, $f'(1) = 1$, $f''(1) = 0$, $f''(x) > 0$ for $x > 1$, and $f(3) = 4$.

9. Sketch the graph of a twice-differentiable function f such that $f'(x) > 0$ for $x > 2$, $f''(x) > 0$ for $x > 2$, $f''(2) = 0$, $f'(x) < 0$, for $x < 2$, and $f''(x) > 0$ for $x < 2$.

10. Do you think that it is possible that, for a twice-differentiable function f, one can have $f(0) = 0$, $f'(0) = 1$, $f''(x) > 0$ for $x > 0$, and $f(1) = 1$? Why?

11. a) Sketch the graph of the function $x^{1/3}$.

b) In terms of concavity, would it be reasonable to say that the graph of $x^{1/3}$ has $(0, 0)$ as an inflection point?

c) Does the second derivative of $x^{1/3}$ at 0 exist?

In Exercises 12 *through* 21,

a) *determine where the function is increasing,*

b) *determine where the function is decreasing,*

c) *find all local maxima,*

d) *find all local minima,*

e) *find all inflection points of the graph,*

f) *sketch the graph, using the information in parts* (a)–(e).

12. $4 - x^2$

13. $x^2 - 6x + 4$

14. $\dfrac{1}{x+1}$

15. $\dfrac{x^3}{3} + x^2 - 3x - 4$

16. $\dfrac{x^3}{3} + x^2 + x - 6$

17. $\dfrac{x^2}{x-1}$

18. $x^4 - 4x + 1$

19. $x^5 - 5x + 1$

20. $x + 2 \sin x$

21. $x - 2 \cos x$

In Exercises 22 *through* 25, *sketch the curve with the given equation.* [*Hint. Regard the equation as giving x as a function g of y, and apply the theory in this section with x and y interchanged.*]

22. $x = y^2 - 2y + 2$

23. $x = y^3 - 3y^2$

24. $x = \dfrac{1}{y^4 + 1}$

25. $x = \dfrac{y}{y^2 + 1}$

26. Find a function f such that $f'(2) = f''(2) = 0$ and such that f is increasing at 2 with $f(2) = 3$.

27. Find a function f such that $f'(2) = f''(2) = 0$ and such that f has a local maximum of -17 at 2.

28. If $f'(x) = g(x) \cdot (x - a)$ where $g(a) \neq 0$ and g is continuous at a, argue that f has a local minimum at a if $g(a) > 0$ and a local maximum at a if $g(a) < 0$.

29. If $f'(x) = g(x) \cdot (x - a)^2$ where $g(a) \neq 0$ and g is continuous at a, argue that f is increasing near a if $g(a) > 0$ and is decreasing near a if $g(a) < 0$.

calculator exercises

Find relative maxima, minima, and inflection points for each of the following functions. You may want to use Newton's method and either algebraic differentiation or the formulas for approximating first, second, and third derivatives (see the Exercises of Section 3.4).

30. $x^3 - 3x^2 + x - 5$ **31.** $x^4 - 2x^3 - x^2 + x$ **32.** $2^x - 5x$

5.6 MAXIMUM AND MINIMUM WORD PROBLEMS

There are many situations in which one wishes to maximize or minimize some quantity. For example, a manufacturer wants to maximize his profit. A builder may want to minimize his materials. Such extremum problems are clearly of great practical importance. One can often solve extremum problems with differential calculus, using the ideas developed in the preceding section.

To find local maxima and minima of a function f, one may proceed as follows. Find all points x where $f'(x) = 0$. These are candidates for points where f has a local maximum or minimum. See if the derivative of f changes sign, or use the second derivative test at each of these points, and determine whether each point corresponds to a local maximum, a local minimum, or neither. Then examine the behavior of f near points which are not inside an interval in which f is differentiable, if any such points exist.

Here is a suggested step-by-step procedure you might wish to use with maximum and minimum word problems.

Suggested outline

STEP 1. Decide what you want to maximize or minimize. (You can't do anything meaningful until you establish your objective.)

STEP 2. Express this quantity which you wish to maximize or minimize as a function f of *one* other quantity. (You may have to draw a figure and use some algebra to do this.)

STEP 3. Find all points x where $f'(x) = 0$.

STEP 4. Decide whether your desired maximum or minimum occurs at one of the points you found in Step 3. Frequently it will be clear that a maximum or minimum exists from the nature of the problem, and if Step 3 gives only one candidate, that is your answer. If there is more than one candidate, or if there are points in the domain of f that are not *inside* an interval where f is differentiable, you may have to make a further examination.

Example 1 Let's find two numbers whose sum is 6 and whose product is as large as possible.

SOLUTION STEP 1. We want to maximize the product P of two numbers.

STEP 2. If the two numbers are x and y, then $P = xy$. Since $x + y = 6$, we have $y = 6 - x$, so $P = x(6 - x) = 6x - x^2$.

STEP 3. We have

$$\frac{dP}{dx} = 6 - 2x.$$

Thus $dP/dx = 0$ when $6 - 2x = 0$, or when $x = 3$.

STEP 4. Since $d^2P/dx^2 = -2 < 0$, we see that P has a maximum at $x = 3$.

Thus the largest value of P occurs when $x = 3$ and $y = 3$, so $xy = 9$. ‖

Example 2 A manufacturer of dog food wishes to package his product in cylindrical metal cans, each of which is to contain a certain volume V_0 of dog food. Let's find the ratio of the height of the can to its radius in order to minimize the amount of metal, assuming that the ends and side of the can are made from metal of the same thickness.

SOLUTION STEP 1. The manufacturer wishes to minimize the surface area S of the can.

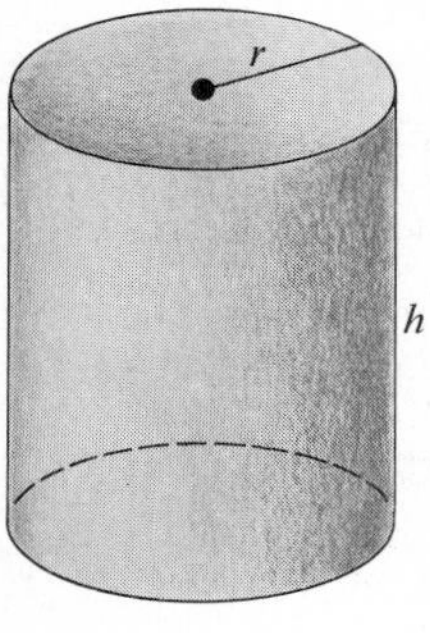

5.16

STEP 2. The surface of the can consists of the two circular disks at the ends and the cylindrical side. If the radius of the can is r and the height is h, the top and bottom disks each have area πr^2, and the cylinder has area $2\pi rh$ (see Fig. 5.16.). Thus

$$S = 2\pi r^2 + 2\pi rh.$$

We would like to find h in terms of r to express S as a function of the single quantity r. From $V_0 = \pi r^2 h$, we obtain $h = V_0/\pi r^2$. Thus

$$S = 2\pi r^2 + 2\pi r\frac{V_0}{\pi r^2} = 2\left(\pi r^2 + \frac{V_0}{r}\right).$$

STEP 3. We easily find that

$$\frac{dS}{dr} = 2\left(2\pi r - \frac{V_0}{r^2}\right).$$

Thus $dS/dr = 0$ when $2\pi r - (V_0/r^2) = 0$, so $2\pi r^3 = V_0$, and

$$r^3 = \frac{V_0}{2\pi}.$$

STEP 4. It is obvious that a minimum for S does exist from the nature of the problem. A can $\frac{1}{32}$ of an inch high and six feet across uses a lot of metal, as does one $\frac{1}{32}$ of an inch in radius and a hundred feet high. Somewhere between these ridiculous measurements there is a can of reasonable dimensions using the least metal. We found only one candidate in Step 3, so we don't have to look further.

We were interested in the ratio h/r. Now $V_0 = \pi r^2 h$, so

$$h = \frac{V_0}{\pi r^2} \quad \text{and} \quad \frac{h}{r} = \frac{V_0}{\pi r^3}.$$

From Step 3, the least metal is used when $r^3 = V_0/2\pi$, so

$$\frac{h}{r} = \frac{V_0}{\pi(V_0/2\pi)} = 2. \quad \|$$

Example 2 illustrates the practical importance of extremum problems. For a cylindrical can of minimal surface area and containing a given volume, we should have $h = 2r$, so the height should equal the diameter. Thus few cylindrical cans on the shelves in our supermarkets represent economical packaging, assuming that the ends and sides are of equally expensive material. The tuna fish cans are usually too short, and the soft drink and beer cans are too high.

SUMMARY 1. *To solve maximum and minimum word problems, you may follow the four-step outline given just before Example 1.*

EXERCISES

1. Find the maximum area a rectangle can have if the perimeter is 20 ft.

2. Generalize Exercise 1 to show that the rectangle of maximum area having a fixed perimeter is a square.

3. Find two positive numbers x and y such that $x + y = 6$ and xy^2 is as large as possible.

4. Find the positive number x such that the sum of x and its reciprocal is minimum.

5. Find two positive numbers whose product is 36 and such that the sum of their cubes is a minimum.

6. Find the point on the parabola $y = x^2$ which is closest to the point (6, 3).

7. A cardboard box of 108 in^3 volume and having a square base and open top is to be constructed. Find the minimum area of cardboard needed. (Neglect waste in construction.)

8. An open box with a reinforced square bottom and volume 96 ft^3 is to be constructed. If material for the bottom costs three times as much per square foot as material for the sides, find the dimensions of the box of minimum cost. (Neglect waste in construction.)

9. A farmer has 1000 rods of fencing with which to fence in three sides of a rectangular pasture; a straight river will form the fourth side of the pasture. Find the dimensions of the pasture of largest area that the farmer can fence.

10. A rancher has 1200 ft of fencing to enclose a double paddock with two rectangular regions of equal areas, as shown in Fig. 5.17. Find the maximum area that the rancher can enclose. (Neglect waste in construction and the need for gates.)

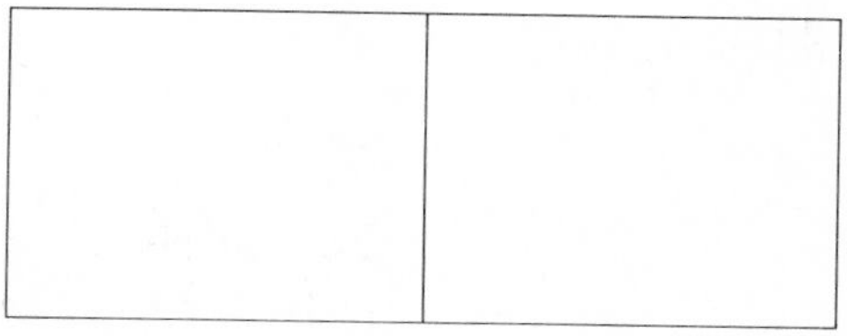

5.17

11. An open gutter of rectangular cross section is to be formed from a long sheet of tin of width 8 in., by bending up the sides of the sheet. Find the dimensions of the cross section of the gutter so formed for maximum carrying capacity.

12. Find the dimensions of the rectangle of maximum area that can be inscribed in a semicircle of radius a. [*Hint.* You may find it easier to maximize the square of the area. Of course, the rectangle having maximum area is the one having the square of its area maximum.]

13. Find the volume of the largest right circular cylinder that can be inscribed in a right circular cone of radius a and altitude b.

14. Find the altitude of the right circular cone of maximum volume that can be inscribed in a sphere of radius a.

15. Find the area of the largest isosceles triangle that can be inscribed in a circle of radius a.

16. A rectangular cardboard poster is to contain 216 in^2 of printed matter with two-inch margins at the sides and three-inch margins at the top and bottom. Find the dimensions of the poster using the least cardboard.

17. The wreck of a plane in a desert is 15 mi from the nearest point A on a straight road. A rescue truck starts for the wreck at a point on the road 30 mi distant from A. If the truck can travel at 80 mph on the road and at 40 mph on a straight path in the desert, how far from A should the truck leave the road in order to reach the wreck in minimum time?

18. A Norman window is to be built in the shape of a rectangle of clear glass surmounted by a semicircle of colored glass. The total perimeter of the window is to be 36 ft.

    a) Find the ratio of the height of the rectangle to the width that maximizes the area of the clear glass in the rectangular portion.

    b) If the clear glass admits twice as much light as the colored glass, find the ratio of the height of the rectangle to the width to admit the maximum amount of light.

    c) If the colored glass costs four times as much as the clear glass and the window is to be at least 4 ft wide, find the width of the window of minimum cost.

19. The strength of a beam of rectangular cross section is proportional to the width and the square of the depth. Find the dimensions of the strongest rectangular beam that can be cut from a circular log of radius 9 in.

20. Ship A travels on a due-north course and passes a buoy at 9:00 A.M. Ship B, traveling twice as fast on a due-east course, passes the same buoy at 11:00 A.M. the same day. At what time are the ships closest together?

21. A silo is to have the form of a cylinder capped with a hemisphere. If the material for the hemisphere is twice as expensive per square foot as the material for the cylinder, find the ratio of the height of the cylinder to its radius for the most economical structure of given volume. (Neglect waste in construction.)

22. An open box is to be formed from a rectangular piece of cardboard by cutting out equal squares at the corners and turning up the resulting flaps. Find the size of the corner squares that should be cut out from a sheet of cardboard of length a and width b in order to obtain a box of maximum volume.

23. The following is a simple economic model for the production of a single, perishable item that must be sold on the day produced if we are to avoid a loss due to spoilage.

    The producer has a basic plant overhead of a dollars per day. Each item produced costs b dollars for ingredients and labor. In addition, if the manufacturer produces x items per day, there is a daily cost of cx^2 dollars, resulting from crowded conditions and inefficient operation as more items are produced. (The value of c is usually quite small, so that cx^2 is of insignificant size until x becomes fairly large.) Each day, if a single item were produced, it could be sold that day for A dollars (the initial demand price). However, the price at which every item produced on a given day can be sold drops B dollars for each item produced on that day. (The number B reflects the degree of saturation of the market per item, and is usually quite small.)

a) Find an algebraic expression giving the daily profit if the manufacturer produces x items per day.

b) Find, in terms of a, b, c, A, and B, the number x of items the manufacturer should produce each day to maximize the daily profit.

24. Suppose that in the economic model in Exercise 23, the government imposes a tax on the manufacturer of t dollars for each item manufactured.

a) Determine the number x of items the manufacturer should produce each day now to maximize the daily profit. [*Hint.* Calculus is not necessary if you worked Exercise 23. Just think how this changes the manufacturer's costs.]

b) Find, in terms of a, b, c, A, and B, the value of t that will maximize the government's return, assuming that the manufacturer maximizes the profit as in part (a).

25. A fence a ft high is located b ft from the side of a house. Find the length of the shortest ladder that will reach over the fence to the house wall from the ground outside the fence.

26. A right circular cone has slant height 10 in. Find the vertex angle θ (see Fig. 5.18) for the cone of maximum volume.

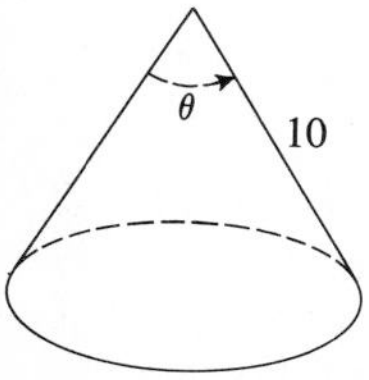

5.18

27. A statue 12 ft high stands on a pedestal 41 ft high. How far from the base of the pedestal on level ground should an observer stand so that the angle θ at his eye subtended by the statue is maximum if the eye of the observer is 5 ft from the ground? [*Hint.* The maximum value of θ occurs when $\tan\theta$ is maximum.]

28. According to Fermat's principle in optics, light travels the path for which the time of travel is minimum. Let light travel with velocity v_1 in medium 1 and velocity v_2 in medium 2, and let the boundary between the media form a plane, as shown in cross section in Fig. 5.19. Show that, according to Fermat's principle, light that travels from A to B in Fig. 5.19 crosses the boundary at a point P such that

$$\frac{\sin\theta_1}{\sin\theta_2} = \frac{v_1}{v_2}.$$

(This is the *law of refraction* or *Snell's law.*)

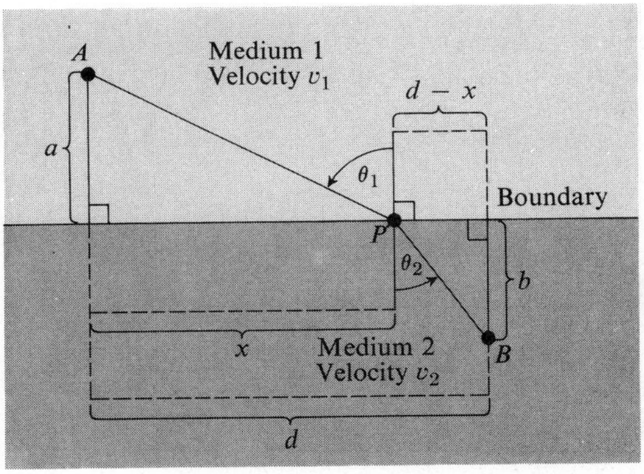

5.19

5.7. CALCULUS IN ECONOMICS AND BUSINESS

At many schools, undergraduate majors in economics who intend to do graduate work are encouraged to take many courses from the mathematics department, including three semesters of calculus. Calculus is an important tool in economic theory.

We shall talk about a very simplified economic situation. Suppose a company is manufacturing some product in a very ideal situation where there is no competition. This gives the company reasonable control over its own destiny. We further assume that the cost of producing an item, the revenue received from its sale, and the profit made are functions of the number x of units of the product manufactured per unit of time (a month, or a year, etc). This is a major assumption. It means that advertising decisions, transportation to market, etc., are all functions of this number x of units produced.

In many economic situations, you are interested only in *integer* values of a variable x. After all, a construction company is not going to build 31.347 houses! While some of our functions, such as cost, may be defined for only integer values of x, we shall assume that there is some differentiable function $C(x)$ defined throughout a whole interval, and giving the cost for all integer values of x in the interval. With this in mind, let

$$C(x) = \text{cost of producing } x \text{ units, per unit time,}$$

$$R(x) = \text{revenue received if } x \text{ units are produced,}$$

$$P(x) = R(x) - C(x) = \text{profit when } x \text{ units are produced.}$$

Economists are interested in the *marginal cost, marginal revenue*, and *marginal profit* when x units are produced. In a low-level economics course where calculus is not used, the marginal cost at x is defined to be the cost of producing one additional unit in the time period, so that

$$\text{Marginal cost} = C(x + 1) - C(x) = \frac{C(x + 1) - C(x)}{1}. \qquad (1)$$

From (1), you see that this marginal cost at x is approximately $C'(x)$, for (1) gives the approximation

$$C'(x) \approx \frac{C(x + \Delta x) - C(x)}{\Delta x}$$

where $\Delta x = 1$. In a higher-level course, the adjective *marginal* generally signifies a derivative.

Definition 5.4 The **marginal cost** is dC/dx, the **marginal revenue** is dR/dx, and the **marginal profit** is dP/dx.

Example 1 A company manufactures a popular pocket calculator. Its annual cost function in dollars is

$$C(x) = 90{,}000 + 500x + 0.01x^2,$$

where x is in hundreds of calculators produced per year. The \$90,000 represents the annual capital outlay for the plant, insurance, and such fixed expenses. The coefficient \$500 might represent the cost, exclusive of fixed expenses, of producing 100 calculators if not too many are produced. The term $0.01x^2$ comes into play only as x becomes fairly large, and might represent problems caused by crowding, storing excessive inventory, and an increase in the cost of materials if many calculators are produced, so material becomes scarce and its cost increases. Suppose the revenue function is

$$R(x) = 1000x - 0.05x^2.$$

Here $1000x$ appears because the first few calculators produced sell for \$10 each. The term $-0.05x^2$ appears because, if x is large, there is a glut on the market, so the sale price falls. Let's find the marginal profit, and find how many calculators the company should manufactuie for maximum profit.

SOLUTION We have

$$\begin{aligned} P(x) &= R(x) - C(x) = (1000x - 0.05x^2) - (90{,}000 + 500x + 0.01x^2) \\ &= -90{,}000 + 500x - 0.06x^2. \end{aligned}$$

Then

$$\text{Marginal profit} = \frac{dP}{dx} = 500 - 0.12x.$$

The profit is maximum when $dP/dx = 0$, or when

$$0.12x = 500 \qquad \text{or} \qquad x = \frac{50{,}000}{12} = 4{,}166.66666\ldots$$

This gives a maximum since $d^2P/dx^2 = -0.12 < 0$. Since x represents hundreds of calculators, we obtain 416,666.66666 . . . calculators. Of course the company isn't going to make $\frac{2}{3}$ of a calculator. A computation on a calculator gives

$$P(4{,}166.66) = P(4{,}166.67) = 951{,}666.6667.$$

From a practical point of view, the company makes the same profit manufacturing 416,666 calculators as manufacturing 416,667 calculators. This profit is \$951,666.67. [Mathematically, $P(4166.67) > P(4166.66)$ by symmetry of the parabola $P(x)$ about its high point where $x = 4166.666666\ldots$] ||

Since $C(x)$ is the cost of producing x units, then the

$$\text{Average cost per unit} = \frac{C(x)}{x}. \tag{2}$$

It is interesting to know how many units should be produced to minimize the average cost. Differentiating to minimize $C(x)/x$, we have

$$\frac{d}{dx}\left(\frac{C(x)}{x}\right) = \frac{xC'(x) - C(x)}{x^2}.$$

This, is zero when

$$xC'(x) - C(x) = 0 \qquad \text{or} \qquad C'(x) = \frac{C(x)}{x}.$$

Thus *minimum average cost occurs when marginal cost equals average cost.* Of course you should check to be sure you have a minimum and not a local maximum.

Example 2 Let's find how many units should be produced to minimize average cost for the company manufacturing calculators described in Example 1.

SOLUTION The cost function is

$$C(x) = 90{,}000 + 500x + 0.01x^2.$$

Thus $C'(x) = 500 + 0.02x$. Setting $C'(x) = C(x)/x$, you obtain

$$500 + 0.02x = \frac{90{,}000 + 500x + 0.01x^2}{x}$$

or

$$500x + 0.02x^2 = 90{,}000 + 500x + 0.01x^2.$$

Then

$$0.01x^2 = 90{,}000,$$

so

$$x^2 = 9{,}000{,}000$$

and

$$x = \sqrt{9{,}000{,}000} = 3000.$$

Since $\lim_{x\to 0^+} C(x)/x = \lim_{x\to\infty} C(x)/x = \infty$, there is a minimum somewhere, and the one candidate we found must be this minimum. This minimum average cost is achieved when 300,000 calculators per year are produced. The minimum average cost is $C(3000)/3000$ = \$560 per hundred calculators, or \$5.60 per calculator. ||

We conclude by giving two examples illustrating calculus applied to business problems.

Example 3 A large home and auto store sells 9,000 auto tires each year. It has *carrying costs* of 50¢ per year for each unsold tire stored (space, insurance, etc.). The *reorder cost* for each lot ordered is 25¢ per tire plus \$10 for the paperwork of the order. Let's find how many times per year the store should reorder, and the size of the order, to minimize the sum of these *inventory costs.*

SOLUTION Let x be the lot size of each order, i.e., the number of tires ordered at a time. Then

$$\text{Number of orders per year} = \frac{9000}{x}.$$

The average number of unsold tires over the year is then $x/2$. Therefore

$$\text{Carrying costs per year} = (0.5)\left(\frac{x}{2}\right) \text{ dollars}$$

and

$$\text{Reorder costs per year} = 10\left(\frac{9000}{x}\right) + (0.25)9000 \text{ dollars.}$$

Consequently, the total inventory costs per year are

$$C(x) = (0.5)\left(\frac{x}{2}\right) + 10\left(\frac{9000}{x}\right) + (0.25)9000$$

$$= \tfrac{1}{4}x + 90{,}000x^{-1} + 2250 \text{ dollars.}$$

Then

$$C'(x) = \frac{1}{4} - 90{,}000x^{-2} = \frac{1}{4} - \frac{90{,}000}{x^2}.$$

You easily see that $C'(x) = 0$ when $x^2 = 360{,}000$, or when $x = \sqrt{360{,}000} = 600$ tires. Since $C''(x) = 180{,}000x^{-3} > 0$, we do have a minimum. Therefore the tires should be ordered in lots of 600, and there should be $9{,}000/600 = 15$ such orders each year. ||

Example 4 A fisherman has exclusive rights to a stretch of clam flats. Suppose that conditions are such that on his flats a population of p bushels of clams this year yields a population of $f(p) = 50p - \tfrac{1}{4}p^2$ bushels next year. How many bushels should the fisherman dig each year to sustain a maximum harvest year after year?

SOLUTION The fisherman can dig (harvest) $h(p) = f(p) - p$ bushels of clams over the year without depleting the original population of p bushels. He wishes to maximize $h(p)$. Now

$$h(p) = f(p) - p = (50p - \tfrac{1}{4}p^2) - p = 49p - \tfrac{1}{4}p^2.$$

Then

$$h'(p) = 49 - \tfrac{1}{2}p$$

and $h'(p) = 0$ when $p = 98$. Since $h''(p) = -\tfrac{1}{2} < 0$, we do have a maximum. Thus the fisherman should first adjust the population of clams to 98 bushels, either by letting them increase without digging or by digging the excess population. Then he can dig

$$h(98) = 50(98) - \tfrac{1}{4}(98)^2 = 2499 \text{ bushels}$$

each year, or about 48 bushels per week, year after year. ||

SUMMARY

1. *If $C(x)$ is the cost, $R(x)$ the revenue, and $P(x) = R(x) - C(x)$ the profit from producing x units of a product per unit of time, then the marginal cost is dC/dx, the marginal revenue is dR/dx, and the marginal profit is dP/dx.*
2. *The average cost is $C(x)/x$. Average revenue and average profit are similarly defined.*
3. *For average cost to be minimum, you must have*

$$\text{Marginal cost} = \text{Average cost.}$$

EXERCISES

1. For the company manufacturing calculators described in Example 1, find the following when 2000 units have been manufactured.
a) The marginal cost
b) The marginal revenue
c) The marginal profit
d) The average profit

2. For the company manufacturing calculators described in Example 1, find how many units should be manufactured for maximum average profit, and the average profit then.

3. A small company manufactures wood stoves. Its cost $C(x)$ and revenue $R(x)$ when x stoves per year are manufactured are given by

$$C(x) = 10{,}000 + 150x + 0.03x^2,$$
$$R(x) = 250x - 0.02x^2.$$

a) Find the marginal profit when $x = 100$.
b) How many stoves should the company make for maximum profit? What is that maximum profit?

4. For the company in Exercise 3, find the minimum average cost and the maximum average profit.

5. Show that at any value $x > 0$ where the average profit achieves a maximum (or minimum) greater than zero, the marginal profit must equal the average profit.

6. Give an intuitive argument, using no calculus, that the average cost can only be minimum at $x > 0$ when it equals the marginal cost.

7. The amount a family saves is a function $S(I)$ of its income I, and the amount it consumes is a function $C(I)$. The *marginal propensity to save* is dS/dI, and the *marginal propensity to consume* is dC/dI. Show that if the entire income is used to either save or consume, then

$$\text{Marginal propensity to consume} = 1 - (\text{marginal propensity to save}).$$

8. The taxes T paid by a family in a certain country are given as a function of the income I by

$$T(I) = \frac{3I^2}{80{,}000 + 5I}.$$

a) Find the marginal tax rate when $I = 10{,}000$.
b) Describe the minimum and maximum average taxes per unit of income, if they exist.

9. An appliance store sells 200 refrigerators each year. It has carrying costs of \$10 per year for each unsold refrigerator stored. The recorder cost for each lot ordered is \$5 per refrigerator plus \$10 for the paperwork of the order. How many times per year should the store reorder, and at what lot size, to minimize these inventory costs?

10. A population of P wild rabbits on a certain farm gives rise to a population of

$$f(P) = 6P - (1/12)p^2$$

next year if left undisturbed.
a) What population will remain constant year after year if left undisturbed?
b) What is the maximum sustainable number that can be trapped year after year? What is the corresponding population size?

5.8 ANTIDERIVATIVES

5.8.1 Antidifferentiation

You have spent a lot of time differentiating, i.e., finding $f'(x)$ if $f(x)$ is known. Of equal importance is *antidifferentiating*, which is finding $f(x)$ if $f'(x)$ is known. We call $f(x)$ an **antiderivative** of $f'(x)$. Here is one indication of the importance of such a computation. You have seen that if you know the position s of a body on a line at time t, then $ds/dt = v$ is the velocity and $d^2s/dt^2 = a$ is the acceleration at time t. But, in practice, you often know the position and velocity only at the initial time $t = 0$, while you know the acceleration at every time $t \geq 0$. This is because you are often applying a controlled force F to produce the motion, perhaps by controlling the thrust of some motor. By Newton's second law of motion, $F = ma$ where m is the mass of the body. Thus if you know the force F, you know the acceleration $a = F/m$. Antidifferentiation of the acceleration will then find the velocity, and antidifferentiation of the velocity will find the position of the body. You are asked to find a well-known formula for the height of a freely-falling body in a vacuum, subject only to gravitational acceleration, in Exercise 18.

We change notation a bit and let $f(x)$ always be the known function, and $F(x)$ the desired antiderivative, so that

$$F'(x) = f(x).$$

Example 1 It is easy to see by differentiating that $x^3/3$ is an antiderivative of x^2. However, $(x^3/3) + 2$ is also an antiderivative of x^2, and indeed, if k is any constant, then $(x^3/3) + k$ is an antiderivative of x^2. ||

An arbitrary constant

As illustrated by Example 1, if $F(x)$ is an antiderivative of $f(x)$, then $F(x) + C$ is also an antiderivative for any constant C. In this context, C is called an *arbitrary constant*. The Mean-Value Theorem can be used to show that *all* antiderivatives of $f(x)$ are of this form $F(x) + C$; there are no others. We show this in two steps.

First, we show that if F is differentiable in $[a, b]$ and $F'(x) = 0$ for all x in $[a, b]$, then $F(x) = F(a)$ for all x in $[a, b]$. By the Mean-Value Theorem applied to $[a, x]$ for any x such that $a < x \leq b$,

$$\frac{F(x) - F(a)}{x - a} = F'(c) = 0,$$

where $a < c < x$. Thus $F(x) - F(a) = 0$, so $F(x) = F(a)$, and $F(x)$ is constant in $[a, b]$.

As the second step, we turn to the general case where $F'(x) = f(x)$ for all x in $[a, b]$. Suppose $G'(x) = f(x)$ also for all x in $[a, b]$. Then

$$\frac{d(G(x) - F(x))}{dx} = G'(x) - F'(x) = 0.$$

By the last paragraph, $G(x) - F(x) = G(a) - F(a)$. If we let $C = G(a) - F(a)$, then $G(x) = F(x) + C$. We use Theorem 5.8 to summarize.

Theorem 5.8 *If $F'(x) = f(x)$, then all antiderivatives of $f(x)$ are of the form $F(x) + C$ for some constant C.*

Example 2 From Example 1, you see that the general antiderivative of x^2 is $(x^3/3) + C$. ||

Example 3 It is easy to check by differentiation that for $n \neq -1$, the general antiderivative of x^n is $(x^{n+1}/(n+1)) + C$. ||

Suppose that f and g are functions with the same domain. If F is an antiderivative of f and G is an antiderivative of g, then it is trivial to check that $F + G$ is an antiderivative of $f + g$ and cF is an antiderivative of cf. One need only differentiate $F + G$ to obtain

$$\frac{d(F(x) + G(x))}{dx} = \frac{d(F(x))}{dx} + \frac{d(G(x))}{dx} = f(x) + g(x)$$

and differentiate cF to obtain

$$\frac{d(c \cdot F(x))}{dx} = c \cdot \frac{d(F(x))}{dx} = c \cdot f(x).$$

Example 4 In view of Example 3 and the preceding remarks, you see that the general antiderivative of a polynomial function

$$a_n x^n + \cdots + a_1 x + a_0$$

is

$$a_n\left(\frac{x^{n+1}}{n+1}\right) + \cdots + a_1\frac{x^2}{2} + a_0 x + C.$$

For example, the general antiderivative of $3x^2 + 4x + 7$ is

$$3\left(\frac{x^3}{3}\right) + 4\left(\frac{x^2}{2}\right) + 7x + C = x^3 + 2x^2 + 7x + C. \quad \|$$

Example 5 By differentiating, you can check that the general antiderivative of $\sin ax$ is $(-1/a) \cos ax + C$. Similarly, the general antiderivative of $\cos ax$ is $(1/a) \sin ax + C$. ||

5.8.2 Differential equations

An equation involving a derivative of an "unknown" function that one desires to find is a **differential equation**. The problem of finding an antiderivative of a function f may be phrased in terms of finding a solution of a differential equation. Namely, to find an antiderivative of f is to find a function F that is a solution of the differential equation

$$F' = f.$$

In Leibniz notation, you want to find a function $y = F(x)$ that is a solution of the differential equation

$$\frac{dy}{dx} = f(x).$$

If f has an antiderivative, that is, if the differential equation $dy/dx = f(x)$ has a solution, then the general antiderivative of f is the **general solution of this differential equation.**

Let's characterize the general solution $F(x) + C$ of the differential equation $dy/dx = f(x)$ geometrically. If $G(x) = F(x) + k$, then the graph of G is simply the graph of F displaced $|k|$ units "upward" if $k > 0$, or $|k|$ units "downward" if $k < 0$. Thus the set $F(x) + C$ of functions may be visualized geometrically as a collection of graphs with the property that any one of them is "congruent" to any other, and may be transformed into the other by sliding up or down.

Example 6 Figure 5.20 shows some of the graphs of solutions of the differential equation

$$\frac{dy}{dx} = x.$$

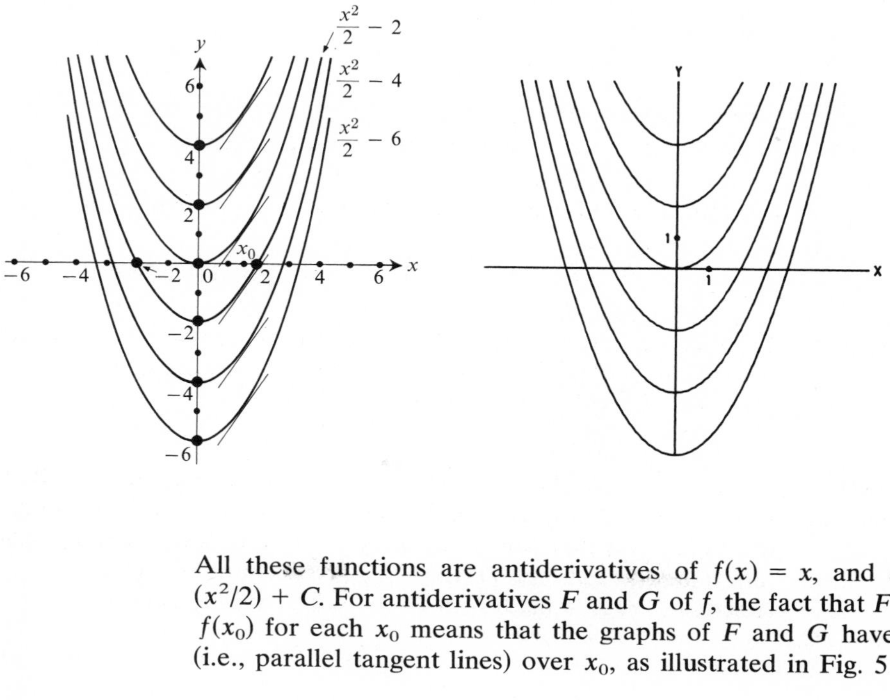

5.20 Artist's figure at left, computer-generated figure at right.

All these functions are antiderivatives of $f(x) = x$, and are of the form $(x^2/2) + C$. For antiderivatives F and G of f, the fact that $F'(x_0) = G'(x_0) = f(x_0)$ for each x_0 means that the graphs of F and G have the same slope (i.e., parallel tangent lines) over x_0, as illustrated in Fig. 5.20. ‖

If f is defined at x_0 and a solution F of $dy/dx = f(x)$ exists, then for any y_0 one can find C such that the solution $G(x) = F(x) + C$ satisfies the *initial condition* expressed by

$$G(x_0) = y_0.$$

Namely, if you desire to have

$$y_0 = G(x_0) = F(x_0) + C,$$

simply take $C = y_0 - F(x_0)$.

Example 7 We find the solution $G(x)$ of $dy/dx = x$ such that $G(1) = 3$.

SOLUTION We take $G(x) = (x^2/2) + C$ and require that

$$3 = G(1) = (1^2/2) + C.$$

You find that $C = 3 - (\frac{1}{2}) = \frac{5}{2}$. Thus $G(x) = (x^2/2) + (\frac{5}{2})$. ||

SUMMARY
1. *If $F'(x) = f(x)$, then $F(x) + C$ is the general antiderivative of $f(x)$, where C is an arbitrary constant.*
2. *The general antiderivative of x^n is $(x^{n+1})/(n + 1) + C$ if $n \neq -1$.*
3. *The antiderivative of a sum is the sum of the antiderivatives, and the antiderivative of a constant times a function is the constant times the antiderivative of the function.*

EXERCISES

In Exercises 1 through 12, find the general antiderivative of the given function.

1. 2 **2.** $x^2 - 3x + 2$ **3.** $8x^3 - 2x^2 + 4$ **4.** $x^{1/2}$

5. $x^{-2/3}$ **6.** $4x + x^{1/2}$ **7.** $(x^2 + 1)^2$ **8.** $\sqrt{x} - 1$

9. $\dfrac{1}{\sqrt{x+1}}$ **10.** $\sin x$ **11.** $\cos 3x$ **12.** $5 \sin 8x$

In Exercises 13 through 17, find the solution $y = F(x)$ of the differential equation that satisfies the given initial condition.

13. $\dfrac{dy}{dx} = 8,\ F(2) = -3$ **14.** $\dfrac{dy}{dx} = 3x^2 + 2,\ F(0) = 1$ **15.** $\dfrac{dy}{dx} = x + \sin x,\ F(0) = 3$

16. $\dfrac{dy}{dx} = 3 \cos 2x,\ F\left(\dfrac{\pi}{4}\right) = -1$ **17.** $\dfrac{dy}{dx} = x - \sqrt{x},\ F(1) = \pi$

18. (*Free fall in a vacuum.*) Suppose a body close to the surface of the earth is subject only to the force of gravity; that is, neglect air resistance. It is known that the acceleration of the body is then directed downward with magnitude $g = 32$ ft/sec^2. Thus if s measures the height of the body above the surface at time t, then you have $d^2s/dt^2 = -32$. Suppose that at time $t = 0$, the body has *initial velocity* v_0 and *initial position* s_0.

a) Find the velocity $v = ds/dt$ of the body as a function of t.

b) Find the position s of the body as a function of t.

exercise sets for chapter 5

review exercise set 5.1

1. If the altitude of a right circular cone remains constant at 10 ft while the volume increases at a steady rate of 8 ft^3/min, find the rate of increase of the radius when the radius is 4 ft. (The volume V of a cone of base radius r and altitude h is $V = \frac{1}{3}\pi r^2 h$).
2. A solution of $f(x) = x^3 - 3x + 1$ is desired. Since $f(1) = -1$ and $f(2) = 3$, there is a solution between 1 and 2. Starting with 2 as a first approximation of the solution, find the next two approximations given by Newton's method.
3. Find the maximum and minimum values assumed by $x^3 - 3x + 1$ in the interval $[0, 3]$.
4. Let f be differentiable for all x. If $f(1) = -2$ and $f'(x) \geq 2$ for x in $[1, 6]$, use the Mean-Value Theorem to show that $f(6) \geq 8$.
5. Find the solution of the differential equation $dy/dx = 3x^2 - 4x + 5$ that satisfies the initial condition $y = -2$ when $x = 1$.
6. Sketch the graph of $y = x^3 - 3x^2 + 2$, finding and labeling all local maxima and minima, and all inflection points.
7. A right circular cone has constant slant height 12. Find the radius of the base for which the cone has maximum volume. (The volume V of a cone of base radius r and altitude h is $V = \frac{1}{3}\pi r^2 h$.)
8. A company has cost and revenue functions

$$C(x) = 5{,}000 + 1{,}500x + 0.02x^2,$$
$$R(x) = 2{,}000x - 0.5^{3/2}.$$

   a) Find the marginal profit when $x = 10{,}000$.

   b) Find the average profit when $x = 10{,}000$.

review exercise set 5.2

1. If the lengths of two sides of a triangle remain constant at 10 ft and 15 ft while the angle θ between them is increasing at the rate of 1/10 radian/min, find the rate of increase of the third side when $\theta = \pi/3$.
2. Use Newton's method to estimate $\sqrt[3]{30}$, starting with 3 as first approximation and finding the next approximation.
3. Find the maximum and minimum values assumed by the function $x^4 - 8x^2 + 4$ in the interval $[-1, 3]$.
4. If $f'(x)$ exists for all x and $f(4) = 12$ while $f'(x) \leq -3$ for x in $[-1, 4]$, use the Mean-Value Theorem to find the smallest possible value for $f(-1)$.
5. A body moving on an s-axis has acceleration $a = 6t - 8$ at time t. If when $t = 1$ the body is at the point $s = 4$ and has velocity $v = -3$, find the position s as a function of time.
6. Sketch the graph of $y = 2x^2 - x^4 + 6$, finding and labeling all relative maxima and minima, and all inflection points.
7. A rectangle has base on the x-axis and upper vertices on the parabola

$$y = 16 - x^2 \quad \text{for} \quad -4 \leq x \leq 4.$$

   Find the maximum area the rectangle can have.
8. A company manufacturing x units of a certain product per year has a cost function $C(x) = 10{,}000 + 500x + 0.05x^2$ dollars, and a marginal revenue of $MR(x) = 900 - 0.04x$ dollars per year. Find the profit if 100 units are manufactured in a year.

more challenging exercises 5

1. A farmer has 1000 ft of fencing. He wishes to use it to fence in one square and one circular enclosure, each having an area of at least 10,000 ft. What size enclosures should he fence to maximize the total area? (Use a calculator.)

2. Show that if f is a twice differentiable function for $a \leq x \leq b$ and $f(x)$ assumes the same value at three distinct points in $[a, b]$, then $f''(c) = 0$ for some c where $a < c < b$.

3. State a generalization of the result in Exercise 2.

4. Suppose f is differentiable for $0 \leq x \leq 10$ and $f(2) = 17$ while $|f'(x)| \leq 3$ for $0 \leq x \leq 10$.

   a) What is the maximum possible value for $f(x)$ for any x in $[0, 10]$?

   b) What is the minimum possible value for $f(x)$ for any x in $[0, 10]$?

5. Maureen is at the edge of a circular lake of radius a miles, and wishes to get to the point directly across the lake. She has a boat which she can row 4 mph, or she can jog along the shore at 8 mph, or she can row to some point and jog the rest of the way. How should she proceed to get across in the shortest time?

6. Town A is located 2 miles back from the bank of a straight river. Town B is located 3 miles from the bank on the same side at a point 15 miles farther downstream. The towns agree to build a pumping station on the bank of the river to supply water to both towns. How far downstream from the point on the bank closest to town A should they build the station to minimize the total length of the water mains to the two towns? (Solve this problem first by using calculus, and then by "moving town A across the river" and using geometry. What have you learned?)

7. Solve the equation $x^2 = \sin x$. (Use a calculator.)

8. Give an example of a function f and a number a_1 such that an attempt to solve $f(x) = 0$ by Newton's method with initial approximation a_1 will lead to oscillation back and forth between two distinct values.

9. Construct an example like that in Exercise 8, but where there is oscillation back and forth between positive and negative numbers of ever increasing magnitude.

the integral

6

The practical importance of integral calculus, as well as of differential calculus, lies in its ability to handle situations in which quantities are varying continuously. Indeed, differential and integral calculus are closely connected, as we shall show in this chapter. Newton (1642–1727) and Leibniz (1646–1716) were both responsible for the development of integral calculus in the form we know it today. However, Archimedes (287?–212 B.C.) used the principles that lie at the heart of the subject in his work on the determination of the areas of certain types of planar regions. For this work alone, which was 2000 years ahead of its time, Archimedes must be regarded as one of the greatest mathematicians of recorded history. When Newton and Leibniz appeared on the scene, it was the natural time for calculus to be developed, as evidenced by their simultaneous but independent achievements in this field. Newton and Leibniz had the analytic geometry of Fermat (1601–1665) and Descartes (1596–1650) on which to build; Archimedes did not.

6.1 THE DEFINITE INTEGRAL

In this chapter, we will often want to consider a sum of quantities represented by subscripted letters, such as

$$a_1 + a_2 + \cdots + a_n.$$

6.1.1 Summation notation

There is a very useful notation for writing such sums. A sum $a_1 + \cdots + a_n$ is denoted in the *summation notation* by

$$\sum_{i=1}^{n} a_i,$$

read, "*the sum of the a_i as i runs from* 1 *to n*." The Greek letter $\sum$ (sigma) stands for *sum*; the value of i under the $\sum$ is the **lower limit of the sum**, and tells where the sum starts. The value above the $\sum$ is the **upper limit of the sum**, and tells where the sum stops. The letter i is the **summation index**. The choice of letter for summation index is of course not significant; the letters i, j, and k are the most commonly used. Thus

$$a_1 + \cdots + a_n = \sum_{i=1}^{n} a_i = \sum_{j=1}^{n} a_j = \sum_{k=1}^{n} a_k.$$

A typical sum that you will run into in this chapter is

$$f(x_1) + f(x_2) + \cdots + f(x_n) = \sum_{i=1}^{n} f(x_i).$$

You will probably best understand the use of the summation notation by studying some examples. In forming the sum, one simply replaces the summation index successively by all integers from the lower limit to the upper limit, and adds the resulting quantities.

Example 1 We have

$$\sum_{i=1}^{2} a_i = a_1 + a_2, \qquad \sum_{i=1}^{3} a_i = a_1 + a_2 + a_3,$$

and

$$\sum_{i=4}^{7} a_i = a_4 + a_5 + a_6 + a_7. \quad \|$$

Example 2 We have

$$\sum_{i=1}^{3} i^2 = 1^2 + 2^2 + 3^2 = 14,$$

$$\sum_{j=2}^{4} (j-1) = (2-1) + (3-1) + (4-1) = 1 + 2 + 3 = 6,$$

and

$$\sum_{k=0}^{2} (k^2+1) = (0^2+1) + (1^2+1) + (2^2+1) = 1 + 2 + 5 = 8. \quad \|$$

Example 3 We have $\sum_{i=1}^{n} (a_i + b_i) = \sum_{i=1}^{n} a_i + \sum_{i=1}^{n} b_i$, for

$$\begin{aligned}\sum_{i=1}^{n} (a_i + b_i) &= (a_1 + b_1) + \cdots + (a_n + b_n)\\ &= (a_1 + \cdots + a_n) + (b_1 + \cdots + b_n)\\ &= \sum_{i=1}^{n} a_i + \sum_{i=1}^{n} b_i. \quad \|\end{aligned}$$

6.1.2 Riemann sums

Let f be a continuous function with domain containing $[a, b]$, and suppose that $f(x) \geq 0$ for all x in $[a, b]$. Since the graph of the *continuous* function f over $[a, b]$ is an unbroken curve, it would seem that the notion of the *area of the region under the graph of f over* $[a, b]$, shown shaded in Fig. 6.1, should be meaningful. Let us denote the area of this region by $A_a^b(f)$.

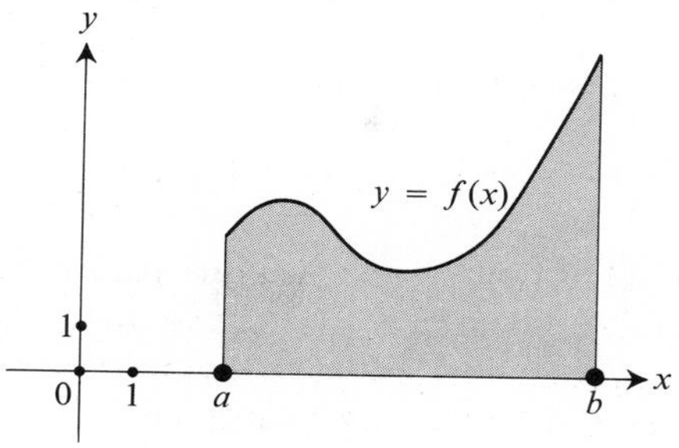

6.1

You can find areas of triangles, rectangles, circles, trapezoids, and certain other familiar plane regions, but the region under the graph of f over $[a, b]$ may not be any of these. Since you can find areas of rectangles, you might approximate $A_a^b(f)$ by areas of rectangles with bottoms on the x-axis and whose tops are horizontal line segments intersecting the graph of f, as shown in Fig. 6.2. The sum of the areas of these rectangles will give an approximate value of $A_a^b(f)$.

The bottoms of the rectangles in Fig. 6.2 divide $[a, b]$ into *subintervals.* To make computations easier, let's divide $[a, b]$ into n subintervals of *equal* lengths. The length of the base of each rectangle is then

$$\text{Length of base} = \Delta x = \frac{b-a}{n}. \tag{1}$$

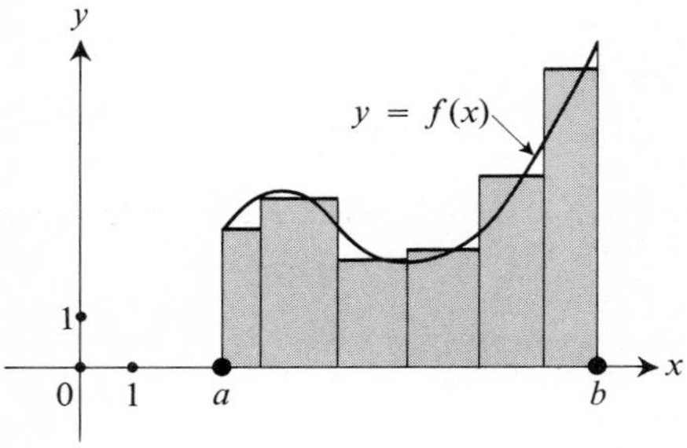

6.2

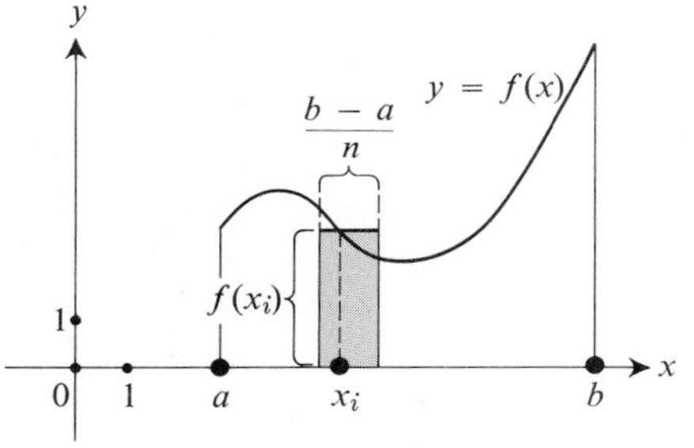

6.3

The height of the ith rectangle is $f(x_i)$ for some point x_i in the ith subinterval, so

$$\text{Area of the } i\text{th rectangle} = f(x_i) \cdot \frac{b-a}{n}.$$

(See Fig. 6.3). If we let $\mathcal{S}_n$ be the *sum* of the areas of these rectangles, then

$$\begin{aligned}\mathcal{S}_n &= \frac{b-a}{n} \cdot f(x_1) + \frac{b-a}{n} \cdot f(x_2) + \cdots + \frac{b-a}{n} \cdot f(x_n) \\ &= \frac{b-a}{n}(f(x_1) + f(x_2) + \cdots + f(x_n)) \\ &= \frac{b-a}{n}\left(\sum_{i=1}^{n} f(x_i)\right). \end{aligned} \tag{2}$$

Riemann sums

This sum $\mathcal{S}_n$ is a *Riemann sum** of order n for f over $[a, b]$.

* Named for the German mathematician, Bernhard Riemann (1826–1866).

Of course, the Riemann sum $\mathcal{S}_n$ in (2) depends not only upon the value of n but also upon the function f, the interval $[a, b]$, and the choice of points $x_1, \ldots, x_n$ in the n subintervals into which $[a, b]$ is divided. However, a notation that reflects all these dependencies would be unwieldy, so we shall simply use $\mathcal{S}_n$.

Example 4 Let's estimate $A_0^2(x^2)$ by computing $\mathcal{S}_4$, taking for the points x_1, x_2, x_3, and x_4 the midpoints of the first, second, third, and fourth intervals, respectively.

SOLUTION The situation is illustrated in Fig. 6.4. We have

$$x_1 = \tfrac{1}{4}, \qquad x_2 = \tfrac{3}{4}, \qquad x_3 = \tfrac{5}{4}, \qquad \text{and} \qquad x_4 = \tfrac{7}{4},$$

while

$$\frac{b-a}{n} = \frac{2-0}{4} = \frac{1}{2}.$$

The estimate $\mathcal{S}_4$ for $A_0^2(x^2)$ given by (2) is then

$$\tfrac{1}{2}(\tfrac{1}{16} + \tfrac{9}{16} + \tfrac{25}{16} + \tfrac{49}{16}) = \tfrac{1}{2}(\tfrac{84}{16}) = \tfrac{21}{8} = 2.625.$$

You will see later that the exact value of $A_0^2(x^2)$ is $\frac{8}{3}$, so our estimate is not too bad. ||

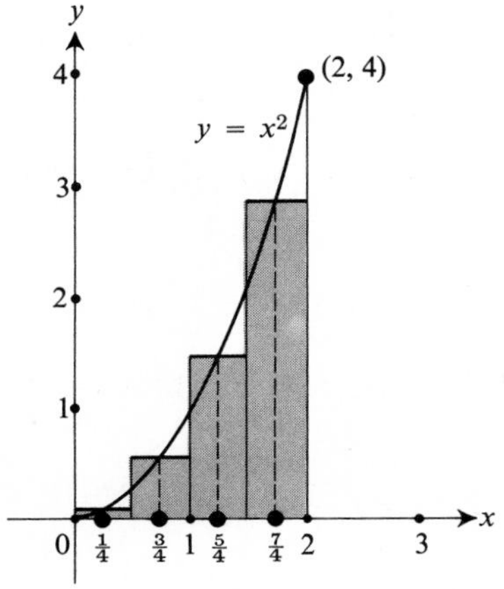

6.4

It is important to note that you can form the sum $\mathcal{S}_n$ in (2) even if $f(x)$ is negative for some x in $[a, b]$; the geometric interpretation of $\mathcal{S}_n$ for such a function will appear later in this section.

There are several natural ways to choose the points x_i in (2). One way is to select x_i so that $f(x_i)$ is the *maximum value* M_i of $f(x)$ for x in the ith subinterval $t_{i-1} \leq x \leq t_i$ (see Fig. 6.5). This gives a Riemann sum that you are sure is $\geq A_a^b(f)$. Another way is to select x_i so that $f(x_i)$ is the *minimum value* m_i of $f(x)$ over the ith subinterval, which gives a Riemann sum that is $\leq A_a^b(f)$ (see Fig. 6.6). To summarize notation, terminology, and relations:

Upper and lower sums

$$S_n = \text{Upper Riemann sum} = \frac{b-a}{n}\left(\sum_{i=1}^{n} M_i\right). \tag{3}$$

$$s_n = \text{Lower Riemann sum} = \frac{b-a}{n}\left(\sum_{i=1}^{n} m_i\right). \tag{4}$$

$$\mathscr{S}_n = \text{General Riemann sum} = \frac{b-a}{n}\left(\sum_{i=1}^{n} f(x_i)\right). \tag{5}$$

$$s_n \leq A_a^b(f) \leq S_n. \tag{6}$$

$$s_n \leq \mathscr{S}_n \leq S_n. \tag{7}$$

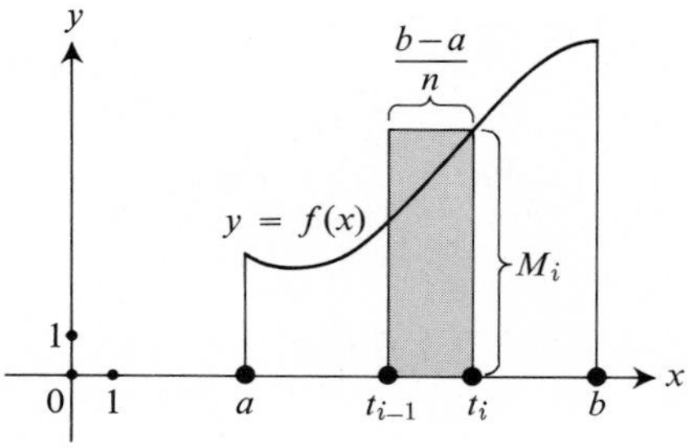

6.5

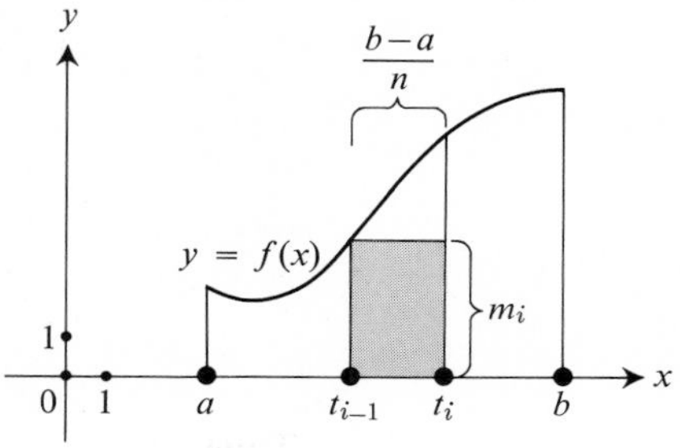

6.6

Example 5 Let us find the upper sum S_2 and lower sum s_2 approximating $A_0^1(x^2)$.

SOLUTION Here $n = 2$, and

$$\frac{b-a}{n} = \frac{1-0}{2} = \frac{1}{2}.$$

It is easily seen from the graph of f in Fig. 6.7(a) that

$$m_1 = 0 \quad \text{and} \quad m_2 = \tfrac{1}{4},$$

while

$$M_1 = \tfrac{1}{4} \quad \text{and} \quad M_2 = 1,$$

as shown in Fig. 6.7(b). Thus

$$s_2 = (\tfrac{1}{2})(0 + \tfrac{1}{4}) \leq A_0^1(x^2) \leq (\tfrac{1}{2})(\tfrac{1}{4} + 1) = S_2,$$

so

$$\tfrac{1}{8} \leq A_0^1(x^2) \leq \tfrac{5}{8}.$$

Of course, these bounds on $A_0^1(x^2)$ are very crude. ||

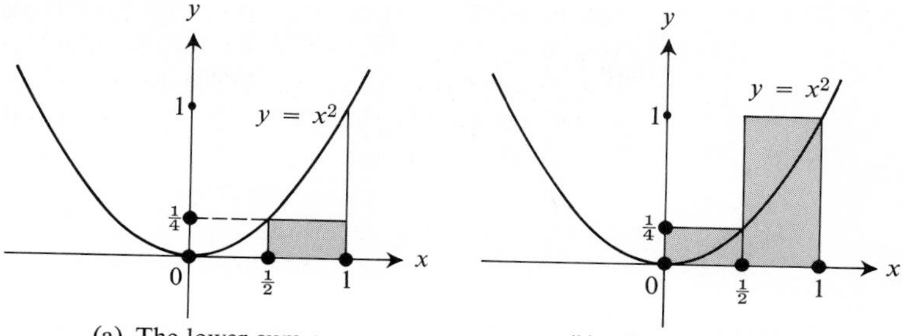

6.7

(a) The lower sum s_2 (b) The upper sum S_2

You can improve the estimate of $A_0^1(x^2)$ in Example 5 by taking a larger value of n.

Example 6 Let's estimate $A_0^1(x^2)$ by finding s_4 and S_4.

SOLUTION As indicated in Fig. 6.8(a) you have

$$m_1 = 0, \quad m_2 = \tfrac{1}{16}, \quad m_3 = \tfrac{1}{4}, \quad m_4 = \tfrac{9}{16},$$

while

$$M_1 = \tfrac{1}{16}, \quad M_2 = \tfrac{1}{4}, \quad M_3 = \tfrac{9}{16}, \quad M_4 = 1,$$

as shown in Fig. 6.8(b). Of course

$$\frac{b-a}{n} = \frac{1-0}{4} = \frac{1}{4}.$$

Thus

$$s_4 = (\tfrac{1}{4})(0 + \tfrac{1}{16} + \tfrac{1}{4} + \tfrac{9}{16}) \le A_0^1(x^2) \le (\tfrac{1}{4})(\tfrac{1}{16} + \tfrac{1}{4} + \tfrac{9}{16} + 1) = S_4.$$

The arithmetic works out to give

$$s_4 = \tfrac{7}{32} \le A_0^1(x^2) \le \tfrac{15}{32} = S_4.$$

This is an improvement over Example 5, which gave $\frac{4}{32} \le A_0^1(x^2) \le \frac{20}{32}$.

You might estimate $A_0^1(x^2)$ by averaging $\frac{7}{32}$ and $\frac{15}{32}$, arriving at $\frac{11}{32}$. We shall show later that the exact value of $A_0^1(x^2)$ is $\frac{1}{3}$. Thus $\frac{11}{32}$ is not a bad estimate. ||

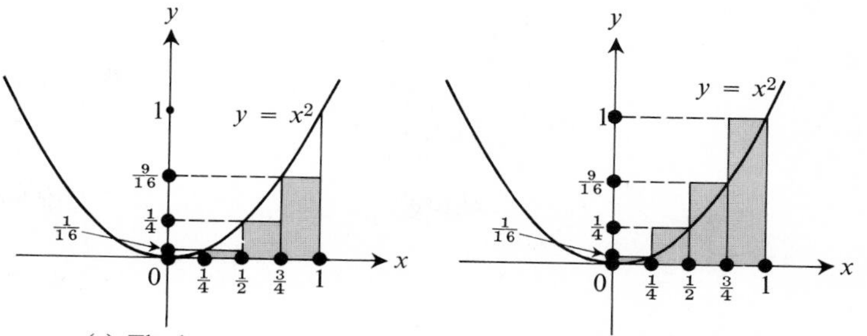

6.8

(a) The lower sum s_4 (b) The upper sum S_4

6.1.3 The definite integral

The accuracy of the bounds s_n and S_n for $A_a^b(f)$ can be measured by $S_n - s_n$. Geometrically, $S_n - s_n$ corresponds to the sum of the areas of the little rectangles along the graph shown shaded in Fig. 6.9. (Figure 6.9(a) illustrates the case $f(x) \geq 0$ for all x in $[a, b]$, and Fig. 6.9(b) illustrates the case where $f(x)$ is negative for some x in $[a, b]$.) Let h_n be the height of the largest of these little rectangles; that is, h_n is the maximum of $M_i - m_i$. By lining up the shaded rectangles in a row, as shown in Fig. 6.9, you find that the sum of the areas of the little rectangles is less than $h_n(b - a)$, so

$$S_n - s_n \leq h_n(b - a). \tag{8}$$

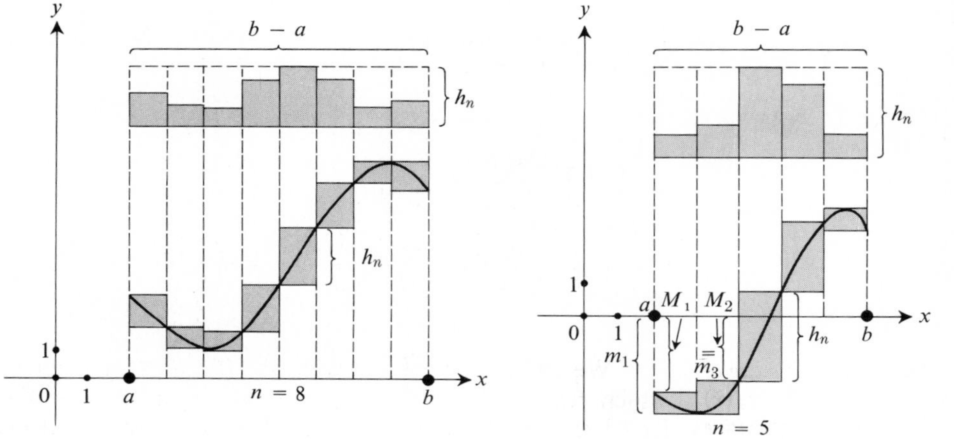

(a) $S_n - s_n$ for $f(x)$ always nonnegative (b) $S_n - s_n$ for $f(x)$ sometimes negative **6.9**

If f is a continuous function, then $f(x)$ is close to $f(x_0)$ for x sufficiently close to x_0, so it is reasonable to expect that, as n becomes large (so that the horizontal dimensions of the little rectangles shaded in Fig. 6.9 become very small), the vertical dimensions become small also. That is, you expect h_n to be close to zero for sufficiently large n, so that

$$\lim_{n \to \infty} h_n = 0.$$

In view of Eq. (8), $\lim_{n \to \infty} h_n = 0$ implies that $S_n - s_n$ is close to zero for sufficiently large n. By Eq. (6), for $f(x) \geq 0$, both S_n and s_n therefore approach $A_a^b(f)$ as n becomes large; that is, the following theorem holds.

Theorem 6.1 *If f is continuous on $[a, b]$, then*

$$\lim_{n \to \infty} s_n = \lim_{n \to \infty} \mathscr{S}_n = \lim_{n \to \infty} S_n. \tag{9}$$

Definition 6.1 The common limit of all the Riemann sums in (9) is the **definite integral of f over** $[a, b]$, written $\int_a^b f(x)\,dx$, so

$$\int_a^b f(x)\,dx = \lim_{n\to\infty} s_n = \lim_{n\to\infty} \mathscr{S}_n = \lim_{n\to\infty} S_n. \tag{10}$$

The Leibniz notation $\int_a^b f(x)\,dx$ may be interpreted as follows. You think of the *integral sign* $\int$ as an elongated letter S standing for "sum." If you let $dx = \Delta x$, an increment in x, then $f(x)\,dx$ is the area of the rectangle shown in Fig. 6.10. The notation for the integral thus suggests the sums in (3), (4), and (5).

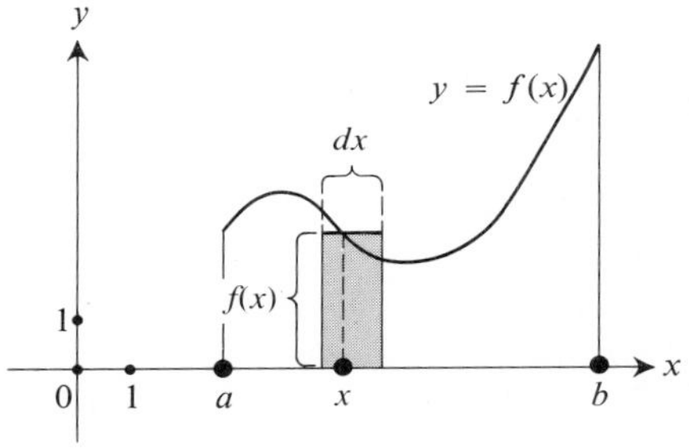

6.10

Geometric meaning of $\int_a^b f(x)\,dx$ for $f(x)$ sometimes negative

We have drawn the graph of f in most of the figures for the case in which $f(x) \geq 0$ for x in $[a, b]$. If $f(x) < 0$ for x in the ith subinterval of $[a, b]$, then both m_i and M_i are negative, so the contributions $[(b - a)/n]m_i$ to s_n and $[(b - a)/n]M_i$ to S_n are both negative. It is obvious that for a function f where $f(x)$ is sometimes negative and sometimes positive $\int_a^b f(x)\,dx$ can be interpreted geometrically as the total area given by the portions of the graph above the x-axis, minus the total area given by the portions of the graph below the x-axis (see Fig. 6.11).

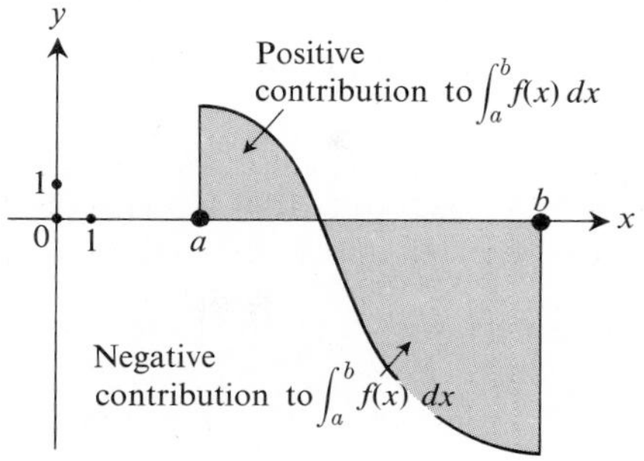

6.11

Example 7 Let's use geometry to find $\int_1^4 (x-2)\,dx$. The graph $y = x - 2$ is shown in Fig. 6.12. Since the small shaded triangle is below the x-axis and the large one is above,

$$\int_1^4 (x-2)\,dx = (\text{area of large triangle}) - (\text{area of small one})$$

$$= \tfrac{1}{2}\cdot 2\cdot 2 - \tfrac{1}{2}\cdot 1\cdot 1 = 2 - \tfrac{1}{2} = \tfrac{3}{2}. \quad \|$$

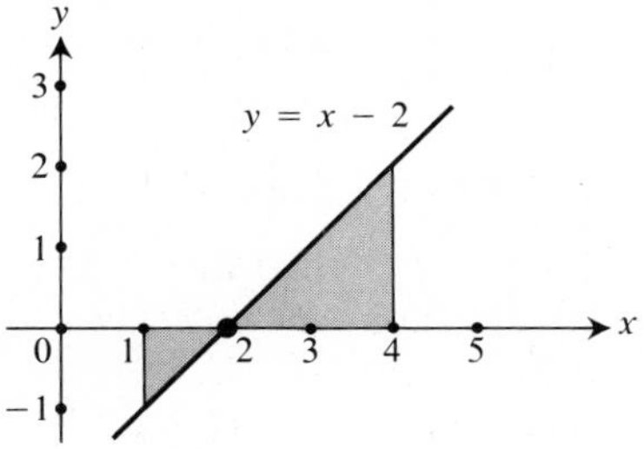

6.12

We have used our intuitive notion of *area* freely in motivating the development of the definite integral. Once Theorem 6.1 was stated and Definition 6.1 given, we may consider the area under a graph of a continuous positive function f over $[a, b]$ as being *defined* by $\int_a^b f(x)\,dx$.

*6.1.4 Computing integrals by limits of sums

The formulas

$$\sum_{i=1}^{n} i = \frac{n(n+1)}{2} \tag{11}$$

and

$$\sum_{i=1}^{n} i^2 = \frac{n(n+1)(2n+1)}{6} \tag{12}$$

can easily be proved using mathematical induction. These formulas can be used to compute definite integrals of linear and quadratic polynomial functions, as shown in Examples 8 and 9 below. We shall find a much easier way to compute the integrals in the next section, so these examples serve merely as illustration of Definition 6.1, and have little practical significance.

Example 8 Let us find the integral $\int_0^4 (5x-3)\,dx$ using (11).

SOLUTION Partition $[0, 4]$ into n subintervals of equal length $4/n$. If you take x_i as the righthand endpoint of the ith subinterval, then

$$x_i = i\cdot\frac{4}{n} = \frac{4i}{n}.$$

---

* This subsection can be omitted without loss of continuity.

Then

$$\mathcal{S}_n = \frac{4}{n}\sum_{i=1}^{n}\left(5\cdot\frac{4i}{n}-3\right) = \frac{4}{n}\left[\frac{20}{n}\left(\sum_{i=1}^{n} i\right)-\sum_{i=1}^{n} 3\right]$$

$$= \frac{4}{n}\left[\frac{20}{n}\cdot\frac{n(n+1)}{2}-3n\right] = \frac{80(n+1)}{2n}-12 = \frac{40(n+1)}{n}-12.$$

Therefore

$$\int_0^4 (5x-3)\,dx = \lim_{n\to\infty}\mathcal{S}_n = \lim_{n\to\infty}\left(\frac{40(n+1)}{n}\right)-12 = 40-12 = 28. \quad \|$$

Example 9 We use (11) and (12) to find $\int_2^4 (x^2+2x)\,dx$.

SOLUTION Partition [2, 4] into n subintervals of equal length $2/n$, and let x_i be the righthand endpoint of the ith subinterval. Then

$$x_i = 2 + i\cdot\frac{2}{n} = 2+\frac{2i}{n}$$

and

$$\mathcal{S}_n = \frac{2}{n}\cdot\sum_{i=1}^{n}\left[\left(2+\frac{2i}{n}\right)^2+2\left(2+\frac{2i}{n}\right)\right] = \frac{2}{n}\cdot\sum_{i=1}^{n}\left[4+\frac{8i}{n}+\frac{4i^2}{n^2}+4+\frac{4i}{n}\right]$$

$$= \frac{2}{n}\left[\sum_{i=1}^{n} 8+\frac{12}{n}\cdot\sum_{i=1}^{n} i+\frac{4}{n^2}\cdot\sum_{i=1}^{n} i^2\right]$$

$$= \frac{2}{n}\left[8n+\frac{12}{n}\cdot\frac{n(n+1)}{2}+\frac{4}{n^2}\cdot\frac{n(n+1)(2n+1)}{6}\right]$$

$$= 16+\frac{12n(n+1)}{n^2}+\frac{4n(n+1)(2n+1)}{3n^3}.$$

Taking the limit as $n\to\infty$, you obtain

$$\int_2^4 (x^2+2x)\,dx = \lim_{n\to\infty}\mathcal{S}_n = 16+12+\tfrac{8}{3} = \tfrac{92}{3}. \quad \|$$

SUMMARY 1. *To write out or compute a sum that is written symbolically using summation notation starting with*

$$\sum_{i=1}^{n},$$

simply replace the summation index i successively by all integers from the lower limit 1 to the upper limit n, and add the resulting quantities. For example,

$$\sum_{i=1}^{5} i^3 = 1^3+2^3+3^3+4^3+5^3.$$

Let f be continuous on $[a, b]$, and let $[a, b]$ be divided into n subintervals of equal lengths. Let M_i be the maximum value and m_i the minimum value of $f(x)$ over the ith subinterval, and let x_i be any point of the ith subinterval.

2. $S_n = \dfrac{b-a}{n}(M_1 + M_2 + \cdots + M_n) = \dfrac{b-a}{n}\left(\sum_{i=1}^{n} M_i\right)$

3. $s_n = \dfrac{b-a}{n}(m_1 + m_2 + \cdots + m_n) = \dfrac{b-a}{n}\left(\sum_{i=1}^{n} m_i\right)$

4. $\mathcal{S}_n = \dfrac{b-a}{n}(f(x_1) + f(x_2) + \cdots + f(x_n)) = \dfrac{b-a}{n}\left(\sum_{i=1}^{n} f(x_i)\right)$

5. $s_n \leq \mathcal{S}_n \leq S_n$

6. $\displaystyle\int_a^b f(x)\,dx = \lim_{n\to\infty} S_n = \lim_{n\to\infty} s_n = \lim_{n\to\infty} \mathcal{S}_n$

7. *For $f(x) > 0$, $s_n \leq$ (area under $y = f(x)$ and over $[a, b]$) $\leq S_n$*

EXERCISES

1. Write out each of the following sums.

a) $\sum_{i=0}^{3} a_i$ b) $\sum_{j=2}^{6} b_j^2$ c) $\sum_{i=1}^{4} a_{2i}$

d) $\sum_{k=4}^{6} (a_{2k} + b_k^2)$ e) $\sum_{i=1}^{5} c^i$ f) $\sum_{i=2}^{4} 2^{a_i}$

2. Compute each of the following sums.

a) $\sum_{i=0}^{3} (i+1)^2$ b) $\sum_{j=2}^{4} 2^j$ c) $\sum_{i=1}^{3} (2i-1)^2$

d) $\sum_{k=1}^{3} (2^k \cdot 3^{k-1})$ e) $\sum_{j=1}^{4} ((-1)^j \cdot j^3)$

3. Express each of the following sums in summation notation. (Many answers are possible).

a) $a_1b_1 + a_2b_2 + a_3b_3$ b) $a_1b_2 + a_2b_3 + a_3b_4$

c) $a_1 + a_2^2 + a_3^3 + a_4^4$ d) $a_1^2 + a_2^3 + a_3^4$

e) $a_1b_2^2 + a_2b_4^2 + a_3b_6^2$ f) $a_1^{b_3} + a_2^{b_6} + a_3^{b_9}$

4. Show, as in Example 3, that $\sum_{i=1}^{n} (c \cdot a_i) = c(\sum_{i=1}^{n} a_i)$.

5. Show, as in Example 3 that

$$\sum_{i=1}^{n} (a_i + b_i)^2 = \sum_{i=1}^{n} a_i^2 + 2\left(\sum_{i=1}^{n} a_ib_i\right) + \sum_{i=1}^{n} b_i^2.$$

6. Estimate $\int_0^4 x^2\,dx$ using the Riemann sum $\mathcal{S}_2$ where x_i is the midpoint of the ith interval.

7. Estimate $\int_1^5 (1/x)\,dx$ using the Riemann sum $\mathcal{S}_4$ where x_i is the midpoint of the ith interval.

8. Estimate $\int_{-1}^{1} x^2\,dx$ using the Riemann sum $\mathcal{S}_2$ where x_i is the midpoint of the ith interval.

9. Find the upper sum S_2 and the lower sum s_2 for x^2 from 0 to 2. (The actual value of $\int_0^2 x^2\,dx$ is $\frac{8}{3}$.)

10. Find the upper sum S_4 and the lower sum s_4 for x^2 from 0 to 2.

11. Estimate $\int_1^2 (1/x)\,dx$ by finding the upper sum S_4 and the lower sum s_4.

12. Give the exact value of $\int_{-2}^{3} 4\,dx$ where 4 is the usual constant function.

13. Let $f(x) = 1 - x$, and consider the interval $[0, 2]$.

a) Sketch the graph of f over this interval.

b) Find S_2 for this interval. (Note that $f(x)$ is sometimes negative for x in $[0, 2]$.)

c) Find s_2 for this interval.

d) What common value do both S_n and s_n approach as n gets large?

14. a) Find the upper sum S_4 and the lower sum s_4 for x^3 from -1 to 1.

b) From the graph of x^3, what common number do both S_n and s_n approach as n gets large?

15. Estimate $\int_0^{\pi} \sin x\,dx$ by finding the upper sum S_4 and the lower sum s_4.

In Exercises 16 through 21, sketch a region whose area is given by the integral. Determine the value of the integral by finding the area of the region.

16. $\int_0^2 x\,dx$ **17.** $\int_1^3 (2x + 3)\,dx$ **18.** $\int_{-1}^1 (x + 1)\,dx$

19. $\int_{-3}^3 \sqrt{9 - x^2}\,dx$ **20.** $\int_0^4 \sqrt{16 - x^2}\,dx$ **21.** $\int_0^4 (3 + \sqrt{16 - x^2})\,dx$

In Exercises 22 through 31, find the exact value of the integral using geometric arguments and the fact (which we shall soon show) that

$$\int_0^\pi \sin x\,dx = 2.$$

22. $\int_0^{\pi/2} \sin x\,dx$ **23.** $\int_{-\pi}^{\pi} \sin x\,dx$ **24.** $\int_0^{\pi/2} \cos x\,dx$ **25.** $\int_{-\pi/2}^{\pi/2} \cos x\,dx$ **26.** $\int_{-2}^2 (x^3 - 3x)\,dx$

27. $\int_0^\pi \cos x\,dx$ **28.** $\int_0^{5\pi} \sin x\,dx$ **29.** $\int_0^{2\pi} |\sin x|\,dx$ **30.** $\int_{-\pi}^{\pi} |\cos x|\,dx$ **31.** $\int_{-1}^1 |2x|\,dx$

32. Sketch the graph of a function f over the interval $[0, 2]$ for which S_2 is much closer to $\int_0^2 f(x)\,dx$ than the average of S_2 and s_2. (This shows that the averaging technique used in Example 6 may not give good results.)

33. Sketch the graph of a function f over the interval $[0, 6]$ for which $S_3 > S_2$. (This shows that while S_n approaches $\int_a^b f(x)\,dx$ for large n, we need not have $S_{n+1} \leq S_n$.)

In Exercises 34 through 37, use formulas (11) and (12) and Definition 6.1 to find the value of the definite integral.

***34.** $\int_0^3 (x - 2)\,dx$ ***35.** $\int_0^2 (4x^2 + 5)\,dx$ ***36.** $\int_3^5 (7 - 2x)\,dx$ ***37.** $\int_2^5 (x^2 - 4x + 2)\,dx$

calculator exercises

Use your calculator to estimate the given integral using $\mathscr{S}_n$ for the indicated value of n, where x_i is the midpoint of the ith subinterval.

38. $\int_1^2 x^x\,dx$ with $n = 8$ **39.** $\int_1^3 2^{\sin x}\,dx$ where $n = 10$ **40.** $\int_0^2 \cos x^2\,dx$ with $n = 16$

6.2 THE FUNDAMENTAL THEOREM OF CALCULUS

Throughout this section, we assume that $f(x)$ and $g(x)$ are continuous in $[a, b]$.

The Fundamental Theorem of Calculus exhibits the relationship between the derivative and the integral. Before proceeding with the fundamen-

6.2.1 Properties of the integral

tal theorem, we prove three properties of the integral, given in Eqs. (1), (2) and (3). Properties (1) and (2) parallel properties of the derivative. When the properties are stated in words, the parallel is striking. You can take your choice of "derivative" or "integral" in the following:

The $\left\{\begin{matrix}\textit{derivative}\\ \textit{integral}\end{matrix}\right\}$ *of a sum is the sum of the* $\left\{\begin{matrix}\textit{derivatives}\\ \textit{integrals}\end{matrix}\right\}$*. The* $\left\{\begin{matrix}\textit{derivative}\\ \textit{integral}\end{matrix}\right\}$ *of a constant times a function is the constant times the* $\left\{\begin{matrix}\textit{derivative}\\ \textit{integral}\end{matrix}\right\}$ *of the function.*

Theorem 6.2 *Three properties of the definite integral are*

$$\int_a^b (f(x) + g(x))\,dx = \int_a^b f(x)\,dx + \int_a^b g(x)\,dx, \tag{1}$$

$$\int_a^b c \cdot f(x)\,dx = c \cdot \int_a^b f(x)\,dx \quad \text{for any constant } c, \tag{2}$$

$$\int_a^b f(x)\,dx = \int_a^h f(x)\,dx + \int_h^b f(x)\,dx \quad \text{for any } h \text{ in } [a, b]. \tag{3}$$

All these properties are evident if you think of the interpretation in terms of areas. Figure 6.13 illustrates (3); surely the area under $f(x)$ from a to b is the area from a to h plus the area from h to b. If you look back at the Riemann sums $\mathcal{S}_n(f)$, $\mathcal{S}_n(g)$, and $\mathcal{S}_n(f + g)$ for $f(x)$, $g(x)$, and their sum, then

$$\begin{aligned}\mathcal{S}_n(f + g) &= \frac{b-a}{n}\left(\sum_{i=1}^{n} (f(x_i) + g(x_i))\right)\\ &= \frac{b-a}{n}\left(\sum_{i=1}^{n} f(x_i)\right) + \frac{b-a}{n}\left(\sum_{i=1}^{n} g(x_i)\right)\\ &= \mathcal{S}_n(f) + \mathcal{S}_n(g).\end{aligned}$$

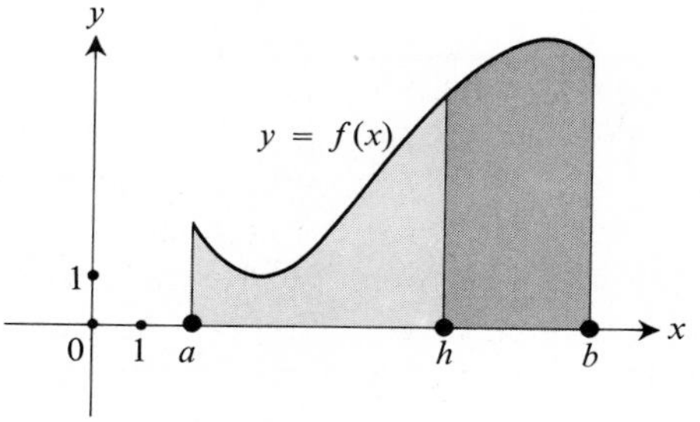

6.13

Therefore

$$\int_a^b (f(x) + g(x))\,dx = \lim_{n\to\infty} \mathcal{S}_n(f+g) = \lim_{n\to\infty} (\mathcal{S}_n(f) + \mathcal{S}_n(g))$$

$$= \lim_{n\to\infty} \mathcal{S}_n(f) + \lim_{n\to\infty} \mathcal{S}_n(g)$$

$$= \int_a^b f(x)\,dx + \int_a^b g(x)\,dx.$$

This establishes (1). Property (2) is established in a similar way.

For another property, note that if $b = a$, then

$$\int_a^a f(x)\,dx = 0. \tag{4}$$

You would like $\int_a^b f(x) = \int_a^h f(x)\,dx + \int_h^b f(x)\,dx$ to hold whether h is between a and b or not. (Naturally we require that a, b, and h all lie in some interval where $f(x)$ is continuous.) Then if you set $b = a$, you would have

$$0 = \int_a^a f(x)\,dx = \int_a^h f(x)\,dx + \int_h^a f(x)\,dx,$$

which leads to

$$\int_h^a f(x)\,dx = -\int_a^h f(x)\,dx. \tag{5}$$

We are motivated by (5) to define $\int_b^a f(x)\,dx$ if $b > a$ by

$$\int_b^a f(x)\,dx = -\int_a^b f(x)\,dx. \tag{6}$$

You will see again at the end of this section that (6) is a convenient definition.

6.2.2 The fundamental theorem

We have defined the definite integral $\int_a^b f(x)\,dx$ and you have had some practice in estimating the integral. You will soon see that the definite integral has a great variety of important applications, so that it is highly desirable to have an easy way to compute $\int_a^b f(x)\,dx$. It turns out that the *exact* value of the integral can be found if you can find an antiderivative $F(x)$ of $f(x)$. This is amazing! What should antiderivatives have to do with areas? Let's state the rule for computation and illustrate it before we see why it works.

Theorem 6.3 *(Fundamental Theorem, Part 2) Let f be continuous in $[a, b]$ and let $F'(x) = f(x)$. Then*

$$\int_a^b f(x)\,dx = F(b) - F(a). \tag{7}$$

Example 1 Let's find the exact value of $\int_0^1 x^2\,dx$, which we estimated several times in Section 6.1.

SOLUTION An antiderivative of the function x^2 is $F(x) = x^3/3$. Thus, by (7),

$$\int_0^1 x^2\,dx = F(1) - F(0) = \tfrac{1}{3} - \tfrac{0}{3} = \tfrac{1}{3}.$$

After all our work trying to get a good estimate for this definite integral in Section 6.1, you should appreciate the elegance and beauty of this easy computation. ||

It is customary to denote $F(b) - F(a)$ by $F(x)]_a^b$. The *upper limit of integration* is b and the *lower limit of integration* is a. For example, you will usually see $\int_0^1 x^2\,dx$ computed as follows:

Computing an integral

$$\int_0^1 x^2\,dx = \frac{x^3}{3}\bigg]_0^1 = \frac{1}{3} - \frac{0}{3} = \frac{1}{3}.$$

Example 2 Let's find the area under one "arch" of the curve $y = \sin x$.

SOLUTION Note that $-\cos x$ is an antiderivative of $\sin x$. The area is given by

$$\int_0^\pi \sin x\,dx = -\cos x\bigg]_0^\pi = -\cos\pi - (-\cos 0)$$

$$= -(-1) - (-1) = 1 + 1 = 2. \quad \|$$

Example 3 Let's sketch the region bounded by the graph of the polynomial function $1 - x^2$ and by the x-axis, and find the area of the region.

SOLUTION The function $1 - x^2$ obviously has a maximum at $x = 0$, and has decreasing values as x gets farther from 0. The graph crosses the x-axis when $1 - x^2 = 0$, that is, when $x = \pm 1$. The graph is shown in Fig. 6.14, where we have shaded the region whose area we wish to find. Clearly the desired

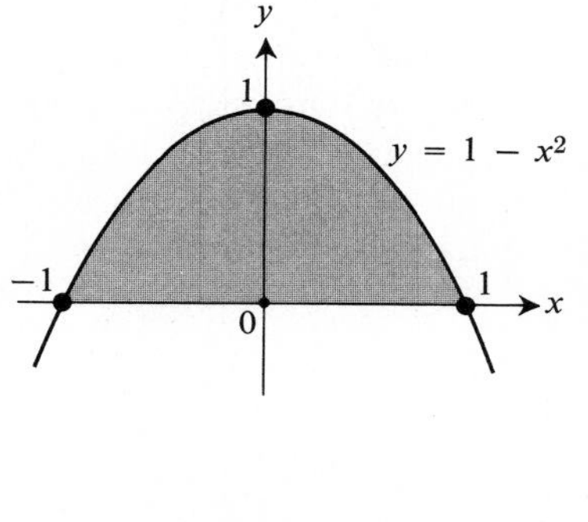

6.14

area is $\int_{-1}^{1}(1 - x^2)\,dx$ square units, and

$$\int_{-1}^{1}(1 - x^2)\,dx = \left(x - \frac{x^3}{3}\right)\Big]_{-1}^{1}$$
$$= \left(1 - \frac{1}{3}\right) - \left(-1 - \frac{-1}{3}\right) = \frac{2}{3} - \left(-\frac{2}{3}\right) = \frac{4}{3}. \quad \|$$

Now we shall show why antidifferentiation enables us to find the area under the graph of a function $f(x)$. The trick is to consider the integral function

$$F(t) = \int_a^t f(x)\,dx, \tag{8}$$

so that if $f(x) \geq 0$, $F(t)$ is the area under the graph of f over $[a, t]$, as shown in Fig. 6.15. We shall show that $F'(t) = f(t)$, and this will remove the mystery. If the derivative of the integral function F is f, then surely to find F one should find an antiderivative of f.

By (3),

$$\int_a^{t+\Delta t} f(x)\,dx = \int_a^t f(x)\,dx + \int_t^{t+\Delta t} f(x)\,dx$$

so

$$F'(t) = \lim_{\Delta t \to 0} \frac{F(t + \Delta t) - F(t)}{\Delta t}$$
$$= \lim_{\Delta t \to 0} \frac{\int_a^{t+\Delta t} f(x)\,dx - \int_a^t f(x)\,dx}{\Delta t} = \lim_{\Delta t \to 0} \frac{\int_t^{t+\Delta t} f(x)\,dx}{\Delta t}. \tag{9}$$

Referring to Fig. 6.16, we see that

$$m \cdot \Delta t \leq \int_t^{t+\Delta t} f(x)\,dx \leq M \cdot \Delta t, \tag{10}$$

where m is the minimum and M the maximum value of $f(x)$ in $[t, t + \Delta t]$.

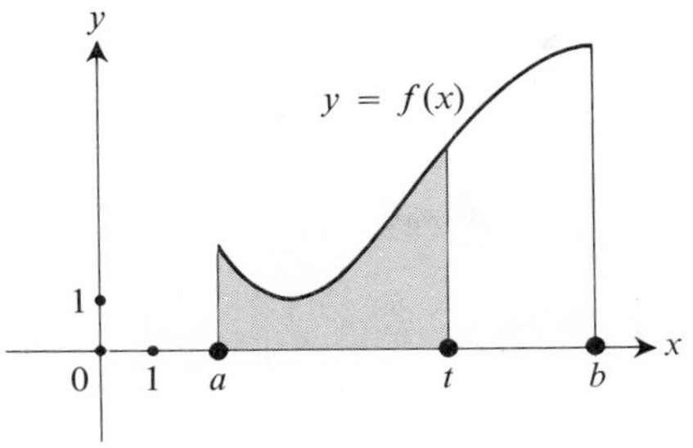

6.15

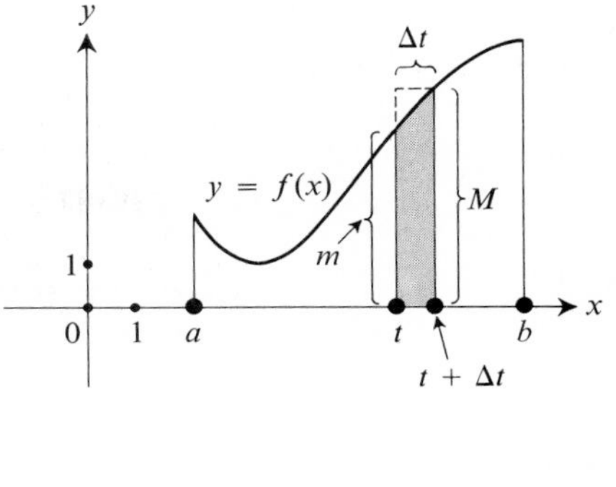

6.16

Therefore

$$m \le \frac{\int_t^{t+\Delta t} f(x)\,dx}{\Delta t} \le M. \tag{11}$$

Now $\lim_{\Delta t \to 0} m = \lim_{\Delta t \to 0} M = f(t)$ since f is continuous. We combine (9) with (11) to obtain

$$F'(t) = \lim_{\Delta t \to 0} \frac{\int_t^{t+\Delta t} f(x)\,dx}{\Delta t} = f(t). \tag{12}$$

This concludes our demonstration that $F'(t) = f(t)$, which is an important result in its own right.

Theorem 6.4 *(Fundamental Theorem, Part* 1). *If f is continuous in $[a, b]$ and $F(t) = \int_a^t f(x)\,dx$, then F is differentiable and $F'(t) = f(t)$. That is,*

$$\frac{d}{dt}\left(\int_a^t f(x)\,dx\right) = f(t).$$

Example 4 By Theorem 6.4,

$$\frac{d}{dt}\left(\int_0^t \sqrt{x^2+1}\,dx\right) = \sqrt{t^2+1},$$

for you can take $f(x) = \sqrt{x^2+1}$ and $F(t) = \int_0^t f(x)\,dx$ in the theorem.

Using properties of the integral as well as Theorem 6.4, you find that

$$\frac{d}{dt}\left(\int_t^{-3} \sin^2 x\,dx\right) = \frac{d}{dt}\left(-\int_{-3}^t \sin^2 x\,dx\right) = -\frac{d}{dt}\left(\int_{-3}^t \sin^2 x\,dx\right) = -\sin^2 t. \quad \|$$

We now continue our explanation of (7) as a means of computing $\int_a^b f(x)\,dx$. Since $F(t) = \int_a^t f(x)\,dx$,

$$\int_a^b f(x)\,dx = F(b). \tag{13}$$

But $F(a) = \int_a^a f(x)\,dx = 0$, so you can write (13) as

$$\int_a^b f(x)\,dx = F(b) - F(a). \tag{14}$$

Now Theorem 6.3 asserted that (14) is true for *any* antiderivative of $f(x)$, not just our integral function F. Let $G(x)$ be any antiderivative, so $G'(x) = f(x)$. Since $F(x)$ and $G(x)$ have the same derivative $f(x)$, you know that $F(x) = G(x) + C$ for some constant C, and

$$\int_a^b f(x)\,dx = F(b) - F(a) = [G(b) + C] - [G(a) + C] = G(b) - G(a).$$

Hence $\int_a^b f(x)\,dx$ can be computed by finding *any* antiderivative of $f(x)$ and subtracting its value at a from its value at b, as stated in (7).

Interchanging a and b in (7) yields

$$\int_b^a f(x)\,dx = F(a) - F(b) = -[F(b) - F(a)] = -\int_a^b f(x)\,dx,$$

which is another reason for our definition in (6).

SUMMARY *Let $f(x)$ and $g(x)$ be continuous in $[a, b]$.*

1. $\int_a^b (f(x) + g(x))\,dx = \int_a^b f(x)\,dx + \int_a^b g(x)\,dx$
2. $\int_a^b c \cdot f(x)\,dx = c \cdot \int_a^b f(x)\,dx$ *for any constant c*
3. $\int_a^b f(x)\,dx = \int_a^h f(x)\,dx + \int_h^b f(x)\,dx$ *for any h in $[a, b]$*
4. $\int_b^a f(x)\,dx = -\int_a^b f(x)\,dx.$
5. *(Fundamental Theorem of Calculus)*
   a) $\dfrac{d}{dt}\left(\int_a^t f(x)\,dx\right) = f(t).$
   b) *If $F'(x) = f(x)$, then* $\int_a^b f(x)\,dx = F(b) - F(a).$

EXERCISES

1. State part 1 of the fundamental theorem of calculus without referring to the text.

2. State the second part of the fundamental theorem of calculus without referring to the text.

In Exercises 3 through 32, use the fundamental theorem, properties of the definite integral, geometry, and the fact that

$$\int_0^\pi \sin^2 x\,dx = \frac{\pi}{2}$$

to evaluate the definite integral.

3. $\int_0^1 x^3\,dx$

4. $\int_{-1}^1 x^4\,dx$

5. $\int_0^2 (x^2 + 3x - 1)\,dx$

6. $\int_4^9 \sqrt{x}\,dx$

7. $\int_0^\pi \sin x\,dx$

8. $\int_0^{\pi/2} 4\cos x\,dx$

9. $\int_0^{\pi/2} (\sin x + 2\cos x)\,dx$

10. $\int_{\pi/4}^{\pi/2} (\sin x - \cos x)\,dx$

11. $\int_{-\pi/4}^{\pi/4} 3\cos x\,dx$

12. $\int_1^0 x^2\,dx$

13. $\int_3^{-2} 4\,dx$

14. $\int_3^{-2} (-4)\,dx$

15. $\int_{-1}^{-2} x\,dx$

16. $\int_4^4 \sqrt{x^3+1}\,dx$

17. $\int_1^{-1} \sqrt{1-x^2}\,dx$

18. $\int_0^1 (x+\sqrt{1-x^2})\,dx$

19. $\int_0^\pi (\sin x + \sin^2 x)\,dx$

20. $\int_{-\pi}^{\pi} \sin^2 x\,dx$

21. $\int_0^{2\pi} (\cos x + \sin^2 x)\,dx$

22. $\int_0^\pi 8\sin^2 x\,dx$

23. $\int_0^\pi (2\sin^2 x + 3\sqrt{\pi^2 - x^2})\,dx$

24. $\int_{-\pi/2}^{\pi/2} \cos^2 x\,dx$

25. $\int_0^\pi (3x - 4\cos^2 x)\,dx$

26. $\int_0^{\pi/2} (2\sin^2 x - \cos^2 x)\,dx$

27. $\int_1^3 \left(\frac{d(x^2)}{dx}\right) dx$

28. $\int_0^1 \left(\frac{d(\sqrt{x^3+1})}{dx}\right) dx$

29. $\int_0^1 \left(\frac{d^2(\sqrt{x^2+1})}{dx^2}\right) dx$

30. $\int_0^2 \left(\frac{d^3(x^2+3x-1)}{dx^3}\right) dx$

31. $\int_0^4 \left[\int_1^x \sqrt{t}\,dt\right] dx$

32. $\int_{-1}^3 \left[\int_x^2 4\,dt\right] dx$

33. Sketch the region bounded by the curve $y = 9 - x^2$ and the x-axis, and find the area of the region.

34. Sketch the region bounded by the curve $y = 2x - x^2$ and the x-axis, and find the area of the region.

35. Sketch the region bounded by the curve $y = x^4 - 16$ and the x-axis, and find the area of the region.

In Exercises 36 *through* 46, *use properties of the definite integral and Theorem* 6.4 *to find the indicated function of t.*

36. $\frac{d}{dt}\left(\int_1^t x^2\,dx\right)$

37. $\frac{d}{dt}\left(\int_{-50}^t \sqrt{x^2+1}\,dx\right)$

38. $\frac{d^2}{dt^2}\left(\int_2^t \sqrt{x^2+1}\,dx\right)$

39. $\frac{d}{dt}\left(\int_t^3 \sqrt{x^2+1}\,dx\right)$

40. $\frac{d}{dt}\left(\int_2^t \sqrt{x^2+1}\,dx + \int_t^{-1} \sqrt{x^2+1}\,dx\right)$

41. $\frac{d}{dt}\left(\int_{-t}^t \sqrt{x^2+1}\,dx\right)$

42. $\frac{d^2}{dt^2}\left(\int_{-t}^t \sqrt{x^2+1}\,dx\right)$

43. $\frac{d}{dt}\left(\int_1^{-t} \sqrt{x^2+1}\,dx\right)$

44. $\frac{d}{dt}\left(\int_1^{3t} \sqrt{x^2+1}\,dx\right)$

45. $\frac{d}{dt}\left(\int_0^{t^2-3} \sqrt{x^2+1}\,dx\right)$

46. $\frac{d}{dt}\left(\int_{t^2}^{t^3} \sqrt{x^2+1}\,dx\right)$

6.3 INTEGRATION AND DIFFERENTIAL EQUATIONS

The fundamental theorem gives a powerful method for computing $\int_a^b f(x)\,dx$. The burden of the computation is placed squarely on finding an antiderivative $F(x)$ of $f(x)$.

6.3.1 The indefinite integral

Definition 6.2. The general antiderivative of $f(x)$ is also called the **indefinite integral** of $f(x)$, and is written

$$\int f(x)\,dx$$

without any limits a and b of integration. Computing an antiderivative is called **integration**.

Example 1 The indefinite integral $\int (x^3 + 2x^2 + 1)\,dx$ is given by

$$\int (x^3 + 2x^2 + 1)\,dx = \frac{x^4}{4} + 2\frac{x^3}{3} + x + C,$$

where C is an arbitrary constant. ||

According to the last section, every function f that has a derivative in $[a, b]$ also has an antiderivative in $[a, b]$. That is, if f is differentiable, it is continuous and by the fundamental theorem, $F(t) = \int_a^t f(x)\,dx$ is an antiderivative of f for t in $[a, b]$. In practice, it is often more difficult to actually compute an antiderivative of a function than to find its derivative. The reason for this is that there are no simple formulas to handle integration of products, quotients, or composite functions, as you have for differentiation. Tables have been prepared that list indefinite integrals of a number of functions frequently encountered. A brief table of this type is found in this text. Such tables are often useful, but no table exists that gives the integral of every continuous function anyone will ever encounter. It can be proved that an antiderivative of an "*elementary function*" (a function formed by algebraic operations from rational, trigonometric, exponential, and logarithmic functions) need not still be an elementary function. This is contrary to the situation for differentiation.

We will discuss integration techniques in considerably more detail later. Integration frequently requires ingenuity, and real facility can be acquired only by practice.

We start your training in integration now. This section presents a small number of integration formulas, which you are expected to remember and to be able to apply faster than you could find the desired formula in a table. For example, you should never have to look up $\int x^2\,dx$ in a table. These formulas are those that arise from the differentiation formulas for basic elementary functions. For example, if u is a differentiable function of x, then

$$\frac{d(\tan u)}{dx} = (\sec^2 u) \cdot \frac{du}{dx}$$

so

$$\int (\sec^2 u) \cdot \frac{du}{dx}\,dx = \tan u + C.$$

Since $du = u'(x)\,dx = (du/dx)\,dx$, the preceding integration formula is usually written in the form

$$\int (\sec^2 u)\, du = \tan u + C.$$

The integration formulas that correspond to our differentiation formulas follow. Formulas 4 through 10 assume that u is a differentiable function of x.

1. $\int (f(x) + g(x))\, dx = \int f(x)\, dx + \int g(x)\, dx$
2. $\int c \cdot f(x)\, dx = c \cdot \int f(x)\, dx$
3. $\displaystyle\int x^n\, dx = \frac{x^{n+1}}{n+1} + C, \quad n \neq -1$
4. $\displaystyle\int u^n\, du = \frac{u^{n+1}}{n+1} + C, \quad n \neq -1$
5. $\int \sin u\, du = -\cos u + C$
6. $\int \cos u\, du = \sin u + C$
7. $\int \sec^2 u\, du = \tan u + C$
8. $\int \csc^2 u\, du = -\cot u + C$
9. $\int \sec u \tan u\, du = \sec u + C$
10. $\int \csc u \cot u\, du = -\csc u + C$

Formula (3) is a special case of Formula (4), which is one of the most often used. The following example illustrates the use of Formula (4).

Example 2 Let us find $\int 2x\sqrt{x^2+1}\, dx$.

SOLUTION If you let $u = x^2 + 1$, then $du = 2x\, dx$ and from Formula (4), you obtain

$$\int 2x\sqrt{x^2+1}\, dx = \int (u^{1/2})\, du = \frac{u^{3/2}}{3/2} + C = \frac{2}{3}(x^2+1)^{3/2} + C. \quad \|$$

Example 3 Let us find $\int 3x^2 \sin x^3\, dx$.

SOLUTION If you let $u = x^3$, then $du = 3x^2\, dx$ and, from Formula (5), you obtain

$$\int 3x^2 \sin x^3\, dx = \int (\sin u)\, du = -\cos u + C = -\cos x^3 + C. \quad \|$$

In practice, you usually do not write out the substitution $u = g(x)$ in simple cases like those in Examples 2 and 3, but rather, realizing that the substitution is appropriate, you compute the integral in one step "in your head." Suppose, for example, you wish to find $\int \cos 2x\, dx$. If you let $u = 2x$, then $du = 2\, dx$ and

$$\int \cos 2x\, dx = \int \tfrac{1}{2} \cdot 2 \cos 2x\, dx = \tfrac{1}{2} \int 2 \cos 2x\, dx.$$

You thus find from Formula (6) that

$$\int \cos 2x\, dx = \tfrac{1}{2} \int 2 \cos 2x\, dx = \tfrac{1}{2} \sin 2x + C.$$

We illustrate this technique further with some more examples.

Example 4 Let's find $\int x^2 \sec^2 x^3\, dx$.

SOLUTION If $u = x^3$, then $du = 3x^2\, dx$ and we would like to have an additional factor 3 in the integral so that Formula (7) would apply. Using Formulas (2) and (7), we obtain

$$\int x^2 \sec^2 x^3\, dx = \tfrac{1}{3}\int 3x^2 \sec^2 x^3\, dx = \tfrac{1}{3}\tan x^3 + C. \quad \|$$

Example 5 Let's find $\int \sin 3x \cos^2 3x\, dx$.

SOLUTION If we let $u = \cos 3x$, then $du = -3\sin 3x\, dx$, and we would like an additional factor (-3) in our integral so that we could apply Formula (4). We obtain from Formulas (2) and (4),

Fixing up integrals

$$\int \sin 3x \cos^2 3x\, dx = -\frac{1}{3}\int (-3\sin 3x)(\cos^2 3x)\, dx = -\frac{1}{3}\frac{\cos^3 3x}{3} + C$$

$$= -\tfrac{1}{9}\cos^3 3x + C. \quad \|$$

As illustrated in the two preceding examples, Formula (2) shows that you can "fix up an integral to supply a desired constant factor." *We warn you that a similar technique to supply a variable factor is not valid.* To illustrate,

$$\int x\, dx \neq \frac{1}{x}\int x \cdot x\, dx = \frac{1}{x}\int x^2\, dx,$$

for $\int x\, dx = x^2/2 + C$, while

$$\frac{1}{x}\int x^2\, dx = \frac{1}{x}\left(\frac{x^3}{3} + C\right) = \frac{x^2}{3} + \frac{C}{x}, \qquad x \neq 0.$$

For example, you can compute $\int x \sin x^2\, dx$, for you can "fix up the desired constant 2" and use Formula (5), but at the moment, you can't compute $\int \sin x^2\, dx$, for there is no way you can "fix up the variable factor $2x$" needed to apply Formula (5).

6.3.2 Differential equations

A differential equation is one containing a derivative or differential. In Chapter 5, we solved some differential equations of the form

$$\frac{dy}{dx} = f(x).$$

Many differential equations cannot be written in this form, where dy/dx equals a function of x alone. (Remember that, when differentiating implicitly, you frequently obtain an expression for dy/dx that involves both x and y.) An example of such a differential equation is

$$\frac{dy}{dx} = \frac{x}{y}.$$

Separating variables

You can solve this differential equation by the following device: Rewrite the equation with all terms involving y (including dy) on the left and all terms involving x (including dx) on the right:

$$y \cdot dy = x \cdot dx.$$

Since the differentials $y \cdot dy$ and $x \cdot dx$ are equal, it appears that you must have

$$\int y \cdot dy = \int x \cdot dx$$

or

$$\frac{y^2}{2} + C_1 = \frac{x^2}{2} + C_2,$$

where C_1 and C_2 are arbitrary constants. This solution may be written as

$$y^2 + 2C_1 = x^2 + 2C_2$$

or

$$y^2 - x^2 = 2C_1 - 2C_2.$$

Now $2C_1 - 2C_2$ may be any constant, so you can express it as a single arbitrary constant C, obtaining

$$y^2 - x^2 = C$$

as solution of the differential equation.

Crucial to this technique is the ability to rewrite the differential equation with all terms involving y on one side and all terms involving x on the other. Such equations are *variables-separable* equations.

Example 6 Let us solve

$$\frac{dy}{dx} = y^2 \sin x.$$

SOLUTION This is a variables-separable equation. The steps of solution are

$$\frac{dy}{y^2} = \sin x\, dx, \qquad \int y^{-2}\, dy = \int \sin x\, dx,$$

$$\frac{y^{-1}}{-1} = -\cos x + C, \qquad \frac{1}{y} = \cos x + C,$$

where we used only one arbitrary constant C and replaced $-C$ by C again in the last step. ||

The equation

$$\frac{dy}{dx} = x + y$$

is *not* a variables-separable equation. The final chapter of the text discusses solution of a few more types of differential equations.

SUMMARY

1. *The general antiderivative of $f(x)$ is also known as the indefinite integral, $\int f(x)\,dx$.*
2. *Ten important formulas for integrating are as follows:*

$$\int (f(x) + g(x))\,dx = \int f(x)\,dx + \int g(x)\,dx$$

$$\int c \cdot f(x)\,dx = c \cdot \int f(x)\,dx$$

$$\int x^n\,dx = \frac{x^{n+1}}{n+1} + C, \quad n \neq -1$$

$$\int u^n\,du = \frac{u^{n+1}}{n+1} + C, \quad n \neq -1$$

$$\int \sin u\,du = -\cos u + C$$

$$\int \cos u\,du = \sin u + C$$

$$\int \sec^2 u\,du = \tan u + C$$

$$\int \csc^2 u\,du = -\cot u + C$$

$$\int \sec u \tan u\,du = \sec u + C$$

$$\int \csc u \cot u\,du = -\csc u + C$$

3. *Suppose all the y terms (including dy) can be put on the left side of a differential equation and all the x terms (including dx) on the right, to become*

$$g(y)\,dy = f(x)\,dx$$

Then the solution of the differential equation is

$$\int g(y)\,dy = \int f(x)\,dx + C,$$

where C is an arbitrary constant. No further arbitrary constants need be included when computing $\int g(y)\,dy$ and $\int f(x)\,dx$.

EXERCISES

In Exercises 1 through 25, find the indicated integral without using tables.

1. $\int (x^3 + 4x^2)\,dx$
2. $\int \left(x + \frac{1}{x^2}\right) dx$
3. $\int (x + 1)^5\,dx$
4. $\int (2x + 1)^4\,dx$
5. $\int x(x^2 + 2)^3\,dx$
6. $\int \frac{x}{(3x^2 + 1)^2}\,dx$
7. $\int \frac{x}{\sqrt{x^2 + 4}}\,dx$
8. $\int \frac{4x + 2}{\sqrt{x^2 + x}}\,dx$
9. $\int x^2(x^3 + 4)^3\,dx$
10. $\int \cos 3x\,dx$
11. $\int \sin x \cos x\,dx$
12. $\int \cos x \sin^2 x\,dx$
13. $\int \sin 4x \cos^3 4x\,dx$
14. $\int x \sec^2 x^2\,dx$
15. $\int \csc 2x \cot 2x\,dx$
16. $\int \tan x \sec^2 x\,dx$
17. $\int \sec^3 2x \tan 2x\,dx$
18. $\int x(\sec^2 x^2)(\tan^3 x^2)\,dx$
19. $\int \frac{\sin x}{(1 + \cos x)^2}\,dx$
20. $\int \frac{\sec^2 2x}{(4 + \tan 2x)^3}\,dx$
21. $\int \frac{1}{\sec 4x}\,dx$
22. $\int \frac{\sin x}{\cos^2 x}\,dx$
23. $\int \frac{\sec^2 x}{\sqrt{1 + \tan x}}\,dx$
24. $\int \frac{\cot 3x}{\sin 3x}\,dx$
25. $\int \csc^5 2x \cot 2x\,dx$
26. Bill is marooned without integral tables on a desert island and is building a ship to escape. At one stage in the construction of his ship, he needs to know $\int \cos^2 x\,dx$.
    a) Show how Bill can find this integral. [*Hint.* Consider the "double-angle formula" for cosine.]
    b) Do you think it likely that you will ever be in Bill's predicament?
27. Sam is also marooned on a desert island (see Exercise 26). Sam is building a larger ship than Bill; consequently, he needs to find $\int \cos^4 x\,dx$. Use Exercise 26(a) to show how Sam can find this integral.

In Exercises 28 through 33, solve the differential equation.

28. $\frac{dy}{dx} = x^2 y^2$
29. $\frac{dy}{dx} = x \cos^2 y$
30. $\frac{ds}{dt} = s^2 t \sqrt{1 + t^2}$
31. $\frac{du}{dx} = x^3 \sqrt{u}$
32. $\frac{dv}{dt} = (\sqrt{5v} - 7)(t \cos t^2)$
33. $\frac{dy}{dx} = (\sin^2 y)(\cos^2 x)(\sin x)$

6.4 INTEGRATION USING TABLES

Integration using tables is a very important technique, and one that should be practiced. You will find a table of integrals on the endpapers of this book. The formulas in the table are given using the variable x, since that is the way they are stated in most tables. For a differentiable function u, it follows, from the chain rule for differentiation and $du = u'(x)\,dx$, that the formulas hold if x is replaced by u and dx by du.

Example 1 Let's compute $\int_0^\pi \sin^2 x\, dx$.

SOLUTION If you worked for a time, you might hit upon the relation

$$\sin^2 x = \tfrac{1}{2}(1 - \cos 2x),$$

which would enable you to find $\int \sin^2 x\, dx$. This indefinite integral is one that at the moment you can probably find more quickly (and more reliably) using a table:

$$\int \sin^2 ax\, dx = \frac{x}{2} - \frac{\sin 2ax}{4a} + C.$$

After looking the formula up a few times, you may remember it. We take $a = 1$, and obtain

$$\int_0^\pi \sin^2 x\, dx = \left(\frac{x}{2} - \frac{\sin 2x}{4}\right)\Bigg]_0^\pi = \left(\frac{\pi}{2} - \frac{0}{4}\right) - \left(0 - \frac{0}{4}\right) = \frac{\pi}{2}. \quad \|$$

Example 2 The integration formula

$$\int \sin^n ax\, dx = \frac{-\sin^{n-1} ax \cos ax}{na} + \frac{n-1}{n}\int \sin^{n-2} ax\, dx$$

is known as a *reduction formula*; the problem of integrating $\sin^n ax$ is reduced to integrating $\sin^{n-2} ax$. One applies this formula repeatedly until the problem is reduced to integrating either $\sin ax$ or $\sin^0 ax = 1$. For example,

$$\begin{aligned}\int \sin^5 2x\, dx &= \frac{-\sin^4 2x \cos 2x}{5 \cdot 2} + \frac{4}{5}\int \sin^3 2x\, dx \\ &= \frac{-\sin^4 2x \cos 2x}{10} + \frac{4}{5}\left(\frac{-\sin^2 2x \cos 2x}{3 \cdot 2} + \frac{2}{3}\int \sin 2x\, dx\right) \\ &= -\frac{\sin^4 2x \cos 2x}{10} - \frac{2\sin^2 2x \cos 2x}{15} - \frac{4}{15}\cos 2x + C. \quad \|\end{aligned}$$

We emphasize again that each integration formula in x given in the table can be modified, using the chain rule, to give a more general result.

Example 3 The formula

$$\int \frac{dx}{x^2\sqrt{a^2 + x^2}} = -\frac{\sqrt{a^2 + x^2}}{a^2 x} + C,$$

together with the chain rule for differentiation, yields the formula

$$\int \frac{du}{u^2\sqrt{a^2 + u^2}} = -\frac{\sqrt{a^2 + u^2}}{a^2 u} + C$$

for any differentiable function u. Thus if $u = \tan x$ so that $du = \sec^2 x\, dx$, and if $a^2 = 3$, you find that

$$\int \frac{\sec^2 x\, dx}{(\tan^2 x)\sqrt{3 + \tan^2 x}}$$

$$= -\frac{\sqrt{3 + \tan^2 x}}{3 \tan x} + C. \quad \|$$

If you are faced with an integration problem such as

$$\int \frac{\sec^2 x}{(\tan^2 x)\sqrt{3 + \tan^2 x}}\, dx,$$

as in the previous example, you should train yourself to spot when the integrand contains, as a factor, some function that is the derivative of some other function u in the integrand. In this case, $\sec^2 x$ should be spotted as a factor that is the derivative of $u = \tan x$. This suggests that we might put the integral into the form

$$\int \frac{du}{u^2\sqrt{3 + u^2}}$$

and therefore hunt in the table for an integral of the form

$$\int \frac{dx}{x^2\sqrt{c + x^2}}$$

for some constant c.

Even if a table of integrals has the formula you need, discovering which formula it is sometimes requires much ingenuity. Facility in integration, as well as facility in differentiation, is developed only by working several hundred examples. The exercises will get you started on this project.

SUMMARY

1. *All formulas $\int f(x)\, dx$ in the tables should be regarded as formulas $\int f(u)\, du$, where u may be any differentiable function of x.*
2. *When faced with a complicated integral that you can't find easily in a table, try to see whether some factor of the integrand can be viewed as the derivative of some other portion u. Then try to write the integral in the form $\int f(u)\, du$ and see if you can find $\int f(x)\, dx$ in the table. See Example 3 and the discussion following the example.*

EXERCISES

In Exercises 1 through 30, find the given integral the fastest way you can, including the use of tables (but excluding looking up the answer in the back of the text).

1. $\int_2^4 x^2\,dx$ **2.** $\int \frac{3}{x^2\sqrt{4+x^2}}\,dx$ **3.** $\int \frac{3x}{(\sqrt{4+x^2})^3}\,dx$ **4.** $\int \frac{1}{x^2\sqrt{x^2-9}}\,dx$

5. $\int \frac{5}{x\sqrt{10x-x^2}}\,dx$ **6.** $\int_0^{\pi} \cos^2 2x\,dx$ **7.** $\int_0^{\pi/4} \sin^2 x\,dx$ **8.** $\int \sin^4 x\,dx$

9. $\int \cos^6 2x\,dx$ **10.** $\int \sin 2x \cos 3x\,dx$ **11.** $\int_0^{\pi} \sin 2x \sin 4x\,dx$ **12.** $\int \sin 4x \cos 4x\,dx$

13. $\int \cos 2x \cos 5x\,dx$ **14.** $\int \sin^4 x \cos^2 x\,dx$ **15.** $\int \frac{3}{1+\cos x}\,dx$ **16.** $\int \frac{x}{1-\cos x^2}\,dx$

17. $\int x \sin 3x\,dx$ **18.** $\int x \cos 4x\,dx$ **19.** $\int x^2 \sin 2x\,dx$ **20.** $\int x^3 \cos x\,dx$

21. $\int \sec^2 4x\,dx$ **22.** $\int \tan^2 3x\,dx$ **23.** $\int x \cot^2 x^2\,dx$ **24.** $\int \tan^4 2x\,dx$

25. $\int x \sec^4 x^2\,dx$ **26.** $\int x^2 \cos^2 x^3\,dx$ **27.** $\int x\,(\sin^2 x^2)(\cos x^2)\,dx$ **28.** $\int \frac{1}{x^3\sqrt{4+x^4}}\,dx$

29. $\int \frac{\cos x}{\sin^2 x\sqrt{4-\sin^2 x}}\,dx$ **30.** $\int \frac{\tan 3x\,dx}{\sqrt{4\sec 3x - \sec^2 3x}}$ [*Hint.* Multiply numerator and denominator by sec $3x$.]

6.5 NUMERICAL METHODS OF INTEGRATION

If you can't find the indefinite integral $\int f(x)\,dx$, even in a table, then the fundamental theorem is of no help in computing $\int_a^b f(x)\,dx$. For example, it can be shown that the integral

$$\int \sqrt{4-\sin^2 x}\,dx$$

can't be expressed in terms of "elementary functions" such as polynomials, trigonometric functions, roots, exponentials, and logarithms. Numerical methods for computing a definite integral such as

$$\int_0^1 \sqrt{4-\sin^2 x}\,dx$$

are therefore very important. With pocket calculators and computers readily available, computation of such integrals is an easy matter.

6.5.1 The rectangular rule

One method of approximating $\int_a^b f(x)\,dx$ is to use a Riemann sum with n subdivisions and midpoints of the intervals for the points x_i where you compute $f(x_i)$. The integral is then computed as a sum of areas of rectangles, as indicated in Fig. 6.17. For obvious reasons, approximation by such Riemann sums is also known as the *rectangular rule.*

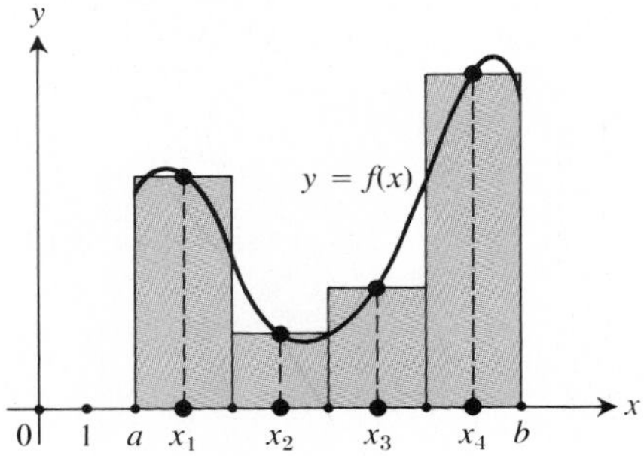

6.17

Theorem 6.5 *(Rectangular Rule) Let $[a, b]$ be divided into n subintervals of equal length and let x_i be the midpoint of the ith subinterval. Then, for f continuous in $[a, b]$,*

$$\int_a^b f(x)\,dx \approx \frac{b-a}{n}[f(x_i) + f(x_2) + \cdots + f(x_n)] \tag{1}$$

and the approximation is very good for sufficiently large n.

It should not be necessary to give an example using the rectangular rule now that you have studied Section 6.1 and done some of the exercises.

6.5.2 The trapezoidal rule

We again divide the interval $[a, b]$ into n subintervals of equal length, but this time let the *endpoints* of the subintervals be

$$a = x_0, \quad x_1, \quad x_2, \quad \ldots, \quad x_n = b,$$

as illustrated in Fig. 6.18 for $n = 4$. Let the (signed) heights to the graph $y = f(x)$ at these endpoints be

$$y_0 = f(x_0), \quad y_1 = f(x_1), \quad y_2 = f(x_2), \quad \ldots, \quad y_n = f(x_n),$$

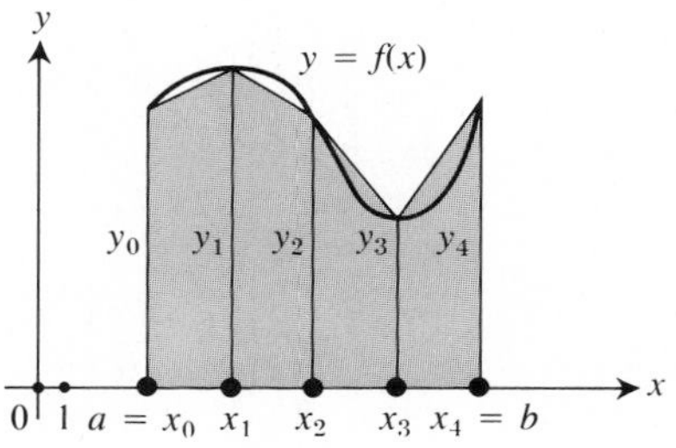

6.18

as in Fig. 6.18. Consider the chord of the curve $y = f(x)$ joining (x_{i-1}, y_{i-1}) and (x_i, y_i), as shown in Fig. 6.19. The shaded regions in Figs. 6.18 and 6.19 are trapezoids. Now the area of a trapezoid is the product of its altitude and the average length of the bases, so the area of the shaded trapezoid in Fig. 6.19 is

$$(x_i - x_{i-1})\frac{y_{i-1} + y_i}{2}.$$

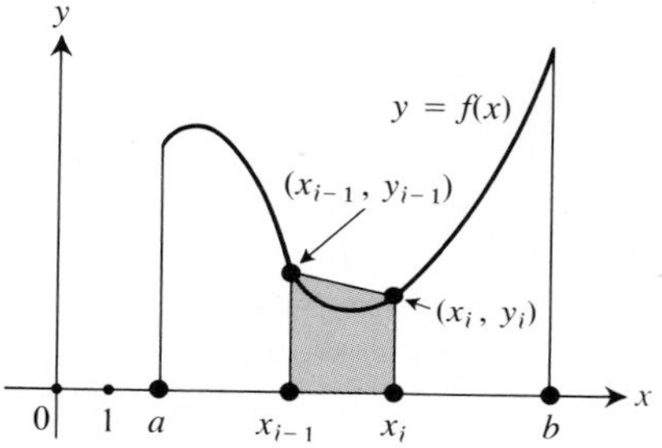

6.19

Since we divide our interval $[a, b]$ into n subintervals of *equal* length,

$$x_i - x_{i-1} = \frac{b-a}{n}.$$

Adding up the area of the trapezoids for $i = 1, \ldots, n$, we obtain

$$\frac{b-a}{n}\left(\frac{y_0 + y_1}{2}\right) + \frac{b-a}{n}\left(\frac{y_1 + y_2}{2}\right) + \frac{b-a}{n}\left(\frac{y_2 + y_3}{2}\right) + \cdots + \frac{b-a}{n}\left(\frac{y_{n-1} + y_n}{2}\right).$$

Factoring out $(b - a)/2n$, we obtain at once the trapezoidal rule.

Theorem 6.6 *(Trapezoidal Rule) For a continuous function f on $[a, b]$,*

$$\int_a^b f(x)\,dx \approx \frac{b-a}{2n}(y_0 + 2y_1 + 2y_2 + \cdots + 2y_{n-1} + y_n) \qquad (2)$$

for sufficiently large values of n.

Example 1 Let's use the trapezoidal rule with $n = 4$ to estimate

$$\int_1^3 \frac{1}{x}\,dx.$$

SOLUTION You have

$$x_0 = 1, \quad x_1 = \tfrac{3}{2}, \quad x_2 = 2, \quad x_3 = \tfrac{5}{2}, \quad \text{and} \quad x_4 = 3.$$

Thus

$$y_0 = 1, \quad y_1 = \tfrac{2}{3}, \quad y_2 = \tfrac{1}{2}, \quad y_3 = \tfrac{2}{5}, \quad \text{and} \quad y_4 = \tfrac{1}{3}.$$

The trapezoidal rule yields

$$\begin{aligned}\int_1^3 \frac{1}{x}\,dx &\approx \frac{3-1}{2\cdot 4}\left(1 + 2\cdot\frac{2}{3} + 2\cdot\frac{1}{2} + 2\cdot\frac{2}{5} + \frac{1}{3}\right)\\ &= \frac{2}{8}\left(1 + \frac{4}{3} + 1 + \frac{4}{5} + \frac{1}{3}\right)\\ &= \frac{1}{4}\left(\frac{15 + 20 + 15 + 12 + 5}{15}\right)\\ &= \frac{1}{4}\left(\frac{67}{15}\right) = \frac{67}{60} \approx 1.12.\end{aligned}$$

Since the graph of $1/x$ is concave up over $[1, 3]$, it is clear that the approximation by chords is too large. The actual value to four decimal places is 1.0986. ‖

6.5.3 Simpson's (parabolic) rule

In the rectangular rule, our approximating rectangles have horizontal lines (constant functions $y = a$) as tops, and the lines usually meet the graph at one point. In the trapezoidal rule, the top of a trapezoid may be any line (linear function $y = ax + b$) and the lines usually meet the graph at two points. One can put a quadratic function $y = ax^2 + bx + c$ through three points of the graph. The graph of $y = ax^2 + bx + c$ is a *parabola*.

For Simpson's (parabolic) rule, we divide the interval $[a, b]$ into an *even number* n of subintervals of equal length having endpoints x_i for $i = 0, 1, \ldots, n$. Let us use the notation $y_i = f(x_i)$ as in the trapezoidal rule. We approximate the graph of f over the first *two* subintervals by a parabola through (x_0, y_0), (x_1, y_1), and (x_2, y_2), over the next two subintervals by a parabola through (x_2, y_2), (x_3, y_3), and (x_4, y_4), etc. (see Fig. 6.20). We then estimate the integral $\int_a^b f(x)\,dx$ by taking the areas under these parabolas.

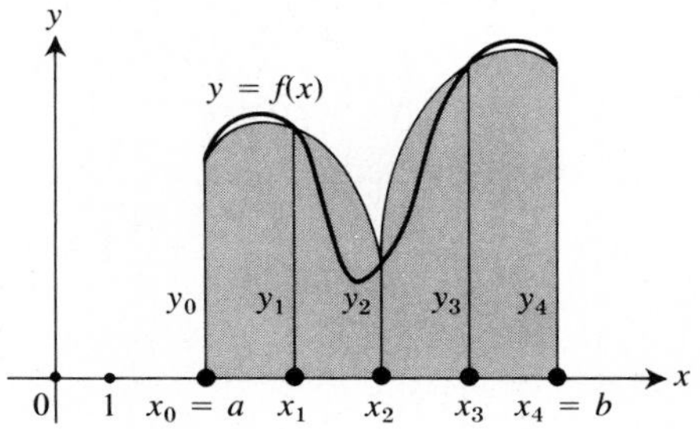

6.20

For our case where $x_1 - x_0 = x_2 - x_1$, there is a convenient formula for the area under a parabola through (x_0, y_0), (x_1, y_1), and (x_2, y_2) in terms of the heights y_0, y_1, and y_2. This formula will allow us to estimate our integral without explicit computation of the equations of the parabolas approximating the graph of f. (Note that you did not have to find the equations of the linear approximations in the trapezoidal rule either.) It can be shown that if $y = ax^2 + bx + c$ is a parabola through the points (x_0, y_0), (x_1, y_1), and (x_2, y_2) where $x_1 - x_0 = x_2 - x_1 = h$, then

$$\int_{x_0}^{x_2} (ax^2 + bx + c)\,dx = \frac{h}{3}(y_0 + 4y_1 + y_2). \tag{3}$$

In a moment, we will show why (3) is true. Of course, the number h is $(b - a)/n$ in the approximation of $\int_a^b f(x)\,dx$ by areas under parabolas. Thus the sum of the areas under the parabolas is

$$\frac{b-a}{3n}(y_0 + 4y_1 + y_2) + \frac{b-a}{3n}(y_2 + 4y_3 + y_4) + \cdots + \frac{b-a}{3n}(y_{n-2} + 4y_{n-1} + y_n)$$
$$= \frac{b-a}{3n}(y_0 + 4y_1 + 2y_2 + 4y_3 + 2y_4 + \cdots + 2y_{n-2} + 4y_{n-1} + y_n).$$

This gives us the parabolic rule.

Theorem 6.7 *(Simpson's Rule) If f is continuous on $[a, b]$ and if n is even, then*

$$\int_a^b f(x)\,dx \approx \frac{b-a}{3n}(y_0 + 4y_1 + 2y_2 + 4y_3 + 2y_4 + \cdots + 2y_{n-2} + 4y_{n-1} + y_n). \tag{4}$$

for n sufficiently large.

Example 2 As in Example 1, let's take $n = 4$ and estimate $\int_1^3 (1/x)\,dx$.

SOLUTION We have

$$x_0 = 1, \quad x_1 = \tfrac{3}{2}, \quad x_2 = 2, \quad x_3 = \tfrac{5}{2}, \quad x_4 = 3$$

and

$$y_0 = 1, \quad y_1 = \tfrac{2}{3}, \quad y_2 = \tfrac{1}{2}, \quad y_3 = \tfrac{2}{5}, \quad y_4 = \tfrac{1}{3}.$$

Simpson's rule yields

$$\int_1^3 \frac{1}{x}\,dx \approx \frac{3-1}{3\cdot 4}\left(1 + 4\cdot\frac{2}{3} + 2\cdot\frac{1}{2} + 4\cdot\frac{2}{5} + \frac{1}{3}\right) = \frac{2}{12}\left(1 + \frac{8}{3} + 1 + \frac{8}{5} + \frac{1}{3}\right)$$
$$= \frac{1}{6}\left(\frac{15 + 40 + 15 + 24 + 5}{15}\right) = \frac{1}{6}\left(\frac{99}{15}\right) = \frac{33}{30} = \frac{11}{10} = 1.1.$$

The correct answer is 1.0986 to four decimal places, and comparison with Example 1 shows that Simpson's rule gives the more accurate estimate for the same value $n = 4$. ||

It can be shown that if $f^{\text{iv}}(x)$ exists for $a < x < b$, then the error in approximating $\int_a^b f(x)\,dx$ by Simpson's rule for a value n is given by

Error in Simpson's rule

$$\left|\frac{(b-a)^5}{180n^4} f^{\text{iv}}(c)\right|$$

where $a < c < b$. For example, our error in estimating $\int_1^3 (1/x)\,dx$ by Simpson's rule, with $n = 4$ in Example 2 is given by

$$\left|\frac{2^5}{180 \cdot n^4} \cdot \frac{24}{c^5}\right| < \frac{2^5}{180 \cdot 4^4} \cdot \frac{24}{1} = \frac{1}{60} \approx 0.0167.$$

The difference between our answer 1.1 in Example 2 and the actual value 1.0986 to four decimal places is 0.0014, and we see that our bound for the error is correct.

We now give the details of the derivation of (3), in case you are interested in seeing them. Recall that (x_0, y_0), (x_1, y_1), and (x_2, y_2) are points on the parabola $y = ax^2 + bx + c$. By translating axes if necessary, you can assume that $x_1 = 0$, so $x_0 = -x_2$ (see Fig. 6.21). Then

$$\begin{aligned}\int_{-x_2}^{x_2} (ax^2 + bx + c)\,dx &= \left(\frac{ax^3}{3} + \frac{bx^2}{2} + cx\right)\Big]_{-x_2}^{x_2} \\ &= \frac{ax_2^3}{3} + \frac{bx_2^2}{2} + cx_2 - \left(-\frac{ax_2^3}{3} + \frac{bx_2^2}{2} - cx_2\right) \\ &= \frac{2}{3}ax_2^3 + 2cx_2. \qquad (5)\end{aligned}$$

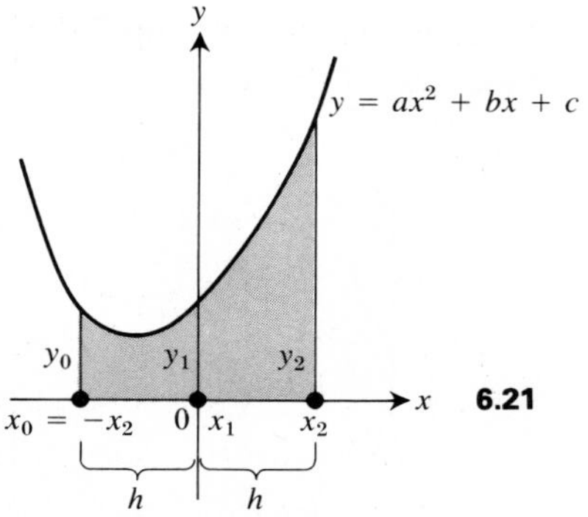

6.21

On the other hand,

$$h = x_1 - x_0 = 0 - (-x_2) = x_2,$$
$$y_0 = f(x_0) = f(-x_2) = ax_2^{\,2} - bx_2 + c,$$
$$y_1 = f(x_1) = f(0) = c,$$
$$y_2 = f(x_2) = ax_2^{\,2} + bx_2 + c.$$

Thus

$$\begin{aligned}\frac{h}{3}(y_0 + 4y_1 + y_2) &= \frac{x_2}{3}(ax_2^{\,2} - bx_2 + c + 4c + ax_2^{\,2} + bx_2 + c)\\ &= \frac{x_2}{3}(2ax_2^{\,2} + 6c)\\ &= \frac{2}{3}ax_2^{\,3} + 2cx_2. \end{aligned} \qquad (6)$$

Comparison of (5) and (6) establishes (3).

Appendix 1 gives a computer program INTRULES, which compares approximations of $\int_a^b f(x)\,dx$ using the rectangular rule, the trapezoidal rule, and Simpson's rule for $n = 10$, 50, 100, and 200. The data supplied is to compute

$$\int_0^1 \sqrt{4 - \sin^2 x}\,dx,$$

which we mentioned earlier as an example of a tough integral. Printout 6.1 indicates the value of this integral is about 1.929752909. You will note that the rectangular and trapezoidal rules have roughly the same accuracy for a given value of n, but that Simpson's rule is more accurate. This is surely to be expected from the geometry of the approximations. Since Simpson's rule is no more difficult to execute than either of the others, it is Simpson's rule you should reach for when you want to approximate an integral. (Of course, there are other approximation methods.)

Printout 6.1

INTRULES

# SUBDIVISIONS	RECTANGULAR	TRAPEZOIDAL	SIMPSON'S
10	1.9298574500E+00	1.929543939E+00	1.929753514E+00
50	1.929757085E+00	1.929744556E+00	1.929752910E+00
100	1.929753953E+00	1.929750820E+00	1.929752909E+00
200	1.929753170E+00	1.929752386E+00	1.929752909E+00

SUMMARY *Let f be continuous in $[a, b]$. Let $[a, b]$ be subdivided into n subintervals of equal length $(b - a)/n$.*

1. *(Rectangular Rule) If x_i is the midpoint of the ith subinterval, then for sufficiently large n,*

$$\int_a^b f(x)\,dx \approx \frac{b-a}{n}[f(x_1) + f(x_2) + \cdots + f(x_n)].$$

2. *(Trapezoidal Rule) If $x_0 = a, x_1, x_2, \ldots, x_n = b$ are the endpoints of the subintervals and $y_0 = f(x_0)$, $y_1 = f(x_1)$, $y_2 = f(x_2), \ldots, y_n = f(x_n)$, then for sufficiently large n,*

$$\int_a^b f(x)\,dx \approx \frac{b-a}{2n}(y_0 + 2y_1 + 2y_2 + \cdots + 2y_{n-1} + y_n).$$

3. *(Simpson's Rule) If n is even and $y_0, y_1, y_2, \ldots, y_n$ are as described in the trapezoidal rule, then, for sufficiently large n,*

$$\int_a^b f(x)\,dx$$

$$\approx \frac{b-a}{3n}(y_0 + 4y_1 + 2y_2 + 4y_3 + 2y_4 + \cdots + 2y_{n-2} + 4y_{n-1} + y_n).$$

EXERCISES

Use the indicated rule and value of n to estimate the integral.

1. $\displaystyle\int_1^4 \frac{1}{x}\,dx$ a) rectangular rule, $n = 3$
b) trapezoidal rule, $n = 3$

2. $\displaystyle\int_1^2 \frac{1}{x}\,dx$ a) trapezoidal rule, $n = 4$
b) Simpson's rule, $n = 4$

3. $\sqrt{3} = \displaystyle\int_0^{\pi/3} 2\cos x\,dx$ (Use a table for cosine.)
a) rectangular rule, $n = 4$,
b) Simpson's rule, $n = 4$

4. $\pi = \displaystyle\int_0^1 \frac{4}{1 + x^2}\,dx$ (Chapter 8 will show this integral equals π.)
a) trapezoidal rule, $n = 4$
b) Simpson's rule, $n = 4$

5. $\displaystyle\int_0^\pi \frac{dx}{2 + \sin^2 x}$
a) trapezoidal rule, $n = 4$
b) Simpson's rule, $n = 4$

calculator exercises

6. $\displaystyle\int_0^\pi \sqrt{4 - \sin^2 x}\,dx$; trapezoidal rule, $n = 10$

7. $\displaystyle\int_0^3 (1 + x)^x\,dx$; Simpson's rule, $n = 12$

8. $\displaystyle\int_0^4 \frac{dx}{\sqrt{6 - \cos^2 x}}$; Simpson's rule, $n = 8$

exercise sets for chapter 6

review exercise set 6.1

1. Find the upper sum S_4 for $f(x) = 1/(x + 1)$ over $[2, 6]$.

2. Use geometry to find $\int_0^3 (2 + \sqrt{9 - x^2})\, dx$.

3. Find

$$\frac{d}{dt}\left(\int_1^{\sqrt{t}} \sqrt{2x^2 + x^4}\, dx\right).$$

4. Find the area of the region under the curve $y = x^3 + 2x$ and over the x-axis between $x = 1$ and $x = 2$.

5. If $\int_1^4 f(x)\, dx = 5$ and $\int_4^2 f(x)\, dx = -7$, find $\int_1^2 f(x)\, dx$.

6. Find:

a) $\displaystyle\int_0^{\pi/2} \sin 2x\, dx$ b) $\displaystyle\int_0^1 x\sqrt{1 - x^2}\, dx$

7. Find:

a) $\int \sin 2x \cos 2x\, dx$ b) $\int \sec^3 x \tan x\, dx$

8. Find the general solution of the differential equation $dy/dx = x^2\sqrt{y}$.

9. Use the formula

$$\int \frac{x\, dx}{\sqrt{ax + b}} = \frac{2(ax - 2b)}{3a^2}\sqrt{ax + b} + C$$

to find

$$\int_0^{\pi/4} \frac{\sin 4x}{\sqrt{3 + \cos 2x}}\, dx.$$

10. Use Simpson's rule with $n = 4$ to estimate

$$\int_0^2 \frac{1}{x + 2}\, dx.$$

review exercise set 6.2

1. Find the Riemann sum $\mathcal{S}_4$ for $\int_0^{2\pi} \sin(x/2)\, dx$, using midpoints of the subintervals.
2. Use geometry to find $\int_{-5}^5 (5 - \sqrt{25 - x^2})\, dx$.
3. Find

$$\frac{d}{dt}\left(\int_{-t}^{2t} \sin^2 x\, dx\right).$$

4. Find the area of the region over the x-axis and under the curve $y = 16 - x^2$ between $x = -4$ and $x = 4$.

5. If $\int_{-1}^4 f(x)\, dx = 4$ and $\int_2^4 (3 - f(x))\, dx = 7$, find $\int_2^{-1} f(x)\, dx$.

6. Find:

a) $\displaystyle\int_{-1}^1 x(x^2 + 1)^3\, dx$

b) $\displaystyle\int_0^{\pi} \sin 3x\, dx$

7. Find:

a) $\displaystyle\int \frac{\cos 3x}{\sin^4 3x}\, dx$ b) $\displaystyle\int \sqrt{\csc x} \cot x\, dx$

8. Find the solution of the differential equation $dy/dx = y^2 \cos 2x$ such that $y = -3$ when $x = \pi/4$.

9. Use the reduction formula

$$\int \cos^n ax\, dx = \frac{\cos^{n-1} ax \sin ax}{na} + \frac{n - 1}{n}\int \cos^{n-2} ax\, dx$$

to find $\int_0^{\pi/8} \cos^4 2x\, dx$

10. Use the trapezoidal rule with $n = 4$ to estimate

$$\int_{-1}^1 \frac{1}{x^2 + 1}\, dx.$$

more challenging exercises

1. Sketch the graph of a function f over the interval $[0, 6]$ for which $s_3 < s_2$. (This shows that while s_n approaches $\int_a^b f(x)\,dx$ as n increases, we need not have $s_{n+1} \geq s_n$.)
2. Show that, for any continuous function f defined on $[a, b]$, we have $S_{2n} \leq S_n$ for each n. (Compare with Exercise 1.)
3. Show that if f is continuous and increasing on $[0, 1]$ with $f(0) = 0$, then $S_n - s_n = f(1)/n$.
4. Exercise 3 can be generalized easily to give a simple formula for $S_n - s_n$ in the case of a continuous increasing function on $[a, b]$. Find this formula for $S_n - s_n$.
5. Give a formula for $S_n - s_n$ analogous to that found in Exercise 4 for the case of a continuous decreasing function on $[a, b]$.
6. Give an example of a function f with domain $[0, 1]$ such that $S_n = 1$ and $s_n = 0$ for *every* positive integer n. (Of course, Theorem 6.1 shows that f could not be continuous, but it is possible to choose f so that it assumes a maximum and a minimum over each subinterval.)

Examples 7 and 8 of Section 6.1 showed how integrals can sometimes be found by evaluating the limit of a sum. Conversely, working backward through the examples, you see how a limit of a sum might be recognized as giving an integral, and then evaluated using the fundamental theorem. In Exercises 7 through 12, use an integral to find each of the limits.

7. $\lim_{n\to\infty} \frac{4}{n} \cdot \sum_{i=1}^{n} \frac{4i^2}{n^2}$

8. $\lim_{n\to\infty} \frac{8}{n^4} \cdot \sum_{i=1}^{n} i^3$

9. $\lim_{n\to\infty} \frac{1}{n^{5/2}} \cdot \sum_{i=1}^{n} i\sqrt{i}$

10. $\lim_{n\to\infty} \frac{1}{n^5} \cdot \sum_{k=1}^{n} (1 + k)^3$

11. $\lim_{n\to\infty} \frac{1}{n^5} \cdot \sum_{k=1}^{n} (2k + 1)^4$

12. $\lim_{n\to\infty} \frac{1}{n^3} \sum_{k=1}^{n} (k - 100)^3$

applications of the integral

7

You have seen that if $f(x)$ is continuous and nonnegative in $[a, b]$, the definite integral $\int_a^b f(x)\,dx$ gives the area of the region under the graph of f from a to b as in Fig. 6.1. Like differential calculus, integral calculus has important applications in situations where quantities vary. For example, the area of a rectangular region is the product of its length and its height, and in the case of a rectangle, these quantities remain constant throughout the region. For the shaded region in Fig. 6.1, however, the height varies as you travel across the region from left to right. This is the type of situation where calculus comes into play.

The definite integral is potentially useful in any type of application where, in the case of quantities that do not vary, you would compute a *product.* (Since any quantity can be viewed as the product of itself with 1, we are talking about a very general situation.) For example, if the length and height of a plane region do not vary over the region its area is the product of these dimensions. We list here several other things that may be given by a product of two quantities.

Volume: For a solid with cross section of constant area A and with constant height h, the volume V is the product Ah.

Work: If a body is moved a distance s by means of a constant force F acting in the direction of the motion, the work W done in moving the body is the product Fs.

Distance: If a body travels a time t with a constant speed v, the distance s traveled is the product vt.

Speed: If a body travels a time t with a constant acceleration a, the speed v attained is the product at.

Force: If the pressure per square unit on a plane region of area A is a constant p throughout the region, then the total force F on the region due to this pressure is the product pA.

Moment: The moment M about an axis of a body of mass m, all points of which are a constant (signed) distance s from the axis, is the product ms.

Moment of inertia: The moment of inertia I about an axis of a body of mass m, all points of which are a constant distance s from the axis, is the product ms^2.

All these notions can be handled by integral calculus if the factors appearing in the products vary continuously. This chapter is devoted to such applications of the integral.

7.1 AREA AND AVERAGE VALUE

This section shows how to find the area of a more general plane region than one under the graph of a continuous function from a to b. Consider, for example, the shaded region in Fig. 7.1 that lies between the graphs of

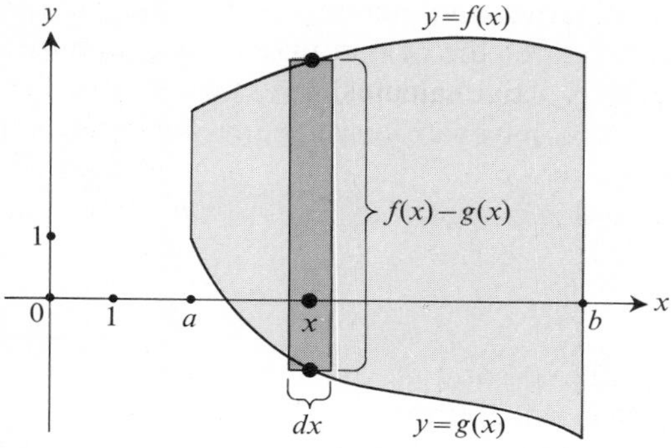

7.1

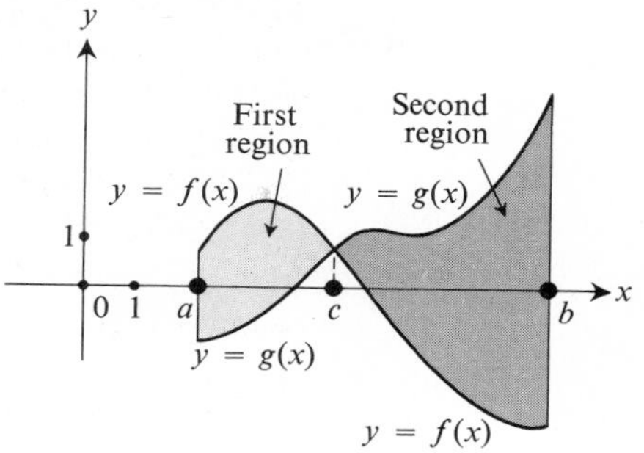

7.2

7.1.1 The area of a plane region

continuous functions f and g from a to b. The area of this region can be estimated by taking thin rectangles like the one shown heavily shaded in Fig. 7.1 and adding up the areas of such rectangles over the region. This rectangle is dx units wide and $f(x) - g(x)$ units high, for x as shown in Fig. 7.1. (Note that $g(x)$ is itself negative. The distance between the graphs of f and g over a point x is always $f(x) - g(x)$ if the graph of f is above that of g at x.) Thus the area of this rectangle is

$$(f(x) - g(x))\, dx.$$

We wish to add up the areas of such rectangles and take the limit of the resulting sum as dx grows smaller and the number of rectangles increases. As we know, the limit of such a sum will be $\int_a^b (f(x) - g(x))\, dx$.

In computing the area of a plane region, care must be taken in finding the correct function to integrate, i.e., in "setting up the integral." The area of the region bounded by the graphs of continuous functions f and g between a and b is not always $\int_a^b (f(x) - g(x))\, dx$. If $g(x) \leq f(x)$ for all x in $[a, b]$, then the correct integral is indeed $\int_a^b (f(x) - g(x))\, dx$. However, if f and g have the graphs shown in Fig. 7.2, then $\int_a^b (f(x) - g(x))\, dx$ is numerically equal to the area of the first region minus the area of the second. The area of the total shaded region in Fig. 7.2 is best obtained by finding the areas of the first and second regions separately, and then adding these areas; a correct expression is

$$\int_a^c (f(x) - g(x))\, dx + \int_c^b (g(x) - f(x))\, dx.$$

The area of the region between the graphs of f and g from a to b can always be expressed as

$$\int_a^b |f(x) - g(x)|\, dx;$$

indeed, as mentioned in Chapter 6, it is appropriate to *define* the area to be this integral. In practice, one does not work with the absolute-value sign, but looks at a figure, and sets up one or more integrals to find the desired area. Setting up the integrals is the interesting part of the problem. Computation of the integrals should be regarded as a mechanical chore, much like adding a column of numbers, although you haven't had as much experience in computing integrals.

We suggest that you use the following steps to find the area of a plane region.

Outline for finding area

STEP 1. Sketch the region, finding the points of intersection of bounding curves.

STEP 2. On your sketch, draw a typical thin rectangle either parallel to the y-axis and of width dx, or parallel to the x-axis and of width dy.

STEP 3. *Looking at your sketch*, write down the area dA of this rectangle as a product of the width (dx or dy) and the length. Express dA entirely in terms of the variable (x or y) appearing in the differential.

STEP 4. Integrate dA between the appropriate (x or y) limits. (By definition of the integral, this amounts to adding up the areas found in Step 3 and taking the limit of the resulting sum.)

We illustrate this procedure with two examples.

Example 1 Let's find the area of the plane region bounded by the curves with equations $y = x^2$ and $y = 3 - 2x$.

SOLUTION STEP 1. The curves are sketched and the region whose area is desired is shown shaded in Fig. 7.3. The points $(-3, 9)$ and $(1, 1)$ of intersection of the

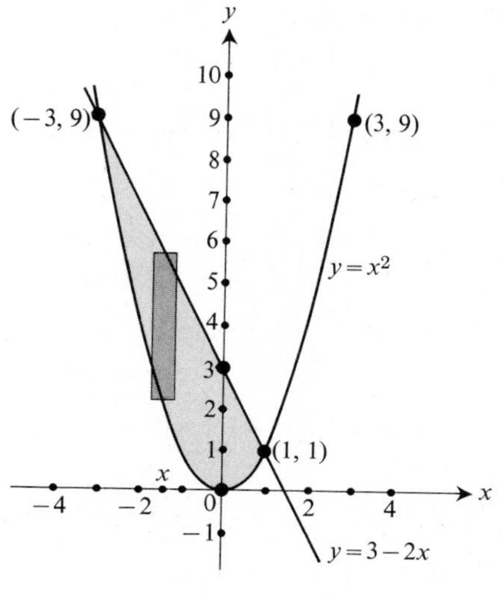

7.3

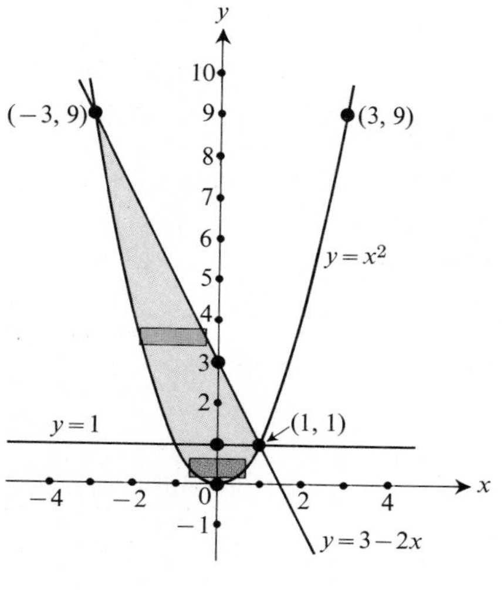

7.4

bounding curves are found by solving the equations $y = x^2$ and $y = 3 - 2x$ simultaneously.

STEP 2. A typical thin vertical rectangle is drawn and heavily shaded in Fig. 7.3,

STEP 3. The graph of the function $3 - 2x$ is above the graph of x^2 over this region, so the height of the thin rectangle over a point x is $(3 - 2x) - x^2$. The area of this thin rectangle is therefore $dA = [(3 - 2x) - x^2]\,dx$.

STEP 4. Since we wish to add up our thin rectangles in the x-direction from $x = -3$ to $x = 1$, the appropriate integral is

$$\int_{-3}^{1} [(3 - 2x) - x^2]\,dx = \left(3x - 2\frac{x^2}{2} - \frac{x^3}{3}\right)\Bigg]_{-3}^{1}$$
$$= 3 - 1 - \tfrac{1}{3} - (-9 - 9 + 9) = \tfrac{32}{3}. \quad \|$$

Example 2 Let's try to solve the same problem that we solved in Example 1 by taking our thin rectangles horizontally.

SOLUTION STEP 1. The appropriate sketch is given in Fig. 7.4.

STEP 2. Note that you can't take just one rectangle as typical for the whole region, for if the rectangle lies above the line given by $y = 1$, it is bounded on the right by the line with equation $y = 3 - 2x$, while a rectangle below the line $y = 1$ is bounded on both ends by the curve given by $y = x^2$. We split the region into two parts by the line with equation $y = 1$, and find the area of each part separately.

STEP 3. To find the horizontal dimensions of our rectangles, you must solve for x in terms of y, obtaining $x = (3 - y)/2$ for the line and $x = \pm\sqrt{y}$ for the curve. The upper rectangle has an area of

$$dA = [\tfrac{1}{2}(3 - y) - (-\sqrt{y})]\,dy,$$

while the lower rectangle has an area of

$$dA = [\sqrt{y} - (-\sqrt{y})]\,dy.$$

STEP 4. The appropriate integrals are

$$\int_{1}^{9} \left[\frac{1}{2}(3 - y) - (-\sqrt{y})\right] dy = \left(\frac{3}{2}y - \frac{y^2}{4} + \frac{2}{3}y^{3/2}\right)\Bigg]_{1}^{9}$$
$$= (\tfrac{27}{2} - \tfrac{81}{4} + \tfrac{2}{3}\cdot 27) - (\tfrac{3}{2} - \tfrac{1}{4} + \tfrac{2}{3}) = \tfrac{28}{3},$$

and

$$\int_{0}^{1} [\sqrt{y} - (-\sqrt{y})]\,dy = \int_{0}^{1} 2\sqrt{y}\,dy = 2\cdot\tfrac{2}{3}y^{3/2}\Big]_{0}^{1} = \tfrac{4}{3} - 0 = \tfrac{4}{3}.$$

Thus the total area is $(\frac{28}{3}) + (\frac{4}{3}) = \frac{32}{3}$ square units. Clearly this computation is not as nice as that in Example 1. $\|$

7.1.2 The average value of a function

Let f be a continuous function with domain containing $[a, b]$, and let's consider the size of $f(x)$ for x in $[a, b]$. This size can vary tremendously over an interval $[a, b]$, even if f is continuous. However, let's see if we can develop some notion of the *average size of* $f(x)$ *over* $[a, b]$.

Perhaps the first natural attempt to find an average for $f(x)$ over $[a, b]$ is to form the average $(f(a) + f(b))/2$. Upon a little consideration, you see that we should reject such a definition of the average size for $f(x)$ over $[a, b]$, for $(f(a) + f(b))/2$ really reflects the size of $f(x)$ only at a and at b. For f with graph shown in Fig. 7.5, this average $(f(a) + f(b))/2$ is zero, even though $f(x) > 0$ for $a < x < b$.

Probably the next natural attempt to define an average for $f(x)$ over $[a, b]$ is to average the maximum size M and the minimum size m of all the $f(x)$ for x in $[a, b]$, if this maximum M and this minimum m exist. We know that M and m do exist if f is continuous in $[a, b]$. For the function f shown in Fig. 7.5, you have $M = 6$ and $m = 0$, so the average is $(M + m)/2 = \frac{6}{2} = 3$, which may seem like a reasonable value for the average of $f(x)$ over $[a, b]$. But for the function with graph shown in Fig. 7.6, the average of $f(x)$ over $[a, b]$ should surely be closer to M than m, reflecting the fact that $f(x)$ is close to M over most of the interval $[a, b]$.

In order to take care of a situation like that shown in Fig. 7.6, let's regard the *average of* $f(x)$ *over* $[a, b]$ as the height h that a rectangle with base $[a, b]$ must have in order for the area of the rectangle to be equal to the area of the region under the graph of f over $[a, b]$. This region under the graph of f over $[a, b]$ is shown shaded in Fig. 7.6, where a rectangle of equal area with height h is also indicated. The area of the rectangle is $h(b-a)$, while the area of the shaded region is of course $\int_a^b f(x)\,dx$, so if the areas are to be equal, you must have

$$h = \frac{1}{b-a}\int_a^b f(x)\,dx.$$

Definition 7.1 Let f be continuous in $[a, b]$. The **average** (or **mean**) **value** of $f(x)$ in $[a, b]$ is

$$\frac{1}{b-a}\int_a^b f(x)\,dx.$$

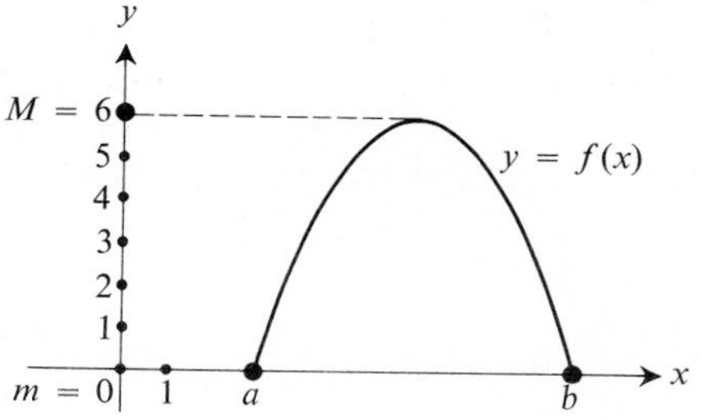

7.5

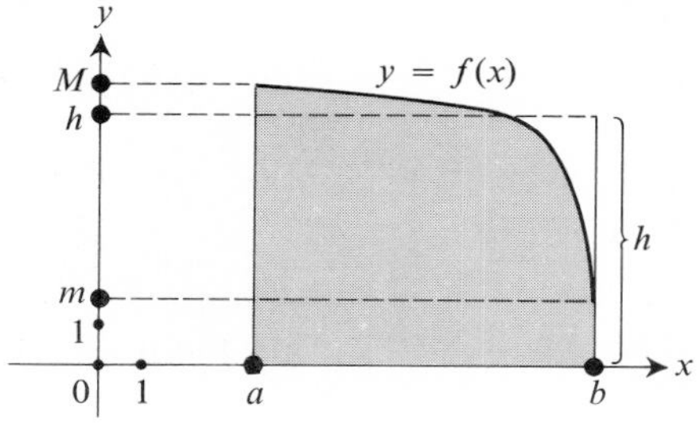

7.6

Example 3 Let's find the average value of x^2 in $[0, 2]$.

SOLUTION We have

$$\int_0^2 x^2\,dx = \frac{x^3}{3}\bigg]_0^2 = \frac{8}{3}.$$

Thus the average value is

$$\frac{1}{2-0}\cdot\frac{8}{3} = \frac{4}{3}. \quad \|$$

If m and M are the minimum value and maximum value for $f(x)$ over $[a, b]$, then you know that

$$s_1 = m(b-a) \le \int_a^b f(x)\,dx \le M(b-a) = S_1.$$

Dividing by $(b - a)$, you obtain

$$m \le \frac{1}{b-a}\int_a^b f(x)\,dx \le M,$$

which proves that the mean value of $f(x)$ in $[a, b]$ is between m and M. Since we are assuming that f is continuous on $[a, b]$, it follows by the Intermediate-Value Theorem that

$$\frac{1}{b-a}\int_a^b f(x)\,dx = f(c)$$

for some c where $a < c < b$. The existence of this c is known as the *Mean-Value Theorem for the Integral.* To illustrate, Example 3 showed that the mean value of x^2 in $[0, 2]$ is $\frac{4}{3}$. In order to have $f(c) = c^2 = \frac{4}{3}$, you must have $c = \pm 2/\sqrt{3}$. The value $c = 2/\sqrt{3}$ does indeed lie inside $[0, 2]$.

SUMMARY

1. *A suggested step-by-step procedure for finding the area of a region bounded by given curves is as follows.*

   STEP 1. *Sketch the region, finding the points of intersection of the bounding curves.*

   STEP 2. *On your sketch, draw a typical thin rectangle either parallel to the y-axis and of width dx, or parallel to the x-axis and of width dy.*

   STEP 3. *Looking at your sketch, write down the area dA of this rectangle as a product of the width (dx or dy) and the length. Express dA entirely in terms of the variable (x or y) appearing in the differential.*

   STEP 4. *Integrate dA between the appropriate (x or y) limits.*

2. *The average value of $f(x)$ in $[a, b]$ is*

$$\frac{1}{b-a}\int_a^b f(x)\,dx.$$

EXERCISES

In Exercises 1 through 14, find the total area of the region or regions bounded by the given curves.

1. $y = x^2, \quad y = 4$
2. $y = x^4, \quad y = 1$
3. $y = x, \quad y = x^2$
4. $y = x, \quad y = x^3$
5. $y = x^4, \quad y = x^2$
6. $y = x^2, \quad y = x^3$
7. $y = x^4 - 1, \quad y = 1 - x^2$
8. $y = x^2 - 1, \quad y = x + 1$
9. $x = y^2, \quad y = x - 2$
10. $y = \sqrt{x}, \quad y = x^2$
11. $x = 2y^2, \quad x = 8$
12. $y = x^2 - 1, \quad y = \sqrt{1 - x^2}$ [*Hint:* Use some known areas.]
13. $y = \sin x, \quad y = 3 \sin x \quad$ for $0 \le x \le \pi$
14. $y = \sin x, \quad y = \cos x \quad$ for $\quad 0 \le x \le \pi/4, \quad$ and $x = 0$
15. Express as an integral the area of the smaller region bounded by the curves $x^2 + y^2 = 4$ and $y = -1$.
16. Express as an integral the area of the region bounded by the curves $x = 2y^2$ and $x^2 + y^2 = 68$.
17. Express as one or more integrals the area of the region in the first quadrant bounded by the curves $x^2 + y^2 = 2$, $y = 0$, and $y = x^2$.
18. Find the value of c such that the region bounded by $y = x^2$ and $y = 4$ is bisected by the line $y = c$.
19. Find the average value of the function $1 - x^2$ in the interval $[-2, 2]$.
20. Find the value of c such that the average value of the function $x^4 - 1$ in the interval $[-c, c]$ is 0.
21. Find the average value of $\sin x$ in the interval $[0, \pi]$.

calculator exercises

Estimate the areas by using Simpson's rule with $n = 10$ to find a numerical approximation of the integral.

22. The area in Exercise 15
23. The area in Exercise 16
24. The area in Exercise 17

7.2 VOLUMES OF REVOLUTION: SLAB METHOD

Let a planar region be revolved in the natural way about a given line (axis) lying in the plane. That is, each point P of the planar region describes a circular orbit, which bounds a disk having the given axis of revolution as perpendicular axis through its center. The three-dimensional region consisting of the points on all such orbits is the *solid of revolution* generated as the planar region is revolved about the axis. Figure 7.7 illustrates these ideas. Using integral calculus, you can frequently find the volume of such a solid of revolution.

Let's consider the case in which a planar region under the graph of a continuous function f from a to b is revolved about the x-axis. Such a region is shown shaded in Fig. 7.8. The treatment for other regions is similar.

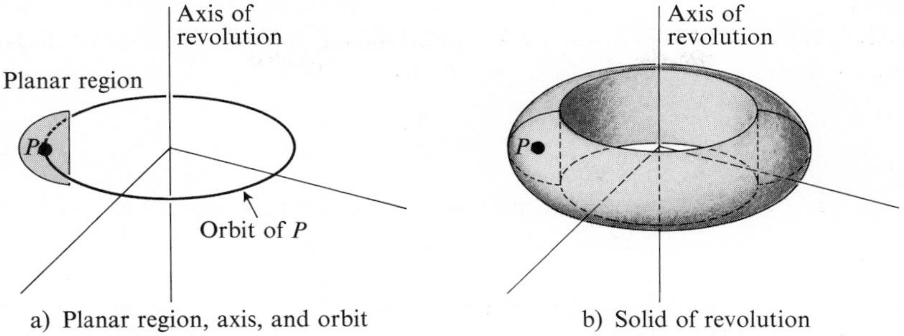

7.7

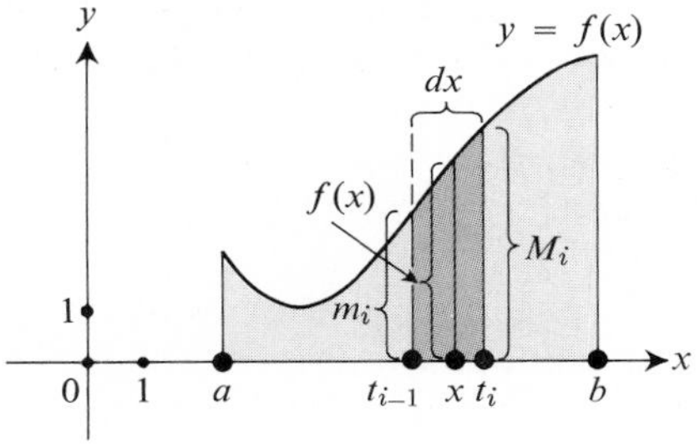

7.8

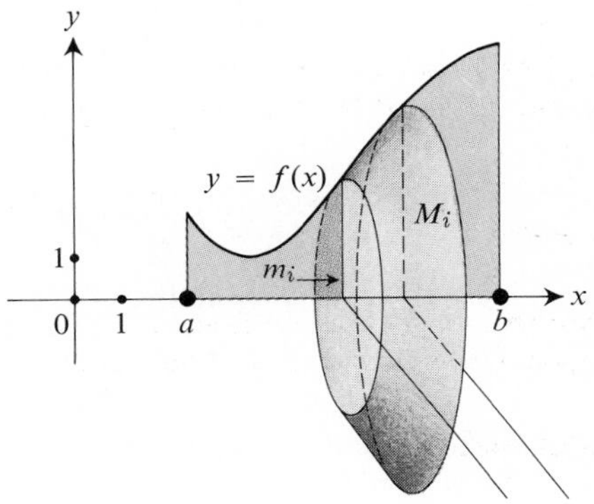

7.9

Let $[a, b]$ be divided into n subintervals of equal lengths, as usual. Consider the contribution to the solid of revolution given by revolving the heavily shaded strip in Fig. 7.8 about the x-axis. The strip sweeps out a slab shown in Fig. 7.9, where the minimum radius is m_i and the maximum radius is M_i. The thickness of the slab is

$$\frac{b-a}{n} = dx.$$

The volume of a circular slab of radius r and thickness h is of course $\pi r^2 h$. Since the radius of our slab varies from m_i to M_i, its volume V_{slab} satisfies

$$\pi m_i^2\left(\frac{b-a}{n}\right) \leq V_{\text{slab}} \leq \pi M_i^2\left(\frac{b-a}{n}\right).$$

Thus for the volume V of our whole solid of revolution, you have the relation

$$\frac{b-a}{n}\left(\sum_{i=1}^{n} \pi m_i^2\right) \leq V \leq \frac{b-a}{n}\left(\sum_{i=1}^{n} \pi M_i^2\right) \tag{1}$$

for all n. Now the extremes of the inequality (1) both approach $\int_a^b \pi f(x)^2\, dx$ as $n \to \infty$, so our desired volume V of revolution is given by

$$V = \int_a^b \pi f(x)^2\, dx. \tag{2}$$

The notation $\int_a^b \pi(f(x))^2\, dx$ is very helpful and suggestive in this situation. For x in $[t_{i-1}, t_i]$ as shown in Fig. 7.8, the radius of the slab in Fig. 7.9 is approximately $f(x)$, so the area of one face of the slab is roughly $\pi f(x)^2$. Since the thickness is dx, the approximate volume of the slab is $\pi f(x)^2\, dx$. You then add up all these little contributions to the volume and take the limit as dx approaches 0 by applying the integral operator $\int_a^b$. You should not memorize Eq. (2), but rather arrive at the correct integral

$$\int_a^b \pi f(x)^2\, dx$$

by such a geometric argument.

We suggest an outline of steps to follow when using this slab method to find the volume of a solid of revolution. The steps are similar to those in our outline in the preceding section for finding the area of a planar region.

Outline for finding volume (slab method)

STEP 1. Sketch the planar region that is to be revolved, finding the points of intersection of bounding curves.

STEP 2. On your sketch, draw a typical thin rectangle perpendicular to the axis of revolution, i.e., either perpendicular to the x-axis and of width dx, or perpendicular to the y-axis and of width dy.

STEP 3. *Looking at your sketch,* write down the volume dV of the slab swept out as the rectangle is revolved about the given axis. Express dV entirely in terms of the variable (x or y) appearing in the differential (dx or dy).

STEP 4. Integrate dV between the appropriate (x or y) limits. (Geometrically, this amounts to adding the volumes found in Step 3 and taking the limit of the resulting sum.)

Example 1 Let the region bounded by the curves with equations $y = x^2$, $x = 0$, and $y = 1$ be revolved about the y-axis, and let's find the volume of the resulting solid.

SOLUTION STEP 1. The planar region to be revolved is shown shaded in Fig. 7.10(a). Revolving the region about the y-axis gives a solid bowl, as shown in Fig. 7.10(b).

STEP 2. A typical rectangle is shown heavily shaded in Fig. 7.10(a). When revolved about the y-axis, the rectangle sweeps out a thin, circular slab.

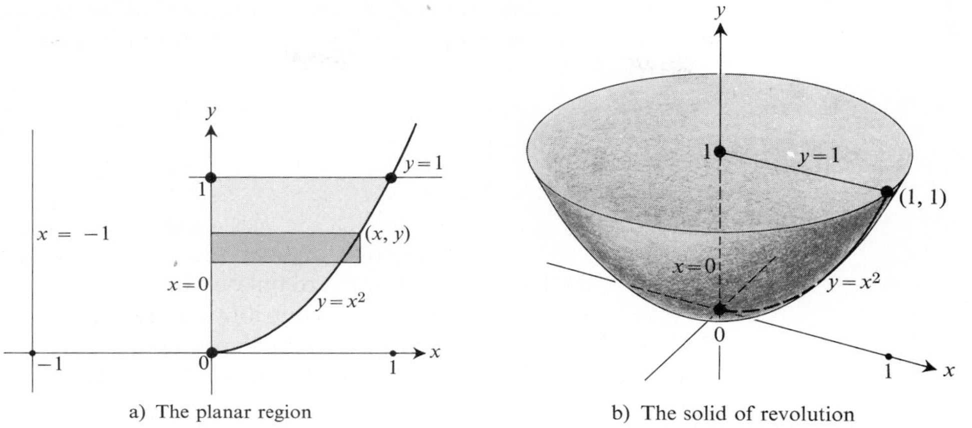

a) The planar region b) The solid of revolution

7.10

STEP 3. The volume of such a slab is the product of the area of the circular face and the thickness (or height) of the slab. The area of a face is πx^2, for the point (x, y) shown in Fig. 7.10(a). The thickness of the slab is dy, so the volume of this slab is $dV = \pi x^2\, dy$. You must express the volume $\pi x^2\, dy$ entirely in terms of y. For the point (x, y) in Fig. 7.10(a), you have $x^2 = y$, so the volume of the slab becomes $dV = (\pi y)\, dy$.

STEP 4. The appropriate integral is

$$\int_0^1 \pi y\, dy = \pi \frac{y^2}{2}\bigg]_0^1 = \frac{\pi}{2} - 0 = \frac{\pi}{2}. \quad \|$$

The axis of revolution need not fall on a boundary of the region, as illustrated in the next example.

Example 2 Let's find the volume generated when the planar region of Example 1 is revolved about the line $x = -1$.

SOLUTION STEP 1. The appropriate sketch is again given in Fig. 7.10(a). This time, the solid generated is shown in Fig. 7.11.

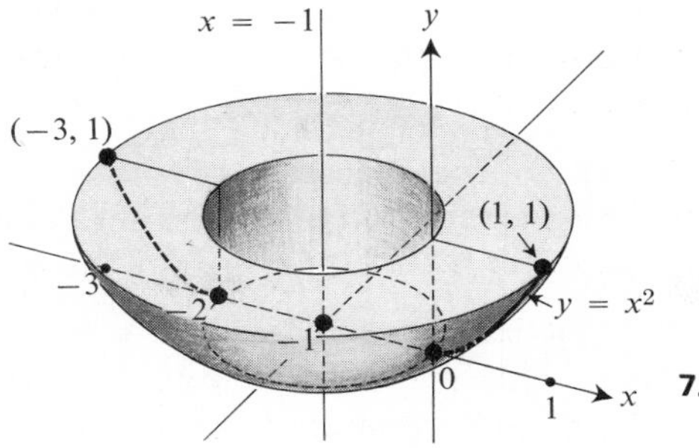

7.11

STEP 2. A typical rectangle is shown heavily shaded in Fig. 7.10(a). When revolved about the line $x = -1$, it sweeps out a thin, annular washer (a disk with a hole in it), as shown in Fig. 7.12.

STEP 3. The volume of an annular washer is the volume of the solid slab, minus the volume of the hole. The radius of our whole disk slab is $x - (-1) = x + 1$ for the point (x, y) shown in Fig. 7.12, so the solid slab has volume $\pi(x + 1)^2\, dy$. On the other hand, the hole has radius 1, and hence volume $\pi(1)^2\, dy$. Thus the volume of the annular washer is

$$dV = \pi(x + 1)^2\, dy - \pi(1)^2\, dy = \pi(x^2 + 2x)\, dy.$$

You must express the volume $\pi(x^2 + 2x)\, dy$ entirely in terms of y. For the point (x, y) in Fig. 7.12, you have $x^2 = y$, so the volume of the annular washer becomes $dV = \pi(y + 2\sqrt{y})\, dy$.

STEP 4. The appropriate integral is

$$\int_0^1 \pi(y + 2\sqrt{y})\, dy = \pi\left(\frac{y^2}{2} + \frac{4}{3}y^{3/2}\right)\Bigg]_0^1 = \pi\left(\frac{1}{2} + \frac{4}{3}\right) - \pi(0) = \frac{11}{6}\pi. \quad \|$$

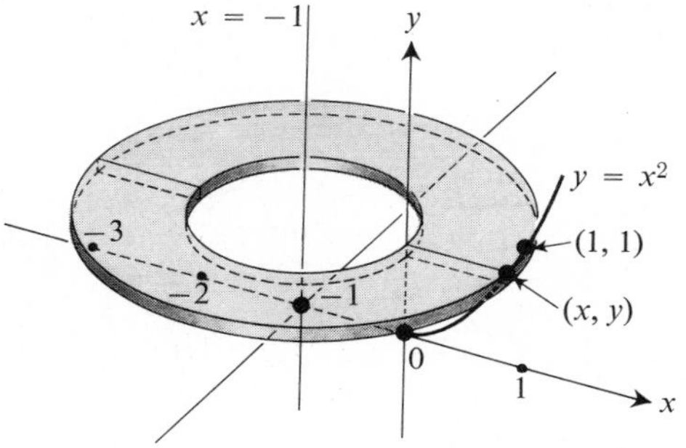

7.12

A slab method can often be used to find the volume of a solid that can be sliced up into parallel slabs whose faces have easily computed areas. A solid of revolution is an example of such a solid, for it can be sliced into parallel slabs having circular faces. The general technique is perhaps best illustrated by an example.

Example 3 A wedge of cheese has a semicircular base of diameter $2a$. If the cheese is cut perpendicular to the diameter of the semicircle, the cross section obtained is an isosceles right triangle with the right angle on the semicircle, as shown in Fig. 7.13(a). Let's find the volume of the cheese.

SOLUTION We may take for the base of the cheese the semicircle bounded by the x-axis and the graph of $\sqrt{a^2 - x^2}$, as shown in Fig. 7.13(b). The slab shown in Fig. 7.13(a) has thickness dx, while its faces are isosceles triangles with legs of lengths y. The area of such a triangle is $y^2/2$, so the volume of the slab is approximately $dV = (y^2/2)\,dx$. Since $y = \sqrt{a^2 - x^2}$, the slab has volume $[(a^2 - x^2)/2]\,dx$. The volumes of these slabs should be added as x goes from $-a$ to a, so the appropriate integral is

$$\int_{-a}^{a} \frac{1}{2}(a^2 - x^2)\,dx = \frac{1}{2}\left(a^2x - \frac{x^3}{3}\right)\Big]_{-a}^{a}$$
$$= \frac{1}{2}\left[\left(a^3 - \frac{a^3}{3}\right) - \left(-a^3 - \frac{(-a)^3}{3}\right)\right]$$
$$= \tfrac{1}{2}[2a^3 - \tfrac{2}{3}a^3] = \tfrac{2}{3}a^3. \quad \|$$

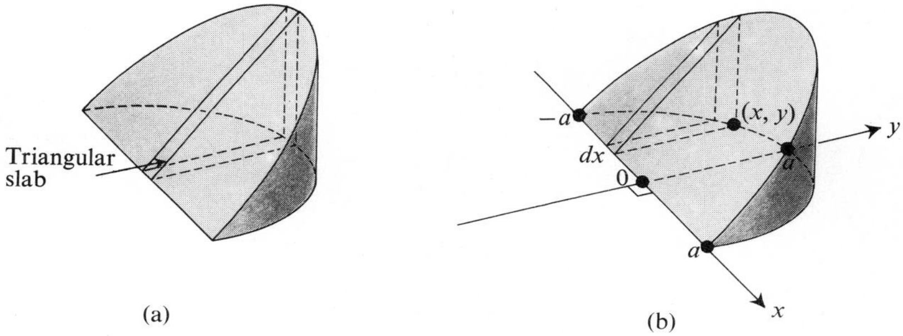

7.13

SUMMARY 1. *A suggested step-by-step outline for finding volumes of revolution by the slab method is as follows:*

STEP 1. *Sketch the planar region that is to be revolved, finding the points of intersection of bounding curves.*

STEP 2. *On your sketch, draw a typical thin rectangle perpendicular to the axis of revolution, i.e., either perpendicular to the x-axis and of width dx, or perpendicular to the y-axis and of width dy.*

STEP 3. *Looking at your sketch, write down the volume dV of the slab swept out as the rectangle is revolved about the given axis. This slab is either a circular slab of volume*

$$dV = \pi r^2 \cdot (dx \text{ or } dy)$$

as in Fig. 17.14(a), *or a circular washer of volume*

$$dV = \pi(R^2 - r^2) \cdot (dx \text{ or } dy)$$

as in Fig. 17.14(b). *In either case, express dV entirely in terms of the variable (x or y) appearing in the differential (dx or dy).*

STEP 4. *Integrate dV between the appropriate (x or y) limits.*

2. *A suggested step-by-step procedure for finding volumes of solids with cross sections of known areas is as follows:*

STEP 1. *Draw a figure; include an axis perpendicular to the cross section of known area. (Say, an x-axis.)*

STEP 2. *Sketch a cross-sectional slab perpendicular to that x-axis.*

STEP 3. *Express the area $A(x)$ of the face of the cross-sectional slab in terms of its position x on the x-axis. The volume of the slab is then*

$$dV = A(x)\,dx,$$

as in Fig. 7.14(c).

STEP 4. *Integrate the expression for dV between appropriate x limits.*

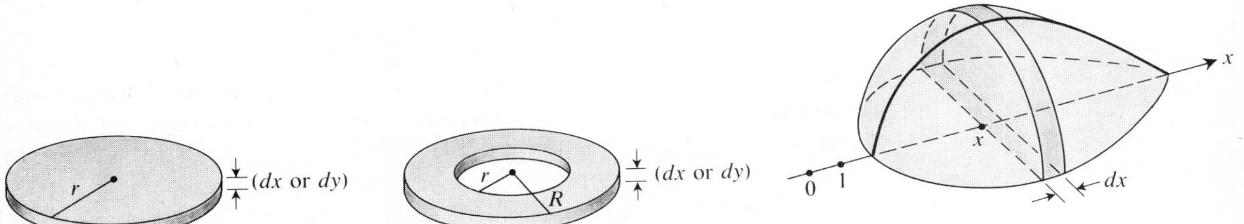

$dV = \pi r^2 \cdot (dx \text{ or } dy)$
(a) Circular slab

$dV = \pi(R^2 - r^2)\cdot(dx \text{ or } dy)$
(b) Circular washer

$dV = A(x)\cdot dx$
(c) Cross section of known area

7.14

EXERCISES

1. Find the volume of the solid generated by revolving the region bounded by the curves with equations $y = x^2$, and $y = 1$ about the line with equation $y = 1$, using circular slabs.

2. Find the volume of the solid generated by revolving the region in Exercise 1 about the line with equation $x = 2$, using the slab method.

3. Verify the formula $V = \frac{1}{3}\pi r^2 h$ for the volume of a right circular cone of height h with base of radius r. [*Hint.* Revolve the region bounded by the lines with equations $y = (r/h)x$, $y = 0$, and $x = h$, about the x-axis.]

4. Verify the formula $V = \frac{4}{3}\pi r^3$ for the volume of a solid ball of radius r. [*Hint.* Revolve the half-disk $x^2 + y^2 \le r^2$, $y \ge 0$, about the x-axis.]

5. Find the volume of the solid generated by revolving the region bounded by the graphs of the functions x and x^2 about the y-axis.

6. Find the volume of the solid generated when the plane region bounded by the curves with equations $y = x^2$ and $y = 3 - 2x$ is revolved about the x-axis.

7. Find the volume of the solid generated when the plane region bounded by the graphs of $\sqrt{x}$ and x^2 is revolved about the x-axis.

8. Find the volume of the solid generated when the plane region bounded by $y = x^2$ and $y = 4$ is revolved about the line $y = -1$.

9. Find the volume of the solid generated when the plane region bounded by $y = \sin x$ and $y = 0$ for $0 \le x \le \pi$ is revolved about the x-axis.

10. Find the volume generated when the region bounded by the curves $y = \csc x$, $x = \pi/4$, $x = 3\pi/4$, and $y = 0$, is revolved about the x-axis.

11. A solid has as base the disk $x^2 + y^2 \le a^2$. Each plane section of the solid cut by a plane perpendicular to the x-axis is an equilateral triangle. Find the volume of the solid.

12. The base of a certain solid is an isosceles right triangle with hypotenuse of length a. Each plane section of the solid cut by a plane perpendicular to the hypotenuse is a square. Find the volume of the solid.

13. Find the volume of the smaller portion of a solid ball of radius a cut off by a plane b units from the center of the ball, where $0 \le b \le a$.

14. A certain solid has as base the plane region bounded by $x = y^2$ and $x = 4$. Each plane section of the solid cut by a plane perpendicular to the x-axis is an isosceles right triangle with the right angle on the graph of $\sqrt{x}$. Find the volume of the solid.

7.3 VOLUMES OF REVOLUTION: SHELL METHOD

Here is another method for finding a volume of revolution. Suppose the region shown in Fig. 7.15 is revolved about the y-axis. This time the heavily shaded strip in Fig. 7.15 sweeps out a cylindrical shell, as shown in Fig. 7.16. Let's try to find an appropriate integral for the volume, using these shells.

The volume of such a cylindrical shell should be approximately the product of the surface area of the cylinder and the thickness of the shell (the wall of the cylinder). The surface area in turn is the product of the perimeter of the circle and the height of the cylinder. For the point (x, y) shown in Fig. 7.15, the perimeter is $2\pi x$ and the height is y. Thus the volume V_{shell} of the cylindrical shell should be about $dV = 2\pi xy\, dx$. Since $y = f(x)$, we would expect, adding up the volumes of these shells and taking the limit of the resulting sum, that the volume of the whole solid of revolution would be

$$\int_a^b 2\pi x f(x)\, dx.$$

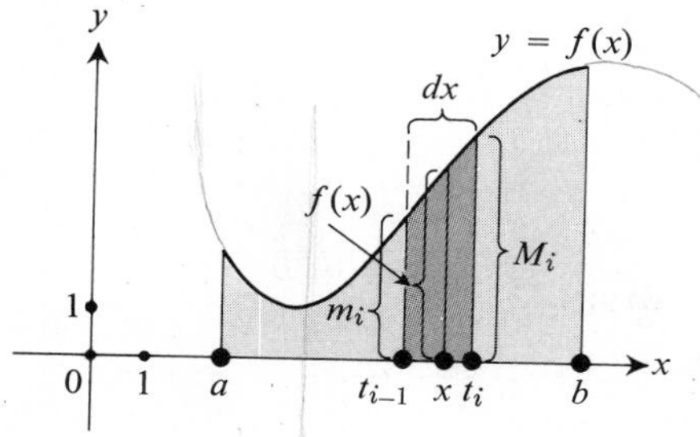

7.15

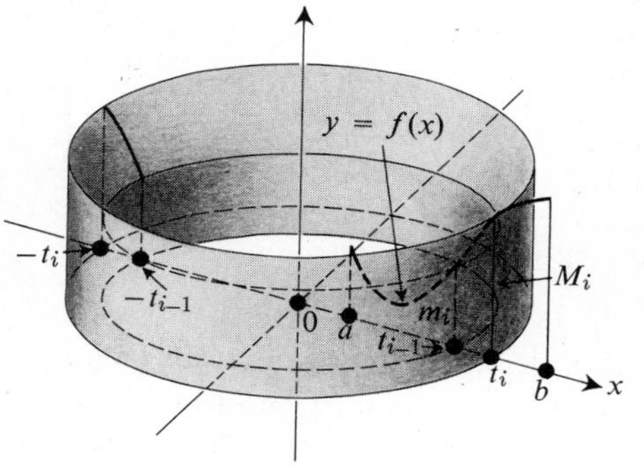

7.16

The preceding integral was set up using rough estimates, in the manner in which one always sets up an integral for the shell method. A justification of the method can be made as follows. Let m_i and M_i be the minimum and maximum values of $f(x)$ over the ith interval of length $(b - a)/n$ as usual. Referring to Fig. 7.16, taking first minimum radius t_{i-1} and minimum height m_i for the cylindrical shell, and then the maximum radius t_i and maximum height M_i, you see that

$$2\pi t_{i-1} m_i \frac{b - a}{n} \leq V_{\text{shell}} \leq 2\pi t_i M_i \frac{b - a}{n}. \tag{1}$$

The expression

$$2\pi t_i M_i \frac{b - a}{n}$$

in (1) is the value of

$$(2\pi x) f(x) \frac{b - a}{n}$$

if you evaluate $2\pi x$ at $x_i = t_i$ and $f(x)$ at the (possibly different) point x_i' where $f(x)$ assumes its maximum value M_i in this ith subinterval. A theorem known as Bliss's theorem, which we shall not prove, shows that the type of Riemann sums for $g(x)f(x)$ where the two functions f and g are evaluated at possibly *different* points x_i and x_i' in the ith interval still approaches the integral. That is,

$$\int_a^b g(x)f(x)\,dx = \lim_{n\to\infty} \frac{b - a}{n}\left(\sum_{i=1}^{n} f(x_i)g(x_i')\right).$$

Using this fact and setting $g(x) = 2\pi x$, you see at once from (1) that

$$V = \int_a^b 2\pi x \cdot f(x)\,dx. \tag{2}$$

Our outline of steps in Section 7.2 needs only slight modification to serve for computing volumes of solids of revolution by the shell method. The outline is revised in the summary.

Example 1 Let's repeat Example 1 of Section 7.2, but let's calculate the volume by taking thin vertical rectangles as shown in Fig. 7.17(a) (Steps 1 and 2).

SOLUTION STEP 3. As the heavily shaded rectangle in Fig. 7.17(a) is revolved about the y-axis, it sweeps out the cylindrical shell shown in Fig. 7.17(b). For the point (x, y) shown in Fig. 7.17(b), the perimeter of the shell is $2\pi x$ and the height is $1 - y$. Thus the volume of the cylindrical shell is $2\pi x(1 - y)\,dx$. With x limits, you must express the volume of the shell entirely in terms of

x. For the point (x, y) of Fig. 7.17(a) you have $y = x^2$, so the volume of the shell is

$$dV = 2\pi x(1 - x^2)\,dx = (2\pi x - 2\pi x^3)\,dx.$$

STEP 4. Thus the appropriate integral is

$$\int_0^1 (2\pi x - 2\pi x^3)\,dx = \left(2\pi\frac{x^2}{2} - 2\pi\frac{x^4}{4}\right)\Big]_0^1$$

$$= \left(\frac{2\pi}{2} - \frac{2\pi}{4}\right) - 0 = \frac{\pi}{2}. \quad \|$$

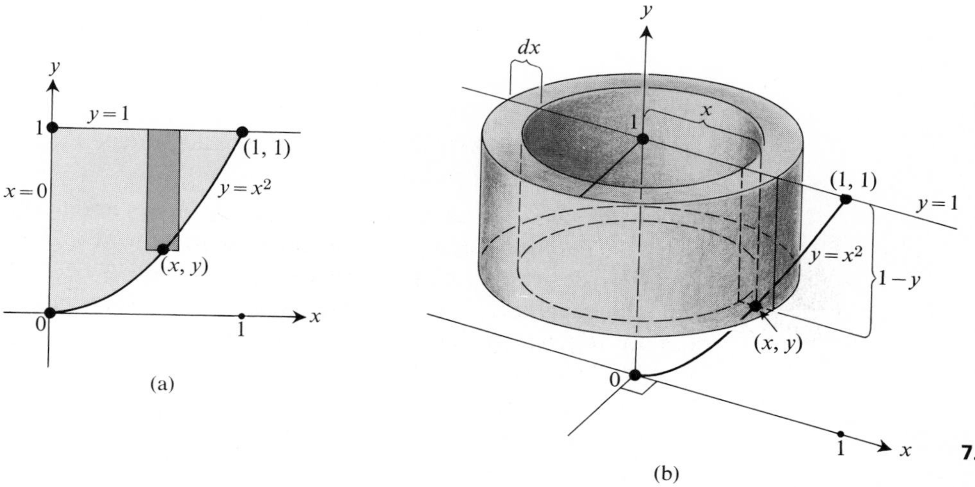

7.17

Example 2 Let's find the volume when the region bounded by $y = 4$ and $y = x^2$ is revolved about the line $y = -1$, using the shell method.

SOLUTION STEPS 1, 2. The sketch is shown in Fig. 7.18.

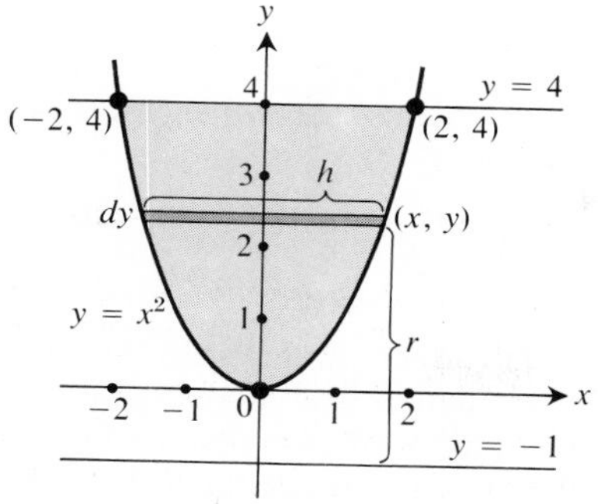

7.18

STEP 3. The volume of the shell is $dV = 2\pi rh(dy)$. We must express r and h in terms of y. In terms of the coordinates of the point (x, y) shown on the curve $y = x^2$, we have $r = y - (-1) = y + 1$, while $h = 2x = 2\sqrt{y}$. Thus

$$dV = 2\pi(y + 1)2\sqrt{y}\,dy = 4\pi(y + 1)\sqrt{y}\,dy.$$

STEP 4. The integral is

$$\int_0^4 4\pi(y + 1)\sqrt{y}\,dy = 4\pi\int_0^4 (y^{3/2} + y^{1/2})\,dy = 4\pi\left(\tfrac{2}{5}y^{5/2} + \tfrac{2}{3}y^{3/2}\right)\Big]_0^4$$

$$= 4\pi\left(\frac{64}{5} + \frac{16}{3}\right) = 64\pi\left(\frac{4}{5} + \frac{1}{3}\right) = 64\pi\frac{17}{15} = \frac{1088\pi}{15}. \quad \|$$

SUMMARY 1. *To find volumes of revolution by the shell method, we suggest the following steps.*

STEP 1. *Sketch the planar region that is to be revolved, finding the points of intersection of bounding curves.*

STEP 2. *On your sketch, draw a typical thin rectangle parallel to the axis of revolution, i.e., either parallel to the x-axis with width dy or parallel to the y-axis with width dx.*

STEP 3. *Looking at your sketch, write down the volume* $dV = 2\pi rh \cdot (dx \text{ or } dy)$ *of the shell swept out as the rectangle is revolved about the given axis. (See Fig. 7.19.) Express r and h in terms of the variable (x or y) in the differential.*

STEP 4. *Integrate the differential volume dV in Step 3 between the appropriate (x or y) limits.*

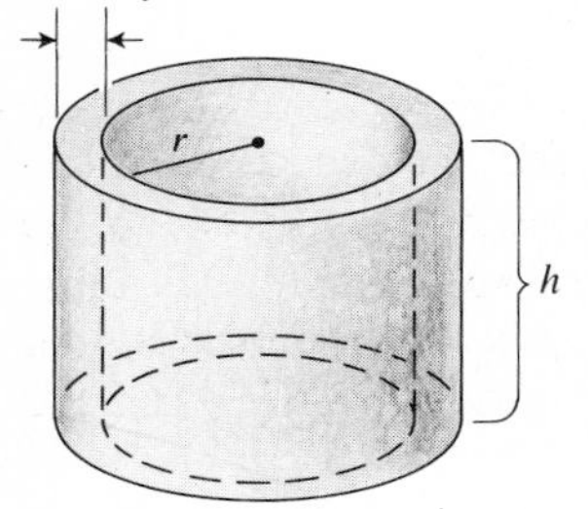

$dV = 2\pi rh(dx \text{ or } dy)$ **7.19**

EXERCISES

Use the method of cylindrical shells in all these exercises.

1. Find the volume of the solid generated by revolving the region bounded by the curves with equations $y = x^2$ and $y = 1$ about the line with equation $y = 1$.

2. Find the volume of the solid generated by revolving the region in Exercise 1 about the line with equation $x = 2$.

3. Find the volume of the solid generated by revolving the region bounded by the graphs of the functions x and x^2 about the y-axis.

4. Find the volume of the solid generated when the plane region bounded by the curves with equations $x = y^2$ and $y = x - 2$ is revolved about the line $y = 2$.

5. Find the volume of the solid generated when the plane region bounded by the graphs of $\sqrt{x}$ and x^2 is revolved about the x-axis.

6. Find the volume of the solid generated when the plane region bounded by $y = \sin x$ and $y = 0$ for $0 \le x \le \pi$ is revolved about the y-axis.

7. Find the volume of the torus (doughnut) generated by revolving the disk $x^2 + y^2 \le a^2$ about the line $x = b$ for $b > a$. [*Hint.* Part of the integral can be evaluated as the area of a familiar figure.]

7.4 ARC LENGTH

7.4.1 Length of a graph

A curve given by a *continuously* differentiable function $y = f(x)$ is a *smooth curve.* Let $f'(x)$ be continuous, so $y = f(x)$ gives a smooth curve. We would like to find the length of the portion of such a curve for x in $[a, b]$ as in Fig. 7.20(a). We present two approaches to this problem.

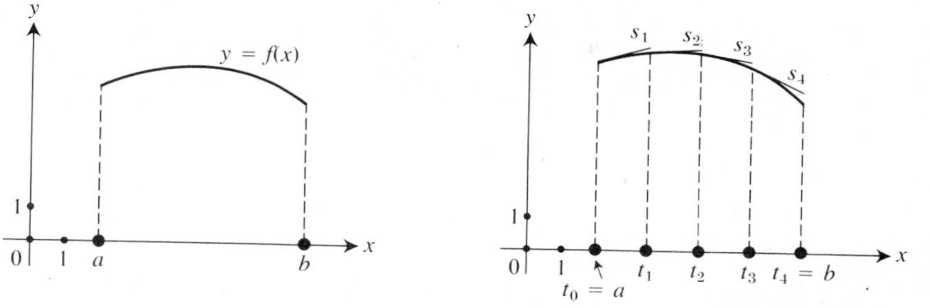

(a)

Approximation by Tangent Line Segments
(b)

7.20

APPROACH 1. Let's partition the interval $[a, b]$ into n subintervals of equal lengths by points $a = t_0 < t_1 < \cdots < t_n = b$. We take as estimate for the length of the curve the sum of the lengths of the line segments tangent to the curve over the points $t_0, t_1, \ldots, t_{n-1}$ and over these subintervals, as illustrated in Fig. 7.20(b) with $n = 4$. That is, if s_i is the length of the tangent segment over the interval $[t_{i-1}, t_i]$, we take as estimate for the length s of the curve

$$s \approx s_1 + s_2 + \cdots + s_n.$$

Figure 7.21 indicates that if n is large, you would expect this sum to be a good approximation to what you intuitively feel is the length of the curve.

Let's compute s_i. Since our subintervals are of equal lengths,

$$t_i - t_{i-1} = (b - a)/n.$$

From Fig. 7.22, we obtain

$$s_i = \sqrt{\left(\frac{b - a}{n}\right)^2 + \left(\frac{b - a}{n} f'(t_{i-1})\right)^2} = \frac{b - a}{n}\sqrt{1 + (f'(t_{i-1}))^2}. \quad (1)$$

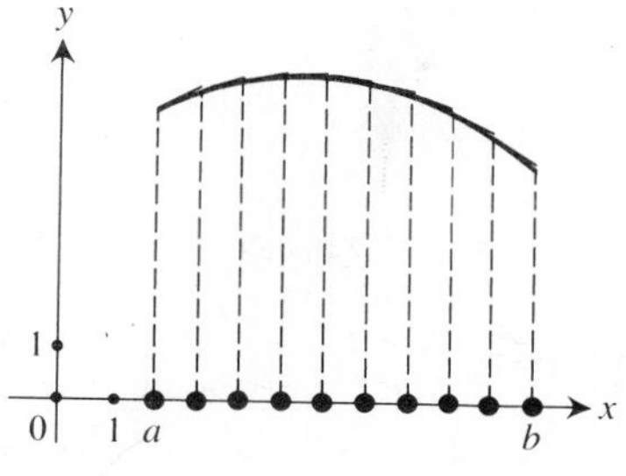

7.21

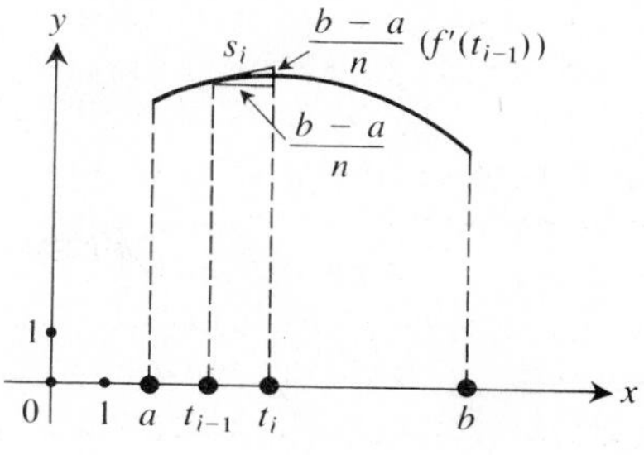

7.22

You then obtain, as an approximation,

$$s \approx \frac{b-a}{n}\left[\sqrt{1+f'(t_0)^2}+\sqrt{1+f'(t_1)^2}+\cdots+\sqrt{1+f'(t_{n-1})^2}\right]. \qquad (2)$$

Now the expression on the righthand side of (2) is a Riemann sum for the function $\sqrt{1+f'(x)^2}$ from a to b. Since you expect the approximation in (2) to become accurate as n becomes large, you obtain

$$s = \int_a^b \sqrt{1+f'(x)^2}\,dx \qquad (3)$$

as the length of the curve. This integral exists, because we are assuming that f' is continuous in $[a, b]$.

The *arc length function* $s(x)$ for the length of the arc $y = f(x)$ from a to x is

$$s(x) = \int_a^x \sqrt{1+f'(t)^2}\,dt.$$

By the fundamental theorem of calculus,

$$\frac{ds}{dx} = \sqrt{1+f'(x)^2},$$

so

$$ds = \sqrt{1+f'(x)^2}\,dx. \qquad (4)$$

This differential ds is the *differential of arc length*, and (3) may be written in the form $s = \int_a^b ds$. Writing $f'(x) = dy/dx$, you have these easily remembered forms of ds:

$$ds = \sqrt{1+(dy/dx)^2}\,dx = \sqrt{(dx)^2+(dy)^2} = \sqrt{(dx/dy)^2+1}\,dy. \qquad (5)$$

The form $ds = \sqrt{(dx)^2+(dy)^2}$ is easy to remember from Fig. 7.23. We think of ds as the length of a short tangent line segment to the curve. This concludes the first approach to arc length.

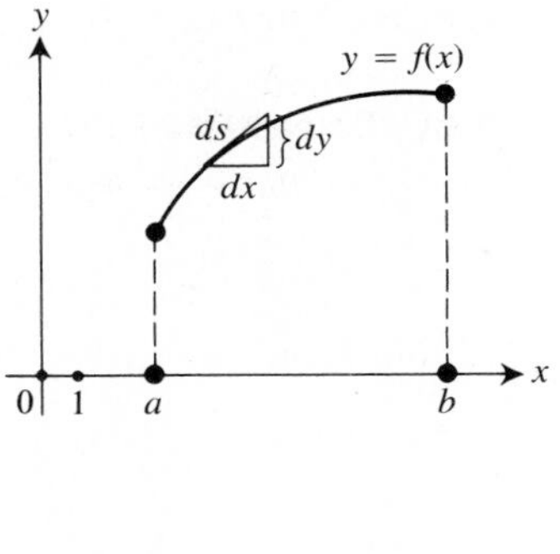

7.23

APPROACH 2. Once again, we partition $[a, b]$ into n subintervals of equal lengths. This time, we take as approximation to s the sum of the lengths of the chords shown in Fig. 7.24 where the ith chord joins $(t_{i-1}, f(t_{i-1}))$ and $(t_i, f(t_i))$. The length of this ith chord is

$$\sqrt{(t_i - t_{i-1})^2 + (f(t_i) - f(t_{i-1}))^2} = (t_i - t_{i-1})\sqrt{1 + \left(\frac{f(t_i) - f(t_{i-1})}{t_i - t_{i-1}}\right)^2}$$

$$= \frac{b-a}{n}\sqrt{1 + \left(\frac{f(t_i) - f(t_{i-1})}{t_i - t_{i-1}}\right)^2}. \qquad (6)$$

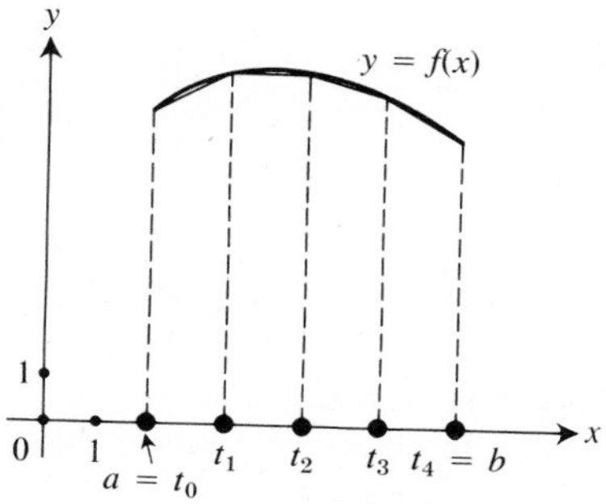

7.24

Approximation by Chords

Our hypotheses on f enable us to apply the Mean-Value Theorem to f on $[t_{i-1}, t_i]$, and we have

$$\frac{f(t_i) - f(t_{i-1})}{t_i - t_{i-1}} = f'(x_i)$$

for some point x_i in $[t_{i-1}, t_i]$. From (6), we obtain, as length of our chord,

$$\frac{b-a}{n}\sqrt{1 + f'(x_i)^2}, \qquad (7)$$

which yields the approximation

$$s \approx \frac{b-a}{n}\left[\sqrt{1 + f'(x_1)^2} + \sqrt{1 + f'(x_2)^2} + \cdots + \sqrt{1 + f'(x_n)^2}\right] \qquad (8)$$

for the length of the arc. Again, the righthand side of (8) is a Riemann sum of the function $\sqrt{1 + f'(x)^2}$ over $[a, b]$, so we again arrive at

$$s = \int_a^b \sqrt{1 + f'(x)^2}\, dx$$

for our arc length.

Example 1 Let's find the length of the curve $y = x^{3/2}$ for $0 \le x \le 1$.

SOLUTION If $y = f(x) = x^{3/2}$, we have

$$\frac{dy}{dx} = f'(x) = \frac{3}{2}x^{1/2},$$

so

$$\sqrt{1 + (f'(x))^2} = \sqrt{1 + \tfrac{9}{4}x}.$$

The length of our curve is therefore

$$\int_0^1 \left(1 + \frac{9}{4}x\right)^{1/2} dx = \frac{4}{9}\cdot\frac{(1 + \frac{9}{4}x)^{3/2}}{3/2}\Bigg]_0^1 = \frac{4}{9}\cdot\frac{2}{3}\left(\left(\frac{13}{4}\right)^{3/2} - 1\right)$$

$$= \frac{8}{27}\cdot\frac{(13)^{3/2}}{8} - \frac{8}{27} = \frac{(13)^{3/2} - 8}{27}. \quad \|$$

The radical that appears in (3) frequently makes evaluation of the integral difficult; examples have to be chosen with care to enable you to compute the integral by taking an antiderivative of $\sqrt{1 + f'(x)^2}$. But you know there are numerical methods (such as Simpson's rule) which you can easily use if integration causes difficulty.

7.4.2 Arc length of a parametric curve

Consider a smooth curve with parametric equations

$$x = h(t), \qquad y = k(t)$$

where $h'(t)$ and $k'(t)$ are continuous functions of t. A formal manipulation of Leibniz notation from (5) leads to

$$ds = \sqrt{(dx)^2 + (dy)^2} = \sqrt{\left(\frac{dx}{dt}\right)^2 + \left(\frac{dy}{dt}\right)^2}\, dt. \tag{9}$$

Thus you expect the total distance traveled on the curve from time $t = a$ to time $t = b$ to be

$$\int_{t=a}^{t=b} \sqrt{\left(\frac{dx}{dt}\right)^2 + \left(\frac{dy}{dt}\right)^2}\, dt. \tag{10}$$

This is the way you *remember* (9) and (10), but it cannot be regarded as a proof.

To prove (10), partition the interval $[a, b]$ on the t-axis into n subintervals of equal lengths, so that $a = t_0 < t_1 < \cdots < t_n = b$. We approximate the arc length by adding the lengths of chords of the curve, as in Approach 2 above. The ith chord joins the points

$$(h(t_{i-1}), k(t_{i-1})) \qquad \text{and} \qquad (h(t_i), k(t_i)).$$

The length of this chord is therefore

$$\sqrt{[h(t_i) - h(t_{i-1})]^2 + [k(t_i) - k(t_{i-1})]^2}$$
$$= \sqrt{\left(\frac{h(t_i) - h(t_{i-1})}{t_i - t_{i-1}}\right)^2 + \left(\frac{k(t_i) - k(t_{i-1})}{t_i - t_{i-1}}\right)^2} \cdot \frac{b - a}{n}.$$

By the Mean-Value Theorem there exist points c_i and c_i' between t_{i-1} and t_i such that this last expression becomes

$$\sqrt{h'(c_i)^2 + k'(c_i')^2} \cdot \frac{b - a}{n}.$$

The total length of the curve from $t = a$ to $t = b$ is therefore approximately

$$\frac{b - a}{n} \cdot \sum_{i=1}^{n} \sqrt{h'(c_i)^2 + k'(c_i')^2}.$$

This sum is *almost* a Riemann sum for the function $\sqrt{h'(t)^2 + k'(t)^2}$ over $[a, b]$. The only problem is the *two* points c_i and c_i' in the ith interval used to evaluate the two summands under the radical. But there is a theorem of Bliss, mentioned in the last section, which also says that such a type of sum approaches the integral of the function as $n \to \infty$. This establishes (10).

Example 2 Let's verify the formula $C = 2\pi a$ for the circumference of a circle having radius a.

SOLUTION We take as parametric equations for the circle

$$x = a \cos t, \qquad y = a \sin t \qquad \text{for} \quad 0 \le t \le 2\pi.$$

Since $dx/dt = -a \sin t$ and $dy/dt = a \cos t$, we see from (10) that the length of the circle is

$$\begin{aligned} s &= \int_0^{2\pi} \sqrt{(-a \sin t)^2 + (a \cos t)^2}\, dt \\ &= \int_0^{2\pi} \sqrt{a^2}\, dt \\ &= a\int_0^{2\pi} (1)\, dt \\ &= 2\pi a. \quad \| \end{aligned}$$

SUMMARY

1. *If $f'(x)$ is continuous, then the length of arc of the graph $y = f(x)$ from $x = a$ to $x = b$ is*

$$s = \int_a^b \sqrt{1 + f'(x)^2}\, dx.$$

2. *If $h'(t)$ and $k'(t)$ are continuous, then the length of the curve $x = h(t)$, $y = k(t)$ traveled from time t_0 to time t_1 is*

$$s = \int_{t_0}^{t_1} \sqrt{(dx/dt)^2 + (dy/dt)^2}\, dt.$$

3. *The differential ds of arc length takes the following forms:*

$$ds = \sqrt{(dx)^2 + (dy)^2} = \sqrt{1 + \left(\frac{dy}{dx}\right)^2}\, dx$$

$$= \sqrt{\left(\frac{dx}{dy}\right)^2 + 1}\, dy$$

$$= \sqrt{\left(\frac{dx}{dt}\right)^2 + \left(\frac{dy}{dt}\right)^2}\, dt.$$

EXERCISES

In Exercises 1 through 8, find the length of the curve with the given x,y-equation or parametric equations.

1. $y^2 = x^3$ from $(0, 0)$ to $(4, 8)$

2. $y = \frac{1}{3}(x^2 - 2)^{3/2}$ from $x = 2$ to $x = 4$

3. $9x^2 = 16y^3$ from $y = 3$ to $y = 6$, $x > 0$

4. $y = x^{2/3}$ from $x = 1$ to $x = 8$

5. $y = x^3 + (1/6x)$ from $x = 1$ to $x = 2$

6. $x^{2/3} + y^{2/3} = a^{2/3}$

7. $x = t^2$, $y = \frac{2}{3}(2t + 1)^{3/2}$ from $t = 0$ to $t = 4$

8. $x = a \cos^3 t$, $y = a \sin^3 t$ from $t = 0$ to $t = \pi/2$

In Exercises 9 through 11, express the arc length as an integral. Do not attempt to perform the integration.

9. The length of $y = x^3$ from $x = 1$ to $x = 5$.

10. The length of arc of the ellipse

$$\frac{x^2}{4} + \frac{y^2}{9} = 1$$

for $|x| \leq 1$, $y > 0$.

11. The total length of arc of the curve

$$x = \sin t, \qquad y = \cos 2t.$$

12. Find the approximate length of the curve $y = \sin x$ from $x = 0$ to $x = 0.03$.

13. Find the approximate length of the arc of the curve $x = 4t$, $y = \sin t$ from $t = 0$ to $t = 0.05$.

14. Find the approximate value of $t_1 > 0$ such that the length of the curve $x = \sin t$, $y = \tan t$ from $t = 0$ to $t = t_1$ is 0.1.

calculator exercises

Use Simpson's rule with $n = 10$ to approximate the arc length described in:

15. Exercise 9

16. Exercise 10

17. Exercise 11

7.5 AREA OF A SURFACE OF REVOLUTION

If the arc of a smooth curve $y = f(x)$ for $a \leq x \leq b$ as shown in Fig. 7.25 is revolved about the x-axis, it generates a *surface of revolution*, as indicated in the figure. If you take a small tangent line segment of length ds at (x, y) and revolve this segment about the x-axis as shown in Fig. 7.26, the surface area generated by this segment would seem to be approximately that of a cylinder of radius $|y|$ and the height ds, that is, approximately $2\pi\,|y|\,ds$. Since $y = f(x)$ and $ds = \sqrt{1 + (f'(x))^2}\,dx$, we arrive at the formula

$$S = \int_a^b 2\pi\,|f(x)|\,\sqrt{1 + (f'(x))^2}\,dx, \tag{1}$$

for the *area of the surface of revolution*.

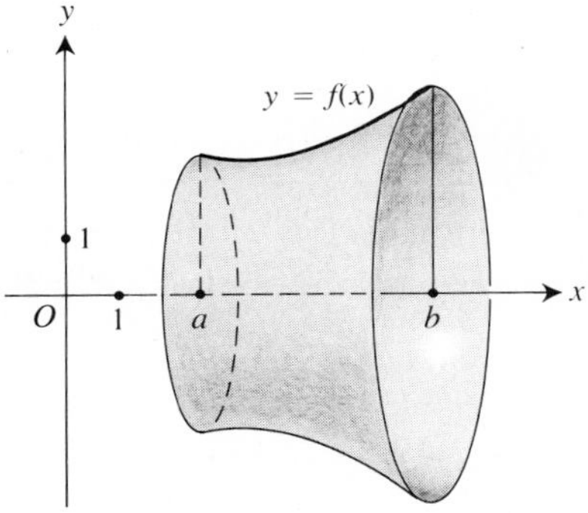

7.25

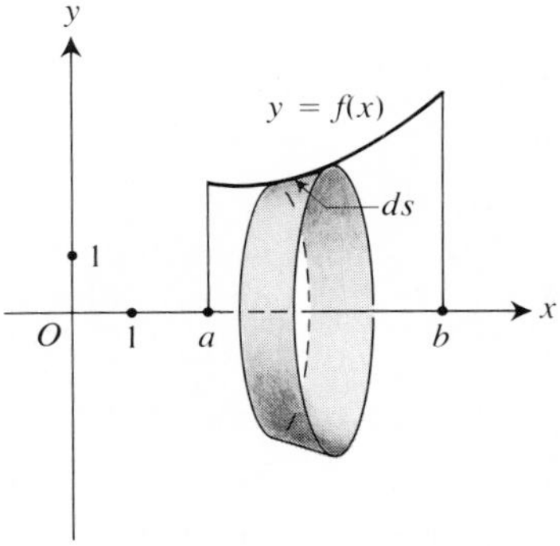

7.26

The preceding paragraph indicates how you can easily remember the formula (1). A justification of the formula can be made along the following lines. Partition the interval $[a, b]$ into n subintervals of equal lengths. Consider the inscribed chords to the curve over the subintervals, as in Approach 2 of the previous section. Equation (7) in the last section showed that the length of the ith inscribed chord of the curve can be written

$$\frac{b - a}{n}\sqrt{1 + f(x_i)^2}$$

for some x_i in the ith subinterval. Let c_i and c_i' be chosen in the ith subinterval so that

$|f(c_i)|$ is the maximum value of $|f(x)|$ over the subinterval

and

$|f(c_i')|$ is the minimum value of $|f(x)|$ over the subinterval.

Then surely the surface area element generated by revolving this ith inscribed chord satisfies

$$2\pi\,|f(c_i')|\,\frac{b - a}{n}\sqrt{1 + f'(x_1)^2} \leq \text{area element} \leq 2\pi\,|f(c_i)|\,\frac{b - a}{n}\sqrt{1 + f'(x_i)^2}.$$

By the theorem of Bliss referred to in Section 7.3, both

$$\frac{b-a}{n}\sum_{i=1}^{n}\left[2\pi\,|f(c_i')|\sqrt{1+f'(x_i)^2}\right]$$

and

$$\frac{b-a}{n}\sum_{i=1}^{n}\left[2\pi\,|f(c_i)|\sqrt{1+f'(x_i)^2}\right]$$

approach the integral in (1) as $n \to \infty$. This justifies (1).

If the curve is given in the parametric form $x = h(t)$, $y = k(t)$ for $t_0 \le t \le t_1$, and if the curve set is traversed just once for t in $[t_0, t_1]$, then (1) takes the form

$$S = \int_{t_0}^{t_1} 2\pi\,|k(t)|\sqrt{(h'(t))^2 + (k'(t))^2}\,dt. \tag{2}$$

If an arc is revolved about an axis other than the x-axis, the formulas (1) and (2) are modified in the obvious way. You "add by integrating" the contributions $2\pi r(ds)$ where ds is the length of a tangent line segment to the curve and r is the radius of the circle through which this tangent segment is revolved.

The general formula for surface area of revolution is thus

$$\text{Surface area} = S = \int_{\ell_1}^{\ell_2} 2\pi(\text{radius of revolution})\,ds, \tag{3}$$

where ℓ_1 and ℓ_2 are appropriate limits and ds is the differential of arc length.

Example 1 Let's find the area of the surface generated by revolving the arc of $y = x^2$ from $(0, 0)$ to $(1, 1)$ about the y-axis.

SOLUTION You have $dy/dx = 2x$, so

$$ds = \sqrt{1 + 4x^2}\,dx.$$

The radius of revolution is x, as indicated in Fig. 7.27. Then (3) becomes

$$S = \int_0^1 2\pi x\,ds = \int_0^1 2\pi x\sqrt{1+4x^2}\,dx = \frac{2\pi}{8}\int_0^1 8x(1+4x^2)^{1/2}\,dx$$

$$= \frac{\pi}{4}\cdot\frac{(1+4x^2)^{3/2}}{3/2}\Bigg]_0^1 = \frac{\pi}{6}(1+4x^2)^{3/2}\Bigg]_0^1 = \frac{\pi}{6}(5\sqrt{5}-1). \quad \|$$

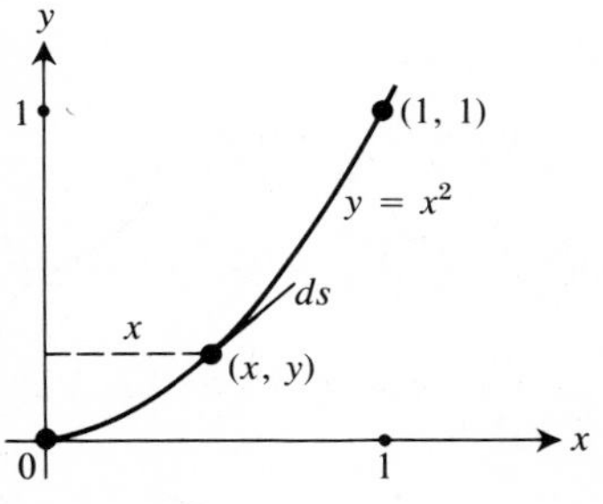

7.27

Example 2 Let's derive the formula $A = 4\pi a^2$ for the surface area of a sphere having radius a.

SOLUTION We view the sphere as the surface generated by revolving the semicircle $x = a \cos t$, $y = a \sin t$ for $0 \leq t \leq \pi$ about the x-axis. Here formula (2) is appropriate, and we obtain

$$A = \int_0^{\pi} 2\pi a \sin t\sqrt{(-a \sin t)^2 + (a \cos t)^2}\, dt$$

$$= 2\pi a\int_0^{\pi} (\sin t)\sqrt{a^2}\, dt = 2\pi a^2\int_0^{\pi} \sin t\, dt$$

$$= 2\pi a^2(-\cos t)\Big]_0^{\pi} = 2\pi a^2(1 - (-1)) = 4\pi a^2. \quad \|$$

SUMMARY 1. *If an arc of a smooth curve is revolved about an axis, then the area of the surface generated is*

$$S = \int_{\ell_1}^{\ell_2} 2\pi(\text{radius of revolution})\, ds$$

for appropriate limits ℓ_1 and ℓ_2.

EXERCISES

In Exercises 1 through 9, find the area of the surface of revolution obtained by revolving the given curve about the indicated axis.

1. $y = x^3$ from $(0, 0)$ to $(1, 1)$ about the x-axis

2. $y = \frac{1}{3}(x^2 - 2)^{3/2}$ from $x = \sqrt{2}$ to $x = 4$ about the y-axis

3. $y = x^4/4 + 1/(8x^2)$ from $x = 1$ to $x = 2$ about the y-axis

4. $y = \sqrt{a^2 - x^2}$ from $x = -a$ to $x = a$ about the x-axis. (Note that this will derive again the formula for the surface area of a sphere of radius a.)

5. $y = \frac{2}{3}x^{3/2}$ from $x = 0$ to $x = 2$ about the y-axis (Use a table.)

6. $x = a \cos^3 t$, $y = a \sin^3 t$ from $t = 0$ to $t = \pi/4$ about the x-axis

7. $x = t$, $y = t^2 - 2$ from $t = 0$ to $t = 3$ about the y-axis

8. $x = t^2/2 + 2t$, $y = t + 1$ from $t = 0$ to $t = 2$ about the line $y = -1$

9. $x = 2t + 1$, $y = t^2 - 3t$ from $t = 3$ to $t = 4$ about the line $x = 4$

In Exercises 10 through 13, express the surface area of revolution as an integral.

10. $y = x^2$ from $x = 0$ to $x = 2$ about the line $y = -4$

11. $y = x^2$ from $x = 0$ to $x = 2$ about the line $x = -3$

12. $x = y^2 + 2y$ from $y = 1$ to $y = 5$ about the line $y = 10$

13. $x = \sin y$ from $y = 0$ to $y = \pi$ about the line $x = 2$

calculator exercises

Use Simpson's rule with $n = 10$ to estimate the area of the surface of revolution given by:

14. $y = x^2$ from $x = 0$ to $x = 2$ about the x-axis

15. $y = \sin x$ from $x = 0$ to $x = \pi$ about the y-axis

16. $x = y^3 + 3y^2$ from $y = 0$ to $y = 4$ about the y-axis

7.6 DISTANCE

If a body travels on a line with constant velocity v, then the distance s traveled after time t is given by the product $|v|\, t$. The absolute-value sign is used since v is negative if the body is moving in the negative direction on the line. Thus $|v|$ is the *speed* of the body.

Example 1 If the velocity at time t sec of a body traveling on a line is $2t + t^2$ ft/sec, let's find how far the body travels from $t = 2$ sec to time $t = 4$ sec, using integral calculus.

SOLUTION Since distance equals the product of speed and time, assuming that speed is constant, we see that at time t, the distance traveled over the next small time interval dt sec is about $(2t + t^2)\, dt$ ft. We wish to add up these distances as t ranges from 2 to 4, and take the limit of the resulting sum as $dt \to 0$. The appropriate integral is

$$\int_2^4 (2t + t^2)\, dt = \left(t^2 + \frac{t^3}{3}\right)\Bigg]_2^4 = \left(16 + \frac{64}{3}\right) - \left(4 + \frac{8}{3}\right) = \frac{92}{3}.$$

Thus the body travels $\frac{92}{3}$ ft. ||

The velocity of a body moving on the x-axis is considered positive if it is moving to the right and negative if the body is moving to the left. Thus the integral of the velocity v from t_1 to t_2 will give the total distance traveled toward the right, minus the distance traveled toward the left between these times. This "resultant distance" may not give a true picture of how far the body has traveled; it does tell you the distance of the body at time t_2 from its position at time t_1. To find the actual distance traveled, you must integrate the speed $|v|$. Thus if a body starts at time t_1, stops at time t_2, and has velocity $v(t)$ for $t_1 \le t \le t_2$, then

$$\text{Distance from start to stop} = \int_{t_1}^{t_2} v(t)\, dt,$$

$$\text{Total distance traveled} = \int_{t_1}^{t_2} |v(t)|\, dt.$$

Example 2 Suppose the velocity at time t of a body traveling on a line is $\cos(\pi t/2)$ ft/sec. Thus at time $t = 0$, the body is moving at the speed of 1 ft/sec in the positive direction, while at time $t = 2$, the body is moving in the negative direction at the speed of 1 ft/sec. Let's find the total distance the body travels from time $t = 0$ to time $t = 2$, and also the distance from start to stop.

SOLUTION We need to compute

$$\int_0^2 \left|\cos\frac{\pi}{2}t\right| dt.$$

Now $\cos(\pi t/2)$ is positive for $0 \leq t < 1$ and negative for $1 < t \leq 2$. Thus we have

$$\begin{aligned}\int_0^2 \left|\cos\frac{\pi}{2}t\right| dt &= \int_0^1 \cos\frac{\pi}{2}t\,dt + \int_1^2 \left(-\cos\frac{\pi}{2}t\right) dt\\ &= \left(\frac{2}{\pi}\sin\frac{\pi}{2}t\right)\Big]_0^1 - \left(\frac{2}{\pi}\sin\frac{\pi}{2}t\right)\Big]_1^2\\ &= \left(\frac{2}{\pi} - 0\right) - \left(0 - \frac{2}{\pi}\right) = \frac{4}{\pi}.\end{aligned}$$

Therefore the body travels $4/\pi$ ft from time $t = 0$ to time $t = 2$ sec.

The distance from start to stop for this body is

$$\int_0^2 \left(\cos\frac{\pi}{2}t\right) dt = \left(\frac{2}{\pi}\sin\frac{\pi}{2}t\right)\Big]_0^2 = \frac{2}{\pi}(0 - 0) = 0.$$

Thus this body returns at $t = 2$ to the point where it started at $t = 0$. ‖

SUMMARY 1. *If a body on a line has velocity $v(t)$ at time t and starts at time t_1 and stops at time t_2, then*

$$\textit{Distance from start to stop} = \int_{t_1}^{t_2} v(t)\,dt,$$

$$\textit{Total distance traveled} = \int_{t_1}^{t_2} |v(t)|\,dt.$$

EXERCISES

In Exercises 1 through 6, the velocity v of a body on a line is given as a function of time t. Find:

i) *the distance from beginning point to ending point, and*

ii) *the total distance traveled in the indicated time interval.*

1. $v = t - 3$	a) $0 \le t \le 2$	b) $1 \le t \le 6$	
2. $v = 2t - t^2$	a) $0 \le t \le 2$	b) $0 \le t \le 4$	
3. $v = t^2 - 3t + 2$	a) $0 \le t \le 1$	b) $0 \le t \le 2$	c) $0 \le t \le 3$
4. $v = \sin t$	a) $0 \le t \le \pi$	b) $0 \le t \le 3\pi$	
5. $v = \sin\left(\frac{\pi t}{2}\right) + \cos\left(\frac{\pi t}{2}\right)$	a) $0 \le t \le 1$	b) $0 \le t \le 2$	c) $0 \le t \le 4$
6. $v = \lvert t - 5\rvert$	a) $0 \le t \le 5$	b) $0 \le t \le 10$	

In Exercises 7 through 10, the acceleration a as a function of time t and the initial velocity v_0 when $t = 0$ are given. Find:

a) *the velocity of the body as a function of t, and*

b) *the total distance the body travels in the indicated time interval.*

7. $a = 3, v_0 = 0, 0 \le t \le 2$

8. $a = 2t - 4, v_0 = 3, 0 \le t \le 3$

9. $a = \sin t, v_0 = 0, 0 \le t \le 3\pi/2$

10. $a = -1/\sqrt{t+1}, v_0 = 2, 0 \le t \le 4$

7.7 WORK AND HYDROSTATIC PRESSURE

7.7.1 Work

If a body is moved a distance s by means of a constant force F in the direction of the motion, then the work W done in moving the body is the product Fs. For example, if a body is being pushed along a line by a constant force of 20 lb directed along the line, then the work done by the force on the body in pushing from position $s = 0$ to position $s = 10$ is $W = 20 \cdot 10 = 200$ ft-lb.

If the force acting on the body does not remain constant, but is a continuous function $F(s)$ of the position s of the body, then the work done over a short interval of length ds is approximately $F(s)\,ds$, for some position s in this interval. Adding all these contributions to the work from starting position $s = a$ to stopping position $s = b$, you obtain

$$W = \int_a^b F(s)\,ds. \tag{1}$$

Example 1 The force F required to stretch (or compress) a coil spring is proportional to the distance x it is stretched (or compressed) from its natural length. That is, $F = kx$ for some constant k, the *spring constant.* Suppose that a spring is such that the force required to stretch it one foot from its natural length is 4 lbs. For this spring, $k = 4$. Let's find the work done in stretching the spring a distance of 4 ft from its natural length.

SOLUTION Work is defined as the product of force and distance, if the force remains constant and is in the direction of the motion. Thus, as our spring is stretched an additional small distance dx at a distance x from its natural length, the work done is approximately $F \cdot dx = 4x \cdot dx$. We wish to add all these little pieces of work from $x = 0$ to $x = 4$, and take the limit of the resulting sum as dx approaches 0. The appropriate integral is

$$\int_0^4 4x\,dx = 4\frac{x^2}{2}\Big]_0^4 = \frac{64}{2} - 0 = 32.$$

Thus 32 ft-lb of work are done. ||

7.7.2 Hydrostatic pressure

If the pressure per square unit on a plane region of area A is a constant p throughout the region, then the total force F on the region due to the pressure is the product pA. If the pressure does not remain constant throughout the region, one attempts to find the total force by integrating $p \cdot dA$ where dA is the area of a small piece of the region throughout which the pressure remains roughly constant.

Example 2 The fluid pressure per square foot at a depth of s ft in water is about $62.4s$ lb. Suppose a dam 16 ft high has the shape of the region bounded by the curves with equations $y = x^2$ and $y = 16$. Let's find the total force due to water pressure on the dam when the water is at the top of the dam.

SOLUTION The region bounded by the curves $y = x^2$ and $y = 16$ is shaded in Fig. 7.28. For small dy, the pressure at a depth s on the strip heavily shaded in Fig. 7.28 is nearly constant at $62.4s$ lb/ft^2. The area of this strip is $2x\,dy$, so the force on it is approximately $62.4s(2x\,dy)$ lb. We want to add up these quantities $62.4s(2x\,dy)$ over the region as y ranges from 0 to 16, and take the limit of the resulting sum as dy approaches 0. We must express this quantity entirely in terms of y. For the point (x, y) in Fig. 7.28, we have $s = 16 - y$, and $y = x^2$, so $x = \sqrt{y}$. Thus

$$62.4s(2x\,dy) = 62.4(16 - y)2\sqrt{y}\,dy = 124.8(16y^{1/2} - y^{3/2})\,dy.$$

The appropriate integral is

$$\begin{aligned}\int_0^{16} 124.8(16y^{1/2} - y^{3/2})dy &= 124.8(16(\tfrac{2}{3})y^{3/2} - \tfrac{2}{5}y^{5/2})\Big]_0^{16} \\ &= 124.8(16(\tfrac{2}{3})64 - (\tfrac{2}{5})1024) - 0 \\ &= 124.8(1024)(\tfrac{2}{3} - \tfrac{2}{5}) \\ &= 124.8(1024)\tfrac{4}{15} = 34{,}078.72.\end{aligned}$$

Thus the force is 34,078.72 lb. ||

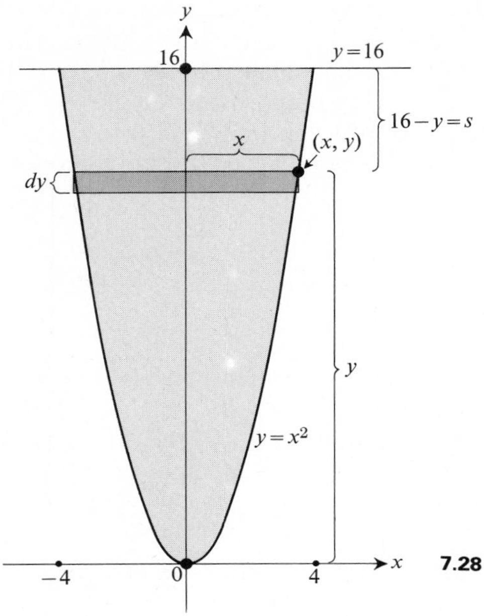

7.28

SUMMARY

1. *If the force acting on a body in the direction of its motion along a line is the function $F(s)$ of the position s of the body, then the work done by the force in moving the body from position $s = a$ to position $s = b$ is*

$$W = \int_a^b F(s)\, ds.$$

2. *The pressure exerted by water at depth s is about $(62.4)s$ $\mathrm{lb/ft^2}$. Let a submerged vertical plate have its top at a depth a and its bottom at depth b. If the area of a thin horizontal strip of the plate at depth s is dA, then*

$$\textit{Total force on the plate} = \int_a^b (62.4)s \cdot dA.$$

EXERCISES

1. Find the work done in stretching the spring of Example 1 from 2 ft longer than its natural length to 6 ft longer than its natural length.

2. If a spring has a natural length of 2 ft and a force of 10 lbs is required to compress the spring 2 in. (to a length of 22 in.), find the work done in stretching the spring from a length of 26 in. to a length of 30 in.

3. A cylindrical tank of radius 2 ft and height 10 ft is full of water and is emptied by dropping in a hose, pumping all the water up to the top of the tank, and letting it spill over. Find the work done.

4. Two electrons a distance s apart repel each other with a force k/s^2 where k is some constant.
   a) If an electron is at the point 2 on the x-axis, find the work done by the force in moving another electron from the point 4 to the point 8.
   b) If an electron is at the point 0 of the x-axis and another is at the point 1, find the work done by the forces in moving a third electron from the point 2 to the point 6.
5. According to Newton's Law of Gravitation, the force of attraction of two bodies of masses m_1 and m_2 is $G(m_1m_2/s^2)$, where s is the distance between the bodies and G is the gravitational constant. If the distance between two bodies of masses m_1 and m_2 is a, find the work done in moving the bodies twice as far apart.
6. Let a body of mass m move in the positive direction on the x-axis subject only to a force F acting in the positive direction along the axis. (The magnitude of the force may vary with the time t.) If the body has velocity v, the *kinetic energy* of the body is $(1/2)mv^2$. Show that the work done by the force from time t_1 to time t_2, for $t_1 < t_2$, is equal to the change in the kinetic energy of the body between these times. [*Hint.* By Newton's Second Law of Motion,

$$F = ma = m\frac{dv}{dt} = m\frac{dv}{dx}\frac{dx}{dt} = m\frac{dv}{dx}v.$$

Consider both F and v as functions of the position x_t of the body at time t, and express the work

$$W = \int_{x_{t_1}}^{x_{t_2}} F(x)\,dx$$

as an integral involving v, using Newton's law.]
7. A vertical dam is in the shape of a semicircle of radius 36 ft, with the diameter of the circle at the top of the dam. Find the force on the dam when the water level is at the top of the dam.
8. A cylindrical drum of diameter 2 ft and length 4 ft is filled with water. Find the force on one end of the drum if the drum is lying on its side.
9. For the drum in Exercise 8, find the force on the cylindrical wall of the drum if the drum is standing on end.

7.8 MASS AND MOMENTS

The *mass* of a body is a numerical measure of the "amount of material" it contains. In this section we are interested in bodies that lie in the plane, such as a thin plate or a piece of wire.

7.8.1 Mass

The *mass density* ρ of a flat plate at a point (x, y) is the mass that a piece of area 1 would have if the plate had everywhere the same composition and thickness as at the point (x, y). A piece of wire has mass density per unit length, rather than per unit area. In general, mass density is a function $\rho(x, y)$ of both x and y, but in this section, we will be concerned only with the case where it is either a function $\rho(x)$ of x alone or $\rho(y)$ of y alone. Three-dimensional bodies and more general mass-density functions are treated in Chapter 18.

To find the total mass of a plate with mass density $\rho(x)$, you multiply $\rho(x)$ by the area dA of a thin *vertical* strip of the plate of width dx, and then integrate the resulting expression:

$$Mass = m = \int_a^b \rho \cdot dA. \qquad (1)$$

Formula (1) also holds for the mass density $\rho(y)$, but then you take a thin *horizontal* strip of height dy.

Example 1 Consider a thin sheet of material exactly covering the plane region bounded by the curves $y = x^2$, $x = 2$, and $y = 0$, as shown in Fig. 7.29. Let's find the mass of the sheet if the mass density at a point (x, y) on the sheet is $\sqrt{x}$.

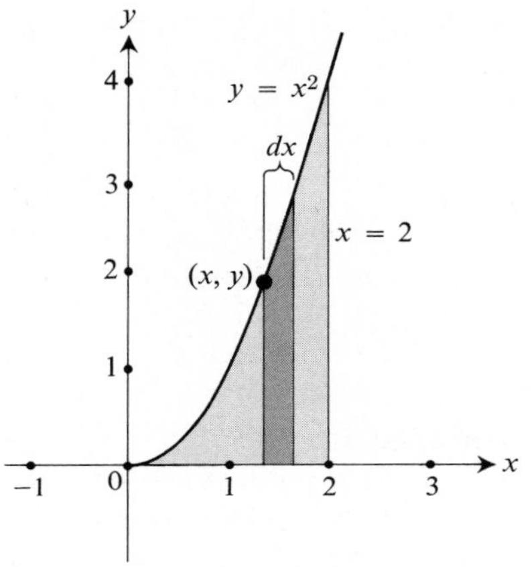

7.29

SOLUTION The mass density is approximately constant with value $\sqrt{x}$ throughout the strip with the heavy shading in Fig. 7.29. The strip has approximate area $x^2\,dx$, and hence approximate mass $(\sqrt{x})(x^2\,dx)$. We wish to add these masses as x goes from 0 to 2 and as dx approaches 0. The total mass is therefore

$$\int_0^2 (\sqrt{x})(x^2\,dx) = \int_0^2 x^{5/2}\,dx = \tfrac{2}{7}x^{7/2}\Big]_0^2 = \tfrac{2}{7}2^{7/2} - 0 = \tfrac{16}{7}\sqrt{2}\text{ units.} \quad \|$$

For a wire of mass density ρ, you integrate $\rho \cdot ds$ to find the total mass.

Example 2 Let a wire lie on the curve $x = y^2$ from $y = 0$ to $y = 2$ and have mass density ky for a constant k. Then

$$ds = \sqrt{(dx/dy)^2 + 1}\,dy = \sqrt{4y^2 + 1}\,dy,$$

so

$$\begin{aligned} m = \int_0^2 ky\sqrt{4y^2+1}\,dy &= \frac{k}{8}\int_0^2 8y(4y^2+1)^{1/2}\,dy \\ &= \frac{k}{8}\cdot\frac{2}{3}(4y^2+1)^{3/2}\Big]_0^2 \\ &= \frac{k}{12}[17\sqrt{17} - 1]. \quad \| \end{aligned}$$

7.8.2 Moments

Let an axis in the plane have one side designated as the positive side and the other as the negative side. For example, points to the right of a vertical axis are on the positive side, while points above a horizontal axis are on the positive side. The *first moment* (or simply the *moment*) about such an axis of a point of mass m is mr, where r is the positive or negative distance from the axis. If the body is a plate, take a thin strip of mass $dm = \rho \cdot dA$ parallel to the axis and a signed distance r from the axis. Then

First moment

$$\text{First moment} = M = \int_a^b r \cdot dm \tag{2}$$

for appropriate limits a and b. If the body is a wire, then $dm = \rho \cdot ds$. We shall often be interested in the moments M_x about the x-axis and M_y about the y-axis.

Example 3 Let's find the moment M_y about the y-axis of the sheet of material described in Example 1.

SOLUTION All points of the heavily shaded strip in Fig. 7.29 are approximately the same distance x from the y-axis if dx is small. Since the mass of this strip is approximately $(\sqrt{x})(x^2\,dx) = x^{5/2}\,dx$, you see that the moment of this strip about the y-axis is approximately $(x)(x^{5/2}\,dx)$. We wish to add the moments of these strips as x goes from 0 to 2 and as dx approaches 0. The moment M_y is therefore given by

$$M_y = \int_0^2 (x)(x^{5/2})\,dx = \int_0^2 x^{7/2}\,dx = \tfrac{2}{9}x^{9/2}\Big]_0^2 = \tfrac{2}{9}2^{9/2} - 0 = \tfrac{32}{9}\sqrt{2}. \quad \|$$

Moment of inertia

The *second moment* (or *moment of intertia*) I of one of these plane bodies about an axis is given by

$$\text{Second moment} = I = \int_a^b r^2\,dm, \tag{3}$$

and in general,

$$n\text{th moment} = \int_a^b r^n\,dm. \tag{4}$$

The second moment I appears in the formula

$$\text{K.E.} = \tfrac{1}{2}I\omega^2 \tag{5}$$

for the kinetic energy of rotation about an axis of a body with moment of inertia I and angular speed of rotation ω. Thus if you wish to double the kinetic energy of flywheel rotating with a fixed angular speed ω, you might modify your flywheel so that the mass is moved $\sqrt{2}$ times as far from the axis.

Example 4 Consider again the sheet of material of Example 1. To find the moment of inertia I_y of this sheet about the y-axis, you should multiply the mass $(\sqrt{x})(x^2\,dx) = x^{5/2}\,dx$ of the strip in Fig. 7.29 by the square of its distance from the y-axis, and add these contributions as dx approaches zero. You arrive at

$$I_y = \int_0^2 (x^2)(x^{5/2}\,dx) = \int_0^2 x^{9/2}\,dx = \tfrac{2}{11}x^{11/2}\Big]_0^2$$

$$= \tfrac{2}{11}2^{11/2} - 0 = \tfrac{64}{11}\sqrt{2}. \quad \|$$

Sometimes it is not convenient to take horizontal strips when finding M_x or vertical strips when finding M_y. If the density ρ is constant, then you can consider the mass to be concentrated at the center of the strip when computing moments, as illustrated in the next example.

Example 5 Let's find the moment M_x about the x-axis of a plate of constant mass density 3 covering the region bounded by $y = 0$ and $y = \sin x$ for $0 \le x \le \pi$, as shown in Fig. 7.30.

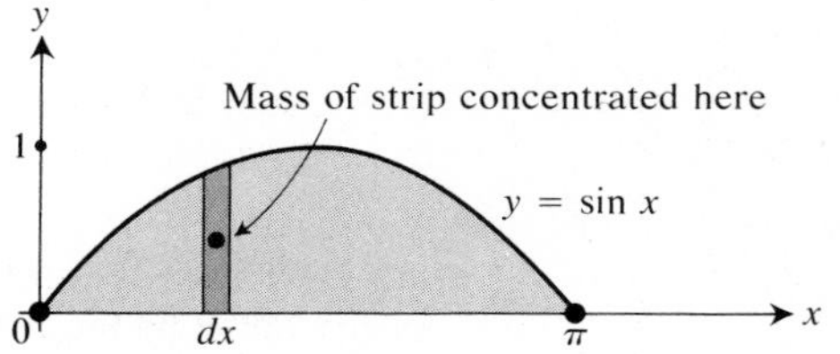

7.30

SOLUTION We consider the mass

$$dm = \rho \cdot dA = 3 \sin x\,dx$$

to be concentrated at the midpoint of the strip, where the distance from the x-axis is $y/2 = (\sin x)/2$. Thus, using a table of integrals,

$$M_x = \int_0^\pi \frac{\sin x}{2}(3 \sin x)\,dx = \frac{3}{2}\int_0^\pi \sin^2 x\,dx = \frac{3}{2}\left(\frac{x}{2} - \frac{\sin 2x}{4}\right)\Big]_0^\pi$$

$$= \frac{3}{2}\left(\frac{\pi}{2} - 0\right) - \frac{3}{2}(0 - 0) = \frac{3\pi}{4}. \quad \|$$

SUMMARY *For mass density ρ and area dA a signed (positive or negative) distance r from an axis,*

1. *Total mass* $= m = \int_a^b \rho \cdot dA$

2. *First moment* $= M = \int_a^b r \cdot \rho \cdot dA$

3. *Second moment (of intertia)* $= I = \int_a^b r^2 \rho \cdot dA$

4. n*th moment* $= \int_a^b r^n \rho \cdot dA$

where a and b are appropriate limits.

EXERCISES

1. The triangular region with vertices $(0, 0)$, $(a, 0)$, and $(0, b)$ for $a > 0$ and $b > 0$ is covered by a sheet of material whose mass density at a point (x, y) is ky^2.

a) Find the mass of the sheet.

b) Find the first moment of the sheet about the x-axis.

2. Let the mass density at a point (x, y) of a plate covering the semicircular disk bounded by the x-axis and $y = \sqrt{1 - x^2}$ be $2(y + 1)$. Find the mass of the plate.

3. A metal plate is cut from a thin sheet of metal of constant mass density 3. If the plate just covers the first quadrant region bounded by the curves $y = x^2$, $y = 1$, and $x = 0$, find the first moment of the plate about the y-axis. (Assume the units are compatible, and don't worry about the units for the answer.)

4. For the plate in Exercise 3, find its first moment about the axis with equation $x = 2$.

5. For the plate in Exercise 3, find its moment of inertia about the x-axis.

6. Consider a thin rod of length a with constant mass density k (mass per unit length).

a) Find the first moment of the rod about an axis through an endpoint and perpendicular to the rod.

b) Find the moment of inertia of the rod about the axis described in (a).

7. Repeat Exercise 6 in case the mass density is kx^2, where x is the distance along the rod from the axis of rotation.

8. Let a wire lie on $y = x^3$ from $x = 0$ to $x = 2$ and have constant mass density $\rho = 5$. Find the moment M_x of the wire about the x-axis.

9. Let a wire lie on $x = y^{3/2}$ from $y = 1$ to $y = 4$ and have mass density $\rho(y) = 3\sqrt{y}$. Express as an integral the first moment M_y of the wire about the y-axis.

10. Find the moment of inertia of a flat disk of radius a and constant mass density k about an axis perpendicular to the disk through its center. [*Hint.* Add the moments of inertia of concentric circular "washers" by integration.]

11. Express as an integral the moment of inertia of a ball of radius a and constant mass density k (mass per unit volume) about a diameter of the ball. [*Hint.* Add the moments of inertia of cylindrical shells with the diameter as axis.]

12. Show that the first moment of a body in the plane about the line $x = -a$ is $M_y + ma$ where M_y is the moment about the y-axis. (This is known as the "*Parallel Axis Theorem.*")

7.9 CENTER OF MASS, CENTROIDS, PAPPUS' THEOREM

It can be proved that for a given body of mass m, there exists a unique point (not necessarily in the body) at which you may consider the entire mass of the body to be concentrated for the computation of first moments about *every* axis. This point is the **center of mass** (or **center of gravity**) of the body. If the (signed) distance from the center of mass of the body to an axis is s, then the first moment of the body about the axis is ms. [*We warn you that there is no "center of inertia" for a body where all the mass can be considered to be concentrated for the computation of moments of inertia about every axis.*]

The **centroid of a plane region** is defined to be the point that would be the location of the center of mass of a thin sheet of material of constant density covering the region. For a thin sheet of material covering a plane region, let the center of mass be $(\bar{x}, \bar{y})$ as shown in Fig. 7.31. If you let the mass of the sheet be m and its moments about the x and y axes be M_x and M_y, respectively, then

$$M_x = m\bar{y},$$

since the (signed) distance from $(\bar{x}, \bar{y})$ to the x-axis is $\bar{y}$. Similarly, $M_y = m\bar{x}$. Thus

$$\bar{x} = \frac{M_y}{m} \quad \text{and} \quad \bar{y} = \frac{M_x}{m}.$$

Since you know how to compute m, M_x, and M_y, you can now compute $(\bar{x}, \bar{y})$.

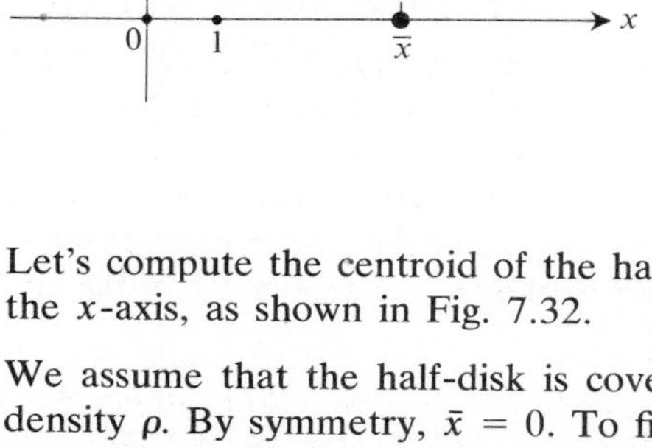

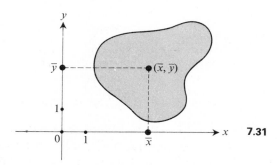

7.31

Example 1 Let's compute the centroid of the half-disk bounded by $y = \sqrt{a^2 - x^2}$ and the x-axis, as shown in Fig. 7.32.

SOLUTION We assume that the half-disk is covered by a thin sheet of constant mass density ρ. By symmetry, $\bar{x} = 0$. To find $\bar{y} = M_x/m$, we need only compute M_x, for we know that the mass m will be the product of the constant density

ρ and the area, so that

$$m = \rho\frac{\pi a^2}{2}.$$

By symmetry, to compute M_x, we need only compute the moment of the quarter-disk in the first quadrant about the x-axis, and then double it. The moment about the x-axis of the strip shown in Fig. 7.32 is approximately $(y)(\rho x\,dy)$. Since $x = \sqrt{a^2 - y^2}$, we arrive at

$$\begin{aligned} M_x &= 2\int_0^a (y)(\rho\sqrt{a^2 - y^2}\,dy) = 2\rho\int_0^a y(a^2 - y^2)^{1/2}\,dy \\ &= \frac{2\rho}{-2}\int_0^a (a^2 - y^2)^{1/2}(-2y\,dy) = -\rho\left(\frac{2}{3}\right)(a^2 - y^2)^{3/2}\Big]_0^a \\ &= -\rho\cdot\tfrac{2}{3}\cdot 0 - (-\rho\cdot\tfrac{2}{3}\cdot a^3) = \tfrac{2}{3}\rho a^3. \end{aligned}$$

Thus

$$\bar{y} = \frac{M_x}{m} = \frac{\frac{2}{3}\rho a^3}{\pi\rho a^2/2} = \frac{4a}{3\pi}.$$

The centroid of the half-disk is therefore $(0, 4a/3\pi)$. ||

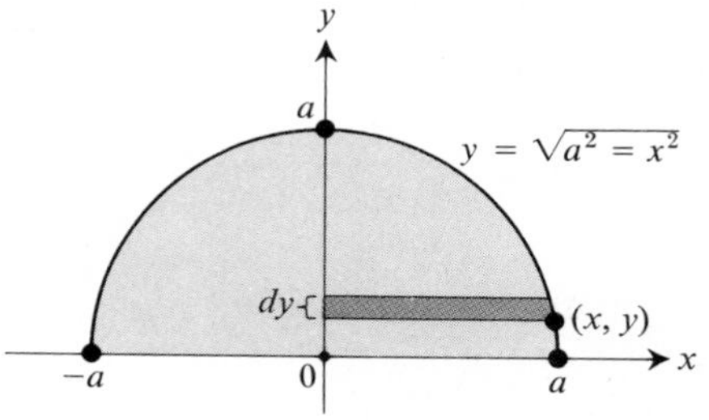

7.32

The following is known as Pappus' theorem.

Theorem 7.1 (Pappus)

1. *If a plane arc of length L is revolved about an axis in the plane but not intersecting the arc, then the area of the surface generated is the product of the length L of the arc and the circumference of the circle described by the centroid of the arc.*
2. *If a plane region of area A is revolved about an axis in the plane but not intersecting the region, then the volume of the solid generated is the product of the area A and the circumference of the circle described by the centroid of the region.*

Obviously these theorems make similar assertions for different dimensional objects. Let's see why the second version is true. Consider the region shown in Fig. 7.33, and suppose it is revolved about the y-axis. Using the shell method, the volume generated is

$$V = \int_a^b 2\pi x \cdot dA = 2\pi \int_a^b x \cdot dA = 2\pi M_y, \tag{1}$$

where dA is the area of the dark-shaded strip. But

$$\bar{x} = \frac{M_y}{A} \tag{2}$$

and from (1), $M_y = V/2\pi$, so $\bar{x} = V/(2\pi A)$ and

$$V = 2\pi\bar{x} \cdot A, \tag{3}$$

which is the assertion of Pappus' theorem.

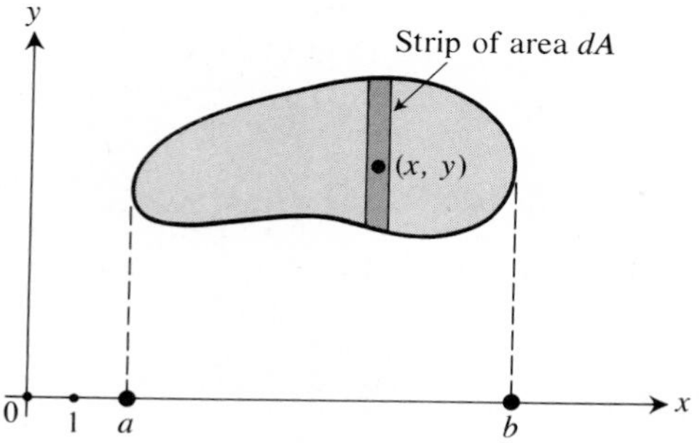

7.33

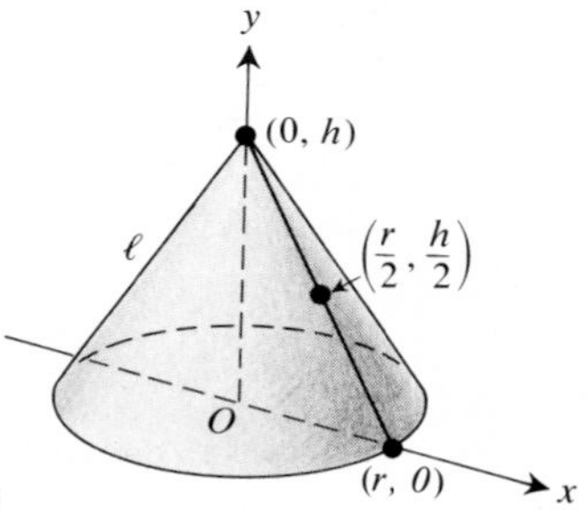

7.34

Example 2 If the line segment joining $(r, 0)$ and $(0, h)$ is revolved about the y-axis, it generates the surface of a cone (see Fig. 7.34). Let

$$\ell = \sqrt{r^2 + h^2}$$

be the slant height of the cone. The centroid of the line segment is at the point $(r/2, h/2)$, so, by Pappus' Theorem, the surface area S of the cone is

$$S = \left(2\pi \cdot \frac{r}{2}\right)\ell = \pi r \ell. \quad \|$$

Example 3 Let's use Pappus' Theorem to find the centroid of the half-disk in Fig. 7.32, using the formula $V = \frac{4}{3}\pi a^3$ for the volume of the generated ball.

SOLUTION You know the area of the half-disk is $\pi a^2/2$. If its centroid is $(0, \bar{y})$, then

$$V = \tfrac{4}{3}\pi a^3 = 2\pi\bar{y} \cdot \tfrac{1}{2}\pi a^2 = \pi^2 a^2 \bar{y}.$$

Therefore $\bar{y} = (4a)/(3\pi)$, as we found in Example 1. $\|$

SUMMARY 1. *The center of mass (centroid) of a plane body (region) is* $(\bar{x}, \bar{y})$ *where*

$$\bar{x} = \frac{M_y}{m} \quad \textit{and} \quad \bar{y} = \frac{M_x}{m}.$$

2. *(Pappus' Theorem) If a plane body is revolved about an axis in the plane of the body but not intersecting the body, then the volume (area) generated is equal to the product of the area (length) of the body and the circumference of the circle described by the centroid of the body.*

EXERCISES

1. Give an example of a body whose center of mass is not a point in the body.

2. Find by integration the centroid of the triangular region whose vertices are $(0, 0)$, $(a, 0)$, $(0, b)$, where $a > 0$ and $b > 0$.

3. Find the centroid of the region bounded by the curves $y = \sqrt{a^2 - x^2}$, $x = a$, and $y = a$, where $a > 0$. [*Hint.* Use symmetry.]

4. Find the center of mass of a sheet of mass density $2y + 3$ that covers the plane region bounded by the curves $y = x^2$ and $y = 4$.

5. Find the centroid of the region bounded by the arch of $y = \sin x$, for $0 \leq x \leq \pi$, and the x-axis.

6. Let the half-disk $0 \leq y \leq \sqrt{1 - x^2}$ be covered by a sheet of material of constant mass density ρ. Find the first moment of the sheet about the line $x + y = 4$. [*Hint.* You know the center of mass of the sheet from Example 1.]

7. Let the portion of the face of a vertical dam that is covered by water be a plane region of area A ft^2 whose centroid is at a depth s ft below the surface of the water. Show that the force on the dam is $(62.4)sA$ lb.

8. Let a flat body of constant mass density k cover the unit square with vertices $(0, 0)$, $(1, 0)$, $(0, 1)$, and $(1, 1)$.

a) Find the moment of inertia of the body about the y-axis.

b) Find the moment of inertia of the body about the line $x = -a$.

c) Find a point (x_1, y_1) in the body such that the moment of inertia of the body about either the x-axis or y-axis is the product of the mass and the square of the distance from (x_1, y_1) to the axis.

d) Find a point (x_2, y_2) in the body such that the moment of inertia of the body about either the line $x = -a$ or the line $y = -a$ is the product of the mass and the square of the distance from (x_2, y_2) to the line.

e) Compare the answers to (c) and (d), and comment on the result.

9. The *radius of gyration* R of a body about an axis is defined by

$$R = \sqrt{I/m},$$

so that $I = mR^2$.

a) From the answer $k/3$ to Exercise 8(a), what is the radius of gyration about the y-axis of a homogeneous flat body covering the unit square?

b) From the answer

$$\frac{k}{3}((a + 1)^3 - a^3)$$

to Exercise 8(b), what is the radius of gyration about the line $x = -a$ of a homogeneous flat body covering the unit square?

10. Use Pappus' Theorem to find the volume generated when the half-disk bounded by $y = \sqrt{1 - x^2}$ and $y = 0$ is revolved about the given line. (The centroid of the disk was found in Example 1.)

a) $y = 1$ b) $y = -1$ c) $y = 2$

d) $x = 2$ e) $x + y = 4$

11. A square of side a is revolved about an axis through a vertex and perpendicular to the diagonal of the square at that vertex. Use Pappus' Theorem to find the volume generated.

12. Given that the volume of a right circular cone of altitude h and base of radius r is $(\frac{1}{3})\pi r^2 h$, use Pappus' Theorem to find the centroid of the triangular region with vertices $(0, 0)$, $(r, 0)$, and $(0, h)$, for $r > 0$ and $h > 0$.

exercise sets for chapter 7

review exercise set 7.1

1. Find the area of the region bounded by the curves $y = \sqrt{2x}$ and $y = \frac{1}{2}x^2$.
2. Find the volume of the solid generated by revolving the region bounded by $y = 4 - x^2$ and $y = 0$ about the line $y = -1$.
3. The work done in stretching a spring one foot from its natural length is 18 ft-lbs. Find the work done in stretching it 4 feet from its natural length.
4. Find the arc length of the curve $y = 2(\frac{5}{3} + x)^{3/2}$ from $x = 0$ to $x = \frac{11}{3}$.
5. Find the area of the surface of revolution generated by rotating the arc $y = x^3/\sqrt{3}$ from $x = 0$ to $x = 1$ about the x-axis.
6. Find the total distance traveled on a line from $t = 1$ to $t = 4$ by a body with velocity $v = \sin \pi t$.
7. A thin plate covers the region bounded by $y = 1 - x^2$ and $y = 0$ and has mass density $y + 3$ at the point (x, y). Express as an integral the moment of inertia of the plate about the x-axis.
8. Find the centroid of the region bounded by the curves $y = \frac{1}{2}x^2$ and $y = \sqrt{2x}$.

review exercise set 7.2

1. Find the area of the region bounded by $x = 4 - y^2$ and $x = 3y$.
2. Find the volume of the solid generated by revolving the region bounded by $y = x^2$ and $y = x + 2$ about the line $x = -1$.
3. The face of a dam is in the form of an isosceles triangle with 40 ft base at the top of the dam and equal 30 ft sides at the sides of the dam. Find the total force of the water on the dam when the water is at the top of the dam.
4. Find the value of x such that the length of arc of the circle $x = 5 \cos t$, $y = 5 \sin t$, measured counterclockwise from $(5, 0)$ to (x, y), is two units.
5. Find the area of the surface of revolution generated by rotating the arc of the circle $x = 3 \cos t$, $y = 3 \sin t$ from $t = 0$ to $t = \pi/4$ about the y-axis.
6. The velocity of a body traveling on a line is given by $v = 4 - t^2$. If the body starts at time $t = 0$, at what time has the body traveled a total distance of 16 units?
7. Find the first moment of a homogeneous semicircular disk of mass m bounded by $y = \sqrt{25 - x^2}$ and $y = 0$ about the x-axis.
8. Find the center of mass of a thin triangular plate with vertices at $(0, 0)$, $(3, 4)$, and $(0, 8)$ if the mass density of the plate at (x, y) is $x + 1$.

more challenging exercises 7

1. Using Pappus' Theorem, express *as a single integral* the volume generated by revolving the region bounded by $y = x^2$ and $y = 4$ about the line $y = x - 2$. Evaluate the integral.
2. Find the area of the region in the first quadrant bounded by $y = x^2$, $y = x^2 + 9$, $x = 0$, and $y = 25$.
3. Express as a sum of integrals the volume generated

when the region in Exercise 2 is revolved about the line $3x + 4y = -12$.

4. Water flows from a pipe at time t minutes at a rate of $40/(t + 1)^2$ gal/min for $t \geq 0$. Find the amount of water that flows from the pipe during the first hour of flow.

The remaining exercises illustrate how an integral may be used to estimate a sum.

5. Estimate $\displaystyle\sum_{k=1}^{800} \frac{k^{3/2}}{100{,}000}$,

6. Estimate $\displaystyle\sum_{k=1}^{4000} \frac{1000}{(1000 + k)^2}$

7. A carpenter has a contract to hang 100 doors in a housing project. It takes him one hour to hang the first door. Using the expertise he continually attains, he finds that the time required to hang the nth door after the first is $3\sqrt{n}$ minutes less than one hour. Estimate the time required for him to hang all 100 doors.

8. A manufacturer finds that a newly hired employee is able to seal $40 + 2\sqrt[3]{n}$ cartons during the nth hour of work, except that, once a level of 60 cartons per hour is reached, there is no further increase. Estimate the number of cartons the employee can seal during the first 1500 hours of work.

other elementary functions

8

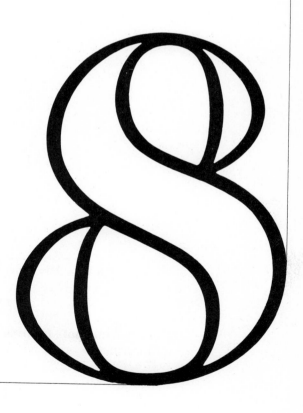

8.1 THE FUNCTION ln *x*

8.1.1 An integration problem

If n is an integer, you know that

$$\int x^n\,dx = \frac{x^{n+1}}{n+1} + C \qquad \text{for} \quad n \neq -1.$$

As yet, we haven't encountered a function that is an antiderivative of the function $1/x$. We would like to find $\int (1/x)\,dx$.

Of course, $\int_1^2 (1/x)\,dx$ exists, since $1/x$ is continuous in $[1, 2]$. Moreover, the fundamental theorem of calculus (Theorem 6.4, Section 6.2) tells us that $1/x$ does have an antiderivative for $x > 0$, namely F, where

$$F(x) = \int_1^x \frac{1}{t}\,dt \quad \text{for} \quad x > 0. \tag{1}$$

The value $F(x)$ is equal to the area of the shaded region in Fig. 8.1. (If a positive number other than 1 is chosen as lower limit for the integral in (1), the resulting function just differs from F by a constant.) Thus the fundamental theorem of calculus enables us to "find" an antiderivative F of $1/x$, at least for $x > 0$. This function F is extremely important. You will see later that F has the formal algebraic properties of the logarithm function you studied in high school, so the notation "ln" for F in the following definition is appropriate.

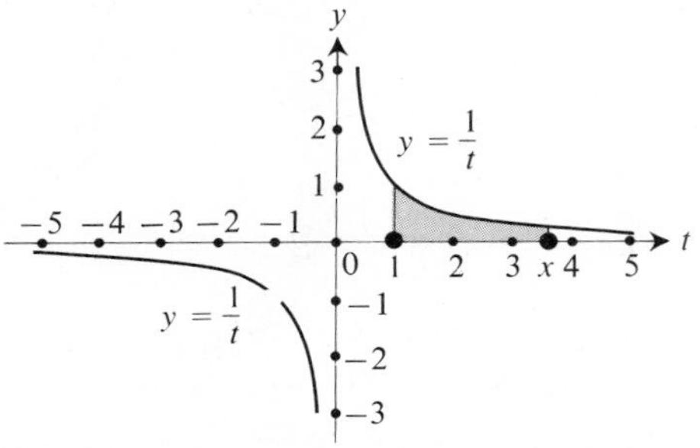

8.1

Definition 8.1 The function ln defined by

$$\ln x = \int_1^x \frac{1}{t}\,dt \quad \text{for} \quad x > 0 \tag{2}$$

is the **(natural) logarithm function**, and $\ln a = \int_1^a (1/t)\,dt$ is the (natural) logarithm of a for any $a > 0$.

From the properties of the integral discussed in Chapter 6, you see at once that

$$\ln 1 = \int_1^1 \frac{1}{t}\,dt = 0 \tag{3}$$

and

$$\ln x \quad \text{is} \quad \begin{cases} >0 & \text{if} \quad x > 1, \\ <0 & \text{if} \quad 0 < x < 1. \end{cases} \tag{4}$$

The graph of $\ln x$ is shown in Fig. 8.2. More justification for this graph appears shortly. A table of some values of $\ln x$ is given in Appendix 3 at the end of the text.

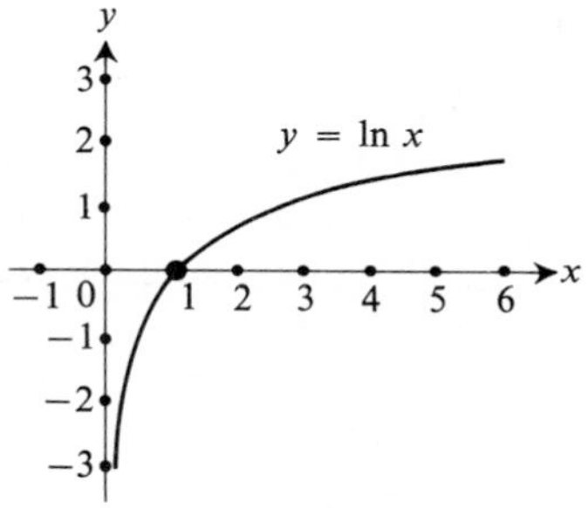

8.2

8.1.2 The calculus of ln x From the definition of $\ln x$ in (2) and the fundamental theorem of calculus, you see that

$$\frac{d(\ln x)}{dx} = \frac{1}{x} \quad \text{for} \quad x > 0. \tag{5}$$

By the chain rule, if $u = g(x)$ where g is a differentiable function, then

$$\frac{d(\ln u)}{dx} = \frac{1}{u} \cdot \frac{du}{dx} \quad \text{for} \quad u > 0. \tag{6}$$

You may wish to memorize formula (6), which we shall use often.

Example 1 Using (6), you have

$$\frac{d(\ln(2x+1))}{dx} = \frac{1}{2x+1} \cdot \frac{d(2x+1)}{dx} = \frac{2}{2x+1}, \quad x > -\tfrac{1}{2}. \quad \|$$

Example 2 Using (6) again, you obtain

$$\frac{d(\ln(\sin x))}{dx} = \frac{1}{\sin x} \cdot \frac{d(\sin x)}{dx} = \frac{\cos x}{\sin x} = \cot x, \quad \sin x > 0. \quad \|$$

From (6), you obtain the integration formula

$$\int \frac{du}{u} = \ln u + C \quad \text{for} \quad u > 0. \tag{7}$$

Let's now consider $\int (1/x)\,dx$ for $x < 0$. If we let $u = -x$, then $u > 0$ and $du = -1\,dx$. Then

$$\int \frac{1}{x}\,dx = \int \frac{-1\,dx}{-x} = \int \frac{du}{u} = \ln u + C = \ln(-x) + C \quad \text{for} \quad x < 0.$$

Thus $\int (1/x)\,dx = \ln(-x) + C$ for $x < 0$, so, by the chain rule,

$$\int \frac{du}{u} = \ln(-u) + C \quad \text{for} \quad u < 0. \tag{8}$$

Formulas (7) and (8) can be collected as one formula, namely

$$\int \frac{du}{u} = \ln|u| + C. \tag{9}$$

Formula (9) is extremely important, and you may wish to memorize it also.

Example 3 We have

$$\int_{-2}^{-1} \frac{1}{x}\,dx = \ln|x|\Big]_{-2}^{-1} = \ln|-1| - \ln|-2|$$
$$= \ln 1 - \ln 2 = -\ln 2. \quad \|$$

Example 4 Let's find $\int x/(x^2 + 1)\,dx$.

SOLUTION If we think of $x^2 + 1$ as u, then $du = 2x\,dx$. Fixing up the integral, we obtain

$$\int \frac{x}{x^2+1}\,dx = \frac{1}{2}\int \frac{2x}{x^2+1}\,dx = \frac{1}{2}\ln|x^2+1| + C.$$

Since $x^2 + 1$ is always positive, the absolute-value sign is really not needed here. ‖

Example 5 You can now find $\int \tan x\,dx$.

SOLUTION You have $\tan x = \sin x/\cos x$, and if you let $u = \cos x$, then $du = -\sin x\,dx$. Thus, fixing up the integral,

$$\int \tan x\,dx = \int \frac{\sin x}{\cos x}\,dx$$
$$= -\int \frac{-\sin x}{\cos x}\,dx = -\ln|\cos x| + C. \quad \|$$

Finally, you would like to be able to integrate ln x. We easily find that

$$\frac{d(x(\ln x) - x)}{dx} = x\frac{1}{x} + \ln x - 1 = \ln x.$$

Thus

$$\int \ln x\,dx = x(\ln x) - x + C. \tag{10}$$

Therefore, by the chain rule,

$$\int (\ln u)\,du = u(\ln u) - u + C, \tag{11}$$

for a differentiable function u. You may well ask how we knew we should try $x(\ln x) - x$ as a candidate for an antiderivative of ln x. You would probably arrive at $x(\ln x) - x$ after a bit of experimentation. There is also a technique of integration called "integration by parts," which can be used to obtain $x(\ln x) - x$ without guessing. Until you learn how to integrate by parts, you should consult a table for formula (11) if you need it.

Example 6 By formula (11),

$$\int x \ln (x^2 + 1)\,dx = \tfrac{1}{2}\int (2x) \ln (x^2 + 1)\,dx$$

$$= \tfrac{1}{2}[(x^2 + 1) \ln (x^2 + 1) - (x^2 + 1)] + C. \quad \|$$

Many integration formulas found in a table involve the logarithm function, and you can now handle these formulas.

8.1.3 Some properties of ln x

We shall now see that the function ln x satisfies properties that relate multiplication with addition. These relationships make ln x a very important function indeed.

Theorem 8.1 *For any $a > 0$ and $b > 0$ and for any rational number (fraction) r, we have*

$$\ln (ab) = \ln a + \ln b; \tag{12}$$

$$\ln (a/b) = \ln a - \ln b; \tag{13}$$

$$\ln (a^r) = r(\ln a). \tag{14}$$

Theorem 8.1 makes our use of the name "logarithm" for this function seem more reasonable. You are familiar (from high school) with a logarithm function that satisfies (12), (13), and (14). You may wonder how we could ever show these properties from the definition of ln x in (2) as an integral. The way it is done is fascinating. To show (12), we show that the functions

$$f(x) = \ln (ax) \qquad \text{and} \qquad g(x) = \ln a + \ln x$$

are identical. We can then obtain (12) by setting $x = b$.

Showing two functions are equal

Here is an outline for a method of showing that two differentiable functions $f(x)$ and $g(x)$ are the same.

A. Show that $f' = g'$.

B. Show that $f(c) = g(c)$ at *one* point c in the domain of the functions.

From (A), you can conclude that

$$(f - g)' = f' - g' = 0,$$

so you must have

$$f(x) - g(x) = k$$

for some constant k and all x in the domain of the functions. Since $f(c) = g(c)$ by (B), you obtain

$$0 = f(c) - g(c) = k,$$

so $k = 0$ and $f(x) = g(x)$ for all x in the domain of the functions.

Let's apply this technique to demonstrate (12), (13), and (14). Let $f(x) = \ln ax$ and let $g(x) = \ln a + \ln x$ for $x > 0$. Then

$$f'(x) = \frac{1}{ax} \cdot a = \frac{1}{x} \qquad \text{and} \qquad g'(x) = 0 + \frac{1}{x} = \frac{1}{x},$$

so $f'(x) = g'(x)$, and condition (A) is satisfied. For condition (B), we find that

$$f(1) = \ln a \qquad \text{and} \qquad g(1) = \ln a + \ln 1 = \ln a,$$

since (3) tells us that $\ln 1 = 0$. Therefore $f(x) = g(x)$ for all $x > 0$; in particular, $f(b) = g(b)$; that is,

$$\ln(ab) = \ln a + \ln b,$$

which establishes (12).

It is convenient to establish (14) next. Let $h(x) = \ln(x^r)$ and $k(x) = r(\ln x)$ for $x > 0$. Then

$$h'(x) = \frac{1}{x^r}(rx^{r-1}) = \frac{r}{x} \qquad \text{and} \qquad k'(x) = r\frac{1}{x} = \frac{r}{x},$$

so condition (A) is satisfied. For condition (B), we have

$$h(1) = \ln(1^r) = \ln 1 = 0 \qquad \text{and} \qquad k(1) = r(\ln 1) = r \cdot 0 = 0.$$

Therefore $h(x) = k(x)$ for all $x > 0$; in particular, taking $x = a$, we have

$$\ln(a^r) = r(\ln a),$$

which is (14).

From (14), we see that

$$\ln\left(\frac{1}{b}\right) = \ln(b^{-1}) = -1(\ln b) = -\ln b.$$

Using (12), we then obtain

$$\ln\left(\frac{a}{b}\right) = \ln\left(a\frac{1}{b}\right) = \ln a + \ln\left(\frac{1}{b}\right) = \ln a - \ln b,$$

which establishes (13).

The properties of ln in Theorem 8.1 can sometimes be used to simplify the differentiation of a function defined in terms of ln.

Example 7 Let's differentiate $\ln(\sqrt{(x^2+1)(2x+3)})$.

SOLUTION Using (12) and (14) we obtain

$$\frac{d[\ln(\sqrt{(x^2+1)(2x+3)})]}{dx} = \frac{d[\frac{1}{2}(\ln(x^2+1) + \ln(2x+3))]}{dx}$$
$$= \frac{1}{2}\left(\frac{1}{x^2+1}\cdot 2x + \frac{1}{2x+3}\cdot 2\right)$$
$$= \frac{x}{x^2+1} + \frac{1}{2x+3}, \quad x > -\frac{3}{2}. \quad \|$$

Since $d(\ln x)/dx = 1/x$ and $(1/x) > 0$ for $x > 0$, we see that $\ln x$ is an increasing function. However, since $\lim_{x\to\infty}(1/x) = 0$, we also see that the graph of $\ln x$ becomes closer and closer to horizontal as x becomes large. We would like to discover whether $\ln x$ becomes large as x becomes large, or whether $\ln x$ remains less than some constant c for all $x > 1$. Similarly, we would like to know the behavior of $\ln x$ near 0. These limits give the answer:

$$\lim_{x\to\infty} \ln x = \infty \quad \text{and} \quad \lim_{x\to 0+} \ln x = -\infty.$$

We obtain these limits easily from (14) as follows. Surely $\ln 2 > 0$, since $\ln 2 = \int_1^2 (1/x)\,dx$. In fact, estimating $\int_1^2 (1/x)\,dx$ by the lower sum s_1, you find that

$$\ln 2 > \tfrac{1}{2}. \tag{15}$$

Therefore by (14),

$$\ln(2^n) = n(\ln 2) > n\left(\frac{1}{2}\right) = \frac{n}{2}.$$

As n approaches ∞, so does $n/2$. Thus for large values 2^n of x, you see that $\ln x$ is also large, so $\lim_{x\to\infty} \ln x = \infty$.

From (13) and (15),

$$\ln\left(\tfrac{1}{2}\right) = \ln 1 - \ln 2 = 0 - \ln 2 < -\tfrac{1}{2}. \tag{16}$$

Using (14), you have

$$\ln\left(\frac{1}{2}\right)^n = n\left(\ln\frac{1}{2}\right) < n\left(-\frac{1}{2}\right) = -\frac{n}{2}.$$

As n gets large, $(\frac{1}{2})^n$ approaches 0 and $-n/2$ approaches $-\infty$. Thus $\lim_{x\to 0+} \ln x = -\infty$. These limits justify our sketch of the graph of $\ln x$ in Fig. 8.2.

SUMMARY

1. $\ln x = \int_1^x \frac{1}{t}\,dt$ *for* $x > 0$.
2. *The graph of* $\ln x$ *is shown in Fig.* 8.2.
3. $\dfrac{d(\ln u)}{dx} = \dfrac{1}{u}\cdot\dfrac{du}{dx}$ 4. $\displaystyle\int \frac{du}{u} = \ln|u| + C$
5. $\ln(ab) = \ln a + \ln b$ *for* $a, b > 0$.
6. $\ln(a/b) = \ln a - \ln b$ *for* $a, b > 0$.
7. $\ln(a^r) = r(\ln a)$ *for* $a > 0$.

EXERCISES

1. Is the function ln continuous? How do you know?

2. Estimate ln 2 by finding the upper sum S_4 for $1/x$ on [1, 2].

3. Find the equation of the line tangent to the graph of $\ln x$ at the point (1, 0).

In Exercises 4 *through* 11, *use properties* (12), (13), *and* (14) *to estimate the quantity, given that* $\ln 2 \approx 0.7$ *and* $\ln 3 \approx 1.1$.

4. $\ln \frac{1}{3}$ **5.** $\ln 6$ **6.** $\ln 8$ **7.** $\ln 12$ **8.** $\ln \frac{3}{4}$ **9.** $\ln 27$ **10.** $\ln\sqrt{3}$ **11.** $\ln\sqrt[3]{4}$

In Exercises 12 *through* 29, *find the derivative of the given function. Use properties of* $\ln x$, *where applicable, to simplify the differentiation.*

12. $\ln(3x + 2)$ **13.** $\ln(\sqrt{x})$ **14.** $\ln(x^3)$ **15.** $\ln(\tan x)$ **16.** $\ln(\cos^2 x)$

17. $\ln\left(\dfrac{2x+3}{x^2+4}\right)$ **18.** $(\ln x)^2$ **19.** $(\ln x)(\sin x)$ **20.** $\ln(\sqrt{3x^3 - 4x})$ **21.** $\dfrac{\ln x}{x^2}$

22. $\ln(\sec x \tan x)$ **23.** $\ln[(x^2 + 4x)^2(3x - 2)^3]$ **24.** $\ln(\cos^2 x \sin^3 2x)$ **25.** $\tan(\ln x)$

26. $\ln(\ln x)$ **27.** $\ln(x\sqrt{2x + 3})$ **28.** $\cos(\ln x)$ **29.** $\ln\left(\dfrac{\cos^2 x}{\sin^3 2x}\right)$

In Exercises 30 through 43, find the given integral without using tables.

30. $\int \frac{1}{x+1}\,dx$ **31.** $\int \frac{1}{2x+3}\,dx$ **32.** $\int \tan 2x\,dx$ **33.** $\int \cot 3x\,dx$ **34.** $\int \frac{x}{x^2+1}\,dx$

35. $\int \frac{1}{(x+1)^2}\,dx$ **36.** $\int \frac{\ln x}{x}\,dx$ **37.** $\int \frac{1}{x(\ln x)}\,dx$ **38.** $\int \frac{1}{x\ln(x^2)}\,dx$ **39.** $\int \frac{\sec^2 x}{\tan x}\,dx$

40. $\int \frac{x}{1-4x^2}\,dx$ **41.** $\int \frac{1}{(3x+1)^2}\,dx$ **42.** $\int \frac{\sin x}{1+\cos x}\,dx$ **43.** $\int \frac{\sin x}{\cos^2 x}\,dx$

In Exercises 44 through 51, find the given integral, using tables if necessary.

44. $\int_0^1 \frac{x}{x+1}\,dx$ **45.** $\int_0^1 \frac{x}{(2x+1)^2}\,dx$ **46.** $\int_1^2 \frac{x+1}{x}\,dx$ **47.** $\int_0^1 \frac{1}{4-x}\,dx$

48. $\int_1^4 \frac{1}{x\sqrt{9+x^2}}\,dx$ **49.** $\int_1^3 \frac{\sqrt{9-x^2}}{x}\,dx$ **50.** $\int_0^{\pi/8} \sec 2x\,dx$ **51.** $\int_{\pi/3}^{\pi/2} \csc\frac{x}{2}\,dx$

52. a) Show that for $t > 1$, we have $\ln t < 2(\sqrt{t} - 1)$. [*Hint.* Compare $\int_1^t (1/x)\,dx$ with $\int_1^t (1/\sqrt{x})\,dx$ for $t > 1$.]

b) Using part (a), find $\lim_{x\to\infty} [(\ln x)/x]$.

c) Using part (b), find $\lim_{x\to 0^+} [x(\ln x)]$. [*Hint.* Let $x = 1/u$ and find the limit as $u \to \infty$.]

calculator exercises

53. Estimate $\ln 5 = \int_1^5 (1/x)\,dx$, using Simpson's rule with $n = 20$. What is the error?

54. Solve $\ln x = x - 2$.

55. Solve $(\ln x)^2 - 4\ln x - 8 = 0$.

8.2 THE FUNCTION e^x

8.2.1 Inverse functions

A curve in the plane is the graph of a function $y = f(x)$ if each vertical line meets the curve in at most one point. (Recall that $y = f(x)$ must *not* assign *two* y-values to *one* x-value.) If each horizontal line meets the curve in at most one point, then the curve defines x as a function of y.

If $y = f(x)$, the graph of f may define x as a function of y also; it will if no horizontal line $y = c$ meets the graph in more than one point. For example, the function $y = f(x) = x^2$ in Fig. 8.3(a) does *not* define x as a function of y; there are actually two natural functions, $x = \sqrt{y}$ and $x = -\sqrt{y}$, given by this graph. On the other hand, $y = x^3$ shown in Fig. 8.3(b) does define x as a function of y, namely, $x = \sqrt[3]{y}$.

If $y = f(x)$ does define x as a function of y also, this function is called the **inverse of** f and is denoted by $x = f^{-1}(y)$. Of course, the domain of f^{-1} is

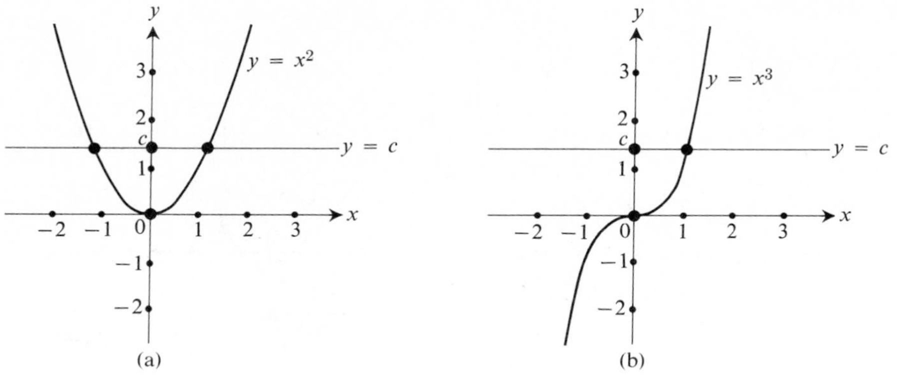

8.3

the range of f, and the range of f^{-1} is the domain of f. Furthermore, since $y = f(x)$ and $x = f^{-1}(y)$, you have upon substituting each relation into the other

$$y = f(f^{-1}(y)) \qquad \text{and} \qquad x = f^{-1}(f(x)).$$

Example 1 The graph of $f(x) = (x - 1)/(x + 2)$ is shown in Fig. 8.4(a), and you see that f has an inverse. Let us find $f^{-1}(x)$, and sketch its graph.

SOLUTION If $y = f(x) = (x - 1)/(x + 2)$, then $x = f^{-1}(y)$. To find $x = f^{-1}(y)$, we simply solve $y = (x - 1)/(x + 2)$ for x. We have

$$\begin{aligned} y &= \frac{x - 1}{x + 2}, \\ (x + 2)y &= x - 1, \\ (y - 1)x &= -2y - 1, \\ x &= -\frac{2y + 1}{y - 1}. \end{aligned}$$

Therefore $f^{-1}(y) = -(2y + 1)/(y - 1)$. We wanted $f^{-1}(x)$, and of course $f^{-1}(x) = -(2x + 1)/(x - 1)$. The graph of $y = f^{-1}(x)$ is shown in Fig. 8.4(b). Note that it can be obtained from the graph in Fig. 8.4(a) by reflecting in the 45° line $y = x$. This reflection interchanges the x-axis and y-axis, which is what happens when you form an inverse function. ||

Suppose $y = f(x)$ is such that the inverse function $x = f^{-1}(y)$ exists, and suppose that $f(x_0) = y_0$. You know that f is differentiable at x_0 if and only if the graph has a *nonvertical* tangent line at (x_0, y_0). By analogy,

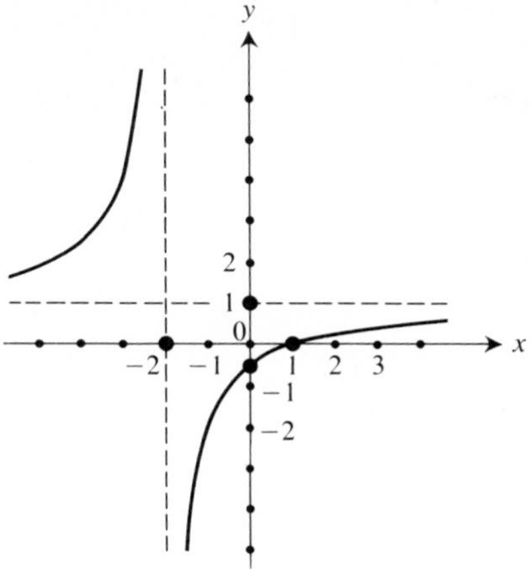

(a) $y = f(x) = \dfrac{x-1}{x+2}$

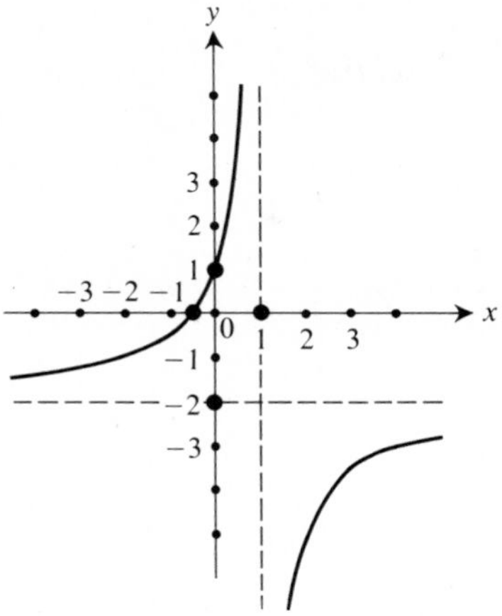

(b) $y = f^{-1}(x) = -\dfrac{2x+1}{x-1}$

8.4

$x = f^{-1}(y)$ is differentiable at y_0 if and only if this same graph has a *nonhorizontal* tangent line at (x_0, y_0). This will be true if $f'(x_0)$ exists (the graph has a tangent line) and is nonzero (the tangent line is not horizontal). This makes the following theorem seem at least plausible.

Theorem 8.2 *Let $y_0 = f(x_0)$ and let $f'(x_0)$ exist and be nonzero. If the inverse function $x = f^{-1}(y)$ exists, then it is differentiable at y_0.*

If $y = f(x)$ is differentiable and defines x as a function f^{-1} of y, then you can find dx/dy by implicit differentiation. From

$$y = f(x)$$

you obtain, differentiating with respect to y,

$$1 = f'(x)\frac{dx}{dy}$$

so if $f'(x) \neq 0$,

$$\frac{dx}{dy} = \frac{1}{f'(x)} \qquad \text{or} \qquad \frac{dx}{dy} = \frac{1}{dy/dx}.$$

Once again, the Leibniz notation for derivatives is very suggestive.

Example 2 Let $f(x) = x^3$. Then f^{-1} exists, and $f^{-1}(y) = y^{1/3}$. Now $f(2) = 8$, so

$$(f^{-1})'(8) = \frac{1}{f'(2)} = \frac{1}{3x^2|_2} = \frac{1}{12}.$$

For an alternate computation, we have $x = y^{1/3}$, so

$$\left.\frac{dx}{dy}\right|_8 = \frac{1}{3}y^{-2/3}\Big|_8 = \frac{1}{3}8^{-2/3} = \frac{1}{3}\cdot\frac{1}{8^{2/3}}$$

$$= \frac{1}{3}\cdot\frac{1}{4} = \frac{1}{12}. \quad \|$$

8.2.2 The exponential function

It is easy to see that if $y = f(x)$ is increasing (or decreasing) for all x in its domain, then the graph does define x as a function of y, for a horizontal line cannot cross an increasing graph twice. The function $\ln x$ is increasing for $x > 0$, since $d(\ln x)/dx = 1/x > 0$. (See Fig. 8.5.) Therefore $y = \ln x$ defines x as a function of y, the inverse function of $\ln x$. We shall denote this function by $\exp(y)$ rather than $\ln^{-1}(y)$ for the moment. It is the *exponential function.*

Definition 8.2 The **(natural) exponential function** is the inverse of the natural logarithm function, $y = \ln x$.

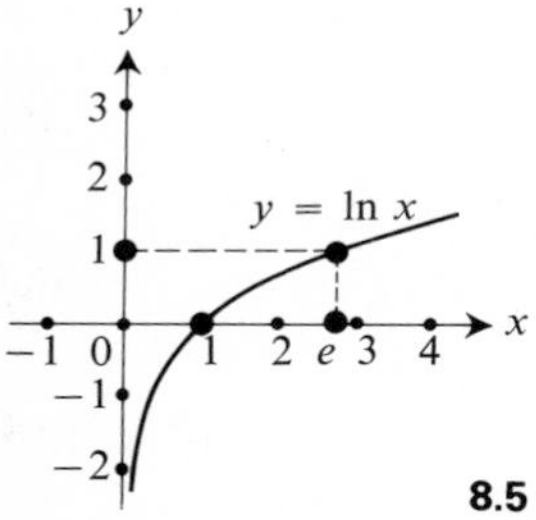

8.5

In the last section, we showed that $\lim_{x\to\infty} \ln x = \infty$ and $\lim_{x\to 0+} \ln x = -\infty$. Since $\ln x$ is continuous, it follows, by the Intermediate-Value Theorem, that its range includes every real number y. Consequently there is a unique number e such that $\ln e = 1$. This number e is one of the most important numbers in mathematics; it has been computed that

$$e \approx 2.71828.$$

The number e, like the number π, is one of those numbers one encounters naturally in mathematics. Both π and e are *irrational numbers*; that is, neither can be expressed as a quotient of integers.

Using one of the properties of $\ln x$, you see that

$$\ln(e^r) = r(\ln e) = r(1) = r \tag{1}$$

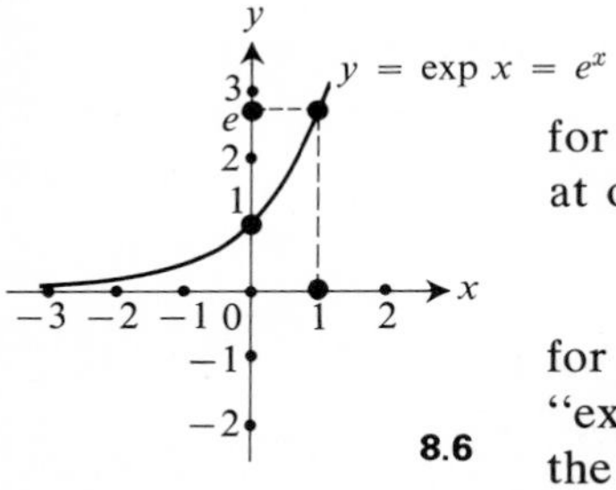

8.6

for any rational number r. Since exp is the inverse of the function ln, you see at once from (1) that

$$\exp r = e^r \tag{2}$$

for any rational number r. Equation (2) explains the use of the name "exponential," for e^r is the number e to the *exponent* r; i.e., it is e raised to the rth power. We have not defined a^x except for x a rational number. Equation (2) suggests that we *define* e^x to be $\exp x$ *for any real number* x; note the labeling of the graph in Fig. 8.6. From now on, we shall use the more intuitive notation e^x to denote the exponential function.

8.2.3 Properties of e^x

As indicated in Fig. 8.6 the exponential function e^x is an increasing function and

$$e^x > 0 \quad \text{for all } x \tag{3}$$

This is a useful fact to remember. Since the equations $y = \ln x$ and $x = e^y$ are inverse relations, you see that $x = e^y = e^{\ln x}$, so

$$e^{\ln x} = x. \tag{4}$$

Similarly,

$$\ln(e^x) = x \tag{5}$$

follows from the inverse relationships $y = e^x$ and $x = \ln y$. You see at once from Eq. (4) that

$\ln x$ is the power to which e must be raised to yield x.

This is sometimes a useful way to think of $\ln x$.

Equation (4) enables us to give a definition of a^b for any $a > 0$ and *any* real number b. Since

$$a = e^{\ln a},$$

we define a^b by

$$a^b = e^{b(\ln a)} \quad \text{for} \quad a > 0. \tag{6}$$

The usual laws of exponents hold for the function e^x, namely,

$$e^a \cdot e^b = e^{a+b}, \tag{7}$$

$$\frac{e^a}{e^b} = e^{a-b}, \tag{8}$$

$$(e^a)^b = e^{ab}. \tag{9}$$

To verify (7) and (8), simply take the natural logarithm of each side of the equation; $\ln x$ is an increasing function, so $\ln x_1 = \ln x_2$ must imply $x_1 = x_2$. Thus

$$\ln(e^a \cdot e^b) = \ln(e^a) + \ln(e^b) = a + b,$$

while

$$\ln(e^{a+b}) = a + b$$

also. Therefore $e^a \cdot e^b = e^{a+b}$. Similarly,

$$\ln\left(\frac{e^a}{e^b}\right) = \ln(e^a) - \ln(e^b) = a - b,$$

and

$$\ln(e^{a-b}) = a - b,$$

which establishes (8). Finally, by our definition in (6),

$$(e^a)^b = e^{b(\ln e^a)} = e^{ba} = e^{ab}.$$

8.2.4 The calculus of e^x

Since $\ln x$ is differentiable and its derivative $1/x$ is never zero, Theorem 8.2 shows that the exponential function is differentiable. If $y = e^x$, then $x = \ln y$. Implicit differentiation of $x = \ln y$ with respect to x yields

$$1 = \frac{1}{y} \cdot \frac{dy}{dx},$$

so

$$\frac{dy}{dx} = y = e^x.$$

We have shown that

$$\frac{d(e^x)}{dx} = e^x.$$

Thus *the exponential function e^x is unchanged by differentiation.* This is one of the reasons the exponential function is so extremely important. If u is a differentiable function of x, then, by the chain rule,

$$\frac{d(e^u)}{dx} = e^u \cdot \frac{du}{dx}. \tag{10}$$

You may wish to memorize formula (10). *We warn you not to confuse* (10), *which gives the "derivative of a constant to a function power," with the formula $d(u^n)/dx = nu^{n-1} \cdot (du/dx)$ for the "derivative of a function to a constant power."*

Example 3 We have

$$\frac{d(e^{\sin x})}{dx} = e^{\sin x} \cdot \frac{d(\sin x)}{dx} = e^{\sin x} \cdot \cos x. \quad \|$$

From $d(e^x)/dx = e^x$, you obtain

$$\int e^x\, dx = e^x + C, \tag{11}$$

and from (10),

$$\int e^u\, du = e^u + C \tag{12}$$

for a differentiable function u. You should learn formula (12) also.

Example 4 Let's find $\int xe^{x^2}\, dx$.

SOLUTION Taking $u = x^2$ so that $du = 2x\, dx$ and fixing up the integral, you find that

$$\int xe^{x^2}\, dx = \frac{1}{2}\int e^{x^2} \cdot 2x\, dx = \frac{1}{2}e^{x^2} + C. \quad \|$$

Finally, we mention that our old formula

$$\frac{d(u^r)}{dx} = r \cdot u^{r-1} \cdot \frac{du}{dx}$$

holds for *every* real number r. We have

$$\frac{d(u^r)}{dx} = \frac{d(e^{r(\ln u)})}{dx} = e^{r(\ln u)} \cdot \frac{d(r(\ln u))}{dx} = e^{r(\ln u)} \cdot r \cdot \frac{1}{u} \cdot \frac{du}{dx}$$

$$= ru^r \cdot \frac{1}{u} \cdot \frac{du}{dx} = ru^{r-1} \cdot \frac{du}{dx}.$$

SUMMARY

1. *If* $y = f(x)$ *also defines* x *as a function of* y, *then* x *is the inverse function* f^{-1} *of* y, *so that* $x = f^{-1}(y)$.
2. *If* $y = f(x)$, *then*

$$\frac{dx}{dy} = \frac{1}{f'(x)} = \frac{1}{dy/dx} \quad \text{if} \quad f'(x) \neq 0.$$

3. e *is the unique number such that* $\ln e = 1$.
4. $y = e^x$ *is the inverse relation to* $x = \ln y$.
5. $e^{\ln x} = x$
6. $\ln e^x = x$
7. $e^a e^b = e^{a+b}$
8. $e^a/e^b = e^{a-b}$
9. $(e^a)^b = e^{ab}$
10. $\dfrac{d(e^u)}{dx} = e^u \cdot \dfrac{du}{dx}$
11. $\int e^u \, du = e^u + C$

EXERCISES

In Exercises 1 through 12, use properties of logarithmic and exponential functions to simplify the given expression.

1. $\ln(e^2)$ **2.** $e^{\ln 3}$ **3.** $\ln(e^{\ln 1})$ **4.** $\ln(e^{1/2})$

5. $\ln\left(\frac{1}{e^2}\right)$ **6.** $e^{(1+\ln 2)}$ **7.** $e^{(\ln 2+\ln 3)}$ **8.** $e^{\ln 3-\ln 4}$

9. $e^{4(\ln 2)}$ **10.** $\ln\left(\frac{\sqrt{e}}{e^2}\right)$ **11.** $(e^2)^{\ln 3}$ **12.** $\ln(\ln e)$

In Exercises 13 through 22, find the derivative of the given function.

13. e^{2x} **14.** xe^x **15.** $e^{2x}\sin x$ **16.** $e^x + e^{-x}$

17. $e^x(\ln 2x)$ **18.** $\tan(e^x)$ **19.** $e^{\sec x}$ **20.** $e^{(x^2+x)}$

21. $e^{1/x}$ **22.** $x^2e^{-x^2}$

In Exercises 23 through 28, find the given integral without the use of tables.

23. $\int_0^{\ln 3} e^{2x}\,dx$ **24.** $\int x^2e^{x^3}\,dx$ **25.** $\int (\sin x)\,e^{\cos x}\,dx$ **26.** $\int_0^{\ln 2} (e^x - e^{-x})\,dx$

27. $\int \frac{e^x}{1+e^x}\,dx$

28. $\int \frac{e^{\sqrt{x}}}{\sqrt{x}}\,dx$

29. Sketch the graph of e^{-x}.

30. Find all values of m such that $y = e^{mx}$ is a solution of the differential equation

$$\frac{d^2y}{dx^2} - 5\frac{dy}{dx} + 6y = 0.$$

calculator exercises

31. Solve $e^{2x} + 7e^x - 3 = 0$. [*Hint.* Substitute $w = e^x$.]

32. Solve $e^{3x} - 9e^{2x} + 10 = 0$.

33. Solve $e^{(ex)} + 3e^x - 10 = 0$.

8.3. OTHER BASES AND LOGARITHMIC DIFFERENTIATION

8.3.1 Other bases

The function e^x is sometimes called the *exponential function with base e.* Using the functions e^x and $\ln x$, we can easily define an exponential function a^x with any base $a > 0$. Since $a = e^{\ln a}$ and $(e^{\ln a})^x = e^{(\ln a)x}$, it is natural to define

$$a^x = e^{(\ln a)x}$$

for any real number x. Now $y = e^x$ is an increasing function with range all $y > 0$. It follows that if $a \neq 1$, then $y = e^{(\ln a)x}$ is an increasing function for $a > 1$ and a decreasing function if $0 < a < 1$, again with range all $y > 0$. Consequently the inverse of the function $a^x = e^{(\ln a)x}$ exists. We summarize these observations.

Definition 8.3 The **exponential function with base** $a > 0$ is defined by

$$a^x = e^{(\ln a)x}. \tag{1}$$

For $a \neq 1$, the inverse of a^x is the **logarithmic function with base** a and is denoted by $\log_a$.

Thus $\ln = \log_e$ is the logarithmic function with base e. For the functions a^x and $\log_a$, we have the relations

$$a^{\log_a x} = x \tag{2}$$

and

$$\log_a (a^x) = x. \tag{3}$$

In high school, you worked with the function $\log_{10} x$. With $a = 10$, Eq. (2) becomes

$$10^{\log_{10} x} = x,$$

so that $\log_{10} x$ is the power to which 10 must be raised to yield x.

For any a_1, and b_1, the three properties

$$a^{a_1} \cdot a^{b_1} = a^{a_1+b_1}, \tag{4}$$

$$\frac{a^{a_1}}{a^{b_1}} = a^{a_1-b_1}, \tag{5}$$

$$(a^{a_1})^{b_1} = a^{a_1 b_1} \tag{6}$$

of the function a^x follow at once from the definition of a^x in (1) and the properties of e^x shown in the last section. From these properties of a^x, one easily obtains, for $a_1 > 0$ and $b_1 > 0$,

$$\log_a (a_1 b_1) = \log_a a_1 + \log_a b_1, \tag{7}$$

$$\log_a \left(\frac{a_1}{b_1}\right) = \log_a a_1 - \log_a b_1, \tag{8}$$

$$\log_a (a_1{}^{b_1}) = b_1(\log_a a_1). \tag{9}$$

If $f(x)$ and $g(x)$ are functions, it is natural to define the function $f(x)^{g(x)}$ for all x such that $f(x) > 0$ by

$$f(x)^{g(x)} = e^{(\ln f(x))\cdot g(x)}.$$

The derivatives of the functions introduced in this article are easily found since they are defined in terms of the exponential function. Indeed, unless you have a good memory, we suggest that you simply express all such functions in terms of the base e when differentiating or integrating, rather than memorize the formulas found below. That is, repeat the derivation of the formula each time. To illustrate,

$$\frac{d(a^x)}{dx} = \frac{d(e^{(\ln a)x})}{dx} = e^{(\ln a)x} \cdot \frac{d((\ln a)x)}{dx}$$
$$= (a^x)(\ln a) = (\ln a)(a^x).$$

This gives the formula

$$\frac{d(a^x)}{dx} = (\ln a)(a^x). \tag{10}$$

Example 1 We have

$$\frac{d(2^x)}{dx} = \frac{d(e^{(\ln 2)x})}{dx} = (\ln 2)e^{(\ln 2)x} = (\ln 2)2^x. \quad \|$$

Example 2 We have

$$\frac{d(x^x)}{dx} = \frac{d(e^{(\ln x)x})}{dx} = (e^{(\ln x)x})\left(\ln x + x \cdot \frac{1}{x}\right)$$
$$= (x^x)(\ln x + 1). \quad \|$$

Example 3 Changing to base e and fixing up the integral,

$$\int a^x\,dx = \int e^{(\ln a)x}\,dx = \frac{1}{\ln a}\int (e^{(\ln a)x} \cdot \ln a)\,dx$$
$$= \frac{1}{\ln a}e^{(\ln a)x} + C = \frac{1}{\ln a}a^x + C.$$

Using the chain rule, you obtain the formula

$$\int a^u\,du = \frac{1}{\ln a}a^u + C. \quad \| \tag{11}$$

The derivative of the function $\log_a$ can be found by implicit differentiation, since (10) gives the derivative of the inverse function a^x. We leave as an exercise the demonstration that

$$\frac{d(\log_a x)}{dx} = \frac{1}{\ln a} \cdot \frac{1}{x} \tag{12}$$

(see Exercise 27).

Probably you will have little occasion to use calculus with exponential or logarithmic functions to bases other than e. With base $a = e$, the annoying constant $\ln a$ becomes $\ln e = 1$, and the calculus is easy to handle.

8.3.2 Logarithmic differentiation

Rather than use the technique of the last article to differentiate a^x, x^x, etc., you might like *logarithmic differentiation.* To differentiate $y = x^x$ by this method, one takes the logarithm of both sides of the equation and then differentiates implicitly:

$$y = x^x,$$
$$\ln y = \ln(x^x),$$
$$\ln y = x(\ln x),$$
$$\frac{1}{y} \cdot \frac{dy}{dx} = x\left(\frac{1}{x}\right) + \ln x = 1 + \ln x,$$
$$\frac{dy}{dx} = y(1 + \ln x) = x^x(1 + \ln x).$$

We have solved our problem very easily. Logarithmic differentiation is handy for such exponentials and also for products and quotients. Here is another illustration.

Example 4 We find dy/dx if $y = 2^x \cdot (\sin x)^{\cos x}$.

SOLUTION The technique of logarithmic differentiation yields

$$\ln y = \ln(2^x) + \ln[(\sin x)^{\cos x}],$$

$$\ln y = x(\ln 2) + (\cos x) \cdot \ln(\sin x),$$

$$\frac{1}{y} \cdot \frac{dy}{dx} = (\ln 2) + (\cos x) \cdot \frac{1}{\sin x} \cdot (\cos x) + (\ln(\sin x))(-\sin x),$$

$$\frac{dy}{dx} = y\left[(\ln 2) + \frac{\cos^2 x}{\sin x} - (\sin x)(\ln(\sin x))\right],$$

$$\frac{dy}{dx} = 2^x(\sin x)^{\cos x}[(\ln 2) + \cos x \cot x - (\sin x)(\ln(\sin x))].$$

The answer isn't very pretty, but it was not hard to find. ||

The summary gives some formulas we proved, and some we didn't.

SUMMARY

1. $a^x = e^{(\ln a)x}$
2. $a^{\log_a x} = x$
3. $\log_a(a^x) = x$
4. $\log_a x = \dfrac{1}{\ln a}(\ln x)$
5. $\dfrac{d(a^u)}{dx} = (\ln a)a^u \cdot \dfrac{du}{dx}$
6. $\displaystyle\int a^u\,du = \frac{1}{\ln a}a^u + C$
7. $\dfrac{d(\log_a u)}{dx} = \dfrac{1}{\ln a} \cdot \dfrac{1}{u} \cdot \dfrac{du}{dx}$

EXERCISES

In Exercises 1 through 6, use properties of logarithmic and exponential functions to simplify the given expression.

1. $2^{3(\log_2 4)}$ **2.** $\log_{10}(2^{\log_2 10})$ **3.** $\log_3(2^{5(\log_2 3)})$ **4.** $2^{(\ln 3)(\log_2 e)}$ **5.** $4^{\log_2 3}$ **6.** $2^{\log_4 2}$

In Exercises 7 through 16, find the derivative of the given function.

7. 2^{3x} **8.** 10^{x^2+2x} **9.** $x(3^{\sin x})$ **10.** x^{2x} **11.** $(\sin x)^x$

12. $x^{\tan x}$ **13.** $\cos(x^x)$ **14.** $(x+1)^{x^2}$ **15.** $\log_{10} 2x$ **16.** $\log_2(x^2+1)$

In Exercises 17 through 21, use logarithmic differentiation to find the derivative of the given function.

17. $2^x \cdot 3^{2x}$ **18.** $(x^2+1)\sqrt{2x+3}(x^3-2x)$ **19.** $5^{x^2}/7^x$

20. $x^{\sin x}/(\cos x)^x$ **21.** $7^x \cdot 8^{-x^2} \cdot 100^x$

In Exercises 22 through 25, find the given integral without the use of tables.

22. $\displaystyle\int \frac{e^{\sqrt{x}}}{\sqrt{x}}\,dx$ **23.** $\displaystyle\int 3^{-x}\,dx$ **24.** $\displaystyle\int x(5^{3x^2})\,dx$ **25.** $\displaystyle\int 2^{\sin x}\cos x\,dx$

26. Discuss the effect of the size of a on the graph of a^x. Distinguish the cases $0 < a < 1$, $a = 1$, and $a > 1$.

27. Show that $d(\log_a x)/dx = 1/[(\ln a)x]$.

calculator exercises

28. Solve $x^{2x} - 5x^x - 6 = 0$.

29. Solve $x^x = \cos x$.

30. Estimate $\int_1^3 x^x\,dx$ using Simpson's Rule with $n = 20$.

8.4 APPLICATIONS TO GROWTH AND DECAY

8.4.1 Growth and the equation $dy/dt = ky$

There are many physical situations in which it is desired to find a function $y = f(t)$ such that

$$\frac{dy}{dt} = ky \qquad \text{or} \qquad f' = k \cdot f. \tag{1}$$

Here are descriptions of three situations where differential equations like (1) occur.

Example 1 It is known that the rate of decay of any particular radioactive element is proportional to the amount of the element present. This means that if $Q = f(t)$ gives the amount of the element present at time t, then

$$\frac{dQ}{dt} = -cQ, \tag{2}$$

where c is a positive constant of proportionality. (The minus sign occurs since the quantity Q present is *decreasing* as time increases.) Since (2) can be written in the form $f' = -c \cdot f$, we see that (2) is an equation of the form (1) where $k = -c$ is a negative constant. ‖

Example 2 Suppose a body traveling through a medium (like air or water) is subject only to a force of retardation proportional to the velocity of the body through the medium. Let $v = f(t)$ be the velocity at time t. By Newton's second law of motion, the force F is ma where m is the mass of the body and $a = dv/dt$ is the acceleration. Thus $m(dv/dt) = -cv$, where c is some positive constant of proportionality, so

$$\frac{dv}{dt} = -\frac{c}{m}v. \tag{3}$$

Since $v = f(t)$, we may write (3) in the form $f' = (-c/m) \cdot f$, which is again an equation of the form (1) with $k = -c/m$. ‖

Example 3 Under "ideal" conditions with no overcrowding, predators, or disease, the rate of growth of a population (whether it be people or bacteria) is proportional to the size of the population. This means that if $Q = f(t)$ is the size of the population at time t, then

$$\frac{dQ}{dt} = cQ, \tag{4}$$

for some constant c. Equation (4) can be written as $f' = c \cdot f$ which is again of the form (1). ‖

8.4.2 Solution of the equation $dy/dt = ky$

You can easily check that, for any constant A, a solution of the equation $dy/dt = ky$ is given by

$$y = f(t) = Ae^{kt}. \tag{5}$$

Namely, from (5)

$$\frac{dy}{dt} = A(e^{kt}) \cdot k = k(Ae^{kt}) = ky.$$

We now show that the functions Ae^{kt} are the only solutions of the differential equation (1). Starting with $dy/dt = ky$, we separate the variables and solve the equation thus:

$$\frac{dy}{dt} = ky, \qquad \frac{dy}{y} = k \cdot dt, \qquad \int \frac{dy}{y} = \int k \cdot dt,$$

$$\ln|y| = kt + C, \qquad |y| = e^{kt+C} = e^{kt} \cdot e^{C},$$

$$y = \pm e^{C} \cdot e^{kt}.$$

Now $\pm e^{C}$ may be any constant except 0. A check shows that $y = 0$ is a solution of $dy/dt = ky$. We lost this solution when we divided by y to form the second line of the solution. Therefore if we let A be any constant, our solution can be written as

$$y = Ae^{kt},$$

which is what we wished to show.

You know now that each of the functions $f(t)$ for the physical situations described in Examples 1 through 3 must be one of the functions Ae^{kt} for some constants A and k. Note that for $f(t) = Ae^{kt}$, we have $f(0) = A$. Thus A has the physical interpretation of the *initial quantity* at time $t = 0$. The two constants A and k can be determined if the value of $f(t)$ is known at two different times t_1 and t_2. The sign of k determines whether $f(t)$ is increasing (if $k > 0$) or decreasing (if $k < 0$) as the time t increases; this is indicated by Figs. 8.7 and 8.8.

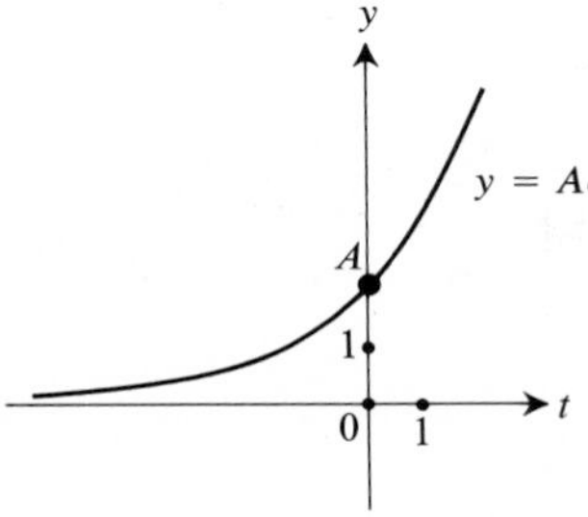

8.7

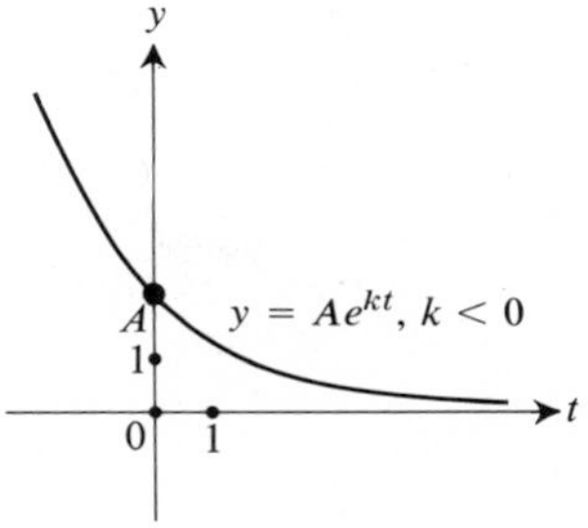

8.8

8.4.3 Applications

We illustrate the preceding discussion with some specific applications.

Example 4 Let's show that the time t_h required for half of an initial amount of a radioactive element to decay is independent of the initial amount of the element present. (This time t_h is the *half-life of the element.*)

SOLUTION If an amount Q of the element is present at time t and if Q_0 is the initial amount present at time $t = 0$, then you know by Example 1 that $dQ/dt = -cQ$ for some positive constant c. Section 8.4.2 then shows that

$$Q = Q_0 e^{-ct}.$$

Since $Q = (\frac{1}{2})Q_0$ when $t = t_h$, we have

$$\tfrac{1}{2}Q_0 = Q_0 e^{-ct_h},$$

so

$$e^{ct_h} = 2.$$

Taking the (natural) logarithm of each side of this equation, we obtain

$$ct_h = \ln 2,$$

so

$$t_h = \frac{1}{c}(\ln 2).$$

Thus t_n depends only upon the constant c, and is independent of the initial amount Q_0 of the element present. ||

Example 5 Suppose a body traveling through a medium is subject only to a force of retardation proportional to the velocity. If the body is traveling at 100 ft/sec at time $t = 0$ and at 20 ft/sec at time $t = 3$ sec, let's find the velocity of the body at time 9 sec, and let's also find the total distance the body travels through the medium.

SOLUTION From Example 2, we know that $dv/dt = -k \cdot v$ for some positive constant k, so by Section 8.4.2,

$$v = Ae^{-kt}$$

for some constant A. From the data, the initial velocity A is 100 ft/sec, so

$$v = 100e^{-kt}.$$

Also, when $t = 3$, we have $v = 20$, so

$$20 = 100e^{-3k}. \tag{6}$$

We can use (6) to find k. We find that $e^{3k} = 5$, so taking logarithms,

$$3k = \ln 5 \qquad \text{and} \qquad k = \tfrac{1}{3}(\ln 5).$$

Thus

$$v = 100e^{-(\ln 5)t/3} = 100 \cdot 5^{-t/3}.$$

When $t = 9$, we obtain

$$v\big|_{t=9} = 100 \cdot 5^{-9/3} = \tfrac{100}{125} = \tfrac{4}{5} \text{ ft/sec.}$$

Now if s is the distance traveled by the body from time $t = 0$, then

$$\frac{ds}{dt} = v = 100e^{-(\ln 5)t/3}.$$

Integrating, we find that

$$s = -\frac{300}{\ln 5} e^{-(\ln 5)t/3} + C$$

for some constant C. Now $s = 0$ when $t = 0$, so we have

$$0 = -\frac{300}{\ln 5} \cdot 1 + C \qquad \text{and} \qquad C = \frac{300}{\ln 5}.$$

Hence, the distance s as a function of t is given by

$$s = -\frac{300}{\ln 5} \cdot 5^{-t/3} + \frac{300}{\ln 5}.$$

Now as t approaches ∞, $5^{-t/3}$ approaches 0 and s approaches $300/(\ln 5)$. Thus, the body travels a total distance of $300/(\ln 5)$ ft through the medium. ||

Example 6 Savings with interest compounded continuously increase at a rate proportional to the amount in the savings account. If $Q(t)$ is the amount in the account after t years, and if interest is compounded continuously at c percent, then

$$\frac{dQ}{dt} = \frac{c}{100} Q.$$

Let's see how long it takes savings to triple at 5 percent, compounded continuously.

SOLUTION The differential equation becomes

$$\frac{dQ}{dt} = \frac{5}{100} Q = \frac{1}{20} Q.$$

Therefore

$$Q = Q_0 e^{t/20}.$$

When $Q = 3Q_0$ so the original amount Q_0 has been tripled, then

$$3Q_0 = Q_0 e^{t/20},$$
$$3 = e^{t/20},$$
$$\ln 3 = \frac{t}{20},$$
$$t = 20(\ln 3) \approx 21.97.$$

Thus money triples in just under 22 years if compounded continuously at 5 percent. ‖

SUMMARY

1. *The general solution of the differential equation $dy/dt = ky$ is $y = Ae^{kt}$. The arbitrary constant A is the initial value of y, when $t = 0$.*
2. *There are several important situations in which the rate of growth (or decay) of some quantity is proportional to the amount $Q(t)$ present at time t. Examples are decay of a radioactive substance, growth of a population, and savings with interest compounded continuously. In all these cases, the differential equation $dQ/dt = kQ$ governs the growth (or decay).*
3. *If a savings account contains $Q(t)$ dollars after t years and interest is compounded continuously at c percent, then*

$$\frac{dQ}{dt} = \frac{c}{100} Q.$$

EXERCISES

1. If the half-life of a radioactive element is 1600 yr, how long does it take for $\frac{1}{3}$ of the original amount to decay?

2. A body traveling in a medium is retarded by a force proportional to its velocity. If the velocity of the body after 4 sec is 80 ft/sec and the velocity after 6 sec is 60 ft/sec, find the initial velocity of the body.

3. In Exercise 2, find the total distance the body travels through the medium.

4. A certain culture of bacteria grows at a rate proportional to the size of the culture. If the culture triples in size in the first two days, find the factor by which the culture increases in size in the first 10 days.

5. A tank initially contains a solution of 100 gallons of brine with a salt concentration of 2 lb/gal. Fresh water is added at a rate of 4 gal/min, and the brine is drawn off at the bottom of the tank at the same rate. Assume that the concentration of salt in the solution is kept homogeneous by stirring.

a) If $f(t)$ is the number of pounds of salt in solution in the tank at time t, show that f satisfies a differential equation of the form (1).

b) Find the amount of salt in solution in the tank after 25 min.

6. Newton's law of cooling states that a body placed in a colder medium will cool at a rate proportional to the difference between its temperature and the temperature of the surrounding medium.

a) Assume the temperature of the surrounding medium remains constant and let $f(t)$ be the difference between the temperature of the body and the temperature of the surrounding medium at time t. Show that f satisfies a differential equation of the form (1).

b) A body having a temperature of 90° is placed in a medium whose temperature is kept constant at 40°. If the temperature of the body after 15 min is 70°, find its temperature after 40 min.

7. Describe all solutions of the differential equation $d^2y/dt^2 = k(dy/dt)$, where k is a constant. [*Hint.* Let $u = dy/dt$.]

8. Find how long it takes savings to double at $5\frac{1}{2}$ percent interest, compounded continuously.

9. Mr. and Mrs. Brice put $10,000 into a savings account in 1970. They plan to leave it there until 1990, when they will withdraw it and all the accrued interest to make a down payment on a new house. During that twenty-year period, their savings will earn 5 percent interest, compounded continuously. How much will the bank pay the Brices in 1990?

10. When you worked Exercise 9, you discovered that the Brices will receive $27,182.82 in 1990. During the time the bank has the Brices' money, the bank loans it continually for home mortgages, and receives what amounts to 9 percent interest, compounded continuously. If it costs the bank an average of $200 per year to handle its investments arising from the Brices' $10,000 deposit, how much profit will the bank have made after paying the Brices in 1990?

11. Answer Exercise 10 if the bank instead continually uses the money generated by the Brices' $10,000 for car loans at 12% interest, compounded continuously. Assume the same expenses as in Exercise 10.

calculator exercises

12. According to the 1971 *World Book Encyclopedia*, the world population in 1971 was about 3,692,000,000 and was increasing at an annual rate of about 1.9 percent. Suppose the rate of increase of population is proportional to the population, and continues to increase at the rate of 1.9 percent annually. If the surface of the earth is a sphere of radius 4000 mi, estimate the year when there will be one person for every square yard of surface area of the earth, including oceans as well as continents.

13. Referring to Exercise 12, suppose it were possible to have people live in space near the earth, say as far out as the moon. Estimate the year in which there would be one person for every three cubic yards of such space, assuming the moon is 250,000 miles from the earth.

8.5 THE INVERSE TRIGONOMETRIC FUNCTIONS

The graphs of the six trigonometric functions are shown in Fig. 8.9. None of these functions has an inverse since, for each graph, a horizontal line $y = c$ may cross it at more than one point. For example, if $-1 \le c \le 1$, there is not a *unique* x such that $\sin x = c$.

Consider the six functions $f_1, \ldots, f_6$, which have as graphs the heavily marked portions of the graphs of the six trigonometric functions in Fig. 8.9. Each of these new functions has the same range as the corresponding trigonometric function, and each new function has an inverse. By abuse

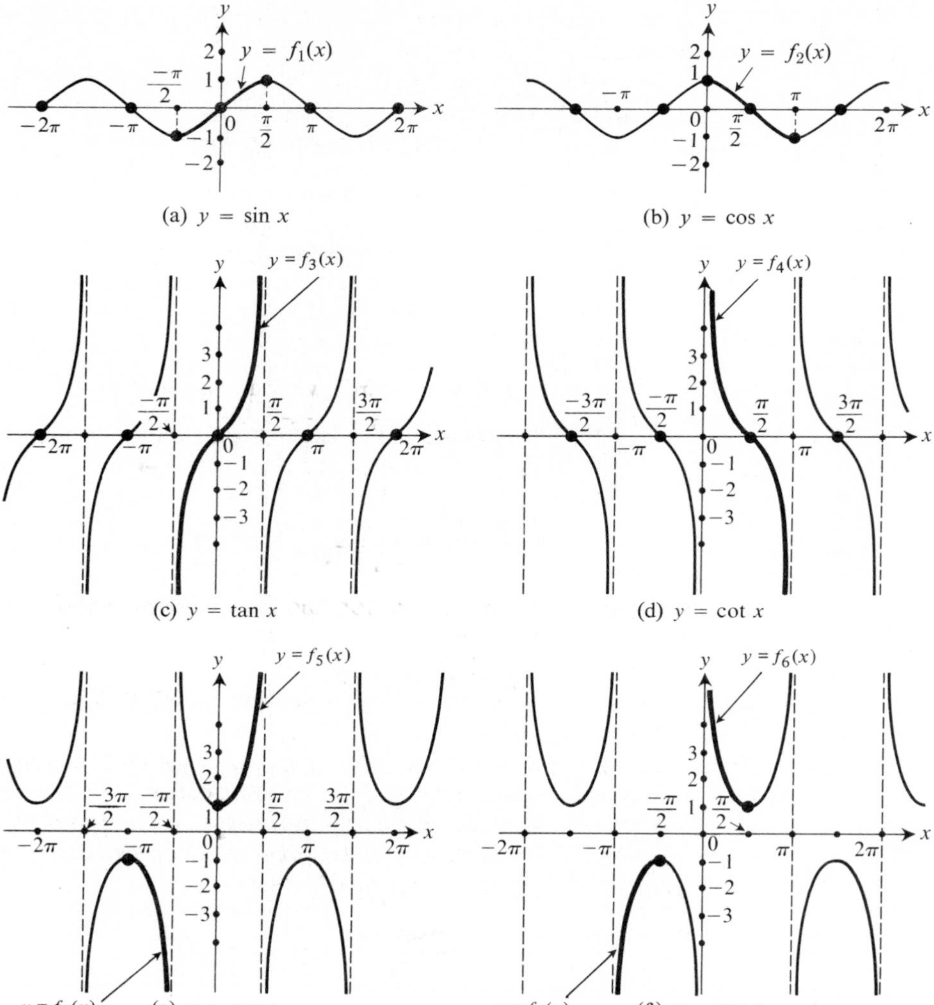

(a) $y = \sin x$ (b) $y = \cos x$
(c) $y = \tan x$ (d) $y = \cot x$
(e) $y = \sec x$ (f) $y = \csc x$

8.9

of terminology, the inverses of $f_1, \ldots, f_6$ are called the **inverse trigonometric functions**, so that f_1^{-1} is the *inverse sine*, f_2^{-1} is the *inverse cosine*, etc. The *inverse sine* is usually denoted by $\sin^{-1}y$, or by arcsin y. Similar notations are used for the other five inverse trigonometric functions. We shall use the first notation given, and *you must remember that the "−1" in* $\sin^{-1}y$ *is* not *an exponent.*

Example 1 We have

$$\sin^{-1}\frac{1}{2} = \frac{\pi}{6}, \qquad \cos^{-1}\frac{-1}{\sqrt{2}} = \frac{3\pi}{4}, \qquad \text{and} \qquad \sec^{-1}2 = \frac{\pi}{3}. \quad \|$$

Values assumed by an inverse trigonometric function are known as the *principal values* for that inverse function.

Principal values

The derivative of f_1 in Fig. 8.9 is the same as the derivative of $\sin x$ at all points in the domain. Hence we can find the derivative dx/dy of $x = \sin^{-1} y$ by differentiating $y = \sin x$ implicitly with respect to y:

$$y = \sin x,$$

$$1 = (\cos x)\frac{dx}{dy},$$

$$\frac{dx}{dy} = \frac{1}{\cos x},$$

if $\cos x \neq 0$. Since $-\pi/2 \leq x \leq \pi/2$, we have $\cos x \geq 0$, so

$$\cos x = \sqrt{1 - \sin^2 x} = \sqrt{1 - y^2}.$$

Thus we obtain

$$\frac{d(\sin^{-1} y)}{dy} = \frac{1}{\sqrt{1 - y^2}} \quad \text{for} \quad -1 < y < 1. \tag{1}$$

It is customary to use x for the independent variable in studying functions, and (1) then yields

$$\frac{d(\sin^{-1} x)}{dx} = \frac{1}{\sqrt{1 - x^2}} \quad \text{for} \quad -1 < x < 1. \tag{2}$$

From now on, we shall not explicitly give the domain of the derivative of an inverse trigonometric function, like the designation "$-1 < x < 1$" in Eq. (2). By Theorem 8.2, the derivative will exist at all points of the domain where the denominator that appears does not become zero.

From Eq. (2), you obtain, by the chain rule,

$$\frac{d(\sin^{-1} u)}{dx} = \frac{1}{\sqrt{1 - u^2}} \cdot \frac{du}{dx} \tag{3}$$

for a differentiable function u.

Example 2 If $y = \sin^{-1} 2x$, then

$$\frac{dy}{dx} = \frac{1}{\sqrt{1 - (2x)^2}} \cdot 2 = \frac{2}{\sqrt{1 - 4x^2}}. \quad \|$$

The graphs of the inverse trigonometric functions as functions of the independent variable x are shown in Fig. 8.10. It is impossible to choose "branches" of the graphs of secant and cosecant so that the inverse functions become continuous. The branches for $\sec^{-1} x$ and $\csc^{-1} x$ are chosen to make the formulas for the derivatives of these functions come out nicely, without ambiguity as to sign (see Exercise 42). Derivatives of all the inverse trigonometric functions can easily be found just as we found the derivative of $\sin^{-1}x$ above (see Exercises 37 through 41). The table in Fig. 8.11 summarizes the data you should remember regarding the inverse trigonometric functions.

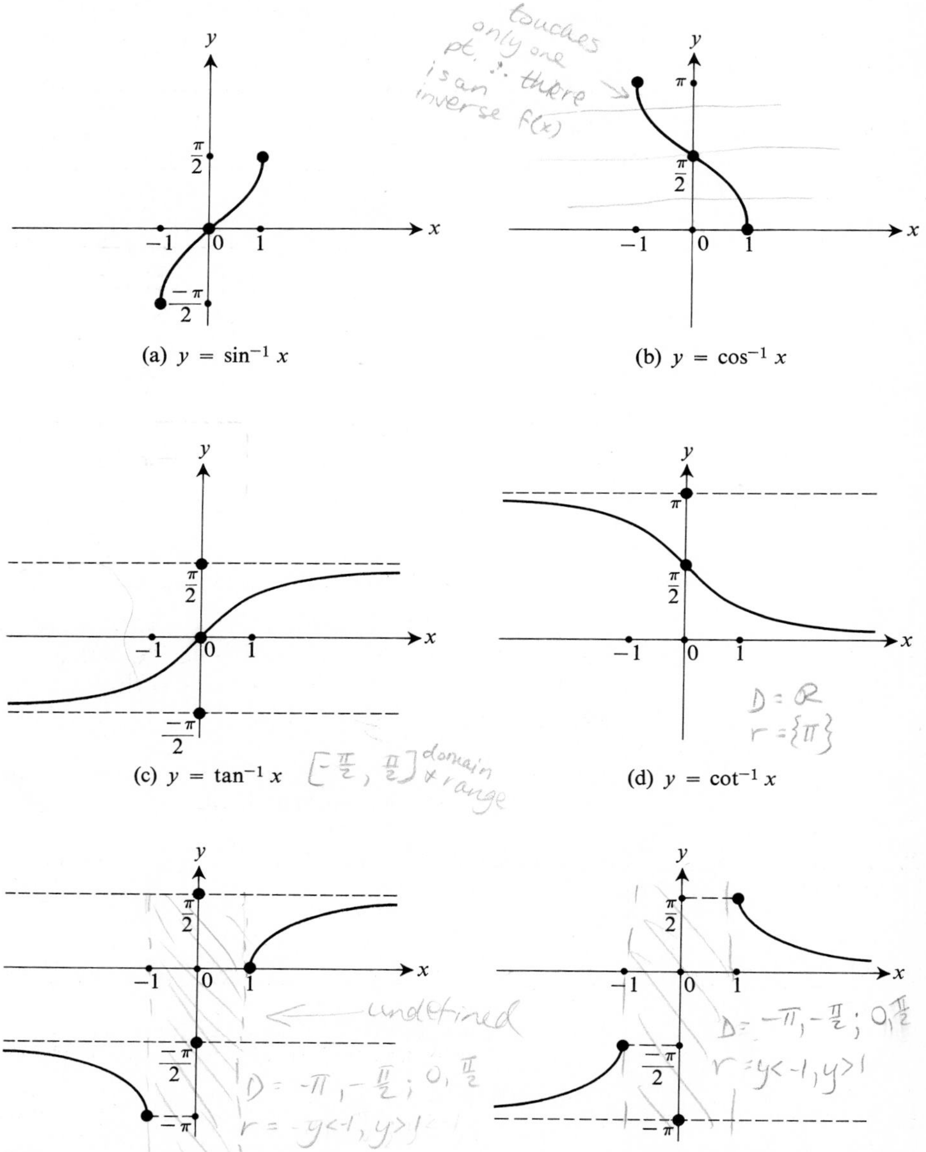

(a) $y = \sin^{-1} x$

touches only one pt. ∴ there is an inverse F(x)

(b) $y = \cos^{-1} x$

(c) $y = \tan^{-1} x$ $\left[-\frac{\pi}{2}, \frac{\pi}{2}\right]$ *domain & range*

(d) $y = \cot^{-1} x$

$D = \mathbb{R}$
$r = \{\pi\}$

8.10   (e) $y = \sec^{-1} x$

← undefined

$D = -\pi, -\frac{\pi}{2}; 0, \frac{\pi}{2}$
$r = -y<-1, y>1$

(f) $y = \csc^{-1} x$

$D = -\pi, -\frac{\pi}{2}; 0, \frac{\pi}{2}$
$r = y<-1, y>1$

take principal values

Function	Domain	Range	Derivative
$\sin^{-1}x$	$[-1, 1]$	$\left[-\frac{\pi}{2}, \frac{\pi}{2}\right]$	$\frac{1}{\sqrt{1-x^2}}$
$\cos^{-1}x$	$[-1, 1]$	$[0, \pi]$	$\frac{-1}{\sqrt{1-x^2}}$
$\tan^{-1}x$	All x	$-\frac{\pi}{2} < y < \frac{\pi}{2}$	$\frac{1}{1+x^2}$
$\cot^{-1}x$	All x	$0 < y < \pi$	$\frac{-1}{1+x^2}$
$\sec^{-1}x$	$x \le -1$ or $x \ge 1$	$-\pi \le y < -\frac{\pi}{2}$ or $0 \le y < \frac{\pi}{2}$	$\frac{1}{x\sqrt{x^2-1}}$
$\csc^{-1}x$	$x \le -1$ or $x \ge 1$	$-\pi < y \le -\frac{\pi}{2}$ or $0 < y \le \frac{\pi}{2}$	$\frac{-1}{x\sqrt{x^2-1}}$

8.11

We now list formulas obtained from those in Fig. 8.11 using the chain rule. These formulas are primarily important for evaluation of certain definite integrals. In fact, this is the main reason for studying the calculus of inverse trigonometric functions.

Differentiation formulas

$$\frac{d(\sin^{-1}u)}{dx} = \frac{1}{\sqrt{1-u^2}} \cdot \frac{du}{dx} \quad (4) \qquad \frac{d(\cos^{-1}u)}{dx} = \frac{-1}{\sqrt{1-u^2}} \cdot \frac{du}{dx} \quad (5) \qquad \frac{d(\tan^{-1}u)}{dx} = \frac{1}{1+u^2} \cdot \frac{du}{dx} \quad (6)$$

$$\frac{d(\cot^{-1}u)}{dx} = \frac{-1}{1+u^2} \cdot \frac{du}{dx} \quad (7) \qquad \frac{d(\sec^{-1}u)}{dx} = \frac{1}{u\sqrt{u^2-1}} \cdot \frac{du}{dx} \quad (8) \qquad \frac{d(\csc^{-1}u)}{dx} = \frac{-1}{u\sqrt{u^2-1}} \cdot \frac{du}{dx} \quad (9)$$

Integration formulas

$$\int \frac{du}{\sqrt{1-u^2}} = \sin^{-1}u + C \quad (10) \qquad \int \frac{du}{1+u^2} = \tan^{-1}u + C \quad (11) \qquad \int \frac{du}{u\sqrt{u^2-1}} = \sec^{-1}u + C \quad (12)$$

Example 3 We have

$$\int_0^{1/2} \frac{1}{\sqrt{1-x^2}}\,dx = \sin^{-1}x\Big]_0^{1/2} = \sin^{-1}\frac{1}{2} - \sin^{-1}0 = \frac{\pi}{6} - 0 = \frac{\pi}{6}. \quad \|$$

Example 4 Using formula 11 with $u = 2x$, you find that

$$\int_0^3 \frac{1}{1+4x^2}\,dx = \frac{1}{2}\int_0^3 \frac{2}{1+(2x)^2}\,dx = \frac{1}{2}\tan^{-1}2x\Big]_0^3$$

$$= \tfrac{1}{2}\tan^{-1}6 - \tfrac{1}{2}\tan^{-1}0 = \tfrac{1}{2}\tan^{-1}6.$$

If you desire a decimal approximation for $\tan^{-1}6$, use a table, and you will find that $\tan^{-1}6 \approx 1.41$. ‖

Example 5 Using a table, you find that

$$\int_0^1 \sqrt{4-x^2}\,dx = \left(\frac{x}{2}\sqrt{4-x^2} + \frac{4}{2}\sin^{-1}\frac{x}{2}\right)\Big]_0^1 = \left(\frac{1}{2}\sqrt{3} + 2\sin^{-1}\frac{1}{2}\right) - 0$$

$$= \frac{\sqrt{3}}{2} + 2\frac{\pi}{6} = \frac{\sqrt{3}}{2} + \frac{\pi}{3}. \quad \|$$

SUMMARY

1. *A summary of the six inverse trigonometric functions is given in Fig. 8.11.*
2. *A summary of differentiation and integration formulas is given in Eqs.* (4) *through* (12).

EXERCISES

In Exercises 1 *through* 12, *find the given quantity.*

1. $\sin^{-1}1$ **2.** $\cos^{-1}\left(\frac{-1}{\sqrt{2}}\right)$ **3.** $\tan^{-1}(-\sqrt{3})$ **4.** $\csc^{-1}2$

5. $\sec^{-1}\left(\frac{-2}{\sqrt{3}}\right)$ **6.** $\sin^{-1}\left(\frac{\sqrt{3}}{-2}\right)$ **7.** $\cot^{-1}1$ **8.** $\sec^{-1}1$

9. $\csc^{-1}(-1)$ **10.** $\cos^{-1}(-1)$ **11.** $\sec^{-1}(-1)$ **12.** $\csc^{-1}\left(\frac{-2}{\sqrt{3}}\right)$

In Exercises 13 *through* 23, *find the derivative of the given function.*

13. $\sin^{-1}(2x)$ **14.** $\cos^{-1}(x^2)$ **15.** $\tan^{-1}(\sqrt{x})$ **16.** $x\sec^{-1}x$

17. $\csc^{-1}\left(\frac{1}{x}\right)$ **18.** $(\sin^{-1}x)(\cos^{-1}x)$ **19.** $(\tan^{-1}2x)^3$ **20.** $\sec^{-1}(x^2+1)$

21. $\frac{1}{\tan^{-1}x}$ **22.** $\sqrt{\csc^{-1}x}$ **23.** $(x + \sin^{-1}3x)^2$

In Exercises 24 through 28, find the integral without the use of tables.

24. $\int_{-1}^{1} \frac{1}{1+x^2}\,dx$ **25.** $\int \frac{1}{\sqrt{1-9x^2}}\,dx$ **26.** $\int_{-1}^{\sqrt{3}} \frac{1}{\sqrt{1-(x^2/4)}}\,dx$ **27.** $\int_{2/\sqrt{3}}^{2} \frac{1}{x\sqrt{x^2-1}}\,dx$ **28.** $\int \frac{1}{3x\sqrt{4x^2-1}}\,dx$

In Exercises 29 through 36, use tables, if necessary, to find the definite integral.

29. $\int_0^1 \frac{1}{\sqrt{4-x^2}}\,dx$ **30.** $\int_{-1}^{1} \sqrt{1-x^2}\,dx$ **31.** $\int_0^2 x^2\sqrt{16-x^2}\,dx$ **32.** $\int_1^{\sqrt{3}} \frac{\sqrt{4-x^2}}{x^2}\,dx$

33. $\int_{-4}^{-4/\sqrt{3}} \frac{\sqrt{x^2-4}}{x}\,dx$ **34.** $\int_2^3 \frac{1}{\sqrt{4x-x^2}}\,dx$ **35.** $\int_0^1 \sqrt{2x-x^2}\,dx$ **36.** $\int_2^3 x\sqrt{4x-x^2}\,dx$

37. Derive the formula for $d(\cos^{-1}x)/dx$

38. Derive the formula for $d(\tan^{-1}x)/dx$.

39. Derive the formula for $d(\cot^{-1}x)/dx$.

40. Derive the formula for $d(\sec^{-1}x)/dx$.

41. Derive the formula for $d(\csc^{-1}x)/dx$.

42. In defining $\sec^{-1}x$, some mathematicians prefer to choose branches of the graph of secant so that the range (set of principal values) of $\sec^{-1}x$ is $0 \le y < \pi/2$ or $\pi/2 < y \le \pi$. With this definition, we see that the relation $\sec^{-1}x = \cos^{-1}(1/x)$ holds.

a) Show by an example that the relation $\sec^{-1}x = \cos^{-1}(1/x)$ does not hold for the definition of $\sec^{-1}x$ in text (Fig. 8.11).

b) Show that if $\sec^{-1}x$ is defined to have range described in this exercise, then the formula for the derivative of the inverse secant changes from that in the table of Fig. 8.11 to

$$\frac{d(\sec^{-1}x)}{dx} = \frac{1}{|x|\sqrt{x^2-1}}.$$

8.6 THE HYPERBOLIC FUNCTIONS

8.6.1 The functions

The hyperbolic functions are defined in terms of exponential functions, and their inverses can be expressed in terms of logarithmic functions. (The inverse of the exponential function is the logarithm function, so this is not surprising!) This section gives just a brief summary of hyperbolic functions.

The hyperbolic functions have many points of similarity with the trigonometric functions. There are six basic hyperbolic functions, just as there are six basic trigonometric functions; the hyperbolic functions are the *hyperbolic sine,* denoted by $\sinh x$, the *hyperbolic cosine* or $\cosh x$, the *hyperbolic tangent* or $\tanh x$, etc. One usually pronounces "$\sinh x$" as though it were spelled "cinch x".

We shall define the hyperbolic functions in terms of exponential functions; this is very different from the way we defined the trigonometric functions. However, if you study complex analysis later, you will discover that the trigonometric functions can also be defined in terms of exponential functions of a "complex variable." The table in Fig. 8.12 gives the definitions of the hyperbolic functions. You have to remember the definitions for $\sinh x$ and $\cosh x$, and then you proceed exactly as for the trigonometric functions.

Function	Definition	Function	Definition
sinh x	$\dfrac{e^x - e^{-x}}{2}$	cosh x	$\dfrac{e^x + e^{-x}}{2}$
tanh x	$\dfrac{\sinh x}{\cosh x}$	coth x	$\dfrac{\cosh x}{\sinh x}$
sech x	$\dfrac{1}{\cosh x}$	csch x	$\dfrac{1}{\sinh x}$

8.12

It is easy to see that

$$\cosh^2 x - \sinh^2 x = 1 \tag{1}$$

for all x, namely,

$$\left(\frac{e^x + e^{-x}}{2}\right)^2 - \left(\frac{e^x - e^{-x}}{2}\right)^2 = \frac{e^{2x} + 2 + e^{-2x}}{4} - \frac{e^{2x} - 2 + e^{-2x}}{4}$$

$$= \frac{2}{4} + \frac{2}{4} = 1.$$

Equation (1) is the basic relation for hyperbolic functions and plays a role similar to the relation $\sin^2 x + \cos^2 x = 1$ for the trigonometric functions. For the trigonometric functions, the point ($\cos t$, $\sin t$) lies on the circle with equation $x^2 + y^2 = 1$ for all t, while for the hyperbolic functions, the point ($\cosh t$, $\sinh t$) lies on the curve with equation $x^2 - y^2 = 1$, which is called a *hyperbola*. This explains the name "hyperbolic functions."

The graphs of the six hyperbolic functions are easily sketched. To sketch the graph of $\sinh x$, we simply take half of the difference in height between the graphs of e^x and e^{-x}, while for $\cosh x$, we average the heights of these two graphs. Since $\lim_{x\to\infty} e^{-x} = 0$, we see that the graphs of both $\sinh x$ and $\cosh x$ are close to the graph of $e^x/2$ for large x. We have sketched the graphs of the six hyperbolic functions in Fig. 8.13.

You see from the graphs that the hyperbolic functions are not periodic functions of a *real* variable. If you study complex analysis later, you will discover that they have imaginary periods; the period of $\sinh x$ is $2\pi i$. A full appreciation of the relationship between the trigonometric functions and the hyperbolic functions comes only with a study of these functions in complex analysis.

Note from the graphs that $\sinh x$, $\tanh x$, $\coth x$, and $\operatorname{csch} x$ all yield a unique x for each y, so the inverse hyperbolic functions $\sinh^{-1} y$, $\tanh^{-1} y$, $\coth^{-1} y$, and $\operatorname{csch}^{-1} y$ exist. While $\cosh x$ and $\operatorname{sech} x$ do not satisfy this condition, we abuse terminology just as we did for trigonometric functions, and pick the branches of their graphs indicated by the darker curves in Fig. 8.13(b) and (e) to define the functions $\cosh^{-1} y$ and $\operatorname{sech}^{-1} y$.

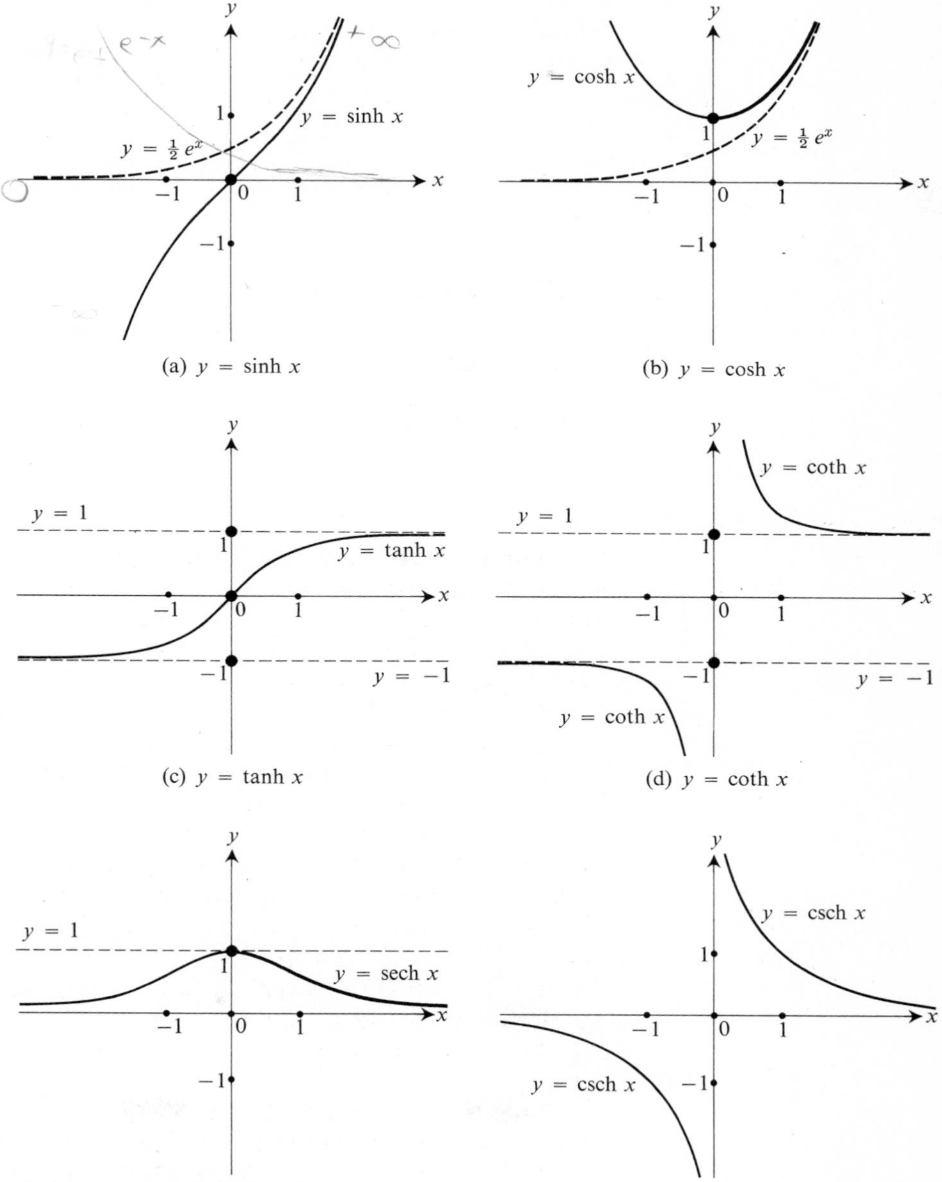

(a) $y = \sinh x$ (b) $y = \cosh x$

(c) $y = \tanh x$ (d) $y = \coth x$

8.13 (e) $y = \operatorname{sech} x$ (f) $y = \operatorname{csch} x$

8.6.2 The calculus of the hyperbolic functions

In this text, the hyperbolic functions and their inverse functions will be used chiefly to supply a few more integration formulas. However, the hyperbolic functions do occur naturally in the physical sciences. For example, a flexible cable suspended between two supports hangs in a "catenary curve" which is the graph of the hyperbolic cosine if appropriate axes and scale are chosen.

Turning to the calculus of these functions, we find that

$$\begin{aligned}\frac{d(\sinh x)}{dx} &= \frac{d}{dx}\left(\frac{e^x - e^{-x}}{2}\right)\\ &= \frac{1}{2}\cdot\frac{d}{dx}(e^x - e^{-x})\\ &= \frac{1}{2}(e^x + e^{-x})\\ &= \cosh x.\end{aligned}$$

Similarly, we find that

$$\frac{d(\cosh x)}{dx} = \sinh x.$$

The derivatives of the remaining four hyperbolic functions can then be found using relation (1) just as for the corresponding trigonometric functions. For example,

$$\begin{aligned}\frac{d(\tanh x)}{dx} &= \frac{d}{dx}\left(\frac{\sinh x}{\cosh x}\right)\\ &= \frac{\cosh x \cosh x - \sinh x \sinh x}{\cosh^2 x}\\ &= \frac{1}{\cosh^2 x}\\ &= \operatorname{sech}^2 x.\end{aligned}$$

The lefthand portion of the table in Fig. 8.14 gives the derivatives of the hyperbolic functions. These differentiation formulas parallel those for the derivatives of the trigonometric functions, except for occasional differences in sign.

The derivatives of the inverse hyperbolic functions can be found from the derivatives of the hyperbolic functions, using relation (1). For example, if $y = f(x) = \sinh x$, then $x = \sinh^{-1} y$ and dx/dy can be obtained by differentiating $y = \sinh x$ implicitly with respect to y:

Function	Derivative	Function	Derivative
$\sinh x$	$\cosh x$	$\sinh^{-1}x$	$\dfrac{1}{\sqrt{1+x^2}}$
$\cosh x$	$\sinh x$	$\cosh^{-1}x$	$\dfrac{1}{\sqrt{x^2-1}}, \quad x > 1$
$\tanh x$	$\text{sech}^2 x$	$\tanh^{-1}x$	$\dfrac{1}{1-x^2}, \quad \|x\| < 1$
$\coth x$	$-\text{csch}^2 x$	$\coth^{-1}x$	$\dfrac{1}{1-x^2}, \quad \|x\| > 1$
$\text{sech}\, x$	$-\text{sech}\, x \tanh x$	$\text{sech}^{-1}x$	$\dfrac{-1}{x\sqrt{1-x^2}}, \quad 0 < x < 1$
$\text{csch}\, x$	$-\text{csch}\, x \coth x$	$\text{csch}^{-1}x$	$\dfrac{-1}{\|x\|\sqrt{1+x^2}}$

8.14

$$y = \sinh x,$$

$$1 = (\cosh x)\frac{dx}{dy},$$

$$\frac{dx}{dy} = \frac{1}{\cosh x} = \frac{1}{\sqrt{1+\sinh^2 x}} = \frac{1}{\sqrt{1+y^2}}.$$

Changing notation so that x is the dependent variable, we obtain

$$\frac{d(\sinh^{-1}x)}{dx} = -\frac{1}{\sqrt{1+x^2}}.$$

The derivatives of the other inverse hyperbolic functions are found similarly, and are given in the righthand portion of the table in Fig. 8.14. These differentiation formulas can, of course, be combined with the chain rule in the usual way.

Example 1 We have

$$\frac{d(\sinh^{-1}(\tan x))}{dx} = \frac{1}{\sqrt{1+\tan^2 x}} \cdot \sec^2 x$$

$$= \frac{\sec^2 x}{|\sec x|} = |\sec x|. \quad \|$$

From the formulas for the derivatives of the inverse hyperbolic functions in the table in Fig. 8.14, we obtain the integration formulas:

$$\int \frac{du}{\sqrt{1+u^2}} = \sinh^{-1} u + C; \tag{2}$$

$$\int \frac{du}{\sqrt{u^2-1}} = \cosh^{-1} u + C, \qquad u > 1; \tag{3}$$

$$\int \frac{du}{1-u^2} = \begin{cases} \tanh^{-1} u + C & \text{for } |u| < 1, \\ \coth^{-1} u + C & \text{for } |u| > 1; \end{cases} \tag{4}$$

$$\int \frac{du}{u\sqrt{1-u^2}} = -\operatorname{sech}^{-1}|u| + C; \tag{5}$$

$$\int \frac{du}{u\sqrt{1+u^2}} = -\operatorname{csch}^{-1}|u| + C. \tag{6}$$

The reason for the split in formula (4) is that $\tanh^{-1} x$ and $\coth^{-1} x$ have algebraically identical derivatives, but the domain of $\tanh^{-1} x$ is $|x| < 1$ while the domain of $\coth^{-1} x$ is $|x| > 1$. Since

$$\frac{1}{1-u^2} = \frac{1/2}{1-u} + \frac{1/2}{1+u},$$

we have, as an alternative to (4),

$$\int \frac{du}{1-u^2} = -\frac{1}{2}\ln|1-u| + \frac{1}{2}\ln|1+u| + C$$

$$= \frac{1}{2}\ln\left|\frac{1+u}{1-u}\right| + C. \tag{7}$$

Example 2 Using formula (5) and "fixing up the integral," we have

$$\int \frac{1}{x\sqrt{4-x^2}}\,dx = \frac{1}{2}\int \frac{1}{x\sqrt{1-(x/2)^2}}\,dx$$

$$= \frac{1}{2}\int \frac{1/2}{(x/2)\sqrt{1-(x/2)^2}}\,dx$$

$$= -\frac{1}{2}\operatorname{sech}^{-1}\left|\frac{x}{2}\right| + C. \quad \|$$

Since the hyperbolic functions can be expressed in terms of exponential functions, it seems reasonable to expect that the inverse hyperbolic functions can be expressed in terms of logarithmic functions, as indicated by the alternative form of (4) computed above in (7). We leave the demonstration of this to the exercises (see Exercises 60 through 65). The integration formulas to which the other alternative forms lead are given in the summary.

SUMMARY

1. *The hyperbolic functions are defined in Fig.* 8.12.
2. *The graphs of the hyperbolic functions are shown in Fig.* 8.13.
3. *Differentiation formulas are in Fig.* 8.14.
4. *Integration formulas given by the inverse hyperbolic functions (and also by logarithm functions) are*

$$\int \frac{du}{\sqrt{1+u^2}} = \sinh^{-1}u + C = \ln\,(u + \sqrt{1+u^2}) + C$$

$$\int \frac{du}{\sqrt{u^2-1}} = \cosh^{-1}u + C = \ln\,(u + \sqrt{u^2-1}) + C$$

$$\int \frac{du}{1-u^2} = \begin{Bmatrix} \tanh^{-1}u + C, & |u| < 1 \\ \coth^{-1}u + C, & |u| > 1 \end{Bmatrix} = \frac{1}{2}\ln\left|\frac{1+u}{1-u}\right| + C$$

$$\int \frac{du}{u\sqrt{1-u^2}} = -\operatorname{sech}^{-1}|u| + C = -\ln\left|\frac{1+\sqrt{1-u^2}}{u}\right| + C$$

$$\int \frac{du}{u\sqrt{1+u^2}} = -\operatorname{csch}^{-1}|u| + C = -\ln\left|\frac{1+\sqrt{1+u^2}}{u}\right| + C$$

EXERCISES

In Exercises 1 through 6, prove the given relation for hyperbolic functions.

1. $1 - \tanh^2 x = \operatorname{sech}^2 x$.

2. $\coth^2 x - 1 = \operatorname{csch}^2 x$.

3. $\sinh(-x) = -\sinh x$.

4. $\cosh(-x) = \cosh x$.

5. $\sinh(x + y) = \sinh x \cosh y + \cosh x \sinh y$.

6. $\cosh(x + y) = \cosh x \cosh y + \sinh x \sinh y$.

7. Use Exercises 5 and 6 to find "double-angle" formulas for $\sinh 2x$ and $\cosh 2x$.

8. If $\sinh a = -\frac{3}{4}$, find $\cosh a$, $\tanh a$, and $\operatorname{csch} a$.

In Exercises 9 through 17, derive the formula for the derivative of the given function. In each exercise, you may use the answer to any previous exercise.

9. $\cosh x$ **10.** $\coth x$ **11.** $\operatorname{sech} x$ **12.** $\operatorname{csch} x$

13. $\cosh^{-1}x$ **14.** $\tanh^{-1}x$ **15.** $\coth^{-1}x$ **16.** $\operatorname{sech}^{-1}x$

17. $\operatorname{csch}^{-1}x$

In Exercises 18 through 35, find the derivative of the given function.

18. $\sinh(2x - 3)$ **19.** $\cosh(x^2)$ **20.** $\tanh^2 3x$ **21.** $\operatorname{sech}(\sqrt{x})$

22. $\coth(e^{2x})$ **23.** $\operatorname{csch}(\ln x)$ **24.** $\operatorname{sech}^3 3x$ **25.** $\sinh^2 x \cosh^2 x$

26. $(e^{-x} + \tanh x)^2$ **27.** $\sinh^{-1}(2x)$ **28.** $\cosh^{-1}(\sec x)$ **29.** $\tanh^{-1}(\sin 2x)$

30. $\coth^{-1}(e^{2x+1})$ **31.** $\operatorname{sech}^{-1}(x^2)$ **32.** $\operatorname{csch}^{-1}(\tan x)$ **33.** $e^x \sinh^{-1}(e^x)$

34. $\cosh^{-1}(x^2 + 1)$ **35.** $\tanh^{-1}(\cot 3x)$

In Exercises 36 through 44, compute the given integral without using tables.

36. $\int \tanh x\,dx$

37. $\int \coth x\,dx$

38. $\int \sinh^2 x \cosh x\,dx$

39. $\int \sinh(3x+2)\,dx$

40. $\int x\cosh(x^2)\,dx$

41. $\int \operatorname{sech}^2 3x\,dx$

42. $\int \frac{1}{(e^x+e^{-x})^2}\,dx$

43. $\int \frac{e^x+e^{-x}}{e^x-e^{-x}}\,dx$

44. $\int (\cosh 2x)(e^{\sinh 2x})\,dx$

In Exercises 45 through 59, use tables, if necessary, to compute the given integral.

45. $\int_0^{1/2} \frac{x}{1-x^2}\,dx$

46. $\int_0^{1/4} \frac{1}{1-4x^2}\,dx$

47. $\int \frac{x}{\sqrt{1+4x^2}}\,dx$

48. $\int \frac{1}{\sqrt{4+x^2}}\,dx$

49. $\int \frac{1}{\sqrt{1-e^{2x}}}\,dx$

50. $\int \frac{\sin 2x}{(\cos 2x)\sqrt{1+\cos^2 2x}}\,dx$

51. $\int_0^1 \sqrt{4+x^2}\,dx$

52. $\int_0^4 x^3\sqrt{9+x^2}\,dx$

53. $\int \frac{\sqrt{16+x^2}}{x}\,dx$

54. $\int \frac{\sqrt{2+x^2}}{3x^2}\,dx$

55. $\int \frac{\sin^2 x\cos x}{\sqrt{9+\sin^2 x}}\,dx$

56. $\int_3^5 x^2\sqrt{x^2-9}\,dx$

57. $\int \frac{e^{3x}}{\sqrt{e^{2x}-16}}\,dx$

58. $\int x\sinh 2x\,dx$

59. $\int \cosh^3 4x\,dx$

60. Let $x = \sinh y = (e^y - e^{-y})/2$. Then $2x = e^y - e^{-y}$, and multiplying by e^y, we obtain

$$2xe^y = e^{2y} - 1 \quad \text{or} \quad e^{2y} - 2xe^y - 1 = 0.$$

This last equation is a quadratic equation

$$(e^y)^2 - 2x(e^y) - 1 = 0$$

in e^y. Solve the equation for e^y in terms of x by the quadratic formula, and then take logarithms to express $y = \sinh^{-1}x$ as a function of x involving logarithms.

61. Follow the outline of Exercise 60 to find a logarithmic formula for $\cosh^{-1}x$.

62. Follow the outline of Exercise 60 to find a logarithmic formula for $\tanh^{-1}x$.

63. Follow the outline of Exercise 60 to find a logarithmic formula for $\coth^{-1}x$.

64. Follow the outline of Exercise 60 to find a logarithmic formula for $\operatorname{sech}^{-1}x$.

65. Follow the outline of Exercise 60 to find a logarithmic formula for $\operatorname{csch}^{-1}x$.

exercise sets for chapter 8

review exercise set 8.1

1. a) Give the definition of $\ln x$ as an integral.

b) Sketch the graph of $\ln(x/2)$.

2. a) Find dy/dx if $y = \ln(x^2+1)$.

b) Find $\int [\sin x/(1+\cos x)]\,dx$.

3. a) Find dy/dx if $y = e^{\tan x}$.

b) Evaluate $\int_0^{\ln 2} e^{3x}\,dx$, simplifying your answer as much as possible.

4. a) Differentiate 2^{3x+4}. b) Differentiate x^{5x}, $x > 0$.

5. a) Solve for x if $(2^x)(3^{x+1}) = 4^{-x}$.
   b) Simplify $2^{\log_4 9}$.
6. If savings are compounded continuously at 5 percent interest, how long does it take for the savings to double?
7. a) Evaluate $\sin^{-1}(-1/2)$.
   b) Evaluate $\tan^{-1}(-\sqrt{3})$.
8. a) Find dy/dx if $y = \sin^{-1}(\sqrt{x})$.
   b) Evaluate $\int_{-1}^{\sqrt{3}} [1/(1 + x^2)]\, dx$.
9. a) Give the definition of $\sinh x$ in terms of the exponential function.
   b) Differentiate $\operatorname{sech}^3 2x$.

review exercise set 8.2

1. a) How is the number e defined?
   b) Sketch the graph of e^{-x^2}.
2. a) Find dy/dx if $y = \ln(\tan x)$.
   b) Find $\int [x^2/(4 + x^3)]\, dx$.
3. a) Find dy/dx if $y = e^{\sin^{-1} x}$.
   b) Find $\int (\sec^2 3x)e^{\tan 3x}\, dx$.
4. a) Differentiate 10^{x^2-x}.
   b) Differentiate $(\cos x)^{\sin x}$.
5. a) Solve for x if $\ln x - 2(\ln x^3) = \ln 10$.
   b) Simplify $25^{\log_5 7}$.
6. The only force on a body traveling on a line is a force of resistance proportional to its velocity. If the velocity is 100 ft/sec at time $t = 0$ and 50 ft/sec at time $t = 10$, find
   a) the velocity of the body as a function of t,
   b) the distance traveled as a function of t.
7. Evaluate: a) $\cos^{-1}(1/2)$ b) $\sec^{-1}(-2)$
8. a) Find dy/dx if $y = \tan^{-1} 4x$.
   b) Evaluate $\int_{-1/4}^{1/4} (1/\sqrt{1 - 4x^2})\, dx$.
9. a) Sketch the graph of $y = \cosh(x + 1)$.
   b) Differentiate $\coth^3(2x + 1)$.

more challenging exercises 8

1. Solve $e^{2x} - 3e^x + 2 = 0$.
2. Outline how it could be proved from our definition of $\ln x$ and of e that
$$2.7 < e < 2.8.$$
3. Show from the definition of $\ln x$ as an integral that, for each integer $n > 1$,
$$\frac{1}{2} + \frac{1}{3} + \cdots + \frac{1}{n} < \ln n < 1 + \frac{1}{2} + \frac{1}{3} + \cdots + \frac{1}{n-1}.$$
4. Show that
$$\lim_{\Delta x \to 0} \frac{a^{\Delta x} - 1}{\Delta x} = \ln a \quad \text{for} \quad a > 0.$$
[*Hint.* You know $d(a^x)/dx$. Use the definition of the derivative for $f(x) = a^x$.]
5. Using Exercise 4, find $\lim_{h\to 0} [(a^{2h} - 1)/h]$.
6. Show that, for any particular integer n_0, e^x is larger than x^{n_0} if x is sufficiently large.
7. How does the size of e^x compare with the size of $f(x)$ for any particular polynomial function f if x is large? [*Hint.* Use Exercise 6.]

technique of integration

9

You had some practice integrating in Sections 6.3 and 6.4. Perhaps you need to review the integration formulas given in Chapters 6 and 8. To save you turning pages, they are collected here for you. We omit those involving the hyperbolic functions and their inverses.

1. $\int (u+v)\,dx = \int u\,dx + \int v\,dx + C$

2. $\int c \cdot u\,dx = c \cdot \int u\,dx + C$

3. $\int u^n\,du = \dfrac{u^{n+1}}{n+1} + C, \quad n \neq -1$

4. $\int \dfrac{1}{u}\,du = \ln|u| + C$

5. $\int \dfrac{du}{\sqrt{1-u^2}} = \sin^{-1}u + C$

6. $\int \dfrac{du}{1+u^2} = \tan^{-1}u + C$

7. $\int \dfrac{du}{u\sqrt{u^2-1}} = \sec^{-1}u + C$

8. $\int \sin u\,du = -\cos u + C$

9. $\int \cos u\,du = \sin u + C$

10. $\int \sec^2 u\,du = \tan u + C$

11. $\int \csc^2 u\,du = -\cot u + C$

12. $\int \sec u \tan u\,du = \sec u + C$

13. $\int \csc u \cot u\,du = -\csc u + C$

14. $\int \sec u\,du = \ln|\sec u + \tan u| + C$

15. $\int \csc u\,du = -\ln|\csc u + \cot u| + C$

16. $\int e^u\,du = e^u + C$

17. $\int a^u\,du = \dfrac{1}{\ln a}a^u + C$

The single most important method of integration is surely the use of a table of integrals. Proper use of a table of integrals is so important that we included a section on this topic when the integral was introduced in Chapter 6. Very large computer programs are now being developed to find formal derivatives, and indefinite as well as definite integrals, and to perform many other tasks of calculus. It may well be that such programs will be easily accessible in the near future, and will largely replace the use of tables as the most efficient and reliable way to find an indefinite integral.

You should learn how to *transform* certain integrals that do not appear in a table into integrals that the table contains. The first three sections of this chapter give three such methods: integration by parts, partial-fraction decomposition, and substitution techniques. Sections 9.4, 9.5, and 9.6 describe how to integrate, without the use of tables, many of the most frequently encountered functions. The functions treated in Section 9.5 and 9.6 especially can be integrated faster and with less chance of error using tables. If time is a factor, you may wish to omit these sections. However, the sections do continue to develop your facility in the general techniques presented in Sections 9.1, 9.2, and 9.3, especially the technique of substitution.

Remember that, if you are unable to integrate a function to find a *definite* integral, it is easy to use Simpson's rule in these days of calculators and computers. The importance of such a nice numerical method of solution cannot be overemphasized.

9.1 INTEGRATION BY PARTS

You know that if $f(x)$ is continuous in an interval containing a, then f has an antiderivative, namely, $F(x) = \int_a^x f(t)\,dt$. Every differentiable function is continuous, so every differentiable function has an indefinite integral. Of course, you know that not every continuous function is differentiable.

Although every differentiable function has an indefinite integral, the problem of computing the indefinite integral of even an elementary function is, in general, much more difficult than differentiating such a function. Let's examine the reason for this. For differentiation, there are certain rules and formulas that enable you to compute derivatives of constant multiples, sums, products, and quotients of functions whose derivatives are known. While it is easy to compute antiderivatives of constant multiples and sums of functions with known antiderivatives, *there are no simple rules or formulas for integrating products or quotients of functions with known antiderivatives.* In this section, we present our only formula for integrating a product of functions. The technique is known as *integration by parts.* The formula is used not only as a tool for integration, but also in some theoretical considerations.

9.1.1 The formula

The formula for integration by parts is easily obtained from the formula for the derivative of a product. Recall that, if $u = f(x)$ and $v = g(x)$ are differentiable functions, then

$$\frac{d(uv)}{dx} = u\frac{dv}{dx} + v\frac{du}{dx}, \tag{1}$$

or, in differential notation,

$$d(uv) = u \cdot dv + v \cdot du. \tag{2}$$

From (2), you obtain

$$u \cdot dv = d(uv) - v \cdot du, \tag{3}$$

so

$$\int u \cdot dv = \int [d(uv) - v \cdot du] = \int d(uv) - \int v \cdot du.$$

Since $\int d(uv) = uv$, you obtain

$$\int u \cdot dv = uv - \int v \cdot du, \tag{4}$$

which is the formula for *integration by parts.*

9.1.2 The technique

The formula $\int u \cdot dv = uv - \int v \cdot du$ for integration by parts is used as follows. Suppose you are faced with the problem of integrating a product of two functions where the antiderivative of at least one of the functions is known. *If integration of the product of the derivative of one of the functions times the antiderivative of the other is easier than your original problem, then integration by parts can be used to advantage.*

To illustrate, suppose you wish to compute $\int (x \sin x)\, dx$. You are faced with the problem of integrating the product of the functions x and $\sin x$. By use of formula (4), you can reduce the problem to that of computing the integral of the derivative of one of these functions times an antiderivative of the other; that is, to computing either

$$\int 1(-\cos x)\, dx \qquad \text{or} \qquad \int \left(\frac{x^2}{2}\right)(\cos x)\, dx. \tag{5}$$

The second of these two integrals in (5) is more complicated than your original problem, but the first one is simpler. Thus you use Eq. (4) with $u = x$ and $dv = \sin x\, dx$. (The alternative substitution $u = \sin x$ and $dv = x\, dx$ would lead to the second integral in (5).) When using Eq. (4), it is customary to write out the substitution in the array shown in Fig. 9.1.

$$\begin{array}{ll} u = x & dv = \sin x\, dx \\ du = dx & v = -\cos x \end{array}$$

9.1

Formula (4) tells you that:

The integral of the product of the functions in the top row of the array equals the product of the functions at the ends of the dashed diagonal minus the integral of the product of the functions in the bottom row.

Thus you obtain

$$\begin{aligned} \int x \sin x\, dx &= x(-\cos x) - \int (1)(-\cos x)\, dx \\ &= -x \cos x + \int \cos x\, dx \\ &= -x \cos x + \sin x + C. \end{aligned}$$

Example 1 Let's compute $\int x^2 e^x\,dx$.

SOLUTION Integration by parts of the product of x^2 and e^x leads to the computation of either $\int 2xe^x\,dx$ or $\int (x^3/3)\,e^x\,dx$. The former integral $\int 2xe^x\,dx$ is simpler than our original integral, and this new integral can obviously be further reduced to $\int 2e^x\,dx$ by an additional application of the formula for integration by parts. Thus we form the array

$$\begin{aligned} u &= x^2 & dv &= e^x\,dx \\ du &= 2x\,dx & v &= e^x \end{aligned}$$

and obtain

$$\int x^2 e^x\,dx = x^2 e^x - \int 2xe^x\,dx. \tag{6}$$

To compute the remaining integral $\int 2xe^x\,dx$, we integrate by parts again. (Do not be confused by the use of u again for the function $2x$ rather than x^2; it is customary.) Our array this time is

$$\begin{aligned} u &= 2x & dv &= e^x\,dx \\ du &= 2\,dx & v &= e^x \end{aligned}$$

and Eq. (6) yields

$$\begin{aligned} \int x^2 e^x\,dx = x^2 e^x - \int 2xe^x\,dx &= x^2 e^x - \left[2xe^x - \int 2e^x\,dx\right] \\ &= x^2 e^x - 2xe^x + 2e^x + C. \quad \| \end{aligned}$$

Example 2 Let's compute $\int e^x \sin x\,dx$.

SOLUTION This time, both substitutions for integration by parts lead to $\pm\int e^x \cos x\,dx$, which seems to be just as hard as our original problem. The following is an effective trick. We form the array

$$\begin{aligned} u &= e^x & dv &= \sin x\,dx \\ du &= e^x\,dx & v &= -\cos x \end{aligned}$$

and obtain

$$\int e^x \sin x\,dx = -e^x \cos x + \int e^x \cos x\,dx. \tag{7}$$

We then integrate by parts again, using the array

$$\begin{aligned} u &= e^x & dv &= \cos x\,dx \\ du &= e^x\,dx & v &= \sin x \end{aligned}$$

and obtain from (7)

$$\begin{aligned} \int e^x \sin x\,dx &= -e^x \cos x + \left[e^x \sin x - \int e^x \sin x\,dx\right] \\ &= (e^x \sin x - e^x \cos x) - \int e^x \sin x\,dx. \end{aligned} \tag{8}$$

We may now solve Eq. (8) for $\int e^x \sin x\, dx$, and we find that

$$\int e^x \sin x\, dx = \tfrac{1}{2}(e^x \sin x - e^x \cos x).$$

This solves our problem. ||

Example 3 Integration by parts is a basic technique for finding reduction formulas. For example, let us derive the reduction formula

$$\int \sin^n x\, dx = -\frac{1}{n}\sin^{n-1}x \cos x + \frac{n-1}{n}\int \sin^{n-2}x\, dx \quad \text{for } n \geq 2$$

SOLUTION We form the array

$$u = \sin^{n-1}x \qquad dv = \sin x\, dx$$

$$du = (n-1)\sin^{n-2}x \cos x\, dx \qquad v = -\cos x$$

which yields

$$\int \sin^n x\, dx = -\sin^{n-1}x \cos x - \int (n-1)(\sin^{n-2}x)(-\cos^2 x)\, dx. \tag{9}$$

From (9) and the identity $-\cos^2 x = \sin^2 x - 1$, we have

$$\begin{aligned}\int \sin^n x\, dx &= -\sin^{n-1}x \cos x - \int (n-1)(\sin^{n-2}x)(\sin^2 x - 1)\, dx \\ &= -\sin^{n-1}x \cos x - (n-1)\int \sin^n x\, dx + (n-1)\int \sin^{n-2}x\, dx. \end{aligned} \tag{10}$$

From (10), we obtain

$$n\int \sin^n x\, dx = -\sin^{n-1}x \cos x + (n-1)\int \sin^{n-2}x\, dx,$$

from which we obtain the reduction formula for $\int \sin^n x\, dx$ upon division by n. ||

Any function can be regarded as the product of itself with 1. Thus, for $\int f(x)\, dx$ we can always try the substitution

$$u = f(x) \qquad dv = 1\, dx$$

$$du = f'(x)\, dx \qquad v = x.$$

Example 4 We shall find $\int \ln x\, dx$ using integration by parts.

SOLUTION The array

$$u = \ln x \qquad dv = 1\, dx$$

$$du = \frac{1}{x}\, dx \qquad v = x$$

yields

$$\begin{aligned}\int \ln x\,dx &= x\ln x - \int x\cdot\frac{1}{x}\,dx\\ &= x\ln x - \int 1\,dx\\ &= x\ln x - x + C. \qquad \|\end{aligned}$$

Example 5 The inverse trigonometric functions can be integrated by parts. Let us find $\int \sin^{-1}x\,dx$.

SOLUTION We set

$$u = \sin^{-1}x \qquad dv = 1\,dx$$
$$du = \frac{1}{\sqrt{1-x^2}}\,dx \qquad v = x$$

and

$$\begin{aligned}\int \sin^{-1}x\,dx &= x\sin^{-1}x - \int \frac{x}{\sqrt{1-x^2}}\,dx\\ &= x\sin^{-1}x + \frac{1}{2}\int -2x(1-x^2)^{-1/2}\,dx\\ &= x\sin^{-1}x + \frac{1}{2}\cdot\frac{(1-x^2)^{1/2}}{1/2} + C\\ &= x\sin^{-1}x + \sqrt{1-x^2} + C. \qquad \|\end{aligned}$$

SUMMARY

1. *Integration by parts is used for finding $\int f(x)\cdot g(x)\,dx$ when the product of the derivative of one of the functions times the antiderivative of the other gives an easier integration problem.*

2. *If you form the array*

$$u = f(x) \qquad dv = g(x)\,dx$$
$$du = f'(x)\,dx \qquad v = \int g(x)\,dx$$

then $\int f(x)\cdot g(x)\,dx$ equals the product of the functions at the ends of the dashed diagonal minus the integral of the product of the terms in the bottom row. In symbols,

$$\int u\cdot dv = uv - \int v\cdot du.$$

EXERCISES

In Exercises 1 through 17, find the indicated indefinite or definite integral without the use of tables.

1. $\int x \cos x \, dx$

2. $\int \ln x^2 \, dx$

3. $\int x^2 e^{-x^3} \, dx$

4. $\int x^2 \sin x \, dx$

5. $\int_0^{1/2} \sin^{-1}(2x) \, dx$

6. $\int \tan^{-1} 3x \, dx$

7. $\int x(\ln x) \, dx$

8. $\int x \sec^2 2x \, dx$

9. $\int e^{ax} \cos bx \, dx$

10. $\int \sin(\ln x) \, dx$

11. $\int x^3 e^x \, dx$

12. $\int x^3 e^{x^2} \, dx$

13. $\int x^5 \sin x^3 \, dx$

14. $\int_1^e x^3(\ln x) \, dx$

15. $\int \ln(1 + x^2) \, dx$

16. $\int (\ln x)^2 \, dx$

17. $\int x \sec^{-1} x \, dx$

In Exercises 18 through 22, derive the given reduction formula.

18. $\int x^n \sin ax \, dx = -\frac{x^n}{a} \cos ax + \frac{n}{a} \int x^{n-1} \cos ax \, dx$

19. $\int x^n e^{ax} \, dx = \frac{1}{a} x^n e^{ax} - \frac{n}{a} \int x^{n-1} e^{ax} \, dx$

20. $\int \cos^n ax \, dx = \frac{\cos^{n-1} ax \sin ax}{na} + \frac{n-1}{n} \int \cos^{n-2} ax \, dx$

21. $\int \sin^n ax \cos^m ax \, dx = -\frac{\sin^{n-1} ax \cos^{m+1} ax}{a(m+n)} + \frac{n-1}{m+n} \int \sin^{n-2} ax \cos^m ax \, dx \quad \text{for } n \neq -m$

22. $\int \sec^n ax \, dx = \frac{\sec^{n-2} ax \tan ax}{a(n-1)} + \frac{n-2}{n-1} \int \sec^{n-2} ax \, dx \quad \text{for } n \neq 1$

9.2 INTEGRATION OF RATIONAL FUNCTIONS BY PARTIAL FRACTIONS

You will not find a formula for

$$\int \frac{x^6 - 2}{x^4 + x^2} \, dx$$

in most tables. This section shows how the integral of $(x^6 - 2)/(x^4 + x^2)$ and any other rational function can be transformed into a sum of integrals that are found in many tables.

9.2.1 The basic integrals

Here are six basic integration formulas which we show can be used to compute the integral of any rational function.

Basic Formulas for Integration of Rational Functions

$$\int \frac{1}{ax+b}\,dx = \frac{1}{a}\ln|ax+b| + C \tag{1}$$

$$\int \frac{1}{(ax+b)^n}\,dx = \frac{-1}{a(n-1)}\cdot\frac{1}{(ax+b)^{n-1}} + C, \qquad n \neq 1 \tag{2}$$

$$\int \frac{1}{ax^2+bx+c}\,dx = \frac{2}{\sqrt{4ac-b^2}}\tan^{-1}\frac{2ax+b}{\sqrt{4ac-b^2}} + C, \quad b^2 < 4ac \tag{3}$$

$$\int \frac{2ax+b}{ax^2+bx+c}\,dx = \ln|ax^2+bx+c| + C \tag{4}$$

$$\int \frac{1}{(ax^2+bx+c)^{n+1}}\,dx = \frac{2ax+b}{n(4ac-b^2)(ax^2+bx+c)^n} + \frac{2(2n-1)a}{n(4ac-b^2)}\int \frac{1}{(ax^2+bx+c)^n}\,dx, \qquad 4ac \neq b^2 \tag{5}$$

$$\int \frac{2ax+b}{(ax^2+bx+c)^n}\,dx = \frac{-1}{(n-1)(ax^2+bx+c)^{n-1}} + C, \quad n \neq 1 \tag{6}$$

Formulas (1) and (4) are easily obtained from the familiar formula

$$\int \frac{du}{u} = \ln|u| + C, \tag{7}$$

and you should not need to look up formulas (1) or (4). For example, using (7),

$$\int \frac{2}{3x+5}\,dx = \frac{2}{3}\int \frac{3}{3x+5}\,dx = \frac{2}{3}\ln|3x+5| + C.$$

The formulas (2) and (6) are instances of the familiar formula

$$\int u^n\,du = \frac{u^{n+1}}{n+1} + C \qquad \text{for } n \neq -1, \tag{8}$$

and you should not need to refer to tables for (2) or (6) either. This leaves just formula (3) and the reduction formula (5), which you may wish to look up when you need them.

9.2.2 Partial fraction decomposition of a rational function

We now indicate how any rational function can be written as the sum of a polynomial function and terms appearing in the integrands in formulas (1) through (6). Let the given rational function be of the form

$$\frac{f(x)}{g(x)}$$

for polynomial functions f and g.

STEP 1. If the polynomial $f(x)$ in the numerator has degree greater than or equal to the degree of the denominator polynomial $g(x)$, perform polynomial long division and write

$$\frac{f(x)}{g(x)} = q(x) + \frac{r(x)}{g(x)} \tag{9}$$

for polynomials $q(x)$ and $r(x)$, where the degree of $r(x)$ is less than the degree of $g(x)$. Since the polynomial function $q(x)$ is easy to integrate, you have essentially reduced the problem of integrating $f(x)/g(x)$ to integrating $r(x)/g(x)$.

Example 1 We illustrate with the rational function

$$\frac{f(x)}{g(x)} = \frac{x^6 - 2}{x^4 + x^2}$$

mentioned at the start of the section. Polynomial long division yields

$$\begin{array}{r|l} & x^2 - 1 \\ \hline x^4 + x^2 & x^6 \qquad\qquad -2 \\ & x^6 + x^4 \\ \hline & \quad - x^4 \\ & \quad - x^4 - x^2 \\ \hline & \qquad\quad x^2 - 2. \end{array}$$

Thus

$$\frac{f(x)}{g(x)} = \frac{x^6 - 2}{x^4 + x^2} = x^2 - 1 + \frac{x^2 - 2}{x^4 + x^2}. \quad \|$$

STEP 2. Factor the denominator polynomial $g(x)$ into a product of (possibly repeated) linear and irreducible quadratic factors, so that the quadratic factors cannot be factored further into a product of real linear factors. *It is a theorem of algebra that such a factorization exists.* Briefly, it can be shown that if a is a root of $g(x) = 0$, then $x - a$ is a factor of $g(x)$. It can be shown that if $g(x)$ is a polynomial of degree n, then $g(x) = 0$ has n (possibly repeated) roots in the complex numbers, which provide a factorization of $g(x)$ into linear factors with complex coefficients. Now the polynomial $g(x)$ has real coefficients, and it can be shown that if $a + bi$ is a root of $g(x) = 0$ with $b \neq 0$, then the conjugate complex number $a - bi$ is also a root. But then $g(x)$ has a factor

$$(x - (a + bi)) \cdot (x - (a - bi)) = x^2 - 2ax + (a^2 + b^2),$$

which is a quadratic polynomial with real coefficients.

Example 2 For the rational function we obtained by long division in Example 1,

$$\frac{r(x)}{g(x)} = \frac{x^2 - 2}{x^4 + x^2} = \frac{x^2 - 2}{x^2(x^2 + 1)}.$$

The linear factor x of $g(x)$ has multiplicity 2, and the irreducible quadratic factor $x^2 + 1$ has multiplicity 1. ||

STEP 3. It is a theorem of algebra (whose proof we shall not attempt to indicate) that $r(x)/g(x)$ can be written as a sum of "partial fractions," which arise from the factors of $g(x)$ as follows.

A linear factor $ax + b$ of $g(x)$ of multiplicity 1 gives rise to a single term

$$\frac{A}{ax + b} \tag{10}$$

for some constant A, while a linear factor with multiplicity n, say $(ax + b)^n$, gives rise to a sum of terms

$$\frac{A_n}{(ax + b)^n} + \frac{A_{n-1}}{(ax + b)^{n-1}} + \cdots + \frac{A_1}{ax + b} \tag{11}$$

for constants $A_n, A_{n-1}, \ldots, A_1$.

An irreducible quadratic factor $ax^2 + bx + c$ of $g(x)$ of multiplicity 1 gives rise to a single term

$$\frac{Ax + B}{ax^2 + bx + c} \tag{12}$$

for constants A and B, while an irreducible quadratic factor of multiplicity n, say $(ax^2 + bx + c)^n$, gives rise to a sum of terms

$$\frac{A_n x + B_n}{(ax^2 + bx + c)^n} + \frac{A_{n-1}x + B_{n-1}}{(ax^2 + bx + c)^{n-1}} + \cdots + \frac{A_1 x + B_1}{ax^2 + bx + c}. \tag{13}$$

As we describe in the following article, the rational function $r(x)/g(x)$ is the sum of these *partial fractions* arising from the various irreducible factors of $g(x)$.

Example 3 For the rational function we obtained in Example 2, we must have

$$\frac{r(x)}{g(x)} = \frac{x^2 - 2}{x^4 + x^2} = \frac{x^2 - 2}{x^2(x^2 + 1)} = \frac{A_2}{x^2} + \frac{A_1}{x} + \frac{Bx + C}{x^2 + 1} \tag{14}$$

for some constants A_2, A_1, B, and C. We show how to compute these constants in the next article. ||

9.2.3 Computation of the partial-fraction decomposition

The numerator constants of a partial-fraction decomposition like (14) can always be found by computing the sum of the partial fractions, and equating the coefficients of powers of x in the resulting numerator with the coefficients of like powers of x in the numerator $r(x)$. This procedure leads to a system of simultaneous linear equations in the "unknown constants." The technique is best illustrated by an example.

Example 4 We continue with the decomposition in Example 3. Adding, we require that

$$\frac{A_2}{x^2} + \frac{A_1}{x} + \frac{Bx + C}{x^2 + 1} = \frac{A_2(x^2 + 1) + A_1 x(x^2 + 1) + (Bx + C)x^2}{x^2(x^2 + 1)}$$

$$= \frac{x^2 - 2}{x^2(x^2 + 1)}. \tag{15}$$

Thus we must determine the constants A_2, A_1, B, and C so that the numerator equation

$$A_2(x^2 + 1) + A_1 x(x^2 + 1) + (Bx + C)x^2 = x^2 - 2 \tag{16}$$

holds. The lefthand side of (16) is of (formal) degree 3. Equating coefficients of x^3 in (16), we find we must have

$$A_1 + B = 0. \tag{17}$$

Equating coefficients of x^2 yields

$$A_2 + C = 1, \tag{18}$$

equating coefficients of x yields

$$A_1 = 0, \tag{19}$$

and, finally, equating constant terms yields

$$A_2 = -2. \tag{20}$$

From the system of four equations (17) through (20), we easily obtain

$$A_1 = 0, \qquad A_2 = -2, \qquad B = 0, \qquad C = 3,$$

so we have arrived at the partial-fraction decomposition

$$\frac{x^2 - 2}{x^4 + x^2} = \frac{x^2 - 2}{x^2(x^2 + 1)} = \frac{-2}{x^2} + \frac{0}{x} + \frac{0x + 3}{x^2 + 1}$$

$$= -\frac{2}{x^2} + \frac{3}{x^2 + 1}.$$

In this case, we "did not need" the constants A_1 and B, but in general, such constants need not be zero.

Let us now finish the integration problem posed at the start of this section. We have, using Example 1 and formula (3),

$$\int \frac{x^6 - 2}{x^4 + x^2}\,dx = \int \left(x^2 - 1 - \frac{2}{x^2} + \frac{3}{x^2 + 1}\right) dx$$
$$= \frac{1}{3}x^3 - x + \frac{2}{x} + 3\tan^{-1}x + C. \quad \|$$

The method of equating coefficients illustrated in Example 4 is a reliable way of finding the numerator constants in a partial-fraction decomposition, although it is not always the easiest method. Some aids have been developed. We mention only one. If the denominator $g(x)$ has a linear factor $x - a$, then setting $x = a$ in the numerator equation, like Eq. (16), gives one of the desired constants at once. For example, in Example 4, we have a linear factor $x = x - 0$ in the denominator. Setting $x = 0$ in (16), we obtain $A_2 = -2$. If $g(x)$ factors into linear factors all of multiplicity 1, then this device gives all the numerator constants quickly.

Example 5 Let us find the constants in the partial-fraction decomposition

$$\frac{x^2 + x - 3}{(x - 2)(x + 1)(x - 1)} = \frac{A}{x - 2} + \frac{B}{x + 1} + \frac{C}{x - 1}.$$

SOLUTION The numerator equation is

$$A(x + 1)(x - 1) + B(x - 2)(x - 1) + C(x - 2)(x + 1) = x^2 + x - 3. \tag{21}$$

Setting $x = 2$ yields

$$3A = 3,$$

setting $x = -1$ yields

$$6B = -3,$$

and setting $x = 1$ yields

$$-2C = -1,$$

so $A = 1$, $B = -\frac{1}{2}$, and $C = \frac{1}{2}$. ‖

9.2.4 Outline and examples

To integrate a rational function, we carry out the following steps.

STEP 1. Perform polynomial long division, if necessary, to reduce the problem to integrating a rational function whose numerator has degree less than the degree of the denominator.

STEP 2. Factor the denominator into linear and irreducible quadratic factors.

STEP 3. Obtain the partial-fraction decomposition of the rational function.

STEP 4. Integrate the summands in the resulting decomposition of the original rational function, using formulas (1) through (6) or other helpful formulas.

We should mention that the actual *execution* of this procedure may break down at Step 2. While our algebraic theory assures us that a polynomial $g(x)$ does have a factorization into linear and irreducible quadratic factors, *finding* such a factorization may be a *very* tough job. But this is the only flaw; all the other steps are mechanical chores.

Concerning Step 4, we should remark that in case $g(x)$ has a quadratic factor $ax^2 + bx + c$, you cannot expect the numerator of a partial fraction having a power of this factor in the denominator to be precisely $2ax + b$ so that you can apply formulas (4) or (6) directly; it may be necessary to fix up the integral, and use formulas (3) and (5) also. To illustrate,

$$\frac{4x - 5}{3x^2 + 4x + 2} = \frac{4}{6} \cdot \frac{6x + 4 - (46/4)}{3x^2 + 4x + 2}$$
$$= \frac{2}{3} \cdot \frac{6x + 4}{3x^2 + 4x + 2} - \frac{23}{3} \cdot \frac{1}{3x^2 + 4x + 2},$$

and formulas (4) and (3) can be used to integrate the two rational functions obtained.

We conclude with two illustrations of the whole technique.

Example 6 Let's find

$$\int \frac{13 - 7x}{(x + 2)(x - 1)^3}\, dx.$$

SOLUTION Since the degree of the numerator is less than that of the denominator, long division as in Step 1 is unnecessary, and the denominator is already factored for Step 2. For the partial-fraction decomposition (Step 3), let

$$\frac{13 - 7x}{(x + 2)(x - 1)^3} = \frac{A}{x + 2} + \frac{B_3}{(x - 1)^3} + \frac{B_2}{(x - 1)^2} + \frac{B_1}{x - 1}.$$

The numerator equation is

$$A(x - 1)^3 + B_3(x + 2) + B_2(x + 2)(x - 1) + B_1(x + 2)(x - 1)^2 = 13 - 7x. \qquad (22)$$

We find as many of the "unknown" constants as possible using the zeros -2 and 1 of the linear factors in the denominator. Setting $x = -2$ in (22) yields $-27A = 27$, so

$$A = -1.$$

Setting $x = 1$ yields $3B_3 = 6$, so

$$B_3 = 2.$$

To find B_2 and B_1, we need two equations containing them found by equating coefficients. The easiest equations to find are often those corresponding to the terms of highest degree and to the constant terms. Computing the coefficient of x^3 (which contains B_1), we obtain

$$A + B_1 = 0$$

so

$$B_1 = -A = 1.$$

Computing the constant terms, we have

$$-A + 2B_3 - 2B_2 + 2B_1 = 13,$$

so

$$2B_2 = -A + 2B_3 + 2B_1 - 13 = 1 + 4 + 2 - 13 = -6,$$

and

$$B_2 = -3.$$

Hence

$$\int \frac{13 - 7x}{(x+2)(x-1)^3}\,dx = \int\left(\frac{-1}{x+2} + \frac{2}{(x-1)^3} + \frac{-3}{(x-1)^2} + \frac{1}{x-1}\right)dx$$
$$= -\ln|x+2| - \frac{1}{(x-1)^2} + \frac{3}{x-1} + \ln|x-1| + C. \quad \|$$

Example 7 Let's find

$$\int \frac{3x^4 + 2x^3 + 8x^2 + x + 2}{x^5 + 2x^3 + x}\,dx.$$

SOLUTION Again, long division is unnecessary, and factoring the denominator yields

$$x^5 + 2x^3 + x = x(x^2 + 1)^2.$$

For the partial-fraction decomposition, we let

$$\frac{3x^4 + 2x^3 + 8x^2 + x + 2}{x^5 + 2x^3 + x} = \frac{A}{x} + \frac{B_2x + C_2}{(x^2+1)^2} + \frac{B_1x + C_1}{x^2 + 1}.$$

The numerator equation is

$$A(x^2+1)^2 + (B_2x + C_2)x + (B_1x + C_1)(x^3 + x)$$
$$= 3x^4 + 2x^3 + 8x^2 + x + 2. \quad (23)$$

Setting $x = 0$, we obtain

$$A = 2.$$

Equating coefficients of x^4 yields $A + B_1 = 3$, so

$$B_1 = 3 - A = 3 - 2 = 1.$$

Equating coefficients of x^3 yields

$$C_1 = 2.$$

Equating coefficients of x^2 yields $2A + B_2 + B_1 = 8$, so

$$B_2 = 8 - 2A - B_1 = 8 - 4 - 1 = 3.$$

Finally, equating coefficients of x yields $C_2 + C_1 = 1$, so

$$C_2 = 1 - C_1 = 1 - 2 = -1.$$

Thus we have

$$\int \frac{3x^4 + 2x^3 + 8x^2 + x + 2}{x(x^2+1)^2}\,dx = \int \left(\frac{2}{x} + \frac{3x - 1}{(x^2+1)^2} + \frac{x+2}{x^2+1}\right)dx. \tag{24}$$

Using formulas (6), (5), and (3), we obtain

$$\begin{aligned}\int \frac{3x - 1}{(x^2+1)^2}\,dx &= \frac{3}{2}\int \frac{2x}{(x^2+1)^2}\,dx - \int \frac{1}{(x^2+1)^2}\,dx \\ &= \frac{3}{2}\cdot\frac{-1}{x^2+1} - \left(\frac{2x}{1(4)(x^2+1)^2} + \frac{2(1)}{1(4)}\int \frac{1}{x^2+1}\,dx\right) \\ &= \frac{-3}{2(x^2+1)} - \frac{x}{2(x^2+1)^2} - \frac{1}{2}\tan^{-1}x + C. \end{aligned} \tag{25}$$

We also find that

$$\begin{aligned}\int \frac{x+2}{x^2+1}\,dx &= \frac{1}{2}\int \frac{2x}{x^2+1}\,dx + 2\int \frac{1}{x^2+1}\,dx \\ &= \frac{1}{2}\ln|x^2+1| + 2\tan^{-1}x + C. \end{aligned} \tag{26}$$

Then (24), (25), and (26) yield

$$\begin{aligned}&\int \frac{3x^4 + 2x^3 + 8x^2 + x + 2}{x(x^2+1)^2}\,dx \\ &\quad = 2\ln|x| - \frac{3}{2(x^2+1)} - \frac{x}{2(x^2+1)^2} - \frac{1}{2}\tan^{-1}x \\ &\quad\quad + \frac{1}{2}\ln|x^2+1| + 2\tan^{-1}x + C \\ &\quad = \frac{1}{2}\ln|x^6 + x^4| - \frac{3}{2(x^2+1)} - \frac{x}{2(x^2+1)^2} + \frac{3}{2}\tan^{-1}x + C. \quad \| \end{aligned}$$

SUMMARY *To integrate a quotient of polynomials, you may follow these steps.*

STEP 1. *Perform polynomial long division, if necessary, to reduce the problem to integrating a rational function whose numerator has degree less than the degree of the denominator.*

STEP 2. *Factor the denominator into linear and irreducible quadratic factors.*

STEP 3. *Obtain the partial-fraction decomposition of the rational function.*

STEP 4. *Integrate the summands in the resulting decomposition of the original rational function, using formulas (1) through (6) in this section, or other helpful formulas.*

EXERCISES

Transform the given integral into a sum of integrals of partial fractions and find the integral.

1. $\int \frac{1}{x^2 - 1}\,dx$
2. $\int \frac{1}{x^2 - x}\,dx$
3. $\int \frac{x^3 + 4}{x^2 - 4}\,dx$
4. $\int \frac{x^4 - 2x^2 + 6}{x^2 - 3x + 2}\,dx$
5. $\int \frac{2x + 1}{x^3 - x^2}\,dx$
6. $\int \frac{x - 1}{x^3 + 2x^2 + x}\,dx$
7. $\int \frac{3x + 4}{x^3 - 2x^2 - 3x}\,dx$
8. $\int \frac{x^4 + 1}{x^4 + 2x^3}\,dx$
9. $\int \frac{x^2 - 3}{(x - 2)(x + 1)^3}\,dx$
10. $\int \frac{2x^2 + 2x - 2}{x^3 + 2x}\,dx$
11. $\int \frac{2x^3 + 5x^2 - 2x + 16}{x^4 - 2x^2 - 8}\,dx$
12. $\int \frac{x^3 + 2x^2 + 3x + 1}{x^4 + 2x^2 + 1}\,dx$
13. $\int \frac{-x^3 + 10x^2 - 19x + 22}{(x - 1)^2(x^2 + 3)}\,dx$
14. $\int \frac{5x^3 - x^2 + 4x - 12}{x^4 - 4x^2}\,dx$
15. $\int \frac{3x^6 + 9x^4 - 4x^3 + 7x^2 + 3}{x^2(x^2 + 1)^3}\,dx$

9.3 SUBSTITUTION

Chapter 6 introduced the type of substitution technique illustrated in the following example.

Example 1 To find $\int x(x^2 + 3)^4\,dx$, note that $x\,dx$ is the differential of $x^2 + 3$, except for a constant factor. We let $u = x^2 + 3$, so $du = 2x\,dx$ and $x\,dx = (\frac{1}{2})\,du$. Then

$$\int x(x^2 + 3)^4\,dx = \int u^4 \cdot \frac{1}{2}\,du = \frac{1}{2} \cdot \frac{u^5}{5} + C = \frac{1}{10}(x^2 + 3)^5 + C. \quad \|$$

Success in this technique depends upon spotting a factor of the integrand to serve as u' for some expression u that already appears in the integral; then $u' \cdot dx = du$. If $f(x)$ can be expressed as $g(u) \cdot u'$ and if $\int g(u)\,du = G(u) + C$, then

$$\frac{d(G(u))}{dx} = \frac{d(G(u))}{du} \cdot \frac{du}{dx}$$
$$= g(u) \cdot u' = f(x),$$

so $G(u)$ is indeed an antiderivative of $f(x)$.

Here is another substitution example.

Example 2 Let's find $\int x\sqrt{x+1}\,dx$ by substitution.

SOLUTION The radical is causing us trouble, so we let $u = \sqrt{x+1}$ to eliminate it. Then $u^2 = x + 1$ so

$$x = u^2 - 1 \quad \text{and} \quad dx = 2u\,du.$$

(*This substitution for dx must be computed. A common error is just to replace dx by du.*) Therefore

$$\int x\sqrt{x+1}\,dx = \int (u^2 - 1)\,u \cdot 2u\,du = 2\int (u^4 - u^2)\,du$$
$$= 2\left(\frac{u^5}{5} - \frac{u^3}{3}\right) + C$$
$$= 2\left(\frac{(\sqrt{x+1})^5}{5} - \frac{(\sqrt{x+1})^3}{3}\right) + C$$
$$= 2(\sqrt{x+1})^3\left(\frac{x+1}{5} - \frac{1}{3}\right) + C$$
$$= 2(\sqrt{x+1})^3\left(\frac{x}{5} - \frac{2}{15}\right) + C. \quad \|$$

In the type of substitution illustrated in Example 2, you let $x = h(u)$, so $dx = h'(u)\,du$. You then write

$$\int f(x)\,dx = \int f(h(u)) \cdot h'(u)\,du = G(u) + C. \tag{1}$$

If $x = h(u)$ has an inverse, so that $u = h^{-1}(x)$, then (1) becomes $G(h^{-1}(x)) + C$. Again,

$$\frac{d(G(u))}{dx} = \frac{d(G(u))}{du} \cdot \frac{du}{dx} = f(h(u)) \cdot h'(u) \cdot \frac{du}{dx} = f(h(u)) \cdot \frac{dx}{du} \cdot \frac{du}{dx}$$
$$= f(h(u)) = f(x),$$

so $G(u) = G(h^{-1}(x))$ is indeed an antiderivative of $f(x)$. You see that the validity of this substitution process is a consequence of the chain rule for differentiation.

A warning As mentioned in Example 2, a common error when substituting $x = h(u)$ is to replace dx by du, instead of replacing dx by $h'(u)\,du$. *Don't forget to compute dx.* Naturally you must choose only substitutions $x = h(u)$ that have continuous derivatives and inverses.

Example 3 Let's try to find

$$\int e^{\sin^{-1}x}\,dx.$$

SOLUTION This integral looks pretty hopeless, but let's try to simplify it by the substitution

$$u = \sin^{-1}x.$$

Then $x = \sin u$, and $dx = \cos u\,du$. The integral becomes

$$\int e^{\sin^{-1}x}\,dx = \int e^{u}\cos u\,du.$$

A table (or integration by parts) yields

$$\int e^{u}\cos u\,du = \frac{e^{u}}{2}(\cos u + \sin u) + C.$$

Since $u = \sin^{-1}x$, we find that

$$\sin u = x \text{ and } \cos u = \sqrt{1 - \sin^2 u} = \sqrt{1 - x^2}.$$

Thus we have

$$\int e^{\sin^{-1}x}\,dx = \frac{e^{\sin^{-1}x}}{2}(x + \sqrt{1 - x^2}) + C. \quad \|$$

An *algebraic substitution* $x = h(u)$ is one where $h(u)$ is a function involving only arithmetic operations and roots (no trigonometric functions, for example).

Example 4 We use an algebraic substitution to find the integral

$$\int x^3\sqrt{4 - x^2}\,dx.$$

SOLUTION To eliminate the radical, we let $u^2 = 4 - x^2$. Implicit differentiation yields $2u(du) = -2x(dx)$, so

$$x\,dx = -u\,du.$$

We thus have

$$\int x^3\sqrt{4-x^2}\,dx = \int x^2\sqrt{4-x^2}(x\,dx) = \int (4-u^2)(u)(-u\,du)$$

$$= \int(-4u^2+u^4)\,du = -\frac{4u^3}{3}+\frac{u^5}{5}+C.$$

Since $u=\sqrt{4-x^2}$, we obtain

$$\int x^3\sqrt{4-x^2}\,dx = -\frac{4(4-x^2)^{3/2}}{3}+\frac{(4-x^2)^{5/2}}{5}+C. \quad \|$$

Example 5 Let's find

$$\int \frac{x^{1/2}}{1+x^{1/3}}\,dx.$$

SOLUTION To eliminate the fractional powers $x^{1/3}$ and $x^{1/2}$ at the same time, we let $x=u^6$, so that $dx=6u^5\,du$ and

$$\int \frac{x^{1/2}}{1+x^{1/3}}\,dx = \int \frac{u^3}{1+u^2}6u^5\,du.$$

Application of the method of partial fractions yields

$$6\int \frac{u^8}{u^2+1}\,du = 6\int\left(u^6-u^4+u^2-1+\frac{1}{u^2+1}\right)du$$

$$= 6\left(\frac{u^7}{7}-\frac{u^5}{5}+\frac{u^3}{3}-u+\tan^{-1}u\right)+C.$$

Since $u=x^{1/6}$, we have

$$\int \frac{x^{1/2}}{1+x^{1/3}}\,dx = 6\left(\frac{x^{7/6}}{7}-\frac{x^{5/6}}{5}+\frac{x^{1/2}}{3}-x^{1/6}+\tan^{-1}x^{1/6}\right)+C. \quad \|$$

You should not think that all integrals with radicals will yield to algebraic substitution. Look at the following example.

Example 6 For the integral $\int x^4\sqrt{9-x^2}\,dx$, it is natural to try the substitution $u^2 = 9-x^2$ so $2u\,du=-2x\,dx$ or $x\,dx=-u\,du$. Then

$$\int x^4\sqrt{9-x^2}\,dx = \int x^3\sqrt{9-x^2}\,x\,dx = \int x^3u(-u\,du).$$

But since $u^2=9-x^2$, we have $x^2=9-u^2$, so

$$x^3 = x^2\cdot x = (9-u^2)\sqrt{9-u^2}$$

and our integral becomes

$$\int x^3u(-u\,du) = \int(9-u^2)\sqrt{9-u^2}\,u(-u\,du) = -\int(9u^2-u^4)\sqrt{9-u^2}\,du.$$

We have not eliminated the radical, and clearly have just as tough an integral. We will see how to find this integral in Section 6. ||

SUMMARY

1. *If $f = h(u)$, then $\int f(x)\,dx$ becomes $\int f(h(u)) \cdot h'(u)\,du$. If the latter integral can be computed as $G(u) + C$, then solve the original substitution for u in terms of x, and substitute in $G(u) + C$ to obtain $\int f(x)\,dx$ in terms of x.*
2. *Integration of expressions involving fractional powers may yield to algebraic substitution designed to eliminate all the fractional powers.*

EXERCISES

1. Integrals that yield easily to one technique often yield easily to another. Show that $\int x\sqrt{x+1}\,dx$ of Example 2 and $\int x^3\sqrt{4-x^2}\,dx$ of Example 4 both can be found using integration by parts.

In Exercises 2 through 18, find the given integral.

2. $\displaystyle\int \frac{x}{\sqrt{1+x^2}}\,dx$
3. $\displaystyle\int \frac{x-3}{\sqrt{4-x^2}}\,dx$
4. $\displaystyle\int \frac{x^3}{\sqrt{1+x^2}}\,dx$

5. $\displaystyle\int \frac{\cos x}{\sqrt{4-\sin^2 x}}\,dx$
6. $\displaystyle\int \frac{x}{\sqrt{1+x}}\,dx$
7. $\displaystyle\int x\sqrt{3-2x}\,dx$

8. $\displaystyle\int x^2\sqrt{4+x}\,dx$
9. $\displaystyle\int \frac{\sqrt{x}-1}{x}\,dx$
10. $\displaystyle\int \frac{\sqrt{x}}{\sqrt{x}+1}\,dx$

11. $\displaystyle\int \frac{\sqrt{x}-1}{\sqrt{x}}\,dx$
12. $\displaystyle\int \frac{x^{1/2}}{4+x^{1/3}}\,dx$
13. $\displaystyle\int \frac{\sqrt{x+2}-1}{\sqrt{x+2}+1}\,dx$

14. $\displaystyle\int \frac{x^5}{\sqrt{1-x^3}}\,dx$
15. $\displaystyle\int x^3(x^2+1)^{3/2}\,dx$
16. $\displaystyle\int \frac{dx}{x^{1/2}+x^{1/3}}$

17. $\displaystyle\int \frac{x^{2/3}}{1+x^{1/2}}\,dx$
18. $\displaystyle\int xe^{\sin^{-1}x}\,dx$

9.4 INTEGRATION OF RATIONAL FUNCTIONS OF sin x AND cos x

We described in Section 9.2 how the integration of any rational function can (theoretically) be accomplished. We show in this article how the integration of any rational expression (quotient of polynomials) in $\sin x$ and $\cos x$ can be reduced to the integration of a rational function by substitution. Since $\tan x$, $\cot x$, $\sec x$, and $\csc x$ can all in turn be written as rational expressions in $\sin x$ and $\cos x$, the technique will actually provide us with a method of integrating any rational expression in the trigonometric functions.

We make the substitution

$$x = 2\tan^{-1}t, \tag{1}$$

so that

$$t = \tan\frac{x}{2}, \tag{2}$$

which was discovered by some ingenious person. Using identities for trigonometric functions, we then obtain

$$\sin x = 2\sin\frac{x}{2}\cos\frac{x}{2} = 2\tan\frac{x}{2}\cos^2\frac{x}{2} = 2\frac{\tan(x/2)}{\sec^2(x/2)}.$$

$$= 2\frac{\tan(x/2)}{1+\tan^2(x/2)} = \frac{2t}{1+t^2}$$

Similarly,

$$\cos x = 2\cos^2\frac{x}{2} - 1 = \frac{2}{\sec^2(x/2)} - 1 = \frac{2}{1+\tan^2(x/2)} - 1$$

$$= \frac{2}{1+t^2} - 1$$

$$= \frac{1-t^2}{1+t^2}.$$

Finally,

$$dx = 2\frac{1}{1+t^2}\,dt.$$

Collecting these formulas in one place, we have

$$\begin{aligned}\sin x &= \frac{2t}{1+t^2},\\ \cos x &= \frac{1-t^2}{1+t^2},\\ dx &= \frac{2}{1+t^2}\,dt.\end{aligned} \tag{3}$$

Clearly the formulas (3) may be used to convert the integration of a rational expression in $\sin x$ and $\cos x$ to the integration of a rational function of t.

You should not blindly make the substitution (3) with every integral of a rational expression in $\sin x$ and $\cos x$. First, look for a shorter method, or try to find the integral in a table. We illustrate with examples.

Example 1 The indefinite integral

$$\int \frac{1-\cos x}{1+\cos x}\,dx$$

can be found using (3), but it is easier to use the "trick"

$$\int \frac{1 - \cos x}{1 + \cos x}\, dx = \int \frac{(1 - \cos x)^2}{(1 + \cos x)(1 - \cos x)}\, dx = \int \frac{(1 - \cos x)^2}{1 - \cos^2 x}\, dx$$

$$= \int \frac{1 - 2\cos x + \cos^2 x}{\sin^2 x}\, dx$$

$$= \int (\csc^2 x - 2\csc x \cot x + \cot^2 x)\, dx$$

$$= \int (\csc^2 x - 2\csc x \cot x + \csc^2 x - 1)\, dx$$

$$= \int (2\csc^2 x - 2\csc x \cot x - 1)\, dx$$

$$= -2\cot x + 2\csc x - x + C. \quad \|$$

Example 2 Using a table like the one in this text, the integral

$$\int \frac{1}{2 + \cos x}\, dx$$

is easily found to be

$$\int \frac{1}{2 + \cos x}\, dx = \frac{2}{\sqrt{3}} \tan^{-1}\left(\frac{1}{\sqrt{3}} \tan \frac{x}{2}\right) + C. \quad \|$$

Example 3 Let us find the integral

$$\frac{1}{3\sin x + 4\cos x}\, dx.$$

SOLUTION This integral is not found in the table in this text. Using the variable transformation in (3), we have

$$\int \frac{1}{3\sin x + 4\cos x}\, dx = \int \frac{1}{3[2t/(1+t^2)] + 4[(1-t^2)/(1+t^2)]} \cdot \frac{2}{1+t^2}\, dt$$

$$= \int \frac{2}{6t + 4 - 4t^2}\, dt = -\int \frac{dt}{2t^2 - 3t - 2}$$

$$= -\int \frac{dt}{(2t+1)(t-2)} = -\int \left(\frac{-2/5}{2t+1} + \frac{1/5}{t-2}\right) dt$$

$$= \tfrac{1}{5} \ln |2t + 1| - \tfrac{1}{5} \ln |t - 2| + C.$$

Since $t = \tan(x/2)$ by Eq. (2), we obtain finally

$$\int \frac{dx}{3\sin x + 4\cos x} = \frac{1}{5} \ln \left| 2\left(\tan \frac{x}{2}\right) + 1 \right| - \frac{1}{5} \ln \left| \left(\tan \frac{x}{2}\right) - 2 \right| + C. \quad \|$$

SUMMARY

1. *A rational function of* sin x *and* cos x *can be integrated using the substitution*

$$\sin x = \frac{2t}{1+t^2}, \qquad \cos x = \frac{1-t^2}{1+t^2}, \qquad dx = \frac{2}{1+t^2}\,dt.$$

Use the method of partial fractions to integrate the resulting rational function of t, and after integration, substitute $t = \tan(x/2)$.

2. *Always look for an easier method before making the substitution just given.*

EXERCISES

1. Find $\int \sec x\,dx = \int (1/\cos x)\,dx$, using the method described in this section.

2. Find $\int \csc x\,dx = \int (1/\sin x)\,dx$, using the method described in this section.

In the remaining exercises, find the indicated indefinite integral by any method, including the use of the tables in the text.

3. $\displaystyle\int \frac{\sin x}{1+\cos x}\,dx$

4. $\displaystyle\int \frac{dx}{1+\sin x}$

5. $\displaystyle\int \frac{dx}{2-\sin x}$

6. $\displaystyle\int \frac{dx}{\sin x+\cos x}$

7. $\displaystyle\int \frac{1-\sin x}{1+\cos x}\,dx$

8. $\displaystyle\int \sqrt{1-\cos 2x}\,dx$

9. $\displaystyle\int \frac{\sin x}{\sin x+\cos x}\,dx$

10. $\displaystyle\int \frac{\cos x-\sin x}{\sin x}\,dx$

11. $\displaystyle\int \frac{dx}{1+\tan x}$

12. $\displaystyle\int \frac{dx}{4\sin x+3\cos x}$

9.5 INTEGRATION OF POWERS OF TRIGONOMETRIC FUNCTIONS

All the integrals treated in this section can be found most easily, and with the least chance for error, by using integral tables. This section may be omitted, especially if time is short. Our feeling toward this section is indicated by Exercises 26 and 27 in Section 6.3.

9.5.1 Integration of odd powers of sin x and cos x

An integral of the form

$$\int \sin^m x \cos^n x\,dx$$

where either m or n is an odd positive integer can be computed by the following device. If m is odd, "save" one factor $\sin x$ and change all other factors of $\sin^2 x$ to $1 - \cos^2 x$. One then obtains an integral of the form

$$\int f(\cos x)\sin x\,dx,$$

where $f(\cos x)$ is a polynomial in $\cos x$. This integral can easily be found. If n

is odd, a similar procedure can be followed, "saving" one factor $\cos x$ and changing all other factors $\cos^2 x$ to $1 - \sin^2 x$.

Example 1 You have

$$\begin{aligned}\int \sin^3 x \cos^2 x\, dx &= \int \sin x(1 - \cos^2 x)\cos^2 x\, dx \\ &= \int (\cos^2 x - \cos^4 x)\sin x\, dx \\ &= -\tfrac{1}{3}\cos^3 x + \tfrac{1}{5}\cos^5 x + C. \quad \|\end{aligned}$$

Example 2 You have

$$\begin{aligned}\int \cos^5 x\, dx &= \int (1 - \sin^2 x)^2 \cos x\, dx \\ &= \int (1 - 2\sin^2 x + \sin^4 x)\cos x\, dx \\ &= \sin x - \tfrac{2}{3}\sin^3 x + \tfrac{1}{5}\sin^5 x + C. \quad \|\end{aligned}$$

9.5.2 Integration of even powers of sin *x* and cos *x*

An integral of the form

$$\int \sin^m x \cos^n x\, dx$$

where both m and n are nonnegative even integers is more difficult to evaluate without tables than the case where either m or n is odd. This time, we use the relation $\sin^2 x + \cos^2 x = 1$, or other trigonometric relations, to express the integrand as a sum of even powers of $\sin x$ only or $\cos x$ only, and then make use of the relations

$$\sin^2 ax = \frac{1 - \cos 2ax}{2} \qquad \text{and} \qquad \cos^2 bx = \frac{1 + \cos 2bx}{2} \tag{1}$$

repeatedly, until we obtain just first powers of cosine functions. The technique is best illustrated by examples.

Example 3 Let's find $\int \sin^4 x\, dx$.

SOLUTION Using (1) repeatedly, we have

$$\begin{aligned}\int \sin^4 x\, dx &= \int (\sin^2 x)^2\, dx = \int \left(\frac{1 - \cos 2x}{2}\right)^2 dx \\ &= \frac{1}{4}\int (1 - 2\cos 2x + \cos^2 2x)\, dx = \frac{1}{4}\int \left(1 - 2\cos 2x + \frac{1 + \cos 4x}{2}\right) dx \\ &= \frac{1}{8}\int (3 - 4\cos 2x + \cos 4x)\, dx = \frac{3}{8}x - \frac{1}{4}\sin 2x + \frac{1}{32}\sin 4x + C. \quad \|\end{aligned}$$

Example 4 Let's find

$$\int \sin^2 x \cos^2 x \, dx.$$

SOLUTION We use the relation $\sin x \cos x = \frac{1}{2} \sin 2x$ followed by (1), and obtain

$$\int \sin^2 x \cos^2 x \, dx = \frac{1}{4} \int \sin^2 2x \, dx = \frac{1}{4} \int \frac{1 - \cos 4x}{2} \, dx$$
$$= \tfrac{1}{8}x - \tfrac{1}{32} \sin 4x + C. \quad \|$$

9.5.3 Integration of expressions such as sin *mx* cos *nx*

The relations

$$\sin mx \sin nx = \tfrac{1}{2}[\cos (m - n)x - \cos (m + n)x] \quad (2)$$

$$\sin mx \cos nx = \tfrac{1}{2}[\sin (m - n)x + \sin (m + n)x] \quad (3)$$

$$\cos mx \cos nx = \tfrac{1}{2}[\cos (m - n)x + \cos (m + n)x] \quad (4)$$

enable us to integrate $\sin mx \sin nx$, $\sin mx \cos nx$, and $\cos mx \cos nx$ without tables. The formulas (2), (3), and (4) are easily derived from the more familiar formulas

$$\sin (mx + nx) = \sin mx \cos nx + \cos mx \sin nx \quad (5)$$

$$\cos (mx + nx) = \cos mx \cos nx - \sin mx \sin nx \quad (6)$$

(see Exercise 1).

Example 5 We have, using (3),

$$\int \sin 2x \cos 3x \, dx = \frac{1}{2} \int [\sin (-x) + \sin 5x] \, dx = \frac{1}{2} \int [-\sin x + \sin 5x] \, dx$$
$$= \tfrac{1}{2} \cos x - \tfrac{1}{10} \cos 5x + C. \quad \|$$

9.5.4 Integration of powers of tan *x* and cot *x*, and even powers of sec *x* and csc *x*

We make use of the identities

$$1 + \tan^2 x = \sec^2 x \qquad \text{or} \qquad \tan^2 x = \sec^2 x - 1 \quad (7)$$

and

$$1 + \cot^2 x = \csc^2 x \qquad \text{or} \qquad \cot^2 x = \csc^2 x - 1. \quad (8)$$

To integrate a power of $\tan x$ or an even power of $\sec x$, we use (7) and the fact that $d(\tan x) = \sec^2 x \, dx$.

Example 6 To integrate an even power of $\sec x$, we "save" one factor $\sec^2 x$ to serve in du and change all the other powers of $\sec^2 x$ to powers of $u = \tan x$. To illustrate,

$$\begin{aligned}\int \sec^6 x\,dx &= \int (1+\tan^2 x)^2 \sec^2 x\,dx\\ &= \int (1 + 2\tan^2 x + \tan^4 x)\sec^2 x\,dx\\ &= \int \sec^2 x\,dx + 2\int \tan^2 x\sec^2 x\,dx + \int \tan^4 x \sec^2 x\,dx\\ &= \tan x + 2\cdot\frac{\tan^3 x}{3} + \frac{\tan^5 x}{5} + C. \quad \|\end{aligned}$$

Example 7 To illustrate the integration of a power of tan x, we have

$$\begin{aligned}\int \tan^5 x\,dx &= \int \tan^3 x(\sec^2 x - 1)\,dx = \int \tan^3 x \sec^2 x\,dx - \int \tan^3 x\,dx\\ &= \frac{\tan^4 x}{4} - \int \tan x(\sec^2 x - 1)\,dx\\ &= \frac{\tan^4 x}{4} - \int \tan x \sec^2 x\,dx + \int \tan x\,dx\\ &= \frac{\tan^4 x}{4} - \frac{\tan^2 x}{2} + \int \frac{\sin x}{\cos x}\,dx\\ &= \frac{\tan^4 x}{4} - \frac{\tan^2 x}{2} - \ln|\cos x| + C. \quad \|\end{aligned}$$

Obviously powers of cot x and even powers of csc x can be integrated in a similar fashion using (8) and $d(\cot x) = -\csc^2 x\,dx$.

9.5.5 Reduction formulas for powers of tan x, cot x, sec x, and csc x

Odd powers of sec x and csc x are tough to integrate without tables. They can be done using integration by parts. We ask you to experiment with $\int \sec^3 x\,dx$ in Exercise 20. Of course the best way to integrate any power of tan x, cot x, sec x, or csc x is to use a table and make repeated use of the appropriate reduction formula

$$\int \tan^n ax\,dx = \frac{\tan^{n-1} ax}{a(n-1)} - \int \tan^{n-2} ax\,dx, \quad n \neq 1 \tag{9}$$

$$\int \cot^n ax\,dx = -\frac{\cot^{n-1} ax}{a(n-1)} - \int \cot^{n-2} ax\,dx, \quad n \neq 1 \tag{10}$$

$$\int \sec^n ax\,dx = \frac{\sec^{n-2} ax \tan ax}{a(n-1)} + \frac{n-2}{n-1}\int \sec^{n-2} ax\,dx, \quad n \neq 1 \tag{11}$$

$$\int \csc^n ax\,dx = -\frac{\csc^{n-2} ax \cot ax}{a(n-1)} + \frac{n-2}{n-1}\int \csc^{n-2} ax\,dx, \quad n \neq 1. \tag{12}$$

This will reduce the problem to the integration of the function to the first power (or to integration of a constant). Then use

$$\int \tan x\, dx = -\ln|\cos x| + C \tag{13}$$

$$\int \cot x\, dx = \ln|\sin x| + C \tag{14}$$

$$\int \sec x\, dx = \ln|\sec x + \tan x| + C \tag{15}$$

$$\int \csc x\, dx = -\ln|\csc x + \cot x| + C \tag{16}$$

which you have already seen. You are asked to derive the reduction formulas (9) through (12) in the exercises, using (7), (8), and integration by parts.

Example 8 We have, using (11),

$$\begin{aligned}\int \sec^5 2x\, dx &= \frac{\sec^3 2x \tan 2x}{2(4)} + \frac{3}{4}\int \sec^3 2x\, dx \\ &= \frac{\sec^3 2x \tan 2x}{8} + \frac{3}{4}\left[\frac{\sec 2x \tan 2x}{2(2)} + \frac{1}{2}\int \sec 2x\, dx\right] \\ &= \frac{\sec^3 2x \tan 2x}{8} + \frac{3}{16}\sec 2x \tan 2x \\ &\quad + \tfrac{3}{16}\ln|\sec 2x + \tan 2x| + C. \quad \|\end{aligned}$$

Finally, remember always to see whether there is an easy way to find an integral before you blindly reach for a "rule." For example,

1. $\displaystyle\int \tan x \sec^2 x\, dx = \int \frac{\sin x}{\cos x}\cdot\frac{1}{\cos^2 x}\, dx = \int \frac{\sin x}{\cos^3 x}\, dx$

   could be done by the $t = \tan(x/2)$ substitution.

   OUUUUUUUUUUUUUUCH!

2. $\displaystyle\int \tan x \sec^2 x\, dx = \int \tan x(1 + \tan^2 x)\, dx = \int (\tan x + \tan^3 x)\, dx$

   could be done using (9) and (13).

   GRRRRRRRRRRRRRRRR!

3. Since $d(\tan x) = \sec^2 x\,dx$, of course $\int \tan x \sec^2 x\,dx = (\tan^2 x)/2 + C$.

YUUUUUMMMMMMMMMMMM!

SUMMARY

1. *Odd powers of* $\sin x$ *can be integrated by "saving" one factor* $\sin x$ *to serve in du and changing all other factors* $\sin^2 x$ *to powers of* $u = \cos x$ *using* $\sin^2 x = 1 - \cos^2 x$. *Odd powers of* $\cos x$ *are integrated similarly.*
2. *Even powers of* $\sin x$ *or* $\cos x$ *can be integrated using*
$$\sin^2 ax = \frac{1 - \cos 2ax}{2} \quad \text{and} \quad \cos^2 bx = \frac{1 + \cos 2bx}{2}$$
repeatedly.
3. *Formulas* (2), (3), *and* (4) *of this section can be used to integrate expressions involving* $\sin mx \sin nx$, $\cos mx \cos nx$, *and* $\sin mx \cos nx$.
4. *Powers of* $\tan x$, $\cot x$, *and even powers of* $\sec x$, $\csc x$ *can be integrated using*
$$1 + \tan^2 x = \sec^2 x \quad \text{and} \quad 1 + \cot^2 x = \csc^2 x.$$
5. *All powers of* $\tan x$, $\cot x$, $\sec x$, *and* $\csc x$ *are easily integrated using the reduction formulas* (9) *through* (12) *and formulas* (13) *through* (16) *of this section.*

EXERCISES

1. Derive Formulas (2), (3), and (4) from (5), (6), and other well-known trigonometric relations.

In Exercises 2 through 24, compute the indicated integrals without the use of tables.

2. $\int \sin^2 x \cos^3 x\,dx$

3. $\int \sin^3 2x \sqrt{\cos 2x}\,dx$

4. $\int \sin^3 x \cos^4 x\,dx$

5. $\int \frac{\cos^3 x}{\sqrt{\sin x}}\,dx$

6. $\int \sin^4 x \cos^5 x\,dx$

7. $\int \frac{\sin^3 x}{\cos^2 x}\,dx$

8. $\int \sin^2 3x\,dx$

9. $\int \sin^2 3x \cos^2 3x\,dx$

10. $\int \cos^4 2x\,dx$

11. $\int \sin^4 x \cos^2 x\,dx$

12. $\int \frac{1}{\sin^2 x}\,dx$

13. $\int \sec^4 x\,dx.$

14. $\int \csc^4 x\,dx$

15. $\int \sec^4 x \tan^2 x\,dx$

16. $\int \frac{\cos^2 x}{\sin^4 x}\,dx$

17. $\int \frac{\sin 3x}{\cos^3 3x}\,dx$

18. $\int \tan^4 x\,dx$

19. $\int \frac{dx}{\cos^6 x}$

20. $\int \sec^3 x\, dx$ [*Hint.* Use integration by parts.]

21. $\int \tan^7 2x\, dx$

22. $\int \tan^3 x \sec^2 x\, dx$

23. $\int \csc^4 x \cot x\, dx$

24. $\int \csc^4 5x \cot^3 5x\, dx$

In Exercises 25 *through* 30, *derive the indicated reduction formula.*

25. The formula in Eq. (9) of this section.

26. The formula in Eq. (10) of this section.

27. The formula in Eq. (11) of this section.

28. The formula in Eq. (12) of this section.

29. $\int \sin^n ax\, dx = \dfrac{-\sin^{n-1} ax \cos ax}{na} + \dfrac{n-1}{n}\int \sin^{n-2} ax\, dx$ [*Note.* This is, the best way to integrate powers of sin x.]

30. $\int \cos^n ax\, dx = \dfrac{\cos^{n-1} ax \sin ax}{na} + \dfrac{n-1}{n}\int \cos^{n-2} ax\, dx$ [*Note.* This is the best way to integrate powers of cos x.]

9.6 TRIGONOMETRIC SUBSTITUTION

9.6.1 Integrals involving $\sqrt{a^2 \pm x^2}$ and $\sqrt{x^2 - a^2}$

The relations

$$a^2 - a^2 \sin^2 t = a^2 \cos^2 t,$$

$$a^2 + a^2 \tan^2 t = a^2 \sec^2 t,$$

$$a^2 \sec^2 t - a^2 = a^2 \tan^2 t$$

are sometimes useful in eliminating the radicals from integrals involving $\sqrt{a^2 - x^2}$, $\sqrt{a^2 + x^2}$, or $\sqrt{x^2 - a^2}$. The resulting integral involves trigonometric functions and is sometimes easy to integrate, or might be found in a table, or (as a last resort) might be integrated by the $t = \tan(x/2)$ substitution. We illustrate with examples.

Example 1 Let us find

$$\int x^3\sqrt{4 - x^2}\, dx$$

which is not found in the little table in this book.

SOLUTION If we let $x = 2 \sin t$, then $dx = 2 \cos t\, dt$, and we obtain

$$\int x^3\sqrt{4 - x^2}\, dx = \int 8 \sin^3 t(\sqrt{4 - 4\sin^2 t})(2 \cos t)\, dt$$

$$= \int 8 \sin^3 t(4 \cos^2 t)\, dt = 32\int \sin t(1 - \cos^2 t)(\cos^2 t)\, dt$$

$$= 32\int (\cos^2 t - \cos^4 t) \sin t\, dt = -\frac{32 \cos^3 t}{3} + \frac{32 \cos^5 t}{5} + C.$$

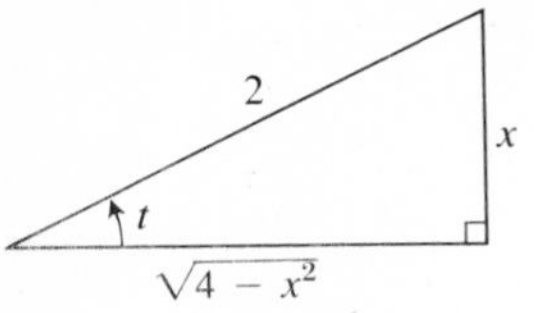

(a) $x = 2 \sin t$ substitution

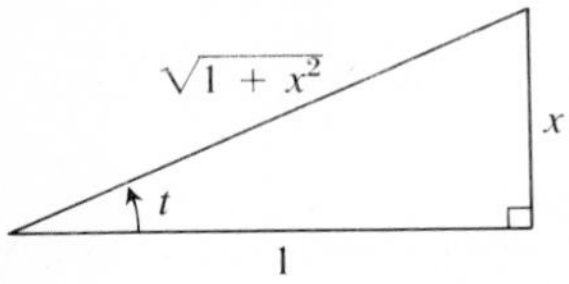

(b) $x = \tan t$ substitution

9.2

From $x = 2 \sin t$, we obtain

$$\begin{aligned}\cos t &= \sqrt{1 - \sin^2 t}\\ &= \sqrt{1 - (x/2)^2} = \tfrac{1}{2}\sqrt{4 - x^2},\end{aligned}$$

so we obtain for our integral

$$\int x^3\sqrt{4 - x^2}\, dx = -\frac{4(4 - x^2)^{3/2}}{3} + \frac{(4 - x^2)^{5/2}}{5} + C. \quad \|$$

When you make a substitution such as $x = 2 \sin t$ in Example 1 and then integrate, the answer obtained is in terms of trigonometric functions of t rather than a function of x. The triangle shown in Fig. 9.2(a) is helpful in figuring out the trigonometric functions of t in terms of x. If $x = 2 \sin t$, then $\sin t = x/2$, so we label the side opposite the angle t with x and the hypotenuse with 2. The remaining side is then found using the Pythagorean theorem. The triangle for the substitution $x = \tan t$, which is used in the following example, is shown in Fig. 9.2(b).

Example 2 We compute

$$\int x^2\sqrt{1 + x^2}\, dx.$$

SOLUTION This time, we let $x = \tan t$, so that $dx = \sec^2 t\, dt$. We obtain

$$\begin{aligned}\int x^2\sqrt{1 + x^2}\, dx &= \int \tan^2 t\sqrt{1 + \tan^2 t}\sec^2 t\, dt\\ &= \int \tan^2 t \sec^3 t\, dt = \int (\sec^2 t - 1)\sec^3 t\, dt\\ &= \int (\sec^5 t - \sec^3 t)\, dt.\end{aligned}$$

We use a reduction formula from the table for powers of secant, and obtain

$$\int \sec^5 t\, dt = \frac{\sec^3 t \tan t}{4} + \frac{3}{4}\int \sec^3 t\, dt.$$

Our integral thus takes the form

$$\begin{aligned}\int (\sec^5 t - \sec^3 t)\, dt &= \frac{\sec^3 t \tan t}{4} - \frac{1}{4}\int \sec^3 t\, dt\\ &= \frac{\sec^3 t \tan t}{4} - \frac{1}{4}\left(\frac{\sec t \tan t}{2} + \frac{1}{2}\int \sec t\, dt\right)\\ &= \frac{\sec^3 t \tan t}{4} - \frac{1}{8}\sec t \tan t - \frac{1}{8}\ln|\sec t + \tan t| + C.\end{aligned}$$

From $x = \tan t$, we obtain $\sec t = \sqrt{1 + \tan^2 t} = \sqrt{1 + x^2}$. Thus we have

$$\int x^2\sqrt{1 + x^2}\, dx = \frac{x(1 + x^2)^{3/2}}{4} - \frac{1}{8} x\sqrt{1 + x^2} - \frac{1}{8} \ln |\sqrt{1 + x^2} + x| + C. \quad \|$$

Example 3 Let's find $\int \sqrt{a^2 - x^2}\, dx$ without the use of tables.

SOLUTION If we let $x = a \sin t$ so that $dx = a \cos t\, dt$, then

$$\begin{aligned}\int \sqrt{a^2 - x^2}\, dx &= \int \sqrt{a^2 - a^2 \sin^2 t}\,(a \cos t)\, dt\\ &= a^2 \int \cos^2 t\, dt = a^2 \int \frac{1 + \cos 2t}{2}\, dt\\ &= \frac{a^2}{2}\left(t + \frac{1}{2} \sin 2t\right) + C = \frac{a^2}{2}(t + \sin t \cos t) + C\\ &= \frac{a^2}{2}\left(\sin^{-1} \frac{x}{a} + \frac{x}{a}\sqrt{1 - \frac{x^2}{a^2}}\right) + C\\ &= \frac{1}{2} a^2 \sin^{-1} \frac{x}{a} + \frac{x}{2}\sqrt{a^2 - x^2} + C. \quad \|\end{aligned}$$

9.6.2 Integrals involving $\sqrt{ax^2 + bx + c}$, where $a \neq 0$

By completing the square, the integral of an expression involving $\sqrt{ax^2 + bx + c}$ may be able to be handled by the method of the preceding article. We illustrate with two examples.

Example 4 Let us find $\int \sqrt{x^2 - 2x}\, dx$.

SOLUTION We have

$$\int \sqrt{x^2 - 2x}\, dx = \int \sqrt{(x - 1)^2 - 1}\, dx.$$

If we let $x - 1 = \sec t$, then we obtain $dx = \sec t \tan t\, dt$, so

$$\begin{aligned}\int \sqrt{(x - 1)^2 - 1}\, dx &= \int \sqrt{\sec^2 t - 1}\, \sec t \tan t\, dt\\ &= \int \tan^2 t \sec t\, dt = \int (\sec^2 t - 1) \sec t\, dt\\ &= \int (\sec^3 t - \sec t)\, dt.\end{aligned}$$

We should know $\int \sec t\, dt$, but pretending we have no tables, we shall integrate $\int \sec^3 t\, dt$ by parts. We let

$$u = \sec t \qquad dv = \sec^2 t\, dt$$
$$du = \sec t \tan t\, dt \qquad v = \tan t$$

so that

$$\int \sec^3 t\,dt = \sec t \tan t - \int \sec t \tan^2 t\,dt$$

$$= \sec t \tan t - \int (\sec^3 t - \sec t)\,dt$$

$$= \sec t \tan t + \ln|\sec t + \tan t| - \int \sec^3 t\,dt.$$

Solving this equation for $\int \sec^3 t\,dt$, we obtain

$$\int \sec^3 t\,dt = \tfrac{1}{2}\sec t \tan t + \tfrac{1}{2}\ln|\sec t + \tan t| + C.$$

We now obtain for our original integral

$$\int \sqrt{x^2 - 2x}\,dx = \int \sec^3 t\,dt - \int \sec t\,dt$$

$$= \tfrac{1}{2}\sec t \tan t - \tfrac{1}{2}\ln|\sec t + \tan t| + C$$

$$= \tfrac{1}{2}(x-1)\sqrt{(x-1)^2 - 1} - \tfrac{1}{2}\ln|(x-1) + \sqrt{(x-1)^2 - 1}| + C$$

$$= \tfrac{1}{2}((x-1)\sqrt{x^2 - 2x} - \ln|x - 1 + \sqrt{x^2 - 2x}|) + C. \quad \|$$

Example 5 Let us find $\int (1/\sqrt{1 + 4x - x^2})\,dx$.

SOLUTION Completing the square, we have

$$\int \frac{dx}{\sqrt{1 + 4x - x^2}} = \int \frac{dx}{\sqrt{5 - (x-2)^2}}.$$

We let

$$x - 2 = \sqrt{5}\sin t,$$

so that

$$dx = \sqrt{5}\cos t\,dt.$$

Then

$$\int \frac{dx}{\sqrt{5 - (x-2)^2}} = \int \frac{\sqrt{5}\cos t}{\sqrt{5 - 5\sin^2 t}}\,dt = \int \frac{\sqrt{5}\cos t}{\sqrt{5}\cos t}\,dt$$

$$= \int 1 \cdot dt = t + C = \sin^{-1}\frac{x-2}{\sqrt{5}} + C. \quad \|$$

SUMMARY

1. *If* $x = a \sin t$, *then* $\sqrt{a^2 - x^2} = a \cos t$. $\sqrt{a^2 - (a \sin t)^2}$
2. *If* $x = a \tan t$, *then* $\sqrt{a^2 + x^2} = a \sec t$. $\sqrt{a^2 + (a \tan t)^2}$
3. *If* $x = a \sec t$, *then* $\sqrt{x^2 - a^2} = a \tan t$. $\sqrt{(a \sec t)^2 - a^2}$
4. *To eliminate the radical from* $\sqrt{ax^2 + bx + c}$ *for* $a \neq 0$, *complete the square and then use the appropriate substitution above.*

EXERCISES

Find these integrals without using tables.

1. $\int \frac{x}{\sqrt{1+x^2}}\,dx$
2. $\int \frac{x-3}{\sqrt{4-x^2}}\,dx$
3. $\int \frac{x^3}{\sqrt{1+x^2}}\,dx$
4. $\int (16-x^2)^{3/2}\,dx$
5. $\int x^3\sqrt{x^2-1}\,dx$
6. $\int \sqrt{4-x^2}\,dx$
7. $\int \frac{x}{\sqrt{5+x^2}}\,dx$
8. $\int \frac{dx}{\sqrt{x^2+1}}$
9. $\int \frac{x-1}{\sqrt{x^2-16}}\,dx$
10. $\int \frac{dx}{x\sqrt{x^2-4}}$
11. $\int \sqrt{1+4x^2}\,dx$
12. $\int \sqrt{x^2+2x+2}\,dx$
13. $\int \sqrt{x^2-6x+8}\,dx$
14. $\int \frac{dx}{\sqrt{2x-x^2}}$
15. $\int \frac{dx}{4x^2+8x+12}$
16. $\int \frac{x}{\sqrt{x^2-4x+5}}\,dx$
17. $\int \frac{3x-2}{x^2+2x+2}\,dx$

9.7 IMPROPER INTEGRALS

Thus far, our discussion of a definite integral $\int_a^b f(x)\,dx$ has been confined to integrals of continuous functions over a closed interval. We now turn to such integrals as

9.7.1 Types of improper integrals

$$\int_a^\infty f(x)\,dx \quad \text{and} \quad \int_a^b f(x)\,dx, \quad \text{where} \lim_{x\to b-} |f(x)| = \infty.$$

Such integrals are *improper integrals.* If $f(x) \geq 0$, the first integral is an attempt to find the area under the graph "from a to ∞," as shown in Fig. 9.3. The second integral is an attempt to find the area of the shaded region in Fig. 9.4.

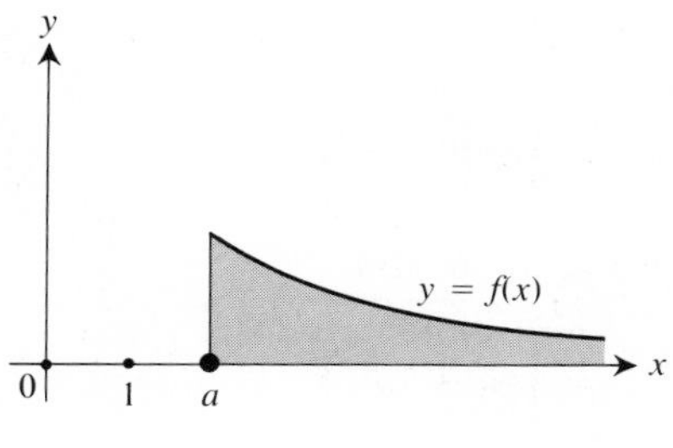

9.3

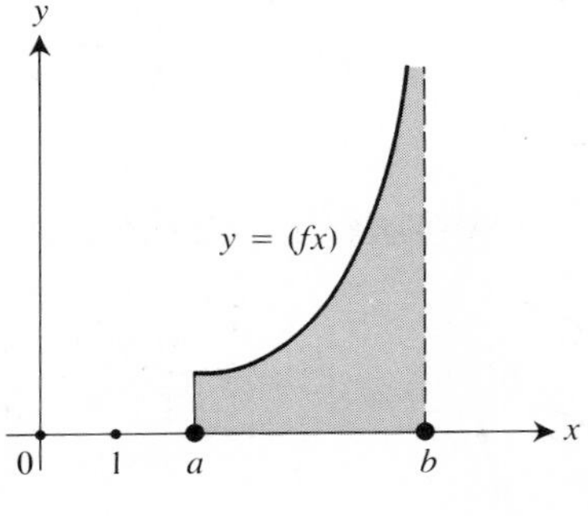

9.4

If $f(x)$ is continuous for all $x > a$, then $\int_a^h f(x)\,dx$ for $h > a$ makes perfectly good sense, and it is natural to let

Convergent and divergent integrals

$$\int_a^\infty f(x)\,dx = \lim_{h\to\infty} \int_a^h f(x)\,dx. \tag{1}$$

If this limit exists, the integral is said to **converge**, while the integral **diverges** if the limit is not a finite number.

Example 1 The improper integral $\int_1^\infty (1/x^2)\,dx$ converges to 1, for

$$\int_1^\infty \frac{1}{x^2}\,dx = \lim_{h\to\infty} \int_1^h \frac{1}{x^2}\,dx = \lim_{h\to\infty} -\frac{1}{x}\bigg]_1^h = \lim_{h\to\infty}\left(-\frac{1}{h} - (-1)\right) = 0 + 1 = 1. \quad \|$$

The nonmathematics major (and even an occasional mathematics major) may tend to compute $\int_1^\infty (1/x^2)\,dx$ as

$$\text{“}\int_1^\infty \frac{1}{x^2}\,dx = -\frac{1}{x}\bigg]_1^\infty = -\frac{1}{\infty} - (-1) = 1,\text{”}$$

rather than write out the limits as in Example 1. While this shorthand rarely causes any difficulty, we shall always write out the limits in the text, for improper integrals are *defined* as limits.

Example 2 The improper integral $\int_1^\infty (1/x)\,dx$ diverges, for

$$\int_1^\infty \frac{1}{x}\,dx = \lim_{h\to\infty} \int_1^h \frac{1}{x}\,dx = \lim_{h\to\infty} \ln|x|\bigg]_1^h = \lim_{h\to\infty} \ln|h| - \ln 1 = \infty. \quad \|$$

Now suppose $\lim_{x\to b-} |f(x)| = \infty$ as in Fig. 9.4. Then we define $\int_a^b f(x)\,dx$ by

$$\int_a^b f(x)\,dx = \lim_{h\to b-} \int_a^h f(x)\,dx. \tag{2}$$

Again, the integral is said to converge or diverge according to whether a finite limit exists or does not exist.

Example 3 The integral $\int_0^1 1/(1-x)\,dx$, diverges, for

$$\int_0^1 \frac{1}{1-x}\,dx = \lim_{h\to 1-} \int_0^h \frac{1}{1-x}\,dx = \lim_{h\to 1-} -\ln|1-x|\bigg]_0^h$$
$$= \lim_{h\to 1-} -\ln|1-h| + \ln 1 = \infty. \quad \|$$

Example 4 The integral $\int_0^1 (1/\sqrt{1-x})\,dx$ converges to 2, for

$$\int_0^1 \frac{1}{\sqrt{1-x}}\,dx = \lim_{h\to 1-} \int_0^1 \frac{1}{\sqrt{1-x}}\,dx = \lim_{h\to 1-} -2\sqrt{1-x}\,\bigg]_0^h$$
$$= \lim_{h\to 1-} -2\sqrt{1-h} + 2 = 2. \quad \|$$

Of course, if $f(x)$ is continuous for $a < x \le b$ but $\lim_{x\to a+} |f(x)| = \infty$, we let

$$\int_a^b f(x)\,dx = \lim_{h\to a+} \int_h^b f(x)\,dx. \tag{3}$$

If $f(x)$ is continuous for all x, then we let

$$\int_{-\infty}^{\infty} f(x)\,dx = \int_{-\infty}^{0} f(x)\,dx + \int_0^{\infty} f(x)\,dx, \tag{4}$$

requiring that both integrals on the righthand side converge.

Example 5 Let us find $\int_{-\infty}^{\infty} dx/(1 + x^2)$.

SOLUTION We have

$$\begin{aligned}\int_{-\infty}^{\infty} \frac{dx}{1+x^2} &= \int_{-\infty}^{0} \frac{dx}{1+x^2} + \int_0^{\infty} \frac{dx}{1+x^2}\\ &= \lim_{h\to-\infty} \int_h^0 \frac{dx}{1+x^2} + \lim_{h\to\infty} \int_0^h \frac{dx}{1+x^2}\\ &= \lim_{h\to-\infty} \tan^{-1}x\Big]_h^0 + \lim_{h\to\infty} \tan^{-1}x\Big]_0^h\\ &= \lim_{h\to-\infty} (0 - \tan^{-1}h) + \lim_{h\to\infty} (\tan^{-1}h - 0)\\ &= 0 - \left(-\frac{\pi}{2}\right) + \left(\frac{\pi}{2} - 0\right) = \pi. \quad \|\end{aligned}$$

Finally, if $f(x)$ is continuous for $a \le x \le b$ except at a point c where $a < c < b$ and if $\lim_{x\to c} |f(x)| = \infty$, then we let

$$\int_a^b f(x)\,dx = \int_a^c f(x)\,dx + \int_c^b f(x)\,dx = \lim_{h\to c-} \int_a^h f(x)\,dx + \lim_{h\to c+} \int_h^b f(x)\,dx.$$

In other words, if we consider ∞, $-\infty$, and points where $f(x)$ becomes infinite as "bad points," then we define an improper integral involving several bad points to be the sum of improper integrals obtained by breaking the interval of integration into pieces, each containing only one bad point at an end of the interval. *Each individual integral must converge in order for the entire integral to converge.*

Example 6 Since $\int_0^1 (1/x)\,dx$ and $\int_{-1}^0 (1/x)\,dx$ do not converge, neither does $\int_{-1}^1 (1/x)\,dx$. The following computation is sometimes found on students' papers, but it is

A warning

WRONG: $$\int_{-1}^1 \frac{1}{x}\,dx = \ln|x|\Big]_{-1}^1 = \ln 1 - \ln 1 = 0 - 0 = 0. \quad \|$$

9.7.2 Convergence of $\int_a^\infty (1/x^p)\,dx$ and $\int_0^b (1/x^p)\,dx$, and a comparison test

The ideas in this subsection provide a preview of things to come in the next chapter on infinite series. We will find Theorem 9.1 very useful in the next chapter. The proof of the theorem is just a question of computation, which we leave to the exercises (see Exercises 24 and 25).

Theorem 9.1

1. *For $a > 0$, the improper integral $\int_a^\infty (1/x^p)\,dx$ converges if $p > 1$ and diverges if $p \le 1$.*
2. *The improper integrals $\int_a^b [1/(b - x)^p]\,dx$ and $\int_a^b [1/(x - a)^p]\,dx$ converge if $p < 1$ and diverge if $p \ge 1$.*

Examples 1 and 2 above illustrate the first statement, and Examples 3 and 4 illustrate the second.

Sometimes we are interested only in whether an integral converges or diverges, rather than its value if it converges. A *comparison test* is often useful. Suppose that $0 \le g(x) \le f(x)$ for $x \ge a$. Then surely

$$\int_a^h g(x)\,dx \le \int_a^h f(x)\,dx \tag{5}$$

for all $h \ge a$. Now $\lim_{h\to\infty} \int_a^h g(x)\,dx$ is either a finite number or approaches ∞, since $g(x) \ge 0$. [This seems obvious, and is proved in more advanced courses.] The same is true for $\int_a^h f(x)\,dx$. Taking the limit as $h \to \infty$ in (5), we see that if $\int_a^\infty f(x)\,dx$ converges, then $\int_a^\infty g(x)\,dx$ must converge also, for if $\int_a^\infty f(x)\,dx$ remains finite, then the smaller integral $\int_a^\infty g(x)\,dx$ must also. Similarly, if $\int_a^\infty g(x)\,dx$ diverges, so does $\int_a^\infty f(x)\,dx$. Let us give a statement of this comparison test not only for integrals at ∞, but also for integrals that have finite "bad points."

Theorem 9.2 (*Comparison Test*) *Let $f(x)$ and $g(x)$ be continuous functions for $x \ge a$ and suppose that $0 \le g(x) \le f(x)$ for $x \ge a$. If $\int_a^\infty f(x)\,dx$ converges, then $\int_a^\infty g(x)$ converges also, while if $\int_a^\infty g(x)\,dx$ diverges, then $\int_a^\infty f(x)\,dx$ diverges also. Similar results hold if ∞ is replaced by b and $x \ge a$ is replaced by $a \le x < b$.*

Example 7 The integral

$$\int_1^\infty \frac{|\sin x|}{x^2}\,dx$$

converges, for

$$\frac{|\sin x|}{x^2} \le \frac{1}{x^2},$$

and $\int_1^\infty (1/x^2)\,dx$ converges by Theorem 9.1 ||

Example 8 Since $\int_1^3 [1/(x - 1)^3]\,dx$ diverges by Theorem 9.1 and

$$\frac{x + 17}{(x - 1)^3} > \frac{1}{(x - 1)^3} \quad \text{for} \quad 1 < x \le 3,$$

Theorem 9.2 shows that $\int_1^3 [(x + 17)/(x - 1)^3]\,dx$ diverges also. ||

SUMMARY

1. $\int_a^\infty f(x)\,dx = \lim_{h\to\infty} \int_a^h f(x)\,dx$

2. $\int_{-\infty}^a f(x)\,dx = \lim_{h\to-\infty} \int_h^a f(x)\,dx$

3. *If* $\lim_{x\to b-} |f(x)| = \infty$, *then* $\int_a^b f(x)\,dx = \lim_{h\to b-} \int_a^h f(x)\,dx.$

4. *If* $\lim_{x\to a+} |f(x)| = \infty$, *then* $\int_a^b f(x)\,dx = \lim_{h\to a+} \int_h^b f(x)\,dx.$

5. $\int_{-\infty}^\infty f(x)\,dx = \int_{-\infty}^0 f(x)\,dx + \int_0^\infty f(x)\,dx$

6. *If* $\lim_{x\to c} |f(x)| = \infty$ *and* $a < c < b$, *then*
$$\int_a^b f(x)\,dx = \int_a^c f(x)\,dx + \int_c^b f(x)\,dx.$$

7. $\int_a^\infty (1/x^p)\,dx$ *where* $a > 0$ *converges if* $p > 1$ *and diverges if* $p \le 1$.

8. $\int_a^b [1/(x-a)^p]\,dx$ *and* $\int_a^b [1/(b-x)^p]\,dx$ *converge if* $p < 1$ *and diverge if* $p \ge 1$.

9. *If* $0 \le g(x) \le f(x)$ *and an improper integral of* $f(x)$ *converges, then the improper integral of* $g(x)$*, having the same limits, converges. If an improper integral of* $g(x)$ *diverges, then the same improper integral of* $f(x)$ *diverges.*

EXERCISES

In Exercises 1 through 16, determine whether or not the improper integral converges, and find its value if it converges.

1. $\int_1^\infty \frac{1}{x^3}\,dx$ **2.** $\int_1^\infty \frac{1}{\sqrt{x}}\,dx$ **3.** $\int_0^\infty \frac{1}{x^2+1}\,dx$ **4.** $\int_{-\infty}^0 \frac{x^2}{x^3+1}\,dx$

5. $\int_{-\infty}^0 e^x\,dx$ **6.** $\int_0^1 \frac{1}{x^{2/3}}\,dx$ **7.** $\int_0^2 \frac{1}{\sqrt{2-x}}\,dx$ **8.** $\int_1^2 \frac{1}{(x-1)^2}\,dx$

9. $\int_{-1}^1 \frac{1}{x^{2/5}}\,dx$ **10.** $\int_0^\infty \frac{1}{\sqrt{x}}\,dx$ **11.** $\int_{-\infty}^\infty |x|e^{-x^2}\,dx$ **12.** $\int_0^1 \frac{1}{\sqrt{2x-x^2}}\,dx$

13. $\int_{-\infty}^\infty e^{-x}\cos x\,dx$ **14.** $\int_0^\infty e^{-x}\sin x\,dx$ **15.** $\int_0^{\pi/2} \tan x\,dx$ **16.** $\int_0^{\pi/2} \sqrt{\cos x}\cot x\,dx$

In Exercises 17 through 23, decide whether the integral converges or diverges. You need not compute its value if it converges.

17. $\int_2^\infty \frac{x^3+3x+2}{x^5-8x}\,dx$ **18.** $\int_1^\infty \frac{|\cos x|}{x^2}\,dx$ **19.** $\int_0^\infty \frac{|\cos x|}{x^2}\,dx$ **20.** $\int_0^\infty \frac{dx}{x^{2/3}+x^2}$

21. $\int_0^{\pi} \frac{|\cos x|}{\sqrt{x}}\, dx$

22. $\int_0^{\pi} \frac{\sin x}{x^{3/2}}\, dx$

23. $\int_0^{\infty} \frac{|\sin x|}{x^{4/3}}\, dx$

24. Show that $\int_1^{\infty} (1/x^p)\, dx$ converges if $p > 1$ and diverges if $p \leq 1$.

25. Show that

$\int_a^b [1/(x-a)^p]\, dx$ and $\int_a^b [1/(b-x)^p]\, dx$

converge if $p<1$ and diverge if $p \geq 1$.

26. Consider the region under the graph of $1/x$ over the half line $x \geq 1$.

a) Does the region have finite area?

b) Show that the unbounded solid obtained by revolving the region about the x-axis has finite volume, and compute this volume.

c) Does the solid of finite volume described in (b) have finite surface area?

27. Consider the region under the graph of $1/\sqrt{x}$ over the half-open interval $0 < x \leq 1$.

a) Does the region have finite area?

b) Show that the unbounded solid obtained by revolving the region about the y-axis has finite volume, and compute this volume.

c) Does the solid of finite volume described in (b) have finite surface area?

exercise sets for chapter 9

review exercise set 9.1

In Problems 1 through 11, find the given integral without using tables.

1. $\int x^2 \cos 3x\, dx$

2. $\int \sin^{-1} 3x\, dx$

3. $\int \frac{5x + 40}{2x^2 - 7x - 15}\, dx$

4. $\int \frac{x}{\sqrt{x+3}}\, dx$

5. $\int \frac{x^{1/3}}{x^{1/6} + 1}\, dx$

6. $\int \frac{1}{\sin x + \cos x}\, dx$

7. $\int \sin^5 2x\, dx$

8. $\int \cos^2 x \sin^2 x\, dx$

9. $\int \sec^4 3x \tan^2 3x\, dx$

10. $\int \cot^3 2x\, dx$

11. $\int \sqrt{16 - 9x^2}\, dx$

12. Find $\int_1^{\infty} [1/(1 + x^2)]\, dx$, if the integral converges.

13. Determine the convergence or divergence of the following improper integrals. Give reasons for your answers.

a) $\int_{-\infty}^{\infty} \frac{1}{x^2}\, dx$

b) $\int_2^{\infty} \frac{x+7}{x^2 - 1}\, dx$

c) $\int_1^{2} \frac{\sqrt{2-x}}{x^2 - 4}\, dx$

review exercise set 9.2

In Problems 1 through 11, find the given integral without using tables.

1. $\int x \ln x\, dx$

2. $\int e^x \sin 2x\, dx$

3. $\int \frac{8x^2 + 3x - 7}{4x^3 - 12x^2 + x - 3}\, dx$

4. $\int \frac{(x-1)^{3/2}}{x}\, dx$

5. $\int \frac{(x-1)^{1/2} - 7}{3 - (x-1)^{1/4}}\, dx$

6. $\int \frac{\sin x}{\sin x - \cos x}\, dx$

7. $\int \cos^3 x \sin^2 x\, dx$

8. $\int \cos^4 2x\, dx$

9. $\int \cot^2 x \csc^4 x\, dx$

10. $\int \cot^4 x\, dx$

11. $\int \frac{x^2}{\sqrt{4 - x^2}}\, dx$

12. Find $\int_0^2 \frac{x}{\sqrt{2 - x}}\, dx$ if the integral converges.

13. Determine the convergence or divergence of the following improper integrals. Give reasons for your answers.

a) $\int_0^\infty \frac{1}{x^3}\, dx$ b) $\int_{-1}^4 \frac{x}{(x - 1)^{2/3}}\, dx$ c) $\int_{-\infty}^\infty \frac{1}{1 + x^3}\, dx$

more challenging exercises 9

1. Let $f''(x)$ be continuous for all x. Show that

$$f(x) = f(0) + f'(0)x + \int_0^x f''(t)(x - t)\, dt.$$

[*Hint.* Show that $\int_0^x f'(t)\, dt = f(x) - f(0)$. Then integrate $\int_0^x f'(t)\, dt$ by parts, taking $u = f'(t)$, $dv = dt$, and letting $v = t - x$.]

2. Continue Exercise 1 to show that if $f'''(x)$ is continuous for all x, then

$$f(x) = f(0) + f'(0)x + \frac{f''(0)}{2}x^2 + \tfrac{1}{2}\int_0^x f'''(t)(x - t)^2\, dt.$$

3. Let $f(x)$ be a continuous function for $x \geq 0$. Show that if $\lim_{x\to\infty} f(x)$ exists and $\int_0^\infty f(x)\, dx$ converges, then $\lim_{x\to\infty} f(x) = 0$.

4. Let $g(x)$ be continuous for $x \geq a$, and let $c \neq 0$. Show that $\int_a^\infty g(x)\, dx$ converges if and only if $\int_a^\infty c \cdot g(x)\, dx$ converges.

5. Let $f(x)$ be continuous for all x. The **Cauchy Principal Value** of $\int_{-\infty}^\infty f(x)\, dx$ is defined to be $\lim_{h\to\infty} \int_{-h}^h f(x)\, dx$, if the limit exists.

a) Give an example of a continuous function f such that the Cauchy principal value of $\int_{-\infty}^\infty f(x)\, dx$ exists, but $\int_{-\infty}^\infty f(x)\, dx$ diverges.

b) Show that if $\int_{-\infty}^\infty f(x)\, dx$ converges, then the Cauchy principal value of $\int_{-\infty}^\infty f(x)\, dx$ exists and is $\int_{-\infty}^\infty f(x)\, dx$.

6. Give an example of a continuous function f such that $f(x) \geq 0$ for $x \geq 0$, and $\int_0^\infty f(x)$ converges, but $\lim_{x\to\infty} f(x)$ does not exist. [*Hint.* Let the graph of f have narrower and narrower upward "spikes" of unit height as $x \to \infty$, and let $f(x) = 0$ between the spikes.]

7. Give an example of a continuous unbounded (assuming arbitrarily large values) function f satisfying the other conditions of the preceding exercise.

infinite series of constants

10

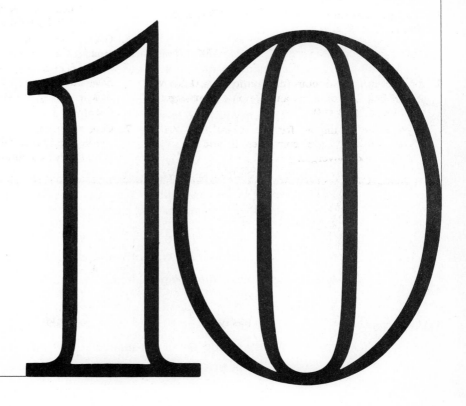

10.1 SEQUENCES

10.1.1 Motivation for series and sequences

It is too bad that not all functions are polynomials. Calculus of the polynomial functions, even integration, is very easy to handle. However, many functions you encounter are not polynomial functions. For example, $\sin x$ is not a polynomial function, for $\sin x$ is always between -1 and 1, while every nonconstant polynomial function is unbounded.

We will see in the next chapter that many elementary functions such as $\sin x$, while not polynomial functions, can be viewed as "infinite polynomial functions," that is, as polynomials with an infinite number of terms. We will see that

$$\sin x = x - \frac{x^3}{3!} + \frac{x^5}{5!} - \frac{x^7}{7!} + \frac{x^9}{9!} - \frac{x^{11}}{11!} + \cdots \tag{1}$$

[Recall that

$$1! = 1, \qquad 2! = 2 \cdot 1 = 2, \qquad 3! = 3 \cdot 2 \cdot 1 = 6,$$

and in general,

$$n! = n(n-1)(n-2) \cdots 3 \cdot 2 \cdot 1.]$$

Then, replacing x by x^2, you obtain

$$\sin x^2 = x^2 - \frac{x^6}{3!} + \frac{x^{10}}{5!} - \frac{x^{14}}{7!} + \frac{x^{18}}{9!} - \frac{x^{22}}{11!} + \cdots \tag{2}$$

It can be shown that you can't find $\int \sin x^2\,dx$ as a combination of elementary functions. However, you can use (2) to integrate $\sin x^2$ as an "infinite polynomial."

If you use (1) to compute sin 1 (the sine of one radian), replacing x by 1, then you have to add up the infinite sum

$$1 - \frac{1}{3!} + \frac{1}{5!} - \frac{1}{7!} + \frac{1}{9!} - \frac{1}{11!} + \cdots \tag{3}$$

to find the answer. This leads us to study infinite sums such as (3), or *infinite series*, as they are called. Of course, we really have this idea embodied in our notation for numbers; for example,

$$\tfrac{1}{3} = 0.33333 \cdots = \tfrac{3}{10} + \tfrac{3}{100} + \tfrac{3}{1000} + \tfrac{3}{10000} + \tfrac{3}{100000} + \cdots$$

Let's illustrate this type of addition of an infinite series by an example.

Example 1 Let's make the obvious attempt to find the "sum" of the infinite series of constants

$$1 + \tfrac{1}{2} + \tfrac{1}{4} + \tfrac{1}{8} + \tfrac{1}{16} + \cdots \tag{4}$$

SOLUTION We try to add up all the numbers in (4), and obtain successively

$$1, \quad 1 + \tfrac{1}{2} = \tfrac{3}{2}, \quad 1 + \tfrac{1}{2} + \tfrac{1}{4} = \tfrac{7}{4}, \quad 1 + \tfrac{1}{2} + \tfrac{1}{4} + \tfrac{1}{8} = \tfrac{15}{8}, \quad \tfrac{15}{8} + \tfrac{1}{16} = \tfrac{31}{16}, \quad \cdots$$

This leads us to consider the *sequence of numbers*

$$1, \quad \tfrac{3}{2}, \quad \tfrac{7}{4}, \quad \tfrac{15}{8}, \quad \tfrac{31}{16}, \quad \cdots \tag{5}$$

It is clear that the numbers in the sequence (5) get very close to 2 as you continue through the sequence. This suggests that the "sum" of the infinite series (4) should be 2. ||

From Example 1, it would seem natural to try to define the sum of an infinite *series* of numbers by examining the *sequence* of numbers obtained by adding more and more terms of the series. In Section 10.1.2 we discuss sequences and their limits. These ideas will then be used to discuss series in Section 10.2.

10.1.2 Sequences

An example of a sequence was given in (5) above. You naively think of a sequence as an endless row of numbers separated by commas, symbolically,

$$a_1, \quad a_2, \quad a_3, \quad \ldots, \quad a_n, \quad \cdots \tag{6}$$

where each a_i is a real number. The sequence in (6) is also written $\{a_n\}$ for brevity. Since a sequence has one term for each positive integer, you can also consider a sequence to be a real-valued function ϕ with domain the set of positive integers. For the sequence (6), you then have

$$\phi(1) = a_1, \qquad \phi(2) = a_2, \qquad \phi(3) = a_3, \qquad \cdots$$

Definition 10.1 A **sequence** is a real-valued function ϕ with domain the set of positive integers. If $\phi(n) = a_n$, the sequence is denoted by $\{a_n\}$ or by $a_1, a_2, \ldots, a_n, \ldots$

Example 2 You may regard (5) as a sequence ϕ where

$$\phi(1) = 1, \qquad \phi(2) = \tfrac{3}{2}, \qquad \phi(3) = \tfrac{7}{4}, \qquad \text{etc.} \quad \|$$

You want to find the limit of a sequence if it exists. Naively, the limit of $\{a_n\}$ is c if the terms a_n get just as close to c as you wish provided that n is large enough. This is a vague statement similar to the one we used to introduce $\lim_{x\to a} f(x) = c$. The more precise definition that follows is very much like the one for the limit of a function.

Definition 10.2 The **limit of a sequence** $\{a_n\}$ is c if for each $\epsilon > 0$, there exists an integer N such that $|a_n - c| < \epsilon$ if $n > N$. We write "$\lim_{n\to\infty} \{a_n\} = c$" or "$\lim_{n\to\infty} a_n = c$," and we say that the sequence $\{a_n\}$ **converges to** c. A sequence that has no limit **diverges**.

We leave to the exercises (see Exercise 19) the easy proof that a sequence can't converge to two different values.

Example 3 For the sequence

$$1, \quad \tfrac{3}{2}, \quad \tfrac{7}{4}, \quad \tfrac{15}{8}, \quad \tfrac{31}{16}, \quad \ldots$$

in (5) you have

$$a_n = \frac{2^n - 1}{2^{n-1}} = \frac{2^n}{2^{n-1}} - \frac{1}{2^{n-1}} = 2 - \frac{1}{2^{n-1}}.$$

Thus

$$|a_n - 2| = \frac{1}{2^{n-1}},$$

so, given $\epsilon > 0$, if you choose N so that $(1/2)^{N-1} < \epsilon$, then

$$|a_n - 2| = \frac{1}{2^{n-1}} < \epsilon$$

for $n > N$. Therefore the sequence converges to 2. ‖

Frequently, one describes a sequence $\{a_n\}$ by giving a formula in terms of n for the "nth term" a_n of the sequence. For example, the sequence discussed in Example 3 is the sequence where $a_n = (2^n - 1)/2^{n-1}$, or, more briefly, the sequence

$$\left\{\frac{2^n - 1}{2^{n-1}}\right\}.$$

Example 3 shows that

$$\lim_{n\to\infty} \frac{2^n - 1}{2^{n-1}} = \lim_{n\to\infty} \left(2 - \frac{1}{2^{n-1}}\right) = 2 - 0 = 2.$$

Example 4 We have

$$\lim_{n\to\infty} \frac{1}{n} = 0;$$

that is, the sequence $1, \frac{1}{2}, \frac{1}{3}, \frac{1}{4}, \ldots, 1/n, \ldots$ converges to 0. ‖

Example 5 The sequence

$$1, \quad -1, \quad 1, \quad -1, \quad 1, \quad -1, \quad \ldots, \quad (-1)^{n+1}, \quad \ldots$$

diverges, since the terms of the sequence do not approach and stay close to any number c as you continue along the sequence. We leave an ϵ,N-proof of this fact to the exercises (see Exercise 23). ‖

Just as for a function, you have the notions of $\lim_{n\to\infty} a_n = \infty$ and $\lim_{n\to\infty} a_n = -\infty$. We ask you to state these definitions for sequences in Exercises 1 and 2. Let us emphasize that a sequence is said to *converge* if and only if it has a *finite* limit. A sequence that does not converge to a finite limit *diverges*.

Example 6 We have

$$\lim_{n\to\infty} n^2 = \infty \qquad \text{while} \qquad \lim_{n\to\infty} \frac{-n^2+2}{n+1} = -\infty.$$

Thus the sequence $\{n^2\}$ diverges to ∞ and the sequence $\{(-n^2+2)/(n+1)\}$ diverges to $-\infty$. ‖

Example 7 The sequence $\{(-1)^n \cdot n\}$ has neither limit ∞ nor limit $-\infty$, since for n even, the terms approach ∞, while for n odd, the terms approach $-\infty$. This sequence diverges. ‖

SUMMARY

1. *A sequence is an endless row of numbers*

$$a_1, \quad a_2, \quad a_3, \quad \ldots, \quad a_n, \quad \ldots$$

separated by commas. Alternatively, it is a real-valued function with domain the set of positive integers.

2. $\text{Lim}_{n\to\infty} a_n = c$ *if for each* $\epsilon > 0$, *there exists a positive integer N depending on* ϵ *such that* $|a_n - c| < \epsilon$ *if* $n > N$.

3. *If* $\lim_{n\to\infty} a_n$ *exists (is a* finite *number c), then the sequence converges; otherwise it diverges.*

EXERCISES

1. Define what is meant by $\lim_{n\to\infty} a_n = \infty$.

2. Define what is meant by $\lim_{n\to\infty} b_n = -\infty$.

In Exercises 3 *through* 18, *find the limit of the sequence if the sequence converges or has limit* ∞ *or* $-\infty$.

3. $\left\{\dfrac{1}{2n}\right\}$

4. $1, 0, \frac{1}{2}, 0, \frac{1}{3}, 0, \frac{1}{4}, 0, \ldots$

5. $\frac{1}{2}, \frac{2}{3}, \frac{3}{4}, \frac{4}{5}, \ldots$

6. $\left\{\dfrac{n^2+1}{3n}\right\}$

7. $\left\{\dfrac{2n-\sqrt{n}}{n}\right\}$

8. $\left\{\dfrac{3n^2-2n}{2n^3}\right\}$

9. $\left\{\dfrac{3n}{\sqrt{n}+100}\right\}$

10 $\left\{\dfrac{n-n^3}{n^2}\right\}$

11. $\left\{\left(-\dfrac{1}{2}\right)^n\right\}$

12. $1, 2, 1, -3, 1, 4, 1, -5, 1, 6, \ldots$

13. $\{e^n\}$

14. $\left\{\dfrac{1}{\ln(n+1)}\right\}$

15. $\left\{\dfrac{\ln n}{n^2}\right\}$

16. $\{\sin n\}$

17. $\left\{\sin\dfrac{1}{n}\right\}$

18. $\{e^{1/\sqrt{n}}\}$

19. Show that a sequence $\{a_n\}$ can't converge to two different limits.

20. Give an ϵ,N-proof that the sequence $1, 1, 1, \ldots, 1, \ldots$ converges.

21. Give an ϵ,N-proof that the sequence $1, -\frac{1}{2}, \frac{1}{3}, -\frac{1}{4}, \frac{1}{5}, -\frac{1}{6}, \ldots$ converges.

22. Give an ϵ,N-proof that the sequence

$$\frac{1}{\sqrt{2}}, \frac{1}{\sqrt{3}}, \frac{1}{\sqrt{4}}, \ldots, \frac{1}{\sqrt{n+1}}, \ldots$$

converges.

23. Give an ϵ,N-proof that the sequence

$$1, \quad -1, \quad 1, \quad -1, \quad 1, \quad \ldots, \quad (-1)^{n+1}, \quad \ldots$$

diverges.

24. Give an ϵ,N-proof that the sequence $1, 2, 3, 4, \ldots, n, \ldots$ diverges.

calculator exercises

25. It can be shown that $\lim_{n\to\infty} (1 + 1/n)^n = e$. Find the smallest integer N such that

$$|(1 + 1/N)^N - e| < \epsilon$$

for the following values of ϵ.

a) 0.1 b) 0.01
c) 0.001 d) 0.0001

In Exercises 26 *through* 28, *find the limit of the sequence, if it exists.*

26. $\left\{\left(1 - \frac{1}{n}\right)^n\right\}$ **27.** $\left\{\left(1 + \frac{1}{2n}\right)^n\right\}$ **28.** $\{n(\sqrt[n]{2} - 1)\}$

10.2 SERIES

10.2.1 The sum of a series

It is mathematically imprecise to define a sequence of constants to be an endless row of numbers

$$a_1, \quad a_2, \quad \ldots, \quad a_n, \quad \ldots \tag{1}$$

separated by commas, although this is the way you probably think of a sequence in your work. In Section 10.1, we defined a sequence to be a *function.* It would be similarly imprecise to define an infinite series of constants to be an endless row of numbers

$$a_1 + a_2 + \cdots + a_n + \cdots \tag{2}$$

with plus signs between them, although this is probably the way you will think of a series in your work. We saw in Section 10.1.1 that in attempting to find the "sum" of the series (2), we are led to consider the limit of the sequence

$$s_1, \quad s_2, \quad \ldots, \quad s_n, \quad \ldots, \tag{3}$$

where $s_1 = a_1$, $s_2 = a_1 + a_2$, and in general, $s_n = a_1 + \cdots + a_n$. From a mathematical standpoint, the precise thing to do is to define the series (2) to *be* the sequence in (3); we have already based the notion of a sequence on the notion of a function.

Definition 10.3 Let $a_1, a_2, \ldots, a_n, \ldots$ be a sequence and let $s_n = a_1 + \cdots + a_n$ for all positive integers n. The **infinite series**

$$\sum_{n=1}^{\infty} a_n = a_1 + a_2 + \cdots + a_n + \cdots$$

is the sequence $\{s_n\}$. The number a_n is the ***n*th term of the series** $\sum_{n=1}^{\infty} a_n$, and s_n is the ***n*th partial sum of the series.**

Definition 10.4 If $\sum_{n=1}^{\infty} a_n$ is a series, then the **sum of the series** is the limit of the sequence $\{s_n\}$ of partial sums, if this limit exists. If $\lim_{n\to\infty} s_n$ is a finite number c, then the series $\sum_{n=1}^{\infty} a_n$ **converges to** c. If $\lim_{n\to\infty} s_n$ is ∞, $-\infty$, or is undefined, then the series **diverges.**

Example 1 The sequence of partial sums for the series

$$0 + 0 + \cdots + 0 + \cdots$$

is

$$0, \quad 0, \quad \ldots, \quad 0, \quad \ldots$$

which converges to 0, so the sum of the series is 0. The series

$$1 - 1 + 1 - 1 + 1 - 1 + \cdots$$

diverges, for the sequence of partial sums is

$$1, \quad 0, \quad 1, \quad 0, \quad 1, \quad 0, \ldots$$

and this sequence diverges. ||

10.2.2 Harmonic and geometric series

For reasons that will become evident in the next section, we are interested in building a "stockpile" of series whose convergence or divergence is known. The series discussed in this article contribute to such a stockpile.

The series

$$1 + \frac{1}{2} + \frac{1}{3} + \frac{1}{4} + \cdots + \frac{1}{n} + \cdots \tag{4}$$

is the **harmonic series**. This series diverges, as the following argument shows. We group together certain terms of the series, using parentheses, so that the series appears as

$$(1) + (\tfrac{1}{2}) + (\tfrac{1}{3} + \tfrac{1}{4}) + (\tfrac{1}{5} + \cdots + \tfrac{1}{8}) + (\tfrac{1}{9} + \cdots + \tfrac{1}{16}) + \cdots \tag{5}$$

The sum of the terms in each parenthesis in (5) is $\geq \frac{1}{2}$; for example,

$$\tfrac{1}{5} + \tfrac{1}{6} + \tfrac{1}{7} + \tfrac{1}{8} > \tfrac{1}{8} + \tfrac{1}{8} + \tfrac{1}{8} + \tfrac{1}{8} = \tfrac{1}{2}.$$

This shows that for the harmonic series, we have

$$s_{2^n} \geq 1 + \tfrac{1}{2}n,$$

and it is then clear that the series diverges to ∞.

We turn now to the series

$$a + ar + ar^2 + \cdots + ar^n + \cdots. \tag{6}$$

The series (6) is the **geometric series** with **initial term** a and **ratio** r, it is a very important series. We have

$$s_n = a + ar + \cdots + ar^{n-1},$$

and consequently

$$rs_n = ar + \cdots + ar^{n-1} + ar^n.$$

Subtracting, we obtain

$$s_n - rs_n = a - ar^n,$$

so if $r \neq 1$,

$$s_n = \frac{a - ar^n}{1 - r} = \frac{a}{1 - r} - \frac{ar^n}{1 - r}.$$

Thus

$$\lim_{n\to\infty} s_n = \lim_{n\to\infty} \left[\frac{a}{1 - r} - \frac{ar^n}{1 - r}\right] = \frac{a}{1 - r} - a\left(\lim_{n\to\infty} \frac{r^n}{1 - r}\right). \tag{7}$$

The value of the limit in (7) depends upon the size of r. We have

$$\lim_{n\to\infty} \frac{r^n}{1 - r} = \begin{cases} 0 & \text{if } -1 < r < 1, \\ -\infty & \text{if } r > 1, \\ \text{undefined} & \text{if } r < -1. \end{cases}$$

If $r = 1$, then the series (6) reduces to

$$a + a + \cdots + a + \cdots \tag{8}$$

which obviously does not converge if $a \neq 0$. Thus the geometric series (6), where $a \neq 0$, converges if $|r| < 1$ and diverges if $|r| \geq 1$. For $|r| < 1$, the sum of the series is found from (7) to be

$$\begin{aligned} \lim_{n\to\infty} s_n &= \frac{a}{1 - r} - a\left(\lim_{n\to\infty} \frac{r^n}{1 - r}\right) \\ &= \frac{a}{1 - r} - 0 = \frac{a}{1 - r}. \end{aligned}$$

We summarize the results of this article in a theorem for easy reference.

Theorem 10.1 *The harmonic series $\sum_{n=1}^{\infty}(1/n)$ diverges to ∞, and the geometric series $\sum_{n=0}^{\infty}(ar^n)$ converges to $a/(1-r)$ if $|r|<1$, and diverges if $|r|\geq 1$ and $a\neq 0$.*

Example 2 The series

$$1+\frac{1}{2}+\frac{1}{4}+\cdots+\frac{1}{2^n}+\cdots$$

discussed in Section 10.1.1 is a geometric series with $a = 1$ and $r = \frac{1}{2}$. By Theorem 10.1 the series converges to

$$\frac{1}{1-r}=\frac{1}{1-\frac{1}{2}}=2,$$

which we guessed to be the case in Section 10.1.1. ‖

In closing this section, we should mention that the harmonic and geometric series are important in their own right, aside from contributing to our stockpile of series whose divergence or convergence is known. Exercises 25 through 30 indicate that geometric series are closely related to questions concerning our representation of numbers in decimal form.

10.2.3 Algebra of sequences and series

There are natural ways to "add" two sequences or two series and to "multiply by a constant" a sequence or a series.

Definition 10.5 Let $\{s_n\}$ and $\{t_n\}$ be sequences and let c be any number. The sequence $\{s_n + t_n\}$ with nth term $s_n + t_n$ is the **sum of** $\{s_n\}$ **and** $\{t_n\}$, while the sequence $\{cs_n\}$ with nth term cs_n is the **product of the constant** c **and** $\{s_n\}$.

Definition 10.6 Let $\sum_{n=1}^{\infty} a_n$ and $\sum_{n=1}^{\infty} b_n$ be series of constants, and let c be any number. The series $\sum_{n=1}^{\infty}(a_n + b_n)$ with nth term $a_n + b_n$ results from **adding the series** $\sum_{n=1}^{\infty} a_n$ **and** $\sum_{n=1}^{\infty} b_n$, while the series $\sum_{n=1}^{\infty}(ca_n)$ with nth term ca_n results from **multiplying the series** $\sum_{n=1}^{\infty} a_n$ **by the constant** c.

It is important to note that if s_n is the nth partial sum of $\sum_{n=1}^{\infty} a_n$ and t_n is the nth partial sum of $\sum_{n=1}^{\infty} b_n$, then the nth partial sum of $\sum_{n=1}^{\infty}(a_n + b_n)$ is $s_n + t_n$, while the nth partial sum of $\sum_{n=1}^{\infty}(ca_n)$ is cs_n.

Throughout this chapter, we are interested primarily in whether sequences (and series) converge or diverge. After the preceding definitions, we at once ask ourselves whether the sequence obtained by adding two convergent sequences still converges, and whether the sequence obtained by multiplying a convergent sequence by a constant still converges. We then ask the analogous questions for series. The answers to these questions are contained in the following theorem and its corollaries; the answers are really intuitively obvious.

Theorem 10.2 *If the sequence $\{s_n\}$ converges to s and the sequence $\{t_n\}$ converges to t, then the sequence $\{s_n + t_n\}$ converges to $s + t$ and the sequence $\{cs_n\}$ converges to cs for all c.*

Proof. Let $\epsilon > 0$ be given. Find an integer N_1 such that $|s_n - s| < \epsilon/2$ for $n > N_1$, and find an integer N_2 such that $|t_n - t| < \epsilon/2$ for $n > N_2$. Let N be the maximum of N_1 and N_2. Then for $n > N$, you have simultaneously

$$-\frac{\epsilon}{2} < s_n - s < \frac{\epsilon}{2}, \qquad -\frac{\epsilon}{2} < t_n - t < \frac{\epsilon}{2}.$$

Adding, you obtain

$$-\epsilon < (s_n + t_n) - (s + t) < \epsilon \qquad \text{or} \qquad |(s_n + t_n) - (s + t)| < \epsilon$$

for $n > N$. Thus the sequence $\{s_n + t_n\}$ converges to $s + t$.

If $c = 0$, then $\{cs_n\}$ is the sequence $0, 0, 0, \ldots$ which clearly converges to $0 = 0 \cdot s$. If $c \neq 0$, you can find N_3 such that

$$|s_n - s| < \frac{\epsilon}{|c|}$$

for $n > N_3$. Then for $n > N_3$, you have

$$-\frac{\epsilon}{|c|} < s_n - s < \frac{\epsilon}{|c|}.$$

Multiplying by c, you obtain for $c < 0$ as well as $c > 0$,

$$-\epsilon < cs_n - cs < \epsilon,$$

so $|cs_n - cs| < \epsilon$ for $n > N_3$. Thus the sequence $\{cs_n\}$ converges to cs. □

Corollary 1 *If $\sum_{n=1}^{\infty} a_n$ converges to a and $\sum_{n=1}^{\infty} b_n$ converges to b, then $\sum_{n=1}^{\infty} (a_n + b_n)$ converges to $a + b$ and $\sum_{n=1}^{\infty} (ca_n)$ converges to ca for all c.*

Proof. The proof is immediate from the preceding theorem and the observation that if $\{s_n\}$ is the sequence of partial sums of $\sum_{n=1}^{\infty} a_n$ and $\{t_n\}$ is the sequence of partial sums of $\sum_{n=1}^{\infty} b_n$, then $\{s_n + t_n\}$ is the sequence of partial sums of $\sum_{n=1}^{\infty} (a_n + b_n)$ and $\{cs_n\}$ is the sequence of partial sums of $\sum_{n=1}^{\infty} (ca_n)$. □

Corollary 2 *If $\sum_{n=1}^{\infty} a_n$ diverges, then for any $c \neq 0$, the series $\sum_{n=1}^{\infty} (ca_n)$ diverges also.*

Proof. If $\sum_{n=1}^{\infty} (ca_n)$ converges, then by Corollary 1, the series

$$\sum_{n=1}^{\infty} \frac{1}{c}(ca_n) = \sum_{n=1}^{\infty} a_n$$

would also converge, contrary to hypothesis. □

Example 3 The series $\sum_{n=1}^{\infty} (1/2n)$ diverges, for this is the divergent harmonic series multiplied by $\frac{1}{2}$. ||

Example 4 The series

$$\sum_{n=0}^{\infty} \left(\frac{1}{2^n} - \frac{2}{3^n}\right) = \sum_{n=0}^{\infty} \frac{1}{2^n} + (-2) \sum_{n=0}^{\infty} \frac{1}{3^n}$$

converges since the two geometric series converge. In fact,

$$\sum_{n=0}^{\infty} \frac{1}{2^n} \text{ converges to } \frac{1}{1 - 1/2} = 2$$

and

$$\sum_{n=0}^{\infty} \frac{1}{3^n} \text{ converges to } \frac{1}{1 - 1/3} = \frac{3}{2},$$

so

$$\sum_{n=0}^{\infty} \left(\frac{1}{2^n} - \frac{2}{3^n}\right)$$

converges to $2 + (-2)(3/2) = 2 - 3 = -1$. ||

SUMMARY

1. *Associated with a series*

$$\sum_{n=1}^{\infty} a_n = a_1 + a_2 + a_3 + \cdots + a_n + \cdots$$

is the sequence of partial sums $\{s_n\}$ *where*

$$s_1 = a_1, \qquad s_2 = a_1 + a_2, \qquad s_3 = a_1 + a_2 + a_3, \qquad \cdots$$

The series converges to the sum c *if* $\lim_{n \to \infty} s_n = c$, *and the series diverges if* $\{s_n\}$ *is a divergent sequence.*

2. *Two sequences (series) may be added term by term, and each may be multiplied by a constant. If each of two sequences (series) converges, then their sum converges to the sum of the limits. A constant multiple of a convergent sequence (series) converges to that multiple of the limit of the original sequence (series).*
3. *The harmonic series* $\sum_{n=1}^{\infty} 1/n$ *diverges.*
4. *The geometric series* $\sum_{n=0}^{\infty} (ar^n)$ *converges to* $a/(1 - r)$ *if* $|r| < 1$ *and diverges if* $|r| \geq 1$ *and* $a \neq 0$.

EXERCISES

1. Consider the series $1 + 0 - 1 + 0 + 1 + 0 - 1 + 0 + \cdots$ Find the indicated partial sums.

a) s_1 b) s_2 c) s_3
d) s_4 e) s_8
f) s_{15} g) s_{122}

2. Find the first four partial sums of the harmonic series $\sum_{n=1}^{\infty} (1/n)$.

3. Find the first five terms $a_1, \ldots, a_5$ of the series having as sequence of partial sums $\{s_n\} = \frac{1}{2}, \frac{1}{3}, \frac{1}{4}, \frac{1}{5}, \frac{1}{6}, \ldots$.

In Exercises 4 through 14, determine whether the series converges or diverges and find the sum of the series if the series converges.

4. $\displaystyle\sum_{n=1}^{\infty} \frac{1}{n}$

5. $\displaystyle\sum_{n=0}^{\infty} \frac{1}{3^n}$

6. $\displaystyle\sum_{n=1}^{\infty} \frac{1}{2^n}$

7. $\displaystyle\sum_{n=2}^{\infty} \frac{1}{2^n}$

8. $\displaystyle\sum_{n=1}^{\infty} \frac{-1}{2n}$

9. $\displaystyle\sum_{n=1}^{\infty} (-1)^n$

10. $\displaystyle\sum_{n=1}^{\infty} \frac{n+1}{n}$

11. $\displaystyle\sum_{n=1}^{\infty} \frac{3}{(10)^n}$

12. $\displaystyle\sum_{n=1}^{\infty} \frac{4}{(-2)^n}$

13. $\displaystyle\sum_{n=1}^{\infty} \frac{1}{n+1}$

14. $\displaystyle\sum_{n=0}^{\infty} e^{-2n}$

15. A ball is dropped from a height of 30 ft. Each time it hits the ground, it rebounds to $\frac{1}{3}$ of the height it attained on the preceding bounce. Find the total distance the ball travels if it is allowed to bounce forever.

16. Give an example of two divergent sequences whose sum converges.

17. Give an example of two divergent series whose sum converges.

18. If $\sum_{n=1}^{\infty} a_n$ converges and $\sum_{n=1}^{\infty} b_n$ diverges, what can be said concerning the convergence or divergence of $\sum_{n=1}^{\infty} (a_n + b_n)$?

In Exercises 19 through 24, find the sum of the given series.

19. $\displaystyle\sum_{n=0}^{\infty} \left(\frac{1}{2^n} + \frac{1}{3^n}\right)$

20. $\displaystyle\sum_{n=0}^{\infty} \left(7\frac{1}{3^n} - 4\frac{1}{2^n}\right)$

21. $\displaystyle\sum_{n=0}^{\infty} \frac{2^n + 3^n}{4^n}$

22. $\displaystyle\sum_{n=0}^{\infty} \frac{3^{n+1} - 7 \cdot 5^n}{10^n}$

23. $\displaystyle\sum_{n=1}^{\infty} \frac{8^n + 9^n}{10^n}$

24. $\displaystyle\sum_{n=1}^{\infty} \frac{2^n}{3^{2n+1}}$

Our representation of a real number as an "unending decimal" is closely related to the notion of an infinite series. Exercises 25 through 30 explain this relationship.

25. Consider the infinite series

$$1 + \frac{3}{10} + \frac{3}{100} + \frac{3}{1000} + \frac{3}{10000} + \cdots$$

a) Express the sum of the series as an "unending decimal." [*Hint.* Think what our decimal notation *means.*]

b) From the decimal found in (a), express the sum of the series as a *rational number* (i.e., as a quotient p/q of two integers p and q where $q \neq 0$).

c) Express the sum of the series as a rational number by noting that the series obtained by deleting the first term is geometric and using Theorem 10.1.

26. With the preceding exercise for illustration, show that every real number ("unending decimal") is the sum of an infinite series of rational numbers with denominators that are powers of 10.

27. Show that the real number 0.222222. . . is a rational number by finding a geometric series that has the decimal as sum, and then using Theorem 10.1 to find the sum of the series.

28. a) Show that the real number 0.12121212. . . is the sum of a geometric series with ratio $r = 1/100$

b) Use (a) and Theorem 10.1 to express the number 0.121212. . . as a quotient of integers.

29. Use the ideas illustrated in the two preceding exercises to show that any "unending decimal" that trails off in a "repeating pattern" such as

$$273.14\,653\,653\,653\,653\,653\ldots$$

represents a rational number.

30. Show the converse of the preceding exercise. That is, show that the decimal representation of any rational number trails off in a repeating pattern, as illustrated in the preceding exercise. [*Hint.* Think of using long division to find the decimal form of a rational number. Argue that some "remainder" obtained in the division must eventually repeat, and that the "quotients" obtained must therefore eventually occur in a repeating pattern.]

10.3 COMPARISON TESTS

10.3.1 Insertion or deletion of terms in a series

The insertion or deletion of a finite number of terms in a series cannot affect whether the series converges or diverges, although if the series converges, the sum of the series may be affected. In particular, suppose the first few terms of a series do not conform to a pattern present in the rest of the series. Then you can neglect those first few terms when studying the convergence or divergence of the series. After an example, we state this property of series as a theorem, and leave the easy proof as an exercise.

Example 1 The geometric series

$$1 + \tfrac{1}{2} + \tfrac{1}{4} + \tfrac{1}{8} + \tfrac{1}{16} + \cdots \tag{1}$$

converges to 2. The series

$$\pi - 3 + 17 + 1 + \tfrac{1}{2} + \tfrac{1}{4} + \tfrac{1}{8} + \tfrac{1}{16} + \cdots,$$

obtained from (1) by inserting three additional terms at the beginning, also converges, and clearly must converge to

$$(\pi - 3 + 17) + 2 = \pi + 16. \quad \|$$

Theorem 10.3 *Suppose that each of two series $\sum_{n=1}^{\infty} a_n$ and $\sum_{n=1}^{\infty} b_n$ contains all but a finite number of terms of the other in the same order. That is, suppose that there exist N and k such that $b_n = a_{n+k}$ for all $n > N$. Then either both the series converge or both the series diverge.*

The condition in the preceding theorem that the common terms in the two series be "in the same order" is very important. Later work will show that the series

$$1 - \tfrac{1}{2} + \tfrac{1}{3} - \tfrac{1}{4} + \tfrac{1}{5} - \tfrac{1}{6} + \tfrac{1}{7} - \tfrac{1}{8} + \tfrac{1}{9} - \tfrac{1}{10} + \tfrac{1}{11} - \cdots$$

converges, and it can be proved that a *divergent* series can be found that contains *exactly the same terms* but in a different order.

10.3.2 If $\lim_{n\to\infty} a_n \neq 0$, then $\sum_{n=1}^{\infty} a_n$ diverges

The following theorem is very important, and is frequently misused. It shows that some series diverge, but it can *never* be used to show convergence of a series.

Theorem 10.4 *If the series $\sum_{n=1}^{\infty} a_n$ converges, then $\lim_{n\to\infty} a_n = 0$. Equivalently, if $\lim_{n\to\infty} a_n \neq 0$, then $\sum_{n=1}^{\infty} a_n$ diverges.*

Proof. Let $\sum_{n=1}^{\infty} a_n$ converge to c, and let $\epsilon > 0$ be given. Then there exists a positive integer N such that for all $n > N$, the nth partial sum s_n satisfies

$$|s_n - c| < \frac{\epsilon}{2}.$$

In particular, if $n > N + 1$, we have

$$|s_{n-1} - c| < \frac{\epsilon}{2} \qquad \text{and} \qquad |s_n - c| < \frac{\epsilon}{2}.$$

Since s_{n-1} and s_n are both within the distance $\epsilon/2$ of c, they must be a distance at most ϵ from each other. That is,

$$|s_n - s_{n-1}| < \epsilon$$

for $n > N + 1$. But

$$s_n - s_{n-1} = (a_1 + \cdots + a_{n-1} + a_n) - (a_1 + \cdots + a_{n-1}) = a_n,$$

so $|a_n| < \epsilon$ for $n > N + 1$. Hence $\lim_{n\to\infty} a_n = 0$. □

Example 2 Theorem 10.4 tells us that the series

$$\sum_{n=1}^{\infty} \frac{n^2}{5n^2 + 100n}$$

diverges, for

$$\lim_{n\to\infty} \frac{n^2}{5n^2 + 100n} = \frac{1}{5} \neq 0. \quad \|$$

Don't confuse the assertion in Theorem 10.4 with the converse assertion: "If $\lim_{n\to\infty} a_n = 0$, then $\sum_{n=1}^{\infty} a_n$ converges." *This converse assertion is not true*; for example, the harmonic series $\sum_{n=1}^{\infty} (1/n)$ has the property that

$\lim_{n\to\infty}(1/n) = 0$, but the harmonic series diverges. It is important that you realize that Theorem 10.4 gives only a *necessary condition* (and not a *sufficient* condition) that a series $\sum_{n=1}^{\infty} a_n$ converges. This means the theorem can *never* be used to demonstrate convergence of a series. It can sometimes be used to show divergence, as in Example 2.

10.3.3 A comparison test for series of nonnegative terms

During the course of this chapter, we shall be introducing certain "tests" for convergence or divergence of a series of constants. Most of these tests depend upon the *comparison* of the series with another series that is either known to converge or known to diverge. (This is one reason why it is important to build a "stockpile" of series whose convergence or divergence is known.) Comparison tests for series with nonnegative terms follow easily from the following fundamental property of the real numbers. A proof of this property is given in more advanced courses where the real numbers are "constructed."

Fundamental Property of the Real Numbers. Let $\{s_n\}$ be a sequence of numbers such that $s_{n+1} \geq s_n$ for $n = 1, 2, 3, \ldots$ (Such a sequence is **monotone increasing**.) Then either $\{s_n\}$ converges to some c or $\lim_{n\to\infty} s_n = \infty$.

Our Fundamental Property asserts that the only way a monotone increasing sequence $\{s_n\}$ can *fail* to converge is to diverge to ∞. Note that if $\sum_{n=1}^{\infty} a_n$ is a series of *nonnegative* terms, then the sequence $\{s_n\}$ of partial sums is monotone increasing.

Theorem 10.5 *(Comparison Test 1) Let $\sum_{n=1}^{\infty} a_n$ and $\sum_{n=1}^{\infty} b_n$ be series of nonnegative terms such that $a_n \leq b_n$ for $n = 1, 2, 3, \ldots$ If $\sum_{n=1}^{\infty} b_n$ converges, then $\sum_{n=1}^{\infty} a_n$ converges also, while if $\sum_{n=1}^{\infty} a_n$ diverges, then $\sum_{n=1}^{\infty} b_n$ diverges also.*

Proof. Let s_n be the nth partial sum of $\sum_{n=1}^{\infty} a_n$ and let t_n be the nth partial sum of $\sum_{n=1}^{\infty} b_n$. Suppose $\sum_{n=1}^{\infty} b_n$ converges to c, so that $\lim_{n\to\infty} t_n = c$. From $a_n \leq b_n$, you see at once that

$$s_n \leq t_n$$

for $n = 1, 2, 3, \ldots$ Since $\lim_{n\to\infty} t_n = c$ and $s_n \leq t_n$, you see that $\lim_{n\to\infty} s_n = \infty$ is impossible, so by the Fundamental Property, the sequence $\{s_n\}$ must converge also.

Suppose that $\sum_{n=1}^{\infty} a_n$ diverges. Since $\{s_n\}$ is monotone increasing and diverges, you must have $\lim_{n\to\infty} s_n = \infty$ by the Fundamental Property. Since $t_n \geq s_n$, obviously

$$\lim_{n\to\infty} t_n = \infty,$$

so $\sum_{n=1}^{\infty} b_n$ diverges also. □

Theorem 10.5 is sometimes summarized by saying that, for series of nonnegative terms, a series "smaller" than a known convergent series also converges, while a series "larger" than a known divergent series must diverge also. If a series is "smaller" than a known divergent series, it may either converge or diverge, depending on how much "smaller" it is. Similarly, a series that is "larger" than a known convergent series may converge or diverge, depending on how much "larger" it is. It is important for you to remember that the comparison test works only in the stated direction.

Example 3 You know that the series

$$1 + \frac{1}{2} + \frac{1}{4} + \frac{1}{8} + \cdots + \frac{1}{2^{n-1}} + \cdots$$

converges. Therefore the "smaller" series

$$\frac{1}{2} + \frac{1}{3} + \frac{1}{5} + \frac{1}{9} + \cdots + \frac{1}{2^{n-1} + 1} + \cdots$$

converges, by Comparison Test 1. ||

Example 4 You know that the harmonic series

$$1 + \frac{1}{2} + \frac{1}{3} + \frac{1}{4} + \cdots + \frac{1}{n} + \cdots$$

diverges. Therefore the "larger" series

$$1 + \frac{1}{\sqrt{2}} + \frac{1}{\sqrt{3}} + \frac{1}{\sqrt{4}} + \cdots + \frac{1}{\sqrt{n}} + \cdots$$

diverges also. ||

Since a finite number of terms can be inserted in a series or deleted from it without affecting its convergence or divergence (Theorem 10.3), we can weaken the hypotheses of Comparison Test 1 and only require that $a_n \le b_n$ for all but a finite number of positive integers n.

10.3.4 Another type of comparison

The series

$$\sum_{n=1}^{\infty} \frac{n^2 + 3n}{2n^3 - n^2}$$

diverges, for

$$\frac{n^2 + 3n}{2n^3 - n^2} > \frac{n^2}{2n^3} = \frac{1}{2n},$$

and $\sum_{n=1}^{\infty} (1/2n)$ diverges since it is "half" the harmonic series. Rather than fuss about such a precise inequality for Comparison Test 1, the mathematician usually says that the given series diverges since $(n^2 + 3n)/(2n^3 - n^2)$ is of order of magnitude $n^2/2n^3 = 1/2n$ for large n, and $\sum_{n=1}^{\infty} (1/2n)$ diverges. The justification for this argument is given in the following theorem.

Theorem 10.6 (*Comparison Test* 2) *Let $\sum_{n=1}^{\infty} a_n$ and $\sum_{n=1}^{\infty} b_n$ be series of nonnegative terms with $a_n \neq 0$ for all sufficiently large n, and suppose that $\lim_{n\to\infty} (b_n/a_n) = c > 0$. Then the two series either both converge or both diverge.*

Proof. Since $\lim_{n\to\infty} (b_n/a_n) = c > 0$, there exists N such that for $n > N$, we have

$$\frac{c}{2} < \frac{b_n}{a_n} < \frac{3c}{2},$$

or

$$\frac{c}{2} a_n < b_n < \frac{3c}{2} a_n. \tag{2}$$

If $\sum_{n=1}^{\infty} a_n$ converges, then $\sum_{n=1}^{\infty} (3c/2)a_n$ converges by Corollary 1 of Theorem 10.2 (Section 10.2). Since (2) provides a "comparison test" for all but a finite number of terms b_n, we see from Comparison Test 1 and the remarks following Example 4 that $\sum_{n=1}^{\infty} b_n$ converges also. Similarly, if $\sum_{n=1}^{\infty} a_n$ diverges, then $\sum_{n=1}^{\infty} (c/2)a_n$ diverges, so by (2) and Comparison Test 1, we see that $\sum_{n=1}^{\infty} b_n$ diverges. □

As indicated for limits as $x \to \infty$ in Chapter 2, for large n, a quotient of polynomials in n "behaves like" the quotient of the monomial terms of highest degree in the numerator and denominator.

Example 5 Let us determine convergence or divergence of the series

$$\sum_{n=1}^{\infty} \frac{2n^3 - 3n^2}{7n^4 + 100n^3 + 7}.$$

SOLUTION For large n, the nth term of the series is of order of magnitude $2n^3/7n^4$ or $2/7n$. More precisely,

$$\lim_{n\to\infty} \frac{(2n^3 - 3n^2)/(7n^4 + 100n^3 + 7)}{2/7n} = \lim_{n\to\infty} \frac{14n^4 - 21n^3}{14n^4 + 200n^3 + 14} = 1.$$

Thus by Comparison Test 2, our given series diverges, since

$$\sum_{n=1}^{\infty} \frac{2}{7n}$$

is $\frac{2}{7}$ times the harmonic series and therefore diverges. ‖

A precise comparison test for all n in the sense of Comparison Test 1 would be a bit messy for the series in Example 5, but Comparison Test 2 handled it easily. We place more than the usual emphasis on developing ability to discover the convergence or divergence of certain series at a glance, and Comparison Test 2 is one of the main tools used for this purpose.

We give one more example of the way in which this theorem is used to determine convergence.

Example 6 The series

$$\sum_{n=1}^{\infty} \frac{2n^2 + 4}{(n^2 + n)2^n}$$

behaves like $\sum_{n=1}^{\infty} (2/2^n)$, which converges since it is a geometric series with ratio 1/2. Thus both series converge. ‖

SUMMARY

1. *The convergence or divergence of a series (or sequence) is not changed by inserting, deleting, or altering any finite number of terms.*
2. *If* $\lim_{n\to\infty} a_n \neq 0$, *then* $\sum_{n=1}^{\infty} a_n$ *diverges.*
3. ***Comparison Test 1:*** *If* $0 \le a_n \le b_n$ *for all sufficiently large* n *and* $\sum_{n=1}^{\infty} b_n$ *converges, then* $\sum_{n=1}^{\infty} a_n$ *converges. On the other hand, if* $\sum_{n=1}^{\infty} a_n$ *diverges, then* $\sum_{n=1}^{\infty} b_n$ *diverges.*
4. ***Comparison Test 2:*** *If* $a_n > 0$ *for sufficiently large* n *and* $b_n \ge 0$ *for all* n *and* $\lim_{n\to\infty} (b_n/a_n) = c > 0$, *then the series* $\sum_{n=1}^{\infty} a_n$ *and* $\sum_{n=1}^{\infty} b_n$ *either both converge or both diverge.*

EXERCISES

In Exercises 1 through 3, determine whether the given series converges or diverges, and if it is convergent, find its sum.

CONVERGES to 1/2

1/2 + 1 − 2 + 1/2 + √5 − √5

1. $1 - 2 + \frac{1}{2} + \sqrt{5} + \frac{1}{4} + \frac{1}{8} + \frac{1}{16} + \cdots + \frac{1}{2^{n-3}} + \cdots$ where $n \ge 5$

2. $1 + \frac{1}{2} + \frac{1}{4} + \frac{1}{8} + \frac{1}{16} + \frac{1}{32} + \frac{2}{32} + \frac{3}{32} + \cdots + \frac{n-5}{32} + \cdots$ where $n \ge 6$

3. $\sqrt{2} + 1 - \frac{1}{3} - \sqrt{3} + \frac{1}{9} - \frac{1}{27} + \frac{1}{81} - \cdots + (-1)^{n+1}\frac{1}{3^{n-3}} + \cdots$ where $n \geq 5$

In Exercises 4 through 24, classify the series as convergent or divergent, and indicate a reason for your answer. (You should do all these exercises, and try to develop facility in ascertaining the convergence or divergence of the series at a glance.)

4. $\sum_{n=1}^{\infty} \frac{1}{10n}$

5. $\sum_{n=1}^{\infty} \frac{\sqrt{n}}{n \cdot 2^n}$

6. $\sum_{n=1}^{\infty} (\ln n)$

7. $\sum_{n=2}^{\infty} \frac{\cos^2 n}{2^n}$

8. $\sum_{n=3}^{\infty} \frac{1}{n-1}$

9. $\sum_{n=1}^{\infty} \frac{1}{n + 2^n}$

10. $\sum_{n=1}^{\infty} \frac{1}{\sqrt[3]{n}}$

11. $\sum_{n=1}^{\infty} \frac{n^3}{3n^3 + n^2}$

12. $\sum_{n=1}^{\infty} \left(\frac{1}{n} - \frac{1}{2^n}\right)$

13. $\sum_{n=1}^{\infty} \frac{|\sin n|}{5^n}$

14. $\sum_{n=1}^{\infty} \frac{\sqrt{n}}{n + 17}$

15. $\sum_{n=1}^{\infty} \frac{8^n + 9^n}{10^n}$

16. $\sum_{n=1}^{\infty} \frac{n^2 - 3n}{4n^3 + n^2}$

17. $\sum_{n=i}^{\infty} \frac{n+2}{n^2 + 3}$

18. $\sum_{n=1}^{\infty} \frac{1}{\sqrt[n]{100}}$

19. $\sum_{n=1}^{\infty} \frac{n^2 e^{-2n}}{3n^2 - 2n}$

20. $\sum_{n=1}^{\infty} \frac{e^n}{n^3 + 4n}$

21. $\sum_{n=1}^{\infty} \frac{1}{\sin^2 n}$

22. $\sum_{n=1}^{\infty} \frac{2^n}{3^{2n+1}}$

23. $\sum_{n=1}^{\infty} \sqrt[n]{n}$

24. $\sum_{n=1}^{\infty} \frac{1}{\sqrt[n]{n}}$

25. Consider the series $\sum_{n=1}^{\infty} \left(\frac{1}{n} - \frac{1}{n+1}\right)$.

a) Compute the first four partial sums of the series.

b) Find a formula for the nth partial sum s_n of the series. (The series is known as a "*telescoping series*." Can you guess why?)

c) Show that the series converges, and find the sum of the series.

26. Consider the series $\sum_{n=1}^{\infty} \left(\ln \frac{n+1}{n}\right)$.

a) Show that the series can be viewed as a telescoping series (see Exercise 25), and compute the nth partial sum s_n. [*Hint.* Use a property of the function ln.]

b) Show that the series diverges.

27. Give an example to show that if we do not require that the terms of our series be nonnegative, it is possible to have $a_n \leq b_n$ for $n = 1, 2, 3, \ldots$ and to have $\sum_{n=1}^{\infty} b_n$ converge while $\sum_{n=1}^{\infty} a_n$ diverges.

10.4 THE INTEGRAL AND RATIO TESTS

You will note that the tests given in this section depend on the comparison test, which can be viewed as the most general test we consider.

10.4.1 The integral test

We discussed convergence of improper integrals of the form $\int_1^\infty f(x)\,dx$ in Section 9.7. The integral test shows that convergence of certain series can be demonstrated by showing the convergence of an integral of this form.

Theorem 10.7 *(The Integral Test) Let $\sum_{n=1}^\infty a_n$ be a series of nonnegative terms. Suppose also that f is a continuous function for $x \geq 1$ such that*

a) *$f(n) = a_n$ for $n = 1, 2, 3, \ldots$*
b) *f is a monotone decreasing for $x \geq 1$; that is, $f(x_i) \geq f(x_j)$ if $x_i \leq x_j$ for $x \geq 1$.*

Then $\sum_{n=1}^\infty a_n$ converges if and only if $\int_1^\infty f(x)\,dx$ converges.

Proof. The proof follows easily from our previous work and from Fig. 10.1. Let s_n be the nth partial sum of the series $\sum_{n=1}^\infty a_n$. Now $\int_1^n f(x)\,dx$ is the area of the region under the graph of f over $[1, n]$. If you approximate $\int_1^n f(x)\,dx$ by the upper and lower sums given by the rectangles with bases of lengths one in Fig. 10.1 you obtain

$$a_2 + \cdots + a_n \leq \int_1^n f(x)\,dx \leq a_1 + a_2 + \cdots + a_n. \tag{1}$$

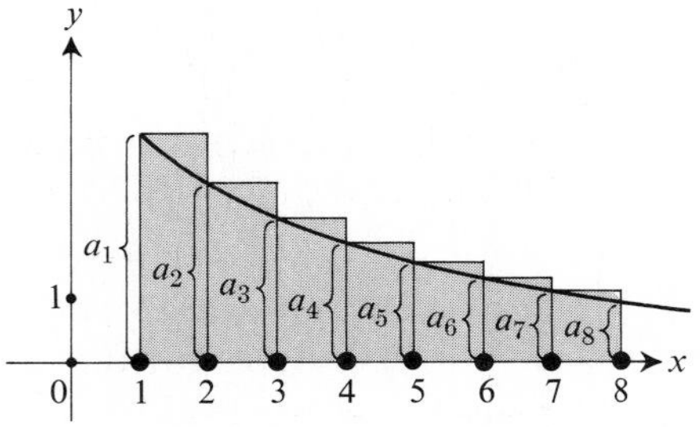

(a) Upper sum approximation

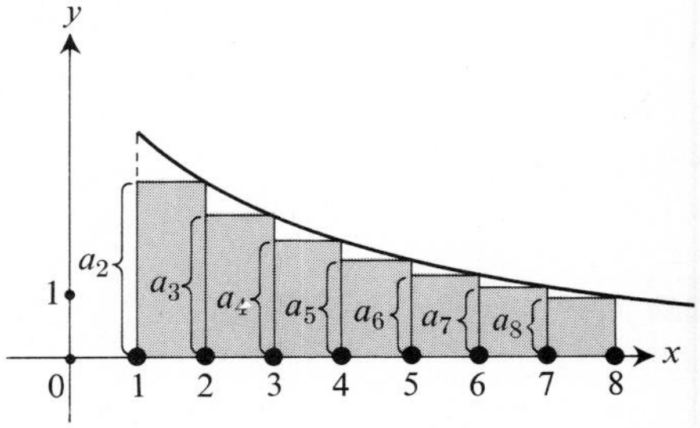

(b) Lower sum approximation

10.1

Note that (1) depends on the fact that f is monotone decreasing for $x \geq 1$. From (1), you obtain

$$s_n - a_1 \leq \int_1^n f(x)\,dx \leq s_n. \tag{2}$$

Since $f(x) \geq 0$ for $x \geq 1$, the sequence $\{\int_1^n f(x)\,dx\}$ is clearly monotone increasing. Therefore, if $\int_1^\infty f(x)\,dx$ converges, then by the Fundamental

Property (Section 10.3), the sequence $\{\int_1^n f(x)\,dx\}$ must converge. From (2), using a comparison test, you find that the sequence $\{s_n - a_1\}$ converges. But then the sequence $\{s_n\}$ converges by Theorem 10.2 (Section 10.2), so the series $\sum_{n=1}^{\infty} a_n$ converges.

On the other hand, if $\{s_n\}$ converges, then the sequence $\{\int_1^n f(x)\,dx\}$ must converge by (2), using a comparison test. Since $F(t) = \int_1^t f(x)\,dx$ is a monotone increasing function of t, it is clear that $\int_1^{\infty} f(x)\,dx$ must converge also. □

You should regard the integral test as asserting that for $\sum_{n=1}^{\infty} a_n$ and f as described, the series "behaves like" $\int_1^{\infty} f(x)\,dx$, as far as convergence or divergence is concerned.

Example 1 The integral test provides another demonstration that the harmonic series $\sum_{n=1}^{\infty} (1/n)$ diverges. The theorem shows that the series behaves like $\lim_{h\to\infty} \int_1^h (1/x)\,dx$. However,

$$\lim_{h\to\infty} \int_1^h \frac{1}{x}\,dx = \lim_{h\to\infty} (\ln x)\Big]_1^h = \lim_{h\to\infty} (\ln h - \ln 1) = \lim_{h\to\infty} (\ln h) = \infty.$$

Thus the integral diverges, so the series diverges also. ‖

One type of series handled by the integral test is so important that we treat it formally in a corollary.

Corollary (p-*series*). *Let p be any real number. The series $\sum_{n=1}^{\infty} (1/n^p)$ converges if $p > 1$, and diverges for $p \le 1$.*

Proof. If $p > 1$, then define f by $f(x) = 1/x^p$. Then f is a continuous function which satisfies the conditions of the Integral Test, and the behavior of the series $\sum_{n=1}^{\infty} (1/n^p)$ is the same as the behavior of $\int_1^{\infty} (1/x^p)\,dx$. We have

$$\int_1^{\infty} \frac{1}{x^p}\,dx = \lim_{h\to\infty} \int_1^h x^{-p}\,dx = \lim_{h\to\infty} \left(\frac{x^{1-p}}{1-p}\Big]_1^h\right)$$

$$= \lim_{h\to\infty} \left(\frac{h^{1-p}}{1-p} - \frac{1}{1-p}\right) = \frac{1}{1-p}\left(\lim_{h\to\infty} \frac{1}{h^{p-1}}\right) + \frac{1}{p-1}.$$

Since we are assuming that $p > 1$, we see that $p - 1 > 0$, so

$$\lim_{h\to\infty} \frac{1}{h^{p-1}} = 0.$$

Thus the integral, and therefore the series, converges if $p > 1$.

If $p \le 1$, then $n^p \le n$, so $1/(n^p) \ge (1/n)$, and the series diverges by comparison with the harmonic series. □

The preceding result is especially useful when combined with Comparison Test 2 (Section 10.3), as illustrated now.

Example 2 By Comparison, Test 2, the series

$$\sum_{n=1}^{\infty} \frac{n+20}{n^3+20n+1}$$

behaves like $\sum_{n=1}^{\infty}(1/n^2)$, which is a "$p$-series" with $p = 2$, and hence converges. ‖

10.4.2 Estimating the sum of a series of monotone-decreasing terms by an integral

The proof of the Integral Test really gives more information than indicated by the statement of the test, for (2) gives a very useful estimate for the partial sum s_n of a series, provided that a function f can be found satisfying the hypotheses of the test. Namely, from (2), we obtain

$$\int_1^n f(x)\,dx \le s_n \le \left(\int_1^n f(x)\,dx\right) + a_1. \tag{3}$$

We illustrate the use of (3) by examples.

Example 3 From (3) with $f(x) = 1/x$, we see that for the partial sum s_n of the harmonic series $1 + (\frac{1}{2}) + (\frac{1}{3}) + \cdots$ we have

$$\int_1^n \frac{1}{x}\,dx \le s_n \le \left(\int_1^n \frac{1}{x}(x)\,dx\right) + 1.$$

Since $\int_1^n (1/x)\,dx = \ln n - \ln 1 = \ln n$, we have

$$\ln n \le s_n \le (\ln n) + 1,$$

or, changing notation slightly,

$$\ln k \le \sum_{n=1}^{k} \frac{1}{n} \le (\ln k) + 1. \quad ‖$$

Example 4 You know that $\sum_{n=1}^{\infty}(1/n^2)$ converges. From (3), you obtain

$$\int_1^n \frac{1}{x^2}\,dx \le s_n \le \left(\int_1^n \frac{1}{x^2}\,dx\right) + 1.$$

Since $\int_1^n (1/x^2)\,dx = (-1/x)]_1^n = (-1/n) + 1 = 1 - (1/n)$, you have

$$1 - \frac{1}{n} \le s_n \le \left(1 - \frac{1}{n}\right) + 1 = 2 - \frac{1}{n}.$$

Therefore

$$1 = \lim_{n\to\infty}\left(1 - \frac{1}{n}\right) \le \lim_{n\to\infty} s_n \le \lim_{n\to\infty}\left(2 - \frac{1}{n}\right) = 2,$$

so $\sum_{n=1}^{\infty}(1/n^2)$ converges to a number in $[1, 2]$. ‖

Exercises 42 and 43 show how an integral can be used to estimate the sum of some convergent series of monotone-decreasing nonnegative terms to any desired accuracy.

10.4.3 The ratio test

Recall that if there is any hope that $\sum_{n=1}^{\infty} a_n$ converges, you must have $\lim_{n\to\infty} a_n = 0$. The ratio test studies the rate at which terms a_n decrease as n increases, and is based on a comparison of the series $\sum_{n=1}^{\infty} a_n$ with a suitable geometric series. The (geometric) decrease from the term a_n to the next term a_{n+1} in $\sum_{n=1}^{\infty} a_n$ can be measured by the *ratio* a_{n+1}/a_n.

Theorem 10.8 (*Ratio Test*) *Let $\sum_{n=1}^{\infty} a_n$ be a series of positive terms and suppose that $\lim_{n\to\infty} (a_{n+1}/a_n)$ exists and is r. Then $\sum_{n=1}^{\infty} a_n$ converges if $r < 1$, and diverges if $r > 1$. (If $r = 1$, further information is necessary to determine the convergence or divergence of the series.)*

Proof. Suppose $\lim_{n\to\infty} (a_{n+1}/a_n) = r < 1$. Find $\delta > 0$ such that

$$r < r + \delta < 1.$$

For example, you could let $\delta = (1 - r)/2$. Since $\lim_{n\to\infty} (a_{n+1}/a_n) = r$, there exists an integer N such that

$$r - \delta < \frac{a_{n+1}}{a_n} < r + \delta$$

for $n > N$. Multiplying by a_n, you find that

$$a_{n+1} < (r + \delta)a_n \tag{4}$$

for $n > N$. Using relation (4) repeatedly, starting with $n = N + 1$, you obtain

$$a_{N+2} < (r + \delta)a_{N+1},$$
$$a_{N+3} < (r + \delta)a_{N+2} < (r + \delta)^2 a_{N+1}$$
$$a_{N+4} < (r + \delta)a_{N+3} < (r + \delta)^3 a_{N+1}, \text{ etc.}$$

Thus each term of the series

$$a_{N+1} + a_{N+2} + a_{N+3} + \cdots + a_{N+k} + \cdots \tag{5}$$

is less than or equal to the corresponding term of the series

$$a_{N+1} + a_{N+1}(r + \delta) + a_{N+1}(r + \delta)^2 + \cdots + a_{N+1}(r + \delta)^{k-1} + \cdots \tag{6}$$

But (6) is a geometric series with ratio $r + \delta < 1$, and converges. By the comparison test, you then see that (5) converges. Since (5) is $\sum_{n=1}^{\infty} a_n$ with a finite number of terms deleted, Theorem 10.3 (Section 10.3) shows that $\sum_{n=1}^{\infty} a_n$ converges also.

Now suppose that $r > 1$. Then for all sufficiently large n,

$$\frac{a_{n+1}}{a_n} > 1;$$

that is,

$$a_{n+1} > a_n$$

for n sufficiently large. This means that after a while, the terms of $\sum_{n=1}^{\infty} a_n$ *increase*. Thus $\lim_{n\to\infty} a_n = 0$ is impossible, so by Theorem 10.4 (Section 10.3), the series $\sum_{n=1}^{\infty} a_n$ must diverge. □

We should comment on the observation in parentheses in the statement of the Ratio Test. For the series $\sum_{n=1}^{\infty} (1/n^2)$, we have

$$r = \lim_{n\to\infty} \frac{a_{n+1}}{a_n} = \lim_{n\to\infty} \frac{1/((n+1)^2)}{1/n^2} = \lim_{n\to\infty} \left(\frac{n}{n+1}\right)^2 = 1,$$

and for the series $\sum_{n=1}^{\infty} (1/n)$, we also have

$$r = \lim_{n\to\infty} \frac{a_{n+1}}{a_n} = \lim_{n\to\infty} \frac{1/(n+1)}{1/n} = \lim_{n\to\infty} \frac{n}{n+1} = 1.$$

In both cases, the *limiting ratio r* exists and is 1, but by the p-series test, the series $\sum_{n=1}^{\infty} (1/n^2)$ converges while the harmonic series $\sum_{n=1}^{\infty} (1/n)$ diverges. This illustrates that for a series having a limiting ratio $r = 1$, more information must be obtained before the convergence or divergence of the series can be determined.

The Ratio Test is especially useful in handling series whose nth term a_n is given by a formula involving a constant to the nth power (e.g., 3^n), or involving a factor $n!$ where

$$n! = (n)(n-1)\cdots(3)(2)(1).$$

It is worth mentioning that for any constant $c > 1$, the exponential c^n increases much faster as $n \to \infty$ than a polynomial in n of any degree s. This is true in the strong sense that

$$\lim_{n\to\infty} \frac{c^n}{n^s} = \infty$$

for any s. You can easily convince yourself that this is true by taking a logarithm and showing that $\lim_{n\to\infty} \ln(c^n/n^s) = \infty$. It is also worth mentioning that $n!$ increases much faster than c^n, once again in the strong sense that

$$\lim_{n\to\infty} \frac{n!}{c^n} = \infty.$$

Write

$$\frac{n!}{c^n} = \frac{n}{c} \cdot \frac{n-1}{c} \cdot \frac{n-2}{c} \cdots \frac{r}{c} \cdots \frac{3}{c} \cdot \frac{2}{c} \cdot \frac{1}{c},$$

where r is the largest integer such that $r < 2c$. Then

$$\frac{n!}{c^n} \geq 2 \cdot 2 \cdot 2 \cdots 2 \cdot \frac{r}{c} \cdot \frac{r-1}{c} \cdots \frac{3}{c} \cdot \frac{2}{c} \cdot \frac{1}{c} = 2^{n-r}\left(\frac{r}{c} \cdots \frac{3}{c} \cdot \frac{2}{c} \cdot \frac{1}{c}\right).$$

Consequently, $\lim_{n\to\infty} (n!/c^n) = \infty$.

Summarizing, for large n, we expect $n!$ to "dominate" c^n for any c, and c^n for $c > 1$ in turn "dominates" any polynomial in n for large n.

We now give some applications of the ratio test that illustrate these ideas.

Example 5 Let's try to find the limiting ratio r of the series

$$\sum_{n=1}^{\infty} \frac{2^n}{n!}.$$

SOLUTION We have

$$r = \lim_{n\to\infty} \frac{a_{n+1}}{a_n} = \lim_{n\to\infty} \frac{[2^{n+1}/(n+1)!]}{2^n/n!} = \lim_{n\to\infty} \left(\frac{2^{n+1}}{(n+1)!} \cdot \frac{n!}{2^n}\right)$$

$$= \lim_{n\to\infty} \frac{2 \cdot n!}{(n+1)n!} = \lim_{n\to\infty} \frac{2}{n+1} = 0.$$

Since r exists and is $0 < 1$, we see the series converges.

Note our use of the relation $(n+1)! = (n+1)n!$; this relation is worth remembering.

This example illustrates that $n!$ increases with n so much faster than 2^n that $\sum_{n=1}^{\infty} (2^n/n!)$ converges. You should try to develop an intuitive feeling for such behavior of a series. Note that the numerator 2^n of the nth term of the series contributed the 2 in the numerator of the ratio. The $n!$ in the denominator contributed the $n + 1$ in the denominator of the ratio, before the limit was taken. ||

Example 6 Let's determine convergence or divergence of the series

$$\sum_{n=1}^{\infty} \frac{n^{25}}{3^n}.$$

SOLUTION We know that 3^n increases more rapidly than n^{25} for large n, and we wonder whether 3^n increases rapidly enough to completely dominate n^{25} and make our series behave like a geometric series with ratio $\frac{1}{3}$.

We find that

$$r = \lim_{n\to\infty} \frac{a_{n+1}}{a_n} = \lim_{n\to\infty} \frac{(n+1)^{25}/3^{n+1}}{n^{25}/3^n} = \lim_{n\to\infty} \frac{(n+1)^{25}}{n^{25}} \cdot \frac{3^n}{3^{n+1}}$$

$$= \lim_{n\to\infty} \left(\frac{n+1}{n}\right)^{25} \frac{1}{3} = 1^{25} \cdot \frac{1}{3} = \frac{1}{3}.$$

Thus r exists and is $\frac{1}{3} < 1$, so our series does indeed converge.

This example shows that if n is large enough, 3^n dominates n^{25} to such an extent that $\sum_{n=1}^{\infty} (n^{25}/3^n)$ converges, and indeed, "behaves" like a geometric series with ratio $\frac{1}{3}$. The example also illustrates that a "polynomial part" of a formula giving the nth term a_n of a series just contributes a factor 1 to the limiting ratio, and hence is never significant in a ratio test. (Note the contribution $1^{25} = 1$ from n^{25} in our computation.) The 3^n in the denominator of the series contributed the 3 in the denominator of the ratio. ‖

Example 7 The series

$$\sum_{n=1}^{\infty} \frac{n!}{n^{100} \cdot 5^n}$$

diverges, for $n!$ dominates both n^{100} and 5^n for large n, and the nth term a_n actually becomes large as n approaches infinity. If you compute the limiting ratio r, the preceding examples show that you will find that $r = \lim_{n\to\infty} [(n + 1)/5] = \infty$. ‖

Example 8 Consider convergence or divergence of the series

$$\sum_{n=1}^{\infty} \frac{(100)^n \cdot (n^3 + 3)}{n!}.$$

SOLUTION Here $(n^3 + 3)$ contributes a factor of 1 to the limiting ratio in a ratio test, and the $n!$ dominates the 100^n to such an extent that the series converges. If you compute the limiting ratio r, the above examples show that you will find that $r = \lim_{n\to\infty} [100/(n + 1)] = 0$. ‖

Example 9 The series

$$\sum_{n=1}^{\infty} \frac{n^5 \cdot 2^{n+1}}{3^n}$$

converges, for the n^5 is dominated by the 2^{n+1} and the 3^n. The series thus behaves like

$$\sum_{n=1}^{\infty} \frac{2^{n+1}}{3^n} = 2\left(\sum_{n=1}^{\infty} \left(\frac{2}{3}\right)^n\right),$$

which converges as a multiple of a geometric series with ratio $\frac{2}{3}$. The limiting ratio of the original series $\sum_{n=1}^{\infty} (n^5 \cdot 2^{n+1}/3^n)$ is

$$r = \lim_{n\to\infty} \frac{2}{3} = \frac{2}{3}. \quad ‖$$

For a series whose nth term a_n is given by a formula involving polynomials, exponentials, and factorials, the mathematician can usually tell whether or not the series converges by examining which things "dominate," and how "fast" the nth term approaches 0. He or she seldom actually writes out a test given in our theorems. You should try to acquire some proficiency in this technique. Practice is provided by the exercises.

SUMMARY

1. ***Integral Test:*** *Let $\sum_{n=1}^{\infty} a_n$ be a series of nonnegative terms. Suppose also that $f(x)$ is a continuous function for $x \geq 1$ such that*
   a) $f(n) = a_n$ *for* $n = 1, 2, 3, \ldots$
   b) *f is monotone decreasing for $x \geq 1$; that is, $f(x_i) \geq f(x_j)$ if $x_i \leq x_j$ for $x_i \geq 1$, $x_j \geq 1$. Then $\sum_{n=1}^{\infty} a_n$ converges if and only if $\int_1^{\infty} f(x)\,dx$ converges.*

2. ***p-Series:*** *Let p be a real number. The series $\sum_{n-1}^{\infty} (1/n^p)$ converges if $p > 1$, and diverges for $p \leq 1$.*

3. ***Ratio Test:*** *Let $\sum_{n=1}^{\infty} a_n$ be a series of positive terms and suppose that $\lim_{n\to\infty} (a_{n+1}/a_n)$ exists and is r. Then $\sum_{n=1}^{\infty} a_n$ converges if $r < 1$, and diverges if $r > 1$. (If $r = 1$, further information is necessary to determine the convergence or divergence of the series.)*

4. *Suppose a_n in $\sum_{n=1}^{\infty} a_n$ is given by a formula involving n, containing factors in either the numerator or denominator of the form b^n, $n!$, or $p(n)$, where $p(n)$ is a polynomial function of n. The individual contributions of these factors in the Ratio Test are as follows.*

   b^n: *Contributes a factor b in the same (numerator or denominator) position.*

   $n!$: *Contributes a factor $n + 1$ in the same position, before $\lim_{n\to\infty}$ is computed.*

   $p(n)$: *Contributes a factor 1 after $\lim_{n\to\infty}$ is computed, and thus may be neglected if there are other factors.*

   Using these ideas, you see at a glance that

$$\sum_{n=1}^{\infty} \frac{2^n \cdot n^4}{n!} \quad \textit{leads to} \quad r = \lim_{n\to\infty} \frac{2}{n+1} = 0$$

   in the Ratio Test, while

$$\sum_{n=1}^{\infty} \frac{n!(n^7 + n^3)}{(n+2)! \cdot 5^n} \quad \textit{leads to} \quad r = \lim_{n\to\infty} \frac{n+1}{(n+3)\cdot 5} = \frac{1}{5}.$$

EXERCISES

While you should be able to ascertain the convergence or divergence of each series in Exercises 1 through 6 at a glance, use the Integral Test, for practice, to discover whether or not the series converges.

1. $\sum_{n=1}^{\infty} \frac{1}{3n}$ **2.** $\sum_{n=1}^{\infty} \frac{1}{n^2+1}$ **3.** $\sum_{n=1}^{\infty} \frac{1}{4n-1}$ **4.** $\sum_{n=1}^{\infty} \frac{n+1}{n^2+2n-2}$ **5.** $\sum_{n=1}^{\infty} \frac{n}{(n+1)^3}$ **6.** $\sum_{n=1}^{\infty} \frac{1}{e^n}$

7. Note that $1/n^2 \le 1/n(\ln n) \le 1/n$ for $n \ge 3$, and $\sum_{n=2}^{\infty}(1/n^2)$ converges while $\sum_{n=2}^{\infty}(1/n)$ diverges. Use the Integral Test to show that $\sum_{n=2}^{\infty}(1/n(\ln n))$ diverges, and try to file this result in your memory with other series that you know diverge.

8. Use the Integral Test to show that the series $\sum_{n=2}^{\infty}(1/n(\ln n)^2)$ converges.

While you should be able to ascertain the convergence or divergence of each series in Exercises 9 through 15 at a glance, write out the Ratio Test, for practice, to determine whether or not the series converges.

9. $\displaystyle\sum_{n=1}^{\infty}\frac{n^2\cdot 2^n}{n!}$

10. $\displaystyle\sum_{n=1}^{\infty}\frac{n^3+3n}{2^n}$

11. $\displaystyle\sum_{n=1}^{\infty}\frac{5^{n+1}}{n^3\cdot 4^{n+2}}$

12. $\displaystyle\sum_{n=1}^{\infty}\frac{(n+1)!}{100^{n+10}}$

13. $\displaystyle\sum_{n=1}^{\infty}\frac{n^2\cdot 5^{n+1}}{3^{2n-1}}$

14. $\displaystyle\sum_{n=1}^{\infty}\frac{(n+3)(n+7)}{n!}$

15. $\displaystyle\sum_{n=1}^{\infty}\frac{(n+5)!}{n^2\cdot n!\cdot 2^n}$

16. Obviously $n^n > n!$ for large n. Let's discover whether n^n dominates $n!$ enough to make $\sum_{n=1}^{\infty} n!/n^n$ converge.

a) Show that if n is even, then
$$\frac{n!}{n^n} < \left(\frac{1}{2}\right)^{n/2} = \left(\frac{1}{\sqrt{2}}\right)^n.$$
[*Hint.* After you are "half-way through the $n!$," the factors in $n!$ are less than $n/2$.]

b) Show that if n is odd, then $n!/n^n \le (\frac{1}{2})^{(n-1)/2} = (1/\sqrt{2})^{n-1}$. [*Hint.* After you are "down to $(n-1)/2$" in the $n!$, the ratio of each remaining factor of $n!$ to n is less than $(n-1)/2n$.]

c) From (a) and (b), conclude that n^n does indeed dominate $n!$ to such an extent that $\sum_{n=1}^{\infty} n!/n^n$ converges.

In Exercises 17 through 40, try to determine by "inspection" (without writing out any computations, as illustrated in Examples 8 and 9, whether or not the series converges. Make use of the results in Exercises 7, 8, and 16 where appropriate. Write out a formal test only if you get stuck.

17. $\displaystyle\sum_{n=1}^{\infty}\frac{3n^2+3n}{n^4+2}$

18. $\displaystyle\sum_{n=1}^{\infty}\frac{n}{\sqrt{n^3+3n}}$

19. $\displaystyle\sum_{n=1}^{\infty}\frac{1}{\sqrt{n+3}}$

20. $\displaystyle\sum_{n=1}^{\infty}\frac{n^{10}}{2^n}$

21. $\displaystyle\sum_{n=1}^{\infty}\frac{\ln n}{n}$

22. $\displaystyle\sum_{n=2}^{\infty}\frac{n-1}{n^2(\ln n)}$

23. $\displaystyle\sum_{n=1}^{\infty}\frac{1}{\ln(n^2+4n)}$

24. $\displaystyle\sum_{n=1}^{\infty}\frac{n^n}{n^3\cdot n!}$

25. $\displaystyle\sum_{n=1}^{\infty}\frac{\sqrt{3n^2+6n-1}}{4n^3-3n}$

26. $\displaystyle\sum_{n=1}^{\infty}\frac{n^3\cdot 4^n}{n!}$

27. $\displaystyle\sum_{n=1}^{\infty}\frac{2^{3n+1}}{n!}$

28. $\displaystyle\sum_{n=1}^{\infty}\frac{3^{2n+4}}{10^n}$

29. $\displaystyle\sum_{n=1}^{\infty}\frac{2^{3n-1}}{8^{n+6}}$

30. $\displaystyle\sum_{n=1}^{\infty}\frac{n!}{(2n)!}$

31. $\displaystyle\sum_{n=1}^{\infty}\frac{n^2\cdot n!}{(n+3)!}$

32. $\displaystyle\sum_{n=1}^{\infty}\frac{n!+3^n}{(n+1)!}$

33. $\displaystyle\sum_{n=1}^{\infty}\frac{n!+2^n}{(n+2)!}$

34. $\displaystyle\sum_{n=1}^{\infty}\frac{(n+3)!-n!}{2^n}$

35. $\displaystyle\sum_{n=1}^{\infty}\frac{(n+3)!-(n+1)!}{2^n\cdot(n+2)!}$

36. $\displaystyle\sum_{n=1}^{\infty}\frac{(n+3)!-(n+1)!}{(n+4)!}$

37. $\displaystyle\sum_{n=1}^{\infty}\frac{(n+3)!-(n+1)!}{(n+5)!}$

38. $\displaystyle\sum_{n=1}^{\infty}\frac{n!}{3^{2n}}$

39. $\sum_{n=1}^{\infty} \frac{n!}{3^{(n^2)}}$

40. $\sum_{n=1}^{\infty} \frac{1}{2^{\ln n}}$

41. Use (3) to estimate $\sum_{n=1}^{\infty} (1/n)^{3/2}$, as illustrated in Example 4.

42. Let f be a continuous monotone-decreasing function for $x \geq 1$ and let $a_n = f(n)$. Use a diagram similar to that in Fig. 10.1 to show that for integers r and s where $r < s$, we have

$$\int_r^s f(x)\,dx \leq \sum_{n=r}^{s} a_n \leq \left(\int_r^s f(x)\,dx\right) + a_r. \qquad (7)$$

[Note the similarity of (7) to (3) in the text.]

43. Let $\sum_{n=1}^{\infty} a_n$ be a convergent series of nonnegative terms where $a_n = f(n)$ for a continuous monotone-decreasing function f for $x \geq 1$. Use (7) in the preceding exercise to show that the following is a valid procedure to estimate the sum of the series with accuracy $\epsilon > 0$.

Let $\epsilon > 0$ be given. Find an integer r such that $a_r < \epsilon$. Then the sum of the series differs from

$$a_1 + a_2 + \cdots + a_{r-1} + \int_r^{\infty} f(x)\,dx$$

by at most ϵ.

44. Use Exercise 43 to estimate the sum of the series $\sum_{n=1}^{\infty} (1/n^2)$ with accuracy 0.1.

10.5 ALTERNATING SERIES; ABSOLUTE CONVERGENCE

10.5.1 Alternating series

We have established several tests for the convergence of a series of nonnegative terms. Analogous tests may be used for series all of whose terms are ≤ 0, since $\sum_{n=1}^{\infty} a_n$ converges to s if and only if $\sum_{n=1}^{\infty} (-a_n)$ converges to $-s$. A finite number of negative (or positive) terms can be neglected in establishing the convergence or divergence of a series. However, if a series contains an infinite number of positive and an infinite number of negative terms, the situation becomes more complicated. One type of series often encountered is an *alternating series*, one in which the terms are alternately positive and negative.

Example 1 The series $1 - 2 + 3 - 4 + 5 - 6 + \cdots$ is an alternating series. This series diverges since the nth term does not approach 0 as $n \to \infty$. ||

The following test establishes the convergence of certain alternating series. While the class of series covered by the test may seem very restrictive, such series occur quite often in practice.

Theorem 10.9 *(Alternating Series Test) Let* $\sum_{n=1}^{\infty} a_n$ *be a series such that*

a) *the series is alternating,*
b) $|a_{n+1}| \leq |a_n|$ *for all* n*, and*
c) $\lim_{n\to\infty} a_n = 0$.

Then the series converges.

Proof. Let s_n be the nth partial sum of the series $\sum_{n=1}^{\infty} a_n$. We may suppose that $a_1 > 0$ so that

$$a_1 > 0, \qquad a_2 < 0, \qquad a_3 > 0, \qquad \text{etc.}$$

(A similar argument holds if $a_1 < 0$.) We have

$$\begin{aligned} s_2 &= (a_1 + a_2), \\ s_4 &= (a_1 + a_2) + (a_3 + a_4), \\ s_6 &= s_4 + (a_5 + a_6), \\ s_8 &= s_6 + (a_7 + a_8), \\ &\ \ \vdots \end{aligned}$$

and the sequence $s_2, s_4, s_6, s_8, \ldots$ is monotone increasing, for each term in parentheses is nonnegative by (b). Since

$$s_{2n} = a_1 + (a_2 + a_3) + \cdots + (a_{2n-2} + a_{2n-1}) + a_{2n} \tag{1}$$

and each term in parentheses in (1) is nonpositive by (b), and since $a_{2n} < 0$, we see that

$$s_{2n} \le a_1.$$

Thus $s_2, s_4, s_6, s_8, \ldots$ is a monotone increasing sequence which is bounded above, and therefore converges to a number c by the Fundamental Property (Section 10.3.3).

We claim that $\sum_{n=1}^{\infty} a_n$ converges to c. Let $\epsilon > 0$ be given and find N such that $|a_n| < \epsilon/2$ for $n > N$ and also $|s_{2m} - c| < \epsilon/2$ for $2m > N$. This is possible by (c) and the preceding paragraph. If n is even and $n > N$, then $|s_n - c| < \epsilon/2 < \epsilon$ by choice of N. If n is odd and $n > N$, then

$$s_n = s_{n+1} - a_{n+1},$$

so

$$|s_n - c| = |(s_{n+1} - c) - a_{n+1}| \le |s_{n+1} - c| + |a_{n+1}| \le \frac{\epsilon}{2} + \frac{\epsilon}{2} = \epsilon$$

by our choice of N. Thus $\{s_n\}$ converges to c, so $\sum_{n=1}^{\infty} a_n$ converges and has sum c. □

Example 2 The *alternating harmonic series*

$$1 - \tfrac{1}{2} + \tfrac{1}{3} - \tfrac{1}{4} + \tfrac{1}{5} - \tfrac{1}{6} + \cdots$$

satisfies all the conditions for the Alternating Series Test, and hence converges. ‖

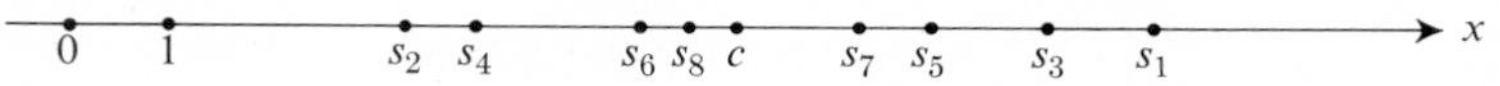

10.2

For an alternating series $\sum_{n=1}^{\infty} a_n$ that satisfies the conditions of Theorem 10.9 and has $a_1 > 0$, we have

$$\begin{aligned} s_1 &= a_1, \\ s_3 &= a_1 + (a_2 + a_3), \\ s_5 &= a_1 + (a_2 + a_3) + (a_4 + a_5), \\ s_7 &= s_5 + (a_6 + a_7), \\ &\ \vdots \end{aligned}$$

and the sequence $s_1, s_3, s_5, s_7, \ldots$ is monotone decreasing since each term in parentheses is nonpositive by (b). Since $s_2, s_4, s_6, s_8, \ldots$ is monotone increasing as shown in the proof of the theorem, we see that the sequence $\{s_n\}$ of partial sums must converge to c in the oscillatory manner indicated in Fig. 10.2. From the figure, it is evident that the error in the approximation s_n for the sum of the series is less than $|a_{n+1}|$. This fact is sometimes useful.

We emphasize that *all three* conditions of the Alternating Series Test must hold before you can conclude that the series converges. We leave as exercises the construction of examples to show that if any one of these three conditions is dropped, a series can be found that satisfies the remaining two conditions, but diverges. (See Exercises 1, 2, and 3.)

10.5.2 Absolute convergence

Let $\sum_{n=1}^{\infty} a_n$ be a series containing both postive and negative terms. The series $\sum_{n=1}^{\infty} |a_n|$ contains only nonnegative terms; we could apply some of the tests developed in the preceding sections to $\sum_{n=1}^{\infty} |a_n|$, and we might be able to establish its convergence or divergence. The next theorem shows that if $\sum_{n=1}^{\infty} |a_n|$ converges, then $\sum_{n=1}^{\infty} a_n$ converges also. It is important to note that if $\sum_{n=1}^{\infty} |a_n|$ diverges, then $\sum_{n=1}^{\infty} a_n$ may diverge or may converge. We illustrate this following the theorem.

Theorem 10.10 *(Absolute Convergence Test) Let $\sum_{n=1}^{\infty} a_n$ be any infinite series. If $\sum_{n=1}^{\infty} |a_n|$ converges, then $\sum_{n=1}^{\infty} a_n$ converges.*

Proof. We define a new series $\sum_{n=1}^{\infty} u_n$ by replacing the negative terms of $\sum_{n=1}^{\infty} a_n$ by zeros, and a new series $\sum_{n=1}^{\infty} v_n$ by replacing the positive terms of $\sum_{n=1}^{\infty} a_n$ by zeros. That is, we let

$$u_n = \begin{cases} a_n & \text{if } a_n \geq 0, \\ 0 & \text{if } a_n < 0, \end{cases} \quad \text{and} \quad v_n = \begin{cases} a_n & \text{if } a_n \leq 0, \\ 0 & \text{if } a_n > 0. \end{cases}$$

Note that $\sum_{n=1}^{\infty} a_n = \sum_{n=1}^{\infty} (u_n + v_n)$.

Let us suppose that $\sum_{n=1}^{\infty} |a_n|$ converges. Now $\sum_{n=1}^{\infty} u_n$ is a series of nonnegative terms and $u_n \leq |a_n|$, so $\sum_{n=1}^{\infty} u_n$ converges by a comparison test. Similarly, $\sum_{n=1}^{\infty} (-v_n)$ is a series of nonnegative terms and $-v_n \leq |a_n|$, so $\sum_{n=1}^{\infty} (-v_n)$ converges by a comparison test also. We then find that the series $(-1) \sum_{n=1}^{\infty} (-v_n) = \sum_{n=1}^{\infty} v_n$ converges, and therefore the series

$$\sum_{n=1}^{\infty} (u_n + v_n) = \sum_{n=1}^{\infty} a_n$$

converges. This is what we wished to prove. □

Example 3 The series

$$1 + \frac{1}{2^2} - \frac{1}{3^2} + \frac{1}{4^2} + \frac{1}{5^2} - \frac{1}{6^2} + \frac{1}{7^2} + \frac{1}{8^2} - \frac{1}{9^2} + \cdots,$$

with two positive terms followed by a negative term, does not satisfy the alternating series test. However, the series does converge, for the corresponding series of absolute values is the series $\sum_{n=1}^{\infty} (1/n^2)$, which we know converges. ‖

We emphasize again that the Absolute Convergence Test does not say anything about the behavior of $\sum_{n=1}^{\infty} a_n$ if $\sum_{n=1}^{\infty} |a_n|$ diverges. To illustrate, both the series $1 - 2 + 3 - 4 + 5 - 6 + \cdots$ and the corresponding series $1 + 2 + 3 + 4 + 5 + 6 + \cdots$ of absolute values diverge since the nth terms do not approach zero. *However, the alternating harmonic series*

$$1 - \tfrac{1}{2} + \tfrac{1}{3} - \tfrac{1}{4} + \tfrac{1}{5} - \tfrac{1}{6} + \cdots;$$

converges (by the Alternating Series Test), while the corresponding series of absolute values is the harmonic series

$$1 + \tfrac{1}{2} + \tfrac{1}{3} + \tfrac{1}{4} + \tfrac{1}{5} + \tfrac{1}{6} + \cdots,$$

which diverges. Let us describe terminology used in this connection.

Definition 10.7 A series $\sum_{n=1}^{\infty} a_n$ **converges absolutely** (or is **absolutely convergent**) if the series $\sum_{n=1}^{\infty} |a_n|$ converges. If $\sum_{n=1}^{\infty} a_n$ converges and $\sum_{n=1}^{\infty} |a_n|$ diverges, then $\sum_{n=1}^{\infty} a_n$ **converges conditionally** (or is **conditionally convergent**).

Example 4 Every convergent series of nonnegative terms is absolutely convergent, since it is identical with the corresponding series of absolute values. ‖

Example 5 The series $\sum_{n=1}^{\infty} (-1)^n (1/n!)$ is absolutely convergent since $\sum_{n=1}^{\infty} (1/n!)$ converges. The alternating harmonic series $\sum_{n=1}^{\infty} (-1)^n (1/n)$ is conditionally convergent, for it converges but the series $\sum_{n=1}^{\infty} (1/n)$ of absolute values diverges. ‖

SUMMARY

1. *An alternating series is one containing alternately positive and negative terms.*
2. ***Alternating Series Test:*** *Let $\sum_{n=1}^{\infty} a_n$ be a series such that*
   a) *the series is alternating,*
   b) $|a_{n+1}| \leq |a_n|$ *for all n, and*
   c) $\lim_{n\to\infty} a_n = 0$.

   Then the series converges.
3. ***Absolute Convergence Test:*** *Let $\sum_{n=1}^{\infty} a_n$ be any infinite series. If $\sum_{n=1}^{\infty} |a_n|$ converges, then $\sum_{n=1}^{\infty} a_n$ converges; the series is said to be absolutely convergent.*
4. *A conditionally convergent series is one that converges, but is not absolutely convergent.*

EXERCISES

1. Give an example of a divergent series that satisfies conditions (a) and (b) of the Alternating Series Test.

2. Give an example of a divergent series that satisfies conditions (a) and (c) of the Alternating Series Test.

3. Give an example of a divergent series that satisfies conditions (b) and (c) of the Alternating Series Test.

4. Show that every convergent series of nonpositive terms converges absolutely.

In Exercises 5 through 20, classify the series as either absolutely convergent, conditionally convergent, or divergent.

5. $\sum_{n=1}^{\infty} (-1)^n \frac{1}{4n}$

6. $\sum_{n=1}^{\infty} (-1)^n \frac{1}{4n^2}$

7. $\sum_{n=1}^{\infty} (-1)^{n+1} \frac{\cos^2 n}{2^n}$

8. $\sum_{n=1}^{\infty} \frac{-1}{\sqrt[3]{n}}$

9. $\sum_{n=4}^{\infty} \frac{2 - n^2}{(n-3)^2}$

10. $\sum_{n=1}^{\infty} (-1)^n \frac{1}{\sin^2 n}$

11. $\sum_{n=2}^{\infty} (-1)^n \frac{1}{\ln n}$

12. $\sum_{n=1}^{\infty} (-1)^n \frac{n^3 \cdot 2^n}{n!}$

13. $\sum_{n=1}^{\infty} (-1)^n \frac{n!}{100^n}$

14. $\sum_{n=1}^{\infty} (-1)^n \frac{\ln n}{n}$

15. $\sum_{n=1}^{\infty} \frac{\ln(1/n)}{n}$

16. $\sum_{n=1}^{\infty} \frac{(-1)^{2n+1}}{\sqrt{n}}$

17. $\sum_{n=1}^{\infty} (-1)^n \frac{\sin^2 n}{n^2}$

18. $\sum_{n=1}^{\infty} (-1)^{n+1} \frac{\sin^2 n}{n^{3/2}}$

19. $\sum_{n=1}^{\infty} (-1)^n \frac{n!}{n^n}$

20. $\sum_{n=2}^{\infty} (-1)^n \frac{\ln(n^n)}{n^2}$

21. Let $\sum_{n=1}^{\infty} a_n$ be a series and let $\sum_{n=1}^{\infty} u_n$ be the series of positive terms (and zeros) and $\sum_{n=1}^{\infty} v_n$ the series of negative terms (and zeros) defined in the proof of Theorem 10.10.

a) Show that if $\sum_{n=1}^{\infty} u_n$ and $\sum_{n=1}^{\infty} v_n$ both converge, then $\sum_{n=1}^{\infty} a_n$ converges.

b) Show that if one of $\sum_{n=1}^{\infty} u_n$ and $\sum_{n=1}^{\infty} v_n$ converges while the other diverges, then $\sum_{n=1}^{\infty} a_n$ diverges.

c) Show by an example that it is possible that $\sum_{n=1}^{\infty} a_n$ converges while both $\sum_{n=1}^{\infty} u_n$ and $\sum_{n=1}^{\infty} v_n$ diverge.

22. Mark each of the following true or false.

a) Every convergent series is absolutely convergent.

b) Every absolutely convergent series is convergent.

c) If $\sum_{n=1}^{\infty} a_n$ is conditionally convergent, then $\sum_{n=1}^{\infty} |a_n|$ diverges.

d) If $\sum_{n=1}^{\infty} |a_n|$ diverges, then $\sum_{n=1}^{\infty} a_n$ is conditionally convergent.

e) Every alternating series converges.

f) Every convergent alternating series is conditionally convergent.

calculator exercises

23. Find the sum of the series $\sum_{n=1}^{\infty} (-1)^n(1/n^3)$ with error less than 0.001.

24. Find the sum of the series $\sum_{n=1}^{\infty} (-1)^n(1/n^2)$ with error less than 0.001.

25. Find the sum of the series $\sum_{n=1}^{\infty} (1/n^2)$ with error at most 0.001. (See Exercise 43 of Section 10.4.)

exercise sets for chapter 10

review exercise set 10.1

1. Define what is meant by $\lim_{n\to\infty} a_n = c$.

2. Find the limit of the given sequence if it converges or has limit ∞ or $-\infty$.

a) $\left\{\dfrac{4n^2 - 2n}{3 - 5n^2}\right\}$

b) $\left\{\dfrac{n^3}{e^n}\right\}$

3. Find the sum of the series

$$\sum_{n=0}^{\infty} \frac{3^{n+1}}{4^{n-1}}$$

if the series converges.

4. Express the repeating decimal 4.731313131. . . as a fraction.

In Problems 5 and 6, classify the series as convergent or divergent, and give a reason for your answer.

5. a) $\displaystyle\sum_{n=1}^{\infty} \frac{n^2 + 3n}{400n + n^2}$ b) $\displaystyle\sum_{n=2}^{\infty} \frac{n^2 - 1}{1 + n^3}$

6. a) $\displaystyle\sum_{n=1}^{\infty} \frac{1}{\sqrt{n + 17}}$ b) $\displaystyle\sum_{n=1}^{\infty} \frac{\sqrt{n + 3}}{n^2 + 4n}$

7. Use the Integral Test to establish the convergence or divergence of

$$\sum_{n=2}^{\infty} \frac{1}{n^2 - n}.$$

8. Write out the Ratio Test to establish the convergence or divergence of

$$\sum_{n=1}^{\infty} \frac{n^2 \cdot 2^n}{e^{n+1}}.$$

In Problems 9 and 10, classify the series as absolutely convergent, conditionally convergent, or divergent, and indicate the reason for your answer.

9. a) $\displaystyle\sum_{n=1}^{\infty} \frac{(-1)^n}{\sqrt{n + 3}}$ b) $\displaystyle\sum_{n=2}^{\infty} \frac{(-1)^n}{n(\ln n)^2}$

10. a) $\displaystyle\sum_{n=1}^{\infty} (-1)^n \frac{\cos(1/n)}{n^{3/2}}$ b) $\displaystyle\sum_{n=1}^{\infty} (-1)^n \frac{3^n}{n^{100}}$

review exercise set 10.2

1. Give an ϵ,N-proof that the sequence $\{(n-1)/n\}$ converges.
2. Find the limit of the given sequence if it converges or has limit ∞ or $-\infty$.

   a) $\left\{\dfrac{\sqrt{n}-n^2}{3n-7}\right\}$

   b) $\left\{\dfrac{1}{n}\cdot\sin\left(\dfrac{1}{n}\right)\right\}$

3. A ball has the property that when it is dropped, it rebounds to $\frac{1}{3}$ of its height on the previous bounce. Find the height from which the ball must be dropped if the total distance it is to travel is 60 ft.
4. If possible, find r such that the sum of the series

$$3-3r+3r^2-3r^3+\cdots+(-1)^n3r^n+\cdots$$

is 7.

In Problems 5 and 6, classify the series as convergent or divergent, and give a reason for your answer.

5. a) $\displaystyle\sum_{n=1}^{\infty}\frac{n^2+3}{n^2\cdot 3^{n-1}}$ b) $\displaystyle\sum_{n=1}^{\infty}\frac{3^n+4^n}{5^n}$

6. a) $\displaystyle\sum_{n=1}^{\infty}\frac{n^2+\sin n}{n^4+3n}$ b) $\displaystyle\sum_{n=1}^{\infty}\frac{1}{\cos^2 n}$

7. Use the Integral Test to establish the convergence or divergence of

$$\sum_{n=2}^{\infty}\frac{n+1}{n^2+1}.$$

8. Write out the Ratio Test to establish the convergence or divergence of

$$\sum_{n=1}^{\infty}\frac{n!}{100^n}.$$

In Problems 9 and 10, classify the series as absolutely convergent, conditionally convergent, or divergent and indicate the reason for your answer.

9. $\displaystyle\sum_{n=2}^{\infty}\frac{(-1)^n}{n(\ln n)}$ b) $\displaystyle\sum_{n=1}^{\infty}(-1)^n\frac{\cos^2 n}{n^{3/2}}$

10. a) $\displaystyle\sum_{n=1}^{\infty}(-1)^n\frac{n^n}{n!}$ b) $\displaystyle\sum_{n=1}^{\infty}(-1)^n\frac{\ln n}{n}$

more challenging exercises 10

1. A sequence $\{t_n\}$ is monotone decreasing if $t_{n+1}\le t_n$ for $n=1,2,3,\ldots$ Assuming the fundamental property of the real numbers stated in the text, show that a monotone decreasing sequence $\{t_n\}$ either converges to a number d or $\lim_{n\to\infty}t_n=-\infty$.
2. Prove Theorem 10.3 [*Hint*. Let s_n be the nth partial sum of $\sum_{n=1}^{\infty}a_n$, and t_n be the nth partial sum of $\sum_{n=1}^{\infty}b_n$. Deduce that there exist an integer N and a number c such that $t_n=s_{n+k}+c$ for $n>N$, and show that if $\{s_n\}$ converges to a, then $\{t_n\}$ converges to $a+c$.]
3. Generalize Theorem 10.6 as follows: Let $\sum_{n=1}^{\infty}a_n$ and $\sum_{n=1}^{\infty}b_n$ be series of nonnegative terms with $a_n\ne 0$ for sufficiently large n. Show that if there exist constants $m>0$ and $M>0$ such that $m<b_n/a_n<M$ for all sufficiently large n, then the two series either both converge or both diverge.
4. A sequence $\{s_n\}$ is a *Cauchy sequence* if for each $\epsilon>0$, there exists an integer N such that

$$|s_n-s_m|<\epsilon$$

provided that both $n>N$ and $m>N$. Prove that every convergent sequence $\{s_n\}$ is a Cauchy sequence. (The converse is also true, but is harder to prove.)

Each of two series is said to be a ***rearrangement*** *of the other if the two series contain exactly the same terms, but the terms do not necessarily appear in the same order. The remaining exercises deal with this concept. It can be shown that rearranging an absolutely convergent series does not change its convergence or divergence, or its sum if it converges. However, let $\sum_{n=1}^{\infty} a_n$ be a series such that $\lim_{n\to\infty} a_n = 0$ and such that the series $\sum_{n=1}^{\infty} u_n$ of nonnegative terms (and zeros) and the series $\sum_{n=1}^{\infty} v_n$ of nonpositive terms (and zeros), defined in the proof of Theorem 10.10, both diverge. Describe how to construct a rearrangement of $\sum_{n=1}^{\infty} a_n$ that has the given behavior.*

5. Converges to 17 **6.** Converges to -50 **7.** Diverges to ∞ **8.** Diverges to $-\infty$

9. Diverges and has partial sums alternately increasing to ≥ 15 and decreasing to ≤ -6.

power series

11

11.1 POWER SERIES

11.1.1 The function represented by a power series

Among the most important series are *power series*; most of our work will be with these series. Power series are precisely the "infinite polynomials" that we mentioned at the start of the last chapter. For example, we shall see that

$$\sin x = x - \frac{x^3}{3!} + \frac{x^5}{5!} - \frac{x^7}{7!} + \frac{x^9}{9!} - \frac{x^{11}}{11!} + \cdots$$

for any value of x. We shall also see that

$$\frac{1}{x} = 1 - (x-1) + (x-1)^2 - (x-1)^3 + (x-1)^4 - (x-1)^5 + \cdots$$

for any x such that $0 < x < 2$. The series for $\sin x$ is in powers of $x = (x - 0)$, and the series for $1/x$ is in powers of $(x - 1)$.

Definition 11.1 **A power series at** x_0 is a series of the form

$$\sum_{n=0}^{\infty} a_n(x - x_0)^n = a_0 + a_1(x - x_0) + \cdots + a_n(x - x_0)^n + \cdots \tag{1}$$

The constants a_i are the **coefficients of the series**; in particular, a_0 is the **constant term of the series**.

It will be convenient to change terminology slightly and to consider the constant term a_0 of (1) to be the *0th term of the series*, and the term $a_n(x - x_0)^n$ to be the *nth term*. With this convention, the nth term of a power series becomes the term with exponent n.

At each value of x, the power series (1) becomes a series of constants that may or may not converge. The sum of such a convergent series of constants generally varies with the value of x, and is a function of x. This function is the *sum function* of the series. The set of all values of x for which the power series converges is the *domain of the sum function defined by the series;* the value of the function at each such point is the sum of the series for that value of x.

The power series (1) should be regarded as an attempt to describe a function *locally*, near x_0; in translated coordinates with $\Delta x = x - x_0$, the series become $\sum_{n=0}^{\infty} a_n (\Delta x)^n$. The series (1) converges for $x = x_0$, for at x_0, the series becomes

$$a_0 + a_1 \cdot 0 + a_2 \cdot 0 + \cdots + a_n \cdot 0 + \cdots,$$

which converges to a_0. It is possible that this is the only point at which the series converges, as the next example shows, but such series are of little importance for us.

Example 1 The series $\sum_{n=0}^{\infty} n!x^n$ converges only for $x = 0$, for the Ratio Test shows that for $a \neq 0$, the series $\sum_{n=0}^{\infty} n!a^n$ diverges. ‖

11.1.2 The radius of convergence of a power series

The next theorem shows that the region of convergence of a power series at x_0 has x_0 as center point. The theorem is stated and proved in the case where $x_0 = 0$, that is, for a series $\sum_{n=0}^{\infty} a_n x^n$. See the remark after the proof of the theorem.

Theorem 11.1 *If a power series $\sum_{n=0}^{\infty} a_n x^n$ converges for $x = c \neq 0$, then the series converges absolutely for all x such that $|x| < |c|$. If the series diverges at $x = d$, then the series diverges for all x such that $|x| > |d|$.*

Proof. Suppose that $\sum_{n=0}^{\infty} a_n c^n$ converges, and let $|b| < |c|$. Since $\sum_{n=0}^{\infty} a_n c^n$ converges, we must have $\lim_{n\to\infty} a_n c^n = 0$; in particular, $|a_n c^n| < 1$ or

$$|a_n| < \frac{1}{|c^n|}$$

for n sufficiently large. We then obtain

$$|a_n b^n| = |a_n| \cdot |b^n| < \left|\frac{b^n}{c^n}\right| = \left|\frac{b}{c}\right|^n$$

for n sufficiently large. Recall that we are assuming $|b/c| < 1$. Thus the series $\sum_{n=0}^{\infty} |a_n b^n|$ is, term for term, less than the convergent geometric series $\sum_{n=0}^{\infty} |b/c|^n$ for n sufficiently large, so $\sum_{n=0}^{\infty} |a_n b^n|$ converges by the comparison test. This shows that $\sum_{n=0}^{\infty} a_n x^n$ converges absolutely for all x such that $|x| < |c|$.

The assertion regarding divergence is really the contrapositive of the assertion we just proved for convergence. Suppose $\sum_{n=0}^{\infty} a_n d^n$ diverges, and let $|h| > |d|$. Convergence of $\sum_{n=0}^{\infty} a_n h^n$ would imply convergence of $\sum_{n=0}^{\infty} a_n d^n$ by the first part of our proof. But this would contradict our hypothesis, so $\sum_{n=0}^{\infty} a_n h^n$ diverges also. □

We stated and proved the theorem for a power series at $x_0 = 0$. It is immediate from the theorem that if $\sum_{n=0}^{\infty} a_n (\Delta x)^n$ converges for $\Delta x = c$, then the series converges for $|\Delta x| < |c|$. Putting $\Delta x = x - x_0$ as usual, you see that if a series $\sum_{n=0}^{\infty} a_n (x - x_0)^n$ converges at $x = x_0 + c$, i.e., for $\Delta x = c$, then it converges for all x such that $|x - x_0| < |c|$.

The important corollary that follows seems very plausible from our theorem. A careful proof depends upon a basic property of the real numbers, and is not given here.

Corollary *For a power series $\sum_{n=0}^{\infty} a_n (x - x_0)^n$, exactly one of the three following alternatives holds.*

a) *The series converges at x_0 only.*

b) *The series converges for all x.*

c) *There exists r such that the series converges for $|x - x_0| < r$, that is, for $x_0 - r < x < x_0 + r$, and diverges for $|x - x_0| > r$.*

The number r that appears in case (c) of the corollary is the **radius of convergence of the series**. In case (a), it is natural to say that the radius of convergence is 0, and in case (b), we say that the radius of convergence is ∞.

In case (c), the behavior of the series at the endpoints of the interval $x_0 - r < x < x_0 + r$ depends upon the individual series; certain series converge at both endpoints, others diverge at both endpoints, and some converge at one endpoint and diverge at the other.

Example 2 The series $\sum_{n=0}^{\infty} x^n$ is the *geometric series*, and our work in Chapter 10 shows that it converges to $1/(1 - x)$ for $|x| < 1$. The radius of convergence of the series is 1. The series diverges for $x = 1$ and $x = -1$, since the nth term does not approach 0 as $n \to \infty$ at these points. ‖

For many power series, the radius of convergence can be computed by using the ratio test. The technique is best illustrated by examples. By Theorem 11.1, if a power series converges at $x - x_0 = c$, it converges absolutely for $|x - x_0| < |c|$, so we try to compute the limit of the absolute value of the ratio.

Example 3 We find the radius of convergence of the series

$$\sum_{n=1}^{\infty} \frac{x^n}{n \cdot 2^n}.$$

SOLUTION The absolute value of the ratio of the $(n + 1)$st term to the nth is

$$\left|\frac{x^{n+1}/((n+1)2^{n+1})}{x^n/(n \cdot 2^n)}\right| = \left|\frac{x^{n+1}}{(n+1)2^{n+1}} \cdot \frac{n \cdot 2^n}{x^n}\right|$$

$$= \left|\frac{nx}{(n+1)2}\right|.$$

The limit as n approaches ∞ is

$$\lim_{n\to\infty} \left|\frac{nx}{(n+1)2}\right| = \left|\frac{x}{2}\right|.$$

Thus the series converges for $|x/2| < 1$ or for $|x| < 2$. The radius of convergence is therefore 2, and the series converges at least for $-2 < x < 2$.

To see what happens at the endpoint 2 of the interval $-2 < x < 2$, examine the series

$$\sum_{n=1}^{\infty} \frac{2^n}{n \cdot 2^n} = \sum_{n=1}^{\infty} \frac{1}{n}.$$

This is the harmonic series, and diverges. At the endpoint -2, you obtain as series

$$\sum_{n=1}^{\infty} \frac{(-2)^n}{n \cdot 2^n} = \sum_{n=1}^{\infty} (-1)^n \cdot \frac{1}{n},$$

which is the convergent alternating harmonic series. Thus the series converges for $-2 \leq x < 2$. This interval is the **interval of convergence of the series**. ||

Example 3 illustrates the usual procedure for finding the radius and interval of convergence for a power series where the limit of the ratio exists. The two series of constants corresponding to the endpoints of the interval must be examined separately. We emphasize that *the Ratio Test never determines the behavior at the endpoints of the interval of convergence, for the limit of the ratio at these endpoints will be* 1.

We illustrate with an example for a power series at a point $x_0 \neq 0$.

Example 4 Let's determine the interval of convergence of the series

$$\sum_{n=1}^{\infty} \frac{(x-3)^{2n}}{n^2 \cdot 5^n},$$

which is a power series at $x_0 = 3$.

SOLUTION For the ratio, we obtain

$$\left| \frac{(x-3)^{2(n+1)}}{(n+1)^2 \cdot 5^{n+1}} \cdot \frac{n^2 \cdot 5^n}{(x-3)^{2n}} \right| = \left| \frac{n^2(x-3)^2}{(n+1)^2 \cdot 5} \right|.$$

We have

$$\lim_{n\to\infty} \left| \frac{n^2(x-3)^2}{(n+1)^2 \cdot 5} \right| = \left| \frac{(x-3)^2}{5} \right|.$$

Thus the series converges if $|(x-3)^2/5| < 1$, or if $|x-3|^2 < 5$. This is equivalent to $|x-3| < \sqrt{5}$. The radius of convergence at $x_0 = 3$ is thus $\sqrt{5}$, and the series converges at least for $3 - \sqrt{5} < x < 3 + \sqrt{5}$.

Turning to the endpoints, at $3 - \sqrt{5}$ our series becomes

$$\sum_{n=1}^{\infty} \frac{(-\sqrt{5})^{2n}}{5^n \cdot n^2} = \sum_{n=1}^{\infty} \frac{5^n}{5^n \cdot n^2}$$

$$= \sum_{n=1}^{\infty} \frac{1}{n^2}.$$

This series converges. The same series is obtained at $3 + \sqrt{5}$, so the series converges at both endpoints, and the interval of convergence is

$$[3 - \sqrt{5}, 3 + \sqrt{5}]. \quad \|$$

With a little practice, you should be able to give the interval of convergence of the series in Example 4 without actually computing the ratio. The nth term is

$$\frac{((x-3)^2)^n}{n^2 \cdot 5^n},$$

and the n^2 is insignificant compared with the nth powers; a polynomial in n always just contributes a factor of 1 in the limit of a ratio. Thus the series will converge for $|x - 3|^2 < 5$, or for $|x - 3| < \sqrt{5}$. The n^2 in the denominator will make the series converge at the endpoints. The exercises give you some opportunity to practice such arguments (see Exercises 9 through 17).

SUMMARY

1. *A power series at x_0 is a series of the form*

$$a_0 + a_1(x - x_0) + a_2(x - x_0)^2 + \cdots + a_n(x - x_0)^n + \cdots$$

The function having as domain all values of x for which the series converges, and having as value at each such x the sum of the series, is called the sum function defined by the series.

2. *The series in* (1) *may converge for $x = x_0$ only, or it may converge for all x, or it may converge if $|x - x_0| < r$ and diverge if $|x - x_0| > r$ for some $r > 0$. Such a number r is the radius of convergence of the series.*
3. *The radius r of convergence of a power series can often be found by:*

   i) *forming the absolute value of the ratio of the $(n + 1)$st term divided by the nth term,*

   ii) *computing the limit of this ratio as $n \to \infty$,*

   iii) *setting the resulting limit <1,*

   iv) *solving the resulting inequality for $|x - x_0|$, obtaining an expression of the form $|x - x_0| < r$. The radius of convergence is then r.*

4. *To determine the interval of convergence of a power series at x_0 after the radius r of convergence has been found, substitute $x = x_0 - r$ and $x = x_0 + r$ to obtain two series of constants. Test these series for convergence to determine which of the endpoints, $x_0 - r$, $x_0 + r$, should be included with $|x - x_0| < r$ to obtain the interval of convergence. The Ratio Test should never be tried at these endpoint series, for the limiting ratio will always be* 1.

EXERCISES

In Exercises 1 through 8, use the Ratio Test to find the radius of convergence of the series. Then find the interval of convergence (including endpoints.)

1. $\sum_{n=1}^{\infty} \frac{x^n}{n}$ **2.** $\sum_{n=1}^{\infty} \frac{x^{2n+1}}{n}$ **3.** $\sum_{n=1}^{\infty} (-1)^n \frac{x^n}{n}$ **4.** $\sum_{n=0}^{\infty} (-1)^n \frac{x^{2n}}{n+3}$

5. $\sum_{n=0}^{\infty} \frac{x^n}{n^2+1}$ **6.** $\sum_{n=0}^{\infty} \frac{x^n}{3^n}$ **7.** $\sum_{n=1}^{\infty} \frac{x^n}{n \cdot 5^{n+2}}$ **8.** $\sum_{n=2}^{\infty} \frac{(-1)^n x^{n+1}}{(\ln n)2^{n-1}}$

In Exercises 9 through 17, proceed as above, but this time try to find the radius of convergence without explicitly writing out the ratio, as illustrated after Example 4 in the text; write out the Ratio Test only if you have to.

9. $\sum_{n=1}^{\infty} \frac{(x-2)^{2n}}{n \cdot 9^n}$ **10.** $\sum_{n=0}^{\infty} \frac{n^2(x+4)^n}{n!}$ **11.** $\sum_{n=0}^{\infty} n^2(x-3)^n$

12. $\sum_{n=0}^{\infty} n!(x+5)^{2n+1}$ **13.** $\sum_{n=1}^{\infty} \frac{(x+4)^{n+1}}{n^2 \cdot 3^n}$ **14.** $\sum_{n=1}^{\infty} \frac{(2x-4)^n}{n^{3/2} \cdot 3^n}$

15. $\sum_{n=1}^{\infty} \frac{(-2)^n(x+3)^{n+1}}{\sqrt{n}}$ **16.** $\sum_{n=0}^{\infty} \frac{3^n(x-2)^{2n+1}}{n!}$ **17.** $\sum_{n=1}^{\infty} \frac{(3x-12)^{2n}}{\sqrt{n}}$

In Exercises 18 through 21, give a power series that has the given interval as interval of convergence, including or excluding endpoints as indicated. (Many answers are possible.)

18. $[0, 6]$ **19.** $1 < x < 4$ **20.** $-2 < x \leq 4$ **21.** $-5 \leq x < -1$

11.2 TAYLOR'S FORMULA

11.2.1 The Taylor polynomial of degree *n*

Let $f(x)$ be a function defined in a neighborhood of x_0, that is, for $x_0 - h < x < x_0 + h$ for some $h > 0$, and let $f^{(n)}(x_0)$ exist. We would like to find the polynomial

$$g(x) = a_0 + a_1(x - x_0) + a_2(x - x_0)^2 + \cdots + a_n(x - x_0)^n$$

of degree n at x_0, which is the best approximation to $f(x)$ near x_0. Surely we want to require that $g(x_0) = f(x_0)$, and that $g'(x_0) = f'(x_0)$, so that the functions have the same slope at x_0. If we also require that $g''(x_0) = f''(x_0)$, then the rates of change of slope will be the same; the graphs will bend or curve at the same rate at x_0. It seems reasonable to require that $g(x)$ have as many as possible of the same derivative values at x_0 as $f(x)$ has. Let's determine the coefficients $a_0, a_1, a_2, \ldots, a_n$ to make this true. Computing, we easily find that

$$\begin{aligned}
g(x) &= a_0 + a_1(x - x_0) + a_2(x - x_0)^2 + \cdots + a_n(x - x_0)^n, \\
g'(x) &= a_1 + 2a_2(x - x_0) + 3a_3(x - x_0)^2 + \cdots + na_n(x - x_0)^{n-1}, \\
g''(x) &= 2a_2 + 3 \cdot 2a_3(x - x_0) + \cdots + n(n-1)a_n(x - x_0)^{n-2}, \\
&\vdots \\
g^{(n)}(x) &= n(n-1)(n-2) \cdots 3 \cdot 2 \cdot 1 \cdot a_n = n!a_n.
\end{aligned}$$

Then

$$\begin{aligned}
g(x_0) &= a_0, \\
g'(x_0) &= a_1, \\
g''(x_0) &= 2a_2, \\
g'''(x_0) &= 3 \cdot 2a_3 = 3!a_3, \\
&\vdots \\
g^{(n)}(x_0) &= n!a_n.
\end{aligned}$$

Setting these equal to the corresponding derivatives of $f(x)$ at x_0, we have

$$f(x_0) = a_0, \quad f'(x_0) = a_1, \quad f''(x_0) = 2a_2, \quad f'''(x_0) = 3!a_3, \quad \ldots, \quad f^{(n)}(x_0) = n!a_n,$$

which leads to

$$\begin{aligned}
a_0 &= f(x_0), \\
a_1 &= f'(x_0), \\
a_2 &= \frac{1}{2!} f''(x_0), \\
a_3 &= \frac{1}{3!} f'''(x_0), \\
&\vdots \\
a_n &= \frac{1}{n!} f^{(n)}(x_0).
\end{aligned}$$

Definition 11.2 The polynomial function

$$\begin{aligned}
g(x) = T_n(x) &= f(x_0) + f'(x_0)(x - x_0) + \frac{f''(x_0)}{2!}(x - x_0)^2 \\
&\quad + \frac{f'''(x_0)}{3!}(x - x_0)^3 + \cdots + \frac{f^{(n)}(x_0)}{n!}(x - x_0)^n \\
&= \sum_{i=0}^{n} \frac{f^{(i)}(x_0)}{i!}(x - x_0)^i
\end{aligned}$$

is the nth **Taylor polynomial for** $f(x)$ **at** x_0. [We define $f^{(0)}(x) = f(x)$ and $0! = 1$, so that we may include the constant term $f(x_0)$ in our formal sum.]

In the preceding definition, we said the "nth Taylor polynomial" rather than the "Taylor polynomial of degree n" since it is possible that $f^{(n)}(x_0) = 0$ so that the degree of $T_n(x)$ is $<n$. These polynomials are named in honor of the English mathematician, Brook Taylor (1685–1731).

Of course, the polynomial $T_n(x)$ depends on f and x_0 as well as n, but a more accurate notation such as "${}^{x_0}T_n^{\,f}(x)$" is simply too cumbersome!

Example 1 Let's find the Taylor polynomial $T_7(x)$ for the function $\sin x$ at $x_0 = 0$.

SOLUTION We have to compute $\sin(0)$ and the derivatives of orders ≤ 7 of $\sin x$ at $x_0 = 0$. For $f(x) = \sin x$, you obtain $f(0) = \sin 0 = 0$, while

$$f'(0) = \cos 0 = 1, \qquad f''(0) = -\sin 0 = 0,$$
$$f'''(0) = -\cos 0 = -1, \qquad f^{\text{iv}}(0) = \sin 0 = 0.$$

At this point, we note that since $f^{\text{iv}}(x) = f(x) = \sin x$, our derivatives will start repeating. Thus the derivatives of $\sin x$ at 0, starting with the first derivative, are

$$1, \quad 0, \quad -1, \quad 0, \quad 1, \quad 0, \quad -1, \quad 0, \quad 1, \quad 0, \quad -1, \quad 0, \ldots$$

for as far as you wish to take them. Since $x_0 = 0$, you have $x - x_0 = x$, and the 7th Taylor polynomial is

$$\begin{aligned} T_7(x) &= 0 + 1 \cdot x + \frac{0}{2!}x^2 + \frac{-1}{3!}x^3 + \frac{0}{4!}x^4 + \frac{1}{5!}x^5 + \frac{0}{6!}x^6 + \frac{-1}{7!}x^7 \\ &= x - \frac{x^3}{3!} + \frac{x^5}{5!} - \frac{x^7}{7!}. \quad \| \end{aligned}$$

Example 2 Let's find the Taylor polynomial $T_8(x)$ for the function $\cos x$ at $x_0 = \pi$.

SOLUTION As in Example 1, you easily find that $\cos \pi = -1$ and that the derivatives of $\cos x$ at π, starting with the first derivative, are

$$0, \quad 1, \quad 0, \quad -1, \quad 0, \quad 1, \quad 0, \quad -1, \quad 0, \quad 1, \quad 0, \quad -1, \ldots$$

for as far as you wish to take them. Dropping the terms with coefficients 0, you obtain the polynomial

$$T_8(x) = -1 + \frac{(x-\pi)^2}{2!} - \frac{(x-\pi)^4}{4!} + \frac{(x-\pi)^6}{6!} - \frac{(x-\pi)^8}{8!}. \quad \|$$

Example 3 Let's find the fifth Taylor polynomial $T_5(x)$ at $x_0 = 0$ for f where $f(x) = 1/(1-x) = (1-x)^{-1}$.

SOLUTION Computing derivatives,

$$\begin{aligned} f'(x) &= (1 - x)^{-2}, \\ f''(x) &= 2(1 - x)^{-3}, \\ f'''(x) &= 3 \cdot 2(1 - x)^{-4} = 3!(1 - x)^{-4}, \\ f^{\text{iv}}(x) &= 4!(1 - x)^{-5}, \\ f^{\text{v}}(x) &= 5!(1 - x)^{-6}, \qquad \text{etc.} \end{aligned}$$

Thus the derivatives of f at 0, starting with the first derivative, are

$$1!, \quad 2!, \quad 3!, \quad 4!, \quad 5!, \quad 6!, \quad 7!, \quad 8!, \ldots$$

for as far as you wish to go. Since $f(0) = 1$, you obtain

$$\begin{aligned} T_5(x) &= 1 + x + \frac{2!}{2!}x^2 + \frac{3!}{3!}x^3 + \frac{4!}{4!}x^4 + \frac{5!}{5!}x^5 \\ &= 1 + x + x^2 + x^3 + x^4 + x^5 \end{aligned}$$

as the fifth Taylor polynomial. ||

11.2.2 Taylor's theorem Let $f(x)$ and x_0 be as described in the last article, so that you can consider the Taylor polynomial $T_n(x)$. You expect to have

$$f(x) \approx T_n(x)$$

for x close to x_0. Naturally, we are interested in the accuracy of this approximation. We would hope that the *error* $E_n(x)$ given by

$$E_n(x) = f(x) - T_n(x) \tag{1}$$

is small if x is close to x_0. The following theorem gives some information on the size of $E_n(x)$, and for this reason, the theorem is extremely important.

Theorem 11.2 *(Taylor's Theorem) Let $f(x)$ be defined for $x_0 - h < x < x_0 + h$ and let the derivatives of orders $\leq n + 1$ exist throughout that interval, and let $E_n(x) = f(x) - T_n(x)$. Then for each x in that interval, there exists a number c depending on x and strictly between x and x_0 (for $x \neq x_0$) such that*

$$E_n(x) = \frac{f^{(n+1)}(c)}{(n + 1)!}(x - x_0)^{n+1}. \tag{2}$$

We defer the proof of Taylor's Theorem to the end of this section. The theorem is really very easy to remember. For f, x, and x_0 as described in the theorem, you have

$$f(x) = T_n(x) + E_n(x) = T_n(x) + \frac{f^{(n+1)}(c)}{(n + 1)!}(x - x_0)^{n+1}.$$

Note that $E_n(x)$ is precisely what you would have for the "next" term of degree $n + 1$ in the Taylor polynomial $T_{n+1}(x)$, *except that the derivative* $f^{(n+1)}$ *must be computed at some* c *between* x_0 *and* x *instead of at* x_0. Note also that in translated coordinates, $E_n(x)$ becomes

$$E_n(x_0 + \Delta x) = \frac{f^{(n+1)}(x_0 + h)}{(n + 1)!}(\Delta x)^{n+1}, \tag{3}$$

where $0 < |h| < |\Delta x|$ if $\Delta x \neq 0$.

The size of the error in the approximation of $f(x)$ by $T_n(x)$ is measured by $|E_n(x)|$. There are two reasons why you would expect $|E_n(x)|$ to be small for large n and for x close to x_0. First of all, if n is large, then the number $(n + 1)!$ in the denominator of $E_n(x)$ is large, and this tends to make $|E_n(x)|$ small. Secondly, if the distance from x to x_0 is less than 1, that is, if $|x - x_0| < 1$, then $|x - x_0|^{n+1}$ also becomes small as n becomes large. The only catch is that $|f^{(n+1)}(c)|$ might become large enough to offset the small size of $|x - x_0|^{n+1}/(n + 1)!$ for large n. However, for many functions, this calamity does not occur. For example, if $f(x) = \sin x$, then $f^{(n+1)}$ is always a sine or cosine function, so $|f^{(n+1)}(c)| \leq 1$ for all n and all numbers c.

We illustrate a few types of applications of Taylor's theorem. We are usually concerned with obtaining a *bound* B on the size of $E_n(x)$, that is, with finding a number $B > 0$ such that

$$|E_n(x)| \leq B.$$

Example 4 Let's use a differential to estimate $\sqrt{101}$, and let's then find a bound for our error.

SOLUTION We let $f(x) = x^{1/2}$, $x_0 = 100$, and $dx = 1$. Our estimate using a differential is simply

$$T_1(x_0 + dx) = f(x_0) + f'(x_0)\,dx.$$

Since $f'(x) = (1/2)x^{-1/2}$, we obtain

$$T_1(101) = \sqrt{100} + \frac{1}{2} \cdot \frac{1}{\sqrt{100}} \cdot 1 = 10 + \frac{1}{20} = 10.05$$

as our estimate for $\sqrt{101}$.

Now $f''(x) = (-1/4)x^{-3/2}$, so by (2) with $n = 1$,

$$|E_1(101)| = \frac{1}{2!} \cdot \left| -\frac{1}{4}c^{-3/2} \right| \cdot 1 = \frac{1}{8} \cdot \frac{1}{c^{3/2}}$$

for some c where $100 < c < 101$. But for such c, the largest values of $1/c^{3/2}$

must occur where $c^{3/2}$ is smallest, that is, for c close to 100, so

$$\frac{1}{c^{3/2}} < \frac{1}{100^{3/2}} = \frac{1}{1000}.$$

Thus

$$|E_1(101)| < \tfrac{1}{8} \cdot \tfrac{1}{1000} = \tfrac{1}{8000} = 0.000125.$$

Therefore our estimate 10.05 for $\sqrt{101}$ is accurate to at least three decimal places. ||

The preceding example illustrates the *typical procedure* in finding a bound on $E_n(x)$. You don't know the *precise value of* c, so you attempt to find the *maximum possible value* that $|f^{(n+1)}(c)|$ can have for any c *over the entire interval* from x_0 to x. If the exact maximum of $|f^{(n+1)}(c)|$ is hard to find, find the best easily computed bound you can for $|f^{(n+1)}(c)|$. The following example illustrates the technique again.

Example 5 Suppose we used the Taylor polynomial $T_7(x)$ for $f(x) = \sin x$ at $x_0 = 0$ which we found in Example 1 to estimate $\sin 2° = \sin(\pi/90)$. Without bothering to compute the actual approximation of sin 2°, let's find a bound for the error.

SOLUTION The error is $E_7(\pi/90)$. Since $f^{(8)}(x) = \sin x$, we need to estimate $|\sin c|$ for $0 < c < \pi/90$. For such c, we have

$$\sin c < \sin\frac{\pi}{90},$$

but of course we don't know $\sin(\pi/90)$; that is what we are trying to estimate. But surely $\sin(\pi/90) < \sin(\pi/6) = \frac{1}{2}$, so $|\sin c| < \frac{1}{2}$. Using also the estimate $\pi/90 < \frac{1}{20}$, we have from (2), with $n = 7$,

$$\left|E_7\left(\frac{\pi}{90}\right)\right| = \left|\frac{\sin c}{8!}\left(\frac{\pi}{90}\right)^8\right| < \frac{1/2}{8!}\left(\frac{1}{20}\right)^8 = \frac{1}{(2^9)(8!)(10^8)}.$$

Thus the error is very small.

We can improve our bound if we note that $T_7(x) = T_8(x)$ for $\sin x$ at 0. Thus $E_7(x) = E_8(x)$, but formula (2) for $E_8(x)$ is different, and permits a better bound. The 9th derivative of $\sin x$ is $\cos x$, and $|\cos c| < 1$, so we have

$$|E_7(x)| = |E_8(x)| < \frac{1}{9!}\left(\frac{\pi}{90}\right)^9 < \frac{1}{9!}\left(\frac{1}{20}\right)^9 = \frac{1}{(9!)(2^9)(10^9)}. \quad \|$$

Example 6 Suppose we wish to approximate sin 46° by using a Taylor polynomial $T_n(x)$ for $f(x) = \sin x$ at $x_0 = \pi/4$. Let's find a value of n such that the error in our approximation will be less than 0.00001.

SOLUTION We need to have $|E_n(46\pi/180)| < 0.00001$. Since the derivatives of $\sin x$ are all either $\pm\sin x$ or $\pm\cos x$, we have $|f^{(n+1)}(c)| < 1$ for any n and c. Since $x - x_0 = \pi/180$, we have from (2),

$$\left|E_n\left(\frac{46\pi}{180}\right)\right| < \frac{1}{(n+1)!}\left(\frac{\pi}{180}\right)^{n+1}.$$

Since $(\pi/180) < (1/50)$, we surely have

$$\left|E_n\left(\frac{46\pi}{180}\right)\right| < \frac{1}{(n+1)!}\left(\frac{1}{50}\right)^{n+1} = \frac{1}{(n+1)!(5)^{n+1}(10)^{n+1}}.$$

We try a few values of n, and discover that for $n = 2$ we obtain

$$\left|E_2\left(\frac{46\pi}{180}\right)\right| < \frac{1}{(3!)(5^3)(10^3)} = \frac{1}{(6)(125)(1000)} < \frac{1}{100000}$$
$$= 0.00001.$$

Thus we may use $n = 2$ with safety to achieve the desired accuracy. ||

The preceding example gives some indication how tables for the trigonometric functions were constructed. Also, rather than use space in the memory of an electronic computer to store tables of trigonometric functions, the manufacturer builds into the computer a program that the machine uses to estimate values of these functions as it needs them, as we estimated values in Examples 5 and 6.

11.2.3 Proof of Taylor's theorem

Proof. For $x = x_0$, the theorem is obvious. Choose $x \neq x_0$ where $x_0 - h < x < x_0 + h$; we think of x as remaining fixed throughout this proof; in particular, we consider x to be a constant in any differentiation. Let

$$A = \left[f(x) - f(x_0) - f'(x_0)(x - x_0) - \frac{f''(x_0)}{2!}(x - x_0)^2 - \cdots - \frac{f^{(n)}(x_0)}{n!}(x - x_0)^n\right]\frac{(n+1)!}{(x - x_0)^{n+1}}.$$

If we let

$$F(t) = f(x) - f(t) - f'(t)(x - t) - \frac{f''(t)}{2!}(x - t)^2 - \cdots - \frac{f^{(n)}(t)}{n!}(x - t)^n - \frac{(x - t)^{n+1}}{(n+1)!}\cdot A,$$

then we see that A was chosen in such a way that $F(x_0) = 0$. Note that $F(x) = 0$ also. We now apply Rolle's Theorem to $F(t)$, and deduce that

$F'(c) = 0$ for some c between x_0 and x. Now

$$\begin{aligned} F'(t) = &- f'(t) - [f'(t)(-1) + (x-t)f''(t)] \\ &- [f''(t)(x-t)(-1) + \frac{f'''(t)}{2!}(x-t)^2] - \cdots \\ &- \left[\frac{f^{(n)}(t)}{(n-1)!}(x-t)^{n-1}(-1) + \frac{f^{(n+1)}(t)}{n!}(x-t)^n\right] \\ &- \frac{(x-t)^n}{n!}(-1)\cdot A. \end{aligned}$$

Many terms cancel in this expression for $F'(t)$; in fact, the first term in each square bracket cancels with the last term of the preceding square bracket. After this cancellation, we are left with only

$$F'(t) = -\frac{f^{(n+1)}(t)}{n!}(x-t)^n + \frac{(x-t)^n}{n!}A;$$

thus

$$F'(c) = -\frac{f^{(n+1)}(c)}{n!}(x-c)^n + \frac{(x-c)^n}{n!}A = 0,$$

and, dividing through by the common factor $(x-c)^n/n!$, we obtain

$$-f^{(n+1)}(c) + A = 0,$$

so

$$A = f^{(n+1)}(c).$$

Equating this expression for A with our original definition of A, we have

$$\begin{aligned} f^{(n+1)}(c) = &\Big[f(x) - f(x_0) - f'(x_0)(x-x_0) - \frac{f''(x_0)}{2!}(x-x_0)^2 - \\ &\cdots - \frac{f^{(n)}(x_0)}{n!}(x-x_0)^n\Big]\frac{(n+1)!}{(x-x_0)^{n+1}}. \end{aligned}$$

Solving for $f(x)$ yields

$$\begin{aligned} f(x) = f(x_0) &+ f'(x_0)(x-x_0) + \frac{f''(x_0)}{2!}(x-x_0)^2 + \\ &\cdots + \frac{f^{(n)}(x_0)}{n!}(x-x_0)^n \\ &+ \frac{f^{(n+1)}(c)}{(n+1)!}(x-x_0)^{n+1}, \end{aligned}$$

which is the assertion of Taylor's Theorem. □

SUMMARY

1. *If $f(x)$ has derivatives of orders $\leq n$ at x_0, then the* nth *Taylor polynomial for $f(x)$ at x_0 is*

$$T_n(x) = f(x_0) + f'(x_0)(x - x_0) + \frac{f''(x_0)}{2!}(x - x_0)^2 + \frac{f'''(x_0)}{3!}(x - x_0)^3 + \cdots + \frac{f^{(n)}(x_0)}{n!}(x - x_0)^n$$

$$= \sum_{i=0}^{n} \frac{f^{(i)}(x_0)}{i!}(x - x_0)^i.$$

The polynomial $T_n(x)$ and the function $f(x)$ have the same derivatives of orders $\leq n$ at x_0.

2. *Taylor's Theorem states that, under suitable conditions,*

$$f(x) = T_n(x) + E_n(x),$$

where the error expression (remainder term) $E_n(x)$ is given by

$$E_n(x) = \frac{f^{(n+1)}(c)}{(n + 1)!}(x - x_0)^{n+1}$$

for some c (depending on x) between x_0 and x.

EXERCISES

In Exercises 1 *through* 6, *find the indicated Taylor polynomial for the function at the point.*

1. $T_{10}(x)$ for $\cos x$ at $x_0 = 0$

2. $T_5(x)$ for $\dfrac{1}{1 + x}$ at $x_0 = 0$

3. $T_6(x)$ for $\sin x$ at $x_0 = \dfrac{\pi}{2}$

4. $T_2(x)$ for $\tan x$ at $x_0 = 0$

5. $T_6(x)$ for $1/x$ at $x_0 = 1$

6. $T_3(x)$ for $\sqrt{x}$ at $x_0 = 4$

7. a) Find the Taylor polynomial $T_7(x)$ for $\sin 2x$ at $x_0 = 0$.
b) Compare your answer in (a) with Example 1. What do you notice?

8. a) Find the Taylor polynomial $T_3(x)$ for $\sin x^2$ at $x_0 = 0$.
b) Compare your answer in (a) with Example 1, as in Exercise 7b.
c) Guess the Taylor polynomial $T_{10}(x)$ for $\sin x^2$ at $x_0 = 0$.

9. a) Find the Taylor polynomial $T_3(x)$ at $x_0 = 0$ for $1/(1 - x^2)$.
b) Compare your answer in (a) with Example 3. What do you notice?
c) Guess the Taylor polynomial $T_8(x)$ at $x_0 = 0$ for $1/(1 - x^2)$.

10. a) Find the translated coordinate form of the Taylor polynomial $T_3(2 + \Delta x)$ for x^2 at $x_0 = 2$.
b) Taking translated coordinates at $x_0 = 2$ so that $x = 2 + \Delta x$, express the polynomial $x^2 = (2 + \Delta x)^2$ as a polynomial in Δx. Compare this answer with your answer in (a).

11. [Parts (a) and (b) of this exercise give a method of finding $T_2(x)$ for $f(x) = x^2 + 3x + 5$ at $x_0 = 1$ without differentiating.]

a) Substitute $1 + \Delta x$ for x in the polynomial $x^2 + 3x + 5$ and expand the resulting polynomial in powers of Δx to obtain the translated coordinate form $T_2(1 + \Delta x)$ at $x_0 = 1$.

b) Substitute $x - 1$ for Δx in the polynomial $T_2(1 + \Delta x)$ found in (a).

c) Find $T_{20}(x)$ for $x^2 + 3x + 5$ at $x_0 = 1$.

12. Show that if f is a polynomial function of degree r, then $T_n(x) = T_r(x)$ at any point x_0 for $n \geq r$.

13. Let f and g be functions which have derivatives of orders $\leq n$ at 0, and let $g(x) = f(cx)$. Show that if $T_n(x)$ is the nth Taylor polynomial for f at 0, then $T_n(cx)$ is the nth Taylor polynomial for g at 0.

14. Let f and g be functions which have derivatives of orders $\leq n$ at x_0. Show that the nth Taylor polynomial for $f + g$ at x is the sum of the Taylor polynomials for f and g at x_0.

15. a) Estimate $(2.98)^3$ using a differential.

b) Find a bound for the error in your estimate in (a).

16. a) Estimate $\sqrt[3]{28}$ using a differential.

b) Find a bound for the error in your estimate in (a).

17. a) Using a differential, estimate the change in volume of a cylindrical silo 20 ft high if the radius is increased from 3 ft to 3 ft 1 in.

b) Find a bound for the error in your estimate in (a).

18. Find a bound for the error if $T_8(x)$ at $x_0 = 0$ is used to estimate $\cos 3°$ by

a) Finding a bound for $E_8(\pi/60)$.

b) Finding a bound for $E_9(\pi/60)$. Why can one use $E_9(\pi/60)$ as a bound for the error in $T_8(\pi/60)$?

19. a) Estimate $\tan 2°$ using the Taylor polynomial $T_2(x)$ at $x_0 = 0$.

b) Find a bound for the error in your estimate in (a).

20. a) Estimate $\sqrt{1.05} + (0.96)^3$ using two Taylor polynomials of degree 2.

b) Find a bound for the error in your estimate in (a).

11.3 TAYLOR SERIES; REPRESENTATION OF A FUNCTION

11.3.1 Taylor series

Taylor's Theorem suggests that, for a function having derivatives of *all* orders in a neighborhood of x_0, we consider the series

$$\sum_{n=0}^{\infty} \frac{f^{(n)}(x_0)}{n!}(x - x_0)^n.$$

This series is the **Taylor series of** $f(x)$ **at** x_0. [If $x_0 = 0$, the series is often called the **Maclaurin series of** $f(x)$.]

When we speak of "the Taylor series of f at x_0," we will understand that f has derivatives of all orders in some neighborhood of x_0. The Taylor series of a function f at x_0 certainly represents f at x_0; that is, it converges to $f(x_0)$ at x_0. Unfortunately, it is not always true that the Taylor series represents f throughout the neighborhood. In extreme cases, the series may represent f only at the point x_0 itself. One can show that the function f

defined by

$$f(x) = \begin{cases} e^{-1/x^2} & \text{for} \quad x \neq 0, \\ 0 & \text{for} \quad x = 0, \end{cases}$$

has derivatives of all orders everywhere; in particular, one can show that $f^{(n)}(0) = 0$ for all n. The Taylor series of f at $x_0 = 0$ is therefore

$$0 + 0x + 0x^2 + 0x^3 + \cdots + 0x^n + \cdots,$$

which represents f only at 0.

A necessary and sufficient condition for the Taylor series of f at x_0 to represent f at a point $x_1 \neq x_0$ is easily obtained from Taylor's theorem.

Theorem 11.3 *Let f have derivatives of all orders in $x_0 - h < x < x_0 + h$. The Taylor series of f at x_0 represents f at x_1 where $x_0 - h < x_1 < x_0 + h$, if and only if $\lim_{n\to\infty} E_n(x_1) = 0$, where*

$$E_n(x_1) = \frac{f^{(n+1)}(c)}{(n+1)!}(x_1 - x_0)^{n+1}$$

is the error term in Taylor's theorem.

Proof. For the Taylor series, the nth partial sum s_n at $x = x_1$ is given by

$$s_n(x_1) = T_n(x_1) = \sum_{i=0}^{n} \frac{f^{(i)}(x_0)}{i!}(x_1 - x_0)^i.$$

By Taylor's theorem, we have

$$f(x_1) - s_n(x_1) = f(x_1) - T_n(x_1) = E_n(x_1).$$

Thus for any $\epsilon > 0$, we have $|f(x_1) - s_n(x_1)| < \epsilon$ if and only if $|E_n(x_1)| < \epsilon$, so the conditions $\lim_{n\to\infty} s_n(x_1) = f(x_1)$ and $\lim_{n\to\infty} E_n(x_1) = 0$ are equivalent. □

The following corollary of Theorem 11.3 will usually suffice for our purposes.

Corollary *Let f have derivatives of all orders in $x_0 - h < x < x_0 + h$. If there is a number $B > 0$ such that $|f^{(n)}(x)| \le B$ for all positive integers n and for all x such that $x_0 - h < x < x_0 + h$, then the Taylor series of f at x_0 represents f throughout the neighborhood $x_0 - h < x < x_0 + h$.*

Proof. By hypothesis, we have, for $x_0 - h < x_1 < x_0 + h$,

$$\lim_{n\to\infty} |E_n(x_1)| = \lim_{n\to\infty} \left| f^{(n+1)}(c)\frac{(x_1 - x_0)^{n+1}}{(n+1)!} \right| \le B\left(\lim_{n\to\infty} \frac{|x_1 - x_0|^{n+1}}{(n+1)!} \right)$$
$$= B \cdot 0 = 0,$$

so by Theorem 11.3, the Taylor series of f represents f at x_1. □

Example 1 All derivatives of the function e^x are again e^x, and the Taylor series of e^x at $x_0 = 0$ is easily found to be $\sum_{n=0}^{\infty} (x^n/n!)$. Since e^x is bounded by e^b in every interval $-b < x < b$, the corollary shows that the series converges to e^x for all x; that is,

$$e^x = 1 + x + \frac{x^2}{2!} + \frac{x^3}{3!} + \cdots + \frac{x^n}{n!} + \cdots \tag{1}$$

for all x. You should remember this series for e^x; it occurs frequently. ||

Example 2 Derivatives of sine and cosine are again just sine or cosine functions, and these are bounded by 1 everywhere. The Taylor series for these functions at any x_0 therefore represent them everywhere. Computing the Taylor series at $x_0 = 0$, you easily find that

$$\sin x = x - \frac{x^3}{3!} + \frac{x^5}{5!} - \frac{x^7}{7!} + \cdots + (-1)^n \frac{x^{2n+1}}{(2n+1)!} + \cdots \tag{2}$$

and

$$\cos x = 1 - \frac{x^2}{2!} + \frac{x^4}{4!} - \frac{x^6}{6!} + \cdots + (-1)^n \frac{x^{2n}}{(2n)!} + \cdots \tag{3}$$

for all x. You should remember these series also. ||

11.3.2 Differentiation and integration of power series

We now turn to the study of functions that are represented in a neighborhood of x_0 by a power series at x_0. Such functions are very important and are called **analytic at** x_0. A function is **analytic** if it is analytic at each point in its domain.

Example 3 Examples 1 and 2 show that the functions e^x, $\sin x$, and $\cos x$ are analytic at $x_0 = 0$. Similar arguments show that the functions are analytic at every point. ||

It can be shown that if f is analytic at x_0 and is represented by a power series throughout $x_0 - r < x < x_0 + r$, then f is analytic at every point in $x_0 - r < x < x_0 + r$. Thus, since the Taylor series for e^x, $\sin x$, and $\cos x$ at $x_0 = 0$ represent these functions for all x, we see again that these three functions are analytic at each point.

The calculus of analytic functions ("infinite polynomial functions") is much like the calculus of the polynomial functions. If f is analytic at x_0, then f has derivatives of all orders at x_0, and these derivatives can be computed by differentiating the series at x_0 representing f just as you would differentiate a polynomial. Antiderivatives can be found similarly. The next theorem states this formally.

Theorem 11.4 *Let f be analytic at x_0 and let $\sum_{n=0}^{\infty} a_n(x - x_0)^n$ represent f in $x_0 - r < x < x_0 + r$. Then f has derivatives of all orders throughout this neighborhood, and the derivatives at any x in the neighborhood may be computed by differentiating the series term by term. For example,*

$$f'(x) = \sum_{n=1}^{\infty} n \cdot a_n(x - x_0)^{n-1}$$

for $x_0 - r < x < x_0 + r$. The indefinite integral of f in $x_0 - r < x < x_0 + r$ is represented by the series

$$C + \sum_{n=0}^{\infty} \frac{a_n}{n+1}(x - x_0)^{n+1},$$

where C is an arbitrary constant function.

Example 4 You can easily check that the series (1) for e^x is unchanged by differentiation, and that the series (2) for sin x becomes the series (3) for cos x upon differentiation. ‖

Example 5 The function $1/(1 - x)$ is analytic at 0, for you know that the geometric series $\sum_{n=0}^{\infty} x^n$ converges to $1/(1-x)$ in $-1<x<1$. Thus

$$\frac{d}{dx}\left(\frac{1}{1-x}\right) = \frac{1}{(1-x)^2}$$

is analytic at 0, and you obtain by differentiation

$$\frac{1}{(1-x)^2} = \sum_{n=1}^{\infty} nx^{n-1}$$
$$= 1 + 2x + 3x^2 + 4x^3 + \cdots + nx^{n-1} + \cdots$$

for $-1 < x < 1$.

Since $\int 1/(1 - x)\,dx = -\ln|1 - x| + C$, you find by integrating the geometric series that

$$-\ln(1 - x) = k + \sum_{n=0}^{\infty} \frac{x^{n+1}}{n+1}$$
$$= k + x + \frac{x^2}{2} + \frac{x^3}{3} + \cdots + \frac{x^n}{n} + \cdots$$

for some constant k and $-1 < x < 1$. Putting $x = 0$, you find that $k = -\ln(1) = 0$, so

$$\ln(1 - x) = -x - \frac{x^2}{2} - \frac{x^3}{3} - \cdots - \frac{x^n}{n} - \cdots \tag{4}$$

for $-1 < x < 1$. ‖

11.3.3 Uniqueness of power series representation

It is an important fact that if f is analytic at x_0, then there is *only one* power series at x_0 that represents f throughout a neighborhood of x_0. This must be the Taylor series, for the coefficients of a power series representing the function are determined by the derivatives of the function to be precisely the coefficients in the Taylor series, as indicated in Section 11.2.1.

Using this uniqueness, you can find the Taylor series for many functions f without differentiating f to compute coefficients. We illustrate with examples.

Example 6 You know that

$$\sin x = x - \frac{x^3}{3!} + \cdots + (-1)^n \frac{x^{2n+1}}{(2n+1)!} + \cdots$$

for all x. Replacing x by x^2,

$$\sin x^2 = x^2 - \frac{x^6}{3!} + \cdots + (-1)^n \frac{x^{4n+2}}{(2n+1)!} + \cdots \tag{5}$$

for all x. Therefore the series (5) must be the Taylor series for $\sin x^2$ at $x_0 = 0$. If you try to find the Taylor series by differentiating $\sin x^2$ repeatedly, you will quickly appreciate the easy way we obtained (5). ‖

Example 7 The series (4) must be the Taylor series at $x_0 = 0$ for the function $\ln(1 - x)$. ‖

Example 8 You know that

$$\frac{1}{1-x} = 1 + x + x^2 + x^3 + \cdots + x^n + \cdots \tag{6}$$

for $-1 < x < 1$. Replacing x by $-x^2$,

$$\frac{1}{1+x^2} = 1 - x^2 + x^4 - x^6 + \cdots + (-1)^n x^{2n} + \cdots \tag{7}$$

for $-1 < x < 1$. Integrating (7), you find that

$$\tan^{-1}x = k + x - \frac{x^3}{3} + \frac{x^5}{5} - \frac{x^7}{7} + \cdots + (-1)^n \frac{x^{2n+1}}{2n+1} + \cdots$$

for some constant k. Putting $x = 0$, you see that $k = 0$, so

$$\tan^{-1}x = x - \frac{x^3}{3} + \frac{x^5}{5} - \frac{x^7}{7} + \cdots + (-1)^n \frac{x^{2n+1}}{2n+1} + \cdots \tag{8}$$

for $-1 < x < 1$. This series (8) must be the Taylor series of $\tan^{-1}x$ at $x_0 = 0$. If you try to compute the series (8) by differentiating $\tan^{-1}x$ repeatedly, you will quickly appreciate the easy way we obtained (8). ‖

The series in Example 8 illustrate a very interesting and perhaps unexpected situation that can occur. The series (6) converges only for

$-1 < x < 1$, but we are not surprised since it represents the function $1/(1 - x)$ which "blows up" at $x = 1$. The function $1/(1 + x^2)$ has derivatives of all orders everywhere, and can be shown to be analytic at every point x_0. However, its Taylor series (7) at $x_0 = 0$ still only converges to $1/(1 + x^2)$ for $-1 < x < 1$. To fully appreciate why this happens, you must study complex analysis. The function $1/(1 + x^2)$ "blows up" at $x = i$ and $x = -i$, and in complex analysis, one sees that the numbers i and $-i$ have distance 1 from $x_0 = 0$. It is for this reason that the radius of convergence of the Taylor series for $1/(1 + x^2)$ at $x_0 = 0$ is only 1.

11.3.4 Multiplication and division of power series

Two power series at x_0 representing functions f and g in $x_0 - r < x < x_0 + r$ can be multiplied and divided as "infinite" polynomials, to yield power series representing the functions fg and f/g in neighborhoods of x_0, with the obvious restriction that $g(x_0) \neq 0$. We state this as a theorem without proof. The functions f and g can be approximated at each point of $x_0 - r < x < x_0 + r$ as closely as you like by polynomial partial sums of the series, so the theorem seems reasonable.

Theorem 11.5 *Let series $\sum_{n=0}^{\infty} (x - x_0)^n$ and $\sum_{n=0}^{\infty} b_n (x - x_0)^n$ converge to functions f and g respectively in $x_0 - r < x < x_0 + r$. Then the product series (called the* ***Cauchy product****)*

$$a_0b_0 + (a_0b_1 + a_1b_0)(x - x_0) + (a_0b_2 + a_1b_1 + a_2b_0)(x - x_0)^2 + \cdots,$$

whose nth coefficient is $\sum_{i=0}^{n} (a_i b_{n-i})$, represents fg throughout $x_0 - r < x < x_0 + r$. Also, if $b_0 = g(x_0) \neq 0$, the series

$$\frac{a_0}{b_0} + \frac{a_1b_0 - a_0b_1}{b_0^{\,2}}(x - x_0) + \cdots,$$

obtained by long division represents f/g in some neighborhood $x_0 - \delta < x < x_0 + \delta$.

You are accustomed to polynomial long division, where you write the polynomials with the terms of highest degree first, and divide only until you obtain a remainder of lower degree than the divisor. *In series long division, you write the terms of lowest degree first and the division may never terminate*; symbolically,

$$\begin{array}{r|l}
 & \dfrac{a_0}{b_0} + \dfrac{a_1b_0 - a_0b_1}{b_0^{\,2}}(x - x_0) + \cdots \\ \hline
b_0 + b_1(x - x_0) + \cdots & a_0 \quad + \quad a_1(x - x_0) + \cdots \\
 & a_0 \quad + \dfrac{a_0b_1}{b_0}(x - x_0) + \cdots \\ \cline{2-2}
 & 0 + \dfrac{a_1b_0 - a_0b_1}{b_0}(x - x_0) + \cdots \\
 & \text{etc.}
\end{array}$$

An illustration of such division appears in Example 10 below.

Example 9 Let's find by series multiplication the first few terms of the Taylor series for $e^x \sin x$ at $x_0 = 0$.

SOLUTION Since

$$e^x = 1 + x + \frac{x^2}{2} + \frac{x^3}{6} + \frac{x^4}{24} + \frac{x^5}{120} + \cdots$$

and

$$\sin x = x - \frac{x^3}{6} + \frac{x^5}{120} - \cdots,$$

we obtain

$$\begin{aligned} e^x \sin x &= x + x^2 + \left(\frac{x^3}{2} - \frac{x^3}{6}\right) + \left(\frac{x^4}{6} - \frac{x^4}{6}\right) + \left(\frac{x^5}{24} - \frac{x^5}{12} + \frac{x^5}{120}\right) + \cdots \\ &= x + x^2 + \frac{x^3}{3} - \frac{x^5}{30} + \cdots \end{aligned}$$

for all x. ||

Example 10 We find the Taylor series for $(1 + x^2)/(1 - x)$ at $x_0 = 0$ in two ways.

SOLUTION The computation using series division is

$$\begin{array}{r|l} & 1 + x + 2x^2 + 2x^3 + \cdots \\ \hline 1 - x & 1 \quad + x^2 \\ & \underline{1 - x} \\ & \quad x + x^2 \\ & \quad \underline{x - x^2} \\ & \qquad 2x^2 \\ & \qquad \underline{2x^2 - 2x^3} \\ & \qquad\quad 2x^3 \\ & \qquad\quad \underline{2x^3 - 2x^4} \\ & \qquad\qquad 2x^4 \end{array}$$

etc.

We thus obtain

$$\frac{1 + x^2}{1 - x} = 1 + x + 2x^2 + 2x^3 + 2x^4 + \cdots + 2x^n + \cdots$$

for $-1 < x < 1$. Alternatively, you could multiply the geometric series $1 + x + x^2 + \cdots + x^n + \cdots$ for $1/(1 - x)$ by $1 + x^2$. The computation is

$$
\begin{array}{r}
1 + x + x^2 + x^3 + \cdots + x^n + x^{n+1} + \cdots \\
\times \qquad 1 + x^2 \\
\hline
1 + x + x^2 + x^3 + \cdots + x^n + x^{n+1} + \cdots \\
x^2 + x^3 + \cdots + x^n + x^{n+1} + \cdots \\
\hline
1 + x + 2x^2 + 2x^3 + \cdots + 2x^n + 2x^{n+1} + \cdots
\end{array}
$$

and you obtain the same series, as you must by uniqueness. If you try to compute the Taylor series for $(1 + x^2)/(1 - x)$ at $x_0 = 0$ by repeated differentiation, you will quickly appreciate the easy ways we found it here. ‖

SUMMARY

1. *If $f(x)$ has derivatives of all orders at x_0, then*
$$\sum_{n=0}^{\infty} \frac{f^{(n)}(x_0)}{n!}(x - x_0)^n$$
is the Taylor series of $f(x)$ at x_0. [Here $0! = 1$ and $f^{(0)}(x) = f(x)$.]
2. *The Taylor series of $f(x)$ at x_0 represents $f(x)$ at x_1 if and only if $\lim_{n\to\infty} E_n(x_1) = 0$. This condition will always hold if all derivatives between x_0 and x_1 are bounded by the same constant B.*
3. *A function is analytic at x_0 if it is represented by some power series in some neighborhood of x_0. It is analytic if it is analytic at each point in its domain.*
4. *If $f(x)$ is represented by a power series in an open interval, then $f'(x)$ is represented by the term-by-term derivative of that power series, and $\int f(x)\,dx$ by the term-by-term antiderivative of that power series, plus an arbitrary constant.*
5. *The only power series at x_0 that can represent $f(x)$ in a neighborhood of x_0 is the Taylor series of $f(x)$.*
6. *Series at x_0 representing $f(x)$ and $g(x)$ in a common neighborhood of x_0 may be multiplied (as infinite polynomials) to obtain the series representing $f(x)g(x)$ in that neighborhood, and divided if $g(x_0) \neq 0$ to represent $f(x)/g(x)$ in* some *neighborhood of x_0.*

EXERCISES

1. Mark each of the following true or false.

a) If f has derivatives of all orders throughout some neighborhood of x_0, then f is analytic at x_0.

b) If f is analytic at x_0, then f has derivatives of all orders throughout some neighborhood of x_0.

c) If f and g are analytic at x_0, then $f + g$ is analytic at x_0.

d) Every power series represents a function which is analytic at every point (except possibly endpoints) of the interval of convergence of the series.

e) There is at most one power series at x_0 which represents a given function f at x_0.

f) There is at most one power series at x_0 which represents a given function f throughout some neighborhood of x_0.

2. Is the function $\sqrt{x}$ analytic at $x_0 = 0$? Why?

In Exercises 3 through 24, find as many terms of the Taylor series of the function at the given point as you conveniently can in the easiest way you can.

3. $x^2 + e^x$ at $x_0 = 0$ **4.** $1 + x^3 - \sin x$ at $x_0 = 0$ **5.** $x \sin x$ at $x_0 = 0$ **6.** $\cos x$ at $x_0 = \pi$

7. $\dfrac{x}{1-x}$ at $x_0 = 0$ **8.** e^{-x^2} at $x_0 = 0$ **9.** $\cos x^3$ at $x_0 = 0$ **10.** $\dfrac{2x+3x^2}{1+4x}$ at $x_0 = 0$

11. $e^x \cos x$ at $x_0 = 0$ **12.** e^{3x} at $x_0 = 0$ **13.** $\dfrac{1}{(1+x)^2}$ at $x_0 = 0$ **14.** $\ln(\cos x)$ at $x_0 = 0$

15. $\sec x$ at $x_0 = 0$ **16.** $\ln x$ at $x_0 = 2$ **17.** $\dfrac{e^x}{1-x}$ at $x_0 = 0$ **18.** $\dfrac{e^x + e^{-x}}{2}$ at $x_0 = 0$

19. $\dfrac{e^x - e^{-x}}{2}$ at $x_0 = 0$ **20.** $\sqrt{x}$ at $x_0 = 1$ **21.** $\dfrac{1}{e^x}$ at $x_0 = 0$ **22.** $\dfrac{x-1}{x^3}$ at $x_0 = 1$

23. $\sec x \tan x$ at $x_0 = 0$ **24.** $\dfrac{1}{2-x}$ at $x_0 = 0$

25. a) Find the terms for $n \le 5$ of the Taylor series of $\sin x \cos x$ at $x_0 = 0$ by series multiplication.

b) Find the Taylor series of $\sin x \cos x_0 = 0$ by use of the identity $\sin x \cos x = (\sin 2x)/2$.

26. Find the Taylor series of $1/x$ at $x_0 = 2$ by

a) differentiating $1/x$ repeatedly to compute the coefficients,

b) using the identity

$$\frac{1}{x} = \frac{1}{2} \cdot \frac{1}{1 - [-(x-2)/2]}$$

and expanding in a geometric series.

27. Use the technique suggested by Exercise 26(b) to find the Taylor series of $1/x$ at $x = -1$; you have to find the appropriate identity.

28. Find the terms for $n \le 3$ of the Taylor series of $\tan x$ at $x_0 = 0$ by dividing the series for $\sin x$ by the series for $\cos x$.

29. a) Obtain the series expansion

$$\ln(1+x) = x - \frac{x^2}{2} + \frac{x^3}{3} - \frac{x^4}{4} + \cdots + (-1)^{n+1}\frac{x^n}{n} + \cdots$$

for $-1 < x < 1$. [*Hint.* You may use (4) or integrate the geometric series for $1/(1+x) = 1/(1-(-x))$.]

b) Show that the series in (a) converges for $x = 1$.

c) Show that the alternating harmonic series

$$1 - \tfrac{1}{2} + \tfrac{1}{3} - \tfrac{1}{4} + \cdots$$

converges to $\ln 2$. [*Hint.* Since 1 is an endpoint of the interval of convergence of the series in (a), Theorem 11.4 cannot be used. You must check $\lim_{n\to\infty} E_n(1)$ for the function $\ln(1+x)$.]

30. Find the series at $x_0 = 0$ representing the function f defined by

$$f(x) = \int_0^x \ln(1-t)\,dt$$

for $-1 < x < 1$.

31. Find the series expansion at $x_0 = 0$ for the indefinite integral of e^{x^2}.

32. Find the series at $x_0 = 0$ representing the function f defined by

$$f(x) = \int_0^x [1/(1-t^3)]\,dt$$

for $-1 < x < 1$.

33. Find the series at $x_0 = 0$ representing the function f defined by $f(x) = \pi + \int_0^x \cos t^2\,dt$ for all x.

34. Find the series at $x_0 = 0$ representing the function f defined by

$$f(x) = \int_0^x \frac{(3+t^2)}{(1+2t)}\,dt$$

for $-\frac{1}{2} < x < \frac{1}{2}$.

35. a) Proceeding purely formally, find the Taylor series at $x_0 = 0$ of e^{ix} and of e^{-ix} where $i^2 = -1$.

b) From part (a), "derive" Euler's formula $e^{ix} = \cos x + i(\sin x)$.

c) From part (a), "derive" the formula $e^{-ix} = \cos x - i(\sin x)$.

d) From (b) and (c), find formulas for $\cos x$ and $\sin x$ in terms of the complex exponential function.

e) Compare the formulas for $\sin x$ and $\cos x$ found in (d) with the formulas for $\sinh x$ and $\cosh x$ in terms of the exponential function.

Exercises 36 through 47 give you practice in series recognition. *The given series represents a familiar elementary function in its interval of convergence. Find the function.*

36. $1 - x + x^2 - x^3 + \cdots + (-1)^n x^n + \cdots$

37. $1 + x + x^2 - x^3 + x^4 - x^5 + \cdots + (-1)^n x^n + \cdots$ for $n \geq 2$

38. $1 - x + \dfrac{x^2}{2!} - \dfrac{x^3}{3!} + \dfrac{x^4}{4!} - \cdots + (-1)^n \dfrac{x^n}{n!} + \cdots$

39. $1 + 3x - \dfrac{x^2}{2!} + \dfrac{x^4}{4!} - \cdots + (-1)^n \dfrac{x^{2n}}{(2n)!} + \cdots$ for $n \geq 1$

40. $1 - \dfrac{x}{2} + \dfrac{x^2}{4} - \dfrac{x^3}{8} + \cdots + \dfrac{(-1)^n}{2^n} x^n + \cdots$

41. $1 + x - \dfrac{x^2}{2!} - \dfrac{x^3}{3!} + \dfrac{x^4}{4!} + \dfrac{x^5}{5!} - \dfrac{x^6}{6!} - \dfrac{x^7}{7!} + \cdots$

42. $1 - x^3 + x^6 - x^9 + x^{12} - \cdots + (-1)^n x^{3n} + \cdots$

43. $1 + 2x + \dfrac{2!+1}{2!} x^2 + \dfrac{3!+1}{3!} x^3 + \cdots + \dfrac{n!+1}{n!} x^n + \cdots$ for $n \geq 2$

44. $-1 + 2x - 3x^2 + 4x^3 - \cdots + (-1)^{n+1}(n+1)x^n + \cdots$ [*Hint*: Integrate the series.]

45. $2 + 3 \cdot 2x + 4 \cdot 3x^2 + 5 \cdot 4x^3 + \cdots + (n+2)(n+1)x^n + \cdots$

46. $x^3 - \dfrac{x^7}{5!} + \dfrac{x^9}{7!} - \cdots + (-1)^n \dfrac{x^{2n+1}}{(2n-1)!} + \cdots$ for $n \geq 3$

47. $4x^4 - 8x^6 + 16x^8 - 32x^{10} + \cdots + (-1)^{n+1} 2^{n+1} x^{2n+2} + \cdots$ for $n \geq 2$

calculator exercises

48. Find the smallest value of n such that the terms of degree $\leq n$ of the series for $\sin x$ about $x_0 = 0$ can be used to find $\sin 1$ with error less than 10^{-6}. How does this value of n compare with that given by the error term in Taylor's formula if all derivatives of $\sin x$ at $x = c$ are replaced by 1 in bounding the error?

49. Repeat Exercise 48, but using the series for e^x about $x_0 = 0$ to estimate $\sqrt{e}$, and answering the second part if all derivatives e^c of e^x are replaced by 2 in bounding the error.

11.4 INDETERMINATE FORMS

11.4.1 Types of indeterminate forms

Note that

$$\lim_{x\to 1}\frac{(x-1)^2}{x-1}=0, \qquad \lim_{x\to 1}\frac{2(x-1)}{x-1}=2, \qquad \text{and} \qquad \lim_{x\to 1}\frac{(x-1)^2}{(x-1)^4}=\infty.$$

In each of these three cases, the function whose limit is being evaluated is undefined at $x = 1$, and a formal substitution of 1 in the numerator and denominator of the quotient leads to the expression "0/0" in each case. These examples show that one could define 0/0 to be 0, 2, or ∞ with equal justification. It is for this reason that one does not attempt to define 0/0; the expression "0/0" is an example of an *indeterminate form*. Similarly, the limits

$$\lim_{x\to 2+}\frac{1/(x-2)}{2/(x-2)}=\frac{1}{2} \qquad \textit{and} \qquad \lim_{x\to 2+}\frac{1/(x-2)}{3/(x-2)^2}=0,$$

Indeterminate quotients

where both the numerator and denominator in each limit approach ∞ as x approaches 2, show that we should consider ∞/∞ to be an indeterminate form. The *indeterminate quotient forms* are

$$\frac{0}{0}, \qquad \pm\frac{\infty}{\infty}.$$

Note that $0/\infty$ is not an indeterminate form, for if $\lim_{x\to a} f(x) = 0$ and $\lim_{x\to a} g(x) = \infty$, then $\lim_{x\to a} f(x)/g(x) = 0$. Also, 2/0 is not an indeterminate form, for if $\lim_{x\to a} f(x) = 2$ and $\lim_{x\to a} g(x) = 0$, then $\lim_{x\to a} f(x)/g(x)$ is always undefined, and the quotient $f(x)/g(x)$ becomes large in absolute value as x approaches a.

Turning to products, the limits

$$\lim_{x\to 1+}(x-1)\frac{3}{x-1}=3 \qquad \text{and} \qquad \lim_{x\to 1+}(x-1)^2\left(\frac{2}{x-1}\right)=0$$

Indeterminate products

show that we should consider $0\cdot\infty$ to be an indeterminate form. The *indeterminate product forms* are

$$0\cdot\infty, \qquad \infty\cdot 0.$$

From

$$\lim_{x\to a+}\left(\frac{1}{x-a}-\frac{1}{x-a}\right)=0$$

and

$$\lim_{x\to a+}\left(\frac{1}{x-a}-\frac{1+2a-2x}{x-a}\right)=\lim_{x\to a+}\frac{2(x-a)}{x-a}=2, \tag{1}$$

Indeterminate sum and difference

you see that $\infty - \infty$ should be considered to be an indeterminate form. The *indeterminate sum and difference forms* are

$$(-\infty) + \infty, \qquad \infty - \infty.$$

Finally, there are indeterminate exponential forms arising from expressions of the form $\lim_{x\to a} f(x)^{g(x)}$. Recall that we have defined the exponential r^s for all s only if $r > 0$; hence we assume that $f(x) > 0$ for $x \neq a$. Since the logarithm function is continuous, and is the inverse of the exponential function, you see that if $\lim_{x\to a} \ln [f(x)^{g(x)}] = b$, then $\lim_{x\to a} f(x)^{g(x)} = e^b$. Now

$$\ln [f(x)^{g(x)}] = g(x) \cdot \ln (f(x)),$$

so for $\lim_{x\to a} f(x)^{g(x)}$ to give rise to an indeterminate exponential form, the product $g(x) \cdot \ln (f(x))$ must give rise to one of the indeterminate product forms $0 \cdot \infty$, $\infty \cdot 0$, $0(-\infty)$, or $(-\infty)0$ at $x = a$. The product $g(x) \cdot \ln (f(x))$ gives rise to $0 \cdot \infty$ at $x = a$ if $\lim_{x\to a} g(x) = 0$ and $\lim_{x\to a} \ln (f(x)) = \infty$, in which case $\lim_{x\to a} f(x) = \infty$ also. Thus $0 \cdot \infty$ gives rise to the indeterminate exponential form ∞^0. Similarly, the product form $\infty \cdot 0$ gives rise to the exponential form 1^∞, while $0(-\infty)$ gives rise to the form 0^0 and $(-\infty)0$ gives rise to $1^{-\infty}$. Thus the *indeterminate exponential forms* are

Indeterminate exponentials

$$0^0, \qquad 1^\infty, \qquad 1^{-\infty}, \qquad \infty^0.$$

11.4.2 Finding limits having indeterminate form by series methods

A limit corresponding to an indeterminate form is usually computed by trying to convert the problem to a limit corresponding to the indeterminate quotient form $0/0$. For example, if $\lim_{x\to a} f(x) = 0$ and $\lim_{x\to a} g(x) = \infty$, then

$$\lim_{x\to a} f(x) \cdot g(x) = \lim_{x\to a} \frac{f(x)}{1/g(x)},$$

and the second limit corresponds to the indeterminate form $0/0$. A mathematically terrifying but mnemonically helpful way to remember how to convert a $0 \cdot \infty$-type problem to a $0/0$-type problem is to write

$$\text{“}0 \cdot \infty = \frac{0}{1/\infty} = \frac{0}{0}.\text{”}$$

Similarly, the mnemonic device

$$\text{“}\frac{\infty}{\infty} = \frac{1/\infty}{1/\infty} = \frac{0}{0}\text{”}$$

enables you to convert an ∞/∞-type problem to a $0/0$-type problem. One usually tries to compute a limit corresponding to an indeterminate sum or difference form such as $\infty - \infty$ by the technique in (1), where the problem was again converted to a 0/0-type limit problem. We saw in Section 11.4.1 that limits having indeterminate exponential form can be reduced to limits having indeterminate product form by taking logarithms, and we have just shown how a $0 \cdot \infty$-type limit problem can be converted to a 0/0-type problem. Thus we concentrate on the 0/0-type problem.

Let f and g be analytic at a, and suppose

$$\lim_{x \to a} f(x) = \lim_{x \to a} g(x) = 0.$$

We wish to find $\lim_{x \to a} f(x)/g(x)$. We assume that neither $f(x)$ nor $g(x)$ is identically 0 throughout an entire neighborhood of a. Since f and g are analytic at a, we have, for x in some sufficiently small neighborhood of a,

$$f(x) = a_r(x-a)^r + a_{r+1}(x-a)^{r+1} + \cdots$$

and

$$g(x) = b_s(x-a)^s + b_{s+1}(x-a)^{s+1} + \cdots,$$

where a_r and b_s are the first nonzero coefficients in the series for f and g, respectively, at a. Then

$$\begin{aligned} \lim_{x \to a} \frac{f(x)}{g(x)} &= \lim_{x \to a} \frac{a_r(x-a)^r + a_{r+1}(x-a)^{r+1} + \cdots}{b_s(x-a)^s + b_{s+1}(x-a)^{s+1} + \cdots} \\ &= \lim_{x \to a} \left\{ \frac{a_r(x-a)^r}{b_s(x-a)^s} \cdot \frac{1 + \dfrac{a_{r+1}}{a_r}(x-a) + \cdots}{1 + \dfrac{b_{s+1}}{b_s}(x-a) + \cdots} \right\} \\ &= \lim_{x \to a} \frac{a_r(x-a)^r}{b_s(x-a)^s}. \end{aligned}$$

We state this result as a theorem.

Theorem 11.6 *Let f and g be analytic at a and let a_r and b_s be the first nonzero coefficients in the series for f and g respectively at a. Then*

$$\lim_{x \to a} \frac{f(x)}{g(x)} = \lim_{x \to a} \frac{a_r(x-a)^r}{b_s(x-a)^s}.$$

You should think of this theorem as showing that the first nonzero terms of series expansions for functions analytic at *a dominate* for x close to a. We give several examples to illustrate this series technique.

Example 1 Let's give a series derivation of the fundamental limit

$$\lim_{x \to 0} \frac{\sin x}{x} = 1.$$

SOLUTION Now $\sin x$ is analytic at 0; in fact,

$$\sin x = x - \frac{x^3}{3!} + \frac{x^5}{5!} - \frac{x^7}{7!} + \cdots$$

for all x. By the theorem

$$\lim_{x\to 0} \frac{\sin x}{x} = \lim_{x\to 0} \frac{x}{x} = 1. \quad \|$$

Example 2 Let's compute

$$\lim_{x\to 0} \frac{\cos x - 1}{x}$$

by series methods.

SOLUTION We have

$$\cos x = 1 - \frac{x^2}{2!} + \frac{x^4}{4!} - \frac{x^6}{6!} + \cdots$$

for all x, so

$$\cos x - 1 = -\frac{x^2}{2!} + \frac{x^4}{4!} - \frac{x^6}{6!} + \cdots$$

for all x. By the theorem,

$$\lim_{x\to 0} \frac{\cos x - 1}{x} = \lim_{x\to 0} \frac{-x^2/2}{x} = \lim_{x\to 0} \frac{-x}{2} = 0. \quad \|$$

Example 3 Let's compute

$$\lim_{x\to 0+} (\cot x)(\ln(1 - x)),$$

which corresponds to the indeterminate form $\infty \cdot 0$.

SOLUTION We convert to a 0/0-type problem by

$$\lim_{x\to 0+} (\cot x)(\ln(1 - x)) = \lim_{x\to 0+} \frac{\ln(1 - x)}{1/(\cot x)} = \lim_{x\to 0+} \frac{\ln(1 - x)}{\tan x}.$$

We saw, in Example 5 (Section 11.3.2), that

$$\ln(1 - x) = -x - \frac{x^2}{2} - \frac{x^3}{3} - \cdots$$

for $-1 < x < 1$, and series division shows that

$$\tan x = \frac{\sin x}{\cos x} = x + \frac{x^3}{3} + \cdots$$

for x sufficiently near 0. Our theorem then yields

$$\lim_{x \to 0+} \frac{\ln(1-x)}{\tan x} = \lim_{x \to 0+} \frac{-x}{x} = -1. \quad \|$$

Example 4 To illustrate an $(\infty - \infty)$-type problem, we compute

$$\lim_{x \to 0+} \left(\frac{2}{x} - \frac{x+1}{x-x^2}\right) = \lim_{x \to 0+} \frac{2(x-x^2) - x(x+1)}{x^2 - x^3}$$

$$= \lim_{x \to 0+} \frac{x - 3x^2}{x^2 - x^3} = \lim_{x \to 0+} \frac{x}{x^2} = \infty. \quad \|$$

Using the fact that $\lim_{x \to \infty} f(x) = \lim_{t \to 0+} f(1/t)$, one can often compute a limit at ∞ by series methods.

Example 5 Let's compute the important limit

$$\lim_{x \to \infty} \left(1 + \frac{1}{x}\right)^x,$$

which corresponds to the indeterminate form 1^∞.

SOLUTION We make use of the relation $\lim_{x \to \infty} f(x) = \lim_{t \to 0+} f(1/t)$. Taking logarithms to handle the exponential indeterminate form, we reduce the computation to

$$\lim_{x \to \infty} \left(x \cdot \ln\left(1 + \frac{1}{x}\right)\right) = \lim_{t \to 0+} \frac{1}{t} \ln(1+t) = \lim_{t \to 0+} \frac{\ln(1+t)}{t}.$$

Example 5 (Section 11.3.2) shows that

$$\ln(1+t) = t - \frac{t^2}{2} + \frac{t^3}{3} - \frac{t^4}{4} + \cdots$$

for $-1 < t < 1$, so

$$\lim_{t \to 0+} \frac{\ln(1+t)}{t} \quad \lim_{t \to 0+} \frac{t}{t} = 1.$$

Thus the limit of the logarithm approaches 1, so

$$\lim_{x \to \infty} \left(1 + \frac{1}{x}\right)^x = e^1 = e.$$

You should remember this limit. ‖

11.4.3 l'Hôpital's rule

The following theorem shows that $\lim_{x \to a} f(x)/g(x)$ can sometimes be computed by taking a limit of quotients of derived functions, assuming that $\lim_{x \to a} f(x)/g(x)$ corresponds to the indeterminate form 0/0.

Theorem 11.7 *(l'Hôpital's Rule) Let f and g be continuous functions of t with derivatives in some neighborhood $a - r < t < a + r$ of a. Suppose, furthermore, that $f(a) = g(a) = 0$ and $g'(t) \neq 0$ for $t \neq a$ in that neighborhood. If $\lim_{t\to a} f'(t)/g'(t)$ exists, then $\lim_{t\to a} f(t)/g(t)$ exists and $\lim_{t\to a} f(t)/g(t) = \lim_{t\to a} f'(t)/g'(t)$.*

Proof. Consider the curve given parametrically by

$$x = g(t), \qquad y = f(t), \qquad \text{for} \quad a - r < t < a + r.$$

The slope of the chord from $(g(a), f(a))$ to $(g(t), f(t))$ is

$$\frac{f(t) - f(a)}{g(t) - g(a)} = \frac{f(t)}{g(t)}.$$

Now the slope of the curve at $(g(t), f(t))$ is

$$\frac{dy}{dx} = \frac{dy/dt}{dx/dt} = \frac{f'(t)}{g'(t)}.$$

It can be shown that the hypothesis $g'(t) \neq 0$ for $t \neq a$ guarantees that $x = g(t)$ for t from a to some t_0 defines a differentiable inverse function $t = g^{-1}(x)$ for x from $g(a) = 0$ to $g(t_0)$. Then the composite function $y = f(g^{-1}(x))$ from $x = 0$ to $x = g(t_0)$ is continuous on $[0, g(t_0)]$ and differentiable inside the interval. This means that the Mean-Value Theorem can be applied to a portion of the parametric curve $x = g(t)$, $y = f(t)$ corresponding to an interval of the parameter from a to a value t. Therefore there exists c between a and t such that the slope of the curve at $(g(c), f(c))$ is equal to the slope of the chord found above. That is,

$$\frac{f(t)}{g(t)} = \frac{f'(c)}{g'(c)} \quad \text{for some } c \text{ betwen } a \text{ and } t. \tag{2}$$

We now take the limit in (2) as $t \to a$. Since c is between t and a, as $t \to a$, we must have $c \to a$ also. Thus

$$\lim_{t\to a} \frac{f(t)}{g(t)} = \lim_{c\to a} \frac{f'(c)}{g'(c)} = \lim_{t\to a} \frac{f'(t)}{g'(t)}$$

as asserted in the theorem. □

The method of proof of Theorem 11.7 led us to state it for function f and g of a variable t. Of course, it may be used for functions of x, u, or any other variable, and we use x as usual in the examples.

Example 6 Computing $\lim_{x\to 0} [(\sin x)/x]$ by l'Hôpital's rule, we have

$$\lim_{x\to 0} \frac{\sin x}{x} = \lim_{x\to 0} \frac{\cos x}{1} = \frac{1}{1} = 1. \quad \|$$

Sometimes $\lim_{x\to a} f'(x)/g'(x)$ is again a 0/0-type problem. If the hypotheses of l'Hôpital's rule are satisfied by the functions f' and g' as well, then

$$\lim_{x\to a} \frac{f(x)}{g(x)} = \lim_{x\to a} \frac{f'(x)}{g'(x)} = \lim_{x\to a} \frac{f''(x)}{g''(x)}.$$

One usually applies l'Hôpital's rule repeatedly *until a limit is obtained which is no longer of indeterminate-form type.*

Example 7 Computation of $\lim_{x\to 0} (x - \sin x)/x^3$ illustrates repeated application of l'Hôpital's rule. We have

$$\lim_{x\to 0} \frac{x - \sin x}{x^3} = \lim_{x\to 0} \frac{1 - \cos x}{3x^2} = \lim_{x\to 0} \frac{\sin x}{6x} = \lim_{x\to 0} \frac{\cos x}{6} = \frac{1}{6}. \quad \|$$

Warning *We emphasize that l'Hôpital's rule must not be applied to compute* $\lim_{x\to a} f(x)/g(x)$ *unless the quotient* $f(a)/g(a)$ *is an indeterminate form.* To illustrate,

$$\lim_{x\to 0} \frac{x^2}{\cos x} = \frac{0}{1} = 0,$$

and the following "l'Hôpital's-rule computation" is

$$\text{WRONG:} \quad \lim_{x\to 0} \frac{x^2}{\cos x} = \lim_{x\to 0} \frac{2x}{-\sin x} = \lim_{x\to 0} \frac{2}{-\cos x} = \frac{2}{-1} = -2.$$

There are many useful variants of l'Hôpital's rule. In particular, if $\lim_{x\to a} f(x) = \infty$ and $\lim_{x\to a} g(x) = \infty$, then

$$\lim_{x\to a} \frac{f(x)}{g(x)} = \lim_{x\to a} \frac{f'(x)}{g'(x)},$$

under suitable conditions on the functions f and g. You can also compute one-sided limits of $f(x)/g(x)$ as well as limits at ∞ or $-\infty$ by the l'Hôpital's-rule procedure with suitable conditions. Finally, if $\lim_{x\to a} f'(x)/g'(x) = \infty$, one can show that $\lim_{x\to a} f(x)/g(x) = \infty$, also under suitable conditions on f and g. We shall use these variants freely, and shall call them all "l'Hôpital's rule."

Example 8 Let's compute $\lim_{x\to\infty} x^{1/x}$.

SOLUTION Taking logarithms, we try to compute

$$\lim_{x\to\infty} \frac{1}{x}(\ln x) = \lim_{x\to\infty} \frac{\ln x}{x}.$$

This is a limit at ∞ of type ∞/∞, and we find by l'Hôpital's rule

$$\lim_{x\to\infty} \frac{\ln x}{x} = \lim_{x\to\infty} \frac{1/x}{1} = \frac{0}{1} = 0.$$

Since $\ln x^{1/x}$ approaches 0 at ∞, we must have

$$\lim_{x\to\infty} x^{1/x} = e^0 = 1. \quad \|$$

We emphasize again that prior to each application of l'Hôpital's rule, you must be sure that the limit to which it is applied corresponds to an indeterminate form of type 0/0 or ∞/∞.

Example 9 Let's find $\lim_{x\to 0+} x^2 e^{1/x}$.

SOLUTION This is a $0 \cdot \infty$-type form, and we write

$$\lim_{x\to 0+} x^2 e^{1/x} = \lim_{x\to 0+} \frac{x^2}{e^{-1/x}}$$

to convert it to a 0/0-type form. Then

$$\lim_{x\to 0+} \frac{x^2}{e^{-1/x}} = \lim_{x\to 0+} \frac{2x}{e^{-1/x} \cdot 1/x^2} = \lim_{x\to 0+} \frac{2x^3}{e^{-1/x}}.$$

This limit is worse; obviously we are going "the wrong way." Starting again and converting to an ∞/∞-type form, we have

$$\begin{aligned}
\lim_{x\to 0+} x^2 e^{1/x} &= \lim_{x\to 0+} \frac{e^{1/x}}{1/x^2} \\
&= \lim_{x\to 0+} \frac{e^{1/x}(-1/x^2)}{-2/x^3} = \lim_{x\to 0+} \frac{e^{1/x}}{2/x} \\
&= \lim_{x\to 0+} \frac{e^{1/x}(-1/x^2)}{-2/x^2} = \lim_{x\to 0+} \frac{e^{1/x}}{2} = \infty. \quad \|
\end{aligned}$$

Example 10 We repeat the previous example, but use the substitution $x = 1/t$.

SOLUTION We have

$$\lim_{x\to 0+} x^2 e^{1/x} = \lim_{t\to\infty} \frac{1}{t^2} e^t = \lim_{t\to\infty} \frac{e^t}{t^2} = \lim_{t\to\infty} \frac{e^t}{2t} = \lim_{t\to\infty} \frac{e^t}{2} = \infty.$$

Clearly, this is easier than the computation in Example 9. $\|$

SUMMARY

1. *Limits of sums, products, quotients, and exponentials involving two functions are called indeterminate forms if they lead formally to an expression of one of the types*

$$\frac{0}{0}, \quad \frac{\infty}{\infty}, \quad 0\cdot\infty, \quad \infty\cdot 0, \quad (-\infty)+\infty,$$
$$\infty-\infty, \quad 0^0, \quad 1^\infty, \quad 1^{-\infty}, \quad \infty^0.$$

2. *Product indeterminate forms are reduced to the quotient type by the algebraic device symbolized by*

$$0\cdot\infty = \frac{0}{1/\infty} = \frac{0}{0} \quad \textit{or} \quad 0\cdot\infty = \frac{\infty}{1/0} = \frac{\infty}{\infty}.$$

Exponential types are reduced to product types by taking a logarithm; as a symbolic illustration, $\ln(0^0) = 0(\ln 0) = 0(-\infty)$.

3. *Let f and g be analytic at a and let a_r and b_s be the first nonzero coefficients in the series for f and g, respectively, at a. Then*

$$\lim_{x\to a}\frac{f(x)}{g(x)} = \lim_{x\to a}\frac{a_r(x-a)^r}{b_s(x-a)^s}.$$

4. l'Hôpital's Rule: *Let f and g be continuous functions with derivatives in some neighborhood of a. Suppose, furthermore, that $f(a) = g(a) = 0$ and $g'(x) \neq 0$ for $x \neq a$ in that neighborhood. If $\lim_{x\to a}[f'(x)/g'(x)]$ exists, then $\lim_{x\to a}[f(x)/g(x)]$ exists, and $\lim_{x\to a}[f(x)/g(x)] = \lim_{x\to a}[f'(x)/g'(x)]$.*

5. *The l'Hôpital's Rule procedure can be used to find limits of ∞/∞-type forms as well as 0/0-type forms.*

EXERCISES

In Exercises 1 through 10, find the limit by series methods.

1. $\displaystyle\lim_{x\to 0}\frac{e^x-1}{x}$ **2.** $\displaystyle\lim_{x\to 0}\frac{\sin x - x}{e^{x^2}-1}$ **3.** $\displaystyle\lim_{x\to 0}\frac{\sin x^2}{e^x-1-x}$ **4.** $\displaystyle\lim_{x\to 0}\frac{x[x^2-\ln(1-x^2)]}{x-\sin x}$

5. $\displaystyle\lim_{x\to 0+}\frac{\ln(1-x)}{x^2}$ **6.** $\displaystyle\lim_{x\to 0}\frac{\sin x - x}{x\cos x - x}$ **7.** $\displaystyle\lim_{x\to 0}\left(\frac{1}{x}-\frac{1}{\sin x}\right)$ **8.** $\displaystyle\lim_{x\to 0}(\cot 2x^2)(\ln(1-x^2))$

9. $\displaystyle\lim_{x\to\infty}\left(1-\frac{1}{x^2}\right)^{\cot(1/x)}$ **10.** $\displaystyle\lim_{x\to 0+}\left(\frac{1}{1-x}\right)^{-1/x^2}$

In Exercises 11 through 20, find the limit by l'Hôpital's rule. Has to be quotient (ratio) a/b

11. $\displaystyle\lim_{x\to\infty}\frac{x}{e^x}$ **12.** $\displaystyle\lim_{x\to\infty}\frac{x^{20}}{e^x}$ **13.** $\displaystyle\lim_{x\to\infty}\frac{\ln x}{x}$ **14.** $\displaystyle\lim_{x\to\infty}\frac{(\ln x)^2}{x}$ **15.** $\displaystyle\lim_{x\to\infty}\frac{(\ln x)^{100}}{x}$

16. $\displaystyle\lim_{x\to 0+}x^x$ **17.** $\displaystyle\lim_{x\to 0+}x(\ln x)$ **18.** $\displaystyle\lim_{x\to 1}\frac{\ln x}{1-x}$ **19.** $\displaystyle\lim_{x\to\pi/2}\frac{\cos x}{x-(\pi/2)}$ **20.** $\displaystyle\lim_{x\to 3}\frac{x-\sqrt{3x}}{27-x^3}$

In Exercises 21 through 40, find the limit by any valid method.

21. $\lim_{x\to 0} \frac{\cos x}{x^2}$

22. $\lim_{x\to 0+} \frac{\ln(\sin x)}{\csc x}$

23. $\lim_{x\to 0} \frac{\sin x}{e^x - e^{-x}}$

24. $\lim_{x\to 0} \frac{\ln(1-x)}{e^x - e^{-x}}$

25. $\lim_{x\to 0} \frac{\sin x^2}{\cos^2 x - 1}$

26. $\lim_{x\to 0} \frac{\ln(\cos x)}{\sin^2 x}$

27. $\lim_{x\to\infty} \frac{e^{2x}}{e^x}$

28. $\lim_{x\to 1+} \frac{\ln(x^2-1)}{\ln(3x^2+3x-6)}$

29. $\lim_{x\to 1+} (1-x)[\ln(\ln x)]$

30. $\lim_{x\to 0+} \frac{\ln(e^x-1)}{\ln x}$

31. $\lim_{x\to(\pi/2)+} (\cos x)\ln\left(x - \frac{\pi}{2}\right)$

32. $\lim_{x\to\infty} \left(1 - \frac{1}{x}\right)^{2x}$

33. $\lim_{x\to\infty} \left(1 + \frac{1}{x}\right)^{x^2}$

34. $\lim_{x\to\pi/2} (\sin x)^{1/(\pi-2x)}$

35. $\lim_{x\to\infty} \left(\frac{3x+2}{2x}\right)^x$

36. $\lim_{x\to\infty} (1+x^2)^{1/x}$

37. $\lim_{x\to\infty} (x+e^x)^{3/x}$

38. $\lim_{x\to 0+} (\sec x)^{\cot x}$

39. $\lim_{x\to 0+} x^3 e^{1/x}$

40. $\lim_{x\to 1+} (x-1)^{\ln x}$

calculator exercises

Find the limit in the indicated exercise above using your calculator, rather than the methods of this section.

41. Exercise 1

42. Exercise 2

43. Exercise 9

44. Exercise 15 (This should be illuminating!)

45. Exercise 23

46. Exercise 37

11.5 THE BINOMIAL SERIES; COMPUTATIONS

11.5.1 The binomial series

The binomial theorem which you learned in high school states that for any numbers a and b and any positive integer n, you have

$$(a+b)^n = a^n + na^{n-1}b + \frac{n(n-1)}{2!}a^{n-2}b^2 + \cdots + \frac{n(n-1)\cdots(n-k+1)}{k!}a^{n-k}b^k + \cdots + b^n. \quad (1)$$

You can write (1) in the form

$$(a+b)^n = \sum_{k=0}^{n} \binom{n}{k} a^{n-k}b^k, \quad (2)$$

where we define the *binomial coefficients* $\binom{n}{k}$ by

$$\binom{n}{k} = \begin{cases} \dfrac{n(n-1)\cdots(n-k+1)}{k!} & \text{for} \quad k > 0, \\ 1 \quad \text{for} \quad k = 0. \end{cases} \tag{3}$$

If you let $a = 1$ and $b = x$ in (2), you obtain

$$(1 + x)^n = \sum_{k=0}^{n} \binom{n}{k} x^k. \tag{4}$$

The sum in (4) must (by uniqueness) be the Taylor series for $(1 + x)^n$ at $x_0 = 0$. You could also verify this directly by differentiation; you easily find that

$$D^k(1 + x)^n = n(n-1)\cdots(n-k+1)(1+x)^{n-k},$$

so

$$D^k(1 + x)^n \big|_{x=0} = n(n-1)\cdots(n-k+1).$$

Since n is a positive *integer*, $(1 + x)^n$ is a polynomial of degree n, so $D^k(1 + x)^n = 0$ for $k > n$. For n a positive integer, you thus expect $\binom{n}{k} = 0$ for $k > n$, and this is easily seen from (3); namely

$$\binom{n}{k} = \frac{n(n-1)\cdots(n-n)\cdots(n-k+1)}{k!} = \frac{0}{k!} = 0$$

for $k > n$.

We generalize (3), and define the *binomial coefficient* $\binom{p}{k}$ for any real number p and integer $k \geq 0$ to be given by

$$\binom{p}{k} = \begin{cases} \dfrac{p(p-1)\cdots(p-k+1)}{k!} & \text{for} \quad k > 0, \\ 1 \quad \text{for} \quad k = 0. \end{cases}$$

Note that if p is not a nonnegative integer, then no factor in the numerator of $\binom{p}{k}$ is zero, even if $k > p$. We are led to consider the series analogue of the sum (4) for any real number p.

Definition 11.3 For any p, the **binomial series for** $(1 + x)^p$ is

$$\sum_{k=0}^{\infty} \binom{p}{k} x^k = 1 + px + \frac{p(p-1)}{2!} x^2 + \cdots$$

$$+ \frac{p(p-1)\cdots(p-k+1)}{k!} x^k + \cdots \tag{5}$$

(We use k rather than n for summation index to avoid confusion with the use of n in Eq. (1).)

It is easily checked that (5) is the Taylor series for $(1+x)^p$ at $x_0 = 0$; you easily find that

$$\begin{aligned} D^k(1+x)^p\big|_{x=0} &= p(p-1)\cdots(p-k+1)(1+x)^{p-k}\big|_{x=0} \\ &= p(p-1)\cdots(p-k+1). \end{aligned}$$

We would like to find the radius of convergence of the binomial series (5), and to determine whether the series does represent $(1+x)^p$ in its interval of convergence. Of course, if p is a nonnegative integer, then the series contains only a finite number of terms with nonzero coefficients, has radius of convergence ∞, and represents $(1+x)^p$ by (4). We now suppose that p is not a nonnegative integer, and compute the ratio

$$\frac{\binom{p}{k+1}x^{k+1}}{\binom{p}{k}x^k} = \frac{([p(p-1)\cdots(p-k)]/(k+1)!)\,x^{k+1}}{([p(p-1)\cdots(p-k+1)]/k!)\,x^k} = \frac{p-k}{k+1}x.$$

We obtain

$$\lim_{k\to\infty}\left|\frac{\binom{p}{k+1}x^{k+1}}{\binom{p}{k}x^k}\right| = \lim_{k\to\infty}\left|\frac{p-k}{k+1}x\right| = |-x| = |x|.$$

The radius of convergence of the binomial series for p not a nonnegative integer is therefore 1; the series converges for $|x| < 1$ or for $-1 < x < 1$.

To show that the series (5) represents $(1+x)^p$ for $-1 < x < 1$, we could try to show that $\lim_{k\to\infty} E_k(x_1) = 0$ for $-1 < x_1 < 1$, but the following argument is less tedious. We set

$$f(x) = \sum_{k=0}^{\infty}\binom{p}{k}x^k \tag{6}$$

for $-1 < x < 1$. Then f is analytic throughout $-1 < x < 1$, and we may differentiate term by term to obtain

$$f'(x) = \sum_{k=0}^{\infty}\binom{p}{k}kx^{k-1} = \sum_{k=1}^{\infty}\frac{p(p-1)\cdots(p-k+1)}{(k-1)!}x^{k-1}. \tag{7}$$

From (6) and (7), you can easily verify that

$$p\cdot f(x) = (1+x)\cdot f'(x) \tag{8}$$

(see Exercise 31). From (8), we obtain

$$\int \frac{f'(x)}{f(x)}\,dx = \int \frac{p}{1+x}\,dx, \tag{9}$$

or, taking the antiderivatives,

$$\ln|f(x)| = p \cdot \ln|1+x| + C = \ln|1+x|^p + C \tag{10}$$

for $-1 < x < 1$. From the series (6), we see that $f(0) = 1$, so putting $x = 0$ in (10), we obtain

$$0 = \ln 1 = \ln(1+0)^p + C = \ln 1 + C = C.$$

Thus $C = 0$ and $\ln|f(x)| = \ln|1+x|^p$. Therefore, $|f(x)| = |1+x|^p$, so $f(x) = \pm(1+x)^p$. However, $f(0) = 1$, so the positive sign is appropriate, and

$$f(x) = (1+x)^p$$

for $-1 < x < 1$, which is what we wished to show. We summarize these results in a theorem.

Theorem 11.8 *If p is not a nonnegative integer, then the binomial series* $\sum_{k=0}^{\infty}\binom{p}{k}x^k$ *has radius of convergence* 1 *and represents* $(1+x)^p$ *for* $-1 < x < 1$. *For a nonnegative integer n, the (finite) binomial series* $\sum_{k=0}^{n}\binom{n}{k}x^k$ *converges for all x to* $(1+x)^n$.

Example 1 We have

$$\begin{aligned}(1+x)^{1/2} &= 1 + \tfrac{1}{2}x + \frac{(\frac{1}{2})(-\frac{1}{2})}{2!}x^2 + \frac{(\frac{1}{2})(-\frac{1}{2})(-\frac{3}{2})}{3!}x^3 + \cdots \\ &= 1 + \tfrac{1}{2}x - \frac{1}{2^2\cdot 2!}x^2 + \frac{3}{2^3\cdot 3!}x^3 - \cdots \\ &\quad + (-1)^{n-1}\frac{(3)(5)(7)\cdots(2n-3)}{2^n\cdot n!}x^n + \cdots,\end{aligned}$$

for $n \ge 1$ and for $-1 < x < 1$. Putting $x = \frac{1}{2}$ and using the terms of exponent ≤ 3, we obtain the estimate

$$\sqrt{\tfrac{3}{2}} \approx 1 + \tfrac{1}{4} - \tfrac{1}{32} + \tfrac{1}{128} \approx 1.2266.$$

When $x = \frac{1}{2}$, our series is alternating with terms of decreasing size, so the error in our estimate is less than the next term $5/2^{11} \approx 0.0024$. The actual value of $\sqrt{\frac{3}{2}}$ to six decimal places is 1.224745. ‖

In finding a decimal estimate as in Example 1, there are two types of errors that may occur. There is a *truncation error* due to estimating a series by a partial sum; this error can be reduced by taking more terms of the series. Also, there is the *round-off error* which occurs when you take decimal estimates for terms in the series or for a partial sum; e.g., you may take 0.333333 as an estimate for $\frac{1}{3}$. The round-off error can be reduced by taking more decimal places. When using decimal estimates for individual terms of a sum, the round-off error can easily accumulate to a point where it actually exceeds the truncation error.

11.5.2 Estimating integrals

Consider a power series $\sum_{n=0}^{\infty} a_n(x - x_0)^n$ with radius of convergence $r > 0$, and let f be the function represented by the series in its interval of convergence. From its very nature as an "infinite polynomial" in $\Delta x = x - x_0$, we should think of our series as describing f *locally*, near x_0. We expect a partial sum of the series to give a good approximation to f provided that we are working sufficiently close to x_0. However, a partial sum s_n for n sufficiently large may be a reasonably good approximation to f in a moderately large interval containing x_0. The accuracy of the approximation throughout an interval might be determined by Taylor's theorem or by an examination of the series.

Suppose we wish to compute $\int_a^b f(x)\,dx$. Perhaps it is hard to find an elementary function that is an antiderivative of f, even if f is a known elementary function. For example, it can be shown that no antiderivative of e^{-x^2} is an elementary function. We might attempt to compute $\int_a^b f(x)\,dx$ by taking a power series $\sum_{n=0}^{\infty} a_n(x - x_0)^n$ that converges to f in $[a, b]$, and computing $\int_a^b [\sum_{n=0}^{\infty} a_n(x - x_0)^n]\,dx$ instead. The theory of power series shows that this integral can be computed by term-by-term integration. As another alternative, we might just use a partial sum s_n of the series to estimate f in $[a, b]$, and estimate $\int_a^b f(x)\,dx$ by $\int_a^b s_n(x)\,dx$. In this case, we would like a bound for our error. If f and g are any continuous functions defined on $[a, b]$, it follows easily from the definition of the definite integral that

$$\text{if} \quad |f(x) - g(x)| < \epsilon \quad \text{for all} \quad x \text{ in } [a, b],$$

$$\text{then} \quad \left|\int_a^b f(x)\,dx - \int_a^b g(x)\,dx\right| < \epsilon(b - a)$$

(see Exercise 32). In the examples that follow, we illustrate both estimation by integrating a series and estimation by integrating a partial sum.

Example 2 Let's estimate $\int_0^1 e^{-x^2}\,dx$ by integrating a series.

SOLUTION Replacing x by $-x^2$ in the well-known series for e^x, we find that

$$e^{-x^2} = 1 - x^2 + \frac{x^4}{2!} - \frac{x^6}{3!} + \cdots + (-1)^n \frac{x^{2n}}{n!} + \cdots$$

for all x. Thus

$$\int_0^1 e^{-x^2}\,dx = \int_0^1 \left(1 - x^2 + \frac{x^4}{2!} - \frac{x^6}{3!} + \frac{x^8}{4!} - \cdots + (-1)^n \frac{x^{2n}}{n!} + \cdots\right) dx$$

$$= \left(x - \frac{x^3}{3} + \frac{x^5}{5\cdot 2!} - \frac{x^7}{7\cdot 3!} + \frac{x^9}{9\cdot 4!} - \cdots + \frac{(-1)^n x^{2n+1}}{(2n+1)n!} + \cdots\right)\Bigg]_0^1$$

$$= 1 - \frac{1}{3} + \frac{1}{5\cdot 2!} - \frac{1}{7\cdot 3!} + \frac{1}{9\cdot 4!} - \cdots + \frac{(-1)^n}{(2n+1)n!} + \cdots$$

Our answer is the sum of an infinite series of constants, and may in turn be approximated by a partial sum of the series. Since the series is alternating with terms of decreasing size, the error in approximating this series by a partial sum is less than the size of the next term. Thus

$$\int_0^1 e^{-x^2}\,dx \approx 1 - \frac{1}{3} + \frac{1}{5\cdot 2!} - \frac{1}{7\cdot 3!} + \frac{1}{9\cdot 4!} \approx 0.7475,$$

with error less than

$$\frac{1}{11\cdot 5!} \approx 0.0008. \quad \|$$

Example 3 We know that

$$\int_0^{1/2} (1 - x^2)^{-1/2}\,dx = \sin^{-1}x\Big]_0^{1/2} = \sin^{-1}\frac{1}{2} - \sin^{-1}0 = \frac{\pi}{6} - 0 = \frac{\pi}{6}.$$

Let's integrate a partial sum of the binomial series representing $(1 - x^2)^{-1/2}$ to estimate $\pi/6$, and then $\pi = 6(\pi/6)$.

SOLUTION We have

$$(1 - x^2)^{-1/2} = 1 + (-\tfrac{1}{2})(-x^2) + \frac{(-\frac{1}{2})(-\frac{3}{2})}{2!}(-x^2)^2$$

$$+ \frac{(-\frac{1}{2})(-\frac{3}{2})(-\frac{5}{2})}{3!}(-x^2)^3 + \cdots$$

$$= 1 + \frac{1}{2}x^2 + \frac{3}{2^2\cdot 2!}x^4 + \frac{3\cdot 5}{2^3\cdot 3!}x^6$$

$$+ \left(\frac{3\cdot 5\cdot 7}{2^4\cdot 4!}x^8 + \frac{3\cdot 5\cdot 7\cdot 9}{2^5\cdot 5!}x^{10} + \cdots\right)$$

for $-1 < x < 1$. The portion of the series which we have placed in parentheses has all coefficients ≤ 1, and is therefore, term for term, less than the series

$$x^8 + x^{10} + \cdots = x^8(1 + x^2 + x^4 + \cdots).$$

Since we will be integrating for $0 \le x \le \frac{1}{2}$, we have

$$x^8(1 + x^2 + x^4 + \cdots) \le \frac{1}{2^8}\left(1 + \frac{1}{4} + \frac{1}{4^2} + \cdots\right)$$
$$= \frac{1}{2^8} \cdot \frac{1}{1 - (\frac{1}{4})} = \frac{1}{2^8} \cdot \frac{4}{3} = \frac{1}{192}.$$

Thus we have

$$\frac{\pi}{6} = \int_0^{1/2} (1 - x^2)^{-1/2}\,dx \approx \int_0^{1/2} \left(1 + \frac{1}{2}x^2 + \frac{3}{8}x^4 + \frac{15}{48}x^6\right) dx$$
$$= \left(x + \tfrac{1}{6}x^3 + \tfrac{3}{40}x^5 + \tfrac{15}{336}x^7\right)\Big]_0^{1/2}$$
$$= \tfrac{1}{2} + \tfrac{1}{48} + \tfrac{3}{1280} + \tfrac{15}{43008} \approx 0.52353,$$

with error at most

$$\tfrac{1}{192}(\tfrac{1}{2} - 0) = \tfrac{1}{384} \approx 0.00260.$$

Therefore,

$$\pi = 6\left(\frac{\pi}{6}\right) \approx 6(0.52353) = 3.14118$$

with error at most $6(0.00260) = 0.01560$. Since the value of π to five decimal places is 3.14159, our actual error in estimating π was only about 0.00041. The reason we obtained such a crude bound for the error is that we estimated the series $x^8 + x^{10} + \cdots$ throughout $[0, \frac{1}{2}]$ by its greatest value at $x = \frac{1}{2}$. ||

11.5.3 More computations

Tables of values for the trigonometric, logarithmic, and exponential functions, with which you are familiar, were computed from series. Indeed, it is more efficient for the large, modern electronic computer to find these values by computing a partial sum of a series than to use space in its memory to store tables. Series have also been used to compute important numbers such as π and e to a large number of decimal places. You may find a few illustrations interesting.

Equation (4) of Section 11.3.2 shows that

$$\ln(1 - x) = -x - \frac{x^2}{2} - \frac{x^3}{3} - \cdots - \frac{x^n}{n} - \cdots$$

for $-1 < x < 1$. Replacing x by $-x$, we obtain

$$\ln(1 + x) = x - \frac{x^2}{2} + \frac{x^3}{3} - \cdots + (-1)^{n+1}\frac{x^n}{n} + \cdots$$

for $-1 < x < 1$. Using the identity

$$\ln \frac{1+x}{1-x} = \ln(1+x) - \ln(1-x),$$

we find that

$$\ln \frac{1+x}{1-x} = 2\left(x + \frac{x^3}{3} + \frac{x^5}{5} + \cdots + \frac{x^{2n-1}}{2n-1} + \cdots\right) \tag{11}$$

for $-1 < x < 1$. If we let $y = (1+x)/(1-x)$, we see from Fig 11.1 that as x ranges through $-1 < x < 1$, y takes on all values >0. Thus (11) may be used to compute tables of the function ln. From $y = (1+x)/(1-x)$, we find that $x = (y-1)/(y+1)$; thus to compute $\ln y$, we may substitute the value $x = (y-1)/(y+1)$ in the series (11). Since the series converges most rapidly for x close to 0, where y is near 1, we could first compute ln 2 by putting $x = \frac{1}{3}$, then $\ln(\frac{3}{2})$, $\ln(\frac{4}{3})$, etc. Once we know $\ln[(n+1)/n]$ for as many values of n as desired, we could easily find $\ln 3 = \ln(\frac{3}{2}) + \ln 2$, $\ln 4 = \ln(\frac{4}{3}) + \ln 3$, and in general

$$\ln(n+1) = \ln \frac{n+1}{n} + \ln n.$$

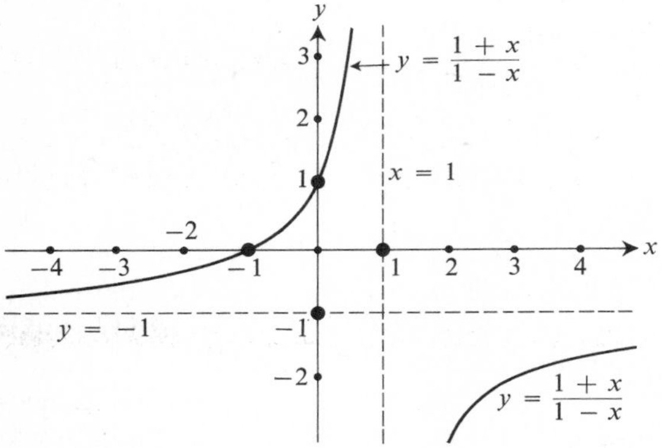

11.1

Since any number r can be approximated by a quotient n/m of integers and $\ln(n/m) = \ln n - \ln m$, we could then estimate $\ln r$ for any $r > 0$.

Turning to the computation of π, Eq. (8) in Section 11.3.3 shows that

$$\tan^{-1} x = x - \frac{x^3}{3} + \frac{x^5}{5} - \frac{x^7}{7} + \cdots + (-1)^n \frac{x^{2n+1}}{2n+1} + \cdots \tag{12}$$

for $-1 < x < 1$. This series converges (by the alternating series test) for $x = 1$, and it can be shown that the sum of the series at $x = 1$ must be

$$\lim_{x\to 1} \tan^{-1}x = \tan^{-1}1 = \frac{\pi}{4}.$$

Thus we obtain the formula

$$\frac{\pi}{4} = 1 - \frac{1}{3} + \frac{1}{5} - \frac{1}{7} + \cdots + (-1)^n \frac{1}{2n+1} + \cdots$$

Of course, π can then be computed by multiplying by 4. The series (12) does not converge very rapidly at $x = 1$. Using identities for circular functions, one can show that

$$\frac{\pi}{4} = 4\tan^{-1}\frac{1}{5} - \tan^{-1}\frac{1}{239}; \tag{13}$$

the series (12) converges quite rapidly at $x = \frac{1}{5}$ and $x = \frac{1}{239}$. Relation (13) has been used to calculate π to 100,000 decimal places. (See Wrench and Shanks, "Calculation of π to 100,000 Decimals," *Mathematics of Computation*, **16,** No. 77, Jan. 1962, pp. 76–99.)

SUMMARY

1. *If p is not a nonnegative integer, then the binomial series $\sum_{k=0}^{\infty} \binom{p}{k} x^k$ has radius of convergence 1 and represents $(1 + x)^p$ for $-1 < x < 1$. For a nonnegative integer n, the (finite length) binomial series $\sum_{k=0}^{n} \binom{n}{k} x^k$ converges for all x to $(1 + x)^n$.*
2. *Integrals $\int_a^b f(x)\,dx$ such as $\int_a^b e^{-x^2}\,dx$, which cannot be found in the usual way, can sometimes be estimated by taking a series expression for $f(x)$ and integrating the series.*

EXERCISES

In Exercises 1 through 12, compute the binomial coefficient.

1. $\binom{4}{2}$ **2.** $\binom{7}{3}$ **3.** $\binom{8}{7}$ **4.** $\binom{10}{0}$ **5.** $\binom{5}{5}$ **6.** $\binom{5}{7}$

7. $\binom{3.5}{3}$ **8.** $\binom{-1.5}{2}$ **9.** $\binom{7.5}{0}$ **10.** $\binom{0.5}{4}$ **11.** $\binom{1/3}{3}$ **12.** $\binom{-4}{4}$

13. What is the Taylor series for $(1 + x)^0$ at $x_0 = 0$? Check that the coefficient of x^k in this series is indeed the binomial coefficient $\binom{0}{k}$ as defined in Section 11.5.1.

In Exercises 14 through 19, find the first five terms of the binomial series which represents the given function in a neighborhood of 0. Give the radius of convergence in each case.

14. $(1 + x)^{3/2}$ **15.** $(1 + x)^{-2}$ **16.** $(1 - x)^{1/2}$

17. $(1 - x^2)^{1/3}$

18. $\left(1 + \dfrac{x}{2}\right)^{2/3}$

19. $(1 + 3x)^{5/3}$

20. Estimate $\sqrt{2}$ by finding a partial sum with $x = \frac{1}{2}$ of a binomial series which represents $(1 - x)^{-1/2}$ in a neighborhood of the origin.

In Exercises 21 *through* 24, *indicate how you could estimate the given quantity by using a suitable binomial series.* (*You need not compute an estimate.*) *To illustrate, Exercise* 20 *shows how* $\sqrt{2}$ *could be estimated* (*rather inefficiently*) *using a binomial series.*

21. $\sqrt{2/3}$

22. $\sqrt{5}$

23. $\sqrt[3]{2}$

24. $\sqrt[4]{17}$

In Exercises 25 *through* 30, *estimate the integral by series methods, and find a bound for your error.* (*You need not simplify your answers, or put them in decimal form.*)

25. $\displaystyle\int_0^1 \sin x^2\, dx$

26. $\displaystyle\int_0^{1/2} e^{x^2}\, dx$

27. $\displaystyle\int_0^1 x^3 \cos x^2\, dx$

28. $\displaystyle\int_0^{1/2} (1 + x^2)^{1/3}\, dx$

29. $\displaystyle\int_0^{0.1} e^{-x^2}\, dx$

30. $\displaystyle\int_0^1 \sqrt{16 - x^4}\, dx$ $\left[\textit{Hint.}\ \sqrt{16 - x^4} = 4\sqrt{1 - \dfrac{x^4}{16}}.\right]$

31. Verify Eq. (8) of Section 11.5.1.

32. Use the definition of $\int_a^b f(x)\, dx$ to show that if f and g are continuous in $[a, b]$ and $|f(x) - g(x)| < \epsilon$ for all x in $[a, b]$, then

$$\left|\int_a^b f(x)\, dx - \int_a^b g(x)\, dx\right| < \epsilon(b - a).$$

calculator exercises

33. Find the absolute numerical difference between the estimate for $\int_0^1 e^{x^2}\, dx$ using Simpson's Rule with $n = 20$ and integrating the first eight nonzero terms of the series for e^{x^2}.

34. Repeat Exercise 33 for $\int_0^1 \cos x^2\, dx$, but using the first five nonzero terms of the series for $\cos x^2$.

35. Using the relation (13) and the first five nonzero terms of the series (12), find the estimate obtained for π.

exercise sets for chapter 11

review exercise set 11.1

1. Find the interval of convergence, including endpoints, of the series

$$\sum_{n=1}^{\infty} \frac{n}{(n^2 + 1) \cdot 3^n}(x + 5)^n.$$

2. Find an expression for the nth term $a_n(x - x_0)^n$ of a power series having as interval of convergence $-1 \le x < 3$. (Many answers are possible.)

3. Find the fourth Taylor polynomial $T_4(x)$ for $f(x) = \sin x$ at $x_0 = \pi/3$.

4. Approximate $\ln(1.04)$ using the Taylor polynomial $T_3(x)$ for $\ln x$ at $x_0 = 1$, and find a bound for the error.

5. a) Define the notion of an analytic function of a real variable.
   b) Find the Taylor series for $x/(1 - x^2)$ at $x_0 = 0$ in the easiest way you can.
6. Use series long division to find the first three terms of the series for $(\sin x)/e^x$ at $x_0 = 0$.
7. Use series methods to find

$$\lim_{x\to 0} \frac{e^{x^3} - 1}{x - \sin x}.$$

8. Find $\lim_{x\to 0+} (1 + (2/\sin x))^x$.
9. Use the binomial series expansion to give the first five nonzero terms of the series for $\sqrt{1 + x^2}$ at $x_0 = 0$.
10. Use series techniques to find $\int_0^1 \sin x^2\, dx$ with error at most 0.001.

review exercise set 11.2

1. Find the interval of convergence, including endpoints, of the series

$$\sum_{n=1}^{\infty} \frac{(2x + 1)^{2n}}{n^2 \cdot 3^n}.$$

2. Find the radius of convergence of the series

$$\sum_{n=1}^{\infty} \frac{(x - 4)^{2n+1}}{3^{4n-3}}.$$

3. Find the fourth Taylor polynomial $T_4(x)$ for $f(x) = \cos 2x$ at $x_0 = \pi/4$.
4. Approximate $\sinh(0.1)$ using the Taylor polynomial $T_4(x)$ at $x_0 = 0$, and find a bound for the error.
5. Find the Taylor series for $x^3 \tan^{-1} x^2$ at $x_0 = 0$ in the easiest way you can.
6. Use series multiplication to find the first four nonzero terms of the Taylor series for $e^x \sin 2x$ at $x_0 = 0$.
7. Use series methods to find

$$\lim_{x\to 0} \frac{\cos x^2 - 1}{x \sin x^3}.$$

8. Find $\lim_{x\to\infty} [(2x + 3)/2x]^x$.
9. Give the first four terms of the binomial series representing $(1 - x/2)^{-1/2}$ in a neighborhood of $x_0 = 0$.
10. Use series methods to estimate

$$\int_0^1 \frac{1 - \cos x^2}{x}\, dx,$$

with error at most 0.001.

more challenging exercises 11

In Exercises 1 through 10, find the radius of convergence of the series.

1. $\displaystyle\sum_{n=1}^{\infty} \frac{n^n}{n!} x^n$
2. $\displaystyle\sum_{n=1}^{\infty} \frac{n^n}{e^{2n} \cdot n!} x^n$
3. $\displaystyle\sum_{n=1}^{\infty} \frac{n^n}{(n!)^2} x^n$
4. $\displaystyle\sum_{n=1}^{\infty} (1 + (-1)^n) x^n$

5. $\sum_{n=1}^{\infty} (1 + (-1)^n)^n x^n$

6. $\sum_{n=1}^{\infty} (2 + (-1)^n 7)^n x^{2n}$

7. $\sum_{n=1}^{\infty} \frac{(2n)!}{n^n} x^n$

8. $\sum_{n=1}^{\infty} \frac{(2n)!}{(n!)^2} x^n$

9. $\sum_{n=1}^{\infty} \frac{2 \cdot 5 \cdot 8 \cdots (3n - 1)}{3 \cdot 7 \cdot 11 \cdot \cdots \cdot (4n - 1)} x^n$

10. $\sum_{n=1}^{\infty} \frac{4 \cdot 7 \cdot 10 \cdots (3n + 1)}{n^n} x^n$

plane curves

12

Sections 12.1 through 12.4 give a treatment of ellipses, hyperbolas, and parabolas, complete with foci, directrices, eccentricities, vertices, major axes, and rotation of axes. Some instructors may feel that this material is not essential for a calculus course. They may decide to assign it as outside reading, if at all, and spend the class time saved on subsequent chapters. We have tried to write these sections to provide as much flexibility as possible, so that you may spend anywhere from zero to four lessons on this material.

1 LESSON:	Do only Section 12.1 on sketching with translation of axes. This is useful for sketching quadric surfaces later.
2 LESSONS:	Do Section 12.1, and either Section 12.2 on synthetic definitions or Section 12.3 on rotation of axes.
3 LESSONS:	Do Sections 12.1, 12.2, and 12.3.
4 LESSONS:	All four sections.

Section 12.4 on applications is especially suitable for outside reading, and we recommend that it be left as such.

12.1 SKETCHING CONIC SECTIONS

Let two congruent right circular cones in space be placed vertex to vertex to form a double cone, as shown in Fig. 12.1. A plane can intersect this cone in three types of curves. Figure 12.1(a) shows the closed-curve type of intersection, which is an *ellipse*. Figure 12.1(b) shows the intersection giving a two-piece, open-ended curve, which is a *hyperbola*. The one-piece, open-ended curve in Fig. 12.1(c) is a *parabola*. The double-lettered figures are computer-drawn surfaces.

In this section, we sketch these curves, starting with their equations with respect to carefully chosen x- and y-axes in the plane of intersection.

(aa)

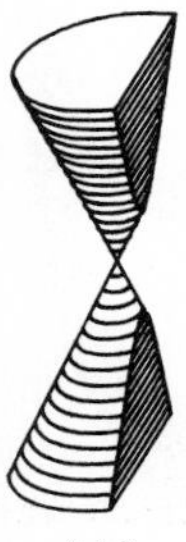

(bb)

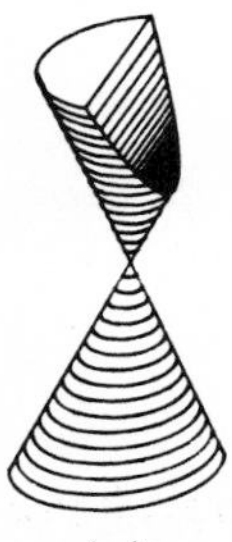

(cc)

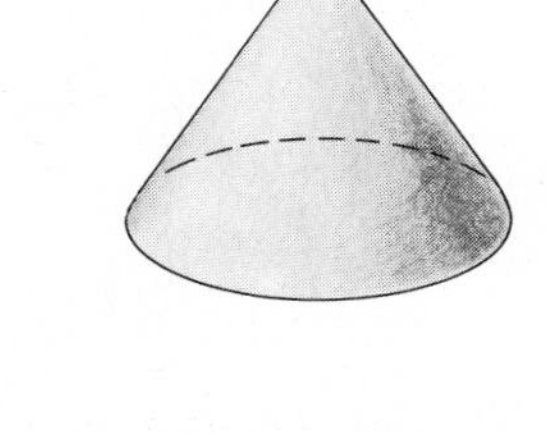

(a) Elliptic section

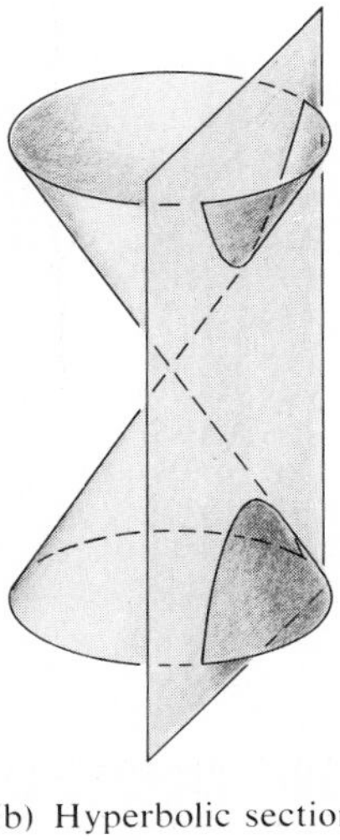

(b) Hyperbolic section

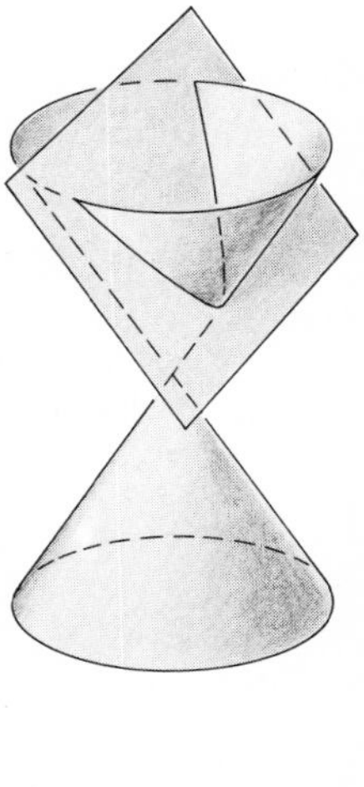

(c) Parabolic section **12.1**

12.1.1 The ellipse The equation

$$\frac{x^2}{a^2} + \frac{y^2}{b^2} = 1 \tag{1}$$

describes an *ellipse*, shown in Fig. 12.2. Setting $x = 0$, you see that the curve meets the y-axis at b and $-b$. Setting $y = 0$, you find that the x-intercepts are a and $-a$. If $a = b$, the ellipse becomes a circle. By translation of axes, you see that

$$\frac{(x-h)^2}{a^2} + \frac{(y-k)^2}{b^2} = 1 \tag{2}$$

is an ellipse with center at (h, k) rather than at the origin.

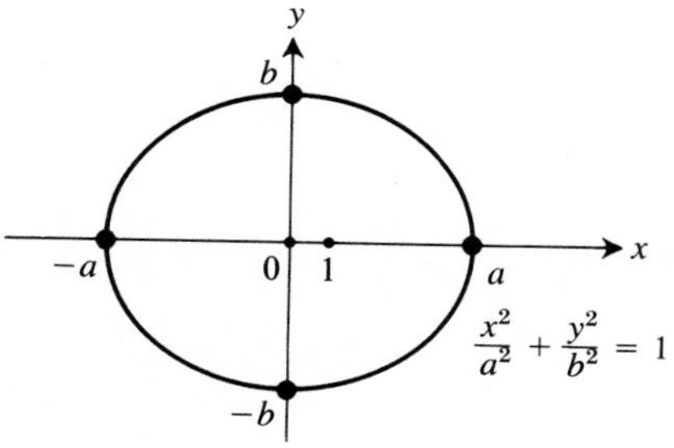

12.2

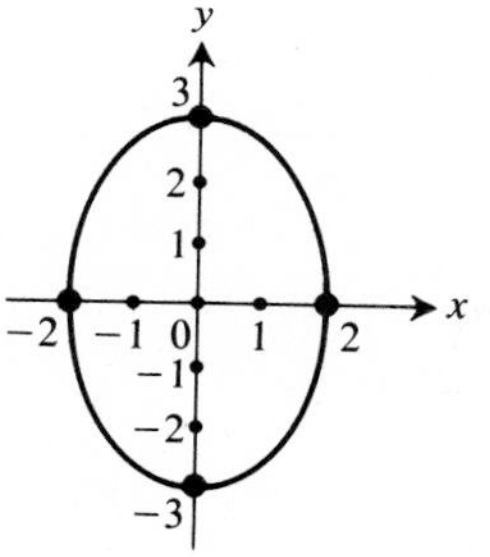

12.3

Example 1 Let's sketch the curve $9x^2 + 4y^2 = 36$.

SOLUTION Dividing through by 36, we obtain

$$\frac{x^2}{4} + \frac{y^2}{9} = 1,$$

which is the ellipse shown in Fig. 12.3. The x-intercepts are $\pm\sqrt{4} = \pm 2$, and the y-intercepts are $\pm\sqrt{9} = \pm 3$. ‖

Example 2 We sketch the curve with equation

$$x^2 + 3y^2 - 4x + 6y = -1.$$

SOLUTION Completing the square we obtain

$$(x^2 - 4x) + 3(y^2 + 2y) = -1,$$

$$(x-2)^2 + 3(y+1)^2 = 4 + 3 - 1 = 6,$$

$$\frac{(x-2)^2}{6} + \frac{(y+1)^2}{2} = 1,$$

which is of the form (2). This is the equation of an ellipse with center $(2, -1)$, as shown in Fig. 12.4. ‖

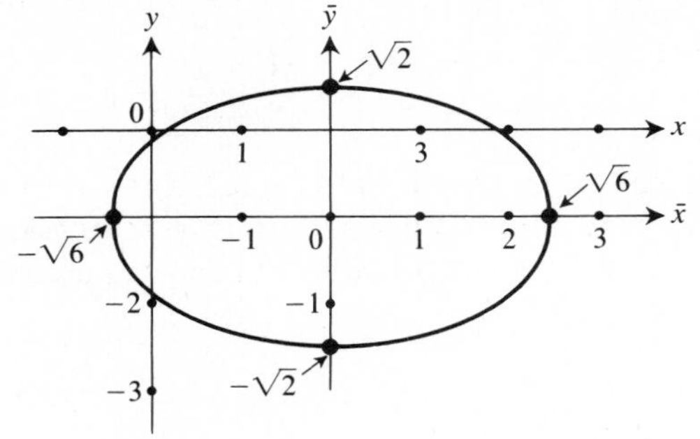

12.4

12.1.2 The hyperbola

The equation

$$\frac{x^2}{a^2} - \frac{y^2}{b^2} = 1 \tag{3}$$

describes a hyperbola, shown in Fig. 12.5. To sketch this hyperbola, proceed as follows. Mark $\pm a$ on the x-axis and $\pm b$ on the y-axis. Then draw the rectangle crossing the axes at those points, and draw and extend the diagonals of that rectangle. The hyperbola has these diagonals

$$y = \pm\frac{b}{a}x$$

as *asymptotes*. The hyperbola crosses the x-axis at $\pm a$, but does not meet the y-axis, for if $x = 0$, then (3) reduces to $y^2 = -b^2$, which has no real solutions. If you solve (3) for y, you obtain

$$y = \pm b\sqrt{(x^2/a^2) - 1}$$

and

$$\lim_{x\to\infty}\left[b\sqrt{(x^2/a^2) - 1} - \frac{b}{a}x\right] = 0,$$

which shows that the lines $y = \pm(b/a)x$ are indeed asymptotes of the curve.

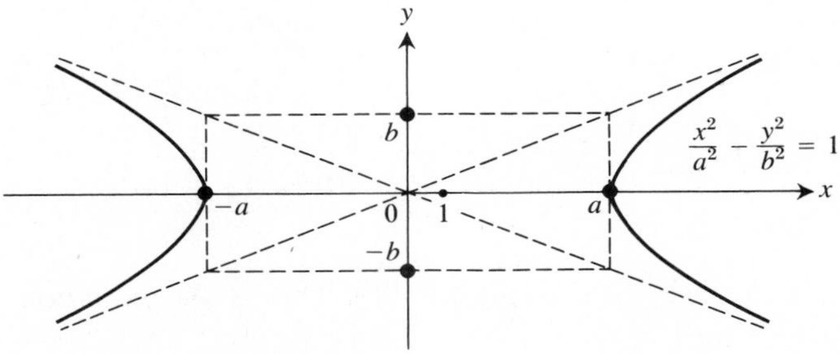

12.5

If the negative sign in (3) appears with the other term x^2/a^2 instead, as in

$$-\frac{x^2}{a^2}+\frac{y^2}{b^2}=1, \tag{4}$$

then the asymptotes are still $y = \pm(b/a)x$, but now the hyperbola crosses the y-axis at $\pm b$ and does not meet the x-axis, as shown in Fig. 12.6.

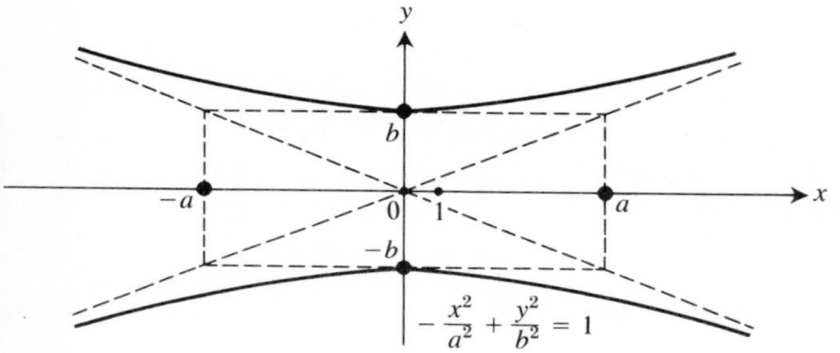

12.6

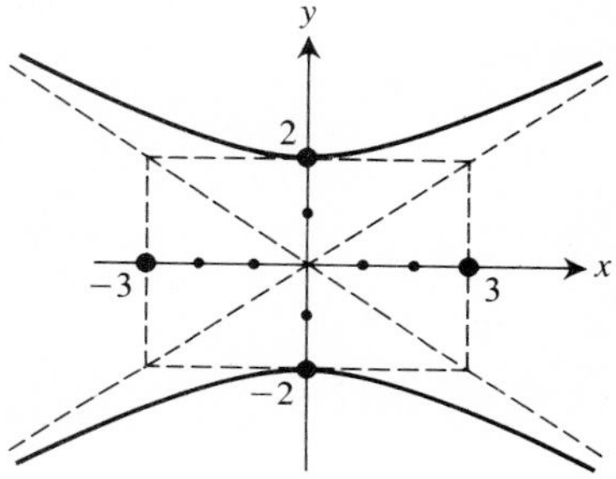

12.7

Example 3 The hyperbola

$$\frac{y^2}{4}-\frac{x^2}{9}=1$$

is shown in Fig. 12.7. ||

The device of completing the square again can be used to sketch a hyperbola whose center is not at (0, 0).

Example 4 Let us sketch $2x^2 - 3y^2 - 4x - 6y = 13$.

SOLUTION Completing the square, we obtain:

$$2(x^2 - 2x) - 3(y^2 + 2y) = 13,$$
$$2(x - 1)^2 - 3(y + 1)^2 = 2 - 3 + 13 = 12,$$
$$\frac{(x-1)^2}{6}-\frac{(y+1)^2}{4}=1,$$

which is the hyperbola with center at $(1, -1)$ and crossing the $\bar{x}$-axis at $\pm\sqrt{6}$, sketched in Fig. 12.8. ||

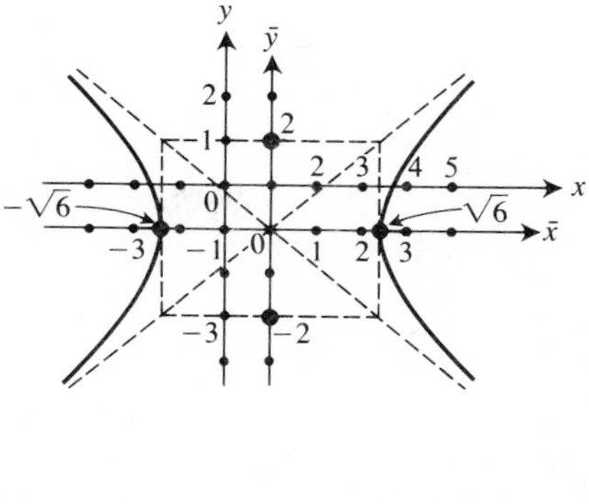

12.8

12.1.3 The parabola

The curves

$$y = 4x^2, \qquad y = -x^2, \qquad x = \tfrac{1}{4}y^2, \qquad x = -4y^2,$$

are parabolas with vertices at the origin, as shown in Fig. 12.9. The size of the coefficient of the quadratic term controls how fast the parabola "opens." If the vertex of $y = 4x^2$ were moved to (h, k), the equation of the curve would become

$$y - k = 4(x - h)^2.$$

As indicated in Chapter 1, any polynomial equation in x and y that is quadratic in one of the variables and linear in the other describes a parabola.

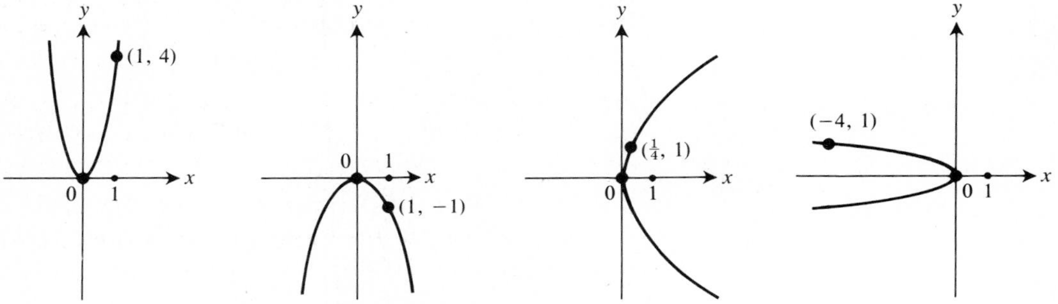

a) $y = 4x^2$ b) $y = -x^2$ c) $x = \frac{1}{4}y^2$ d) $x = -4y^2$

12.9

Example 5 Let us sketch

$$3y + 4x^2 + 8x = 5.$$

SOLUTION Completing the square yields

$$\begin{aligned} 3y + 4(x^2 + 2x) &= 5, \\ 3y + 4(x + 1)^2 &= 4 + 5 = 9, \\ 3y - 9 &= -4(x + 1)^2, \\ 3(y - 3) &= -4(x + 1)^2, \\ y - 3 &= -\tfrac{4}{3}(x + 1)^2. \end{aligned}$$

This is the parabola with vertex at $(-1, 3)$ shown in Fig. 12.10. ||

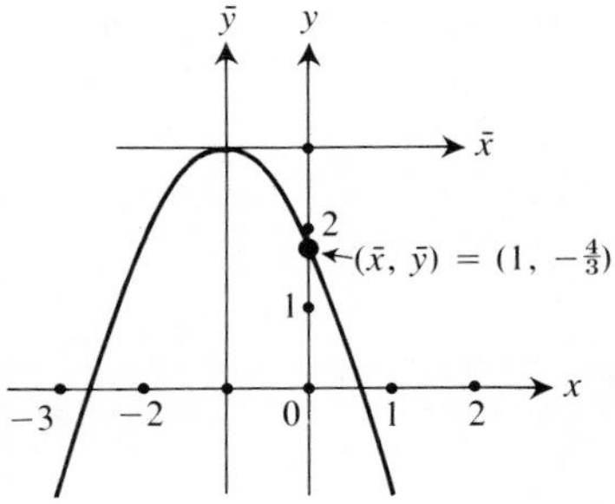

12.10

SUMMARY

1. *An equation of the form*

$$\frac{(x-h)^2}{a^2}+\frac{(y-k)^2}{b^2}=1$$

describes an ellipse with center at (h, k), meeting the $\bar{x}$-axis at $\pm a$ and the $\bar{y}$-axis at $\pm b$, where $\bar{x} = x - h$ and $\bar{y} = y - k$.

2. *Equations of the form*

$$\frac{(x-h)^2}{a^2}-\frac{(y-k)^2}{b^2}=1$$

or

$$-\frac{(x-h)^2}{a^2}+\frac{(y-k)^2}{b^2}=1$$

describe hyperbolas with center at (h, k). To sketch them, draw the rectangle with center (h, k) crossing at right angles the $\bar{x}$-axis at $\pm a$ and the $\bar{y}$-axis at $\pm b$. Draw and extend the diagonals of the rectangle. The first hyperbola crosses the $\bar{x}$-axis at $\pm a$ while the second hyperbola crosses the $\bar{y}$-axis at $\pm b$. Both hyperbolas have the diagonals of the rectangle as asymptotes.

3. *Equations of the form*

$$y-k=c(x-h)^2$$

and

$$x-h=c(y-k)^2$$

describe parabolas with vertices at (h, k). The magnitude $|c|$ controls how fast the parabola "opens," and the sign of c determines the direction in which it opens.

EXERCISES

In Exercises 1 through 12, sketch the given curve.

1. $x^2 + 4y^2 = 16$

2. $x^2 - 4y^2 = 16$

3. $x^2 = -4y$

4. $x = 4y^2$

5. $4x^2 - 9y^2 = -36$

6. $5x^2 + 2y^2 = 50$

7. $2x^2 - 3y^2 + 4x + 12y = 0$

8. $4x^2 - 8x + 2y = 5$

9. $x^2 + y^2 - 4y = 9$

10. $4x^2 + y^2 + 6y = -5$

11. $3x^2 - 4y^2 - 6x - 8y = 0$

12. $-x^2 + 2y^2 + 2x + 8y = -1$

12.2 SYNTHETIC DEFINITIONS OF CONIC SECTIONS

12.2.1 The parabola

Definition 12.1 A **parabola** is the locus in the plane of all points equidistant from a fixed line ℓ (the **directrix**) and a fixed point F (the **focus**) not on the line. The point V on the parabola closest to the directrix is the **vertex** of the parabola.

Figure 12.11 shows a parabola with focus F, directrix ℓ, and vertex V. To describe the locus analytically, we choose axes through the vertex, as indicated in Fig. 12.12, so that the focus F has coordinates $(p, 0)$ and the directrix is the line $x = -p$. If (x, y) is on the parabola, then the definition states that

$$\sqrt{(x-p)^2 + y^2} = x + p.$$

Simplifying, we obtain

$$(x-p)^2 + y^2 = (x+p)^2,$$
$$x^2 - 2xp + p^2 + y^2 = x^2 + 2xp + p^2,$$
$$y^2 = 4px. \qquad (1)$$

Equation (1) is the *standard form* for the equation of the parabola.

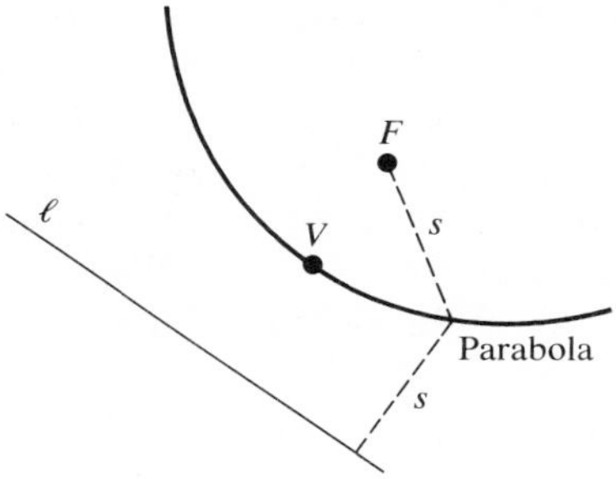

12.11

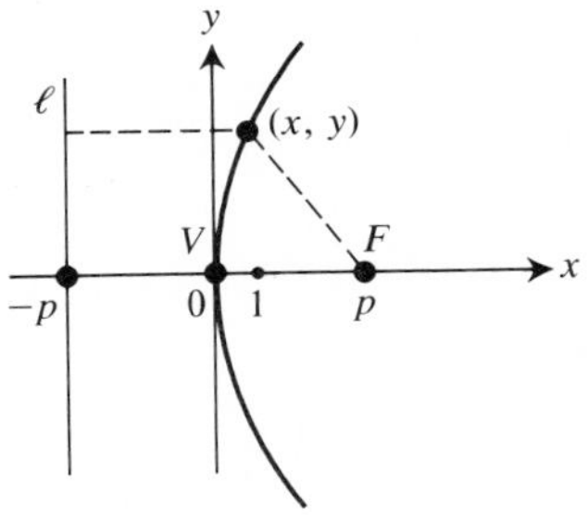

12.12

Example 1 Let's find the focus and directrix for $y^2 = 10x$.

SOLUTION Here, $4p = 10$ so $p = 5/2$. The focus of the parabola is $(5/2, 0)$ and the directrix is the line $x = -5/2$. We do not draw a sketch, for you learned to sketch a parabola in Section 12.1. ||

Of course $x^2 = 4py$ is the parabola "opening upward" with focus $(0, p)$ and directrix $y = -p$. Similarly, $x^2 = -4py$ is the parabola "opening downward" with focus $(0, -p)$ and directrix $y = p$.

Our work in Section 12.1 shows that an equation in x and y that is quadratic in one variable and linear in the other has as locus a parabola. It may be necessary to complete a square and translate axes to bring the equation into standard form.

Example 2 We shall sketch $x^2 + 4x + 3y = 2$.

SOLUTION Completing the square, we obtain

$$(x+2)^2 + 3y = 2 + 4 = 6, \qquad (x+2)^2 = -3y + 6,$$
$$(x+2)^2 = -3(y-2), \qquad \bar{x}^2 = -4(\tfrac{3}{4})\bar{y}.$$

This parabola has vertex at $(-2, 2)$. This time, $p = \frac{3}{4}$. With $\bar{x},\bar{y}$-axes at $(-2, 2)$, the equation became $\bar{x}^2 = -4p\bar{y}$. The focus is at $(x, y) = (-2, \frac{5}{4})$, and the directrix is $y = \frac{11}{4}$. The parabola is sketched in Fig. 12.13. ‖

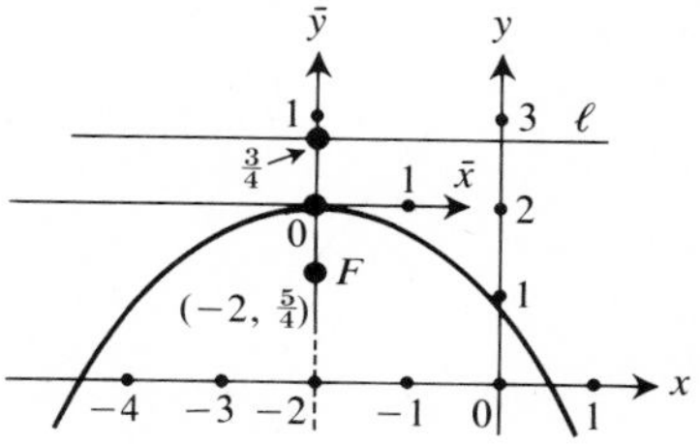

12.13

12.2.2 The ellipse

Definition 12.2 An **ellipse** is the locus in the plane of all points the sum of whose distances from two fixed points F_1 and F_2 (the **foci**) remains a constant value $2a$ (greater than the distance between the foci).

Figure 12.14 shows an ellipse with foci F_1 and F_2. This ellipse may be drawn as follows: Take a piece of string a bit more than $2a$ units long and fasten near the ends at F_1 and F_2 with staples, so that the length of string between the staples is exactly $2a$. A pencil placed to keep the string taut as in Fig. 12.14 will then trace out the ellipse.

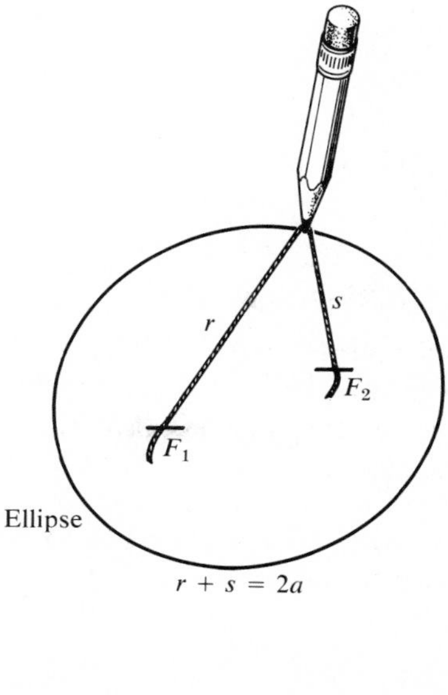

12.14

To describe an ellipse analytically, we choose axes as shown in Fig. 12.15, so that the foci are $(-c, 0)$ and $(c, 0)$. Then, for (x, y) on the ellipse, we obtain

$$\sqrt{(x-c)^2+y^2}+\sqrt{(x+c)^2+y^2}=2a, \tag{2}$$

$$\begin{aligned}\sqrt{(x-c)^2+y^2}&=2a-\sqrt{(x+c)^2+y^2},\\ x^2-2cx+c^2+y^2&=4a^2-4a\sqrt{(x+c)^2+y^2}\\ &\quad+x^2+2cx+c^2+y^2,\\ a\sqrt{(x+c)^2+y^2}&=a^2+cx,\\ a^2(x^2+2cx+c^2+y^2)&=a^4+2a^2cx+c^2x^2,\\ (a^2-c^2)x^2+a^2y^2&=a^4-a^2c^2=a^2(a^2-c^2). \qquad (3)\end{aligned}$$

From Fig. 12.15, it is clear that we must have $a > c$, as stated in the definition. Therefore we may set

$$b^2 = a^2 - c^2. \tag{4}$$

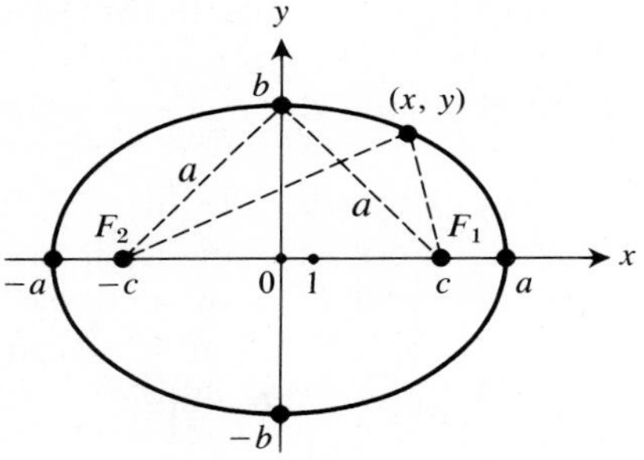

12.15

Then Eq. (3) can be simplified to become

$$b^2x^2 + a^2y^2 = a^2b^2,$$

$$\frac{x^2}{a^2}+\frac{y^2}{b^2}=1. \tag{5}$$

Equation (5) is the *standard form* of the equation of the ellipse. The line segment from $(-a, 0)$ to $(a, 0)$ in Fig. 12.15 is the **major axis** of the ellipse, and the segment from $(0, -b)$ to $(0, b)$ is the **minor axis**.

Example 3 Let us find the foci and the lengths of major and minor axes for the ellipse

$$\frac{x^2}{25}+\frac{y^2}{9}=1.$$

SOLUTION Here $a^2 = 25$ and $b^2 = 9$. From Eq. (4), you obtain

$$c^2 = a^2 - b^2. \tag{6}$$

Thus $c^2 = 25 - 9 = 16$, so $c = 4$. The foci are $(-4, 0)$ and $(0, 4)$. The length of the major axis is $2a = 10$, while the length of the minor axis is $2b = 6$. ||

Example 4 We study the ellipse $9x^2 + 4y^2 - 54x + 16y = -61$.

SOLUTION Completing the square, we obtain

$$9(x^2 - 6x) + 4(y^2 + 4y) = -61,$$
$$9(x - 3)^2 + 4(y + 2)^2 = -61 + 81 + 16 = 36,$$
$$\frac{(x - 3)^2}{4} + \frac{(y + 2)^2}{9} = 1,$$
$$\frac{\bar{x}^2}{4} + \frac{\bar{y}^2}{9} = 1,$$

where we chose translated $\bar{x},\bar{y}$-axes at $(x,y) = (3, -2)$. *This time the major axis of the ellipse is vertical.* We should always take as a^2 the *larger* of the two numbers in the denominator when the equation is brought into standard form. (We did not bother to do this in Section 12.1.) Thus $a^2 = 9$, $b^2 = 4$, and $c^2 = a^2 - b^2 = 5$, so $c = \sqrt{5}$. The foci are then $(3, -2 - \sqrt{5})$ and $(3, -2 + \sqrt{5})$. The major axis is of length $2a = 6$, and the minor axis is of length $2b = 4$. ||

12.2.3 The hyperbola

Definition 12.3 A **hyperbola** is the locus in the plane of all points the difference of whose distances from two fixed points F_1 and F_2 (the **foci**) remains at a constant value of $2a$ (which must be less than the distance between the foci).

Figure 12.16 shows a hyperbola with foci F_1 and F_2. To determine the locus analytically, we choose axes as indicated in Fig. 12.17, so that the foci

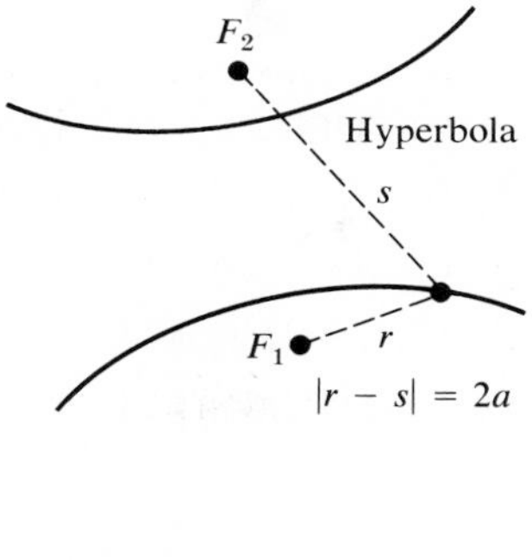

12.16

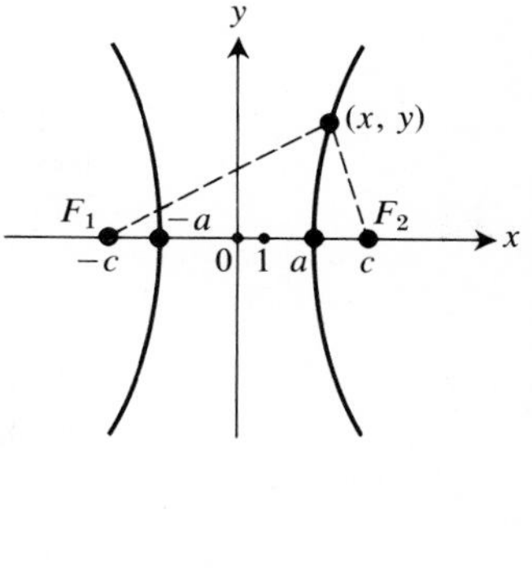

12.17

are at $(-c, 0)$ and $(c, 0)$. This time, $a < c$, which we ask you to prove from Definition 12.3 in Exercise 1. For a point (x, y) on the hyperbola, we then have

$$|\sqrt{(x-c)^2+y^2} - \sqrt{(x+c)^2+y^2}| = 2a. \tag{7}$$

The simplification of (7) is similar to that of (2); there is just a difference in sign, which disappears the second time the two sides of the equation are squared. In Exercise 2, we ask you to show that Eq. (3) is again obtained. This time $c > a$, so we set

$$b^2 = c^2 - a^2 \tag{8}$$

and obtain from (3)

$$-b^2x^2 + a^2y^2 = a^2(-b^2)$$

or

$$\frac{x^2}{a^2} - \frac{y^2}{b^2} = 1. \tag{9}$$

Equation (9) is the *standard form* of the equation of a hyperbola. We discussed the sketching of the hyperbola, and in particular its *asymptotes*, in Section 12.1.

Example 5 The hyperbola

$$\frac{x^2}{16} - \frac{y^2}{9} = 1$$

has asymptotes $y = \pm(3/4)x$. From Eq. (8), we obtain

$$c^2 = a^2 + b^2 \tag{10}$$

so $c = \sqrt{16+9} = 5$. The foci of the hyperbola are then $(-5, 0)$ and $(5, 0)$. ‖

Example 6 The hyperbola

$$-\frac{x^2}{144} + \frac{y^2}{25} = 1$$

has foci on the y-axis. For the hyperbola, you should think of a^2 as the number under the quadratic variable appearing with the *positive sign* in the standard form. (We did not bother to do this in Section 12.1.) So this time $a^2 = 25$ and $b^2 = 144$. From Eq. (8),

$$c = \sqrt{a^2+b^2} = \sqrt{25+144} = \sqrt{169} = 13.$$

Thus the foci are $(0, -13)$ and $(0, 13)$. The asymptotes are $y = \pm(5/12)x$. ‖

12.2.4 Eccentricity and directrices For the ellipse and hyperbola, each focus has an associated line as directrix, just as for the parabola. Figure 12.18 shows the foci F_1 and F_2 with their associated directrices ℓ_1 and ℓ_2, respectively. These **directrices** are the lines $x = \pm a^2/c$.

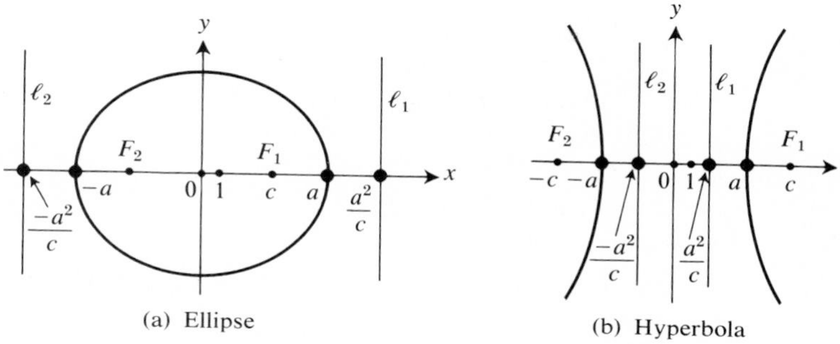

(a) Ellipse (b) Hyperbola

12.18

We leave to Exercise 3 the demonstration of the following theorem.

Theorem 12.1 *If P is a point on an ellipse or a hyperbola, then*

$$\frac{\textit{Distance from P to a focus}}{\textit{Distance from P to the associated directrix}} = \frac{c}{a}. \quad (11)$$

Definition 12.4 The number

$$e = \frac{c}{a} \quad (12)$$

is called the **eccentricity** of the ellipse or hyperbola.

Example 7 Let's find the eccentricity and directrices for the hyperbola

$$\frac{x^2}{16} - \frac{y^2}{9} = 1.$$

SOLUTION We have $a = 4$ and $c = 5$, as in Example 5. Thus $e = 5/4$, and the directrices are the lines $x = \pm 16/5$. ||

Example 8 For the ellipse

$$\frac{x^2}{25} + \frac{y^2}{9} = 1,$$

$a = 5$ and $c = 4$, so $e = c/a = 4/5$. The directrices are the lines $x = \pm a^2/c = \pm 25/4$. ||

In view of (11), (12), and Definition 12.1, you see that all three conic sections can be defined concisely as follows.

Definition 12.5 Let F be a point in the plane and ℓ a line not containing F. Let e be any positive number. The locus in the plane of all points P such that

$$\frac{\text{Distance from } P \text{ to } F}{\text{Distance from } P \text{ to } \ell} = e$$

is

1. a **parabola** if $e = 1$,
2. an **ellipse** if $e < 1$,
3. a **hyperbola** if $e > 1$.

Example 9 Let's find the equation of the hyperbola with a focus at $(-1, 2)$, associated directrix $y = 1$, and eccentricity $3/2$.

SOLUTION We use Definition 12.5, and we see that a point (x, y) lies on the hyperbola if and only if

$$\frac{\sqrt{(x+1)^2 + (y-2)^2}}{|y-1|} = \frac{3}{2}.$$

Squaring, we obtain

$$(x+1)^2 + (y-2)^2 = \frac{9}{4}(y-1)^2$$

or

$$4x^2 - 5y^2 + 8x + 2y + 11 = 0. \quad \|$$

SUMMARY

1. *An ellipse is the plane locus of points such that the sum of their distances from two fixed points (foci) remains constant.*
2. *A hyperbola is the plane locus of points such that the difference of their distances from two fixed points (foci) remains constant.*
3. *Let a fixed point F (focus) and a line ℓ (directrix) not through F be given in the plane. For a point P in the plane, let $|PF|$ be the distance from P to F and $|PD|$ the distance from P to ℓ. Let $e > 0$ (eccentricity) be given. The locus of all points P such that $|PF|/|PD| = e$ is*

   a) *a parabola if $e = 1$,* b) *an ellipse if $e < 1$,*

   c) *a hyperbola if $e > 1$.*
4. *The chart below gives other information about the conic sections.*

Type of curve	Standard forms	Focus	Directrix	Eccentricity
Parabola	$y^2 = 4px$, $y^2 = -4px$, $x^2 = 4py$, $x^2 = -4py$	$(p, 0)$ $(-p, 0)$ $(0, p)$ $(0, -p)$	$x = -p$, $x = p$, $y = -p$, $y = p$	$e = 1$
Ellipse a^2 is the *larger* number in the denominator. $c = \sqrt{a^2 - b^2}$	$\frac{x^2}{a^2} + \frac{y^2}{b^2} = 1$, $\frac{x^2}{b^2} + \frac{y^2}{a^2} = 1$	$(c. 0)$ $(-c, 0)$ $(0, c)$ $(0, -c)$	$x = a^2/c$, $x = -a^2/c$, $y = a^2/c$, $y = -a^2/c$	$e = \frac{c}{a} < 1$
Hyperbola a^2 is the number in the denominator of the *plus* term. $c = \sqrt{a^2 + b^2}$	$\frac{x^2}{a^2} - \frac{y^2}{b^2} = 1$, $-\frac{x^2}{b^2} + \frac{y^2}{a^2} = 1$	$(c, 0)$ $(-c, 0)$ $(0, c)$ $(0, -c)$	$x = a^2/c$, $x = -a^2/c$, $y = a^2/c$, $y = -a^2/c$	$e = \frac{c}{a} > 1$

EXERCISES

1. Show from Definition 12.3 that $a < c$ for a hyperbola. [*Hint.* Use the fact that the sum of the lengths of two sides of a triangle is greater than the length of the third side.

2. Show that Eq. (7) of the text again leads to Eq. (3), as asserted in the text.

3. Starting from the equation

$$\frac{x^2}{a^2} + \frac{y^2}{b^2} = 1$$

for an ellipse, prove that Eq. (11) of the text holds. That is, prove Theorem 12.1. [*Hint.* Using the equation of the ellipse and $b^2 = a^2 - c^2$, express everything on the left in Eq. (11) in terms of x, a, and c. Square, and show that it simplifies to c^2/a^2.] Observe that the same computation is valid for the hyperbola, since two sign changes cancel each other.

In Exercises 4 through 9, find the foci and directrices of the conic section with the given equation, and sketch the curve, showing the foci and directrices.

4. $\frac{x^2}{9} + \frac{y^2}{25} = 1$

5. $\frac{x^2}{25} - \frac{y^2}{16} = 1$

6. $x^2 = 12y$

7. $x^2 + 2x + 2y^2 - 8y + 5 = 0$

8. $3y^2 + 12y - 4x^2 + 3 = 0$

9. $y^2 - 4x + 14y + 57 = 0$

10. Find the x-intercept of the line tangent to the parabola $y^2 = 4px$ at a point (x_0, y_0) on the parabola.

11. Find the area of the region bounded by an ellipse with major axis of length $2a$ and minor axis of length $2b$.

12. Find the volume generated by revolving the region in Exercise 11 about the major axis of the ellipse.

In Exercises 13 *through* 18, *find the equation of the second-degree curve with the given focus, directrix, and eccentricity e.*

13. Focus $(0, 0)$, directrix $x = -2$, $e = 3$
14. Focus $(0, 0)$, directrix $y = 4$, $e = 1/2$
15. Focus $(0, 0)$, directrix $x = 3$, $e = 1$
16. Focus $(-1, 2)$, directrix $x = 1$, $e = 2$
17. Focus $(-2, 2)$, directrix $y = 3$, $e = 1$
18. Focus $(-2, 3)$, directrix $x - y = 6$, $e = 3/2$

19. Find the equation of the ellipse with center at the origin, foci at $(\pm 3, 0)$, and eccentricity $1/2$.
20. Find the equation of the ellipse with center at the origin, foci at $(0, \pm 1)$, and directrices $y = \pm 4$.
21. Find the equation of the hyperbola with center at the origin, foci at $(\pm 4, 0)$, and directrices $x = \pm 2$.
22. Find the equation of the hyperbola with center at the origin, foci at $(0, \pm 6)$, and eccentricity 2.
23. Find the equation of the parabola with vertex at the origin and focus at $(3, 0)$.
24. Find the equation of the parabola with vertex at the origin and directrix $y = 5$.
25. Find the equation of the ellipse with center at $(-1, 2)$, a focus at $(-1, 4)$, and eccentricity $1/2$.
26. Find the equation of the hyperbola with center at $(1, -3)$, a focus at $(1, 0)$, and a directrix at $(1, -2)$.
27. Find the equation of the parabola with vertex at $(-5, 2)$ and focus at $(-5, 0)$.
28. Find the equation of the parabola with vertex at the origin and focus at $(1, 1)$.
29. The chord of a parabola that passes through the focus and is parallel to the directrix is the *latus rectum* of the parabola. Show that the latus rectum of the parabola $y^2 = 4px$ has length $4p$.

12.3 CLASSIFICATION OF SECOND-DEGREE CURVES

In this section, we shall classify and sketch the various types of plane curves that are described by a second-degree equation of the form

$$Ax^2 + Bxy + Cy^2 + Dx + Ey + F = 0, \tag{1}$$

where at least one of A, B, or C is nonzero. We shall see that each such curve can be characterized as either an *ellipse*, a *hyperbola*, or a *parabola*.

12.3.1 Rotation of axes

If the coefficient B of xy in Eq. (1) is 0, then the curve can be sketched quite easily by completing the squares on x and on y and choosing translated coordinates; we shall do this in Section 12.3.2. In the present section, we show that by *rotating our axes*, we can "get rid" of the term Bxy.

Let's choose a new set of perpendicular axes, an x'-axis and a y'-axis, through the origin O of the Euclidean plane. Let the x'-axis make an angle θ with the x-axis, as shown in Fig. 12.19. We take the same scale on our new axes as on our old axes. Each point of our Euclidean plane then has two

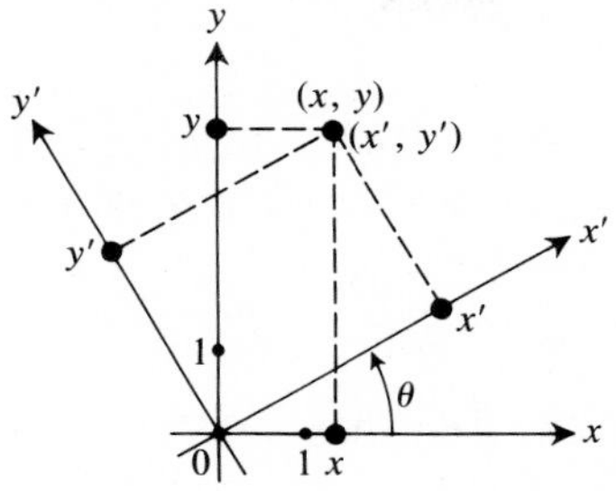

12.19

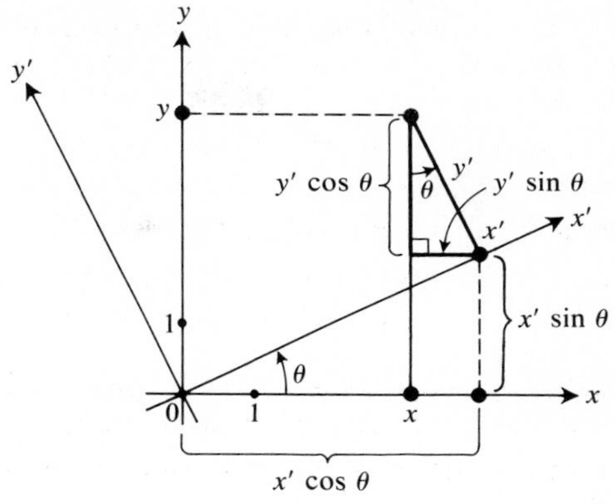

12.20

types of rectangular coordinates, (x, y) and (x',y'), as shown in Fig. 12.19. We would like to obtain formulas for the x,y-coordinates of a point in terms of its x',y'-coordinates, and vice versa. The following equations are easily obtained from Fig. 12.20:

$$\begin{aligned} x &= x' \cos \theta - y' \sin \theta, \\ y &= x' \sin \theta + y' \cos \theta. \end{aligned} \tag{2}$$

For a curve with Eq. (1), we can use Eqs. (2) to obtain the equation of the same curve in terms of the x',y'-coordinate system. Substitution yields

$$\begin{aligned} A(x' \cos \theta - y' \sin \theta)^2 &+ B(x' \cos \theta - y' \sin \theta)(x' \sin \theta + y' \cos \theta) \\ &+ C(x' \sin \theta + y' \cos \theta)^2 + D(x' \cos \theta - y' \sin \theta) \\ &+ E(x' \sin \theta + y' \cos \theta) + F = 0. \end{aligned} \tag{3}$$

If the terms of (3) are multiplied out and collected, you obtain an equation of the form

$$A'x'^2 + B'x'y' + C'y'^2 + D'x' + E'y' + F' = 0, \tag{4}$$

where, in particular,

$$\begin{aligned} A' &= A \cos^2 \theta + B \sin \theta \cos \theta + C \sin^2 \theta, \\ B' &= A(-2 \sin \theta \cos \theta) + B(\cos^2 \theta - \sin^2 \theta) + C(2 \sin \theta \cos \theta) \\ &= (C - A) \sin 2\theta + B \cos 2\theta, \\ C' &= A \sin^2 \theta + B(-\sin \theta \cos \theta) + C \cos^2 \theta. \end{aligned} \tag{5}$$

Several interesting results can be obtained from the relations (5). In particular, we see that $B' = 0$ if and only if

$$(C - A) \sin 2\theta + B \cos 2\theta = 0,$$

or if and only if (assuming $B \neq 0$)

$$\cot 2\theta = \frac{A - C}{B}. \tag{6}$$

Equation (6) shows that to rotate axes so that the x',y'-form of Eq. (1) has zero coefficient of $x'y'$, you should rotate through an angle θ such that

$$2\theta = \cot^{-1}\frac{A - C}{B}.$$

If $B \neq 0$, there will be exactly one such angle θ where $0 < \theta < \pi/2$.

From the relations (5), it is easy to compute directly (see Exercises 2 and 3) that

$$A' + C' = A + C \qquad \text{and} \qquad B'^2 - 4A'C' = B^2 - 4AC;$$

that is, these quantities are *invariants* under rotation of axes.

Let's summarize:

Theorem 12.2 *If the axes of the Euclidean plane are rotated through an angle θ as shown in Fig. 12.19 then the x',y'-form of Eq. (1) has zero as coefficient B' of $x'y'$ if*

$$\cot 2\theta = \frac{A - C}{B}.$$

Furthermore, $A + C$ and $B^2 - 4AC$ remain invariant: that is $A + C = A' + C'$ and $B^2 - 4AC = B'^2 - 4A'C'$.

To employ rotation of axes, you need to find Eqs. (2). First you find $\cot 2\theta = (A - C)/B$. Sometimes 2θ and θ are angles such that $\sin\theta$ and $\cos\theta$ are familiar. If this is not the case, find $\cos 2\theta$ from $\cot 2\theta$, and then use the relations

$$\sin\theta = \sqrt{\frac{1 - \cos 2\theta}{2}} \qquad \text{and} \qquad \cos\theta = \sqrt{\frac{1 + \cos 2\theta}{2}}. \tag{7}$$

Example 1 Let's eliminate the "crossterm xy" in the equation $x^2 - xy + y^2 - 4x + 5 = 0$ by rotation of axes.

SOLUTION Here $A = 1$, $B = -1$, and $C = 1$, so

$$\cot 2\theta = \frac{1-1}{-1} = 0.$$

Therefore $2\theta = \pi/2$ and $\theta = \pi/4$. The transformation equations (2) then become

$$x = \frac{1}{\sqrt{2}}x' - \frac{1}{\sqrt{2}}y', \qquad y = \frac{1}{\sqrt{2}}x' + \frac{1}{\sqrt{2}}y'.$$

Substituting, we obtain

$$\frac{1}{2}(x' - y')^2 - \frac{1}{2}(x' - y')(x' + y') + \frac{1}{2}(x' + y')^2 - 2\sqrt{2}(x' - y') + 5 = 0.$$

We now obtain

$$\frac{1}{2}x'^2 + \frac{3}{2}y'^2 - 2\sqrt{2}x' + 2\sqrt{2}y' + 5 = 0$$

or

$$x'^2 + 3y'^2 - 4\sqrt{2}x' + 4\sqrt{2}y' + 10 = 0$$

as an x',y'-equation for our curve with 0 for coefficient of $x'y'$. ‖

12.3.2 The classification Section 12.3.1 has shown that in studying the curve given by Eq. (1), you can assume (by rotation of axes if necessary) that the equation is of the form

$$Ax^2 + Cy^2 + Dx + Ey + F = 0 \tag{8}$$

where not both A and C are zero.

CASE 1. If $A = 0$ and $D \neq 0$, then upon completing the square on the terms involving y, you obtain the equation of a parabola. Namely, you obtain

$$C\left(y + \frac{E}{2C}\right)^2 = -D\left(x + \frac{F - (E^2/4C)}{D}\right). \tag{9}$$

Taking translated $\bar{x},\bar{y}$-coordinates at the point

$$\left(\frac{-E}{2C}, \frac{(E^2/4C) - F}{D}\right),$$

we write Eq. (9) as

$$(\bar{y})^2 = -\frac{D}{C}\bar{x},$$

which describes a parabola having vertex at our translated origin and opening parallel to the x-axis. A similar analysis can be made if $C = 0$ and $E \neq 0$.

If A and D are both zero, then Eq. (8) either has no locus, or has a straight line parallel to the x-axis as locus, or has two such straight lines as locus. If C and E are both zero, similar cases may occur. It is natural to consider straight lines that occur in this fashion to be *degenerate parabolas*.

Example 2 Consider the curve $x^2 + 2x + y - 3 = 0$. Completing the square, you obtain

$$(x + 1)^2 = -(y - 4),$$

which is the equation of a parabola having vertex at $(-1, 4)$ and opening in the negative y-direction, as shown in Fig. 12.21.

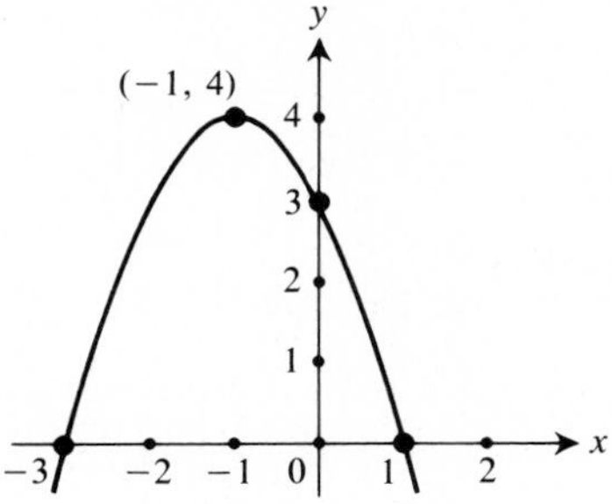

12.21

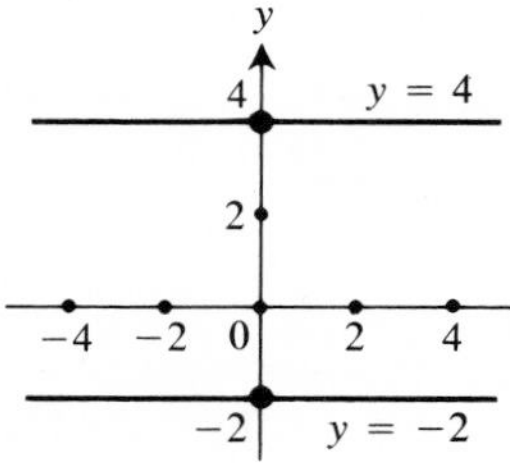

12.22

The plane locus described by the equation $y^2 - 2y - 8 = (y - 4)(y + 2) = 0$ consists of the lines $y = 4$ and $y = -2$, as shown in Fig. 12.22. There is no real plane of the equation $x^2 + 2x + 2 = 0$, for completion of the square yields $(x + 1)^2 = -1$, and a square of a number is nonnegative. ‖

CASE 2. Referring again to Eq. (8), if A and C are either both positive or both negative, then you obtain either no locus, a single point locus, or an ellipse. Completing squares, you obtain

$$A\left(x + \frac{D}{2A}\right)^2 + C\left(y + \frac{E}{2C}\right)^2 = \frac{D^2}{4A} + \frac{E^2}{4C} - F. \tag{10}$$

Taking translated coordinates at $(-D/2A, -E/2C)$, Eq. (10) can be written in the form

$$A(\bar{x})^2 + C(\bar{y})^2 = G. \tag{11}$$

By changing all signs if necessary, we can assume that A and C are both positive. If $G < 0$, there is no locus. If $G = 0$, the only locus is the translated origin $(\bar{x}, \bar{y}) = (0, 0)$. If $G > 0$, then (11) is an ellipse, which may be written in the form

$$\frac{(\bar{x})^2}{a^2} + \frac{(\bar{y})^2}{b^2} = 1 \tag{12}$$

Example 3 We classify the curve $x^2 + 4y^2 + 2x - 24y + 33 = 0$.

SOLUTION Completing the square, you obtain

$$(x + 1)^2 + 4(y - 3)^2 = 4$$

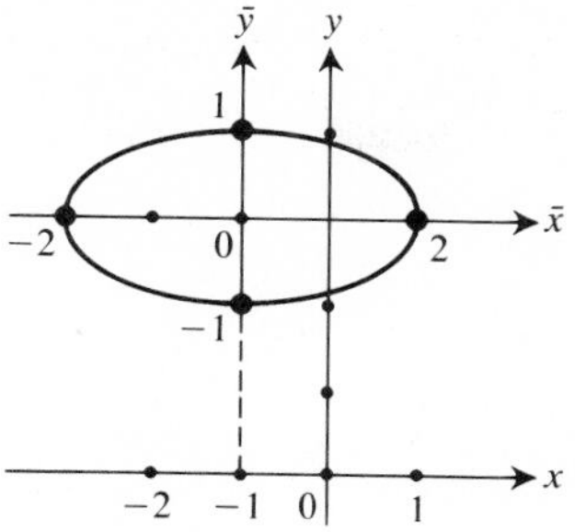

12.23

or

$$\frac{(x+1)^2}{2^2}+\frac{(y-3)^2}{1^2}=1.$$

This is the equation of an ellipse with *center* at $(-1, 3)$, as shown in Fig. 12.23. ‖

CASE 3. Referring again to Eq. (8), if A and C have opposite sign, then upon completing squares, choosing a translated origin and renaming constants as in Case 2, you can simplify your equation to either

$$\frac{(\bar{x})^2}{a^2}-\frac{(\bar{y})^2}{b^2}=1 \tag{13}$$

or

$$\frac{(\bar{y})^2}{a^2}-\frac{(\bar{x})^2}{b^2}=1 \tag{14}$$

or

$$\frac{(\bar{x})^2}{a^2}-\frac{(\bar{y})^2}{b^2}=0. \tag{15}$$

Equations (13) and (14) give hyperbolas. Equation (15) has as locus the two lines $\bar{y}=(b/a)\bar{x}$ and $\bar{y}=-(b/a)\bar{x}$. These two lines are considered to form a *degenerate hyperbola.*

Example 4 We classify the curve

$$x^2-4y^2+4x+24y-48=0.$$

SOLUTION Completing the square, you obtain

$$(x+2)^2-4(y-3)^2=16$$

or

$$\frac{(x+2)^2}{4^2}-\frac{(y-3)^2}{2^2}=1.$$

This curve is the hyperbola sketched in Fig. 12.24. ‖

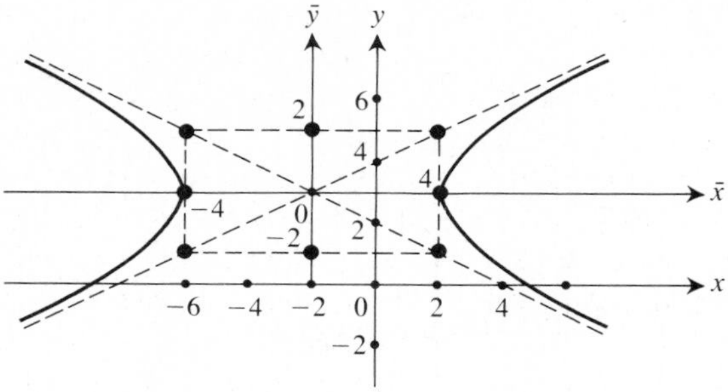

12.24

12.3.3 Classification without reduction to standard form

You have seen that if either A or C is 0 in Eq. (8), you have a parabola, while you have an ellipse if A and C have the same sign and a hyperbola if A and C have the opposite sign. These cases are easily separated by considering $B^2 - 4AC$. Recall that $B^2 - 4AC$ is invariant under rotation of axes (Theorem 12.2). If $B = 0$ as in Eq. (8) so that $B^2 - 4AC = -4AC$, you see that

$$\text{if } -4AC = 0, \quad \text{then } A \text{ or } C = 0;$$
$$\text{if } -4AC < 0, \quad \text{then } A \text{ and } C \text{ have the same sign;}$$
$$\text{if } -4AC > 0, \quad \text{then } A \text{ and } C \text{ have opposite sign.}$$

We obtain at once:

Theorem 12.3 (*Classification*) *The second-degree equation*

$$Ax^2 + Bxy + Cy^2 + Dx + Ey + F = 0$$

represents a (possibly imaginary or degenerate)

i) *parabola if* $B^2 - 4AC = 0$,
ii) *ellipse if* $B^2 - 4AC < 0$,
iii) *hyperbola if* $B^2 - 4AC > 0$.

Example 5 We classify the equation

$$3x^2 + 4xy + y^2 - 8x + 7y - 7 = 0.$$

SOLUTION Since $B^2 - 4AC = 4^2 - 4(3)(1) = 16 - 12 = 4 > 0$, you see that the equation describes a hyperbola. ||

12.3.4 A projective view of the conic sections

The Euclidean plane can be enlarged to form a *projective plane*, and this projective plane provides an intriguing view of the distinction between an ellipse, a hyperbola, and a parabola. To form the projective plane, we adjoin to our Euclidean plane a *line* ℓ_∞ *at infinity*. It is postulated that each line ℓ in

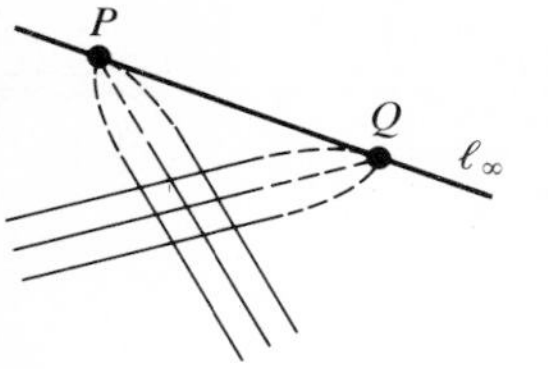

12.25

our old Euclidean plane intersects ℓ_∞ at exactly one point, the *point at infinity on the line ℓ*. (You should think of the point at infinity on a line as being a common point at both "ends" of the line. That is, you may visualize a line in the projective plane as a "huge circle," which looks straight to you nearby, where you can see it, but which is in reality closed at "infinity.") It is further postulated that any two parallel lines of our old Euclidean plane have the same point at infinity, so there are no such things as "parallel lines" in a projective plane. We have tried to illustrate this in Fig. 12.25, where we show some parallel lines in the Euclidean plane that meet at P on the line ℓ_∞, and another set of parallel lines in the Euclidean plane that meet at Q on ℓ_∞.

In real projective geometry, all conic sections (second-degree curves) look alike, and the different appearance in the old Euclidean part of the plane is just due to the way the curves meet the line ℓ_∞, as shown in Fig. 12.26. The ellipse does not meet the line at infinity; it lies totally within the old Euclidean plane, and can be seen in its entirety by a Euclidean bug. The hyperbola is cut into two pieces by the line at infinity, and a Euclidean bug can see only these two pieces. The parabola is tangent to the line at infinity, and a Euclidean bug sees only one piece whose extremities are approaching parallel.

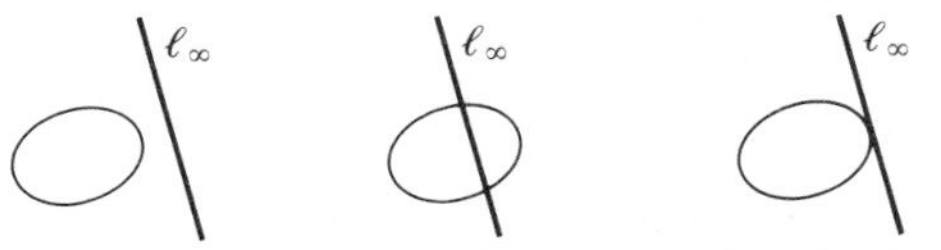

a) An ellipse b) A hyperbola c) A parabola **12.26**

SUMMARY

1. *If new x',y'-axes are obtained by rotating the old x,y-axes counterclockwise through the angle θ, then*

$$x = x'\cos\theta - y'\sin\theta, \qquad y = x'\sin\theta + y'\cos\theta.$$

2. *Starting with* $Ax^2 + Bxy + Cy^2 + Dx + Ey + F = 0$, *rotation of axes through θ such that*

$$\cot 2\theta = \frac{A - C}{B}$$

gives an x',y'-equation in which the coefficient of $x'y'$ is zero.

3. *For the equation in No. 2, $A+C$ and B^2-4AC are invariants under rotation of axes.*

4. *The equation in No. 2 has as locus*

   a) *an ellipse if* $B^2 - 4AC < 0$, b) *a hyperbola if* $B^2 - 4AC > 0$,

   c) *a parabola if* $B^2 - 4AC = 0$.

EXERCISES

1. Express x' and y' in terms of x and y for the case of rotation of axes shown in Fig. 12.19. [*Hint.* You can solve Eqs. (2) for x' and y', but it is easier to think of a rotation through the angle $-\theta$.]

2. Show from Eqs. (5) that $A + C$ is invariant under rotation, as stated in Theorem 12.2.

3. Show from Eqs. (5) that $B^2 - 4AC$ is invariant under rotation, as stated in Theorem 12.2.

In Exercises 4 *through* 9, *obtain an equation of the given locus in* x',y'*-coordinates formed by rotating axes so that the coefficient of* $x'y'$ *is* 0.

4. $xy = -3$
5. $x^2 - 3xy + y^2 = 5$
6. $2x^2 + \sqrt{3}xy + y^2 - 2x = 20$
7. $x^2 + \sqrt{3}xy + 2y^2 - 4y = 18$
8. $x^2 + xy = 10$. [*Hint.* Use Eqs. (7).]
9. $5x^2 + 3xy + y^2 + x = 12$. [*Hint.* Use Eqs. (7).]

In Exercises 10 *through* 15, *use Theorem* 12.3 *to classify the curve with the given equation as a (possibly imaginary or degenerate) ellipse, hyperbola, or parabola.*

10. $2x^2 - 3xy + y^2 - 8x + 5y = 30$
11. $x^2 + 6xy + 9y^2 - 2x + 14y = 10$
12. $4x^2 - 2xy - 3y^2 + 8x - 5y = 17$
13. $8x^2 + 6xy + 2y^2 - 5x = 25$
14. $x^2 - 2xy + 4x - 5y = 6$
15. $2x^2 - 3xy + 2y^2 - 8y = 15$
16. Can you think of a "use" for the invariant $A + C$ given in Theorem 12.2?

In Exercises 17 *through* 19, *rotate axes to simplify the equation, complete squares, and sketch the curve, showing the* x,y*-axes, the* x',y'*-axes, and the* $\bar{x},\bar{y}$*-axes. That is, do the complete job.*

17. $x^2 + 2xy + y^2 - 2\sqrt{2}x + 6\sqrt{2}y = 6$
18. $2x^2 + 4xy - y^2 - 12\sqrt{5}x = -6$
19. $3x^2 + 4xy + 3y^2 + 2\sqrt{2}x - 12\sqrt{2}y = -29$
20. For every real number t, the point $(\cos t, \sin t)$ lies on the circle $x^2 + y^2 = 1$. The functions sine and cosine are *circular functions.* Can you think of a reason why the hyperbolic sine and the hyperbolic cosine are called "*hyperbolic functions*"?

12.4 WHY STUDY CONIC SECTIONS?

You may quite properly ask why we should bother to classify all the second-degree plane curves. After all, it is clearly pretty hopeless to attempt to classify successively all plane curves of degree three, then all of degree four, etc. The plane curves of degree one (the lines) are certainly important; they are the graphs of linear functions, and calculus deals with approximation using these functions. The classification of the second-degree curves is classical and quite elegant, but that is not in itself sufficient reason to include it in this text. However, the second-degree plane curves do have important physical applications.

12.4.1 Orbits in central force fields

Consider a moving body that is subject only to a force of attraction toward a single fixed point. (The fixed point is considered to be the *center* of the system, and an attraction of this type constitutes a *central force field*. It can be shown from the laws of Newtonian mechanics that the orbit (path) on which the body travels due to such a force field is a second-degree plane curve, i.e., either an ellipse, a hyperbola, or a parabola, having the center of the force field system as focus. For example, if you neglect the force upon the earth due to celestial bodies other than the sun, then you are in a central force field situation with center at the sun. The earth has an elliptical orbit about the sun, with the sun at one focus of the ellipse. Orbits of all the planets about the sun are ellipses with a focus at the sun. Also, a satellite that we send up about our earth travels in an elliptical orbit. The smaller the eccentricity of an ellipse the more nearly circular the ellipse, for $b^2 = a^2 - c^2$ yields $(b/a)^2 = 1-(c/a)^2 = 1-e^2$. The table in Fig. 12.27 gives the eccentricities of the orbits of the planets.

12.27

Planet	Eccentricity	Period	Mean distance from sun*
Mercury	0.2056234	87.967 days	36.0
Venus	0.0067992	224.701 days	67.3
Earth	0.0167322	365.256 days	93.0
Mars	0.0933543	1.881 years	141.7
Jupiter	0.0484108	11.862 years	483.9
Saturn	0.0557337	29.458 years	887.1
Uranus	0.0471703	84.015 years	1785.0
Neptune	0.0085646	164.788 years	2797.0
Pluto	0.2485200	247.697 years	3670.0

* Millions of miles.

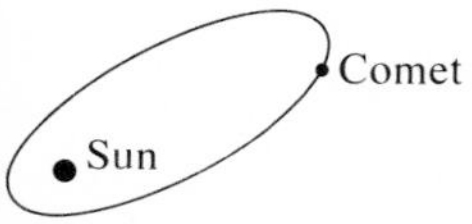

12.28

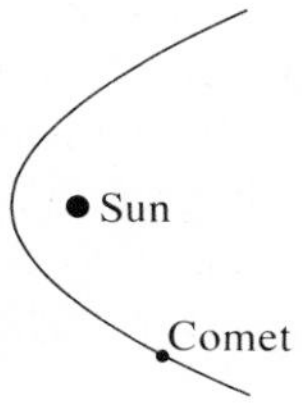

12.29

The orbits of comets about our sun are of two types. Some comets have "narrow" elliptical orbits with the sun at a focus that is comparatively close to one "end" of the ellipse, as illustrated in Fig. 12.28 (see Exercise 1). One such comet is Halley's comet, which takes about 75 or 76 years to travel once around its orbit, and which was last seen, when it came to this "end" of its orbit near the sun, in 1910 Another type of comet enters our solar system from "outer space," is attracted by our sun in accordance with the central force-field principle, and has such velocity that when it has passed the sun, it continues on and escapes into outer space again. Such a comet is traveling on a hyperbolic or parabolic path while in our solar system, see Fig. 12.29. (The path will be a hyperbola with probability 1, for in order for it to be a parabola, the comet would have to enter our solar system in precisely the right way to get a solar orbit of eccentricity *exactly* 1.)

12.4.2 Applications to optics and sound

The second-degree plane curves have important applications to reflected energy waves (e.g., light and sound). It has been found that when such a wave meets a reflecting barrier, the directions of the wave as it meets the barrier and as it is reflected make equal angles with the normal to the barrier (see Fig. 12.30). If the barrier is curved, then the normal is considered to be perpendicular to the tangent line (or plane) of the barrier (see Fig. 12.31).

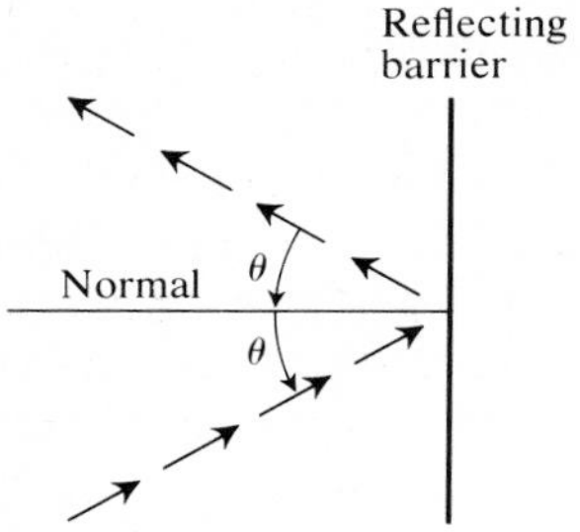

12.30

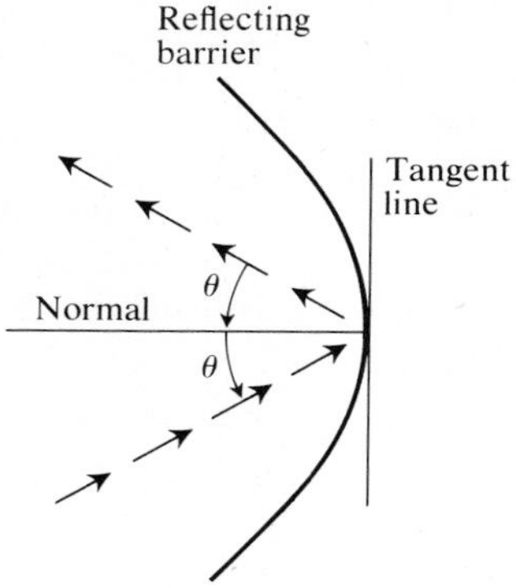

12.31

The conic sections have special geometric properties that yield important applications for these physical situations. We state these properties without proof. Proofs are not difficult, and are good exercises if you are interested.

Theorem 12.4 (*Reflecting Properties of Conic Sections*)

1. (*Ellipse and hyperbola*) *The lines joining a point* (x, y) *on an ellipse or hyperbola to the foci make equal angles with the normal to the curve at* (x, y). (*See Figs.* 12.32 *and* 12.33.)

2. (*Parabola*) *The line joining a point* (x, y) *on a parabola to the focus and the line through* (x, y) *parallel to the axis of the parabola make equal angles with the normal to the parabola.* (*See Fig.* 12.34.)

12.32

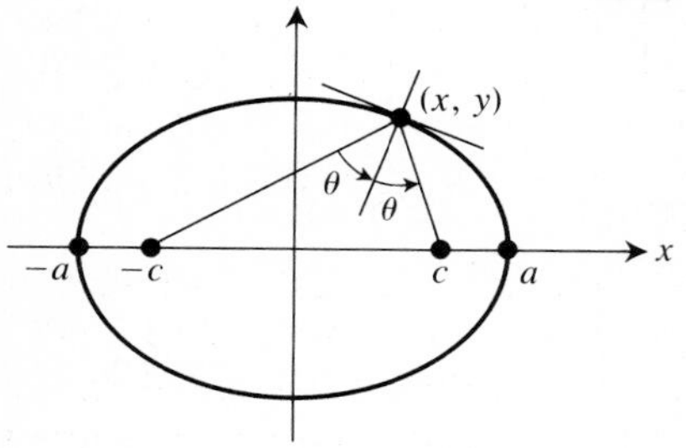

12.33

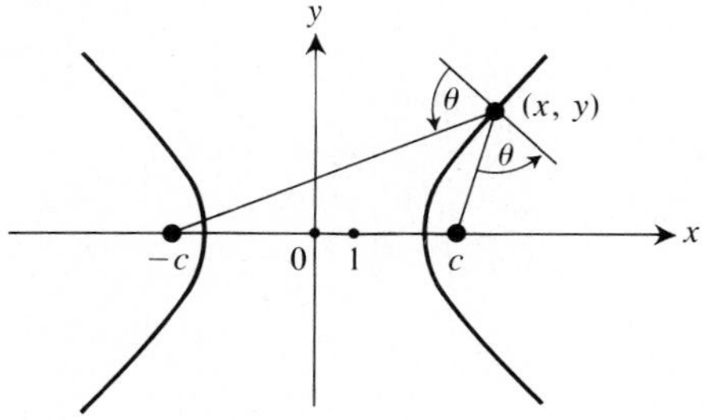

12.34

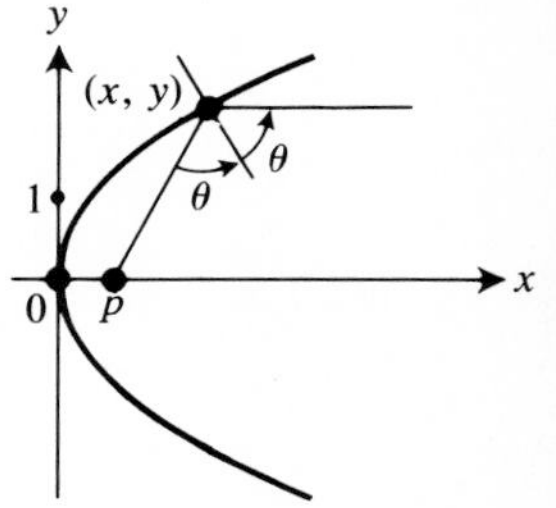

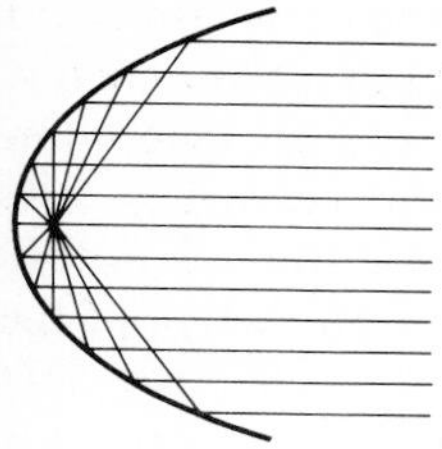
12.35

Consider the *paraboloid* formed by revolving the parabola $y^2 = 4px$ about the x-axis. If the inside of this paraboloid is silvered to reflect light, then the beams from a light source at the focus of the paraboloid will be reflected in rays parallel to the axis of the paraboloid (see Fig. 12.35). This principle is used in a searchlight. Conversely, light emanating from a source at a great distance out from the axis enters the silvered paraboloid in essentially parallel rays, and is all reflected to the focus. This principle is used in telescopes, such as the one at Mount Polamar in California, which has a parabolic mirror of diameter 200 in. The large antennas used in radio astronomy are also constructed in the shape of a paraboloid, and small parabolic reflectors are used in sending and receiving radio and television signals over short distances.

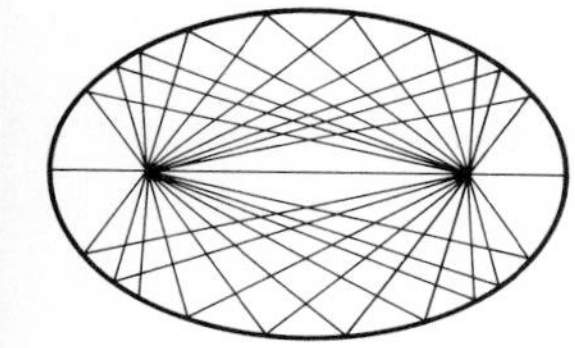
12.36

Let an *ellipsoid* be formed by revolving an ellipse about its major axis. If the inside of the ellipsoid is silvered, then light emanating in all directions from a source at one focus will all be reflected to the other focus (see Fig. 12.36). Definition 12.2 of Section 12.2 shows that the distance traveled from one focus to the other via reflection is always the length $2a$ of the major axis, independent of the point of reflection. Thus, sound that emanates in all directions from one focus of an ellipsoid will not only all be reflected to the other focus, but will all reach the other focus *at the same time.* This is the principle behind a "whisper gallery" with walls and ceiling forming a portion of an ellipsoid; a whisper at one focus can be heard distinctly at a great distance at the other focus.

12.4.3 Application to navigation

In World War I, the following scheme was sometimes used to determine the location of an enemy cannon. Three observers would synchronize their watches, and then move to observation points A, B, and C with known coordinate locations. These observers at A, B, and C would note the precise time the cannon fired. Knowing the speed of sound, they could then calculate the difference of the distances from A and B to the cannon. For example, if the observer at A heard the sound three seconds before the observer at B, then the cannon must have been about 3300 ft farther from B than from A, for sound travels at about 1100 ft/sec. Regarding A and B as foci, the cannon must lie on a hyperbola with foci A and B and $2a = 3300$. Similarly, if the observer at C heard the same firing two seconds before the observer at B, the cannon was also on a hyperbola having B and C as foci, and where $2a = 2200$. The cannon then must have been at the intersection of these two hyperbolas.

The same principle used in locating the cannon just described is currently used in LORAN (Long-Range Navigation). This time, radio waves, which travel at 186,000 mi/sec, are used rather than sound waves traveling only 1100 ft/sec. Consequently, time differences must be measured very accurately, using sensitive equipment, in microseconds (millionths of seconds). A master station at A transmits a signal pattern, and two remote stations at B and C transmit the same pattern, but with a known time delay measured in microseconds. Equipment on board a ship measures the number of microseconds delay in the signal patterns received from A and B. This locates the ship on a hyperbola with foci at A and B. The delay in the signals received from A and C provide another hyperbola, and the ship must lie on the intersection of these two hyperbolas. If all the equipment is in good order, a fisherman can determine his position using LORAN within about 50 yards.

SUMMARY

1. *Bodies traveling subject only to gravitational attraction of a single much larger body move in conic-section orbits.*
2. *The reflective properties of conic sections* (given in Theorem 12.4) *make them useful in optics and acoustics.*
3. *The hyperbola has important applications to navigation.*

EXERCISES

1. Show that if the major axis of an ellipse is very large compared with the minor axis, then the foci of the ellipse are comparatively near the "ends."
2. Show that the closest point on an ellipse to a focus is the point nearest the focus on the end of the major axis.
3. The *apogee* of an earth satellite is its maximum altitude above the surface of the earth during orbit, and its *perigee* is its minimum altitude during orbit. Using Exercise 2, argue that for a satellite in elliptical earth orbit with major axis of length $2a$, we have
$$2a = (\text{diameter of earth}) + (\text{apogee}) + (\text{perigee}).$$
4. Prove the reflecting property of the hyperbola.
5. Prove the reflecting property of the parabola.
6. An ellipse and a hyperbola are *confocal* if they have the same foci. Using the reflecting properties, give a synthetic proof that an ellipse is perpendicular to a confocal hyperbola at each point of intersection. [*Hint.* Show that the normals to the curves are perpendicular at a point of intersection.]

12.5 PARAMETRIC CURVES REVIEWED

In Section 3.4.2, we discussed parametric equations $x = h(t)$, $y = k(t)$, which can be viewed as giving the location of a body on a curve at time t. In this section, we discuss sketching parametric curves, and then we review the formulas for the slope and arc length of such curves. We always assume that $h(t)$ and $k(t)$ are continuous functions. In discussing the slope of the curve and arc length, we assume further that $h'(t)$ and $k'(t)$ exist and are continuous; such a curve is a *smooth curve.*

12.5.1 Sketching and parametrizing

Curves given parametrically can sometimes be sketched by eliminating the parameter t to obtain a (we hope, familiar) x,y-equation.

Example 1 Let us sketch the plane curve given parametrically by

$$x = \cos t, \qquad y = \sin t.$$

SOLUTION Squaring and adding these equations, you find that each point (x, y) of the curve lies on the circle $x^2 + y^2 = 1$ shown in Fig. 12.37. You may view these parametric equations as picking up the t-axis and wrapping it around and around this circle, with $t = 0$ placed at $(1, 0)$. ||

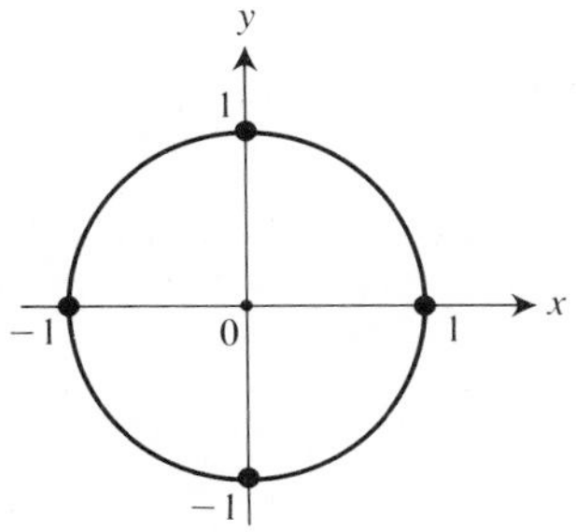

12.37

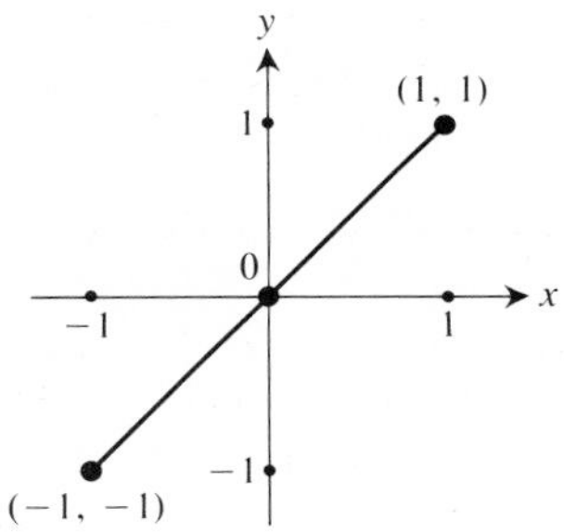

12.38

Warning Our next example shows that you have to be careful in sketching a curve by eliminating the parameter to obtain an x,y-equation. The nature of the functions $x = h(t)$ and $y = k(t)$ may have placed restrictions on x and y that are not apparent from the x,y-equation.

Example 2 Elimination of t from the parametric equations

$$x = \sin t, \qquad y = \sin t,$$

yields $y = x$. However, since $-1 \le \sin t \le 1$ for all values of t, we see that the points of the curve given by these parametric equations all lie on the segment of the line $y = x$ joining $(-1, -1)$ and $(1, 1)$, as shown in Fig. 12.38. An object whose position in the plane at time t is given by these equations travels back and forth along this line segment. ||

Sometimes a curve given by an equation in x and y can be parametrized in terms of some quantity (parameter) arising naturally from the curve or from some physical consideration. For such a parametrization to be accomplished, there must be a *unique* point on the curve corresponding to each value of the parameter. The next two examples illustrate this idea.

Example 3 Let's parametrize the parabola $y = x^2$, taking as parameter the slope m of the parabola at each point.

SOLUTION At a point (x, y) on the parabola, the slope is given by $m = y' = 2x$, so $x = m/2$. Since $y = x^2$, we obtain the parametrization

$$x = \tfrac{1}{2}m,$$
$$y = \tfrac{1}{4}m^2,$$

of the parabola. ‖

Example 4 Consider the plane curve traced by a point P on a circle of radius a as the circle rolls along the x-axis. This curve is *cycloid.* We assume that the point P on the circle touches the x-axis at the points 0 and $\pm 2n\pi a$ as the circle rolls along, as shown in Fig. 12.39.

Let's parametrize this cycloid in terms of the angle θ through which the circle has rolled, starting with $\theta = 0$ when P is at the origin, as shown in Fig. 12.39.

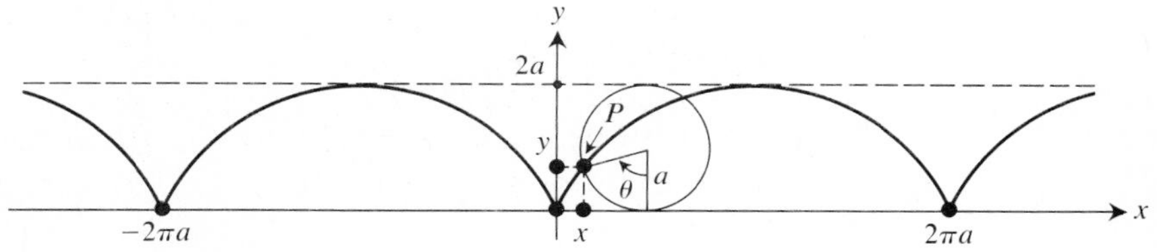

12.39

SOLUTION We assume that θ is taken to be positive as the circle rolls to the right. From Fig. 12.39, you easily obtain

$$x = a\theta - a \sin \theta,$$
$$y = a - a \cos \theta,$$

as the desired parametric equations. ‖

The cycloid discussed in Example 4 has some very fascinating physical properties. Let a point Q be given in the fourth quadrant, and imagine a drop of water placed at (0, 0), which slides (without friction) along a curve from the origin (0, 0) to Q, subject only to a force mg downward due to gravity, as shown in Fig. 12.40. What shape should the curve have so that the drop slides from (0, 0) to Q in the least possible time? (This is the *brachistochrone problem*; the name comes from two Greek words meaning "shortest time.") Initially, you might think that the drop should slide along the straight line segment joining (0, 0) and Q, but after a little thought, it might seem reasonable that the curve should drop more steeply at first, and allow the drop to gain speed more quickly. It can be shown that the "smooth" curve (with no sharp corners) corresponding to the shortest time is a portion of a single arc of an inverted cycloid

$$x = a\theta - a\sin\theta,$$
$$y = a\cos\theta - a,$$

generated by a point P on a circle rolling on the *under* side of the x-axis, with P starting at the origin. The cycloid is thus the solution to the brachistochrone problem.

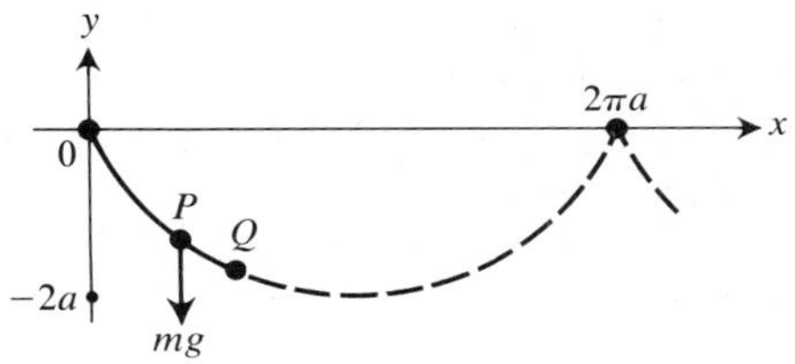

12.40

It can be shown that the inverted cycloid is also the solution of the *tautochrone problem* (meaning "same time"), for if a drop of water is placed at a point other than the low point on an arc of an inverted cycloid, the time required for it to slide to the low point of the arc is independent of the initial point where the drop is placed.

12.5.2 Review of slope

Now assume that $x = h(t)$ and $y = k(t)$ both have continuous derivatives. Recall from our previous work that the slope of the parametric curve at a point corresponding to t_1 is given by

$$\left.\frac{dy}{dx}\right|_{t_1} = \left.\frac{dy/dt}{dx/dt}\right|_{t_1} \tag{1}$$

if $dx/dt \neq 0$ at t_1.

Example 5 Let us find the tangent and normal lines to the smooth curve $x = \cos t$, $y = \sin t$ at the point $(\sqrt{3}/2, 1/2)$ corresponding to $t_1 = \pi/6$.

SOLUTION We have

$$\left.\frac{dy}{dx}\right|_{t=\pi/6} = \left.\frac{dy/dt}{dx/dt}\right|_{t=\pi/6} = \frac{\cos(\pi/6)}{-\sin(\pi/6)} = \frac{\sqrt{3}/2}{-1/2} = -\sqrt{3}.$$

The tangent line thus has x,y-equation $y = -\sqrt{3}x + 2$ and the normal line has equation $y = (1/\sqrt{3})x$. ‖

If dx/dt is zero and dy/dt nonzero at $t = t_1$, then the curve has a vertical tangent at the corresponding point (x_1, y_1). It may happen that both dx/dt and dy/dt are zero at $t = t_1$. In that case, the slope is given by the limit of the quotient (1) as $t \to t_1$ if the limit exists.

Example 6 If $x = \cos t$ and $y = t^2$, then

$$\frac{dy}{dx} = \frac{dy/dt}{dx/dt} = \frac{-\sin t}{2t}.$$

The slope of the curve when $t = 0$ is given by

$$\lim_{t\to 0} \frac{-\sin t}{2t} = \lim_{t\to 0} -\frac{1}{2} \cdot \frac{\sin t}{t} = -\frac{1}{2} \cdot 1 = -\frac{1}{2}. \quad \|$$

If $h(t)$ and $k(t)$ have derivatives of sufficiently high order, then it is a simple (but often tedious) matter to compute the higher-order derivatives d^2y/dx^2, d^3y/dx^3, etc. For example,

$$\begin{aligned}
\frac{d^2y}{dx^2} &= \frac{d(dy/dx)}{dx} = \frac{d(dy/dx)}{dt} \cdot \frac{dt}{dx} \\
&= \frac{d\left(\dfrac{dy/dt}{dx/dt}\right)}{dt} \cdot \frac{dt}{dx} = \frac{\dfrac{dx}{dt}\left(\dfrac{d^2y}{dt^2}\right) - \dfrac{dy}{dt}\left(\dfrac{d^2x}{dt^2}\right)}{(dx/dt)^2} \cdot \frac{1}{dx/dt} \\
&= \frac{\dfrac{dx}{dt}\left(\dfrac{d^2y}{dt^2}\right) - \dfrac{dy}{dt}\left(\dfrac{d^2x}{dt^2}\right)}{(dx/dt)^3}.
\end{aligned}$$

One does not memorize this result, but simply applies the chain rule as indicated in the first line of its derivation when the result is needed. We illustrate with an example.

Example 7 Let us find $^{d2}y/dx^2$ the curve $x = \cos t$, $y = \sin t$.

SOLUTION We have

$$\frac{dy}{dx} = \frac{dy/dt}{dx/dt} = \frac{\cos t}{-\sin t} = -\cot t.$$

Then

$$\begin{aligned}
\frac{d^2y}{dx^2} &= \frac{d(dy/dx)}{dt} \cdot \frac{dt}{dx} \\
&= \frac{d(-\cot t)}{dt} \cdot \frac{1}{dx/dt} \\
&= \frac{\csc^2 t}{dx/dt} = \frac{\csc^2 t}{-\sin t} = -\csc^3 t. \quad \|
\end{aligned}$$

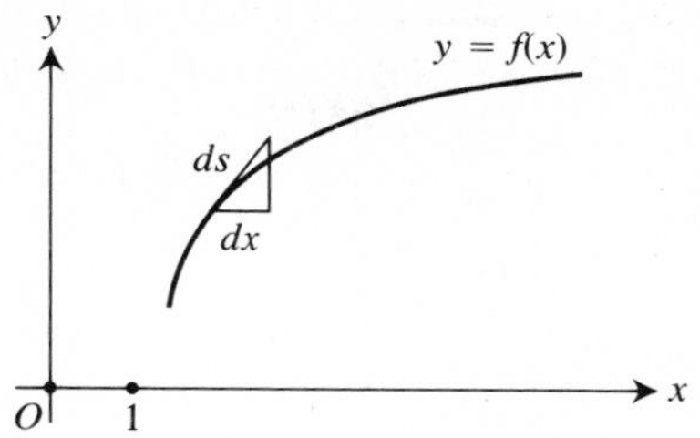

12.41

12.5.3 Review of arc length Recall that if $x = h(t)$ and $y = k(t)$ have continuous derivatives, then the arc length of the curve from $t = t_1$ to $t = t_2$ is given by

$$\int_{t_1}^{t_2} \sqrt{(dx/dt)^2 + (dy/dt)^2}\, dt, \tag{2}$$

and $ds = \sqrt{(dx/dt)^2 + (dy/dt)^2}\, dt$ is the *differential of arc length.* We may consider ds to be the length of a short piece of tangent line segment to the curve, corresponding to a change dx in x, as shown in Fig. 12.41.

Example 8 Let's verify the formula $C = 2\pi a$ for the circumference of a circle whose radius is a.

SOLUTION We take as parametric equation for the circle

$$x = a \cos t, \qquad y = a \sin t, \qquad \text{for} \quad 0 \le t \le 2\pi.$$

Since $dx/dt = -a \sin t$ and $dy/dt = a \cos t$, we see from (2) that the length of the circle is

$$C = \int_0^{2\pi} \sqrt{(-a \sin t)^2 + (a \cos t)^2}\, dt$$

$$= \int_0^{2\pi} \sqrt{a^2}\, dt = a\int_0^{2\pi} (1)\, dt = 2\pi a. \quad \|$$

SUMMARY

1. *A parametric curve can sometimes be sketched by eliminating the parameter to obtain an x,y-equation, but care must be taken to note any restrictions on x and y imposed by the parametric equations.*
2. *If $x = h(t)$ and $y = k(t)$, then*

$$\frac{dy}{dx} = \frac{dy/dt}{dx/dt}.$$

3. *The differential of arc length is $ds = \sqrt{(dx/dt)^2 + (dy/dt)^2}\, dt$ and*

$$\text{Arc length} = \int_{t_1}^{t_2} ds = \int_{t_1}^{t_2} \sqrt{(dx/dt)^2 + (dy/dt)^2}\, dt.$$

EXERCISES

In Exercises 1 through 12, sketch the curve having the given parametric representation.

1. $x = \sin t,\ y = \cos t$ for $0 \le t \le \pi$

2. $x = 2\cos t,\ y = 3\sin t$ for all t

3. $x = \tan t,\ y = \sec t$ for $-\frac{\pi}{2} < t < \frac{\pi}{2}$

4. $x = t^2,\ y = t + 1$ for all t

5. $x = t^2,\ y = t^3$ for all t

6. $x = \sin t,\ y = \cos 2t$ for $0 \le t \le 2\pi$

7. $x = t - 3,\ y = t^2 + 1$ for all t

8. $x = 3\cosh t,\ y = 2\sinh t$ for all t

9. $x = t - 1,\ y = \ln t$ for $0 < t < \infty$

10. $x = \sin t,\ y = 1 + \sin^2 t$ for all t

11. $x = t^2,\ y = t^2 + 1$ for all t

12. $x = t^2,\ y = t^4$ for all t

13. Parametrize the curve $y = e^x$ taking as parameter the slope m of the curve at (x, y).

14. Parametrize the curve $y = \sqrt{x}$ taking as parameter the slope s of the normal to the curve at (x, y).

15. Parametrize the curve $y = \sqrt{x}$ taking as parameter the distance d from (x, y) to $(0, 0)$.

16. Can you parametrize the curve $y = x^2$ by taking as parameter the distance from (x, y) to $(0, 0)$? Why?

17. A circular disk of radius a rolls along the x-axis. Find parametric equations of the curve traced by a point P on the disk a distance b from the center of the disk, where $0 \le b \le a$. Assume that the point P falls on the interval $[0, a]$ on the y-axis when the disk touches the origin, and take as parameter the angle θ through which the disk rolls from its position at the origin. (This curve is a *trochoid.* For $b = a$, we obtain the cycloid, while $b = 0$ yields the line $y = a$.)

In Exercises 18 through 22, find the slope of the curve, with the given parametric representation, at the point on the curve corresponding to the indicated value of the parameter.

18. The cycloid $x = a\theta - a\sin\theta,\ y = a - a\cos\theta = \pi/4$

19. $x = t^2 - 3t,\ y = \sin 2t$ where $t = 0$

20. $x = \sinh t,\ y = \cosh 2t$ where $t = 0$

21. $x = e^t,\ y = \ln(t + 1)$ where $t = 0$

22. $x = t^3 - 3t^2 + 3t - 5,\ y = 4(t - 1)^3$ where $t = 1$

23. Find all points where the curve $x = \cos 2t,\ y = \sin t$ has a vertical tangent.

In Exercises 24 through 26, the parametric equations represent a curve which is the graph of a function in a neighborhood of the point corresponding to the indicated value of the parameter. Find d^2y/dx^2 at this point.

24. $x = t^2,\ y = t^3 - 2t^2 + 5$ where $t = 1$

25. $x = \sin 3t,\ y = e^t$ where $t = 0$

26. $x = \ln(t + 3),\ y = \cos 2t$ where $t = 0$

In Exercises 27 through 29, find the length of the curve with the given parametric representation.

27. $x = t^2,\ y = \frac{2}{3}(2t + 1)^{3/2}$ from $t = 0$ to $t = 4$

28. $x = t,\ y = \ln(\cos t)$ from $t = 0$ to $t = \pi/4$

29. $x = a\cos^3 t,\ y = a\sin^3 t$ from $t = 0$ to $t = \pi/2$

12.6 CURVATURE

12.6.1 What is curvature?

We are interested in the rate at which a plane curve bends (or "curves") as we travel along the curve. We shall attempt to give a numerical measure of such a rate of turning at a point on the curve; this number will be the *curvature of the curve at the point.* The more the curve bends at a point, the larger the curvature there will be.

Example 1 A straight line does not bend at all. We shall see that it is convenient to let the curvature of a straight line be zero at each point. ‖

Example 2 A circle "curves" at a uniform rate; we would like the curvature of a circle at one point to be the same as the curvature of the circle at every other point on the circle, and the same as the curvature at each point of any other circle of the same radius. This would allow us to speak of the *curvature of a circle of radius a*. ‖

Example 3 The smaller the radius of a circle, the more the circle "curves" at each point. We would like the curvature of a small circle to be greater than the curvature of a circle of larger radius. ‖

Let's consider a smooth curve having a tangent line at each point. Let's choose a point (x_0, y_0) on our curve from which we measure arc length s along the curve, so that (x_0, y_0) corresponds to $s = 0$, as shown in Fig. 12.42. Let ϕ be the angle shown in the figure, measured from the positive $\bar{x}$-axis at (x, y) to the direction of increasing s along the tangent line. The rate at which the curve bends at (x, y) can be described in terms of the rate at which the tangent line is turning as you travel along the curve at (x, y) and this in turn can be measured by the rate at which the angle ϕ is increasing at (x, y). We want our notion of curvature to be intrinsic to the curve set, and not dependent upon the rate at which you may be traveling along the curve. Thus it would not be appropriate to let the curvature be the rate of change of ϕ per unit change in *time* as you travel along the curve. It is intuitively apparent that the notion of arc length is intrinsic to the curve set, and we let the curvature of the curve at a point be the rate of change of ϕ per unit change in arc length along the curve at that point.

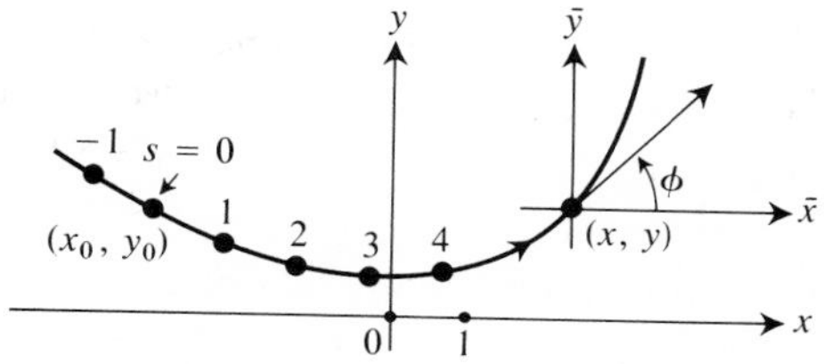

12.42

Definition 12.6 The **curvature** κ of a curve at (x, y) is $|d\phi/ds|$ where ϕ is the angle in Fig. 12.42 and s is arc length measured along the curve.*

* Some texts define the curvature to be the signed quantity $d\phi/ds$, but we take a definition that specializes the notion of curvature for a space curve, which will be introduced in Chapter 15. The interpretation of the sign of $d\phi/ds$ is explained later in this section.

Example 4 Since the angle ϕ along a straight line is a constant function of the arc length s (see Fig. 12.43), we have $d\phi/ds = 0$, and the curvature of a straight line at each point is 0. ||

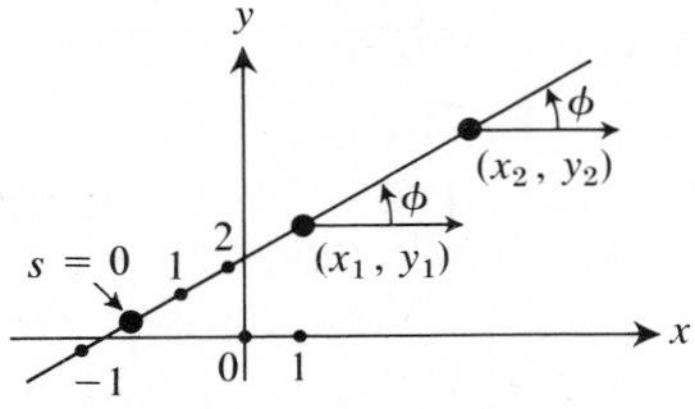

12.43

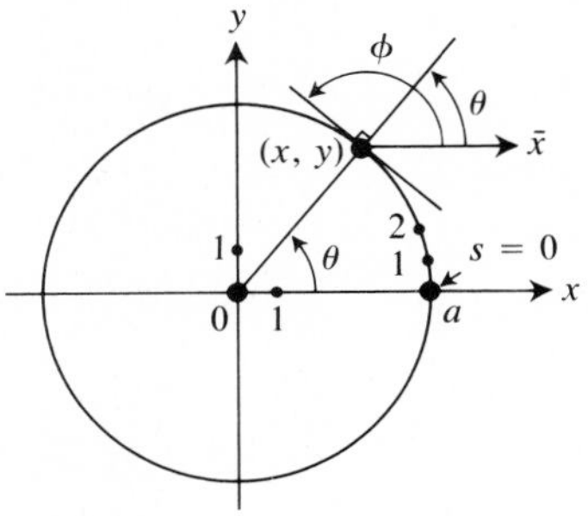

12.44

Example 5 Let's measure arc length on the circle $x^2 + y^2 = a^2$ in the counterclockwise direction, starting with $s = 0$ at $(a, 0)$ as shown in Fig. 12.44.

From this figure, you see that at a point (x, y) on the circle corresponding to a central angle of θ and hence to an arc length $s = a\theta$, you have $\phi = \theta + n(\pi/2)$. (The value of n depends upon the quadrant of θ; for θ in the first quadrant as in Fig. 12.44, you have $n = 1$.) Therefore

$$\phi = \theta + n\frac{\pi}{2} = \frac{1}{a}s + n\frac{\pi}{2},$$

so

$$\kappa = \frac{d\phi}{ds} = \frac{1}{a}.$$

Thus the curvature at any point of a circle of radius a is the reciprocal $1/a$ of the radius. ||

The preceding example supports the following terminology. If a curve has curvature $\kappa \neq 0$ at a point, then the **radius of curvature** of the curve at this point is $\rho = 1/\kappa$.

12.6.2 The formula for the curvature of $y = f(x)$

Let a curve be the graph $y = f(x)$ for a *twice*-differentiable function f, and let the direction of increasing arc length measured from (x_0, y_0) be the direction of increasing x, so that

$$s = \int_{x_0}^{x} \sqrt{1 + f'(t)^2}\, dt. \qquad (1)$$

For the angle ϕ shown in Fig. 12.42 we have

$$\tan \phi = f'(x)$$

at any point (x, y) on the graph. Thus in a neighborhood of (x, y),

$$\phi = b + \tan^{-1} f'(x) \tag{2}$$

where the constant b occurs since ϕ may not fall in the principal-value range of the inverse tangent function. Since f is a twice-differentiable function of x, you see from (2) that ϕ is a differentiable function of x. From (1) you see that s is a differentiable function of x; in a neighborhood of a point where $ds/dx \neq 0$ it can then be shown that x appears as a differentiable function of s. Under these conditions, ϕ then appears as a differentiable function of s, and by the chain rule,

$$\frac{d\phi}{ds} = \frac{d\phi}{dx} \cdot \frac{dx}{ds}. \tag{3}$$

From Eq. (2),

$$\frac{d\phi}{dx} = \frac{1}{1 + (f'(x))^2} \cdot f''(x), \tag{4}$$

and from (1),

$$\frac{ds}{dx} = \sqrt{1 + (f'(x))^2}.$$

Thus

$$\frac{dx}{ds} = \frac{1}{\sqrt{1 + (f'(x))^2}}, \tag{5}$$

and from Eqs. (3), (4), and (5),

$$\kappa = \left|\frac{d\phi}{ds}\right| = \left|\frac{f''(x)}{[1 + (f'(x))^2]^{3/2}}\right|. \tag{6}$$

You can easily verify that $d\phi/ds > 0$ where, when traveling along the curve in the direction of increasing s, the curve bends to the left (ϕ is increasing), and $d\phi/ds < 0$ where the curve bends to the right (ϕ is decreasing).

Theorem 12.5 *The curvature of the graph of a twice-differentiable function f at a point (x, y) on the graph is given by*

$$\kappa = \left|\frac{d\phi}{ds}\right| = \left|\frac{f''(x)}{[1 + (f'(x))^2]^{3/2}}\right|. \tag{7}$$

(Geometrically, $d\phi/ds$ is positive at a point if, when traveling along the curve in the direction of increasing s, the curve bends to the left, and $d\phi/ds$ is negative if the curve bends to the right.)

Example 6 Let's find the curvature and radius of curvature of the parabola $y = x^2$ at the origin, taking as direction of increasing s the direction of increasing x.

SOLUTION For $f(x) = x^2$, we have

$$\kappa = \left|\frac{f''(x)}{[1 + (f'(x))^2]^{3/2}}\right| = \frac{2}{(1 + 4x^2)^{3/2}},$$

so

$$\kappa|_{(0,0)} = 2.$$

The radius of curvature is $\rho = 1/\kappa = 1/2$. ||

Let a curve have radius of curvature ρ at a point (x, y). The **osculating circle of the curve at** (x, y) is the circle of radius ρ through (x, y) having the same tangent as the curve at (x, y) and having center on the concave side of the curve. The center of this circle is the **center of curvature of the curve at** (x, y).

Example 7 From Example 6, you see that the center of curvature of the parabola $y = x^2$ at (0, 0) is (0, 1/2), and the osculating circle has equation

$$x^2 + (y - \tfrac{1}{2})^2 = \tfrac{1}{4}.$$

This is illustrated in Fig. 12.45. ||

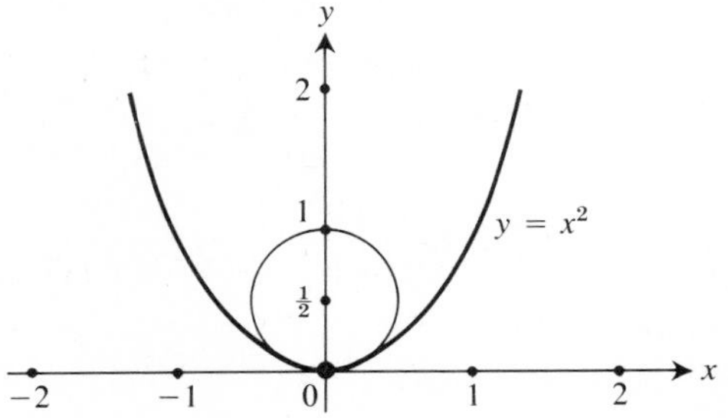

12.45

12.6.3 The curvature formula in parametric form

Let a curve be given parametrically by $x = h(t)$, $y = k(t)$ where h and k are twice differentiable functions of t. We have previously defined the arc length function

$$s(t) = \int_{t_0}^{t} \sqrt{h'(u)^2 + k'(u)^2}\, du. \tag{8}$$

If $ds/dt \neq 0$, then it can be shown that t appears as a differentiable function of s with

$$\frac{dt}{ds} = \frac{1}{\sqrt{(h'(t))^2 + (k'(t))^2}}. \tag{9}$$

For the angle ϕ shown in Fig. 12.42 you obtain

$$\phi = b + \tan^{-1}\frac{k'(t)}{h'(t)} \tag{10}$$

as analogue of Eq. (2), so that ϕ is a differentiable function of t. Under these conditions,

$$\begin{aligned}\frac{d\phi}{ds} &= \frac{d\phi}{dt}\cdot\frac{dt}{ds}\\ &= \frac{1}{1 + (k'(t)/h'(t))^2}\cdot\frac{h'(t)k''(t) - k'(t)h''(t)}{(h'(t))^2}\cdot\frac{1}{\sqrt{(h'(t))^2 + (k'(t))^2}}\\ &= \frac{h'(t)k''(t) - k'(t)h''(t)}{[(h'(t))^2 + (k'(t))^2]^{3/2}}.\end{aligned} \tag{11}$$

Formula (11) looks less cumbersome if we let

$$\dot{x} = \frac{dx}{dt} = h'(t) \quad \text{and} \quad \dot{y} = \frac{dy}{dt} = k'(t),$$

and denote second derivatives with respect to t by $\ddot{x}$ and $\ddot{y}$. You then obtain Formula (12) given in the following theorem.

Theorem 12.6 *Let h and k be twice-differentiable functions. The curvature of the parametric curve $x = h(t)$, $y = k(t)$ is given at a point corresponding to a value t by*

$$\kappa = \left|\frac{d\phi}{ds}\right| = \left|\frac{\dot{x}\ddot{y} - \dot{y}\ddot{x}}{(\dot{x}^2 + \dot{y}^2)^{3/2}}\right|. \tag{12}$$

Example 8 Let's find the curvature of the ellipse $x = 3\cos t$, $y = 4\sin t$ at the point $(3, 0)$ corresponding to $t = 0$, taking the direction of increasing s counterclockwise to coincide with the direction of increasing t.

SOLUTION From (12), you obtain

$$\begin{aligned}\kappa &= \left|\frac{\dot{x}\ddot{y} - \dot{y}\ddot{x}}{(\dot{x}^2 + \dot{y}^2)^{3/2}}\right|\\ &= \left|\frac{12\sin^3 t - (-12\cos^3 t)}{(9\sin^2 t + 16\cos^2 t)^{3/2}}\right|.\end{aligned}$$

For $t = 0$ corresponding to the point $(3, 0)$, you obtain

$$\kappa\big|_{t=0} = \frac{12}{16^{3/2}} = \frac{12}{64} = \frac{3}{16}. \quad \|$$

SUMMARY

1. *Curvature $\kappa = |d\phi/ds|$ where ϕ is the angle (measured counterclockwise) from the horizontal to the tangent to the curve, and s is arc length.*
2. *If $y = f(x)$, then*

$$\kappa = \left|\frac{d^2y/dx^2}{(1 + (dy/dx)^2)^{3/2}}\right|.$$

3. *If $x = h(t)$ and $y = k(t)$, then*

$$\kappa = \left|\frac{\dot{x}\ddot{y} - \dot{y}\ddot{x}}{(\dot{x}^2 + \dot{y}^2)^{3/2}}\right|,$$

where $\dot{x} = dx/dt$, $\ddot{x} = d^2x/dt^2$, and $\dot{y}$ and $\ddot{y}$ are similarly defined.

4. *The radius of curvature ρ at a point is equal to $1/\kappa$.*
5. *The osculating circle to a curve at a point is the circle with center on the concave side of the curve, radius $\rho = 1/\kappa$, and tangent to the curve.*

EXERCISES

In Exercises 1 through 10, find the curvature κ at the indicated point of the curve having the given x,y-equation or parametric equations.

1. $y = \sin x$ at $(\pi/2, 1)$

2. $xy = 1$ at $(1, 1)$

3. $y = \ln x$ at $(1, 0)$

4. $x^2y + 2xy^2 = 3$ at $(1, 1)$

5. $\dfrac{x^2}{16} - \dfrac{y^2}{25} = 1$ at $(4, 0)$ [*Hint.* The hyperbola defines x as a function of y near $(4, 0)$.]

6. $x = 4 \sin t$, $y = 5 \cos t$ where $t = 0$

7. The cycloid $x = a(\theta - \sin \theta)$, $y = a(1 - \cos \theta)$ where $\theta = \pi$

8. $x = e^t$, $y = t^2$ where $t = 0$

9. $x = \sqrt{25 - y^2}$ at $(3, -4)$

10. $x = \cosh t$, $y = \sinh t$ where $t = 0$

11. Give rate of change arguments that $d\phi/ds$ at a point of a curve is *positive* if the curve bends to the left as you travel in the direction of increasing s at the point, and is *negative* if the curve bends to the right.

12. Discuss the curvature at the origin of the graphs of the monomial functions ax^n for integers $n \geq 1$.

13. What is the curvature of the graph of a twice-differentiable function at an inflection point on the graph?

14. The osculating circle to a given curve at a point (x, y) has the same tangent line and the same radius of curvature as the given curve at (x, y).

a) Show that the given curve and the circle have the same curvature κ at (x, y).

b) Argue from (a) that if both the given curve and the circle are graphs of functions in a neighborhood of (x, y), then these two functions have the same first derivatives and the same second derivatives for this value x.

15. Prove that the only twice-differentiable functions whose graphs have curvature zero at each point are those functions having straight lines as graphs.

16. Find the equation of the osculating circle to the curve $y = \ln x$ at $(1, 0)$.

17. Find the equation of the osculating circle to the hyperbola $(x^2/4) - (y^2/9) = 1$ at the point $(-2, 0)$.

18. Let the graph $y = f(x)$ of a twice-differentiable function f have nonzero curvature at a point (x, y). Show that the coordinates (α, β) of the center of curvature of the graph at (x, y) are given by

$$\alpha = x - y'\frac{1 + (y')^2}{y''}, \qquad \beta = y + \frac{1 + (y')^2}{y''}.$$

The locus of centers of curvature of a given curve is the ***evolute of the curve****. The given curve is an* ***involute*** *of this locus of centers of curvature. (While a curve has only one evolute, a single locus may have many involutes, i.e., it may be the evolute of many curves, as illustrated by the next exercise.)*

19. a) What is the evolute of a circle?

b) What are the involutes of a point?

20. Find parametric equations for the evolute of the parabola $y = x^2$. [*Hint.* Use the formulas given in Exercise 18.]

exercise sets for chapter 12

review exercise set 12.1

[*If you did Section* 12.1 *but not Sections* 12.2, 12.3, *and* 12.4, *replace Problems* 1, 2, *and* 3 *by these.*

1.1. Sketch the curve $x^2 - 4y^2 = 16$.

1.2. Sketch the curve $x + 4y^2 - 8y = -12$.

1.3. Sketch the curve $3x^2 + y^2 - 12x + 4y = -4$.]

1. Find the equation of the conic section with focus $(3, 2)$, directrix $x = -1$, and eccentricity 2.

2. Find the foci and directrices of the conic section with equation $2x^2 + 4x + y^2 - 2y + 1 = 0$, and sketch the curve.

3. Classify the curve as a (possibly imaginary or degenerate) ellipse, hyperbola, or parabola.

a) $x^2 - 3xy + 2y^2 - 3x + 4y = 7$

b) $x^2 + 2xy + y^2 - 2x + 4y = 11$

c) $x^2 + 3xy + 3y^2 - 8x + 6y = 0$

4. Sketch the curve described parametrically by $x = t^2$, $y = t^2 - 1$.

5. Find dy/dx and d^2y/dx^2 at $t = 1$ if $x = t^2$, $y = t^3 - 2t$.

6. Paramatrize the curve $y = x^3$, taking as parameter the y-coordinate h at each point on the curve.

7. Express as an integral the length of the curve $x = \sin t$, $y = t^2 - 2$ from $t = 0$ to $t = 3$.

8. Find the curvature of the parabola $y = x^2$ at the point $(2, 4)$.

9. Find the curvature of the curve $x = \sin 2t$, $y = \cos 3t$ at the point $t = \pi/6$.

10. Find the equation of the osculating circle to the curve $y = \cos x$ at the point $(0, 1)$.

review exercise set 12.2

[*If you did Section* 12.1 *but not Sections* 12.2, 12.3, *and* 12.4, *replace Problems* 1, 2, *and* 3 *by these.*

1.1. Sketch the curve $x^2 + 10x + 4y = 3$.

1.2. Sketch the curve $2x^2 - y^2 - 4x + 6y = 1$.

1.3. Sketch the curve $4x^2 + y^2 - x + 3y = \frac{27}{16}$.]

1. Find the equation of the conic section with focus $(-1, 1)$, directrix $y = 2$, and eccentricity $\frac{1}{3}$.
2. Find the foci, directrices, and asymptotes of the hyperbola $4x^2 - 16x - y^2 - 2y = -11$, and sketch the curve.
3. Rotate axes, complete squares, and then sketch the curve $x^2 - 2xy + y^2 + \sqrt{2}x - 3\sqrt{2}y = 4$.
4. Sketch the curve described parametrically by $x = 1 + 3 \sin t$, $y = 2 - \cos t$.
5. Find dy/dx and d^2y/dx^2 at $t = \pi/3$ if $x = \sin t$, $y = \cos 2t$.
6. Paramatrize the curve $y = e^{-x/2}$, taking as parameter the slope m of the curve at each point.
7. Find the point (x_0, y_0) on the curve $x = t$, $y = \frac{2}{3}t^{3/2}$ such that the distance from the origin to (x_0, y_0) measured along the curve is 42.
8. Find the curvature of the curve $y = \sin x$ at the point where $x = \pi/4$.
9. Find the curvature of the curve $x = t^2$, $y = t^3 + 2t$ at the point where $t = 1$.
10. Find the equation of the osculating circle to the curve $x = 2 \cos t$, $y = 3 \sin t$ at the point $t = \pi/2$.

more challenging exercises 12

1. Consider ellipses of the form $(x^2/a^2) + (y^2/b^2) = 1$, where a is held constant and b approaches a.
   a) What does the limiting locus of such ellipses approach as b approaches a?
   b) What is the limiting position of the foci as b approaches a?
   c) What is the limiting position of the directrices as b approaches a?
   d) What is the limiting value of the eccentricity?
2. Answer Exercise 1 if $a \to \infty$ while b is held constant, rather than $b \to a$.
3. Answer Exercise 1 for the hyperbolas $(x^2/a^2) - (y^2/b^2) = 1$ if a is held constant and $b \to \infty$.
4. Answer Exercise 1 for the hyperbolas $(x^2/a^2) - (y^2/b^2) = 1$ if b is held constant and $a \to \infty$.
5. Answer Exercise 1 for the hyperbolas $(x^2/a^2) - (y^2/b^2) = 1$ if a is held constant and $b \to 0$.
6. Answer Exercise 1 for the hyperbolas $(x^2/a^2) - (y^2/b^2) = 1$ if b is held constant and $a \to 0$.
7. Let f be a twice differentiable function. Show that $|\kappa(x)| \leq |f''(x)|$, where $\kappa(x)$ is the curvature of the graph at $(x, f(x))$.
8. Let $f(x) = 1 + x + (x/2)^2 + \cdots + (x/n)^n + \cdots$ Find the curvature of the graph where $x = 0$.
9. Find the length of arc of the curve $x = a \cos^3 t$, $y = a \sin^3 t$ for $0 \leq t \leq 2\pi/3$.

polar coordinates

13

13.1 THE POLAR COORDINATE SYSTEM

13.1.1 Polar coordinates

We have used cartesian x,y-coordinates so far for all our work in the plane. Let's look at a different coordinate system, and its relation to x,y-coordinates. Choose any point O in the Euclidean plane as the *pole* for our coordinate system, and any half-line emanating from O as the *polar axis.* It is conventional in sketching to let the polar axis be a horizontal half-line, extending to the right, as shown in Fig. 13.1. We choose a scale on the polar axis, as indicated in the figure.

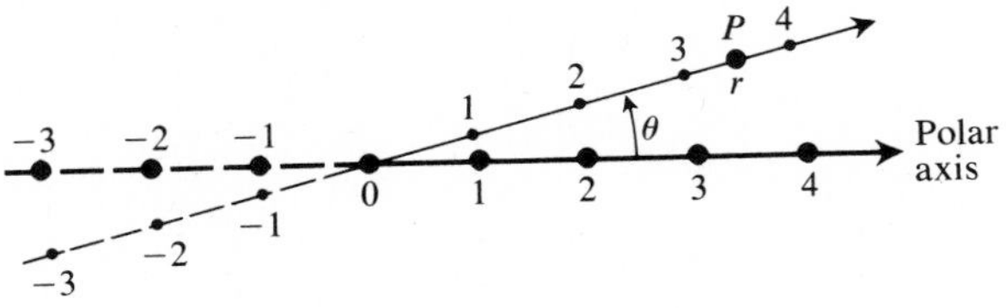

13.1

Let P be any point of the plane, and rotate the polar axis through an angle θ so that the rotated axis passes through the point P, as indicated in Fig. 13.1. (We let positive values of θ correspond to counterclockwise rotation). The point P falls at a number r on the scale of the rotated axis, and the ordered pair (r, θ) of numbers constitutes **polar coordinates** for the point P.

In a cartesian coordinate system, each point of the plane corresponds to a *unique* ordered pair (x, y) of numbers. A disadvantage of a polar coordinate system is that each point has an *infinite* number of polar coordinate pairs. As indicated in Fig. 13.2, if a point P has polar coordinates (r, θ) it also has polar coordinates $(r, \theta + 2n\pi)$ for each integer n. Figure 13.3 shows that the same point P also has polar coordinates $(-r, \theta + \pi + 2n\pi)$ for all integers n. All polar coordinates of a point (r, θ) different from O are of the form

$$(r, \theta + 2n\pi) \qquad \text{or} \qquad (-r, \theta + \pi + 2n\pi).$$

Note that the pole O has coordinates $(0, \theta)$ for every real number θ.

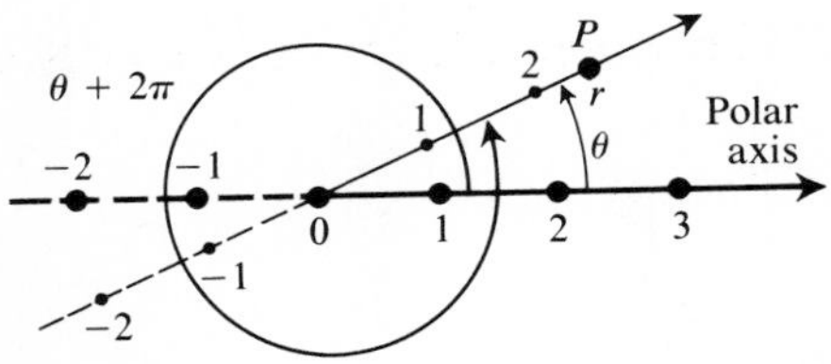

13.2

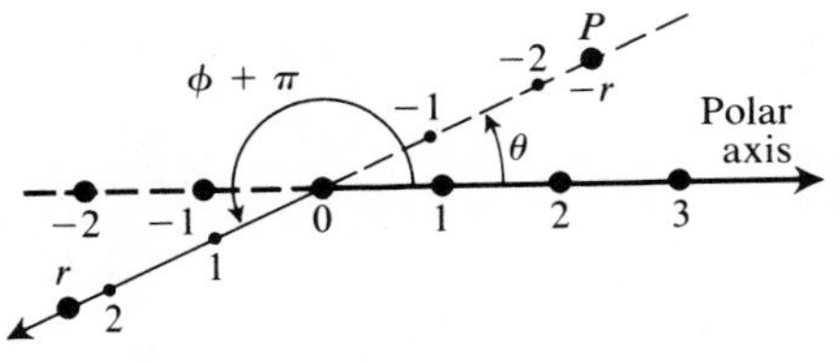

13.3

Let's simultaneously consider a cartesian and a polar coordinate system for the plane. It is conventional to let the pole be at the Euclidean origin (0, 0) with the polar axis falling on the positive x-axis. (We shall always

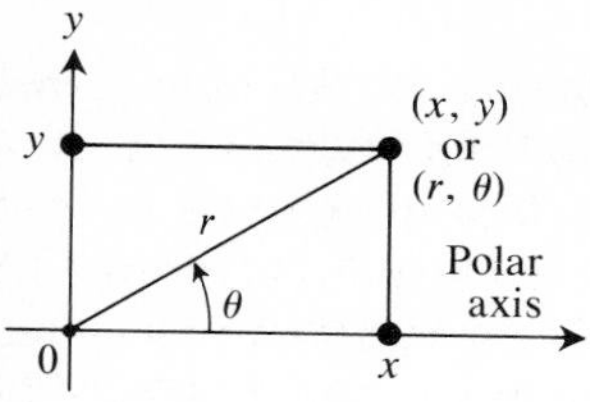

13.4

follow this convention.) From Fig. 13.4, you see that the cartesian x,y-coordinates can be expressed in terms of polar r,θ-coordinates by the equations

$$\begin{aligned} x &= r\cos\theta, \\ y &= r\sin\theta. \end{aligned} \tag{1}$$

It also follows at once from the figure that

$$\begin{aligned} r^2 &= x^2 + y^2, \\ \theta &= \tan^{-1}\frac{y}{x}, \end{aligned} \tag{2}$$

provided that you can select θ such that $-\pi/2 < \theta < \pi/2$. Equations (1) and (2) are useful in changing from one coordinate system to the other.

Example 1 From Eqs. (1), you find that the point with polar coordinates $(r, \theta) = (2, \pi/3)$ has cartesian coordinates

$$(x, y) = \left(2\cos\frac{\pi}{3}, 2\sin\frac{\pi}{3}\right) = (1, \sqrt{3}).$$

From Eqs. (2), the point with cartesian coordinates $(x, y) = (-\sqrt{3}, 1)$ has polar coordinates (r, θ) such that

$$r^2 = x^2 + y^2 = 4$$

and

$$\theta = \tan^{-1}\frac{y}{x} = \tan^{-1}\left(-\frac{1}{\sqrt{3}}\right) = -\frac{\pi}{6}.$$

Since $(-\sqrt{3}, 1)$ lies in the second cartesian quadrant, you easily see that $r = -2$, and the point has polar coordinates $(r, \theta) = (-2, -\pi/6)$. Of course, the point also has polar coordinates

$$\left(-2, -\frac{\pi}{6} + 2n\pi\right) \quad \text{and} \quad \left(2, \frac{5\pi}{6} + 2n\pi\right)$$

for all integers n. This also may be seen graphically, as shown in Fig. 13.5. ‖

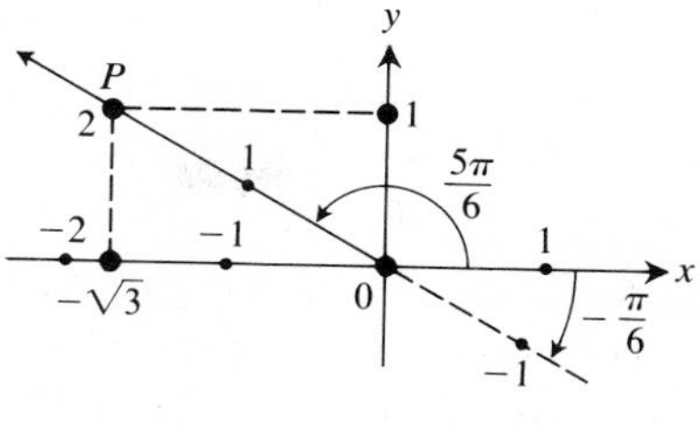

13.5

In such a coordinate system for the plane, among the first things you examine are the *level coordinate curves* of the system, obtained by putting the coordinate variables equal to constants. In a cartesian system, a level curve is either a vertical line $x = c$ or a horizontal line $y = k$. In a polar coordinate system, a level curve is either a locus $r = a$, which is a circle about the pole of radius a, or a locus $\theta = \beta$, which is a line through the pole, as indicated in Fig. 13.6. This suggests that polar coordinates might be useful in handling circles and lines through the origin.

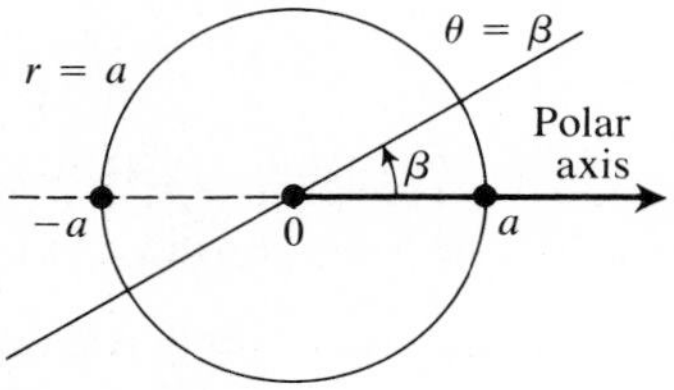

13.6

The Eqs. (1) and (2) can be used to express a cartesian characterization of a curve in polar form, and vice versa. For example, you see immediately from Eqs. (1) that the vertical line $x = c$ has polar equation $r \cos \theta = c$, or $r = c \sec \theta$.

Example 2 Let's sketch the polar curve

$$r = a \sin \theta$$

by changing it to cartesian form.

SOLUTION If $r \neq 0$, our equation is equivalent to $r^2 = ar \sin \theta$. Since $r = 0$ if $\theta = 0$ in our original equation, we see that our new equation actually gives exactly the original locus. By Eqs. (1) and (2), the equation $r^2 = ar \sin \theta$ has cartesian form

$$x^2 + y^2 = ay,$$

or

$$x^2 + \left(y - \frac{a}{2}\right)^2 = \frac{a^2}{4}.$$

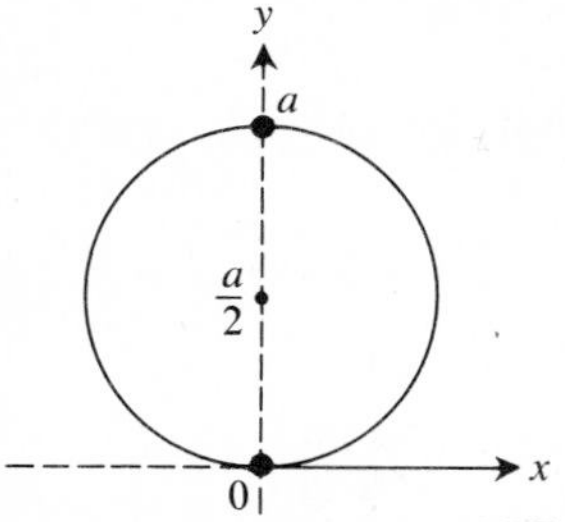

13.7

We see that the locus is the circle with center at the point $(x, y) = (0, a/2)$ and radius $a/2$, as shown in Fig. 13.7. ‖

Example 3 Let's find the polar coordinate equation of the ellipse with x,y-equation $x^2 + 3y^2 = 10$.

SOLUTION We use Eqs. (1) and obtain

$$r^2 \cos \theta + 3(r^2 \sin^2\theta) = 10, \qquad r^2(\cos^2\theta + 3 \sin^2\theta) = 10,$$

$$r^2(1 + 2 \sin^2\theta) = 10.$$

Thus

$$r^2 = \frac{10}{1 + 2\sin^2\theta}. \quad \|$$

***13.1.2 Conic sections in polar coordinates**

We describe equations of the conic sections in polar coordinates. Let a conic section of eccentricity e have focus at the pole O and have the line $x = -q$ as directrix, as shown in Fig. 13.8. Using the notation of that figure, we have

$$FP = e(DP).$$

Now

$$FP = r \qquad \text{and} \qquad DP = q + (FP)\cos\theta = q + r\cos\theta.$$

Thus the polar equation of our conic section is

$$r = e(q + r\cos\theta)$$

or

$$r = \frac{qe}{1 - e\cos\theta}. \tag{3}$$

From Eq. (3), you see that every equation of the form

$$r = \frac{c}{1 - e\cos\theta}, \tag{4}$$

where $c > 0$ and $e > 0$, has as locus a conic section of eccentricity e and focus at the pole, for if you let $q = c/e$, you obtain Eq. (3).

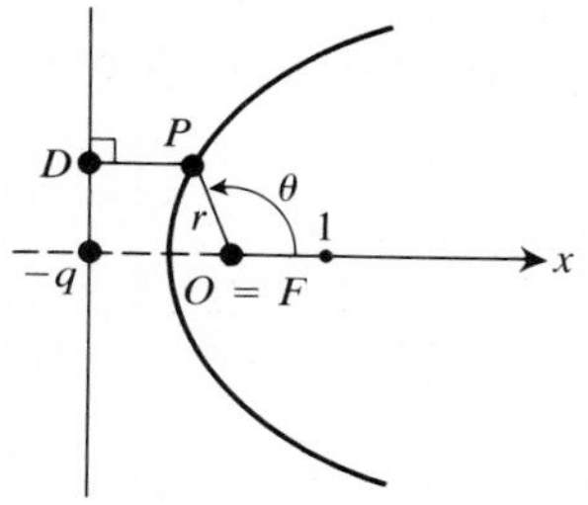

13.8

Example 4 The equation

$$r = \frac{4}{1 - \cos\theta}$$

* Omit this section if you did not study Section 12.2.

has as locus a parabola with focus at O and directrix at $x = -4$.

The equation

$$r = \frac{4}{1 - 2\cos\theta} = \frac{2 \cdot 2}{1 - 2\cos\theta}$$

has as locus a hyperbola with a focus at O and corresponding directrix $x = -2$. ‖

SUMMARY

1. *A point with the polar coordinates (r, θ) also has as polar coordinates $(r, \theta + 2n\pi)$ and $(-r, \theta + \pi + 2n\pi)$ for every integer n.*
2. *The equations for changing from x,y-coordinates to polar r,θ-coordinates and vice versa are*

$$x = r\cos\theta, \qquad r^2 = x^2 + y^2,$$

$$y = r\sin\theta, \qquad \theta = \tan^{-1}\frac{y}{x}.$$

3. *The polar equation*

$$r = \frac{c}{1 - e\cos\theta} \quad \text{for} \quad c > 0,\ e > 0$$

has as locus a conic section of eccentricity e and focus at the pole. The directrix is the line $x = -q$, where $q = c/e$.

EXERCISES

In Exercises 1 through 8, find the cartesian coordinates of the points with the given polar coordinates.

1. $(4, \pi/4)$ **2.** $(0, 5\pi/7)$ **3.** $(6, -\pi/2)$ **4.** $(3, 5\pi/6)$

5. $(-2, \pi/4)$ **6.** $(-1, 2\pi/3)$ **7.** $(-4, 3\pi)$ **8.** $(-2, 11\pi/4)$

In Exercises 9 through 12, find all polar coordinates of the points with the given cartesian coordinates.

9. $(2, 2)$ **10.** $(-1, \sqrt{3})$ **11.** $(-2, -3)$ **12.** $(0, -5)$

13. Show that the distance d in the plane between points having polar coordinates (r_1, θ_1) and (r_2, θ_2) is

$$\sqrt{r_1^2 - 2r_1r_2\cos(\theta_1 - \theta_2) + r_2^2}.$$

[*Hint.* Use the distance formula in cartesian form and Eqs. (1).]

14. Find the polar equation of the circle with $(r, \theta) = (3, \pi/4)$ as center and radius 5.

15. Find the polar equation of the line with x,y-equation $2x + 3y = 5$.

16. Find the polar equation of the parabola $y^2 = 8x$.

17. Find the x,y-equation of the curve with polar equation $r = 2a\cos\theta$.

***18.** Find the polar coordinate equation of the parabola with focus at the pole and directrix $x = -3$.

***19.** Find the eccentricity and directrix of the conic section with polar equation

$$r = \frac{5}{2 - 3\cos\theta}.$$

*20. Find the length of the major axis of the ellipse with polar equation

$$r = \frac{8}{1 - \frac{1}{2}\cos\theta}.$$

*21. Find the length of the minor axis of the ellipse in Exercise 20.

13.2 SKETCHING CURVES IN POLAR COORDINATES

13.2.1 Sketching curves

We turn now to the problem of sketching a curve given in polar form without changing back to cartesian form. We limit ourselves to simple polar equations of the form $r = f(\theta)$, where $f(\theta)$ usually involves just a sine or cosine function to the first power. You would like to plot points corresponding to local maximum or local minimum values of r, to find where the curve is nearest to or farthest from the origin. The technique is best illustrated by examples.

Example 1 Let's sketch the *cardioid*

$$r = a(1 + \cos\theta).$$

SOLUTION Since θ appears in the equation only in "$\cos\theta$", we plot points (r, θ) "every 90°," starting with $\theta = 0$. We mark in the "90° lines" with dashes in Fig. 13.9. Of course, our curve $r = a(1 + \cos\theta)$ repeats itself after θ runs through 2π radians.

We find that for $\theta = 0$, we have $r = a(1 + \cos 0) = 2a$; we mark this point on the polar axis. As θ increases to $\pi/2$, $\cos\theta$ decreases from 1 to 0 and r decreases from $2a$ to a. This enables us to sketch the first-quadrant portion of our curve, as shown in Fig. 13.9. To establish the actual shape of the curve in the first quadrant, you could plot (r, θ) for a few more values of θ, say $\theta = \pi/6$, $\pi/4$, and $\pi/3$. We shall not bother to plot these points, but will content ourselves with rough sketches. Note that $r \geq 0$ for all θ since $\cos\theta \geq -1$. Thus we may always measure r along our *positive* rotated polar axis.

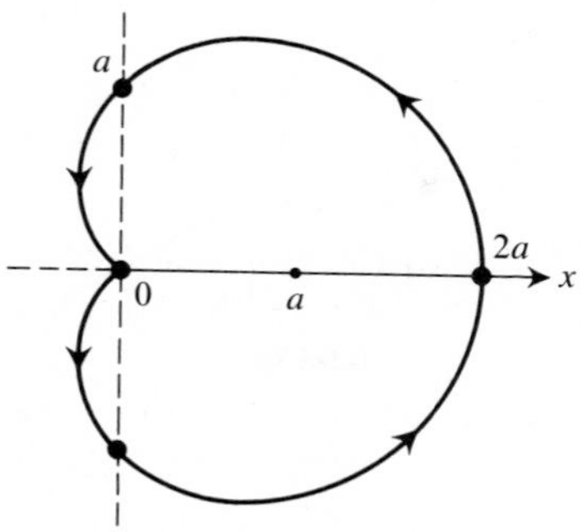

13.9

Cardioid: $r = a(1 + \cos\theta)$

As θ increases from $\pi/2$ to π, we see that $\cos\theta$ decreases from 0 to -1, so r decreases from a to 0; this enables us to sketch the second-quadrant portion of the curve. In a similar fashion, we see that r increases from 0 to a through the third quadrant, and from a to $2a$ through the fourth quadrant. The origin of the name *cardioid* for this curve is clear from the shape of the curve. The arrows in Fig. 13.9 indicate the direction of increasing θ along the curve. ||

Example 2 Let's sketch the *four-leaved rose*

$$r = a \sin 2\theta.$$

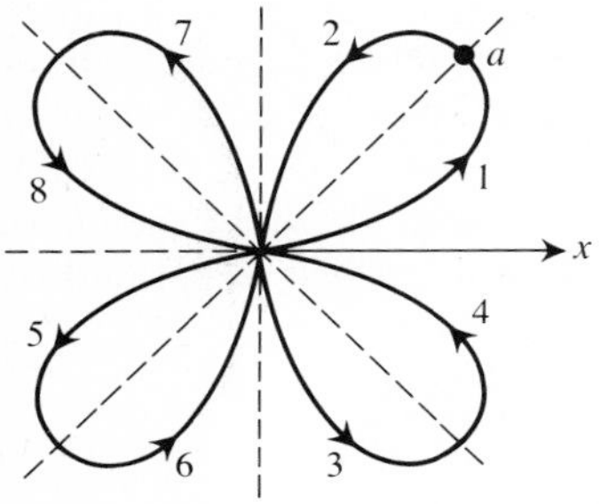

13.10

Four-leaved rose: $r = \sin 2\theta$

SOLUTION We are interested in plotting points where r assumes maximum and minimum values, or becomes zero; this occurs every 90° for 2θ, or every 45° for θ. Thus we plot the curve "every 45°," and start by marking the dashed "45° lines" in Fig. 13.10.

As θ increases from 0 to $\pi/4$, we see that r increases from 0 to a; this gives the arc of the curve we have numbered 1 in Fig. 13.10. As θ increases from $\pi/4$ to $\pi/2$, r decreases from a to 0, giving the arc numbered 2.

Now as θ increases from $\pi/2$ to $3\pi/4$, we see that r runs through the *negative* values from 0 to $-a$, and our curve therefore lies in the "diagonally opposite wedge" and forms the arc numbered 3 in our figure. You can check that as θ continues to increase up to 2π, you obtain in succession the arcs numbered 4 through 8 for each 45° increment in θ. The arrows on the arcs indicate the direction of increasing θ. ||

Example 3 The curve with polar equation

$$r = \theta$$

is a double spiral. This time the curve does *not* repeat after θ runs through 2π radians, for r increases without bound as θ increases. The double spiral is shown in Fig. 13.11. Again, the arrows indicate the direction of increasing θ. ||

Example 4 The curve

$$r = a(1 + 2\cos\theta)$$

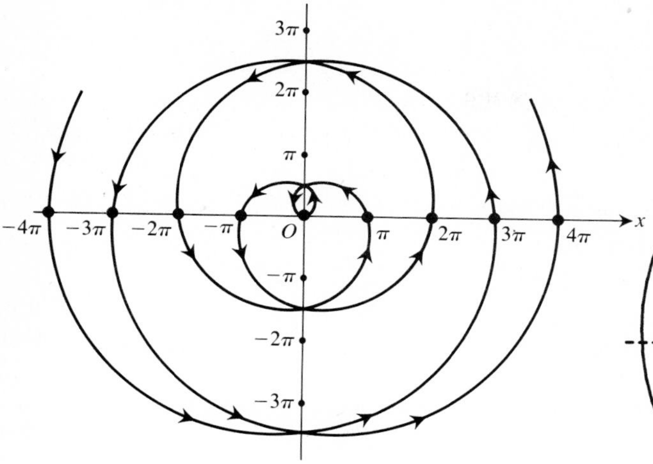

Spiral of Archimedes: $r = \theta$
(a) Artist's Figure

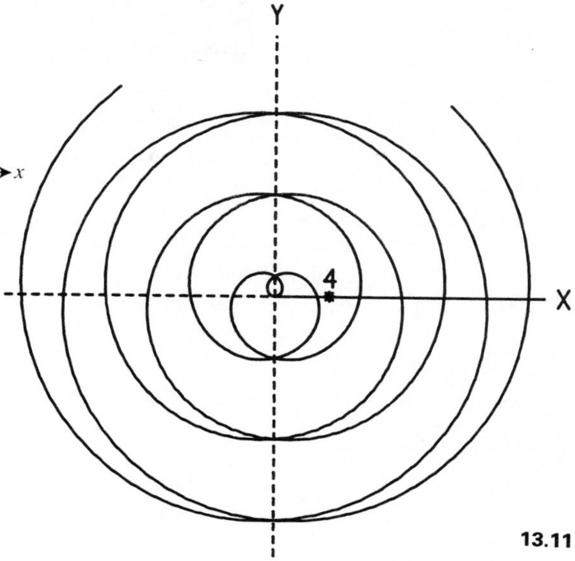

13.11

(b) Computer generated figure

is sketched in Fig. 13.12. Note that $r = 0$ when $\cos\theta = -1/2$, or when $\theta = 2\pi/3$ and $\theta = 4\pi/3$. At these values of θ, r changes sign from positive to negative and from negative to positive respectively; our curve goes through the pole O tangent to the rays $\theta = 2\pi/3$ and $\theta = 4\pi/3$. ‖

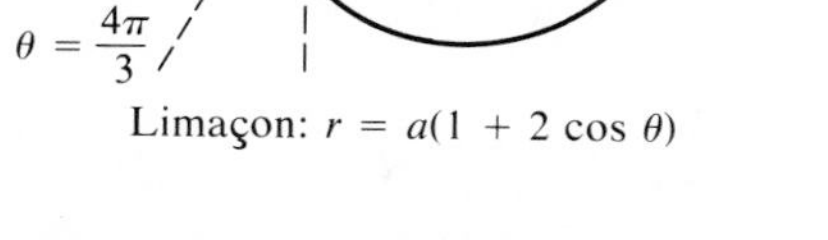

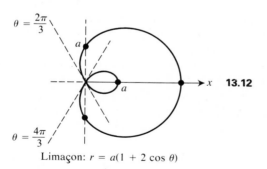

13.12

Limaçon: $r = a(1 + 2\cos\theta)$

13.2.2 Intersections of curves in polar coordinates

The problem of finding intersections of curves in polar coordinates is more complicated than in cartesian coordinates, since a point may have many polar coordinates. For example, the point with cartesian coordinates $(x, y) = (0, -1)$ lies on the polar curve $r = 3 + 2\sin\theta$, for the polar coordinate pair $(1, 3\pi/2)$ satisfies this equation. This same point also lies on the polar curve $r = \cos 2\theta$, since the polar pair $(-1, \pi/2)$ satisfies this equation. Note that the polar coordinate pair $(-1, \pi/2)$ does *not* satisfy the first equation $r = 3 + 2 \sin \theta$. Thus if we simply solve $r = 3 + 2 \sin \theta$ and $r = \cos 2\theta$ simultaneously for r and θ, we would "miss" this point of intersection. We illustrate how one should find the point of intersection of these curves in the following example.

Example 5 Let's find all points of intersection of

$$r = 3 + 2 \sin \theta \qquad \text{and} \qquad r = \cos 2\theta.$$

SOLUTION First, we check whether the pole O lies on both curves. Now $3 + 2 \sin \theta$ is never zero, since $-1 \leq \sin \theta \leq 1$, so the pole does not lie on even the first curve.

We now try to find coordinates (r, θ) which satisfy the first equation while coordinates $(r, \theta + 2n\pi)$ or $(-r, \theta + \pi + 2n\pi)$ satisfy the second equation; that is, we find solutions of either

$$3 + 2 \sin \theta = \cos 2(\theta + 2n\pi) \tag{1}$$

or

$$3 + 2 \sin \theta = -\cos 2(\theta + \pi + 2n\pi). \tag{2}$$

From Eq. (1), we obtain

$$\begin{aligned} 3 + 2 \sin \theta &= \cos (2\theta + 4n\pi) \\ &= \cos 2\theta = 1 - 2 \sin^2 \theta. \end{aligned}$$

This yields

$$\sin^2 \theta + \sin \theta + 1 = 0,$$

or, solving by the quadratic formula,

$$\sin \theta = \frac{-1 \pm \sqrt{1 - 4}}{2},$$

which has no real solutions.

Turning to Eq. (2), we obtain

$$\begin{aligned} 3 + 2 \sin \theta &= -\cos (2\theta + 2\pi + 4n\pi) = -\cos 2\theta \\ &= 2 \sin^2 \theta - 1, \end{aligned}$$

which yields

$$\sin^2 \theta - \sin \theta - 2 = 0$$

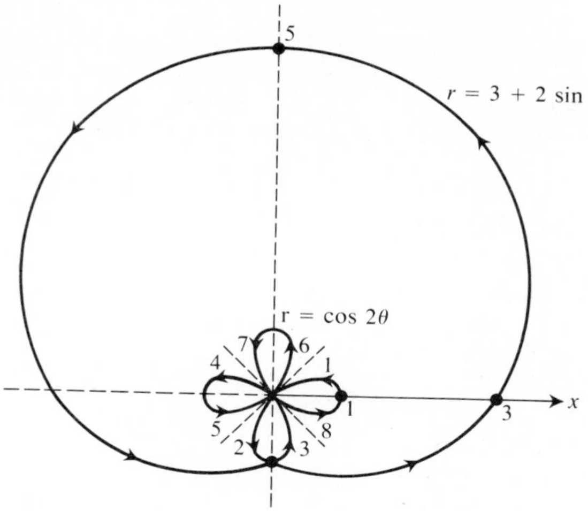

13.13

or

$$(\sin \theta - 2)(\sin \theta + 1) = 0.$$

Since $\sin \theta = 2$ is impossible, we are left with $\sin \theta = -1$ or $\theta = 3\pi/2$. Thus the point $(1, 3\pi/2)$ is the only point of intersection of the curves, which are sketched in Fig. 13.13. ‖

SUMMARY

1. *When plotting a polar curve $r = f(\theta)$, plot those points where r assumes relative maximum or minimum values, or becomes zero. For example, $r = 4 \cos 3\theta$ should be plotted in increments of 30° in θ, starting with $\theta = 0$.*
2. *To find points of intersection of two polar curves, find (r, θ) satisfying the first equation for which some points $(r, \theta + 2n\pi)$ or $(-r, \theta + \pi + 2n\pi)$ satisfy the second equation. Check separately to see if the origin lies on both curves; that is, if r can be zero.*

EXERCISES

In Exercises 1 through 15, sketch the curve having the given polar equation.

1. $r = a\theta$ (Spiral of Archimedes)

2. $r\theta = a$ (Hyperbolic spiral)

3. $r = a \sin 3\theta$ (Three-leaved rose)

4. $r = 2a \cos \theta$

5. $r = 2a \sin \theta$

6. $r^2 = a^2 \cos 2\theta$ (Lemniscate of Bernoulli)

7. $r^2 = a^2 \sin 2\theta$

8. $r = a(1 - \cos \theta)$

9. $r = ae^{\theta}$ (Logarithmic spiral)

10. $r = 3 \csc \theta$

11. $r = 1 + 2 \sin \theta$

12. $r = 3 + 2 \cos \theta$

13. $r = 2 + 3 \cos \theta$

14. $r = a \sin \dfrac{\theta}{2}$

15. $r = a\left(1 + \cos \dfrac{\theta}{2}\right)$

16. Find all points of intersection of the polar curves $r = a \sin \theta$ and $r = a \cos \theta$.

17. Find all points of intersection of the polar curves $r = a$ and $r^2 = 2a^2 \sin 2\theta$.

18. Find all points of intersection of the polar curves $r = a \cos 2\theta$ and $r = a(1 + \cos \theta)$.

13.3 AREA IN POLAR COORDINATES

Let's attempt to find the area A of a region bounded by a polar curve $r = f(\theta)$, where f is a continuous function, and by two rays $\theta = \theta_1$ and $\theta = \theta_2$, as shown in Fig. 13.14.

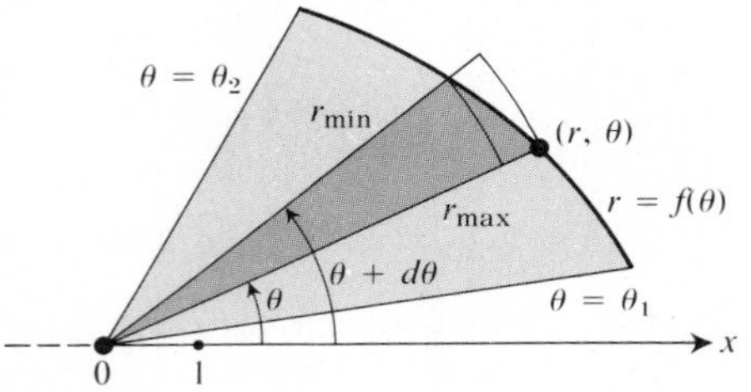

13.14

From Fig. 13.14, we see that the area of the dark-shaded wedge corresponding to the interval $[\theta, \theta + d\theta]$ is greater than the area of a sector of a circle having central angle $d\theta$ and radius r_{min} which is the minimum value of $f(\theta)$ over the interval $[\theta, \theta + d\theta]$. On the other hand, this area is less than the area of a sector of a circle having central angle $d\theta$ and radius r_{max}. Since the area of a sector with central angle $d\theta$ and radius a is

$$\frac{d\theta}{2\pi}\pi a^2 = \frac{1}{2}a^2\,d\theta,$$

you see that you obtain from Fig. 13.14

$$\tfrac{1}{2}(r_{min})^2\,d\theta \le \text{area of the dark wedge} \le \tfrac{1}{2}(r_{max})^2\,d\theta.$$

Since $r = f(\theta)$ is a continuous function, the theory of the integral shows that the area A of the whole shaded region in Fig. 13.14 is

$$A = \int_{\theta_1}^{\theta_2} \frac{1}{2} r^2\,d\theta = \int_{\theta_1}^{\theta_2} \frac{1}{2}(f(\theta))^2\,d\theta. \qquad (1)$$

Example 1 Let's find the total area of the regions bounded by the lemniscate $r^2 = a^2 \cos 2\theta$ shown in Fig. 13.15.

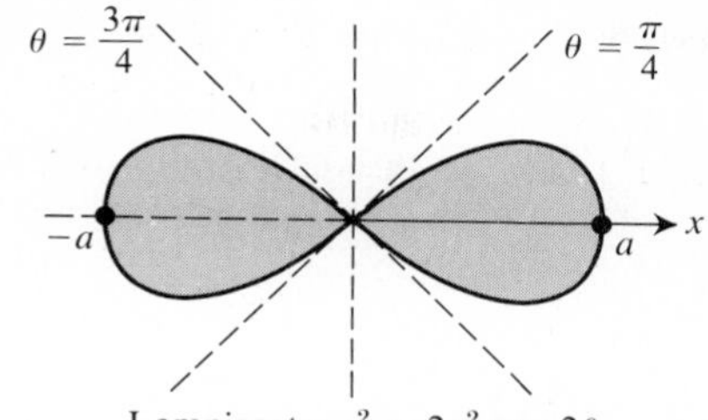

13.15

Lemniscate: $r^2 = 2a^2 \cos 2\theta$

SOLUTION By symmetry, we may find the area of the first-quadrant portion of our region, and multiply by four. We thus obtain from (1)

$$4\int_0^{\pi/4} \frac{1}{2} r^2 \, d\theta = 4\int_0^{\pi/4} \frac{1}{2} a^2 \cos 2\theta \, d\theta = 2a^2 \int_0^{\pi/4} \cos 2\theta \, d\theta$$

$$= 2a^2 \frac{1}{2} \sin 2\theta \Big]_0^{\pi/4} = a^2\left(\sin \frac{\pi}{2}\right) - a^2 (\sin 0) = a^2$$

as our desired area. ‖

Example 2 Let's find the area of the region that is inside the cardioid $r = a(1 + \cos \theta)$ but outside the circle $r = a$, shown shaded in Fig. 13.16.

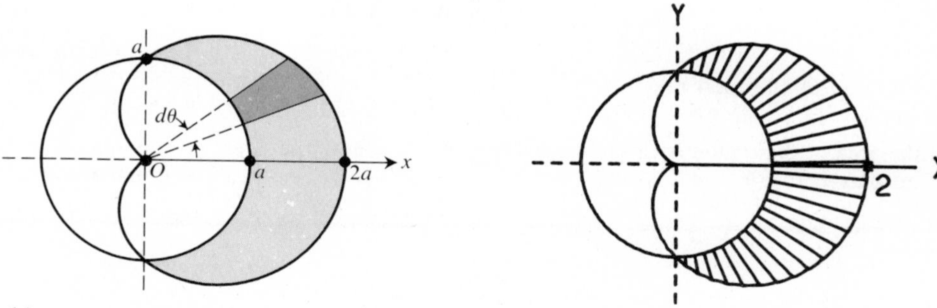

(a) Artist's figure (b) Computer generated figure

13.16

SOLUTION We may double the area of the first-quadrant portion of our region. The area of the dark partial wedge of central angle $d\theta$ shown in Fig. 13.16 is approximately

$$\tfrac{1}{2}(r_{\text{cardioid}})^2 \, d\theta - \tfrac{1}{2}(r_{\text{circle}})^2 \, d\theta,$$

and you obtain

$$2 \cdot \frac{1}{2}\int_0^{\pi/2} [a^2(1 + \cos\theta)^2 - a^2]\, d\theta = \int_0^{\pi/2} (a^2 + 2a^2\cos\theta + a^2\cos^2\theta - a^2)\, d\theta$$

$$= a^2\int_0^{\pi/2} (2\cos\theta + \cos^2\theta)\, d\theta = a^2\left(2\sin\theta + \frac{\theta}{2} + \frac{\sin 2\theta}{4}\right)\Bigg]_0^{\pi/2}$$

$$= a^2\left(2 + \frac{\pi}{4}\right) - a^2(0) = 2a^2 + \frac{\pi a^2}{4}$$

as the desired area. ‖

SUMMARY 1. *To find the area of a region bounded by polar curves:*

STEP 1. *Draw a figure.*

STEP 2. *Draw polar rays corresponding to a small increment $d\theta$ in θ.*

STEP 3. *Write down the area of the resulting wedge. A wedge with vertex at the origin, small central angle $d\theta$, and going out to $r = f(\theta)$ has approximate area $dA = \frac{1}{2}r^2\, d\theta = \frac{1}{2}f(\theta)^2\, d\theta$.*

STEP 4. *Integrate between the appropriate limits.*

EXERCISES

1. Use integration in polar coordinates to find the area of a circle of radius a.
2. Find the area of the total region enclosed by the cardioid $r = a(1 + \sin\theta)$.
3. Find the area of the total region enclosed by the polar curve $r^2 = a^2 \sin 2\theta$.
4. Find the area of the region enclosed by one leaf of the four-leaved rose $r = a\cos 2\theta$.
5. Find the total area of the regions inside the four-leaved rose $r = 2a\cos 2\theta$ but outside the circle $r = a$.
6. Find the area of the region bounded by the portion of the hyperbolic spiral $r\theta = 1$ where $\pi/2 \le \theta \le \pi$ and by the rays $\theta = \pi/2$ and $\theta = \pi$.
7. Find the area of the region in common to the circles $r = 2a\cos\theta$ and $r = 2a\sin\theta$.

13.4 THE ANGLE ψ AND ARC LENGTH

13.4.1 The angle ψ from the radius to the tangent

We would like to find the direction of the tangent line to a smooth curve. From Fig. 13.17, we see that if we can find the angle ψ from the radius vector to the tangent, then we can find the angle ϕ that the tangent makes with the x-axis, for by plane geometry,

$$\phi = \theta + \psi. \tag{1}$$

We shall show that if $r = f(\theta)$, where f is a differentiable function, then

$$\tan\psi = \frac{r}{dr/d\theta} = \frac{f(\theta)}{f'(\theta)} \tag{2}$$

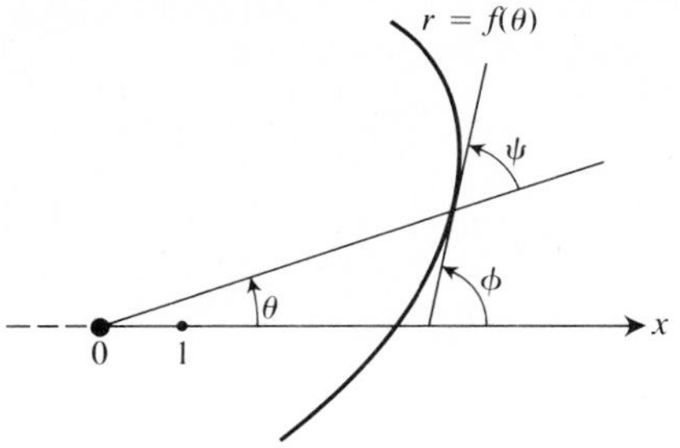

13.17

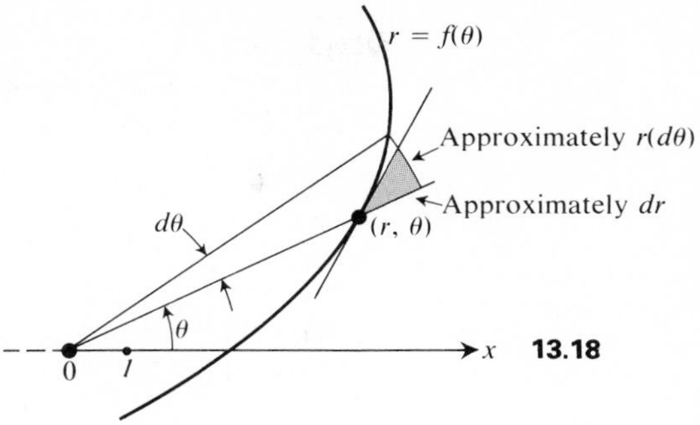

13.18

if $f'(\theta) \neq 0$. Figure 13.18 indicates how formula (2) can be remembered. At a point (r, θ) on the curve, take a small change in r as θ is held constant, and then a small change in θ as r is held constant; this gives the shaded "differential right triangle" with hypotenuse along the tangent to the curve. (Actually, one leg of this "triangle" is a circular arc.) For small $d\theta$, the legs of the "right triangle" are approximately of lengths dr and $r(d\theta)$, which at once suggests that

$$\tan \psi = \frac{r(d\theta)}{dr} = \frac{r}{dr/d\theta}.$$

For a careful derivation of Eq. (2), note from Eq. (1) that

$$\tan \psi = \tan (\phi - \theta) = \frac{\tan \phi - \tan \theta}{1 + \tan \phi \tan \theta}. \tag{3}$$

Now $\tan \phi = dy/dx$, and from the parametric equations

$$x = r \cos \theta = f(\theta) \cos \theta, \qquad y = r \sin \theta = f(\theta) \sin \theta,$$

you obtain

$$\tan \phi = \frac{dy}{dx} = \frac{dy/d\theta}{dx/d\theta} = \frac{r \cos \theta + (dr/d\theta) \sin \theta}{-r \sin \theta + (dr/d\theta) \cos \theta}. \tag{4}$$

Substituting in (3) the value for $\tan \phi$ found in (4), and putting $\tan \theta = (\sin \theta)/(\cos \theta)$, you obtain a compound quotient which you can easily show yields

$$\tan \psi = \frac{r \cos^2\theta + (dr/d\theta) \sin \theta \cos \theta + r \sin^2\theta - (dr/d\theta) \sin \theta \cos \theta}{-r \sin \theta \cos \theta + (dr/d\theta) \cos^2\theta + r \sin \theta \cos \theta + (dr/d\theta) \sin^2\theta}$$

$$= \frac{r}{dr/d\theta}.$$

Example 1 Let's find the acute angle β that the cardioid $r = a(1 + \cos \theta)$ makes with the y-axis at the point $(r, \theta) = (a, \pi/2)$.

SOLUTION From Fig. 13.19, we see that we can reduce our problem to finding the angle ψ, for the angle β shown in the figure is then given by $\beta = \pi - \psi$.

We have

$$\tan \psi = \left.\frac{r}{dr/d\theta}\right|_{\theta=\pi/2} = \left.\frac{a(1+\cos\theta)}{-a\sin\theta}\right|_{\pi/2} = \frac{a}{-a} = -1.$$

Thus $\psi = 3\pi/4$, and $\beta = \pi - 3\pi/4 = \pi/4$. ‖

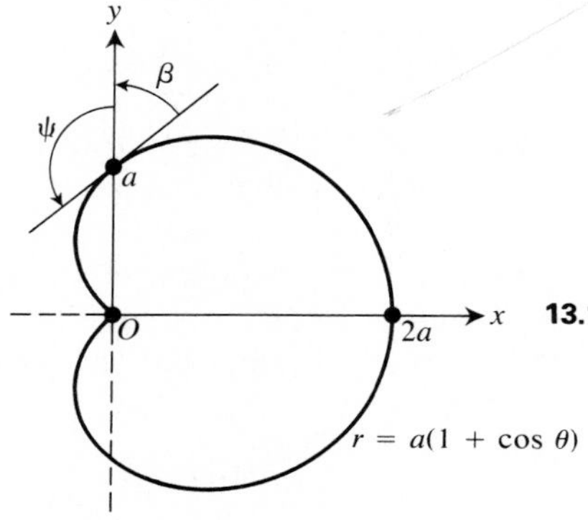

13.19

From Fig. 13.20, you see that the angle β between curves $r = f_1(\theta)$ and $r = f_2(\theta)$ can be computed by finding

$$\tan \beta = \tan(\psi_2 - \psi_1) = \frac{\tan\psi_2 - \tan\psi_1}{1 + \tan\psi_2 \tan\psi_1}.$$

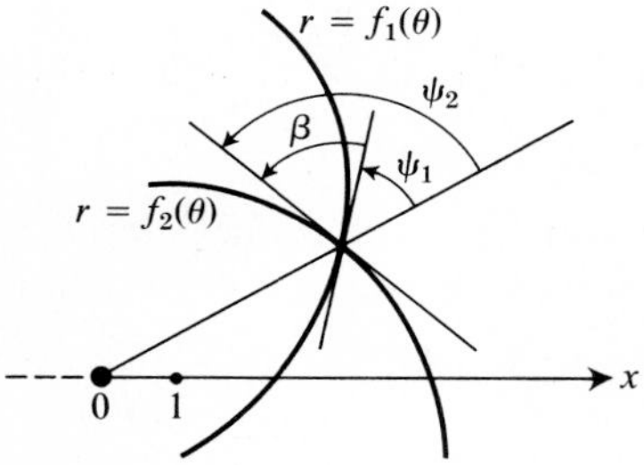

13.20

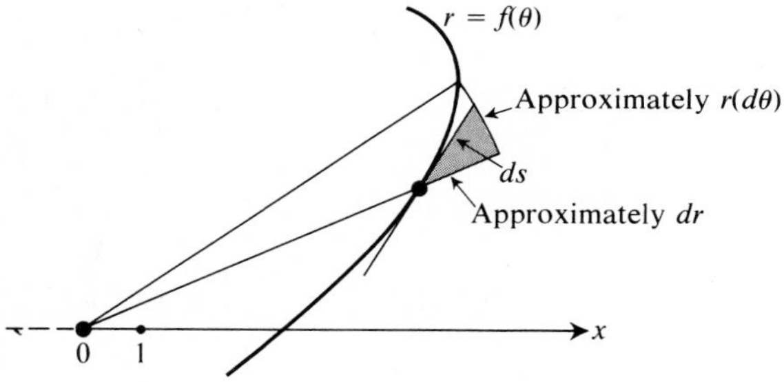

13.21

13.4.2 Arc length in polar coordinates In Fig. 13.21, we shade again the "differential right triangle" shown in Fig. 13.18. From this triangle, you obtain the estimate

$$ds = \sqrt{(dr)^2 + (r\,d\theta)^2} = \sqrt{(dr/d\theta)^2 + r^2}\,d\theta \tag{5}$$

for the length of the tangent line segment to the curve. To find arc length, you may add lengths of tangent line segments, and from (5), you expect the arc length of a smooth polar curve $r = f(\theta)$ from (r_1, θ_1) to (r_2, θ_2) to be given by

$$\int_{\theta_1}^{\theta_2} \sqrt{(dr/d\theta)^2 + r^2}\, d\theta. \tag{6}$$

For a careful derivation of (6), simply note that the polar curve $r = f(\theta)$ is defined parametrically by

$$x = r\cos\theta = f(\theta)\cos\theta,$$
$$y = r\sin\theta = f(\theta)\sin\theta,$$

and use the parametric formula

$$ds = \sqrt{(dx/d\theta)^2 + (dy/d\theta)^2}\, d\theta.$$

We have

$$\frac{dx}{d\theta} = -f(\theta)\sin\theta + f'(\theta)\cos\theta,$$

$$\frac{dy}{d\theta} = f(\theta)\cos\theta + f'(\theta)\sin\theta,$$

and you obtain,

$$\begin{aligned} ds &= \sqrt{(dx/d\theta)^2 + (dy/d\theta)^2}\, d\theta \\ &= \sqrt{(f(\theta))^2(\sin^2\theta + \cos^2\theta) + (f'(\theta))^2(\sin^2\theta + \cos^2\theta)}\, d\theta \\ &= \sqrt{f((\theta))^2 + (f'(\theta))^2}\, d\theta \\ &= \sqrt{r^2 + (dr/d\theta)^2}\, d\theta. \end{aligned}$$

Example 2 Let's find the length of the spiral $r = \theta$ shown in Fig. 13.11 from $\theta = 0$ to $\theta = 2\pi$.

SOLUTION The arc length is given by the integral

$$\begin{aligned} \int_0^{2\pi} \sqrt{r^2 + (dr/d\theta)^2}\, d\theta &= \int_0^{2\pi} \sqrt{\theta^2 + 1}\, d\theta \\ &= \left(\frac{\theta}{2}\sqrt{1 + \theta^2} + \frac{1}{2}\ln\left(\theta + \sqrt{1 + \theta^2}\right)\right)\Big]_0^{2\pi} \\ &= \pi\sqrt{1 + 4\pi^2} + \frac{1}{2}\ln\left(2\pi + \sqrt{1 + 4\pi^2}\right). \quad \| \end{aligned}$$

Example 3 Let's find the area of the surface generated when the cardioid $r = a(1 + \cos\theta)$ shown in Fig. 13.19 is revolved about the x-axis.

SOLUTION Our surface area is given by

$$\begin{aligned}
\int_0^{\pi} 2\pi y\, ds &= \int_0^{\pi} 2\pi(r \sin\theta)\sqrt{r^2 + (dr/d\theta)^2}\, d\theta \\
&= \int_0^{\pi} 2\pi a(1 + \cos\theta)(\sin\theta)\sqrt{a^2(1 + \cos\theta)^2 + a^2 \sin^2\theta}\, d\theta \\
&= 2\pi a^2 \int_0^{\pi} (1 + \cos\theta)(\sin\theta)\sqrt{2 + 2\cos\theta}\, d\theta \\
&= 2\sqrt{2}\pi a^2 \int_0^{\pi} (1 + \cos\theta)^{3/2}(\sin\theta)\, d\theta \\
&= -2\sqrt{2}\pi a^2 \frac{2}{5}(1 + \cos\theta)^{5/2}\Big]_0^{\pi} = \frac{32}{5}\pi a^2. \quad \|
\end{aligned}$$

SUMMARY

1. *If $r = f(\theta)$ is differentiable and ψ is the angle from the radius vector to the tangent to the polar curve $r = f(\theta)$, then*

$$\tan\psi = \frac{r}{dr/d\theta} = \frac{f(\theta)}{f'(\theta)}.$$

2. *The angle β between polar curves $r = f_1(\theta)$ and $r = f_2(\theta)$ at a point of intersection is given by*

$$\tan\beta = \frac{\tan\psi_2 - \tan\psi_1}{1 + \tan\psi_1 \tan\psi_2}.$$

3. *If $r = f(\theta)$ is continuously differentiable, then the arc length of the polar curve from θ_1 to θ_2 is*

$$s = \int_{\theta_1}^{\theta_2} \sqrt{(dr/d\theta)^2 + r^2}\, d\theta.$$

EXERCISES

1. Find the acute angle that the hyperbolic spiral $r\theta = a$ makes with the y-axis at the point $(r, \theta) = (2a/\pi, \pi/2)$.
2. Find the angle that the polar curve $r = 2 + 3\sin\theta$ makes with the x-axis at the point $(r, \theta) = (2, 0)$.
3. Find all points on the cardioid $r = a(1 - \cos\theta)$ where the tangent line is horizontal. [*Hint.* Use Eq. (4).]
4. Find all points on the cardioid $r = a(1 - \cos\theta)$ where the tangent line is vertical. [*Hint.* Use Eq. (4).]
5. Find the angle between the circles $r = 2a\cos\theta$ and $r = 2a\sin\theta$ at the point $(r, \theta) = (a\sqrt{2}, \pi/4)$ of intersection.
6. Find the angle between the circle $r = a$ and the four-leaved rose $r = 2a\cos 2\theta$ at the point of intersection $(r, \theta) = (a, \pi/6)$.
7. Find the length of the parabolic spiral $r = a\theta^2$ from $\theta = 0$ to $\theta = 2\pi$.
8. Find the total length of the cardioid $r = a(1 + \sin\theta)$. [*Hint.* Evaluate the integral by multiplying the integrand by $\sqrt{2 - 2\sin\theta}/\sqrt{2 - 2\sin\theta}$.]
9. Express as an integral the length of the polar curve $r = a\cos\theta/2$ from $\theta = 0$ to $\theta = \pi$.
10. Express as an integral the total length of the three-leaved rose $r = a\sin 3\theta$.

11. Find the area of the surface generated when the circle $r = 2a \sin \theta$ is revolved about the x-axis.

12. Express as an integral the area of the surface generated when the arc of the spiral $r = \theta$ from $\theta = 0$ to $\theta = \pi$ is revolved about the y-axis.

13. Let f be a twice-differentiable function. Show that the curvature κ of the curve $r = f(\theta)$ at a point (r, θ) is given by the formula

$$\kappa = \left| \frac{(f(\theta))^2 + 2(f'(\theta))^2 - f(\theta)f''(\theta)}{[(f(\theta))^2 + (f'(\theta))^2]^{3/2}} \right|.$$

exercise sets for chapter 13

review exercise set 13.1

1. Find *all* polar coordinates of the point $(-\sqrt{3}, 1)$.

2. Find the polar coordinate equation of the ellipse $4x^2 + 9y^2 = 1$.

3. Find the x,y-equation of the polar curve $r = \sin \theta + \cos \theta$.

4. Sketch the curve with polar-coordinate equation $r = a(1 + 2 \sin \theta)$.

5. Find all points of intersection of $r^2 = a^2 \sin \theta$ and $r = a/\sqrt{2}$.

6. Find the area inside $r = a(1 + \frac{1}{2} \sin \theta)$ and outside the circle $r = a$.

7. Find the angle between $r^2 = a^2 \sin \theta$ and $r = a/\sqrt{2}$ at their first-quadrant point of intersection.

8. Find the length of arc of the spiral $r = e^{2\theta}$ from $\theta = 0$ to $\theta = 2\pi$.

review exercise set 13.2

1. Find *all* polar coordinates of the point $(1, -1)$.

2. Find the polar coordinate equation of the hyperbola $x^2 - y^2 + 4x = 9$.

3. Find the x,y-equation of the polar curve $r^2 = 2 + \sin 2\theta$.

4. Sketch the curve with polar coordinate equation $r = a \sin 2\theta$.

5. Find all points of intersection of $r^2 = a^2 \cos 2\theta$ and $r = a/\sqrt{2}$.

6. Find the total area inside the curve $r^2 = a^2 \sin 2\theta$.

7. Find the angle between the cardioids $r = a(1 + \cos \theta)$ and $r = -a(1 + \cos \theta)$ at their point of intersection in the upper half-plane.

8. Find the length of arc of the curve $r = a \cos^2(\theta/2)$ from $\theta = 0$ to $\theta = \pi/2$.

a more challenging exercise 13

1. Fly A is located at $(x, y) = (1, 1)$, while fly B is at $(-1, 1)$, fly C is at $(-1, -1)$, and fly D is at $(1, -1)$. The flies all crawl at the same rate of one unit distance per unit time. The flies all start crawling at the same instant, with fly A always crawling directly toward B, B directly toward C, C directly toward D, and D directly toward A.

a) Find the point at which the flies meet.

b) How long do the flies crawl before they meet?

c) Find the polar coordinate equation of the path traveled by fly A. Sketch the path. [*Hint.* Find the angle ψ at a point on this path, and solve the differential equation $r = (\tan \psi)\, dr/d\theta$.]

d) Find the length of the path traveled by fly A.

e) What physiological problem will fly A encounter as he travels the path found in (c) in the time found in (b)?

space geometry and vectors

14

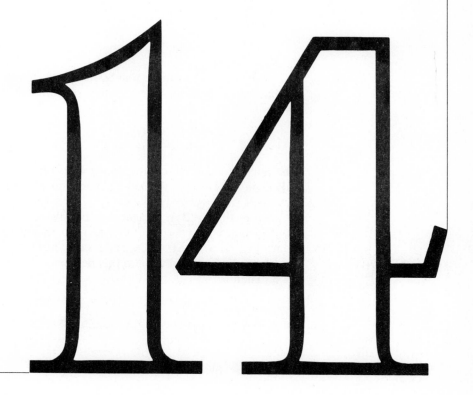

14.1 COORDINATES IN SPACE

14.1.1 Rectangular coordinates

You know how to describe the location of a point in the plane using an ordered pair (x, y) of real numbers. You can describe the location of a point in space using an ordered triple (x, y, z) of numbers. We set up a rectangular (or cartesian) system of coordinates as follows. Select some point of space as *origin*, and imagine three coordinate axes, any two of which are perpendicular, through this point. Figure 14.1 shows only half of each of these x-, y-, and z-axes for clarity. There is some difficulty in trying to sketch a space picture on a piece of paper. The plane containing the x- and y-coordinate axes is the *x,y-coordinate plane*, and the x,z-coordinate plane and y,z-coordinate plane are similarly defined.

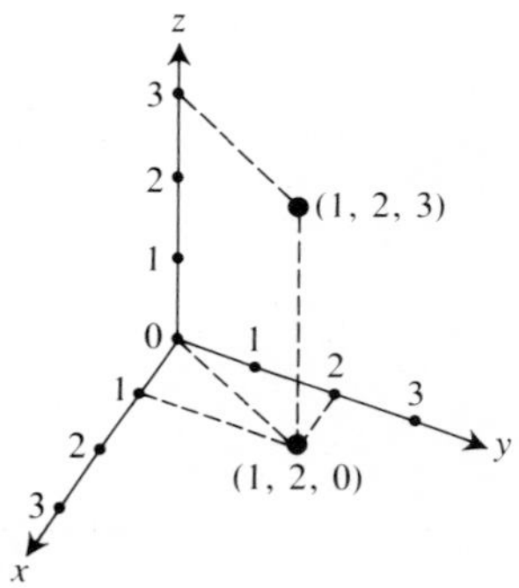

14.1

The three coordinate planes naturally divide space into eight parts or *octants* according to whether the coordinates are positive or negative. Symbolically

$$(+,+,+),\quad (+,+,-),\quad (+,-,+),\quad (+,-,-),$$
$$(-,+,+),\quad (-,+,-),\quad (-,-,+),\quad (-,-,-).$$

The portion where all coordinates are positive, that is, the $(+,+,+)$ part, is called the *first octant*. We have never seen any attempt to number the other octants.

From Fig. 14.2, it is clear that the distance from the origin to the point (x, y, z) is $\sqrt{x^2 + y^2 + z^2}$. Let's turn to the distance from (x_1, y_1, z_1) to a point (x_2, y_2, z_2). At (x_1, y_1, z_1), take new Δx, Δy, Δz-axes which are translations of our old axes to this new origin at (x_1, y_1, z_1), as shown in Fig. 14.3. Then

$$x_2 = x_1 + \Delta x, \qquad y_2 = y_1 + \Delta y, \qquad z_2 = z_1 + \Delta z.$$

Of course, the distance from $(\Delta x, \Delta y, \Delta z)$ to the new translated origin is

$$d = \sqrt{(\Delta x)^2 + (\Delta y)^2 + (\Delta z)^2}.$$

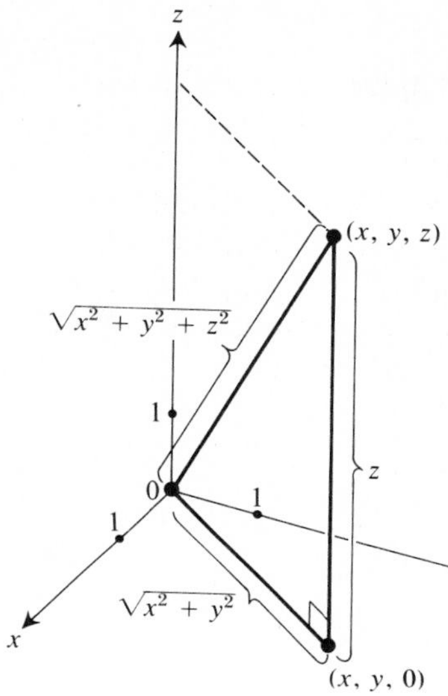

14.2

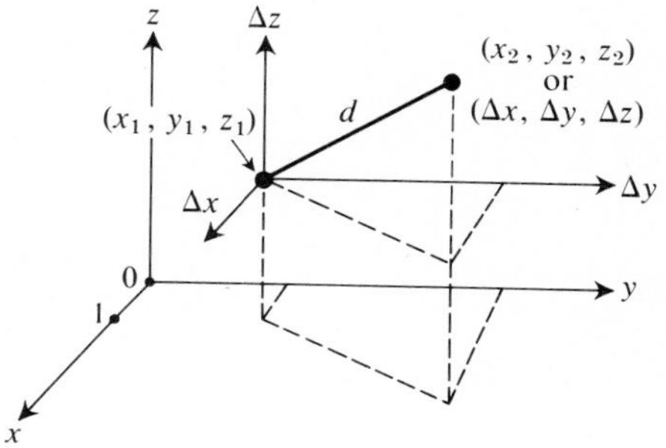

14.3

Therefore the distance in terms of the old coordinates is

$$d = \sqrt{(x_2 - x_1)^2 + (y_2 - y_1)^2 + (z_2 - z_1)^2}. \tag{1}$$

This is an easily remembered generalization of the formula for distance between points (x_1, y_1) and (x_2, y_2) in the plane.

The locus of all points (x, y, z) a fixed distance r from a point (x_1, y_1, z_1) is a *sphere of radius r with center at* (x_1, y_1, z_1). From the distance formula, you see that (x, y, z) lies on this sphere if and only if

$$\sqrt{(x - x_1)^2 + (y - y_1)^2 + (z - z_1)^2} = r$$

or

$$(x - x_1)^2 + (y - y_1)^2 + (z - z_1)^2 = r^2. \tag{2}$$

Equation (2) is thus the equation of a sphere. By completing the square, it is easily seen that the locus of any equation

$$x^2 + y^2 + z^2 + ax + by + cz = d$$

is a sphere, if the equation has a real locus in space.

Example 1 Let's find the center and radius of the sphere

$$x^2 + y^2 + z^2 - 6x + 4y = -9$$

and then sketch the sphere.

SOLUTION The steps for completing the square are

$$(x^2 - 6x) + (y^2 + 4y) + z^2 = -9$$

$$(x - 3)^2 + (y + 2)^2 + (z - 0)^2 = -9 + 9 + 4 = 4.$$

Thus the sphere has center $(3, -2, 0)$ and radius 2, and is shown in Fig. 14.4 ‖

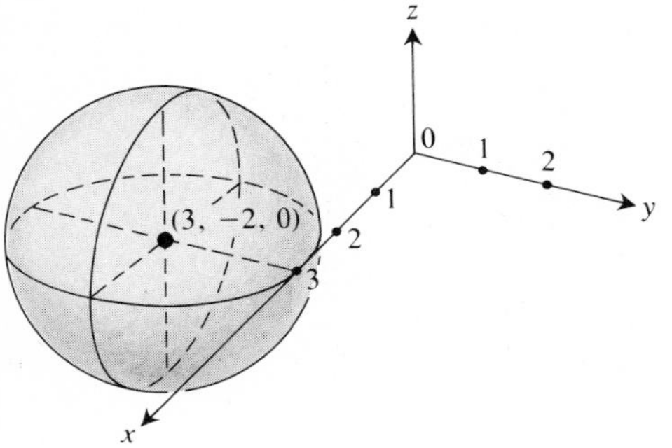

14.4

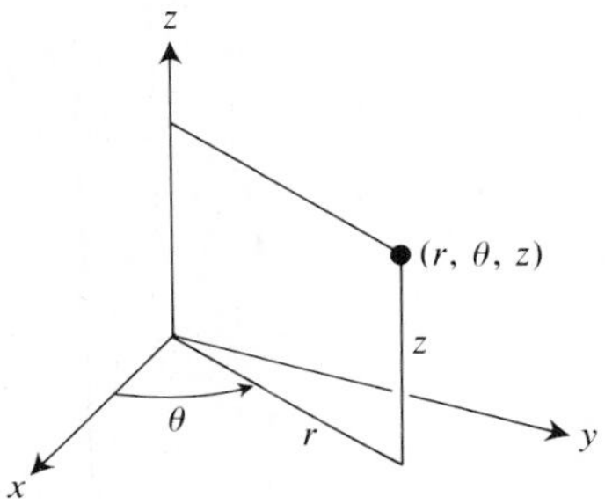

14.5

14.1.2 Cylindrical coordinates

You can also locate a point in space by specifying the x,y-coordinates of its position using polar r,θ-coordinates, and specifying its height by the z-coordinate. The point then has coordinates (r, θ, z) as well as coordinates (x, y, z). Of course, the coordinates r amd θ are not unique, being our usual polar coordinates. Such coordinates are shown in Fig. 14.5. As shown in Fig. 14.6, the locus of $r = a$ is a cylinder about the z-axis, for $r = a$ has a circle in the x,y-plane as polar locus, and there is no restriction placed on z. Consequently (a, θ, z) is on the locus for all θ and all z. For this reason these r,θ,z-coordinates are called *cylindrical coordinates.* Since you know how to change from polar r,θ-coordinates to x,y-coordinates in the plane, you know how to change from cylindrical to rectangular coordinates in space. Namely,

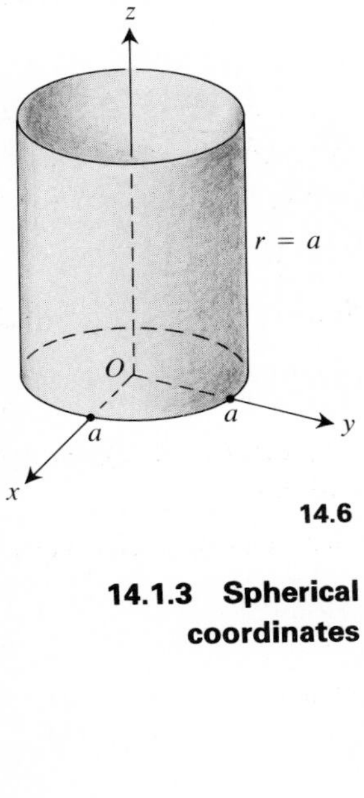

14.6

$$x = r\cos\theta, \qquad r^2 = x^2 + y^2,$$

$$y = r\sin\theta, \qquad \theta = \tan^{-1}(y/x),$$

$$z = z, \qquad z = z.$$

14.1.3 Spherical coordinates

Another coordinate system in space that is sometimes useful is the *spherica' coordinate system*, where a point has coordinates (ρ, ϕ, θ) as indicated in Fig. 14.7.

The coordinate ρ is the length of the line segment joining the point and the origin, ϕ is the angle from the z-axis to this line segment, and θ is the same angle as in cylindrical coordinates. Note that the locus of $\rho = a$ is a sphere with center at the origin and radius a, as indicated in Fig. 14.8. This is the reason for the term, "spherical coordinates."

Since ρ is the distance from the point to the origin, it is obvious that

$$\rho^2 = x^2 + y^2 + z^2.$$

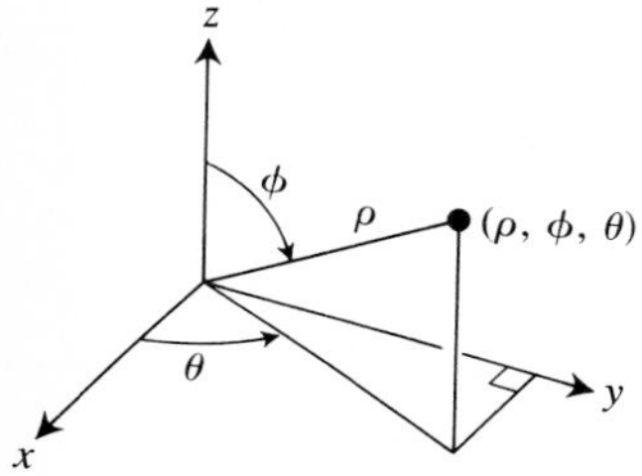

14.7

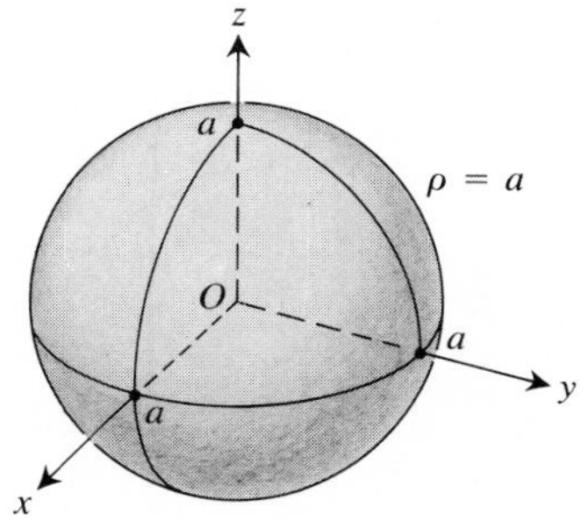

14.8

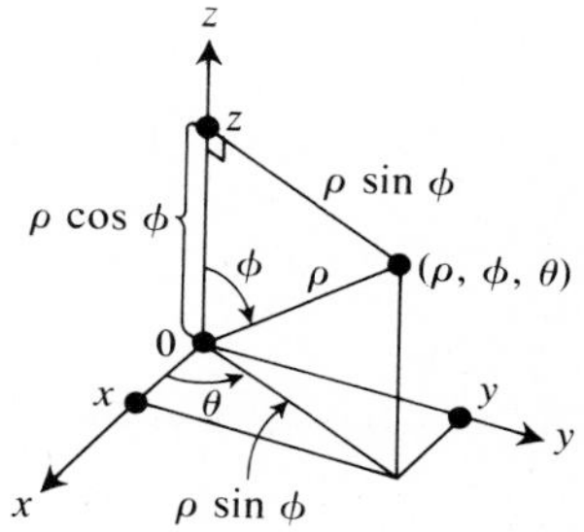

14.9

We need to express x, y, and z in terms of spherical ρ,ϕ,θ-coordinates. From Fig. 14.9, you easily see that

$$x = \rho \sin \phi \cos \theta, \qquad y = \rho \sin \phi \sin \theta, \qquad z = \rho \cos \phi.$$

We shall allow only values of ρ where $\rho \geq 0$ and only values of ϕ where $0 \leq \phi \leq \pi$.

SUMMARY

1. *In rectangular coordinates for space, a point has coordinates* (x, y, z).
2. *The distance from* (x_1, y_1, z_1) *to* (x_2, y_2, z_2) *in rectangular coordinates is*

$$d = \sqrt{(x_2 - x_1)^2 + (y_2 - y_1)^2 + (z_2 - z_1)^2}.$$

3. *The rectangular equation of a sphere in space with center* (x_1, y_1, z_1) *and radius* r *is*

$$(x - x_1)^2 + (y - y_1)^2 + (z - z_1)^2 = r^2.$$

4. *In cylindrical coordinates for space, a point has coordinates* (r, θ, z), *where* r *and* θ *are the usual polar coordinates in the* x,y*-plane.*
5. *Transformation from rectangular to cylindrical coordinates and vice versa is accomplished using the formulas*

$$\begin{aligned} x &= r \cos \theta, & \qquad & r = \sqrt{x^2 + y^2}, \\ y &= r \sin \theta, & \quad \text{and} \quad & \theta = \tan^{-1}(y/x), \\ z &= z, & & z = z. \end{aligned}$$

6. *In spherical ρ,ϕ,θ-coordinates for a point in space,*

ρ *is the distance to the origin,*
ϕ *is the angle from the positive z-axis to the ray from the origin, where $0 \leq \phi \leq \pi$,*
θ *is the same angle as for cylindrical coordinates.*

7. *Transformation from rectangular to spherical coordinates is accomplished using the formulas*

$$x = \rho \sin\phi \cos\theta,$$

$$y = \rho \sin\phi \sin\theta,$$

$$z = \rho \cos\phi.$$

EXERCISES

1. Sketch in space all (x, y, z) satisfying the given equation.

a) $x = 2$ b) $z = 3$ c) $x = y$

d) $y^2 = 1$ e) $x = y = z$

2. While we have defined neither a line nor a plane in space, use your geometric intuition to find the desired point.

a) The point such that the line segment joining it to $(-2, 1, -4)$ is bisected by and perpendicular to the plane $x = 0$.

b) The point such that the line segment joining it to $(-1, \pi, \sqrt{2})$ has the origin as midpoint.

c) The point such that the line segment joining it to $(-1, 4, -3)$ has $(-1, 2, -3)$ as midpoint.

d) The point in the plane $y = 2$ that is closest to the point $(-1, -5, 2)$.

3. Let $(-1, 2, 1)$ be chosen as origin for $\Delta x,\Delta y,\Delta z$-coordinates. Express each point in terms of these new translated coordinates.

a) $(1, -2, 1)$ b) $(-3, 4, 0)$ c) $(5, -1, 2)$

4. Find the distance between the given points.

a) $(-1, 0, 4)$ and $(1, 1, 6)$

b) $(2, -1, 3)$ and $(0, 1, 7)$

5. Find the equation of the sphere with center $(-1, 2, 4)$ and passing through the point $(2, -1, 5)$.

6. Find the equation of the sphere having $(-1, 2, 6)$ and $(1, 6, 0)$ as endpoints of a diameter.

7. Find the center and radius of the given sphere.

a) $x^2 + y^2 + z^2 - 2x + 2y = 0$
b) $x^2 + y^2 + z^2 - 6x - 4y + 8z = -4$

8. Find *all* cylindrical coordinates of the point $(1, 1, 1)$.

9. Sketch in space the locus of $\theta = \pi/4$ in cylindrical coordinates.

10. Sketch in space the locus of $r = 2$ in cylindrical coordinates.

11. Sketch in space the locus of $x^2 + y^2 = 9$.

12. Sketch in space the locus of $x^2 + z^2 = 4$.

13. Find x,y,z-coordinates of the point with the following spherical coordinates. Use a figure rather than the transformation equations.

a) $(2, \pi/4, -\pi)$ b) $(0, 3\pi/4, \pi/6)$
c) $(4, \pi/2, \pi/3)$

14. Find ρ,ϕ,θ-coordinates for the point with the following x,y,z-coordinates. Use a figure to find the answers.

a) $(1, 0, 0)$ b) $(0, 0, -4)$
c) $(1, 1, 1)$ d) $(-3, -4, 5)$

15. Find transformation equations for spherical coordinates ρ, ϕ, and θ in terms of rectangular coordinates x, y, and z. [*Hint.* Use a figure.]

16. a) Express the cylindrical coordinates r, θ, and z in terms of the spherical coordinates ρ, ϕ and θ.

b) Express the spherical coordinates ρ, ϕ, and θ in terms of the cylindrical coordinates r, θ, and z.

17. Sketch in space the locus of $\phi = \pi/4$ in spherical coordinates.

18. Find the volume of the region described in spherical coordinates by $2 \le \rho \le 5$ and $0 \le \phi \le \pi/2$.

19. Describe in terms of x,y,z-coordinates all points in space where the cylindrical r-coordinate is equal to the spherical ρ-coordinate.

20. Try to visualize the surface described in spherical coordinates by $\rho = \theta$ for $0 \le \theta \le 2\pi$. It's not realistic to ask you to sketch this surface.

21. Find the spherical coordinate equation for $x^2 + y^2 + z^2 = 25$ by using the transformation equations.

14.2 QUADRIC SURFACES

A quadric surface in space is the locus of a polynomial equation in x, y, and z of degree two. As an aid in sketching such surfaces, examine the curves in which they intersect planes parallel to the coordinate planes. Note that $x = x_0$ has as locus in space a plane parallel to the y,z-coordinate plane. Similarly $y = y_0$ is a plane parallel to the x,z-coordinate plane, and $z = z_0$ is a plane parallel to the x,y-coordinate plane. The coordinate planes themselves, $x = 0$, $y = 0$, and $z = 0$, are especially useful. Each such plane meeting a quadric surface intersects it in an ellipse, hyperbola, or parabola.

Cylinders

If one of the variables x, y, or z is missing from the equation, then the locus is a cylinder. A *cylinder* in space is a surface that can be generated by a line that moves along a plane curve, keeping a fixed direction. The parallel lines on the cylinder corresponding to positions of the generating line are the *elements of the cylinder.* To illustrate, suppose the variable x is missing in an equation $F(x, y, z) = 0$. If (a, b, c) lies on the locus, then so will (x, b, c) for all x; i.e., the line through (a, b, c) parallel to the x-axis. Consequently, the locus is a cylinder with elements parallel to the x-axis and meeting the y,z-plane in the curve $F(0, y, z) = 0$. Similar results hold if y or z is missing.

Example 1 The cylinder

$$\frac{x^2}{a^2} + \frac{y^2}{b^2} = 1$$

in space meets the x,y-plane in an ellipse and has elements parallel to the z-axis. This *elliptic cylinder* is shown in Fig. 14.10. ‖

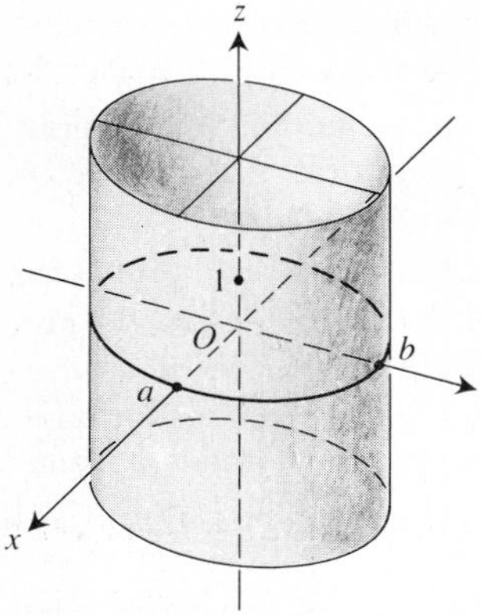

Elliptic Cylinder $\frac{x^2}{a^2} + \frac{y^2}{b^2} = 1$ **14.10**

Parabolic Cylinder $z^2 = 4py$

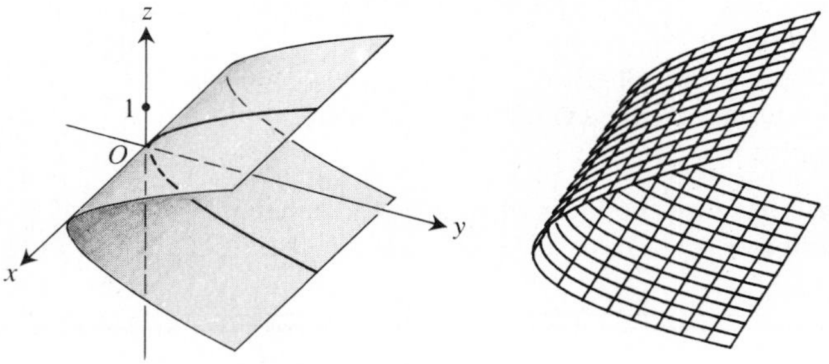

14.11

(a) Artist's figure (b) Computer generated figure

Example 2 The *parabolic cylinder* $z^2 = 4py$ is sketched in Fig. 14.11 ‖

As for second-degree plane curves, the device of completing squares and choosing a new origin can often be used to simplify the sketching of a quadric surface. We assume that this now causes you no difficulty, and the examples that follow start with equations where the completion of squares is unnecessary. These examples exhibit some types of quadric surfaces.

Example 3 Consider the surface with equation

$$\frac{x^2}{a^2} + \frac{y^2}{b^2} + \frac{z^2}{c^2} = 1.$$

If this surface is cut by a plane $x = x_0$ for $-a < x_0 < a$, an ellipse is obtained (or a circle if $b = c$). This is clear upon substitution of x_0 for x in the equation. The closer x_0 is to $-a$ or a, the smaller the elliptical section obtained. Similar results hold for a plane $y = y_0$ if $-b < y_0 < b$, and for a plane $z = z_0$ if $-c < z_0 < c$. This surface is an *ellipsoid*, and is sketched in Fig. 14.12. ‖

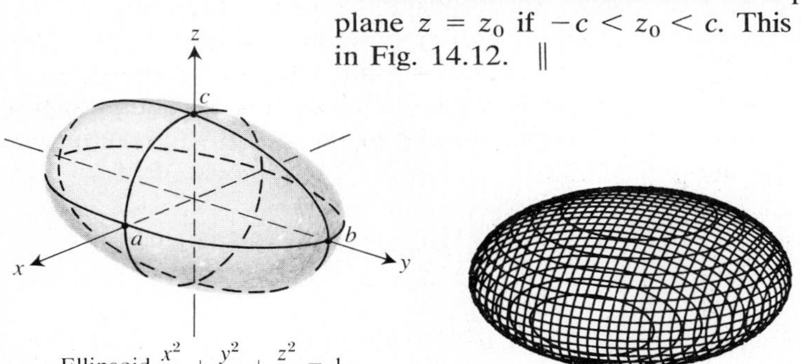

Ellipsoid $\frac{x^2}{a^2} + \frac{y^2}{b^2} + \frac{z^2}{c^2} = 1$ **14.12**

(a) Artist's figure (b) Computer generated figure

Example 4 The surface with equation

$$z = \frac{x^2}{a^2} + \frac{y^2}{b^2}$$

is an *elliptic paraboloid* if $a \neq b$ and a *circular paraboloid* if $a = b$. The plane $z = z_0$ does not meet the surface if $z_0 < 0$, meets the surface in a point if $z_0 = 0$, and meets the surface in an ellipse if $z_0 > 0$. The planes $x = x_0$ and $y = y_0$ meet the surface in parabolas. This surface is sketched in Fig. 14.13. ||

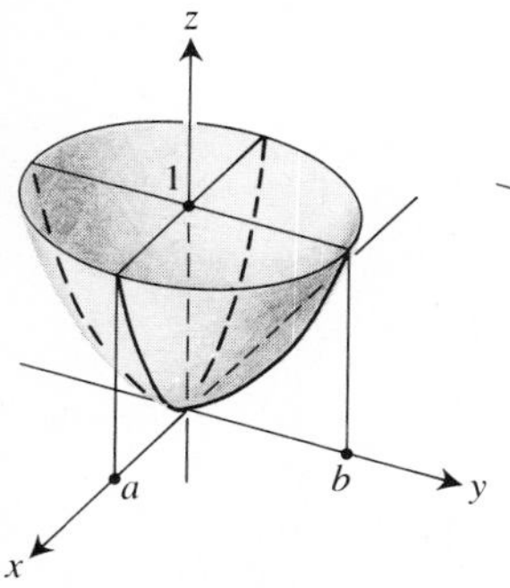

Elliptic Paraboloid $z = \frac{x^2}{a^2} + \frac{y^2}{b^2}$

14.13

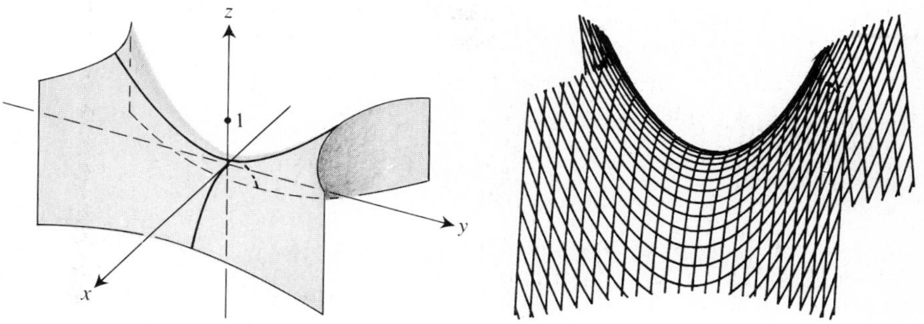

Hyperbolic Paraboloid $z = \frac{y^2}{b^2} - \frac{x^2}{a^2}$

(a) Artist's figure

(b) Computer generated figure

14.14

Example 5 The surface with equation

$$z = \frac{y^2}{b^2} - \frac{x^2}{a^2}$$

is a *hyperbolic paraboloid*, and is sketched in Fig. 14.14. This is not an easy surface for a person without artistic ability to sketch. The plane $z = z_0$ meets the surface in a hyperbola that "opens" in the y-direction if $z_0 > 0$ and in the x-direction if $z_0 < 0$, while the plane $z = 0$ meets the surface in the degenerate hyperbola consisting of two intersecting lines. A plane $x = x_0$ meets the surface in a parabola "opening upward" while a plane $y = y_0$ meets the surface in a parabola "opening downward." ||

Example 6 The surface with equation

$$z^2 = \frac{x^2}{a^2} + \frac{y^2}{b^2}$$

is an *elliptic cone* (a *circular cone* if $a = b$). Putting $z = z_0$, you obtain an elliptical section, while the planes $x = x_0$ and $y = y_0$ yield hyperbolic sections. The surface is sketched in Fig. 14.15. ||

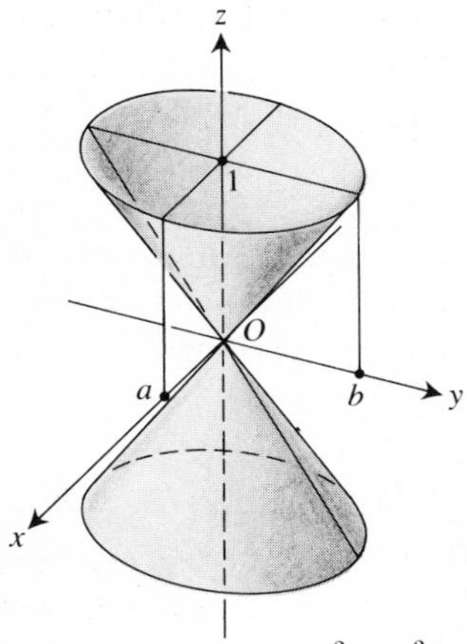

Elliptic Cone $z^2 = \frac{x^2}{a^2} + \frac{y^2}{b^2}$

14.15

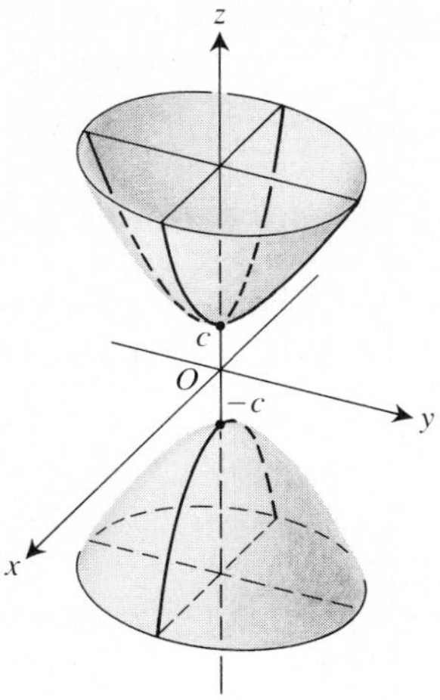

Hyperboloid of two sheets $\frac{z^2}{c^2} = 1 + \frac{x^2}{a^2} + \frac{y^2}{b^2}$

14.16

Example 7 The surface with equation

$$\frac{z^2}{c^2} = 1 + \frac{x^2}{a^2} + \frac{y^2}{b^2}$$

is a *hyperboloid of two sheets* and is sketched in Fig. 14.16. We leave as an exercise the discussion of sections by planes parallel to the coordinate planes (see Exercise 1). ‖

Example 8 The surface with equation

$$1 + \frac{z^2}{c^2} = \frac{x^2}{a^2} + \frac{y^2}{b^2}$$

is a *hyperboloid of one sheet,* and is sketched in Fig. 14.17. Again, we leave the discussion of planar sections to the exercises (see Exercise 2). ‖

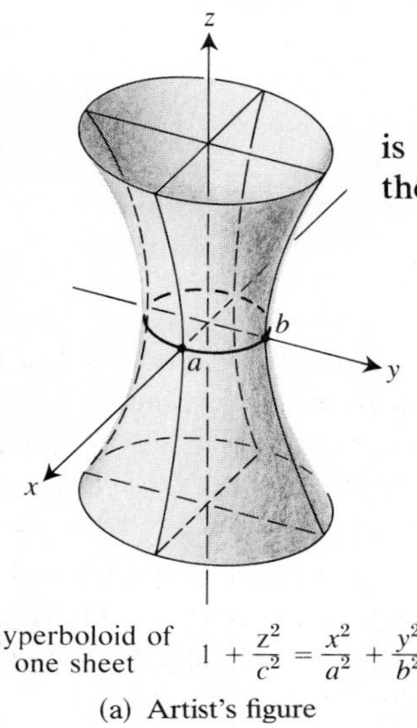

Hyperboloid of one sheet $1 + \frac{z^2}{c^2} = \frac{x^2}{a^2} + \frac{y^2}{b^2}$

(a) Artist's figure

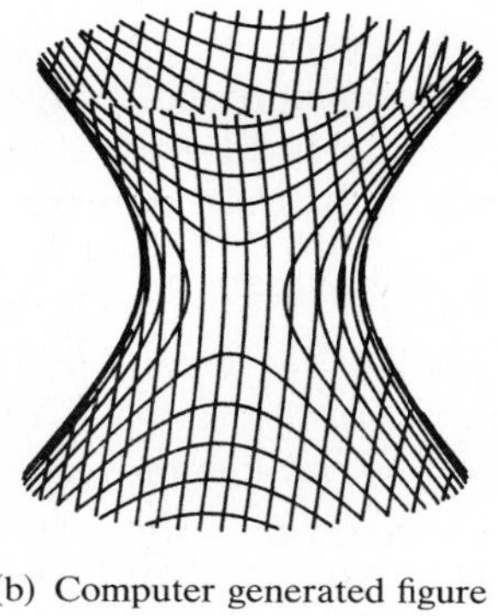

(b) Computer generated figure

14.17

We conclude with an example with specific numbers rather than a, b, and c, and where it is necessary to complete squares.

Example 9 Let's sketch the surface $-16x^2 + 4y^2 - z^2 - 8y + 4z = 0$.

SOLUTION We complete squares and arrange our equation as follows:

$$-16x^2 + 4(y^2 - 2y) - (z^2 - 4z) = 0$$

$$-16x^2 + 4(y - 1)^2 - (z - 2)^2 = 4 - 4 = 0$$

$$4(y - 1)^2 = 16x^2 + (z - 2)^2$$

$$4\bar{y}^2 = 16\bar{x}^2 + \bar{z}^2 \quad \text{where } \bar{x} = x,\ \bar{y} = y - 1,\ \bar{z} = z - 2.$$

In Fig. 14.18, we take $\bar{x},\bar{y},\bar{z}$-axes at $(0, 1, 2)$. Setting $\bar{x} = 0$, you see that the $\bar{y},\bar{z}$-plane meets the surface in the two lines $\bar{z} = \pm 2\bar{y}$. Setting $\bar{y} = 0$, you obtain only the locus $(\bar{x}, \bar{y}, \bar{z}) = (0, 0, 0)$. Setting $\bar{z} = 0$, you obtain the two lines $\bar{y} = \pm 2\bar{x}$ in the $\bar{x},\bar{y}$-plane. Planes $\bar{y} = c$ meet the surface in elliptical sections. The surface is the double elliptic cone shown in Fig. 14.18. ‖

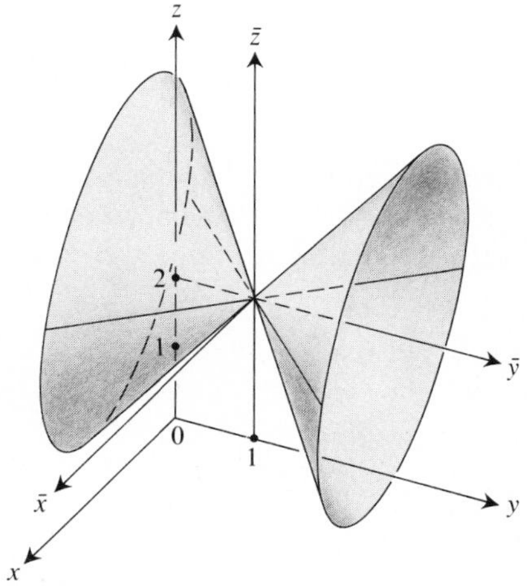

14.18

SUMMARY

1. *An equation containing only two of the three space variables x, y, z has as locus a cylinder with the axis of the cylinder parallel to the axis of the missing variable. To illustrate, suppose x and y are present in the equation. The cylinder intersects the x,y-coordinate plane in the curve given by the equation and is parallel to the z-axis.*
2. *Quadric surfaces are best sketched by determining the curves of intersection of the surface with planes $x = x_0$, $y = y_0$, or $z = z_0$. Start by sketching the trace curves in the coordinate planes, $x = 0$, $y = 0$, and $z = 0$. See Figs. 14.10 through 14.17 for the possible types of surfaces.*

EXERCISES

1. Describe the curves of intersection of the hyperboloid of two sheets in Example 7 with planes parallel to the coordinate planes.

2. Describe the curves of intersection of the hyperboloid of one sheet in Example 8 with planes parallel to the coordinate planes.

In Exercises 3 *through* 17, *sketch the quadric surface in space having the given equation, and give the descriptive name of the surface as in Figs.* 14.10 *through* 14.17.

3. $y^2 + z^2 - 4 = 0$
4. $x^2 + 2x + y^2 = 0$
5. $y^2 - z = 0$
6. $xz - 1 = 0$
7. $4x - y^2 + 2y + 3 = 0$
8. $y^2 - x^2 - z^2 = 0$
9. $36x - 9y^2 - 16z^2 = 0$
10. $\dfrac{x^2}{4} - \dfrac{y^2}{9} + z^2 + 1 = 0$
11. $\dfrac{x^2}{4} - \dfrac{y^2}{25} - \dfrac{z^2}{9} + 4 = 0$
12. $\dfrac{x^2}{4} + \dfrac{y^2}{9} + z - 3 = 0$
13. $\dfrac{x^2}{4} - \dfrac{y^2}{9} + z - 1 = 0$
14. $2x^2 + 3y^2 + 4z^2 - 24 = 0$
15. $x^2 - 4y^2 + 16z^2 = 0$
10. $\dfrac{x^2}{4} - \dfrac{y^2}{25} - \dfrac{z^2}{9} - 4 = 0$
17. $x^2 - 4y + z^2 - 8 = 0$

14.3 VECTORS AND THEIR ALGEBRA

14.3.1 Vector notation and terminology

We start out with notation considerations. In this chapter, you will want to be thinking in terms of the *first, second,* or *third* coordinate of a point. This suggests changing notation for coordinates to index the coordinate position. We often write

$$\begin{aligned} (a_1, a_2) &\text{ in place of } (a, b),\\ (x_1, x_2) &\text{ in place of } (x, y),\\ (a_1, a_2, a_3) &\text{ in place of } (a, b, c),\\ (x_1, x_2, x_3) &\text{ in place of } (x, y, z). \end{aligned}$$

As you work with more coordinates, lengthy notations such as (a_1, a_2, a_3) are time-consuming to write, and may cause printing problems if a number of such notations appear in a single formula. We shall often use a single boldface letter $\boldsymbol{a}$ to denote such a point (a_1, a_2, a_3). The number of coordinates will always be either explicitly stated or clear from the context. For example, the point $\boldsymbol{a}$ in space is (a_1, a_2, a_3), the point $\boldsymbol{b}$ in the plane is (b_1, b_2), the point $\boldsymbol{x}$ in space is (x_1, x_2, x_3), etc. We suggest that in your written work, you use $\vec{a}$ with an arrow over the letter in place of the boldfaced letter. This is *vector notation,* and points correspond to vectors, as we shall explain. We use a boldface zero for the origin; for example, in space $\mathbf{0} = (0, 0, 0)$.

Each point $\boldsymbol{a}$ in space (or the plane) yields a numerical *magnitude*, the distance $\sqrt{a_1^2 + a_2^2 + a_3^2}$ from $\mathbf{0}$ to $\boldsymbol{a}$, and a *direction*, the direction from $\mathbf{0}$ to $\boldsymbol{a}$. In the terminology of classical mechanics, any quantity that has associated with it a magnitude and a direction is called a *vector*. Let's use this classical terminology, and consider (a_1, a_2, a_3) to be a **vector** in space as well as a point in space. In vector terminology, $\mathbf{0}$ is the **zero vector**. The vectors

$$\boldsymbol{i} = (1, 0) \qquad \text{and} \qquad \boldsymbol{j} = (0, 1)$$

are the **unit coordinate vectors** in the plane, while

$$\boldsymbol{i} = (1, 0, 0), \qquad \boldsymbol{j} = (0, 1, 0), \qquad \text{and} \qquad \boldsymbol{k} = (0, 0, 1)$$

are the **unit coordinate vectors** in space.

We emphasize that the mathematical definition of a *vector* is identical with the definition of a *point;* each is an ordered collection of real numbers. The names "vector" and "point" indicate different geometric interpretations for such a collection. If you ask mathematicians to show pictorially the *vector* (1, 2) in the plane, they will draw the arrow indicating length and direction shown in Fig. 14.19. On the other hand, if you ask them to show pictorially the *point* (3, −2), they will make the large dot shown in the figure, just indicating a position.

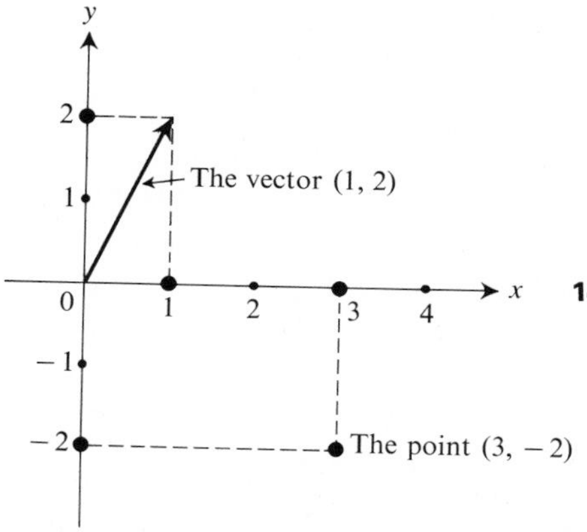

14.19

Vector terminology

In summary, each triple $\boldsymbol{a} = (a_1, a_2, a_3)$ is both a point in space and a vector in space. The **length** of vector $\boldsymbol{a}$, denoted by $|\boldsymbol{a}|$, is $\sqrt{a_1^2 + a_2^2 + a_3^2}$. The number a_i is the **ith component** of the vector. Any vector of length 1 is a **unit vector**; in particular, $\boldsymbol{i} = (1, 0, 0)$, $\boldsymbol{j} = (0, 1, 0)$, $\boldsymbol{k} = (0, 0, 1)$ are the unit coordinate vectors in space. The vector $\mathbf{0} = (0, 0, 0)$ is the **zero vector**. Similar terminology is used for a vector $\boldsymbol{a} = (a_1, a_2)$ in the plane.

Example 1 The vectors $\boldsymbol{a} = (0, 1)$ and $\boldsymbol{b} = (1/2, \sqrt{3}/2)$ are both unit vectors in the plane. ||

14.3.2 The algebra of vectors

You are aware of the importance of the notion of addition of real numbers. Addition of numbers on the line has a generalization to addition in the plane and in space, which is of basic importance in analysis. Addition in space or the plane is usually phrased in the language of vectors. We shall follow this convention.

Definition 14.1 Let $\boldsymbol{a}$ and $\boldsymbol{b}$ be vectors in space. The **sum** of $\boldsymbol{a}$ and $\boldsymbol{b}$ is the vector in space defined by

$$\boldsymbol{a} + \boldsymbol{b} = (a_1 + b_1, a_2 + b_2, a_3 + b_3).$$

The analogous notion of sum holds for vectors in the plane. Note that vector addition is defined only for two vectors with the *same number of components.*

Example 2 You have $(0, -1, 3) + (4, 2, -6) = (4, 1, -3)$. ||

Let's see how to interpret vector addition geometrically, in terms of arrows. We make our sketches in the plane, but similar arguments hold for vectors in space.

There are two ways that you can arrive at the arrow representing $\boldsymbol{a} + \boldsymbol{b}$ from arrows representing $\boldsymbol{a}$ and representing $\boldsymbol{b}$. One way is to find the diagonal of the parallelogram that has $(0, 0)$ as a vertex and the arrows represented by $\boldsymbol{a}$ and $\boldsymbol{b}$ as sides emanating from $(0, 0)$. As indicated in Fig. 14.20, the diagonal arrow of the parallelogram that starts at $(0, 0)$ represents the vector $\boldsymbol{a} + \boldsymbol{b}$.

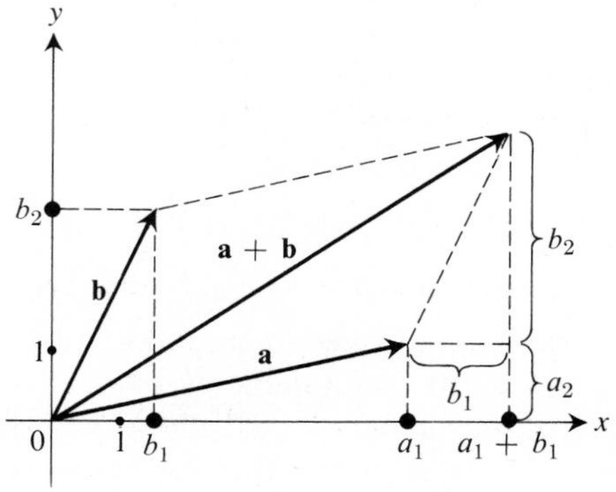

14.20

For an alternative method to represent $\boldsymbol{a} + \boldsymbol{b}$, choose the point (a_1, a_2) as new origin, and draw the arrow representing the *translated vector* $\boldsymbol{b}$. This arrow starts at the tip of the original vector $\boldsymbol{a}$ and goes to the point with translated coordinates (b_1, b_2). The tip of this translated vector $\boldsymbol{b}$ then falls at the same point as the tip of the desired vector $\boldsymbol{a} + \boldsymbol{b}$ emanating from the original origin, $(0, 0)$, as shown in Fig. 14.21. It is worth noting that the vector $\boldsymbol{b}$ in Fig. 14.20 and the translated vector $\boldsymbol{b}$ in Fig. 14.21 have the *same length* and the *same direction.*

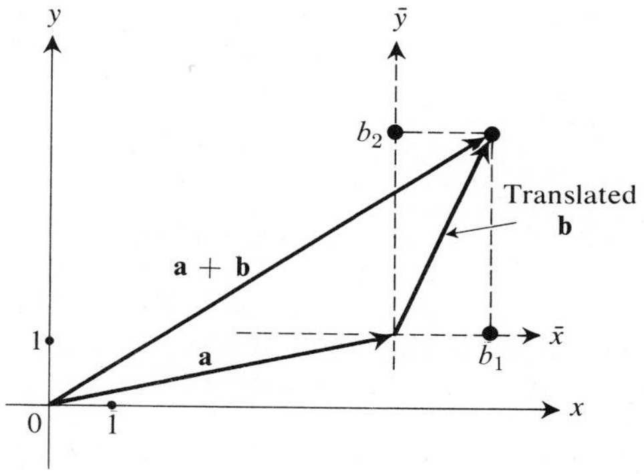

14.21

Scalars Let's turn now to another operation in the vector algebra. When dealing with vectors in space (or the plane), one often refers to a real number as a **scalar** to distinguish it from the vectors. As our second algebraic operation involving vectors, we define the product of a scalar r and a vector $\boldsymbol{a} = (a_1, a_2, a_3)$ in space.

Definition 14.2 The **product** $r\boldsymbol{a}$ of the scalar r and the vector $\boldsymbol{a}$ is the vector

$$r\boldsymbol{a} = (ra_1, ra_2, ra_3).$$

Again, the analogous notion holds for the product of a vector in the plane by the scalar r.

Example 3 In space, $2(3, -1, 4) = (6, -2, 8)$. ‖

Note that for every vector $\boldsymbol{a} = (a_1, a_2, a_3)$ in space, you have

$$\begin{aligned} \boldsymbol{a} &= (a_1, a_2, a_3) \\ &= a_1(1, 0, 0) + a_2(0, 1, 0) + a_3(0, 0, 1) \\ &= a_1\boldsymbol{i} + a_2\boldsymbol{j} + a_3\boldsymbol{k}. \end{aligned}$$

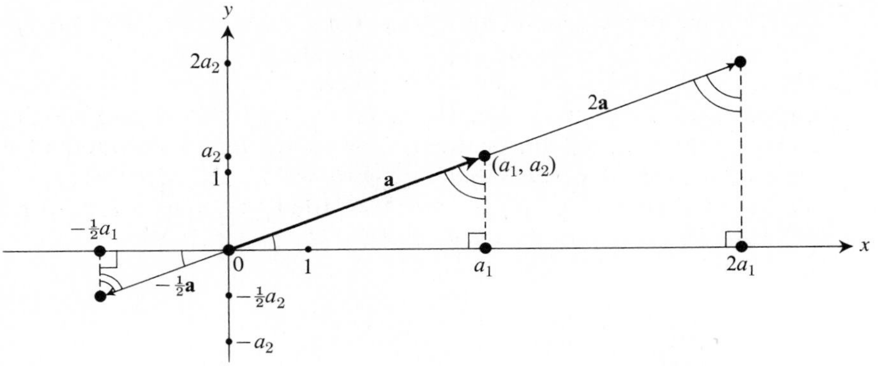

14.22

Similarly, in the plane

$$\boldsymbol{b} = (b_1, b_2) = b_1(1, 0) + b_2(0, 1) = b_1\boldsymbol{i} + b_2\boldsymbol{j}.$$

Such $\boldsymbol{i}$, $\boldsymbol{j}$, $\boldsymbol{k}$ expressions for vectors are used extensively. We shall use this notation, wherever it is not too cumbersome, to indicate where we are thinking in terms of the "vector interpretation" of the ordered triple or pair of numbers.

From Definition 14.2, you see at once that for any real number r and vector $\boldsymbol{a} = a_1\boldsymbol{i} + a_2\boldsymbol{j} + a_3\boldsymbol{k}$, you have

$$|r\boldsymbol{a}| = |ra_1\boldsymbol{i} + ra_2\boldsymbol{j} + ra_3\boldsymbol{k}| = \sqrt{(ra_1)^2 + (ra_2)^2 + (ra_3)^2}$$

$$= \sqrt{r^2} \cdot \sqrt{a_1{}^2 + a_2{}^2 + a_3{}^2} = |r| \cdot |\boldsymbol{a}|. \qquad (1)$$

Thus if you wish to describe $r\boldsymbol{a}$ in terms of *length* and *direction*, Eq. (1) shows that the length of the product $r\boldsymbol{a}$ is $|r|$ times the length of $\boldsymbol{a}$. Figure 14.22 indicates that you should consider $r\boldsymbol{a}$ to have the same direction as $\boldsymbol{a}$ if $r > 0$, and opposite direction if $r < 0$.

Parallel vectors

Note that nonzero vectors $\boldsymbol{a}$ and $\boldsymbol{b}$ are *parallel* if and only if $\boldsymbol{b} = r\boldsymbol{a}$ for some real number r. Thus $r\boldsymbol{a}$ is a vector parallel to $\boldsymbol{a}$ of length $|r| \cdot |\boldsymbol{a}|$ and having the *same* direction as $\boldsymbol{a}$ if $r > 0$, and the *opposite* direction if $r < 0$.

Example 4 The vectors $\boldsymbol{a} = \boldsymbol{i} - 3\boldsymbol{j}$ and $\boldsymbol{b} = 2\boldsymbol{i} - 6\boldsymbol{j}$ in the plane are parallel and have the same direction, for $2\boldsymbol{i} - 6\boldsymbol{j} = 2(\boldsymbol{i} - 3\boldsymbol{j})$. However, $\boldsymbol{c} = 4\boldsymbol{i} - 3\boldsymbol{j}$ and $\boldsymbol{d} = 2\boldsymbol{i} - 7\boldsymbol{j}$ are *not* parallel. ||

Example 5 Let $|\boldsymbol{a}| = 5$. Then

$$|3\boldsymbol{a}| = |3| \cdot |\boldsymbol{a}| = 3 \cdot 5 = 15,$$

and $3\boldsymbol{a}$ has the same direction as $\boldsymbol{a}$. However, $-7\boldsymbol{a}$ has the opposite direction to $\boldsymbol{a}$, and

$$|-7\boldsymbol{a}| = |-7| \cdot |\boldsymbol{a}| = 7 \cdot 5 = 35. \quad ||$$

We may now define the *difference* $\boldsymbol{a} - \boldsymbol{b}$ of vectors $\boldsymbol{a}$ and $\boldsymbol{b}$ by

$$\boldsymbol{a} - \boldsymbol{b} = \boldsymbol{a} + (-1)\boldsymbol{b};$$

we have already defined vector addition and the product $(-1)\boldsymbol{b}$. Since $\boldsymbol{b} + (\boldsymbol{a} - \boldsymbol{b}) = \boldsymbol{a}$, you see that $\boldsymbol{a} - \boldsymbol{b}$ is the vector which, when added to $\boldsymbol{b}$, yields $\boldsymbol{a}$. A translated coordinate representation of $\boldsymbol{a} - \boldsymbol{b}$, starting from the tip of $\boldsymbol{b}$ as translated origin, therefore ends at (a_1, a_2), as shown in Fig. 14.23.

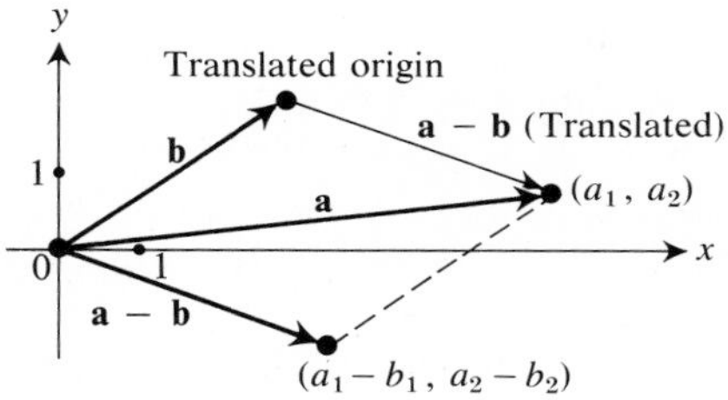

14.23

In summary, vector addition, subtraction, and multiplication by a scalar are easy to compute; you simply compute the corresponding numerical operations in each component.

We list some algebraic laws that hold for vector algebra. These laws are easy to prove. Their demonstration is left to the exercises (see Exercise 12).

Theorem 14.1 *For all vectors* **a**, **b**, *and* **c** *in space or the plane, and all scalars r and s, the following laws hold.*

a) $(\boldsymbol{a} + \boldsymbol{b}) + \boldsymbol{c} = \boldsymbol{a} + (\boldsymbol{b} + \boldsymbol{c})$ (*associativity of addition*)

b) $\boldsymbol{a} + \boldsymbol{b} = \boldsymbol{b} + \boldsymbol{a}$ (*commutativity of addition*)

c) $r(s\boldsymbol{a}) = (rs)\boldsymbol{a}$ (*associativity of multiplication by scalars*)

d) $(r + s)\boldsymbol{a} = r\boldsymbol{a} + s\boldsymbol{a}$ (*a right distributive law*)

e) $r(\boldsymbol{a} + \boldsymbol{b}) = r\boldsymbol{a} + r\boldsymbol{b}$ (*a left distributive law*)

14.3.3 A physical model for vectors

When studying motion, a physicist may use a vector to represent a *force*. Suppose, for example, that you are moving some object by pushing it. The motion of the object is influenced by the *direction* in which you are pushing and how *hard* you are pushing. Thus the force with which you are pushing can be conveniently represented by a vector, which you regard as having the *direction* in which you are pushing, and a *length* representing how hard you are pushing. If you double the force with which you push, the force vector doubles in length; this corresponds to the multiplication of the force vector by the scalar 2.

Suppose that two people are pushing on an object with forces that correspond to vectors $\boldsymbol{a}$ and $\boldsymbol{b}$ as shown in Fig. 14.20. It can be shown that the motion of the object due to these combined forces is the same as the motion that would result if only one person were pushing with a force expressed by a vector that is the diagonal of the parallelogram with arrow vectors $\boldsymbol{a}$ and $\boldsymbol{b}$ as adjacent sides (see Fig. 14.20). Thus this *resultant force vector* is precisely the vector $\boldsymbol{a} + \boldsymbol{b}$.

14.3.4 Perpendicular vectors

It is very important for us to know when two directions are perpendicular. There is a very easy criterion for this in terms of vectors. Any three points in space not all on the same line determine a plane. Imagine that Fig. 14.24 gives a picture of such a plane determined by three points, $(0, 0, 0)$, (a_1, a_2, a_3), and (b_1, b_2, b_3). The vectors $\boldsymbol{a}$ and $\boldsymbol{b}$ of Fig. 14.24 are perpendicular if and only if the Pythagorean relation holds, that is, if and only if

$$|\boldsymbol{a}|^2 + |\boldsymbol{b}|^2 = d^2 = |\boldsymbol{a} - \boldsymbol{b}|^2. \tag{2}$$

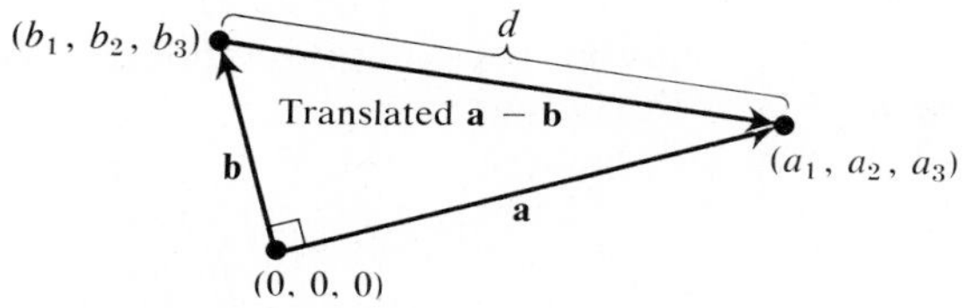

14.24

By definition of the length of a vector, you see that this is true when

$$a_1^2 + a_2^2 + a_3^2 + b_1^2 + b_2^2 + b_3^2 = (a_1 - b_1)^2 + (a_2 - b_2)^2 + (a_3 - b_3)^2. \tag{3}$$

Squaring the terms on the righthand side of (3) and cancelling, you obtain the condition

$$0 = -2a_1b_1 - 2a_2b_2 - 2a_3b_3$$

or

$$a_1b_1 + a_2b_2 + a_3b_3 = 0.$$

Perpendicular condition

Thus the vectors $\boldsymbol{a}$ and $\boldsymbol{b}$ in space are *perpendicular* if and only if

$$a_1b_1 + a_2b_2 + a_3b_3 = 0. \tag{4}$$

Of course, the corresponding result with just two components holds for vectors in the plane.

Example 6 The unit coordinate vectors $\boldsymbol{i} = 1\boldsymbol{i} + 0\boldsymbol{j}$ and $\boldsymbol{j} = 0\boldsymbol{i} + 1\boldsymbol{j}$ in the plane are perpendicular, for

$$1 \cdot 0 + 0 \cdot 1 = 0.$$

Also, $-\boldsymbol{i} + 3\boldsymbol{j} + 2\boldsymbol{k}$ and $5\boldsymbol{i} - \boldsymbol{j} + 4\boldsymbol{k}$ are perpendicular in space, for

$$-1 \cdot 5 + 3 \cdot (-1) + 2 \cdot 4 = -5 - 3 + 8 = 0. \quad \|$$

Example 7 According to condition (4), the zero vector $\mathbf{0}$ is perpendicular to *every* vector in space. This is why it is convenient to think of $\mathbf{0}$ as having *all directions*, rather than *no direction*. ‖

SUMMARY *Formulas are given for vectors in space, but are equally valid for vectors in the plane.*

1. *Each point $\boldsymbol{a} = (a_1, a_2, a_3)$ can also be considered as a vector having direction from the origin to the point (a_1, a_2, a_3) and length $|\boldsymbol{a}| = \sqrt{a_1{}^2 + a_2{}^2 + a_3{}^2}$. A unit vector has length 1. The zero vector is $\mathbf{0} = (0, 0, 0)$.*
2. *Addition of vectors is given by adding corresponding coordinates, that is, $\boldsymbol{a} + \boldsymbol{b} = (a_1, a_2, a_3) + (b_1, b_2, b_3) = (a_1 + b_1, a_2 + b_2, a_3 + b_3)$.*
3. *Multiplication of a vector $\boldsymbol{a}$ by a scalar (real number) r is given by $r\boldsymbol{a} = r(a_1, a_2, a_3) = (ra_1, ra_2, ra_3)$.*
4. *Two nonzero vectors $\boldsymbol{a}$ and $\boldsymbol{b}$ are parallel if $\boldsymbol{b} = r\boldsymbol{a}$ for some scalar r.*
5. *Two vectors $\boldsymbol{a}$ and $\boldsymbol{b}$ are perpendicular if $a_1b_1 + a_2b_2 + a_3b_3 = 0$.*
6. *In space, the unit coordinate vectors are often written*

$$\boldsymbol{i} = (1, 0, 0), \qquad \boldsymbol{j} = (0, 1, 0), \qquad \boldsymbol{k} = (0, 0, 1),$$

so $(a_1, a_2, a_3) = a_1\boldsymbol{i} + a_2\boldsymbol{j} + a_3\boldsymbol{k}$. In the plane, one uses just $\boldsymbol{i} = (1, 0)$ and $\boldsymbol{j} = (0, 1)$.

EXERCISES

1. Let $\boldsymbol{a} = 2\boldsymbol{i} - \boldsymbol{j}$ and $\boldsymbol{b} = -3\boldsymbol{i} - 2\boldsymbol{j}$ be vectors in the plane. Sketch, using arrows, the vectors $\boldsymbol{a}$, $\boldsymbol{b}$, $\boldsymbol{a} + \boldsymbol{b}$, $\boldsymbol{a} - \boldsymbol{b}$, and $-(4/3)\boldsymbol{a}$.

2. Let $\boldsymbol{a} = -\boldsymbol{i} + 3\boldsymbol{j} - 2\boldsymbol{k}$, $\boldsymbol{b} = 4\boldsymbol{i} - \boldsymbol{k}$, and $\boldsymbol{c} = -3\boldsymbol{i} - \boldsymbol{j} + 2\boldsymbol{k}$ be vectors in space. Compute the following.

a) $3\boldsymbol{a}$ b) $-2\boldsymbol{c}$ c) $\boldsymbol{a} + \boldsymbol{b}$ d) $3\boldsymbol{b} - 2\boldsymbol{c}$

e) $\boldsymbol{a} + 2(\boldsymbol{b} - 3\boldsymbol{c})$ f) $3(\boldsymbol{a} - 2\boldsymbol{b})$ g) $4(3\boldsymbol{a} + 5\boldsymbol{b})$

3. Let $\boldsymbol{a} = 3\boldsymbol{i} - 2\boldsymbol{j} + 2\boldsymbol{k}$ and $\boldsymbol{b} = -\boldsymbol{i} + 4\boldsymbol{j} + \boldsymbol{k}$. Compute each of the following.

a) $|\boldsymbol{a}|$ b) $|\boldsymbol{a} + \boldsymbol{b}|$ c) $|-2\boldsymbol{a}|$ d) $|\boldsymbol{b} - 3\boldsymbol{a}|$

4. Determine whether each of the following pairs of vectors is parallel, perpendicular, or neither. If two vectors are parallel, state whether they have the same or opposite directions.

a) $3\boldsymbol{i} - \boldsymbol{j}$ and $4\boldsymbol{i} + 12\boldsymbol{j}$ in the plane

b) $-2\boldsymbol{i} + 6\boldsymbol{j}$ and $4\boldsymbol{i} - 12\boldsymbol{j}$ in the plane

c) $3\boldsymbol{i} - \boldsymbol{j}$ and $4\boldsymbol{i} + 3\boldsymbol{j} + 2\boldsymbol{k}$ in space

d) $2\boldsymbol{i} - 3\boldsymbol{j} + \boldsymbol{k}$ and $8\boldsymbol{i} + 2\boldsymbol{j} - 10\boldsymbol{k}$ in space

e) $\sqrt{2}\boldsymbol{i} + \sqrt{18}\boldsymbol{j} - \sqrt{8}\boldsymbol{k}$ and $2\boldsymbol{i} + 6\boldsymbol{j} - 4\boldsymbol{k}$ in space

5. If possible, determine c so that the vector $2\boldsymbol{i} + c\boldsymbol{j}$ in the plane is parallel to the given vector.

a) $4\boldsymbol{i} + 6\boldsymbol{j}$ b) $-5\boldsymbol{i} + 3\boldsymbol{j}$ c) $3\boldsymbol{i}$ d) $3\boldsymbol{j}$

6. If possible, determine c so that the vector $c\boldsymbol{i} + 2\boldsymbol{j} - \boldsymbol{k}$ is perpendicular to the given vector.

a) $\boldsymbol{j} - 4\boldsymbol{k}$ b) $\boldsymbol{i} - 3\boldsymbol{k}$ c) $-5\boldsymbol{i} + \boldsymbol{j} + 2\boldsymbol{k}$

7. Find a unit vector in space parallel to $\boldsymbol{i} - \boldsymbol{j} + 3\boldsymbol{k}$ and having the same direction.

8. Find two unit vectors in the plane perpendicular to $3\boldsymbol{i} - 4\boldsymbol{j}$.

9. Find two unit vectors in space that are not parallel and each of which is perpendicular to $-2\boldsymbol{i} + \boldsymbol{j} + 2\boldsymbol{k}$.

10. Show that $(1, -5)$, $(9, -11)$, and $(4, -1)$ are vertices of a right triangle in the plane.

11. Show that $(1, -1, 4)$, $(3, -2, 4)$, $(-4, 2, 6)$, and $(-2, 1, 6)$ are vertices of a parallelogram in space.

12. Show that, for all vectors $\boldsymbol{a} = a_1\boldsymbol{i} + a_2\boldsymbol{j} + a_3\boldsymbol{k}$, $\boldsymbol{b} = b_1\boldsymbol{i} + b_2\boldsymbol{j} + b_3\boldsymbol{k}$, and $\boldsymbol{c} = c_1\boldsymbol{i} + c_2\boldsymbol{j} + c_3\boldsymbol{k}$, and all scalars r and s, the following relations hold.

a) $(\boldsymbol{a} + \boldsymbol{b}) + \boldsymbol{c} = \boldsymbol{a} + (\boldsymbol{b} + \boldsymbol{c})$

b) $\boldsymbol{a} + \boldsymbol{b} = \boldsymbol{b} + \boldsymbol{a}$

c) $r(s\boldsymbol{a}) = (rs)\boldsymbol{a}$

d) $(r + s)\boldsymbol{a} = r\boldsymbol{a} + s\boldsymbol{a}$

e) $r(\boldsymbol{a} + \boldsymbol{b}) = r\boldsymbol{a} + r\boldsymbol{b}$

14.4 THE DOT PRODUCT OF VECTORS

14.4.1 The dot product

Let $\boldsymbol{a} = a_1\boldsymbol{i} + a_2\boldsymbol{j} + a_3\boldsymbol{k}$ and $\boldsymbol{b} = b_1\boldsymbol{i} + b_2\boldsymbol{j} + b_3\boldsymbol{k}$ be nonparallel and nonzero vectors in space. You can view $\boldsymbol{a}$ and $\boldsymbol{b}$ geometrically as arrows emanating from the origin, as shown in Fig. 14.25. The vector $\boldsymbol{a}$ gives the direction for a line through the origin, labeled "the line along $\boldsymbol{a}$" in Fig. 14.25. Similarly, the vector $\boldsymbol{b}$ gives the direction for the line along $\boldsymbol{b}$. These two intersecting lines in space determine a plane, as indicated in Fig. 14.26. The points in this plane are precisely those points $\boldsymbol{x} = (x_1, x_2, x_3)$ that can be expressed in the form $\boldsymbol{x} = t\boldsymbol{a} + s\boldsymbol{b}$ for some scalars t and s.

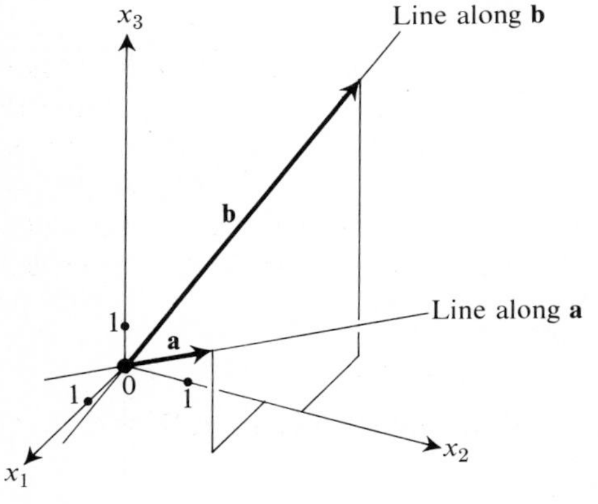

14.25

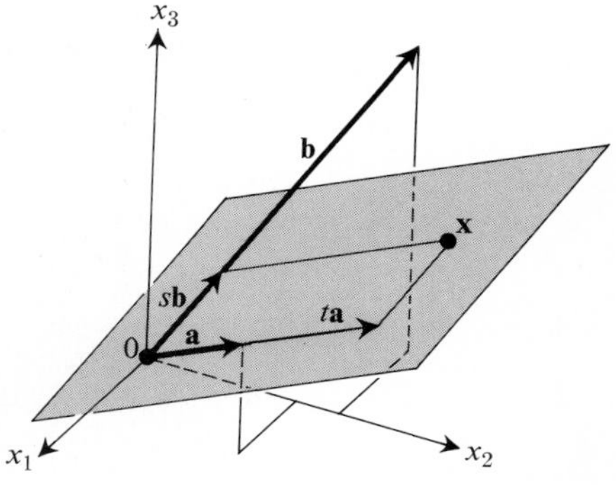

14.26

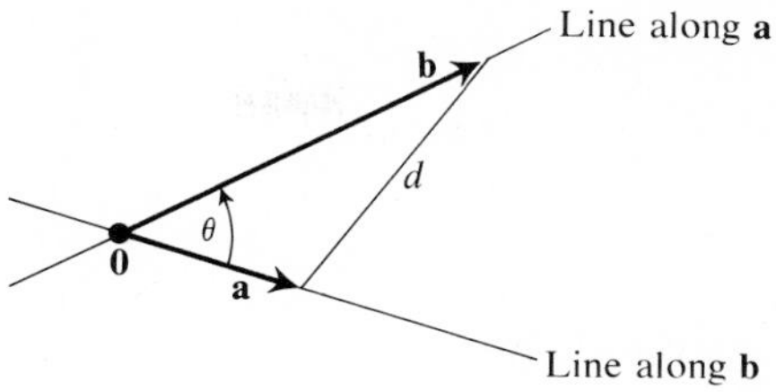

14.27

In Fig. 14.27, we take the plane in Fig. 14.26 as the plane of our page. We are interested in finding the angle θ between $\boldsymbol{a}$ and $\boldsymbol{b}$ as shown in the figure. To find θ, we apply the law of cosines to the triangle shown in that figure, and obtain

$$d^2 = |\boldsymbol{a}|^2 + |\boldsymbol{b}|^2 - 2|\boldsymbol{a}|\,|\boldsymbol{b}| \cos \theta. \tag{1}$$

You can easily compute d^2, $|\boldsymbol{a}|$, and $|\boldsymbol{b}|$ in terms of the components a_i and b_i of the vectors $\boldsymbol{a}$ and $\boldsymbol{b}$, so you can use (1) to find $\cos \theta$ in terms of these components. You have

$$d^2 = (b_1 - a_1)^2 + (b_2 - a_2)^2 + (b_3 - a_3)^2,$$

while

$$|\boldsymbol{a}|^2 = a_1{}^2 + a_2{}^2 + a_3{}^2$$

and

$$|\boldsymbol{b}|^2 = b_1{}^2 + b_2{}^2 + b_3{}^2.$$

Substituting these quantities in (1), squaring out the terms $(b_i - a_i)^2$ in d^2, and cancelling the terms a_i^2 and b_i^2 from both sides of the resulting equation, you obtain

$$-2a_1b_1 - 2a_2b_2 - 2a_3b_3 = -2|\boldsymbol{a}|\,|\boldsymbol{b}| \cos \theta, \tag{2}$$

so that

$$\cos \theta = \frac{a_1b_1 + a_2b_2 + a_3b_3}{|\boldsymbol{a}|\,|\boldsymbol{b}|}. \tag{3}$$

The numerator in (3) is familiar; you saw in Section 14.3.4 that vectors $\boldsymbol{a}$ and $\boldsymbol{b}$ are perpendicular if and only if $a_1b_1 + a_2b_2 + a_3b_3 = 0$. Note that this is consistent with (3); the vectors should be perpendicular if and only if $\theta = \pi/2$ or $\theta = 3\pi/2$ so that $\cos \theta = 0$. The number $a_1b_1 + a_2b_2 + a_3b_3$ appearing in (3) is so important that it is given a special name, the *dot* (or *scalar,* or *inner*) *product of* $\boldsymbol{a}$ *and* $\boldsymbol{b}$. (The result of this product of $\boldsymbol{a}$ and $\boldsymbol{b}$ is a *scalar.*) From (3), you obtain

$$a_1b_1 + a_2b_2 + a_3b_3 = |\boldsymbol{a}|\,|\boldsymbol{b}| \cos \theta. \tag{4}$$

Note that if either $\boldsymbol{a}$ or $\boldsymbol{b}$ is $\mathbf{0}$ so that θ is undefined, then $|\boldsymbol{a}|$ or $|\boldsymbol{b}|$ is zero, and Eq. (4) still holds formally.

Definition 14.3 The **dot product** $\boldsymbol{a} \cdot \boldsymbol{b}$ of $\boldsymbol{a}$ and $\boldsymbol{b}$ is

$$\boldsymbol{a} \cdot \boldsymbol{b} = a_1b_1 + a_2b_2 + a_3b_3. \tag{5}$$

Equation (4) shows that $\boldsymbol{a} \cdot \boldsymbol{b} = |\boldsymbol{a}|\,|\boldsymbol{b}| \cos\theta$, where θ is the angle between $\boldsymbol{a}$ and $\boldsymbol{b}$.

Of course, this same work could be done with vectors in the plane having only two components. The notion of the dot product is defined for any two vectors having the same number of components.

Example 1 We compute $\boldsymbol{a} \cdot \boldsymbol{b}$ for the vectors

$$\boldsymbol{a} = \boldsymbol{i} - 4\boldsymbol{j} + 3\boldsymbol{k} \quad \text{and} \quad \boldsymbol{b} = 6\boldsymbol{i} - 2\boldsymbol{j} - \boldsymbol{k}.$$

SOLUTION You have

$$\boldsymbol{a} \cdot \boldsymbol{b} = (1)(6) + (-4)(-2) + (3)(-1) = 6 + 8 - 3 = 11. \quad \|$$

Example 2 Let's find the angle between the vectors $\boldsymbol{a} = \boldsymbol{i} - 4\boldsymbol{j}$ and $\boldsymbol{b} = 3\boldsymbol{i} + 2\boldsymbol{j}$ in the plane.

SOLUTION You have

$$\theta = \cos^{-1}\frac{\boldsymbol{a} \cdot \boldsymbol{b}}{|\boldsymbol{a}|\,|\boldsymbol{b}|} = \cos^{-1}\frac{3 - 8}{\sqrt{17}\sqrt{13}} = \cos^{-1}\frac{-5}{\sqrt{17}\sqrt{13}} \approx 109.65°. \quad \|$$

Example 3 Let's find the angle θ that the diagonal of a cube in space makes with an edge of the cube.

SOLUTION We may take as our cube one with a vertex at the origin and with edges falling on the positive coordinate axes, as shown in Fig. 14.28. Then $\boldsymbol{i}$, $\boldsymbol{j}$, and $\boldsymbol{k}$ are vectors along edges of the cube, while a vector along a diagonal is

$$\boldsymbol{d} = \boldsymbol{i} + \boldsymbol{j} + \boldsymbol{k}.$$

You have

$$\theta = \cos^{-1}\frac{\boldsymbol{i} \cdot \boldsymbol{d}}{|\boldsymbol{i}|\,|\boldsymbol{d}|} = \cos^{-1}\frac{1}{1 \cdot \sqrt{3}}$$

$$= \cos^{-1}\frac{1}{\sqrt{3}} \approx 54.74°. \quad \|$$

14.4.2 Algebraic properties of the dot product

Theorem 14.2 below lists some of the algebraic properties of the dot product. We observe the usual convention that an algebraic operation written in multiplicative notation is performed before one written in additive notation, in the absence of parentheses. For example,

$$\boldsymbol{a} \cdot \boldsymbol{b} + \boldsymbol{a} \cdot \boldsymbol{c} = (\boldsymbol{a} \cdot \boldsymbol{b}) + (\boldsymbol{a} \cdot \boldsymbol{c}).$$

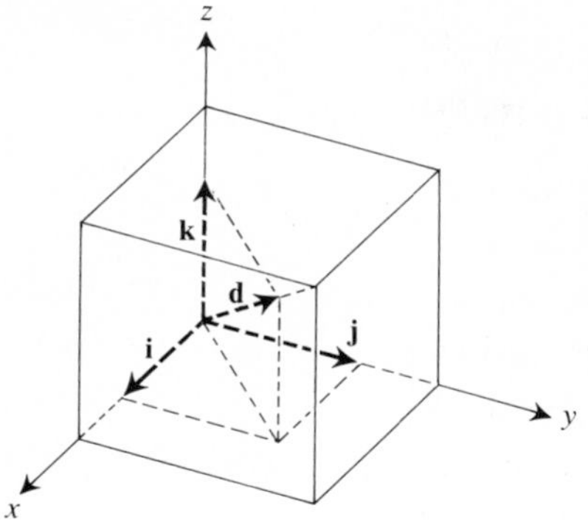

14.28

Theorem 14.2 *(Properties of the dot product) Let **a**, **b** and **c** be vectors with the same number of components and let r be a scalar. Then*

a) $\boldsymbol{a} \cdot \boldsymbol{a} \geq 0$ *and* $\boldsymbol{a} \cdot \boldsymbol{a} = 0$ *if and only if* $\boldsymbol{a} = \boldsymbol{0}$ *(nonnegative property)*

b) $\boldsymbol{a} \cdot \boldsymbol{b} = \boldsymbol{b} \cdot \boldsymbol{a}$ *(commutative property)*

c) $\boldsymbol{a} \cdot (\boldsymbol{b} + \boldsymbol{c}) = \boldsymbol{a} \cdot \boldsymbol{b} + \boldsymbol{a} \cdot \boldsymbol{c}$ *(distributive property)*

d) $(r\boldsymbol{a}) \cdot \boldsymbol{b} = \boldsymbol{a} \cdot (r\boldsymbol{b}) = r(\boldsymbol{a} \cdot \boldsymbol{b})$ *(homogeneous property)*

e) $\boldsymbol{a} \cdot \boldsymbol{a} = |\boldsymbol{a}|^2$ *(length property)*

f) $\boldsymbol{a} \cdot \boldsymbol{b} = 0$ *if and only if* ***a** and **b** are perpendicular* *(perpendicular property)*

Properties (a), (b), (c), and (d) are proved easily from the formula in (5) for $\boldsymbol{a} \cdot \boldsymbol{b}$ in terms of the components of $\boldsymbol{a}$ and of $\boldsymbol{b}$. Illustrating with the proof of (a), you have, for vectors in space,

$$\boldsymbol{a} \cdot \boldsymbol{a} = a_1a_1 + a_2a_2 + a_3a_3 = a_1{}^2 + a_2{}^2 + a_3{}^2 \geq 0,$$

and this sum of squares is 0 if and only if each $a_i = 0$, that is, if and only if $\boldsymbol{a} = \boldsymbol{0}$. We leave the proofs of (b), (c), and (d) to the exercises (see Exercises 14, 15, and 16).

Properties (e) and (f) are really restatements of previous definitions in the notation of the dot product. We defined the length of a vector $\boldsymbol{a}$ in space to be $\sqrt{a_1{}^2 + a_2{}^2 + a_3{}^2} = \sqrt{\boldsymbol{a} \cdot \boldsymbol{a}}$, and we also defined $\boldsymbol{a}$ and $\boldsymbol{b}$ to be perpendicular vectors if and only if $\boldsymbol{a} \cdot \boldsymbol{b} = a_1b_1 + a_2b_2 + a_3b_3 = 0$. Recall that we defined the vector $\boldsymbol{0}$ to be perpendicular to every vector.

The properties of the dot product are very important, and all sorts of consequences can be easily derived from them. We shall give a geometric illustration.

Example 4 Let's show that the sum of the squares of the lengths of the diagonals of a parallelogram is equal to the sum of the squares of the lengths of the sides. (This is the *parallelogram relation*.)

SOLUTION We take our parallelogram with a vertex at the origin and vectors $\boldsymbol{a}$ and $\boldsymbol{b}$ as coterminous sides, as shown in Fig. 14.29. The lengths of the diagonals are then $|\boldsymbol{a}+\boldsymbol{b}|$ and $|\boldsymbol{a}-\boldsymbol{b}|$. Using Theorem 14.2, you have

$$\begin{aligned}|\boldsymbol{a}+\boldsymbol{b}|^2+|\boldsymbol{a}-\boldsymbol{b}|^2 &= (\boldsymbol{a}+\boldsymbol{b})\cdot(\boldsymbol{a}+\boldsymbol{b})+(\boldsymbol{a}-\boldsymbol{b})\cdot(\boldsymbol{a}-\boldsymbol{b})\\ &= \boldsymbol{a}\cdot\boldsymbol{a}+2\boldsymbol{a}\cdot\boldsymbol{b}+\boldsymbol{b}\cdot\boldsymbol{b}+\boldsymbol{a}\cdot\boldsymbol{a}-2\boldsymbol{a}\cdot\boldsymbol{b}+\boldsymbol{b}\cdot\boldsymbol{b}\\ &= 2(\boldsymbol{a}\cdot\boldsymbol{a})+2(\boldsymbol{b}\cdot\boldsymbol{b})\\ &= 2|\boldsymbol{a}|^2+2|\boldsymbol{b}|^2,\end{aligned}$$

which is what we wished to prove. You may think that we have used only property (e) of Theorem 14.2, but we also used properties (b), (c), and (d) as we ask you to show in Exercise 17. ||

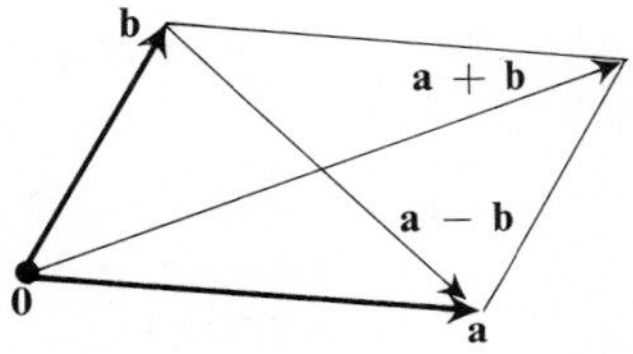

14.29

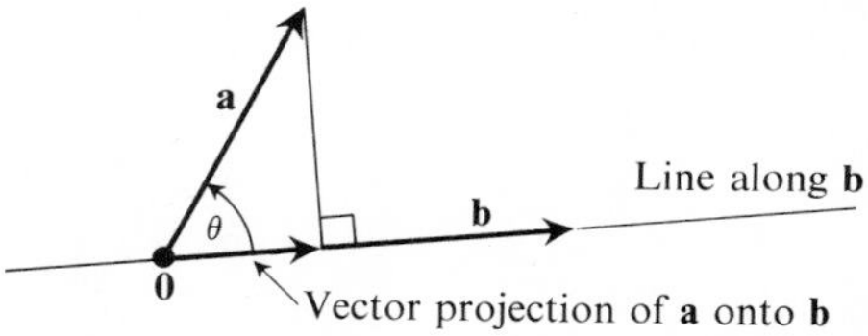

14.30

14.4.3 Vector projection Let $\boldsymbol{a}$ and $\boldsymbol{b}$ be vectors in space with $\boldsymbol{b} \neq \boldsymbol{0}$. Then

$$\frac{1}{|\boldsymbol{a}|}\boldsymbol{a} \quad \text{and} \quad \frac{1}{|\boldsymbol{b}|}\boldsymbol{b}$$

are unit vectors in the directions of $\boldsymbol{a}$ and $\boldsymbol{b}$, respectively. We shall write such vectors as

$$\frac{\boldsymbol{a}}{|\boldsymbol{a}|} \quad \text{and} \quad \frac{\boldsymbol{b}}{|\boldsymbol{b}|}$$

in what follows, although we have not formally defined division of a vector by a scalar.

In Fig. 14.30, we again imagine that the plane containing both $\boldsymbol{a}$ and $\boldsymbol{b}$ is the plane of the page, and show the angle θ between $\boldsymbol{a}$ and $\boldsymbol{b}$. In the figure, we have also labeled the *vector projection of* $\boldsymbol{a}$ *onto* $\boldsymbol{b}$. This vector has the direction of $\boldsymbol{b}$ if θ is an acute angle, and the direction of $-\boldsymbol{b}$ if θ is an obtuse angle. The length of this projection vector is the distance from the origin $\boldsymbol{0}$ to the foot of the perpendicular dropped to the line along $\boldsymbol{b}$ from the tip of the vector $\boldsymbol{a}$. For the figure, this length should be $|\boldsymbol{a}|\cos\theta$, so the vector projection should be

$$\begin{aligned}(|\boldsymbol{a}|\cos\theta)\frac{\boldsymbol{b}}{|\boldsymbol{b}|} &= |\boldsymbol{a}|\,|\boldsymbol{b}|\,(\cos\theta)\frac{\boldsymbol{b}}{|\boldsymbol{b}|^2}\\ &= \frac{\boldsymbol{a}\cdot\boldsymbol{b}}{|\boldsymbol{b}|^2}\boldsymbol{b}.\end{aligned}$$

The number

$$|\boldsymbol{a}|\,(\cos\theta) = \frac{\boldsymbol{a}\cdot\boldsymbol{b}}{|\boldsymbol{b}|}$$

is the (signed) length of the vector projection. In summary:

The vector projection of $\boldsymbol{a}$ *onto* $\boldsymbol{b}$ *is* $[(\boldsymbol{a}\cdot\boldsymbol{b})/|\boldsymbol{b}|^2]\boldsymbol{b}$, *for* $\boldsymbol{b}\neq\boldsymbol{0}$. *The component of* $\boldsymbol{a}$ *along* $\boldsymbol{b}$ *is the number* $(\boldsymbol{a}\cdot\boldsymbol{b})/|\boldsymbol{b}|$, *for* $\boldsymbol{b}\neq\boldsymbol{0}$.

Example 5 Let's find the vector projection of $\boldsymbol{a} = 2\boldsymbol{i} - \boldsymbol{j} + 3\boldsymbol{k}$ onto $\boldsymbol{b} = \boldsymbol{i} - 3\boldsymbol{j} - \boldsymbol{k}$ and the component of $\boldsymbol{a}$ along $\boldsymbol{b}$.

SOLUTION The vector projection of $\boldsymbol{a}$ onto $\boldsymbol{b}$ is

$$\frac{\boldsymbol{a}\cdot\boldsymbol{b}}{|\boldsymbol{b}|^2}\boldsymbol{b} = \frac{2+3-3}{1+9+1}(\boldsymbol{i} - 3\boldsymbol{j} - \boldsymbol{k}) = \frac{2}{11}(\boldsymbol{i} - 3\boldsymbol{j} - \boldsymbol{k}) = \frac{2}{11}\boldsymbol{i} - \frac{6}{11}\boldsymbol{j} - \frac{2}{11}\boldsymbol{k}.$$

The component of $\boldsymbol{a}$ along $\boldsymbol{b}$ is

$$\frac{\boldsymbol{a}\cdot\boldsymbol{b}}{|\boldsymbol{b}|} = \frac{2}{\sqrt{11}}. \quad \|$$

If $\boldsymbol{b}$ is a unit vector so that $|\boldsymbol{b}| = 1$, then the vector projection of $\boldsymbol{a}$ onto $\boldsymbol{b}$ takes the simpler form

$$(\boldsymbol{a}\cdot\boldsymbol{b})\boldsymbol{b}.$$

Example 6 A physicist may be interested in finding the components of a force vector in the coordinate directions. For a force vector $\boldsymbol{F}$ in space and the directions given by $\boldsymbol{i}$, $\boldsymbol{j}$, and $\boldsymbol{k}$, these components of $\boldsymbol{F}$ along $\boldsymbol{i}$, $\boldsymbol{j}$, and $\boldsymbol{k}$, are

$$(\boldsymbol{F}\cdot\boldsymbol{i}), \qquad (\boldsymbol{F}\cdot\boldsymbol{j}), \qquad \text{and} \qquad (\boldsymbol{F}\cdot\boldsymbol{k}). \quad \|$$

SUMMARY

1. *Let* $\boldsymbol{a}$ *and* $\boldsymbol{b}$ *be vectors in space or the plane. The dot (or scalar) product of* $\boldsymbol{a}$ *and* $\boldsymbol{b}$ *is the number*

$$\boldsymbol{a}\cdot\boldsymbol{b} = a_1b_1 + a_2b_2 + a_3b_3.$$

2. *The dot product of* $\boldsymbol{a}$ *and* $\boldsymbol{b}$ *is described geometrically by*

$$\boldsymbol{a}\cdot\boldsymbol{b} = |\boldsymbol{a}|\,|\boldsymbol{b}|\cos\theta,$$

where θ *is the angle between* $\boldsymbol{a}$ *and* $\boldsymbol{b}$.

3. *Algebraic properties of the dot product are listed in Theorem* 14.2 *of Section* 14.4.2.

4. *The vector projection of* $\boldsymbol{a}$ *onto* $\boldsymbol{b}$ *if* $\boldsymbol{b}\neq\boldsymbol{0}$ *is the vector*

$$\frac{\boldsymbol{a}\cdot\boldsymbol{b}}{|\boldsymbol{b}|^2}\boldsymbol{b},$$

and the number $(\boldsymbol{a}\cdot\boldsymbol{b})/|\boldsymbol{b}|$ *is the component of* $\boldsymbol{a}$ *along* $\boldsymbol{b}$.

EXERCISES

In Exercises 1 through 4, find the angle between the vectors.

1. $\boldsymbol{i} + 4\boldsymbol{j}$ and $-8\boldsymbol{i} + 2\boldsymbol{j}$

2. $3\boldsymbol{i} + 2\boldsymbol{j} - 2\boldsymbol{k}$ and $4\boldsymbol{j} + \boldsymbol{k}$

3. $\boldsymbol{a} = \boldsymbol{k}$ and $\boldsymbol{b} = \boldsymbol{i} - \boldsymbol{k}$

4. $\boldsymbol{a} = 3\boldsymbol{i} + 4\boldsymbol{j}$ and $\boldsymbol{b} = -\boldsymbol{i}$

5. Find the angle BAC of the triangle with vertices $A(2, -3)$, $B(-1, 1)$, and $C(5, -9)$.

6. Find the angle ACB of the triangle with vertices $A(0, 1, 6)$, $B(2, 3, 0)$, and $C(-1, 3, 4)$.

7. Find the angle between the line through $(-1, 2, 4)$ and $(3, 4, 0)$ and the line also through $(-1, 2, 4)$ that passes through $(5, 7, 2)$.

8. Find two lines in the plane through $(1, -4)$ that intersect the line $y = -2x + 7$ at an angle of 45°.

9. Use vector methods to show that the diagonals of a rhombus (parallelogram with equal sides) are perpendicular. [*Hint.* Use a figure like Fig. 14.29, and show that $(\boldsymbol{a} + \boldsymbol{b}) \cdot (\boldsymbol{a} - \boldsymbol{b}) = 0$.]

10. Use vector methods to show that the midpoint of the hypotenuse of a right triangle is equidistant from the three vertices. [*Hint.* See Fig. 14.31.

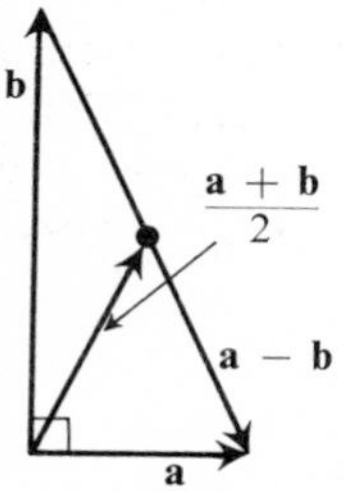

14.31

Show that

$$\left|\frac{\boldsymbol{a} + \boldsymbol{b}}{2}\right| = \left|\frac{\boldsymbol{a} - \boldsymbol{b}}{2}\right|.]$$

11. Show that the vectors $|\boldsymbol{a}|\,\boldsymbol{b} + |\boldsymbol{b}|\,\boldsymbol{a}$ and $|\boldsymbol{a}|\boldsymbol{b} - |\boldsymbol{b}|\,\boldsymbol{a}$ are perpendicular.

12. Show that the vector

$$\frac{|\boldsymbol{a}|\boldsymbol{b} + |\boldsymbol{b}|\boldsymbol{a}}{|\boldsymbol{a}| + |\boldsymbol{b}|}$$

bisects the angle between $\boldsymbol{a}$ and $\boldsymbol{b}$.

13. Show that for $\boldsymbol{a}, \boldsymbol{b}, \boldsymbol{c}$, the equation

$$\boldsymbol{a} \cdot \boldsymbol{b} = \boldsymbol{a} \cdot \boldsymbol{c},$$

where $\boldsymbol{a} \neq \boldsymbol{0}$, need not imply $\boldsymbol{b} = \boldsymbol{c}$.

14. Using the formula for the dot product, show that

$$\boldsymbol{a} \cdot \boldsymbol{b} = \boldsymbol{b} \cdot \boldsymbol{a}$$

for all $\boldsymbol{a}$ and $\boldsymbol{b}$ in space.

15. Using the formula for the dot product, show that

$$\boldsymbol{a} \cdot (\boldsymbol{b} + \boldsymbol{c}) = \boldsymbol{a} \cdot \boldsymbol{b} + \boldsymbol{a} \cdot \boldsymbol{c}$$

for all $\boldsymbol{a}, \boldsymbol{b}, \boldsymbol{c}$ in space.

16. Using the formula for the dot product, show that $(r\boldsymbol{a}) \cdot \boldsymbol{b} = \boldsymbol{a} \cdot (r\boldsymbol{b}) = r(\boldsymbol{a} \cdot \boldsymbol{b})$ for all $\boldsymbol{a}$ and $\boldsymbol{b}$ in space and all scalars r.

17. Find where properties (b), (c), and (d) given in Theorem 14.2 were used in the proof of the parallelogram relation in Example 4.

In Exercises 18 through 22, find the vector projection of the first vector on the second vector, and the component of the first vector along the second.

18. $\boldsymbol{i} + 3\boldsymbol{j} + 4\boldsymbol{k}$ on $\boldsymbol{j}$

19. $\boldsymbol{j}$ on $\boldsymbol{i} + 3\boldsymbol{j} + 4\boldsymbol{k}$

20. $2\boldsymbol{i} - \boldsymbol{j}$ on $-2\boldsymbol{i} + 3\boldsymbol{j}$

21. $3\boldsymbol{i} + \boldsymbol{j} - 2\boldsymbol{k}$ on $4\boldsymbol{i} + 2\boldsymbol{j} + 7\boldsymbol{k}$

22. $\boldsymbol{a} = -\boldsymbol{i} + \boldsymbol{j} + 3\boldsymbol{k}$ on $\boldsymbol{b} = 3\boldsymbol{i} - 2\boldsymbol{j} + \boldsymbol{k}$

23. Let $\boldsymbol{a}$ and $\boldsymbol{b}$ be vectors with $\boldsymbol{b} \neq \boldsymbol{0}$ and let $\boldsymbol{c}$ be the vector projection of $\boldsymbol{a}$ onto $\boldsymbol{b}$. Show that $\boldsymbol{a} - \boldsymbol{c}$ is perpendicular to $\boldsymbol{b}$.

14.5 THE CROSS PRODUCT AND TRIPLE PRODUCTS

14.5.1 Review of 2 × 2 and 3 × 3 determinants

A *square matrix* is a square array of numbers. For example,

$$\begin{pmatrix} -3 & 4 \\ 2 & 6 \end{pmatrix}$$

is a 2 × 2 (read "two by two") matrix, and

$$\begin{pmatrix} -1 & 0 & 4 \\ 2 & 1 & 0 \\ 3 & -4 & 5 \end{pmatrix}$$

is a 3 × 3 matrix. Each square matrix has associated with it a number, called the *determinant* of the matrix. The determinant is denoted by vertical lines rather than large parentheses on the sides of the array. The determinant of a 2 × 2 matrix is defined to be

$$\begin{vmatrix} a_1 & a_2 \\ b_1 & b_2 \end{vmatrix} = a_1b_2 - a_2b_1. \tag{1}$$

The determinant of a 3 × 3 matrix is defined in terms of the determinants of 2 × 2 matrices as follows:

$$\begin{vmatrix} a_1 & a_2 & a_3 \\ b_1 & b_2 & b_3 \\ c_1 & c_2 & c_3 \end{vmatrix} = a_1\begin{vmatrix} b_2 & b_3 \\ c_2 & c_3 \end{vmatrix} - a_2\begin{vmatrix} b_1 & b_3 \\ c_1 & c_3 \end{vmatrix} + a_3\begin{vmatrix} b_1 & b_2 \\ c_1 & c_2 \end{vmatrix}. \tag{2}$$

Formula (2) is easily remembered. The coefficients of the three determinants on the righthand side of (2) are the entries of the first row of the original 3 × 3 matrix, with alternate plus and minus signs. The first determinant on the righthand side in (2) is the determinant of the 2 × 2 matrix obtained by crossing out the row and column in which the coefficient a_1 appears in the 3 × 3 matrix. The second determinant is obtained by crossing out the row and column in which a_2 appears, etc.

Example 1 You have

$$\begin{vmatrix} 7 & -2 \\ 4 & 3 \end{vmatrix} = 7 \cdot 3 - (-2)4 = 21 + 8 = 29.$$

For a 3 × 3 illustration,

$$\begin{aligned}\begin{vmatrix} 2 & 3 & 5 \\ -4 & 2 & 6 \\ 1 & 0 & 3 \end{vmatrix} &= 2\begin{vmatrix} 2 & 6 \\ 0 & 3 \end{vmatrix} - 3\begin{vmatrix} -4 & 6 \\ 1 & 3 \end{vmatrix} + 5\begin{vmatrix} -4 & 2 \\ 1 & 0 \end{vmatrix} \\ &= 2(6 - 0) - 3(-12 - 6) + 5(0 - 2) \\ &= 12 + 54 - 10 = 56. \quad \|\end{aligned}$$

The only facts you need to know about determinants for your work with this text are given in this theorem.

Theorem 14.3 *If either the second row or the third row of a* 3×3 *matrix is the same as the first row, then the determinant of the matrix is zero. If the second and third rows of a* 3×3 *matrix are interchanged, the determinant of the new matrix differs from the determinant of the original matrix only in sign.*

You are asked to prove Theorem 14.3 in Exercises 7 and 8 by computing that

$$\begin{vmatrix} a_1 & a_2 & a_3 \\ a_1 & a_2 & a_3 \\ c_1 & c_2 & c_3 \end{vmatrix} = 0, \qquad \begin{vmatrix} a_1 & a_2 & a_3 \\ b_1 & b_2 & b_3 \\ a_1 & a_2 & a_3 \end{vmatrix} = 0,$$

and

$$\begin{vmatrix} a_1 & a_2 & a_3 \\ b_1 & b_2 & b_3 \\ c_1 & c_2 & c_3 \end{vmatrix} = - \begin{vmatrix} a_1 & a_2 & a_3 \\ c_1 & c_2 & c_3 \\ b_1 & b_2 & b_3 \end{vmatrix}.$$

14.5.2 The cross product of vectors

Let $\boldsymbol{a} = a_1\boldsymbol{i} + a_2\boldsymbol{j} + a_3\boldsymbol{k}$ and $\boldsymbol{b} = b_1\boldsymbol{i} + b_2\boldsymbol{j} + b_3\boldsymbol{k}$ be vectors in space.

Definition 14.4 The **cross product** $\boldsymbol{a} \times \boldsymbol{b}$ of $\boldsymbol{a}$ and $\boldsymbol{b}$ is the vector found by computing a symbolic determinant as follows:

$$\boldsymbol{a} \times \boldsymbol{b} = \begin{vmatrix} \boldsymbol{i} & \boldsymbol{j} & \boldsymbol{k} \\ a_1 & a_2 & a_3 \\ b_1 & b_2 & b_3 \end{vmatrix} = \begin{vmatrix} a_2 & a_3 \\ b_2 & b_3 \end{vmatrix}\boldsymbol{i} - \begin{vmatrix} a_1 & a_3 \\ b_1 & b_3 \end{vmatrix}\boldsymbol{j} + \begin{vmatrix} a_1 & a_2 \\ b_1 & b_2 \end{vmatrix}\boldsymbol{k}. \tag{3}$$

This cross product is also known as the **vector product**, for $\boldsymbol{a} \times \boldsymbol{b}$ is a vector quantity.

Example 2 If $\boldsymbol{a} = 3\boldsymbol{i} - 2\boldsymbol{j} + \boldsymbol{k}$ and $\boldsymbol{b} = -2\boldsymbol{i} + 3\boldsymbol{j} + 4\boldsymbol{k}$, then

$$\boldsymbol{a} \times \boldsymbol{b} = \begin{vmatrix} \boldsymbol{i} & \boldsymbol{j} & \boldsymbol{k} \\ 3 & -2 & 1 \\ -2 & 3 & 4 \end{vmatrix} = \begin{vmatrix} -2 & 1 \\ 3 & 4 \end{vmatrix}\boldsymbol{i} - \begin{vmatrix} 3 & 1 \\ -2 & 4 \end{vmatrix}\boldsymbol{j} + \begin{vmatrix} 3 & -2 \\ -2 & 3 \end{vmatrix}\boldsymbol{k}$$

$$= -11\boldsymbol{i} - 14\boldsymbol{j} + 5\boldsymbol{k}. \quad \|$$

Example 3 You easily find that

$$\boldsymbol{i} \times \boldsymbol{j} = \begin{vmatrix} \boldsymbol{i} & \boldsymbol{j} & \boldsymbol{k} \\ 1 & 0 & 0 \\ 0 & 1 & 0 \end{vmatrix} = \begin{vmatrix} 0 & 0 \\ 1 & 0 \end{vmatrix}\boldsymbol{i} - \begin{vmatrix} 1 & 0 \\ 0 & 0 \end{vmatrix}\boldsymbol{j} + \begin{vmatrix} 1 & 0 \\ 0 & 1 \end{vmatrix}\boldsymbol{k}$$

$$= 0\boldsymbol{i} + 0\boldsymbol{j} + \boldsymbol{k} = \boldsymbol{k}$$

and that $\boldsymbol{j} \times \boldsymbol{k} = \boldsymbol{i}$ and $\boldsymbol{k} \times \boldsymbol{i} = \boldsymbol{j}$. $\|$

The important geometric properties of any vector are its *length* and its *direction.* In Fig. 14.32, we take the plane determined by the lines along $\boldsymbol{a}$ and $\boldsymbol{b}$ as the plane of the page of the text, and shade the parallelogram having $\boldsymbol{a}$ and $\boldsymbol{b}$ as two edges. We claim that the length $|\boldsymbol{a} \times \boldsymbol{b}|$ is equal to the area of this shaded parallelogram. Let's compute the area to verify this. Referring to Fig. 14.32,

$$\begin{aligned}\text{Area} &= (\text{length of base})(\text{altitude})\\ &= |\boldsymbol{a}| \cdot h = |\boldsymbol{a}| \cdot |\boldsymbol{b}| \sin \theta.\end{aligned}$$

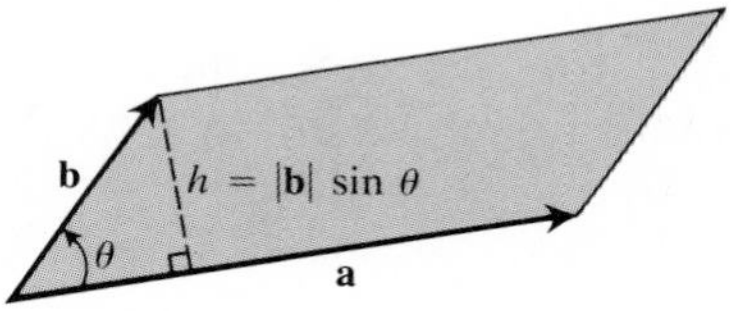

14.32

Therefore

$$\begin{aligned}(\text{Area})^2 &= |\boldsymbol{a}|^2\ |\boldsymbol{b}|^2 \sin^2\theta\\ &= |\boldsymbol{a}|^2\,|\boldsymbol{b}|^2\,(1 - \cos^2\theta)\\ &= |\boldsymbol{a}|^2\,|\boldsymbol{b}|^2 - (|\boldsymbol{a}| \cdot |\boldsymbol{b}| \cos\theta)^2\\ &= |\boldsymbol{a}|^2\,|\boldsymbol{b}|^2 - (\boldsymbol{a} \cdot \boldsymbol{b})^2\\ &= (a_1{}^2 + a_2{}^2 + a_3{}^2)(b_1{}^2 + b_2{}^2 + b_3{}^2) - (a_1b_1 + a_2b_2 + a_3b_3)^2.\end{aligned}$$

A bit of straightforward pencil-pushing shows that this all boils down to

$$\begin{aligned}(\text{Area})^2 &= (a_2b_3 - a_3b_2)^2 + (a_1b_3 - a_3b_1)^2 + (a_1b_2 - a_2b_1)^2\\ &= \begin{vmatrix} a_2 & a_3\\ b_2 & b_3\end{vmatrix}^2 + \begin{vmatrix} a_1 & a_3\\ b_1 & b_3\end{vmatrix}^2 + \begin{vmatrix} a_1 & a_2\\ b_1 & b_2\end{vmatrix}^2.\end{aligned}$$

But this is the square of the length of the cross product $\boldsymbol{a} \times \boldsymbol{b}$, so $(\text{area})^2 = |\boldsymbol{a} \times \boldsymbol{b}|^2$ and area $= |\boldsymbol{a} \times \boldsymbol{b}|$. This shows that

Length of $a \times b$

$$|\boldsymbol{a} \times \boldsymbol{b}| = (\text{Area of parallelogram}) = |\boldsymbol{a}| \cdot |\boldsymbol{b}| \sin\theta. \tag{4}$$

Finally, you want to know the direction of $\boldsymbol{a} \times \boldsymbol{b}$. First, let's see that $\boldsymbol{a} \times \boldsymbol{b}$ is perpendicular to both $\boldsymbol{a}$ and $\boldsymbol{b}$, and therefore perpendicular to the plane containing the parallelogram shaded in Fig. 14.32. We need only show that $\boldsymbol{a} \cdot (\boldsymbol{a} \times \boldsymbol{b}) = 0$ and $\boldsymbol{b} \cdot (\boldsymbol{a} \times \boldsymbol{b}) = 0$. Referring to Eq. (3), where $\boldsymbol{a} \times \boldsymbol{b}$ is defined, you see that

$$\boldsymbol{a} \cdot (\boldsymbol{a} \times \boldsymbol{b}) = a_1\begin{vmatrix} a_2 & a_3\\ b_2 & b_3\end{vmatrix} - a_2\begin{vmatrix} a_1 & a_3\\ b_1 & b_3\end{vmatrix} + a_3\begin{vmatrix} a_1 & a_2\\ b_1 & b_2\end{vmatrix}.$$

But this is equal to the determinant

$$\begin{vmatrix} a_1 & a_2 & a_3 \\ a_1 & a_2 & a_3 \\ b_1 & b_2 & b_3 \end{vmatrix} = 0,$$

which is zero since the first and second rows are the same (Theorem 14.3). Thus $\boldsymbol{a} \cdot (\boldsymbol{a} \times \boldsymbol{b}) = 0$ and $\boldsymbol{a}$ is perpendicular to $\boldsymbol{a} \times \boldsymbol{b}$. A similar argument shows that $\boldsymbol{b} \cdot (\boldsymbol{a} \times \boldsymbol{b}) = 0$; this time the determinant has its first and third rows the same.

You now know that $\boldsymbol{a} \times \boldsymbol{b}$ has length $|\boldsymbol{a}| \cdot |\boldsymbol{b}| \sin \theta$ and direction perpendicular to the plane determined by $\boldsymbol{a}$ and $\boldsymbol{b}$. There are two vectors of this length perpendicular to the plane; one is the negative of the other. One of them is $\boldsymbol{a} \times \boldsymbol{b}$, and the other one is $-(\boldsymbol{a} \times \boldsymbol{b})$, which we claim equals $\boldsymbol{b} \times \boldsymbol{a}$. Let's see why this is so.

The symoblic determinant used to find $\boldsymbol{b} \times \boldsymbol{a}$ is the one used to find $\boldsymbol{a} \times \boldsymbol{b}$ with the second and third rows interchanged. Theorem 14.3 thus shows at once that

$$\boldsymbol{b} \times \boldsymbol{a} = -(\boldsymbol{a} \times \boldsymbol{b}). \tag{5}$$

Let's summarize what we have done and a bit more in a theorem.

Theorem 14.4 *The vector $\boldsymbol{a} \times \boldsymbol{b}$ has length given by*

$$|\boldsymbol{a} \times \boldsymbol{b}| = |\boldsymbol{a}| \cdot |\boldsymbol{b}| \sin \theta,$$

where θ satisfying $0 \leq \theta \leq \pi$ is the angle between $\boldsymbol{a}$ and $\boldsymbol{b}$. The direction of $\boldsymbol{a} \times \boldsymbol{b}$ is perpendicular to both $\boldsymbol{a}$ and $\boldsymbol{b}$ in the direction in which the thumb of the right hand points as the fingers curl through θ from $\boldsymbol{a}$ to $\boldsymbol{b}$. This manner of describing the direction of $\boldsymbol{a} \times \boldsymbol{b}$ is known as the "righthand rule," and is illustrated in Fig. 14.33.

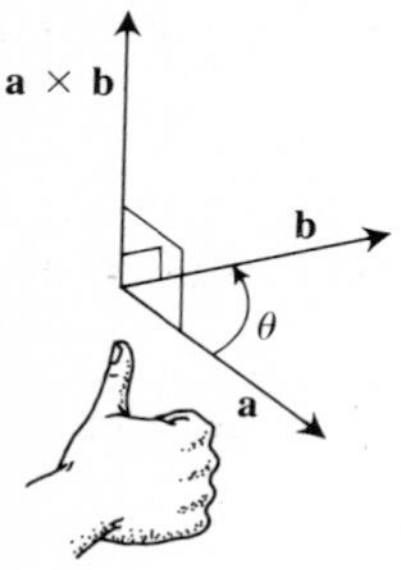

14.33

The only part of the theorem that we have not proved is the "righthand rule" part. We shall not prove this; you can easily illustrate it using

$$\boldsymbol{i} \times \boldsymbol{j} = \boldsymbol{k}, \qquad \boldsymbol{j} \times \boldsymbol{k} = \boldsymbol{i}, \qquad \text{and} \qquad \boldsymbol{k} \times \boldsymbol{i} = \boldsymbol{j},$$

which were shown in Example 3.

Example 4 In Example 3, you saw that

$$\boldsymbol{i} \times \boldsymbol{j} = \boldsymbol{k}, \qquad \boldsymbol{j} \times \boldsymbol{k} = \boldsymbol{i}, \qquad \text{and} \qquad \boldsymbol{k} \times \boldsymbol{i} = \boldsymbol{j}. \tag{6}$$

Therefore, from (5),

$$\boldsymbol{j} \times \boldsymbol{i} = -\boldsymbol{k}, \qquad \boldsymbol{k} \times \boldsymbol{j} = -\boldsymbol{i}, \qquad \text{and} \qquad \boldsymbol{i} \times \boldsymbol{k} = -\boldsymbol{j}. \tag{7}$$

You can remember (6) and (7) by writing the sequence

$$\boldsymbol{i}, \ \boldsymbol{j}, \ \boldsymbol{k}, \ \boldsymbol{i}, \ \boldsymbol{j}, \ \boldsymbol{k}.$$

The cross product of two consecutive vectors in left-to-right order is the next one to the right, while the cross product in right-to-left order is the negative of the next vector to the left. ||

Equation (5) shows that taking cross products is not a commutative operation. However, it is true that

$$\boldsymbol{a} \times (\boldsymbol{b} + \boldsymbol{c}) = \boldsymbol{a} \times \boldsymbol{b} + \boldsymbol{a} \times \boldsymbol{c} \tag{8}$$

and

$$(k\boldsymbol{a}) \times \boldsymbol{b} = \boldsymbol{a} \times (k\boldsymbol{b}) = k(\boldsymbol{a} \times \boldsymbol{b}) \tag{9}$$

for any vectors $\boldsymbol{a}$, $\boldsymbol{b}$, $\boldsymbol{c}$ in space and any scalar k. You can easily prove (8) and (9) as exercises.

14.5.3 Triple products

You now know two ways of taking a product of vectors $\boldsymbol{a}$ and $\boldsymbol{b}$ in space. You can find $\boldsymbol{a} \cdot \boldsymbol{b}$, which is a scalar, or $\boldsymbol{a} \times \boldsymbol{b}$, which is a vector. It is natural to try to multiply three vectors, $\boldsymbol{a}$, $\boldsymbol{b}$, and $\boldsymbol{c}$, in space. The product $\boldsymbol{a} \cdot (\boldsymbol{b} \cdot \boldsymbol{c})$ makes no sense, for $\boldsymbol{a}$ is a vector and $\boldsymbol{b} \cdot \boldsymbol{c}$ is a scalar. However, $\boldsymbol{a} \times (\boldsymbol{b} \times \boldsymbol{c})$ makes sense, for both $\boldsymbol{a}$ and $\boldsymbol{b} \times \boldsymbol{c}$ are vectors. This product $\boldsymbol{a} \times (\boldsymbol{b} \times \boldsymbol{c})$ is the *triple vector product*. It is most easily computed using the formula

$$\boldsymbol{a} \times (\boldsymbol{b} \times \boldsymbol{c}) = (\boldsymbol{a} \cdot \boldsymbol{c})\boldsymbol{b} - (\boldsymbol{a} \cdot \boldsymbol{b})\boldsymbol{c}, \tag{10}$$

which we ask you to establish in Exercise 24. In Exercise 29, you are asked to convince yourself that

$$\boldsymbol{a} \times (\boldsymbol{b} \times \boldsymbol{c}) \neq (\boldsymbol{a} \times \boldsymbol{b}) \times \boldsymbol{c}$$

unless $\boldsymbol{a}$, $\boldsymbol{b}$, and $\boldsymbol{c}$ are carefully chosen. That is, the triple cross product is not associative.

Example 5 If $\boldsymbol{a} = 2\boldsymbol{i} - 3\boldsymbol{j} + 4\boldsymbol{k}$ while $\boldsymbol{b} = 3\boldsymbol{i} - \boldsymbol{j} - 2\boldsymbol{k}$ and $\boldsymbol{c} = -3\boldsymbol{i} - 5\boldsymbol{j} - \boldsymbol{k}$, then

$$\boldsymbol{a} \times (\boldsymbol{b} \times \boldsymbol{c}) = (\boldsymbol{a} \cdot \boldsymbol{c})\boldsymbol{b} - (\boldsymbol{a} \cdot \boldsymbol{b})\boldsymbol{c} = 5\boldsymbol{b} + \boldsymbol{c} = 12\boldsymbol{i} - 10\boldsymbol{j} - 11\boldsymbol{k}. \quad \|$$

The product $\boldsymbol{a} \cdot (\boldsymbol{b} \times \boldsymbol{c})$ also makes sense, but the answer this time is a scalar. Consequently $\boldsymbol{a} \cdot (\boldsymbol{b} \times \boldsymbol{c})$ is called the *triple scalar product.* Let $\boldsymbol{a}$, $\boldsymbol{b}$, and $\boldsymbol{c}$ be as shown in Fig. 14.34. The vectors are coterminous edges of a box, shown shaded in the figure. For such a box, you have

$$\text{Volume} = (\text{area of base})\,(\text{altitude}).$$

Using the vectors and angles shown in Fig. 14.34, you see that

$$\text{Volume} = (\text{area of base})(|\boldsymbol{a}| \cos \phi).$$

Now the area of the base is $|\boldsymbol{b} \times \boldsymbol{c}|$, by the last article. Therefore

$$\text{Volume} = |\boldsymbol{b} \times \boldsymbol{c}| \cdot |\boldsymbol{a}| \cos \phi.$$

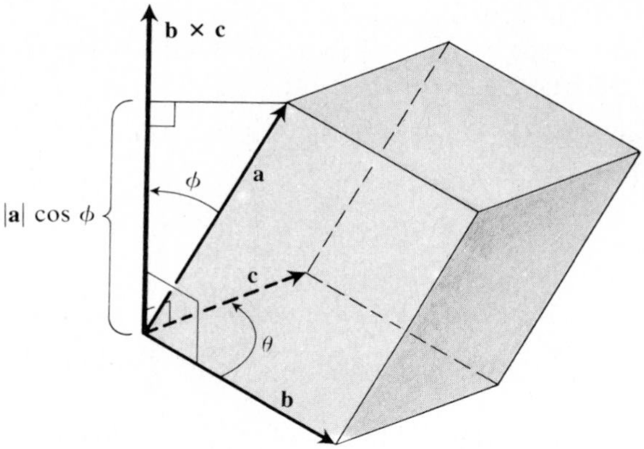

14.34

But $\boldsymbol{b} \times \boldsymbol{c}$ is perpendicular to the base of the box, and ϕ is the angle between $\boldsymbol{b} \times \boldsymbol{c}$ and $\boldsymbol{a}$. Therefore

$$|\boldsymbol{b} \times \boldsymbol{c}| \cdot |\boldsymbol{a}| \cos \phi = \boldsymbol{a} \cdot (\boldsymbol{b} \times \boldsymbol{c}).$$

If the order of $\boldsymbol{b}$ and $\boldsymbol{c}$ is reversed, you obtain $\boldsymbol{a} \cdot (\boldsymbol{c} \times \boldsymbol{b}) = -\boldsymbol{a} \cdot (\boldsymbol{b} \times \boldsymbol{c})$, but of course the box given by $\boldsymbol{a}$, $\boldsymbol{c}$, $\boldsymbol{b}$ is the same one as that given by $\boldsymbol{a}$, $\boldsymbol{b}$, $\boldsymbol{c}$. Therefore you have the formula

$$\text{Volume} = |\boldsymbol{a} \cdot (\boldsymbol{b} \times \boldsymbol{c})|. \tag{11}$$

There is a very easy way to compute $\boldsymbol{a} \cdot (\boldsymbol{b} \times \boldsymbol{c})$. Form the matrix having the vectors $\boldsymbol{a}$, $\boldsymbol{b}$, and $\boldsymbol{c}$ as the first, second, and third rows, respectively. Then $\boldsymbol{a} \cdot (\boldsymbol{b} \times \boldsymbol{c})$ is the determinant of this matrix. To see why this is true, note that, if $\boldsymbol{a} = a_1\boldsymbol{i} + a_2\boldsymbol{j} + a_3\boldsymbol{k}$ and $\boldsymbol{d} = d_1\boldsymbol{i} + d_2\boldsymbol{j} + d_3\boldsymbol{k}$, then $\boldsymbol{a} \cdot \boldsymbol{d} = d_1a_1 + d_2a_2 + d_3a_3$ can be found by formally replacing the $\boldsymbol{i}$, $\boldsymbol{j}$, and $\boldsymbol{k}$ in the expression for $\boldsymbol{d}$ by a_1, a_2, and a_3, respectively. If you do this for

$$\boldsymbol{d} = \boldsymbol{b} \times \boldsymbol{c} = \begin{vmatrix} \boldsymbol{i} & \boldsymbol{j} & \boldsymbol{k} \\ b_1 & b_2 & b_3 \\ c_1 & c_2 & c_3 \end{vmatrix}$$

to form $\boldsymbol{a} \cdot \boldsymbol{d} = \boldsymbol{a} \cdot (\boldsymbol{b} \times \boldsymbol{c})$, you obtain

$$\boldsymbol{a} \cdot (\boldsymbol{b} \times \boldsymbol{c}) = \begin{vmatrix} a_1 & a_2 & a_3 \\ b_1 & b_2 & b_3 \\ c_1 & c_2 & c_3 \end{vmatrix}. \tag{12}$$

Example 6 Let us find the volume of the box in space having as coterminous edges the vectors $\boldsymbol{a} = \boldsymbol{i} - 2\boldsymbol{j} + \boldsymbol{k}$, $\boldsymbol{b} = 2\boldsymbol{i} + 3\boldsymbol{j} - 2\boldsymbol{k}$, and $\boldsymbol{c} = -\boldsymbol{i} + 3\boldsymbol{j} - 2\boldsymbol{k}$.

SOLUTION We have:

$$\boldsymbol{a}\cdot(\boldsymbol{b}\times\boldsymbol{c}) = \begin{vmatrix} 1 & -2 & 1 \\ 2 & 3 & -2 \\ -1 & 3 & -2 \end{vmatrix} = 1(0) - (-2)(-6) + 1(9) = -3.$$

Therefore,

$$\text{Volume} = |\boldsymbol{a}\cdot(\boldsymbol{b}\times\boldsymbol{c})| = |-3| = 3. \quad \|$$

SUMMARY

1. $\begin{vmatrix} a_1 & a_2 \\ b_1 & b_2 \end{vmatrix} = a_1b_2 - a_2b_1$

$$\begin{vmatrix} a_1 & a_2 & a_3 \\ b_1 & b_2 & b_3 \\ c_1 & c_2 & c_3 \end{vmatrix} = a_1\begin{vmatrix} b_2 & b_3 \\ c_2 & c_3 \end{vmatrix} - a_2\begin{vmatrix} b_1 & b_3 \\ c_1 & c_3 \end{vmatrix} + a_3\begin{vmatrix} b_1 & b_2 \\ c_1 & c_2 \end{vmatrix}$$

2. $\boldsymbol{a}\times\boldsymbol{b} = \begin{vmatrix} \boldsymbol{i} & \boldsymbol{j} & \boldsymbol{k} \\ a_1 & a_2 & a_3 \\ b_1 & b_2 & b_3 \end{vmatrix}$

3. *The cross product* $\boldsymbol{a}\times\boldsymbol{b}$ *has length* $|\boldsymbol{a}|\cdot|\boldsymbol{b}|\sin\theta$, *where* θ *is the angle between* $\boldsymbol{a}$ *and* $\boldsymbol{b}$. *This length equals the area of the parallelogram having* $\boldsymbol{a}$ *and* $\boldsymbol{b}$ *as adjacent sides.*

4. *The direction of* $\boldsymbol{a}\times\boldsymbol{b}$ *is perpendicular to the plane of* $\boldsymbol{a}$ *and* $\boldsymbol{b}$ *and in the direction given by the righthand rule, illustrated in Fig.* 14.33.

5. *For any vectors* $\boldsymbol{a}$, $\boldsymbol{b}$, $\boldsymbol{c}$ *in space and any scalar* k,

$$\boldsymbol{a}\times\boldsymbol{b} = -\boldsymbol{b}\times\boldsymbol{a},$$
$$\boldsymbol{a}\times(\boldsymbol{b}+\boldsymbol{c}) = \boldsymbol{a}\times\boldsymbol{b} + \boldsymbol{a}\times\boldsymbol{c},$$
$$(k\boldsymbol{a})\times\boldsymbol{b} = \boldsymbol{a}\times(k\boldsymbol{b}) = k(\boldsymbol{a}\times\boldsymbol{b}).$$

6. $\boldsymbol{i}\times\boldsymbol{j} = \boldsymbol{k}$, $\boldsymbol{j}\times\boldsymbol{k} = \boldsymbol{i}$, $\boldsymbol{k}\times\boldsymbol{i} = \boldsymbol{j}$, while $\boldsymbol{j}\times\boldsymbol{i} = -\boldsymbol{k}$, $\boldsymbol{k}\times\boldsymbol{j} = -\boldsymbol{i}$, $\boldsymbol{i}\times\boldsymbol{k} = -\boldsymbol{j}$.

7. *The triple scalar product* $\boldsymbol{a}\cdot(\boldsymbol{b}\times\boldsymbol{c})$ *and the triple vector product* $\boldsymbol{a}\times(\boldsymbol{b}\times\boldsymbol{c})$ *are most easily computed using*

$$\boldsymbol{a}\cdot(\boldsymbol{b}\times\boldsymbol{c}) = \begin{vmatrix} a_1 & a_2 & a_3 \\ b_1 & b_2 & b_3 \\ c_1 & c_2 & c_3 \end{vmatrix}$$

and

$$\boldsymbol{a}\times(\boldsymbol{b}\times\boldsymbol{c}) = (\boldsymbol{a}\cdot\boldsymbol{c})\boldsymbol{b} - (\boldsymbol{a}\cdot\boldsymbol{b})\boldsymbol{c}.$$

8. $|(\boldsymbol{a}\cdot\boldsymbol{b}\times\boldsymbol{c})|$ *is the volume of the box having* $\boldsymbol{a}$, $\boldsymbol{b}$, *and* $\boldsymbol{c}$ *as coterminous edges.*

EXERCISES

In Exercises 1 through 6, find the indicated determinant.

1. $\begin{vmatrix} -1 & 3 \\ 5 & 0 \end{vmatrix}$

2. $\begin{vmatrix} -1 & 0 \\ 0 & 7 \end{vmatrix}$

3. $\begin{vmatrix} 0 & -3 \\ 5 & 0 \end{vmatrix}$

4. $\begin{vmatrix} 21 & -4 \\ 10 & 7 \end{vmatrix}$

5. $\begin{vmatrix} 1 & 4 & -2 \\ 3 & 13 & 0 \\ 2 & -1 & 3 \end{vmatrix}$

6. $\begin{vmatrix} 2 & -5 & 3 \\ 1 & 3 & 4 \\ -2 & 3 & 7 \end{vmatrix}$

7. Show by direct computation that

a) $\begin{vmatrix} a_1 & a_2 & a_3 \\ a_1 & a_2 & a_3 \\ c_1 & c_2 & c_3 \end{vmatrix} = 0,$ b) $\begin{vmatrix} a_1 & a_2 & a_3 \\ b_1 & b_2 & b_3 \\ a_1 & a_2 & a_3 \end{vmatrix} = 0.$

8. Show by direct computation that

$$\begin{vmatrix} a_1 & a_2 & a_3 \\ b_1 & b_2 & b_3 \\ c_1 & c_2 & c_3 \end{vmatrix} = -\begin{vmatrix} a_1 & a_2 & a_3 \\ c_1 & c_2 & c_3 \\ b_1 & b_2 & b_3 \end{vmatrix}.$$

In Exercises 9 through 11, find $\boldsymbol{a} \times \boldsymbol{b}$.

9. $\boldsymbol{a} = 2\boldsymbol{i} - \boldsymbol{j} + 3\boldsymbol{k}$, $\boldsymbol{b} = \boldsymbol{i} + 2\boldsymbol{j}$

10. $\boldsymbol{a} = -5\boldsymbol{i} + \boldsymbol{j} + 4\boldsymbol{k}$, $\boldsymbol{b} = 2\boldsymbol{i} + \boldsymbol{j} - 3\boldsymbol{k}$

11. $\boldsymbol{a} = -\boldsymbol{i} + 2\boldsymbol{j} + 4\boldsymbol{k}$, $\boldsymbol{b} = 2\boldsymbol{i} - 4\boldsymbol{j} - 8\boldsymbol{k}$

12. Find the area of the parallelogram in space with edges $\boldsymbol{a} = \boldsymbol{i} - 4\boldsymbol{j} + \boldsymbol{k}$ and $\boldsymbol{b} = 2\boldsymbol{i} + 3\boldsymbol{j} - 2\boldsymbol{k}$ by computing $|\boldsymbol{a} \times \boldsymbol{b}|$.

In Exercises 13 through 16, find the area of the parallelogram having the given vectors as edges. If the vectors are in the plane, regard them as being in space with the coefficient of $\boldsymbol{k}$ *being zero.*

13. $-\boldsymbol{i} + 4\boldsymbol{j}$ and $2\boldsymbol{i} + 3\boldsymbol{j}$

14. $-5\boldsymbol{i} + 3\boldsymbol{j}$ and $\boldsymbol{i} + 7\boldsymbol{j}$

15. $\boldsymbol{i} + 3\boldsymbol{j} - 5\boldsymbol{k}$ and $2\boldsymbol{i} + 4\boldsymbol{j} - \boldsymbol{k}$

16. $2\boldsymbol{i} - \boldsymbol{j} + \boldsymbol{k}$ and $\boldsymbol{i} + 3\boldsymbol{j} - \boldsymbol{k}$

17. Find the area of the triangle in the plane with vertices $(-1, 2)$, $(3, -1)$, and $(4, 3)$. [*Hint.* Take translated coordinates with translated origin $(-1, 2)$. The triangle may be viewed as half a parallelogram.]

18. Find the area of the parallelogram in the plane bounded by the lines $x - 2y = 3$, $x - 2y = 8$, $2x + 3y = -1$, and $2x + 3y = -5$.

In Exercises 19 and 20, find $\boldsymbol{a} \cdot (\boldsymbol{b} \times \boldsymbol{c})$ *and* $\boldsymbol{a} \times (\boldsymbol{b} \times \boldsymbol{c})$.

19. $\boldsymbol{a} = \boldsymbol{i} + 2\boldsymbol{j} - 3\boldsymbol{k}$, $\boldsymbol{b} = 4\boldsymbol{i} - \boldsymbol{j} + 2\boldsymbol{k}$, $\boldsymbol{c} = 3\boldsymbol{i} + \boldsymbol{k}$

20. $\boldsymbol{a} = -\boldsymbol{i} + \boldsymbol{j} + 2\boldsymbol{k}$, $\boldsymbol{b} = \boldsymbol{i} + \boldsymbol{k}$, $\boldsymbol{c} = 3\boldsymbol{i} - 2\boldsymbol{j} + 5\boldsymbol{k}$

In Exercises 21 and 22, find the volume of the box having the given vectors as coterminous edges.

21. $-\boldsymbol{i} + 4\boldsymbol{j} + 7\boldsymbol{k}$, $3\boldsymbol{i} - 2\boldsymbol{j} - \boldsymbol{k}$, and $4\boldsymbol{i} + 2\boldsymbol{k}$

22. $2\boldsymbol{i} + \boldsymbol{j} - 4\boldsymbol{k}$, $3\boldsymbol{i} - \boldsymbol{j} + 2\boldsymbol{k}$, and $\boldsymbol{i} + 3\boldsymbol{j} - 10\boldsymbol{k}$

23. Find the volume of the tetrahedron in space with vertices at

$$(-1, 2, 4), \quad (2, -3, 0), \quad (-4, 2, -1), \quad \text{and} \quad (0, 3, -2).$$

24. Let $\boldsymbol{a} = a_1\boldsymbol{i} + a_2\boldsymbol{j} + a_3\boldsymbol{k}$, $\boldsymbol{b} = b_1\boldsymbol{i} + b_2\boldsymbol{j} + b_3\boldsymbol{k}$, and $\boldsymbol{c} = c_1\boldsymbol{i} + c_2\boldsymbol{j} + c_3\boldsymbol{k}$. Verify that $\boldsymbol{a} \times (\boldsymbol{b} \times \boldsymbol{c}) = (\boldsymbol{a} \cdot \boldsymbol{c})\boldsymbol{b} - (\boldsymbol{a} \cdot \boldsymbol{b})\boldsymbol{c}$. (This dull problem involves merely a lot of tedious algebra.)

25. Use the properties of the cross product to find a formula, similar to that in the preceding exercise, for the computation of $(\boldsymbol{a} \times \boldsymbol{b}) \times \boldsymbol{c}$.

26. Use the results of Exercises 24 and 25 to express $(\boldsymbol{a} \times \boldsymbol{b}) \times (\boldsymbol{c} \times \boldsymbol{d})$ in each of the forms $h\boldsymbol{a} + k\boldsymbol{b}$ and $r\boldsymbol{c} + s\boldsymbol{d}$ for scalars h, k, r, and s.

27. Show that, for any vectors $\boldsymbol{a}$, $\boldsymbol{b}$ in space, you have $\boldsymbol{a} \cdot (\boldsymbol{a} \times \boldsymbol{b}) = 0$.

28. Show that $\boldsymbol{i} \times (\boldsymbol{i} \times \boldsymbol{k}) = -\boldsymbol{k}$, while $(\boldsymbol{i} \times \boldsymbol{i}) \times \boldsymbol{k} = \boldsymbol{0}$, illustrating that the cross product is not associative.

29. Consider the triple vector products $\boldsymbol{a} \times (\boldsymbol{b} \times \boldsymbol{c})$ and $(\boldsymbol{a} \times \boldsymbol{b}) \times \boldsymbol{c}$ for vectors $\boldsymbol{a}$, $\boldsymbol{b}$, and $\boldsymbol{c}$ in space.

a) Argue geometrically that $\boldsymbol{a} \times (\boldsymbol{b} \times \boldsymbol{c})$ is a vector in the plane containing $\boldsymbol{b}$ and $\boldsymbol{c}$.

b) Argue geometrically that $(\boldsymbol{a} \times \boldsymbol{b}) \times \boldsymbol{c}$ is a vector in the plane containing $\boldsymbol{a}$ and $\boldsymbol{b}$.

c) Use (a) and (b) to argue that the cross product is not associative.

14.6 LINES

14.6.1 Parametric equations for a line

You are probably accustomed to thinking of a line as being determined by two points. While this is perfectly correct, *it will also be useful for us to think of a line as being determined by one point on the line and the direction of the line.* Of course, the direction of a line can be specified in terms of a nonzero vector. Figure 14.35 shows a line in the plane that passes through the origin and has a direction given by the direction vector $\boldsymbol{d} = d_1\boldsymbol{i} + d_2\boldsymbol{j}$. Clearly, for any vector $\boldsymbol{x} = x_1\boldsymbol{i} + x_2\boldsymbol{j}$ along this line, you must have

$$(x_1, x_2) = t(d_1, d_2)$$

for some scalar t. (We are using subscripted variables x_1 for x and x_2 for y, since it makes the structure of the analysis clearer.) Conversely, for each real number t, the point (td_1, td_2) is on the line.

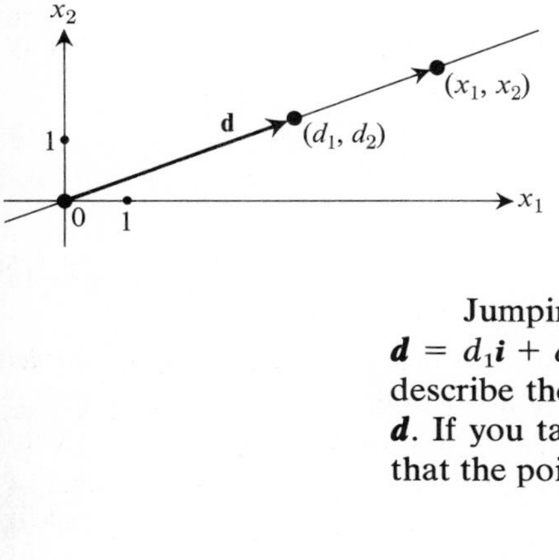

14.35

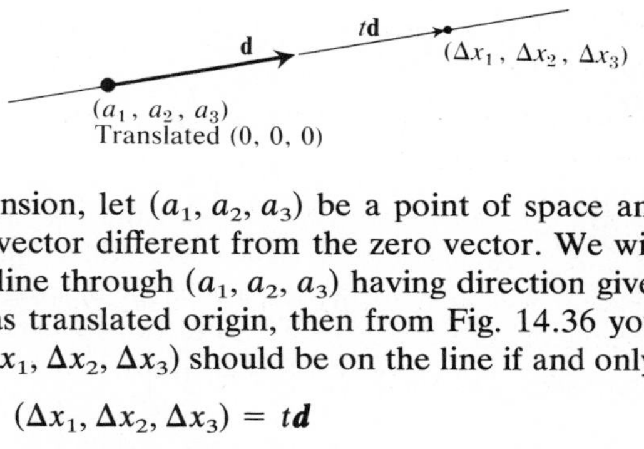

14.36

Jumping up one dimension, let (a_1, a_2, a_3) be a point of space and let $\boldsymbol{d} = d_1\boldsymbol{i} + d_2\boldsymbol{j} + d_3\boldsymbol{k}$ be a vector different from the zero vector. We wish to describe the points on the line through (a_1, a_2, a_3) having direction given by $\boldsymbol{d}$. If you take (a_1, a_2, a_3) as translated origin, then from Fig. 14.36 you see that the point translated $(\Delta x_1, \Delta x_2, \Delta x_3)$ should be on the line if and only if

$$(\Delta x_1, \Delta x_2, \Delta x_3) = t\boldsymbol{d}$$

for some t. This vector equation can be rewritten by components as

$$\begin{aligned} \Delta x_1 &= td_1, \\ \Delta x_2 &= td_2, \\ \Delta x_3 &= td_3. \end{aligned} \qquad (1)$$

Recall that if translated $(\Delta x_1, \Delta x_2, \Delta x_3)$ is old (x_1, x_2, x_3), then $\Delta x_i = x_i - a_i$ for $i = 1, 2,$ and 3. Equations (1) then take the form

$$\begin{aligned} x_1 - a_1 &= d_1 t, \\ x_2 - a_2 &= d_2 t, \\ x_3 - a_3 &= d_3 t. \end{aligned} \tag{2}$$

Let's summarize this discussion in a definition.

Definition 14.5 Let (a_1, a_2, a_3) be a point and let $\boldsymbol{d} = d_1\boldsymbol{i} + d_2\boldsymbol{j} + d_3\boldsymbol{k}$ be a vector. The **line** through (a_1, a_2, a_3) with direction $\boldsymbol{d}$ is the set of all points (x, y, z) such that, for some scalar t,

$$\begin{aligned} x &= a_1 + d_1 t, \\ y &= a_2 + d_2 t, \\ z &= a_3 + d_3 t. \end{aligned} \tag{3}$$

The Eqs. (3) are **parametric equations** for the line, and t is the **parameter**.

Example 1 Parametric equations for the line through $(-1, 0, 2)$ with direction $\boldsymbol{d} = 2\boldsymbol{i} - 3\boldsymbol{j} - \boldsymbol{k}$ are

$$x = -1 + 2t, \qquad y = -3t, \qquad z = 2 - t.$$

For example, putting $t = -2$, you see that $(-5, 6, 4)$ is on the line. ||

Of course, if a line has direction $\boldsymbol{d}$, then it also has direction $r\boldsymbol{d}$ for any nonzero scalar r.

Remember, if you want to find parametric equations of a line, you should always find a point on the line and a direction for the line, and use Eqs. (3).

Of course, for parametric equations of a line in the plane, one simply uses (3) with the last equation dropped.

It is sometimes helpful to think of the parametric equations (3) as giving a rule for picking up each point t of a Euclidean line (a t-axis; see Fig. 14.37), and putting it down in space. The whole t-axis, that is, the whole Euclidean line, is picked up and put down, with the origin 0 of the line put down at (a_1, a_2, a_3). While the Eqs. (3) do not "bend" the line as they pick it up and put it down, they may "stretch" it if $|\boldsymbol{d}| \neq 1$, for the point 1 is put down at

$$(a_1 + d_1, a_2 + d_2, a_3 + d_3),$$

a distance of $|\boldsymbol{d}|$ from (a_1, a_2, a_3).

14.37

A physicist usually thinks of the parameter t as representing *time*. Equations (3) can then be viewed as giving the location at time t of a body traveling along a line in space. For this reason, we sometimes refer to (a_1, a_2, a_3) as the point on the line *when* $t = 0$.

14.6.2 The angle between intersecting lines

Parallel and perpendicular lines

One nice thing about *parametric* equations for a line is that the direction of the line is easily found from the equations. Naturally two lines are *parallel* if they have parallel direction vectors, and are *perpendicular* if they meet and have perpendicular direction vectors. Of course, two lines in the plane either are parallel or intersect, but this need not be true in space.

Definition 14.6 The **angle between intersecting lines** is the angle between their direction vectors.

Example 2 Let's find the angle between the lines

$$\begin{cases} x = 3 + 4t, \\ y = -2 + t, \end{cases} \quad \text{and} \quad \begin{cases} x = -1 - 3t, \\ y = 5 + 12t, \end{cases}$$

in the plane.

SOLUTION Direction vectors for these lines are $\boldsymbol{d} = 4\boldsymbol{i} + \boldsymbol{j}$ and $\boldsymbol{d}' = -3\boldsymbol{i} + 12\boldsymbol{j}$. Now $\boldsymbol{d} \cdot \boldsymbol{d}' = -12 + 12 = 0$, so the vectors, and therefore the lines, are perpendicular. ‖

Example 3 Let's find the acute angle between the lines

$$\begin{cases} x = -2 + 2t, \\ y = 3 - 4t, \\ z = -4 + t, \end{cases} \quad \text{and} \quad \begin{cases} x = -2 - t, \\ y = 3 + 2t, \\ z = -4 + 3t, \end{cases}$$

which meet at the point $(-2, 3, -4)$ in space.

SOLUTION Direction vectors for these lines are

$$\boldsymbol{d} = 2\boldsymbol{i} - 4\boldsymbol{j} + \boldsymbol{k} \quad \text{and} \quad \boldsymbol{d}' = -\boldsymbol{i} + 2\boldsymbol{j} + 3\boldsymbol{k}.$$

Now $\boldsymbol{d} \cdot \boldsymbol{d}' = -2 - 8 + 3 = -7$. We change the direction vector of the first line to $-\boldsymbol{d} = -2\boldsymbol{i} + 4\boldsymbol{j} - \boldsymbol{k}$ to obtain a positive dot product, corresponding to an acute angle θ. Then

$$\cos\theta = \frac{(-\boldsymbol{d}) \cdot \boldsymbol{d}'}{|-\boldsymbol{d}| \cdot |\boldsymbol{d}'|} = \frac{7}{\sqrt{21}\sqrt{14}} = \frac{1}{\sqrt{6}}.$$

Therefore,

$$\theta = \cos^{-1}\left(\frac{1}{\sqrt{6}}\right) \approx 65.91°. \quad ‖$$

If (a_1, a_2, a_3) and (b_1, b_2, b_3) are two distinct points, then taking (a_1, a_2, a_3) as translated origin, you see that

$$\boldsymbol{d} = \boldsymbol{b} - \boldsymbol{a} = (b_1 - a_1)\boldsymbol{i} + (b_2 - a_2)\boldsymbol{j} + (b_3 - a_3)\boldsymbol{k}$$

is a direction vector for the line joining these points. Therefore

$$\begin{aligned} x &= a_1 + (b_1 - a_1)t, \\ y &= a_2 + (b_2 - a_2)t, \\ z &= a_3 + (b_3 - a_3)t, \end{aligned} \tag{4}$$

are parametric equations of a line that passes through (a_1, a_2, a_3) when $t = 0$ and (b_1, b_2, b_3) when $t = 1$. This shows how to find parametric equations for a line given by two points. Of course, the first two equations in (4) are used for a line in the plane.

Example 4 Parametric equations for the line in the plane determined by the points $(2, -1)$ and $(-1, 0)$ are

$$x = 2 - 3t, \qquad y = -1 + t,$$

or, equivalently,

$$x = -1 + 3t, \qquad y = -t. \quad \|$$

14.6.3 Line segments We now describe analytically the *line segment* between two points (a_1, a_2, a_3) and (b_1, b_2, b_3) in space (or the plane). Again, let's take (a_1, a_2, a_3) as translated origin, so that (b_1, b_2, b_3) becomes after translation $(b_1 - a_1, b_2 - a_2, b_3 - a_3)$. The point which is translated $(\Delta x, \Delta y, \Delta z)$ should be on the line segment if and only if

$$(\Delta x, \Delta y, \Delta z) = t(\boldsymbol{b} - \boldsymbol{a}) = t(b_1 - a_1, b_2 - a_2, b_3 - a_3)$$

for some t *such that* $0 \le t \le 1$, as indicated in Fig. 14.38. Since $\Delta x = x - a_1$, etc., you see that (x, y, z) should be on this line segment if and only if, for some t where $0 \le t \le 1$,

$$\begin{aligned} x - a_1 &= t(b_1 - a_1), \\ y - a_2 &= t(b_2 - a_2), \\ z - a_3 &= t(b_3 - a_3). \end{aligned} \tag{5}$$

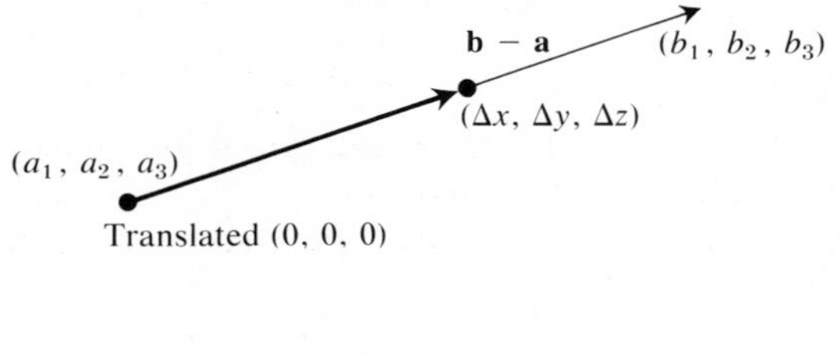

14.38

In particular, $t = \frac{1}{2}$ should give the point halfway between (a_1, a_2, a_3) and (b_1, b_2, b_3). Let's summarize in a definition.

Definition 14.7 Let (a_1, a_2, a_3) and (b_1, b_2, b_3) be two points in space (or the plane). The **line segment** joining them consists of all points (x, y, z) such that, for some value of t, where $0 \le t \le 1$,

$$\begin{aligned} x &= a_1 + (b_1 - a_1)t, \\ y &= a_2 + (b_2 - a_2)t, \\ z &= a_3 + (b_3 - a_3)t. \end{aligned} \tag{6}$$

The **midpoint** of the line segment is the point

$$\left(\frac{a_1 + b_1}{2}, \frac{a_2 + b_2}{2}, \frac{a_3 + b_3}{2}\right).$$

Example 5 The midpoint of the line segment in space joining $(-1, 3, 2)$ and $(3, 1, -1)$ is $(1, 2, \frac{1}{2})$. The point (x, y, z) one-third of the way from $(-1, 3, 2)$ to $(3, 1, -1)$ is obtained by putting $t = \frac{1}{3}$ in Eqs. (6) for the given points. For example,

$$x = -1 + \tfrac{1}{3}(3 - (-1)) = -1 + \tfrac{4}{3} = \tfrac{1}{3}.$$

Computing y and z similarly, you find that the desired point is $(\frac{1}{3}, \frac{7}{3}, 1)$. ‖

SUMMARY

1. *A line in the plane or space is determined by a point on the line and a vector* $\boldsymbol{d}$ *parallel to the line.*
2. *If a line goes through a point* (a_1, a_2, a_3) *and has direction given by the nonzero vector* $\boldsymbol{d} = d_1\boldsymbol{i} + d_2\boldsymbol{j} + d_3\boldsymbol{k}$, *then parametric equations for the line are*

$$\begin{aligned} x &= a_1 + d_1t, \\ y &= a_2 + d_2t, \\ z &= a_3 + d_3t. \end{aligned}$$

   All values of the parameter t give all points on the line.
3. *Parametric equations of the line through* (a_1, a_2, a_3) *and* (b_1, b_2, b_3) *are*

$$\begin{aligned} x &= a_1 + (b_1 - a_1)t, \\ y &= a_2 + (b_2 - a_2)t, \\ z &= a_3 + (b_3 - a_3)t. \end{aligned}$$

   The first two equations are used for a line in the plane. The points obtained by restricting t to the interval $0 \le t \le 1$ *make up the line segment joining* (a_1, a_2, a_3) *and* (b_1, b_2, b_3). *The midpoint is obtained when* $t = \frac{1}{2}$.

EXERCISES

1. Give parametric equations for the line in the plane through $(3, -2)$ with direction $\boldsymbol{d} = -8\boldsymbol{i} + 4\boldsymbol{j}$. Sketch the line in an appropriate figure.

2. Give parametric equations for the line through $(-1, 3, 0)$ with direction $\boldsymbol{d} = -2\boldsymbol{i} - \boldsymbol{j} + 4\boldsymbol{k}$. Sketch in an appropriate figure.

3. Give parametric equations for the line through $(2, -1, 0)$ with direction $\boldsymbol{d} = -\boldsymbol{i} + 4\boldsymbol{k}$.

4. Consider the line in the plane with parametric equations $x = -3 + 4t$, $y = 2 - 3t$.

a) Find a direction vector for the line.

b) Find the point on the line whose x-coordinate is 1.

c) Find the point on the line whose y-coordinate is 8.

5. Find parametric equations for the line in the plane through $(5, -1)$ and perpendicular to the line with parametric equations $x = 4 - 2t$, $y = 7 + t$.

6. For each pair of points, find parametric equations of the line containing them.

a) $(-2, 4)$ and $(3, -1)$ in the plane

b) $(3, -1, 6)$ and $(0, -3, -1)$ in space

7. Find the acute angle between the following pairs of intersecting lines.

a) $\begin{cases} x = 3 - 4t \\ y = 2 + 3t \end{cases}$ and $\begin{cases} x = 5 - t \\ y = 7 + 2t \end{cases}$ in the plane

b) $\begin{cases} x = 2 - 4t \\ y = -1 - t \\ z = 3 + 5t \end{cases}$ and $\begin{cases} x = 2 - t \\ y = -1 + 2t \\ z = 3 \end{cases}$ in space

8. Find the acute angle that the line $x = t$, $y = t$, $z = t$ in space makes with the coordinate axes.

9. Let a line in space pass through the origin and make angles of α, β, and γ with the three positive coordinate axes. Show that $\cos^2\alpha + \cos^2\beta + \cos^2\gamma = 1$.

10. Find parametric equations for the line in space through $(-1, 2, 3)$ that is perpendicular to each of the two lines having parametric equations

$$\begin{cases} x = -1 + 3t, \\ y = 2, \\ z = 3 - t, \end{cases} \quad \text{and} \quad \begin{cases} x = -1 - t, \\ y = 2 + 3t, \\ z = 3 + t. \end{cases}$$

11. Find the midpoint of the line segment joining the pair of points.

a) $(-2, 4)$ and $(3, -1)$ in the plane

b) $(3, -1, 6)$ and $(0, -3, -1)$ in space

12. Find the point in the plane on the line segment joining $(-1, 3)$ and $(2, 5)$ that is twice as close to $(-1, 3)$ as to $(2, 5)$.

13. Find the point in space on the line segment joining $(-2, 1, 3)$ and $(0, -5, 2)$ that is one fourth of the way from $(-2, 1, 3)$ to $(0, -5, 2)$.

14. Let (a_1, a_2, a_3) and (b_1, b_2, b_3) be any two points in space. Show that the line segment joining the points consists of all points (x_1, x_2, x_3), where $x_i = (1 - t)a_i + tb_i$ for $0 \le t \le 1$, $i = 1, 2, 3$.

15. Find the cosines of the angles that the direction vector of the line

$$x = a_1 + d_1 t,$$
$$y = a_2 + d_2 t,$$
$$z = a_3 + d_3 t$$

makes with the vectors $\boldsymbol{i}$, $\boldsymbol{j}$, and $\boldsymbol{k}$ along the positive coordinate axes. These cosines are called the *direction cosines* of the line. They are usually denoted by $\cos\alpha$, $\cos\beta$, and $\cos\gamma$, respectively. (See Exercise 9.)

14.7 PLANES

14.7.1 The equation of a plane

Let (a_1, a_2, a_3) be chosen as translated origin, and let's, for the moment, use subscripted variable notation. The line with parametric equations

$$x_1 = a_1 + d_1t, \qquad x_2 = a_2 + d_2t, \qquad x_3 = a_3 + d_3t$$

can be characterized as the set of all points (x_1, x_2, x_3) such that the vector

$$\boldsymbol{x} - \boldsymbol{a} = (x_1 - a_1)\boldsymbol{i} + (x_2 - a_2)\boldsymbol{j} + (x_3 - a_3)\boldsymbol{k}$$

is *parallel* to the direction vector $\boldsymbol{d} = d_1\boldsymbol{i} + d_2\boldsymbol{j} + d_3\boldsymbol{k}$ of the line. Let's consider now the set of all (x_1, x_2, x_3) such that the vector

$$\boldsymbol{x} - \boldsymbol{a} = (x_1 - a_1)\boldsymbol{i} + (x_2 - a_2)\boldsymbol{j} + (x_3 - a_3)\boldsymbol{k}$$

is *perpendicular* to a vector $\boldsymbol{d} = d_1\boldsymbol{i} + d_2\boldsymbol{j} + d_3\boldsymbol{k}$, where not all d_i are zero. These two vectors are perpendicular if and only if

$$d_1(x_1 - a_1) + d_2(x_2 - a_2) + d_3(x_3 - a_3) = 0. \tag{1}$$

Let's try to see whether this concept gives some familiar geometric configuration in space, and let's also look at the similarly defined locus with equation $d_1(x_1 - a_1) + d_2(x_2 - a_2) = 0$ in the plane. From Fig. 14.39, you see that all such points (x_1, x_2) in the plane lie on a line through (a_1, a_2) that is perpendicuar to $\boldsymbol{d}$, and from Fig. 14.40, you see that all such points (x_1, x_2, x_3) in space lie in a plane through (a_1, a_2, a_3). This suggests the following definition.

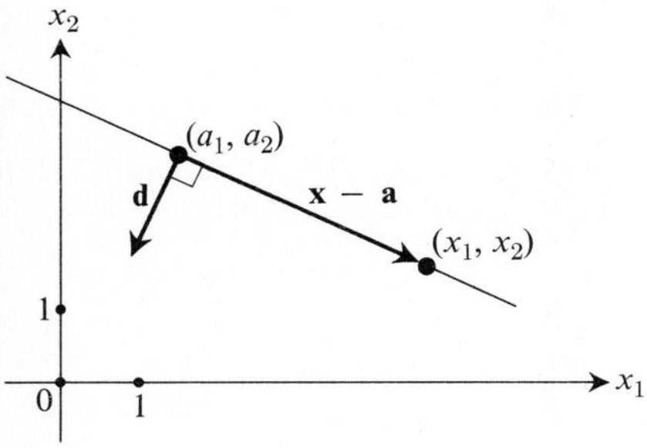

14.39

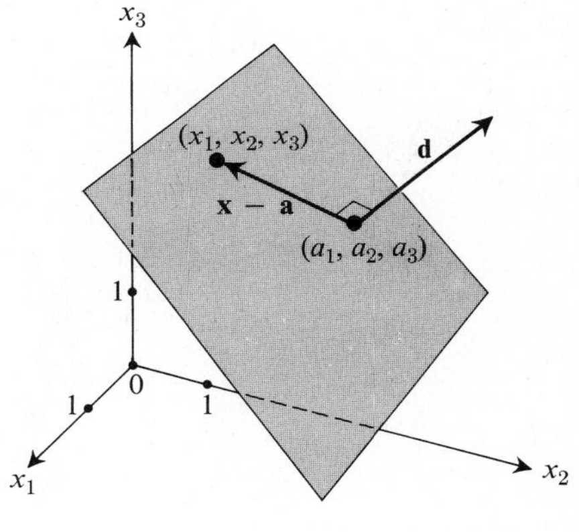

14.40

Definition 14.8 Let (a_1, a_2, a_3) be a point in space and $\boldsymbol{d} = d_1\boldsymbol{i} + d_2\boldsymbol{j} + d_3\boldsymbol{k}$ a nonzero vector. The **plane** through (a_1, a_2, a_3) with direction vector $\boldsymbol{d}$ is the set of all points (x, y, z) satisfying

$$d_1(x - a_1) + d_2(y - a_2) + d_3(z - a_3) = 0. \tag{2}$$

We want to emphasize that the *direction* associated with the plane in Fig. 14.40 is the direction of a *normal* (perpendicular) vector to the plane. Please remember:

a) *To find parametric equations of a line, you want to know a point on the line and a (parallel) direction vector for the line.*

b) *To find an equation of a plane, you want to know a point in the plane and a (normal) direction vector for the plane.*

Example 1 Let's find an equation of the plane in space that passes through $(-1, 2, 1)$ and has the normal direction vector $\boldsymbol{d} = \boldsymbol{i} - 3\boldsymbol{j} + 2\boldsymbol{k}$.

SOLUTION Equation (2) becomes

$$1(x - (-1)) + (-3)(y - 2) + 2(z - 1) = 0,$$

or

$$x - 3y + 2z = -5. \quad \|$$

Of course, if $\boldsymbol{d}$ is a direction vector for a plane, then $r\boldsymbol{d}$ is also a direction vector for the same plane for all nonzero r.

Just as in Example 1, Eq. (2) can always be rewritten in the form

$$d_1x + d_2y + d_3z = c,$$

where $c = d_1a_1 + d_2a_2 + d_3a_3$ is a real number. Let's show now that every linear equation of the form $d_1x + d_2y + d_3z = c$, where not all d_i are zero, indeed has as locus some plane in space. We may suppose that $d_1 \neq 0$. Choose any numbers a_2 and a_3, and let

$$a_1 = \frac{c - d_2a_2 - d_3a_3}{d_1}.$$

Then $d_1a_1 + d_2a_2 + d_3a_3 = c$. Now for any point (x, y, z) such that $d_1x + d_2y + d_3z = c$, you have the two equations

$$d_1x + d_2y + d_3z = c,$$
$$d_1a_1 + d_2a_2 + d_3a_3 = c.$$

Subtracting, you obtain

$$d_1(x - a_1) + d_2(y - a_2) + d_3(z - a_3) = 0.$$

Thus (x, y, z) lies on the plane through (a_1, a_2, a_3) with direction vector $\boldsymbol{d} = d_1\boldsymbol{i} + d_2\boldsymbol{j} + d_3\boldsymbol{k}$.

Example 2 Dropping back to two dimensions, the set of all (x, y) in the plane such that

$$0x + 1y = 0$$

should be a line in the plane. This line contains $(0, 0)$ and has as normal direction vector $\boldsymbol{d} = 0\boldsymbol{i} + 1\boldsymbol{j} = \boldsymbol{j}$. Thus this line is simply the x-axis. Of course, it is clear that the solutions of $y = 0$ are precisely the points $(a, 0)$. ‖

Example 3 As in Example 2, it is easy to see that $z = 0$ is the equation of the plane in space through the origin that contains the x- and y-coordinate axes, that is, the x,y-coordinate plane. ‖

You can easily check that the following definitions coincide with your intuitive ideas of parallel and perpendicular.

Definition 14.9 Two planes in space are **parallel** if direction vectors for the two planes are parallel. The planes are **perpendicular** if their direction vectors are perpendicular. The **angle between** two planes is the angle between their direction vectors. A line and a plane are **perpendicular** if a direction vector for the line is also a (normal) direction vector of the plane.

14.7.2 Computations

The following examples illustrate a few of the many types of problems you can now solve. Remember:

a) *To find parametric equations for a line, you want to know a point on the line and a (parallel) direction vector of the line.*

b) *To find an equation of a plane, you want to know a point in the plane and a (normal) direction vector for the plane.*

Example 4 Let's find parametric equations for the line in space passing through the point $(1, 2, -1)$ and perpendicular to the plane with equation

$$3x + 5y - z = 6.$$

SOLUTION You know a point $(1, 2, -1)$ on the line, and you need a direction vector for the line. Since the line is to be perpendicular to the plane given by $3x + 5y - z = 6$, you see that $\boldsymbol{d} = 3\boldsymbol{i} + 5\boldsymbol{j} - \boldsymbol{k}$ is a direction vector for the line. Thus the line has parametric equations

$$x = 1 + 3t, \qquad y = 2 + 5t, \qquad z = -1 - t. \quad \|$$

Example 5 Let's find all points of intersection in space of the plane with equation

$$3x + 5y - z = -2$$

and the line with parametric equations

$$x = -3 + 2t, \qquad y = 4 + t, \qquad z = -1 - 3t.$$

SOLUTION If (x, y, z) lies on both the line and the plane, then, substituting, you must have

$$3(-3 + 2t) + 5(4 + t) - (-1 - 3t) = -2,$$

so

$$14t = -14 \qquad \text{and} \qquad t = -1.$$

Thus the only point of intersection is $(-5, 3, 2)$. ‖

Example 6 Let's find the acute angle θ between the planes $x - 3y + z = 4$ and $3x + 2y + 4z = 6$.

SOLUTION Direction vectors for the planes are

$$\boldsymbol{d} = \boldsymbol{i} - 3\boldsymbol{j} + \boldsymbol{k} \qquad \text{and} \qquad \boldsymbol{d}' = 3\boldsymbol{i} + 2\boldsymbol{j} + 4\boldsymbol{k}.$$

Now $\boldsymbol{d} \cdot \boldsymbol{d}' = 3 - 6 + 4 = 1$, so

$$\cos\theta = \frac{\boldsymbol{d} \cdot \boldsymbol{d}}{|\boldsymbol{d}| \cdot |\boldsymbol{d}|} = \frac{1}{\sqrt{11}\sqrt{29}}.$$

Therefore

$$\theta = \cos^{-1}\left(\frac{1}{\sqrt{11}\sqrt{29}}\right) \approx 86.79°. \quad ‖$$

Example 7 Let's find the equation of the plane in space that contains the points $(1, -1, 1)$, $(2, 3, -4)$, and $(0, 1, -2)$.

SOLUTION To find the equation of a plane, you want to know a vector normal to the plane. Taking a translated origin at $(1, -1, 1)$, you see that

$$(2, 3, -4) - (1, -1, 1) = (1, 4, -5)$$

and

$$(0, 1, -2) - (1, -1, 1) = (-1, 2, -3)$$

are vectors in the desired plane. A normal vector to the plane must be perpendicular to both of the vectors; their cross product will be such a vector. The symbolic determinant

$$\begin{vmatrix} \boldsymbol{i} & \boldsymbol{j} & \boldsymbol{k} \\ 1 & 4 & -5 \\ -1 & 2 & -3 \end{vmatrix} = -2\boldsymbol{i} + 8\boldsymbol{j} + 6\boldsymbol{k}$$

shows that $\boldsymbol{i} - 4\boldsymbol{j} - 3\boldsymbol{k}$ is a direction vector for the plane. Since the plane contains the point $(1, -1, 1)$,

$$x - 4y - 3z = 2$$

is the desired equation. ‖

14.7.3 The distance from a point to a plane

Let $d_1x + d_2y + d_3z = c$ be the equation of a plane in space, and let (a_1, a_2, a_3) be a point in space. We would like to find the distance from (a_1, a_2, a_3) to the plane. This distance is measured along the line through (a_1, a_2, a_3) and perpendicular to the plane, since this gives the shortest distance. This line has direction vector $\boldsymbol{d} = d_1\boldsymbol{i} + d_2\boldsymbol{j} + d_3\boldsymbol{k}$, so parametric equations of the line are

$$x = a_1 + d_1t, \qquad y = a_2 + d_2t, \qquad z = a_3 + d_3t.$$

Substituting, you find that the line meets the plane when

$$d_1(a_1 + d_1t) + d_2(a_2 + d_2t) + d_3(a_3 + d_3t) = c,$$

or when $t = t_0$, where

$$t_0 = \frac{c - d_1a_1 - d_2a_2 - d_3a_3}{d_1^2 + d_2^2 + d_3^2}.$$

(Recall that not all d_i can be zero, so $d_1^2 + d_2^2 + d_3^2 \neq 0$.) Thus the line intersects the plane at the point $(a_1 + d_1t_0, a_2 + d_2t_0, a_3 + d_3t_0)$, and the distance from (a_1, a_2, a_3) to this point is

$$\sqrt{d_1^2t_0^2 + d_2^2t_0^2 + d_3^2t_0^2} = |t_0|\sqrt{d_1^2 + d_2^2 + d_3^2}.$$

Substituting the value you found for t_0, you obtain

$$\frac{|c - d_1a_1 - d_2a_2 - d_3a_3|}{d_1^2 + d_2^2 + d_3^2}\sqrt{d_1^2 + d_2^2 + d_3^2} = \frac{|d_1a_1 + d_2a_2 + d_3a_3 - c|}{\sqrt{d_1^2 + d_2^2 + d_3^2}}$$

as the distance from the point to the plane.

Of course, the same derivation using two components gives the formula

$$\frac{|d_1a_1 + d_2a_2 - c|}{\sqrt{d_1^2 + d_2^2}}$$

as the distance from (a_1, a_2) to the line $d_1x + d_2y = c$.

Example 8 Let's find the distance from the point $(-9, 6, 2)$ to the plane with equation $5x - 2y + z = 2$.

SOLUTION The preceding shows that this distance is

$$\frac{|-45 - 12 + 2 - 2|}{\sqrt{25 + 4 + 1}} = \frac{|-57|}{\sqrt{30}} = \frac{57}{\sqrt{30}}. \quad ‖$$

SUMMARY

1. *A plane is determined by a point in the plane and a normal (perpendicular) vector.*
2. *The plane through* (a_1, a_2, a_3) *and having direction vector* $d_1\boldsymbol{i} + d_2\boldsymbol{j} + d_3\boldsymbol{k}$ *normal to the plane has as equation*

$$d_1(x - a_1) + d_2(y - a_2) + d_3(z - a_3) = 0.$$

3. *The equation* $d_1x + d_2y + d_3z = c$, *where not all* d_i *are zero, has as locus in space a plane with* $d_1\boldsymbol{i} + d_2\boldsymbol{j} + d_3\boldsymbol{k}$ *as normal vector.*
4. *The line* $d_1x + d_2y = c$ *in the plane has* $d_1\boldsymbol{i} + d_2\boldsymbol{j}$ *as normal vector.*
5. *Planes are parallel if their direction vectors are parallel, and perpendicular if these vectors are perpendicular. The angle between two planes is the angle between their direction vectors.*
6. *The distance from a point* (a_1, a_2, a_3) *to the plane* $d_1x + d_2y + d_3z = c$ *is*

$$\frac{|d_1a_1 + d_2a_2 + d_3a_3 - c|}{\sqrt{d_1^2 + d_2^2 + d_3^2}}$$

Similarly, the distance from (a_1, a_2) *to the line* $d_1x + d_2y = c$ *in the plane is*

$$\frac{|d_1a_1 + d_2a_2 - c|}{\sqrt{d_1^2 + d_2^2}}.$$

EXERCISES

1. Find the equation of the plane passing through the point $(-1, 4, 2)$ and having direction vector $\boldsymbol{i} - 2\boldsymbol{j} + \boldsymbol{k}$.

2. Find an equation for the plane through $(1, -3, 0)$ and with direction vector $\boldsymbol{j} + 4\boldsymbol{k}$.

3. Find an equation of the plane passing through $(3, -1, 4)$ and parallel to the plane with equation $3x - 2y + 7z = 14$.

4. Find an equation of the plane passing through $(-1, 4, -3)$ and perpendicular to the line with parametric equations

$$x = 3 - 7t, \qquad y = 4 + t, \qquad z = 2t.$$

5. Find an equation of the plane passing through the origin and perpendicular to the line through the points $(-1, 3, 0)$ and $(2, -4, 3)$.

6. Find the angles between the following pairs of lines or planes.

a) $x - 3y = 7$ and $2x + 4y = 1$ in the plane

b) $3x + 2y = -7$ and $6x + 4y = 2$ in the plane

c) $3x + 4y - z = 1$ and $x - 2y = 3$ in space

d) $4x - 7y + z = 3$ and $3x + 2y + 2z = 17$ in space

7. Find parametric equations of the line that is perpendicular to the plane with equation $x - 2y + 4z = 3$ and passes through the point $(-2, 1, 0)$.

8. Find the intersection of the line

$$x = 5 + t, \qquad y = -3t, \qquad z = -2 + 4t$$

and the plane with equation

$$x - 3y + 2z = -35.$$

9. As in Example 7, find the equation of the plane that contains $(1, 0, 1)$, $(-1, 2, 0)$, and $(0, 1, 3)$.

10. Find an equation of the plane that passes through

the unit coordinate points $(1, 0, 0)$, $(0, 1, 0)$, and $(0, 0, 1)$.

11. Find the equation of the plane through $(-1, 4, 2)$, $(1, 2, -1)$, and $(3, -2, 0)$.

12. Find parametric equations of the line of intersection of the planes $x - 3y + z = 7$ and $3x + 2y + z = -1$.

13. Find the equation of the plane through $(-1, 4, 2)$ that contains the lines $x = -1 + 3t$, $y = 4 + t$, $z = 2 - 2t$, and $x = -1 + t$, $y = 4$, $z = 2 + 7t$.

14. Show that the equation of the line in the plane through two distinct points (a_1, a_2) and (b_1, b_2) is given by

$$\begin{vmatrix} x & y & 1 \\ a_1 & a_2 & 1 \\ b_1 & b_2 & 1 \end{vmatrix} = 0.$$

[*Hint.* Do not expand the determinant, but argue that a linear equation in x and y is obtained, and that the equation is satisfied if $(x, y) = (a_1, a_2)$ or $(x, y) = (b_1, b_2)$.]

15. Use the method suggested by Exercise 14 to find the equation of the line in the plane through $(1, -4)$ and $(2, 3)$.

16. Find the distance from the point $(-1, 3)$ to the line with equation $3x - 4y = 5$.

17. Find the distance from the point $(1, 3, -1)$ to the plane with equation $2x + y + z = 4$.

18. Find the distance in the plane from the point $(2, -1)$ to the line with parametric equations $x = 3 - t$, $y = 2 + 4t$. [*Hint.* Obtain the x,y-equation for the line by eliminating t from the parametric equations.]

19. Find parametric equations for the line of intersection of the two planes $x + y - 2z = 7$ and $2x + y - z = -2$.

20. Give an alternative derivation of the formula for the distance from the point (a_1, a_2, a_3) to the plane $d_1x + d_2y + d_3z = c$, by finding the absolute value of the component of the vector from (a_1, a_2, a_3) to a point (x, y, z) in the plane along the normal vector $d_1\boldsymbol{i} + d_2\boldsymbol{j} + d_3\boldsymbol{k}$.

exercise sets for chapter 14

review exercise set 14.1

1. a) Find the distance between $(-1, 2, 4)$ and $(3, 8, -2)$.

b) Find the equation of the sphere having $(1, -3, 5)$ and $(-3, 5, 7)$ as endpoints of a diameter.

2. a) Sketch the locus of the cylindrical coordinate equation $r = 4$.

b) Find spherical coordinates of the point $(x, y, z) = (0, 1, -1)$.

3. Sketch the surface

$$\frac{x^2}{4} + \frac{y^2}{9} = 1 + z^2.$$

4. a) Find the length of the vector $2\boldsymbol{i} - 3\boldsymbol{j} + \boldsymbol{k}$.

b) Find c_2 so that the vectors $\boldsymbol{i} - c_2\boldsymbol{j} + 3\boldsymbol{k}$ and $4\boldsymbol{i} + 2\boldsymbol{j} - 7\boldsymbol{k}$ are perpendicular.

5. Find the angle between the vectors $2\boldsymbol{i} - 3\boldsymbol{j} + \boldsymbol{k}$ and $-\boldsymbol{i} + 2\boldsymbol{j} + 3\boldsymbol{k}$.

6. Find the vector projection of

$$2\boldsymbol{i} - 3\boldsymbol{j} + \boldsymbol{k}$$

on $3\boldsymbol{i} - 4\boldsymbol{k}$.

7. Find $\boldsymbol{a} \times \boldsymbol{b}$ if $\boldsymbol{a} = \boldsymbol{i} + 2\boldsymbol{j} - 3\boldsymbol{k}$ and $\boldsymbol{b} = 4\boldsymbol{i} - \boldsymbol{j} + \boldsymbol{k}$.

8. Find the volume of the box in space having the vectors $\boldsymbol{i} - 3\boldsymbol{j} + \boldsymbol{k}$, $2\boldsymbol{i} - 3\boldsymbol{j} + 2\boldsymbol{k}$, $-3\boldsymbol{i} + 5\boldsymbol{j} - 2\boldsymbol{k}$ as coterminous edges.

9. If $\boldsymbol{a} = \boldsymbol{i} - 2\boldsymbol{j} + \boldsymbol{k}$, $\boldsymbol{b} = -2\boldsymbol{i} + 3\boldsymbol{j} - \boldsymbol{k}$, and $\boldsymbol{c} = 4\boldsymbol{i} - 2\boldsymbol{j} + 2\boldsymbol{k}$, compute $(\boldsymbol{a} \times \boldsymbol{b}) \times \boldsymbol{c}$.

10. Find parametric equations of the line through $(-5, 1, 2)$ and $(2, -1, 7)$.

11. Find parametric equations of the line perpendicular to the plane $x - 2y + 3z = 8$ and passing through the origin.

12. Find the equation of the plane passing through $(-1, 1, 4)$ and perpendicular to the line $x = 2 + t$, $y = 3 - 4t$, $z = -7 + 2t$.

13. Find the distance from $(-1, 1, 3)$ to the plane $2x + y - 2z = 4$.

14. Classify the given planes as parallel, perpendicular, or neither parallel nor perpendicular.

a) $x - 3y + 4z = 7, \quad -3x + y + 2z = 11$

b) $2x + y + 3z = 8, \quad 4x + 7y - 5z = -3$

c) $-4x + 6y - 12z = 7, \quad 2x - 3y + 6z = 5$

15. Find the point of intersection of the line $x = 2 - t$, $y = 4 + t$, $z = 1 - 3t$, with the plane $x - 3y + 4z = -1$.

review exercise set 14.2

1. a) Find c so that the distance from $(c, -2, 3)$ to $(1, 4, -5)$ is 15.

b) Find the center and radius of the sphere $x^2 + y^2 + z^2 - 2x + 4z = 4$.

2. a) Find cylindrical coordinates of $(x, y, z) = (1, -\sqrt{3}, -2)$.

b) Find rectangular coordinates of $(\rho, \phi, \theta) = (3, 5\pi/6, 3\pi/4)$.

3. a) Sketch the surface $z = y^2 + 1$.

b) Sketch the surface

$$y/4 = x^2/4 + z^2/9.$$

4. Let $\boldsymbol{a} = \boldsymbol{i} - 2\boldsymbol{j} + 3\boldsymbol{k}$, $\boldsymbol{b} = 4\boldsymbol{i} - \boldsymbol{j} + 2\boldsymbol{k}$, and $\boldsymbol{c} = 2\boldsymbol{i} - 3\boldsymbol{k}$. Find s such that $\boldsymbol{a} + s\boldsymbol{b}$ is perpendicular to $\boldsymbol{c}$.

5. Find the angle between the vectors $\boldsymbol{i} - 3\boldsymbol{j} + \boldsymbol{k}$ and $2\boldsymbol{i} - \boldsymbol{j} + 2\boldsymbol{k}$.

6. Find the vector projection of $3\boldsymbol{i} - 4\boldsymbol{j} + 2\boldsymbol{k}$ onto $\boldsymbol{i} - 2\boldsymbol{j} + \boldsymbol{k}$.

7. Let $\boldsymbol{a}$ and $\boldsymbol{b}$ be perpendicular unit vectors in space. Simplify each of the following as much as possible from this data.

a) $(\boldsymbol{a} \times \boldsymbol{b}) \cdot \boldsymbol{a}$ b) $(\boldsymbol{a} \times \boldsymbol{b}) \times \boldsymbol{a}$

c) $\boldsymbol{a} \times (\boldsymbol{b} \times \boldsymbol{b})$

8. Find the area of the parallelogram in space having as adjacent edges $\boldsymbol{i} + 2\boldsymbol{j} - \boldsymbol{k}$ and $2\boldsymbol{i} - 3\boldsymbol{j} + 4\boldsymbol{k}$.

9. Find the volume of the box in space with one corner at $(-1, 3, 4)$ and adjacent corners at $(0, -1, 2)$, $(3, -1, 4)$, and $(-1, 2, -1)$.

10. Find all points on the line $x = 1 - 2t$, $y = 3 + 4t$, $z = 2 - t$, whose distance from $(1, 6, 3)$ is $\sqrt{18}$.

11. Find parametric equations of the line through $(-1, 5, 2)$ and $(3, -1, 4)$.

12. Find the equation of the plane through $(-2, 5, 4)$ and parallel to the plane $3x - 4y + 7z = 0$.

13. Find the distance from $(-2, 1)$ to the line $x = 4 + t$, $y = 2 - 3t$ in the plane.

14. Find the point of intersection of the line $x = 4 + t$, $y = 3 - t$, $z = 2 + 3t$, with the plane $2x + y + 3z = -3$.

15. Find the equation of the plane containing the two lines

$$x = 2 + t, \quad y = -1 + 2t, \quad z = 3 - 4t,$$

and

$$x = 2 - 3t, \quad y = -1 + 4t, \quad z = 3 - t$$

that intersect at $(2, -1, 3)$.

more challenging exercises 14

1. Find the distance between the planes $x - 2y + 3z = 10$ and $4x - 8y + 12z = -7$.

2. Find the (shortest) distance between the lines $x = 1 - t$, $y = 2 + 3t$, $z = -1 + 2t$, and $x = 4 + 2t$, $y = -3 + t$, $z = -2 + 5t$.

3. Find the equation of the plane containing the intersection of the sphere with center $(-1, 2, 4)$ and radius 5 with the sphere with center $(1, -1, 3)$ and radius 3.

4. Find the shortest distance from the point $(-1, 4, 8)$ to the line $x = 7 + t$, $y = 4 - 3t$, $z = -2 - 2t$.

5. Find the shortest distance from the line $x = 7 - 3t$, $y = -5 + 4t$, $z = 6 + 2t$, to the sphere $x^2 + y^2 + z^2 = 9$.

6. Consider the parallelogram in the plane having $\boldsymbol{a} = a_1\boldsymbol{i} + a_2\boldsymbol{j}$ and $\boldsymbol{b} = b_1\boldsymbol{i} + b_2\boldsymbol{j}$ as coterminous edges emanating from the origin, as in Fig. 14.32. Show that the area of this parallelogram is the absolute value of the determinant

$$\begin{vmatrix} a_1 & a_2 \\ b_1 & b_2 \end{vmatrix}.$$

7. We motivated consideration of $\boldsymbol{a} \cdot \boldsymbol{b}$ by finding the cosine of the angle between two vectors $\boldsymbol{a}$ and $\boldsymbol{b}$. For an alternative approach, suppose we had simply *defined*, with no motivation,

$$\begin{aligned}\boldsymbol{a} \cdot \boldsymbol{b} &= (a_1\boldsymbol{i} + a_2\boldsymbol{j} + a_3\boldsymbol{k}) \cdot (b_1\boldsymbol{i} + b_2\boldsymbol{j} + b_3\boldsymbol{k}) \\ &= a_1b_1 + a_2b_2 + a_3b_3. \qquad (1)\end{aligned}$$

The properties (a)–(d) of Theorem 14.2 (Section 14.4) are easy to show from the definition (1). One then attempts to *define* the angle θ between the vectors $\boldsymbol{a}$ and $\boldsymbol{b}$ by

$$\theta = \cos^{-1}\frac{\boldsymbol{a} \cdot \boldsymbol{b}}{\sqrt{\boldsymbol{a} \cdot \boldsymbol{a}}\sqrt{\boldsymbol{b} \cdot \boldsymbol{b}}}.$$

But one must first know that

$$\left|\frac{\boldsymbol{a} \cdot \boldsymbol{b}}{\sqrt{\boldsymbol{a} \cdot \boldsymbol{a}}\sqrt{\boldsymbol{b} \cdot \boldsymbol{b}}}\right| \leq 1 \qquad (2)$$

in order for this definition to be meaningful. This inequality is known as the *Schwarz inequality*, and is usually written as

$$|\boldsymbol{a} \cdot \boldsymbol{b}| \leq \|\boldsymbol{a}\|\,\|\boldsymbol{b}\|, \qquad (3)$$

where $\|\boldsymbol{a}\|$ is simply another notation for $|\boldsymbol{a}| = \sqrt{\boldsymbol{a} \cdot \boldsymbol{a}}$, used to avoid confusing the length of a vector with the absolute value of a scalar. Prove the Schwarz inequality (3), using just (1) and properties (a)–(d) of Theorem 14.2. [*Hint.* Use the fact that $(\boldsymbol{a} - x\boldsymbol{b}) \cdot (\boldsymbol{a} - x\boldsymbol{b}) \geq 0$ for all values x by (a) of Theorem 14.2, and use the quadratic formula.]

8. Continuing Exercise 7, prove from the Schwarz inequality the *triangle inequality*

$$\|\boldsymbol{a} + \boldsymbol{b}\| \leq \|\boldsymbol{a}\| + \|\boldsymbol{b}\|$$

for all vectors $\boldsymbol{a}$ and $\boldsymbol{b}$ in space. [If you draw a figure, you will see that this inequality can be viewed as asserting that the length of one side of a triangle is less than or equal to the sum of the lengths of the other two sides; hence the name.]

vector analysis of curves

15

15.1 VELOCITY AND ACCELERATION VECTORS

15.1.1 The position and velocity vectors

Let $x = h(t)$ and $y = k(t)$ be a parametric representation of a smooth curve, so that h' and k' exist and are continuous. You may view these parametric equations as giving the position of a body in the plane at time t. The vector

$$\boldsymbol{r} = x\boldsymbol{i} + y\boldsymbol{j} = h(t)\boldsymbol{i} + k(t)\boldsymbol{j} \tag{1}$$

is the **position vector** of the body at time t. This position $\boldsymbol{r}$ is illustrated in Fig. 15.1.

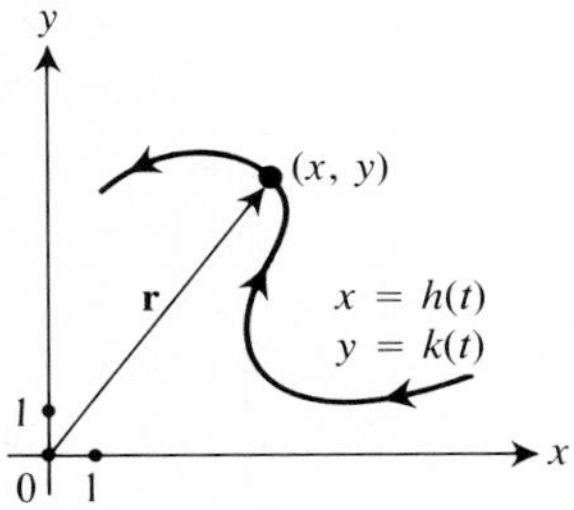

15.1

You can view Eq. (1) as describing a function f that carries a portion of a t-axis into the x,y-plane, so that

$$(x, y) = f(t) = (h(t), k(t)).$$

Such a function, having as values points (which may also be regarded as vectors) in a space of dimension greater than one, is often called a *vector-valued function.*

A change Δt in t produces a change $\Delta\boldsymbol{r}$ in $\boldsymbol{r}$, where $\Delta\boldsymbol{r}$ is as shown in Fig. 15.2. The vector $\Delta\boldsymbol{r}$ is directed along a chord of the curve. Let the notation $\Delta\boldsymbol{r}/\Delta t$ be understood to mean the scalar $1/\Delta t$ times $\Delta\boldsymbol{r}$. Then

$$\frac{\Delta\boldsymbol{r}}{\Delta t} = \frac{\Delta x}{\Delta t}\boldsymbol{i} + \frac{\Delta y}{\Delta t}\boldsymbol{j}.$$

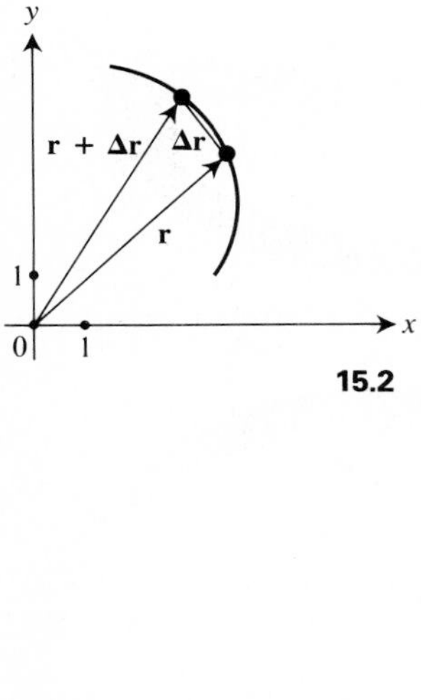

15.2

Taking the limit as $\Delta t \to 0$, we define the **vector derivative** by

$$\frac{d\boldsymbol{r}}{dt} = \lim_{t\to 0}\frac{\Delta\boldsymbol{r}}{\Delta t} = \frac{dx}{dt}\boldsymbol{i} + \frac{dy}{dt}\boldsymbol{j}. \tag{2}$$

Note that dx/dt and dy/dt exist, since we are assuming that $x = h(t)$ and $y = k(t)$ are differentiable functions.

Let's study the vector $d\boldsymbol{r}/dt$. The first things you want to know about a vector are its length and direction. From Eq. (2),

$$\left|\frac{d\boldsymbol{r}}{dt}\right| = \sqrt{\left(\frac{dx}{dt}\right)^2 + \left(\frac{dy}{dt}\right)^2}. \tag{3}$$

From the formula

$$s(t) = \int_{t_0}^{t}\sqrt{\left(\frac{dx}{dt}\right)^2 + \left(\frac{dy}{dt}\right)^2}\,dt$$

for the arc length function, you obtain

$$\frac{ds}{dt} = \sqrt{\left(\frac{dx}{dt}\right)^2 + \left(\frac{dy}{dt}\right)^2}, \tag{4}$$

so (3) yields

$$\left|\frac{d\boldsymbol{r}}{dt}\right| = \frac{ds}{dt}. \tag{5}$$

Now ds/dt has the physical interpretation of the instantaneous rate of change of arc length per unit change in time as the body travels along the curve, so ds/dt is the *speed* of the body along the curve. Turning to the direction of $d\boldsymbol{r}/dt$, the limiting direction of $\Delta\boldsymbol{r}$ as $\Delta t \to 0$ is tangent to the curve. Consequently $d\boldsymbol{r}/dt$ has direction tangent to the curve in the direction given by increasing t. You can also see this from Eq. (2), since the "slope" of the vector arrow $d\boldsymbol{r}/dt$ is

$$\frac{dy/dt}{dx/dt} = \frac{dy}{dx},$$

which is the slope of the tangent to the curve. Thus, at any instant, the vector $d\boldsymbol{r}/dt$ has the same direction as that in which the body is moving, and magnitude equal to the speed of the body.

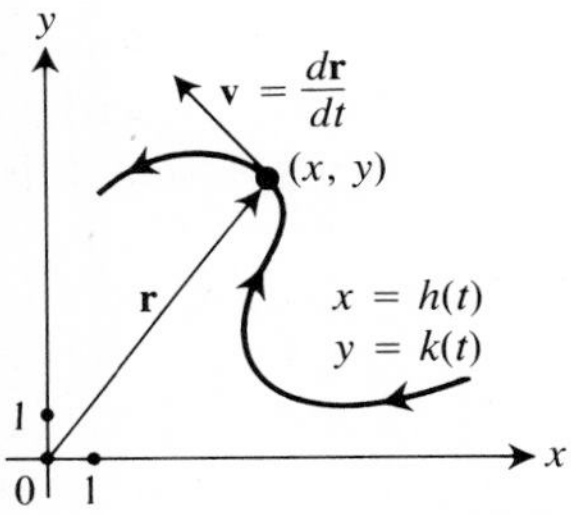

15.3

Definition 15.1 If $\boldsymbol{r} = h(t)\boldsymbol{i} + k(t)\boldsymbol{j}$ is the position vector of a body traveling on a smooth curve, then

$$\boldsymbol{v} = \frac{d\boldsymbol{r}}{dt} = h'(t)\boldsymbol{i} + k'(t)\boldsymbol{j} \tag{6}$$

is the **velocity vector** (or simply the **velocity**) of the body at time t.

Figure 15.3 illustrates the situation just described.

Example 1 If the position of a body at time t on the circle $x^2 + y^2 = 1$ is given by

$$\boldsymbol{r} = (\cos t)\boldsymbol{i} + (\sin t)\boldsymbol{j},$$

then

$$\boldsymbol{v} = \frac{d\boldsymbol{r}}{dt} = (-\sin t)\boldsymbol{i} + (\cos t)\boldsymbol{j}.$$

Note that the speed of the body is constant, namely

$$\text{Speed} = |\boldsymbol{v}| = \left|\frac{d\boldsymbol{r}}{dt}\right| = \sqrt{\sin^2 t + \cos^2 t} = 1. \quad \|$$

Since $\boldsymbol{v}$ is tangent to the curve $x = h(t)$, $y = k(t)$ in the direction corresponding to increasing t, you see that, for $|\boldsymbol{v}| \neq 0$,

$$\boldsymbol{t} = \frac{1}{|\boldsymbol{v}|}\,\boldsymbol{v} \tag{7}$$

is a *unit* vector tangent to the curve in the direction of increasing t. Since $|\boldsymbol{v}| = ds/dt$, you obtain, from (7),

$$\boldsymbol{v} = \frac{ds}{dt}\boldsymbol{t}, \tag{8}$$

which is a compact description of the velocity vector $\boldsymbol{v}$ as tangent to the curve and of magnitude ds/dt.

Exactly the same analysis can be made for a smooth space curve

$$x = h(t), \qquad y = k(t), \qquad z = g(t),$$

where $h(t)$, $k(t)$, and $g(t)$ all have continuous derivatives. The position vector is

$$\boldsymbol{r} = h(t)\boldsymbol{i} + k(t)\boldsymbol{j} + g(t)\boldsymbol{k}, \tag{9}$$

and the velocity vector is

$$\boldsymbol{v} = \frac{d\boldsymbol{r}}{dt} = \frac{dx}{dt}\boldsymbol{i} + \frac{dy}{dt}\boldsymbol{j} + \frac{dz}{dt}\boldsymbol{k}. \tag{10}$$

Since the direction of $\boldsymbol{v}$ is determined by the limiting direction of a chord of the curve corresponding to a vector $\Delta\boldsymbol{r}$ as $\Delta t \to 0$, you again see that $\boldsymbol{v}$ is tangent to the curve and points in the direction given by increasing t. A small change dt in t produces approximate changes dx in x, dy in y, and dz in z, so the approximate change in arc length is:

$$ds = \sqrt{(dx)^2 + (dy)^2 + (dz)^2} = \sqrt{(dx/dt)^2 + (dy/dt)^2 + (dz/dt)^2}\, dt. \tag{11}$$

(A rigorous demonstration of (11) can be given just as for arc length of parametrically described plane curves, which was treated in Chapter 7.) Thus (11) gives the *differential of arc length* for a space curve; the length of arc as a function of t, measured from a point where $t = t_0$, is

$$s(t) = \int_{t_0}^{t} \sqrt{(dx/dt)^2 + (dy/dt)^2 + (dz/dt)^2}\, dt. \tag{12}$$

Once again, the length of the velocity vector is the speed, for

$$|\boldsymbol{v}| = \sqrt{(dx/dt)^2 + (dy/dt)^2 + (dz/dt)^2} = \frac{ds}{dt}. \tag{13}$$

The unit tangent vector $\boldsymbol{t}$ is again given by (7), and (8) is still valid.

Example 2 The space curve

$$x = a\cos t, \qquad y = a\sin t, \qquad z = t,$$

lies on the cyclinder $x^2 + y^2 = a^2$. This curve, which is a *helix*, is sketched in Fig. 15.4. For this helix,

$$\boldsymbol{r} = (a \cos t)\boldsymbol{i} + (a \sin t)\boldsymbol{j} + t\boldsymbol{k},$$

$$\boldsymbol{v} = -(a \sin t)\boldsymbol{i} + (a \cos t)\boldsymbol{j} + \boldsymbol{k},$$

$$\frac{ds}{dt} = |\boldsymbol{v}| = \sqrt{a^2 \cos^2 t + a^2 \sin^2 t + 1} = \sqrt{a^2 + 1}.$$

The length of one "turn" of the helix is therefore

$$\int_0^{2\pi} \sqrt{a^2 + 1}\, dt = 2\pi\sqrt{a^2 + 1}. \quad \|$$

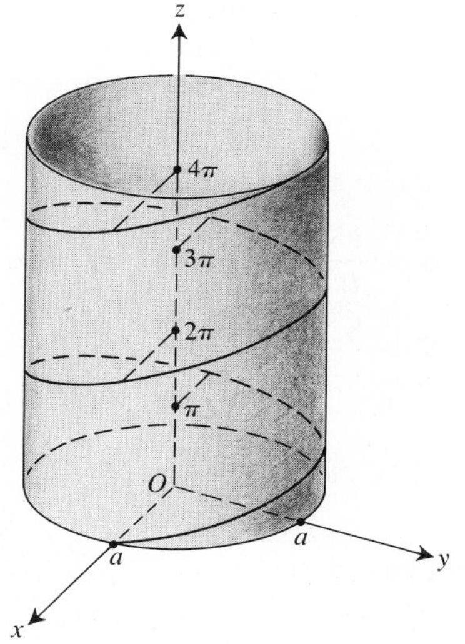

15.4

15.1.2 The acceleration vector

Definition 15.2 If the position vector of a body at time t is $\boldsymbol{r} = h(t)\boldsymbol{i} + k(t)\boldsymbol{j}$, where h and k are twice-differentiable functions of t, then

$$\boldsymbol{a} = \frac{d\boldsymbol{v}}{dt} = \frac{d^2\boldsymbol{r}}{dt^2} = \frac{d^2x}{dt^2}\boldsymbol{i} + \frac{d^2y}{dt^2}\boldsymbol{j} \tag{14}$$

is the **acceleration vector** (or simply the **acceleration**) of the body at time t.

As in the case of the velocity vector, you would like to know the magnitude and direction of the acceleration vector. From Eq. (14), you obtain

$$|\boldsymbol{a}| = \sqrt{(h''(t))^2 + (k''(t))^2}. \tag{15}$$

We leave discussion of the direction of the acceleration vector $\boldsymbol{a}$ to the next section. It depends on the speed and rate of change of speed of the body, and also on the curvature of the curve.

Example 3 For a body traveling on the unit circle with position vector

$$\boldsymbol{r} = (\cos t)\boldsymbol{i} + (\sin t)\boldsymbol{j},$$

you have

$$\boldsymbol{v} = (-\sin t)\boldsymbol{i} + (\cos t)\boldsymbol{j},$$

while

$$\boldsymbol{a} = (-\cos t)\boldsymbol{i} - (\sin t)\boldsymbol{j}.$$

Note that

$$\boldsymbol{v} \cdot \boldsymbol{a} = \sin t \cos t - \cos t \sin t = 0.$$

Since $\boldsymbol{v}$ is tangent to the circle, this shows that $\boldsymbol{a}$ must be normal to the circle. Exercise 2 shows that, whenever $|\boldsymbol{v}|$ remains constant, as in this example where $|\boldsymbol{v}| = 1$, the acceleration vector $\boldsymbol{a}$ has direction normal to the curve. ||

Of course, for a space curve,

$$\boldsymbol{a} = \frac{d^2\boldsymbol{r}}{dt^2} = \frac{d^2x}{dt^2}\boldsymbol{i} + \frac{d^2y}{dt^2}\boldsymbol{j} + \frac{d^2z}{dt^2}\boldsymbol{k}. \tag{16}$$

Example 4 For the helix in Example 2,

$$\boldsymbol{a} = -(a \cos t)\boldsymbol{i} - (a \sin t)\boldsymbol{j},$$

so

$$|\boldsymbol{a}| = \sqrt{a^2 \cos^2 t + a^2 \sin^2 t} = a. \quad ||$$

15.1.3 Differentiation of products of vectors

The previous sections have shown that the differentiation of a vector can be accomplished by just differentiating each component of the vector. If $\boldsymbol{a}$ and $\boldsymbol{b}$ are differentiable vector functions of t in space, then you can form the products $\boldsymbol{a} \cdot \boldsymbol{b}$ and $\boldsymbol{a} \times \boldsymbol{b}$. Using properties of the dot product $\boldsymbol{a} \cdot \boldsymbol{b}$, you see that a change Δt in t produces a change

$$\begin{aligned}(\boldsymbol{a} + \Delta\boldsymbol{a}) \cdot (\boldsymbol{b} + \Delta\boldsymbol{b}) - \boldsymbol{a} \cdot \boldsymbol{b} &= \boldsymbol{a} \cdot \boldsymbol{b} + \boldsymbol{a} \cdot \Delta\boldsymbol{b} + \Delta\boldsymbol{a} \cdot \boldsymbol{b} + \Delta\boldsymbol{a} \cdot \Delta\boldsymbol{b} - \boldsymbol{a} \cdot \boldsymbol{b} \\ &= \boldsymbol{a} \cdot \Delta\boldsymbol{b} + \Delta\boldsymbol{a} \cdot \boldsymbol{b} + \Delta\boldsymbol{a} \cdot \Delta\boldsymbol{b}\end{aligned}$$

in $\boldsymbol{a} \cdot \boldsymbol{b}$. Thus the derivative $d(\boldsymbol{a} \cdot \boldsymbol{b})/dt$ is

$$\begin{aligned}\frac{d(\boldsymbol{a} \cdot \boldsymbol{b})}{dt} &= \lim_{\Delta t \to 0} \left(\boldsymbol{a} \cdot \frac{\Delta\boldsymbol{b}}{\Delta t} + \frac{\Delta\boldsymbol{a}}{\Delta t} \cdot \boldsymbol{b} + \frac{\Delta\boldsymbol{a}}{\Delta t} \cdot \Delta\boldsymbol{b}\right) \\ &= \boldsymbol{a} \cdot \frac{d\boldsymbol{b}}{dt} + \frac{d\boldsymbol{a}}{dt} \cdot \boldsymbol{b} + \frac{d\boldsymbol{a}}{dt} \cdot \boldsymbol{0} \\ &= \boldsymbol{a} \cdot \frac{d\boldsymbol{b}}{dt} + \frac{d\boldsymbol{a}}{dt} \cdot \boldsymbol{b},\end{aligned} \tag{17}$$

which is the familiar formula for differentiation of a product. Similarly,

$$\frac{d(\boldsymbol{a} \times \boldsymbol{b})}{dt} = \boldsymbol{a} \times \frac{d\boldsymbol{b}}{dt} + \frac{d\boldsymbol{a}}{dt} \times \boldsymbol{b}. \tag{18}$$

Since the cross product is not commutative, you should note that while

$$\frac{d(\boldsymbol{a} \cdot \boldsymbol{b})}{dt} = \boldsymbol{a} \cdot \frac{d\boldsymbol{b}}{dt} + \boldsymbol{b} \cdot \frac{d\boldsymbol{a}}{dt},$$

this formula with $\cdot$ replaced by $\times$ is *not* valid for the cross product, for $(d\boldsymbol{a}/dt) \times \boldsymbol{b} = -\boldsymbol{b} \times (d\boldsymbol{a}/dt)$. We ask you to differentiate some products in the exercises.

SUMMARY *Assume that you have parametric equations $x = h(t)$, $y = k(t)$ of a plane curve, or $x = h(t)$, $y = k(t)$, $z = g(t)$ for a space curve, where all the functions of t are twice differentiable.*

1. *The position vector to the curve is*

$$\boldsymbol{r} = h(t)\boldsymbol{i} + k(t)\boldsymbol{j} \quad \textit{for a plane curve,}$$

$$\boldsymbol{r} = h(t)\boldsymbol{i} + k(t)\boldsymbol{j} + g(t)\boldsymbol{k} \quad \textit{for a space curve.}$$

2. *The velocity vector is*

$$\boldsymbol{v} = \frac{d\boldsymbol{r}}{dt} = \frac{dx}{dt}\boldsymbol{i} + \frac{dy}{dt}\boldsymbol{j} \quad \textit{for a plane curve,}$$

$$\boldsymbol{v} = \frac{d\boldsymbol{r}}{dt} = \frac{dx}{dt}\boldsymbol{i} + \frac{dy}{dt}\boldsymbol{j} + \frac{dz}{dt}\boldsymbol{k} \quad \textit{for a space curve.}$$

3. *The length of the velocity vector is*

$$|\boldsymbol{v}| = \frac{ds}{dt} = \textit{Speed along the curve.}$$

4. *The distance traveled along a space curve from t_0 to t is*

$$s(t) = \int_{t_0}^{t} \sqrt{\left(\frac{dx}{dt}\right)^2 + \left(\frac{dy}{dt}\right)^2 + \left(\frac{dz}{dt}\right)^2}\, dt.$$

5. *The direction of $\boldsymbol{v}$ is tangent to the curve pointing in the direction corresponding to increasing t.*

6. *The acceleration vector is*

$$\boldsymbol{a} = \frac{d^2\boldsymbol{r}}{dt^2} = \frac{d\boldsymbol{v}}{dt} = \frac{d^2x}{dt^2}\boldsymbol{i} + \frac{d^2y}{dt^2}\boldsymbol{j} \quad \textit{for a plane curve,}$$

$$\boldsymbol{a} = \frac{d^2\boldsymbol{r}}{dt^2} = \frac{d\boldsymbol{v}}{dt} = \frac{d^2x}{dt^2}\boldsymbol{i} + \frac{d^2y}{dt^2}\boldsymbol{j} + \frac{d^2z}{dt^2}\boldsymbol{k} \quad \textit{for a space curve.}$$

7. *For differentiable vector functions* $\boldsymbol{a}$ *and* $\boldsymbol{b}$ *of* t *in space,*

$$\frac{d(\boldsymbol{a}\cdot\boldsymbol{b})}{dt} = \boldsymbol{a}\cdot\frac{d\boldsymbol{b}}{dt} + \frac{d\boldsymbol{a}}{dt}\cdot\boldsymbol{b} \quad \textit{and} \quad \frac{d(\boldsymbol{a}\times\boldsymbol{b})}{dt} = \boldsymbol{a}\times\frac{d\boldsymbol{b}}{dt} + \frac{d\boldsymbol{a}}{dt}\times\boldsymbol{b}.$$

EXERCISES

1. Let $\boldsymbol{r}(t)$ be the position vector from the origin to a point $(x, y) = (h(t), k(t))$ on a smooth curve. Show that if s is the arc length along the curve defined as usual, then $d\boldsymbol{r}/ds = \boldsymbol{t}$, where $\boldsymbol{t}$ is the unit tangent vector to the curve in the direction of increasing t.

In Exercise 2, use the fact that, if a body is traveling subject to a force vector $\mathbf{F}$, *then* $\mathbf{F} = m\boldsymbol{a}$ *where* $\boldsymbol{a}$ *is the acceleration vector and* m *is the mass of the body.*

2. Let the path of a body, moving subject to a force, be a "twice-differentiable" curve. Show that the force is always perpendicular to the direction of motion if and only if the speed of the body is constant. [*Hint.* The speed is constant if and only if $\boldsymbol{v}\cdot\boldsymbol{v} = |\boldsymbol{v}|^2$ is constant. Differentiate $\boldsymbol{v}\cdot\boldsymbol{v}$ as a product of vectors.]

In Exercises 3 through 13, the position vector $\boldsymbol{r}$ *at time* t *of a body moving on a curve is given. Find the following at the indicated time* t_0:

a) *the velocity vector of the body,* b) *the speed of the body,* c) *the acceleration vector of the body.*

3. $\boldsymbol{r} = 2t\boldsymbol{i} + (3t - 1)\boldsymbol{j}$ at $t_0 = 0$

4. $\boldsymbol{r} = (3t + 1)\boldsymbol{i} + t^2\boldsymbol{j}$ at $t_0 = 1$

5. $\boldsymbol{r} = (\sin t)\boldsymbol{i} + (\cos 2t)\boldsymbol{j}$ at $t_0 = \pi$

6. $\boldsymbol{r} = e^t\boldsymbol{i} + t^2\boldsymbol{j}$ at $t_0 = 0$

7. $\boldsymbol{r} = (\ln t)\boldsymbol{i} + (\cosh(t - 1))\boldsymbol{j}$ at $t_0 = 1$

8. $\boldsymbol{r} = (e^t \sin t)\boldsymbol{i} + (e^t \cos t)\boldsymbol{j}$ at $t_0 = 0$

9. $\boldsymbol{r} = (1/t)\boldsymbol{i} + (1/t^2)\boldsymbol{j}$ at $t_0 = 1$

10. $\boldsymbol{r} = (\ln(\sin t))\boldsymbol{i} + (\ln(\cos t))\boldsymbol{j}$ at $t_0 = \pi/4$

11. $\boldsymbol{r} = t^2\boldsymbol{i} + t^3\boldsymbol{j} - (t + 1)\boldsymbol{k}$ at $t_0 = 2$

12. $\boldsymbol{r} = e^t\boldsymbol{i} - (\sin t)\boldsymbol{j} + (\cos t)\boldsymbol{k}$ at $t_0 = 0$

13. $\boldsymbol{r} = \dfrac{t+1}{t}\boldsymbol{i} + \dfrac{t-1}{t}\boldsymbol{j} + t^2\boldsymbol{k}$ at $t_0 = 1$

14. Find the length of the space curve $x = 2t$, $y = t^2$, $z = -t^2$, from $t = 0$ to $t = 2$.

15. Find the length of the twisted cubic $x = 3t^2$, $y = 2t^3$, $z = 3t$, from $t = 0$ to $t = 4$.

16. Let $\boldsymbol{a} = t^2\boldsymbol{i} - (3t + 1)\boldsymbol{j}$ and $\boldsymbol{b} = (2t)\boldsymbol{i} - t^3\boldsymbol{j}$ in the plane. Illustrate Eq. (17) of the text by computing $d(\boldsymbol{a}\cdot\boldsymbol{b})/dt$ in two ways.

17. Let $\boldsymbol{a} = 3t\boldsymbol{i} - (4t + 1)\boldsymbol{j} + t^2\boldsymbol{k}$ and $\boldsymbol{b} = (t^2 - 2)\boldsymbol{i} + 5t\boldsymbol{j} - 6t\boldsymbol{k}$ in space. Illustrate Eq. (18) of the text by computing $d(\boldsymbol{a}\times\boldsymbol{b})/dt$ in two ways.

15.2 NORMAL AND TANGENTIAL COMPONENTS OF ACCELERATION

Let $\boldsymbol{r} = h(t)\boldsymbol{i} + k(t)\boldsymbol{j}$ be the position vector of a plane curve and let $h(t)$ and $k(t)$ be twice differentiable functions. It is fruitful to examine the components of the acceleration vector $\boldsymbol{a}$ at a point (x, y) in the direction tangent to the curve and in the direction normal to the curve. First we do a bit of preliminary work.

15.2.1 The direction of the acceleration vector

At a point (x, y) on a smooth curve, let ϕ be the angle from the positive dx-axis to the positive tangent ds-axis, as shown in Fig. 15.5. For the unit tangent vector $\boldsymbol{t}$ to the curve at (x, y), you obtain from the figure,

$$\boldsymbol{t} = (\cos\phi)\boldsymbol{i} + (\sin\phi)\boldsymbol{j}. \tag{1}$$

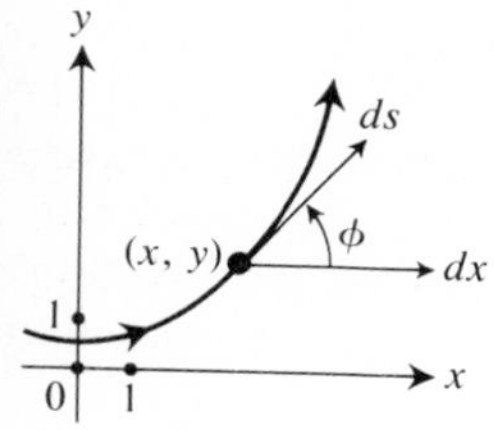

15.5

Thus $\boldsymbol{t}$ is a differentiable function of ϕ, and

$$\frac{d\boldsymbol{t}}{d\phi} = (-\sin\phi)\boldsymbol{i} + (\cos\phi)\boldsymbol{j}. \tag{2}$$

From (1) and (2), you see that $\boldsymbol{t}$ and $d\boldsymbol{t}/d\phi$ are perpendicular for

$$\boldsymbol{t}\cdot\frac{d\boldsymbol{t}}{d\phi} = (\cos\phi)(-\sin\phi) + (\sin\phi)(\cos\phi) = 0.$$

Furthermore,

$$\left|\frac{d\boldsymbol{t}}{d\phi}\right| = \sqrt{(-\sin\phi)^2 + \cos^2\phi} = 1.$$

Hence $d\boldsymbol{t}/d\phi$ is a *unit vector normal to the curve.* [You can easily see from (2) that $d\boldsymbol{t}/d\phi$ may be obtained by rotating $\boldsymbol{t}$ counterclockwise through an angle of 90°.] It follows that

$$\frac{d\boldsymbol{t}}{ds} = \frac{d\phi}{ds}\frac{d\boldsymbol{t}}{d\phi} \tag{3}$$

is also normal to the curve. We let the *unit normal vector* $\boldsymbol{n}$ be the unit vector having the direction of $d\boldsymbol{t}/ds$, if $d\boldsymbol{t}/ds \neq \boldsymbol{0}$. If $d\boldsymbol{t}/ds = \boldsymbol{0}$, then take $\boldsymbol{n} = d\boldsymbol{t}/d\phi$. From the definition of $d\boldsymbol{t}/ds$ as

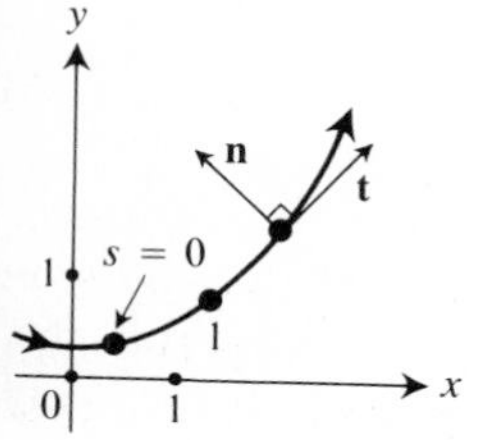

15.6

$$\frac{d\boldsymbol{t}}{ds} = \lim_{\Delta s\to 0}\frac{\boldsymbol{t}(s+\Delta s) - \boldsymbol{t}(s)}{\Delta s},$$

you see at once that $\boldsymbol{n}$ points toward the *concave* side of the curve if $d\boldsymbol{t}/ds \neq 0$ (see Fig. 15.6).

From Eq. (3) and the fact that $|d\boldsymbol{t}/d\phi| = 1$,

$$\left|\frac{d\boldsymbol{t}}{ds}\right| = \left|\frac{d\phi}{ds}\right| = \kappa,$$

so

$$\frac{d\boldsymbol{t}}{ds} = \kappa\boldsymbol{n}. \tag{4}$$

Equation (4) could be used to *define* curvature κ, and will be so used when we study differential geometry of space curves. We prefer the more intuitive definition $\kappa = |d\phi/ds|$ at this time.

We have now described an *orthogonal frame of unit vectors* $\boldsymbol{t}$ and $\boldsymbol{n}$ at each point of our curve. These unit vectors might be regarded as "forward" and "sideways" vectors by a two-dimensional bug living on the curve; the curve may appear locally "straight" to him. You may think of $\boldsymbol{t}$ and $\boldsymbol{n}$ as local unit coordinate vectors on the curve for such a bug.

Now assume that our curve possesses a curvature κ at (x, y) so that the angle ϕ shown in Fig. 15.5 is a differentiable function of the arc length s along the curve in a neighborhood of (x, y). The product and chain rules for differentiation hold for products such as $(ds/dt)\boldsymbol{t}$ and for vector functions such as $\boldsymbol{t}$; you can check by computing components. Using these rules and (4), differentiate

$$\boldsymbol{v} = \frac{ds}{dt}\boldsymbol{t}$$

and you will obtain

$$\begin{aligned}\boldsymbol{a} = \frac{d\boldsymbol{v}}{dt} = \frac{d^2s}{dt^2}\boldsymbol{t} + \frac{ds}{dt}\frac{d\boldsymbol{t}}{dt} &= \frac{d^2s}{dt^2}\boldsymbol{t} + \frac{ds}{dt}\left(\frac{ds}{dt}\frac{d\boldsymbol{t}}{ds}\right)\\ &= \frac{d^2s}{dt^2}\boldsymbol{t} + \kappa\left(\frac{ds}{dt}\right)^2\boldsymbol{n}.\end{aligned} \tag{5}$$

Thus if you resolve the acceleration vector into tangential and normal components so that

$$\boldsymbol{a} = a_{\mathrm{t}}\boldsymbol{t} + a_{\mathrm{n}}\boldsymbol{n},$$

then

$$a_{\mathrm{t}} = \frac{d^2s}{dt^2} \quad \text{and} \quad a_{\mathrm{n}} = \kappa\left(\frac{ds}{dt}\right)^2. \tag{6}$$

The relations in (6) have a nice physical interpretation. By Newton's second law of motion, the force vector $\boldsymbol{F}$ governing a body of mass m moving on a plane curve with acceleration $\boldsymbol{a}$ is given by $\boldsymbol{F} = m\boldsymbol{a}$. From (6), you see that the component of the force tangential to the path of the body controls the rate of change of the speed of the body, for

$$ma_{\mathrm{t}} = m\frac{d^2s}{dt^2} = m\frac{d(ds/dt)}{dt}.$$

On the other hand, the component of the force normal to the direction of a body controls the curvature of the path on which the body travels. This seems intuitively reasonable. The formula for a_{n} in (6) shows that, when you are driving a car around an unbanked curve, the force normal to the curve exerted by the road on the wheels must be proportional to the product of the curvature of the curve and the square of the speed of the car. The larger the curvature, the more force is required. If the speed of the car is doubled, the force required normal to the curve to "hold the car on the road" is quadrupled. If you take a sharp unbanked curve too fast, then the available force due to friction normal to the wheels is not sufficient for the car to make the curve; you skid and go off the road.

15.2.2 Computations

Let's summarize here the terminology and formulas developed in our study of a curve $x = h(t)$, $y = k(t)$ by vector methods. Let's use the notation of a dot over a variable to denote the derivative "with respect to t," so that $\dot{x} = dx/dt$, $\ddot{x} = d^2x/dt^2$, etc. Then

$$\boldsymbol{r} = \text{position vector} = x\boldsymbol{i} + y\boldsymbol{j}, \tag{7}$$

$$\boldsymbol{v} = \text{velocity vector} = \dot{x}\boldsymbol{i} + \dot{y}\boldsymbol{j} = \dot{s}\boldsymbol{t}, \tag{8}$$

$$|\boldsymbol{v}| = \text{speed} = \dot{s} = \sqrt{\dot{x}^2 + \dot{y}^2}, \tag{9}$$

$$\boldsymbol{a} = \text{acceleration} = \ddot{x}\boldsymbol{i} + \ddot{y}\boldsymbol{j} = a_t\boldsymbol{t} + a_n\boldsymbol{n} = \ddot{s}\boldsymbol{t} + \kappa(\dot{s}^2)\boldsymbol{n}, \tag{10}$$

$$|\boldsymbol{a}| = \sqrt{\ddot{x}^2 + \ddot{y}^2} = \sqrt{a_t^2 + a_n^2} = \sqrt{\ddot{s}^2 + \kappa^2\dot{s}^4}, \tag{11}$$

$$a_t = \ddot{s}, \tag{12}$$

$$a_n = \kappa(\dot{s})^2, \tag{13}$$

$$\kappa = \left|\frac{d\boldsymbol{t}}{ds}\right| = \left|\frac{d\phi}{ds}\right|. \tag{14}$$

The final expression for $|\boldsymbol{a}|$ in (11) follows from relation (10), using the Pythagorean theorem and the fact that $\boldsymbol{t}$ and $\boldsymbol{n}$ are *perpendicular unit vectors.*

The use of relations (7) through (14) is best illustrated by an example.

Example 1 Suppose the position (x, y) of a body in the plane at time t is given by

$$x = 1 + \cos t - \sin t, \qquad y = \sin t + \cos t.$$

Let's find the velocity $\boldsymbol{v}$, speed s, acceleration $\boldsymbol{a}$, magnitudes $|a_t|$ and $|a_n|$ of the tangential and normal components of the acceleration, and the radius of curvature of the curve at any time t.

SOLUTION The position vector at time t is

$$\boldsymbol{r} = (1 + \cos t - \sin t)\boldsymbol{i} + (\sin t + \cos t)\boldsymbol{j}.$$

From (8) and (10),

$$\boldsymbol{v} = (-\sin t - \cos t)\boldsymbol{i} + (\cos t - \sin t)\boldsymbol{j}$$

and

$$\boldsymbol{a} = (-\cos t + \sin t)\boldsymbol{i} + (-\sin t - \cos t)\boldsymbol{j}.$$

For speed, you obtain, from (9),

$$\dot{s} = \sqrt{(-\sin t - \cos t)^2 + (\cos t - \sin t)^2} = \sqrt{2}.$$

Therefore

$$a_t = \ddot{s} = 0.$$

From (11) and the fact that $a_t = 0$,

$$\begin{aligned} |\boldsymbol{a}| &= \sqrt{(-\cos t + \sin t)^2 + (-\sin t - \cos t)^2} \\ &= \sqrt{2} = \sqrt{a_t^{\,2} + a_n^{\,2}} = \sqrt{a_n^{\,2}}. \end{aligned}$$

Thus $a_n = \sqrt{2}$. From Formula (13), you now obtain

$$\kappa \dot{s}^2 = \sqrt{2},$$

so

$$\kappa = \frac{\sqrt{2}}{2} = \frac{1}{\sqrt{2}},$$

and $\rho = 1/\kappa = \sqrt{2}$. Consequently, the curve is a circle of radius $\sqrt{2}$. ||

SUMMARY *Let $\boldsymbol{t}$ and $\boldsymbol{n}$ be tangent and normal unit vectors to a twice-differentiable plane curve at a point, with $\boldsymbol{t}$ pointing in the direction corresponding to increasing t, and $\boldsymbol{n}$ pointing toward the concave side of the curve.*

1. $d\boldsymbol{t}/ds = \kappa \boldsymbol{n}$, *where κ is the curvature of the curve*
2. $\boldsymbol{v} = (ds/dt)\boldsymbol{t}$
3. $\boldsymbol{a} = d^2\boldsymbol{r}/dt^2 = (d^2s/dt^2)\boldsymbol{t} + \kappa(ds/dt)^2\boldsymbol{n}$,
   so the tangential and normal components of acceleration are

$$a_t = \frac{d^2s}{dt^2} \quad \text{and} \quad a_n = \kappa\left(\frac{ds}{dt}\right)^2.$$

EXERCISES

1. Let the path of a body, moving in the plane subject to a force, be a twice-differentiable curve. Show that, if the speed of the body at time $t = t_0$ is zero, then the force on the body at time t_0 has direction tangent to the curve.

In Exercises 2 through 9, find, at the indicated time t_0:

a) *the tangential and normal components of the acceleration, and*

b) *the curvature of the curve.*

2. $\boldsymbol{r} = 2t\boldsymbol{i} + (3t - 1)\boldsymbol{j}$ at $t_0 = 0$

3. $\boldsymbol{r} = (3t + 1)\boldsymbol{i} + t^2\boldsymbol{j}$ at $t_0 = 1$

4. $\boldsymbol{r} = (\sin t)\boldsymbol{i} + (\cos 2t)\boldsymbol{j}$ at $t_0 = \pi$

5. $\boldsymbol{r} = e^t\boldsymbol{i} + t^2\boldsymbol{j}$ at $t_0 = 0$

6. $\boldsymbol{r} = (\ln t)\boldsymbol{i} + (\cosh (t - 1))\boldsymbol{j}$ at $t_0 = 1$

7. $\boldsymbol{r} = (e^t \sin t)\boldsymbol{i} + (e^t \cos t)\boldsymbol{j}$ at $t_0 = 0$

8. $\boldsymbol{r} = (1/t)\boldsymbol{i} + (1/t^2)\boldsymbol{j}$ at $t_0 = 1$

9. $\boldsymbol{r} = (\ln (\sin t))\boldsymbol{i} + (\ln (\cos t))\boldsymbol{j}$ at $t_0 = \pi/4$

*15.3 POLAR VECTOR ANALYSIS AND KEPLER'S LAWS

*15.3.1 Velocity and acceleration in polar coordinates

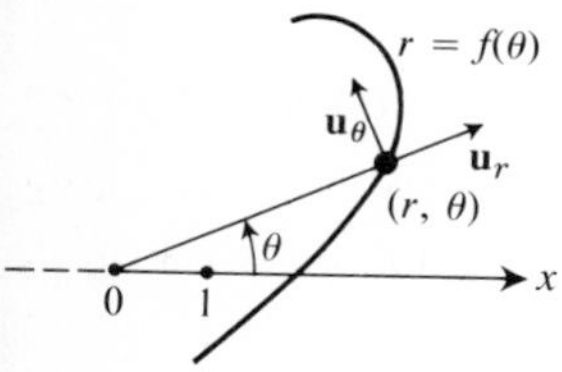

15.7

Let $r = f(\theta)$ where f is a twice-differentiable function, and consider a body moving along the curve $r = f(\theta)$. We wish to express the velocity and acceleration vectors of the body at time t in terms of the unit vector $\boldsymbol{u}_r$ directed away from the origin and the perpendicular unit vector $\boldsymbol{u}_\theta$ in the direction of increasing θ, as shown in Fig. 15.7.

Note that

$$\boldsymbol{u}_r = (\cos \theta)\boldsymbol{i} + (\sin \theta)\boldsymbol{j}, \tag{1}$$

while

$$\boldsymbol{u}_\theta = (-\sin \theta)\boldsymbol{i} + (\cos \theta)\boldsymbol{j}. \tag{2}$$

From (3) and (4), you at once obtain

$$\frac{d\boldsymbol{u}_r}{d\theta} = (-\sin \theta)\boldsymbol{i} + (\cos \theta)\boldsymbol{j} = \boldsymbol{u}_\theta \tag{3}$$

and

$$\frac{d\boldsymbol{u}_\theta}{d\theta} = (-\cos \theta)\boldsymbol{i} + (-\sin \theta)\boldsymbol{j} = -\boldsymbol{u}_r. \tag{4}$$

You can now obtain the desired formulas for the velocity vector $\boldsymbol{v}$ and acceleration vector $\boldsymbol{a}$ by differentiating the position vector

$$\boldsymbol{r} = r\boldsymbol{u}_r \tag{5}$$

with respect to t. It is easily checked by writing out components, that the usual "product rule for differentiation" holds for differentiating a scalar function times a vector. Using the chain rule and (3), you obtain from (5)

$$\boldsymbol{v} = \frac{d\boldsymbol{r}}{dt} = r\frac{d\boldsymbol{u}_r}{dt} + \frac{dr}{dt}\boldsymbol{u}_r = r\frac{d\theta}{dt}\frac{d\boldsymbol{u}_r}{d\theta} + \frac{dr}{dt}\boldsymbol{u}_r = r\frac{d\theta}{dt}\boldsymbol{u}_\theta + \frac{dr}{dt}\boldsymbol{u}_r,$$

so

$$\boldsymbol{v} = \dot{r}\boldsymbol{u}_r + r\dot{\theta}\boldsymbol{u}_\theta, \tag{6}$$

* This section may be omitted without loss of continuity. Some may want to cover just the one subsection on vectors in polar coordinates.

where a dot over a variable denotes a derivative with respect to time. Differentiating (6), using chain rules and (3) and (4),

$$\begin{aligned} \boldsymbol{a} = \frac{d\boldsymbol{v}}{dt} &= \dot{r}\frac{d\boldsymbol{u}_r}{dt} + \ddot{r}\boldsymbol{u}_r + r\dot{\theta}\frac{d\boldsymbol{u}_\theta}{dt} + (r\ddot{\theta} + \dot{r}\dot{\theta})\boldsymbol{u}_\theta \\ &= \dot{r}\frac{d\theta}{dt}\frac{d\boldsymbol{u}_r}{d\theta} + \ddot{r}\boldsymbol{u}_r + r\dot{\theta}\frac{d\theta}{dt}\frac{d\boldsymbol{u}_\theta}{d\theta} + (r\ddot{\theta} + \dot{r}\dot{\theta})\boldsymbol{u}_\theta \\ &= \dot{r}\dot{\theta}\boldsymbol{u}_\theta + \ddot{r}\boldsymbol{u}_r + r\dot{\theta}^2(-\boldsymbol{u}_r) + (r\ddot{\theta} + \dot{r}\dot{\theta})\boldsymbol{u}_\theta, \end{aligned}$$

which yields

$$\boldsymbol{a} = (\ddot{r} - r\dot{\theta}^2)\boldsymbol{u}_r + (r\ddot{\theta} + 2\dot{r}\dot{\theta})\boldsymbol{u}_\theta. \tag{7}$$

Equation (7) is the one needed to study central force fields. Particular illustrations of Eqs. (6) and (7) are routine, and are left to the exercises.

*15.3.2 Central force fields and Kepler's laws

In this section, we work in polar coordinates and show how one can derive the planetary laws of Johannes Kepler (1571–1630) from Newton's law of gravitation. These laws are as follows:

Newton's Law of Universal Gravitation. *The force F of attraction between two bodies of masses m and M is given by*

$$F = \frac{GmM}{d^2},$$

where G is a "universal" constant of gravitational attraction and d is the distance between the bodies.

Kepler's Laws

1. *The orbit of each planet is an ellipse with the sun at a focus.*
2. *A line segment joining the sun and a particular planet sweeps out regions of equal areas during equal time intervals.*
3. *Let the elliptical orbit of a planet have a major axis of length $2a$ and let T be the* ***period*** *of the planet (i.e., the time required for one complete revolution about the sun). Then the ratio a^3/T^2 is the same for all planets of our sun.*

We derive Kepler's Laws from Newton's Law. This reverses the historical order. Newton actually derived his inverse square law from Kepler's laws. Kepler, in turn, formulated his laws on the basis of his analysis of astronomical observations by Tycho Brahe (1546–1607). After some exercises illustrating the material in the text, we give a sequence of exercises that constitute a detailed outline for the derivation of Newton's inverse square law from Kepler's three laws.

Let's consider a body of mass M to be fixed at the pole O of a polar coordinate system, and examine the motion of a body of mass m, which is subject only to the force of gravitational attraction toward the body at O, as specified by Newton's law. Using Newton's second law of motion,

$$\boldsymbol{F} = m\boldsymbol{a}.$$

Since the force is directed entirely in the direction $-\boldsymbol{u}_r$ toward O, you see from (7) that

$$r\ddot{\theta} + 2\dot{r}\dot{\theta} = 0. \tag{8}$$

You obtain, from (7) and Newton's universal law of gravitation,

$$m(\ddot{r} - r\dot{\theta}^2) = -\frac{mMG}{r^2}$$

or

$$\ddot{r} - r\dot{\theta}^2 = -\frac{MG}{r^2} = -\frac{k}{r^2}, \tag{9}$$

where $k = MG > 0$ is a constant independent of the mass m of the moving body.

We derive Kepler's laws from (8) and (9). Kepler's second law is the easiest one to derive.

Derivation of Kepler's Second Law. From (8), we see that

$$\frac{d}{dt}\left(\frac{1}{2}r^2\dot{\theta}\right) = \frac{1}{2}r^2\ddot{\theta} + r\dot{r}\dot{\theta} = \frac{1}{2}r(r\ddot{\theta} + 2\dot{r}\dot{\theta}) = 0.$$

Thus

$$r^2\dot{\theta} = K \tag{10}$$

for some constant K. Now the area A of the region swept out by the line segment joining O and the moving body from time $t = t_0$ where $\theta = \theta_0$ to time t is given by

$$A = \int_{\theta_0}^{\theta(t)} \tfrac{1}{2}r^2\,d\theta = \int_{\theta_0}^{\theta(t)} \tfrac{1}{2}(f(\theta))^2\,d\theta.$$

Thus

$$\frac{dA}{dt} = \frac{dA}{d\theta}\frac{d\theta}{dt} = \frac{1}{2}r^2\frac{d\theta}{dt} = \frac{1}{2}r^2\dot{\theta} = \frac{1}{2}K. \tag{11}$$

From (11),

$$A = \tfrac{1}{2}Kt + C,$$

from which it follows at once that equal time intervals correspond to equal increments in A.

Derivation of Kepler's First Law. From (10),

$$r\dot{\theta}^2 = \frac{(r^2\dot{\theta})^2}{r^3} = \frac{K^2}{r^3}. \tag{12}$$

Using (9), you then have

$$\ddot{r} - \frac{K^2}{r^3} = -\frac{k}{r^2}. \tag{13}$$

Multiplication of (13) by $\dot{r}$ yields

$$\dot{r}\ddot{r} - K^2\frac{\dot{r}}{r^3} = -k\frac{\dot{r}}{r^2}. \tag{14}$$

Integrating (13) with respect to t and multiplying by 2,

$$\dot{r}^2 + K^2\frac{1}{r^2} = 2k\frac{1}{r} + C. \tag{15}$$

If you now let $p = 1/r$, then $r = 1/p$ and $\dot{r} = -\dot{p}/p^2$. You may then write (15) as

$$\frac{\dot{p}^2}{p^4} + K^2p^2 = 2kp + C. \tag{16}$$

From $r^2\dot{\theta} = K$ in (10), you obtain $\dot{\theta}/p^2 = K$, so

$$K^2\left(\frac{dp}{d\theta}\right)^2 = \frac{1}{p^4}\left(\frac{d\theta}{dt}\right)^2\left(\frac{dp}{d\theta}\right)^2 = \frac{1}{p^4}\left(\frac{dp}{dt}\right)^2 = \frac{\dot{p}^2}{p^4}. \tag{17}$$

From (16) and (17),

$$K^2\left(\left(\frac{dp}{d\theta}\right)^2 + p^2\right) = 2kp + C. \tag{18}$$

Differentiation of (18) with respect to θ yields

$$K^2\left(2\left(\frac{dp}{d\theta}\right)\frac{d^2p}{d\theta^2} + 2p\frac{dp}{d\theta}\right) = 2k\frac{dp}{d\theta},$$

or

$$2K^2\frac{dp}{d\theta}\left(\frac{d^2p}{d\theta^2} + p - \frac{k}{K^2}\right) = 0. \tag{19}$$

If $dp/d\theta = 0$, then $p = 1/r = a$ for some constant a, so $r = 1/a$ and the orbit of our body is a circle with center O, which is in accordance with

Kepler's first law. Suppose

$$\frac{d^2p}{d\theta^2} + p - \frac{k}{K^2} = 0. \tag{20}$$

In Section 20.6 you will see that the solution of the differential equation

$$\frac{d^2p}{d\theta^2} + p = \frac{k}{K^2} \tag{21}$$

describing the motion of our body must be given by

$$\begin{aligned} p &= A\cos\theta + B\sin\theta + \frac{k}{K^2} \\ &= \sqrt{A^2+B^2}\left(\frac{A}{\sqrt{A^2+B^2}}\cos\theta + \frac{B}{\sqrt{A^2+B^2}}\sin\theta\right) + \frac{k}{K^2} \end{aligned} \tag{22}$$

for some constants A and B. You can find an angle γ such that

$$\frac{A}{\sqrt{A^2+B^2}} = -\cos\gamma \qquad \text{and} \qquad \frac{B}{\sqrt{A^2+B^2}} = -\sin\gamma. \tag{23}$$

Let $E = \sqrt{A^2+B^2}$. From (22) and (23), we obtain

$$p = -E\cos(\theta - \gamma) + \frac{k}{K^2}. \tag{24}$$

By choosing a new polar axis, if necessary, you can assume that $\gamma = 0$, so that (24) becomes

$$\frac{1}{r} = -E\cos\theta + \frac{k}{K^2}, \tag{25}$$

which yields

$$\begin{aligned} r &= \frac{1}{(k/K^2) - E\cos\theta} \\ &= \frac{K^2/k}{1 - (K^2E/k)\cos\theta}. \end{aligned} \tag{26}$$

Note that $K^2E/k > 0$. Setting $e = K^2E/k$, you obtain finally

$$r = \frac{e(1/E)}{1 - e\cos\theta}, \tag{27}$$

which is the equation of a conic section with focus at the pole O. In particular, if the orbit is a closed path (as for the planets orbiting our sun) the orbit must be an ellipse.

Derivation of Kepler's Third Law. By integration, you easily find that the area of the region enclosed by the ellipse $(x^2/a^2) + (y^2/b^2) = 1$ of major axis $2a$ and minor axis $2b$ is πab. If T is the period of the orbit of our planet, then (11) shows that the area swept out by the line segment joining O and the planet during one revolution is given by

$$\int_0^T \frac{dA}{dt}\,dt = \int_0^T \frac{1}{2}K\,dt = \frac{1}{2}Kt\Big]_0^T = \frac{1}{2}KT.$$

Thus we have

$$\tfrac{1}{2}KT = \pi ab. \tag{28}$$

Now refer to Fig. 15.8. For an ellipse $r = f(\theta)$ with focus at the pole and major axis of length $2a$ along the x-axis as shown, you see that $2a = r|_{\theta=0} + r|_{\theta=\pi}$. From Eq. (26),

$$2a = \frac{K^2}{k - EK^2} + \frac{K^2}{k + EK^2} = \frac{2kK^2}{k^2 - E^2K^4}. \tag{29}$$

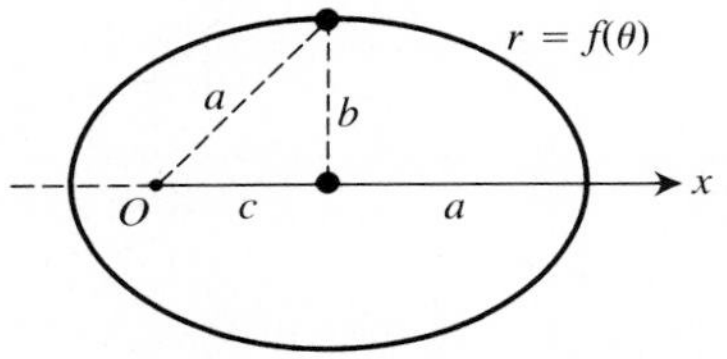

15.8

Letting b and c be as indicated in Fig. 15.8,

$$c = a - r|_{\theta=\pi} = \frac{kK^2}{k^2 - E^2K^4} - \frac{K^2}{k + EK^2} = \frac{kK^2}{k^2 - E^2K^4} - \frac{K^2(k - EK^2)}{k^2 - E^2K^4}$$
$$= \frac{EK^4}{k^2 - E^2K^4}. \tag{30}$$

A computation then yields

$$\frac{b^2}{K^2} = \frac{a^2 - c^2}{K^2} = \frac{k^2K^4 - E^2K^8}{K^2(k^2 - E^2K^4)^2} = \frac{K^2}{k^2 - E^2K^4} = \frac{a}{k}. \tag{31}$$

Equations (28) and (31) now yield

$$T^2 = 4\pi^2 a^2 \frac{b^2}{K^2} = 4\pi^2 \frac{a^3}{k},$$

so

$$\frac{a^3}{T^2} = \frac{k}{4\pi^2} = \frac{GM}{4\pi^2}, \tag{32}$$

which yields Kepler's third law, for $GM/(4\pi^2)$ is independent of the mass m of the planet orbiting the sun of mass M.

SUMMARY 1. *For unit vectors* $\boldsymbol{u}_r$ *and* $\boldsymbol{u}_\theta$ *in polar coordinates,*

$$\text{Position vector} = \boldsymbol{r} = r\boldsymbol{u}_r,$$

$$\text{Velocity vector} = \boldsymbol{v} = \dot{r}\boldsymbol{u}_r + r\dot{\theta}\boldsymbol{u}_\theta$$

$$\text{Acceleration vector} = \boldsymbol{a} = (\ddot{r} - r\dot{\theta}^2)\boldsymbol{u}_r + (r\ddot{\theta} + 2\dot{r}\dot{\theta})\boldsymbol{u}_\theta.$$

2. *Newton's and Kepler's Laws are given at the start of Section* 15.3.2.

EXERCISES

*__1.__ If the polar coordinate position of a body in the plane at time t is given by $(r, \theta) = (t^3 - 2t, t - 1)$, find the velocity and acceleration vectors of the body in terms of the unit vectors $\boldsymbol{u}_r$ and $\boldsymbol{u}_\theta$ at time $t = 1$.

*__2.__ Express the velocity and acceleration vectors found in Exercise 1 in terms of the cartesian unit vectors $\boldsymbol{i}$ and $\boldsymbol{j}$.

*__3.__ The value of the universal gravitational constant G was calculated by Cavendish in 1798. Explain how one might use Eq. (32) to calculate the mass of the sun.

*__4.__ Argue from Eq. (32) that the period of a body in an elliptical orbit in a given central force field can be found if the length of the major axis of the ellipse is known.

*__5.__ The *apogee* of an earth satellite is its maximum distance from the surface of the earth, and the *perigee* is the minimum distance from the surface. Use Eq. (32) to argue that the period of an earth satellite is completely determined by the apogee and perigee of the satellite. [*Hint.* Let the radius of the earth be R and the length of the major axis of the elliptical orbit be $2a$. Find a relation between R, a, the apogee, and the perigee.]

*__6.__ Would you expect the ratio a^3/T^2 of Eq. (32) to be the same for the earth in revolution about the sun as for a moon in revolution about the planet Jupiter? Explain.

The remaining exercises deal with the derivation of the inverse-square property and the formulation of Newton's law of universal gravitation from Kepler's laws. Suppose that planets describe planar orbits about the sun in accordance with Kepler's laws, and let the position of a certain planet with respect to a polar coordinate system with pole at the sun be given by $r = r(t)$, $\theta = \theta(t)$.

*__7.__ Deduce from Kepler's second law that if $A = A(t)$ is the area of the region swept out by the line segment joining the sun and the planet from time t_0 to time t, then $dA/dt = \beta$ for some constant β.

*__8.__ Deduce from Exercise 7 and the formula for area in polar coordinates that $r^2\dot{\theta} = 2\beta$.

*__9.__ Deduce from Exercise 8 that $r\ddot{\theta} + 2\dot{r}\dot{\theta} = 0$.

*__10.__ Deduce from Exercise 9 and Section 15.3.1 that the acceleration of the planet is toward the sun at all times. (This shows you are indeed in a "central force field" situation.)

*__11.__ Let orbit of the planet be an ellipse

$$r = \frac{B}{1 - e\cos\theta}$$

for $B > 0$. (This is Kepler's first law.) Show that

$$\dot{r}(1 - e\cos\theta) + re\dot{\theta}\sin\theta = 0.$$

*__12.__ Multiply the result of Exercise 11 by r and use Exercise 8 to show that

$$B\dot{r} + 2\beta e\sin\theta = 0.$$

*13. Deduce by differentiating the result in Exercise 12 and using Exercise 8 that

$$\ddot{r} = -\frac{4\beta^2}{r^2}\frac{e\cos\theta}{B} = \frac{4\beta^2}{r^2}\left(\frac{1}{r} - \frac{1}{B}\right).$$

*14. Deduce from Exercise 8 that $r\dot{\theta}^2 = 4\beta^2/r^3$, and conclude from Exercise 13 that

$$\ddot{r} - r\dot{\theta}^2 = -\frac{4\beta^2}{B}\cdot\frac{1}{r^2}.$$

*15. Use the result of Exercise 10 and Section 15.3.1 to argue that the force on a planet of mass m is of magnitude

$$\frac{4m\beta^2/B}{r^2}$$

and is directed toward the sun. (This gives the inverse-square property.)

*16. Let $2a$ be the length of the major axis and $2b$ the length of the minor axis of the ellipse in Exercise 11 and show that

$$B = a(1 - e^2) \quad\text{and}\quad b = a\sqrt{1 - e^2}.$$

*17. Show from Exercise 7 that if the planet has elliptical orbit of period T, then the area of the ellipse is βT.

*18. Deduce from Exercise 17 that $\beta = \pi ab/T$.

*19. Deduce from Exercises 16 and 18 that the force in Exercise 15 may be written as

$$\frac{4\pi^2 a^3 m}{T^2 r^2}.$$

*20. Conclude from Exercise 19 and Kepler's third law that the force of attraction of the sun per unit mass of planet is independent of the planet ("universal"), depending only on the distance from the planet to the sun.

*21. The ratio a^3/T^2 can be measured astronomically for various situations (a planet orbiting the sun or a moon orbiting a planet) in our solar system. Show that astronomical indications that a^3/T^2 is proportional to the mass at the center of the central force field, together with Exercises 19 and 20, would lead to prediction of Newton's universal law of gravitation.

*15.4 NORMAL VECTORS AND CURVATURE FOR SPACE CURVES

Let's assume that we are working with a smooth space curve with parametrizing functions having continuous derivatives of all orders that we desire to take.

*15.4.1 The principal normal and the curvature

Let $\boldsymbol{t}$ be the unit vector tangent to a smooth space curve in the direction of increasing t. The fact that $\boldsymbol{v}$ has length ds/dt and direction tangent to the curve may be summarized by

$$\boldsymbol{v} = \frac{ds}{dt}\boldsymbol{t}. \tag{1}$$

The unit tangent vector $\boldsymbol{t}$ does not necessarily have constant direction, but $\boldsymbol{t}$ does have constant length 1, so that

$$\boldsymbol{t}\cdot\boldsymbol{t} = 1. \tag{2}$$

Viewing the vector $\boldsymbol{t}$ as a function of the arc-length parameter s and

* This section may be omitted without loss of continuity.

differentiating (2) with respect to s,

$$\boldsymbol{t} \cdot \frac{d\boldsymbol{t}}{ds} + \frac{d\boldsymbol{t}}{ds} \cdot \boldsymbol{t} = 0,$$

so that $2\boldsymbol{t} \cdot (d\boldsymbol{t}/ds) = 0$ and

$$\boldsymbol{t} \cdot \frac{d\boldsymbol{t}}{ds} = 0. \tag{3}$$

Thus $d\boldsymbol{t}/ds$ is perpendicular to the unit tangent vector $\boldsymbol{t}$. If $d\boldsymbol{t}/ds \neq \mathbf{0}$, we let $\boldsymbol{n}$ be a unit vector in the direction of $d\boldsymbol{t}/ds$; then

$$\frac{d\boldsymbol{t}}{ds} = \kappa\boldsymbol{n} \tag{4}$$

for some positive constant κ. The vector $\boldsymbol{n}$ is the **principal normal** to the curve at this point, and by analogy with Eq. (4) of Section 15.2.1, we consider the constant κ in Eq. (4) to be the **curvature** of the curve at this point.

At each point of the curve, the perpendicular vectors $\boldsymbol{t}$ and $\boldsymbol{n}$ lie in a plane through the point; this is the **osculating plane** of the curve at this point. It can be shown that the osculating plane of the curve at each point is the plane that "most nearly contains" the curve in a neighborhood of the point. Note that the vector $\boldsymbol{n}$ points to the left or right as you face along $\boldsymbol{t}$ in the osculating plane, depending on whether the curve (regarded as being "in" the osculating plane near the point) bends to the left or right respectively.

Example 1 Let's turn again to the helix with position vector

$$\boldsymbol{r} = (a \cos t)\boldsymbol{i} + (a \sin t)\boldsymbol{j} + t\boldsymbol{k}.$$

Then

$$\boldsymbol{v} = \frac{d\boldsymbol{r}}{dt} = (-a \sin t)\boldsymbol{i} + (a \cos t)\boldsymbol{j} + \boldsymbol{k},$$

so

$$\frac{ds}{dt} = \sqrt{a^2 + 1}.$$

Thus we see that

$$\boldsymbol{t} = \frac{1}{\sqrt{a^2 + 1}}[-(a \sin t)\boldsymbol{i} + (a \cos t)\boldsymbol{j} + \boldsymbol{k}] \tag{5}$$

and

$$\frac{d\boldsymbol{t}}{ds} = \frac{dt}{ds}\frac{d\boldsymbol{t}}{dt} = \frac{1}{a^2 + 1}[-(a \cos t)\boldsymbol{i} - (a \sin t)\boldsymbol{j}]$$

$$= \frac{-a}{a^2 + 1}[(\cos t)\boldsymbol{i} + (\sin t)\boldsymbol{j}]. \tag{6}$$

From (6), you obtain

$$\kappa = \left|\frac{d\boldsymbol{t}}{ds}\right| = \frac{a}{a^2 + 1}$$

and

$$\boldsymbol{n} = -(\cos t)\boldsymbol{i} - (\sin t)\boldsymbol{j}. \tag{7}$$

The vector in (7) is directed toward the z-axis, parallel to the x,y-plane. ||

*15.4.2 The binormal vector and torsion

The *binormal vector* $\boldsymbol{b}$ at a point on a space curve where $\kappa \neq 0$ is defined by

$$\boldsymbol{b} = \boldsymbol{t} \times \boldsymbol{n}.$$

Therefore

$$|\boldsymbol{b}| = |\boldsymbol{t}|\,|\boldsymbol{n}| \sin 90° = 1,$$

so $\boldsymbol{b}$ is a unit vector perpendicular to both $\boldsymbol{t}$ and $\boldsymbol{n}$. The sequence

$$\boldsymbol{t}, \quad \boldsymbol{n}, \quad \boldsymbol{b}$$

of vectors is a righthand triple of vectors at each point on the space curve, and may be regarded as a local-coordinate 3-frame at each point on the curve (see Fig. 15.9). To a bug crawling in the curve in the direction of increasing t so that $\boldsymbol{t}$ points "straight ahead" and $\boldsymbol{n}$ points "left," it appears that $\boldsymbol{b}$ points "up." Of course,

$$\boldsymbol{b} = \boldsymbol{t} \times \boldsymbol{n}, \qquad \boldsymbol{t} = \boldsymbol{n} \times \boldsymbol{b}, \qquad \text{and} \qquad \boldsymbol{n} = \boldsymbol{b} \times \boldsymbol{t}. \tag{8}$$

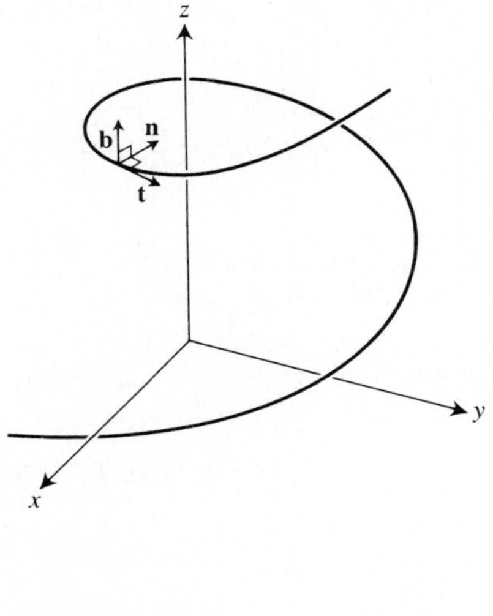

15.9

Example 2 For the helix in Example 1, you see from the example that

$$b = t \times n = \frac{1}{\sqrt{a^2+1}}\begin{vmatrix} i & j & k \\ -a\sin t & a\cos t & 1 \\ -\cos t & -\sin t & 0 \end{vmatrix}$$

$$= \frac{1}{\sqrt{a^2+1}}[(\sin t)i - (\cos t)j + ak]. \quad \|$$

Viewing t, n, and b as functions of the arc length s and differentiating $b = t \times n$ with respect to s, using the product rule,

$$\frac{db}{ds} = \frac{d(t \times n)}{ds} = t \times \frac{dn}{ds} + \frac{dt}{ds} \times n$$

$$= t \times \frac{dn}{ds} + (\kappa n) \times n = t \times \frac{dn}{ds}. \tag{9}$$

By differentiating the equation $n \cdot n = 1$, you obtain, just as we obtained Eq. (3),

$$\frac{dn.}{ds} \cdot n = 0,$$

so n is perpendicular to dn/ds. From (9), you see that db/ds is perpendicular to both t and dn/ds, and is therefore parallel to n. Hence

$$\frac{db}{ds} = -\tau n \tag{10}$$

for some (positive or negative) constant τ. The vector b is direction vector for the osculating plane, so db/ds measures the rate at which the osculating plane twists per unit change in arc length; the number τ is therefore called the *torsion* of the curve at the point. The minus sign is introduced in (10) so that positive torsion corresponds to the vector b turning toward the right in the direction of $-n$ (like a right-threaded screw) as you travel the curve in the direction of increasing t.

Example 3 For the helix in Examples 1 and 2, you obtain

$$\frac{db}{ds} = \frac{dt}{ds}\frac{db}{dt} = \frac{1}{a^2+1}[(\cos t)i + (\sin t)j] = -\frac{1}{a^2+1}n.$$

Thus the torsion of the helix is constant, independent of the point on the curve, and is given by

$$\tau = \frac{1}{a^2+1}. \quad \|$$

***15.4.3 Formulas for computing κ and τ**

The sequence of Exercises 18 through 21 leads to the formulas

$$\kappa = \frac{|\dot{\boldsymbol{r}} \times \ddot{\boldsymbol{r}}|}{|\dot{\boldsymbol{r}}|^3} \quad \text{and} \quad \tau = \frac{(\dot{\boldsymbol{r}} \times \ddot{\boldsymbol{r}}) \cdot \dddot{\boldsymbol{r}}}{|\dot{\boldsymbol{r}} \times \ddot{\boldsymbol{r}}|^2}. \tag{11}$$

For the helix in the previous examples, we found κ and τ easily from their definitions, but for some more complicated examples, the formulas (11) may be much easier.

Example 4 Let's find the curvature κ and torsion τ for the curve with position vector

$$\boldsymbol{r} = t^2\boldsymbol{i} - (3t + 1)\boldsymbol{j} + t^3\boldsymbol{k}$$

at the point where $t = 1$.

SOLUTION Now

$$\begin{aligned} \dot{\boldsymbol{r}} &= 2t\boldsymbol{i} - 3\boldsymbol{j} + 3t^2\boldsymbol{k}, \\ \ddot{\boldsymbol{r}} &= 2\boldsymbol{i} + 6t\boldsymbol{k}, \\ \dddot{\boldsymbol{r}} &= 6\boldsymbol{k}. \end{aligned}$$

Thus at $t = 1$,

$$\begin{aligned} \dot{\boldsymbol{r}} &= 2\boldsymbol{i} - 3\boldsymbol{j} + 3\boldsymbol{k}, \\ \ddot{\boldsymbol{r}} &= 2\boldsymbol{i} + 6\boldsymbol{k}, \\ \dddot{\boldsymbol{r}} &= 6\boldsymbol{k}; \end{aligned}$$

$$\dot{\boldsymbol{r}} \times \ddot{\boldsymbol{r}} = \begin{vmatrix} \boldsymbol{i} & \boldsymbol{j} & \boldsymbol{k} \\ 2 & -3 & 3 \\ 2 & 0 & 6 \end{vmatrix} = -18\boldsymbol{i} - 6\boldsymbol{j} + 6\boldsymbol{k},$$

$$|\dot{\boldsymbol{r}} \times \ddot{\boldsymbol{r}}| = 6\sqrt{9 + 1 + 1} = 6\sqrt{11},$$

$$|\dot{\boldsymbol{r}}| = \sqrt{4 + 9 + 9} = \sqrt{22}.$$

Hence

$$\kappa = \frac{|\dot{\boldsymbol{r}} \times \ddot{\boldsymbol{r}}|}{|\dot{\boldsymbol{r}}|^3} = \frac{6\sqrt{11}}{22\sqrt{22}} = \frac{3}{11\sqrt{2}}$$

and

$$\begin{aligned} \tau = \frac{(\dot{\boldsymbol{r}} \times \ddot{\boldsymbol{r}}) \cdot \dddot{\boldsymbol{r}}}{|\dot{\boldsymbol{r}} \times \ddot{\boldsymbol{r}}|^2} &= \frac{(-18)(0) + (-6)(0) + (6)(6)}{36 \cdot 11} \\ &= \frac{36}{36 \cdot 11} = \frac{1}{11}. \quad \| \end{aligned}$$

***15.4.4 The Frenet formulas**

From $\boldsymbol{n} = \boldsymbol{b} \times \boldsymbol{t}$, you obtain, upon differentiating,

$$\frac{d\boldsymbol{n}}{ds} = \boldsymbol{b} \times \frac{d\boldsymbol{t}}{ds} + \frac{d\boldsymbol{b}}{ds} \times \boldsymbol{t}. \tag{12}$$

Making use of Eqs. (4), (8), and (10), you obtain

$$\frac{d\boldsymbol{n}}{ds} = \boldsymbol{b} \times (\kappa\boldsymbol{n}) - (\tau\boldsymbol{n}) \times \boldsymbol{t} = \kappa(\boldsymbol{b} \times \boldsymbol{n}) - \tau(\boldsymbol{n} \times \boldsymbol{t}) = -\kappa\boldsymbol{t} + \tau\boldsymbol{b}. \tag{13}$$

Equations (4), (10), and (13) are known as the "*Frenet formulas.*" Let's collect them in one place.

Frenet Formulas

$$\frac{d\boldsymbol{t}}{ds} = \kappa\boldsymbol{n}, \qquad \frac{d\boldsymbol{n}}{ds} = -\kappa\boldsymbol{t} + \tau\boldsymbol{b}, \qquad \frac{d\boldsymbol{b}}{ds} = -\tau\boldsymbol{n}.$$

SUMMARY

1. *For a space curve with unit tangent vector* $\boldsymbol{t}$ *so that* $\boldsymbol{v} = (ds/dt)\boldsymbol{t}$, *the principal normal (unit) vector* $\boldsymbol{n}$ *and curvature* κ *are defined by*

$$\frac{d\boldsymbol{t}}{ds} = \kappa\boldsymbol{n}.$$

2. *The (unit) binormal vector is* $\boldsymbol{b} = \boldsymbol{t} \times \boldsymbol{n}$.
3. *The torsion* τ *of the curve is defined by*

$$\frac{d\boldsymbol{b}}{ds} = -\tau\boldsymbol{n}.$$

4. *Curvature* κ *and torsion* τ *may also be found from the position vector* $\boldsymbol{r}$ *using*

$$\kappa = \frac{|\dot{\boldsymbol{r}} \times \ddot{\boldsymbol{r}}|}{|\dot{\boldsymbol{r}}|^3} \quad \text{and} \quad \tau = \frac{(\dot{\boldsymbol{r}} \times \ddot{\boldsymbol{r}}) \cdot \dddot{\boldsymbol{r}}}{|\dot{\boldsymbol{r}} \times \ddot{\boldsymbol{r}}|^2}.$$

5. *The Frenet formulas are*

$$\frac{d\boldsymbol{t}}{ds} = \kappa\boldsymbol{n},$$

$$\frac{d\boldsymbol{n}}{ds} = -\kappa\boldsymbol{t} + \tau\boldsymbol{b},$$

$$\frac{d\boldsymbol{b}}{ds} = -\tau\boldsymbol{n}.$$

EXERCISES

*1. Find the curvature and torsion of a straight line in space.

Exercises 2 through 9 are concerned with the space curve

$$x = 2t, \quad y = t^2, \quad z = -t^2.$$

*2. Find the velocity and acceleration vectors when $t = 1$.

*3. Find the speed of a body on the curve, thinking of t as time parameter, when $t = 2$.

*4. Find the length of the curve from $t = 0$ to $t = 2$.

*5. Find the vectors $\boldsymbol{t}$, $\boldsymbol{n}$, and $\boldsymbol{b}$ at a point on the curve in terms of the parameter t.

*6. Find the curvature κ of the curve in terms of the parameter t.

*7. Find the equation of the osculating plane to the curve for a value t_0 of the parameter.

*8. Find the torsion τ of the curve in terms of the parameter t.

*9. From the answers to the two preceding exercises, it appears that the curve lies in a plane. Deduce this directly from the equations of the curve.

Exercises 10 through 16 are concerned with the twisted cubic

$$x = 3t^2, \quad y = 2t^3, \quad z = 3t.$$

*10. Find the velocity and acceleration vectors in terms of the parameter t.

*11. Find the speed of a body on the curve when $t = 2$, regarding t as time parameter.

*12. Find the length of the curve from $t = 0$ to $t = 4$.

*13. Find the vectors $\boldsymbol{t}$ and $\boldsymbol{n}$ at the point on the curve where $t = 2$.

*14. Find the equation of the osculating plane to the curve where $t = 2$.

*15. Find the curvature κ of the curve where $t = 2$.

*16. Find the binormal vector $\boldsymbol{b}$ and torsion τ of the curve where $t = 2$.

*17. Show that if we let $\boldsymbol{\delta} = \tau\boldsymbol{t} + \kappa\boldsymbol{b}$ (the *Darboux vector*), then the Frenet formulas take the symmetric form

$$\frac{d\boldsymbol{t}}{ds} = \boldsymbol{\delta} \times \boldsymbol{t}, \quad \frac{d\boldsymbol{n}}{ds} = \boldsymbol{\delta} \times \boldsymbol{n}, \quad \frac{d\boldsymbol{b}}{ds} = \boldsymbol{\delta} \times \boldsymbol{b}.$$

In Exercises 18 through 21, we denote a derivative with respect to t by a dot over the vector or variable. We assume we are working with a space curve whose coordinate functions are differentiable as often as we like.

*18. Show that $\ddot{\boldsymbol{r}} = \ddot{s}\boldsymbol{t} + \dot{s}^2\kappa\boldsymbol{n}$. [*Hint.* Differentiate $\dot{\boldsymbol{r}} = \dot{s}\boldsymbol{t}$.]

*19. Show that $\dddot{\boldsymbol{r}} = (\dddot{s} - \dot{s}^3\kappa^2)\boldsymbol{t} + (3\dot{s}\ddot{s}\kappa + \dot{s}^2\dot{\kappa})\boldsymbol{n} + \dot{s}^3\kappa\tau\boldsymbol{b}$. [*Hint.* Differentiate the result in Exercise 18.]

*20. Deduce from the preceding two exercises that

a) $\dot{\boldsymbol{r}} \times \ddot{\boldsymbol{r}} = \dot{s}^3\kappa\boldsymbol{b}$, b) $(\dot{\boldsymbol{r}} \times \ddot{\boldsymbol{r}}) \cdot \dddot{\boldsymbol{r}} = \dot{s}^6\kappa^2\tau$.

[*Hint.* Since $\boldsymbol{t}$, $\boldsymbol{n}$, $\boldsymbol{b}$, form a righthand perpendicular unit 3-frame at each point, they may be used in the role of $\boldsymbol{i}$, $\boldsymbol{j}$, and $\boldsymbol{k}$ in computations of the cross product.]

*21. Deduce from the preceding exercise that

a) $\kappa = \dfrac{|\dot{\boldsymbol{r}} \times \ddot{\boldsymbol{r}}|}{|\dot{\boldsymbol{r}}|^3}$, b) $\tau = \dfrac{(\dot{\boldsymbol{r}} \times \ddot{\boldsymbol{r}}) \cdot \dddot{\boldsymbol{r}}}{|\dot{\boldsymbol{r}} \times \ddot{\boldsymbol{r}}|^2}$.

*22. Use the results in the preceding exercise to find the curvature κ and torsion τ of the curve

$$x = \cos t, \quad y = e^{2t}, \quad z = (t + 1)^3$$

when $t = 0$.

*23. Show that if the curvature of a smooth space curve is zero at each point, then the curve is a straight line.

*24. Show that a space curve with torsion zero at each point lies in a plane. [*Hint.* Deduce that $\boldsymbol{b}$ is a constant vector. Show that $d(\boldsymbol{b} \cdot \boldsymbol{r})/ds = 0$, and conclude that $\boldsymbol{b} \cdot (\boldsymbol{r}(t) - \boldsymbol{r}(t_0)) = 0$, so that the curve lies in the plane through the point where $t = t_0$ having orthogonal vector $\boldsymbol{b}$.]

*25. (*Fundamental Theorem*) Show that two space curves having the same curvature and torsion for each value of the arc-length parameter s are congruent. [*Hint.* You may suppose that the position vectors of the curves are given by $\boldsymbol{r}(s)$ and $\bar{\boldsymbol{r}}(s)$, and that

$$\boldsymbol{r}(0) = \bar{\boldsymbol{r}}(0), \quad \boldsymbol{t}(0) = \bar{\boldsymbol{t}}(0), \quad \boldsymbol{n}(0) = \bar{\boldsymbol{n}}(0),$$
and
$$\boldsymbol{b}(0) = \bar{\boldsymbol{b}}(0).$$

You must then show that $\boldsymbol{r}(s) = \bar{\boldsymbol{r}}(s)$. Set $w = \boldsymbol{t} \cdot \bar{\boldsymbol{t}} + \boldsymbol{n} \cdot \bar{\boldsymbol{n}} + \boldsymbol{b} \cdot \bar{\boldsymbol{b}}$, and show that $dw/ds = 0$. Show that $w(0) = 3$, and conclude that

$$\boldsymbol{t} = \bar{\boldsymbol{t}}, \quad \boldsymbol{n} = \bar{\boldsymbol{n}}, \quad \text{and} \quad \boldsymbol{b} = \bar{\boldsymbol{b}}$$

for all values of s. From $\boldsymbol{t} = \bar{\boldsymbol{t}}$, deduce that $\boldsymbol{r} = \bar{\boldsymbol{r}} + \mathbf{c}$, and, taking $s = 0$, that $\mathbf{c} = \mathbf{0}$, so that $\boldsymbol{r}(s) = \bar{\boldsymbol{r}}(s)$.]

exercise sets for chapter 15

review exercise set 15.1

1. Let the position of a body on a plane curve at time t be given parametrically by $x = \sin 2t$, $y = \cos t$.

a) Give the position vector $\boldsymbol{r}$ of the body at time t.

b) Find the velocity vector $\boldsymbol{v}$ at time $t = \pi/4$.

c) Find the speed at time $t = \pi/4$.

d) Find the acceleration vector of the body at time $t = \pi/4$.

2. For the body in Problem 1, find:

a) The normal and tangential components of the acceleration at time $t = \pi/4$.

b) The curvature of the path at time $t = \pi/4$.

*3. If the polar coordinate position of a body in the plane at time t is given by $(r, \theta) = (\sqrt{t}, \sqrt{3t + 4})$, find the velocity and acceleration vectors in terms of the unit vectors $\boldsymbol{u}_r$ and $\boldsymbol{u}_\theta$ at time $t = 4$.

*4. State Kepler's Laws.

*5. What type of curve is traveled by a body moving subject to a central force field?

*6. Explain how the curvature of a space curve is defined.

7. Consider a body in space whose position at time t is given by $x = t$, $y = 3 \sin t$, $z = -3 \cos t$.

a) Find the velocity vector $\boldsymbol{v}$ at time $t = \pi/4$.

b) Find the speed at time $t = \pi/4$.

c) Find the acceleration at time $t = \pi/4$.

*8. For the body in Problem 7, find:

a) The unit normal vector $\boldsymbol{n}$ at time $t = \pi/4$.

b) The unit binormal vector $\boldsymbol{b}$ at time $t = \pi/4$.

*9. For the body in Problem 7, find:

a) The curvature κ of the curve at time $t = \pi/4$.

b) The torsion τ of the curve at time $t = \pi/4$.

review exercise set 15.2

1. Let the position vector of a body in the plane be given by

$$\boldsymbol{r} = \sqrt{2t + 1}\boldsymbol{i} + \sqrt{t}\boldsymbol{j}.$$

a) Find the velocity vector $\boldsymbol{v}$ at time $t = 4$.

b) Find the speed at time $t = 4$.

c) Find the acceleration vector $\boldsymbol{a}$ at time $t = 4$.

2. Let a body be moving in the plane subject to some force. Show that if

$$\frac{d^2s/dt^2}{\kappa(ds/dt)^2} = 1$$

at all times t, then the force must be directed at an angle of 45° to the direction of motion of the body at every time t.

3. For a body traveling a circular track at 20 ft/sec, a force of 500 lbs perpendicular to the track is required to keep the body from leaving the track. If the maximum force the track can exert against the body in this perpendicular direction is 4500 lbs, how fast can the body travel without leaving the track?

*4. Give the polar formula for $\boldsymbol{u}_r$ and $\boldsymbol{u}_\theta$ components of acceleration for a body traveling a curve in the plane.

*5. State Newton's Law of Universal Gravitation.

*6. Two satellites A and B are traveling elliptic orbits about the same body. If the major axis of A's orbit is four times the length of the major axis of B's orbit, and if A's period is 48 hours, find B's period.

*7. Explain how the torsion of a space curve is defined.

8. Consider a body in space whose position at time t is given by

$$x = t^3, \qquad y = 2t, \qquad z = t^2.$$

a) Find the velocity vector $\boldsymbol{v}$ at time $t = -1$.

b) Find the speed at time $t = -1$.

c) Find the acceleration vector $\boldsymbol{a}$ at time $t = -1$.

*9. For the body in Problem 8, find the curvature κ and torsion τ at time $t = -1$.

differential calculus of several variables

16

16.1 PARTIAL DERIVATIVES

16.1.1 Graphs of functions of two variables

Continuity

Let $z = f(x, y)$ be a function of two independent variables x and y. For example, perhaps $z = f(x, y) = x^2 - 3xy$. All points (x, y, z) in space such that $z = f(x, y)$ form the graph of the function. Such a function is **continuous at a point** (x_0, y_0) in its domain if sufficiently small changes Δx in x and Δy in y produce only very small changes Δz in z. More precisely, given $\epsilon > 0$, you should be able to find a $\delta > 0$ such that $|\Delta z| < \epsilon$ provided that $\sqrt{(\Delta x)^2 + (\Delta y)^2} < \delta$. Of course, $\sqrt{(\Delta x)^2 + (\Delta y)^2}$ is simply the distance from the changed point $(x_0 + \Delta x, y_0 + \Delta y)$ to the old point (x_0, y_0), while $\Delta z = f(x_0 + \Delta x, y_0 + \Delta y) - f(x_0, y_0)$. The function is **continuous** if it is continuous at each point in its domain. Once again, all the elementary functions (rational functions, trigonometric functions, exponential and logarithmic functions) are continuous.

We shall not talk any more about continuity, but simply state that, if a function is continuous, then its graph can be regarded as a surface lying over (or sometimes under, depending on the sign of z) its domain in the x,y-plane. A picture of such a surface is shown in Fig. 16.1.

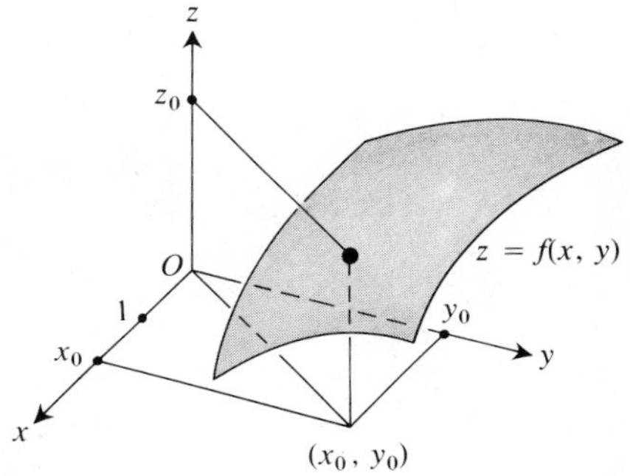

16.1

Example 1 Since $x^2 + y^2 + z^2 = a^2$ is the equation of a sphere with center at the origin and radius a you see that $z = f(x, y) = \sqrt{a^2 - x^2 - y^2}$ has the upper hemisphere as its graph. ||

Example 2 Of course, $z = f(x, y) = 2x - 3y$ has as graph a plane through the origin. Since the plane can be written $2x - 3y - z = 0$, you see it has as normal vector $2\boldsymbol{i} - 3\boldsymbol{j} - \boldsymbol{k}$. Of course, $z = ax + by + c$ always has as graph a plane in space. ||

16.1.2 Partial derivatives

If $y = f(x)$, the derivative dy/dx gives the rate of change of y with respect to x. For a function of one variable, there are only two directions you can travel from a point in the domain, in the direction of increasing x or in the direction of decreasing x. If $z = f(x, y)$, you can also attempt to find some rates of change. This time there are many directions you can go from a point (x_0, y_0) in the domain, and we shall eventually show how to find the rate at which z changes in each of these directions. Let's first find the rate of change in the direction parallel to the x-axis given by increasing x as y is held constant. That is, let's find the rate of change at (x_0, y_0) in the direction given by the vector $\boldsymbol{i}$ at that point. Suppose $z = f(x, y)$ has as graph the surface shown in Fig. 16.2. The portion of the surface over the line $y = y_0$ in the x,y-plane is a curve on the surface, as shown in Fig. 16.2. On the curve, the height z appears as a function $g(x)$ of x only, as indicated in Fig. 16.3. This function $g(x)$ is, of course, found by setting $y = y_0$ in $z = f(x, y)$, that is,

$$z = g(x) = f(x, y_0).$$

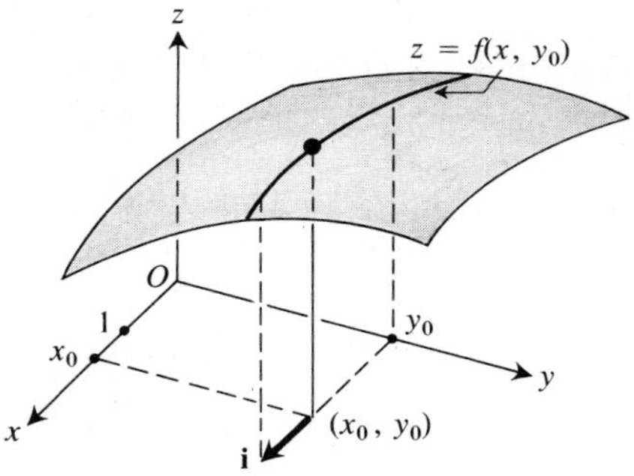

16.2

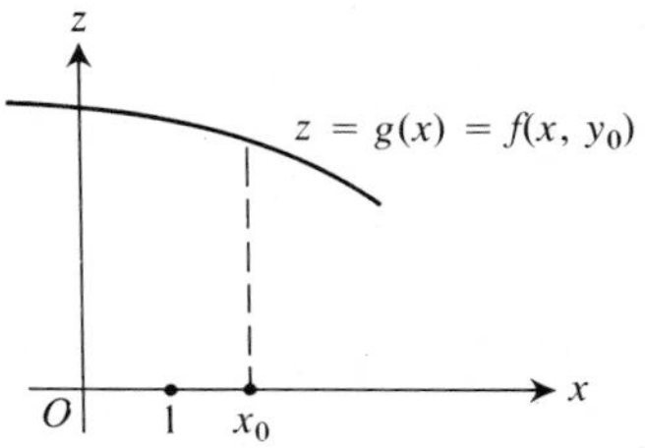

16.3

Partial derivatives

Since you have a function of one variable $z = g(x)$, you can easily find the rate of change $g'(x)$ of z with respect to x. This time we denote the rate of change by $\partial z/\partial x$ rather than dz/dx. The "round d's" remind us that z is actually a function of more than one variable. We call $\partial z/\partial x$ the **partial derivative** of z with respect to x. The value of this derivative at $x = x_0$ (recall that $y = y_0$) is written

$$\left.\frac{\partial z}{\partial x}\right|_{(x_0,y_0)} \quad \text{or} \quad \left.\frac{\partial f}{\partial x}\right|_{(x_0,y_0)} \quad \text{or} \quad f_x(x_0, y_0).$$

You use the function $z = h(y) = f(x_0, y)$ obtained by holding x constant at x_0 to find the rate of change $h'(y)$ of $z = f(x, y)$ at (x_0, y_0) in the direction given by the vector $\boldsymbol{j}$ at that point. The notations are

$$\left.\frac{\partial z}{\partial y}\right|_{(x_0,y_0)} \quad \text{or} \quad \left.\frac{\partial f}{\partial y}\right|_{(x_0,y_0)} \quad \text{or} \quad f_y(x_0, y_0).$$

Of course the derivative of $g(x) = f(x, y_0)$ is defined using a limit. This is the way the limit appears in the definition of the partial derivatives.

Definition 16.1 Let $f(x, y)$ be defined everywhere inside some sufficiently small circle with (x_0, y_0) as center. Then

$$\left.\frac{\partial f}{\partial x}\right|_{(x_0,y_0)} = f_x(x_0, y_0) = \lim_{\Delta x \to 0} \frac{f(x_0 + \Delta x, y_0) - f(x_0, y_0)}{\Delta x}$$

and

$$\left.\frac{\partial f}{\partial y}\right|_{(x_0,y_0)} = f_y(x_0, y_0) = \lim_{\Delta y \to 0} \frac{f(x_0, y_0 + \Delta y) - f(x_0, y_0)}{\Delta y}$$

if these limits exist.

16.1.3 Computation of partial derivatives

We can regard $f_x(x_0, y_0)$ as the derivative at x_0 of the function g given by $g(x) = f(x, y_0)$, that is, the function of one real variable x obtained from f by keeping $y = y_0$ and allowing only x to vary. This means that partial derivatives can be computed using the techniques for finding derivatives of functions of one variable. This is illustrated in the following examples.

Example 3 Let f be the polynomial function given by $f(x, y) = x^2 + 3xy + 2y^3$. Let's find

$$f_x(1, 2) = \left.\frac{\partial f}{\partial x}\right|_{(1,2)} \quad \text{and} \quad f_y(1, 2) = \left.\frac{\partial f}{\partial y}\right|_{(1,2)}.$$

SOLUTION Now $f_x(1, 2)$ can be viewed as the derivative at $x = 1$ of the function g obtained by setting $y = 2$ in the polynomial $x^2 + 3xy + 2y^3$. That is,

$$g(x) = x^2 + 3x(2) + 2(2^3) = x^2 + 6x + 16.$$

The derivative of $x^2 + 6x + 16$ is $2x + 6$, and the value of $2x + 6$ when $x = 1$ is 8. Hence

$$f_x(1, 2) = \left.\frac{\partial f}{\partial x}\right|_{(1,2)} = 8.$$

Similarly, putting $x = 1$, you find that $f_y(1, 2)$ is the derivative, when $y = 2$, of $1 + 3y + 2y^3$. Since.

$$\frac{d(1 + 3y + 2y^3)}{dy} = 3 + 6y^2,$$

you obtain

$$f_y(1, 2) = \frac{\partial f}{\partial y}(1, 2) = 3 + 6 \cdot 2^2 = 27. \quad \|$$

Example 4 You could simplify the computation of the partial derivatives in Example 3 by noting that you can compute $f_x(x, y)$ for any point (x, y) by differentiating $x^2 + 3xy + 2y^3$ "with respect to x only," treating y as a constant. The notation "$\partial f/\partial x$" is a practical way to indicate this derivative "with respect to x." Thus

$$\frac{\partial f}{\partial x} = \frac{\partial(x^2 + 3xy + 2y^3)}{\partial x} = 2x + 3y,$$

for $\partial(2y^3)/\partial x = 0$, since you think of y as a constant. Similarly,

$$\frac{\partial f}{\partial y} = \frac{\partial(x^2 + 3xy + 2y^3)}{\partial y} = 3x + 6y^2.$$

Therefore

$$\frac{\partial f}{\partial x}(1, 2) = 2 \cdot 1 + 3 \cdot 2 = 8 \quad \text{and} \quad \frac{\partial f}{\partial y}(1, 2) = 3 \cdot 1 + 6 \cdot 2^2 = 27. \quad \|$$

In practice, you always find partial derivatives by the technique illustrated in Example 4, and the ∂-notation is obviously suggestive here; you differentiate "with respect to x" or "with respect to y." For instance, to find $\partial f/\partial x$, just differentiate f with respect to x only, pretending that y is constant. Practice is given in the exercises.

16.1.4 Functions of more variables

If (x_0, y_0, z_0) is a point in the domain of a function $w = f(x, y, z)$, then you can try to find the partial derivatives of w in the directions of $\boldsymbol{i}$, $\boldsymbol{j}$, and $\boldsymbol{k}$ at (x_0, y_0, z_0). To find the derivative in the direction given by $\boldsymbol{i}$, you should set $y = y_0$ and $z = z_0$, allowing only x to vary. That is, you hold y and z constant and differentiate just with respect to x to find

$$\left.\frac{\partial w}{\partial x}\right|_{(x_0, y_0, z_0)} \quad \text{or} \quad f_x(x_0, y_0, z_0).$$

Similarly, you can find $\partial w/\partial y$ and $\partial w/\partial z$.

Example 5 We find the three partial derivatives for $w = x \sin yz$.

SOLUTION You easily see that

$$\frac{\partial w}{\partial x} = \sin yz, \qquad \frac{\partial w}{\partial y} = xz \cos yz, \qquad \frac{\partial w}{\partial z} = xy \cos yz. \quad \|$$

For functions of two or more variables, it is sometimes handy to use the subscripted notation presented in Chapter 14, where a point in the plane is (x_1, x_2) and a point in space is (x_1, x_2, x_3). The vector notation $\boldsymbol{x}$ may be used to represent either of these points; the dimension should be clear from the context. Then a function of three variables may be written

$$y = f(\boldsymbol{x}) = f(x_1, x_2, x_3).$$

Of course, then you can compute all the partial derivatives

$$\frac{\partial y}{\partial x_1}, \quad \frac{\partial y}{\partial x_2}, \quad \frac{\partial y}{\partial x_3},$$

and $\partial y/\partial x_i$ is also written $\partial f/\partial x_i$ or $f_{x_i}(x_1, x_2, x_3)$.

16.1.5 Higher-order derivatives

Let $z = f(x, y)$. Then $\partial z/\partial x = f_x(x, y)$ is again a function of two variables and you can attempt to compute its partial derivatives with respect to either x or y. These are *second-order derivatives* of our original function $f(x, y)$. The notations are

$$\frac{\partial}{\partial x}\left(\frac{\partial z}{\partial x}\right) = \frac{\partial^2 z}{\partial x^2} = f_{xx}(x, y) \quad \text{and} \quad \frac{\partial}{\partial y}\left(\frac{\partial z}{\partial x}\right) = \frac{\partial^2 z}{\partial y\, \partial x} = f_{xy}(x, y).$$

Of course, you could equally well find first $\partial z/\partial y$ and then the second-order derivatives

$$\frac{\partial}{\partial y}\left(\frac{\partial z}{\partial y}\right) = \frac{\partial^2 z}{\partial y^2} = f_{yy}(x, y) \quad \text{and} \quad \frac{\partial}{\partial x}\left(\frac{\partial z}{\partial y}\right) = \frac{\partial^2 z}{\partial x\, \partial y} = f_{yx}(x, y).$$

Note that f_{xy} means $(f_x)_y$ while f_{yx} means $(f_y)_x$.

Equality of mixed partial derivatives

It is a theorem that, if the function has continuous second partial derivatives, then the "mixed" partial derivatives are equal, that is,

$$\frac{\partial^2 z}{\partial y\, \partial x} = \frac{\partial^2 z}{\partial x\, \partial y}.$$

This hypothesis of continuity is true for all elementary functions with which we shall work.

Of course, you can keep taking derivatives. Thus $\partial^3 f/\partial y^2\, \partial x$ means the third partial derivative of $f(x, y)$, first with respect to x, and then twice more with respect to y. We have

$$\frac{\partial^3 f}{\partial y^2\, \partial x} = \frac{\partial^3 f}{\partial y\, \partial x\, \partial y} = \frac{\partial^3 f}{\partial x\, \partial y^2}$$

for functions with continuous partial derivatives of order 3. Similar notations are used for functions of more than two variables, and again,

mixed partial derivatives of the same order are equal if all partial derivatives of that order are continuous and the total number of differentiations with respect to each variable is the same.

Example 6 We illustrate that $f_{xy} = f_{yx}$ for $f(x, y) = \sin(x^2y)$.

SOLUTION Now

$$f_x(x, y) = 2xy \cos(x^2y),$$

so

$$f_{xy}(x, y) = -2x^3y \sin(x^2y) + 2x \cos(x^2y).$$

Differentiating in the other order,

$$f_y(x, y) = x^2 \cos(x^2y),$$

so

$$f_{yx}(x, y) = -2x^3y \sin(x^2y) + 2x \cos(x^2y).$$

Thus $f_{xy} = f_{yx}$. ||

SUMMARY

1. *If $z = f(x, y)$, then $\partial z/\partial x = f_x(x, y)$ is the partial derivative of $f(x, y)$ in the direction* **i** *corresponding to increasing x as y is held constant. It is computed by the usual differentiation methods, regarding y as a constant. The partial derivative $\partial z/\partial y = f_y(x, y)$ is similarly defined and computed.*
2. *If $y = f(\mathbf{x}) = f(x_1, x_2, x_3)$, then $\partial y/\partial x_i$ is computed by differentiating with respect to x_i only, treating all other variables as though they were constants in the differentiation.*
3. *If $z = f(x, y)$, then second-order partial derivatives are*

$$\frac{\partial}{\partial x}\left(\frac{\partial z}{\partial x}\right) = \frac{\partial^2 z}{\partial x^2} = f_{xx}(x, y),$$

$$\frac{\partial}{\partial y}\left(\frac{\partial z}{\partial y}\right) = \frac{\partial^2 z}{\partial y^2} = f_{yy}(x, y),$$

together with the mixed partial derivatives

$$\frac{\partial}{\partial y}\left(\frac{\partial z}{\partial x}\right) = \frac{\partial^2 z}{\partial y\, \partial x} = f_{xy}(x, y)$$

and

$$\frac{\partial}{\partial x}\left(\frac{\partial z}{\partial y}\right) = \frac{\partial^2 z}{\partial x\, \partial y} = f_{yx}(x, y).$$

For all the functions that we shall encounter, we have $\partial^2 z/\partial x\, \partial y = \partial^2 z/\partial y\, \partial x$. Similar notations are used for more variables and higher-order derivatives. For all the functions we shall use, mixed partials of the same order are equal if differentiation with respect to each variable occurs the same total number of times. For example, if $w = f(x, y, z)$, we have $\partial^4 w/\partial x\, \partial y\, \partial x\, \partial z = \partial^4 w/\partial z\, \partial y\, \partial x^2$.

EXERCISES

In Exercises 1 through 21, find $f_x(x, y)$ and $f_y(x, y)$ for the given function $f(x, y)$. You need not simplify the answers.

1. $3x + 4y$ **2.** xy **3.** $x^2 + y^2$

4. $xy^3 + x^2y^2$ **5.** $e^{x/y}$ **6.** $\dfrac{x^2 + 3x + 1}{y}$

7. $xy^2 + \dfrac{3x^2}{y^3}$ **8.** $(3xy^2 - 2x^2y)^3$ **9.** $(x^2 + 2xy)(y^3 + x^2)$

10. $(xy)^3(x^2 - y^3)^2$ **11.** $\dfrac{x^2 + y^2}{x^2 + y}$ **12.** $\sin xy$

13. $\tan(x^2 + y^2)$ **14.** $e^{xy} \cos x^2$ **15.** $e^{xy^2} \sec(x^2y)$

16. $\ln(2x + 3y)$ **17.** $\ln(2x + y) \cdot \cot y^2$ **18.** $(\sin x^2 + \cos y^2)^5$

19. $y \sec^3 x + xy^2$ **20.** $\ln(\sin(xy))$ **21.** $\tan^{-1}(xy^2)$

In Exercises 22 through 27, find the indicated partial derivative of the given function f. You need not simplify the answers.

22. xyz; $\partial^2 f/\partial z^2$ **23.** x^2yz; $\partial^2 f/\partial z\, \partial x$ **24.** $\dfrac{xy}{z}$; f_{xyz} **25.** xe^{yz}; f_{zx}

26. $\dfrac{x^3 + y}{y^3 + yz}$; $\partial^4 f/\partial x^4$ **27.** $\ln[\cos(2x + y - 3z)]$; $\partial^3 f/\partial z\, \partial x^2$

28. Let $f(x, y) = \sin xy$. Show that $x^2 \cdot f_{xx} = y^2 \cdot f_{yy}$.

29. Verify that $f_{xy} = f_{yx}$ for the given function f.

a) x^2y b) $x^3y^2 + \dfrac{x^2}{y}$ c) $\ln(xy^2)$

30. Verify that $f_{xxy} = f_{xyx}$ if $f(x, y) = x^3y^2 + (x^2/y^3)$.

16.2 TANGENT PLANES AND APPROXIMATIONS

16.2.1 Tangent planes

For a function $y = f(x)$ of one variable, its derivative has geometric significance as the slope of the tangent line to the graph. The translated vector

$$\boldsymbol{i} + \frac{dy}{dx}\boldsymbol{j}$$

emanating from a point (x, y) on the curve is tangent to the curve.

For a function of two variables $z = f(x, y)$, the partial derivatives are related to the tangent plane to the graph. Since $\partial z/\partial x$ gives the rate of increase in z with respect to x as y is held constant, the translated vector

$$\boldsymbol{i} + \frac{\partial z}{\partial x}\boldsymbol{k}$$

emanating from a point (x, y, z) on the surface is tangent to the surface. That is, the vector is tangent to the curve on the surface obtained by holding y constant, as shown in Fig. 16.4. Similarly,

$$\boldsymbol{j} + \frac{\partial z}{\partial y}\boldsymbol{k}$$

is tangent to the surface there also.

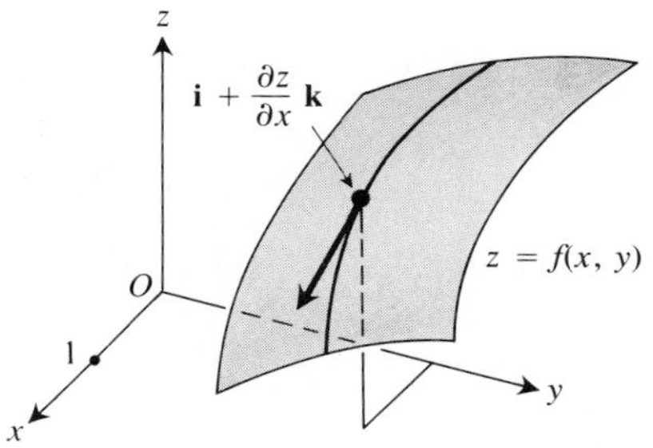

16.4

To find the equation of a tangent plane to the graph $z = f(x, y)$ at a point (x, y, z) on the surface, you need to find a vector normal to the plane. Since the translated vectors

$$\boldsymbol{i} + \frac{\partial z}{\partial x}\boldsymbol{k}$$

and

$$\boldsymbol{j} + \frac{\partial z}{\partial y}\boldsymbol{k}$$

are in the tangent plane, their cross product

$$\begin{vmatrix} \boldsymbol{i} & \boldsymbol{j} & \boldsymbol{k} \\ 1 & 0 & \dfrac{\partial z}{\partial x} \\ 0 & 1 & \dfrac{\partial z}{\partial y} \end{vmatrix} = -\frac{\partial z}{\partial x}\boldsymbol{i} - \frac{\partial z}{\partial y}\boldsymbol{j} + \boldsymbol{k}$$

is normal to the plane there. We multiply by -1, and remember that

Normal vector to a graph

$(\partial z/\partial x)\boldsymbol{i} + (\partial z/\partial y)\boldsymbol{j} - \boldsymbol{k}$ *is normal to the graph of* $z = f(x, y)$ *at each point on the graph.*

Example 1 Let $z = f(x, y) = x^2 + 3xy + 2y^3$ and let's find the equation of the tangent plane at $(1, 2, 23)$.

SOLUTION You have

$$\frac{\partial z}{\partial x} = 2x + 3y \qquad \text{and} \qquad \frac{\partial z}{\partial y} = 3x + 6y^2,$$

so at (1, 2, 23), you have

$$\left.\frac{\partial z}{\partial x}\right|_{(1,2)} = 2 + 6 = 8 \qquad \text{and} \qquad \left.\frac{\partial z}{\partial y}\right|_{(1,2)} = 3 + 24 = 27.$$

Therefore a vector normal to the plane is $8\boldsymbol{i} + 27\boldsymbol{j} - \boldsymbol{k}$. The equation of the plane is

$$8(x - 1) + 27(y - 2) - (z - 23) = 0 \qquad \text{or} \qquad 8x + 27y - z = 39. \quad \|$$

Of course, you can also find parametric equations of the normal line to a surface if you know a point and a vector parallel to the line, i.e., normal to the surface.

Example 2 Parametric equations of the normal line to the surface in Example 1 at (1, 2, 23) are

$$\begin{aligned} x &= 1 + 8t, \\ y &= 2 + 27t, \\ z &= 23 - t. \quad \| \end{aligned}$$

16.2.2 Approximations

In Chapter 3, you saw how to approximate $f(x_0 + \Delta x)$, where $f(x_0)$ is easily computed and $\Delta x = dx$ is small. Geometrically, you computed the height over $x_0 + \Delta x$ to the line tangent to the graph at $(x_0, f(x_0))$. We now have a plane tangent to the surface graph of $z = f(x, y)$ at $(x_0, y_0, f(x_0, y_0))$. To strengthen your geometric intuition in preparation for the differential in the next section, we treat geometrically the analogous approximation of $f(x_0 + \Delta x, y_0 + \Delta y)$, where $f(x_0, y_0)$ is easily computed and Δx and Δy are small. This approximation will be treated again in the language of differentials in the next section, where we shall describe conditions on f that guarantee that the approximation is a good one for sufficiently small values Δx and Δy.

Let $z = f(x, y)$ and let's take $\Delta x, \Delta y, \Delta z$-axes at a point (x_0, y_0, z_0) on the graph, as shown in Fig. 16.5. The equation of the tangent plane is

$$f_x(x_0, y_0)(x - x_0) + f_y(x_0, y_0)(y - y_0) - (z - z_0) = 0.$$

Therefore the change $\Delta z_{\tan} = z - z_0$ in the tangent plane to the surface corresponding to changes $\Delta x = x - x_0$ and $\Delta y = y - y_0$ is

$$\Delta z_{\tan} = f_x(x_0, y_0)\,\Delta x + f_y(x_0, y_0)\,\Delta y. \tag{1}$$

Recall that for a function $y = f(x)$ of one variable, we had $\Delta y_{\tan} = f'(x_0)\,\Delta x$, which led to the approximation

$$f(x_0 + \Delta x) \approx f(x_0) + f'(x_0)\,\Delta x$$

given in Chapter 3. Correspondingly, you will now be able to approximate $f(x_0 + \Delta x, y_0 + \Delta y)$ by taking the height to the tangent plane as an approximation to the height of the surface. Thus

$$f(x_0 + \Delta x, y_0 + \Delta y) \approx f(x_0, y_0) + f_x(x_0, y_0)\,\Delta x + f_y(x_0, y_0)\,\Delta y, \tag{2}$$

where the approximation is best for small values of Δx and Δy.

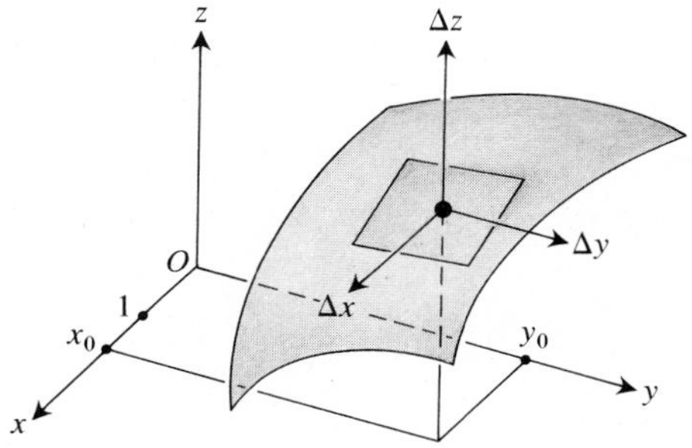

16.5

Example 3 Let's estimate $(1.05)^2(2.99)^3$.

SOLUTION Let $f(x, y) = x^2y^3$. You know $f(1, 3) = 1^2 \cdot 3^3 = 27$. Let $\Delta x = 0.05$ and $\Delta y = -0.01$. Now

$$f_x = 2xy^3 \quad \text{and} \quad f_y = 3x^2y^2.$$

Then by (2),

$$\begin{aligned} f(1.05, 2.99) &= f(1 + \Delta x, 3 + \Delta y) \\ &\approx 27 + f_x(1, 3)\cdot(0.05) + f_y(1, 3)\cdot(-0.01) \\ &= 27 + 54(0.05) + 27(-0.01) = 27 + 2.70 - 0.27 \\ &= 29.43. \end{aligned}$$

Actually, $(1.05)^2(2.99)^3 = 29.4708161475$. ||

Similar results hold for functions of more variables. For example, if $w = f(x, y, z)$, then the analogue of the approximation (2) is

$$f(x_0 + \Delta x, y_0 + \Delta y, z_0 + \Delta z) \approx f(x_0, y_0, z_0) + f_x(x_0, y_0, z_0)\,\Delta x + f_y(x_0, y_0, z_0)\,\Delta y + f_z(x_0, y_0, z_0)\,\Delta z. \tag{3}$$

Such formulas can be written more compactly using vector notation with subscripted variables. For example, the approximation near $\boldsymbol{a} = (a_1, a_2, a_3)$ of

$$y = f(\boldsymbol{x}) = f(x_1, x_2, x_3)$$

corresponding to a change $\boldsymbol{\Delta x} = (\Delta x_1, \Delta x_2, \Delta x_3)$ is given by

$$f(\boldsymbol{a} + \boldsymbol{\Delta x}) \approx f(\boldsymbol{a}) + f_{x_1}(\boldsymbol{a})\,\Delta x_1 + f_{x_2}(\boldsymbol{a})\,\Delta x_2 + f_{x_3}(\boldsymbol{a})\,\Delta x_3, \tag{4}$$

for a vector $\boldsymbol{\Delta x}$ of sufficiently small magnitude. The use of vector notation and subscripted variables often makes the structure of formulas easier to follow. Formulas just like (4) hold for approximation of functions of one or two variables also; it just depends on the subscript of the "last x."

SUMMARY

1. *A vector normal to the surface given by $z = f(x, y)$ at any point (x, y, z) on the surface is*

$$\frac{\partial z}{\partial x}\boldsymbol{i} + \frac{\partial z}{\partial y}\boldsymbol{j} - \boldsymbol{k}.$$

2. *In view of the vector just described, it is easy to find the tangent plane and normal line to the graph of the function, for these loci are determined by the point on the graph and a vector perpendicular to the graph.*
3. *The approximate value of $f(x_0 + \Delta x, y_0 + \Delta y)$ given by the height of the tangent plane over $(x_0 + \Delta x, y_0 + \Delta y)$ is*

$$f(x_0 + \Delta x, y_0 + \Delta y) \approx f(x_0, y_0) + f_x(x_0, y_0)\,\Delta x + f_y(x_0, y_0)\,\Delta y$$

for sufficiently small Δx and Δy.

4. *If $y = f(\boldsymbol{x}) = f(x_1, x_2, x_3)$ while $\boldsymbol{a} = (a_1, a_2, a_3)$ and $\boldsymbol{\Delta x} = (\Delta x_1, \Delta x_2, \Delta x_3)$, then*

$$f(\boldsymbol{a} + \boldsymbol{\Delta x}) \approx f(\boldsymbol{a}) + f_{x_1}(\boldsymbol{a})\,\Delta x_1 + f_{x_2}(\boldsymbol{a})\,\Delta x_2 + f_{x_3}(\boldsymbol{a})\,\Delta x_3$$

for $\boldsymbol{\Delta x}$ of sufficiently small magnitude.

EXERCISES

1. Find the equation of the tangent plane to the graph of the polynomial function $xy + 3y^2$ at the point $(-2, 3, 21)$.

2. Find the equation of the tangent plane to the graph of the function $\sin xy$ at the point having coordinates $(1, \pi/2, 1)$.

3. Find parametric equations for the normal line to the graph of $x^3y + xy^2$ at the point $(1, -1, 0)$.

4. Find parametric equations of the normal line to the graph of $\ln(x^2 + y^2)$ at the point $(3, 4, \ln 25)$.

5. Find the equation of the tangent plane and parametric equations for the normal line to the surface $x_3 = x_1{}^2x_2 - 4x_1x_2{}^3$ at the point having coordinates $(-2, 1, 12)$.

In Exercises 6 through 9, use calculus to estimate the indicated quantity.

6. $\sqrt{(4.04)(0.95)}$ **7.** $(2.01)(1.98)^3 + (2.01)^2(1.98)$ **8.** $(\cos 1°)(\tan 44°)$ **9.** $\sqrt{(1.97)^2 + (2.02)^2 + (1.05)^2}$

10. A rectangular box has inside measurements of 14 in. wide, 20 in. long, and 8 in. high. Estimate the volume of material used in construction of the box if the sides and bottom are $\frac{1}{8}$ in. thick, and the box has no top.

11. A cylindrical silo with a hemispherical cap has volume
$$V = \pi r^2 h + \tfrac{2}{3}\pi r^3,$$
where h is the height and r the radius of the cylinder. Estimate the change in volume of a silo of 6-ft radius and 30-ft height if the radius is increased by 4 in. and the height is decreased by 6 in.

16.3 THE DERIVATIVE AND DIFFERENTIAL

16.3.1 The derivative

If $z = f(x, y)$, then $\partial f/\partial x$ and $\partial f/\partial y$ give the rates of change of $f(x, y)$ in the coordinate directions. The vector

$$\left(\frac{\partial f}{\partial x}, \frac{\partial f}{\partial y}\right)$$

summarizes both these rates of change, and we consider this vector to be the **derivative** of $f(x, y)$, and denote it by $\mathbf{f}'(x, y)$.

Example 1 If $f(x, y) = x^2 - 3xy + y^3$, then

$$\mathbf{f}'(x, y) = (2x - 3y, -3x + 3y^2). \quad \|$$

More generally, illustrating subscripted notation, if $y = f(\mathbf{x}) = f(x_1, x_2, x_3)$, then

$$\mathbf{f}'(\mathbf{x}) = \left(\frac{\partial y}{\partial x_1}, \frac{\partial y}{\partial x_2}, \frac{\partial y}{\partial x_3}\right). \tag{1}$$

The vector notation $y = f(\mathbf{x})$ is reminiscent of the single-variable calculus in the first part of this text. It is *very* tempting to introduce the notation $\partial y/\boldsymbol{\partial}\mathbf{x}$ (which you would write $\partial y/\partial\vec{x}$) for this vector derivative. Just as in the single-variable case, the notation is very handy for remembering the chain rule presented in the next section. We must give you one word of caution: This time, there will not be any interpretation of $\partial y/\boldsymbol{\partial}\mathbf{x}$ as a quotient of differential quantities. The notation $\partial y/\boldsymbol{\partial}\mathbf{x}$ is purely a happy symbolism, useful for remembering formulas, and is *not* to be regarded as a quotient. Perhaps someone someday will come along and make quotient sense out of it. Who knows? That is one of the ways that mathematics develops.

Example 2 We find $\mathbf{f}'(\mathbf{x}) = \partial y/\boldsymbol{\partial}\mathbf{x}$ at $\mathbf{x} = (1, -1, 2)$ for $y = f(\mathbf{x}) = x_1{}^2x_3 - 2x_2{}^3x_3{}^2$.

SOLUTION Now

$$\frac{\partial y}{\partial x_1} = 2x_1x_3, \qquad \frac{\partial y}{\partial x_2} = -6x_2^2x_3^2, \qquad \frac{\partial y}{\partial x_3} = x_1^2 - 4x_2^3x_3.$$

Therefore

$$\mathbf{f}'(\mathbf{x}) = \frac{\partial y}{\partial \mathbf{x}} = (2x_1x_3, -6x_2^2x_3^2, x_1^2 - 4x_2^3x_3).$$

Consequently,

$$\mathbf{f}'(1, -1, 2) = \left.\frac{\partial y}{\partial \mathbf{x}}\right|_{(1,-1,2)} = (4, -24, 9). \quad \|$$

16.3.2 The differential

Recall from Chapter 3 that if $y = f(x)$ is a differentiable function of one variable, the differential dy is given by

$$dy = f'(x)\,dx \tag{2}$$

where dx is an independent variable. If $dx = \Delta x$ is an increment in x, then $dy = \Delta y_{\tan}$, which is the change in the y-height of the tangent line corresponding to the change Δx in x. Also, dy is a good approximation to the change Δy in y since there is a function ϵ, depending on Δx, such that

$$y = f'(x) \cdot \Delta x + \epsilon \cdot \Delta x \qquad \text{where} \qquad \lim_{\Delta x \to 0} \epsilon = 0. \tag{3}$$

Relation (3) was used to prove the chain rule for differentiation.

Now let's parrot these ideas for $z = f(x, y)$. The analogue of $\Delta y_{\tan}$ is

$$\Delta z_{\tan} = f_x(x, y)\,\Delta x + f_y(x, y)\,\Delta y, \tag{4}$$

so by analogy we define the differential dz as follows:

Definition 16.2 Let $z = f(x, y)$ have partial derivatives f_x and f_y at (x_0, y_0). Let dx and dy be independent variables. The **differential** dz (or df) **of** f **at** (x_0, y_0) is

$$dz = f_x(x_0, y_0)\,dx + f_y(x_0, y_0)\,dy.$$

If f_x and f_y exist at all points in the domain of f, then the **differential** dz (or df) is

$$dz = f_x(x, y)\,dx + f_y(x, y)\,dy. \tag{5}$$

The expression (5) can be found as soon as f_x and f_y exist. However, examples have been found showing that $dz = \Delta z_{\tan}$ may not be a good approximation to Δz for small $dx = \Delta x$ and $dy = \Delta y$ if f_x and f_y are not continuous. Consequently, we shall say that a function $f(x, y)$ is **differenti-**

Differentiable function **able** only if the partial derivatives f_x and f_y *exist* and are *continuous.* This is contrary to the case for a function $y = f(x)$ of one variable, where the mere existence of the derivative always guarantees that $dy = \Delta y_{\tan}$ is a good approximation to Δy for small $dx = \Delta x$.

We now develop the analogue of (3). Assume that f_x and f_y exist and are continuous. Then

$$\begin{aligned}\Delta z &= f(x + \Delta x, y + \Delta y) - f(x, y)\\ &= [f(x + \Delta x, y + \Delta y) - f(x, y + \Delta y)] + [f(x, y + \Delta y) - f(x, y)]. \quad (6)\end{aligned}$$

[Relation (6) is known as the old subtract-and-add trick.] Now hold x and y fixed, and let the function ϵ_1 of the variables Δx and Δy be given by

$$\epsilon_1 = \begin{cases} \dfrac{f(x + \Delta x, y + \Delta y) - f(x, y + \Delta y)}{\Delta x} - f_x(x, y) & \text{for} \quad \Delta x \neq 0,\\ 0 \quad \text{for} \quad \Delta x = 0, \end{cases}$$

so that

$$f(x + \Delta x, y + \Delta y) - f(x, y + \Delta y) = f_x(x, y) \cdot \Delta x + \epsilon_1 \cdot \Delta x. \quad (7)$$

Using the Mean-Value Theorem on $g(x) = f(x, y + \Delta y)$ for y and Δy held fixed, you can write

$$\epsilon_1 = f_x(c, y + \Delta y) - f_x(x, y) \quad \text{where } c \text{ is between } x \text{ and } x + \Delta x.$$

Since f_x is assumed to be continuous, you see that

$$\lim_{\Delta x, \Delta y \to 0} \epsilon_1 = f_x(x, y) - f_x(x, y) = 0.$$

Similarly, if

$$\epsilon_2 = \begin{cases} \dfrac{f(x, y + \Delta y) - f(x, y)}{\Delta y} - f_y(x, y) & \text{for} \quad \Delta y \neq 0,\\ 0 \quad \text{for} \quad \Delta y = 0, \end{cases}$$

then

$$f(x, y + \Delta y) - f(x, y) = f_y(x, y) \cdot \Delta y + \epsilon_2 \cdot \Delta y. \quad (8)$$

An argument similar to the one for ϵ_1 shows that $\lim_{\Delta x, \Delta y \to 0} \epsilon_2 = 0$. Substituting the expressions in (7) and (8) for the bracketed expressions in (6), you have

$$\Delta z = f_x(x, y) \cdot \Delta x + f_y(x, y) \cdot \Delta y + \epsilon_1 \cdot \Delta x + \epsilon_2 \cdot \Delta y \quad (9)$$

where

$$\lim_{\Delta x, \Delta y \to 0} \epsilon_1 = \lim_{\Delta x, \Delta y \to 0} \epsilon_2 = 0.$$

Relation (9) is the desired analogue of (3) for $z = f(x, y)$. Of course, similar relations hold for still more variables. In differential notation, (9) assures us that when the approximation

$$f(x + dx, y + dy) \approx f(x, y) + f_x(x, y)\,dx + f_y(x, y)\,dy \tag{10}$$

is used, the approximation will be a good one for sufficiently small dx and dy, in the sense that the error

$$\epsilon_1 \cdot dx + \epsilon_2 \cdot dy$$

is small in comparison with the size of dx and dy, since $\epsilon_1 \to 0$ and $\epsilon_2 \to 0$ as $dx \to 0$ and $dy \to 0$.

Relation (9) is so important that we state it as a theorem.

Theorem 16.1 *Let $f(x, y)$ have continuous partial derivatives f_x and f_y inside a circle of radius $r > 0$ with center (x_0, y_0). Then there exist functions $\epsilon_1(\Delta x, \Delta y)$ and $\epsilon_2(\Delta x, \Delta y)$, defined for $(\Delta x)^2 + (\Delta y)^2 < r^2$, such that*

$$\lim_{\Delta x, \Delta y \to 0} \epsilon_1 = \lim_{\Delta x, \Delta y \to 0} \epsilon_2 = 0$$

and such that

$$f(x_0 + \Delta x, y_0 + \Delta y) = f(x_0, y_0) + f_x(x_0, y_0)\,\Delta x + f_y(x_0, y_0)\,\Delta y + \epsilon_1(\Delta x, \Delta y) \cdot \Delta x + \epsilon_2(\Delta x, \Delta y) \cdot \Delta y.$$

Of course, if $w = f(x, y, z)$, then you let

$$dw = f_x(x, y, z)\,dx + f_y(x, y, z)\,dy + f_z(x, y, z)\,dz.$$

Once again, vector notation with subscripted variables makes everything look very much like the single-variable case. If $y = f(\boldsymbol{x}) = f(x_1, x_2, x_3)$, then

$$dy = f_{x_1}(\boldsymbol{x})\,dx_1 + f_{x_2}(\boldsymbol{x})\,dx_2 + f_{x_3}(\boldsymbol{x})\,dx_3. \tag{11}$$

Recall that

$$\boldsymbol{f}'(\boldsymbol{x}) = (f_{x_1}(\boldsymbol{x}), f_{x_2}(\boldsymbol{x}), f_{x_3}(\boldsymbol{x})).$$

If you let $\boldsymbol{dx}$ be the vector $\boldsymbol{dx} = (dx_1, dx_2, dx_3)$, then (11) takes the simple form of a dot product,

$$dy = \boldsymbol{f}'(\boldsymbol{x}) \cdot \boldsymbol{dx}. \tag{12}$$

By using vector notation with a dot product, we have retained the same simple formula for our differential as for a function of one variable.

Example 3 We find dy if $y = f(\boldsymbol{x}) = f(x_1, x_2, x_3) = x_1^{\,2} + x_2 e^{x_3}$.

SOLUTION Now

$$\begin{aligned} \boldsymbol{f}'(\boldsymbol{x}) &= (f_{x_1}, f_{x_2}, f_{x_3}) \\ &= (2x_1, e^{x_3}, x_2 e^{x_3}), \end{aligned}$$

so

$$dy = \mathbf{f}'(\mathbf{x}) \cdot \mathbf{dx} = (2x_1, e^{x_3}, x_2e^{x_3}) \cdot (dx_1, dx_2, dx_3)$$
$$= 2x_1\,dx_1 + e^{x_3}\,dx_2 + x_2e^{x_3}\,dx_3. \quad \|$$

Let $y = f(x)$ be a differentiable function of one variable. We saw in Chapter 3 that the estimate $f(x + \Delta x) \approx f(x) + f'(x)\,\Delta x$ becomes, in differential notation,

$$f(x + dx) \approx f(x) + f'(x)\,dx \tag{13}$$

for sufficiently small dx. Now let $y = f(\mathbf{x}) = f(x_1, x_2)$ be a differentiable function of two variables. Theorem 16.1 shows that if $|\mathbf{dx}|$ is sufficiently small, then

$$f(\mathbf{x} + \mathbf{dx}) \approx f(\mathbf{x}) + \mathbf{f}'(\mathbf{x}) \cdot \mathbf{dx}. \tag{14}$$

This time $\mathbf{dx} = (dx_1, dx_2)$ and $\mathbf{f}'(\mathbf{x}) = (f_{x_1}, f_{x_2})$. Of course, (14) also holds for a differentiable function $y = f(\mathbf{x}) = f(x_1, x_2, x_3)$.

In summary, *the familiar formulas from differential calculus of one variable are still valid if you use subscripted variables, vector notation, and dot products.*

Example 4 Let $y = f(\mathbf{x}) = f(x_1, x_2, x_3) = x_1^2 + x_2e^{x_3}$, as in Example 3. Let's use (14) to estimate $f(3.9, 6.05, 0.2)$.

SOLUTION We let $\mathbf{x} = (4, 6, 0)$ and $\mathbf{dx} = (-0.1, 0.05, 0.2)$. Example 3 showed that

$$\mathbf{f}'(\mathbf{x}) = (2x_1, e^{x_3}, x_2e^{x_3}).$$

Therefore,

$$f(3.9, 6.05, 0.2) \approx f(\mathbf{x}) + \mathbf{f}'(\mathbf{x}) \cdot \mathbf{dx} = (16 + 6) + (8, 1, 6) \cdot (-0.1, 0.05, 0.2)$$
$$= 22 + (-0.8 + 0.05 + 1.2) = 22 + 0.45 = 22.45. \quad \|$$

SUMMARY *If* $z = f(x, y)$, *then we let* $\mathbf{f}'(x, y) = (\partial z/\partial x, \partial z/\partial y)$. *If* $y = f(\mathbf{x}) = f(x_1, x_2, x_3)$, *then*

$$\mathbf{f}'(\mathbf{x}) = \frac{\partial y}{\partial \mathbf{x}} = \left(\frac{\partial y}{\partial x_1}, \frac{\partial y}{\partial x_2}, \frac{\partial y}{\partial x_3}\right) = (f_{x_1}, f_{x_2}, f_{x_3}).$$

2. *If* $z = f(x, y)$, *we define the differential* dz *by*

$$dz = f_x(x, y)\,dx + f_y(x, y)\,dy.$$

3. *If* f_x *and* f_y *are continuous, then*

$$f(x + \Delta x, y + \Delta y) - f(x, y)$$
$$= f_x(x, y)\,\Delta x + f_y(x, y)\,\Delta y + \epsilon_1 \cdot \Delta x + \epsilon_2 \cdot \Delta y,$$

where $\epsilon_1 \to 0$ *and* $\epsilon_2 \to 0$ *as* $\Delta x \to 0$ *and* $\Delta y \to 0$.

4. *If* $y = f(\mathbf{x}) = f(x_1, x_2, x_3)$, *then*

$$dy = f_{x_1}(\mathbf{x})\, dx_1 + f_{x_2}(\mathbf{x})\, dx_2 + f_{x_3}(\mathbf{x})\, dx_3 = \mathbf{f}'(\mathbf{x}) \cdot \mathbf{dx},$$

where

$$\mathbf{dx} = (dx_1, dx_2, dx_3).$$

5. *If* $y = f(\mathbf{x}) = f(x_1, x_2, x_3)$ *is differentiable, then the formula for estimating using differentials takes the form*

$$f(\mathbf{x} + \mathbf{dx}) \approx f(\mathbf{x}) + \mathbf{f}'(\mathbf{x}) \cdot \mathbf{dx}$$

for $|\mathbf{dx}|$ *sufficiently small.*

EXERCISES

In Exercises 1 through 6, find the vector derivative $\mathbf{f}'$ *and the differential of the indicated function at the given point.*

1. $f(x, y, z) = x^2 + 2yz$ at $(4, -1, 2)$

2. $f(x, y) = (x^3y^2 + 2x^2y)$ at $(2, 3)$

3. $f(x, y, z) = \ln(xy) + e^{yz} + \sin(xz)$ at $(2, 4, \pi)$

4. $f(x, y) = x^2 - 2y^2 + \tan(xy) + (1/x)$ at $(-2, 0)$

5. $f(x) = 3x^2 + \sec x + \ln x$ at π

6. $f(x, y) = 2x + x\cos y + x\sin y$ at $(3, \pi)$

7. Let f be defined by $f(x, y, z) = xy + \sin z$. Use a differential to estimate $f(0.98, 2.03, 0.05)$.

8. Let f be defined by $f(x, y) = e^{xy} + \sin xy + 4y$. Use a differential to estimate $f(0.02, 4.97)$.

9. Illustrate relation (9) of the text by explicitly finding the functions ϵ_1 and ϵ_2 for $f(x, y) = x^2 - 2xy + 3x - y^2$. Observe that $\epsilon_1 \to 0$ and $\epsilon_2 \to 0$ as $\Delta x \to 0$ and $\Delta y \to 0$.

In Exercises 10 through 12, use relation (14) of the text to estimate the function at the indicated point.

10. $f(x_1, x_2) = x_1^2x_2 + x_2^3$ at $(3.9, 2.05)$.

11. $f(x_1, x_2, x_3) = (x_1/x_3) + 2x_1^3x_2$ at $(2.95, 5.1, 1.2)$.

12. $f(x_1, x_2, x_3) = \sin(x_1x_3) + \ln(4x_1x_2^3)$ at $(0.95, 2.1, -0.03)$.

16.4 CHAIN RULES

Suppose $z = f(x, y)$ and $x = g_1(t)$, $y = g_2(t)$ so t determines z. Suppose also that the derivatives $\partial z/\partial x$, $\partial z/\partial y$, dx/dt, and dy/dt all exist and are continuous. We want to find dz/dt, since z appears as a function of the one variable t. The rate of change of z with respect to t can be expressed in terms of the rates at which z changes with respect to x and to y and the rates at which x and y change with respect to t. This is surely not too surprising. The total rate of change of z is the sum of the rates of change due to the changing quantities x and y. These rates of change due to x and y individually are, by the chain rule in Chapter 3,

$$\frac{\partial z}{\partial x} \cdot \frac{dx}{dt} \quad \text{and} \quad \frac{\partial z}{\partial y} \cdot \frac{dy}{dt}.$$

Thus, although we have not proved it yet, the valid formula

$$\frac{dz}{dt} = \frac{\partial z}{\partial x} \cdot \frac{dx}{dt} + \frac{\partial z}{\partial y} \cdot \frac{dy}{dt} \tag{1}$$

should seem reasonable to you.

We can prove (1) from our work in the last section. Theorem 16.1 shows that

$$\Delta z = f_x(x, y)\,\Delta x + f_y(x, y)\,\Delta y + \epsilon_1 \cdot \Delta x + \epsilon_2 \cdot \Delta y \tag{2}$$

where $\epsilon_1 \to 0$ and $\epsilon_2 \to 0$ as $\Delta x \to 0$ and $\Delta y \to 0$. Thus

$$\begin{aligned} \frac{dz}{dt} = \lim_{\Delta t \to 0} \frac{\Delta z}{\Delta t} &= \lim_{\Delta t \to 0} \left[f_x(x, y) \frac{\Delta x}{\Delta t} + f_y(x, y) \frac{\Delta y}{\Delta t} + \epsilon_1 \frac{\Delta x}{\Delta t} + \epsilon_2 \frac{\Delta y}{\Delta t} \right] \\ &= f_x(x, y) \frac{dx}{dt} + f_y(x, y) \frac{dy}{dt} + 0 \cdot \frac{dx}{dt} + 0 \cdot \frac{dy}{dt} \\ &= \frac{\partial z}{\partial x} \frac{dx}{dt} + \frac{\partial z}{\partial y} \frac{dy}{dt}, \end{aligned}$$

which substantiates (1).

A similar argument shows that, if $w = f(x, y, z)$ and $x = g_1(t)$, $y = g_2(t)$, and $z = g_3(t)$, then

$$\frac{dw}{dt} = \frac{\partial w}{\partial x} \frac{dx}{dt} + \frac{\partial w}{\partial y} \frac{dy}{dt} + \frac{\partial w}{\partial z} \frac{dz}{dt}.$$

In subscripted notation, if $y = f(\boldsymbol{x}) = f(x_1, x_2, x_3)$ and $x_1 = g_1(t)$, $x_2 = g_2(t)$, $x_3 = g_3(t)$, then

$$\frac{dy}{dt} = \frac{\partial y}{\partial x_1} \frac{dx_1}{dt} + \frac{\partial y}{\partial x_2} \frac{dx_2}{dt} + \frac{\partial y}{\partial x_3} \frac{dx_3}{dt}.$$

Example 1 The volume V of a right circular cylinder of radius r and height h is given by $V = \pi r^2 h$. If the volume is increasing at a rate of 72π in.^{3}/min while the height is decreasing at a rate of 4 in./min, let us find the rate of increase of the radius when the height is 3 in. and the radius is 6 in.

SOLUTION We have

$$\frac{dV}{dt} = \frac{\partial V}{\partial r} \cdot \frac{dr}{dt} + \frac{\partial V}{\partial h} \cdot \frac{dh}{dt},$$

so

$$\frac{dV}{dt} = 2\pi r h \frac{dr}{dt} + \pi r^2 \left(\frac{dh}{dt} \right).$$

You know that $dV/dt = 72\pi$ and $dh/dt = -4$. (The negative sign occurs because h is decreasing.) You want to find dr/dt when $r = 6$ and $h = 3$.

Substituting,

$$72\pi = 2\pi(6)(3)\frac{dr}{dt} + \pi(6^2)(-4),$$

so

$$36\pi\frac{dr}{dt} = 216\pi.$$

Hence $dr/dt = 216\pi/36\pi = 6$ in./min. ||

Now let $z = f(x, y)$ and $x = g_1(s, t)$ and $y = g_2(s, t)$. This time z appears as the composite function of *two* variables, s and t, and you are interested in the *partial* derivatives $\partial z/\partial s$ and $\partial z/\partial t$. But (2) is still valid, and you divide by the increment Δt and take the limit as s is held constant, to find $\partial z/\partial t$. That is, you obtain from (2)

$$\begin{aligned}\frac{\partial z}{\partial t} = \lim_{\Delta t \to 0}\frac{\Delta z}{\Delta t} &= f_x(x, y)\frac{\partial x}{\partial t} + f_y(x, y)\frac{\partial y}{\partial t} + 0\cdot\frac{\partial x}{\partial t} + 0\cdot\frac{\partial y}{\partial t}\\ &= \frac{\partial z}{\partial x}\frac{\partial x}{\partial t} + \frac{\partial z}{\partial y}\frac{\partial y}{\partial t}.\end{aligned}$$

Thus the derivatives dx/dt and dy/dt in (1) simply become partial derivatives in this case.

Example 2 Consider the situation where

$$w = f(x, y, z) = xy^2 + ze^{x^2}$$

while $x = u$, $y = v - 1$, and $z = uv$. Let's compute $\partial w/\partial u$ at $(0, 2)$ in two ways: by expressing w directly as a function of u and v, and by using the chain rule.

SOLUTION Expressing w directly as a function of u and v,

$$\begin{aligned} w &= f(u, v - 1, uv)\\ &= u(v - 1)^2 + uve^{u^2}.\end{aligned}$$

Thus

$$\frac{\partial w}{\partial u} = (v - 1)^2 + 2u^2ve^{u^2} + ve^{u^2}.$$

Therefore

$$\begin{aligned}\left.\frac{\partial w}{\partial u}\right|_{(0,2)} &= 1 + 0 + 2\cdot e^0\\ &= 1 + 2 = 3.\end{aligned}$$

To use the chain rule, note that when $(u, v) = (0, 2)$,

$$(x, y, z) = (0, 1, 0).$$

If $w = xy^2 + ze^{x^2}$, then

$$\frac{\partial w}{\partial u} = \frac{\partial w}{\partial x} \cdot \frac{\partial x}{\partial u} + \frac{\partial w}{\partial y} \cdot \frac{\partial y}{\partial u} + \frac{\partial w}{\partial z} \cdot \frac{dy}{\partial u}$$
$$= (y^2 + 2xze^{x^2})(1) + (2xy)(0) + (e^{x^2})(v).$$

Thus

$$\left.\frac{\partial w}{\partial u}\right|_{(u,v)=(0,2)} = (1 + 0)(1) + 0 + (1)(2) = 1 + 2 = 3. \quad \|$$

Let's take a case involving subscripted variables. Suppose that $y = f(\mathbf{x}) = f(x_1, x_2, x_3)$ and $x_1 = g_1(t_1, t_2)$, $x_2 = g_2(t_1, t_2)$, and $x_3 = g_3(t_1, t_2)$. Then

$$\frac{\partial y}{\partial t_1} = \frac{\partial y}{\partial x_1}\frac{\partial x_1}{\partial t_1} + \frac{\partial y}{\partial x_2}\frac{\partial x_2}{\partial t_1} + \frac{\partial y}{\partial x_3}\frac{\partial x_3}{\partial t_1}$$
$$= \left(\frac{\partial y}{\partial x_1}, \frac{\partial y}{\partial x_2}, \frac{\partial y}{\partial x_3}\right) \cdot \left(\frac{\partial x_1}{\partial t_1}, \frac{\partial x_2}{\partial t_1}, \frac{\partial x_3}{\partial t_1}\right), \qquad (3)$$

where the last expression is the dot product of the vectors. Now we are going to succumb as before to the temptation to introduce Leibniz-type vector notations,

$$\frac{\partial y}{\partial \mathbf{x}} = \left(\frac{\partial y}{\partial x_1}, \frac{\partial y}{\partial x_2}, \frac{\partial y}{\partial x_3}\right) \quad \text{and} \quad \frac{\partial \mathbf{x}}{\partial t_1} = \left(\frac{\partial x_1}{\partial t_1}, \frac{\partial x_2}{\partial t_1}, \frac{\partial x_3}{\partial t_1}\right).$$

Equation (3) then becomes

$$\frac{\partial y}{\partial t_1} = \frac{\partial y}{\partial \mathbf{x}} \cdot \frac{\partial \mathbf{x}}{\partial t_1}. \qquad (4)$$

Once again, we have retained our old chain-rule formula by using vector notations, subscripted variables, and a dot product. Using these subscripted variables illuminates the structure of these formulas for functions of more than one variable.

Example 3 Let $y = f(x) = x_1{}^2 x_2{}^3$ and

$$x_1 = t_1{}^2 - 2t_2 + t_3, \qquad x_2 = t_1 t_2 - t_3{}^2.$$

Then

$$\frac{\partial y}{\partial t_1} = \frac{\partial y}{\partial \mathbf{x}} \cdot \frac{\partial \mathbf{x}}{\partial t_1} = (2x_1 x_2{}^3, 3x_1{}^2 x_2{}^2) \cdot (2t_1, t_2).$$

When $(t_1, t_2, t_3) = (-1, 1, 2)$, then $(x_1, x_2) = (1, -5)$, so

$$\left.\frac{\partial y}{\partial t_1}\right|_{(-1,1,2)} = (-250, 75) \cdot (-2, 1) = 575.$$

For another illustration,

$$\frac{\partial y}{\partial t_3} = \frac{\partial y}{\partial \mathbf{x}} \cdot \frac{\partial \mathbf{x}}{\partial t_3} = (2x_1x_2^3, 3x_1^2x_2^2) \cdot (1, -2t_3),$$

so

$$\left.\frac{\partial y}{\partial t_3}\right|_{(-1,1,2)} = (-250, 75) \cdot (1, -4) = -550. \quad \|$$

SUMMARY

1. *If* $z = f(x, y)$ *and* $x = g(t)$, $y = g(t)$, *then*

$$\frac{dz}{dt} = \frac{\partial z}{\partial x}\frac{dx}{dt} + \frac{\partial z}{\partial y}\frac{dy}{dt}.$$

Turning to subscripted variables, if $y = f(\mathbf{x}) = f(x_1, x_2, x_3)$ *and* $\mathbf{x} = g(t)$, *then*

$$\frac{dy}{dt} = \frac{\partial y}{\partial x_1}\frac{dx_1}{dt} + \frac{\partial y}{\partial x_2}\frac{dx_2}{dt} + \frac{\partial y}{\partial x_3}\frac{dx_3}{dt}.$$

2. *If* $z = f(x, y)$ *and* $x = g_1(s, t)$, $y = g_2(s, t)$, *then*

$$\frac{\partial z}{\partial t} = \frac{\partial z}{\partial x}\frac{\partial x}{\partial t} + \frac{\partial z}{\partial y}\frac{\partial y}{\partial t}.$$

3. *Turning to vector notation, if* $y = f(\mathbf{x}) = f(x_1, x_2, x_3)$ *while* $x_1 = g_1(t_1, t_2)$, $x_2 = g_2(t_1, t_2)$, *and* $x_3 = g_3(t_1, t_2)$, *then*

$$\frac{\partial y}{\partial t_k} = \frac{\partial y}{\partial x_1}\frac{\partial x_1}{\partial t_k} + \frac{\partial y}{\partial x_2}\frac{\partial x_2}{\partial t_k} + \frac{\partial y}{\partial x_3}\frac{\partial x_3}{\partial t_k}.$$

The partial derivative $\partial y/\partial t_k$ *is given by the dot product*

$$\frac{\partial y}{\partial t_k} = \frac{\partial y}{\partial \mathbf{x}} \cdot \frac{\partial \mathbf{x}}{\partial t_k},$$

where

$$\frac{\partial y}{\partial \mathbf{x}} = \left(\frac{\partial y}{\partial x_1}, \frac{\partial y}{\partial x_2}, \frac{\partial y}{\partial x_3}\right) \quad \text{and} \quad \frac{\partial \mathbf{x}}{\partial t_k} = \left(\frac{\partial x_1}{\partial t_k}, \frac{\partial x_2}{\partial t_k}, \frac{\partial x_3}{\partial t_k}\right).$$

EXERCISES

1. Let $z = x^2 + (1/y^2)$, $x = t^2$, and $y = t + 1$.

a) Find x, y, and z when $t = 1$.

b) Find $dz/dt|_{t=1}$, using a chain rule.

c) Express z as a function of t by substitution.

d) Find $dz/dt|_{t=1}$ by differentiating your answer to part (c). Compare with your answer in (b).

2. If $z = x^2 - 2xy + xy^3$, $x = t^3 + 1$, and $y = 1/t$, use a chain rule to find dz/dt when $t = 1$.

3. If $w = x \sin(yz)$, $x = 2t + 1$, $y = 3t^2$, and $z = (\pi/2)t$, use a chain rule to find dw/dt when $t = -1$.

4. If $w = x^2 + yz$, $x = uv$, $y = u - v$, and $z = 2u^2v$, use a chain rule to find $\partial w/\partial u$ when $u = -1$ and $v = 2$.

5. If $w = y_1^2 + \sin y_2 y_3 - e^{y_1}$ while $y_1 = x_1 x_3$, $y_2 = \ln(x_3^2 + 1)$, and $y_3 = x_2 \cos x_3$, use a chain rule to find $\partial w/\partial x_2|_{(-1,2,0)}$.

6. If $w = 2u - 3v$, $u = xy$, $v = ye^x$, $x = t^3$, and $y = 2 \sin t$, find $dw/dt|_{t=0}$.

7. If the radius of a circular cylinder is increasing at a rate of 4 in./min while the length is increasing at a rate of 8 in./min, find the rate of change of the volume of the cylinder when the radius is 10 in. and the length is 50 in.

8. The voltage drop V, measured in volts, across a certain conductor of variable resistance R ohms is IR, where I, measured in amperes, is the current flowing through the conductor. The current increases at a constant rate of 2 amperes per second while the voltage drop is kept constant by decreasing the resistance as the current increases. Find the rate of change of the resistance when $I = 5$ amperes and $R = 1000$ ohms.

9. The moment of inertia I about an axis of a body of mass m and distance s from the axis is given by $I = ms^2$. Find the rate of change of the moment of inertia about the axis when $s = 50$ if the mass remains constant while the distance s is decreasing at a rate of 3 units length per unit time.

10. Answer Exercise 9 if the body is gaining mass at a rate of 2 units of mass per unit time, and has mass 20 when $s = 50$, while the other data remain the same.

11. The pressure P in lb/ft^2 of a certain gas at temperature T degrees in a container of variable volume V ft^3 is given by $P = 8T/V$. The temperature is increased at a rate of 5°/min while the volume of the container is increased at a rate of 2 ft^3/min. If, at time t_0, the temperature was 20° and the volume was 10 ft^3, find the rate of change of the pressure 5 min later.

12. By Newton's Law of Gravitation, the force of attraction between two bodies of masses m_1 and m_2 is Gm_1m_2/s^2 where s is the distance between the bodies and G is the universal gravitational constant. Find the rate of change of the force of attraction for two bodies of constant masses of $(10)^4$ and $(10)^7$ units that are $(10)^4$ units distance apart and are separating at a rate of $(10)^2$ units distance per unit time. (Assume the given units are compatible, and don't worry about the value of G or the name of the units in the answer.)

13. Repeat Exercise 12 if the first body is gaining mass at the rate of 30 units mass per unit time, and the second body is losing mass at the rate of 80 units mass per unit time, while the other data remain the same.

In Exercises 14 through 23, assume that all functions encountered satisfy enough differentiability conditions to enable you to use any chain rule you wish.

14. If $w = f(u)$ and $u = ax + by$, show that $b(\partial w/\partial x) = a(\partial w/\partial y)$.

15. Obtain a result similar to that in Exercise 14 for $w = f(u)$ and $u = ax + by + cz$.

16. If $w = f(u, v)$, $u = x + y$, and $c = 2x - 2y$, show that

$$\frac{\partial w}{\partial x} \cdot \frac{\partial w}{\partial y} = \left(\frac{\partial f}{\partial u}\right)^2 - 4\left(\frac{\partial f}{\partial v}\right)^2.$$

17. If $w = f(u)$ and $u = xy^2$, show that $2x(\partial w/\partial x) - y(\partial w/\partial y) = 0$.

18. If $w = f(u) + g(v)$, $u = ax + by$, and $v = ax - by$, show that $b^2(\partial^2 w/\partial x^2) = a^2(\partial^2 w/\partial y^2)$.

19. If $w = f(u)$ and $u = x/y$, show that $x(\partial w/\partial x) + y(\partial w/\partial y) = 0$.

20. If f is a differentiable function of two variables and $f(tx, ty) = t^2 f(x, y)$ for all t, show that $x \cdot f_x(x, y) + y \cdot f_y(x, y) = 2 \cdot f(x, y)$. (Such a function is *homogeneous of degree* 2.) [*Hint.* Differentiate the equation $f(tx_0, ty_0) = t^2 f(x_0, y_0)$ with respect to t, and put $t = 1$.]

21. Generalize the conclusion in Exercise 20 in the case that $f(tx, ty) = t^k f(x, y)$ for all t. (Such a function is *homogeneous of degree k*.)

22. Show that the result in Exercise 19 is a special case of your generalization in Exercise 21.

23. Generalize the conclusion in Exercise 21 for a differentiable function of n variables.

16.5 THE DIRECTIONAL DERIVATIVE AND GRADIENT

The partial derivatives of $z = f(x, y)$ give the rates of change of z in the positive coordinate directions. Let's find the rates of change of z in directions other than the coordinate directions.

Let $\boldsymbol{a} = (a_1, a_2)$ be a point in the x,y-plane in the domain of $f(x, y)$. We want to find the instantaneous rate at which $f(x, y)$ increases per unit change in a direction given by a *unit* vector $\boldsymbol{u} = u_1\boldsymbol{i} + u_2\boldsymbol{j}$ at $\boldsymbol{a}$, as indicated in Fig. 16.6. Let's think of taking an s-axis, with origin at (a_1, a_2) in the plane, in the direction given by $\boldsymbol{u}$, as shown in the figure. The curve on the surface $z = f(x, y)$ lying directly over this s-axis is the graph of a function $z = h(s)$. The rate of change of z with respect to s at (a_1, a_2) is the derivative $h'(s) = dz/ds$ evaluated at $s = 0$. Clearly, this derivative gives the rate of change of z in the direction given by the vector $\boldsymbol{u}$ at (a_1, a_2); it is called the **directional derivative** of f at (a_1, a_2) in the direction given by $\boldsymbol{u}$.

You can compute dz/ds using a chain rule, for z is a function of x and y, given by $z = f(x, y)$, and it is easy to express x and y in terms of s. Remember that $\boldsymbol{u}$ is a *unit* vector. Consequently, for the point (x, y) on the s-axis corresponding to a value s, shown in Fig. 16.6, you have, in vector notation,

$$\begin{aligned} x\boldsymbol{i} + y\boldsymbol{j} &= (a_1\boldsymbol{i} + a_2\boldsymbol{j}) + s\boldsymbol{u} \\ &= a_1\boldsymbol{i} + a_2\boldsymbol{j} + s(u_1\boldsymbol{i} + u_2\boldsymbol{j}) \\ &= (a_1 + u_1 s)\boldsymbol{i} + (a_2 + u_2 s)\boldsymbol{j}. \end{aligned}$$

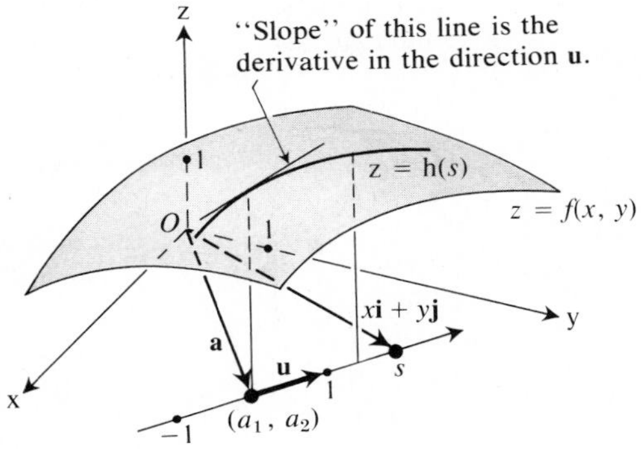

16.6

Therefore,

$$x = a_1 + u_1 s,$$
$$y = a_2 + u_2 s.$$

Then the chain rule shows that

$$\frac{dz}{ds} = \frac{\partial z}{\partial x}\frac{dx}{ds} + \frac{\partial z}{\partial y}\frac{dy}{ds} = \frac{\partial z}{\partial x}\cdot u_1 + \frac{\partial z}{\partial y}\cdot u_2. \tag{1}$$

When $s = 0$, you have

$$\left.\frac{dz}{ds}\right|_{s=0} = u_1 f_x(a_1, a_2) + u_2 f_y(a_1, a_2). \tag{2}$$

If $w = f(x, y, z)$, then the directional derivative of w at $\boldsymbol{a} = (a_1, a_2, a_3)$ in the direction given by the *unit* vector $\boldsymbol{u} = u_1\boldsymbol{i} + u_2\boldsymbol{j} + u_3\boldsymbol{k}$ is, of course,

$$\left.\frac{dw}{ds}\right|_{s=0} = u_1 f_x(\boldsymbol{a}) + u_2 f_y(\boldsymbol{a}) + u_3 f_z(\boldsymbol{a}). \tag{3}$$

Let's also denote the directional derivative given in (2) by $f_{\boldsymbol{u}}(\boldsymbol{a})$, the derivative of $f(x, y)$ at $\boldsymbol{a} = (a_1, a_2)$ in the direction given by the unit vector $\boldsymbol{u} = u_1\boldsymbol{i} + u_2\boldsymbol{j}$. Thus

$$f_{\boldsymbol{u}}(\boldsymbol{a}) = u_1 \frac{\partial f}{\partial x}(\boldsymbol{a}) + u_2 \frac{\partial f}{\partial y}(\boldsymbol{a}). \tag{4}$$

Using subscripted variables, if $y = f(\boldsymbol{x}) = f(x_1, x_2, x_3)$, while $\boldsymbol{a} = (a_1, a_2, a_3)$ and $\boldsymbol{u} = u_1\boldsymbol{i} + u_2\boldsymbol{j} + u_3\boldsymbol{k}$, then

$$f_{\boldsymbol{u}}(\boldsymbol{a}) = u_1 \frac{\partial f}{\partial x_1}(\boldsymbol{a}) + u_2 \frac{\partial f}{\partial x_2}(\boldsymbol{a}) + u_3 \frac{\partial f}{\partial x_2}(\boldsymbol{a}). \tag{5}$$

Example 1 If f is differentiable at $\boldsymbol{a}$, then the directional derivative $f_{\boldsymbol{i}}(\boldsymbol{a})$ in the direction of $\boldsymbol{u} = \boldsymbol{i}$ at $\boldsymbol{a}$ is simply $f_x(\boldsymbol{a})$, for $u_1 = 1$ and $u_2 = 0$. Similarly, $f_{\boldsymbol{j}}(\boldsymbol{a}) = f_y(\boldsymbol{a})$. ‖

It is important that you remember that $\boldsymbol{u}$ must be a *unit* vector in the direction of differentiation.

Example 2 Let f be defined by

$$f(x, y) = x^2 + 3xy^2$$

and let's find the directional derivative of f at $(1, 2)$ in the direction toward the origin.

SOLUTION A vector in the direction from $(1, 2)$ to $(0, 0)$ is $-\boldsymbol{i} - 2\boldsymbol{j}$, so a *unit* vector in this direction is therefore

$$\boldsymbol{u} = -\frac{1}{\sqrt{5}}\boldsymbol{i} - \frac{2}{\sqrt{5}}\boldsymbol{j}.$$

You find that

$$\frac{\partial f}{\partial x}(1, 2) = (2x + 3y^2)\Big|_{(1,2)} = 14$$

and

$$\frac{\partial f}{\partial y}(1, 2) = 6xy\Big|_{(1,2)} = 12.$$

Consequently,

$$f_{\boldsymbol{u}}(1, 2) = -\frac{1}{\sqrt{5}} \cdot 14 + \left(-\frac{2}{\sqrt{5}}\right) \cdot 12 = -\frac{38}{\sqrt{5}}. \quad \|$$

Recall that if $z = f(x, y)$, then the derivative $\boldsymbol{f}'(\boldsymbol{a})$ of f at $\boldsymbol{a} = (a_1, a_2)$ is

$$\boldsymbol{f}'(\boldsymbol{a}) = (f_x(\boldsymbol{a}), f_y(\boldsymbol{a})).$$

You see from (4) that the directional derivative $f_{\boldsymbol{u}}(\boldsymbol{a})$ appears as the dot product of the vectors

$$\boldsymbol{f}'(\boldsymbol{a}) = f_x(\boldsymbol{a})\boldsymbol{i} + f_y(\boldsymbol{a})\boldsymbol{j} \qquad \text{and} \qquad \boldsymbol{u} = u_1\boldsymbol{i} + u_2\boldsymbol{j};$$

for

$$f_{\boldsymbol{u}}(\boldsymbol{a}) = u_1 f_x(\boldsymbol{a}) + u_2 f_y(\boldsymbol{a}) = \boldsymbol{f}'(\boldsymbol{a}) \cdot \boldsymbol{u}.$$

But

$$\boldsymbol{f}'(\boldsymbol{a}) \cdot \boldsymbol{u} = |\boldsymbol{f}'(\boldsymbol{a})| \cdot |\boldsymbol{u}| \cos\theta = |\boldsymbol{f}'(\boldsymbol{a})| \cdot 1 \cos\theta = |\boldsymbol{f}'(\boldsymbol{a})| \cos\theta,$$

where θ is the angle between $\boldsymbol{f}'(\boldsymbol{a})$ and $\boldsymbol{u}$. Now $|\boldsymbol{f}'(\boldsymbol{a})| \cos\theta$ depends only on θ at the point $\boldsymbol{a}$, and has maximum value $|\boldsymbol{f}'(\boldsymbol{a})|$ when $\cos\theta = 1$, that is, when $\theta = 0$ corresponding to $\boldsymbol{u}$ aimed in the direction of the vector $\boldsymbol{f}'(\boldsymbol{a}) = f_x(\boldsymbol{a})\boldsymbol{i} + f_y(\boldsymbol{a})\boldsymbol{j}$. This gives a nice geometric interpretation of the derivative vector $\boldsymbol{f}'(\boldsymbol{a}) = f_x(\boldsymbol{a})\boldsymbol{i} + f_y(\boldsymbol{a})\boldsymbol{j}$. It points in the direction of maximum rate of increase of $f(x, y)$, and has magnitude equal to that maximum rate of increase. For this reason the vector $\boldsymbol{f}'(\boldsymbol{a})$ is also called the **gradient vector** of $f(x, y)$ at $\boldsymbol{a} = (a_1, a_2)$; it points in the direction of maximum steepness (grade) of the surface $z = f(x, y)$, and has length equal to that maximum slope. The notation $\nabla f(\boldsymbol{a})$ is used in this gradient vector interpretation, so

Gradient vector

$$\nabla f(\boldsymbol{a}) = \boldsymbol{f}'(\boldsymbol{a}) = f_x(\boldsymbol{a})\boldsymbol{i} + f_y(\boldsymbol{a})\boldsymbol{j}.$$

Of course, if $y = f(\boldsymbol{x}) = f(x_1, x_2, x_3)$, then

$$\nabla f(\boldsymbol{a}) = \boldsymbol{f}'(\boldsymbol{a}) = f_{x_1}(\boldsymbol{a})\boldsymbol{i} + f_{x_2}(\boldsymbol{a})\boldsymbol{j} + f_{x_3}(\boldsymbol{a})\boldsymbol{k},$$

and again

$$f_{\boldsymbol{u}}(\boldsymbol{a}) = \boldsymbol{f}'(\boldsymbol{a}) \cdot \boldsymbol{u} = \nabla f(\boldsymbol{a}) \cdot \boldsymbol{u}.$$

Note that this directional derivative is the component of $\nabla f(\boldsymbol{a})$ along $\boldsymbol{u}$.

Example 3 Let us find the maximum rate of increase of $f(x, y, z) = x^2 e^y + y \tan^{-1} z$ at $(1, 0, 0)$.

SOLUTION The gradient vector of f at $(1, 0, 0)$ is given by

$$\nabla f(1, 0, 0) = f_x(1, 0, 0)\boldsymbol{i} + f_y(1, 0, 0)\boldsymbol{j} + f_z(1, 0, 0)\boldsymbol{k} = 2\boldsymbol{i} + \boldsymbol{j}.$$

Thus the direction of maximum rate of increase of $f(x, y, z)$ at $(1, 0, 0)$ is given by the vector $2\boldsymbol{i} + \boldsymbol{j}$ and this maximum rate of increase is

$$|\nabla f(1, 0, 0)| = |2\boldsymbol{i} + \boldsymbol{j}| = \sqrt{5}. \quad \|$$

Finally, note that $-\nabla f(\boldsymbol{a})$ points in the direction of maximum rate of *decrease* of $f(\boldsymbol{x})$ at $\boldsymbol{a}$, and the directional derivative in that direction is $-|\nabla f(\boldsymbol{a})|$.

SUMMARY

1. *The directional derivative of $z = f(x, y)$ at $\boldsymbol{a} = (a_1, a_2)$ in the direction of the unit vector $\boldsymbol{u} = u_1\boldsymbol{i} + u_2\boldsymbol{j}$ is*

$$\frac{dz}{ds} = f_{\boldsymbol{u}}(\boldsymbol{a}) = u_1 f_x(a_1, a_2) + u_2 f_y(a_1, a_2).$$

2. *Restating No. 1 with three subscripted variables, the directional derivative of $y = f(\boldsymbol{x}) = f(x_1, x_2, x_3)$ at $\boldsymbol{a} = (a_1, a_2, a_3)$ in the direction of the unit vector $\boldsymbol{u} = u_1\boldsymbol{i} + u_2\boldsymbol{j} + u_3\boldsymbol{k}$ is*

$$\frac{dy}{ds} = f_{\boldsymbol{u}}(\boldsymbol{a}) = u_1 f_{x_1}(\boldsymbol{a}) + u_2 f_{x_2}(\boldsymbol{a}) + u_3 f_{x_3}(\boldsymbol{a}).$$

3. *If $z = f(x, y)$, the derivative $\boldsymbol{f}'(x, y) = (\partial f/\partial x)\boldsymbol{i} + (\partial f/\partial y)\boldsymbol{j}$ is also called the gradient vector $\nabla f(x, y)$. The directional derivative $f_{\boldsymbol{u}}(\boldsymbol{a})$ is then*

$$f_{\boldsymbol{u}}(\boldsymbol{a}) = \nabla f(\boldsymbol{a}) \cdot \boldsymbol{u}.$$

The same formula holds for functions of more than two variables.

4. *The gradient vector $\nabla f(\boldsymbol{a})$ points in the direction of maximum rate of increase of the function at $\boldsymbol{a}$, and has magnitude equal to that rate of increase.*

EXERCISES

In Exercises 1 through 3, find the directional derivative of the function at the given point in the indicated direction.

1. $f(x, y) = x^2 - 3xy^3$ at $(-2, 1)$ toward the origin

2. $f(x, y) = \sin^{-1}(x/y)$ at $(3, 5)$ toward $(4, 4)$

3. $f(x_1, x_2, x_3) = x_1 x_2^2 e^{x_3}$ at $(1, 3, 0)$ toward $(1, 3, -1)$

4. Let $f(x, y, z) = x^2y + \tan^{-1}xz$. Find the direction at the point $(1, -2, 1)$ in which $f(x, y, z)$ increases most rapidly, and find this maximum rate of increase. In what direction is the rate of increase of $f(x, y, z)$ minimum, and what is this minimum rate?

5. Repeat Exercise 4 for the function $f(x, y) = x^2 + x(\ln y)$ at the point $(x, y) = (2, 1)$.

6. Let f be differentiable at $\boldsymbol{a}$ and have a local maximum at $\boldsymbol{a}$. What can be said concerning the directional derivatives of f in all directions at $\boldsymbol{a}$?

7. Give an example to show that it is possible to have $f(x, y)$ such that $f_{\boldsymbol{u}}(0, 0) = 0$ for all unit vectors $\boldsymbol{u}$, and still have neither a local maximum nor a local minimum for f at $(0, 0)$.

You know that for a suitable function f, the gradient vector $\nabla f(\boldsymbol{a})$ points in the direction of maximum rate of change of $f(\mathbf{x})$ at $\boldsymbol{a}$ if $\nabla f(\boldsymbol{a}) \neq \mathbf{0}$. If $\nabla f(\boldsymbol{a}) = \boldsymbol{f}'(\boldsymbol{a}) = \mathbf{0}$, then the rate of increase of $f(\mathbf{x})$ in all directions at $\boldsymbol{a}$ is zero. In Exercises 8 through 14, find the direction or directions in which you should travel from the given point to attain maximum increase $f(\mathbf{x})$ over a short distance. The text gives no instruction in this; you just think about it in each case.

8. $f(x) = x^2$ at 0

9. $f(x) = x^3$ at 0

10. $f(x, y) = x^2 + y^2$ at $(0, 0)$

11. $f(x, y) = x^2 - y^2$ at $(0, 0)$

12. $f(x, y) = x^2 + 2xy + y^2$ at $(0, 0)$

13. $f(x, y) = x^3 - x^2 + y^2$ at $(0, 0)$

14. $f(x, y, z) = 5$ at $(1, -2, 7)$

16.6 DIFFERENTIATION OF IMPLICIT FUNCTIONS

Chapter 3 treated differentiation of implicitly defined functions of one variable, finding dy/dx given an equation of the form $G(x, y) = c$. We refresh your memory with an example.

Example 1 Let $x^2y^3 + 2xy^2 - x^3 = 3$ and find dy/dx by differentiating implicitly.

SOLUTION We have

$$x^2 \cdot 3y^2 \frac{dy}{dx} + 2xy^3 + 2x \cdot 2y \frac{dy}{dx} + 2y^2 - 3x^2 = 0,$$

so

$$\frac{dy}{dx} = \frac{3x^2 - 2xy^3 - 2y^2}{3x^2y^2 + 4xy}. \quad \|$$

An equation of the form $G(x, y, z) = c$ may define z implicitly as one or more functions of both x and y. For example, $x^2 + y^2 + z^2 = 16$ gives rise to the functions

$$z = \sqrt{16 - x^2 - y^2} \quad \text{and} \quad z = -\sqrt{16 - x^2 - y^2}.$$

We shall not describe exactly when $G(x, y, z) = c$ gives such implicitly defined functions. But for functions you will use here, the locus of $G(x, y, z) = c$ is a surface in space, and a piece of such a surface containing a point (x_0, y_0, z_0) can often be regarded as the graph of a function $z = f(x, y)$ such that $z_0 = f(x_0, y_0)$. In this case, you would like to find $\partial z/\partial x$ and $\partial z/\partial y$. This can again be accomplished by implicit differentiation.

Example 2 Let $x^2z + yz^3 - 2xy^2 = -9$, and let's find $\partial z/\partial x$ at the point $(1, -2, 1)$ on the surface.

SOLUTION Differentiate implicitly with respect to x, thinking of y as a constant, obtaining

$$x^2 \frac{\partial z}{\partial x} + z \cdot 2x + y \cdot 3z^2 \frac{\partial z}{\partial x} - 2y^2 = 0,$$

so

$$\frac{\partial z}{\partial x} = \frac{2y^2 - 2xz}{x^2 + 3yz^2}$$

and

$$\left.\frac{\partial z}{\partial x}\right|_{(1,-2,1)} = \frac{8 - 2}{1 - 6} = -\frac{6}{5}. \quad \|$$

There is an easy formula that avoids the technique of implicit differentiation. Suppose $G(x, y, z) = c$ defines $z = f(x, y)$. Then the equations

$$w = G(x, y, z), \tag{1}$$

$$x = x, \qquad y = y, \qquad z = f(x, y), \tag{2}$$

define w as a composite function of the two variables x and y. Furthermore,

$$w = G(x, y, z) = G(x, y, f(x, y)) = c$$

for all x and y, since $z = f(x, y)$ was chosen so that $G(x, y, z) = c$. Thus $\partial w/\partial x = 0$ and $\partial w/\partial y = 0$. But, by the chain rule for the sequence of functions given by (1) and (2),

$$\frac{\partial w}{\partial x} = \frac{\partial G}{\partial x}\frac{\partial x}{\partial x} + \frac{\partial G}{\partial y}\frac{\partial y}{\partial x} + \frac{\partial G}{\partial z}\frac{\partial z}{\partial x} = \frac{\partial G}{\partial x} \cdot 1 + \frac{\partial G}{\partial y} \cdot 0 + \frac{\partial G}{\partial z}\frac{\partial z}{\partial x}. \tag{3}$$

[Note that $\partial y/\partial x = 0$ from (2) where y is regarded as a function of x and y.] Since $\partial w/\partial x = 0$, you obtain from (3)

$$\frac{\partial G}{\partial x} + \frac{\partial G}{\partial z}\frac{\partial z}{\partial x} = 0,$$

so

$$\frac{\partial z}{\partial x} = -\frac{\partial G/\partial x}{\partial G/\partial z}. \tag{4}$$

Formula (4) for $\partial z/\partial x$ is valid wherever $\partial G/\partial z \neq 0$, and it can be shown that, for a nice function G, this is a condition that guarantees that $G(x, y, z) = c$ does define z implicitly as a function of x and y.

Example 3 Let's do Example 2 again using the formula (4).

SOLUTION From

$$G(x, y, z) = x^2z + yz^3 - 2xy^2 = -9,$$

you obtain

$$\frac{\partial G}{\partial x} = 2xz - 2y^2 \quad \text{and} \quad \frac{\partial G}{\partial z} = x^2 + 3yz^2,$$

so

$$\frac{\partial z}{\partial x} = -\frac{2xz - 2y^2}{x^2 + 3yz^2} = \frac{2y^2 - 2xz}{x^2 + 3yz^2}.$$

The derivative is then computed at $(1, -2, 1)$ as in Example 2, but the messy implicit differentiation is gone. ‖

The formula

$$\frac{\partial z}{\partial x} = -\frac{\partial G/\partial x}{\partial G/\partial z}$$

is fairly easy to remember in this Leibniz notation; the ∂G's "cancel" to give what you want, *but you must remember the minus sign also.* This is one place where the Leibniz notation lets you down just a bit.

Of course, a similar argument shows that

$$\frac{\partial z}{\partial y} = -\frac{\partial G/\partial y}{\partial G/\partial z}.$$

Indeed, if $G(x_1, x_2, x_3) = c$ and you solve for x_i in terms of the other x's, similar arguments show that

$$\frac{\partial x_i}{\partial x_j} = -\frac{\partial G/\partial x_j}{\partial G/\partial x_i}, \qquad i \neq j. \tag{5}$$

You can even use the formula to solve the Chapter 3 implicit differentiation problems, finding dy/dx if $G(x, y) = c$, for

$$\frac{dy}{dx} = -\frac{\partial G/\partial x}{\partial G/\partial y}.$$

To illustrate, we solve Example 1 again, finding dy/dx if

$$G(x, y) = x^2y^3 + 2xy^2 - x^3 = 3.$$

Then

$$\frac{\partial G}{\partial x} = 2xy^3 + 2y^2 - 3x^2 \qquad \text{and} \qquad \frac{\partial G}{\partial y} = 3x^2y^2 + 4xy,$$

so

$$\frac{dy}{dx} = -\frac{2xy^3 + 2y^2 - 3x^2}{3x^2y^2 + 4xy} = \frac{3x^2 - 2xy^3 - 2y^2}{3x^2y^2 + 4xy},$$

as obtained in Example 1.

Recall that a vector normal to a surface $z = f(x, y)$ given implicitly by $G(x, y, z) = c$ is

$$\frac{\partial z}{\partial x}\boldsymbol{i} + \frac{\partial z}{\partial y}\boldsymbol{j} - \boldsymbol{k} = -\frac{\partial G/\partial x}{\partial G/\partial z}\boldsymbol{i} - \frac{\partial G/\partial y}{\partial G/\partial z}\boldsymbol{j} - \boldsymbol{k}. \tag{6}$$

Multiplying the vector in (6) by $-\partial G/\partial z$, you obtain, as a vector normal to the surface $G(x, y, z) = c$,

$$\frac{\partial G}{\partial x}\boldsymbol{i} + \frac{\partial G}{\partial y}\boldsymbol{j} + \frac{\partial G}{\partial z}\boldsymbol{k}. \tag{7}$$

But (7) is the gradient vector ∇G. The surface $G(x, y, z) = c$ is called a *level surface* of the function $G(x, y, z)$, for it consists of all points (x, y, z) where the function has the "level" c. Summarizing,

the gradient vector of a function at a point is perpendicular to the level surface of the function through that point.

Of course, for a function $G(x, y)$ of just two variables, $G(x, y) = c$ is a curve in the plane, called a *level curve* of G. Again, the gradient vector ∇G is perpendicular to the level curve at each point. This follows at once from the statement above if you consider the locus of $G(x, y) = c$ in space, which is a vertical cylinder intersecting the x,y-plane in the curve $G(x, y) = c$.

Example 4 The curve $y = x^2 - 1$ is a level curve in the plane of $G(x, y) = y - x^2$. Thus a perpendicular vector to the curve at (2, 3) is given by $\nabla G(2, 3)$. We obtain

$$\frac{\partial G}{\partial x} = -2x \qquad \text{and} \qquad \frac{\partial G}{\partial y} = 1,$$

so $\nabla G = -2x\boldsymbol{i} + \boldsymbol{j}$, and $\nabla G(2, 3) = -4\boldsymbol{i} + \boldsymbol{j}$. The curve and this vector are shown in Fig. 16.7. ||

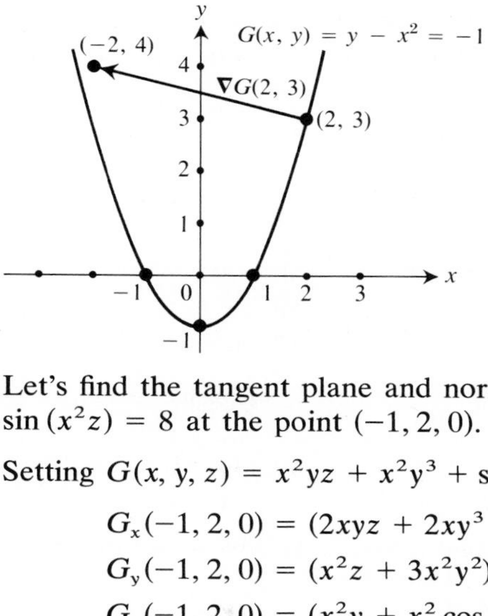

16.7

Example 5 Let's find the tangent plane and normal line to the surface $x^2yz + x^2y^3 + \sin(x^2z) = 8$ at the point $(-1, 2, 0)$.

SOLUTION Setting $G(x, y, z) = x^2yz + x^2y^3 + \sin(x^2z)$, we have

$$G_x(-1, 2, 0) = (2xyz + 2xy^3 + 2xz\cos(x^2z))|_{(-1,2,0)} = -16,$$
$$G_y(-1, 2, 0) = (x^2z + 3x^2y^2)|_{(-1,2,0)} = 12,$$
$$G_z(-1, 2, 0) = (x^2y + x^2\cos(x^2z))|_{(-1,2,0)} = 3.$$

Therefore the normal vector (7) is $-16\boldsymbol{i} + 12\boldsymbol{j} + 3\boldsymbol{k}$. The tangent plane at $(-1, 2, 0)$ has equation

$$16x - 12y - 3z = -40,$$

and the normal line has parametric equations

$$x = -1 - 16t, \qquad y = 2 + 12t, \qquad z = 3t. \quad \|$$

SUMMARY

1. *If $G(x, y, z) = c$ defines $z = f(x, y)$, then $\partial z/\partial x$ and $\partial z/\partial y$ can be found using either implicit differentiation or the formulas*

$$\frac{\partial z}{\partial x} = -\frac{\partial G/\partial x}{\partial G/\partial z} \quad \text{and} \quad \frac{\partial z}{\partial y} = -\frac{\partial G/\partial y}{\partial G/\partial z}.$$

2. *If $G(\mathbf{x}) = G(x_1, x_2, x_3) = c$ defines x_i as a function of the other x's, then $\partial x_i/\partial x_j$ for $i \neq j$ can be found by implicit differentiation or using the formula*

$$\frac{\partial x_i}{\partial x_j} = -\frac{\partial G/\partial x_j}{\partial G/\partial x_i}.$$

3. *Given a level surface $G(x, y, z) = c$, the gradient vector*

$$\nabla G = \frac{\partial G}{\partial x}\mathbf{i} + \frac{\partial G}{\partial y}\mathbf{j} + \frac{\partial G}{\partial z}\mathbf{k}$$

is normal to the surface at each point on the surface.

EXERCISES

In Exercises 1 through 8, find the desired derivative using formulas like those in Eqs. (4) and (5).

1. $dy/dx|_{(1,-1)}$ if $x^2y - x\sin(\pi y) - y^3 = 0$

2. $dx/dy|_{(0,2)}$ if $e^{xy} - (3xy + 2y)^3 = -63$

3. $\partial z/\partial x|_{(-1,2,0)}$ if $x\sin(yz) - 3x^2z + ye^z = 2$

4. $\partial x_3/\partial x_2|_{(1,-1,1)}$ if $x_1^2 - 2x_2x_3^4 = 3x_1x_3^2$

5. $\partial y/\partial z|_{(0,1,3)}$ if $\tan^{-1}(x + y) + \ln(xz + y) = \pi/4$

6. $\partial x_1/\partial x_3|_{(1,2,\pi)}$ if $\sin(x_1x_3) - x_1^3x_2^2 = -4$

7. $\partial x/\partial z|_{(3,0,0)}$ if $\sin^{-1}z + x^2e^{y^2} = 12 - xe^y$

8. $\partial y/\partial x|_{(2,0,2)}$ if $\ln(xy + 1) + 3xz^2 = 24 + \tan(x - z)$

In Exercises 9 through 12, use implicit differentiation to find the desired derivative.

9. $\partial x/\partial z|_{(-1,0,1)}$ if $\cos(x^2y) - yz^3 + xz^2 = 0$

10. $\partial x_2/\partial x_3|_{(0,1,-1)}$ if $x_1x_2^3 - 3x_2x_3 + x_1x_3^4 = 3$

11. $\partial z/\partial y|_{(-1,2,0)}$ if $e^{xyz} - \ln(xz + 1) = 1$

12. $\partial x_3/\partial x_1|_{(1,2,0)}$ if $x_2^2 - 4x_1x_2^2 = 7x_1x_3 - 12x_1^3$

In Exercises 13 through 15, find the equation of the tangent line or plane and parametric equations of the normal line to the given curve or surface at the given point.

13. $x^3y - 3y^2 = -2$ at $(1, 1)$

14. $3xe^y + xy^3 = 2 + x$ at $(1, 0)$

15. $\sin(x^2y) + (3x + 2z)^5 = x$ at $(-1, 0, 1)$

16. Show that if $G(x, y)$ has continuous second partial derivatives and if the curve $G(x, y) = c$ defines y as a twice-differentiable function of x, then at a point (x, y) where $G_y(x, y) \neq 0$, we have

$$\frac{d^2y}{dx^2} = -\frac{G_y^2G_{xx} - 2G_xG_yG_{xy} + G_x^2G_{yy}}{G_y^3}.$$

17. Use the result in Exercise 16 to compute $d^2y/dx^2|_{(0,1)}$ for y defined implicitly as a function of x by $x^3 - 3xy^2 + 4y^3 = 4$.

18. Show that the curves

$$5x^4y - 10x^2y^3 + y^5 = 4$$

and

$$x^5 - 10x^3y^2 + 5xy^4 = -4$$

are orthogonal at all points of intersection.

19. Show that the surfaces

$$x^2 - 2y^2 + z^2 = 0 \quad \text{and} \quad xyz = 1$$

are orthogonal at all points of intersection. (Surfaces are *orthogonal* at a point of intersection if their normal lines at that point are orthogonal.)

20. Show that the surfaces

$$x + y^2 + 2z^3 = 4$$

and

$$12x - 3(\ln y) + z^{-1} = 13$$

are orthogonal at all points of intersection.

exercise sets for chapter 16

review exercise set 16.1

1. If $w = f(x, y, z) = xz^2 + y^2e^{xz}$, find:
a) $\partial w/\partial x$ b) $\partial^2 w/\partial x^2$ c) $\partial^2 w/\partial x\,\partial y$

2. Find the equation of the plane tangent to the surface

$$z = \frac{x + y}{x - y}$$

at the point (2, 1, 3).

3. Let $w = f(x, y, z) = x^2y - xy^2z^3$. Find the vector derivative $\boldsymbol{f}'(1, -1, 2)$.

4. Use a differential to estimate

$$\frac{(2.05)^4 - (3.97)^2}{(1.08)^5}.$$

5. Let $z = f(x, y) = e^{x^2y}$ while $x = 2st$ and $y = s^2 + t^3$. Use the chain rule to find $\partial z/\partial t$ at the point $(s, t) = (1, -1)$.

6. Let $z = f(x_1, x_2, x_3)$ while $x_1 = g_1(t_1, t_2)$, $x_2 = g_2(t_1, t_2)$, and $x_3 = g_3(t_1, t_2)$. Suppose that when $\boldsymbol{t} = \boldsymbol{a}$, you obtain a vector $\boldsymbol{x} = \boldsymbol{b}$, and that

$$\frac{\partial z}{\partial t_1}(\boldsymbol{a}) = -4, \qquad \frac{\partial z}{\partial \boldsymbol{x}}(\boldsymbol{b}) = (3, 0, c),$$

and

$$\frac{\partial \boldsymbol{x}}{\partial t_1}(\boldsymbol{a}) = (-4, 3, 2).$$

Find the value of c.

7. Find the directional derivative of $z = f(x, y) = x^2y^3 - 3y^2$ at $(-3, 4)$ in the direction toward the origin.

8. Find the direction for maximum rate of increase of $f(x, y, z) = y^2 \sin xz^2$ at the point $(0, 3, -1)$, and the magnitude of this maximum rate of increase.

9. Let y be defined implicitly as a function of x and z by $x^2y - 3xz^2 + y^3 = -13$. Find $\partial y/\partial x$ at the point $(1, -2, 1)$.

10. Find parametric equations of the line normal to the surface $x^2y - 3xz^2 + y^3 = -13$ at the point $(1, -2, 1)$.

review exercise set 16.2

1. If $y = x_1 \sin x_2x_3 - x_2{}^2x_1{}^3$, find:
a) $\partial y/\partial x_2$ b) $\partial^2 y/\partial x_1\partial x_3$

2. Find the equation of the line normal to the surface

$$z = \frac{x^2 - y}{x + y}$$

at the point (2, 1, 1).

3. Let $y = f(x_1, x_2) = x_2{}^3 + x_1 \cos x_2$. Find the vector derivative $\boldsymbol{f}'(1, 0)$.

4. Use a differential to estimate $(2.03)^2 \cos(-0.05)$.

5. Let $z = f(x, y)$ while $x = g_1(t)$ and $y = g_2(t)$. If $g_1(1) = -3$ and $g_2(1) = 4$, while

$$\left.\frac{dz}{dt}\right|_{t=1} = 2, \qquad \left.\frac{dx}{dt}\right|_{t=1} = 4,$$

$$\left.\frac{dy}{dt}\right|_{t=1} = -1, \qquad \left.\frac{\partial z}{\partial x}\right|_{(-3,4)} = 3,$$

find $f_y(-3, 4)$.

6. Let $x_1 = g_1(t_1, t_2)$ and $x_2 = g_2(t_1, t_2)$ be differentiable functions. Suppose $g_1(-1, 2) = -1$ while $g_2(-1, 2) = 2$. Let

$$\left.\frac{\partial x_1}{\partial t}\right|_{(-1,2)} = (3, -4) \quad \text{and} \quad \left.\frac{\partial x_2}{\partial t}\right|_{(-1,2)} = (1, 2).$$

If $f(t_1, t_2) = g_2(g_1(t_1, t_2), g_2(t_1, t_2))$, find $f_{t_2}(-1, 2)$.

7. Find the directional derivative of $x^3y + (x/y)$ at $(-1, 1)$ in the direction toward $(2, -3)$.

8. Find the direction of the maximum rate of increase of $f(x, y) = x^2e^{xy}$ at the point $(3, 0)$, and then find the magnitude of this rate of change.

9. If x is defined implicitly as a function of y and z by $y^2 \cos x + xz^3 - 3y^2z = -5$, find $\partial x/\partial z$ at $(0, 1, 2)$.

10. Find the equation of the plane tangent to the surface $x^3y + y^3z + z^3x = -1$ at the point $(1, 1, -1)$.

more challenging exercises 16

1. Find the appropriate "subtract-and-add trick" that could be used to prove the 3-variable analogue of Theorem 16.1 for a function $w = f(x, y, z)$. That is, give the analogue of Eq. (6) of Section 16.3 for $w = f(x, y, z)$.

The remaining exercises introduce power-series representation for a function of two variables.

2. Consider the polynomial function

$$\begin{aligned} P(x, y) = {} & a_{00} + a_{10}(x - x_0) + a_{01}(y - y_0) + \cdots \\ & + a_{ij}(x - x_0)^i(y - y_0)^j + \cdots \\ & + a_{0n}(y - y_0)^n \end{aligned}$$

for all nonnegative integers i and j where $i + j \le n$. Let $f(x, y)$ have continuous partial derivatives of all orders $\le n$ at (x_0, y_0). Find a formula for a_{ij} if $P(x, y)$ has the same partial derivatives as $f(x, y)$ of all orders $\le n$ at (x_0, y_0); that is, if

$$\left.\frac{\partial^{i+j}P}{\partial x^i\,\partial y^j}\right|_{(x_0,y_0)} = \left.\frac{\partial^{i+j}f}{\partial x^i\partial y^j}\right|_{(x_0,y_0)}$$

We define the **nth Taylor polynomial** $T_n(x, y)$ *for* $f(x, y)$ *at* (x_0, y_0) *to be*

$$T_n(x, y) = \sum_{i+j\le n}\left[\frac{1}{i!\,j!}\cdot\left.\frac{\partial^{i+j}}{\partial x^i\,\partial y^j}\right|_{(x_0,y_0)}\cdot(x - x_0)^i(y - y_0)^j\right]. \tag{1}$$

3. Find the Taylor polynomial $T_3(x, y)$ for $\sin(x + y)$ at $(0, 0)$.

4. Explain how the answer to Exercise 3 is predictable in terms of Taylor series for a function of one variable.

5. Following the idea of Exercise 4, predict the polynomial $T_4(x, y)$ for e^{xy} at $(0, 0)$. Then verify your answer by computing (1) with $n = 4$.

6. Find the Taylor polynomial $T_4(x, y)$ for e^{xy} at $(0, 1)$.

Taylor's Theorem *Let $f(x, y)$ have continuous partial derivatives of all orders $\le(n + 1)$ in some disk with center at (x_0, y_0). Then for each $(x, y) \ne (x_0, y_0)$ in the disk, there exists (c_1, c_2), depending on x and y and strictly between (x_0, y_0) and (x, y) on the line segment joining them, such that*

$$E_n(x, y) = \sum_{\substack{i,j \\ i+j=n+1}}\left[\frac{1}{i!\,j!}\cdot\left.\frac{\partial^{i+j}f}{\partial x^i\,\partial y^j}\right|_{(c_1,c_2)}\cdot(x - x_0)^i(y - y_0)^j\right],$$

where $E_n(x, y) = f(x, y) - T_n(x, y)$.

7. Estimate $(1.02)^2 \ln(0.97)$ using a differential, and then use Taylor's Theorem to find a bound for the error.

8. Use the Taylor Polynomial $T_2(x, y)$ at $(0, 1)$ for $y^3 \cos x$ to estimate $(1.03)^3 \cos(-0.02)$, and then find a bound for the error.

9. Let f be a function of two variables with continuous coordinate derivatives of all orders $\leq n$. Let us introduce the *operation notations*

$$\left(\frac{\partial}{\partial x} h + \frac{\partial}{\partial y} k\right) f = \frac{\partial f}{\partial x} h + \frac{\partial f}{\partial y} k,$$

$$\left(\frac{\partial}{\partial x} h + \frac{\partial}{\partial y} k\right)^2 f = \frac{\partial^2 f}{\partial x^2} h^2 + 2 \frac{\partial^2 f}{\partial x\, \partial y} hk + \frac{\partial^2 f}{\partial y^2} k^2,$$

$$\vdots \qquad\qquad \vdots$$

$$\left(\frac{\partial}{\partial x} h + \frac{\partial}{\partial y} k\right)^n f = \sum_{i=0}^{n} \binom{n}{i} \frac{\partial^n f}{\partial x^i\, \partial y^{n-i}} h^i k^{n-i}.$$

Show that the nth Taylor polynomial for f at (x_0, y_0) is given by

$$T_n(x, y) = f(x_0, y_0)$$

$$+ \sum_{i=1}^{n} \frac{1}{i!} \left(\frac{\partial}{\partial x}\bigg|_{(x_0, y_0)} (x - x_0) + \frac{\partial}{\partial y}\bigg|_{(x_0, y_0)} (y - y_0)\right)^i f.$$

applications of partial derivatives

17

17.1 MAXIMA AND MINIMA

Let $z = f(x, y)$ and suppose $f(x, y) \leq f(x_0, y_0)$ for all (x, y) inside some sufficiently small circle with center at (x_0, y_0). Then $f(x, y)$ has a **local maximum** or **relative maximum** of $f(x_0, y_0)$ at the point (x_0, y_0). Of course, if the inequality were reversed so that $f(x, y) \geq f(x_0, y_0)$ for all such (x_0, y_0), then $f(x_0, y_0)$ would be a **local minimum** or **relative minimum**. This is an obvious generalization of the same notions for a function of one variable; and still further generalizations to functions of three variables are clear.

Of course you want to find such local maxima and minima of $z = f(x, y)$. Suppose that $f_x(x, y)$ and $f_y(x, y)$ exist. If $f(x_0, y_0)$ is a local maximum, then the function $g(x) = f(x, y_0)$ has a local maximum $g(x_0)$ at x_0, and consequently

$$g'(x_0) = f_x(x_0, y_0) = 0.$$

Also, you must have a local maximum at y_0 of $h(y) = f(x_0, y)$, so

$$h'(y_0) = f_y(x_0, y_0) = 0.$$

A similar argument shows that first partial derivatives must be zero if $f(x_0, y_0)$ is a local minimum, and indeed, the same results hold for functions of more than two variables.

Theorem 17.1 *If a function has first-order partial derivatives, then these derivatives will be zero at any point where the function has a local maximum or a local minimum.*

Consequently you can find all *candidates* for local extrema of differentiable functions by finding those points where all first-order partial derivatives are zero simultaneously.

So far the situation is just the same as for a function of one variable. From here on, more complicated things can happen for functions of two or more variables, as the following example shows.

Example 1 Let f be defined by $f(x, y) = 1 - x^2 + y^2$. Then $f_x(0, 0) = f_y(0, 0) = 0$. Clearly f has neither a local maximum nor a local minimum at $(0, 0)$, for $f(0, 0) = 1$, but $f(x_1, 0) < 1$ and $f(0, y_1) > 1$ for nonzero x_1 and y_1 close to 0. The graph of f is shown in Fig. 17.1. ‖

For a function of one variable with first derivative 0 at x_0, the sign of a nonzero second derivative at x_0 determines whether the function has a local minimum or a local maximum at x_0. The situation is not so simple for functions of more than one variable, as we show in the next example.

Example 2 Let f be defined by $f(x, y) = x^2 + 4xy + y^2$. Then $f_x(0, 0) = f_y(0, 0) = 0$, and all partial derivatives of order two are positive at $(0, 0)$; namely,

$$f_{xx}(0, 0) = 2, \qquad f_{xy}(0, 0) = 4, \qquad f_{yy}(0, 0) = 2.$$

However, f still does not have a local maximum or a local minimum at

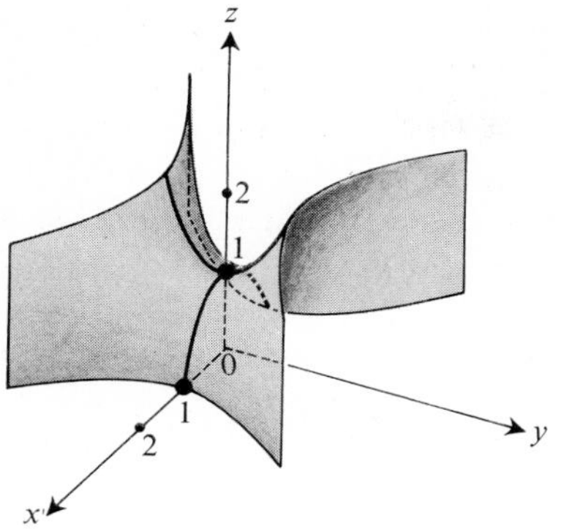

17.1

(0, 0). To see this, note that

$$f(t, mt) = (1 + 4m + m^2)t^2$$

and that $1 + 4m + m^2$ assumes values of opposite signs for $m = 1$ and $m = -1$. Thus on the line $x = t$, $y = t$ corresponding to $m = 1$, the function f assumes positive values, while on the line $x = t$, $y = -t$ corresponding to $m = -1$, the function assumes negative values. Since $f(0, 0) = 0$, you see that f has neither a local maximum nor a local minimum at (0, 0). ‖

We state without proof a second-order derivative test for a local maximum or minimum of a function of two variables. A proof can be found in any advanced calculus text.

Theorem 17.2 *Let $f(x, y)$ be a function of two variables with continuous partial derivatives of orders ≤ 2 in some disk with center (x_0, y_0), and suppose that*

$$f_x(x_0, y_0) = f_y(x_0, y_0) = 0,$$

while not all second-order partial derivatives are zero at (x_0, y_0). Let

$$A = f_{xx}(x_0, y_0), \qquad B = f_{xy}(x_0, y_0), \qquad \text{and} \qquad C = f_{yy}(x_0, y_0).$$

Then the function f has either a local maximum or a local minimum at (x_0, y_0) if $AC - B^2 > 0$. In this case, the function has a local minimum if $A > 0$ or $C > 0$, and has a local maximum if $A < 0$ or $C < 0$. If $AC - B^2 < 0$, then f has neither a local maximum nor a local minimum at (x_0, y_0), but rather a saddle point like that in Fig. 17.1.

Referring to Example 2, you see that there $A = 2$, $B = 4$, and $C = 2$, so $AC - B^2 = 4 - 16 = -12 < 0$. Consequently $f(x, y) = x^2 + 4xy + y^2$ had neither a local maximum nor a local minimum at (0, 0).

We illustrate with two more examples.

Example 3 Let's find all local minima and maxima of the function f where

$$f(x, y) = x^2 - 2xy + 2y^2 - 2x + 2y + 4.$$

SOLUTION We have

$$f_x(x, y) = 2x - 2y - 2 \qquad \text{and} \qquad f_y(x, y) = -2x + 4y + 2.$$

In order for both partial derivatives to be zero, you must have

$$2x - 2y - 2 = 0,$$
$$-2x + 4y + 2 = 0.$$

Adding these equations, you find that

$$2y = 0$$

or

$$y = 0.$$

Then you must have $x = 1$. Therefore $(1, 0)$ is the only candidate for a point where f can have a local maximum or minimum. Computing second partial derivatives,

$$A = f_{xx}(1, 0) = 2, \qquad B = f_{xy}(1, 0) = -2, \qquad C = f_{yy}(1, 0) = 4.$$

Therefore $AC - B^2 = 8 - 4 = 4 > 0$. Since $A > 0$, you know by Theorem 17.2 that f has a local minimum of $f(1, 0) = 3$ at $(1, 0)$. ||

Example 4 Let's find all local maxima and minima of the function f where

$$f(x, y) = x^3 - y^3 - 3xy + 4.$$

SOLUTION You have

$$f_x(x, y) = 3x^2 - 3y \qquad \text{and} \qquad f_y(x, y) = -3y^2 - 3x.$$

Setting these derivatives equal to 0, you obtain from the first one $y = x^2$, and then from the second one you have $x^4 + x = 0$. Now

$$\begin{aligned} x^4 + x &= x(x^3 + 1) \\ &= x(x + 1)(x^2 - x + 1) = 0 \end{aligned}$$

has only $x = 0$ and $x = -1$ as real solutions.

At $x = 0$, you obtain $y = 0$, and

$$A = f_{xx}(0, 0) = 0, \qquad B = f_{xy}(0, 0) = -3, \qquad C = f_{yy}(0, 0) = 0.$$

In this case, $AC - B^2 = -9 < 0$, so f has neither a local maximum nor a local minimum at $(0, 0)$.

At $x = -1$, you have $y = 1$, and

$$A = f_{xx}(-1, 1) = -6, \qquad B = f_{xy}(-1, 1) = -3, \qquad C = f_{yy}(-1, 1) = -6,$$

so $AC - B^2 = 36 - 9 = 27 > 0$. Since $A < 0$, you see in this case that f has a local maximum of $f(-1, 1) = 5$ at $(-1, 1)$. ||

SUMMARY

1. *If a function has a local maximum or minimum at a point where the first-order partial derivatives exist, then these first-order partial derivatives must be zero there.*
2. *Let $f(x, y)$ be a function of two variables with continuous partial derivatives of orders ≤ 2 in a neighborhood of (x_0, y_0), and suppose that*

$$f_x(x_0, y_0) = f_y(x_0, y_0) = 0,$$

while not all second-order partial derivatives are zero at (x_0, y_0). Let

$$A = f_{xx}(x_0, y_0), \qquad B = f_{xy}(x_0, y_0), \qquad \text{and} \qquad C = f_{yy}(x_0, y_0).$$

Then the function $f(x, y)$ has either a local maximum or a local minimum at (x_0, y_0) if $AC - B^2 > 0$. In this case, the function has a local minimum if $A > 0$ or $C > 0$, and has a local maximum if $A < 0$ or $C < 0$. If $AC - B^2 < 0$, then $f(x, y)$ has neither a local maximum nor a local minimum at (x_0, y_0), but rather a saddle point there.

EXERCISES

In Exercises 1 through 12, find all local maxima and minima of the function.

1. $\sin xy$

2. $x^2 + y^2 - 4$

3. $e^{x^2+y^2}$

4. $x^2 + y^2 + 4x - 2y + 3$

5. $x^2 - y^2 + 2x + 8y - 7$

6. $x^2 + y^2 + 4xy - 2x + 6y$

7. $3x^2 + y^2 - 3xy + 6x - 4y$

8. $\ln(x^2 + 2xy + 2y^2 - 2x - 8y + 20)$. [*Hint.* Use the fact that $\ln u$ is an increasing function of u.]

9. $x^3 + 2y^3 - 3x^2 - 24y + 16$

10. $x^3 + y^3 + 3xy - 6$

11. $x^4 + 2y^2 + 3z^2 - 2x^2 + 4y - 12z + 3$

12. $2x^2 - 2y^2 + 4yz - 3z^2 - x^4 + 5$

13. Consider a function f of two variables having continuous partial derivatives of orders ≤ 2 with $f_x(0, 0) = f_y(0, 0) = 0$, and let

$$A = f_{xx}(0, 0), \qquad B = f_{xy}(0, 0),$$

and

$$C = f_{yy}(0, 0).$$

Suppose that $AC - B^2 = 0$ with A, B, and C not all zero.

a) Give an example of such a function f with a local maximum $(0, 0)$.

b) Give an example of such a function f with a local minimum at $(0, 0)$.

c) Give an example of such a function f with neither a local maximum nor a local minimum at $(0, 0)$.

d) What is the significance of this exercise?

In Exercises 14 through 18, use common sense to find a point at which the function assumes its maximum value on the square where $-1 \leq x \leq 1$ and $-1 \leq y \leq 1$. Then find a point where it assumes its minimum value on this square.

14. $x^2 + y^2$ **15.** xy **16.** $y - 2x$ **17.** $x^2 + y^2 - xy$ **18.** $x^2 - y^2 + y$

17.2 LAGRANGE MULTIPLIERS

Sometimes you want to maximize or minimize a function $f(x, y)$ subject to some relation $g(x, y) = 0$. The relation $g(x, y) = 0$ is called a *side condition* or a *constraint*. You already met this problem in the maximum and minimum word problems back in Chapter 5.

Example 1 Let's describe a method to find the dimensions of the rectangle of maximum area that can be inscribed in a semicircle of radius a.

SOLUTION From Fig. 17.2, you see that the problem is to maximize $f(x, y) = 2xy$ subject to the constraint $x^2 + y^2 = a^2$. In Chapter 5, we would have solved this problem by using the constraint to express the area as a function of *one* variable as follows:

$$\text{Area} = 2xy,$$

$$x^2 + y^2 - a^2 = 0, \quad \text{so} \quad y = \sqrt{a^2 - x^2},$$

$$\text{Area} = 2x\sqrt{a^2 - x^2}.$$

We then took the derivative of this area function, set it equal to zero, solved that equation, etc. We continue this example in a moment. ||

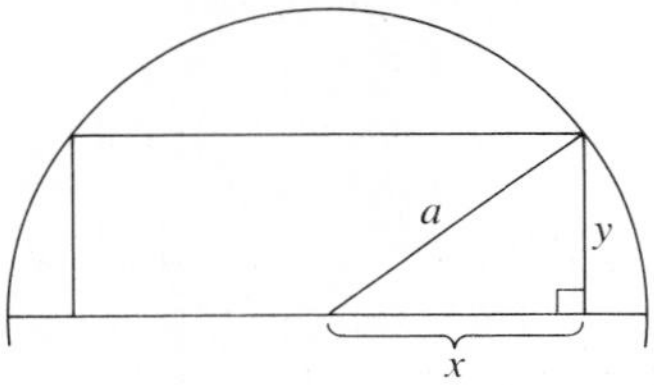

17.2

This section presents an alternative method for finding extrema, subject to constraints. This other method, the technique of Lagrange multipliers, is frequently just as much work as the method described in the preceding example. The constraint $g(x, y) = 0$ is a curve in the plane. Since this curve is a level curve of $g(x, y)$, you know that

$$\nabla g = \frac{\partial g}{\partial x}\boldsymbol{i} + \frac{\partial g}{\partial y}\boldsymbol{j} \tag{1}$$

is perpendicular to the curve at each point on the curve. Turning to the function $f(x, y)$ to be maximized or minimized, you know that

$$\nabla f = \frac{\partial f}{\partial x}\boldsymbol{i} + \frac{\partial f}{\partial y}\boldsymbol{j} \tag{2}$$

points in the direction of maximum increase of the function $f(x, y)$ at each point. Furthermore, if $\boldsymbol{u}$ is a unit vector,

$$\nabla f \cdot \boldsymbol{u} = \text{Directional derivative of } f \text{ in the direction } \boldsymbol{u}. \tag{3}$$

Now at a point (x_0, y_0) on the curve $g(x, y) = 0$ where $f(x, y)$ has a local extremum *when considered only on the curve*, the directional derivative of f along the curve must be zero. That is ∇f must be normal to the curve at that point. Consequently, ∇f and ∇g must be *parallel* at such a point, so there must exist λ such that

$$\nabla f = \lambda(\nabla g). \tag{4}$$

Equation (4), together with $g(x, y) = 0$, leads to the three conditions

$$\frac{\partial f}{\partial x} = \lambda\frac{\partial g}{\partial x}, \qquad \frac{\partial f}{\partial y} = \lambda\frac{\partial g}{\partial y}, \qquad g(x, y) = 0. \tag{5}$$

With luck, these three conditions (5) in the three unknowns x, y, and λ, can be solved. Equations (5) are the conditions of the method of Lagrange multipliers. The method itself is nothing more than a handy device for obtaining the conditions (5). Let

$$L(x, y, \lambda) = f(x, y) - \lambda g(x, y). \tag{6}$$

The conditions (5) are then equivalent to these conditions, in the same order:

$$\frac{\partial L}{\partial x} = 0, \qquad \frac{\partial L}{\partial y} = 0, \qquad \frac{\partial L}{\partial \lambda} = 0. \tag{7}$$

The variable λ is called the *Lagrange multiplier.*

A point satisfying (7) is a *candidate* for a point where $f(x, y)$ has a local maximum or minimum value, subject to $g(x, y) = 0$. If you know that such a local maximum or minimum exists, and if you can find all solutions of (7), then computation of $f(x, y)$ at those points may indicate which is the desired extremum.

Example 2 Let's continue Example 1, and solve the problem of maximizing the function $f(x, y) = 2xy$, subject to $x^2 + y^2 - a^2 = 0$.

SOLUTION First, we form

$$L(x, y, \lambda) = 2xy - \lambda(x^2 + y^2 - a^2).$$

Then we set

$$\frac{\partial L}{\partial x} = 2y - 2x\lambda = 0, \tag{8}$$

$$\frac{\partial L}{\partial y} = 2x - 2y\lambda = 0, \tag{9}$$

$$\frac{\partial L}{\partial \lambda} = -x^2 - y^2 + a^2 = 0. \tag{10}$$

Substituting $y = x\lambda$ and $x = y\lambda$ in (10), you obtain

$$-y^2\lambda^2 - x^2\lambda^2 + a^2 = 0, \qquad \text{or} \qquad -(x^2 + y^2)\lambda^2 + a^2 = 0.$$

From (10), you then know that

$$-a^2\lambda^2 + a^2 = 0 \qquad \text{or} \qquad a^2(-\lambda^2 + 1) = 0,$$

so

$$\lambda^2 = 1 \qquad \text{and} \qquad \lambda = \pm 1.$$

The value $\lambda = -1$ would give $y = -x$, which is impossible for our geometric problem. Thus $\lambda = 1$, and $y = x$. From (10), you obtain

$$2x^2 = a^2 \qquad \text{so} \qquad x = \frac{a}{\sqrt{2}},$$

and $(x, y) = (a/\sqrt{2}, a/\sqrt{2})$ is the desired maximum. That is, you know from geometry that a maximum exists, and we found only one candidate for it. ||

We can make analogous use of Lagrange multipliers to handle situations with more variables or more constraints. For example, to maximize $f(x, y, z)$ subject to $g(x, y, z) = 0$, the gradient

$$\nabla f = \frac{\partial f}{\partial x}\boldsymbol{i} + \frac{\partial f}{\partial y}\boldsymbol{j} + \frac{\partial f}{\partial z}\boldsymbol{k}$$

must be perpendicular to the level surface $g(x, y, z) = 0$, and consequently, ∇f must be parallel to ∇g, so again $\nabla f = \lambda(\nabla g)$. The four conditions

$$\frac{\partial f}{\partial x} = \lambda\frac{\partial g}{\partial x}, \qquad \frac{\partial f}{\partial y} = \lambda\frac{\partial g}{\partial y}, \qquad \frac{\partial f}{\partial z} = \lambda\frac{\partial g}{\partial z}, \qquad g(x, y, z) = 0$$

in x, y, z, and λ can again be concisely expressed as

$$\frac{\partial L}{\partial x} = 0, \qquad \frac{\partial L}{\partial y} = 0, \qquad \frac{\partial L}{\partial z} = 0, \qquad \frac{\partial L}{\partial \lambda} = 0,$$

where $L(x, y, z, \lambda) = f(x, y, z) - \lambda g(x, y, z)$.

For another case, suppose you wish to maximize or minimize $f(x, y, z)$ subject to *two* constraints $g(x, y, z) = 0$ and $h(x, y, z) = 0$. The locus of the two constraints $g(x, y, z) = 0$ and $h(x, y, z) = 0$ is the curve of intersection of the two level surfaces. At a point where $f(x, y, z)$ is maximum or minimum, subject to these constraints, the directional derivative along this curve must be zero. Consequently, ∇f must be perpendicular to the curve. Since both ∇g and ∇h are perpendicular to the curve, it must be that ∇f lies in the plane determined by the vectors* ∇g and ∇h. This time there are two Lagrange multipliers. You must have

$$\nabla f = \lambda_1(\nabla g) + \lambda_2(\nabla h)$$

---

* One hopes these will not be parallel.

for some λ_1 and λ_2. This leads to

$$\frac{\partial f}{\partial x} + \lambda_1\frac{\partial g}{\partial x} + \lambda_2\frac{\partial h}{\partial x}, \qquad \frac{\partial f}{\partial y} = \lambda_1\frac{\partial g}{\partial y} + \lambda_2\frac{\partial h}{\partial y}, \qquad \frac{\partial f}{\partial z} = \lambda_1\frac{\partial g}{\partial z} + \lambda_2\frac{\partial h}{\partial z},$$

and

$$g(x, y, z) = 0, \qquad h(x, y, z) = 0.$$

These are five equations in five unknowns, and can be written as

$$\frac{\partial L}{\partial x} = 0, \quad \frac{\partial L}{\partial y} = 0, \quad \frac{\partial L}{\partial z} = 0, \quad \frac{\partial L}{\partial \lambda_1} = 0, \quad \frac{\partial L}{\partial \lambda_2} = 0,$$

where

$$L(x, y, z, \lambda_1, \lambda_2) = f(x, y, z) - \lambda_1 g(x, y, z) - \lambda_2 h(x, y, z).$$

Of course, solving these equations can be very messy. We conclude with two more examples.

Example 3 Let's use Lagrange multipliers to find the point on the plane $2x - 2y + z = 4$ that is closest to the origin.

SOLUTION We want to minimize $\sqrt{x^2 + y^2 + z^2}$ subject to $2x - 2y + z - 4 = 0$. To make things easier, we minimize the square of the distance $x^2 + y^2 + z^2$ subject to $2x - 2y + z - 4 = 0$.

Let $L(x, y, z, \lambda) = x^2 + y^2 + z^2 - \lambda(2x - 2y + z - 4)$. Then the conditions are

$$\frac{\partial L}{\partial x} = 2x - 2\lambda = 0,$$

$$\frac{\partial L}{\partial y} = 2y + 2\lambda = 0,$$

$$\frac{\partial L}{\partial z} = 2z - \lambda = 0,$$

$$\frac{\partial L}{\partial \lambda} = -2x + 2y - z + 4 = 0.$$

Substituting the values for $2x$, $2y$, and z from the first three conditions into the fourth condition yields

$$-2\lambda - 2\lambda - \frac{\lambda}{2} + 4 = 0, \qquad \text{or} \qquad -\frac{9}{2}\lambda + 4 = 0, \qquad \text{or} \qquad \lambda = \frac{8}{9}.$$

Therefore $x = 8/9$, $y = -8/9$, and $z = 4/9$, so $(8/9, -8/9, 4/9)$ is the desired point, and the distance from this point to the origin is

$$\sqrt{\frac{64 + 64 + 16}{81}} = 4 \cdot \sqrt{\frac{4 + 4 + 1}{81}} = 4 \cdot \sqrt{\frac{1}{9}} = \frac{4}{3}. \quad \|$$

Example 4 To illustrate Lagrange multipliers when there are two constraints, let's find the point on the line of intersection of the planes $x - y = 2$ and $x - 2z = 4$ that is closest to the origin.

SOLUTION This time we want to minimize $x^2 + y^2 + z^2$ subject to the side conditions $x - y - 2 = 0$ and $x - 2z - 4 = 0$. We form

$$L(x, y, z, \lambda_1, \lambda_2) = x^2 + y^2 + z^2 - \lambda_1(x - y - 2) - \lambda_2(x - 2z - 4).$$

The conditions are

$$\frac{\partial L}{\partial x} = 2x - \lambda_1 - \lambda_2 = 0,$$

$$\frac{\partial L}{\partial y} = 2y + \lambda_1 = 0,$$

$$\frac{\partial L}{\partial z} = 2z + 2\lambda_2 = 0,$$

$$\frac{\partial L}{\partial \lambda_1} = -x + y + 2 = 0,$$

$$\frac{\partial L}{\partial \lambda_2} = -x + 2z + 4 = 0.$$

The second and third conditions give

$$\lambda_1 = -2y \qquad \text{and} \qquad \lambda_2 = -z,$$

so the first condition becomes $2x + 2y + z = 0$. We then have

$$\begin{aligned} 2x + 2y + z &= 0, \\ -x + y &= -2, \\ -x + 2z &= -4. \end{aligned}$$

The last two equations may be written as $y = x - 2$ and $z = (x - 4)/2$. Substitution of these values into the first equation gives

$$2x + 2(x - 2) + \frac{x - 4}{2} = 0$$

or

$$\frac{9}{2}x - 6 = 0$$

or

$$x = \frac{4}{3}.$$

Consequently, $y = -\frac{2}{3}$ and $z = -\frac{4}{3}$. The desired point is therefore $(\frac{4}{3}, -\frac{2}{3}, -\frac{4}{3})$. ‖

SUMMARY

1. *To maximize or minimize a function $f(x, y)$ subject to a constraint $g(x, y) = 0$, form $L(x, y, \lambda) = f(x, y) - \lambda g(x, y)$. Candidates for points (x, y) where extrema occur are such that x, y, and λ satisfy the three conditions*

$$\frac{\partial L}{\partial x} = 0,$$

$$\frac{\partial L}{\partial y} = 0,$$

$$\frac{\partial L}{\partial \lambda} = 0.$$

For a function $f(x, y, z)$ subject to a constraint $g(x, y, z) = 0$, form $L(x, y, z, \lambda) = f(x, y, z) - \lambda g(x, y, z)$ and solve the four conditions

$$\frac{\partial L}{\partial x} = 0, \quad \frac{\partial L}{\partial y} = 0, \quad \frac{\partial L}{\partial z} = 0, \quad \frac{\partial L}{\partial \lambda} = 0, \quad \text{etc.}$$

2. *If more than one constraint is present, use Lagrange multipliers, λ_1, λ_2, etc., equal in number to the number of constraints. To illustrate with $f(x, y, z)$ subject to constraints $g(x, y, z) = 0$ and $h(x, y, z) = 0$, form $L(x, y, z, \lambda_1, \lambda_2) = f(x, y, z) - \lambda_1 g(x, y, z) - \lambda_2 h(x, y, z)$. The conditions are then*

$$\frac{\partial L}{\partial x} = 0, \quad \frac{\partial L}{\partial y} = 0, \quad \frac{\partial L}{\partial z} = 0, \quad \frac{\partial L}{\partial \lambda_1} = 0, \quad \frac{\partial L}{\partial \lambda_2} = 0.$$

EXERCISES

Use the method of Lagrange multipliers to solve the following problems.

1. Maximize $x + y$ on the circle

$$x^2 + y^2 = 4.$$

2. Maximize xy on the circle

$$x^2 + y^2 = 4.$$

3. Find the point on the plane $2x - 3y + 6z = 5$ that is closest to the origin.

4. A box has a square base and open top. Find the dimensions for minimum surface area if the volume is to be 108 cubic inches.

5. Find the maximum possible volume of a right circular cone inscribed in a sphere of radius a.

6. Maximize xy^2 subject to $x + y = 6$ if x and y are both positive.

7. Find the point on the plane $4x - 8y + 20z = 3$ where $x^2 + 4y^2 + 2z^2$ is minimum.

8. Find the point on the curve of intersection of $x^2 + z^2 = 4$ and $x - y = 8$ that is farthest from the origin.

9. Find the point on the line of intersection of the planes $x - y = 4$ and $y + 3z = 6$ that is closest to $(-1, 3, 2)$.

10. Minimize $x^2 + y^2 - 3z^2$ subject to $x - y + z = 4$ and $2x - y = 6$.

17.3 EXACT DIFFERENTIALS

In Section 6.3 of Chapter 6, we saw how to solve some differential equations by separating variables. For example, we can solve the equation $dy/dx = x^2/y^2$ as follows:

$$\frac{dy}{dx} = \frac{x^2}{y^2},$$

$$y^2\,dy = x^2\,dx,$$

$$\frac{y^3}{3} = \frac{x^3}{3} + C.$$

However, if we try that technique to solve the differential equation $dy/dx = -3x^2y/(x^3 + 6y)$, we run into a problem. We have

$$\frac{dy}{dx} = \frac{-3x^2y}{x^3 + 6y},$$

$$(x^3 + 6y)\,dy = -3x^2y\,dx.$$

We cannot "separate" the variables. However, let us write the equation in the form

$$3x^2y\,dx + (x^3 + 6y)\,dy = 0,$$

and let $F(x, y) = x^3y + 3y^2$. If you just *happen* to notice that

$$dF = \frac{\partial F}{\partial x}\,dx + \frac{\partial F}{\partial y}\,dy = 3x^2y\,dx + (x^3 + 6y)\,dy,$$

you see that the differential equation becomes

$$dF = 0.$$

This equation has as solution

$$F(x, y) = C$$

or

$$x^3y + 3y^2 = C,$$

and the differential equation is solved. This discussion motivates the following definition.

Definition 17.1 A differential expression

$$M(x, y)\,dx + N(x, y)\,dy \tag{1}$$

is called an **exact differential** if there exists a function $F(x, y)$ such that the differential (1) is equal to dF.

Since

$$dF = \frac{\partial F}{\partial x}\,dx + \frac{\partial F}{\partial y}\,dy, \tag{2}$$

you see that, in order for (2) to be the same differential as (1), you must have

$$\frac{\partial F}{\partial x} = M(x, y) \qquad \text{and} \qquad \frac{\partial F}{\partial y} = N(x, y). \tag{3}$$

Now if $M(x, y)$ and $N(x, y)$ have continuous partial derivatives, so that $F(x, y)$ would have to have continuous second partial derivatives, you would have

$$\frac{\partial^2 F}{\partial y\,\partial x} = \frac{\partial^2 F}{\partial x\,\partial y}. \tag{4}$$

From (3) and (4), you see that, in order for (1) to be an exact differential, you must have

$$\frac{\partial M}{\partial y} = \frac{\partial^2 F}{\partial y\,\partial x} = \frac{\partial^2 F}{\partial x\,\partial y} = \frac{\partial N}{\partial x}.$$

Theorem 17.3 *If $M(x, y)$ and $N(x, y)$ have continuous partial derivatives, then the differential $M(x, y)\,dx + N(x, y)\,dy$ can be exact only if*

$$\frac{\partial M}{\partial y} = \frac{\partial N}{\partial x}. \tag{5}$$

Example 1 Let us test the differential $x^2y\,dx + (x^2 - y^2)\,dy$ for exactness.

SOLUTION Here $M(x, y) = x^2y$ and $N(x, y) = x^2 - y^2$. Now

$$\frac{\partial M}{\partial y} = \frac{\partial(x^2y)}{\partial y} = x^2 \qquad \text{and} \qquad \frac{\partial N}{\partial x} = \frac{\partial(x^2 - y^2)}{\partial x} = 2x.$$

Thus $\partial M/\partial y \neq \partial N/\partial x$, so (5) is not satisfied. Consequently the differential cannot be exact. ||

Example 2 Let us test the differential $(2xy + y^2)\,dx + (x^2 + 2xy)\,dy$.

SOLUTION This time

$$\frac{\partial M}{\partial y} = \frac{\partial(2xy + y^2)}{\partial y} = 2x + 2y \qquad \text{and} \qquad \frac{\partial N}{\partial x} = \frac{\partial(x^2 + 2xy)}{\partial x} = 2x + 2y.$$

Thus $\partial M/\partial y = \partial N/\partial x$, so (5) is satisfied. In this case, a little experimentation shows that our differential is indeed exact; it is dF for $F(x, y) = x^2y + xy^2$, for then

$$dF = \frac{\partial F}{\partial x}\,dx + \frac{\partial F}{\partial y}\,dy = (2xy + y^2)\,dx + (x^2 + 2xy)\,dy. \quad \|$$

The preceding example suggests that perhaps (5) is not only a necessary condition for $M(x, y)\,dx + N(x, y)\,dy$ to be exact, but is also a sufficient condition, at least if $M(x, y)$ and $N(x, y)$ have continuous derivatives. This is *not* always true, but it is true if the domain in the plane where $M(x, y)$ and $N(x, y)$ are both defined has no "holes" in it. It can be demonstrated that

$$\frac{y}{x^2 + y^2}\,dx - \frac{x}{x^2 + y^2}\,dy,$$

which does satisfy (5), is not exact in its entire domain, which consists of the plane minus the origin. (See the final exercise in this chapter.) Here, the origin is a "hole" in the domain.

We shall not concern ourselves with the niceties of finding as large regions as possible where a differential that satisfies (5) is exact. Let's just show that, if $M(x, y)\,dx + N(x, y)\,dy$ is defined throughout some neighborhood of a point and $M(x, y)$ and $N(x, y)$ are differentiable there with $\partial M/\partial y = \partial N/\partial x$, then the differential is exact in that neighborhood. Our demonstration will indicate a four-step procedure to find all functions $F(x, y)$ such that $M(x, y)\,dx + N(x, y)\,dy = dF$.

Since you must have $\partial F/\partial x = M(x, y)$, we let $G(x, y)$ be some antiderivative of $M(x, y)$ with respect to x only, treating y as a constant. (This is partial antidifferentiation with respect to x.) Now an indefinite integral is defined only up to a constant. Since y is treated as a constant in this integration with respect to x, you see that the most general partial antiderivative of $M(x, y)$ with respect to x is of the form $G(x, y) + h(y)$ for an arbitrary function $h(y)$ of y only.

STEP 1. Compute

$$F(x, y) = \int M(x, y)\,dx = G(x, y) + h(y) \tag{6}$$

where $G(x, y)$ is any computed antiderivative of $M(x, y)$ with respect to x only, treating y as a constant.

Our problem is to determine $h(y)$ such that $\partial F/\partial y = N(x, y)$. Now $\partial F/\partial y = \partial G/\partial y + h'(y)$, so $\partial G/\partial y + h'(y) = N(x, y)$, or

$$h'(y) = N(x, y) - \frac{\partial G}{\partial y}. \tag{7}$$

If $N(x, y) - \partial G/\partial y$ is a continuous function of y *only*, you can then set $h'(y)$ equal to it and integrate to find the desired $h(y) = \int h'(y)\,dy$. Now (7) is a function of y only provided that

$$\frac{\partial}{\partial x}\left(N(x, y) - \frac{\partial G}{\partial y}\right) = 0, \tag{8}$$

that is, if

$$\frac{\partial N}{\partial x} - \frac{\partial^2 G}{\partial x\,\partial y} = 0. \tag{9}$$

But if $G(x, y)$ has continuous second-order partial derivatives throughout the neighborhood where you are working, then

$$\frac{\partial^2 G}{\partial x\, \partial y} = \frac{\partial^2 G}{\partial y\, \partial x} = \frac{\partial}{\partial y}\left(\frac{\partial G}{\partial x}\right) = \frac{\partial}{\partial y}(M(x, y)) = \frac{\partial M}{\partial y}.$$

Under these conditions, (9) is satisfied, for the lefthand side becomes

$$\frac{\partial N}{\partial x} - \frac{\partial M}{\partial y},$$

which is zero by assumption.

STEP 2. Compute $\partial F/\partial y = \partial G/\partial y + h'(y)$, set it equal to $N(x, y)$, and solve for $h'(y)$.

STEP 3. Integrate to find $h(y) = \int h'(y)\, dy$.

The final answer is:

STEP 4. $F(x, y) = G(x, y) + h(y) + C$.

Example 3 Let's illustrate this procedure to show that

$$(2xy^3 + 6x)\, dx + (3x^2y^2 + 4y^3)\, dy$$

is an exact differential dF, and to find the function $F(x, y)$.

SOLUTION Note that

$$\frac{\partial M}{\partial y} = \frac{\partial(2xy^3 + 6x)}{\partial y} = 6xy^2 \quad \text{and} \quad \frac{\partial N}{\partial x} = \frac{\partial(3x^2y^2 + 4y^3)}{\partial x} = 6xy^2,$$

so the differential is indeed exact. Then

STEP 1. $F(x, y) = \int M(x, y)\, dx = \int (2xy^3 + 6x)\, dx = x^2y^3 + 3x^2 + h(y)$.

STEP 2. $\partial F/\partial y = 3x^2y^2 + h'(y)$, so

$$3x^2y^2 + h'(y) = 3x^2y^2 + 4y^3, \quad \text{and} \quad h'(y) = 4y^3.$$

STEP 3. $h(y) = \int 4y^3\, dy = y^4 + C$.

STEP 4. $F(x, y) = x^2y^3 + 3x^2 + y^4 + C$. ‖

For a differential in three variables

$$M(x, y, z)\, dx + N(x, y, z)\, dy + P(x, y, z)\, dz$$

to be exact, the corresponding conditions are

$$\frac{\partial M}{\partial y} = \frac{\partial N}{\partial x}, \quad \frac{\partial M}{\partial z} = \frac{\partial P}{\partial x}, \quad \frac{\partial N}{\partial z} = \frac{\partial P}{\partial y}. \tag{10}$$

Using the notation of subscripted variables, if $\boldsymbol{x} = (x_1, x_2, x_3)$, then

$$f_1(\boldsymbol{x})\, dx_1 + f_2(\boldsymbol{x})\, dx_2 + f_3(\boldsymbol{x})\, dx_3$$

is exact in some neighborhood of each point if

$$\frac{\partial f_i}{\partial x_j} = \frac{\partial f_j}{\partial x_i} \quad \text{for all } i \text{ and } j. \tag{11}$$

Again, the necessity of these relations follows from equality of mixed second-order partial derivatives. Here is one illustration of the computation of F corresponding to an exact differential dF for three variables.

Example 4 Consider the differential

$$(yz^2 - 6x \sin z)\, dx + (xz^2 - 3y^2 \cos z)\, dy + (2xyz - 3x^2 \cos z + y^3 \sin z)\, dz$$

which can easily be checked to satisfy (10). Let us find $F(x, y, z)$ so that the differential is dF.

SOLUTION Now from $\partial F/\partial x = yz^2 - 6x \sin z$, you have

$$\begin{aligned} F(x, y, z) &= \int (yz^2 - 6x \sin z)\, dx \\ &= xyz^2 - 3x^2 \sin z + h(y, z). \end{aligned}$$

Thus, $\partial F/\partial y = xz^2 + \partial h/\partial y = N(x, y)$ yields

$$xz^2 + \frac{\partial h}{\partial y} = xz^2 - 3y^2 \cos z,$$

so

$$\frac{\partial h}{\partial y} = -3y^2 \cos z,$$

and

$$\begin{aligned} h(y, z) &= \int -3y^2 \cos z\, dy \\ &= -y^3 \cos z + k(z). \end{aligned}$$

You are now down to

$$F(x, y, z) = xyz^2 - 3x^2 \sin z - y^3 \cos z + k(z).$$

Finally, $P(x, y, z) = \partial F/\partial z = 2xyz - 3x^2 \cos z + y^3 \sin z + k(z)$ yields

$$2xyz - 3x^2 \cos z + y^3 \sin z = 2xyz - 3x^2 \cos z + y^3 \sin z + k'(z),$$

so $k'(z) = 0$ and $k(z) = C$. Thus

$$F(x, y, z) = xyz^2 - 3x^2 \sin z - y^3 \cos z + C. \quad \|$$

SUMMARY

1. *A differential $M(x, y)\,dx + N(x, y)\,dy$ is exact if it is the differential of some function $F(x, y)$. A differential $f_1(\mathbf{x})\,dx_1 + f_2(\mathbf{x})\,dx_2 + f_3(\mathbf{x})\,dx_3$ is exact if it is the differential of some function $F(\mathbf{x})$.*
2. *If M and N are continuously differentiable, then $M(x, y)\,dx + N(x, y)\,dy$ is exact in a neighborhood of a point if and only if $\partial M/\partial y = \partial N/\partial x$. Similarly, $f_1(\mathbf{x})\,dx_1 + f_2(\mathbf{x})\,dx_2 + f_3(\mathbf{x})\,dx_3$ is exact if and only if $\partial f_i/\partial x_j = \partial f_j/\partial x_i$ for each choice of i and j.*
3. *If $M(x, y)\,dx + N(x, y)\,dy = dF$, then $F(x, y)$ is found as follows:*

   STEP 1. *Compute $F(x, y) = \int M(x, y)\,dx = G(x, y) + h(y)$ where $G(x, y)$ is any partial antiderivative of $M(x, y)$ with respect to x, treating y as a constant.*

   STEP 2. *Compute $\partial F/\partial y = \partial G/\partial y + h'(y)$, set it equal to $N(x, y)$, and solve for $h'(y)$.*

   STEP 3. *Integrate to find $h(y) = \int h'(y)\,dy$.*

   STEP 4. *Then $F(x, y) = G(x, y) + h(y) + C$.*

   A similar technique is valid for differentials involving more variables.

EXERCISES

In Exercises 1 through 15, determine whether the given differential is exact, and if it is, find a function F such that the differential is dF.

1. $x^2dx - y\,dy$
2. $x^2z\,dx - yz\,dy$
3. $2xy\,dx + x^2\,dy$
4. $2xz\,dx + x^2dz$
5. $(3x - 2y)\,dx + (2x - 3y)\,dy$
6. $\cos y\,dx + (1 - x\sin y)\,dy$
7. $2xy\,dx + (x^2 - e^{-y})\,dy$
8. $y^2dx + \left(\frac{1}{y} + 2xy\right)dy$
9. $(y\sec^2 xy)\,dx + (1 + x\sec^2 xy)\,dy$
10. $(e^y - y\cos xy)\,dx + (xe^y - x\cos xy)\,dy$
11. $(2xy^3 - 3)\,dx + (3x^2y^2 + 4y)\,dy$
12. $\dfrac{-y}{x^2 + z^2}\,dy + \dfrac{x}{x^2 + z^2}\,dz$ for $x > 0$
13. $\dfrac{-z}{y^2 + z^2}\,dy + \dfrac{y}{y^2 + z^2}\,dz$ for $z > 0$
14. $(2xyz - 3y^2 + 2z^3)\,dx + (x^2z - 6xy)\,dy + (x^2y + 8z + 6xz^2)\,dz$
15. $(yz\cos xyz - 3z^2)\,dx + (xz\cos xyz + 3y^2)\,dy + (xy\cos xyz - 6xz)\,dz$

17.4 LINE INTEGRALS

This section describes integrals over a smooth curve in the plane. Such integrals are called "line integrals". Generalizations to integrals over a smooth curve in space will be obvious, and we shall feel free to employ such integrals with no additional discussion.

Let a smooth curve γ in the plane be given parametrically by

$$x = h(t), \qquad y = k(t), \qquad \text{for } a \le t \le b, \tag{1}$$

where the coordinate functions h and k have continuous derivatives. The point A on γ where $t = a$ is the *initial point* of γ and the point B where $t = b$ is the *terminal point*. Figure 17.3 shows a curve γ with initial point A and terminal point B.

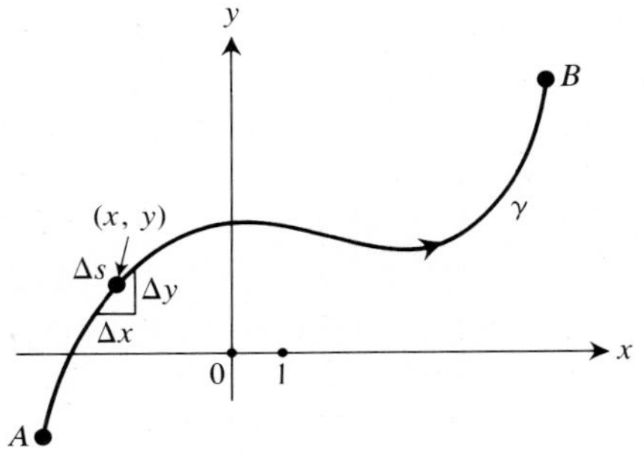

17.3

Imagine γ to be partitioned into little short pieces, corresponding to changes Δx in x and Δy in y. One such short piece of length Δs is shown in the figure. Let $f(x, y)$ be a continuous function with domain containing γ, and let (x, y) be a point on the short piece of γ of length Δs. The integrals in which we are interested arise from summing, over a partition of γ, terms of the form

$$f(x, y)\,\Delta s \qquad \text{or} \qquad f(x, y)\,\Delta x \qquad \text{or} \qquad f(x, y)\,\Delta y,$$

and then taking the limit of the sums as $\Delta s \to 0$, and hence $\Delta x \to 0$ and $\Delta y \to 0$. These integrals are written as:

$$\int_\gamma f(x, y)\,ds, \quad \text{coming from the terms } f(x, y)\,\Delta s,$$

$$\int_\gamma f(x, y)\,dx, \quad \text{coming from the terms } f(x, y)\,\Delta x,$$

$$\int_\gamma f(x, y)\,dy, \quad \text{coming from the terms } f(x, y)\,\Delta y.$$

To evaluate such integrals, simply use the parametric equations

$$x = h(t), \qquad y = k(t), \qquad \text{for } a \le t \le b$$

for γ, and substitute as follows:

$$\begin{aligned} f(x, y) &= f(h(t), k(t)), \\ dx &= h'(t)\,dt, \\ dy &= k'(t)\,dt, \\ ds &= \sqrt{\left(\frac{dx}{dt}\right)^2 + \left(\frac{dy}{dt}\right)^2}\,dt. \end{aligned}$$

Then integrate the resulting expression from $t = a$ to $t = b$.

It can be shown that evaluation of the integrals $\int_\gamma f(x, y)\,ds$, $\int_\gamma f(x, y)\,dx$, and $\int_\gamma f(x, y)\,dy$, using parametric equations to describe γ, really depends only on the continuous function $f(x, y)$ and the path γ from the initial point A to the terminal point B in the plane. That is, two different parametrizations leading to the same smooth path going from A to B will yield the same values for the integrals.

This invariance of parametrization seems reasonable from our introduction of these integrals as limits of sums of terms $f(x, y)\,\Delta s$, $f(x, y)\,\Delta x$, and $f(x, y)\,\Delta y$. To illustrate, the top semicircular arc γ of $x^2 + y^2 = a^2$ from $(a, 0)$ to $(-a, 0)$ can be parametrized as:

$$x = a \cos t, \qquad y = a \sin t \qquad \text{for } 0 \le t \le \pi,$$

and also as

$$x = a \sin\left(\frac{\pi}{2} t\right), \qquad y = -a \cos\left(\frac{\pi}{2} t\right) \qquad \text{for } 1 \le t \le 3.$$

An integral of the form $\int_\gamma f(x, y)\,ds$ could be computed using either parametrization; the same answer will be obtained.

The integral $\int_\gamma f(x, y)\,ds$ is called *the integral of $f(x, y)$ over γ with respect to arc length.* Here are some examples illustrating how such integrals may arise.

Mass

Suppose a wire is in the shape of a smooth plane curve γ described by (1), and suppose the mass density (mass per unit length) of the wire at a point (x, y) is $\sigma(x, y)$. Then the mass covering a little piece of γ of length Δs is approximately $\sigma(x, y)\,\Delta s$. Summing such contributions to the mass of the wire and taking the limit as $\Delta s \to 0$ leads to the formula

$$m = \text{Mass of the wire} = \int_\gamma \sigma(x, y)\,ds.$$

Example 1 Let's find the mass of a wire covering the circle $x^2 + y^2 = 1$ with mass density $\sigma(x, y) = ky^2$.

SOLUTION The equations

$$x = h(t) = \cos t, \qquad y = k(t) = \sin t \qquad \text{for } 0 \le t \le 2\pi$$

parametrized the circle γ. Thus

$$m = \int_\gamma \sigma(x, y)\, ds = \int_\gamma ky^2\, ds = \int_0^{2\pi} k \sin^2 t\sqrt{(-\sin t)^2 + (\cos t)^2}\, dt$$

$$= k\int_0^{2\pi} \sin^2 t\, dt = k\left(\frac{t}{2} - \frac{\sin 2t}{4}\right)\Big]_0^{2\pi} = k\pi. \quad \|$$

Example 2 *The total curvature* of a smooth curve γ is the integral with respect to arc length of the curvature κ of γ, that is, $\int_\gamma \kappa\, ds$. Let's find the total curvature of the portion γ of the helix

Total curvature

$$x = a \cos t, \qquad y = a \sin t, \qquad z = t,$$

where $0 \le t \le 2\pi$, as shown in Fig. 17.4.

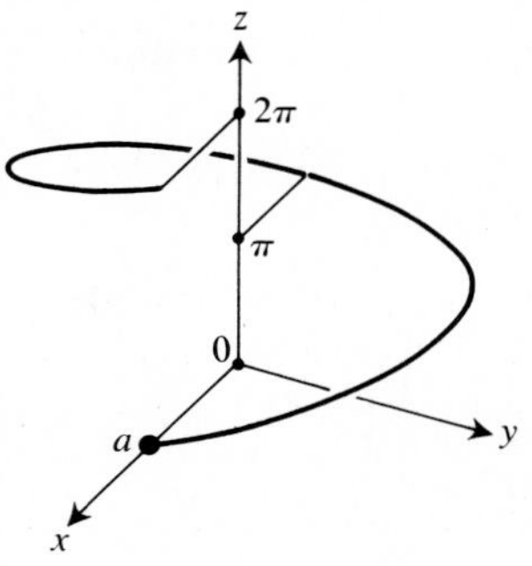

17.4

SOLUTION In Example 1 of Section 15.4, you saw that the curvature of the helix is the constant

$$\kappa = \frac{a}{a^2 + 1}.$$

Thus the total curvature of our "single turn" of the helix is

$$\int_\gamma \frac{a}{a^2 + 1}\, ds = \int_0^{2\pi} \frac{a}{a^2 + 1}\sqrt{(-a \sin t)^2 + (a \cos t)^2 + 1}\, dt$$

$$= \int_0^{2\pi} \frac{a}{a^2 + 1}\sqrt{a^2 + 1}\, dt = \frac{at}{\sqrt{a^2 + 1}}\Big]_0^{2\pi}$$

$$= \frac{2\pi a}{\sqrt{a^2 + 1}}. \quad \|$$

Example 3 Let a body be moved along a plane curve γ by a force, and let the component of the force *acting in the direction of the motion* at a point (x, y) on the curve be $F(x, y)$. Then since

Work

$$\text{Work} = (\text{force})(\text{distance}),$$

you see that the work W done by the force in moving the body along the curve γ is given by

$$W = \int_\gamma F(x, y)\, ds.$$

We will return to this application in the next section. ||

Now we illustrate the computation of integrals of the form $\int_\gamma f(x, y)\, dx$ and $\int_\gamma f(x, y)\, dy$.

Example 4 Let us find $\int_\gamma f(x, y)\, dx$ and $\int_\gamma f(x, y)\, dy$ if $f(x, y) = xy^2$ and γ is the curve joining $(0, 0)$ and $(1, 1)$ defined by

$$x = t, \qquad y = t^2 \qquad \text{for } 0 \le t \le 1.$$

SOLUTION Now $f(x, y) = xy^2$ so that $f(t, t^2) = t^5$. Therefore

$$\int_\gamma f(x, y)\, dx = \int_0^1 t^5 \cdot 1\, dt = \frac{t^6}{6}\Big]_0^1 = \frac{1}{6}$$

while

$$\int_\gamma f(x, y)\, dy = \int_0^1 t^5 \cdot 2t\, dt = \frac{2t^7}{7}\Big]_0^1 = \frac{2}{7}. \quad ||$$

Finally, it is often useful to relax the condition that γ be smooth. (Remember that a curve is not smooth where it has a sharp point.) We will allow γ to be any curve that consists of a finite number of smooth arcs joined together, as illustrated in Fig. 17.5. Such a curve is called *piecewise smooth.* Let γ be a curve with smooth pieces γ_1, γ_2, and γ_3, as shown in the figure. It is natural to write $\gamma = \gamma_1 + \gamma_2 + \gamma_3$. We define

$$\int_\gamma f(x, y)\, ds = \int_{\gamma_1} f(x, y)\, ds + \int_{\gamma_2} f(x, y)\, ds + \int_{\gamma_3} f(x, y)\, ds,$$

and the integrals with respect to dx and dy are similarly defined.

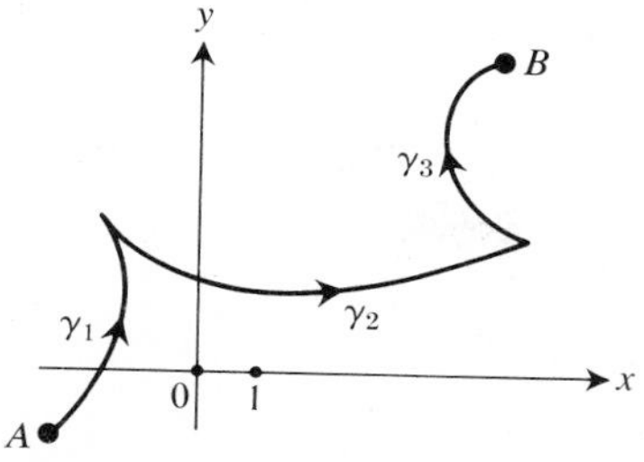

17.5

Example 5 Let's find $\int_\gamma xy^2\,dx$ and $\int_\gamma xy^2\,dy$ if $\gamma = \gamma_1 + \gamma_2$ is the piecewise-smooth curve joining $A(0, 0)$ and $B(1, 1)$ shown in Fig. 17.6.

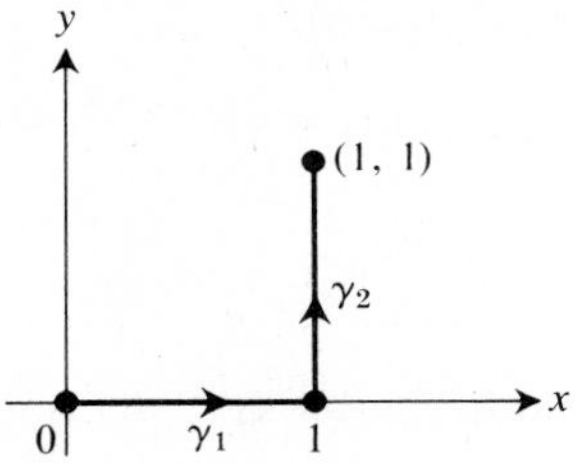

17.6

SOLUTION Here, γ_1 is the straight line segment from $(0, 0)$ to $(1, 0)$, and γ_2 the vertical line segment from $(1, 0)$ to $(1, 1)$.

Since the value of xy^2 on γ_1 is zero at each point, you see that

$$\int_{\gamma_1} xy^2\,dx = 0.$$

On γ_2, every short piece corresponds to $\Delta x = 0$, so also

$$\int_{\gamma_2} xy^2\,dx = 0.$$

Consequently,

$$\int_\gamma xy^2\,dx = \int_{\gamma_1} xy^2\,dx + \int_{\gamma_2} xy^2\,dx = 0 + 0 = 0.$$

Note that we discovered that this integral was zero by thinking in terms of the contributions to the typical sums. We did not even bother to parametrize γ_1 or γ_2.

Turning to $\int_\gamma xy^2\,dy$, a little piece of γ_1 corresponds to a change $\Delta y = 0$, so

$$\int_{\gamma_1} xy^2\,dy = 0.$$

But xy^2 is nonzero on most of γ_2, and $\Delta y \neq 0$ there, so there will be nonzero terms contributing to $\int_{\gamma_2} xy^2\,dy$. We may take as parametrization of γ_2

$$x = 1, \qquad y = t \qquad \text{for} \quad 0 \le t \le 1.$$

Then $dy = dt$, and

$$\int_{\gamma_2} xy^2\,dy = \int_{\gamma_2} 1 \cdot t^2\,dt = \frac{t^3}{3}\bigg]_0^1 = \frac{1}{3},$$

so

$$\int_{\gamma} xy^2\,dy = \int_{\gamma_1} xy^2\,dy + \int_{\gamma_2} xy^2\,dy = 0 + \frac{1}{3} = \frac{1}{3}. \quad \|$$

SUMMARY *Throughout the following, γ is a path in the plane given by $x = h(t)$, $y = k(t)$ for $a \le t \le b$, where $h(t)$ and $k(t)$ have continuous derivatives. Also, $f(x, y)$ is continuous.*

1. *$\int_\gamma f(x, y)\,ds$ is computed as $\int_a^b f(h(t), k(t))\sqrt{(dx/dt)^2 + (dy/dt)^2}\,dt$.*
2. *$\int_\gamma f(x, y)\,dx$ is computed as $\int_a^b f(h(t), k(t))(dx/dt)\,dt$.*
3. *$\int_\gamma f(x, y)\,dy$ is computed as $\int_a^b f(h(t), k(t))(dy/dt)\,dt$.*
4. *The value obtained in computing a line integral of a continuous function along a path from A to B is independent of the (smooth) parametrization used for the path.*
5. *The length of a curve γ is $\int_\gamma ds$.*
6. *The mass of a wire covering γ with density function $\sigma(x, y)$ is $\int_\gamma \sigma(x, y)\,ds$.*
7. *The total curvature of a curve γ is $\int_\gamma \kappa\,ds$.*
8. *A line integral over a piecewise-smooth curve is the sum of the integrals over the smooth pieces.*

EXERCISES

In Exercises 1 through 5, the curves are

$$\gamma_1 \text{ given by } x = t, \quad y = t^2 \qquad \text{for } 0 \le t \le 1$$

and

$$\gamma_2 \text{ given by } x = t + 1, \quad y = 2t + 1 \qquad \text{for } 0 \le t \le 1.$$

Compute the indicated integrals.

1. $\int_{\gamma_1} x\,ds$ **2.** $\int_{\gamma_2} \sqrt{y}\,ds$ **3.** $\int_{\gamma_2} (x + y)\,dx$ **4.** $\int_{\gamma_1} (x - y)\,dy$ **5.** $\int_{\gamma_2} (x^2 dx - y\,dx)$

6. Let γ be the space curve given by $x = 3t^2$, $y = 2t^3$, $z = 3t$ for $0 \le t \le 1$, and let $F(x, y, z) = 3x^2yz$.

a) Compute $\int_\gamma F(x, y, z)\,dx$.

b) Compute $\int_\gamma F(x, y, z)\,dy$.

7. Repeat Exercise 6 for the space curve γ given by $x = \sin t$, $y = \cos t$, $z = t$, for $0 \le t \le \pi/4$ but this time compute

a) $\int_\gamma F(x, y, z)\,ds$ b) $\int_\gamma F(x, y, z)\,dx$.

8. Find $\int_\gamma (x^2 + y^2)\,ds$ if γ is the straight line segment from $(1, 0)$ to $(3, 4)$.

9. Find $\int_\gamma (x^2 + xy)\,dx$ if $\gamma = \gamma_1 + \gamma_2$, where γ_1 is the straight line segment from $(0, 0)$ to $(2, 0)$ and γ_2 the straight line segment from $(2, 0)$ to $(4, 2)$.

10. Find $\int_\gamma (x + 2y)\,dy$ if $\gamma = \gamma_1 + \gamma_2$, where γ_1 is the arc of the parabola $y = x^2$ from $(0, 0)$ to $(1, 1)$ and γ_2 is the arc of the parabola $y = 2 - x^2$ from $(1, 1)$ to $(0, 2)$.

11. Find $\int_\gamma (x + 2y + z)\,ds$ if $\gamma = \gamma_1 + \gamma_2 + \gamma_3$, where γ_1 is the shorter arc of the circle with center at the origin from $(0, -2, 0)$ to $(2, 0, 0)$, γ_2 is the vertical line segment from $(2, 0, 0)$ to $(2, 0, 1)$, and γ_3 is the shorter arc of the circle that has its center at $(0, 0, 1)$, from $(2, 0, 1)$ to $(0, 2, 1)$.

12. Let the mass density of a thin wire covering the portion of the parabola $y = x^2$ from $(0, 0)$ to $(2, 4)$ by $\sigma(x, y) = xy$. Find the total mass of the wire.

13. Let the mass density of a thin wire covering the semicircle $y = \sqrt{4 - x^2}$ be $\sigma(x, y) = y$. Find the mass of the wire.

14. Find the total curvature of the circle

$$x = a \cos t,$$
$$y = a \sin t$$

for $0 \le t \le 2\pi$.

15. Find the total curvature of the plane curve $y = x^2$ from $(0, 0)$ to $(2, 4)$.

17.5 INTEGRATION OF VECTOR FIELDS ALONG CURVES

17.5.1 Vector fields

A **vector field** on a region G in the plane assigns, to each point (x, y) in G, a vector in the plane, which you usually visualize as emanating from (x, y). A vector field

$$\boldsymbol{F}(x, y) = F_1(x, y)\boldsymbol{i} + F_2(x, y)\boldsymbol{j}$$

is *continuous* if F_1 and F_2 are continuous functions. We use a boldface letter $\boldsymbol{F}$ to indicate that we are considering a vector field; of course, you will use $\vec{F}$ in your written work.

Example 1 You encountered the notion of a vector field in the plane in Chapter 16, when we discussed the gradient ∇f of a differentiable function f of two variables. To each point (x, y) in the domain of f, you assign the gradient vector

$$\nabla f = \frac{\partial f}{\partial x}\boldsymbol{i} + \frac{\partial f}{\partial y}\boldsymbol{j}.$$

Recall that the direction at (x, y) of the gradient vector is normal to the level curve of f through (x, y), and the gradient vector points in the direction of maximum increase of $f(x, y)$ at (x, y). ||

Example 2 An electrical charge at the origin in the plane exerts a force of repulsion on a like charge at any other point (x, y). This force at (x, y) may be represented by a vector of length equal to the magnitude of the force and having the direction of the force; namely, away from the origin. The vector field of these vectors for $(x, y) \neq (0, 0)$ is the *force field of the charge.* This vector field is continuous, and grows weaker (shorter vectors) as the distance from the origin increases (see Fig. 17.7). ||

Example 3 If a region G in the plane is covered with a flowing liquid (or gas), then you can associate with each (x, y) in G and time t the velocity vector of the fluid flow at (x, y) at the time t. This vector has the direction of the flow, and length equal to the speed of the flow. This vector field is the *velocity field of the flow at time t.* If the field does not vary with time, then the flow is called *steady-state.* ||

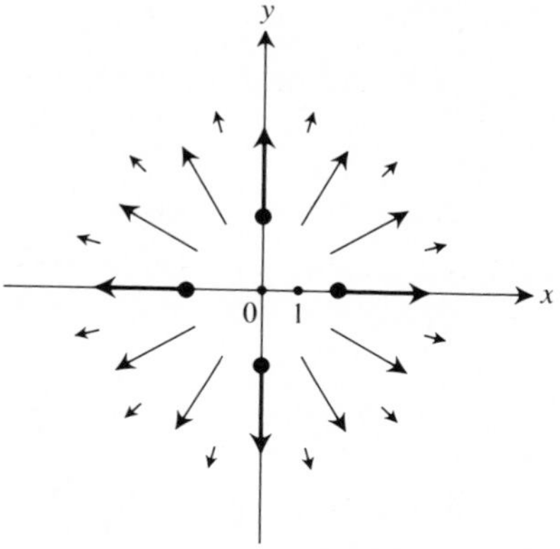

17.7

17.5.2 Integral of a vector field along a curve

Let $\boldsymbol{F}$ be a continuous vector field on a region G of the plane, and let

$$x = h(t), \qquad y = k(t), \qquad \text{for} \quad a \le t \le b$$

be a smooth curve γ lying in G. Recall that the position vector to a point on γ at time t is

$$\boldsymbol{r} = x\boldsymbol{i} + y\boldsymbol{j} = h(t)\boldsymbol{i} + k(t)\boldsymbol{j}. \tag{1}$$

The *unit* tangent vector (in the direction of increasing t) to the curve at time t is

$$\boldsymbol{t} = \frac{d\boldsymbol{r}/dt}{|d\boldsymbol{r}/dt|} = \frac{d\boldsymbol{r}/dt}{ds/dt} = \frac{d\boldsymbol{r}}{ds} = \frac{dx}{ds}\boldsymbol{i} + \frac{dy}{ds}\boldsymbol{j}. \tag{2}$$

The component of the vector field $\boldsymbol{F}$ *tangent to the curve* at a point (x, y) is thus given by

$$\begin{aligned}\boldsymbol{F}\cdot\boldsymbol{t} &= (F_1(x, y)\boldsymbol{i} + F_2(x, y)\boldsymbol{j})\cdot\left(\frac{dx}{ds}\boldsymbol{i} + \frac{dy}{ds}\boldsymbol{j}\right)\\ &= F_1(x, y)\frac{dx}{ds} + F_2(x, y)\frac{dy}{ds}.\end{aligned} \tag{3}$$

The **integral of the vector field F along** γ is the integral with respect to arc length of this tangential component (3). This integral is

$$\begin{aligned}\int_\gamma \left(F_1(x, y)\frac{dx}{ds} + F_2(x, y)\frac{dy}{ds}\right) ds &= \int_\gamma [F_1(x, y)\, dx + F_2(x, y)\, dy]\\ &= \int_\gamma F_1(x, y)\, dx + \int_\gamma F_2(x, y)\, dy.\end{aligned} \tag{4}$$

Now the vector

$$d\boldsymbol{r} = dx\boldsymbol{i} + dy\boldsymbol{j} \tag{5}$$

has direction tangent to the curve and length $ds = \sqrt{(dx)^2 + (dy)^2}$. Note that

$$\begin{aligned} \boldsymbol{F} \cdot d\boldsymbol{r} &= (F_1(x, y)\boldsymbol{i} + F_2(x, y)\boldsymbol{j}) \cdot (dx\boldsymbol{i} + dy\boldsymbol{j}) \\ &= F_1(x, y)\, dx + F_2(x, y)\, dy. \end{aligned} \tag{6}$$

Comparing (4) and (6), we write the integral of the vector field $\boldsymbol{F}$ along γ as

$$\int_\gamma \boldsymbol{F} \cdot d\boldsymbol{r}. \tag{7}$$

In practice, (7) is calculated by expressing both $\boldsymbol{F}$ and $d\boldsymbol{r}$ in terms of the parameter t and dt.

Work

If $\boldsymbol{F}$ is a force field, then the component of $\boldsymbol{F}$ tangent to a smooth curve γ is the portion of the force that acts to move a body along γ. The *work* done by this force field in moving a body along γ is then the integral with respect to arc length of this component, which we have seen is $\int_\gamma \boldsymbol{F} \cdot d\boldsymbol{r}$. Thus,

$$\text{Work} = \int_\gamma \boldsymbol{F} \cdot d\boldsymbol{r}. \tag{8}$$

Example 4 Let's find the work done by the force field

$$\boldsymbol{F}(x, y) = x^2\boldsymbol{i} + y^2\boldsymbol{j}$$

in moving a body from (0, 0) to (1, 1) if the position (x, y) of the body at time t is given by

$$x = t, \qquad y = t^2 \qquad \text{for} \quad 0 \le t \le 1.$$

SOLUTION In terms of t, you have

$$\boldsymbol{F}(x, y) = \boldsymbol{F}(t, t^2) = t^2\boldsymbol{i} + t^4\boldsymbol{j},$$

and

$$d\boldsymbol{r} = dx\boldsymbol{i} + dy\boldsymbol{j} = (\boldsymbol{i} + 2t\boldsymbol{j})\, dt.$$

Thus

$$\begin{aligned} \boldsymbol{F} \cdot d\boldsymbol{r} &= \int_0^1 (t^2\boldsymbol{i} + t^4\boldsymbol{j}) \cdot (\boldsymbol{i} + 2t\boldsymbol{j})\, dt \\ &= \int_0^1 (t^2 + 2t^5)\, dt = \left(\frac{t^3}{3} + \frac{2t^6}{6}\right)\Bigg]_0^1 = \frac{1}{3} + \frac{1}{3} = \frac{2}{3}. \quad \| \end{aligned}$$

Example 5

Circulation and $\oint$ notation

Let $\boldsymbol{F}$ be the velocity field of a steady-state flow in the plane, and let $\boldsymbol{F}$ be defined on a smooth curve γ that is *closed*, so that its initial and terminal points coincide. Then $\oint_\gamma \boldsymbol{F} \cdot d\boldsymbol{r}$ is the integral with respect to arc length of the tangential component of $\boldsymbol{F}$, and is called the *circulation of the flow around* γ. One uses an integral sign $\oint$ rather than $\int$ to denote an integral such as this around a *closed* curve. ||

17.5.3 Exact differentials and conservative force fields

If γ is a smooth curve joining A and B in the plane and if $M(x, y)\,dx + N(x, y)\,dy$ is an *exact* differential, then

$$\int_\gamma M(x, y)\,dx + N(x, y)\,dy$$

depends only on A and B and is independent of which smooth curve γ is chosen joining A and B. To see that this is so, let the curve γ be given by $x = h(t)$, $y = k(t)$ for $a \le t \le b$, so $(x, y) = (h(t), k(t))$. Also, let $G(x, y)$ be such that $dG = M(x, y)\,dx + N(x, y)\,dy$. Then

$$\begin{aligned}\int_\gamma M(x, y)\,dx + N(x, y)\,dy &= \int_a^b \left[M(x, y)\frac{dx}{dt} + N(x, y)\frac{dy}{dt}\right] dt\\ &= \int_a^b \frac{d}{dt}(G(h(t), k(t)))\,dt\\ &= G(h(b), k(b)) - G(h(a), k(a)).\end{aligned}$$

Thus the integral depends only on the endpoints $A = (h(a), k(a))$ and $B = (h(b), k(b))$. In summary,

the line integral of an exact differential is independent of the path.

It can be shown that independence of the path for $\int_\gamma M(x, y)\,dx + N(x, y)\,dy$ implies that the differential is exact; i.e., independence of path holds if and only if the differential is exact.

As an exercise, we ask you to show that independence of path for a line integral is equivalent to the assertion that the integral is zero around any *closed* path, whose terminal point is the same as its initial point. Then for a force field $\boldsymbol{F}$ where $\boldsymbol{F} \cdot d\boldsymbol{r}$ is an exact differential, the work done around any path that comes back to the starting point is equal to zero; for this reason, such a force field is called *conservative.* For another example, if the velocity field of a fluid gives rise to an exact differential, the *circulation* around any path that comes back to the starting point is equal to zero; for this reason such a velocity field is called *irrotational.*

Conservative and irrotational fields

Example 6 The force field $\boldsymbol{F}(x, y) = x^2\boldsymbol{i} + y^2\boldsymbol{j}$ is conservative, for $\partial(x^2)/\partial y = 0 = \partial(y^2)/\partial x$, so $x^2\,dx + y^2\,dy$ is an exact differential. Of course, $x^2\,dx +$

$y^2\,dy = d(\frac{1}{3}(x^3 + y^3))$. Thus the work in moving a body from the point (1, 1) to the point (2, 3) is

$$\left.\frac{x^3 + y^3}{3}\right]_{(1,1)}^{(2,3)} = \frac{8 + 27}{3} - \frac{1 + 1}{3} = \frac{33}{3} = 11.$$

This result will be obtained by integrating over *any* path from (1, 1) to (2, 3). For example, if γ_1 is given by $x = 1 + t$, $y = 1 + 2t$ for $0 \le t \le 1$, then

$$\begin{aligned}\int_{\gamma_1} x^2\,dx + y^2\,dy &= \int_0^1 [(1 + t)^2 \cdot 1 + (1 + 2t)^2 \cdot 2]\,dt\\ &= \left(\frac{(1 + t)^3}{3} + \frac{(1 + 2t)^3}{3}\right)\Bigg]_0^1 = \frac{8}{3} + \frac{27}{3} - \frac{1}{3} - \frac{1}{3}\\ &= \frac{33}{3} = 11.\end{aligned}$$

Again, if γ_2 consists of the straight-line path from (1, 1) to (2, 1) followed by the straight-line path from (2, 1) to (2, 3), then

$$\begin{aligned}\int_{\gamma_2} x^2\,dx + y^2\,dy &= \int_{(1,1)}^{(2,1)} x^2\,dx + y^2\,dy + \int_{(2,1)}^{(2,3)} x^2\,dx + y^2\,dy\\ &= \int_{x=1}^{x=2} (x^2\,dx + 1 \cdot 0) + \int_{y=1}^{y=3} (4 \cdot 0 + y^2\,dy)\\ &= \left.\frac{x^3}{3}\right]_1^2 + \left.\frac{y^3}{3}\right]_1^3\\ &= \frac{8}{3} - \frac{1}{3} + \frac{27}{3} - \frac{1}{3} = \frac{7}{3} + \frac{26}{3} = \frac{33}{3} = 11. \quad \|\end{aligned}$$

Everything we have done is equally valid for integrals over curves in space. The extensions of the definitions and formulas are obvious.

SUMMARY *Throughout the following, γ is a piecewise-smooth curve and $\mathbf{F}(x, y) = F_1(x, y)\mathbf{i} + F_2(x, y)\mathbf{j}$ is a continuously differentiable vector field.*

1. *The integral of the vector field $\mathbf{F}(x, y)$ along γ is*

$$\int_\gamma \mathbf{F}(x, y) \cdot d\mathbf{r} = \int_\gamma [F_1(x, y)\,dx + F_2(x, y)\,dy].$$

2. *If $\mathbf{F}$ is a force field, then $\int_\gamma \mathbf{F} \cdot d\mathbf{r}$ is the work done by the force in moving a body along γ.*
3. *If $\mathbf{F}$ is the velocity field of a steady-state flow in the plane and γ is a closed curve (initial point coincides with terminal point), then $\oint_\gamma \mathbf{F} \cdot d\mathbf{r}$ is the circulation of the flow around γ.*

4. *$\int_\gamma \boldsymbol{F}(x, y) \cdot d\boldsymbol{r}$ is independent of the path γ from A to B if and only if $F_1(x, y)\,dx + F_2(x, y)\,dy$ is an exact differential. If the differential is dG, then*

$$\int_\gamma \boldsymbol{F}(x, y) \cdot d\boldsymbol{r} = \int_\gamma dG = G(B) - G(A).$$

EXERCISES

In Exercises 1 through 4, the curves are

$$\gamma_1 \text{ given by } x = t, \quad y = t^2 \quad \text{for } 0 \le t \le 1,$$

$$\gamma_2 \text{ given by } x = t + 1, \quad y = 2t + 1 \quad \text{for } 0 \le t \le 1,$$

and the vector fields are

$$\boldsymbol{F}(x, y) = x^2\boldsymbol{i} + y^2\boldsymbol{j} \quad \text{and} \quad \boldsymbol{G}(x, y) = xy\boldsymbol{i} - x^2y\boldsymbol{j}.$$

Find the indicated integral.

1. $\int_{\gamma_1} \boldsymbol{F} \cdot d\boldsymbol{r}$ **2.** $\int_{\gamma_2} \boldsymbol{G} \cdot d\boldsymbol{r}$ **3.** $\int_{\gamma_1} (\boldsymbol{F} - \boldsymbol{G}) \cdot d\boldsymbol{r}$ **4.** $\int_{\gamma_2} (2\boldsymbol{F} + \boldsymbol{G}) \cdot d\boldsymbol{r}$

5. Let $\boldsymbol{F}(x, y) = xy\boldsymbol{i} + x^2\boldsymbol{j}$ and let γ be the shorter arc from (3, 0) to (0, 3) of a circle with center at the origin. Find $\int_\gamma \boldsymbol{F} \cdot d\boldsymbol{r}$.

6. Let $F(x, y, z) = (x^2 + z)\boldsymbol{i} + (x + y^2)\boldsymbol{j} + xz\boldsymbol{k}$ and let $\gamma = \gamma_1 + \gamma_2 + \gamma_3$ be the path consisting of three straight-line segments from (0, 0, 0) to (1, 0, 0) to (1, 1, 1) to (0, 1, 2). Find $\int_\gamma \boldsymbol{F} \cdot d\boldsymbol{r}$.

7. Let γ be the space curve given by $x = 3t^2$, $y = 2t^3$, $z = 3t$ for $0 \le t \le 1$ and let $F(x, y, z) = 3x^2yz$.

a) Compute $\int_\gamma dF$ by integration.

b) Compute the integral again, using the fact that dF is an exact differential, and the result in (4) of the Summary.

8. Repeat Exercise 7(b) for the space curve γ given by $x = \sin t$, $y = \cos t$, $z = t$ for $0 \le t \le \pi/4$.

9. Find a function whose differential is $xy^2\,dx + x^2y\,dy$ and use (4) of the Summary to find $\int_\gamma (xy^2\,dx + x^2y\,dy)$ for γ given by $x = 3t^2$, $y = 2t^3$, where $0 \le t \le 1$.

10. Let the position of a body moving in the plane be $(\cos t, \sin 2t)$ at time t, and let the body be subject to the force field $x\boldsymbol{i} + y\boldsymbol{j}$. Find the work done by the force field on the body from time $t = 0$ to time $t = \pi/2$.

11. Let the position of a moving body in space be $(3t^2, 2t^3, 3t)$ at time t, and let the body be subject to the force field $x\boldsymbol{i} + z\boldsymbol{j} + y\boldsymbol{k}$. Find the work done by the force field on the body from time $t = 1$ to time $t = 2$.

12. Let the velocity field of a steady-state fluid flow in the plane be $x\boldsymbol{i} - y\boldsymbol{j}$. Find the circulation of the flow around the circle $x = \cos t$, $y = \sin t$ for $0 \le t \le 2\pi$.

13. Let the velocity field of a steady-state fluid flow in the plane be $xy\boldsymbol{i} - x\boldsymbol{j}$. Find the circulation of the flow around the ellipse $x = 2\cos t$, $y = 3\sin t$ for $0 \le t \le 2\pi$.

14. Let $\boldsymbol{F}$ be a continuous vector field defined throughout a region G of the plane. Show that $\int_\gamma \boldsymbol{F}(x, y) \cdot d\boldsymbol{r}$ is independent of the path joining the endpoints of γ, for all choices of the endpoints and all smooth paths γ joining them, if and only if the integral around any closed path is zero. [*Hint.* Show that if γ_1 is given by $x = h(t)$, $y = k(t)$ for $a \le t \le b$ and γ_2 is given by $x = h(b - (t - a))$, $y = k(b - (t - a))$ for $a \le t \le b$, then γ_2 is γ_1 traveled backwards; we say that $\gamma_2 = -\gamma_1$. Then show that $\int_{\gamma_1} \boldsymbol{F} \cdot d\boldsymbol{r} = -\int_{\gamma_2} \boldsymbol{F} \cdot d\boldsymbol{r}$, and use this in your proof of the main result.]

exercise sets for chapter 17

review exercise set 17.1

1. Find all local maxima and minima of the function tion

$$f(x, y) = 2x^2 - xy + y - y^2 - 7x + 3.$$

2. Use Lagrange multipliers to maximize $x + y^2$ over the ellipse $x^2 + 4y^2 = 4$.
3. Use the method of Lagrange multipliers to find the point on the intersection of the spheres $x^2 + y^2 + z^2 = 4$ and $(x - 2)^2 + y^2 + z^2 = 12$ that is closest to $(1, 1, 0)$. Find also the point farthest from $(1, 1, 0)$.
4. Let $M(x, y)$ and $N(x, y)$ have continuous partial derivatives in a region without any "holes" in it. State a criterion for $M(x, y)\,dx + N(x, y)\,dy$ to be an exact differential.
5. Test whether

$$(y^2e^x + 3x^2y)\,dx + (2ye^x + x^3 + \sin y)\,dy$$

is an exact differential, and if it is, find $F(x, y)$ such that the differential is dF.

6. Find the total curvature of a circle of radius a.
7. Evaluate $\int_\gamma x\,ds$ if γ is the plane curve given by $x = 8t$, $y = 3t^2$ for $0 \le t \le 1$.
8. Consider the force field in space $\boldsymbol{F}(x, y, z) = xy\boldsymbol{i} + (z/x)\boldsymbol{j} + y^2z\boldsymbol{k}$. Find the work done by the field in moving a body along the curve $x = t$, $y = t$, $z = t - 1$, from $t = 1$ to $t = 2$.
9. Let γ be a smooth curve from $(-1, 2)$ to $(3, 4)$. When is $\int_\gamma M(x, y)\,dx + N(x, y)\,dy$ independent of the choice of path joining these points?

review exercise set 17.2

1. Find all relative maxima and minima of the function tion

$$f(x, y) = xy - x^2 - 2y^2 + 3x - 5y - 6.$$

2. Use the method of Lagrange multipliers to maximize $x^2 + y + z^2$ over the ellipsoid $x^2 + y^2 + 4z^2 = 4$.
3. Use the method of Lagrange multipliers to find the point on the line of intersection of the planes

$$x - 2y + z = 4 \quad \text{and} \quad 2x + y - z = 8$$

that is closest to the origin.

4. Find c and k such that the differential

$$(y^2 + kxy)\,dx + (3x^2 + cxy)\,dy$$

is exact.

5. Find $F(x, y)$ such that dF is

$$(2x \sin y - 3x^2y)\,dx + (x^2 \cos y - x^3 + 4y^2)\,dy.$$

6. A wire of variable density $\rho(x, y) = 2x$ covers the curve

$$x = t^2 - 1, \qquad y = 3t + 1 \qquad \text{for } 1 \le t \le 2.$$

Express as an integral the moment of the wire about the y-axis.

7. If γ is the curve $x = \sin t$, $y = 2 \cos t$ for $0 \le t \le \pi/4$, and $\boldsymbol{F}(x, y)$ is the vector field $xy\boldsymbol{i} - (x/y)\boldsymbol{j}$, find $\int_\gamma \boldsymbol{F} \cdot d\boldsymbol{r}$.
8. Note that

$$\boldsymbol{F}(x, y) = (x^2 - 4xy)\boldsymbol{i} + (y^2 - 2x^2)\boldsymbol{j}$$

is a conservative force field. Find $\int_\gamma \boldsymbol{F} \cdot d\boldsymbol{r}$ where γ is any smooth curve from $(-1, 1)$ to $(1, -2)$.

9. If the velocity field of a fluid is $xy\boldsymbol{i} + 2y\boldsymbol{j}$, find the circulation of the fluid about the unit circle $x^2 + y^2 = 1$ in the counterclockwise direction. [*Hint.* Take the parametrization $x = \cos t$, $y = \sin t$ for $0 \le t \le 2\pi$.]

more challenging exercises 17

Exercises 1 through 3 are designed to shed a little light on the criterion $AC - B^2 > 0$ given in Theorem 17.2 for a local maximum or local minimum of $f(x, y)$.

1. Taylor's Theorem for functions $f(x, y)$ of two variables was stated in the last exercise set at the end of Chapter 16.

a) Write the expression for $E_1(x, y)$ given in the theorem.

b) Suppose $f_x(x_0, y_0) = f_y(x_0, y_0) = 0$. Show that:

i) $f(x_0, y_0)$ is a local minimum if $E_1(x, y) \geq 0$ for all (x, y) in some small disk with center at (x_0, y_0).

ii) $f(x_0, y_0)$ is a local maximum if $E_1(x, y) \leq 0$ for all (x, y) in some small disk with center at (x_0, y_0).

iii) $f(x_0, y_0)$ is neither a local minimum nor a local maximum if $E_1(x, y)$ assumes both positive and negative values in every small disk with center at (x_0, y_0).

2. Let $f(x, y)$ have continuous second partial derivatives, and let $A = f_{xx}(x_0, y_0)$, $B = f_{xy}(x_0, y_0)$, and $C = f_{yy}(x_0, y_0)$. Convince yourself that if $A(\Delta x)^2 + 2B(\Delta x)(\Delta y) + C(\Delta y)^2 > 0$ for all choices of $(\Delta x, \Delta y) \neq (0, 0)$, then $E_1(x, y) \geq 0$ for all (x, y) in some small disk with center at (x_0, y_0). [*Hint.* Think of $\Delta x = x - x_0$, $\Delta y = y - y_0$, and set $\Delta y = t(\Delta x)$.] What would be true if $A(\Delta x)^2 + 2B(\Delta x)(\Delta y) + C(\Delta y)^2 > 0$ for $(\Delta x, \Delta y) \neq (0, 0)$? If it were sometimes positive and sometimes negative?

3. Show that $A(\Delta x)^2 + 2B(\Delta x)(\Delta y) + C(\Delta y)^2 > 0$ for all $(\Delta x, \Delta y) \neq (0, 0)$ if $AC - B^2 > 0$ and $A > 0$ or $C > 0$. [*Hint.* Note that

$$A(A(\Delta x)^2 + 2B(\Delta x)(\Delta y) + C(\Delta y)^2) = (A(\Delta x) + B(\Delta y))^2 + (AC - B^2)(\Delta y)^2.]$$

4. Suppose $f(x, y)$ has continuous partial derivatives of all orders $\leq n$. Suppose, further, that all partial derivatives of order $< n$ are zero at (x_0, y_0), but some partial derivative of order n is nonzero at (x_0, y_0).

a) Show that if n is odd, then $f(x, y)$ has neither a local minimum nor a local maximum at (x_0, y_0).

b) If n is even, show that Exercise 1(b) holds with $E_{n-1}(x, y)$ in place of $E_1(x, y)$.

5. Let P, Q, R, and S be four functions of the four variables w, x, y, and z, all having continuous partial derivatives. State the criteria for $P\,dw + Q\,dx + R\,dy + S\,dz$ to be an exact differential in some region of four-space having no holes in it.

6. Using subscript notation, let $\boldsymbol{x} = (x_1, \ldots, x_n)$ and let $F_1, F_2, \ldots, F_n$ be n functions of $\boldsymbol{x}$ having continuous partial derivatives. Repeat Exercise 5, giving criteria for $F_1dx_1 + F_2dx_2 + \cdots + F_ndx_n$ to be exact in a suitable region. (It is much easier in this notation.)

7. Consider the differential

$$\frac{y}{x^2 + y^2}\,dx - \frac{x}{x^2 + y^2}\,dy, \qquad (x, y) \neq (0, 0).$$

a) Show that this differential satisfies the criterion $\partial M/\partial y = \partial N/\partial x$ for all $(x, y) \neq (0, 0)$.

b) Show that, if the given differential were dF for $F(x, y)$ and all $(x, y) \neq (0, 0)$, then you would have to have

$$F(x, y) = \tan^{-1}\left(\frac{x}{y}\right) + A \qquad \text{for } y > 0 \text{ and some } A$$

and

$$F(x, y) = \tan^{-1}\left(\frac{x}{y}\right) + B \qquad \text{for } y < 0 \text{ and some } B.$$

Show that there are no choices for A and B and no way of defining $F(x, y)$ on the positive and negative portions of the x-axis that would give a differentiable function $F(x, y)$ for all $(x, y) \neq (0, 0)$. (This illustrates that the region should have no holes in it in order for the partial-derivative criterion for exactness to guarantee exactness in the entire region.) [*Hint.* Show that you would have to have $B =$

$A + \pi$, and $F(x, 0) = A + \pi/2$ for $x > 0$. Then show that it becomes impossible to define $F(x, 0)$ for $x < 0$ to make F even continuous on the negative x-axis, to say nothing of differentiable.]

multiple integrals

18

18.1 INTEGRALS OVER A RECTANGLE

18.1.1 The integral as a limit of sums

Let a rectangular region R where $a \leq x \leq b$ and $c \leq y \leq d$ be given in the plane, as illustrated in Fig. 18.1. We partition R into n^2 nonoverlapping subrectangles as follows: Partition $[a, b]$ into n subintervals of equal lengths with endpoints

$$a = t_0 < t_1 < \cdots < t_n = b$$

and $[c, d]$ into n subintervals of equal lengths with endpoints

$$c = s_0 < s_1 < \cdots < s_n = d.$$

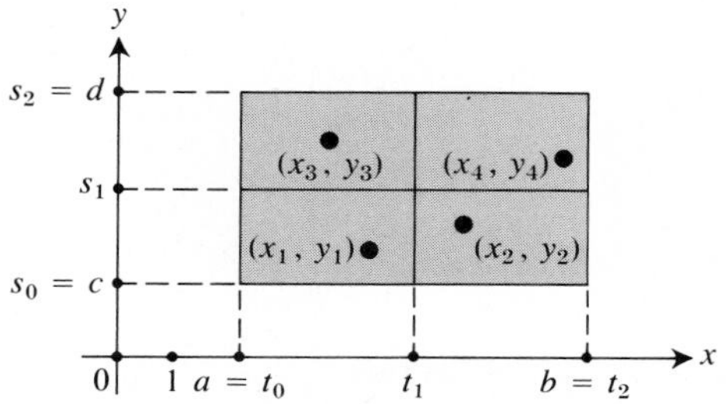

18.1

The n^2 rectangular subregions where $t_{i-1} \leq x \leq t_i$ and $s_{j-1} \leq y \leq s_j$ for $i = 1, \ldots, n$ and $j = 1, \ldots, n$ then have equal areas and constitute a partition of R into n^2 nonoverlapping subrectangles. Suppose that we have numbered the n^2 subrectangles in some convenient fashion, and let (x_k, y_k) be a point in the kth subrectangle of the partition (see Fig. 18.1).

It can be shown that a continuous function f of two variables assumes a maximum value and a minimum value on each closed* rectangular region contained in the domain of f. Let m_k and M_k be the minimum and maximum values, respectively, of f on the kth of the n^2 subrectangles in our partition of R. The area of each subrectangle in our partition is of course

$$\frac{b-a}{n} \cdot \frac{d-c}{n} = \frac{(b-a)(d-c)}{n^2},$$

and we have, for each k,

$$\frac{(b-a)(d-c)}{n^2} m_k \leq \frac{(b-a)(d-c)}{n^2} f(x_k, y_k) \leq \frac{(b-a)(d-c)}{n^2} M_k. \quad (1)$$

(Relation (1) is analogous to the relation

$$\frac{b-a}{n} m_i \leq \frac{b-a}{n} f(x_i) \leq \frac{b-a}{n} M_i$$

---

* To say that a rectangular region is *closed* is to say that the region includes the boundary of the rectangle.

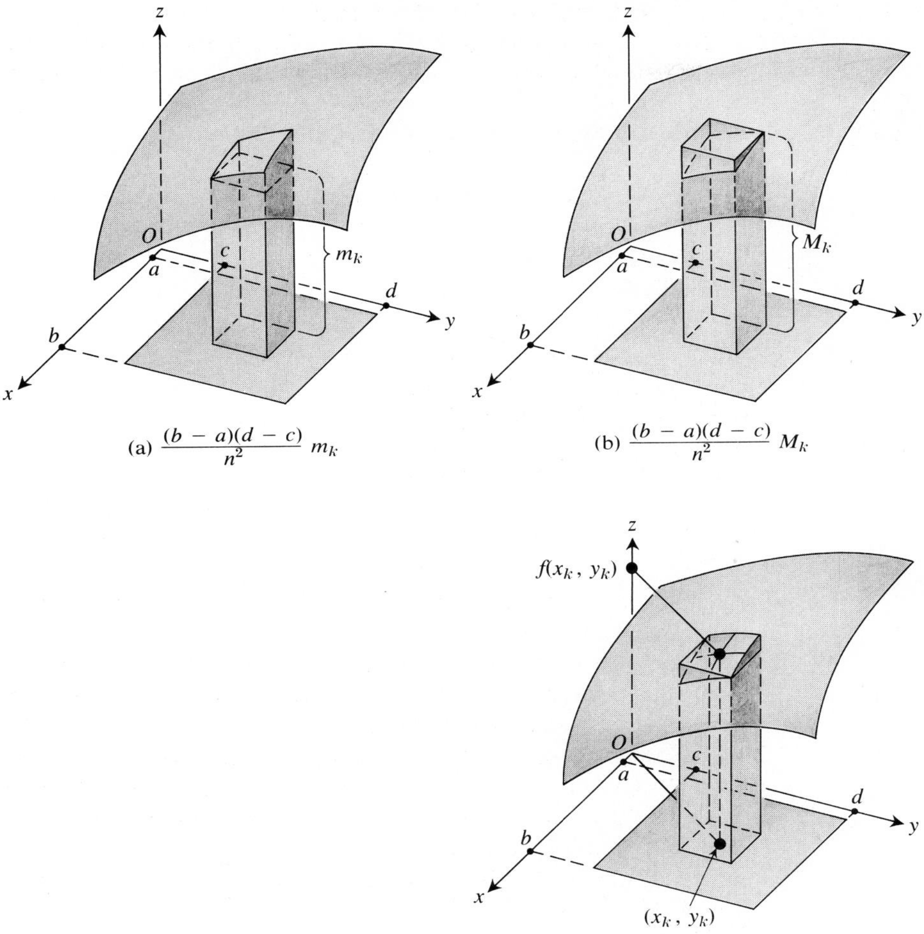

(a) $\frac{(b-a)(d-c)}{n^2} m_k$

(b) $\frac{(b-a)(d-c)}{n^2} M_k$

(c) $\frac{(b-a)(d-c)}{n^2} f(x_k, y_k)$

18.2

for a function of one variable.) If $f(x, y)$ is nonnegative for (x, y) in R, each of the terms in (1) can be viewed geometrically as the volume of a prism of rectangular cross section, as indicated in Fig. 18.2. The first term is the volume of the rectangular prism with base the kth of the n^2 subrectangles and height the minimum height m_k to the surface $z = f(x, y)$ over this kth subrectangle, etc. Then

$$s_n = \frac{(b-a)(d-c)}{n^2} \cdot \sum_{k=1}^{n^2} m_k$$

is the ***n*th lower sum for** f **over** R,

$$S_n = \frac{(b-a)(d-c)}{n^2} \cdot \sum_{k=1}^{n^2} M_k$$

is the ***n*th upper sum for** f **over** R, *and*

$$\mathcal{S}_n = \frac{(b-a)(d-c)}{n^2} \cdot \sum_{k=1}^{n^2} f(x_k, y_k)$$

is an ***n*th Riemann sum for** f **over** R.

From (1), you see that

$$s_n \leq \mathcal{S}_n \leq S_n \tag{2}$$

for all $n > 0$. We should point out that the notation "$\mathcal{S}_n$" is unjustifiably simple; it does not reflect the function f, the region R, or the choice of (x_k, y_k) as a point in the kth subrectangle of the partition.

Let h_n be the maximum of the quantities $M_k - m_k$ for $k = 1, \ldots, n^2$, and you easily see that

$$S_n - s_n \leq h_n \cdot \sum_{k=1}^{n^2} \frac{(b-a)(d-c)}{n^2} = h_n(b-a)(d-c). \tag{3}$$

Geometrically, $(b - a)(d - c)$ is the area of the region R, and $h_n(b - a) \times (d - c)$ is the volume of a box of height h_n and base of area $(b - a)(d - c)$. The illustration for $n = 2$ in Fig. 18.3 is the two-dimensional analogue of Fig. 6.9. Since f is continuous, it seems reasonable that as $n \to \infty$, the height h_n should approach zero. As in Chapter 6 we then see that $\lim_{n\to\infty} s_n$ and $\lim_{n\to\infty} S_n$ exist and are equal. Now (2) tells us that $\mathcal{S}_n$ is between s_n and S_n, so $\lim_{n\to\infty} \mathcal{S}_n$ must exist and have this same value.

Theorem 18.1 *Let f be a continuous function of two variables with domain containing a rectangular region R. Then the limits of s_n, S_n, and $\mathcal{S}_n$ defined above exist as $n \to \infty$, and*

$$\lim_{n\to\infty} s_n = \lim_{n\to\infty} \mathcal{S}_n = \lim_{n\to\infty} S_n.$$

Definition 18.1 The **definite integral** $\iint_R f(x, y)\,dx\,dy$ **of** f **over** R *is*

$$\iint_R f(x, y)\,dx\,dy = \lim_{n\to\infty} s_n = \lim_{n\to\infty} \mathcal{S}_n = \lim_{n\to\infty} S_n. \tag{4}$$

Example 1 Let's estimate the integral of $x^2 + y^2$ over R where $0 \leq x \leq 4$ and $0 \leq y \leq 4$ by computing s_2 and S_2.

SOLUTION The partition of R for $n = 2$ is shown in Fig. 18.4, which also indicates a numbering for the rectangles. Since $x^2 + y^2$ is the square of the distance

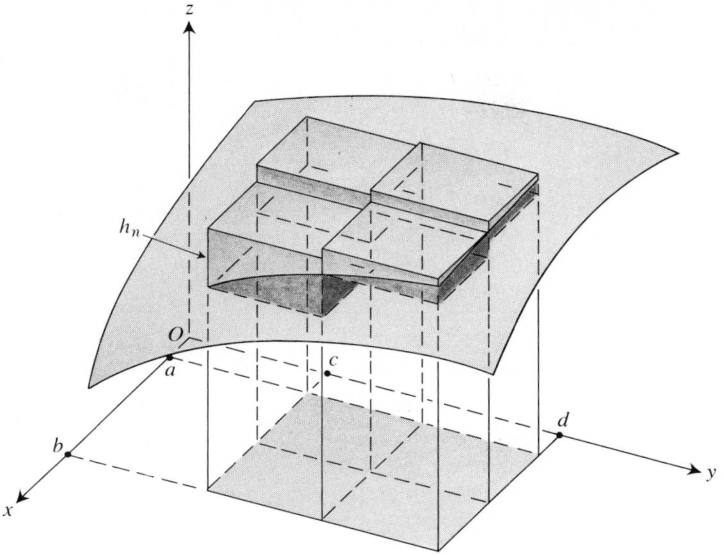

18.3

from (x, y) to the origin, you see that the points where f assumes minimum values for computation of s_2 are as indicated in (a) of the figure. The points where f assumes maximum values are shown in (b). Thus

$$\begin{aligned} s_2 &= \frac{(4-0)(4-0)}{4}[(0^2+0^2)+(2^2+0^2)+(0^2+2^2)+(2^2+2^2)] \\ &= 4[0+4+4+8] = 4\cdot 16 = 64. \end{aligned}$$

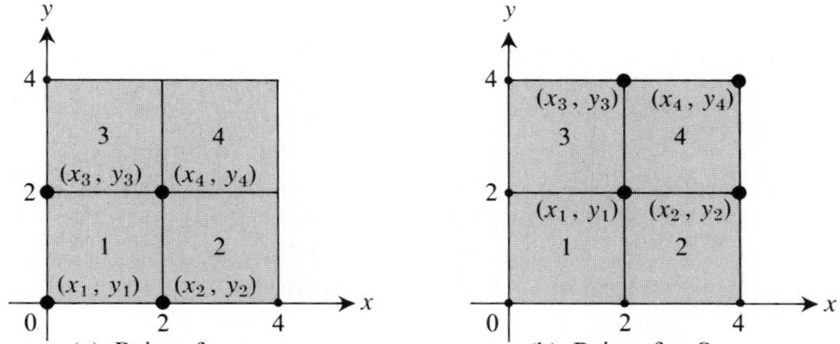

(a) Points for s_2

(b) Points for S_2

18.4

On the other hand,

$$\begin{aligned} S_2 &= \frac{(4-0)(4-0)}{4}[(2^2+2^2)+(4^2+2^2)+(2^2+4^2)+(4^2+4^2)] \\ &= 4[8+20+20+32] = 4\cdot 80 = 320. \end{aligned}$$

Thus

$$64 \le \iint_R (x^2 + y^2)\,dx\,dy \le 320.$$

Of course, these estimates are very crude indeed. ||

The Leibniz notation

$$\iint_R f(x, y)\,dx\,dy$$

may be interpreted geometrically as follows: Choose translated axes at a vertex of a rectangle in the partition, as shown in Fig. 18.5. The area of the rectangle is regarded as $dx\,dy$, as indicated in the figure. This area is multiplied by the value of the function at a point (x, y) in the rectangle. The integral then sums these products, and takes the limit of such sums as dx and dy approach zero.

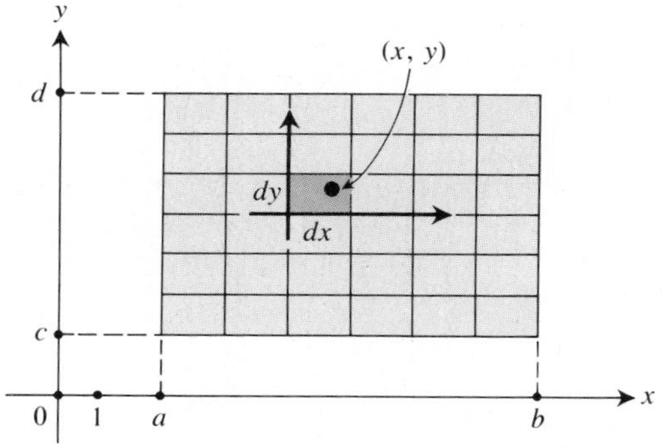

18.5

Finally, we mention that, just as for the integral of a function of one variable, it is not necessary that the region R be partitioned into subrectangles of *equal* areas in forming Riemann sums; one can form a partition of the type shown in Fig. 18.6. Indeed, it is not even necessary that the region be subdivided into *rectangular* pieces. However, in order to be sure that the limit of a sequence of Riemann sums indeed approaches the integral $\iint_R f(x, y)\,dx\,dy$, it is essential that not only the areas of subregions in the partitions of the rectangle approach zero, but also that the *maximum span* of the subregions approaches zero as $n \to \infty$ (see Figs. 18.6 and 18.7). For example, a subdivision of R into thin vertical strips as in Fig. 18.7 does not, in general, suffice to approximate $\iint_R f(x, y)\,dx\,dy$, even though the areas of the strips approach zero as $n \to \infty$. The maximum span of each strip is greater than $d - c$, which does not approach zero. (See Exercise 3.)

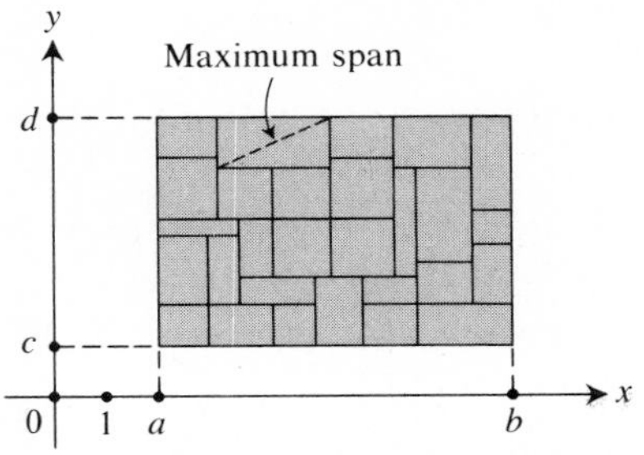

18.6

A good subdivision for Riemann sums

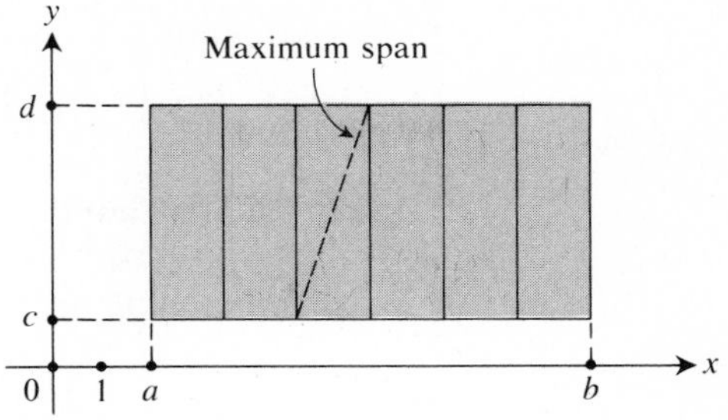

18.7

A poor subdivision for Riemann sums

18.1.2 The three-variable case It is easy to generalize the work in the preceding article and construct the integral of a continuous function f of three variables x, y, z over a rectangular box R in space given by:

$$a_1 \leq x \leq b_1, \qquad a_2 \leq y \leq b_2, \qquad a_3 \leq z \leq b_3.$$

We partition the box into n^3 subboxes by partitioning the three intervals $[a_i, b_i]$ into n subintervals of equal lengths. The volume of each subbox is easily seen to be

$$V_n = \frac{(b_1 - a_1)(b_2 - a_2)(b_3 - a_3)}{n^3}. \tag{5}$$

Let M_k be the maximum value and m_k the minimum value of $f(x, y, z)$ on the kth subbox (for some convenient numbering), and let (x_k, y_k, z_k) be a point in the kth subbox. We then let

$$s_n = V_n \cdot \sum_{k=1}^{n^3} m_k, \qquad \mathcal{S}_n = V_n \cdot \sum_{k=1}^{n^3} f(x_k, y_k, z_k),$$

and

$$S_n = V_n \cdot \sum_{k=1}^{n^3} M_k. \tag{6}$$

Obviously

$$s_n \leq \mathcal{S}_n \leq S_n. \tag{7}$$

The integral $\iiint_R f(x, y, z)\, dx\, dy\, dz$ is defined to be the common limit of s_n, $\mathcal{S}_n$, and S_n as $n \to \infty$.

18.1.3 Iterated integrals Let f be a continuous function on a rectangular region $a \leq x \leq b$, $c \leq y \leq d$ in the plane. We define the **iterated integral** $\int_c^d \int_a^b f(x, y)\, dx\, dy$ as follows. Find a function $F(x, y)$ such that

$$\frac{\partial F}{\partial x} = f(x, y), \tag{8}$$

and define

$$\int_c^d \int_a^b f(x, y)\,dx\,dy = \int_c^d (F(b, y) - F(a, y))\,dy, \tag{9}$$

where this latter integral is of a function of one variable over an interval $[c, d]$. In Exercise 4, we ask you to show that this result does not depend on the choice of the function F satisfying (8). If $\partial G/\partial y = f(x, y)$, we define

$$\int_a^b \int_c^d f(x, y)\,dy\,dx = \int_a^b (G(x, d) - G(x, c))\,dx. \tag{10}$$

Note that these iterated integrals are computed from the "inside outward." That is, you think of $\int_c^d \int_a^b f(x, y)\,dx\,dy$ as

$$\int_c^d \left(\int_a^b f(x, y)\,dx\right) dy,$$

where the inside integral is computed with respect to x only, thinking of y as a constant, and the final integral is computed with respect to y.

Clearly, the order of the limits on the integrals makes a difference; for example

$$\int_c^d \int_a^b f(x, y)\,dx\,dy = -\int_c^d \int_b^a f(x, y)\,dx\,dy$$

$$= \int_d^c \int_b^a f(x, y)\,dx\,dy.$$

You are asked to examine this matter further in Exercises 6 and 7.

Example 2 Let's compute

$$\int_0^1 \int_1^3 xy^2\,dx\,dy.$$

SOLUTION We have

$$\int_0^1 \int_1^3 xy^2\,dx\,dy = \int_0^1 \left(\frac{x^2}{2} y^2\Big]_{x=1}^{x=3}\right) dy = \int_0^1 \left(\frac{9}{2} y^2 - \frac{1}{2} y^2\right) dy$$

$$= \int_0^1 4y^2\,dy = \frac{4}{3} y^3\Big]_0^1 = \frac{4}{3}. \quad \|$$

Example 3 Let's compute the iterated integral

$$\int_1^3 \int_0^1 xy^2\,dy\,dx,$$

which is the integral in Example 2 in the "reverse order."

SOLUTION We have

$$\int_1^3 \int_0^1 xy^2\,dy\,dx = \int_1^3 \left(\frac{xy^3}{3} \Big]_{y=0}^{y=1} \right) dx = \int_1^3 \left(\frac{x}{3} - 0 \right) dx = \frac{x^2}{6} \Big]_1^3$$

$$= \frac{9}{6} - \frac{1}{6} = \frac{4}{3}. \quad \|$$

Note that the iterated integrals in Examples 2 and 3 are equal; the *order* of the integration does not seem to matter. This is an example of a general theorem.

Theorem 18.2 *Let $f(x, y)$ be continuous for (x, y) in the rectangular region R where $a \le x \le b$ and $c \le y \le d$. Then*

$$\int_c^d \int_a^b f(x, y)\,dx\,dy = \int_a^b \int_c^d f(x, y)\,dy\,dx = \iint_R f(x, y)\,dx\,dy.$$

While we shall not give a rigorous proof of this, we do give analytic and geometric explanations why it is true.

Analytic Explanation. Our numbering of the subrectangles in a partition of R into n^2 subrectangles of equal area is arbitrary. Two different styles of numbering are suggested by Fig. 18.8. The first "horizontal numbering" indicates that you may view $\mathcal{S}_n$ as a sum over n horizontal strips, each of which is, in turn, a sum over n rectangles. That is, if you let (x_i, y_j) be the midpoint of the rectangle in the ith column from the left and the jth row from the bottom in a horizontal numbering such as Fig. 18.8(a), then

$$\mathcal{S}_n = \sum_{j=1}^{n} \left(\sum_{i=1}^{n} f(x_i, y_j) \frac{(b-a)(d-c)}{n^2} \right) = \sum_{j=1}^{n} \left(\sum_{i=1}^{n} f(x_i, y_j) \frac{b-a}{n} \right) \frac{d-c}{n}. \quad (11)$$

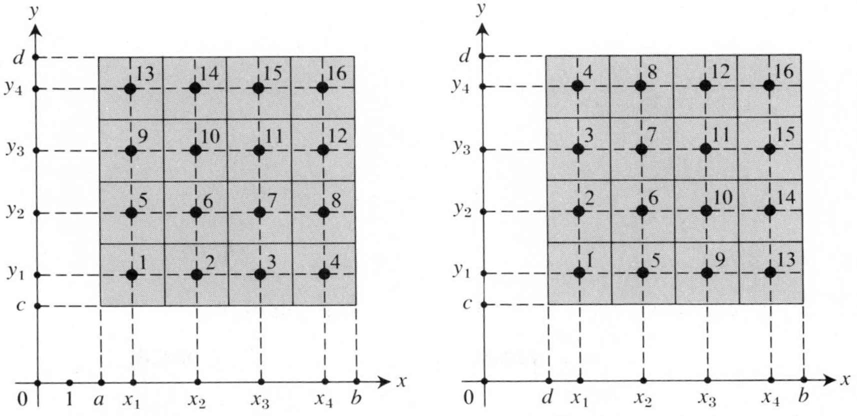

(a) Horizontal numbering $n = 4$ (b) Vertical numbering $n = 4$

18.8

By the fundamental theorem of calculus, the inside sum over i in the last expression in (11) approaches $\int_a^b f(x, y_j)\,dx$ as $n \to \infty$. It seems reasonable that the whole sum therefore approaches $\int_c^d (\int_a^b f(x, y)\,dx)\,dy$ as $n \to \infty$. A similar argument, starting with the vertical numbering in Fig. 18.8(b), leads to the iterated integral $\int_a^b (\int_c^d f(x, y)\,dy)\,dx$. This indicates analytically that the iterated integrals are both equal to $\iint_R f(x, y)\,dx\,dy$.

Geometric Explanation. The Leibniz notation suggests the following interpretation of $\int_c^d \int_a^b f(x, y)\,dx\,dy$. If you form $\int_a^b f(x, y_j)\,dx$, thinking of y_j as a constant, you obtain the area of the vertical shaded plane region shown in Fig. 18.9, which lies in the plane $y = y_j$ and under the surface $z = f(x, y)$ between $x = a$ and $x = b$. Multiplying by dy, you obtain the volume of the slab shown in Fig. 18.10, and the iterated integral

$$\int_c^d \left(\int_a^b f(x, y)\,dx\right) dy$$

thus adds up the volumes of the slabs from $y = c$ to $y = d$ as $dy \to 0$. Clearly the result for $f(x, y) \geq 0$ is the volume of the three-dimensional region under the surface $z = f(x, y)$ and over the rectangle R, which is precisely the geometric interpretation of $\iint_R f(x, y)\,dx\,dy$. Similar geometric considerations indicate that $\int_a^b \int_c^d f(x, y)\,dy\,dx$ must also be equal to $\iint_R f(x, y)\,dx\,dy$. We leave the sketching of figures like Figs. 18.9 and 18.10 to the exercises (see Exercise 5).

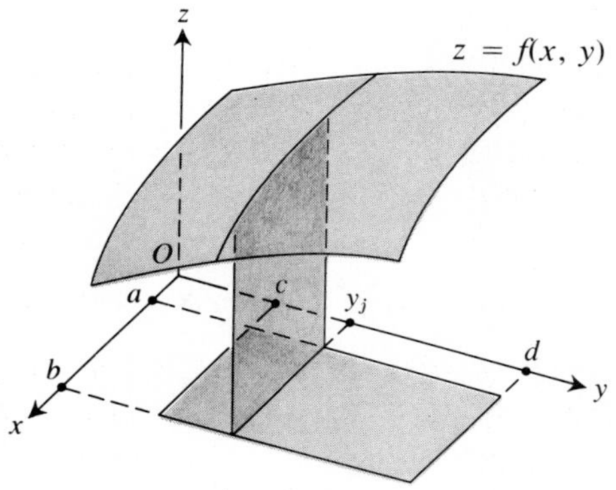

18.9

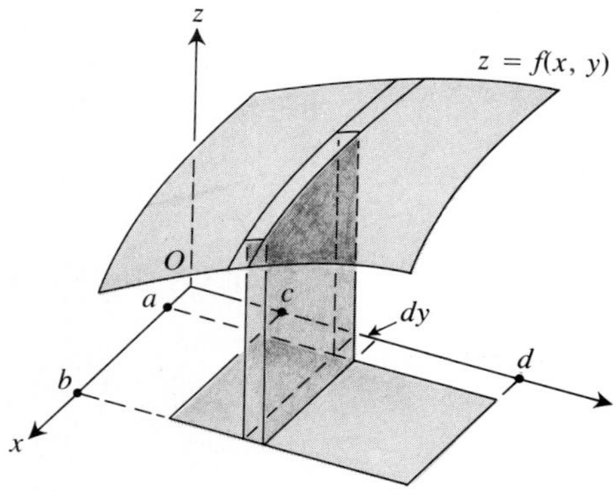

18.10

Since $\iint_R f(x, y)\,dx\,dy$ and a corresponding notation for an iterated integral contain two integral signs, integrals of functions of two variables are often called **double integrals**. Iteration provides a technique for computing a double integral, just as the fundamental theorem of calculus provides a technique for computing an integral of a function of one variable. In fact, the computation of a double integral is reduced to the computation of two, successive integrals of functions of a single variable.

Example 4 In Example 1, we estimated $\iint_R (x^2 + y^2)\,dx\,dy$ for R given by $0 \le x \le 4$, $0 \le y \le 4$, using upper and lower sums with $n = 2$. Let's compute its exact value using an iterated integral.

SOLUTION The exact value of the integral is given by

$$\iint_R (x^2 + y^2)\,dx\,dy = \int_0^4 \int_0^4 (x^2 + y^2)\,dx\,dy = \int_0^4 \left(\frac{x^3}{3} + y^2 x\right)\Bigg]_{x=0}^{x=4} dy$$

$$= \int_0^4 \left(\frac{64}{3} + 4y^2\right) dy = \left(\frac{64}{3} y + \frac{4y^3}{3}\right)\Bigg]_0^4 = \frac{256}{3} + \frac{256}{3} = \frac{512}{3}. \quad \|$$

18.1.4 Three-dimensional iterated integrals If f is a function of three variables that is continuous on a rectangular box R where $a_1 \le x \le b_1$, $a_2 \le y \le b_2$, and $a_3 \le z \le b_3$, then you can form several different iterated integrals, such as

$$\int_{a_1}^{b_1} \int_{a_2}^{b_2} \int_{a_3}^{b_3} f(x, y, z)\,dz\,dy\,dx$$

and

$$\int_{a_2}^{b_2} \int_{a_3}^{b_3} \int_{a_1}^{b_1} f(x, y, z)\,dx\,dz\,dy.$$

The iterated integrals are all to be computed "from the inside outward." Any order of integration is possible; there are six possible orders in all (see Exercise 7). If f is continuous over R, then all these iterated integrals with a_i's as lower limits and b_i's as upper limits are equal, and equal the multiple integral $\iiint_R f(x, y, z)\,dx\,dy\,dz$.

SUMMARY *Let $f(x, y)$ be continuous in the rectangle R where $a \le x \le b$ and $c \le y \le d$. Partition R into n^2 subrectangles by subdividing both intervals $[a, b]$ and $[c, d]$ into n subintervals of equal lengths. Number the subrectangles from 1 to n^2 and let m_k be the minimum value and M_k the maximum value of $f(x, y)$ in the kth subrectangle. Let (x_k, y_k) be any point in the kth subrectangle.*

1. $s_n = \dfrac{(b-a)(d-c)}{n^2} \cdot \sum_{k=1}^{n} m_k$ *is the nth lower sum.*

2. $S_n = \dfrac{(b-a)(d-c)}{n^2} \cdot \sum_{k=1}^{n} M_k$ *is the nth upper sum.*

3. $\mathscr{S}_n = \dfrac{(b-a)(d-c)}{n^2} \cdot \sum_{k=1}^{n} f(x_k, y_k)$ *is an* nth *Riemann sum.*

4. $s_n \le \mathscr{S}_n \le S_n$ *and*

$$\lim_{n\to\infty} s_n = \lim_{n\to\infty} S_n = \lim_{n\to\infty} \mathscr{S}_n = \iint_R f(x, y)\, dx\, dy.$$

5. *If* $\partial F/\partial x = f(x, y)$, *then*

$$\int_c^d \int_a^b f(x, y)\, dx\, dy = \int_c^d F(x, y)\Big]_{x=a}^{x=b} dy = \int_c^d [F(b, y) - F(a, y)]\, dy.$$

6. *If* $\partial G/\partial y = f(x, y)$, *then*

$$\int_a^b \int_c^d f(x, y)\, dy\, dx = \int_a^b G(x, y)\Big]_{y=c}^{y=d} dx = \int_a^b [G(x, d) - G(x, c)]\, dx.$$

7. $$\iint_R f(x, y)\, dx\, dy = \int_c^d \int_a^b f(x, y)\, dx\, dy = \int_a^b \int_c^d f(x, y)\, dy\, dx.$$

8. *Analogous treatments hold for more than two variables.*

EXERCISES

1. Give a geometric interpretation of $\iint_R f(x, y)\, dx\, dy$ where f is a function of two variables, continuous over a rectangular region R, but possibly assuming both positive and negative values on R.

2. We have treated only a very restrictive case of multiple integrals of continuous functions in this section. Can you guess the restriction to which we are referring?

3. Let R be given by $a \le x \le b$, $c \le x \le d$ and let $f(x, y) = y$ for (x, y) in R. Consider the partition of R into n vertical strips of equal areas, as in Fig. 18.7 where $n = 6$.

a) Compute the upper sums s_n and the lower sums s_n for these partitions of R.

b) Is it true that $\lim_{n\to\infty} s_n = \lim_{n\to\infty} S_n$ for these partitions of R?

c) What point is illustrated by this exercise?

4. Show that the definition of $\int_c^d \int_a^b f(x, y)\, dx\, dy$ as given by Eq. (9) in the text is *well-defined*, that is, independent of the choice of "partial antiderivative" $F(x, y)$.

5. Sketch figures similar to Figs. 18.9 and 18.10 to illustrate geometrically that $\int_a^b \int_c^d f(x, y)\, dy\, dx = \iint_R f(x, y)\, dx\, dy$ where R is the rectangle $a \le x \le b$, $c \le y \le d$, and f is continuous over R.

6. In Section 18.1.3 we defined iterated integrals $\int_c^d \int_a^b f(x, y)\, dx\, dy$ and $\int_a^b \int_c^d f(x, y)\, dy\, dx$ over R where $a \le x \le b$, $c \le y \le d$ for f continuous on R. We could also consider

$$\int_d^c \int_a^b f(x, y)\, dx\, dy,$$

where, again, the integral is to be computed from the "inside outward." List all such iterated integrals of f arising from R, and compare the values of these integrals. [*Hint.* There are eight of them.]

7. Let f be a continuous function of three variables with domain containing a rectangular box $a_1 \le x \le b_1$, $a_2 \le y \le b_2$, $a_3 \le z \le b_3$.

a) Show that there are six possible "orders of integration" for an iterated integral, where each integral has some a_i for lower limit and b_i for upper limit.

b) How many iterated integrals can you form for f over the box if the a_i's are not restricted to lower limits and the b_i's are not restricted to upper limits? (See Exercise 6.)

c) Compare the values of these integrals described in (b).

8. Let R be the rectangle $0 \le x \le 2$, $1 \le y \le 5$. Estimate $\iint_R (x + y)\,dx\,dy$ by s_2 and S_2.

9. Repeat Exercise 8, but estimate $\iint_R (x + y)\,dx\,dy$ using $\mathscr{S}_2$ and the midpoints of the subrectangles.

10. Let R be given by $0 \le x \le 2$, $0 \le y \le 4$, $2 \le z \le 4$. Estimate $\iiint_R x^2yz\,dx\,dy\,dz$ using $\mathscr{S}_2$ and the midpoints of the regions in the partitions.

11. Let R be given by $0 \le x \le 4$, $1 \le y \le 5$, $-4 \le z \le 4$. Estimate $\iiint_R xy^2z^2\,dx\,dy\,dz$ using $\mathscr{S}_2$ and the midpoints of the regions in the partition.

12. Compute the integral in Exercise 8, using an iterated integral.

13. Compute the integral in Exercise 10, using an iterated integral.

14. Compute the integral in Exercise 11, using an iterated integral.

In Exercises 15 *through* 24, *compute the given iterated integral.*

15. $\displaystyle\int_1^4\int_0^2 (x + y^2)\,dx\,dy$

16. $\displaystyle\int_1^3\int_{-1}^2 x^2y\,dy\,dx$

17. $\displaystyle\int_0^{\pi}\int_0^{\pi/2} x\sin y\,dx\,dy$

18. $\displaystyle\int_0^{\pi}\int_0^{\pi/2} x\sin y\,dy\,dx$

19. $\displaystyle\int_0^2\int_0^{\pi} x\sin^2 y\,dy\,dx$

20. $\displaystyle\int_1^{e^2}\int_1^{e} \ln(xy)\,dx\,dy$

21. $\displaystyle\int_0^1\int_2^3\int_{-1}^4 xy^2z\,dx\,dy\,dz$

22. $\displaystyle\int_0^1\int_{-1}^1\int_1^2 (x^2 + yz)\,dz\,dx\,dy$

23. $\displaystyle\int_{-1}^3\int_0^{\ln 2}\int_0^4 xze^y\,dx\,dy\,dz$

24. $\displaystyle\int_0^2\int_0^1\int_0^1 xyz\sqrt{2 - x^2 - y^2}\,dx\,dy\,dz$

18.2 INTEGRALS OVER A REGION

18.2.1 Riemann sums over regions

A region G in the plane is *bounded* if it is contained in some sufficiently large rectangle in the plane. Naively, bounded regions are those that do not go off to infinity in any direction. We will consider "nice" bounded regions like the one in Fig. 18.11, where we have a natural idea of the *boundary* (bounding curve) of the region and the *interior* (inside) of the region. A region is *closed* if the boundary is considered to be part of the region.

We wish to integrate a continuous function f over a bounded region G of the plane contained in the domain of f. We shall restrict ourselves to the important case where the boundary of G is a smooth curve of finite length.

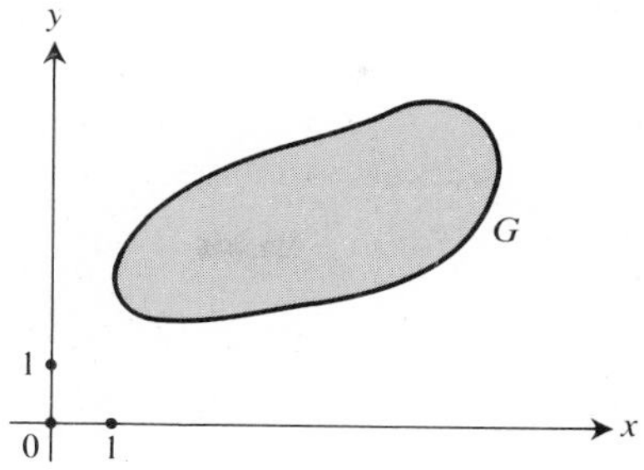

18.11

Let R be some rectangle with sides parallel to the coordinate axes and containing G. Divide R into n^2 subrectangles of equal area as in Section 18.1. This gives a grid of rectangles of equal size superimposed over the plane region G, as illustrated in Fig. 18.12.

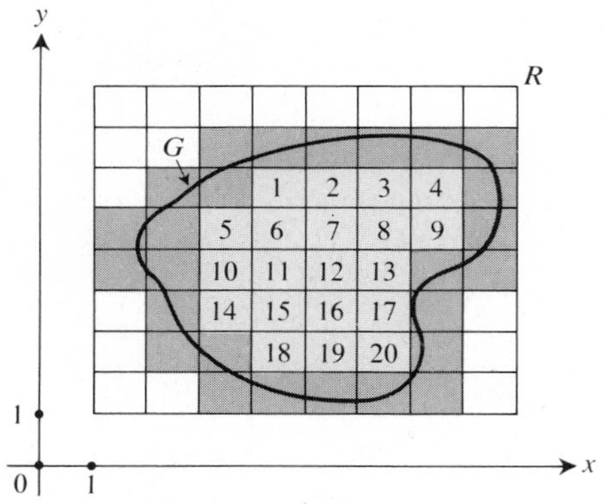

18.12

Consider now only those little rectangles that lie totally *inside* the region G. In Fig. 18.12, they are the twenty lightly shaded rectangles. In general, there may be r such rectangles, which we number from 1 to r in some convenient fashion. It can be proved that if f is continuous on the kth such rectangle, then f assumes a maximum value M_k and a minimum value m_k on that rectangle. Using only those r rectangles lying totally within G, we define the *lower sum* s_n, the *upper sum* S_n, and a *Riemann sum* $\mathcal{S}_n$ for $f(x, y)$ over G as follows: If each rectangle in the grid has area ΔA and if (x_k, y_k) is any point in the kth inside rectangle, we let

$$s_n = (\Delta A)\left[\sum_{k=1}^{r} m_k\right],$$

$$\mathcal{S}_n = (\Delta A)\left[\sum_{k=1}^{r} f(x_k, y_k)\right],$$

$$S_n = (\Delta A)\left[\sum_{k=1}^{r} M_k\right].$$

Note that the dark-shaded rectangles in the partition of R in Fig. 18.12, which contain the boundary curve of G, are not considered at all in this formation of the sums s_n, $\mathcal{S}_n$, and S_n. This is permissible for the following reason. We assumed that the boundary curve of G has finite length, and it can then be shown that the sum of the areas of those rectangles containing the boundary approaches zero as $n \to \infty$. Consequently, if we did attempt to include their contributions in s_n, $\mathcal{S}_n$, or S_n, those contributions would approach zero as $n \to \infty$, and in the next paragraph, we are going to let $n \to \infty$.

For the kth little rectangle lying inside G, you have

$$(M_k - m_k) \to 0 \quad \text{as} \quad n \to \infty,$$

precisely as in Section 18.1. If h_n is the maximum of $M_k - m_k$ over these r rectangles, then

$$S_n - s_n \leq h_n \text{ (area of } G).$$

Consequently,

$$\lim_{n\to\infty} (S_n - s_n) = 0.$$

As in Section 18.1, it seems reasonable that $\lim_{n\to\infty} s_n$ exists and equals $\lim_{n\to\infty} S_n$. These limits can be shown to be independent of the choice of the rectangle R containing G.

Definition 18.2 Let f be a continuous function over a closed bounded region G of the plane whose boundary is a smooth curve of finite length. The **integral** $\iint_G f(x, y)\, dx\, dy$ **of** f **over** G is the common value

$$\iint_G f(x, y)\, dx\, dy = \lim_{n\to\infty} s_n = \lim_{n\to\infty} \mathcal{S}_n = \lim_{n\to\infty} S_n,$$

where s_n, $\mathcal{S}_n$ and S_n are as described above.

It is obvious that this work can be extended to integrals of continuous functions of three variables over bounded regions in space, whose boundaries are smooth surfaces of finite area. We need only replace our rectangular grid by a grid of boxes.

18.2.2 Iterated integrals over regions The integrals defined in the preceding article are usually computed using an iterated integral, much as in Section 18.1. Consider the plane region G shown in Fig. 18.13, where the "lower" portion of the boundary is the curve $y = h_1(x)$ and the "top" portion is the curve $y = h_2(x)$, where h_1 and h_2 are continuous functions. If f is continuous on G, we form the iterated integral

$$\int_a^b \int_{h_1(x)}^{h_2(x)} f(x, y)\, dy\, dx = \int_a^b \left(\int_{h_1(x)}^{h_{2}(x)} f(x, y)\, dy\right) dx. \tag{1}$$

The computation of (1) is again to be made from the "inside outward," as indicated by the parentheses. The first integration is to be performed "with

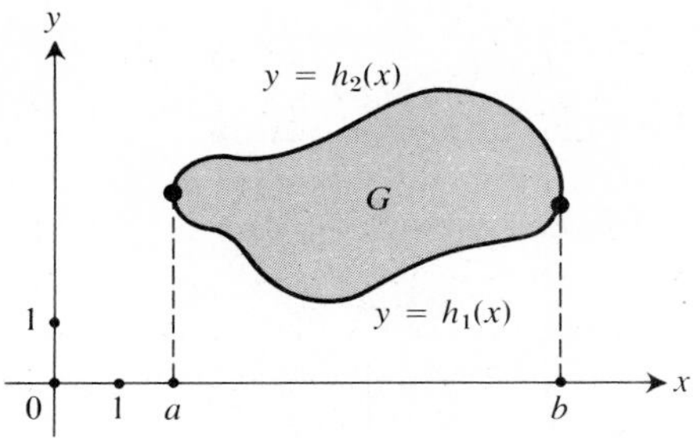

18.13

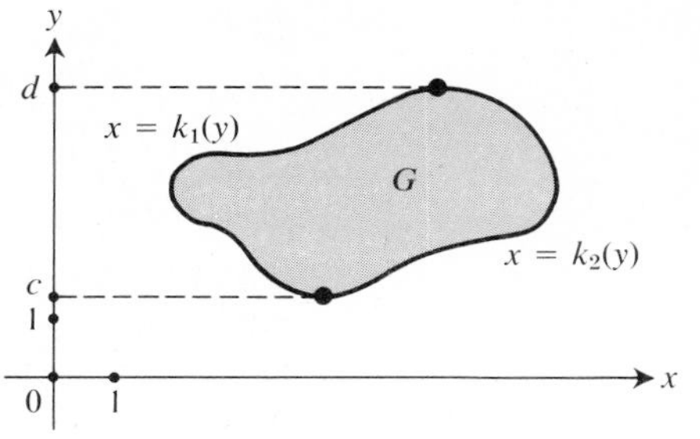

18.14

respect to y only," and the first limits depend upon x. A function of x results upon substituting the limits after the first integration, and this function is then integrated "with respect to x" from a to b. Analytic and geometric arguments like those in Section 18.1 easily convince us that this iterated integral is equal to $\iint_G f(x, y)\,dx\,dy$, as defined above.

We could also integrate in the reverse order, in which case we consider the "left" and "right" boundary curves of G to be given by $x = k_1(y)$ and $x = k_2(y)$ as shown in Fig. 18.14. Our integral then takes the form

$$\iint_G f(x, y)\,dx\,dy = \int_c^d \int_{k_1(y)}^{k_2(y)} f(x, y)\,dx\,dy = \int_c^d \left(\int_{k_1(y)}^{k_2(y)} f(x, y)\,dx \right) dy.$$

Analogous iterated integrals can be formed for continuous functions of three variables over a suitable region in space.

Example 1 For illustration, we evaluate the iterated integral $\int_0^2 \int_{x^2}^4 x^2 y\,dy\,dx$.

SOLUTION We have

$$\int_0^2 \int_{x^2}^4 x^2 y\,dy\,dx = \int_0^2 \left(\int_{x^2}^4 x^2 y\,dy \right) dx = \int_0^2 x^2 \frac{y^2}{2} \bigg]_{y=x^2}^{y=4} dx$$

$$= \int_0^2 (8x^2 - \tfrac{1}{2}x^6)\,dx$$

$$= \left(\frac{8}{3}x^3 - \frac{1}{14}x^7 \right) \bigg]_0^2$$

$$= \frac{64}{3} - \frac{64}{7} = \frac{256}{21}. \quad \|$$

It is important to be able to change the order of integration in an iterated integral. An outline of a procedure by which this may be done for double integrals is as follows.

Changing order of integration

STEP 1. *Draw a sketch,* and shade the region of the plane over which integration takes place. Draw a little "differential rectangle" of dimensions dx by dy in the shaded region.

STEP 2. Convert your given iterated integral into one with the order of integration reversed *by looking at your sketch* and writing down the appropriate limits. The limits on the inside integral sign are functions of the remaining variable of integration, while the limits on the outside integral sign are always constants.

Similar steps may be followed to change the order of integration in a triple iterated integral.

Example 2 Let's reverse the order of integration for the integral

$$\int_0^2 \int_{x^2}^4 x^2 y \, dy \, dx$$

considered in Example 1.

SOLUTION STEP 1. Starting with the limits on the inside integral, the first integration with respect to y, goes from $y = x^2$ to $y = 4$, so we sketch these curves in the plane. See Fig. 18.15(a). Since this first integration was with respect to y, we think of $y = x^2$ and $y = 4$ as forming the bottom and top boundaries of our region. Now the final integration with respect to x goes only from $x = 0$ to $x = 4$. Thus the region is the one in the first quadrant only, shown shaded in Fig. 18.15(a).

STEP 2. To reverse the order of integration, we wish to integrate first with respect to x. Pushing our differential rectangle as far to the left (negative x-direction) as it will go, it is always stopped by the line $x = 0$. Pushing it to the right (positive x-direction), it is stopped by the curve $y = x^2$. Since these inside limits must be expressed as functions of y, we write $y = x^2$ as $x = \sqrt{y}$; the positive square root is clearly appropriate from the sketch. So

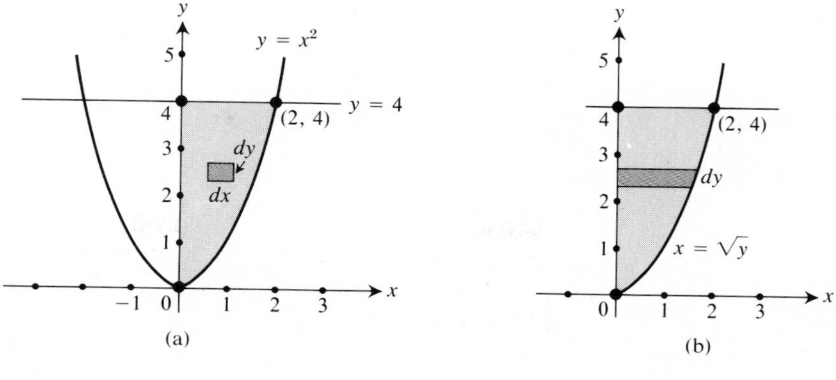

18.15

our inside integral becomes

$$\int_0^{\sqrt{y}} x^2 y \, dx.$$

This corresponds to adding the contributions to the integral over the thin horizontal strip shown in Fig. 18.15(b). Now we must add the contributions of these strips from a strip at the bottom of the region, where $y = 0$, to a strip at the top, where $y = 4$. Thus we arrive at

$$\int_0^4 \int_0^{\sqrt{y}} x^2 y \, dx \, dy$$

as the desired integral. A computation yields

$$\int_0^4 \int_0^{\sqrt{y}} x^2 y \, dx \, dy = \int_0^4 \frac{x^3}{3} y \Big]_{x=0}^{x=\sqrt{y}} dy = \int_0^4 \frac{1}{3} y^{5/2} \, dy = \frac{1}{3} \cdot \frac{2}{7} y^{7/2} \Big]_0^4$$

$$= \frac{2}{21}(4^{7/2}) = \frac{2}{21} \cdot 2^7 = \frac{256}{21}.$$

Of course, this is the same value as we obtained in Example 1. ||

Example 3 Let us convert

$$\int_0^1 \int_{2x-2}^0 \int_0^{3-3x+(3y/2)} xyz^2 \, dz \, dy \, dx$$

to the order

$$\iiint xyz^2 \, dx \, dz \, dy.$$

SOLUTION STEP 1. The inside limits with respect to z show that the bottom of the region is the plane $z = 0$ and the top is the plane $z = 3 - 3x + (3y/2)$, or $6x - 3y + 2z = 6$. We sketch these planes in Fig. 18.16. The remaining limits of integration then show that the region of integration is the tetrahedron shaded in the figure.

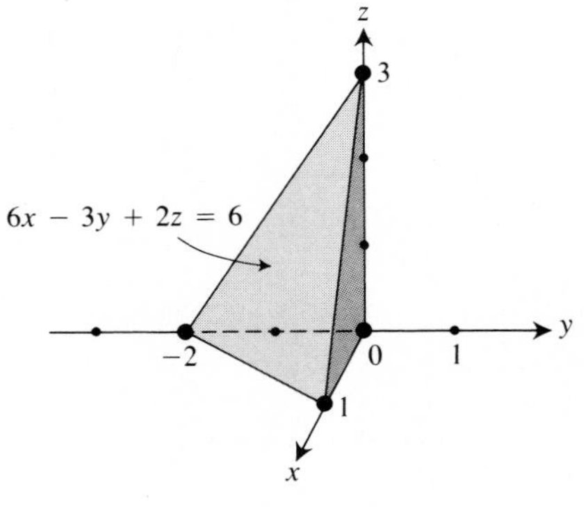

18.16

STEP 2 In the new order, we wish to integrate first in the x-direction, from $x = 0$ out to the plane $6x - 3y + 2z = 6$, where $x = 1 + (y/2) - (z/3)$. Thus we start with

$$\int_0^{1+(y/2)-(z/3)} xyz^2\,dx.$$

The next z-limits may be found from the back triangle of the region, in the yz-plane, which has as base the line where $z = 0$ and as top the line where $-3y + 2z = 6$, or $z = 3 + (3y/2)$, obtained by setting $x = 0$ in $6x - 3y + 2z = 6$. We have arrived so far at

$$\int_0^{3+(3y/2)} \int_0^{1+(y/2)-(z/3)} xyz^2\,dx\,dz.$$

Finally, the constant y-limits go from the minimum y-value of -2 to the maximum y-value of 0, so the desired integral is

$$\int_{-2}^{0} \int_0^{3+(3y/2)} \int_0^{1+(y/2)-(z/3)} xyz^2\,dx\,dz\,dy. \quad \|$$

18.2.3 Areas and volumes by multiple integration

Let G be a closed, bounded region in the plane. It is geometrically clear that for the constant function 1, the integral

$$\iint_G 1\,dx\,dy$$

gives the area of the region. The integral is usually computed using an iterated integral. We give two examples.

Example 4 We start by solving, using "double integrals," an area problem we could easily have solved in Chapter 7. Let's find the area of the plane region in the first quadrant of the plane bounded by the curves $y = x^3$ and $y = \sqrt{x}$.

SOLUTION The region is sketched in Fig. 18.17. Using Leibniz notation, we think of

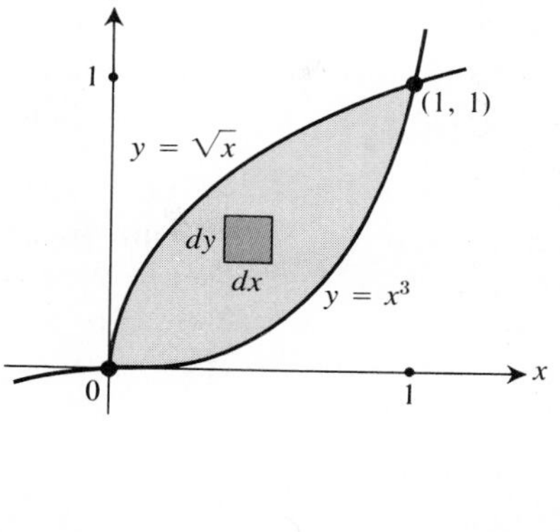

18.17

$dx\,dy = dy\,dx$ as the area of a small rectangle in the region, and add up the areas of such rectangles over the region as $dx \to 0$ and $dy \to 0$ with an integral. The "lower" and "upper" bounding curves of the region are $y = x^3$ and $y = \sqrt{x}$, respectively, so our iterated integral is

$$\int_0^1 \int_{x^3}^{\sqrt{x}} dy\,dx = \int_0^1 \left(\int_{x^3}^{\sqrt{x}} dy\right) dx = \int_0^1 y\Big]_{x^3}^{\sqrt{x}} dx = \int_0^1 (\sqrt{x} - x^3)\,dx$$

$$= \left(\frac{2}{3}x^{3/2} - \frac{x^4}{4}\right)\Big]_0^1 = \frac{2}{3} - \frac{1}{4} = \frac{5}{12}.$$

Note that the integral $\int_0^1 (\sqrt{x} - x^3)\,dx$ that occurs in the middle of the computation above is the one we would have started with in Chapter 7 to compute the area.

If we wish to compute the iterated integral in the other order, we must find the "left" and "right" bounding curves of our region, and we obtain $x = y^2$ and $x = y^{1/3}$, respectively. The iterated integral in this order is therefore

$$\int_0^1 \int_{y^2}^{y^{1/3}} dx\,dy = \int_0^1 x\Big]_{y^2}^{y^{1/3}} dy = \int_0^1 (y^{1/3} - y^2)\,dy$$

$$= \left(\frac{3}{4}y^{4/3} - \frac{y^3}{3}\right)\Big]_0^1 = \frac{3}{4} - \frac{1}{3} = \frac{5}{12}.$$

Of course, we obtained the same answer. ||

People often have trouble finding the appropriate limits of integration in setting up an iterated integral. Note in particular that the *final* limits (on the lefthand integral sign) are always constant, and other limits may be functions of only those variables with respect to which integration will be performed *later.*

$$\text{WRONG:} \qquad \int_y^{y^2}\int_1^3 dy\,dx, \qquad \int_1^3\int_x^{x+5}\int_z^{z-y} yz^2\,dx\,dy\,dz.$$

Example 5 Let's find the volume of the region in space bounded above by the surface $z = 1 - x^2 - y^2$, on the sides by the planes $x = 0$, $y = 0$, $x + y = 1$, and below by the plane $z = 0$. The region is sketched in Fig. 18.18.

SOLUTION We think of adding the volumes of small rectangular boxes with edges having lengths dx, dy, and dz. An attempt to find our iterated integral by integrating in the x-direction first leads to problems in the x-limits, for the "lower" boxes would have to be summed from the plane $x = 0$ forward to the plane $x = 1 - y$, while the "higher" boxes would have to be summed from $x = 0$ forward to the surface $x = \sqrt{1 - y^2 - z}$, as indicated in Fig.

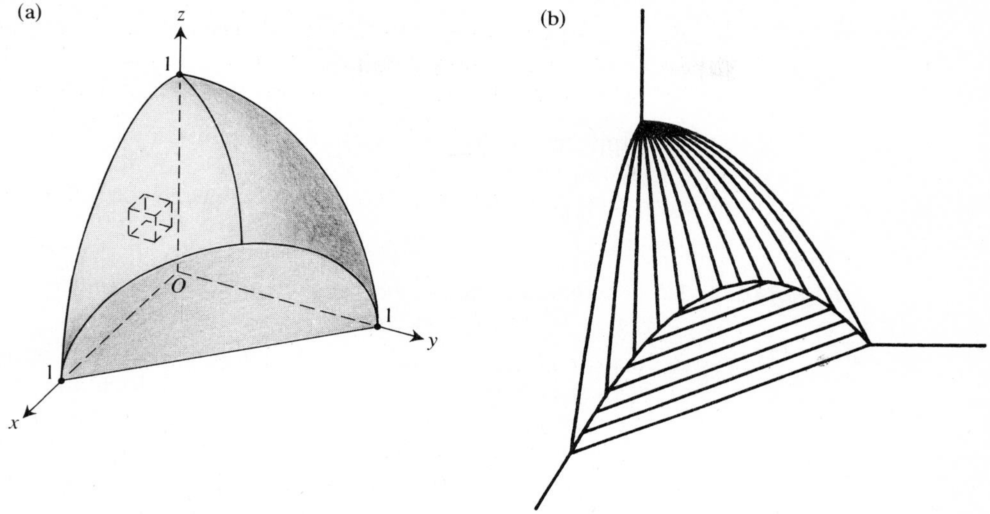

18.18 (a) The region of Example 5. (b) Computer-plotted representation.

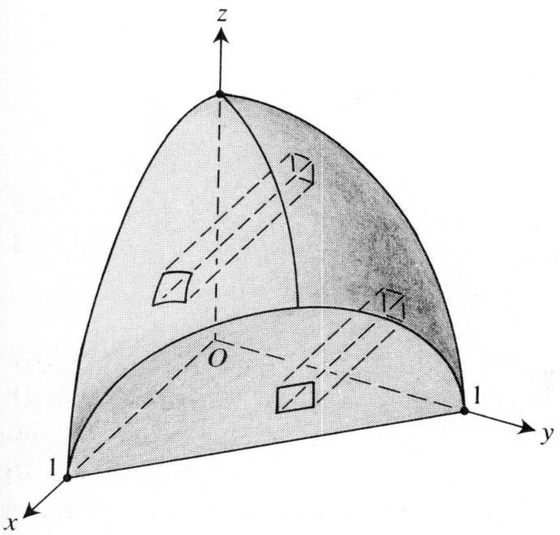

18.19

18.19. The same problem occurs if we integrate first in the y-direction. Thus we integrate in the order z, x, y, and obtain the iterated integral

$$\begin{aligned}\int_0^1 \int_0^{1-y} \int_0^{1-x^2-y^2} dz\,dx\,dy &= \int_0^1 \int_0^{1-y} z\Big]_0^{1-x^2-y^2} dx\,dy \\ &= \int_0^1 \int_0^{1-y} (1 - x^2 - y^2)\,dx\,dy \\ &= \int_0^1 \left(x - \frac{x^3}{3} - y^2x\right)\Big]_0^{1-y} dy \\ &= \int_0^1 \left((1-y) - \frac{(1-y)^3}{3} - y^2(1-y)\right) dy \\ &= \int_0^1 \left(1 - y - y^2 + y^3 - \frac{(1-y)^3}{3}\right) dy \\ &= \left(y - \frac{y^2}{2} - \frac{y^3}{3} + \frac{y^4}{4} + \frac{(1-y)^4}{12}\right)\Big]_0^1 \\ &= 1 - \frac{1}{2} - \frac{1}{3} + \frac{1}{4} - \frac{1}{12} \\ &= 1 - \frac{6 + 4 - 3 + 1}{12} = 1 - \frac{8}{12} = \frac{1}{3}.\end{aligned}$$

Note that we could have started with the integral

$$\int_0^1 \int_0^{1-y} (1 - x^2 - y^2)\, dx\, dy$$

that appears in the computation if we just thought of our region as lying under the surface $z = 1 - x^2 - y^2$ and over the triangular region in the x,y-plane bounded by the coordinate axes and the line $x + y = 1$. ||

SUMMARY

1. *A bounded plane region G is one that can be enclosed in a rectangle. The region is closed if the region includes its boundary.*
2. *Let a rectangle containing G be partitioned into n^2 subrectangles of equal size as in Section 18.1. Upper, lower, and Riemann sums taken over just those rectangles of the partition that lie totally inside G approach a common value $\iint_G f(x, y)\, dx\, dy$ as $n \to \infty$, for a continuous function $f(x, y)$ over a bounded region G with smooth boundary curve of finite length.*
3. *Iterated integrals are used for computation of an integral over a region. See Section 18.2.2 for a description of the limits.*
4. *Draw a sketch and work from that sketch when changing the order of integration.*
5. *The area of a region G is found by integrating the constant function 1 over G.*
6. *Analogous integration can be done in space.*

EXERCISES

In Exercises 1 through 8, compute the given iterated integral.

1. $\int_0^1 \int_0^{\sqrt{1-y^2}} 4xy\, dx\, dy$

2. $\int_0^{\pi} \int_0^{\sin x} x\, dy\, dx$

3. $\int_1^e \int_0^{\ln x} \frac{y}{x}\, dy\, dx$

4. $\int_0^1 \int_{-\sqrt{y}}^{y} (x + y^2)\, dx\, dy$

5. $\int_0^1 \int_0^{1-y} \int_0^{x^2+y^2} y\, dz\, dx\, dy$

6. $\int_0^a \int_0^{\sqrt{a^2-y^2}} \int_0^{\sqrt{a^2-x^2-y^2}} x\, dz\, dx\, dy$

7. $\int_0^2 \int_0^{\sqrt{4-z^2}} \int_{y^2+z^2-4}^{4-y^2-z^2} 1\, dx\, dy\, dz$

8. $\int_0^1 \int_0^z \int_0^{y+z} yz\, dx\, dy\, dz$

In Exercises 9 through 16, sketch the region of integration for the given iterated integral and then write an equal iterated integral (or an equal sum of iterated integrals) with the integration in the "reverse order." (You are not asked to compute the integral.)

9. $\int_0^1 \int_{-\sqrt{1-y^2}}^{\sqrt{1-y^2}} 4xy\, dx\, dy$

10. $\int_0^2 \int_0^{x^2} (\sin xy)\, dy\, dx$

11. $\int_0^{\pi} \int_0^{\sin x} x^2 y\, dy\, dx$

12. $\int_0^1 \int_1^{e^y} x^2\, dx\, dy$

13. $\int_0^1 \int_{x^2}^{x} y \cos x \, dy \, dx$ **14.** $\int_{-3}^{1} \int_{-\sqrt{1-y}}^{\sqrt{1-y}} e^{xy} \, dx \, dy$ **15.** $\int_0^1 \int_{2y-2}^{1-y} x \cos^2 y \, dx \, dy$ **16.** $\int_{-1}^{1} \int_{1-y^2}^{y^2-1} x^2 y^3 \, dx \, dy$

17. Find the appropriate limits for $\iiint xyz^2 \, dx \, dy \, dz$ in order for it to equal the iterated integral

$$\int_{-1}^{1} \int_{-\sqrt{1-x^2}}^{\sqrt{1-x^2}} \int_0^{1-x^2-y^2} xyz^2 \, dz \, dy \, dx.$$

Do not evaluate the integral.

18. Find the appropriate limits for $\iiint x \sin yz \, dy \, dz \, dx$ in order for it to equal the iterated integral

$$\int_{-2}^{1} \int_0^{\pi/2} \int_0^{\cos z} x \sin yz \, dx \, dz \, dy.$$

Do not evaluate the integral.

In Exercises 19 *through* 24, *use an iterated integral to find the area or volume of the indicated region.*

19. The region of the plane bounded by the curves $y = 0$, $y = 1 + x$, and $y = \sqrt{1 - x}$.

20. The region of the plane bounded by the curves $y = \sin x$, $x = \pi/2$, and $y = x$.

21. The region of the plane bounded by $y = \ln x$, $y = 1 - x$, and $y = 1$.

22. The region of space bounded by $z = \cos x$ for $-\pi/2 \leq x \leq \pi/2$, $z = 0$, $y = -1$, and $y = 2$.

23. The region of space bounded by $z = 4 - x^2 - y^2$ and $z = x^2 + y^2 - 4$.

24. The region of space bounded by $z = 1 + x^2 + y^2$, $y = 1 - x^2$, $y = 0$, and $z = 0$.

18.3 MULTIPLE INTEGRATION IN POLAR AND CYLINDRICAL COORDINATES

Consider the polar r,θ-coordinates in the plane. Recall that the transformation from polar r,θ-coordinates to rectangular x,y-coordinates is given by

$$x = r \cos \theta, \quad y = r \sin \theta, \tag{1}$$

as indicated in Fig. 18.20.

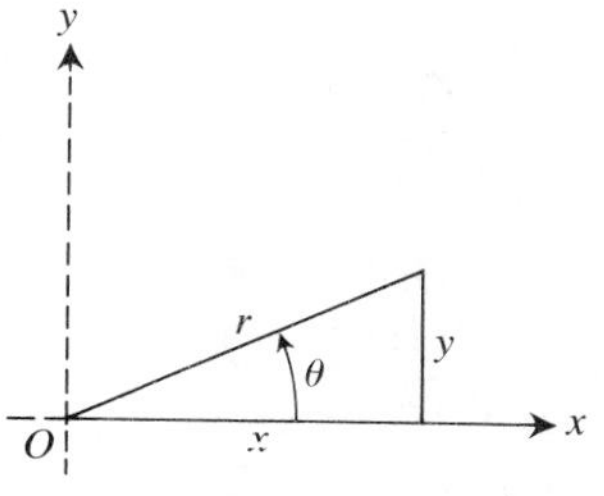

18.20

18.3.1 Double integrals in polar coordinates

Suppose we wish to integrate a continuous function over a region G in the plane. We may often express our function either in terms of the independent variables x and y, or in terms of r and θ. If the function is expressed by $f(x, y)$ in terms of x and y, then in terms of r,θ-coordinates it is

$$h(r, \theta) = f(r \cos \theta, r \sin \theta).$$

The differential area element, $dx\,dy$, should be replaced by one in polar coordinates. Now the rectangular area element, $dx\,dy$, is obtained by starting at a point in the region, increasing x by dx, increasing y by dy, then decreasing x by dx, and finally decreasing y by dy. For r,θ-coordinates, we start at a point, increase r by dr, increase θ by $d\theta$, decrease r by dr, and finally decrease θ by $d\theta$. The result is shown in Fig. 18.21.

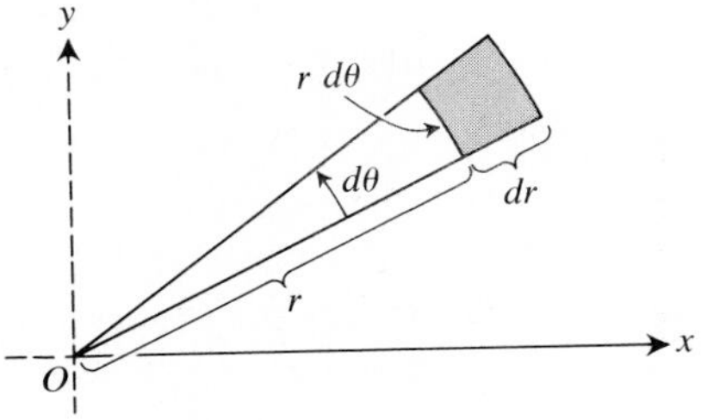

18.21

Note that the area of the element is approximately $(r\,d\theta)(dr)$, *not* $dr\,d\theta$. Then

$$\text{Differential area element} = r\,dr\,d\theta, \tag{2}$$

and the integral of $f(x, y)$ over G becomes

$$\iint_G f(r\cos\theta, r\sin\theta)r\,dr\,d\theta. \tag{3}$$

A rigorous derivation of (3) is left to a more advanced course. The integral (3) "adds up, over G, these areas, multiplied by values of the function, and takes the limit as $d\theta \to \theta$ and $dr \to 0$."

Example 1 Let's find the area of the region G bounded by the cardioid $r = 1 + \cos\theta$, shown in Fig. 18.22, by integrating the constant function 1 over this region.

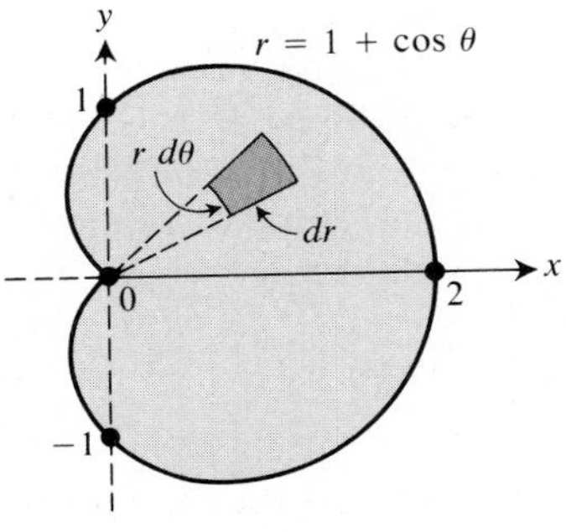

18.22

SOLUTION By symmetry, our desired area is

$$2\int_0^{\pi}\int_0^{1+\cos\theta} 1\cdot r\,dr\,d\theta = 2\int_0^{\pi}\frac{1}{2}r^2\bigg]_0^{1+\cos\theta} d\theta = 2\int_0^{\pi}\frac{1}{2}(1+\cos\theta)^2\,d\theta$$

$$= \int_0^{\pi}(1 + 2\cos\theta + \cos^2\theta)\,d\theta$$

$$= \left(\theta + 2\sin\theta + \frac{\theta}{2} + \frac{\sin 2\theta}{4}\right)\bigg]_0^{\pi}$$

$$= \pi + \frac{\pi}{2} = \frac{3\pi}{2}.$$

Note that the integral $2\int_0^{\pi}(1/2)(1+\cos\theta)^2\,d\theta$ in our computation is the one we would have started with in Chapter 13 to find our area. ||

18.3.2 Cylindrical coordinates Recall that cylindrical r,θ,z-coordinates for space are formed by taking polar r,θ-coordinates in the x,y-plane and the usual rectangular z-coordinate, as indicated in Fig. 18.23. The locus of the equation $r = a$ is the cylinder shown in Fig. 18.24; hence the name "cylindrical coordinates."

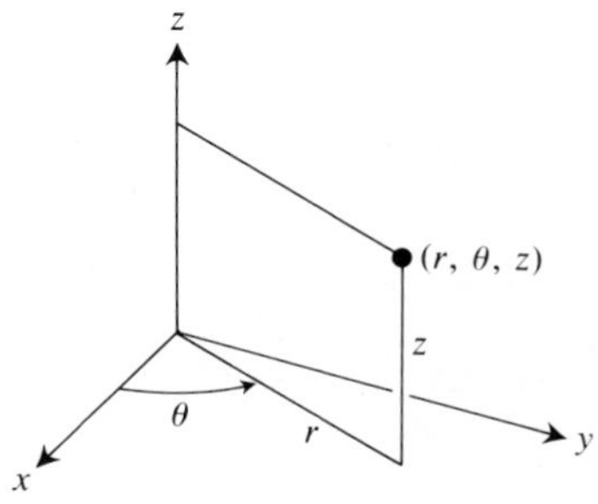

18.23

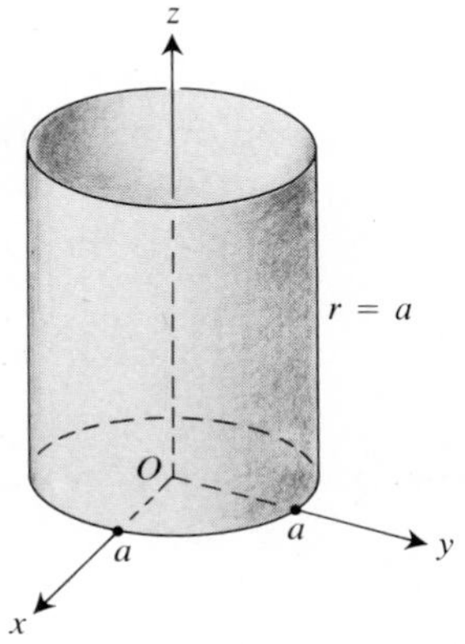

18.24

Let $h(r, \theta, z)$ be a continuous cylindrical-coordinate function defined on a region G in space; we are interested in the integral of $h(r, \theta, z)$ over G. The integral is

$$\iiint_G h(r, \theta, z)\cdot r\,dr\,d\theta\,dz, \tag{4}$$

which is the analogue of the polar coordinate formula (3). We think of

$r\,dr\,d\theta\,dz$ as a little "volume element" in cylindrical coordinates, as shown in Fig. 18.25.

Since $x^2 + y^2 = r^2$, transformation to cylindrical coordinates is especially useful in integrating a Cartesian expression involving $x^2 + y^2$, or in integrating over regions bounded by surfaces with simple cylindrical-coordinate equations. If an integral involves $x^2 + z^2$, we may take "cylindrical r,θ,y-coordinates," corresponding to polar r,θ-coordinates in the x,z-plane.

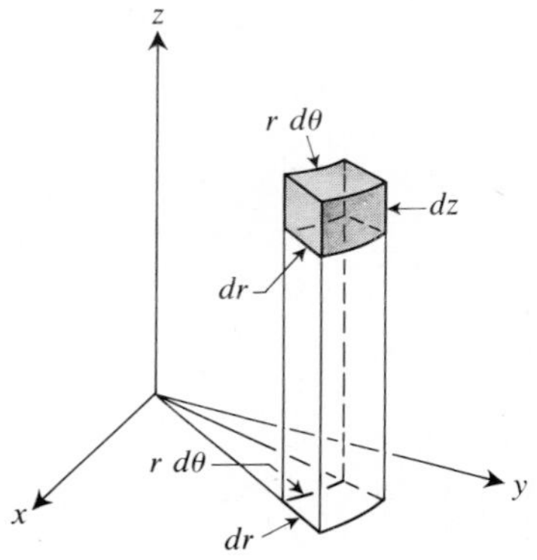

18.25

Example 2 Let G be the region bounded above by $z = 1 + x^2 + y^2$, below by $z = 0$, and on the sides by $x^2 + y^2 = 4$, as indicated in Fig. 18.26. Let's integrate f over G, where $f(x, y, z) = x - y + z^2$.

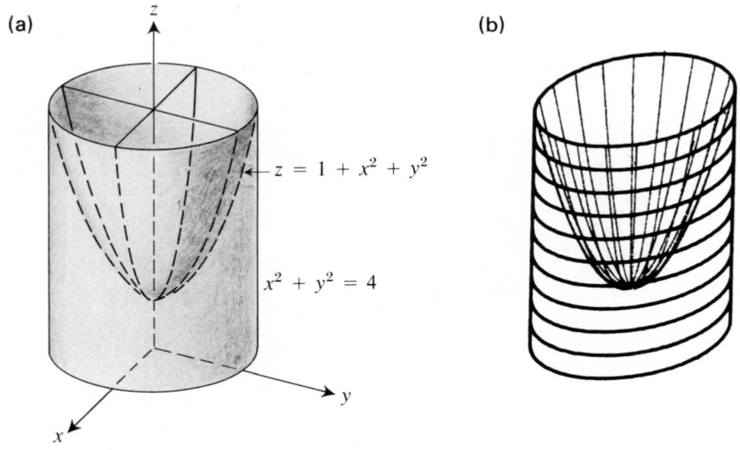

18.26 (a) The bounded region of Example 2. (b) Computer-plotted representation.

SOLUTION Our x,y,z-integral would be

$$\int_{-2}^{2} \int_{-\sqrt{4-y^2}}^{\sqrt{4-y^2}} \int_{0}^{1+x^2+y^2} (x - y + z^2)\, dz\, dx\, dy,$$

which is not too pleasant to evaluate. However, changing to cylindrical coordinates, our integral becomes

$$\int_0^{2\pi} \int_0^2 \int_0^{1+r^2} (r \cos \theta - r \sin \theta + z^2) r\, dz\, dr\, d\theta$$

$$= \int_0^{2\pi} \int_0^2 \left[(\cos \theta - \sin \theta) r^2 z + r \frac{z^3}{3} \right]_{z=0}^{z=1+r^2} dr\, d\theta$$

$$= \int_0^{2\pi} \int_0^2 [(\cos \theta - \sin \theta)(r^2)(1 + r^2) + \tfrac{1}{3} r(1 + r^2)^3]\, dr\, d\theta$$

$$= \int_0^{2\pi} \left[(\cos \theta - \sin \theta)\left(\frac{r^3}{3} + \frac{r^5}{5}\right) + \frac{1}{6} \cdot \frac{(1 + r^2)^4}{4} \right]_{r=0}^{r=2} d\theta$$

$$= \int_0^{2\pi} \left[(\cos \theta - \sin \theta)\left(\frac{8}{3} + \frac{32}{5}\right) + \frac{5^4 - 1}{24} \right] d\theta$$

$$= \int_0^{2\pi} \left[(\cos \theta - \sin \theta) \frac{136}{15} + \frac{624}{24} \right] d\theta$$

$$= \left[\frac{136}{15} (\sin \theta + \cos \theta) + \frac{624}{24} \theta \right]_0^{2\pi}$$

$$= \frac{136}{15}(1 - 1) + \frac{624}{24} 2\pi = \frac{624}{12} \pi = 52\pi.$$

You probably think this computation is bad enough! ||

In many cases, the order dz, dr, $d\theta$ seems the most natural order for integration in cylindrical coordinates. However, other orders are possible.

Example 3 Let's express the volume bounded by the sphere $x^2 + y^2 + z^2 = a^2$ as integrals that illustrate various orders of integration in cylindrical coordinates.

SOLUTION The cylindrical coordinate equation of the sphere is of course $r^2 + z^2 = a^2$. Then

$$\text{Volume} = \int_0^{2\pi} \int_0^a \int_{-\sqrt{a^2-r^2}}^{\sqrt{a^2-r^2}} r\, dz\, dr\, d\theta = \int_0^a \int_0^{2\pi} \int_{-\sqrt{a^2-r^2}}^{\sqrt{a^2-r^2}} r\, dz\, d\theta\, dr$$

$$= \int_0^{2\pi} \int_{-a}^a \int_0^{\sqrt{a^2-z^2}} r\, dr\, dz\, d\theta = \int_{-a}^a \int_0^{2\pi} \int_0^{\sqrt{a^2-z^2}} r\, dr\, d\theta\, dz$$

$$= \int_0^a \int_{-\sqrt{a^2-r^2}}^{\sqrt{a^2-r^2}} \int_0^{2\pi} r\, d\theta\, dz\, dr = \int_{-a}^a \int_0^{\sqrt{a^2-z^2}} \int_0^{2\pi} r\, d\theta\, dr\, dz.$$

You should see from a figure how each of these integrals "adds up the volume contributions" in such a way as to give the volume of the entire ball. ||

SUMMARY

1. *An integral $\iint f(x, y)\, dx\, dy$ with appropriate x,y-limits becomes in polar coordinates $\iint f(r \cos \theta, r \sin \theta) \cdot r\, dr\, d\theta$ with appropriate polar limits for the region.*
2. *An integral $\iiint f(x, y, z)\, dx\, dy\, dz$ with appropriate x,y,z-limits becomes, in cylindrical coordinates,*

$$\iiint f(r \cos \theta, r \sin \theta, z) \cdot r\, dz\, dr\, d\theta$$

with appropriate cylindrical-coordinate limits for the region.

EXERCISES

In Exercises 1 through 4, find the area of the plane region using double integration in polar coordinates.

1. The region inside the cardioid $r = a(1 + \cos \theta)$ and outside the circle $r = a$.
2. The region inside one loop of the four-leaved rose $r = a \sin 2\theta$.
3. The region in the first quadrant bounded by $x^2 + y^2 = a^2$, $y = 0$, and $x = a/2$.
4. The region inside the larger loop and outside the smaller loop of the limacon $r = a(1 + 2 \cos \theta)$.
5. Find the integral of the polar function $h(r, \theta) = r \sin^2\theta$, $r \geq 0$, over the closed disk bounded by $r = a$.
6. Find the integral of the polar function $h(r, \theta) = \cos \theta$, $r \geq 0$, over the region bounded by the cardioid $r = a(1 + \sin \theta)$.
7. In cylindrical coordinates, the level coordinate surface $r = a$ is a cylinder (see Fig. 18.24). Describe the level coordinate surfaces

   a) $\theta = \theta_0$ and b) $z = b$.

In Exercises 8 through 11, find the volume of the given region in space, using triple integration in cylindrical coordinates.

8. The region bounded by the paraboloid $z = 4 - x^2 - y^2$ and the plane $z = 0$
9. The region bounded by the paraboloid $z = x^2 + y^2$, the plane $z = 0$, and the cylinder $x^2 + y^2 = 2x$
10. The region bounded by the hemisphere $z = \sqrt{a^2 - x^2 - y^2}$ and the plane $z = b$ for $0 \leq b < a$
11. The region inside the semicircular cylinder bounded by $x = \sqrt{4 - z^2}$ and $x = 0$, and bounded on the ends by $y = 0$ and the hemisphere $y = \sqrt{16 - x^2 - z^2}$. [*Hint.* Use "cylindrical coordinates" (r, y, θ).]
12. Find the integral of the cylindrical-coordinate function $h(r, \theta, z) = rz \cos^2 \theta$ for $r \geq 0$ over the region in space bounded by $r = a$, $z = 0$, and $z = 4$.

13. Find the integral of the cylindrical-coordinate function $h(r, \theta, z) = rz^2$ for $r \geq 0$ over the region in space bounded by the cone $z^2 = x^2 + y^2$ and the plane $z = 4$.

18.4 INTEGRATION IN SPHERICAL COORDINATES

Recall now the spherical-coordinate system, where a point has coordinates (ρ, ϕ, θ) as indicated in Fig. 18.27. The coordinate ρ is the length of the line segment joining the point and the origin, ϕ is the angle from the z-axis to this line segment, and θ is the same angle as in cylindrical coordinates. Note that the locus of $\rho = a$ is a sphere with center at the origin and radius a, as indicated in Fig. 18.28. This is the reason for the term, "spherical coordinates."

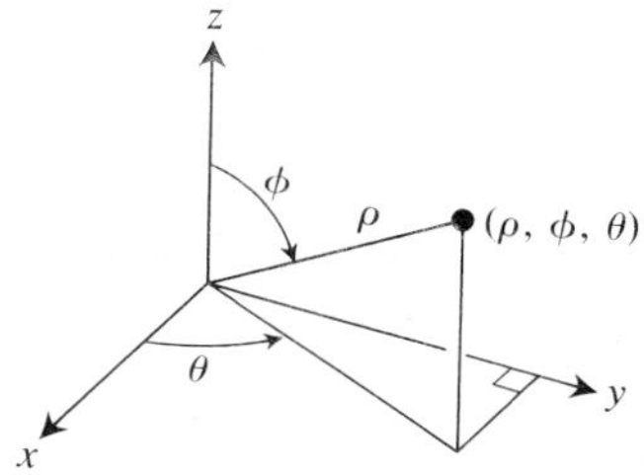

18.27

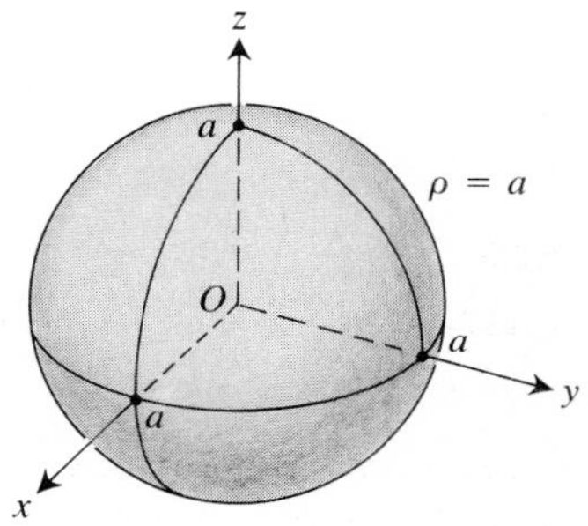

18.28

Since ρ is the distance from the point to the origin, it is obvious that

$$\rho^2 = x^2 + y^2 + z^2, \tag{1}$$

and consequently, transformation to spherical coordinates is useful in the triple integration of cartesian expressions involving $x^2 + y^2 + z^2$, or integrals over regions bounded in part by spherical surfaces.

We need to express x, y, and z in terms of spherical ρ,ϕ,θ-coordinates, so that we can express an integral $\iiint_G f(x, y, z)\, dx\, dy\, dz$ in terms of spherical coordinates. From Fig. 18.29, we easily recall that

$$\begin{aligned} x &= \rho \sin \phi \cos \theta, \\ y &= \rho \sin \phi \sin \theta, \\ z &= \rho \cos \phi. \end{aligned} \tag{2}$$

Increasing and decreasing ρ, ϕ, and θ by amounts $d\rho$, $d\phi$, and $d\theta$, you generate the differential element shown in Fig. 18.29(b). The volume of this element is approximately $(\rho \sin \phi\, d\theta)(d\rho)(\rho\, d\phi)$, as shown in the figure. Therefore

$$\text{Differential volume element} = \rho^2 \sin \phi\, d\rho\, d\phi\, d\theta. \tag{3}$$

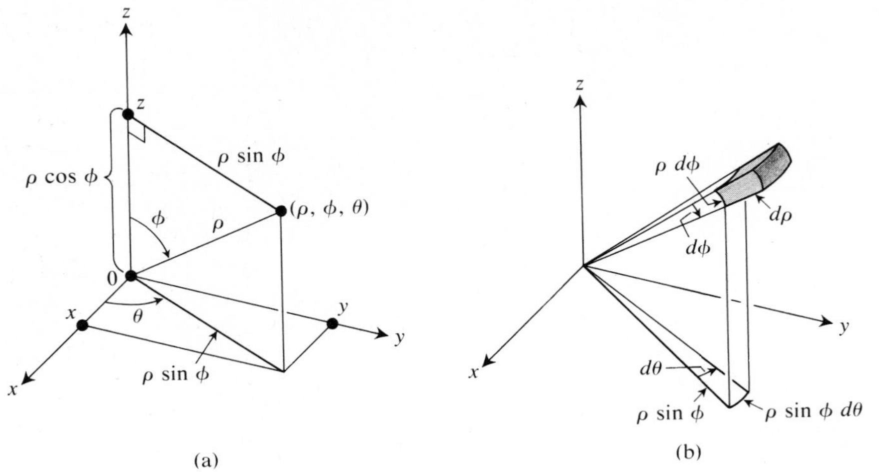

(a) (b) **18.29**

A real proof that (3) is appropriate is left to an advanced calculus course. Remember that we restrict ϕ to the range $0 \leq \phi \leq \pi$; thus $\sin \phi \geq 0$.

Example 1 Let's find the volume of the ball bounded by the sphere $x^2 + y^2 + z^2 = a^2$, which has spherical-coordinate equation $\rho = a$.

SOLUTION We integrate the constant function 1 over this region, using the volume element (3) and limits in spherical coordinates. We shall integrate in the order $d\rho$, $d\phi$, $d\theta$. We think of the first integration with respect to ρ as adding up our volume elements to give the spike shown in Fig. 18.30(a). The next integration with respect to ϕ adds up the volumes of these spikes to give the volume of the wedge in Fig. 18.30(b), and the final integration with respect to

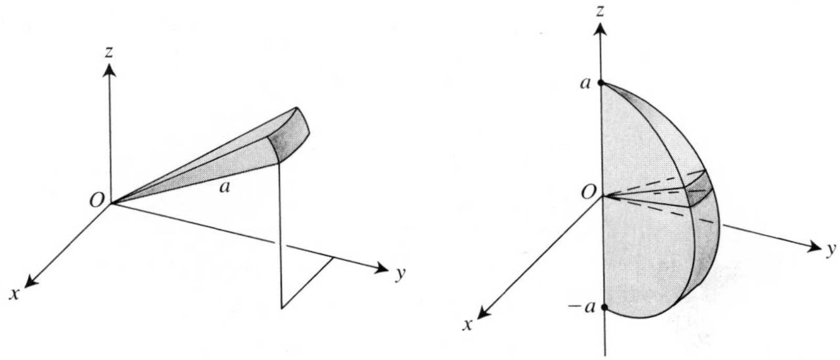

(a) Integration with respect to ρ (b) Subsequent integration with respect to ϕ **18.30**

θ from 0 to 2π adds up the volumes of these wedges to give the volume of the entire ball. Computing, we have

$$\int_0^{2\pi}\int_0^{\pi}\int_0^{a}(1)\rho^2\sin\phi\,d\rho\,d\phi\,d\theta = \int_0^{2\pi}\int_0^{\pi}\frac{\rho^3}{3}\sin\phi\bigg]_{\rho=0}^{\rho=a}d\phi\,d\theta$$

$$= \int_0^{2\pi}\int_0^{\pi}\frac{a^3}{3}\sin\phi\,d\phi\,d\theta$$

$$= \frac{a^3}{3}\int_0^{2\pi} -\cos\phi\bigg]_0^{\pi}d\theta$$

$$= \frac{a^3}{3}\int_0^{2\pi}[-(-1)+1]\,d\theta$$

$$= \frac{a^3}{3}2\theta\bigg]_0^{2\pi} = \frac{a^3}{3}4\pi = \frac{4}{3}\pi a^3. \quad ||$$

Example 2 Let's integrate the function $f(x, y, z) = z$ over the half-ball bounded above by $z = \sqrt{1 - x^2 - y^2}$ and below by $z = 0$, as indicated in Fig. 18.31.

SOLUTION In spherical coordinates, $z = \rho\cos\phi$ by (2), so our integral in spherical coordinate form is

$$\int_0^{2\pi}\int_0^{\pi/2}\int_0^{1}(\rho\cos\phi)(\rho^2\sin\phi)\,d\rho\,d\phi\,d\theta$$

$$= \int_0^{2\pi}\int_0^{\pi/2}\frac{\rho^4}{4}\cos\phi\sin\phi\bigg]_{\rho=0}^{\rho=1}d\phi\,d\theta$$

$$= \int_0^{2\pi}\int_0^{\pi/2}\frac{1}{4}\cos\phi\sin\phi\,d\phi\,d\theta$$

$$= \frac{1}{4}\int_0^{2\pi}\frac{\sin^2\phi}{2}\bigg]_0^{\pi/2}d\theta$$

$$= \frac{1}{4}\int_0^{2\pi}\frac{1}{2}\,d\theta = \frac{1}{8}\int_0^{2\pi}d\theta = \frac{1}{8}\theta\bigg]_0^{2\pi} = \frac{1}{8}2\pi = \frac{\pi}{4}.$$

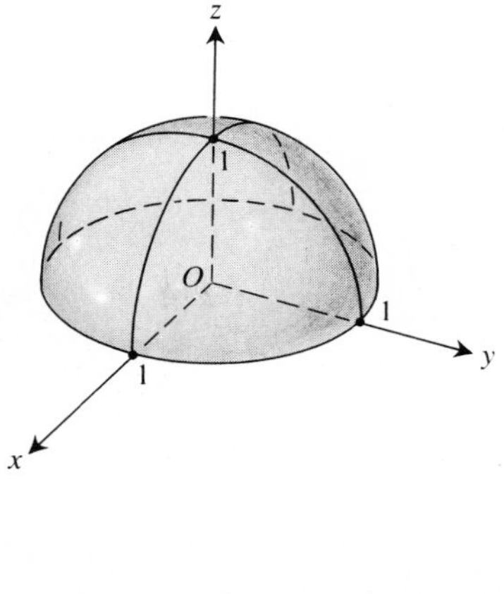

18.31

Note that the rectangular-coordinate integral of z over this half-ball is

$$\int_0^1 \int_{-\sqrt{1-y^2}}^{\sqrt{1-y^2}} \int_0^{\sqrt{1-x^2-y^2}} z\, dz\, dx\, dy.$$

This integral is less pleasant to compute! ||

Example 3 Let us integrate in spherical coordinates to derive the formula $V = (1/3)\pi a^2 h$ for the volume of a solid right-circular cone of altitude h and radius of base a.

SOLUTION The solid bounded by $a^2z^2 = h^2(x^2 + y^2)$ and $z = h$ is such a cone, as shown in Fig. 18.32. The plane $z = h$ becomes $\rho \cos \phi = h$, and the integral is

$$\int_0^{2\pi} \int_0^{\tan^{-1}(a/h)} \int_0^{h/\cos\phi} p^2 \sin\phi\, d\rho\, d\phi\, d\theta = \int_0^{2\pi} \int_0^{\tan^{-1}(a/h)} \frac{\rho^3}{3} \sin\phi \Big]_{\rho=0}^{\rho=h/\cos\phi} d\phi\, d\theta$$

$$= \int_0^{2\pi} \int_0^{\tan^{-1}(a/h)} \frac{h^3}{3} (\cos\phi)^{-3} \sin\phi\, d\phi\, d\theta$$

$$= \int_0^{2\pi} -\frac{h^3}{3} \cdot \frac{\sec^2\phi}{-2} \Big]_0^{\tan^{-1}(a/h)} d\theta$$

$$= \int_0^{2\pi} -\frac{h^3}{3} \left(\frac{a^2 + h^2}{-2h^2} + \frac{1}{2}\right) d\theta$$

$$= 2\pi\left(-\frac{h^3}{3}\right)\left(\frac{a^2 + h^2 - h^2}{-2h^2}\right)$$

$$= 2\pi\left(-\frac{h^3}{3}\right)\left(\frac{a^2}{-2h^2}\right) = \frac{1}{3}\pi a^2 h. \quad ||$$

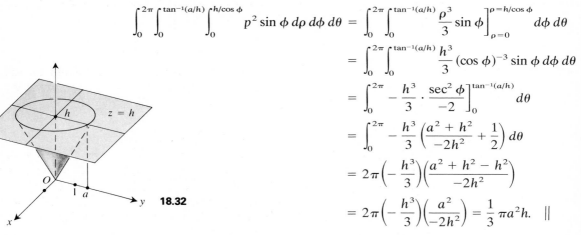

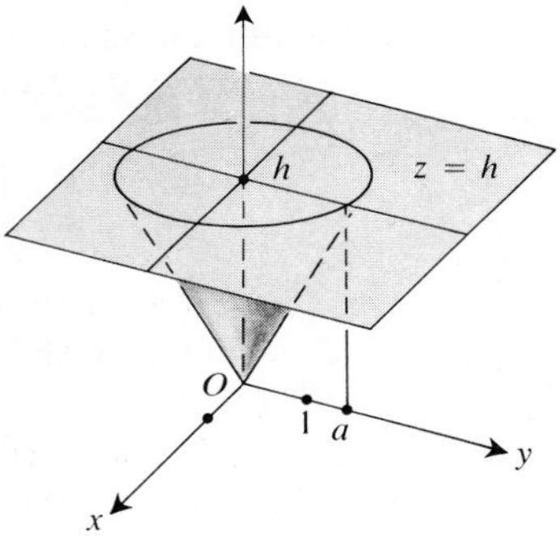

18.32

SUMMARY

1. *The meaning of spherical ρ,ϕ,θ-coordinates is given in Fig. 18.27.*
2. *Transformation from x,y,z-coordinates to ρ,ϕ,θ-coordinates is accomplished by*

$$x = \rho \sin\phi \cos\theta,$$
$$y = \rho \sin\phi \sin\theta,$$
$$z = \rho \cos\phi.$$

3. *An integral $\iiint f(x, y, z)\, dx\, dy\, dz$ with appropriate x,y,z-limits becomes, in spherical coordinates,*

$$\iiint f(\rho \sin\phi \cos\theta, \rho \sin\phi \sin\theta, \rho \cos\phi) \cdot \rho^2 \sin\phi\, d\rho\, d\phi\, d\theta,$$

with appropriate spherical coordinate limits for the region chosen so that $0 \le \phi \le \pi$.

EXERCISES

In Exercises 1 *through* 4 *find the volume of the given region in space by triple integration in spherical coordinates.*

1. The region bounded by the hemisphere $z = \sqrt{a^2 - x^2 - y^2}$ and the plane $z = b$ for $0 \leq b < a$.

2. The region bounded by the cone $z^2 = x^2 + y^2$ and the hemisphere $z = \sqrt{16 - x^2 - y^2}$.

3. The region bounded by the hemisphere $y = \sqrt{4 - x^2 - z^2}$ and the planes $y = x$ and $y = \sqrt{3}x$.

4. The region between the cones $z^2 = x^2 + y^2$ and $3z^2 = x^2 + y^2$ and below the hemisphere $z = \sqrt{4 - x^2 - y^2}$.

5. Find the integral of the spherical coordinate function $h(\rho, \phi, \theta) = \rho^2$ over the ball bounded by the sphere $x^2 + y^2 + z^2 = a^2$.

6. Find the integral of the spherical coordinate function $h(\rho, \phi, \theta) = \rho^2 \cos \phi$ over the region bounded by the cone $z^2 = x^2 + y^2$ and the hemisphere $z = \sqrt{4 - x^2 - y^2}$.

18.5 MOMENTS AND CENTERS OF MASS

We introduced moments and centroids in Chapter 7, treating only the special cases that can be handled with an integral of a function of one variable. We can give a better presentation now, using multiple integrals.

18.5.1 Mass

Imagine a physical body to occupy some region G in space. The *mass m* of the body is a numerical measure of the "amount of material" it contains. Near the surface of the earth, the *weight* of a body is mg, where g is the gravitational acceleration; one slug of mass weighs about 32 pounds.

The *mass density* of the body is the *mass per unit volume.* If the body is not homogeneous, the mass density may vary and be a function of the position within the body. To say that the mass density at a point (x_0, y_0, z_0) is $\sigma(x_0, y_0, z_0)$ is to say that, if the body had the same composition everywhere that it has at (x_0, y_0, z_0), then its mass would be

$$\sigma(x_0, y_0, z_0) \cdot (\text{volume of } G).$$

Suppose the mass density $\sigma(x, y, z)$ is a *continuous* function for (x, y, z) in G. If (x, y, z) is a point of a small box with edges of lengths dx, dy, and dz, then the approximate mass of the material in this box is

$$\sigma(x, y, z)\, dx\, dy\, dz.$$

If you add all these small amounts of mass with an integral as dx, dy, and dz approach zero, you obtain the **mass of the body**

$$m = \iiint_G \sigma(x, y, z)\, dx\, dy\, dz. \tag{1}$$

Of course, in cylindrical and spherical coordinates, our volume elements are $r\,dz\,dr\,d\theta$ and $\rho^2 \sin\phi\,d\rho\,d\phi\,d\theta$, respectively.

Example 1 Let the mass density of a ball of radius a be proportional to the distance from the center of the ball, and let's find the mass of the ball if the mass density at a distance of one unit from the center is k.

SOLUTION If we take the center of the ball as origin, then the mass density is given by $k\sqrt{x^2 + y^2 + z^2}$. It is natural to use spherical coordinates in integrating over a ball; in terms of spherical coordinates, the mass density is given by

$$k\sqrt{x^2 + y^2 + z^2} = k\rho.$$

The mass is then

$$\begin{aligned} m &= \int_0^{2\pi}\int_0^{\pi}\int_0^{a} k\rho \cdot \rho^2 \sin\phi\,d\rho\,d\phi\,d\theta \\ &= \int_0^{2\pi}\int_0^{\pi} k\frac{\rho^4}{4}\bigg]_0^a \sin\phi\,d\phi\,d\theta \\ &= \int_0^{2\pi}\int_0^{\pi} k\frac{a^4}{4}\sin\phi\,d\phi\,d\theta = \frac{ka^4}{4}\int_0^{2\pi} -\cos\phi\bigg]_0^{\pi} d\theta \\ &= \frac{ka^4}{4}\int_0^{2\pi} [-(-1) + 1]\,d\theta \\ &= \frac{2ka^4}{4}\int_0^{2\pi} d\theta = \frac{ka^4}{2}\theta\bigg]_0^{2\pi} = k\pi a^4. \quad \| \end{aligned}$$

If we are dealing with a flat sheet of material of constant thickness that is homogeneous in the direction perpendicular to the sheet, we often use mass per unit *area* as mass density.

Example 2 Let a flat sheet of material of constant thickness cover the region bounded by the cardioid $r = a(1 + \sin\theta)$ shown in Fig. 18.33, and let the area mass density of the sheet be proportional to the distance from the y-axis. Let's find the mass of the body.

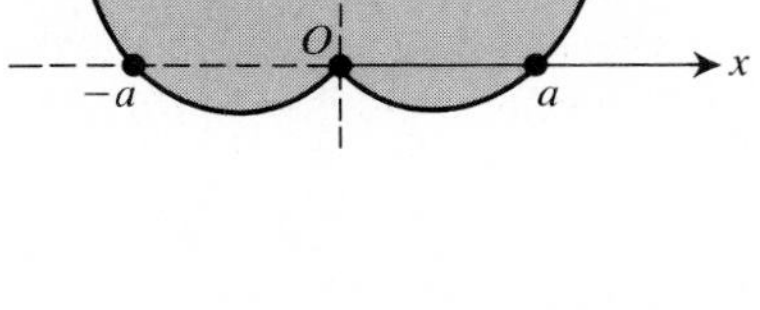

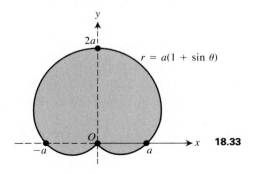

18.33

SOLUTION The mass density is then

$$\sigma(x, y) = k|x| = k|r \cos \theta|,$$

where k is a constant of proportionality. The mass is given by the integral

$$\begin{aligned} m &= 2\int_{-\pi/2}^{\pi/2} \int_0^{a(1+\sin\theta)} k(r\cos\theta) r\, dr\, d\theta \\ &= 2k\int_{-\pi/2}^{\pi/2} \frac{r^3}{3}\Big]_0^{a(1+\sin\theta)} \cos\theta\, d\theta \\ &= 2k\int_{-\pi/2}^{\pi/2} \frac{a^3}{3}(1+\sin\theta)^3 \cos\theta\, d\theta \\ &= \frac{2}{3}ka^3 \frac{(1+\sin\theta)^4}{4}\Big]_{-\pi/2}^{\pi/2} \\ &= \frac{2}{3}ka^3\frac{2^4}{4} - \frac{2}{3}ka^3(0) \\ &= \frac{8ka^3}{3}. \quad \| \end{aligned}$$

18.5.2 First moments

The **first moment** (or simply the moment) about an axis in a plane of a "point mass" in the plane is the product of the mass and the *signed* distance from the axis. If the point mass is in space, we consider the first moment about a plane, which is the product of the mass and the signed distance from the plane.

Now consider a body whose mass is not concentrated at one point (which is usually the case). We compute the first moment by adding up products of the masses of small pieces and the signed distances of the pieces from the axis (or plane), and take the limit as the pieces become smaller and smaller. Of course this leads to an integral. For a flat sheet of material in the plane, we will let M_x and M_y be the first moments about the x-axis and y-axis, respectively, while M_{xy}, M_{yz}, and M_{xz} are first moments about the x,y-plane, the y,z-plane, and the x,z-plane, respectively, for a body in space.

We illustrate with two examples.

Example 3 Let's find the first moments about the x-axis and y-axis of a flat sheet of material covering the region bounded by the cardioid $r = a(1 + \sin\theta)$ shown in Fig. 18.33, if the area mass density is the constant k.

SOLUTION By symmetry, the moment about the y-axis is zero, since the mass of a small piece is multiplied by the *signed* distance from the axis; a positive contribution of a piece on the righthand side of the y-axis is counterbalanced by the negative contribution of the symmetric piece on the lefthand side. Since the signed distance from a point (x, y) to the x-axis is $y = r\sin\theta$, we obtain

$$\begin{aligned}
M_x &= 2\int_{-\pi/2}^{\pi/2}\int_0^{a(1+\sin\theta)} (r\sin\theta)kr\,dr\,d\theta \\
&= 2k\int_{-\pi/2}^{\pi/2} \frac{r^3}{3}\Big]_0^{a(1+\sin\theta)} \sin\theta\,d\theta \\
&= 2k\int_{-\pi/2}^{\pi/2} \frac{a^3(1+\sin\theta)^3}{3}\sin\theta\,d\theta \\
&= \frac{2ka^3}{3}\int_{-\pi/2}^{\pi/2} (\sin\theta + 3\sin^2\theta + 3\sin^3\theta + \sin^4\theta)\,d\theta.
\end{aligned}$$

Since $\sin\theta = -\sin(-\theta)$ and $\sin^3\theta = -\sin^3(-\theta)$, their integrals over $[-\pi/2, \pi/2]$ are zero. Our integral reduces to

$$\begin{aligned}
\frac{2ka^3}{3}\int_{-\pi/2}^{\pi/2} (3\sin^2\theta + \sin^4\theta)\,d\theta \\
&= \frac{2ka^3}{3}\left(\frac{3\theta}{2} - \frac{3\sin 2\theta}{4} + \frac{3\theta}{8} - \frac{\sin 2\theta}{4} + \frac{\sin 4\theta}{32}\right)\Big]_{-\pi/2}^{\pi/2} \\
&= \frac{2ka^3}{3}\left(\frac{3\pi}{4} + \frac{3\pi}{16} - \left(-\frac{3\pi}{4} - \frac{3\pi}{16}\right)\right) \\
&= \frac{2ka^3}{3}\left(\frac{3\pi}{2} + \frac{3\pi}{8}\right) = \frac{2ka^3}{3}\cdot\frac{15\pi}{8} = \frac{5k\pi a^3}{4}. \quad \|
\end{aligned}$$

Example 4 Let a solid in space be bounded by the cylinder $x^2 + y^2 = a^2$ and the planes $z = 0$ and $z = b$. If the mass density at a height z above the x,y-plane is kz, let us find the first moments of the solid about the coordinate planes.

SOLUTION Symmetry shows at once that

$$M_{xz} = M_{yz} = 0.$$

We use cylindrical coordinates and obtain

$$\begin{aligned}
M_{xy} &= \int_0^{2\pi}\int_0^a\int_0^b (z)(kz)r\,dz\,dr\,d\theta = k\int_0^{2\pi}\int_0^a \frac{z^3}{3}\Big]_0^b r\,dr\,d\theta \\
&= k\int_0^{2\pi}\int_0^a \frac{b^3}{3} r\,dr\,d\theta = \frac{kb^3}{3}\int_0^{2\pi} \frac{r^2}{2}\Big]_0^a d\theta \\
&= \frac{kb^3}{3}\int_0^{2\pi} \frac{a^2}{2}\,d\theta = \frac{ka^2b^3}{6}\,\theta\Big]_0^{2\pi} = \frac{k\pi a^2b^3}{3}. \quad \|
\end{aligned}$$

18.5.3 Second moments The **second moment** I (or **moment of inertia**) about an axis of a "point mass" is the product of the mass and the *square* of the distance from the axis. A moment of inertia is used in computing kinetic energy of rotation, which is given by the formula

$$\text{K.E.} = \tfrac{1}{2}I\omega^2,$$

where ω is the angular speed of rotation. Computation of a moment of inertia often is accomplished by integration. We illustrate with an example.

Example 5 Let's find the moment of inertia of a homogeneous ball of radius a and constant mass density k about a diameter.

SOLUTION We take the center of the ball at the origin, and let the z-axis be the diameter about which the moment of inertia is to be computed. The distance from a point (ρ, ϕ, θ) in spherical coordinates to the z-axis is easily seen to be $\rho \sin \phi$ (see Fig. 18.29(a) in Section 18.4. Thus

$$\begin{aligned} I &= \int_0^{2\pi}\int_0^{\pi}\int_0^{a} (\rho \sin \phi)^2 k\rho^2 \sin \phi \, d\rho \, d\phi \, d\theta \\ &= k\int_0^{2\pi}\int_0^{\pi} \frac{\rho^5}{5}\Big]_0^a \sin^3\phi \, d\phi \, d\theta = k\int_0^{2\pi}\int_0^{\pi} \frac{a^5}{5} \sin^3\phi \, d\phi \, d\theta \\ &= \frac{ka^5}{5}\int_0^{2\pi}\int_0^{\pi} (1 - \cos^2\phi) \sin \phi \, d\phi \, d\theta = \frac{ka^5}{5}\int_0^{2\pi} \left(-\cos \phi + \frac{\cos^3\phi}{3}\right)\Big]_0^{\pi} d\theta \\ &= \frac{ka^5}{5}\int_0^{2\pi} \left[-(-1) - \frac{1}{3} - \left(-1 + \frac{1}{3}\right)\right] d\theta = \frac{ka^5}{5}\int_0^{2\pi} \frac{4}{3}\, d\theta = \frac{4ka^5}{15}\,\theta\Big]_0^{2\pi} \\ &= \frac{8\pi ka^5}{15}. \quad \| \end{aligned}$$

18.5.4 Centers of mass and centroids

Consider a flat sheet of material in the plane. The **center of mass** of the sheet is the point at which you can consider all the mass to be concentrated for computation of first moments about the coordinate axes (see Fig. 18.34). Thus if the center of mass is $(\bar{x}, \bar{y})$ and the body has mass m, we must have

$$M_x = m\bar{y} \qquad \text{and} \qquad M_y = m\bar{x}.$$

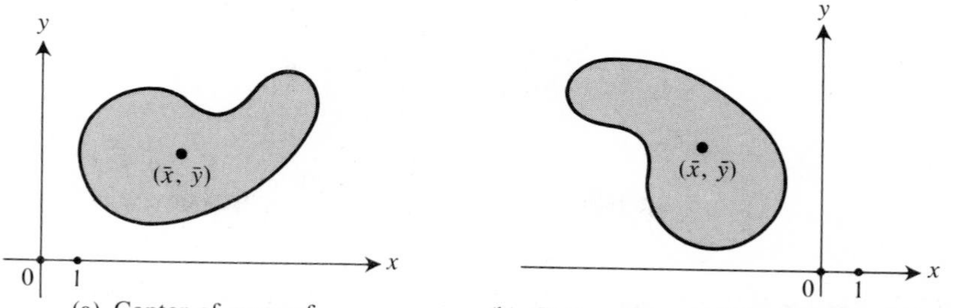

(a) Center of mass of a flat body

(b) Center of mass of the same body in a different position

18.34

Hence

$$\bar{x} = \frac{M_y}{m} \quad \text{and} \quad \bar{y} = \frac{M_x}{m}. \tag{2}$$

It is a fact that

the location (2) *of the center of mass in relation to the body is independent of the position of the body in the plane*

(see Fig. 18.34). We give some exercises that indicate the reason for this at the end of the section (see Exercises 12 and 13). Also

the first moment of the body about any *axis is the product of its mass and the* (*signed*) *distance from its center of mass to the axis.*

We should warn you that, in general, there is *no* single point in a body at which the mass can be considered to be concentrated for computation of moments of intertia about *every* axis (see Exercise 14).

For a body in space, coordinates of the center of mass $(\bar{x}, \bar{y}, \bar{z})$ are given by

$$\bar{x} = \frac{M_{yz}}{m}, \quad \bar{y} = \frac{M_{xz}}{m}, \quad \text{and} \quad \bar{z} = \frac{M_{xy}}{m} \tag{3}$$

in analogy with (2).

If a body is homogeneous with constant mase density, the center of mass is also called the **centroid of the body**, or the **centroid of the region** which the body occupies.

To compute the center of mass, we simply form the quotients of the first moments by the mass, and we have illustrated how to compute mass and first moments.

Example 6 Consider a solid bounded by the cylinder $x^2 + y^2 = a^2$ and the planes $z = 0$ and $z = b$. Suppose the mass density of the solid at the point (x, y, z) is kz. Let's find the center of mass of the solid.

SOLUTION In Example 4, we found that $M_{xz} = M_{yz} = 0$ and $M_{xy} = k\pi a^2 b^3/3$. It only remains to compute the mass, which is given by the integral

$$\begin{aligned} m &= \int_0^{2\pi} \int_0^a \int_0^b kzr\,dz\,dr\,d\theta = k\int_0^{2\pi} \int_0^a \left[\frac{z^2}{2}\right]_0^b r\,dr\,d\theta = k\int_0^{2\pi} \int_0^a \frac{b^2}{2} r\,dr\,d\theta \\ &= \frac{kb^2}{2} \int_0^{2\pi} \left[\frac{r^2}{2}\right]_0^a d\theta = \frac{kb^2}{2} \int_0^{2\pi} \frac{a^2}{2}\,d\theta = \frac{ka^2b^2}{4}\,\theta\Big]_0^{2\pi} = \frac{k\pi a^2 b^2}{2}. \end{aligned}$$

Hence

$$\bar{z} = \frac{M_{xy}}{m} = \frac{(k\pi a^2 b^3/3)}{(k\pi a^2 b^2/2)} = \frac{2}{3}b,$$

so the center of mass is at the point $(0, 0, 2b/3)$. ||

SUMMARY *Let a body in space occupy a region G.*

1. *The mass of the body is* $m = \iiint_G \sigma(x, y, z)\, dx\, dy\, dz$ *where* $\sigma(x, y, z)$ *is the mass density of the body at* (x, y, z).
2. *The first moment of the body about a plane is*

$$\iiint_G (\text{signed distance to plane}) \cdot \sigma(x, y, z)\, dx\, dy\, dz,$$

where the distance is from the differential element of volume $dx\, dy\, dz$ *to the plane.*

3. *The second moment or moment of inertia of the body about an axis is*

$$\iiint_G (\text{distance from axis})^2 \cdot \sigma(x, y, z)\, dx\, dy\, dz,$$

where the distance is from the differential element of volume $dx\, dy\, dz$ *to the axis.*

4. *The center of mass of the body is* $(\bar{x}, \bar{y}, \bar{z})$, *where*

$$\bar{x} = \frac{M_{yz}}{m}, \qquad \bar{y} = \frac{M_{xz}}{m}, \qquad \bar{z} = \frac{M_{xy}}{m}.$$

Here M_{yz} *is the first moment about the* y,z*-plane, etc.*

EXERCISES

1. Let the area mass density at a point (x, y) of a flat body covering the square $0 \le x \le 1$, $0 \le y \le 1$, be xy^2.
   a) Find the mass of the body.
   b) Find the center of mass of the body.

2. Consider a flat body covering the plane region bounded by $y = x^2$ and $x = y^2$, and let the area mass density of the body at a point (x, y) be xy. Find the center of mass of the body. [*Hint.* Use symmetry.]

3. Let a flat body cover the closed disk $x^2 + y^2 \le a$ in the plane, and let the area mass density be proportional to the distance from the center of the disk, with an area density of k at a distance of one unit from the center.
   a) Find the mass of the body.
   b) Find the first moment of the body about the line $x = -a$.
   c) Find the absolute value of the first moment of the body about the line $x + y = 2a$.
   d) Find the moment of inertia of the body about an axis perpendicular to the plane through the origin.

4. Find the centroid of the plane region inside the cardioid $r = a(1 + \cos \theta)$ and outside the circle $r = a$.

5. Let the mass density of a solid cone bounded by $x^2 + y^2 = z^2$ and the plane $z = a$ be proportional to the distance from the z-axis, with the mass density of k one unit away from the z-axis.

a) Find the mass of the solid.

b) Find the center of mass of the solid.

c) Find the absolute value of the first moment of the solid about the plane $x = a$.

d) Find the absolute value of the first moment of the solid about the plane $z = -a$.

e) Find the moment of inertia of the solid about the z-axis.

6. Let a solid in space be bounded by the cylinder $y = a^2 - z^2$ and the planes $y = 0$, $x = 0$, and $x = b$. Let the mass density of the solid at a point (x, y, z) be ky.

a) Find the mass of the solid.

b) Find the centroid of the solid.

c) Find the absolute value of the first moment of the solid about the plane $x + y - 2z = 4$.

7. Find the moment of inertia of a solid ball in space of radius a and constant mass density k about a line tangent to the ball.

8. Find the centroid of the hemispherical region in space bounded by $z = \sqrt{a^2 - x^2 - y^2}$ and $z = 0$.

9. Find the centroid of the region in space bounded by $z = 0$, $x^2 + y^2 = 4$, and $z = 1 + x^2 + y^2$.

10. Find the centroid of the region in space bounded by $x = 0$ and $x = 4 - y^2 - z^2$.

11. Find the centroid of the region in space bounded above by the hemisphere $z = \sqrt{a^2 - x^2 - y^2}$ and below by the cone $z = \sqrt{x^2 + y^2}$.

12. a) Show that the first moment of a body in the plane about the line $x = -a$ is $M_y + ma$. (This is known as the *Parallel Axis Theorem.*)

b) Let a new origin (h, k) be chosen in the plane and let the x'-axis be the line $y = k$ and the y'-axis the line $x = h$. Argue from (a) that the same location for the center of mass of a body in the plane, relative to the body, is obtained whether one computes coordinates of the center using the x-axis and y-axis or using the x'-axis and y'-axis.

13. State the analogue for space of Exercise 12(a).

14. Let a flat body of constant mass density k cover the unit square $0 \le x \le 1$, $0 \le y \le 1$, in the plane.

a) Find the moment of inertia of the body about the y-axis.

b) Find the moment of inertia of the body about the line $x = -a$.

c) Find a point (x_1, y_1) in the body such that the moment of inertia of the body about either the x-axis or the y-axis is the product of the mass and the square of the distance from (x_1, y_1) to the axis.

d) Find a point (x_2, y_2) in the body such that the moment of inertia of the body about either the line $x = -a$ or the line $y = -a$ is the product of the mass and the square of the distance from (x_2, y_2) to the line.

e) Compare the answers to (c) and (d), and comment on the result.

15. The **radius of gyration** R of a body about an axis is defined by

$$R = \sqrt{I/m},$$

so that $I = mR^2$.

a) From the answer $k/3$ to Exercise 14(a), what is the radius of gyration about the y-axis of a homogeneous flat body covering the square $0 \le x \le 1$, $0 \le y \le 1$?

b) From the answer

$$\frac{k}{3}((a + 1)^3 - a^3)$$

to Exercise 14(b), what is the radius of gyration about the line $x = -a$ of a homogeneous flat body covering this square?

18.6 SURFACE AREA

Let $z = f(x, y)$ be a function of two variables with continuous partial derivatives. Let G be a closed bounded region in the domain of f having boundary of finite length. The graph of f over G is then a smooth surface in space, as indicated in Fig. 18.35. We attempt to find the area of this surface.

The situation is analogous to finding the length of a curve lying over an interval $[a, b]$ on the x-axis, as shown in Fig. 18.36. In that case, we approximated the arc length over a differential element of the interval by the length ds of the tangent line segment lying over the differential element of length dx. For surfaces, we approximate the area of the surface over a differential element of G by the portion of the tangent plane lying over the differential element of area $dx\,dy$.

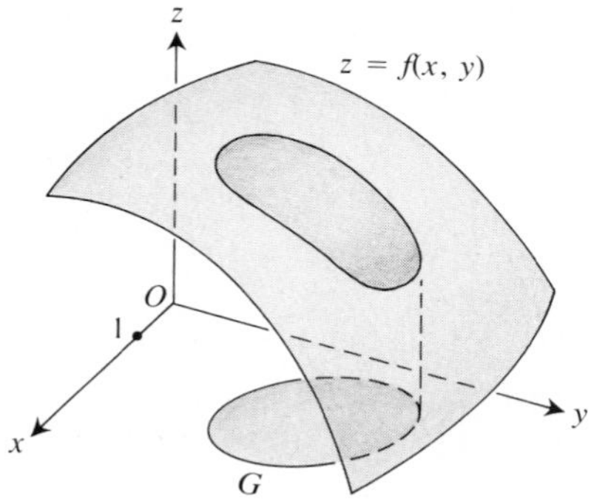

18.35

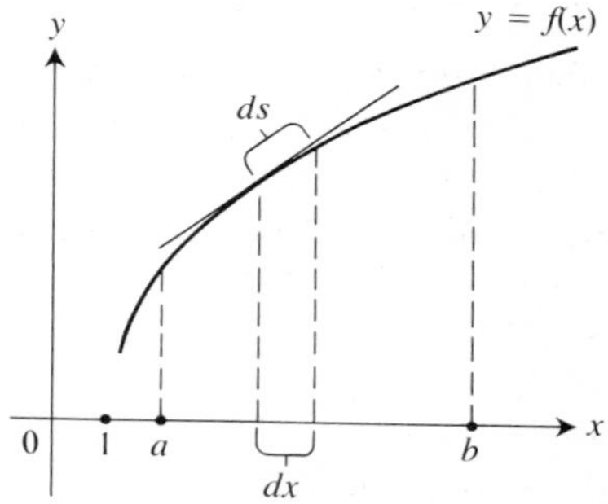

18.36

As shown in Fig. 18.37, vectors along the edges of the parallelogram lying over the differential element of area $dx\,dy$ are

$$(dx)\boldsymbol{i} + \left(\frac{\partial z}{\partial x}\,dx\right)\boldsymbol{k} \qquad \text{and} \qquad (dy)\boldsymbol{j} + \left(\frac{\partial z}{\partial y}\,dy\right)\boldsymbol{k}.$$

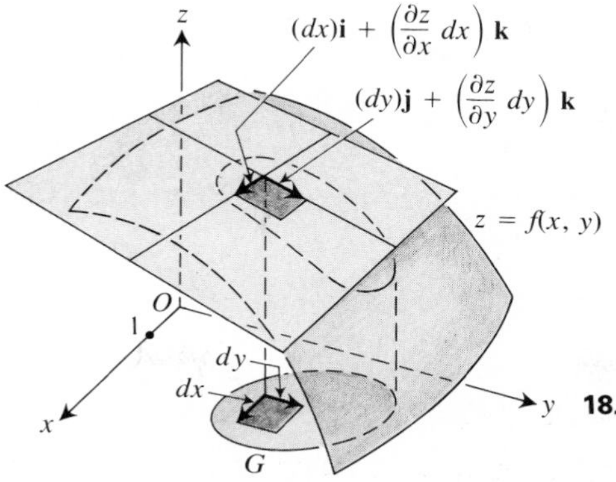

18.37

The area of this parallelogram is the magnitude of the cross product of the vectors. Computing, we obtain the cross product

$$\begin{vmatrix} \boldsymbol{i} & \boldsymbol{j} & \boldsymbol{k} \\ dx & 0 & \dfrac{\partial z}{dx}\,dx \\ 0 & dy & \dfrac{\partial z}{\partial y}\,dy \end{vmatrix} = -\left(\frac{\partial z}{dx}\,dx\,dy\right)\boldsymbol{i} - \left(\frac{\partial z}{\partial y}\,dx\,dy\right)\boldsymbol{j} + (dx\,dy)\boldsymbol{k}.$$

The length of this cross product is the *differential element dS of surface area,* so

$$dS = \sqrt{\left(\frac{\partial z}{\partial x}\right)^2 + \left(\frac{\partial z}{\partial y}\right)^2 + 1}\,dx\,dy.$$

Consequently, we have

$$\text{Surface area} = \iint_G \sqrt{\left(\frac{\partial z}{\partial x}\right)^2 + \left(\frac{\partial z}{\partial y}\right)^2 + 1}\,dx\,dy. \tag{1}$$

Example 1 Let's find the area of the sphere $x^2 + y^2 + z^2 = a^2$.

SOLUTION We find the area of the top hemisphere $z = \sqrt{a^2 - x^2 - y^2}$, and then double the result for our final answer. Computing, we find that

$$\frac{\partial z}{\partial x} = \frac{-x}{\sqrt{a^2 - x^2 - y^2}} \qquad \text{and} \qquad \frac{\partial z}{\partial y} = \frac{-y}{\sqrt{a^2 - x^2 - y^2}}.$$

Hence

$$\begin{aligned} \sqrt{\left(\frac{\partial z}{\partial x}\right)^2 + \left(\frac{\partial z}{\partial y}\right)^2 + 1} &= \sqrt{\frac{x^2 + y^2}{a^2 - x^2 - y^2} + 1} \\ &= \sqrt{\frac{x^2 + y^2 + a^2 - x^2 - y^2}{a^2 - x^2 - y^2}} \\ &= \sqrt{\frac{a^2}{a^2 - x^2 - y^2}} = \frac{a}{\sqrt{a^2 - x^2 - y^2}}. \end{aligned}$$

This computation suggests that we change to cylindrical coordinates, where

$$\frac{a}{\sqrt{a^2 - x^2 - y^2}} = \frac{a}{\sqrt{a^2 - r^2}}.$$

We thus form the integral

$$\int_0^{2\pi}\int_0^a \frac{a}{\sqrt{a^2 - r^2}}\,r\,dr\,d\theta.$$

We should point out that our integrand is undefined for $r = a$, so this is an improper integral of two variables, which we really have not discussed. (Geometrically, this happens because our surface is perpendicular to the x,y-plane at $r = a$, so that $\partial z/\partial x$ and $\partial z/\partial y$ are undefined there.) In straightforward analogy with our improper integrals of a function of one variable, we compute

$$\begin{aligned}\int_0^{2\pi}\left(\lim_{h\to a-}\int_0^h \frac{a}{\sqrt{a^2-r^2}}\,r\,dr\right)d\theta &= \int_0^{2\pi}\lim_{h\to a-}\left(-a\sqrt{a^2-r^2}\right]_0^h d\theta\\ &= \int_0^{2\pi}\left[\lim_{h\to a-}\left(-a\sqrt{a^2-h^2}+a^2\right]d\theta\\ &= \int_0^{2\pi} a^2\,d\theta = a^2\theta\Big]_0^{2\pi} = 2\pi a^2.\end{aligned}$$

Doubling, we obtain $4\pi a^2$ as the area of our sphere. ||

Sometimes a surface is given in the form $F(x, y, z) = 0$, rather than in the form $z = f(x, y)$. Recall that if F has continuous partial derivatives and $\partial F/\partial z$ does not assume the value zero in a neighborhood of a point, then the surface does define an implicit function $z = f(x, y)$ in a neighborhood of the point, and furthermore,

$$\frac{\partial z}{\partial x} = -\frac{\partial F/\partial x}{\partial F/\partial z} \quad \text{and} \quad \frac{\partial z}{\partial y} = -\frac{\partial F/\partial y}{\partial F/\partial z}.$$

We then obtain

$$\begin{aligned}\sqrt{\left(\frac{\partial z}{\partial x}\right)^2+\left(\frac{\partial z}{\partial y}\right)^2+1} &= \sqrt{\left(\frac{\partial F/\partial x}{\partial F/\partial z}\right)^2+\left(\frac{\partial F/\partial y}{\partial F/\partial z}\right)^2+1}\\ &= \frac{\sqrt{(\partial F/\partial x)^2+(\partial F/\partial y)^2+(\partial F/\partial z)^2}}{|\partial F/\partial z|}.\end{aligned} \tag{2}$$

In case our surface is given in the form $F(x, y, z) = 0$, we may use this last form in (2) for our integrand in surface area. Illustrating with the sphere $x^2 + y^2 + z^2 - a^2 = 0$ in Example 1, we could have computed our integrand as

$$\begin{aligned}\frac{\sqrt{(2x)^2+(2y)^2+(2z)^2}}{|2z|} &= \frac{2\sqrt{x^2+y^2+z^2}}{|2z|}\\ &= \frac{2\sqrt{a^2}}{|2z|} = \frac{a}{\sqrt{a^2-x^2-y^2}}.\end{aligned}$$

Illustrations of further computations of surface area are really unnecessary; one simply computes

$$\sqrt{\left(\frac{\partial z}{\partial x}\right)^2 + \left(\frac{\partial z}{\partial y}\right)^2 + 1} \qquad \text{or} \qquad \frac{\sqrt{(\partial F/\partial x)^2 + (\partial F/\partial y)^2 + (\partial F/\partial z)^2}}{|\partial F/\partial z|},$$

or analogous expressions if the surface is projected on the x,z-plane or y,z-plane, and then evaluates the integral.

SUMMARY

1. *The area of a surface consisting of part of a graph $z = f(x, y)$ is equal to the integral*

$$\iint_G \sqrt{\left(\frac{\partial z}{\partial x}\right)^2 + \left(\frac{\partial z}{\partial y}\right)^2 + 1}\, dx\, dy$$

evaluated over the region G in the x,y-plane under the surface.

2. *The area of a surface consisting of part of a locus $F(x, y, z) = c$ is equal to the integral*

$$\iint_G \frac{\sqrt{(\partial F/\partial x)^2 + (\partial F/\partial y)^2 + (\partial F/\partial z)^2}}{|\partial F/\partial z|}\, dx\, dy$$

evaluated over the region G in the x,y-plane under the surface.

3. *The differential element of surface area is*

$$dS = \sqrt{\left(\frac{dz}{\partial x}\right)^2 + \left(\frac{dz}{dy}\right)^2 + 1}\, dx\, dy \quad \text{for the graph of } z = f(x, y)$$

and

$$dS = \frac{\sqrt{(\partial F/\partial x)^2 + (\partial F/\partial y)^2 + (\partial F/\partial z)^2}}{|\partial F/\partial z|}\, dx\, dy = \frac{|\nabla F|}{|\partial F/\partial z|} \quad \text{for } F(x, y, z) = c.$$

EXERCISES

1. Find the area of the portion of the surface $z = \frac{2}{3}(x^{3/2} + y^{3/2})$ over the rectangle $0 \le x \le 1$, $0 \le y \le 2$.

2. Find the area of the surface $z = x^2 + y^2$ inside the cylinder $x^2 + y^2 = a^2$.

3. Find the area of the portion of the surface of the sphere $x^2 + y^2 + z^2 = a^2$ which lies inside the cone $z = \sqrt{x^2 + y^2}$.

4. Find the area of the portion of the sphere $x^2 + y^2 + z^2 = a^2$ inside the cylinder $x^2 + z^2 = az$.

5. Find the area of the portion of the surface $ax = z^2 - y^2$ which is inside the cylinder $y^2 + z^2 = a^2$.

6. Find the area of the surface of the solid bounded above by $z = 4 - x^2 - y^2$ and below by $z = -4 + x^2 + y^2$.

exercise sets for chapter 18

review exercise set 18.1

1. Find the upper sum S_2 and lower sum s_2 approximating the integral of the function $f(x, y) = 2x - 3y$ over the rectangle $1 \leq x \leq 3$, $-1 \leq y \leq 3$.
2. Compute $\int_{-1}^{2} \int_{1}^{4} (3xy^2 - 2y)\, dx\, dy$.
3. Express $\int_{-2}^{2} \int_{x^2}^{4} (x^2 - 3xy)\, dy\, dx$ in the form $\iint (x^2 - 3xy)\, dx\, dy$, reversing the order of integration. Do not compute either integral.
4. Compute, using an iterated integral, the area of the region in the plane bounded by $y = x^2$ and $y = x$.
5. Express as an iterated integral the volume of the region in space bounded below by $z = x^2 + y^2$ and above by the hemisphere $z - 4 = \sqrt{4 - x^2 - y^2}$. Do not compute the integral.
6. Find the integral of the polar function $h(r, \theta) = r \cos \theta$, $r \geq 0$, over the circle $r = 2a \cos \theta$.
7. Use triple integration in cylindrical coordinates to find the volume of the solid bounded by the paraboloids $z = x^2 + y^2$ and $z = 8 - x^2 - y^2$.
8. Find the integral of the spherical coordinate function $h(p, \phi, \theta) = \rho \cos^2 \theta$ over the ball $0 \leq \rho \leq a$.
9. Find the moment of inertia of a solid ball $x^2 + y^2 + z^2 \leq a^2$ about the z-axis if the mass density of the ball at (x, y, z) is given by $|z|$.
10. Find the area of the portion of the surface of the sphere $x^2 + y^2 + z^2 = a^2$ lying above the plane $z = b$ for $0 \leq b \leq a$.

review exercise set 18.2

1. Find the Riemann sum $\mathscr{S}_2$, using midpoints of the subrectangles, approximating the integral of the function $f(x, y) = 3x - 2y$ over the rectangle $-1 \leq x \leq 3$, $1 \leq y \leq 3$.
2. Compute $\int_{4}^{-2} \int_{-1}^{3} (2xy - 3y^2)\, dy\, dx$.
3. Express $\int_{0}^{2} \int_{0}^{\sqrt{4-y^2}} \int_{x^2+y^2}^{4} x^2 z\, dz\, dx\, dy$ in the form $\iiint x^2 z\, dx\, dy\, dz$, changing the order of integration. Do not evaluate either integral.
4. Compute, using an iterated integral, the area of the region in the plane bounded by $y = 1/x$ and $x + y = \frac{5}{2}$.
5. Compute, using an iterated integral, the volume of the region in space bounded below by $z = 0$, on the sides by $x = 0$, $y = 0$, and $x + y = 2$, and above by $z = x^2 + y^2$.
6. Using double integration in polar coordinates, find the area of one loop of the rose $r = \cos 3\theta$.
7. Find the integral of the cylindrical coordinate function $h(r, \theta, z) = rz \sin^2\theta$, $r \geq 0$, over the region bounded above by $z = 4 + x^2 + y^2$, below by $z = 0$, and on the sides by $x^2 + y^2 = 4$.
8. Use triple integration in spherical coordinates to find the volume of the solid bounded by the cone $z^2 = 3x^2 + 3y^2$ and the hemisphere $z = \sqrt{16 - x^2 - y^2}$.
9. Find the centroid of the region in space bounded by $z = -1$ and $z = 3 - x^2 - y^2$.
10. Find the area of the surface $z = 16 - x^2 - y^2$, which lies above the plane $z = 12$.

more challenging exercises 18

1. Use integration to find the four-dimensional "volume" of the "4-ball of radius a" consisting of all (x, y, z, w) such that $x^2 + y^2 + z^2 + w^2 \leq a^2$. [*Hint.* Use (ρ, ϕ, θ, w) coordinates where ρ, ϕ, and θ are the usual spherical coordinates replacing x, y, and z. This is analogous to cylindrical coordinates (r, θ, z) in space compared with the polar coordinates (r, θ) in the plane.]

2. Work Exercise 1 again, but this time use Pappus' Theorem and the fact that the centroid of the half-ball $x^2 + y^2 + z^2 \leq a^2$, where $z \geq 0$, is $(0, 0, 3a/8)$. (See Exercise 8 of Section 18.5.)

3. Find the three-dimensional volume of the "3-sphere of radius a" consisting of all (x, y, z, w) such that $x^2 + y^2 + z^2 + w^2 = a^2$. [*Hint*. Recall that for a 3-ball $x^2 + y^2 + z^2 \leq a^2$ of volume $(4/3)\pi a^3$, the area of the surface of the 2-sphere $x^2 + y^2 + z^2 = a^2$ is $4\pi a^2$. Consider the relation between $V = (4/3)\pi a^3$ and $A = 4\pi a^2$ in terms of approximation of the volume V by a differential for a small change in radius. Then jump up one dimension and answer the new problem with essentially no work, using the answer to Exercise 1.]

If we partition the rectangle $0 \leq x \leq 4$, $-2 \leq y \leq 6$ into n^2 subrectangles of equal areas, then each subrectangle has area $32/n^2$. Using Riemann sums with the upper right corner of each subrectangle as the place to evaluate the function, you see that

$$\int_{-2}^{6}\int_{0}^{4} x^2 y\, dx\, dy = \lim_{n\to\infty} \sum_{i,j=1}^{n} \left[\left(\frac{4i}{n}\right)^2\left(-2 + \frac{8j}{n}\right)\left(\frac{32}{n^2}\right)\right].$$

Exercises 4 through 7 give you some practice in writing double integrals as limits of double sums, and in estimating some double sums using integrals.

4. Write a sum whose limit as $n \to \infty$ is equal to

$$\int_1^7 \int_{-3}^0 (x + 4y)^3\, dx\, dy.$$

5. Repeat Exercise 4 for $\int_5^{10}\int_{-2}^{2} (x^2 + 3xy)\, dy\, dx$.

6. Use an integral to estimate

$$\sum_{i,j=1}^{100} \left[\frac{3i}{100}\left(-1 + \left(\frac{2j}{100}\right)^2\right)\frac{6}{10{,}000}\right].$$

7. Use an integral to estimate $(8/10^7)\sum_{i,j=1}^{10} i^2 j^3$.

The remaining exercises deal with a simple algebraic device that is useful in multivariable calculus. We define a type of product of differential expressions, which is written using the symbol $\wedge$. All the properties of regular multiplication hold for this $\wedge$-product except that

$$dx \wedge dy = -dy \wedge dx$$

while

$$dx \wedge dx = dy \wedge dy = 0.$$

With three variables, you have the logical extension of such relations. For example,

$$dx \wedge dy \wedge dz = -dx \wedge dz \wedge dy$$

and

$$dy \wedge dz \wedge dy = 0.$$

Roughly speaking, a term containing the $\wedge$-product of differentials like dx, dy, etc., is zero if two differentials are the same, and changes sign if two of the differentials are interchanged. For a further illustration,

$$\begin{aligned}(x^2\, dx + y\, dy) \wedge (2y\, dx - x^2 y\, dy) &= 2x^2 y\, dx \wedge dx - x^4 y\, dx \wedge dy + 2y^2\, dy \wedge dx - x^2 y^2\, dy \wedge dy \\ &= -x^4 y\, dx \wedge dy + 2y^2\, dy \wedge dx \\ &= -x^4 y\, dx \wedge dy - 2y^2\, dx \wedge dy \\ &= -(x^4 y + 2y^2)\, dx \wedge dy.\end{aligned}$$

8. Compute

$$(x\,dx + x^2y\,dy - xz\,dz) \wedge (yz\,dz + xz\,dy - z^2\,dz),$$

as in the preceding illustration, simplifying as much as possible.

The remaining exercises indicate that this new multiplication is useful in changing variables in multiple integrals.

9. Recall that for polar coordinates, $x = r\cos\theta$ and $y = r\sin\theta$.

a) Compute dx and dy in terms of the polar-coordinate variables.

b) Using (a), compute $dx \wedge dy$ in terms of polar coordinates, simplifying as much as possible.

10. Recall that, for spherical coordinates, you have

$$x = \rho\sin\phi\cos\theta, \quad y = \rho\sin\phi\sin\theta,$$

$$z = \rho\cos\phi.$$

a) Compute dx, dy, and dz in terms of spherical coordinates.

b) Using (a), compute $dx \wedge dy \wedge dz$ in terms of spherical coordinates, simplifying as much as possible.

11. Consider $\iint_G (x - y)^4(3x + 2y)^5\,dx\,dy$ where G is the parallelogram bounded by $x - y = 1$, $x - y = 3$, $3x + 2y = -1$, and $3x + 2y = 2$. This integral is tough to evaluate in x,y-coordinates. The last two exercises should give you the clue as to how to proceed. Make the variable substitution $u = x - y$, $v = 3x + 2y$. Solve for x and y in terms of u and v, and compute dx and dy. Then compute $dx\,dy = dx \wedge dy$ in terms of du and dv. Form the new integral in terms of u,v-coordinates; be sure you change to u,v-limits. Evaluate the integral.

divergence, Green's, and Stokes' theorems

19

This chapter gives an intuitive introduction to some fundamental integral theorems of vector calculus. Precise statements and proofs of the most general cases of these theorems are not appropriate for a first calculus course.

19.1 PHYSICAL MODELS FOR GREEN'S THEOREM AND THE DIVERGENCE THEOREM

The theorems to be studied in this chapter are all of the same nature. They all assert:

The integral of some quantity over the boundary of a region is equal to the integral of a related quantity over the region itself. (1)

Weeks could be spent discussing what is meant by a region and by its boundary. Your calculus course is almost over, and you have only a few days to spend on this material. We shall be very intuitive, and hope it doesn't get us into any trouble. Throughout this section, we assume that all the functions considered have continuous partial derivatives, so that we can consider integrals of these partial derivatives.

19.1.1 A new look at the fundamental theorem of calculus

We shall start off with a new look at the fundamental theorem of calculus in Theorem 19.1. We then describe a physical demonstration of Theorem 19.1. This demonstration will generalize to give us Green's Theorem in Section 19.1.2, and the Divergence Theorem in Section 19.1.3.

Theorem 19.1 *(Fundamental Theorem of Calculus) If $f(x)$ has a continuous derivative $f'(x)$ for all x in the one-dimensional region $[a, b]$, then*

$$f(b) - f(a) = \int_a^b f'(x)\,dx.$$

When stated this way, the fundamental theorem is of the type described in (1). Surely $f(x)$ and $f'(x)$ are "related quantities," and the endpoints a and b can be viewed as the boundary of the one-dimensional region $[a, b]$. By suitable definition, $f(b) - f(a)$ can be considered to be the "integral" of $f(x)$ over this boundary.

For a physical demonstration of Theorem 19.1, imagine a gas flowing through a long cylinder *with cross section area* 1 and reaching from a to b on the x-axis, as shown in Fig. 19.1. In this idealistic situation, we shall suppose that the velocity and mass density of the gas depend only on the location x along the cylinder, and are independent of the up-down and front-back locations within a cross section of the cylinder. This is a model for *one-dimensional flow.* Both the velocity and mass density may vary with time, but we shall concentrate on one particular instant, and consider how the

One-dimensional flow

total mass of gas in the cylinder would change if it were always to flow just as it did at that instant.

The velocity could be represented by a vector function

$$\boldsymbol{V} = v(x)\boldsymbol{i}$$

and the mass density by a scalar function $\rho(x)$. The vector function

$$\boldsymbol{F} = \rho(x)v(x)\boldsymbol{i} = f(x)\boldsymbol{i}$$

is known as the *flux vector* of the flow. If the flow did not vary with time, then, since the cylinder has cross section of area one, $f(x)$ would be a (signed) measure of the mass of gas that would go past x in one unit of time. Thus $f(b)$ is the mass of gas leaving the right end of the cylinder in one unit time, and $f(a)$ the mass entering the left end per unit time. Consequently

$$f(b) - f(a) = \begin{cases}\text{Decrease of mass of gas in the}\\ \text{cylinder per unit of time.}\end{cases} \tag{2}$$

The same reasoning applied to the short element of cylinder from x to $x + \Delta x$, shown in Fig. 19.1, shows that

$$f(x + \Delta x) - f(x) = \begin{cases}\text{Decrease of mass of gas in this}\\ \text{cylindrical element per unit time.}\end{cases}$$

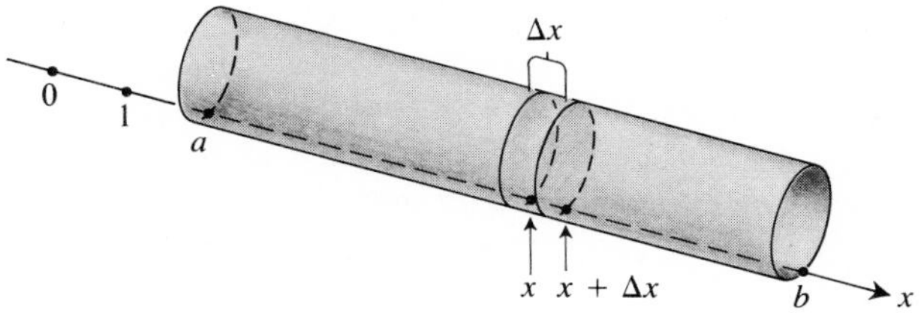

19.1

Therefore

$$\frac{f(x + \Delta x) - f(x)}{\Delta x} = \begin{cases}\text{Average decrease in mass of gas per}\\ \text{unit length of cylinder, per unit time,}\end{cases}$$

and

$$f'(x) = \lim_{\Delta x \to 0} \frac{f(x + \Delta x) - f(x)}{\Delta x} = \begin{cases}\text{Decrease of mass of gas measured at}\\ x \text{ per unit length, per unit time.}\end{cases}$$

Consequently $\int_a^b f'(x)\,dx$ has the following interpretation: The product $f'(x) \cdot dx$ may be viewed as the decrease in one unit time of mass of the gas over a short length dx of cylinder at x. Thus

$$\int_a^b f'(x)\,dx = \begin{cases}\text{Decrease of mass of the gas over the}\\ \text{entire cylinder in one unit of time.}\end{cases} \tag{3}$$

Comparison of (2) and (3) shows that you must have

$$f(b) - f(a) = \int_a^b f'(x)\,dx,$$

which is, of course, the fundamental theorem (Theorem 19.1).

19.1.2 A physical demonstration of Green's theorem

Two-dimensional flow

The same ideas as those of the last article, but with a two-dimensional flow, lead us to Green's Theorem. Imagine gas to be flowing between two identical parallel plates, placed *with one unit between them,* with the lower plate in the x,y-plane, as in Fig. 19.2. This time we suppose the *gas flow is two-dimensional,* so its velocity vector

$$\mathbf{V} = v_1(x, y)\boldsymbol{i} + v_2(x, y)\boldsymbol{j}$$

has no $\boldsymbol{k}$-component, and also does not depend on the position $0 \le z \le 1$ between the plates. Also, we assume that the mass density $\rho(x, y)$ does not depend on z. The flux vector

$$\boldsymbol{F} = \rho(x, y)\mathbf{V} = P(x, y)\boldsymbol{i} + Q(x, y)\boldsymbol{j},$$

where $P(x, y) = \rho(x, y)v_1(x, y)$ and $Q(x, y) = \rho(x, y)v_2(x, y)$, again measures the flow of mass of gas at each point per unit time. This flux vector has as direction the direction of the flow. To interpret the magnitude of $\boldsymbol{F}$, imagine a small square placed perpendicular to $\boldsymbol{F}$ at a point, with the flux vector at the center of the square. A certain mass of gas flows through this square per unit time. The magnitude of the flux vector $\boldsymbol{F}$ is the limit of the quotients of these masses of gas divided by the areas of the squares, as the areas approach zero. That is, the magnitude of the flux is the mass of gas per unit time flowing per unit area of cross section perpendicular to the flow.

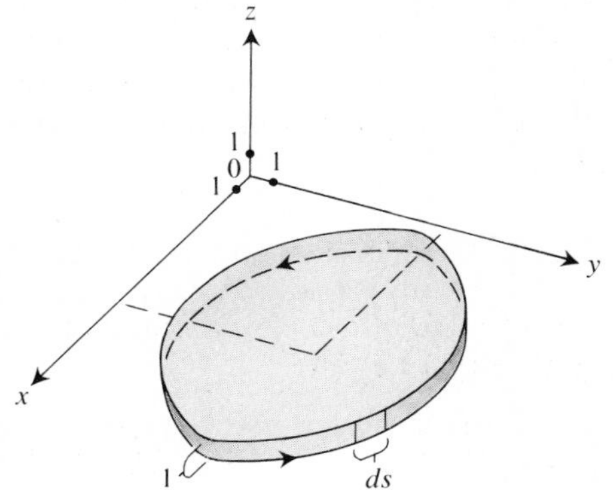

19.2

Figure 19.3 shows the region G in the x,y-plane occupied by the lower of the two plates. The boundary of the region is denoted by ∂G, read "*the boundary of G.*" Imagine that you travel along this curve, ∂G, so that the region G lies on your lefthand side, as indicated by the arrows on the curve. From previous work, you know that

$$\boldsymbol{t} = \frac{dx}{ds}\boldsymbol{i} + \frac{dy}{ds}\boldsymbol{j}$$

is a unit vector tangent to this curve at each point. It is then easy to see that

$$\boldsymbol{n} = \frac{dy}{ds}\boldsymbol{i} - \frac{dx}{ds}\boldsymbol{j}$$

is a unit vector normal to the curve, and directed *outward* from the region.

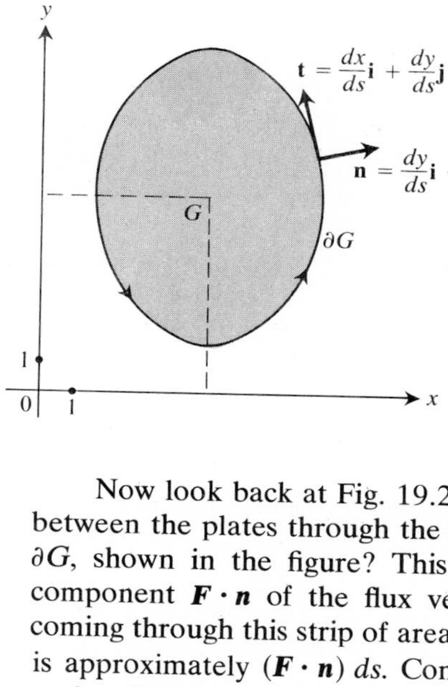

19.3

Now look back at Fig. 19.2. How much gas is flowing out of the region between the plates through the little strip of height one and width ds along ∂G, shown in the figure? This outward flow is measured by the *normal* component $\boldsymbol{F} \cdot \boldsymbol{n}$ of the flux vector. Thus the mass of gas per unit time coming through this strip of area ds (recall that the plates are one unit apart) is approximately $(\boldsymbol{F} \cdot \boldsymbol{n})\, ds$. Consequently the total mass of gas leaving the region between the plates per unit time is

$$\oint_{\partial G} (\boldsymbol{F} \cdot \boldsymbol{n})\, ds, \tag{4}$$

where $\oint_{\partial G}$ denotes the line integral once around the boundary of G in the direction given by the arrows in Fig. 19.3. But

$$\boldsymbol{F} \cdot \boldsymbol{n} = [P(x, y)\boldsymbol{i} + Q(x, y)\boldsymbol{j}] \cdot \left(\frac{dy}{ds}\boldsymbol{i} - \frac{dx}{ds}\boldsymbol{j}\right) = P(x, y)\frac{dy}{ds} - Q(x, y)\frac{dx}{ds}.$$

Therefore (4) becomes

$$\oint_{\partial G} [P(x, y)\, dy - Q(x, y)\, dx] = \begin{cases}\text{Mass of gas leaving the region} \\ \text{between the plates per unit time.}\end{cases} \tag{5}$$

Let's now compute the mass of gas leaving in another way, namely, by considering separately the contributions of $P(x, y)\boldsymbol{i}$ and $Q(x, y)\boldsymbol{j}$ to the flux vector $\boldsymbol{F}$. The vector $P(x, y)\boldsymbol{i}$ is the horizontal component of the flux. The region between the two plates over the strip of width dy shown in Fig. 19.4 is a cylinder having area of cross section $(dy)(1) = dy$. Referring back to (3) in Section 19.1.1, where we discussed flow along a cylinder, and considering just the component $P(x, y)$,

$$\int_{g_1(y)}^{g_2(y)} \frac{\partial P}{\partial x}(x, y)\, dx = \begin{cases}\text{Mass of gas leaving the ends of the cylinder} \\ \text{per unit area cross section, per unit time.}\end{cases}$$

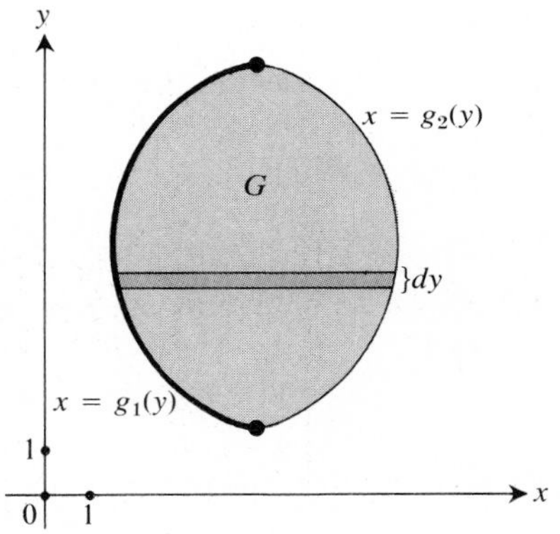

19.4

Since the cylinder has cross section area dy (the plates are still one unit apart), you have

$$\left(\int_{g_1(y)}^{g_2(y)} \frac{\partial P}{\partial x}(x, y)\, dx\right) dy = \begin{cases}\text{Mass of gas leaving the cylinder} \\ \text{at the ends per unit time.}\end{cases}$$

Therefore

$$\int_c^d \left(\int_{g_1(y)}^{g_2(y)} \frac{\partial P}{\partial x}(x, y)\, dx\right) dy = \iint_G \frac{\partial P}{\partial x}(x, y)\, dx\, dy$$

$$= \begin{cases}\text{Mass of gas leaving the region between} \\ \text{the plates per unit time due to } P(x, y)\boldsymbol{i}.\end{cases}$$

A similar computation with a vertical cylindrical strip of width dx and the component $Q(x, y)\boldsymbol{j}$ of $\boldsymbol{F}$ shows that

$$\iint_G \frac{\partial Q}{\partial y}(x, y)\, dx\, dy = \begin{cases}\text{Mass of gas leaving the region between} \\ \text{the plates per unit time due to } Q(x, y)\boldsymbol{j}.\end{cases}$$

Thus

$$\iint_G \left[\frac{\partial P}{\partial x}(x, y) + \frac{\partial Q}{\partial y}(x, y)\right] dx\, dy = \begin{cases}\text{Mass of gas leaving the region} \\ \text{between the plates per unit time.}\end{cases} \tag{6}$$

Comparing (5) and (6), you have

$$\iint_G \left[\frac{\partial P}{\partial x}(x, y) + \frac{\partial Q}{\partial y}(x, y)\right] dx\, dy = \oint_{\partial G} [P(x, y)\, dy - Q(x, y)\, dx], \tag{7}$$

which is again a relation of the form (1). Equation (7) is known as Green's Theorem.

Theorem 19.2 *(Green's Theorem) For a suitable plane region G, and for functions $P(x, y)$ and $Q(x, y)$ with continuous partial derivatives,*

$$\oint_{\partial G} (P\, dy - Q\, dx) = \iint_G \left(\frac{\partial P}{\partial x} + \frac{\partial Q}{\partial y}\right) dx\, dy. \tag{8}$$

In the line integral, ∂G is traced in the direction that keeps G on the left.

Equation (8) is often stated in vector form, reflecting the physical demonstration we gave for it. Let $\boldsymbol{F} = P(x, y)\boldsymbol{i} + Q(x, y)\boldsymbol{j}$ be a flux vector. A *symbolic operator*

$$\boldsymbol{\nabla} = \frac{\partial}{\partial x}\boldsymbol{i} + \frac{\partial}{\partial y}\boldsymbol{j}$$

is introduced, where $\boldsymbol{\nabla}$ is read "del." Symbolically, you have

$$\begin{aligned}\boldsymbol{\nabla} \cdot \boldsymbol{F} &= \left(\frac{\partial}{\partial x}\boldsymbol{i} + \frac{\partial}{\partial y}\boldsymbol{j}\right) \cdot (P(x, y)\boldsymbol{i} + Q(x, y)\boldsymbol{j}) \\ &= \frac{\partial}{\partial x} P(x, y) + \frac{\partial}{\partial y} Q(x, y) \\ &= \frac{\partial P}{\partial x} + \frac{\partial Q}{\partial y}.\end{aligned}$$

Also, you have seen that if $\boldsymbol{n}$ is a unit normal vector to the boundary, then

$$\oint_{\partial G} (P\, dy - Q\, dx) = \iint_G (\boldsymbol{F} \cdot \boldsymbol{n})\, ds.$$

Thus (8) becomes

$$\oint_{\partial G} (\boldsymbol{F} \cdot \boldsymbol{n})\, ds = \iint_G (\boldsymbol{\nabla} \cdot \boldsymbol{F})\, dx\, dy. \tag{9}$$

Divergence The scalar $\boldsymbol{\nabla} \cdot \boldsymbol{F}$ is the *divergence of* $\boldsymbol{F}$. It measures the rate at which the gas diverges (leaves) at each point.

In the next section, we shall give a mathematical, rather than physical, demonstration of Green's Theorem. In the exercises of this section, you are asked to illustrate the theorem. Here is an illustration to use as a model.

Example 1 Let's illustrate Green's Theorem for the functions

$$P(x, y) = x^3 \qquad \text{and} \qquad Q(x, y) = 2x + y^3$$

over the region G bounded by the circle $x^2 + y^2 = a^2$.

SOLUTION Now

$$\frac{\partial P}{\partial x} = 3x^2 \qquad \text{and} \qquad \frac{\partial Q}{\partial y} = 3y^2,$$

so

$$\iint_G \left(\frac{\partial P}{\partial x} + \frac{\partial Q}{\partial y}\right) dx\, dy = \iint_G 3(x^2 + y^2)\, dx\, dy.$$

Changing to polar coordinates,

$$\iint_G 3(x^2 + y^2)\, dx\, dy = \int_0^{2\pi} \int_0^a 3r^2 \cdot r\, dr\, d\theta$$

$$= \int_0^{2\pi} 3\frac{r^4}{4}\Big]_0^a d\theta = 3\frac{a^4}{4}\theta\Big]_0^{2\pi} = \frac{3a^4\pi}{2}.$$

The boundary ∂G may be parametrized by $x = a \cos\theta$, $y = a \sin\theta$ for $0 \le \theta \le 2\pi$. The integral around the boundary becomes

$$\oint_{\partial G} (P\, dy - Q\, dx) = \oint_{\partial G} [x^3\, dy - (2x + y^3)\, dx]$$

$$= \int_0^{2\pi} [a^3 \cos^3\theta \cdot a \cos\theta\, d\theta - (2a\cos\theta + a^3 \sin\theta)(-a \sin\theta\, d\theta)]$$

$$= \int_0^{2\pi} [a^4 (\cos^4\theta + \sin^4\theta) + 2a^2 \sin\theta\cos\theta]\, d\theta$$

$$= \int_0^{2\pi} \left[a^4\left(\left(\frac{1 + \cos 2\theta}{2}\right)^2 + \left(\frac{1 - \cos 2\theta}{2}\right)^2\right) + 2a^2 \sin\theta\cos\theta\right] d\theta$$

$$= \int_0^{2\pi} \left[a^4\left(\frac{\cos^2 2\theta}{2} + \frac{1}{2}\right) + 2a^2 \sin\theta\cos\theta \right] d\theta$$

$$= \int_0^{2\pi} \left[a^4\left(\frac{1 + \cos 4\theta}{4} + \frac{1}{2}\right) + 2a^2 \sin\theta\cos\theta \right] d\theta$$

$$= \left[a^4\left(\frac{1}{4}\theta + \frac{\sin 4\theta}{16} + \frac{1}{2}\theta\right) + a^2 \sin^2\theta \right]_0^{2\pi}$$

$$= a^4\left(\frac{2\pi}{4} + \frac{2\pi}{2}\right) - 0 = a^4 \cdot \frac{3\pi}{2} = \frac{3\pi a^4}{2}.$$

The same answer was obtained, illustrating Green's Theorem. Obviously, the area integral was much easier to evaluate than the line integral in this case. ||

19.1.3 The divergence theorem

If you repeat the gas-flow arguments we made earlier, but in a three-dimensional setting, you obtain the divergence theorem. Let G now be a region in space with boundary ∂G, as illustrated in Fig. 19.5. This time ∂G is a *surface*. Imagine gas flowing in space, and able to flow in and out of G without impediment. (You can think of the boundary of G as not being physically present, or, if you prefer, consider it to be a netting through which gas can flow unhindered.) Let

$$\boldsymbol{F} = P(x, y, z)\boldsymbol{i} + Q(x, y, z)\boldsymbol{j} + R(x, y, z)\boldsymbol{k}$$

be the flux vector of the flow at (x, y, z) in the region. Let $\boldsymbol{n}$ be a unit normal vector outward from the boundary surface at a point and let a little "differential" piece of surface at that point have area dS. Then the mass of gas flowing out of the region G per unit time is given by

$$\iint_{\partial G} (\boldsymbol{F} \cdot \boldsymbol{n})\, dS, \tag{10}$$

by reasoning like that in the last article.

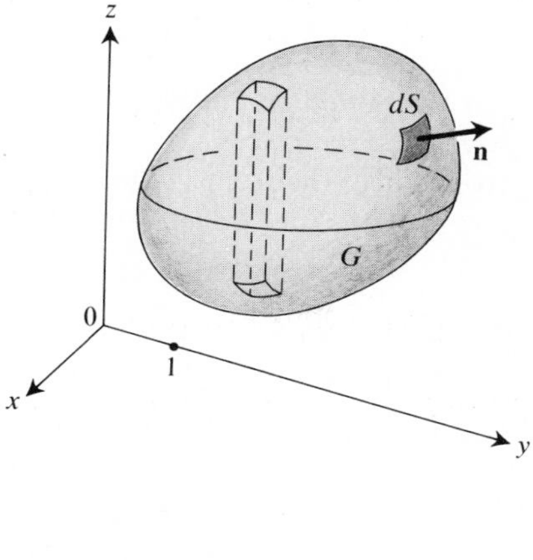

19.5

The contribution of the z-component $R(x, y, z)\boldsymbol{k}$ of $\boldsymbol{F}$ to the mass leaving G can be computed by finding the contribution in the cylinder of cross-section area $dx\,dy$, shown in Fig. 19.5, and then adding these contributions over the entire region G. Obviously you obtain

$$\iiint_G \frac{\partial R}{\partial z}\,dx\,dy\,dz$$

by reasoning just like that in Section 19.1.2. The total mass of gas leaving the region per unit time is thus

$$\iiint_G \left(\frac{\partial P}{\partial x} + \frac{\partial Q}{\partial y} + \frac{\partial R}{\partial z}\right) dx\,dy\,dz, \tag{11}$$

which may also be written

$$\iiint_G \boldsymbol{\nabla} \cdot \boldsymbol{F}\,dx\,dy\,dz \tag{12}$$

where this time

$$\boldsymbol{\nabla} = \frac{\partial}{\partial x}\boldsymbol{i} + \frac{\partial}{\partial y}\boldsymbol{j} + \frac{\partial}{\partial z}\boldsymbol{k}.$$

This gives the divergence theorem; $\boldsymbol{\nabla} \cdot \boldsymbol{F}$ is called the *divergence of* $\boldsymbol{F}$.

Theorem 19.3 (*Divergence Theorem*) *If* $\boldsymbol{F} = P(x, y, z)\boldsymbol{i} + Q(x, y, z)\boldsymbol{j} + R(x, y, z)\boldsymbol{k}$ *is a vector field with continuously differentiable components over a suitable region* G *in space, then*

$$\iint_{\partial G} (\boldsymbol{F} \cdot \boldsymbol{n})\,dS = \iiint_G (\boldsymbol{\nabla} \cdot \boldsymbol{F})\,dx\,dy\,dz. \tag{13}$$

In the exercises, you are asked to illustrate the divergence theorem for a box region with faces parallel to the coordinate planes. Illustrations for regions with more general curved boundaries will have to wait until we have discussed integrals over such surfaces. We wanted to take this opportunity to present the divergence theorem, while you have the gas-flow argument well in mind.

SUMMARY

1. *The theorems to be studied in this chapter all relate the integral of a quantity over the boundary of a region to the integral of a related quantity over the region itself.*

2. *(Green's Theorem) If $P(x, y)$ and $Q(x, y)$ are continuously differentiable throughout a suitable plane region G with boundary ∂G, then*

$$\oint_{\partial G} (P\,dy - Q\,dx) = \iint_G \left(\frac{\partial P}{\partial x} + \frac{\partial Q}{\partial y}\right) dx\,dy.$$

In the line integral, ∂G is traced in the direction that keeps G on the left.

3. *(Vector Form of Green's Theorem) If $\boldsymbol{F} = P(x, y)\boldsymbol{i} + Q(x, y)\boldsymbol{j}$ is a continuously differentiable vector field throughout a suitable plane region G with boundary ∂G, then*

$$\oint_{\partial G} (\boldsymbol{F} \cdot \boldsymbol{n})\,ds = \iint_G (\boldsymbol{\nabla} \cdot \boldsymbol{F})\,dx\,dy,$$

where $\boldsymbol{n}$ is an outward unit normal vector to ∂G and $\boldsymbol{\nabla}$ is the symbolic operator

$$\boldsymbol{\nabla} = \frac{\partial}{\partial x}\boldsymbol{i} + \frac{\partial}{\partial y}\boldsymbol{j}.$$

4. *(Divergence Theorem) Let $\boldsymbol{F} = P(x, y, z)\boldsymbol{i} + Q(x, y, z)\boldsymbol{j} + R(x, y, z)\boldsymbol{k}$ be a continuously differentiable vector field over a suitable region G in space with boundary surface ∂G. Then*

$$\iint_{\partial G} (\boldsymbol{F} \cdot \boldsymbol{n})\,dS = \iiint_G (\boldsymbol{\nabla} \cdot \boldsymbol{F})\,dx\,dy\,dz$$

where $\boldsymbol{n}$ is a unit normal vector outward from ∂G, and dS is a differential element of surface area, and $\boldsymbol{\nabla}$ is the symbolic operator

$$\boldsymbol{\nabla} = \frac{\partial}{\partial x}\boldsymbol{i} + \frac{\partial}{\partial y}\boldsymbol{j} + \frac{\partial}{\partial z}\boldsymbol{k}.$$

EXERCISES

1. Gas is flowing in the cylinder $y^2 + z^2 = 1$ for $0 \le x \le 8$, with flux vector at a certain instant equal to $\boldsymbol{F} = (4 + x^{2/3})\boldsymbol{i}$. Find the rate of decrease of mass of the gas in the cylinder at that instant.

2. Repeat Exercise 1 if the flux vector is

$$\boldsymbol{F} = \frac{16}{2 + x^{1/3}}\boldsymbol{i}.$$

In Exercises 3 through 6, illustrate Green's Theorem by computing

$$\oint_{\partial G} (P\,dy - Q\,dx) \quad \text{and} \quad \iint_G \left(\frac{\partial P}{\partial x} + \frac{\partial Q}{\partial y}\right) dx\,dy$$

for the given $P(x, y)$, $Q(x, y)$, and the region G.

3. $P(x, y) = x^2y^2$, $Q(x, y) = 2x - 3y$, where G is the square region bounded by $x = 0$, $x = 1$, $y = 0$, and $y = 1$.

4. $P(x, y) = x^2y$, $Q(x, y) = xy^2$, where G is the disk $x^2 + y^2 \leq 4$.

5. $P(x, y) = y$, $Q(x, y) = xy$, where G is the region bounded by $y = x^2$ and $y = x$.

6. $P(x, y) = xe^y$, $Q(x, y) = xy^2$, where G is the triangular region bounded by $x = 0$, $y = 0$, and $x + y = 2$.

As indicated by Example 1, it may be the case that one of the integrals appearing in Green's Theorem is much easier to compute than the other. In Exercises 7 through 10, find the indicated quantity by computing whichever integral is easier.

7. Two parallel identical plates are three units apart with the lower one having as boundary the circle $(x - 3)^2 + y^2 = 16$ in the xy-plane. Gas flowing between the plates has as flux vector

$$(2x + y^3)\boldsymbol{i} + (e^x - 4y)\boldsymbol{j}$$

at a certain instant. Find the rate of decrease of mass of gas between the plates at that instant.

8. Find the integral of the normal component of the vector field

$$\boldsymbol{F} = (x^2y + xy^2)\boldsymbol{i} + (xy)\boldsymbol{j}$$

counterclockwise around the triangle bounded by

$$x = 0, \qquad y = 0, \qquad \text{and} \qquad y = 1 - x.$$

9. Find the integral of the divergence of the vector field $x\boldsymbol{i} + y\boldsymbol{j}$ over the ellipse $(x^2/a^2) + (y^2/b^2) = 1$. [*Hint.* A parametrization of the ellipse is

$$x = a \cos t, \qquad y = b \sin t$$

for $0 \leq t \leq 2\pi$.]

10. Find the integral of the normal component of the gradient vector field of the function $f(x, y) = x^2/y^2$ counterclockwise around the boundary of the rectangular region bounded by

$$x = 0, \qquad x = 4, \qquad y = 1, \qquad \text{and} \qquad y = 3.$$

11. Show that for a region G where Green's Theorem applies,

$$\text{Area of } G = \frac{1}{2}\oint_{\partial G} (x\, dy - y\, dx).$$

12. Suppose that $\boldsymbol{F} = P(x, y)\boldsymbol{i} + Q(x, y)\boldsymbol{j}$ is a vector field such that $\partial P/\partial x = -\partial Q/\partial y$. Show that the integral of the normal component of $\boldsymbol{F}$ around ∂G is zero for any region G in the domain of $\boldsymbol{F}$ for which Green's Theorem applies.

13. Let G be a region where Green's Theorem applies. If $f(x, y)$ has G in its domain, find a line integral around ∂G equal to $\iint_G (\nabla \cdot \nabla f)\, dx\, dy$.

14. Follow the steps indicated to illustrate the divergence theorem for the vector field $\boldsymbol{F} = x^2z^2\boldsymbol{i} + xz\boldsymbol{j} + xy^2(z + 1)\boldsymbol{k}$ and the unit cube G with one vertex at the origin and the diagonally opposite vertex at $(1, 1, 1)$.

a) i) Compute $\iint (\boldsymbol{F} \cdot \boldsymbol{n})\, dS$ over the square face of G in the x,y-coordinate plane, where $z = 0$. Here

$$\boldsymbol{n} = -\boldsymbol{k} \qquad \text{and} \qquad dS = dx\, dy.$$

ii) Compute $\iint (\boldsymbol{F} \cdot \boldsymbol{n})\, dS$ over the square face of G in the plane $z = 1$. Here

$$\boldsymbol{n} = \boldsymbol{k} \qquad \text{and} \qquad dS = dx\, dy.$$

iii) Proceed as in (i) and (ii) to find the integrals $\iint (\boldsymbol{F} \cdot \boldsymbol{n})\, dS$ over the square faces of G in the planes $y = 0$ and $y = 1$.

iv) Proceed as in (i) and (ii) to find the integrals $\iint (\boldsymbol{F} \cdot \boldsymbol{n})\, dS$ over the square faces of G in the planes $x = 0$ and $x = 1$.

v) Add up the six answers found in (i) through (iv) to give $\iint_{\partial G} (\boldsymbol{F} \cdot \boldsymbol{n})\, dS$.

b) Compute $\iiint_G (\nabla \cdot \boldsymbol{F})\, dx\, dy\, dz$; the answer should be the same as in (v) of part (a).

19.2 GREEN'S THEOREM AND APPLICATIONS

Again, we just assume you have the correct notion of a plane region G and its boundary ∂G, traced so that G is on the left. A region is *bounded* if it all lies inside some sufficiently large square, and is *closed* if the boundary is considered to be part of the region.

19.2.1 Proof of Green's theorem

Simple regions

It is usual to prove Green's theorem first for a special type of region G, and then extend to more general regions by decomposing them into regions of the special type. We shall call a region G *simple* if any line parallel to one of the coordinate axes *crosses* the boundary of G in at most two points. (We allow such a line to *coincide* with the boundary for a whole interval.) The regions shown in Fig. 19.6 are simple.

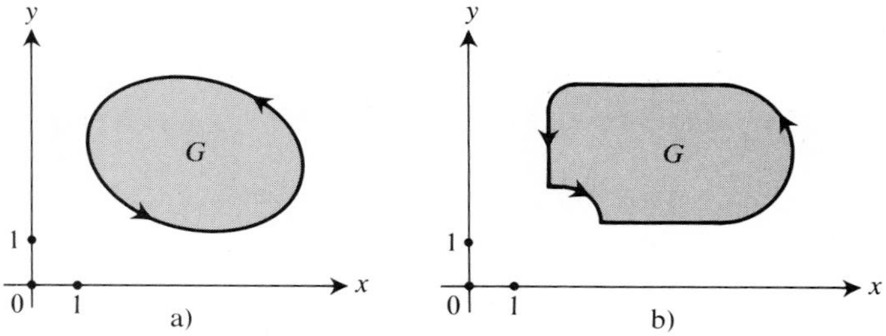

19.6

You will note that Green's Theorem in the following form differs slightly in notation from Theorem 19.2 of the last section. The substitution $Q(x, y) = -P(x, y)$ and $P(x, y) = Q(x, y)$ in Theorem 19.2 gives (1) below in Theorem 19.4. The reason for our change of notation appears later in this section.

Theorem 19.4 (*Green's Theorem for Simple Regions*) *Let G be a bounded, simple, closed region in the plane with boundary ∂G. If $P(x, y)$ and $Q(x, y)$ are continuously differentiable functions defined on G, then*

$$\oint_{\partial G} [P(x, y)\, dx + Q(x, y)\, dy] = \iint_G \left(\frac{\partial Q}{\partial x} - \frac{\partial P}{\partial y}\right) dx\, dy. \tag{1}$$

Proof. Since G is a simple region, ∂G can be split into a "top curve" and a "bottom curve," possibly separated by straight-line "sides," as shown in Fig. 19.7. The top and bottom curves may be parametrized by the parameter x; we let the equations of these curves be

$$y = u(x) \quad \text{for the bottom curve}$$

and

$$y = v(x) \quad \text{for the top curve,}$$

as shown in Fig. 19.7. The integral $\int P(x, y)\, dx$ over any portion of ∂G

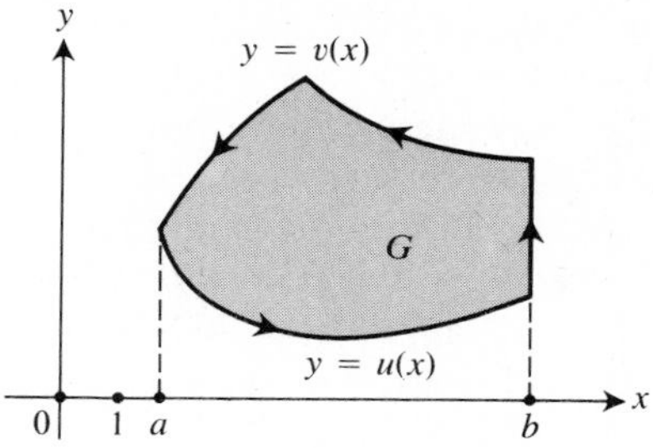

19.7

consisting of vertical line segments is zero, since x remains constant, so $dx = 0$ on any such segment. Consequently, from Fig. 19.7, you obtain

$$\oint_{\partial G} P(x, y)\,dx = \int_a^b P(x, u(x))\,dx + \int_b^a P(x, v(x))\,dx, \tag{2}$$

where the right-to-left integral sign $\int_b^a$ occurs since the top curve is traced from right to left. From (2),

$$\oint_{\partial G} P(x, y)\,dx = \int_a^b [P(x, u(x)) - P(x, v(x))]\,dx = \int_a^b -P(x, y)\Big]_{y=u(x)}^{y=v(x)} dx$$

$$= \int_a^b \int_{u(x)}^{v(x)} -\frac{\partial P}{\partial y}\,dy\,dx = \iint_G -\frac{\partial P}{\partial y}\,dx\,dy. \tag{3}$$

A similar argument, which we ask you to give in Exercise 5, shows that

$$\oint_{\partial G} Q(x, y)\,dy = \iint_G \frac{\partial Q}{\partial x}\,dx\,dy. \tag{4}$$

From (3) and (4),

$$\oint_{\partial G} [P(x, y)\,dx + Q(x, y)\,dy] = \iint_G \left(\frac{\partial Q}{\partial x} - \frac{\partial P}{\partial y}\right) dx\,dy,$$

which is the assertion of the theorem. □

The result in Theorem 19.4 can easily be extended to a region that can be decomposed into a finite number of simple regions. Consider, for example, the region shown in Fig. 19.8. This region G is not simple, but it may be decomposed into the two simple regions G_1 and G_2 separated by the curve γ_3 shown in the figure. Let's introduce some convenient algebraic notation. If γ_3 is the curve traced in the direction of the arrow in Fig. 19.8, let $-\gamma_3$ be the curve traced in the opposite direction. From line integrals, you know that

$$\int_{-\gamma_3} (P\,dx + Q\,dy) = -\int_{\gamma_3} (P\,dx + Q\,dy) \tag{5}$$

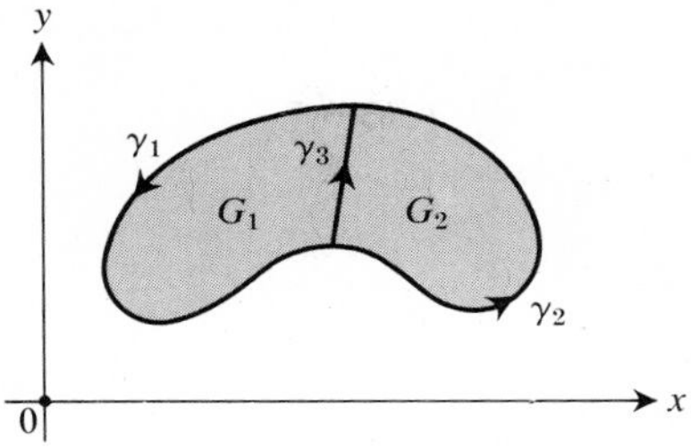

19.8

(see Chapter 17, Section 17.5, Exercise 14). Symbolically, we write $\partial G_1 = \gamma_1 + \gamma_3$ and $\partial G_2 = \gamma_2 - \gamma_3$. In view of (5), we have symbolically

$$\oint_{\partial G_1} + \oint_{\partial G_2} = \oint_{\gamma_1+\gamma_3} + \oint_{\gamma_2-\gamma_3} = \int_{\gamma_1} + \int_{\gamma_3} + \int_{\gamma_2} - \int_{\gamma_3}$$

$$= \oint_{\gamma_1+\gamma_2} = \oint_{\partial G}, \tag{6}$$

where we have omitted the integrand $P\,dx + Q\,dy$ for brevity. Theorem 19.4 applied to the simple regions G_1 and G_2 tells us that

$$\oint_{\partial G_1} (P\,dx + Q\,dy) = \iint_{G_1} \left(\frac{\partial Q}{\partial x} - \frac{\partial P}{\partial y}\right) dx\,dy \tag{7}$$

and

$$\oint_{\partial G_2} (P\,dx + Q\,dy) = \iint_{G_2} \left(\frac{\partial Q}{\partial x} - \frac{\partial P}{\partial y}\right) dx\,dy. \tag{8}$$

Adding Eqs. (7) and (8), you see from (6) that

$$\oint_{\partial G} (P\,dx + Q\,dy) = \iint_{G} \left(\frac{\partial Q}{\partial x} - \frac{\partial P}{\partial y}\right) dx\,dy,$$

which is Green's Theorem for the region G. Similar arguments can be used for any decomposition of a region into a finite number of simple regions, and we state the result as a corollary.

Corollary. (*Green's Theorem for More General Regions*) *Let G be a bounded, closed region in the plane that can be decomposed into a finite number of simple regions. If $P(x, y)$ and $Q(x, y)$ are continuously differentiable on G, then*

$$\oint_{\partial G} (P\,dx + Q\,dy) = \iint_{G} \left(\frac{\partial Q}{\partial x} - \frac{\partial P}{\partial y}\right) dx\,dy. \tag{9}$$

Example 1 Let G be the annular region between the circles $x^2 + y^2 = 4$ and $x^2 + y^2 = 16$ shown in Fig. 19.9. The direction in which the boundary circles should be traced to keep G on the left is indicated in the figure. Now G is decomposed into four simple regions by the coordinate axes. Let's illustrate Green's Theorem for G using $P(x, y) = xy$ and $Q(x, y) = -x$.

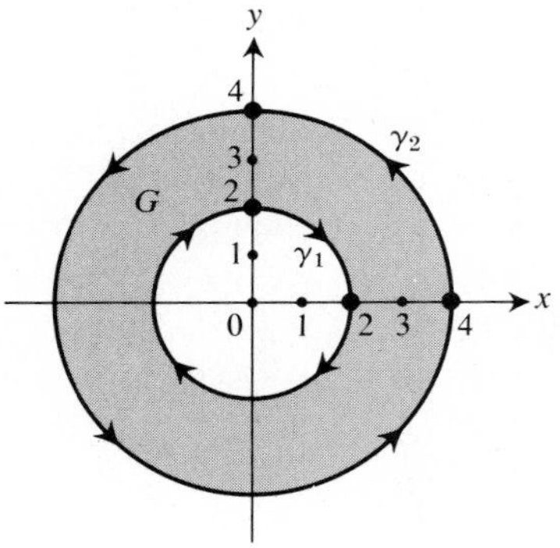

19.9

SOLUTION Let γ_1 be the clockwise-traced inner circle parametrized by

$$x = h_1(t) = 2\cos t, \qquad y = k_1(t) = -2\sin t \quad \text{for } 0 \le t \le 2\pi,$$

and let γ_2 be the counterclockwise-traced outer circle parametrized by

$$x = h_2(t) = 4\cos t, \qquad y = k_2(t) = 4\sin t \quad \text{for } 0 \le t \le 2\pi.$$

Then

$$\begin{aligned}
\int_{\gamma_1+\gamma_2} (xy\,dx - x\,dy) &= \int_{\gamma_1} (xy\,dx - x\,dy) + \int_{\gamma_2} (xy\,dx - x\,dy)\\
&= \int_0^{2\pi} [8\cos t\sin^2 t + 4\cos^2 t]\,dt\\
&\quad + \int_0^{2\pi} [-64\cos t\sin^2 t - 16\cos^2 t]\,dt\\
&= \int_0^{2\pi} [-56\cos t\sin^2 t - 12\cos^2 t]\,dt\\
&= \left[-56\frac{\sin^3 t}{3} - 12\left(\frac{t}{2} + \frac{\sin 2t}{4}\right)\right]_0^{2\pi}\\
&= -12\frac{2\pi}{2} = -12\pi.
\end{aligned}$$

On the other hand, we obtain, using polar coordinates,

$$\begin{aligned}\iint_G \left(\frac{\partial(-x)}{\partial x} - \frac{\partial(xy)}{\partial y}\right) dx\,dy &= \iint_G (-1 - x)\,dx\,dy \\ &= \int_0^{2\pi}\int_2^4 (-1 - r\cos\theta)r\,dr\,d\theta \\ &= \int_0^{2\pi} \left(-\frac{r^2}{2} - \frac{r^3}{3}\cos\theta\right)\Bigg]_{r=2}^{r=4} d\theta \\ &= \int_0^{2\pi} (-6 - \tfrac{56}{3}\cos\theta)\,d\theta \\ &= (-6\theta - \tfrac{56}{3}\sin\theta)\Bigg]_0^{2\pi} = -12\pi.\end{aligned}$$

The same answers were obtained for each integral, illustrating Green's theorem. ‖

19.2.2 Independence of path

In Chapter 17, Section 17.5, we showed that

> *if $P\,dx + Q\,dy$ is an exact differential, then $\int_\gamma (P\,dx + Q\,dy)$ depends only on the endpoints of γ.*

Such *independence of path* is equivalent to the assertion that the integral about any closed path is zero. For if γ_1 and γ_2 are two paths joining A to B, then $\gamma_1 - \gamma_2$ joins A to A. Then

$$\oint_{\gamma_1-\gamma_2} = \int_{\gamma_1} - \int_{\gamma_2} = 0 \qquad \text{if and only if} \qquad \int_{\gamma_1} = \int_{\gamma_2}$$

We also stated in Chapter 17, Section 17.3, that

> *in a region with no holes in it, $P\,dx + Q\,dy$ is exact if and only if $\partial P/\partial y = \partial Q/\partial x$.*

Green's theorem provides an easy demonstration that if $\int_\gamma (P\,dx + Q\,dy)$ depends only on the endpoints of γ for all choices of γ in the region, then $\partial P/\partial y = \partial Q/\partial x$. (Theorem 19.5 below proves this.) Combined with the statements above, this yields the following.

> *In a region G with no holes in it, the integral $\int_\gamma (P\,dx + Q\,dy)$ is independent of the path if and only if $\partial P/\partial y = \partial Q/\partial x$. This in turn is true if and only if $P\,dx + Q\,dy$ is an exact differential.*

We now proceed to show independence of path implies $\partial P/\partial y = \partial Q/\partial x$.

Theorem 19.5 *Let $P(x, y)$ and $Q(x, y)$ be continuously differentiable, and suppose that $\int_\gamma (P\,dx + Q\,dy) = 0$ for every closed path γ in a region G. Then $\partial P/\partial y = \partial Q/\partial x$ throughout G.*

Proof. Let (x_0, y_0) be a point in G that is not on a boundary edge of G. Suppose

$$\left.\frac{\partial P}{\partial y}\right|_{(x_0, y_0)} \neq \left.\frac{\partial Q}{\partial x}\right|_{(x_0, y_0)};$$

we shall derive a contradiction to the independence of path hypothesis. Then

$$\left.\frac{\partial Q}{\partial x}\right|_{(x_0, y_0)} - \left.\frac{\partial P}{\partial y}\right|_{(x_0, y_0)} = c \neq 0.$$

Suppose $c > 0$; a similar argument will work if $c < 0$. By continuity of $\partial Q/\partial x - \partial P/\partial y$, there exists some small circular disk D of radius $\epsilon > 0$ inside G with center at (x_0, y_0) throughout which $(\partial Q/\partial x - \partial P/\partial y) > c/2$. Let γ be the closed curve consisting of the boundary circle of this disk, traced counterclockwise. Then

$$\iint_D \left(\frac{\partial Q}{\partial x} - \frac{\partial P}{\partial y}\right) dx\,dy > \frac{\pi\epsilon^2 c}{2} > 0;$$

so by Green's Theorem,

$$\oint_\gamma (P\,dx + Q\,dy) > \frac{\pi\epsilon^2 c}{2} > 0,$$

contradicting the hypothesis that such an integral over every closed curve is zero. So the assumption made above is false, and

$$\left.\frac{\partial P}{\partial y}\right|_{(x_0, y_0)} = \left.\frac{\partial Q}{\partial x}\right|_{(x_0, y_0)}$$

for every (x_0, y_0) in G not on a boundary edge. But then $\partial P/\partial y = \partial Q/\partial x$ at all points on the boundary edges also, since these partial derivatives are continuous. □

19.2.3 Application to circulation of a flow

Let $\boldsymbol{F} = P(x, y)\boldsymbol{i} + Q(x, y)\boldsymbol{j}$ be the flux vector of a flow over a region G. You have seen a physical interpretation of Green's Theorem in terms of the divergence of the flow, which measures the rate that mass is leaving G. This interpretation arises from integrating the outward *normal* component of the flux over the boundary of G.

The integral of the *tangential* component of the flux over the boundary measures the rotation or *circulation* of the flow around the boundary of G.

A unit tangent vector is

$$\boldsymbol{t} = \frac{dx}{ds}\boldsymbol{i} + \frac{dy}{ds}\boldsymbol{j}.$$

The circulation around ∂G is

$$\oint_{\partial G} (\boldsymbol{F} \cdot \boldsymbol{t})\, ds = \oint_{\partial G} \left(P\frac{dx}{ds} + Q\frac{dy}{ds}\right) ds = \oint_{\partial G} (P\,dx + Q\,dy).$$

By Theorem 19.4,

$$\text{Circulation of flow around } \partial G = \oint_{\partial G} (\boldsymbol{F} \cdot \boldsymbol{t})\, ds = \oint_{\partial G} (P\,dx + Q\,dy)$$
$$= \iint_G \left(\frac{\partial Q}{\partial x} - \frac{\partial P}{\partial y}\right) dx\,dy.$$

(The notation in Theorem 19.4 was changed from that in Theorem 19.2 precisely to illustrate this application of Green's Theorem.) The scalar quantity $\partial Q/\partial x - \partial P/\partial y$ thus measures the tendency of the flow to rotate or *curl* at each point (x, y). This is seen if you take a very small disk with center (x, y) as your region G. For this reason, we write

$$\text{curl}\,\boldsymbol{F} = \frac{\partial Q}{\partial x} - \frac{\partial P}{\partial y}.$$

We then obtain

$$\oint_{\partial G} (\boldsymbol{F} \cdot \boldsymbol{t})\, ds = \iint_G (\text{curl}\,\boldsymbol{F})\, dx\,dy. \tag{10}$$

Irrotational flow

If $\text{curl}\,\boldsymbol{F} = 0$ at all (x, y) in G, then the integral of the tangential component of $\boldsymbol{F}$ is zero about any closed path bounding a suitable region lying within G. That is, the total rotation or circulation around any such closed path is zero. Such a flow is called *irrotational.*

19.2.4 Application to work

If $\boldsymbol{F} = P(x, y)\boldsymbol{i} + Q(x, y)\boldsymbol{j}$ is regarded as a force field, then the integral of the tangential component of $\boldsymbol{F}$ along a curve is the work done by the force in moving a body along the curve. Once again, if $\partial Q/\partial x - \partial P/\partial y = 0$ for all (x, y), then the work done by the force in moving a body around any closed path bounding a suitable region is zero. In this context, the force field $\boldsymbol{F}$ is said to be *conservative.* The work done by the force field along a curve depends only on the endpoints of the curve; i.e., you have independence of path. Note that $P\,dx + Q\,dy$ is then the exact differential of some function $-u(x, y)$. Then $u(x, y)$ is called a *potential function* of the force field. The potential function gives the *potential energy* of the field at each point (x, y). By our work in Section 17.5, the work done by the force in moving a body from A to B is then $-u(B) - (-u(A)) = u(A) - u(B)$.

Conservative force field

Potential function

19.2.5 Incompressible flow

Think back again to the divergence interpretation of Green's Theorem (Theorem 19.2). You see that if $\boldsymbol{F} = P(x, y)\boldsymbol{i} + Q(x, y)\boldsymbol{j}$ is a flux vector such that the divergence $\nabla \cdot \boldsymbol{F} = \partial P/\partial x + \partial Q/\partial y = 0$, for all (x, y), then the total flux of mass outward (or inward) through the boundary of any suitable region is zero. Such a flow is said to be *incompressible.*

SUMMARY

1. *(Green's Theorem) If $P(x, y)$ and $Q(x, y)$ are continuously differentiable, and if G can be decomposed into a finite number of simple regions, then*

$$\oint_{\partial G} (P\,dx + Q\,dy) = \iint_G \left(\frac{\partial Q}{\partial x} - \frac{\partial P}{\partial y}\right) dx\,dy.$$

2. *If $P(x, y)$ and $Q(x, y)$ are continuously differentiable and $\partial Q/\partial x = \partial P/\partial y$, then $\int_{\gamma_1} (P\,dx + Q\,dy) = \int_{\gamma_2} (P\,dx + Q\,dy)$ if γ_1 and γ_2 join the same points, and if $\gamma_1 - \gamma_2$ is the boundary of a region G that can be decomposed into a finite number of simple regions.*
3. *If $\boldsymbol{F} = P(x, y)\boldsymbol{i} + Q(x, y)\boldsymbol{j}$ is the flux vector of a flow, then*

$$\oint_{\partial G} (\boldsymbol{F} \cdot \boldsymbol{t})\,ds = \iint_G (\text{curl}\,\boldsymbol{F})\,dx\,dy$$

where $\text{curl}\,\boldsymbol{F} = \partial Q/\partial x - \partial P/\partial y$. *If* $\text{curl}\,\boldsymbol{F} = 0$, *then the flow is called irrotational.*

4. *The work done by a force field $\boldsymbol{F}$ along γ is $\int_\gamma (\boldsymbol{F} \cdot \boldsymbol{t})\,ds$.*
5. *A force field $\boldsymbol{F} = P(x, y)\boldsymbol{i} + Q(x, y)\boldsymbol{j}$ is conservative if $\partial Q/\partial x - \partial P/\partial y = 0$. For a conservative force field in a region with no holes in it, the work done is independent of the path, and $u(x, y)$ such that $d(-u) = P\,dx + Q\,dy$ is a potential function of the force field. The work done by the force field in moving a body from A to B is then $u(A) - u(B)$.*
6. *A flow with flux vector $\boldsymbol{F} = P(x, y)\boldsymbol{i} + Q(x, y)\boldsymbol{j}$ is incompressible if $\nabla \cdot \boldsymbol{F} = \partial P/\partial x + \partial Q/\partial y = 0$.*

EXERCISES

1. Let G be the region between the two squares with center at the origin and sides of lengths 2 and 4. Illustrate Green's Theorem for $P(x, y) = y^3$ and $Q(x, y) = -x^3$; i.e., for the vector field $\boldsymbol{F} = y^3\boldsymbol{i} - x^3\boldsymbol{j}$.

2. Let $\boldsymbol{F}$ be a vector field in the plane. Suppose the integrals $\oint (\boldsymbol{F} \cdot \boldsymbol{t})\,ds$ taken counterclockwise about the circles $x^2 + y^2 = 100$ and $x^2 + y^2 = 25$ are 35 and -24, respectively. Find $\iint_G (\text{curl}\,\boldsymbol{F})\,dx\,dy$, where G is the region between the two circles.

3. Let $\boldsymbol{E}$ and $\boldsymbol{F}$ be two vector fields on a simple, closed, bounded region G. Suppose that $\boldsymbol{E} = \boldsymbol{F}$ at each point on ∂G.

a) Show that

$$\iint_G (\nabla \cdot \boldsymbol{E})\,dx\,dy = \iint_G (\nabla \cdot \boldsymbol{F})\,dx\,dy.$$

b) Show that

$$\iint_G (\text{curl } \boldsymbol{E})\, dx\, dy = \iint_G (\text{curl } \boldsymbol{F})\, dx\, dy.$$

4. Let G be a simple, closed, bounded region, and let $\boldsymbol{F}$ be a vector field on G.

a) Suppose $\oint_\gamma (\boldsymbol{F} \cdot \boldsymbol{t})\, ds = 0$ for every closed curve γ lying in G. Argue as best you can that curl $\boldsymbol{F} = 0$ at each (x, y) in G.

b) State a result similar to that in part (a) under the hypothesis $\oint_\gamma (\boldsymbol{F} \cdot \boldsymbol{n})\, ds = 0$.

5. Prove Eq. (4) of the proof of Green's Theorem in the text.

6. Let a plane region G, which can be decomposed into a finite number of simple regions, be bounded by a piecewise smooth curve γ. Using Green's Theorem, show that the line integral $\oint_\gamma (-y\, dx + x\, dy)$ is equal to twice the area of G.

7. Let $\boldsymbol{F}$ be a vector field defined at all points except a point A of the plane, and suppose that curl $\boldsymbol{F} = 0$. Show that for any two closed piecewise-smooth curves γ_1 and γ_2, bounding simple regions and each going around A once in the same direction, we have

$$\oint_{\gamma_1} (\boldsymbol{F} \cdot \boldsymbol{t})\, ds = \oint_{\gamma_2} (\boldsymbol{F} \cdot \boldsymbol{t})\, ds.$$

[*Hint.* Consider $\int_{\gamma_2+\gamma_3-\gamma_1-\gamma_3}$ for γ_3 shown in Fig. 19.10.]

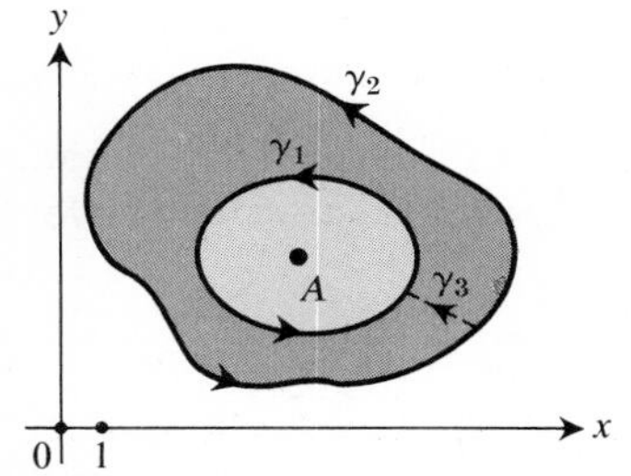

19.10

8. Consider the vector field

$$\boldsymbol{F} = \frac{y}{x^2 + y^2}\boldsymbol{i} - \frac{x}{x^2 + y^2}\boldsymbol{j} \quad \text{for } (x, y) \neq (0, 0).$$

a) Show that curl $\boldsymbol{F} = 0$.

b) Let γ be the ellipse about $(0, 0)$ given by $x^2/9 + y^2/4 = 1$. Use (a) and the result of the previous exercise to compute $\int_\gamma (\boldsymbol{F} \cdot \boldsymbol{t})\, ds$ by computing, in its place, the integral of the tangential component of $\boldsymbol{F}$ around a more convenient curve that encircles the origin once counterclockwise. [*Hint.* Try a circle or a square.]

9. Show that if $f(x, y)$ has continuous second partial derivatives, then curl $\nabla f = 0$.

10. Show that if ∇f is the flux vector of an incompressible flow, then f satisfies Laplace's equation $f_{xx} + f_{yy} = 0$.

11. Use Green's Theorem to find the work done by the force field $\boldsymbol{F}(x, y) = (x^2 + y^2)\boldsymbol{i} - 2xy\boldsymbol{j}$ in moving a body counterclockwise around the square with vertices at $(0, 0)$, $(1, 0)$, $(1, 1)$, and $(0, 1)$.

12. Find all functions $g(x, y)$ such that the force field $\boldsymbol{F}(x, y) = 3xy^2\boldsymbol{i} + g(x, y)\boldsymbol{j}$ is conservative.

13. Let the flux vector of a plane flow be $\boldsymbol{F}(x, y) = (x^2 + y^2)\boldsymbol{i} + 2xy\boldsymbol{j}$.

a) Find the divergence of $\boldsymbol{F}$.

b) Use Green's Theorem to find the flux of the flow across the border of the square with vertices $(0, 0)$, $(1, 0)$, $(1, 1)$, and $(0, 1)$.

14. For the flow in Exercise 13, use Green's Theorem to find the circulation of the flow counterclockwise around the circle $x^2 + y^2 = 4$.

15. Show that every constant force field is conservative. What can be said concerning the work done by a constant force field in moving a body from a point A to a point B?

16. Show that every constant flux field in the plane gives a flow that is both irrotational and incompressible. Give physical interpretations of this result.

17. Let $\boldsymbol{E}$ and $\boldsymbol{F}$ be conservative force fields in a plane region G.

a) Show that for any numbers a and b, the force field $a\boldsymbol{E} + b\boldsymbol{F}$ is conservative.

b) If u is a potential function of $\boldsymbol{E}$ and v a potential function of $\boldsymbol{F}$ in G, describe all potential functions of $a\boldsymbol{E} + b\boldsymbol{F}$.

18. Consider the force field $\boldsymbol{F} = 2xy\boldsymbol{i} + x^2\boldsymbol{j}$ in the plane.

a) Show that the field is conservative.

b) Find a potential function for the field.

c) Find the potential energy $u(x, y)$ of the field such that $u(0, 0) = 5$.

d) Find the work done by the field in moving a body from the point $(1, -1)$ to the point $(2, 1)$.

19. Let $\boldsymbol{F}$ be a force field in the plane, and let the position of a body in the plane at time t be given by $x = h_1(t)$, $y = h_2(t)$ for $a \le t \le b$. Let the position be A when $t = a$ and B when $t = b$. Let $W(A, B)$ be the work done by the field in moving the body from A to B for $a \le t \le b$. Show that

$$W(A, B) = \tfrac{1}{2}m|\boldsymbol{v}(b)|^2 - \tfrac{1}{2}m|\boldsymbol{v}(a)|^2 = k(b) - k(a),$$

where $k(t) = \frac{1}{2}m\,|\boldsymbol{v}(t)|^2$ is the *kinetic energy* of the body at time t. [*Hint.* Use $\boldsymbol{F}(t) = m\boldsymbol{a}(t) = m\boldsymbol{v}'(t)$ and $d\boldsymbol{r}/dt = \boldsymbol{v}(t)$ to express $W(A, B) = \int_a^b (\boldsymbol{F} \cdot d\boldsymbol{r})$ in terms of $\boldsymbol{v}$.]

20. Continuing Exercise 19, show that if $\boldsymbol{F}$ is the force field for a potential function u, then $u(A) + k(A) = u(B) + k(B)$, so the sum of the potential and kinetic energies remains constant. (Note that it is permissible to write u and k as functions of positions in the plane since $\boldsymbol{F}$ is a conservative force field, so the work in Exercise 19 is independent of the path.) This is the reason why such a force field is called *conservative*.

21. A positive electric charge in the plane at the origin exerts a force of attraction on a negative unit charge in the plane that is inversely proportional to the square of the distance between the charges. Let the force of attraction on a negative unit charge at $(1, 0)$ be k units.

a) Find the force field created by the charge at the origin.

b) Show that the force field in (a) is conservative by finding a potential function; the function is the *Newtonian potential.*

c) Using the answer to (b), find the work done by the field in moving a unit negative charge from $(0, 2)$ to $(2, 3)$.

22. Let u be a potential function of a conservative force field $\boldsymbol{F}$ in the plane. Level curves of u are known as *equipotential curves.*

a) What geometric relation exists between the field $\boldsymbol{F}$ and its equipotential curves?

b) If γ_1 and γ_2 are equipotential curves, show that the work done by the field $\boldsymbol{F}$ in moving a body from a point P_1 on γ_1 to a point P_2 on γ_2 is independent of the choices of P_1 on γ_1 and P_2 on γ_2.

19.3 STOKES' THEOREM

19.3.1 Integration over a surface

Recall the Divergence Theorem from Section 19.1, which states that, if $\boldsymbol{F} = P\boldsymbol{i} + Q\boldsymbol{j} + R\boldsymbol{k}$ is a continuously differentiable vector field over a region G in space with boundary surface ∂G, then

$$\iint_{\partial G} (\boldsymbol{F} \cdot \boldsymbol{n})\, dS = \iiint_G \left(\frac{\partial P}{\partial x} + \frac{\partial Q}{\partial y} + \frac{\partial R}{\partial z}\right) dx\, dy\, dz. \tag{1}$$

Here $\boldsymbol{n}$ is the *outward* unit normal vector to the surface ∂G. We did not illustrate the Divergence Theorem (1) in Section 19.1. We must first explain how to integrate the normal component of a vector field $\boldsymbol{F} = P\boldsymbol{i} + Q\boldsymbol{j} + R\boldsymbol{k}$ over a surface. You can compute $\boldsymbol{F} \cdot \boldsymbol{n}$ if you know $\boldsymbol{n}$; presumably $\boldsymbol{F}$ is given. Now if the surface is given in the form $z = f(x, y)$, a normal vector is

$$\frac{\partial f}{\partial x}\boldsymbol{i} + \frac{\partial f}{\partial y}\boldsymbol{j} - \boldsymbol{k},$$

so

$$\boldsymbol{n} = \pm\frac{1}{\sqrt{(f_x)^2 + (f_y)^2 + 1}}(f_x\boldsymbol{i} + f_y\boldsymbol{j} - \boldsymbol{k}). \tag{2}$$

It is usually best to see which sign is appropriate for the *outward* normal by sketching. From Section 18.6, you know that

$$dS = \sqrt{(f_x)^2 + (f_y)^2 + 1}\, dx\, dy$$

is the differential surface area.

Suppose, on the other hand, that the surface ∂G is given in terms of $f(x, y, z) = c$. Then a normal vector is $f_x\boldsymbol{i} + f_y\boldsymbol{j} + f_z\boldsymbol{k}$, so

$$\boldsymbol{n} = \pm\frac{1}{\sqrt{(f_x)^2 + (f_y)^2 + (f_z)^2}}(f_x\boldsymbol{i} + f_y\boldsymbol{j} + f_z\boldsymbol{k}). \tag{3}$$

This time

$$dS = \frac{\sqrt{(f_x)^2 + (f_y)^2 + (f_z)^2}}{|f_z|}\, dx\, dy.$$

Example 1 Let's illustrate the Divergence Theorem (1) for the vector field

$$\boldsymbol{F} = x\boldsymbol{i} + y\boldsymbol{j} + 2z\boldsymbol{k}$$

over the region G, which is the tetrahedron with vertices $(0, 0, 0)$, $(1, 0, 0)$, $(0, 2, 0)$, and $(0, 0, 1)$ shown in Fig. 19.11.

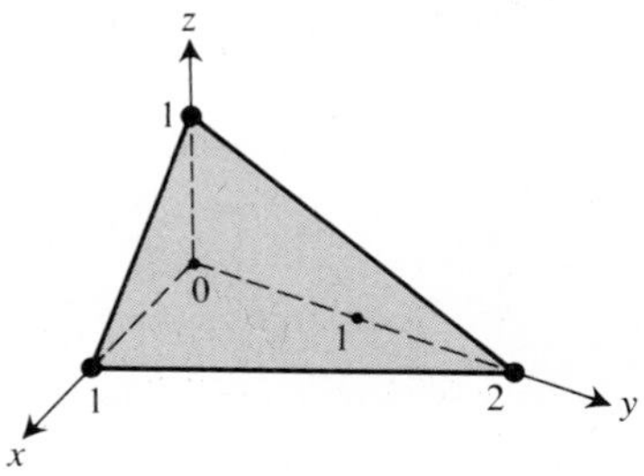

19.11

SOLUTION The volume integral is

$$\iiint_G \left(\frac{\partial P}{\partial x} + \frac{\partial Q}{\partial y} + \frac{\partial R}{\partial z}\right) dx\, dy\, dz = \iiint_G (1 + 1 + 2)\, dx\, dy\, dz$$

$$= 4(\text{volume of tetrahedron})$$

$$= 4 \cdot \tfrac{1}{3} \cdot 1 \cdot 1 = \tfrac{4}{3}.$$

Now we turn to the integral $\iint_{\partial G} (\boldsymbol{F} \cdot \boldsymbol{n})\, dS$ over the four triangles that form the surface of the tetrahedron. For the triangle in the x,y-plane, $\boldsymbol{n} = -\boldsymbol{k}$, $z = 0$, and $dS = dx\, dy$, so the integral becomes

$$\int_0^1 \int_0^{2-2x} [(x\boldsymbol{i} + y\boldsymbol{j}) \cdot (-\boldsymbol{k})]\, dy\, dx = \int_0^1 \int_0^{2-2x} 0\, dy\, dx = 0.$$

The integrals over the triangles in the x,z-plane and in the y,z-plane are similarly zero.

We must now find the equation of the front plane of the region, determined by the points $(1, 0, 0)$, $(0, 2, 0)$ and $(0, 0, 1)$. A vector perpendicular to the plane is

$$\begin{vmatrix} \boldsymbol{i} & \boldsymbol{j} & \boldsymbol{k} \\ -1 & 2 & 0 \\ -1 & 0 & 1 \end{vmatrix} = 2\boldsymbol{i} + \boldsymbol{j} + 2\boldsymbol{k},$$

so the equation of the plane is

$$2x + y + 2z = 2.$$

For this front triangle, you see from the equation $2x + y + 2z = 2$ of its plane that

$$\boldsymbol{n} = \tfrac{2}{3}\boldsymbol{i} + \tfrac{1}{3}\boldsymbol{j} + \tfrac{2}{3}\boldsymbol{k}$$

and

$$dS = (\sqrt{4 + 1 + 4}/2)\, dx\, dy = (\tfrac{3}{2})\, dx\, dy.$$

Therefore this integral becomes

$$\begin{aligned}
\int_0^1 \int_0^{2-2x} & \left[\frac{2}{3}x + \frac{1}{3}y + \frac{4}{3}\left(1 - x - \frac{y}{2}\right)\right]\frac{3}{2}\, dy\, dx \\
&= \int_0^1 \int_0^{2-2x} \left(2 - x - \frac{y}{2}\right) dy\, dx \\
&= \int_0^1 \left[2(2 - 2x) - x(2 - 2x) - \frac{1}{4}(2 - 2x)^2\right] dx \\
&= \int_0^1 (3 - 4x + x^2)\, dx \\
&= \left(3x - 2x^2 + \frac{x^3}{3}\right)\Big]_0^1 \\
&= 3 - 2 + \frac{1}{3} = \frac{4}{3}.
\end{aligned}$$

Thus both sides of (5) are equal to 4/3, illustrating the Divergence Theorem. ||

19.3.2 The curl of a vector field Let $\boldsymbol{F} = P\boldsymbol{i} + Q\boldsymbol{j} + R\boldsymbol{k}$ be a vector field in space. The **curl of $\boldsymbol{F}$** is the *vector* defined by the symbolic determinant

$$\boldsymbol{curl\,F} = \nabla \times \boldsymbol{F} = \begin{vmatrix} \boldsymbol{i} & \boldsymbol{j} & \boldsymbol{k} \\ \dfrac{\partial}{\partial x} & \dfrac{\partial}{\partial y} & \dfrac{\partial}{\partial z} \\ P & Q & R \end{vmatrix}.$$

Note that ***curl F*** is a *vector.* Our definition of curl $\boldsymbol{F}$ as a scalar quantity in the preceding section was a temporary expedient in discussing rotation of a flow in the plane, and is explained in Example 3 below.

Example 2 If $\boldsymbol{F} = xy\boldsymbol{i} + y^2\boldsymbol{j} + yz^2\boldsymbol{k}$, then

$$\boldsymbol{curl\,F} = \nabla \times \boldsymbol{F} = \begin{vmatrix} \boldsymbol{i} & \boldsymbol{j} & \boldsymbol{k} \\ \dfrac{\partial}{\partial x} & \dfrac{\partial}{\partial y} & \dfrac{\partial}{\partial z} \\ xy & y^2 & yz^2 \end{vmatrix} = z^2\boldsymbol{i} + 0\boldsymbol{j} - x\boldsymbol{k}$$

$$= z^2\boldsymbol{i} - x\boldsymbol{k}. \quad \|$$

Example 3 If $R(x, y, z) = 0$ so $\boldsymbol{F} = P\boldsymbol{i} + Q\boldsymbol{j}$, then

$$\boldsymbol{curl\,F} = \nabla \times \boldsymbol{F} = \begin{vmatrix} \boldsymbol{i} & \boldsymbol{j} & \boldsymbol{k} \\ \dfrac{\partial}{\partial x} & \dfrac{\partial}{\partial y} & \dfrac{\partial}{\partial z} \\ P & Q & 0 \end{vmatrix} = -\frac{\partial Q}{\partial z}\boldsymbol{i} + \frac{\partial P}{\partial z}\boldsymbol{j} + \left(\frac{\partial Q}{\partial x} - \frac{\partial P}{\partial y}\right)\boldsymbol{k}.$$

You see that what we called curl $\boldsymbol{F}$ in Section 19.2 for $\boldsymbol{F} = P\boldsymbol{i} + Q\boldsymbol{j}$ was really the $\boldsymbol{k}$-component of ***curl F***, viewed as a vector in space. ‖

19.3.3 Stokes' theorem

We make no attempt to prove Stokes' Theorem, but shall, roughly, state it. It is the generalization of Green's Theorem to a two-dimensional surface G in space. For example, G might be the surface shown in Fig. 19.12. (The theorem is not true for some "one-sided" surfaces, which we shall not describe.) Figure 19.12 also shows a unit normal vector $\boldsymbol{n}$ at a point on G, and indicates by arrows a direction around ∂G. The directions for $\boldsymbol{n}$ and around ∂G are always to be related so that as the fingers of the right hand curve in the direction given by the direction around ∂G, the thumb points in the direction of $\boldsymbol{n}$ (a righthand rule). This is illustrated in Fig. 19.12. To tie this in with G and ∂G for Green's Theorem in the plane, note that this means that when walking along ∂G with your head in the direction of $\boldsymbol{n}$, you have the region G on your left.

Theorem 19.6 *(Stokes' Theorem) For a suitable surface G in space and a continuously differentiable vector field $\boldsymbol{F} = P\boldsymbol{i} + Q\boldsymbol{j} + R\boldsymbol{k}$, you have*

$$\oint_{\partial G} (\boldsymbol{F} \cdot \boldsymbol{t})\, ds = \iint_G [(\nabla \times \boldsymbol{F}) \cdot \boldsymbol{n}]\, dS. \tag{4}$$

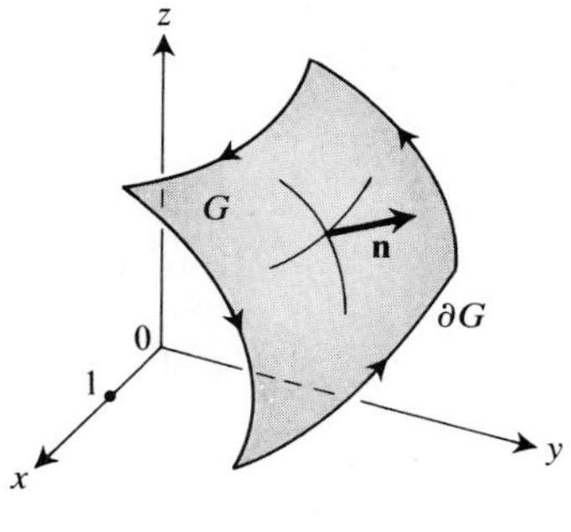

19.12

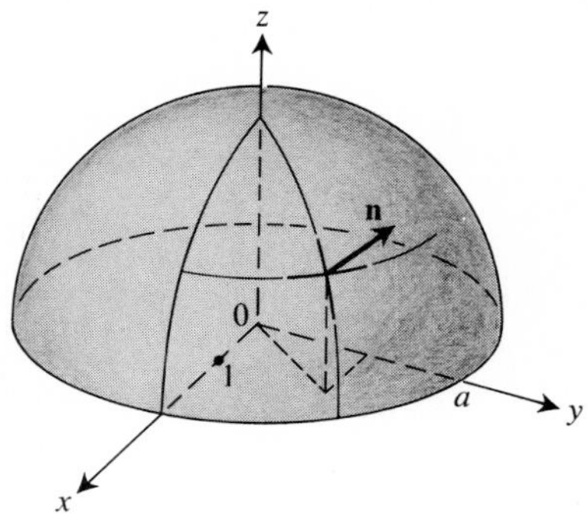

19.13

In words, the integral of the tangential component of the vector field around ∂G is equal to the integral of the normal component of ***curl F*** over the surface G. If $\boldsymbol{F}$ is the flux field of a flow, this has the interpretation that the circulation around ∂G equals the integral of the normal component of the curl over G. The curl measures the rotation of the flow at each point, and the normal component measures the portion of this rotation that acts *tangent* to the surface. Remember that the direction of a *tangent* plane to a surface is specified by a *normal* direction to the surface.

Example 4 If G lies in the x,y-plane, then $\boldsymbol{n} = \boldsymbol{k}$ and $dS = dx\,dy$. If $R(x, y, z) = 0$ so $\boldsymbol{F} = P\boldsymbol{i} + Q\boldsymbol{j}$, then as shown in Example 3,

$$\nabla \times \boldsymbol{F} = -\frac{\partial Q}{\partial z}\boldsymbol{i} + \frac{\partial P}{\partial z}\boldsymbol{j} + \left(\frac{\partial Q}{\partial x} - \frac{\partial P}{\partial y}\right)\boldsymbol{k}.$$

Equation (4) then becomes

$$\oint_{\partial G} (\boldsymbol{F} \cdot \boldsymbol{t})\, ds = \iint_G \left(\frac{\partial Q}{\partial x} - \frac{\partial P}{\partial y}\right) dx\,dy.$$

This is Green's Theorem, which is thus a special case of Stokes' Theorem. ||

Example 5 Let us illustrate Stokes' Theorem if the surface G is the hemisphere $z = \sqrt{a^2 - x^2 - y^2}$, shown in Fig. 19.13, and

$$\boldsymbol{F} = yz\boldsymbol{i} - xz\boldsymbol{j} + 3\boldsymbol{k}.$$

SOLUTION Now

$$\boldsymbol{n} = \frac{1}{a}(x\boldsymbol{i} + y\boldsymbol{j} + z\boldsymbol{k})$$

and Fig. 19.14 shows that we could use spherical coordinates and take

$$dS = (a \sin\phi\, d\theta)(a\, d\phi) = a^2 \sin\phi\, d\phi\, d\theta.$$

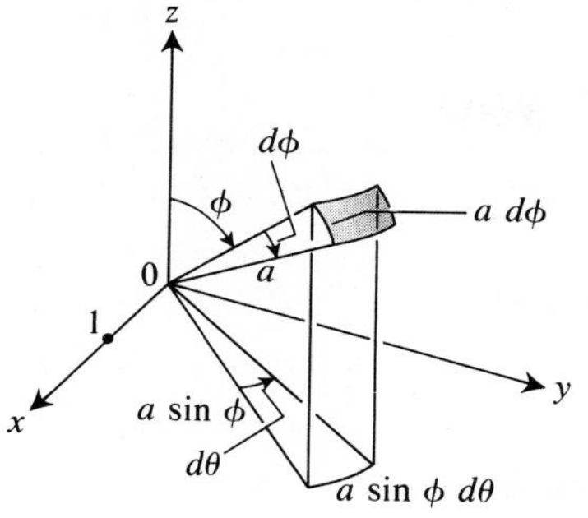

19.14

Also

$$\nabla \times \boldsymbol{F} = \begin{vmatrix} \boldsymbol{i} & \boldsymbol{j} & \boldsymbol{k} \\ \dfrac{\partial}{\partial x} & \dfrac{\partial}{\partial y} & \dfrac{\partial}{\partial z} \\ yz & -xz & 3 \end{vmatrix} = x\boldsymbol{i} + y\boldsymbol{j} - 2z\boldsymbol{k}.$$

Consequently,

$$\begin{aligned}(\nabla \times \boldsymbol{F}) \cdot \boldsymbol{n} &= \frac{1}{a}(x^2 + y^2 - 2z^2) = \frac{1}{a}(x^2 + y^2 + z^2 - 3z^2) \\ &= \frac{1}{a}(a^2 - 3z^2) = \frac{1}{a}(a^2 - 3a^2 \cos^2 \phi) \\ &= a(1 - 3\cos^2 \phi).\end{aligned}$$

Then

$$\begin{aligned}\iint_G [(\nabla \times \boldsymbol{F}) \cdot \boldsymbol{n}]\, dS &= \int_0^{2\pi} \int_0^{\pi/2} a(1 - 3\cos^2 \phi) a^2 \sin \phi \, d\phi \, d\theta \\ &= a^3 \int_0^{2\pi} (-\cos \phi + \cos^3 \phi) \Big]_0^{\pi/2} d\theta \\ &= a^3 \int_0^{2\pi} 0 \, d\theta = 0.\end{aligned}$$

Turning to $\oint_{\partial G} (\boldsymbol{F} \cdot \boldsymbol{t})\, ds$, we parametrize ∂G by $x = a \cos t$ and $y = a \sin t$ for $0 \leq t \leq 2\pi$. Then

$$\boldsymbol{t} = \frac{dx}{ds}\boldsymbol{i} + \frac{dy}{ds}\boldsymbol{j} + 0\boldsymbol{k}.$$

Since ∂G lies in the plane $z = 0$, you see that there $\boldsymbol{F} = 0\boldsymbol{i} + 0\boldsymbol{j} + 3\boldsymbol{k}$. Therefore $\boldsymbol{F} \cdot \boldsymbol{t} = 0$, so $\oint_{\partial G} (\boldsymbol{F} \cdot \boldsymbol{t})\, ds = 0$ also, illustrating Stokes' Theorem. ‖

SUMMARY *Let $\boldsymbol{F} = P\boldsymbol{i} + Q\boldsymbol{j} + R\boldsymbol{k}$ be a continuously differentiable vector field.*

1. ***Curl F*** $= \nabla \times \boldsymbol{F}$.
2. *(Divergence Theorem) If G is a suitable three-dimensional region with boundary surface ∂G and* $\boldsymbol{n}$ *is a unit outward normal vector, then*

$$\iint_{\partial G} (\boldsymbol{F} \cdot \boldsymbol{n})\, dS = \iiint_G (\nabla \cdot \boldsymbol{F})\, dx\, dy\, dz.$$

3. *(Stokes' Theorem) If G is a suitable surface in space, and if the directions of the unit normal vector* $\boldsymbol{n}$ *and the boundary ∂G are related by the righthand rule, then*

$$\oint_{\partial G} (\boldsymbol{F} \cdot \boldsymbol{t})\, ds = \iint_G [(\nabla \times \boldsymbol{F}) \cdot \boldsymbol{n}]\, dS.$$

EXERCISES

In Exercises 1 through 3, find ***curl F***.

1. $\boldsymbol{F} = x^2\boldsymbol{i} + y^2\boldsymbol{j} + z^2\boldsymbol{k}$ **2.** $\boldsymbol{F} = xyz\boldsymbol{i} - 2y^2z\boldsymbol{j} + 3z^4\boldsymbol{k}$ **3.** $\boldsymbol{F} = (\sin xy)\boldsymbol{i} + xe^y\boldsymbol{j} + (\ln yz)\boldsymbol{k}$

4. Illustrate the Divergence Theorem for the ball G where $x^2 + y^2 + z^2 \le a^2$ and $\boldsymbol{F} = x\boldsymbol{i} + y\boldsymbol{j} + z\boldsymbol{k}$.

5. Illustrate Stokes' Theorem if G is the portion of the sphere $x^2 + y^2 + z^2 = 5$ on or above the plane $z = 1$, and if $\boldsymbol{F} = 2y\boldsymbol{i} + 3x\boldsymbol{j} + xz\boldsymbol{k}$.

6. Illustrate Stokes' Theorem if G is the triangular region consisting of the portion of the plane $2x + y + 2z = 2$. in the first octant, and $\boldsymbol{F} = yz\boldsymbol{i} + y^2\boldsymbol{j} + xy\boldsymbol{k}$.

In Exercises 7 through 9, use Stokes' Theorem or the Divergence Theorem to make an easier computation of the indicated surface integral.

7. $\iint_G [(\nabla \times \boldsymbol{F}) \cdot \boldsymbol{n}]\, dS$ where G is the cylindrical surface $x^2 + y^2 = a^2$ between $z = 0$ and $z = b$, $\boldsymbol{n}$ is the outward normal, and $\boldsymbol{F} = xz\boldsymbol{i} + (4 + z^2)\boldsymbol{j} + xe^{yz}\boldsymbol{k}$.

8. $\iint_G (\boldsymbol{F} \cdot \boldsymbol{n})\, dS$ where $\boldsymbol{F} = 4x\boldsymbol{i} + 5y\boldsymbol{j} - 3z\boldsymbol{k}$, G is the sphere $x^2 + y^2 + z^2 = a^2$, and $\boldsymbol{n}$ is the outward normal.

9. $\iint_G [(\nabla \times \boldsymbol{F}) \cdot \boldsymbol{n}]\, dS$ where G is the portion of the sphere $x^2 + y^2 + z^2 = 16$ *below* the plane $z = 2$, $\boldsymbol{n}$ is the outward normal, and $\boldsymbol{F} = yz\boldsymbol{i} + 3xz\boldsymbol{j} + z^2\boldsymbol{k}$.

10. Let G be a three-dimensional region where the Divergence Theorem applies. Use the theorem to show that

$$\text{Volume of } G = \frac{1}{3}\iint_{\partial G} [(x\boldsymbol{i} + y\boldsymbol{j} + z\boldsymbol{k}) \cdot \boldsymbol{n}]\, dS.$$

11. Let $\boldsymbol{F}$ be a continuously differentiable vector field on a surface G in space.

a) Show that if $\boldsymbol{F}$ is normal to the unit tangent vector $\boldsymbol{t}$ at each point on ∂G, then $\iint_G [(\textbf{curl}\, \boldsymbol{F}) \cdot \boldsymbol{n}]\, dS = 0$.

b) Show that if the vector **curl F** is tangent to G at each point of G, then $\oint_{\partial G} (\boldsymbol{F} \cdot \boldsymbol{t})\, ds = 0$.

12. Use the Divergence Theorem for the region $a^2 \le x^2 + y^2 + z^2 \le b^2$ to show that the flux of the gradient field $\boldsymbol{F}$ of the Newtonian potential function $(x^2 + y^2 + z^2)^{-1/2}$ across a sphere with center at the origin is independent of the radius of the sphere.

13. A vector field $\boldsymbol{F}$ in a region of space is *irrotational* if $\textbf{\textit{curl F}} = \mathbf{0}$ thoughout the region. Show that the circulation of an irrotational vector field about every piecewise-smooth closed path bounding a suitable surface G lying within the region is zero.

14. A vector field $\boldsymbol{F}$ on a region of space is *incompressible* if $\nabla \cdot \boldsymbol{F} = 0$ throughout the region. Show that the flux of an incompressible field $\boldsymbol{F}$ across the boundary of each ball inside the region is zero.

15. a) Show that $\nabla \cdot (\textbf{\textit{curl F}}) = 0$ for a twice continuously differentiable vector field in space.

b) Show that $\nabla \times (\nabla f) = \mathbf{0}$ for a twice continuously differentiable function of three variables.

16. Often $\nabla \cdot (\nabla f)$ is written as $\nabla^2 f$. Show that for a suitable region G in space, we have

$$\iiint_G (\nabla^2 f)\, dx\, dy\, dz = \iint_{\partial G} (\nabla f \cdot \boldsymbol{n})\, dS$$

for a twice continuously differentiable function f of three variables.

17. Argue as best you can from Stokes' Theorem that if G is a surface that is the entire boundary of a suitable three-dimensional region, then

$$\iint_G [(\nabla \times \boldsymbol{F}) \cdot \boldsymbol{n}]\, dS = 0$$

for every continuously differentiable vector field $\boldsymbol{F}$.

18. Let $\boldsymbol{F}$ be a twice continuously differentiable vector field and let G be a three-dimensional region in space where the Divergence Theorem can be applied. Show that

$$\iint_{\partial G} [(\textbf{\textit{curl F}}) \cdot \boldsymbol{n}]\, dS = \iiint_G \nabla \cdot (\textbf{\textit{curl F}})\, dx\, dy\, dz.$$

19. Use the results of the last two exercises to show that for $\boldsymbol{F}$ and G as described in Exercise 18,

$$\iiint_G \nabla \cdot (\textbf{\textit{curl F}})\, dx\, dy\, dz = 0.$$

20. Argue as best you can from the preceding exercise, with no computation, that $\nabla \cdot (\nabla \times \boldsymbol{F}) = 0$ for every twice continuously differentiable vector field $\boldsymbol{F}$. You were asked to show this by computation in Exercise 15a.

exercise sets for chapter 19

review exercise set 19.1

1. State Green's Theorem.

2. Use Green's Theorem to compute

$$\int_\gamma [xy\, dx + (x^3 + y^3)\, dy]$$

where γ is the boundary of the region G bounded by $y = x^2$ and $y = 4$.

3. State the vector form of the Divergence Theorem, and explain its interpretation for the flux vector $\boldsymbol{F}$ of a flow.

4. Let $\boldsymbol{F}$ be the vector field

$$\boldsymbol{F} = x^3\boldsymbol{i} + y^3\boldsymbol{j} + z^2\boldsymbol{k}.$$

Use the Divergence Theorem to compute $\iint_{\partial G} (\boldsymbol{F} \cdot \boldsymbol{n})\, dS$ where G is the region bounded by $z = x^2 + y^2$ and $z = 4$.

5. Let $\boldsymbol{F} = P\boldsymbol{i} + Q\boldsymbol{j}$ be the continuously differentiable flux vector for a flow in the plane. Give the conditions for $\boldsymbol{F}$ to be

a) irrotational, b) incompressible.

6. Let $\boldsymbol{F} = xyz^2\boldsymbol{i} + 2yz\boldsymbol{j} - x^3\boldsymbol{k}$.

a) Find $\textbf{\textit{curl F}}$. b) Find the divergence of $\boldsymbol{F}$.

7. Let $\boldsymbol{F} = 4\boldsymbol{i} + xz\boldsymbol{j} - xy\boldsymbol{k}$. Use Stokes' Theorem to find $\iint_G [(\textbf{\textit{curl F}}) \cdot \boldsymbol{n}]\, dS$, where G is the surface $x = y^2 + z^2$ for $0 \le x \le 9$, while vectors $\boldsymbol{n}$ are chosen so that their $\boldsymbol{i}$-components are positive.

review exercise set 19.2

1. State a vector form of Green's Theorem.

2. Use Green's Theorem to find the circulation of the flow with flux vector

$$\boldsymbol{F} = x\boldsymbol{i} + 3xy\boldsymbol{j}$$

about the boundary of the plane region bounded by

$$x = y^2 \quad \text{and} \quad y = x - 2.$$

3. Let a ball of mass density 1 be bounded by the spherical surface with equation

$$x^2 + y^2 + z^2 = a^2.$$

Show that the moment of inertia of the ball about the z-axis is equal to

$$\iint_G \left[\left(\frac{x^3}{3}\boldsymbol{i} + \frac{y^3}{3}\boldsymbol{j} + 0\boldsymbol{k}\right) \cdot \boldsymbol{n}\right] dS,$$

where $\boldsymbol{n}$ is the outward normal to the sphere.

4. a) Give the condition for a force field $\boldsymbol{F} = P\boldsymbol{i} + Q\boldsymbol{j}$ in the plane to be conservative.

b) Show the force field $\boldsymbol{F} = y^2\boldsymbol{i} + 2xy\boldsymbol{j}$ is conservative, and find a potential function u.

c) Find the work done by the force field in (b) in moving a body from $(-1, 2)$ to $(3, -1)$.

5. Find $\textbf{curl}\,\boldsymbol{F}$ if $F = x\boldsymbol{i} + yz\boldsymbol{j} + xyz\boldsymbol{k}$.

6. Let G be the surface $z = 4 + \sqrt{9 - x^2 - y^2}$ with normal vectors $\boldsymbol{n}$ having positive $\boldsymbol{k}$-component. Let G' be the portion of the sphere $x^2 + y^2 + z^2 = 25$, where $z \leq 4$, with outward normal vectors $\boldsymbol{n}$. Let $\boldsymbol{F}$ be a vector field. What relation holds between

$$\iint_G [(\textbf{curl}\,\boldsymbol{F}) \cdot \boldsymbol{n}]\, dS$$

and

$$\iint_{G'} [(\textbf{curl}\,\boldsymbol{F}) \cdot \boldsymbol{n}]\, dS?$$

Why?

7. Use Stokes' Theorem to find $\iint_G [(\textbf{curl}\,\boldsymbol{F}) \cdot \boldsymbol{n}]\, dS$ where $\boldsymbol{F} = xy\boldsymbol{i} + y\boldsymbol{j} + yz\boldsymbol{k}$ and G is the surface of the cubical box with no top and having diagonally opposite vertices at $(0, 0, 0)$ and $(1, 1, 1)$. Let $\boldsymbol{n}$ be the outward normal.

more challenging exercises 19

1. Let G be a suitable surface in space and let $P(x, y, z)$, $Q(x, y, z)$, and $R(x, y, z)$ be continuously differentiable functions. Stokes' Theorem is often stated in the form

$$\int_{\partial G} (P\,dx + Q\,dy + R\,dz)$$

$$= \iint_G \left[\left(\frac{\partial R}{\partial y} - \frac{\partial Q}{\partial z}\right) dy\,dz + \left(\frac{\partial P}{\partial z} - \frac{\partial R}{\partial x}\right) dz\,dx + \left(\frac{\partial Q}{\partial x} - \frac{\partial P}{\partial y}\right) dx\,dy\right].$$

Convince yourself that this is equivalent to the statement

$$\int_{\partial G} (\boldsymbol{F} \cdot \boldsymbol{t})\, ds = \iint_G [(\boldsymbol{\nabla} \times \boldsymbol{F}) \cdot \boldsymbol{n}]\, dS$$

for $\boldsymbol{F} = P\boldsymbol{i} + Q\boldsymbol{j} + R\boldsymbol{k}$.

The remaining exercises use the notation of the $\wedge$-product introduced in the last exercise set of Chapter 18. *Review that material before proceeding.*

The table below indicates what is meant by a differential form ω, *the* order *of the form, and its* exterior derivative $d\omega$. *Here P. Q, and R are differentiable functions of two or three variables (we give the table for three variables), and for* $P(x, y, z)$,

$$dP = \frac{\partial P}{\partial x}dx + \frac{\partial P}{\partial y}dy + \frac{\partial P}{\partial z}dz$$

as usual.

Differential forms ω

Order of ω	ω	$d\omega$
0	$P(x, y, z)$	$dP = \frac{\partial P}{\partial x}dx + \frac{\partial P}{\partial y}dy + \frac{\partial P}{\partial z}dz$
1	$P\,dx + Q\,dy + R\,dz$	$dP \wedge dx + dQ \wedge dy + dR \wedge dz$
2	$P\,dy \wedge dz + Q\,dz \wedge dx + R\,dx \wedge dy$	$dP \wedge dy \wedge dz + dQ \wedge dz \wedge dx + dR \wedge dx \wedge dy$
3	$P\,dx \wedge dy \wedge dz$	$dP \wedge dx \wedge dy \wedge dz$

2. Show that if $\omega = P\,dx \wedge dy \wedge dz$, then $d\omega = 0$.

3. Note that if ω has order r, then $d\omega$ has order $r + 1$. Compute $d\omega$, simplifying as much as possible.

a) $\omega = xy\,dx - x^2\,dy$

b) $\omega = xz\,dx \wedge dy - yz^3\,dx \wedge dz$

All the main theorems in this chapter can be expressed in the following form, as the remaining exercises ask you to show.

Generalized Stokes' Theorem *If ω is a differential form of order r with continuously differentiable coefficients P, Q, etc., and if G is a suitable (r + 1)-dimensional region, then*

$$\int_{\partial G} \omega = \int_G d\omega. \tag{1}$$

In (1), a single integral sign is used for each integral, rather than using double or triple integral signs as we have in the past. The integrals of differential forms are defined so that

$$\int_G P(x, y)\,dx \wedge dy = \iint_G P(x, y)\,dx\,dy = -\int_G P(x, y)\,dy \wedge dx,$$

$$\int_G P(x, y, z)\,dx \wedge dy \wedge dz = \iiint_G P(x, y, z)\,dx\,dy\,dz = -\int_G P(x, y, z)\,dz \wedge dy \wedge dx,$$

etc.

4. If G is a one-dimensional region consisting of a curve joining point A to point B, then ∂G is the symbolic expression $B - A$. The integral of a function f (a form of order 0) over ∂G is defined to be

$f(B) - f(A)$. Show that if $\omega = f(x)$ and G is the line segment $[a, b]$ from a to b, then (1) reduces to the fundamental theorem of calculus.

5. State (1) for the special case of $\omega = f(x, y, z)$ and G a curve γ joining $A = (a_1, a_2, a_3)$ to $B = (b_1, b_2, b_3)$ in space. (See the preceding exercise.)

6. State (1) for $\omega = P(x, y)\,dx + Q(x, y)\,dy$ in the plane, and G a plane region. Simplify the statement as much as possible. What familiar theorem do you obtain?

7. State (1) for $\omega = P\,dx + Q\,dy + R\,dz$ and G a surface in space. Simplify the statement as much as possible. What theorem do you obtain?

8. a) State (1) for $\omega = P\,dy \wedge dz + Q\,dz \wedge dx + R\,dx \wedge dy$, simplifying as much as possible.

b) Convince yourself that the statement obtained in (a) is equivalent to the divergence theorem.

differential equations

20

20.1 INTRODUCTION

20.1.1 The notion of a differential equation

A *differential equation* is an equation that involves derivatives (or differentials) of an "unknown function" f. To solve the differential equation is to find all possible unknown functions f for which the equation is true. You first encountered differential equations in Chapter 5, under the guise of an antidifferentiation problem. For example, the *general solution* of the equation

$$\frac{dy}{dx} = x^2$$

is

$$y = f(x) = \frac{1}{3}x^3 + C,$$

where C is an arbitrary constant, for this expression embodies *all* solutions of $dy/dx = x^2$. This equation is of *order* 1, since only derivatives of order 1 (i.e., first derivatives) appear in the equation. The *order* of an equation is r if the equation involves an rth derivative, such as d^ry/dx^r, but no derivatives of higher order. Thus the equation

$$\frac{d^2y}{dx^2} = x^2$$

is of second order. Solving the equation, we have

$$\frac{dy}{dx} = \frac{1}{3}x^3 + C_1$$

so

$$y = f(x) = \frac{1}{12}x^4 + C_1x + C_2$$

is the general solution. Without going into detail, we state that one expects the general solution of a differential equation of order n to contain n arbitrary constants.

A differential equation of order 1 need not be of the form $dy/dx = g(x)$. Recall that we considered equations such as

$$\frac{dy}{dx} = y^2x$$

in Chapter 6, and solved them by separating the variables:

$$\frac{dy}{y^2} = x\,dx, \qquad \int y^{-2}\,dy = \int x\,dx, \qquad \frac{y^{-1}}{-1} = \frac{x^2}{2} + C,$$

$$y = f(x) = \frac{-1}{(x^2/2) + C} = \frac{-2}{x^2 + 2C}.$$

We ask you to review this technique in Section 20.2.

20.1.2 Geometric interpretation of the equation $y' = F(x, y)$

Let a function F of two variables be given and consider the differential equation $y' = F(x, y)$. If $y = h(x)$ is a solution of this equation, then for each x in the domain of h, you must have

$$h'(x) = F(x, y) = F(x, h(x)).$$

Now $h'(x)$ can be interpreted geometrically as the slope of the tangent line to the graph of h at the point $(x, y) = (x, h(x))$. Thus if F is a known function, the equation $y' = F(x, y)$ allows you to compute slopes of tangents to solutions at points (x, y) in the domain of F. If you place a short line segment of slope $m = F(x, y)$ at each such point (x, y), you obtain the *direction field* of the equation $y' = F(x, y)$. By actually drawing a few of these line segments in a direction field, you may be able to obtain graphically some information about the solutions of the differential equation. We illustrate with two examples.

Example 1 Let's sketch the direction field of the differential equation $y' = -x/y$.

SOLUTION A useful device in sketching direction fields is to find all points (x, y) where the segments in the field have a particular slope c. For an equation $y' = F(x, y)$, the locus of such points is the curve

$$F(x, y) = c,$$

and for simple functions F, this curve may be easy to sketch. In our case,

$$F(x, y) = \frac{-x}{y},$$

and the equation $F(x, y) = c$ takes the form

$$\frac{-x}{y} = c, \qquad \text{or} \qquad y = \frac{-1}{c}x \qquad \text{where} \quad y \neq 0. \tag{1}$$

This is an equation of a line through the origin with the origin omitted. Thus at all points but $(0, 0)$ on the line $y = -x/c$, the segments in the direction field have slope $m = c$. For example, all segments have slope -1 on the line $y = x$; all segments have slope 1 on the line $y = -x$; all segments have slope 2 on the line $y = (-1/2)x$, etc. Putting $c = 0$ in the first equation in (1), you see that the segments at all points but $(0, 0)$ on the line $x = 0$ (that is, on the y-axis), have slope 0. We have sketched the direction field of $y' = -x/y$ in Fig. 20.1, labeling some lines through the origin and indicating the slopes m of the segments of the direction field on these lines.

From Fig. 20.1, we would guess that solutions of the equation $y' = -x/y$ have as graphs portions of certain curves about the origin; we might even guess that they are the circles

$$x^2 + y^2 = r^2,$$

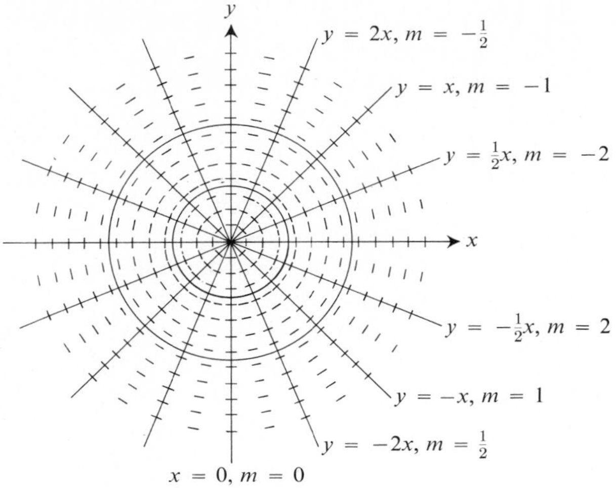

20.1

which are level curves of the function

$$G(x, y) = x^2 + y^2.$$

We can easily check our conjecture. We find that for y defined implicitly as a function of x by

$$G(x, y) = x^2 + y^2 = a,$$

we have

$$y' = \frac{dy}{dx} = -\frac{G_x(x, y)}{G_y(x, y)} = -\frac{2x}{2y} = -\frac{x}{y},$$

so our guess was correct. ||

Example 2 The direction field of the differential equation $y' = x$ is sketched in Fig. 20.2. This time, segments of equal slope $m = c$ lie on the vertical line $x = c$. Again, we have estimated a couple of graphs of solutions from the direction field; this time, the solution curves look like parabolas. Of course, we can easily solve $y' = x$, and obtain as general solution

$$y = \frac{x^2}{2} + C,$$

which is indeed a collection of parabolas. ||

20.1.2 An existence theorem

We state without proof an existence theorem for solutions of a differential equation $y' = F(x, y)$. The hypotheses of the theorem could be weakened somewhat and the conclusion would still hold; however, the statement suffices for our needs.

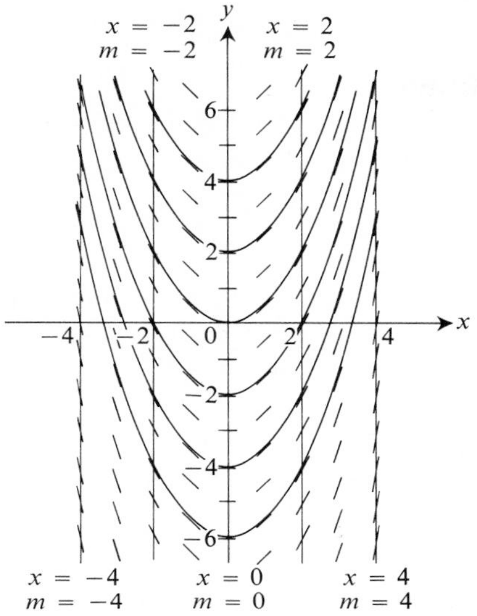

20.2

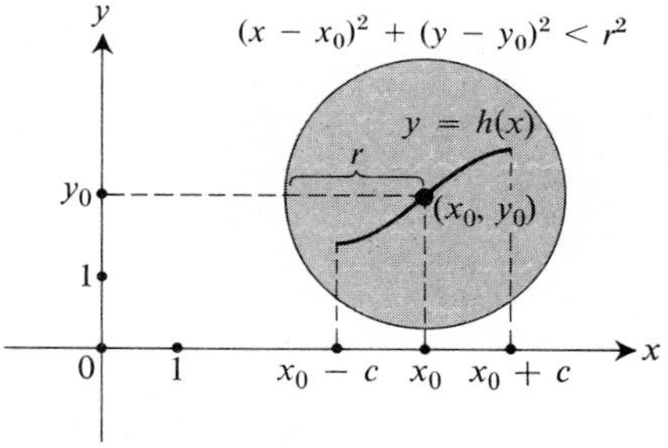

20.3

Theorem 20.1 *Let F be a continuous function of two variables with domain containing a neighborhood $(x - x_0)^2 + (y - y_0)^2 < r^2$ of (x_0, y_0) in which F_y exists and is continuous. Then there exists a number $c > 0$ and a differentiable function $h(x)$ for $x_0 - c < x < x_0 + c$ such that $y = h(x)$ is a solution of the differential equation $y' = F(x, y)$, and such that $y_0 = h(x_0)$. Furthermore, h is the unique such function with domain $x_0 - c < x < x_0 + c$.*

Theorem 20.1 essentially asserts the existence of a (unique) solution of $y' = F(x, y)$ through any point (x_0, y_0), provided that F is sufficiently well behaved in some neighborhood of (x_0, y_0). A sketch illustrating the theorem is given in Fig. 20.3, and the direction fields shown in Figs. 20.1 and 20.2 also illustrate the theorem.

In finding the solution $y = h(x)$ of a differential equation $y' = F(x, y)$ such that $y_0 = h(x_0)$, one often attempts to find the *general solution* of the differential equation. The general solution frequently can be expressed in the form

$$G(x, y) = C,$$

where C is an arbitrary constant. (Recall that you would expect the general solution of a first-order differential equation to contain an arbitrary constant.) Each individual solution then appears as a level curve of the function G, and the solution

$$y = h(x) \qquad \text{such that} \qquad h(x_0) = y_0$$

is given by the level curve of G through (x_0, y_0), namely,

$$G(x, y) = G(x_0, y_0).$$

Thus the *initial condition* $y_0 = h(x_0)$ can be used to determine the value of the arbitrary constant in the general solution to give the desired *particular solution* through (x_0, y_0).

SUMMARY

1. *A differential equation is one involving derivatives (or differentials) of an unknown function f. The order of the equation is that of the highest-order derivative that appears. The general solution is the expression, containing as many arbitrary constants as the order of the equation, which yields functions $f(x)$ satisfying the equation as the constants assume all values.*
2. *The equation $y' = F(x, y)$ can be viewed geometrically as specifying the slope at each point (x, y) of any solution of the equation that passes through that point. If a short line segment of slope $F(x, y)$ is placed at (x, y), we obtain the direction field of the equation.*
3. *Let F be a continuous function of two variables with domain containing a neighborhood of (x_0, y_0) in which F_y exists and is continuous. Then there exists a number $c > 0$ and a differentiable function $h(x)$ for $x_0 - c < x < x_0 + c$ such that $y = h(x)$ is a solution of the differential equation $y' = F(x, y)$ and such that $y_0 = h(x_0)$. Furthermore, h is the unique such function with domain $x_0 - c < x < x_0 + c$.*

EXERCISES

In Exercises 1 through 8, sketch the direction field of the differential equation, and estimate a few solution curves, as in Figs. 20.1 and 20.2.

1. $y' = y$

2. $y' = -y$

3. $y' = x^2 + y^2$

4. $y' = \dfrac{1}{x^2 + y^2}$

5. $y' = xy$

6. $y' = x + y$

7. $y' = \dfrac{x + y}{x - y}$

8. $y' = y^2$

9. (This exercise shows that the number c in our existence theorem (Theorem 20.1) may be very small compared with the radius r of the neighborhood of (x_0, y_0) in which F_y exists and is continuous.) Consider the differential equation

$$y' = 1 + y^2 = F(x, y).$$

a) For what values of r is it true that F and F_y are continuous within the neighborhood $x^2 + y^2 < r^2$ of $(0, 0)$?

b) What is the largest value of c such that there exists a differentiable function $h(x)$ for $-c < x < c$ that is a solution of $y' = 1 + y^2$ with $h(0) = 0$? [*Hint.* Solve the equation $y' = 1 + y^2$, and examine the solution through $(0, 0)$.]

10. a) Check that

$$y = h_1(x) = -x^2/4$$

and

$$y = h_2(x) = 1 - x$$

are both solutions of the differential equation

$$y' = \frac{-x + (x^2 + 4y)^{1/2}}{2}$$

for $x \geqslant 2$, and that $h_1(2) = h_2(2) = -1$.

b) Why doesn't the result in (a) contradict the uniqueness statement in Theorem 20.1?

20.2 VARIABLES SEPARABLE AND HOMOGENEOUS EQUATIONS

Recall from Chapter 6 that, if a differential equation $y' = F(x, y)$ can be expressed in the form

$$y' = \frac{dy}{dx} = \frac{f(x)}{g(y)},$$

20.2.1 Variables separable equations

then the variables can be separated and the equation solved as follows:

$$g(y)\,dy = f(x)\,dx,$$

$$\int g(y)\,dy = \int f(x)\,dx.$$

That is, one puts all terms involving y (including dy) on the lefthand side of the equation and all terms involving x (including dx) on the righthand side; the variables are then "separated." The equation is then solved by finding two indefinite integrals of a function of a single variable.

Example 1 Let's solve

$$\frac{dy}{dx} = \frac{\ln x}{xy^2}.$$

SOLUTION The equation is of variables separable type, and

$$y^2\,dy = \frac{\ln x}{x}\,dx,$$

so

$$\int y^2\,dy = \int \frac{\ln x}{x}\,dx$$

and

$$\frac{1}{3}y^3 = \frac{(\ln x)^2}{2} + C$$

is the general solution. We may also express this as

$$2y^3 = 3(\ln x)^2 + 6C,$$

or, since $6C$ may again be any arbitrary constant, it is acceptable to simply write

$$2y^3 = 3(\ln x)^2 + C.$$

Such informality with arbitrary constants is conventional. ||

20.2.2 Homogeneous equations The differential equation $y' = F(x, y)$ is *homogeneous* if $F(x, y)$ can be expressed in the form

$$F(x, y) = g\left(\frac{y}{x}\right).$$

A criterion for this to be the case is that

$$F(kx, ky) = F(x, y) \quad \text{for all } k \neq 0,$$

for, if this is true, then taking $k = 1/x$ you obtain

$$F(x, y) = F\left(\frac{1}{x}x, \frac{1}{x}y\right) = F\left(1, \frac{y}{x}\right) = g\left(\frac{y}{x}\right).$$

Conversely, if $F(x, y) = g(y/x)$, then

$$F(kx, ky) = g(ky/kx) = g(y/x) = F(x, y).$$

Such a function $F(x, y)$ is called *homogeneous of degree* 0. (If $F(kx, ky) = k^r F(x, y)$, then $F(x, y)$ is homogeneous of degree r.)

Example 2 The equation

$$\frac{dy}{dx} = F(x, y) = \frac{x^2 + y^2}{2x^2}$$

is not of variables separable type, but is homogeneous, for it can be written

$$\frac{dy}{dx} = \frac{x^2}{2x^2} + \frac{y^2}{2x^2} = \frac{1}{2} + \frac{1}{2}\left(\frac{y}{x}\right)^2.$$

Note that

$$F(kx, ky) = \frac{(kx)^2 + (ky)^2}{2(kx)^2} = \frac{k^2x^2 + k^2y^2}{2k^2x^2}$$

$$= \frac{x^2 + y^2}{2x^2} = F(x, y). \quad ||$$

If

$$\frac{dy}{dx} = g\left(\frac{y}{x}\right), \tag{1}$$

then the substitution $y = vx$ yields an equation in v and x of variables separable type. For then, from $y = vx$,

$$\frac{dy}{dx} = v \cdot 1 + x\frac{dv}{dx} \quad \text{and} \quad \frac{y}{x} = v.$$

Then the Eq. (1) becomes

$$v + x\frac{dv}{dx} = g(v) \quad \text{or} \quad x\frac{dv}{dx} = g(v) - v$$

or

$$\frac{dv}{g(v) - v} = \frac{dx}{x}, \tag{2}$$

and the variables are separated. Equation (2) can be integrated to find the solution relation in x and v, and then v replaced by y/x to find the solution in terms of x and y.

Example 3 Let us solve the equation of Example 2,

$$\frac{dy}{dx} = \frac{x^2 + y^2}{2x^2} = \frac{1}{2} + \frac{1}{2}\left(\frac{y}{x}\right)^2.$$

SOLUTION The substitution $y = vx$ yields

$$v + x\frac{dv}{dx} = \frac{1}{2} + \frac{1}{2}v^2,$$

so

$$x\frac{dv}{dx} = \frac{1 + v^2}{2} - v = \frac{v^2 - 2v + 1}{2}.$$

Then

$$\frac{dv}{v^2 - 2v + 1} = \frac{dx}{2x}, \quad \text{so} \quad (v - 1)^{-2}\, dv = \frac{1}{2} \cdot \frac{dx}{x}.$$

Hence

$$\int (v - 1)^{-2}\, dv = \frac{1}{2}\int \frac{dx}{x}, \quad \text{so} \quad \frac{(v - 1)^{-1}}{-1} = \frac{1}{2}\ln|x| + C.$$

Substituting $v = y/x$, we obtain

$$\frac{-1}{(y/x) - 1} = \frac{1}{2}\ln|x| + C$$

or

$$\frac{-2x}{y - x} = \ln|x| + C,$$

as our general solution relation. ||

20.2.3 Applications to geometry

Let G be a function of two variables with continuous partial derivatives. The relation $G(x, y) = C$ for an arbitrary constant C gives a *family of curves*, one curve for each value of C. Another family of curves $H(x, y) = K$ is **orthogonal** to the first family (or consists of **orthogonal trajectories** of the first family) if every curve of the first family is perpendicular to every curve of the second family at all points of intersection.

The problem of finding a family of curves orthogonal to a given family $G(x, y) = C$ is a problem in differential equations, and can sometimes be solved by the techniques described in this section. We illustrate with an example.

Example 4 Let's find the family of orthogonal trajectories to the curves $x^2 - y^2 = C$.

SOLUTION Let $G(x, y) = x^2 - y^2$. The slope of a curve in $x^2 - y^2 = C$ at a point (x, y) is given by

$$\frac{dy}{dx} = -\frac{\partial G/\partial x}{\partial G/\partial y} = -\frac{2x}{-2y} = \frac{x}{y}.$$

The slope of a curve in the orthogonal family at (x, y) is the negative reciprocal, $-y/x$. Thus our orthogonal family of curves consists of the solutions of the differential equation

$$\frac{dy}{dx} = -\frac{y}{x} = \frac{1/x}{-1/y}.$$

The solution of this equation is

$$\int \frac{1}{x}\,dx - \int \frac{-1}{y}\,dy = K$$

or

$$\ln|x| + \ln|y| = K. \tag{3}$$

We may write (3) in the form

$$\ln|xy| = K.$$

Then $e^{\ln|xy|} = |xy| = e^K$, so $xy = \pm e^K$. Now as K runs through all constants, $\pm e^K$ runs through all constants except 0. A special examination shows that the curves $x = 0$ and $y = 0$ are also orthogonal to the curves $x^2 - y^2 = C$; changing notation, we obtain, as orthogonal family,

$$xy = K.$$

Figure 20.4 shows some of the curves of these two orthogonal families. ||

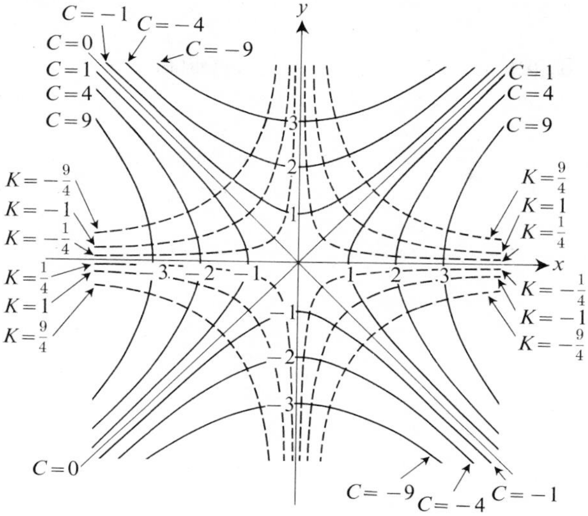

20.4

SUMMARY

1. *A differential equation $dy/dx = F(x, y)$ is of variables separable type if all the terms involving y, including dy, can be placed on the lefthand side and all the terms involving x, including dx, on the righthand side. The equation is then solved by integrating each side of this separated equation.*
2. *The equation $dy/dx = F(x, y)$ is homogeneous if $F(x, y)$ can be written in the form $g(y/x)$. This is the case if and only if $F(kx, ky) = F(x, y)$ for all $k \neq 0$.*
3. *If the substitution $y = vx$ is made in a homogeneous equation $dy/dx = g(y/x)$, so that*

$$\frac{dy}{dx} = v + x\frac{dv}{dx} \qquad \textit{and} \qquad g\left(\frac{y}{x}\right) = g(v),$$

then the new equation in x and v is of variables-separable type. Solving the new equation, and then replacing v by y/x yields the solution of the original equation.

4. *For a family of curves $G(x, y) = C$, the family of orthogonal trajectories is found as follows:*

STEP 1. *Compute*

$$\frac{dy}{dx} = -\frac{\partial G/\partial x}{\partial G/\partial y}$$

to find the slope of a curve of the first family at (x, y).

STEP 2. *Solve the differential equation*

$$\frac{dy}{dx} = \frac{\partial G/\partial y}{\partial G/\partial x};$$

the solution is the family of orthogonal trajectories.

EXERCISES

In Exercises 1 *through* 10, *find the general solution of the differential equation.*

1. $\dfrac{dy}{dx} = xy^2$ **2.** $\dfrac{dy}{dx} = e^x \tan y$

3. $\dfrac{dy}{dx} = x^2 + x^2y^2$ **4.** $\dfrac{dy}{dx} = \sin x \cos^2 y$

5. $\dfrac{dy}{dx} = x(1 - y^2)^{1/2}$ **6.** $\dfrac{dy}{dx} = y\ (\ln y) \cos^2 x$

7. $\dfrac{dy}{dx} = \dfrac{x^2 + y^2}{2xy}$ **8.** $(x + y)\, dy = (y - x)\, dx$

9. $x\, dy = (x + y)\, dx$ **10.** $x\, dy = (y + x \sin (y/x))\, dx$

In Exercises 11 *through* 16, *find the particular solution of the differential equation that satisfies the indicated initial condition.*

11. $y' = 1 + x,\ y(1) = -1$ **12.** $y' = x \sin y,\ y(2) = \dfrac{\pi}{2}$

13. $y' = \dfrac{x}{y},\ y(2) = -3$ **14.** $y' = x^2y + 2xy,\ y(2) = 1$

15. $y' = \dfrac{1 + y^2}{xy},\ y(1) = 5$ **16.** $y' = y^3e^{2x},\ y(0) = 2$

In Exercises 17 *through* 20, *find the family of orthogonal trajectories of the given family of curves. Sketch both families as in Fig.* 20.4.

17. $y - x = C$ **18.** $y - x^2 = C$ **19.** $y^2 + x = C$ **20.** $x^2 + y^2 = C$

20.3 EXACT EQUATIONS

Consider the differential equation $y' = F(x, y)$. Suppose

$$F(x, y) = \frac{F_1(x, y)}{F_2(x, y)} \tag{1}$$

20.3.1 Solution of an exact differential equation

for functions F_1 and F_2, and suppose we can find a function G of two variables with continuous derivatives such that

$$\frac{\partial G}{\partial x} = F_1(x, y) \qquad \text{and} \qquad \frac{\partial G}{\partial y} = -F_2(x, y). \tag{2}$$

Then a function $y = h(x)$ defined implicitly by a level curve of G will be a solution of $y' = F(x, y)$, for the derivative of such a function h will be

$$\frac{dy}{dx} = h'(x) = -\frac{\partial G/\partial x}{\partial G/\partial y} = -\frac{F_1(x, y)}{-F_2(x, y)} = F(x, y).$$

The general solution of $y' = F(x, y)$ will therefore be $G(x, y) = C$.

We have reduced our problem of solving $y' = F(x, y)$ to expressing $F(x, y)$ in the form (1) and solving the two *partial differential equations* (2). Of course, (1) can always be achieved; you can take

$$F_1 = F \qquad \text{and} \qquad F_2(x, y) = 1.$$

In fact, (1) can usually be achieved in a number of ways; for example, if $F(x, y) = x^2/y^2$, then

$$F(x, y) = \frac{x^2/y^2}{1} = \frac{x^2}{y^2} = \frac{x^2/y}{y} = \frac{x}{y^2/x},$$

so there are many possibilities for F_1 and F_2. The assertion that the partial differential equations (2) have a simultaneous solution is precisely the assertion that

$$F_1(x, y)\, dx - F_2(x, y)\, dy \tag{3}$$

is an exact differential. Note that (3) can be obtained from

$$\frac{dy}{dx} = \frac{F_1(x, y)}{F_2(x, y)} \tag{4}$$

by formal algebraic manipulation; namely, we obtain from (4)

$$F_2(x, y)\, dy = F_1(x, y)\, dx$$

or

$$F_1(x, y)\, dx - F_2(x, y)\, dy = 0. \tag{5}$$

Once more, Leibniz notation works beautifully. We shall view (5) as an equivalent form of Eq. (1). We now replace "F_1" by "M" and "F_2" by "$-N$" so that our discussion from here on will use the notation classically employed in treating exact differential equations.

Definition 20.1 A differential equation

$$M(x, y)\,dx + N(x, y)\,dy = 0 \tag{6}$$

is **exact** if $M(x, y)\,dx + N(x, y)\,dy$ is an exact differential.

Our work in Chapter 17 shows that if M and N have continuous partial derivatives in a suitable region of the plane, then the differential equation (6) is exact if and only if

$$\frac{\partial M}{\partial y} = \frac{\partial N}{\partial x}. \tag{7}$$

If (7) is satisfied and if G is a function of two variables such that

$$dG = M(x, y)\,dx + N(x, y)\,dy,$$

then the differential equation (6) has as general solution

$$G(x, y) = C. \tag{8}$$

The technique used to solve an exact differential equation $M\,dx + N\,dy = 0$ is, of course, precisely the technique employed to find a function $G(x, y)$ such that $dG = M(x, y)\,dx + N(x, y)\,dy$, described in Section 17.3 of Chapter 17. We illustrate with an example.

Example 1 Let us solve $3x^2y\,dx + (x^3 - y^2)\,dy = 0$.

SOLUTION The differential equation

$$M\,dx + N\,dy = 3x^2y\,dx + (x^3 - y^2)\,dy = 0$$

is exact since

$$\frac{\partial M}{\partial y} = \frac{\partial(3x^2y)}{\partial y} = 3x^2 \quad \text{and} \quad \frac{\partial N}{\partial x} = \frac{\partial(x^3 - y^2)}{\partial x} = 3x^2.$$

Setting $\partial G/\partial x = 3x^2y$, you find that

$$G(x, y) = x^3y + h(y). \tag{9}$$

Then

$$\frac{\partial G}{\partial y} = x^3 + h'(y) = x^3 - y^2.$$

Consequently,

$$h'(y) = -y^2 \quad \text{and} \quad h(y) = \frac{-y^3}{3}.$$

This shows that

$$G(x, y) = x^3y - \frac{y^3}{3} \tag{10}$$

has differential $3x^2y\,dx + (x^3 - y)\,dy$, so the general solution of our equation is

$$x^3y - \frac{y^3}{3} = C. \quad \|$$

20.3.2 Integrating factors

Let a differential equation

$$y' = F(x, y) = \frac{-M(x, y)}{N(x, y)}$$

be written in the form

$$M(x, y)\,dx + N(x, y)\,dy = 0, \tag{11}$$

and let (11) have general solution

$$G(x, y) = C. \tag{12}$$

From (12) you see that, at any point on a solution curve $y = h(x)$, we have

$$y' = -\frac{\partial G/\partial x}{\partial G/\partial y}.$$

Thus we must have

$$\frac{\partial G/\partial x}{\partial G/\partial y} = \frac{M(x, y)}{N(x, y)}. \tag{13}$$

From (13), we obtain

$$\frac{\partial G/\partial x}{M(x, y)} = \frac{\partial G/\partial y}{N(x, y)} = \mu(x, y), \tag{14}$$

where we let $\mu(x, y)$ be the common ratio in (14). From (14), we obtain

$$\frac{\partial G}{\partial x} = \mu(x, y)M(x, y) \quad \text{and} \quad \frac{\partial G}{\partial y} = \mu(x, y)N(x, y). \tag{15}$$

The Eqs. (15) show that the equation

$$\mu(x, y)M(x, y)\,dx + \mu(x, y)N(x, y)\,dy = 0, \tag{16}$$

obtained by multiplying (11) by $\mu(x, y)$, is exact.

Definition 20.2 A function $\mu(x, y)$ is an **integrating factor** for a differential equation $M(x, y)\,dx + N(x, y)\,dy = 0$ if

$$\mu(x, y)M(x, y)\,dx + \mu(x, y)N(x, y)\,dy = 0$$

is an exact differential equation.

Our work prior to this definition shows that if the differential equation (11) has a general solution $G(x, y) = C$, then the equation has an integrating factor $\mu(x, y)$.

Integrating factors are by no means unique as the following example shows.

Example 2 The differential equation

$$3xy\,dx + x^2\,dy = 0$$

has x as an integrating factor, for

$$\begin{aligned} x(3xy\,dx + x^2\,dy) &= 3x^2y\,dx + x^3\,dy \\ &= d(x^3y). \end{aligned}$$

Another integrating factor is $1/x^2y$, for

$$\frac{1}{x^2y}(3xy\,dx + x^2\,dy) = \frac{3}{x}\,dx + \frac{1}{y}\,dy = d(3\ln|x| + \ln|y|). \quad \|$$

Exercise 15 indicates that, in general, the Eq. (11) can be expected to have an infinite number of integrating factors.

In Exercises 16 and 17, we give a partial differential equation involving $M(x, y)$ and $N(x, y)$ which a function μ must satisfy to be an integrating factor of (11). From this differential equation, one can characterize the differential equations (11) that have certain types of integrating factors, such as factors that are functions of just x or of just y. Some results and illustrations along these lines are given in Exercises 18 through 20. In the article that follows, we indicate how one may sometimes find an integrating factor by "inspection."

20.3.3 Finding integrating factors by inspection

Integrating factors for certain differential equations of the form (11) can be found by inspection; facility requires a certain amount of practice. One useful technique is to watch for expressions such as $y\,dx + x\,dy$ that are themselves exact or have obvious integrating factors; note that

$$d(xy) = y\,dx + x\,dy.$$

Thus the presence of $y\,dx + x\,dy$ suggests that a function of xy might be an integrating factor. Similarly, the differentials

$$d\left(\frac{x}{y}\right) = \frac{1}{y^2}(y\,dx - x\,dy) \qquad \text{and} \qquad d\left(\frac{y}{x}\right) = \frac{1}{x^2}(x\,dy - y\,dx)$$

suggest watching for the expressions $y\,dx - x\,dy$ and $x\,dy - y\,dx$, whose presence would lead us to try integrating factors of the form

$$\frac{1}{y^2} f(x/y) \qquad \text{and} \qquad \frac{1}{x^2} g(y/x),$$

respectively. We illustrate with some examples.

Example 3 Let's solve the differential equation

$$y\,dx + (x + x^2 y)\,dy = 0.$$

SOLUTION The expression $y\,dx + x\,dy$ contained in this equation suggests an integrating factor that is a function of xy. If we divide through the equation by $(xy)^2$, then the term $x^2 y\,dy$ becomes $(1/y)\,dy$, which can be integrated. Thus we take as integrating factor $1/(xy)^2$, and obtain the equation

$$\frac{y\,dx + x\,dy}{(xy)^2} + \frac{x^2 y}{(xy)^2}\,dy = 0$$

or

$$(xy)^{-2}\,d(xy) + \frac{1}{y}\,dy = 0.$$

Integration of the last equation yields

$$-\frac{1}{xy} + \ln|y| = C$$

as general solution. ‖

Example 4 Let's solve the differential equation

$$(xy^4 + y)\,dx - x\,dy = 0.$$

SOLUTION The presence of $y\,dx - x\,dy$ suggests an integrating factor of the form $(1/y^2)f(x/y)$. To "eliminate" the y^4 from $xy^4\,dx$ for integration, we use $(1/y^2)(x/y)^2$ as integrating factor, and obtain

$$\frac{1}{y^2}\left(\frac{x}{y}\right)^2 xy^4\,dx + \frac{1}{y^2}\left(\frac{x}{y}\right)^2 (y\,dx - x\,dy) = 0$$

or

$$x^3\,dx + \left(\frac{x}{y}\right)^2 d\left(\frac{x}{y}\right) = 0.$$

Integrating this last equation, we obtain

$$\frac{x^4}{4} + \frac{1}{3}\left(\frac{x}{y}\right)^3 = C$$

as general solution. ‖

SUMMARY

1. *The differential equation* $M(x, y)\,dx + N(x, y)\,dy = 0$ *is exact if and only if* $\partial M/\partial y = \partial N/\partial x$, *and* $M(x, y)$ *and* $N(x, y)$ *satisfy suitable conditions. In this case, find* $G(x, y)$ *such that*

$$dG = M(x, y)\,dx + N(x, y)\,dy$$

as described in Section 17.3 *of Chapter* 17, *and the solution of the given equation is* $G(x, y) = C$.

2. *If* $M(x, y)\,dx + N(x, y)\,dy = 0$ *has a solution* $G(x, y) = C$, *then there exists a function (integrating factor)* $\mu(x, y)$ *such that*

$$\mu(x, y)M(x, y)\,dx + \mu(x, y)N(x, y)\,dy = 0$$

is an exact equation, and can be solved as in (1) *above. Integrating factors can sometimes be found by inspection; attempt to create obvious exact differentials such as*

$$d(xy) = x\,dy + y\,dx,$$

$$d\left(\frac{x}{y}\right) = \frac{1}{y^2}(y\,dx - x\,dx),$$

$$d\left(\frac{y}{x}\right) = \frac{1}{x^2}(x\,dy - y\,dx).$$

See Examples 3 *and* 4 *in the text for illustration.*

EXERCISES

In Exercises 1 *through* 6, *verify that the equation is exact and find the general solution.*

1. $\cos y\,dx + (1 - x\sin y)\,dy = 0$

2. $2xy\,dx + (x^2 - e^{-y})\,dy = 0$

3. $y^2\,dx + \left(\frac{1}{y} + 2xy\right)dy = 0$

4. $(y\sec^2 xy)\,dx + (1 + x\sec^2 xy)\,dy = 0$

5. $(e^y - y\cos xy)\,dx + (xe^y - x\cos xy)\,dy = 0$

6. $(2xy^3 - 3)\,dx + (3x^2y^2 + 4y)\,dy = 0$

In Exercises 7 *through* 14, *find an integrating factor by inspection, and solve the differential equation.*

7. $(xy^2 + y)\,dx + (x - 3x^2)\,dy = 0$

8. $(x^2y^2 + y^2)\,dx + (2xy + x)\,dy = 0$

9. $(xy^2 + y)\,dx - (x + y)\,dy = 0$

10. $(4 - y)\,dx + (x + 3x^2)\,dy = 0$

11. $(4 + y)\,dx - (x + 3x^2)\,dy = 0$

12. $\left(\frac{x^2}{y} + y\right)dx + (2x - e^y)\,dy = 0$

13. $(1 + 2x^2y^3)\,dx + (xy^2 + 3x^3y^2)\,dy = 0$

14. $(x^2y^2 + 2xy)\,dx - (x^2 + 3)\,dy = 0$

15. Let $M(x, y)\,dx + N(x, y)\,dy = 0$ have general solution $G(x, y) = C$ and let μ be an integrating factor such that $dG = \mu M\,dx + \mu N\,dy$. Show that if f is any continuous function of one variable, then $\mu(x, y)\cdot f(G(x, y))$ is an integrating factor. (This indicates that the differential equation (11) can be expected to have an infinite number of integrating factors.)

16. Show that μ is an integrating factor of a differential equation $M\,dx + N\,dy = 0$, where M and N are continuously differentiable functions in a suitable region, if and only if μ is a solution of the partial differential equation

$$N\frac{\partial u}{\partial x} - M\frac{\partial \mu}{\partial y} = \mu\left(\frac{\partial M}{\partial y} - \frac{\partial N}{\partial x}\right).$$

[*Hint.* Apply the condition for $\mu M\,dx + \mu N\,dy$ to be exact.]

17. Show that the partial differential equation in Exercise 16 can be written in the form

$$N\frac{\partial(\ln|\mu|)}{\partial x} - M\frac{\partial(\ln|\mu|)}{\partial y} = \frac{\partial M}{\partial y} - \frac{\partial N}{\partial x}.$$

18. Using Exercises 16 and 17, show that if M and N are continuously differentiable in a suitable region, then $M\,dx + N\,dy = 0$ has an integrating factor μ which is a function of x only (that is, $\partial\mu/\partial y = 0$) if and only if $(1/N)\cdot(\partial M/\partial y - \partial N/\partial x)$ is a function of x only, say $f(x)$, and that the integrating factor μ is then of the form $\mu(x) = Ke^{\int f(x)dx}$

19. Use the result in Exercise 18 to solve the differential equations

a) $(xy^2 + x + y^2 + 1)\,dx + (2x^2y - 3xy^2 + 2xy - 3y^2)\,dy = 0$

b) $(xy - y)\,dx + (x^2 + x\cos y - x - \cos y)\,dy = 0.$

20. State a result analogous to that in Exercise 18 in the case that $M\,dx + N\,dy = 0$ has an integrating factor that is a function of y only.

20.4 FIRST-ORDER LINEAR EQUATIONS

In this section, we are concerned with finding all solutions in a neighborhood of some point x_0 of the first-order linear differential equation

$$p_0(x)y' + p_1(x)y = q(x) \tag{1}$$

where $p_0(x)$, $p_1(x)$, and $q(x)$ are known functions defined in some neighborhood of x_0. We shall restrict our discussion to the case where the coefficient $p_0(x)$ of y' in Eq. (1) takes on only nonzero values throughout some neighborhood of x_0. We may then divide Eq. (1) by $p_0(x)$ and set $p(x) = p_1(x)/p_0(x)$ and $g(x) = q(x)/p_0(x)$ to write Eq. (1) in the more simple form

$$y' + p(x)y = g(x). \tag{2}$$

We assume that the functions $p(x)$ and $g(x)$ in Eq. (2) are continuous near x_0 so that we can integrate them. This differential equation (2) is one whose solutions can be found easily; we can actually obtain a formula for the general solution.

The trick is to find a continuous function $\mu(x)$ that is nonzero in the neighborhood, and which has the property

$$\mu(x)y' + \mu(x)p(x)y = v' \tag{3}$$

for some function v. (Such a function μ is an *integrating factor.*) Upon multiplication by $\mu(x)$, the Eq. (2) would then reduce to the equivalent equation

$$v' = \mu(x)g(x), \tag{4}$$

which can be solved by a single integration. It has been found that

$$\mu(x) = e^{\int p(x)\,dx}, \tag{5}$$

for any choice of antiderivative of $p(x)$, is such an integrating factor. (Observe that $\mu(x)$ is continuous and $\mu(x) \neq 0$.) To see that μ is indeed an integrating factor, note that if

$$v = \mu(x) \cdot y, \tag{6}$$

then

$$\begin{aligned} v' &= \frac{d}{dx}(\mu(x)) \cdot y + \mu(x) \cdot y' = \frac{d}{dx}(e^{\int p(x)dy} \cdot y + [\mu(x) \cdot y'] \\ &= \left[\left(\frac{d}{dx}\right)\int p(x)\,dx \cdot e^{\int p(x)dx} \cdot y\right] + [\mu(x) \cdot y'] \\ &= p(x)\mu(x)y + \mu(x)y', \end{aligned}$$

which is Eq. (3). Thus multiplying Eq. (2) by our $\mu(x)$, we obtain

$$v' = (\mu(x)y)' = \mu(x)g(x), \tag{7}$$

which yields

$$\mu(x)y = \int \mu(x)g(x)\,dx. \tag{8}$$

From (8), we obtain

$$y = \frac{1}{\mu(x)} \int \mu(x)g(x)\,dx. \tag{9}$$

If we take $\int_{x_0}^{x} \mu(t)g(t)\,dt$ as a particular antiderivative of $\mu(x)g(x)$, then Eq. (9) becomes

$$y = \frac{1}{\mu(x)}\left[\int_{x_0}^{x} \mu(t)g(t)\,dt + C\right]. \tag{10}$$

We have almost proved the following theorem.

Theorem 20.2 *Let $p(x)$ and $g(x)$ be continuous in a neighborhood of x_0, and let y_0 be any real number. Then there exists a* unique *solution $y = f(x)$ of the differential equation $y' + p(x)y = g(x)$ such that $y(x_0) = y_0$, and this solution satisfies the differential equation throughout the neighborhood. Furthermore, the general solution of the differential equation is*

$$y = \frac{1}{\mu(x)} \int \mu(x)g(x)\,dx$$

where

$$\mu(x) = e^{\int p(x)\,dx}.$$

Proof. We have already seen in Eq. (9) that the general solution is as stated in the theorem, and that these solutions are valid for all x in the neighborhood. It remains only to demonstrate the existence and uniqueness of a particular solution through the point (x_0, y_0). But putting $x = x_0$ and $y = y_0$ in the form of the general solution given in Eq. (10), we obtain the relation

$$y_0 = \frac{1}{\mu(x_0)}C,$$

and hence $C = y_0\mu(x_0)$ yields the only solution through (x_0, y_0). □

Example 1 We showed in Section 8.4 of Chapter 8 that the general solution of the differential equation $y' = ky$ is $y = Ae^{kx}$, where A is an arbitrary constant that controls $y(0)$. Let us obtain it again by using the theorem.

SOLUTION Our equation is $y' - ky = 0$, so we have $p(x) = -k$ and $g(x) = 0$. Our integrating factor is thus

$$\mu(x) = e^{\int -k\,dx} = e^{-kx}.$$

The general solution is therefore

$$y = \frac{1}{e^{-kx}}\int e^{-kx}\cdot 0\,dx = \frac{1}{e^{-kx}}C = Ce^{kx}.$$

This coincides with our previous result. ||

Example 2 Let's find the particular solution of the differential equation

$$y' + 3xy = x,$$

which passes through the point (0, 4).

SOLUTION Here $p(x) = 3x$ and our integrating factor is

$$\mu(x) = e^{\int 3x\,dx} = e^{3x^2/2}.$$

By Theorem 20.2, the general solution is

$$y = \frac{1}{e^{3x^2/2}}\int xe^{3x^2/2}\,dx = \frac{1}{e^{3x^2/2}}\left[\frac{1}{3}e^{3x^2/2} + C\right] = \frac{1}{3} + Ce^{-3x^2/2}.$$

Putting $y = 4$ and $x = 0$, we find that

$$4 = \frac{1}{3} + C\cdot 1,$$

so $C = 11/3$, and the desired particular solution is

$$y = \frac{1}{3} + \frac{11}{3} e^{-3x^2/2}. \quad \|$$

The only problem in using Theorem 20.2 to solve $y' + p(x)y = g(x)$ is that the solution contains two integrals, namely

$$\mu(x) = e^{\int p(x)\,dx} \qquad \text{and} \qquad \int \mu(x)g(x)\,dx.$$

Sometimes it is impossible to evaluate one of these integrals in terms of elementary functions, even if $p(x)$ and $g(x)$ are themselves elementary functions. For example, if $p(x) = -x$ and $g(x) = 1$, then for $x_0 = 0$, we have

$$\mu(x) = e^{-x^2/2} \qquad \text{and} \qquad \int \mu(x)g(x)\,dx = \int e^{-x^2/2}\,dx.$$

This last integral cannot be evaluated in terms of elementary functions. However we have seen how this integral can be expressed as an infinite series. (This particular integral is so important in the theory of probability that $\int_0^x e^{-t^2/2}\,dt$ has actually been tabulated for many values of x.) There are many numerical methods for estimating an integral, so the presence of integrals in the general solution of the first-order linear differential equation is not really a serious problem in practical applications.

SUMMARY

1. *The general solution of the differential equation*

$$y' + p(x)y = g(x),$$

where $p(x)$ and $g(x)$ satisfy suitable conditions described in the text, can be found as follows:

STEP 1. *Compute the integrating factor*

$$\mu(x) = e^{\int p(x)\,dx},$$

where $\int p(x)\,dx$ is any particular antiderivative of $p(x)$.

STEP 2. *Upon multiplication by $\mu(x)\,dx$, the equation $y' + p(x)y = g(x)$ becomes*

$$d(\mu(x) \cdot y) = \mu(x)g(x)\,dx.$$

STEP 3. *The solution is then*

$$\mu(x) \cdot y = \int \mu(x)g(x)\,dx$$

or

$$y = \frac{1}{\mu(x)} \int \mu(x)g(x)\,dx.$$

EXERCISES

In Exercises 1 *through* 8, *find the general solution of the differential equation.*

1. $y' - xy = 0$

2. $y' - 3y = 2$

3. $y' + y = 3e^x$

4. $y' + 2y = x + e^{-3x}$

5. $y' + 2y = xe^{-2x} + 3$

6. $xy' + y = 2x \sin x$; $x > 0$. [*Hint.* Reduce to the form of Eq. (2) by dividing by x.]

7. $y' + (\cot x)y = 3x + 1$, $0 < x < \pi$

8. $y' + (\sin x)y = 3 \sin x$

In Exercises 9 *through* 12, *find the solution* $y = f(x)$ *of the differential equation having the indicated value for* $y(x_0)$.

9. $y' - 3y = x + 2$, $y(0) = -1$

10. $xy' - y = x^3$, $y(1) = 5$

11. $(1 + x^2)y' + y = 3$, $y(0) = 2$

12. $y' - (\cos 2x)y = \cos 2x$, $y(\pi/2) = 3$

13. Find the general solution of the differential equation $y' - xy = x^2$ by using Theorem 20.2 and expressing the integral in series form.

14. Find the general solution of the differential equation

$$y' + 2x(\cot x^2)y = 3, 0 < x < \sqrt{\pi}$$

by using Theorem 20.2 and expressing the integral in series form.

15. If $i(t)$ is the current at time t in an electrical circuit with constant resistance R, constant inductance L, and variable electromotive force $E(t)$, then it can be shown that

$$L\frac{di}{dt} + Ri = E(t).$$

a) Let the current at time $t = 0$ be i_0, and let $E(t)$ be a continuous function of the time t. Find an expression for i at time t.

b) Show that if E is constant, then for large values of t, we have $i \approx E/R$ (so that Ohm's law is approximately true after a long time).

c) Describe the current after a long period of time if E diminishes exponentially, that is, if $E(t) = E_0e^{-kt}$ where

$$E_0 = E(0) \quad \text{and} \quad k > 0.$$

d) Describe the behavior of the current as $t \to \infty$ if E is abruptly cut off at time t_0, so that $E(t) = 0$ for $t > t_0$.

16. According to Newton's law of cooling, the rate at which a body in a medium changes temperature is proportional to the difference between its temperature and the temperature of the medium.

a) Assuming that the temperature of the medium remains constant at a degrees and that the temperature T_0 of the body at time $t = 0$ is higher than that of the medium, express the temperature T of the body as a function of the time t for $t > 0$. (Let the constant of proportionality in Newton's law be $-k$.)

b) For $k > 0$, what is the approximate temperature of the body as $t \to \infty$.

17. If n is a constant different from 0 or 1, then the equation

$$y' + p(x)y = g(x)y^n$$

is known as *Bernoulli's equation.* (For both $n = 0$ and $n = 1$, the equation is linear and can be solved as described in this section.)

a) Show that the substitution $v = y^{1-n}$ enables us to reduce the solution of Bernoulli's equation to a differential equation for v which is linear.

b) Use the result of (a) and Theorem 20.2 to solve the differential equation

$$y' - 2xy = 5xy^3.$$

20.5 HOMOGENEOUS LINEAR EQUATIONS WITH CONSTANT COEFFICIENTS

20.5.1 The existence theorem

We shall state the main existence theorem for solutions in a neighborhood of x_0 of a linear differential equation of order n, and then restrict ourselves to the case of a homogeneous linear equation with constant coefficients for the rest of the section. Suppose that the coefficient of $y^{(n)}$ assumes only nonzero values throughout a neighborhood of x_0, so that we may divide by this coefficient and assume that the equation is of the form

$$y^{(n)} + p_1(x)y^{(n-1)} + \cdots + p_{n-1}(x)y' + p_n(x)y = g(x). \tag{1}$$

We state the main existence theorem without proof.

Theorem 20.3 *If $p_1(x), \ldots, p_n(x)$, $g(x)$ are continuous in a neighborhood of x_0 and if a_0, $a_1, \ldots, a_{n-1}$ are any constants, then there is a unique solution $y = f(x)$ of Eq. (1) that is valid throughout the neighborhood and has the property*

$$y(x_0) = a_0,\ y'(x_0) = a_1, \ldots, y^{(n-1)}(x_0) = a_{n-1}.$$

In the case to be considered in the remainder of this section where the coefficient functions $p_1(x), \ldots, p_n(x)$ are constant functions and where $g(x) = 0$, Eq. (1) takes the form

$$y^{(n)} + b_1 y^{(n-1)} + \cdots + b_{n-1}y' + b_n y = 0. \tag{2}$$

Equation (2) is a linear homogeneous differential equation with constant coefficients. In this section, we shall see how the general solution of Eq. (2) can be obtained in terms of elementary functions by very simple algebraic means.

20.5.2 Polynomials in the operator D

We may write Eq. (2) in the form

$$D^n y + b_1 D^{n-1} y + \cdots + b_{n-1} D y + b_n y = 0, \tag{3}$$

where of course $Dy = y'$, $D^2 y = y''$, etc. Proceeding purely formally, it is natural to factor out y on the lefthand side of Eq. (3) and write the equation in the form

$$(D^n + b_1 D^{n-1} + \cdots + b_{n-1} D + b_n) y = 0 \tag{4}$$

or, more briefly,

$$P(D)y = 0, \tag{5}$$

where $P(D)$ is the polynomial $D^n + b_1 D^{n-1} + \cdots + b_{n-1} D + b_n$ in D.

Suppose that the polynomial $P(D)$ factors (in the sense of polynomial factorization) so that $P(D) = Q_1(D)Q_2(D)$ for polynomials $Q_1(D)$ and $Q_2(D)$ in D. It is not difficult to show that

$$P(D)y = Q_1(D)(Q_2(D)y), \tag{6}$$

where $y = f(x)$ has derivatives of all orders $\leq n$. A careful proof of (6) can be given using mathematical induction; you will probably find it sufficiently convincing if we compute a special case to illustrate what we mean.

Example 1 As a polynomial, we have

$$D^2 - 3D + 2 = (D - 1)(D - 2).$$

Computing, we find that for a twice differentiable function $y = f(x)$, we have

$$\begin{aligned}(D - 1)((D - 2)y) &= (D - 1)(Dy - 2y)\\ &= D(Dy - 2y) - 1(Dy - 2y)\\ &= D^2y - D(2y) - Dy + 2y\\ &= D^2y - 2Dy - Dy + 2y\\ &= D^2y - 3Dy + 2y\\ &= (D^2 - 3D + 2)y. \quad \|\end{aligned}$$

The result in (6) can easily be extended to more than two factors. Since polynomial multiplication is commutative (i.e., does not depend on the order of multiplication), we see at once from (6) that

$$Q_1(D)(Q_2(D)y) = Q_2(D)(Q_1(D)y) \tag{7}$$

for polynomials $Q_1(D)$ and $Q_2(D)$. Equation (7) can also be extended to any number of factors in any orders.

20.5.3 Case 1: *P(D)* factors into distinct linear factors

We first consider a differential equation of the form

$$P(D)y = 0, \tag{8}$$

where $P(D)$ is a product of *distinct* linear factors. In this case, we have

$$P(D) = (D - r_1)(D - r_2) \cdots (D - r_n), \tag{9}$$

where $r_i \neq r_j$ for $i \neq j$. Such an equation can be solved by repeated application of Theorem 20.2. The technique is best illustrated by an example.

Example 2 Let's solve the homogeneous equation $y'' - 3y' + 2 = 0$.

SOLUTION Our equation can be written in the form

$$(D^2 - 3D + 2)y = (D - 1)(D - 2)y = 0. \tag{10}$$

If we let $u = (D - 2)y$, then we must have

$$(D - 1)u = 0. \tag{11}$$

Equation (11) is a linear first-order equation for u, and by Section 20.4 we have

$$u = \frac{1}{e^{-x}}\int e^{-x} \cdot 0\, dx = C_1e^x.$$

We then solve the equation

$$(D - 2)y = u = C_1e^x,$$

and obtain

$$\begin{aligned} y &= \frac{1}{e^{-2x}}\int e^{-2x}(C_1e^x)\,dx \\ &= \frac{1}{e^{-2x}}(-C_1e^{-x} + C_2) \\ &= -C_1e^x + C_2e^{2x}. \end{aligned}$$

If we write our linear factors in the reverse order so that Eq. (10) becomes

$$(D - 2)(D - 1)y = 0$$

and make the substitution $(D - 1)y = v$, we obtain first the e^{2x} part of the solution, namely $v = C_1e^{2x}$. The equation $(D - 1)y = v$ then yields the same solution $y = C_1e^{2x} + C_2e^x$. ‖

It is clear that if $D - r$ is a factor of $P(D)$, then some solutions of $P(D)y = 0$ are given by $y = Ce^{rx}$, since

$$(D - r)(Ce^{rx}) = rCe^{rx} - rCe^{rx} = 0.$$

The argument in Example 2 can obviously be extended to give the following theorem.

Theorem 20.4 *Let $P(D) = (D - r_1)(D - r_2)\cdots(D - r_n)$ where $r_i \neq r_j$ for $i \neq j$. Then the general solution of the differential equation $P(D)y = 0$ is*

$$y = C_1e^{r_1x} + C_2e^{r_2x} + \cdots + C_ne^{r_nx}.$$

Example 3 Let's solve the differential equation

$$3y''' - 2y'' - y' = 0.$$

SOLUTION We write our equation in the form

$$(3D^3 - 2D^2 - D)y = D(3D + 1)(D - 1)y = 0.$$

While Theorem 20.4 treats the case where the coefficient of D^n is 1, the solutions of our differential equation remain the same if we divide through by 3. We see that the numbers r_i in Theorem 20.4 can be characterized as solutions of the polynomial equation

$$P(r) = 0, \tag{12}$$

whether the coefficient of D^n is 1 or not. This polynomial equation (12) is the *auxiliary* or *characteristic equation* of the differential equation $P(D)y = 0$. Our characteristic equation in this example is

$$3r^3 - 2r^2 - r = r(3r + 1)(r - 1) = 0,$$

and has as solutions $r_1 = 0$, $r_2 = -1/3$, and $r_3 = 1$. Thus our general solution is

$$y = C_1e^{0x} + C_2e^{-x/3} + C_3e^x = C_1 + C_2e^{-x/3} + C_3e^x. \quad \|$$

20.5.4 Case 2: Repeated linear factors

We now take up the case where the polynomial $P(D)$ factors into a product of the form

$$P(D) = (D - r_1)^{n_1}(D - r_2)^{n_2} \cdots (D - r_m)^{n_m}, \tag{13}$$

where, of course, $n_1 + n_2 + \cdots + n_m = n$.

The general solution of $P(D)y = 0$ can again be obtained by repeated application of Theorem 20.2. To see what form the solution now takes, we consider a simple case where $P(D) = (D - r)^2$. Let $u = (D - r)y$, so that the equation $(D - r)^2y = 0$ becomes $(D - r)u = 0$. We then obtain $u = C_1e^{rx}$. Therefore

$$(D - r)y = C_1e^{rx}$$

and

$$y = \frac{1}{e^{-rx}}\int e^{-rx}C_1e^{rx}\,dx = \frac{1}{e^{-rx}}\int C_1\,dx = \frac{1}{e^{-rx}}(C_1x + C_2)$$
$$= e^{rx}(C_1x + C_2).$$

It is easy to check that the general solution of $(D - r)^3y = 0$ is

$$y = e^{rx}(C_1x^2 + C_2x + C_3).$$

The factors in (13) can be written in any order. For the given order, the first n_1 iterations to solve $P(D)y = 0$ show that

$$e^{r_1x}(C_1x^{n_1-1} + C_2x^{n_1-2} + \cdots + C_{n_1})$$

forms a portion of the general solution. We consider that we have proved the following theorem.

Theorem 20.5 *Let $P(D) = (D - r_1)^{n_1} \cdots (D - r_m)^{n_m}$ where $n_1 + \cdots + n_m = n$. Then the general solution of $P(D)y = 0$ is*

$$y = e^{r_1x}(C_1x^{n_1-1} + C_2x^{n_1-2} \cdots + C_{n_1}) + \cdots$$
$$+ e^{r_mx}(C_{n-n_m+1}x^{n_m-1} + C_{n-n_m+2}x^{n_m-2} + \cdots + C_n).$$

Example 4 Let's solve $y''' + 2y'' + y' = 0$.

SOLUTION The characteristic equation of $y''' + 2y'' + y' = 0$ is

$$r^3 + 2r^2 + r = r(r + 1)^2 = 0,$$

and we see that the general solution of the equation is

$$y = C_1 e^{0x} + e^{-x}(C_2 x + C_3) = C_1 + e^{-x}(C_2 x + C_3). \quad \|$$

20.5.5 Case 3: *P(D)* Contains quadratic factors

It is a theorem of algebra that every polynomial with real coefficients can be factored into a product of linear and quadratic factors with real coefficients. Namely, it can be shown that a polynomial can be factored into a product of linear factors if one allows complex coefficients. (This is the *Fundamental Theorem of Algebra.*) If $D - (a + bi)$ is a factor, then $D - (a - bi)$ can be shown to be a factor also, and the product $D^2 - 2aD + (a^2 + b^2)$ is therefore a quadratic factor of $P(D)$ with real coefficients. We are interested in discovering what such a quadratic factor contributes to the general solution of $P(D)y = 0$.

Proceeding purely formally with complex numbers, we might expect the general solution of

$$((D^2 - 2aD + (a^2 + b^2))y = (D - (a + bi))(D - (a - bi))y = 0 \quad (14)$$

to be

$$\begin{aligned} y &= C_1 e^{(a+bi)x} + C_2 e^{(a-bi)x} \\ &= C_1 e^{ax} e^{bix} + C_2 e^{ax} e^{-bix} \\ &= e^{ax}(C_1 e^{i(bx)} + C_2 e^{-i(bx)}). \end{aligned} \quad (15)$$

In Exercise 35 of Section 11.3, we asked you to formally verify Euler's formula

$$e^{ix} = \cos x + i \sin x.$$

Using this formula and proceeding formally from (15), we obtain

$$\begin{aligned} y &= e^{ax}[C_1(\cos bx + i \sin bx) + C_2(\cos(-bx) + i \sin(-bx))] \\ &= e^{ax}[(C_1 + C_2) \cos bx + (C_1 i - C_2 i) \sin bx]. \end{aligned} \quad (16)$$

Replacing the arbitrary constant $C_1 + C_2$ by C_1 and replacing $C_1 i - C_2 i$ by C_2, we obtain from (16)

$$y = e^{ax}(C_1 \cos bx + C_2 \sin bx). \quad (17)$$

We conjecture that (17) is the general solution of (14). The preceding use of complex numbers can be justified, and our conjecture is indeed correct. If you are not satisfied, you may compute directly that

$$[D^2 - 2aD + (a^2 + b^2)](e^{ax}(C_1 \cos bx + C_2 \sin bx)) = 0$$

(see Exercise 1). From our work with repeated linear factors, we would guess that a repeated quadratic factor $(D_2 - 2aD + (a^2 + b^2))^2$ would give rise to a repetition of (17) with an additional factor x in the general solution. This can also be verified.

We can now solve any homogeneous linear differential $P(D)y = 0$ with constant coefficients, provided we are able to factor the polynomial $P(D)$ into quadratic and linear factors. We conclude with two examples.

Example 5 Let's solve the differential equation

$$D(D - 1)^2(D + 2D + 4)y = 0.$$

SOLUTION The characteristic equation is $r(r - 1)^2(r^2 + 2r + 4) = 0$, which has a root $r_1 = 0$, a double root $r_2 = 1$, and complex roots $-1 \pm i\sqrt{3}$ that are obtained by solving $r^2 + 2r + 4 = 0$ by the quadratic formula. The general solution of our equation is therefore

$$y = C_1 + e^x(C_2x + C_3) + e^{-x}(C_4 \cos \sqrt{3}x + C_5 \sin \sqrt{3}x). \quad \|$$

Example 6 Let's solve the differential equation $(D^6 + 4D^4 + 4D^2)y = 0$

SOLUTION We turn to the characteristic equation

$$r^6 + 4r^4 + 4r^2 = r^2(r^2 + 2)^2 = 0.$$

Here $r_1 = 0$, $r_2 = \sqrt{2}i$, and $r_3 = -\sqrt{2}i$ are all double roots. The general solution of our differential equation is

$$y = C_1x + C_2 + (\cos \sqrt{2}x)(C_3x + C_4) + (\sin \sqrt{2}x)(C_5x + C_6). \quad \|$$

SUMMARY

1. *The differential equation*

$$y^{(n)} + b_1y^{(n-1)} + b_2y^{(n-2)} + \cdots + b_{n-1}y' + b_ny = 0$$

is a linear homogeneous differential equation with constant coefficients b_i. *We let*

$$P(D) = D^n + b_1D^{n-1} + b_2D^{n-2} + \cdots + b_{n-1}D + b_n$$

be the polynomial in the differential operator D (standing for differentiation). The equation then is symbolically written

$$P(D)y = 0.$$

2. *If* $P(D) = (D - r_1)(D - r_2)\cdots(D - r_n)$, *where* $r_i \neq r_j$ *for* $i \neq j$, *then the solution of the equation is*

$$y = C_1e^{r_1x} + C_2e^{r_2x} + \cdots + C_ne^{r_nx}.$$

3. *Let* $P(D) = (D - r_1)^{n_1} \cdots (D - r_m)^{n_m}$, *where* $n_1 + \cdots + n_m = n$. *Then the general solution of* $P(D)y = 0$ *is*

$$y = e^{r_1 x}(C_1 x^{n_1 - 1} + C_2 x^{n_1 - 2} + \cdots + C_{n_1}) + \cdots$$
$$+ e^{r_m x}(C_{n - n_m + 1} x^{n_m - 1} + C_{n - n_m + 2} x^{n_m - 2} + \cdots + C_n).$$

4. *An irreducible quadratic factor in* $P(D)$ *having as complex roots* $a + bi$ *and* $a - bi$ *gives rise to a summand of the general solution of* $P(D)y = 0$ *of the form*

$$e^{ax}(C_1 \cos bx + C_2 \sin bx).$$

EXERCISES

1. Convince yourself by direct computation that

$$[D^2 - 2aD + (a^2 + b^2)](e^{ax}(C_1 \cos bx + C_2 \sin bx)) = 0.$$

In Exercises 2 *through* 16, *find the general solution of the given differential equation.*

2. $y' + 3y = 0$

3. $2y' + 4y = 0$

4. $y'' + 4y' + 3y = 0$

5. $4y'' + 12y' + 5y = 0$

6. $y'' - 6y' + 9y = 0$

7. $4y''' + 4y'' + y' = 0$

8. $y''' - 3y'' = 0$

9. $y'' + 3y = 0$

10. $y'' + 2y' + 6y = 0$

11. $y''' - y = 0$

12. $D(D - 3)^2(D^2 + 1)y = 0$

13. $D^2(D + 2)(D^2 + 2)y = 0$

14. $D^3(D^2 + 1)^2 y = 0$

15. $(D + 1)^2(D^2 + D + 2)y = 0$

16. $D^2(D + 5)(D^2 + 3D + 5)^2 y =$

17. Find the particular solution $y = f(x)$ of $y'' - 5y' + 6y = 0$ such that $y(0) = 1$ and $y'(0) = -1$.

18. Find the particular solution $y = f(x)$ of $y'' + y = 0$ such that $y(\pi/2) = 3$ and $y'(\pi/2) = -2$.

19. Find the particular solution $y = f(x)$ of $y''' - 8y = 0$ such that $y(0) = 2$, $y'(0) = 0$, and $y''(0) = 4$.

20.6 THE NONHOMOGENEOUS CASE; APPLICATIONS

In the first two articles of this section, we consider the problem of finding the solutions of a linear differential equation with constant coefficients of the form

$$y^{(n)} + b_1 y^{(n-1)} + \cdots + b_{n-1} y' + b_n y = g(x)$$

20.6.1 The form of the solution

for a continuous function $g(x)$. The existence theorem was stated in the preceding section. If $g(x) = 0$, the equation is called **homogeneous**. Writing the equation in the form

$$P(D)y = g(x), \tag{1}$$

suppose that $C_1y_1(x) + \cdots + C_ny_n(x)$ is the general solution of the *homogeneous* equation obtained from Eq. (1) by replacing $g(x)$ by 0. Suppose also that $f(x)$ is any particular solution of $P(D)y = g(x)$. Then

$$y = C_1y_1(x) + \cdots + C_ny_n(x) + f(x) \tag{2}$$

gives solutions of (1), for

$$\begin{aligned} P(D)y &= P(D)(C_1y_1(x) + \cdots + C_ny_n(x)) + P(D)f(x) \\ &= 0 + g(x) = g(x). \end{aligned}$$

Also, if $h(x)$ is any solution of (1), then

$$P(D)(h(x) - f(x)) = P(D)h(x) - P(D)f(x) = g(x) - g(x) = 0,$$

so $h(x) - f(x)$ is a solution of the homogeneous equation obtained by setting $g(x) = 0$. Thus

$$h(x) = [\text{A solution of } P(D)y = 0] + f(x),$$

so any solution $h(x)$ is of the form (2). Thus (2) is the general solution of (1).

In this section 20.5, we showed how to find solutions of the homogeneous equation. We therefore focus our attention on finding a particular solution $y = f(x)$ of Eq. (1). Two methods are presented in Sections 20.6.2 and 20.6.3. In Section 20.6.4 we turn to applications involving second-order linear differential equations with constant coefficients.

20.6.2 Successive reduction to first-order equations

Let $P(D)$ be a polynomial in D that factors into (not necessarily distinct) linear factors, so that

$$P(D) = (D - r_1)(D - r_2) \cdots (D - r_n).$$

Consider an equation of the form

$$P(D)y = (D - r_1)(D - r_2) \cdots (D - r_n)y = g(x). \tag{3}$$

We can solve Eq. (3) by the method illustrated in Example 2 of Section 20.5. This technique consists of setting

$$(D - r_2) \cdots (D - r_n)y = u, \tag{4}$$

so that Eq. (3) becomes

$$(D - r_1)u = g(x). \tag{5}$$

Equation (5) is linear in u and can be solved using the method of Section 20.4. We have thus reduced our problem to the solution of Eq. (4), which is a differential equation of order $n - 1$. Repetition of this process n times enables us to find y.

Actually, the procedure just outlined yields the general solution of Eq. (3). Since it is easy for us to write down "most" of this general solution, namely the part in Eq. (2) involving the arbitrary constants, it is more efficient when using this technique to take all arbitrary constants = 0 and obtain just a particular solution of Eq. (3).

Example 1 Let's solve the differential equation

$$y'' - 3y' + 2y = x + 1.$$

SOLUTION Our equation can be written in the form

$$(D - 1)(D - 2)y = x + 1.$$

We let $(D - 2)y = u$, and find a particular solution of $(D - 1)u = x + 1$. Integrating by parts or using tables, we obtain as particular solution, taking zero for all constants of integration,

$$\begin{aligned} u &= \frac{1}{e^{-x}}\int e^{-x}(x + 1)\,dx = \frac{1}{e^{-x}}(-xe^{-x} - 2e^{-x}) \\ &= -x - 2. \end{aligned}$$

Solving $(D - 2)y = u = -x - 2$ by the same method, we obtain as a particular solution of our original equation

$$\begin{aligned} y &= \frac{1}{e^{-2x}}\int -e^{-2x}(x + 2)\,dx = \frac{1}{e^{-2x}}\left(\frac{1}{2}xe^{-2x} + \frac{5}{4}e^{-2x}\right) \\ &= \frac{1}{2}x + \frac{5}{4}. \end{aligned}$$

Our general solution is therefore

$$y = C_1e^x + C_2e^{2x} + \frac{1}{2}x + \frac{5}{4}. \quad \|$$

20.6.3 The method of undetermined coefficients

Sometimes a particular solution of $P(D)y = g(x)$ can be determined by inspection. Equations for which this is true include those where $g(x)$ is a polynomial function.

Example 2 Obviously a particular solution of $y'' + 4y' + 4y = 12$ is $y = 3$; just check that $y = 3$ satisfies the equation. Since the characteristic equation is $r^2 + 4r + 4 = (r + 2)^2 = 0$, we see that the general solution is

$$y = e^{-2x}(C_1x + C_2) + 3. \quad \|$$

Example 3 Let us find by inspection a particular solution of the equation $y'' - 3y' + 2y = x + 1$, which we solved in Example 1.

SOLUTION We try to find $y = f(x)$ such that we obtain $x + 1$ upon computing $y'' - 3y' + 2y$. Clearly $y = x/2$ will give the desired amount of x; namely for $y = x/2$, we have

$$y'' - 3y' + 2y = D^2\left(\frac{x}{2}\right) - 3D\left(\frac{x}{2}\right) + 2\left(\frac{x}{2}\right) = 0 - \left(\frac{3}{2}\right) + x.$$

To get the desired constant 1 rather than $-3/2$, we need to add to our $x/2$ a constant which when *doubled* yields $1 - (3/2) = 5/2$, since our equation contains $2y$. Thus the needed constant is $5/4$, and a particular solution is $y = (x/2) + (5/4)$, as obtained in Example 1. ||

A more systematic attack suggested by Example 3 leads to the method of undetermined coefficients.

Example 4 Let's solve the differential equation

$$y'' + 2y' - 3y = 2x - 17.$$

SOLUTION From Example 3, we would guess that $y = ax + b$ should be a solution for some a and b. For $y = ax + b$, we have

$$y = ax + b, \qquad y' = a, \qquad y'' = 0.$$

Multiplying the first equation by -3, the second by 2, the last by 1, and adding, we see from our original differential equation that we must have

$$2x - 17 = -3ax - 3b + 2a.$$

Equating the coefficients of x and the constant terms, we obtain

$$-3a = 2 \qquad \text{and} \qquad 2a - 3b = -17.$$

Thus $a = -2/3$ and $b = 47/9$. A particular solution is therefore

$$y = -\frac{2}{3}x + \frac{47}{9}.$$

The general solution is then easily found to be

$$y = C_1e^{-3x} + C_2e^x - \frac{2}{3}x + \frac{47}{9}. \quad \|$$

The *method of undetermined coefficients* illustrated in Example 4 works well when the function $g(x)$ on the righthand side of Eq. (1) and derivatives of all orders of $g(x)$ involve sums of only a finite number of different functions, except for constant factors. For example, a polynomial of degree n and its derivatives of all orders involve only sums of constant multiples of the finite number of functions 1, x, $x^2, \ldots, x^n$. Also, derivatives of $\sin ax$ or $\cos ax$ are just constant multiples of these same trigonometric functions. Derivatives of the exponential function e^{ax} involve only constant multiples of e^{ax}. If $g(x)$ contains only sums and products of these functions, then the method of undetermined coefficients is often useful. If $g(x)$ is a function of this type, and if none of the summands in $g(x)$ or any of their derivatives is a solution of the homogeneous equation where $g(x) = 0$, then one takes as a trial particular solution a sum of all the summands in $g(x)$ and derivatives of all orders of these summands, with coefficients a, b, c, etc., to be determined. *If some summand of $g(x)$ or a derivative of any order of a summand is a solution of the homogeneous equation corresponding to an s-fold root of the characteristic equation, then one should start with a trial particular solution in which that summand of $g(x)$ is multiplied by x^s before taking derivatives of it.* This seemingly complicated procedure is quite simple (although frequently tedious) in practice, and is best understood by further examples.

Example 5 Let's solve the differential equation

$$y'' - 3y' + 2y = 2 \cos 4x.$$

SOLUTION Now $\cos 4x$ is not part of the solution of the homogeneous equation $y'' - 3y' + 2y = 0$, so we try as particular solution

$$y = a \cos 4x + b \sin 4x,$$

which is a sum with coefficients a and b of $\cos 4x$ and all different types of derivatives of $\cos 4x$. We then obtain

$$\begin{aligned} y &= a \cos 4x + b \sin 4x, \\ y' &= 4b \cos 4x - 4a \sin 4x, \\ y'' &= -16a \cos 4x - 16b \sin 4x. \end{aligned}$$

Multiplying the first equation by 2, the second by -3, the last by 1, and adding, we see from our original differential equation that we must have

$$\begin{aligned} 2 \cos 4x &= (2a - 12b - 16a) \cos 4x + (2b + 12a - 16b) \sin 4x \\ &= (-14a - 12b) \cos 4x + (12a - 14b) \sin 4x. \end{aligned}$$

We therefore must have

$$\begin{aligned} -14a - 12b &= 2, \\ 12a - 14b &= 0, \end{aligned}$$

which yields upon solution $a = -7/85$ and $b = -6/85$. Our general solution is therefore

$$y = C_1 e^x + C_2 e^{2x} - \frac{7}{85} \cos 4x - \frac{6}{85} \sin 4x. \quad \|$$

Example 6 Let's solve the differential equation

$$y'' - 2y' + y = e^x.$$

SOLUTION In this case, e^x is part of the solution of the homogeneous equation corresponding to the double root $r = 1$ of the characteristic equation $r^2 - 2r + 1 = (r - 1)^2 = 0$. We therefore multiply by x^2, and start with x^2e^x and functions obtained by differentiating it; that is, we first take as a trial particular solution

$$y = ax^2e^x + bxe^x + ce^x.$$

But since e^x and xe^x are solutions of the homogeneous equation, the bxe^x and ce^x portions will contribute 0 when we compute $y'' - 2y' + y$. We thus simply take $y = ax^2e^x$ as trial solution, and we find that

$$y = ax^2e^x,$$
$$y' = ax^2e^x + 2axe^x,$$
$$y'' = ax^2e^x + 4axe^x + 2ae^x.$$

Multiplying the first equation by 1, the second by -2, the last by 1, and adding, we obtain from our original differential equation

$$e^x = 0(x^2e^x) + 0(xe^x) + 2ae^x = 2ae^x,$$

so we must have $2a = 1$ or $a = 1/2$. Our general solution is therefore

$$y = e^x(C_1x + C_2) + \tfrac{1}{2}x^2e^x. \quad \|$$

20.6.4 Applications Let us consider motion of a body of mass m along a straight line, which we consider to be an s-axis. By Newton's second law of motion, we have

$$F(t) = m\frac{d^2s}{dt^2},$$

where $F(t)$ is the force at time t acting on the body and directed along the line. Frequently, one denotes a derivative with respect to time t by a dot over a variable, rather than by a prime. Thus we let $\dot{s} = ds/dt$, $\ddot{s} = d^2s/dt^2$, etc.

There are many physical situations in which the force on the body has a certain basic component $g(t)$ at time t together with components due to the velocity and position of the body. If these additional components due to velocity and position are constant multiples of the velocity and position, then by Newton's law, we have

$$\ddot{s} = k_1\dot{s} + k_2s + \frac{1}{m}g(t) \qquad (6)$$

for constants k_1 and k_2. The differential equation (6) is linear with constant coefficients and might be solved by the methods we have presented. We illustrate several particular cases.

Example 7 (*Free Fall in a Vacuum*) We consider a body falling freely near the surface of a planet without atmosphere and with gravitational acceleration g, which is essentially constant near the surface. The motion of the body is then governed by the differential equation

$$\ddot{s} = -g, \tag{7}$$

where s is measured upwards from the surface of the planet. The general solution of Eq. (7) is easily found to be

$$s = C_1 t + C_2 - \frac{1}{2} g t^2,$$

and the constants C_1 and C_2 can be determined if the position s and velocity $\dot{s}$ of the body are known at a particular time t_0. ||

Example 8 (*Free Fall in a Medium*) Let a body be falling freely through a medium (atmosphere) near the surface of a planet with gravitational acceleration g near its surface. Suppose that, due to the medium, the motion of the body is retarded by a force proportional to its velocity. The motion of the body is then governed by the differential equation

$$\ddot{s} = -k\dot{s} - g, \tag{8}$$

where $k > 0$. The characteristic equation of (8) is $r^2 + kr = 0$, and a particular solution is easily found to be $s = -(g/k)t$. The general solution is therefore

$$s = C_1 + C_2 e^{-kt} - \frac{g}{k} t. \tag{9}$$

Again, C_1 and C_2 can be determined by the position s and velocity $\dot{s}$ at a particular time t_0.

Differentiating (9), we obtain for the velocity

$$\dot{s} = -kC_2 e^{-kt} - \frac{g}{k}.$$

As t becomes large, the term $-kC_2e^{-kt}$ becomes very small; the *terminal velocity* of the body is thus $-g/k$. Note that the body approaches this terminal velocity exponentially (quite rapidly). ||

Example 9 (*Undamped vibrating spring*) Consider a body of mass m hanging on a spring, shown in natural position of rest at $s = 0$ on the vertical s-axis in Fig. 20.5. If the spring is stretched (or compressed) from this natural position, then the spring exerts a restoring force proportional to the dis-

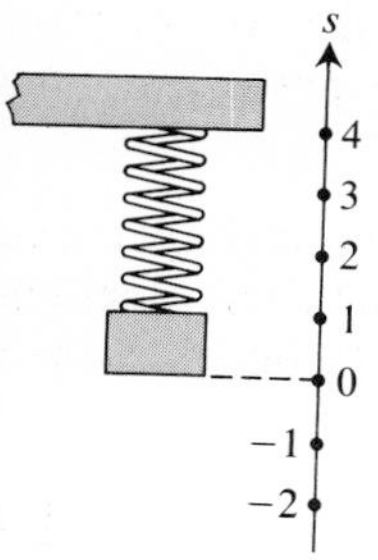

20.5

placement of the body, assuming that the elastic limit of the spring is not exceeded. If the body is set in vertical motion and this restoring force is the only force on the system, then the motion of the body must satisfy the differential equation

$$m\ddot{s} = -ks \tag{10}$$

for a spring constant $k > 0$. The characteristic equation for (10) is $mr^2 + k = 0$, and we see that the general solution is

$$s = C_1 \cos\left(\sqrt{\frac{k}{m}}\,t\right) + C_2 \sin\left(\sqrt{\frac{k}{m}}\,t\right), \tag{11}$$

where C_1 and C_2 can be determined by the position and velocity of the body at a particular time t_0.

If we set $A = \sqrt{C_1{}^2 + C_2{}^2}$, then (11) can be written

$$s = A\left(\frac{C_1}{A} \cos\left(\sqrt{\frac{k}{m}}\,t\right) + \frac{C_2}{A} \sin\left(\sqrt{\frac{k}{m}}\,t\right)\right).$$

Since $(C_1/A)^2 + (C_2/A)^2 = 1$, there exists θ where $0 \le \theta < 2\pi$ and where $\sin\theta = C_1/A$ and $\cos\theta = C_2/A$. Thus

$$\begin{aligned} s &= A\left(\sin\theta \cos\sqrt{\frac{k}{m}}\,t + \cos\theta \sin\sqrt{\frac{k}{m}}\,t\right) \\ &= A \sin\left(\sqrt{\frac{k}{m}}\,t + \theta\right). \end{aligned} \tag{12}$$

Equation (12) shows that our vibratory motion is sinusoidal, with amplitude A and phase angle θ. Such sinusoidal undamped vibratory motion is frequently referred to as "*simple harmonic motion.*" ‖

Example 10 (*Damped vibrating spring*) Suppose the spring in Fig. 20.5 has a damping mechanism attached that exerts a force against the direction of motion and proportional to the velocity of the body. The differential equation then becomes

$$m\ddot{s} = -ks - c\dot{s} \tag{13}$$

for $k > 0$ and $c > 0$. The character of the general solution of this equation $m\ddot{s} + c\dot{s} + ks = 0$ depends upon the relative sizes of m, c, and k, and upon the initial conditions. In the exercises, we ask you to describe the nature of particular motions in the *overdamped case* where $c^2 > 4km$, the *critically damped case* where $c^2 = 4km$, and the *underdamped case* where $c^2 < 4km$ (see Exercises 17, 18, and 19). ‖

SUMMARY

1. *The general solution of a linear equation with constant coefficients*

$$P(D)y = g(x)$$

is of the form

$$y = [\textit{General solution of } P(D)y = 0] + f(x),$$

where $f(x)$ is any particular solution of $P(D)y = g(x)$.

2. *If $P(D) = (D - r_1)(D - r_2) \cdots (D - r_n)$, then $P(D)y = g(x)$ can be solved by setting $u = (D - r_2) \cdots (D - r_n)y$, so $P(D)y = g(x)$ becomes $(D - r_1)u = g(x)$. Solve for u, and apply the same technique again to $(D - r_2) \cdots (D - r_n)y = u$, etc., to find the general solution.*

3. *The method of undetermined coefficients to find a particular solution $f(x)$ of $P(D)y = g(x)$ is too complicated to describe in a summary. See Section 20.6.3 and the examples.*

EXERCISES

In Exercises 1 through 4, find the general solution of the differential equation by finding a particular solution by the method of successive reduction to first order equations, as in Example 1.

1. $y'' - 2y' - 3y = 4x$

2. $y'' - y = \cos x$

3. $y'' + 2y' + y = e^{-x}$

4. $y''' - y'' = x^2$

In Exercises 5 through 14, find the general solution of the differential equation by the method of undetermined coefficients.

5. $y'' - 3y' + 2y = x - 3$

6. $y'' + 4y' = x^2$

7. $y'' + 4y = \sin x$

8. $y'' - 4y' - 5y = x^2 + 2e^{-x}$

9. $y''' + 3y'' = x + e^{3x}$

10. $y''' - y'' = x^2 + e^x$.

11. $y'' + 4y' + 4y = e^{2x} \cos x$

12. $y'' - 4y = x \sin 2x$

13. $(D^4 - 2D^3)y = x^3 + 3x^2$

14. $(D - 1)^3 y = 4 - 3e^x$

15. A body of mass 2 slugs is dropped from an altitude of 3000 ft above the Atlantic Ocean. Suppose the motion of the body is retarded by a force due to air resistance and of magnitude $v/2$ lbs where v is the velocity of the body measured in ft/sec. Find the height s of the body above the ocean as a function of time t, and find the terminal velocity of the body. (Take $g =$ 32 ft/sec^2.)

16. A body of mass one slug (weighing 32 lbs) is attached to a vertical spring, and stretches the spring two feet. The body is then raised one foot and released.

a) Find the displacement s of the body from its natural position at rest at the end of the spring as a function of the time after it was released. (For example, $s = 1$ when $t = 0$.)

b) What is the amplitude of this oscillatory motion?

c) What is the period of this oscillatory motion? (The period is the time required for one complete oscillation.)

17. For a damped vibrating spring with spring constant k and with a weight of mass m attached, suppose the weight is raised a height h from its position at rest and then released. Show that if

$c^2 > 4km$, the weight eases back to its original position of rest without crossing this position.

18. With reference to Exercise 17, suppose now that $c^2 = 4km$ and that the weight is given an initial velocity $v_0 < -ch/2m$ toward the position of rest. Show that the weight crosses its original position of rest, and then eases back to this position.

19. With reference to Exercise 17, show now that if $c^2 < 4km$, then the motion of the weight about its position of rest is oscillatory, but has amplitude decreasing exponentially to 0 as time increases.

20. It can be shown that if an electric circuit has a (constant) resistance R, a (constant) inductance L, (constant) capacitance C, and variable impressed electromotive force $E(t)$, then the charge $Q(t)$ on the capacitor at time t is governed by the differential equation

$$L\ddot{Q} + R\dot{Q} + \frac{1}{C}Q = E(t).$$

Suppose the electromotive force $E(t)$ is abruptly cut off, so that $E(t) = 0$ for $t > t_0$. Use Exercises 17, 18, and 19 to discuss the possibilities for the behavior of the charge Q after time t_0.

21. Some possibilities for the behavior of the solutions of a differential equation $a\ddot{s} + b\dot{s} + cs = 0$ where $a,\ b,\ c > 0$ are indicated in Exercises 17, 18, and 19. This exercise exhibits a phenomenon that may appear in the solutions of the differential equation $a\ddot{s} + b\dot{s} + cs = \sin kt$. A physical model for such an equation is given by the electric circuit described in Exercise 20 where the impressed electromotive force $E(t)$ is sinusoidal.

Suppose the "damping factor" b in our equation is 0. Show that if $k = 0$, the motion described by the resulting (homogeneous) equation is oscillatory with period $2\pi\sqrt{a}/\sqrt{c}$. Then show that if the impressed sinusoidal force $\sin kt$ has the same period so that $k = \sqrt{c/a}$, the amplitude of the motion increases without bound as time increases. (If the damping factor b is nonzero but quite small, the amplitude can still get quite large if $k = \sqrt{4ac - b^2}/2a$. This phenomenon is known as "*resonance*" and can be very destructive. A group of men marching in step should be instructed to break step when crossing a bridge, on the outside chance that the frequency of their step might be the same as the natural frequency of vibration of the bridge.)

20.7 SERIES SOLUTIONS: THE LINEAR HOMOGENEOUS CASE

20.7.1 The nature of a series solution

Consider the differential equation

$$y' = g(x). \tag{1}$$

If the function g in Eq. (1) is analytic at x_0 and if we can find a power series in $x - x_0$ that represents g in a neighborhood of x_0, then we can find the general solution of Eq. (1) near x_0 by integrating the series term by term. Since some elementary functions g do not have elementary functions as antiderivatives, series solutions of Eq. (1) can be very useful. You should think of a series solution in powers of $x - x_0$ as describing the behavior of y for x near x_0.

Example 1 We have stated that the function e^{x^2} does not have an elementary function as antiderivative. Let's find a series solution of the differential equation $y' = e^{x^2}$.

SOLUTION If we replace x by x^2 in the well-known series for e^x, our differential equation becomes

$$y' = e^{x^2} = 1 + x^2 + \frac{x^4}{2!} + \frac{x^6}{3!} + \cdots + \frac{x^{2n}}{n!} + \cdots$$

The series converges to e^{x^2} for all x. Integrating, we obtain as general solution,

$$y = C + x + \frac{x^3}{3} + \frac{x^5}{5 \cdot 2!} + \frac{x^7}{7 \cdot 3!} + \cdots + \frac{x^{2n+1}}{(2n+1)n!} + \cdots$$

for all x. By adjusting the arbitrary constant C, we can find a particular solution having any desired value at the origin. For example, $C = 0$ gives a solution f_1 where $f_1(0) = 0$, while $C = -5$ gives a solution f_2 where $f_2(0) = -5$. ‖

Example 2 From Example 1, we see that, starting with the equation $y'' = e^{x^2}$, we obtain, by integrating once,

$$y' = C_1 + x + \frac{x^3}{3} + \frac{x^5}{5 \cdot 2!} + \frac{x^7}{7 \cdot 3!} + \cdots + \frac{x^{2n+1}}{(2n+1)n!} + \cdots,$$

for all x. Another integration yields the general solution

$$y = C_2 + C_1 x + \frac{x^2}{2} + \frac{x^4}{4 \cdot 3} + \frac{x^6}{6 \cdot 5 \cdot 2!} + \frac{x^8}{8 \cdot 7 \cdot 3!} + \cdots + \frac{x^{2n+2}}{(2n+2)(2n+1)n!} + \cdots,$$

for all x. If $y = f(x)$, then $C_1 = f'(0)$ and $C_2 = f(0)$; both the value of a particular solution at 0 and the value of its derivative at 0 are controlled by the two arbitrary constants C_2 and C_1. ‖

We should mention that not every differential equation has a power-series solution in $x - x_0$ for general solution, even if the equation has a general solution in a neighborhood of x_0. For example, the differential equation $y' = g(x)$, where

$$g(x) = \begin{cases} e^{-1/x^2} & \text{for } x \neq 0, \\ 0 & \text{for } x = 0, \end{cases}$$

has a general solution, namely

$$y = \int_0^x g(x)\,dx + C,$$

which is valid for all x. (The function g is continuous at 0.) However, the function g is not analytic at 0, so of course no antiderivative could be analytic at 0 either. Hence there is no power-series solution at $x_0 = 0$ for the equation $y' = g(x)$.

20.7.2 Homogeneous linear equations

The equation

$$y^{(n)} + p_1(x)y^{(n-1)} + \cdots + p_{n-1}(x)y' + p_n(x)y = 0, \tag{3}$$

where $p_1, \ldots, p_n$ are known functions, is a homogeneous linear differential equation of order n with coefficients $p_1(x), \ldots, p_n(x)$. We state without proof the main theorem concerning the existence of analytic solutions $y = f(x)$ of Eq. (3) in a neighborhood of a point x_0. (See also Theorem 20.3 in Section 20.5.) Since Eq. (3) is of order n, we would expect the general solution to have n arbitrary constants, which could be adjusted to control the values

$$f(x_0), \quad f'(x_0), \quad \ldots, \quad f^{(n-1)}(x_0).$$

It is a convenient notational device to let

$$y(x_0) = f(x_0), \qquad y'(x_0) = f'(x_0), \qquad \ldots, \qquad y^{(n-1)}(x_0) = f^{(n-1)}(x_0).$$

Theorem 20.6 *Let the coefficients $p_1(x), \ldots, p_n(x)$ of Eq. (3) be analytic at x_0 with power-series expansions in $x - x_0$ which represent them in a neighborhood $x_0 - r < x < x_0 + r$. Then every solution of Eq. (3) in this neighborhood is analytic at x_0, and for any constants $a_0, \ldots, a_{n-1}$, there exist unique constants a_n, $a_{n+1}, \ldots$ such that the series $\sum_{k=0}^{\infty} a_k(x - x_0)^k$ represents a solution of Eq. (3) in this neighborhood. Furthermore, if we regard $a_0, \ldots, a_{n-1}$ as arbitrary constants, there exist n functions $y_1(x), \ldots, y_n(x)$ such that the general solution of Eq. (3) in this neighborhood is*

$$a_0 y_1(x) + \cdots + a_{n-1} y_n(x). \tag{4}$$

In the following article, we present a technique for finding series solutions of Eq. (3). Another technique and a discussion of the nonhomogeneous case (where the righthand side of Eq. (3) is nonzero) appear in Section 20.8.

20.7.3 Solving Eq. (3) by differentiating

Let $y = f(x)$ give a solution of Eq. (3) in $x_0 - r < x < x_0 + r$ such that

$$y(x_0) = a_0, \quad y'(x_0) = a_1, \quad \ldots, \quad y^{(n-1)}(x_0) = a_{n-1},$$

for constants $a_0, \ldots, a_{n-1}$. By Theorem 20.6, if the coefficients $p_i(x)$ are analytic in $x_0 - r < x < x_0 + r$, then there exist constants $a_n, a_{n+1}, \ldots$ such that

$$y = \sum_{k=0}^{\infty} a_k(x - x_0)^k$$

for $x_0 - r < x < x_0 + r$. By our uniqueness theorem for series expansions, the coefficient a_k in the series must be the kth Taylor coefficient, so

$$a_k = \frac{y^{(k)}(x_0)}{k!} \tag{5}$$

for $k = 0, 1, \ldots$ We are assuming that the coefficients $p_1(x), \ldots, p_n(x)$ in Eq. (3) are known functions that are analytic at x_0, and that $a_0, \ldots, a_{n-1}$ are known constants. Our job is to find

$$a_n = y^{(n)}(x_0)/n!, \qquad a_{n+1} = y^{(n+1)}(x_0)/(n+1)!, \qquad \ldots.$$

Obviously it suffices to find the derivatives $y^{(i)}(x_0)$ for $i \geq n$. From (3), we obtain

$$y^{(n)} = -p_1(x)y^{(n-1)} - \cdots - p_n(x)y, \tag{6}$$

and we can use (6) to compute $y^{(n)}(x_0)$, since we know the values at x_0 of all the functions that appear on the righthand side of (6). We then differentiate (6) to obtain

$$y^{(n+1)} = -[p_1(x)y^{(n)} + p_1'(x)y^{(n-1)}] - \cdots - [p_n(x)y' + p_n'(x)y]. \tag{7}$$

Since we have computed $y^{(n)}(x_0)$, we now know the values at x_0 of all the functions that appear on the righthand side of (7), so we can compute $y^{(n+1)}(x_0)$. By differentiating (7), we can compute $y^{(n+2)}(x_0)$, etc., and find our series solution.

If all the $p_i(x)$ are constant functions so that $p_i'(x) = 0$ for $i = 1, \ldots, n$, then the differentiation as in (7) is quite easy. We know that if Eq. (3) has constant coefficients, the general solution can be expressed in terms of elementary functions, and we know how to compute these functions quite easily. Thus the series technique is important chiefly in case the $p_i(x)$ are not all constant functions. However, constant-coefficient cases give nice, easy illustrations of the differentiation technique and of Theorem 20.6.

Example 3 We showed in Section 20.4 that the general solution of the differential equation

$$y' = ky \tag{8}$$

is $y = Ae^{kx}$, where A is an arbitrary constant that controls $y(0)$. Let's derive this general solution by series methods. Note that Eq. (8) is of the form (3) where $n = 1$ and $p_1(x) = -k$.

SOLUTION We try to find a series $y = \sum_{n=0}^{\infty} a_n x^n$ at $x_0 = 0$ that is a solution of (8) such that $y(0) = a_0$. We give the equations obtained from (8) and repeated differentiation of (8) in a column at the left; the values $y(0)$, $y'(0)$, $y''(0)$, etc., are given in a column at the right.

$$\begin{array}{ll}
 & y(0) = a_0 \\
y' = ky & y'(0) = ka_0 \\
y'' = ky' & y''(0) = k(ka_0) = k^2a_0 \\
y''' = ky'' & y'''(0) = k(k^2a_0) = k^3a_0 \\
\vdots & \vdots \\
y^{(n)} = ky^{(n-1)} & y^{(n)}(0) = k^na_0 \\
\vdots & \vdots
\end{array}$$

From (5), we have $a_n = y^{(n)}(0)/n!$, so we obtain for our solution

$$\begin{aligned} y &= a_0 + ka_0x + \frac{k^2a_0}{2!}x^2 + \frac{k^3a_0}{3!}x^3 + \cdots + \frac{k^na_0}{n!}x^n + \cdots \\ &= a_0\left(1 + kx + \frac{(kx)^2}{2!} + \frac{(kx)^3}{3!} + \cdots + \frac{(kx)^n}{n!} + \cdots\right) \\ &= a_0e^{kx}. \end{aligned}$$

Thus we have, as general solution, $y = a_0e^{kx}$ for all x, where a_0 may be any constant. Note that we have given the general solution in the form described in the last sentence of Theorem 20.6, where $y_1(x) = e^{kx}$. ‖

Example 4 Let's find the general solution of the equation $y'' + y = 0$ in series form at $x_0 = 0$.

SOLUTION For this second-degree equation, we let $a_0 = y(0)$ and $a_1 = y'(0)$ be "arbitrary," and compute further derivatives of y at 0 using two columns as in the preceding example.

$$\begin{array}{rcl} & & y(0) = a_0 \\ & & y'(0) = a_1 \\ y'' = -y & & y''(0) = -a_0 \\ y''' = -y' & & y'''(0) = -a_1 \\ y^{iv} = -y'' & & y^{iv}(0) = a_0 \\ y^{v} = -y''' & & y^{v}(0) = a_1 \\ y^{vi} = -y^{iv} & & y^{vi}(0) = -a_0 \\ y^{vii} = -y^{v} & & y^{vii}(0) = -a_1 \\ \vdots & & \vdots \end{array}$$

The general solution is therefore

$$\begin{aligned} y &= a_0 + a_1x - \frac{a_0}{2!}x^2 - \frac{a_1}{3!}x^3 + \frac{a_0}{4!}x^4 + \frac{a_1}{5!}x^5 - \frac{a_0}{6!}x^6 - \frac{a_1}{7!}x^7 + \cdots \\ &= a_0\left(1 - \frac{x^2}{2!} + \frac{x^4}{4!} - \frac{x^6}{6!} \cdots\right) + a_1\left(x - \frac{x^3}{3!} + \frac{x^5}{5!} - \frac{x^7}{7!} + \cdots\right) \\ &= a_0 \cos x + a_1 \sin x, \end{aligned}$$

for all x. Note that we have expressed our solution in the form $a_0y_1(x) + a_1y_2(x)$ described in Theorem 20.6, where $y_1(x) = \cos x$ and $y_2(x) = \sin x$. ‖

We give an example that illustrates the differentiation method when the functions $p_i(x)$ are not all constant.

Example 5 Let's find the general solution at $x_0 = 0$ of the differential equation

$$y'' - xy = 0$$

SOLUTION We let $y(0) = a_0$ and $y'(0) = a_1$, and compute in two columns as before.

$$
\begin{array}{ll}
 & y(0) = a_0 \\
 & y'(0) = a_1 \\
y'' = xy & y''(0) = 0 \cdot a_0 = 0 \\
y''' = xy' + y & y'''(0) = a_0 \\
y^{\text{iv}} = xy'' + y' + y' = xy'' + 2y' & y^{\text{iv}}(0) = 2a_1 \\
y^{\text{v}} = xy''' + 3y'' & y^{\text{v}}(0) = 0 \\
y^{\text{vi}} = xy^{\text{iv}} + 4y''' & y^{\text{vi}}(0) = 4a_0 \\
y^{\text{vii}} = xy^{\text{v}} + 5y^{\text{iv}} & y^{\text{vii}}(0) = 5(2a_1) = 10a_1 \\
\vdots & \vdots \\
y^{(n)} = xy^{(n-2)} + (n-2)y^{(n-3)} &
\end{array}
$$

Here we find that

$$y^{(n)}(0) = \begin{cases} (4)(7)(10)\cdots(n-2)a_0 & \text{if } n = 3m, \\ (2)(5)(8)\cdots(n-2)a_1 & \text{if } n = 3m+1, \\ 0 & \text{if } n = 3m+2. \end{cases}$$

The general solution is therefore

$$y = a_0 + a_1 x + 0x^2 + \frac{a_0}{3!}x^3 + \frac{2a_1}{4!}x^4 + 0x^5 + \frac{4a_0}{6!}x^6 + \frac{2\cdot 5a_1}{7!}x^7 + \cdots$$

$$= a_0\left(1 + \frac{x^3}{3!} + \frac{4x^6}{6!} + \frac{4\cdot 7x^9}{9!} + \cdots + \frac{(4)(7)(10)\cdots(3n-2)}{(3n)!}x^{3n} + \cdots\right)$$

$$= a_1\left(x + \frac{2x^4}{4!} + \frac{2\cdot 5x^7}{7!} + \frac{2\cdot 5\cdot 8x^{10}}{10!} + \cdots + \frac{(2)(5)(8)\cdots(3n-1)}{(3n+1)!}x^{3n+1} + \cdots\right),$$

for all x. Here we have again expressed the solution in the form $a_0y_1(x) + a_1y_2(x)$ described in Theorem 20.6. ‖

SUMMARY 1. *A series solution of*

$$y^{(n)} + p_1(x)y^{(n-1)} + \cdots + p_{n-1}(x)y' + p_n(x)y = 0$$

in a neighborhood of x_0 *is of the form*

$$\sum_{k=0}^{\infty} a_k(x - x_0)^k,$$

where $a_0, \ldots, a_{n-1}$ *may be arbitrary constants, and*

$$a_k = \frac{y^{(k)}(x_0)}{k!}.$$

Find a_n *by solving the equation for* $y^{(n)}$,

$$y^{(n)} = -p_1(x)y^{(n-1)} - \cdots - p_{n-1}(x)y' - p_n(x)y,$$

and evaluate at x_0, *and then divide by* $n!$. *Then differentiate this equation for* $y^{(n)}$ *to obtain an expression for* $y^{(n+1)}$ *and evaluate it at* x_0 *and divide by* $(n + 1)!$ *to find* a_{n+1}. *Continue this differentiation and evaluation as far as desired to obtain successive terms of the series solution.*

EXERCISES

1. Let $y = f(x)$ be the solution of the differential equation $y'' - x^2y' + 2y = 0$ such that $y(0) = 1$ and $y'(0) = -2$. Find $y^{iv}(0)$.

2. Let $y = f(x)$ be the solution of the differential equation $y''' + (\sin x)y' - 3(\cos x)y = 0$ such that $y(0) = -1$, $y'(0) = 2$, and $y''(0) = 0$. Find $y^{iv}(0)$.

In Exercises 3 through 9, find, by the method of repeated differentiation, as many terms as you conveniently can of the general solution of the differential equation in a neighborhood of the indicated point x_0. *Express your answer in the form* $a_0y_1(x) + \cdots + a_{n-1}y_n(x)$ *described in Theorem 20.6, and express the functions* $y_i(x)$ *in terms of elementary functions wherever you can.*

3. $y' + xy = 0;\ x_0 = 0$

4. $y'' - y' = 0;\ x_0 = 1$

5. $y'' - 3y' + 2y = 0;\ x_0 = -1$

6. $y'' - xy' = 0;\ x_0 = 0$

7. $y'' + xy' - y = 0;\ x_0 = 0$

8. $y'' + (\sin x)y' + (\cos x)y = 0;\ x_0 = 0$

9. $(x + 1)y' - y = 0,\ x_0 = 0$

20.8 SERIES SOLUTIONS: THE NONHOMOGENEOUS CASE

We consider again the homogeneous linear differential equation

$$y^{(n)} + p_1(x)y^{(n-1)} + \cdots + p_{n-1}(x)y' + p_n(x)y = 0, \tag{1}$$

20.8.1 The method of undetermined coefficients

with coefficients that are analytic in $x_0 - r < x < x_0 + r$. The method of undetermined coefficients for solving Eq. (1) consists of substituting $y = \sum_{k=0}^{\infty} a_k(x - x_0)^k$ in the lefthand side and determining the constants a_n, $a_{n+1}, \ldots$ in terms of $a_0, \ldots, a_{n-1}$, so that the resulting expression is indeed 0. The technique is best illustrated by an example. We take $x_0 = 0$. The same procedure can be used at any point x_0 to obtain a series in $\Delta x = x - x_0$.

Example 1 Let's find the general power-series solution at $x_0 = 0$ of the equation

$$y'' - 2xy' + y = 0. \tag{2}$$

SOLUTION We let

$$y = \sum_{n=0}^{\infty} a_n x^n = a_0 + a_1 x + a_2 x^2 + \cdots + a_n x^n + \cdots$$

We wish to compute $y'' - 2xy' + y$. To compute $-2xy'$, we differentiate to find

$$y' = a_1 + 2a_2 x + 3a_3 x^2 + \cdots + na_n x^{n-1} + \cdots$$

and multiply by $-2x$. We then easily compute the sum $y'' + (-2xy) + y$ by writing out the summands, *keeping like powers of x aligned* as when adding polynomials:

$$\begin{array}{rlll} y &= a_0 & + a_1 x + \cdots & + a_n x^n + \cdots \\ -2xy' &= & -2a_1 x - \cdots & -2na_n x^n + \cdots \\ y'' &= 2a_2 & + 3\cdot 2a_3 x + \cdots & +(n+2)(n+1)a_{n+2}x^n + \cdots \\ \text{(add)} & & & \\ \hline \end{array}$$

$$0 = (a_0 + 2a_2) + (6a_3 - a_1)x + \cdots + [(n+2)(n+1)a_{n+2} - (2n-1)a_n]x^n + \cdots$$

The uniqueness of power series shows that, in order for the series we obtained by adding to represent the constant function 0, we must have

$$a_0 + 2a_2 = 0, \qquad 6a_3 - a_1 = 0, \qquad \ldots, \qquad (n+2)(n+1)a_{n+2} - (2n-1)a_n = 0, \qquad \ldots$$

or

$$a_2 = -\frac{a_0}{2}, \qquad a_3 = \frac{a_1}{6}, \qquad \ldots, \qquad a_{n+2} = \frac{2n-1}{(n+2)(n+1)}a_n, \qquad \ldots$$

The relation

$$a_{n+2} = \frac{2n - 1}{(n + 2)(n + 1)} a_n \tag{3}$$

is a *recursion relation*, and can be used to compute all the coefficients in terms of a_0 and a_1. We obtain as general solution

$$\begin{aligned} y &= a_0 + a_1 x - \frac{a_0}{2} x^2 + \frac{a_1}{3 \cdot 2} x^3 + \frac{3}{4 \cdot 3}\left(-\frac{a_0}{2}\right)x^4 + \frac{5}{5 \cdot 4}\left(\frac{a_1}{3 \cdot 2}\right)x^5 + \cdots \\ &= a_0 + a_1 x - \frac{a_0}{2!} x^2 + \frac{a_1}{3!} x^3 - \frac{3a_0}{4!} x^4 + \frac{5a_1}{5!} x^5 \\ &\quad - \frac{7 \cdot 3a_0}{6!} x^6 + \frac{9 \cdot 5a_1}{7!} x^7 - \cdots \\ &= a_0\left(1 - \frac{1}{2!} x^2 - \frac{3}{4!} x^4 - \frac{7 \cdot 3}{6!} x^6 - \frac{11 \cdot 7 \cdot 3}{8!} x^8 - \cdots\right) \\ &\quad + a_1\left(x + \frac{1}{3!} x^3 + \frac{5}{5!} x^5 + \frac{9 \cdot 5}{7!} x^7 + \frac{13 \cdot 9 \cdot 5}{9!} x^9 + \cdots\right) \end{aligned}$$

for all x. Again, we have expressed our solution in the form $a_0 y_1(x) + a_1 y_2(x)$ given in Theorem 20.6 (Section 20.7). ||

20.8.2 The nonhomogeneous case

The equation

$$y^{(n)} + p_1(x)y^{(n-1)} + \cdots + p_{n-1}(x)y' + p_n(x)y = g(x) \tag{4}$$

for a nonzero function g is a nonhomogeneous linear differential equation of order n with coefficients $p_1(x), \ldots, p_n(x)$. Equation (4) differs from the homogeneous Eq. (1) only in that the righthand side of Eq. (4) is not 0. As in Theorem 20.6 (Section 20.7), it is true that if $p_1(x), \ldots, p_n(x)$ and $g(x)$ are all analytic at x_0 with power-series expansions in $x - x_0$ representing them in some neighborhood of x_0, than every solution of Eq. (4) in that neighborhood is analytic at x_0, and for any constants $a_1, \ldots, a_{n-1}$, there exist *unique* constants $a_n, a_{n+1}, \ldots$ such that the series $y = \sum_{k=0}^{\infty} a_k (x - x_0)^k$ represents a solution of Eq. (4) in that neighborhood. As in Section 20.6, it is easy to see that the solution of (4) is of the form

$$y = [\text{Solution of homogeneous equation where } g(x) = 0] + f(x),$$

where $f(x)$ is any particular solution of (4).

Both the technique of differentiation described in Section 20.7 and the technique of undetermined coefficients can be used to find series solutions of Eq. (4). We give an illustration using the method of undetermined coefficients.

Example 2 Let us find the general series solution at $x_0 = 0$ of

$$y'' - 2xy' + y = x, \tag{5}$$

which is similar to Eq. (2) of Example 1, but with x in place of 0 on the righthand side.

SOLUTION We let $y = \sum_{n=0}^{\infty} a_n x^n$ as in Example 1, and we have exactly the same computations, except that this time when we add to compute $y'' - 2xy' + y$, we must have

$$x = (a_0 + 2a_2) + (6a_3 - a_1)x + \cdots + [(n+2)(n+1)a_{n+2} - (2n-1)a_n]x^n + \cdots$$

This time, we obtain

$$0 = a_0 + 2a_2, \qquad 1 = 6a_3 - a_1, \qquad \ldots, \qquad 0 = (n+2)(n+1)a_{n+2} - (2n-1)a_n, \qquad \ldots$$

so $a_2 = -a_0/2$ and $a_3 = (a_1 + 1)/6$, while for $n > 1$, we have again the recursion relation

$$a_{n+2} = \frac{2n-1}{(n+2)(n+1)} a_n.$$

Our general solution is therefore

$$\begin{aligned} y &= a_0 + a_1 x - \frac{a_0}{2}x^2 + \frac{a_1+1}{3\cdot 2}x^3 + \frac{3}{4\cdot 3}\left(-\frac{a_0}{2}\right)x^4 + \frac{5}{5\cdot 4}\left(\frac{a_1+1}{3\cdot 2}\right)x^5 + \cdots \\ &= a_0 + a_1 x - \frac{a_0}{2!}x^2 + \frac{a_1+1}{3!}x^3 - \frac{3a_0}{4!}x^4 + \frac{5(a_1+1)}{5!}x^5 - \cdots \\ &= a_0\left(1 - \frac{1}{2!}x^2 - \frac{3}{4!}x^4 - \frac{7\cdot 3}{6!}x^6 - \frac{11\cdot 7\cdot 3}{8!}x^8 - \cdots\right) \\ &\quad + a_1\left(x + \frac{1}{3!}x^3 + \frac{5}{5!}x^5 + \frac{9\cdot 5}{7!}x^7 + \frac{13\cdot 9\cdot 5}{9!}x^9 + \cdots\right) \\ &\quad + \left(\frac{1}{3!}x^3 + \frac{5}{5!}x^5 + \frac{9\cdot 5}{7!}x^7 + \frac{13\cdot 9\cdot 5}{9!}x^9 + \cdots\right), \end{aligned}$$

for all x. Note that this is the solution found in Example 1 to the homogeneous equation $y'' - 2xy' + y = 0$ plus the particular solution

$$y = \frac{1}{3!}x^3 + \frac{5}{5!}x^5 + \frac{9\cdot 5}{7!}x^7 + \frac{13\cdot 9\cdot 5}{9!}x^9 + \cdots \quad \|$$

SUMMARY

1. *A series solution in a neighborhood of 0 for a linear differential equation may be able to be found by substituting*

$$y = \sum_{k=0} a_k x^k$$

in the equation, and equating coefficients of like powers of x on the two sides of the equation. (All coefficients and functions in the equation must first be expressed in power series at $x = 0$.) Hopefully, one may obtain a recursion relation to determine a_n in terms of earlier coefficients in the series.

2. *It may be possible to obtain a series solution using the differentiation technique described in the last section for this nonhomogeneous case also.*

EXERCISES

In Exercises 1 through 4, find, by the method of undetermined coefficients, as many terms as you conveniently can of the general series solution of the differential equation in a neighborhood of the indicated point x_0. Express your answer in the form $a_0 y_1(x) + \cdots + a_{n-1} y_n(x)$ described in Theorem 20.6.

1. $y'' + xy' + 2y = 0;\ x_0 = 0$

2. $y^{iv} - x^2 y'' = 0;\ x_0 = 0$

3. $y'' + 2xy' - xy = 0;\ x_0 = 0$

4. $y'' + 3(x - 1)y' - 2y = 0;\ x_0 = 1$

In Exercises 5 through 12, use any method you please to find as many terms as you conveniently can of the general series solution of the differential equation in a neighborhood of x_0.

5. $y'' = \sin x^2;\ x_0 = 0$

6. $y'' = x \cos x^2;\ x_0 = 0$

7. $y'' = x^2 \tan^{-1} x;\ x_0 = 0$

8. $(x^2 + 1)y'' + 2xy' - y = 0;\ x_0 = 0$

9. $xy''' - y'' = 0;\ x_0 = 1$

10. $y'' - y = x^2;\ x_0 = 0$

11. $y'' + y = \sin x;\ x_0 = 0$

12. $y'' - xy' + 2y = x^2 + 2;\ x_0 = 0$

exercise sets for chapter 20

review exercise set 20.1

1. Sketch the direction field of the differential equation

$$y' = y - x,$$

and sketch a few solution curves.

2. Find the family of orthogonal trajectories of $x^2 - y^2 = C$.

3. Find the particular solution of $dy/dx = x \tan y$ such that $y = \pi/4$ when $x = 0$.

4. Find the general solution of $dy/dx = (x + y)/x$.

5. Find the general solution of $(x^2 + y^2)\,dx + 2xy\,dy = 0$.

6. Find the general solution of $(x + y^2 + y)\,dy - y\,dx = 0$.

7. Find the general solution of $y' + (\sin x)y = \sin x$.

8. Find the general solution of $y''' - 6y'' + 9y' = 0$.

9. Find the general solution of $y'' - 2y' + 5y = e^{3x}$.

10. Find the first eight terms of the series solution about $x_0 = 0$ of $y'' - 2xy' + y = 0$.

review exercise set 20.2

1. State the existence theorem for solutions of a differential equation $y' = F(x, y)$ through a point (x_0, y_0).
2. Find the family of orthogonal trajectories of $y = 4x^2 + C$.
3. Find the general solution of $dy/dx = (y - x)/(y + x)$.
4. Find the solution of
$$(2x \sin y)\, dx + (x^2 \cos y)\, dy = 0$$
such that $y = \pi/2$ when $x = 3$.
5. Find the general solution of $(x + \cos y)\, dy + (y + \sin x)\, dx = 0$.
6. Find the general solution of $y' + (\cot x)y = x$.
7. Find the general solution of $y''' - 2y'' = 0$.
8. Find the general solution of
$$(D^2 - 4)(D^2 + 3)y = 0.$$
9. Find the general solution of
$$y'' - 3y' + 2y = \sin x.$$
10. Find the first five terms of the series solution about $x_0 = 0$ of
$$y'' - 2xy' + xy = 1 - \sin x.$$

appendixes

1. BASIC programs

2. Brief summary of algebra and geometry assumed

3. Tables of functions

BASIC programs

APPENDIX 1

PRINTOUT 1 XYVALUES

```
100  REM   THIS PROGRAM PRINTS A TABLE OF X AND Y VALUES
110  REM   FOR A FUNCTION.  THERE ARE  N  EQUALLY SPACED
120  REM   X VALUES OVER AN INTERVAL  [A, B]
130 REM
140  REM   ENTER DATA
150  DEF FNF(X) = X**X*X + 10*X*X + 8*X - 50
160  READ A,B,N
170  DATA -10,3,27
180 REM
190  REM   PRINT LABELS
200  PRINT "X-VALUE", "Y-VALUE"
210 REM
220  REM   FIND AND PRINT Y VALUES
230  PRINT A, FNF(A)
240    FOR I = 1 TO N-1
250    LET X = A + I*(B-A)/(N-1)
260    PRINT X, FNF(X)
270    NEXT I
280  END
```

PRINTOUT 2 PLOT

```
100  REM   SEND TO END OF PROGRAM FOR INSTRUCTIONS
110  GO TO 960
120 REM
130 REM
140  REM   DIMENSION ARRAY AND ENTER DATA
150  DIM Y(100)
160  LET Z = 1
170  IF Z = 0 THEN 220
180  PRINT "YOU DID NOT ENTER THE FUNCTION TO BE GRAPHED IN LINE 160."
190  PRINT "TYPE      160 DEF FNF(X) = <FUNCTION TO BE GRAPHED>"
200  PRINT "BEFORE RUNNING THE PROGRAM."
210  GO TO 1210
220  PRINT "INPUT ENDPOINTS  A,B  OF INTERVAL OF X VALUES."
230  INPUT A,B
240  PRINT "INPUT NUMBER <= 100 OF POINTS TO BE PLOTTED."
250  INPUT N
260  PRINT
270  PRINT
280 REM
290 REM
300  REM   COMPUTE FIRST Y VALUE;  INITIALIZE SMALLEST (S) AND
310  REM   LARGEST (L) VALUES
320  LET D = (B-A)/(N-1)
330  LET X = A
340  LET I = 1
350  LET Y = FNF(X)
360  LET Y(I) = Y
370  LET S, L = Y
380 REM
390 REM
400  REM   COMPUTE OTHER Y VALUES, S, AND L
410  LET I = I + 1
420  IF I > N THEN 530
430  LET X = X + D
440  LET Y = FNF(X)
450  LET Y(I) = Y
460  IF S <= Y THEN 480
470  LET S = Y
480  IF L >= Y THEN 500
490  LET L = Y
500  GO TO 410
510 REM
```

```
520 REM
530  REM   HANDLE SPECIAL CASE OF CONSTANT FUNCTION
540  IF L - S <> 0 THEN 580
550  PRINT "CONSTANT FUNCTION WITH VALUE "S; "OVER THE WHOLE"
560  PRINT "INTERVAL.  GRAPH NOT DRAWN."
570  GO TO 1210
580 REM
590 REM
600  REM   PRINT SCALE INFORMATION
610  PRINT "THE SMALLEST Y VALUE IS "; S
620  PRINT "THE LARGEST Y VALUE IS "; L
630  LET Q = (L-S)/50
640  PRINT "ONE Y-AXIS MARK EQUALS"; Q; "UNITS"
650  PRINT
660  PRINT
670  PRINT " ",S," ","                    ";L
680  IF S*L>0 THEN 770
690  PRINT " ",
700    FOR I = 1 TO 50
710    IF (S + (I-1)*Q)*(S + I*Q) > 0 THEN 740
720    PRINT "0"
730    GO TO 760
740    PRINT " ";
750    NEXT I
760  PRINT
770  PRINT "Y-AXIS SCALE",
780  PRINT "+....+....+....+....+....+....+....+....+....+....+"
790  PRINT
800  PRINT "X VALUES"
810 REM
820 REM
830  REM   PLOT GRAPH
840    FOR K = 1 TO N
850    LET P = (Y(K)-S)/Q
860    PRINT A + (K-1)*D,
870      FOR J = 0 TO 50
880      IF J < INT(P) THEN 910
890      PRINT "*"
900      GO TO 930
910      PRINT " ";
920      NEXT J
930    NEXT K
940  GO TO 1210
950 REM
960  REM   INSTRUCTIONS FOR USERS OF THE PROGRAM
970  PRINT "WANT INSTRUCTIONS?  TYPE  YES  OR  NO , THEN RETURN CARRIAGE."
980  INPUT A$
990  IF A$ = "NO" THEN 140
1000  PRINT
1010  PRINT "THIS PROGRAM PLOTS UP TO 100 POINTS OF A GRAPH"
1020  PRINT "WITH EQUALLY SPACED X VALUES OVER AN INTERVAL"
1030  PRINT "[A, B]  OF X VALUES WHICH YOU WILL SPECIFY."
1040  PRINT
1050  PRINT "TYPE      WIDTH 90      AND RETURN THE CARRIAGE."
1060  PRINT
1070  PRINT "TYPE      160 DEF FNF(X) = <FUNCTION TO BE GRAPHED>      AND"
1080  PRINT "RETURN THE CARRIAGE. THEN RUN THE PROGRAM."
1090  PRINT
1100  PRINT "DURING THE RUN, THE ENDPOINTS  A,B  OF THE INTERVAL"
1110  PRINT "OF X VALUES WILL BE REQUESTED.  ENTER THE VALUES,"
1120  PRINT "SEPARATED BY A COMMA, AND RETURN THE CARRIAGE."
1130  PRINT
1140  PRINT "DURING THE RUN, ENTER THE NUMBER OF POINTS TO BE PLOTTED"
1150  PRINT "WHEN REQUESTED, AND RETURN THE CARRIAGE."
1160  PRINT
1170  PRINT "TO SUPPRESS THE OFFER OF INSTRUCTIONS, TYPE      110"
1180  PRINT "BEFORE RUNNING THE PROGRAM, AND RETURN THE CARRIAGE."
1190  PRINT
1200  PRINT "YOU ARE NOW READY TO RUN THE PROGRAM."
1210  END
```

PRINTOUT 3 SECHORD

```
100  REM   THIS PROGRAM PRINTS 20 VALUES OF THE SLOPES OF SECANTS
110  REM   AND CHORDS TO A GRAPH FOR X = X1 AND INCREMENTS  H =
120  REM   1/2, 1/4, 1/8, ..., 1/2**20 .
130 REM
140  REM  GIVE DATA
150  DEF FNF(X) = SQR(X*X + 16)
160  READ X1
170  DATA 2
180  LET H = 0.5
190 REM
200  REM   COMPUTE AND PRINT SLOPES
210  PRINT " INCREMENT", " SEC. SLOPE", " CHORD SLOPE"
220  PRINT
230    FOR I = 1 TO 20
240    LET S = (FNF(X1+H)-FNF(X1))/H
250    LET C = (FNF(X1+H)-FNF(X1-H))/(2*H)
260    PRINT USING 290, H,S,C
270    LET H = H/2
280    NEXT I
290  : .##########       .##########       .##########
300  END
```

PRINTOUT 4 LIMIT

```
100  REM   THIS PROGRAM PRINTS A SEQUENCE OF FUNCTION VALUES F(X)
110  REM   FOR A RANDOM SEQUENCE OF 20 X-VALUES APPROACHING C.
120 REM
130  REM   ENTER DATA
140  DEF FNF(X) = (X*X - 4)/(X - 2)
150  PRINT "ENTER VALUE  C "
160  INPUT C
170 REM
180  REM   FIND AND PRINT X-VALUES AND VALUES F(X)
190    FOR I = 1 TO 20
200    LET R = RND
210    LET X = C + (2*R - 1)*(1/2)**I
220    PRINT USING 240, X, FNF(X)
230    NEXT I
240  : #.##########     #.##########
250  END
```

PRINTOUT 5 DERIVE

```
100  REM   THIS PROGRAM FINDS THE DERIVATIVE OF FNF(X)
110  REM   AT  X = C  TO  S  SIGNIFICANT FIGURES.
120 REM
130  REM   ENTER DATA
140  DEF FNF(X) = (X*X - 4*X)/(X + 6)
150  READ S
160  DATA 6
170  PRINT 'ENTER VALUE C'
180  INPUT C
190 REM
200  REM   FIND AND PRINT APPROXIMATIONS TO THE DERIVATIVE
210  REM   UNTIL  S  FIGURE ACCURACY IS ACHIEVED.
220  PRINT '    INCREMENT         APPROXIMATION'
230    FOR I = 1 TO 30
240    LET X1 = C - (1/2)**I
250    LET Y1 = FNF(X1)
260    LET X2 = C + (1/2)**I
270    LET Y2 = FNF(X2)
280    LET D = (Y2 - Y1)/(X2 - X1)
290    PRINT USING 370, (1/2)**I, D
300   REM
310    REM   TEST FOR  S  FIGURE ACCURACY
320    IF I = 1 THEN 340
330    IF ABS(D/L - 1) < .5*10**(-S) THEN 380
340    IF ABS(D) < .5*10**(-S) THEN 380
350    LET L = D
360    NEXT I
370  : #.##########     ####.##########
380  END
```

PRINTOUT 6 NEWTON

```
100  REM   THIS PROGRAM ATTEMPTS TO FIND A ZERO OF A FUNCTION USING
110  REM   NEWTON'S METHOD, STARTING WITH INITIAL APPROXIMATION  A .
120  REM   COMPUTATION IS TO  S  SIGNIFICANT FIGURES.
130 REM
140  REM   ENTER DATA
150  DEF FNF(X) = X*X*X + 10*X*X + 8*X - 50
160  READ S
170  DATA 8
180  PRINT "INITIAL APPROXIMATION";
190  INPUT A
200  PRINT
210 REM
220  REM   APPLY NEWTON'S METHOD
230  PRINT "APPROXIMATION X", "F(X)"
240  PRINT
250  PRINT A, FNF(A)
260    FOR  J = 1 TO 40
270    GOSUB 450
280    LET Z = A - FNF(A)/D
290    PRINT Z, FNF(Z)
300    IF A = 0 THEN 320
310    IF ABS(Z/A - 1) < .5*10**(-S) THEN 350
320    IF ABS(Z) < .5*10**(-S)  THEN 350
330    LET A = Z
340    NEXT J
350 REM
360  REM   PRINT RESULTS
370  PRINT
380  IF J = 40 THEN 410
390  PRINT "NEWTON'S METHOD CONVERGED TO A SOLUTION AT X = ";Z
400  STOP
410  PRINT "ACCURACY NOT ACHIEVED"
420  STOP
430 REM
440 REM
450  REM   SUBROUTINE TO COMPUTE DERIVATIVE OF FNF(X) AT X = A
460  LET H = 0.5
470    FOR I = 1 TO 40
480    LET D = (FNF(A+H)-FNF(A-H))/(2*H)
490    IF I = 1 THEN 510
500    IF ABS(D/L - 1) < 10**(-S-1) THEN 550
510    IF ABS(D) < 10**(-S-1) THEN 550
520    LET L = D
530    LET H = H/2
540    NEXT I
550  RETURN
560  END
```

PRINTOUT 7 INTRULES

```
100  REM   THIS PROGRAM COMPARES VALUES OF A DEFINITE INTEGRAL
110  REM   USING RECTANGULAR, TRAPEZOIDAL, AND SIMPSON'S RULES.
120 REM
130  REM   PRINT LABELS AND ENTER DATA
140  PRINT "# SUBDIVISIONS"," RECTANGULAR"," TRAPEZOIDAL"," SIMPSON'S"
150  PRINT
160  DEF FNF(X) = SQR(4 - SIN(X)*SIN(X))
170  READ A,B
180  DATA 0,1
190  READ N
200  DATA 10,50,100,200,-1
210  IF N = -1 THEN 430
220 REM
230  REM   FIND FIRST TERMS OF SUMS
240  LET H = (B-A)/N
250  LET R = FNF(A + H/2)
260  LET T = FNF(A)
270  LET S = FNF(A)
280 REM
290  REM   COMPUTE SUMS EXCEPT FOR LAST TERM OF T AND S
300    FOR I = 1 TO N-1
310    LET R = R + FNF(A + (I+1/2)*H)
320    LET T = T + 2*FNF(A + I*H)
330    LET C = 2 + 2*(I - 2*INT(I/2))
340    LET S = S + C*FNF(A + I*H)
350    NEXT I
360 REM
370  REM   ADD LAST TERMS OF SUMS AND PRINT ANSWERS
380  LET T = T + FNF(B)
390  LET S = S + FNF(B)
400  PRINT N, H*R, (H/2)*T, (H/3)*S
410  PRINT
420  GO TO 190
430  END
```

brief summary of algebra and geometry assumed

APPENDIX 2

A. ALGEBRA

1. Inequalities

Notation	*Read*
$a < b$	a is less than b
$a > b$	a is greater than b
$a \leq b$	a is less than or equal to b
$a \geq b$	a is greater than or equal to b

Laws

$a \leq a$

If $a \leq b$ and $b \leq a$, then $a = b$

If $a \leq b$ and $b \leq c$, then $a \leq c$

If $a \leq b$, then $a + c \leq b + c$

If $a \leq b$ and $c \leq d$, then $a + c \leq b + d$

If $a \leq b$ and $c > 0$, then $ac \leq bc$

If $a \leq b$ and $c < 0$, then $bc \leq ac$

2. Absolute value

Notation	*Read*
$\lvert a \rvert$	The absolute value of a

Properties

$$|a| = \begin{cases} a & \text{if} \quad a \geq 0 \\ -a & \text{if} \quad a < 0 \end{cases}$$

$|a| \geq 0$, and $|a| = 0$ if and only if $a = 0$

$|ab| = |a| \cdot |b|$

$|a + b| \leq |a| + |b|$

$|a - b| \geq |a| - |b|$

$|a - b| =$ distance from a to b on the number line

3. Arithmetic of rational numbers

$$\frac{a}{b} + \frac{c}{d} = \frac{ad + bc}{bd}, \qquad \frac{a}{b} \cdot \frac{c}{d} = \frac{ac}{bd}, \qquad \frac{a/b}{c/d} = \frac{ad}{bc}$$

4. Laws of signs

$(-a)(b) = a(-b) = -(ab), \qquad (-a)(-b) = ab$

5. Distributive laws

$a(b + c) = ab + ac, \qquad (a + b)c = ac + bc$

6. Laws of exponents

$a^m a^n = a^{m+n}, \qquad (ab)^m = a^m b^m, \qquad (a^m)^n = a^{mn}$

$a^{m/n} = \sqrt[n]{a^m} = (\sqrt[n]{a})^m, \qquad a^{-n} = 1/a^n,$

$a^m / a^n = a^{m-n}$

7. Arithmetic involving zero

$a \cdot 0 = 0 \cdot a = 0$ for any number a

$a + 0 = 0 + a = a$ for any number a

$\dfrac{0}{a} = 0$

$a^0 = 1$ and $0^a = 0$ if $a > 0$

8. Arithmetic involving one

$1 \cdot a = a \cdot 1 = a$ for any number a

$\dfrac{a}{1} = a^1 = a$ for any number a

$1^n = 1$ for any integer n

9. Binomial theorem

$$(a + b)^n = a^n + na^{n-1}b + \frac{n(n-1)}{1 \cdot 2} a^{n-2}b^2 + \frac{n(n-1)(n-2)}{1 \cdot 2 \cdot 3} a^{n-3}b^3 + \cdots + nab^{n-1} + b^n$$

where n is a positive integer

10. Quadratic formula

If $a \neq 0$, then the solutions of the quadratic equation $ax^2 + bx + c = 0$ are given by the formula

$$x = \frac{-b \pm \sqrt{b^2 - 4ac}}{2a}.$$

B. GEOMETRY

In the formulas that follow, we let

A = area
s = slant height
C = circumference
b = length of base
r = radius
S = surface area or lateral area
h = altitude
V = volume

1. Triangle:

$$A = \tfrac{1}{2}bh$$

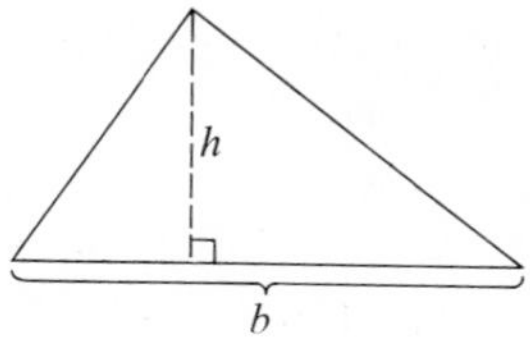

2. Similar triangles:

$$\frac{a'}{a} = \frac{b'}{b} = \frac{c'}{c}$$

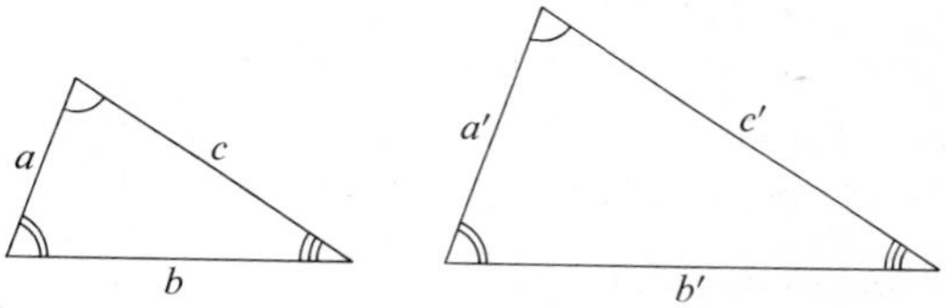

3. Pythagorean theorem:

$$c^2 = a^2 + b^2$$

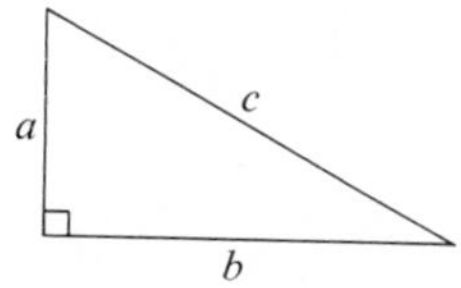

4. Parallelogram:

$$A = bh$$

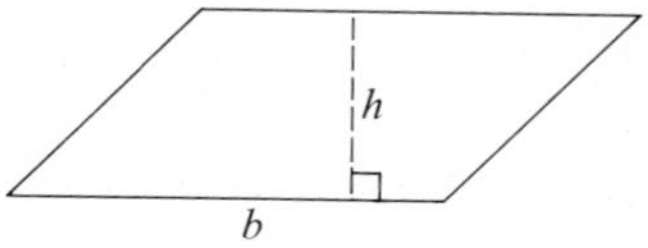

5. Trapezoid:

$$A = \tfrac{1}{2}(b_1 + b_2)h$$

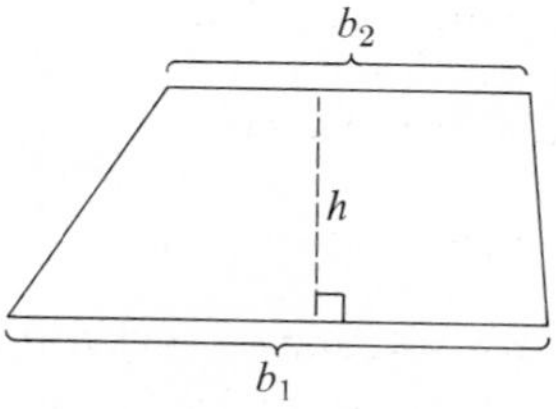

6. Circle:

$$A = \pi r^2, \qquad C = 2\pi r$$

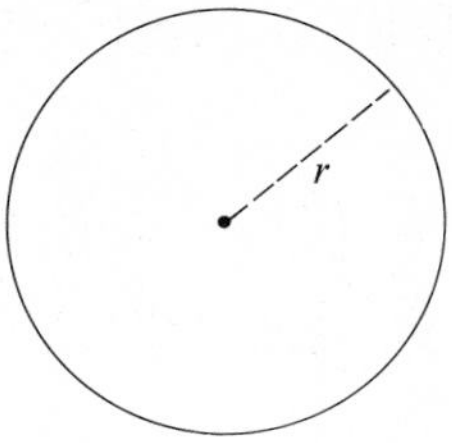

7. Right circular cylinder:

$$V = \pi r^2 h, \qquad S = 2\pi r h$$

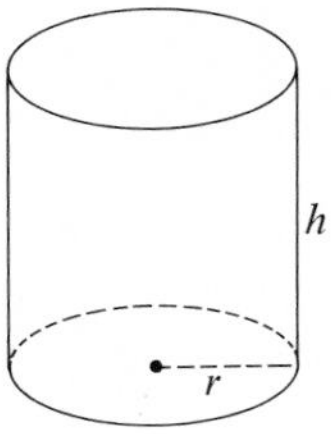

8. Right circular cone:

$$V = \tfrac{1}{3}\pi r^2 h, \qquad S = \pi r s$$

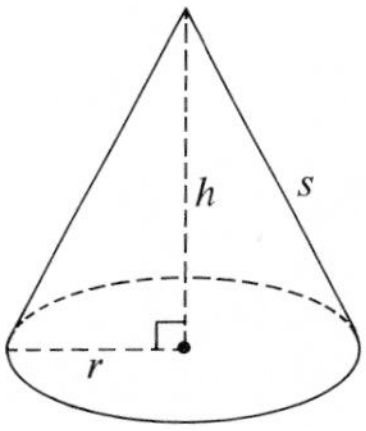

9. Sphere:

$$V = \tfrac{4}{3}\pi r^3, \qquad S = 4\pi r^2$$

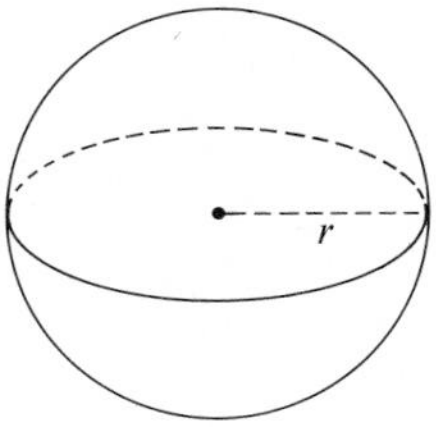

tables of functions

APPENDIX 3

TABLE 1. Natural trigonometric functions

Angle					Angle				
Degree	Radian	Sine	Cosine	Tangent	Degree	Radian	Sine	Cosine	Tangent
0°	0.000	0.000	1.000	0.000					
1°	0.017	0.017	1.000	0.017	46°	0.803	0.719	0.695	1.036
2°	0.035	0.035	0.999	0.035	47°	0.820	0.731	0.682	1.072
3°	0.052	0.052	0.999	0.052	48°	0.838	0.743	0.669	1.111
4°	0.070	0.070	0.998	0.070	49°	0.855	0.755	0.656	1.150
5°	0.087	0.087	0.996	0.087	50°	0.873	0.766	0.643	1.192
6°	0.105	0.105	0.995	0.105	51°	0.890	0.777	0.629	1.235
7°	0.122	0.122	0.993	0.123	52°	0.908	0.788	0.616	1.280
8°	0.140	0.139	0.990	0.141	53°	0.925	0.799	0.602	1.327
9°	0.157	0.156	0.988	0.158	54°	0.942	0.809	0.588	1.376
10°	0.175	0.174	0.985	0.176	55°	0.960	0.819	0.574	1.428
11°	0.192	0.191	0.982	0.194	56°	0.977	0.829	0.559	1.483
12°	0.209	0.208	0.978	0.213	57°	0.995	0.839	0.545	1.540
13°	0.227	0.225	0.974	0.231	58°	1.012	0.848	0.530	1.600
14°	0.244	0.242	0.970	0.249	59°	1.030	0.857	0.515	1.664
15°	0.262	0.259	0.966	0.268	60°	1.047	0.866	0.500	1.732
16°	0.279	0.276	0.961	0.287	61°	1.065	0.875	0.485	1.804
17°	0.297	0.292	0.956	0.306	62°	1.082	0.883	0.469	1.881
18°	0.314	0.309	0.951	0.325	63°	1.100	0.891	0.454	1.963
19°	0.332	0.326	0.946	0.344	64°	1.117	0.899	0.438	2.050
20°	0.349	0.342	0.940	0.364	65°	1.134	0.906	0.423	2.145
21°	0.367	0.358	0.934	0.384	66°	1.152	0.914	0.407	2.246
22°	0.384	0.375	0.927	0.404	67°	1.169	0.921	0.391	2.356
23°	0.401	0.391	0.921	0.424	68°	1.187	0.927	0.375	2.475
24°	0.419	0.407	0.914	0.445	69°	1.204	0.934	0.358	2.605
25°	0.436	0.423	0.906	0.466	70°	1.222	0.940	0.342	2.748
26°	0.454	0.438	0.899	0.488	71°	1.239	0.946	0.326	2.904
27°	0.471	0.454	0.891	0.510	72°	1.257	0.951	0.309	3.078
28°	0.489	0.469	0.883	0.532	73°	1.274	0.956	0.292	3.271
29°	0.506	0.485	0.875	0.554	74°	1.292	0.961	0.276	3.487
30°	0.524	0.500	0.866	0.577	75°	1.309	0.966	0.259	3.732
31°	0.541	0.515	0.857	0.601	76°	1.326	0.970	0.242	4.011
32°	0.559	0.530	0.848	0.625	77°	1.344	0.974	0.225	4.332
33°	0.576	0.545	0.839	0.649	78°	1.361	0.978	0.208	4.705
34°	0.593	0.559	0.829	0.675	79°	1.379	0.982	0.191	5.145
35°	0.611	0.574	0.819	0.700	80°	1.396	0.985	0.174	5.671
36°	0.628	0.588	0.809	0.727	81°	1.414	0.988	0.156	6.314
37°	0.646	0.602	0.799	0.754	82°	1.431	0.990	0.139	7.115
38°	0.663	0.616	0.788	0.781	83°	1.449	0.993	0.122	8.144
39°	0.681	0.629	0.777	0.810	84°	1.466	0.995	0.105	9.514
40°	0.698	0.643	0.766	0.839	85°	1.484	0.996	0.087	11.43
41°	0.716	0.656	0.755	0.869	86°	1.501	0.998	0.070	14.30
42°	0.733	0.669	0.743	0.900	87°	1.518	0.999	0.052	19.08
43°	0.750	0.682	0.731	0.933	88°	1.536	0.999	0.035	28.64
44°	0.768	0.695	0.719	0.966	89°	1.553	1.000	0.017	57.29
45°	0.785	0.707	0.707	1.000	90°	1.571	1.000	0.000	

TABLE 2. Exponential functions

x	e^x	e^{-x}	x	e^x	e^{-x}
0.00	1.0000	1.0000	2.5	12.182	0.0821
0.05	1.0513	0.9512	2.6	13.464	0.0743
0.10	1.1052	0.9048	2.7	14.880	0.0672
0.15	1.1618	0.8607	2.8	16.445	0.0608
0.20	1.2214	0.8187	2.9	18.174	0.0550
0.25	1.2840	0.7788	3.0	20.086	0.0498
0.30	1.3499	0.7408	3.1	22.198	0.0450
0.35	1.4191	0.7047	3.2	24.533	0.0408
0.40	1.4918	0.6703	3.3	27.113	0.0369
0.45	1.5683	0.6376	3.4	29.964	0.0334
0.50	1.6487	0.6065	3.5	33.115	0.0302
0.55	1.7333	0.5769	3.6	36.598	0.0273
0.60	1.8221	0.5488	3.7	40.447	0.0247
0.65	1.9155	0.5220	3.8	44.701	0.0224
0.70	2.0138	0.4966	3.9	49.402	0.0202
0.75	2.1170	0.4724	4.0	54.598	0.0183
0.80	2.2255	0.4493	4.1	60.340	0.0166
0.85	2.3396	0.4274	4.2	66.686	0.0150
0.90	2.4596	0.4066	4.3	73.700	0.0136
0.95	2.5857	0.3867	4.4	81.451	0.0123
1.0	2.7183	0.3679	4.5	90.017	0.0111
1.1	3.0042	0.3329	4.6	99.484	0.0101
1.2	3.3201	0.3012	4.7	109.95	0.0091
1.3	3.6693	0.2725	4.8	121.51	0.0082
1.4	4.0552	0.2466	4.9	134.29	0.0074
1.5	4.4817	0.2231	5	148.41	0.0067
1.6	4.9530	0.2019	6	403.43	0.0025
1.7	5.4739	0.1827	7	1096.6	0.0009
1.8	6.0496	0.1653	8	2981.0	0.0003
1.9	6.6859	0.1496	9	8103.1	0.0001
2.0	7.3891	0.1353	10	22026	0.00005
2.1	8.1662	0.1225			
2.2	9.0250	0.1108			
2.3	9.9742	0.1003			
2.4	11.023	0.0907			

TABLE 3. Natural logarithms

n	$\log_e n$	n	$\log_e n$	n	$\log_e n$
0.0	*	4.5	1.5041	9.0	2.1972
0.1	7.6974	4.6	1.5261	9.1	2.2083
0.2	8.3906	4.7	1.5476	9.2	2.2192
0.3	8.7960	4.8	1.5686	9.3	2.2300
0.4	9.0837	4.9	1.5892	9.4	2.2407
0.5	9.3069	5.0	1.6094	9.5	2.2513
0.6	9.4892	5.1	1.6292	9.6	2.2618
0.7	9.6433	5.2	1.6487	9.7	2.2721
0.8	9.7769	5.3	1.6677	9.8	2.2824
0.9	9.8946	5.4	1.6864	9.9	2.2925
1.0	0.0000	5.5	1.7047	10	2.3026
1.1	0.0953	5.6	1.7228	11	2.3979
1.2	0.1823	5.7	1.7405	12	2.4849
1.3	0.2624	5.8	1.7579	13	2.5649
1.4	0.3365	5.9	1.7750	14	2.6391
1.5	0.4055	6.0	1.7918	15	2.7081
1.6	0.4700	6.1	1.8083	16	2.7726
1.7	0.5306	6.2	1.8245	17	2.8332
1.8	0.5878	6.3	1.8405	18	2.8904
1.9	0.6419	6.4	1.8563	19	2.9444
2.0	0.6931	6.5	1.8718	20	2.9957
2.1	0.7419	6.6	1.8871	25	3.2189
2.2	0.7885	6.7	1.9021	30	3.4012
2.3	0.8329	6.8	1.9169	35	3.5553
2.4	0.8755	6.9	1.9315	40	3.6889
2.5	0.9163	7.0	1.9459	45	3.8067
2.6	0.9555	7.1	1.9601	50	3.9120
2.7	0.9933	7.2	1.9741	55	4.0073
2.8	1.0296	7.3	1.9879	60	4.0943
2.9	1.0647	7.4	2.0015	65	4.1744
3.0	1.0986	7.5	2.0149	70	4.2485
3.1	1.1314	7.6	2.0281	75	4.3175
3.2	1.1632	7.7	2.0412	80	4.3820
3.3	1.1939	7.8	2.0541	85	4.4427
3.4	1.2238	7.9	2.0669	90	4.4998
3.5	1.2528	8.0	2.0794	95	4.5539
3.6	1.2809	8.1	2.0919	100	4.6052
3.7	1.3083	8.2	2.1041		
3.8	1.3350	8.3	2.1163		
3.9	1.3610	8.4	2.1282		
4.0	1.3863	8.5	2.1401		
4.1	1.4110	8.6	2.1518		
4.2	1.4351	8.7	2.1633		
4.3	1.4586	8.8	2.1748		
4.4	1.4816	8.9	2.1861		

* Subtract 10 from these entries.

answers to odd-numbered exercises

CHAPTER 1

Section 1.1

1. a) 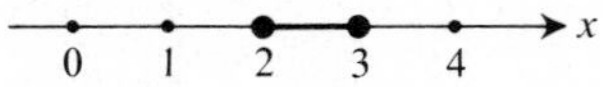

b) 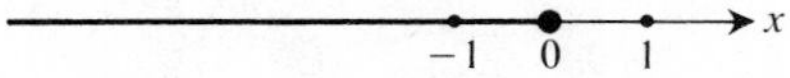

c) 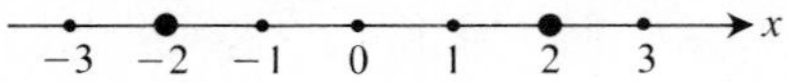

d) 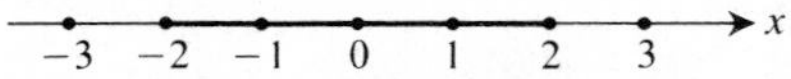

e) 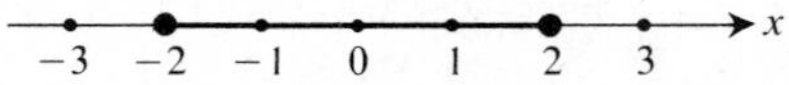

f) 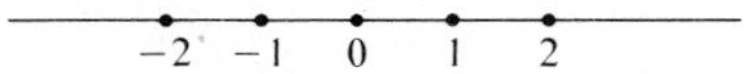 (No locus)

3. a) 8.5 b) 7/3 c) $3\sqrt{2}$ d) $\pi - \sqrt{2}$

5. a) 0 b) 1.5 c) −4.5 d) −5/12 e) $-\sqrt{2}/2$

f) $\dfrac{\sqrt{2} + \pi}{2}$

7. a) 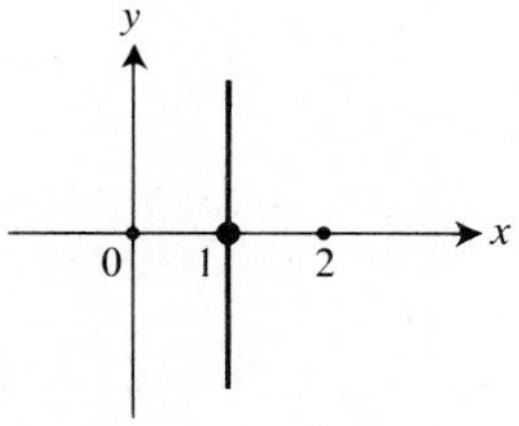

b) 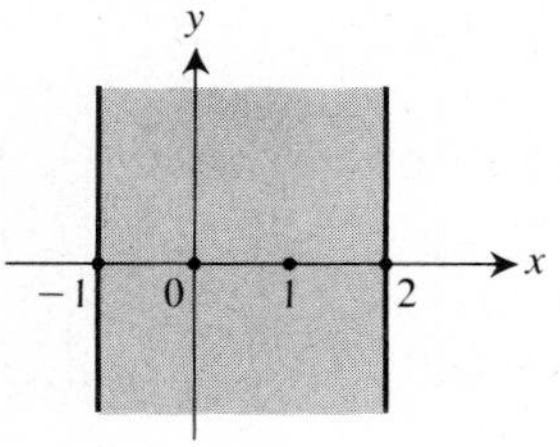

c)

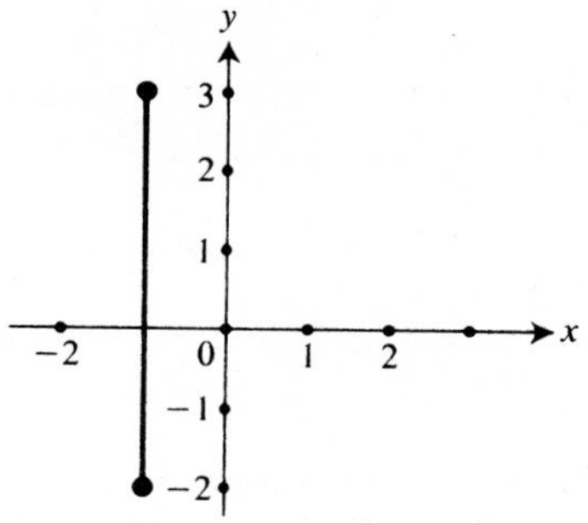

d)

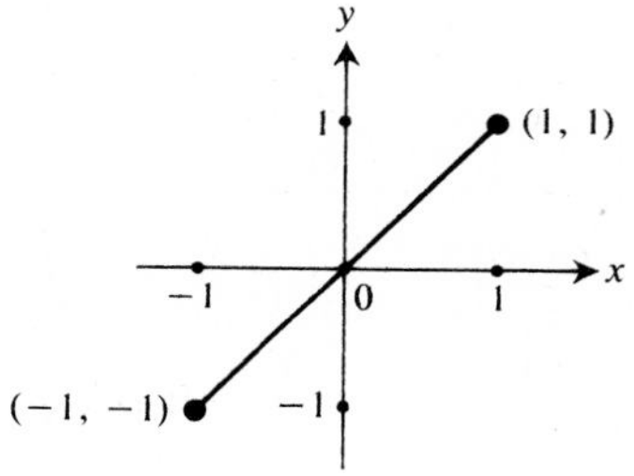

9. a) (2, 1) b) (3, 2) c) (1, −3) d) (2, 6)

11. $\sqrt{29}$

13. 4.88233

15. 4.47214

17. 29.84445

Section 1.2

1. a) $x^2 + y^2 = 25$ b) $(x + 1)^2 + (y - 2)^2 = 9$
c) $(x - 3)^2 + (y + 4)^2 = 30$

3. a) Center (2, −3), radius 4 b) Center (−4, 0), radius 5
c) Center (3/2, 3), radius $9\sqrt{2}/4$

5. $(x - 2)^2 + (y + 2)^2 = 25$

7. a) −3/5 b) 1/11 c) Undefined d) 4/5 e) 0

9. −8

11. 13/2

13. Slope = 9/5 = increase in Fahrenheit temperature per degree increase in Celsius temperature.

15. a) Center (3.14159, 1.77245), radius 1.65409
b) Center (−1.5788, 0.6177), radius 2.49256
c) Center (10.96237, −4.69151), radius 12.27534

Section 1.3

1. a) $y - 4 = 5(x + 1)$ or $y = 5x + 9$ b) $y = 5$
c) $y + 5 = -\frac{6}{5}(x - 4)$ or $5y + 6x = -1$ d) $x = -3$

3. $2x + 3y = -1$

5. No. The first has slope −3/4 and the second −4/3, and $(-3/4)(-4/3) = 1 \neq -1$.

7. $3y + 2x = 26$

9. 2

11. $x^2 + y^2 + 4x - 2y = 20$

13. $d = 13 + \frac{3}{2}(t - 3)$ inches

Section 1.4

1. $V = x^3, \quad x > 0$

3. $A = s^2/4\pi, \quad s > 0$

5. $V = d^3/3\sqrt{3}, \quad d > 0$

7. $s = (\sqrt{34})t, \quad t \geq 0$

9. a) 1 b) 1/2 c) −3/2

11. a) 0 b) Undefined c) −1/2 d) $1/(1 + \Delta t), \quad \Delta t \neq -1, 0$

13. a) $u \leq -1$ or $u \geq 1$ b) $t \geq 2$ but $t \neq 4$
c) $x \geq 4$ d) $-3 < v < 3$

15. a)

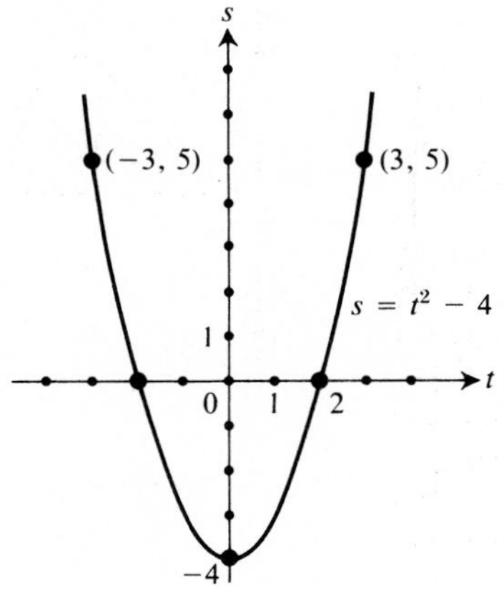

b)

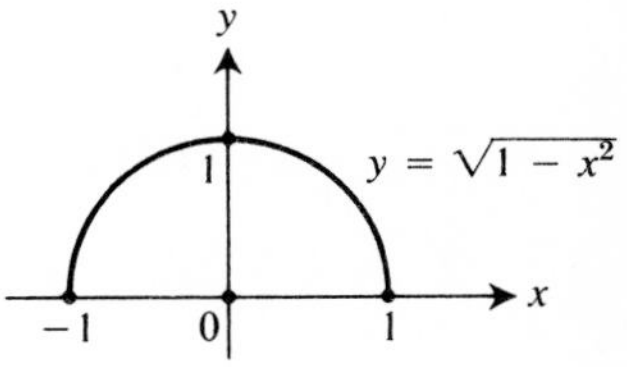

c)

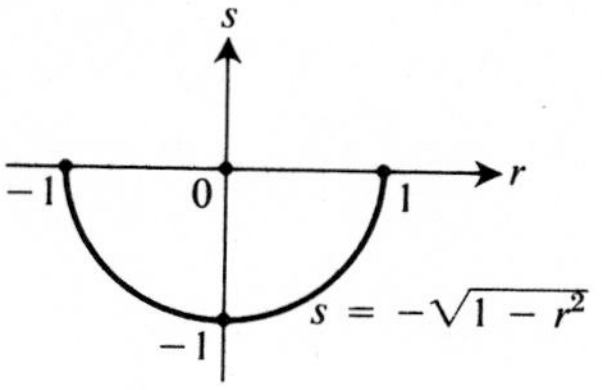

17.

x	y
−1	0
−1/2	−1/3
0	−1
1/2	−3
3/4	−7
7/8	−15
9/8	17
5/4	9
3/2	5
2	3
5/2	7/3
3	2

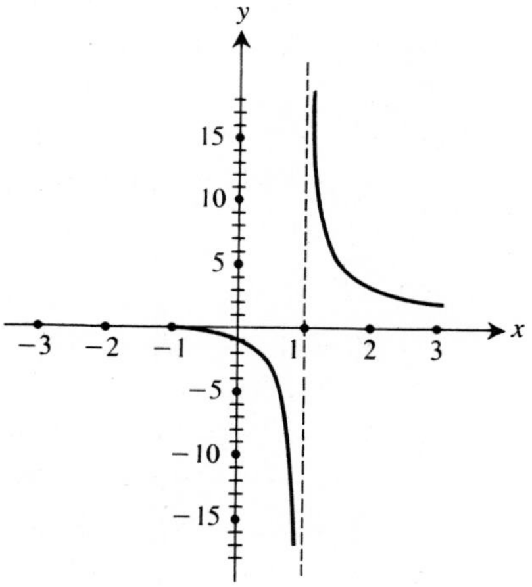

19.

x	$\sin x^2$
0.0	0.0
0.25	0.062246
0.5	0.2474
0.75	0.5333
1.0	0.84147
1.25	0.99997
1.5	0.77807
1.75	0.07901
2.0	−0.7568
2.25	−0.93933
2.5	−0.03318
2.75	0.95782
3.0	0.41212

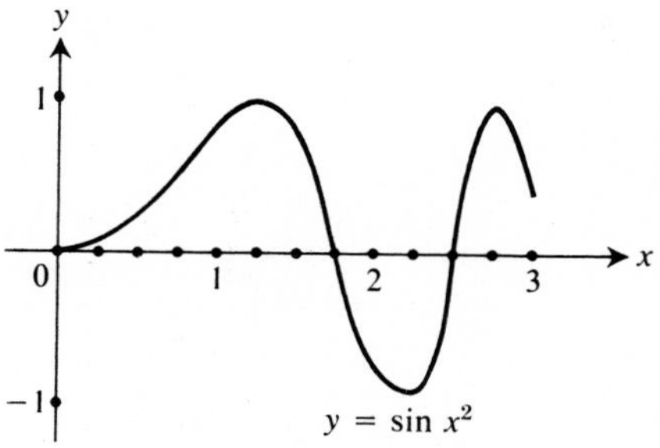

Section 1.5

1.
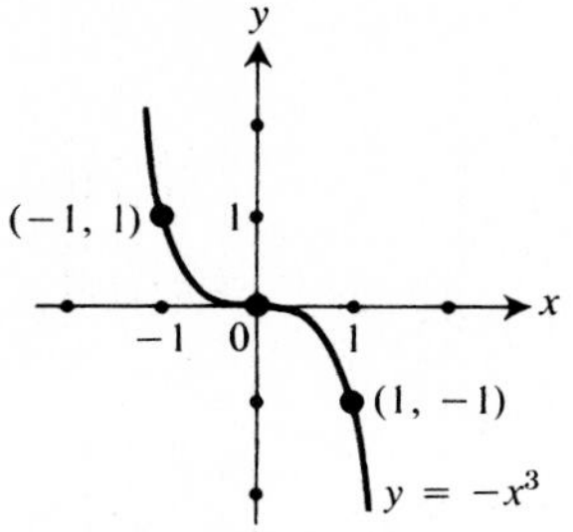

3.
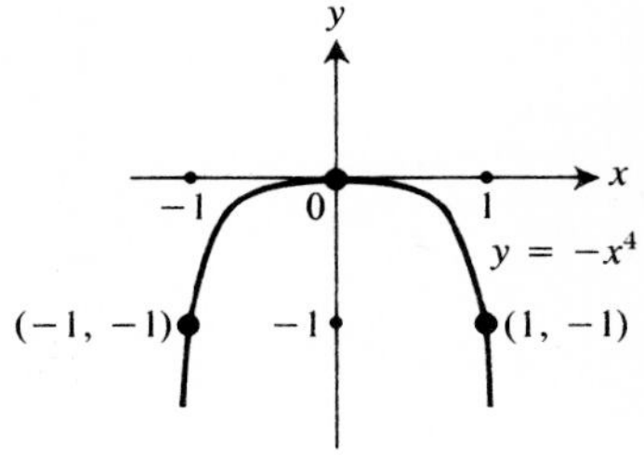

5.
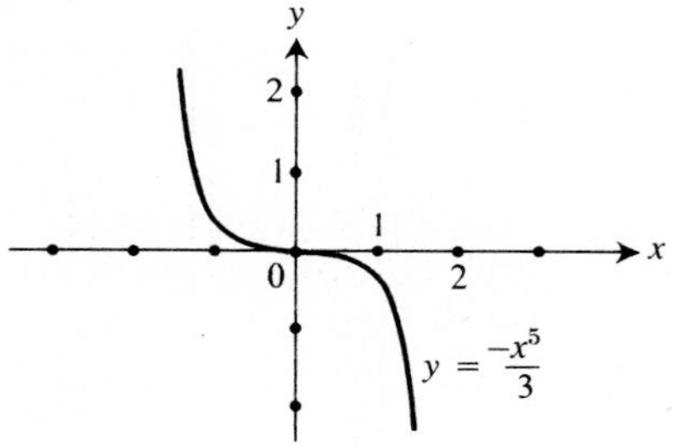

7.
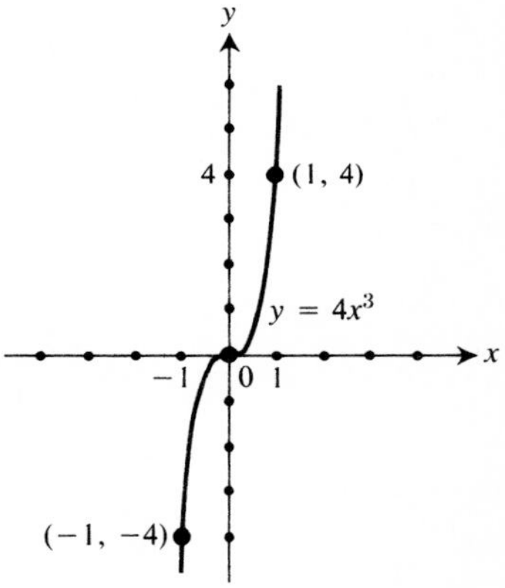

9.
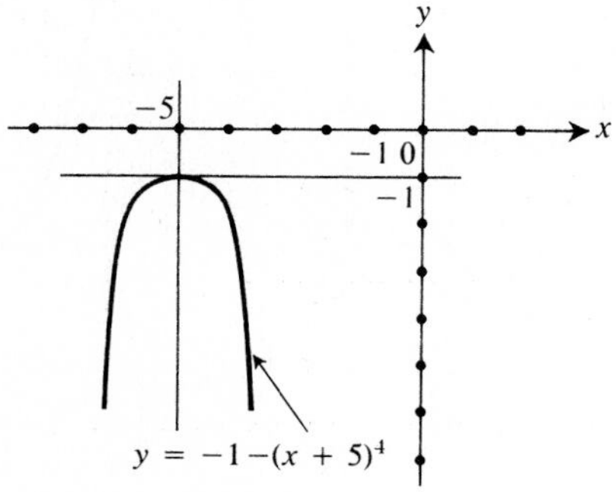

11.
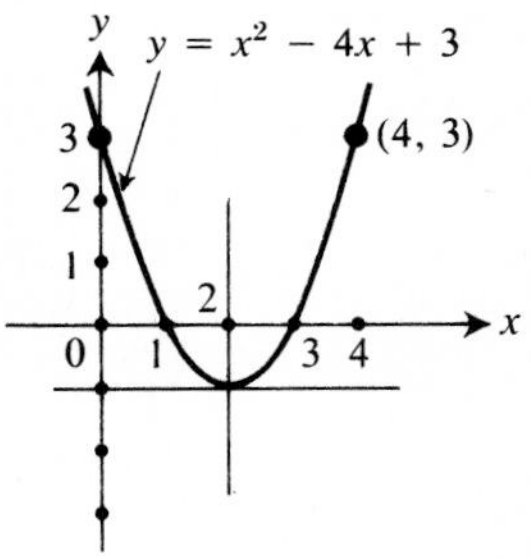

13.

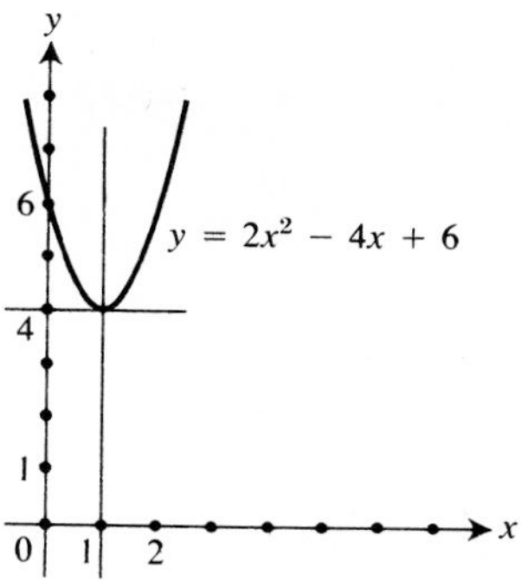

15. No

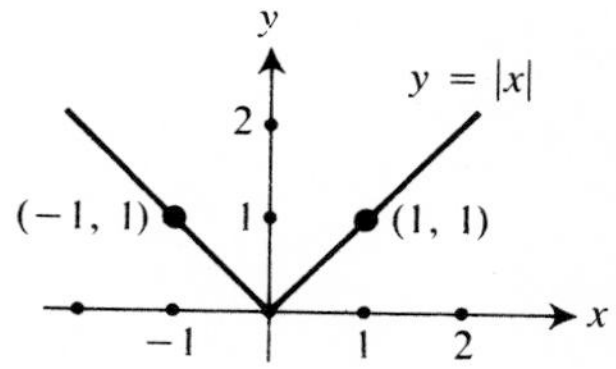

Review Exercise Set 1.1

1. a) 7

b)

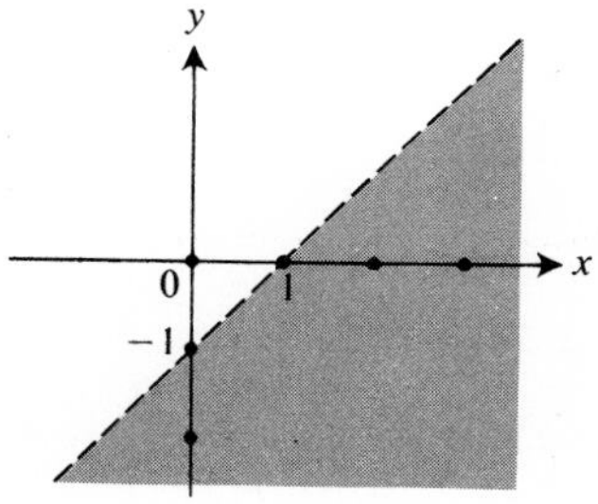

3. a) $(x - 2)^2 + (y + 1)^2 = 53$

b)

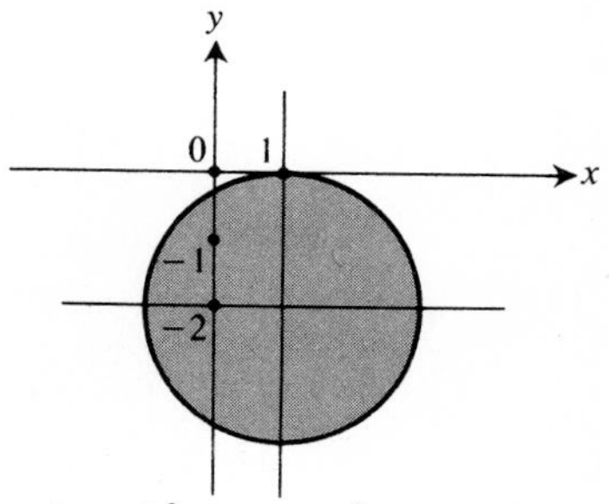

5. a) $x = -4$ b) $x - 3y = -7$

7. a) $x \neq 0, 5$ b) 6/7

9.

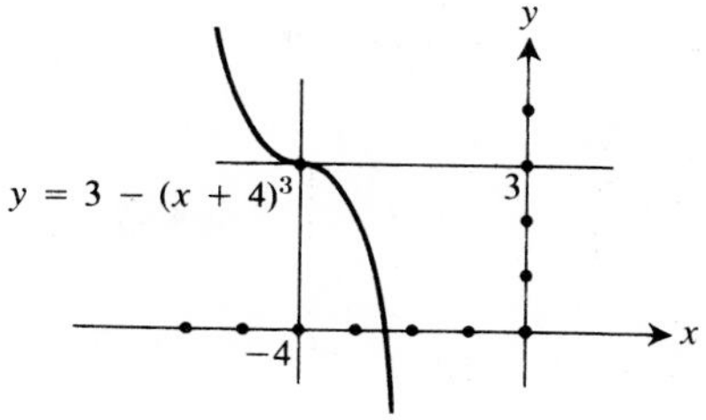

Review Exercise Set 1.2

1. a) 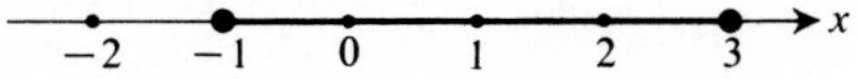

b) −1.6

3. a) $(x - 1)^2 + (y - 5)^2 = 10$ b) Center $(3, -4)$, radius 6
5. a) $4y = x + 17$ b) $x = 3$
7. a) $-5 \le x \le 5$

b) 4

c) 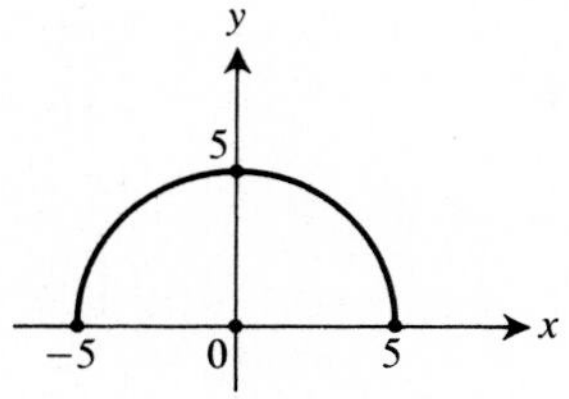

9. 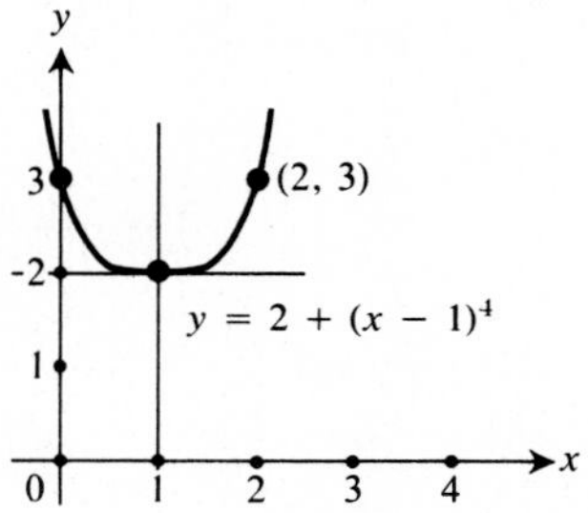

More Challenging Exercises 1

1. a)

$$\begin{aligned} -|a| \le a &\le |a| \\ \text{Adding:} \quad -|b| \le b &\le |b| \\ \hline -(|a| + |b|) \le a + b &\le |a| + |b| \end{aligned}$$

so $|a + b| \le |a| + |b|$.

b) From (a), you have $|a| = |(a - b) + b| \le |a - b| + |b|$, so $|a - b| \ge |a| - |b|$.

3. From Exercise 2,

$$-2(a_1a_2 + b_1b_2) \le 2\sqrt{a_1{}^2 + b_1{}^2} \cdot \sqrt{a_2{}^2 + b_2{}^2}.$$

Adding $a_1{}^2 + a_2{}^2 + b_1{}^2 + b_2{}^2$ to both sides, you obtain

$$(a_2 - a_1)^2 + (b_2 - b_1)^2 \le (\sqrt{a_1{}^2 + b_1{}^2} + \sqrt{a_2{}^2 + b_2{}^2})^2.$$

The desired relation follows upon taking square roots.

5. 5
7. $x^2 + y^2 + 6x - 6y + 9 = 0$
9. $8 - \sqrt{34}$

CHAPTER 2

Section 2.1

1. $m_{\text{sec}} = 8.01,\ m_{\text{chord}} = 8$

3. $m_{\text{sec}} = 0.31,\ m_{\text{chord}} = 0.01$

5. 8

7. 0

9. $\frac{1}{4}$

11. 20 mph

13. 1.34164

15. 2.582

Section 2.2

1. $\delta = \epsilon$

3. $\delta = \epsilon/5$

5. $\lim_{x\to 0} \left|\dfrac{x^3 + x^2 + 2}{x}\right| = \infty$

7. 0

9. $\lim_{r\to 0} \left|\dfrac{2r^2 - 3r}{r^3 + 4r^2}\right| = \infty$

11. 0

13. 0

15. 4/3

17. 2

19. 4

21. Does not exist

23. 2

25. 0.166667

27. -0.5

29. 1

Section 2.3

1. $2x - 3$

3. $-2/(2x + 3)^2$

5. $1/(x + 1)^2$

7. 3

9. $14x^6 + 8x$

11. $x - \frac{3}{2}$

13. $324x^3 - 160x^4$

15. $4x^3 + 12x^2 + 8x$

17. $36x^2 + 40x$

19. $x^2 - \frac{1}{3}$

21. $x + 12y = 50$

23. a) $1/2\sqrt{x}$

b)
$$\lim_{\Delta x\to 0} \frac{\sqrt{x+\Delta x} - \sqrt{x}}{\Delta x} = \lim_{\Delta x\to 0} \frac{(\sqrt{x+\Delta x} - \sqrt{x})(\sqrt{x+\Delta x} + \sqrt{x})}{\Delta x(\sqrt{x+\Delta x} + \sqrt{x})}$$
$$= \lim_{\Delta x\to 0} \frac{x + \Delta x - x}{\Delta x(\sqrt{x+\Delta x} + \sqrt{x})}$$
$$= \lim_{\Delta x\to 0} \frac{1}{\sqrt{x+\Delta x} + \sqrt{x}} = \frac{1}{2\sqrt{x}}$$

c) $\dfrac{3}{2\sqrt{x}} - 4x$

d) $\dfrac{\sqrt{5}}{2\sqrt{x}} - \dfrac{\sqrt{7}}{2\sqrt{x}}$

25. a) 12 in^3/sec b) 75 in^3/sec

27. a) 256π in^2 b) 256π in^2/sec

29. $$\lim_{\Delta x\to 0}\frac{f(x_1+\Delta x)-f(x_1-\Delta x)}{2\cdot\Delta x}$$

$$= \lim_{\Delta x\to 0}\left(\frac{1}{2}\cdot\frac{f(x_1+\Delta x)-f(x_1)}{\Delta x}+\frac{1}{2}\cdot\frac{f(x_1+(-\Delta x))-f(x_1)}{-\Delta x}\right)$$

$$= \lim_{\Delta x\to 0}\frac{1}{2}\cdot\frac{f(x_1+\Delta x)-f(x_1)}{\Delta x}$$

$$+ \lim_{-\Delta x\to 0}\frac{1}{2}\cdot\frac{f(x_1+(-\Delta x))-f(x_1)}{-\Delta x}$$

$$= \tfrac{1}{2}\cdot f'(x_1)+\tfrac{1}{2}\cdot f'(x_1) = f'(x_1)$$

31. 2

33. 14.1017

35. 0.74522

Section 2.4

1. $\lim_{x\to 2}\left|\dfrac{1}{2-x}\right| = \infty$

3. ∞

5. ∞

7. $-\infty$

9. $-\infty$

11. 3/2

13. ∞

15. 0

17. $-\infty$

19. ∞

21. a)

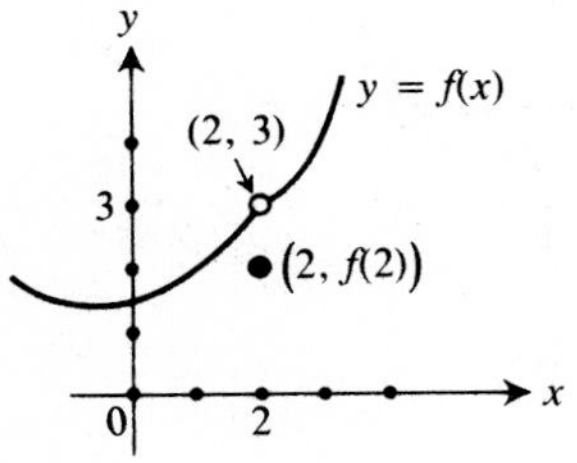

b)

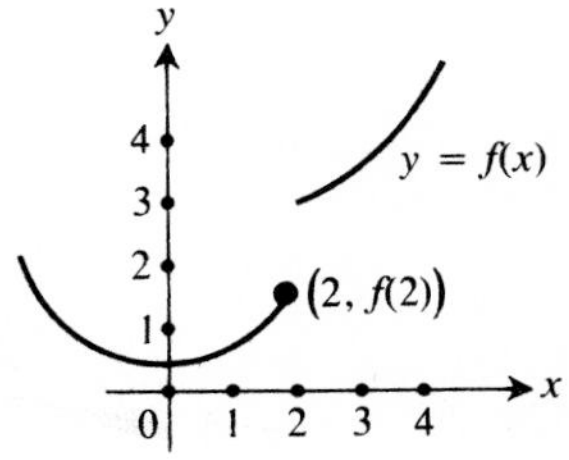

c) 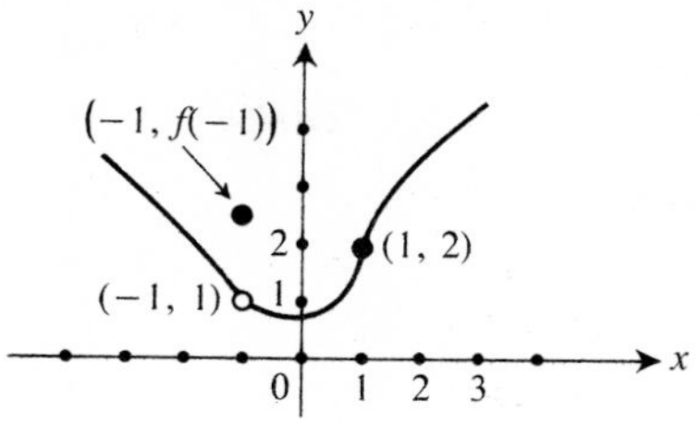

23. No. $\lim_{x\to 2} f(x) = -8 \neq 8$

25. $\dfrac{8\sqrt{4000}}{\sqrt{5280}} = 6.9631$ mi/sec

27. 0

29. 7.38906

31. ∞

Section *2.5

1. a) 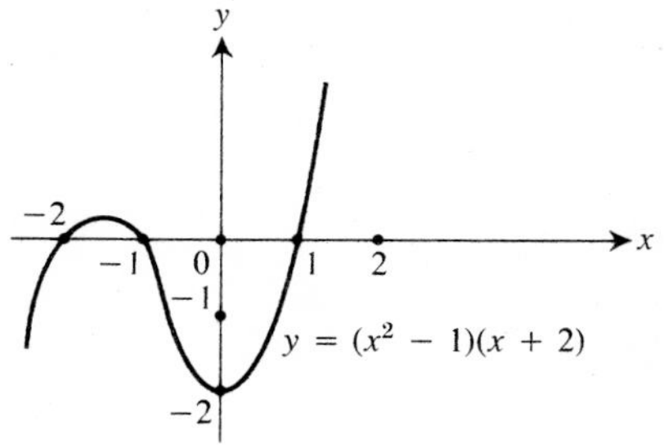

b) 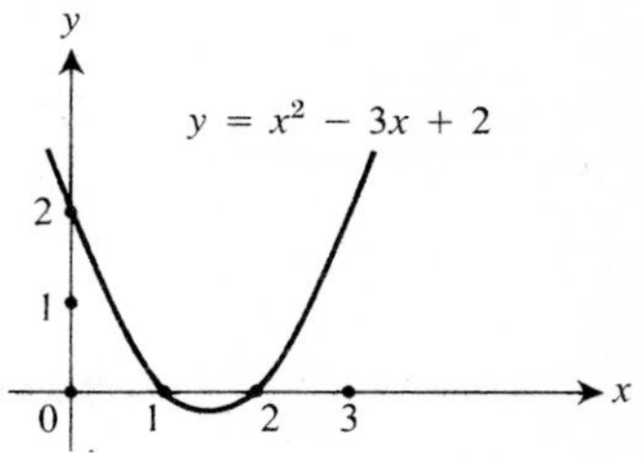

c) 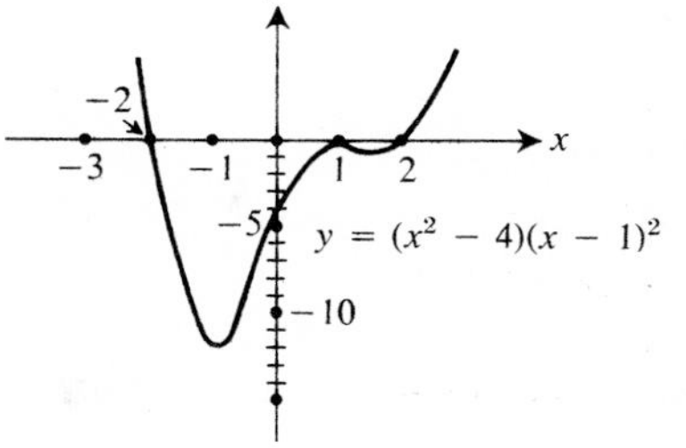

d) 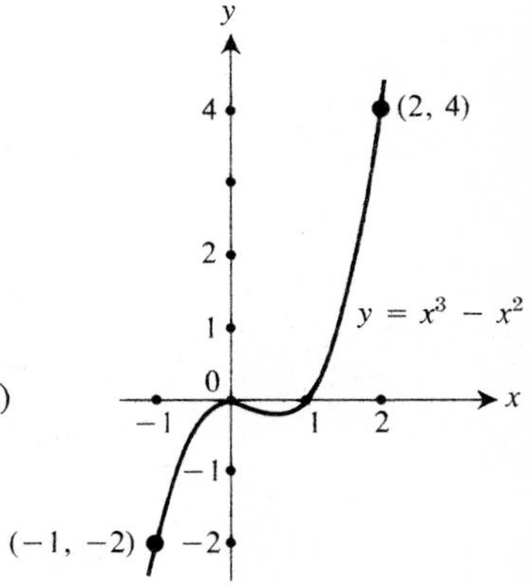

e) 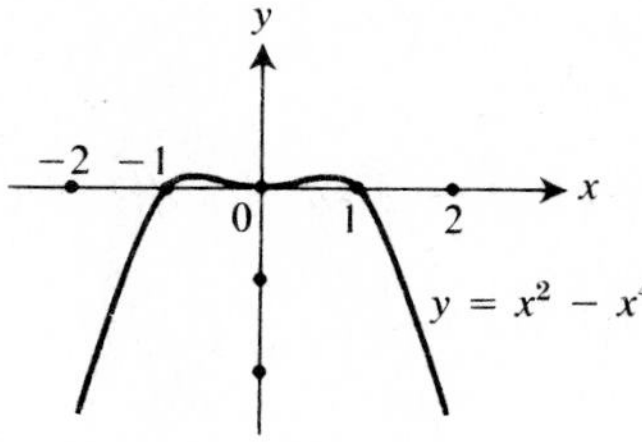

f) 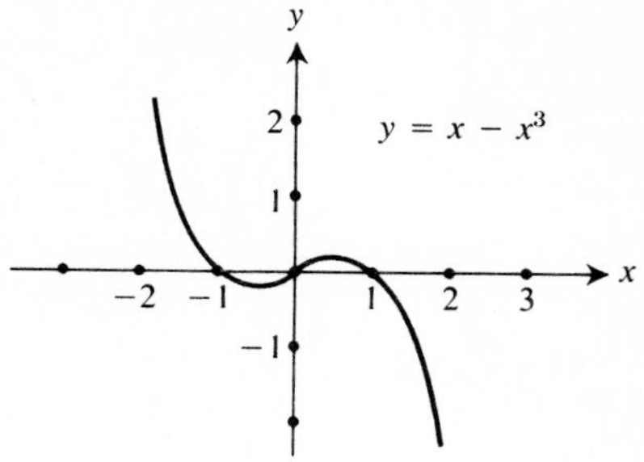

3. a) 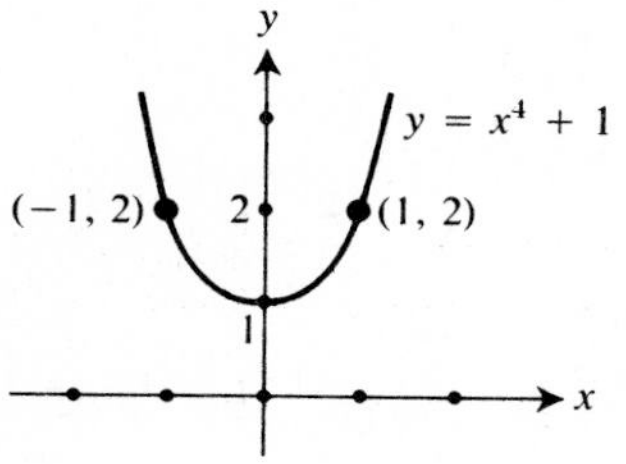

b) 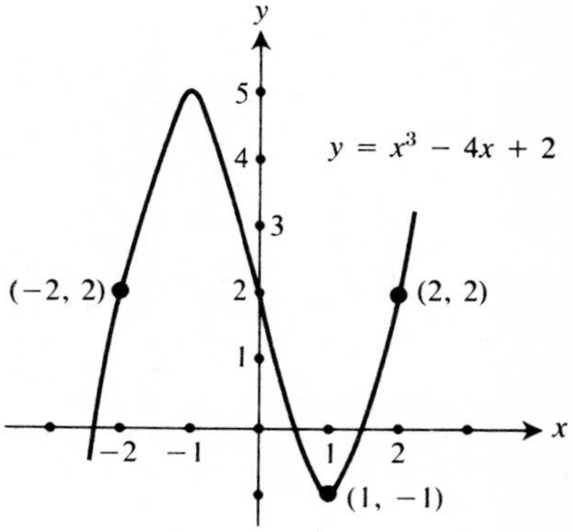

c) 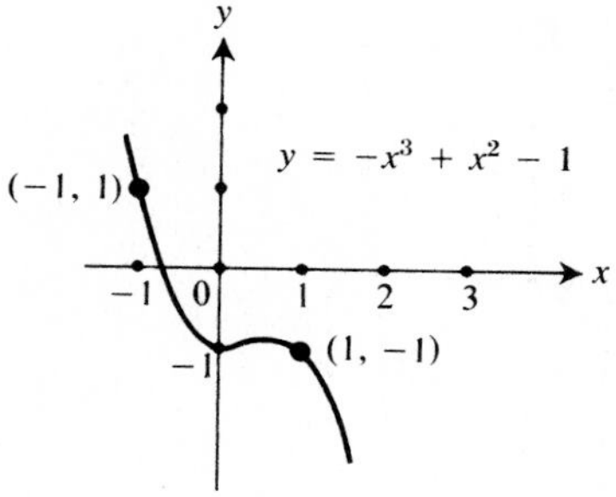

d) 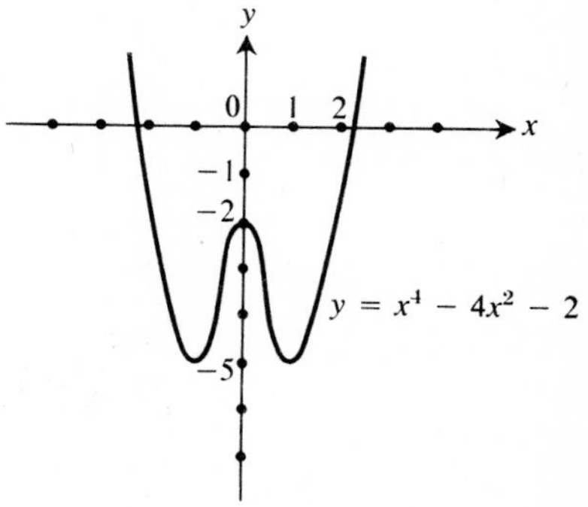

5. a) $x = 1$, sign change b) $x = 0$, sign change; $x = -1$, no sign change

c) $x = 3$, no sign change; $x = -1$, sign change

7. a)

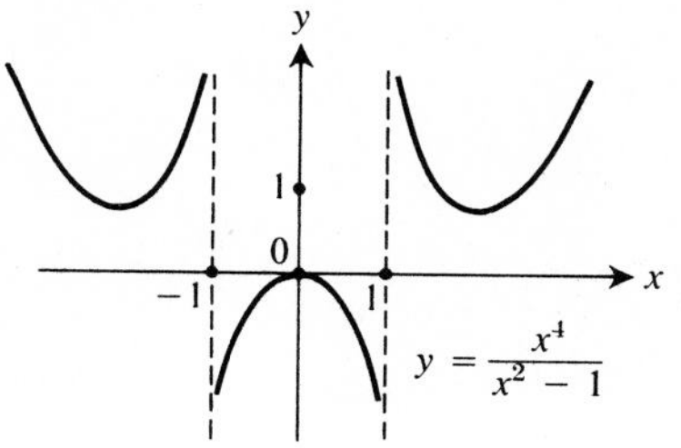
y
1
0
−1
1
x
$y = \frac{x^4}{x^2 - 1}$

b)

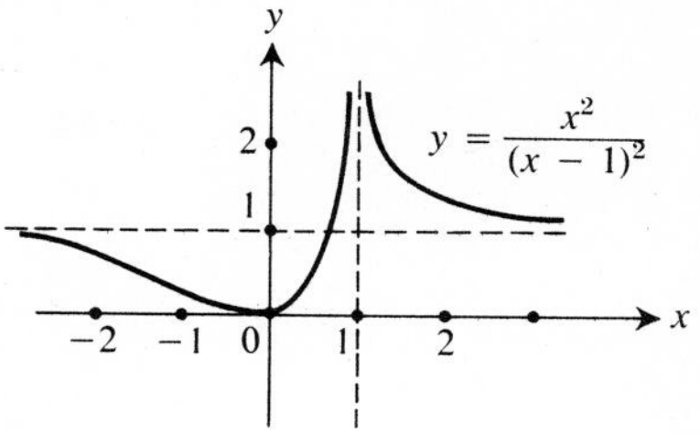
y
2
1
−2
−1
0
1
2
x
$y = \frac{x^2}{(x - 1)^2}$

c)

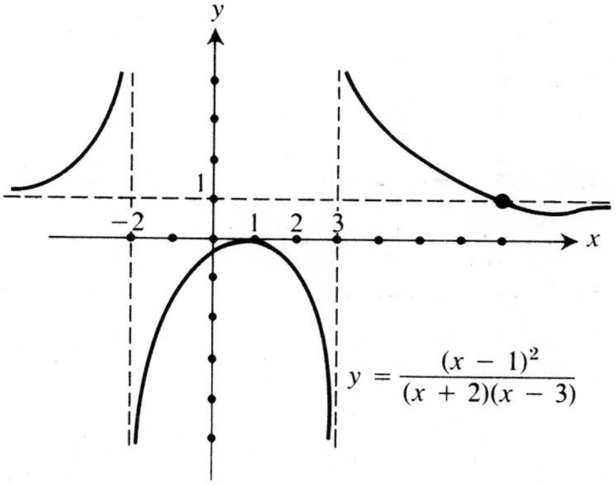
y
1
−2
1
2
3
x
$y = \frac{(x - 1)^2}{(x + 2)(x - 3)}$

d)

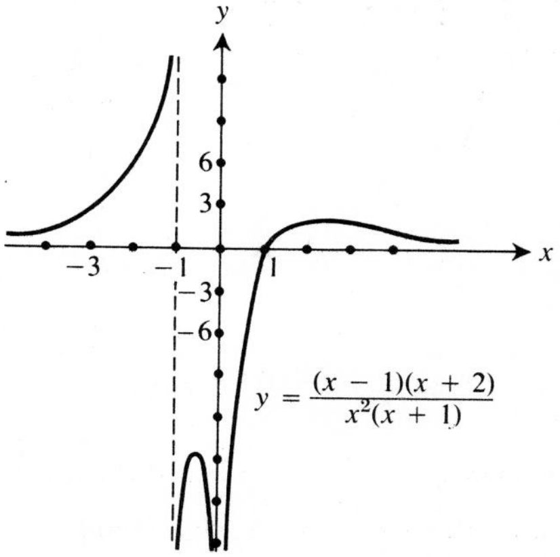
y
6
3
−3
−1
1
−3
−6
x
$y = \frac{(x - 1)(x + 2)}{x^2(x + 1)}$

e)

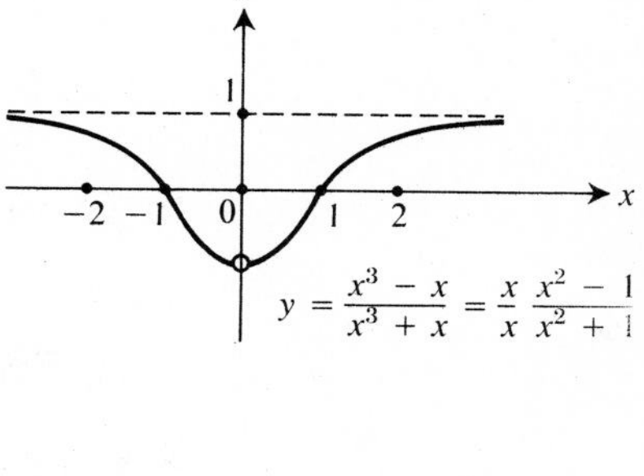
y
1
−2
−1
0
1
2
x
$y = \frac{x^3 - x}{x^3 + x} = \frac{x}{x}\,\frac{x^2 - 1}{x^2 + 1}$

9. a) $y = x$

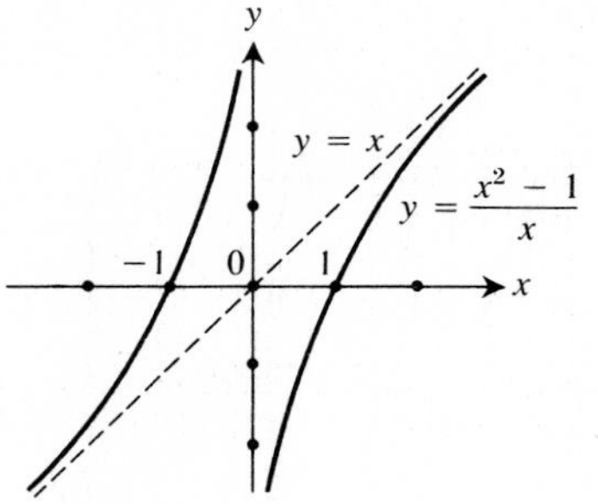

b) $y = x - 2$

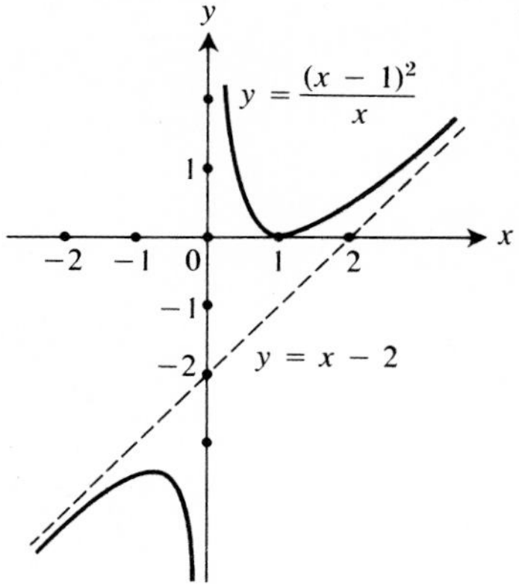

c) $y = x - 1$

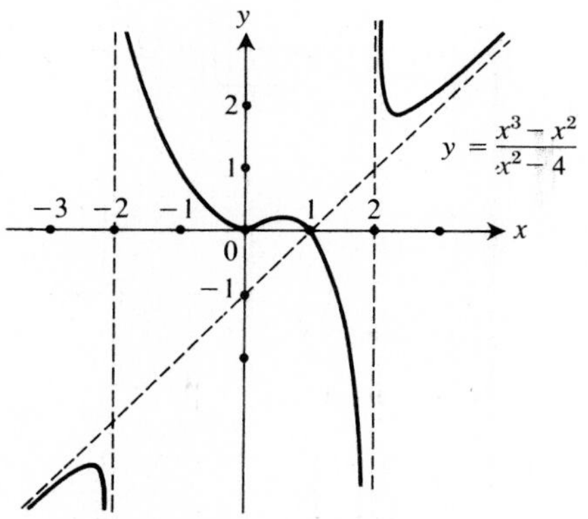

Review Exercise Set 2.1

1. a) $-2/3$ b) $-4/3$

3. a) $f'(x_1) = \lim_{\Delta x \to 0} \dfrac{f(x_1 + \Delta x) - f(x_1)}{\Delta x}$

b) $f'(x) = \lim_{\Delta x \to 0} \dfrac{(x + \Delta x)^2 - 3(x + \Delta x) - (x^2 - 3x)}{\Delta x}$

$= \lim_{\Delta x \to 0} \dfrac{x^2 + 2x \cdot \Delta x + (\Delta x)^2 - 3x - 3 \cdot \Delta x - x^2 + 3x}{\Delta x}$

$= \lim_{\Delta x \to 0} \dfrac{\Delta x(2x - 3 + \Delta x)}{\Delta x} = \lim_{\Delta x \to 0} (2x - 3 + \Delta x) = 2x - 3$

5. a) i) $-1/3$ ii) ∞

b) No, for $\lim_{x \to -3} f(x) = \lim_{x \to -3} \dfrac{(x + 3)(x - 3)}{x + 3} = -6 \neq f(-3)$.

Review Exercise Set 2.2

1. a) ∞ b) $1/4$ c) $-1/10$ d) 0

3. $f'(x) = \lim_{\Delta x \to 0} \dfrac{[1/(2(x + \Delta x) + 1)] - [1/(2x + 1)]}{\Delta x}$

$= \lim_{\Delta x \to 0} \dfrac{(2x + 1 - 2x - 2 \cdot \Delta x - 1)/([2(x + \Delta x) + 1][2x + 1])}{\Delta x}$

$= \lim_{\Delta x \to 0} \dfrac{-2}{[2(x + \Delta x) + 1][2x + 1]} = \dfrac{-2}{(2x + 1)^2}$

5. a) 0

b) 7/4

*7. ∞

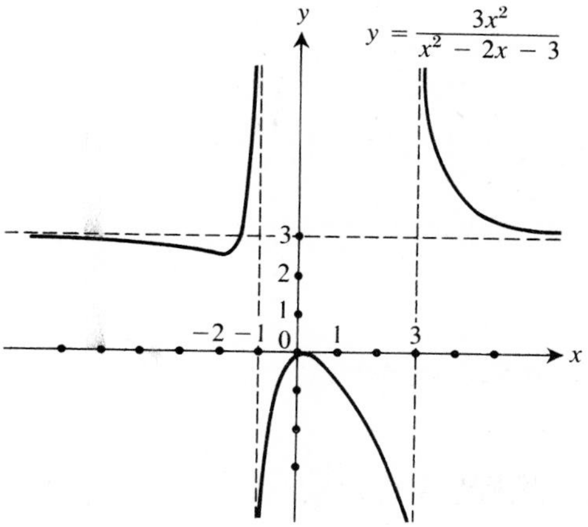

More Challenging Exercises 2

1. 1

3. 2

5. 0

7. ∞

9. 1

11. a) For some $\epsilon > 0$, there does not exist $\delta > 0$.

b) For some apple blossom, there exists no apple.

d) Find one $\epsilon > 0$ such that for every $\delta > 0$, there is an x_δ such that $0 < |x_\delta - a| < \delta$ but $|f(x_\delta) - c| \geq \epsilon$.

13. Let $\epsilon > 0$ be given. Find $\delta_1 > 0$ such that

$$L - \frac{\epsilon}{2} < f(x) < L + \frac{\epsilon}{2}$$

if

$$0 < |x - a| < \delta_1$$

and $\delta_2 > 0$ such that

$$M - \frac{\epsilon}{2} < g(x) < M + \frac{\epsilon}{2}$$

if

$$0 < |x - a| < \delta_2.$$

Let δ be the minimum of δ_1 and δ_2, so both inequalities hold if $0 < |x - a| < \delta$. Adding the inequalities,

$$L + M - \epsilon < f(x) + g(x) < L + M + \epsilon$$

if

$$0 < |x - a| < \delta.$$

15. a) Incorrect; replace "$|f(x) - a|$" by "$|f(x) - f(a)|$".

b) Incorrect; delete "$0<$".

c) Correct

d) Incorrect; replace "some" by "each".

e) Incorrect; replace "$\leq$" by "$<$".

f) Correct

17. Let

$$f(x) = \begin{cases} 1 & \text{for } x \geq 2, \\ 0 & \text{for } x < 2, \end{cases} \quad \text{and} \quad g(x) = \begin{cases} 0 & \text{for } x \geq 2, \\ 10 & \text{for } x < 2. \end{cases}$$

Then $f(x)g(x) = 0$ for all x.

CHAPTER 3

Section 3.1

1. $6x + 17$

3. $2x/3$

5. $-3/x^2$

7. $12x^2 + (4/x^3)$

9. $(x^2 - 1)(2x + 1) + (x^2 + x + 2)(2x)$

11. $(x^2 + 1)[(x - 1)3x^2 + (x^3 + 3)] + [(x - 1)(x^3 + 3)](2x)$

13. $4 + (3/x^2)$

15. $[(x + 3)2x - (x^2 - 2)]/(x + 3)^2$

17. $(x^2 + 2)((x^2 + 9) + (x - 3)2x) - (x^2 + 9)(x - 3)2x]/(x^2 + 2)^2$

19. $$\frac{(x - 1)(4x^2 + 5)[(2x + 3)2x + (x^2 - 4)2] - (2x + 3)(x^2 - 4)[(x - 1)8x + (4x^2 + 5)]}{(x - 1)^2(4x^2 + 5)^2}$$

21. $x \cos x + \sin x$

23. $2 \sin x \cos x$

25. $$\frac{(\cos x)^2 + (\sin x)^2}{(\cos x)^2}$$

27. $$\frac{3x^2 \sin x - x^3 \cos x}{(\sin x)^2}$$

29. $$\frac{(x^2 - 4x) \cos x - (2x - 4) \sin x}{(x^2 - 4x)^2}$$

31. Tangent line: $y = 10x - 12$; normal line: $10y + x = 82$

33. Tangent line: $y = -5x - 3$; normal line: $5y - x = -15$

Section 3.2

1. $dy = \dfrac{1}{(x + 1)^2}\, dx$

3. $dA = 2\pi r\, dr$

5. $dx = \dfrac{-4t}{(t^2 - 1)^2}\, dt$

7. 0.00000975

9. 10.05

11. 3.975

13. a) $3/\pi$ ft

b) The estimate is exact, for the circumference of the earth is a linear function of the radius.

15. $7/24\ \text{ft}^2$

17. 6%

19. 0.5%

21. $\epsilon = \Delta x$; $\lim_{\Delta x \to 0} \epsilon = \lim_{\Delta x \to 0} \Delta x = 0$

23. 2.022148

Section 3.3

1. a) 3 b) 3
3. $(2x + 1)^{-1/2}$
5. $8x - 2 + 5x^{2/3}$
7. $-\frac{1}{2}x^{-3/2}$
9. $\frac{2}{3}x^{-1/3} + \frac{1}{5}x^{-4/5}$
11. $x(x^2 + 1)^{-1/2}$
13. $\dfrac{\frac{1}{2}(x + 1)x^{-1/2} - \sqrt{x}}{(x + 1)^2}$
15. $12(3x + 2)^3$
17. $9x^2(x^2 + 3x)^2(x^3 - 1)^2 + 2(x^3 - 1)^3(x^2 + 3x)(2x + 3)$
19. $\dfrac{16x(4x^2 + 1)^2 - 16x(8x^2 - 2)(4x^2 + 1)}{(4x^2 + 1)^4}$
21. $\sqrt{2x + 1}\left(\dfrac{2(2x + 5)(4x^2 - 3x)(8x - 3) - 2(4x^2 - 3x)^2}{(2x + 5)^2}\right) + \dfrac{(4x^2 - 3x)^2}{2x + 5}(2x + 1)^{-1/2}$
23. $2 \cos 2x$
25. $3 \sin^2 x \cos x$
27. $\frac{1}{2}(x + \sin x)^{-1/2}(1 + \cos x)$
29. $4y + 3x = 25$
31. $3y - x = 5$
33. 70.4875
35. 8.6875

Section 3.4

1. $y' = 5x^4 - 12x^3$, $y'' = 20x^3 - 36x^2$, $y''' = 60x^2 - 72x$
3. $y' = (1/\sqrt{5})(-\frac{1}{2})x^{-3/2}$, $y'' = (1/\sqrt{5})(3/4)x^{-5/2}$, $y''' = (1/\sqrt{5})(-15/8)x^{-7/2}$
5. $y' = x(x^2+1)^{-1/2}$, $y'' = -x^2(x^2 + 1)^{-3/2} + (x^2 + 1)^{-1/2}$,
$y''' = 3x^3(x^2 + 1)^{-5/2} - 3x(x^2 + 1)^{-3/2}$
7. $y' = (x + 1)^{-2}$, $y'' = -2(x + 1)^{-3}$, $y''' = 6(x + 1)^{-4}$
9. Velocity 2; acceleration 18
11. a) $v = -32t + 48$ ft/sec b) $a = -32$ ft/sec^2 c) 48 ft/sec
d) $t = \frac{3}{2}$ sec e) 36 ft f) $0 \le t \le 3$
13. Speed $= |t|\sqrt{4 + 9t^2}$, slope $= 2/3t$
15. Speed $= 1/\sqrt{2}$, slope 0
17. $3y - 8x = 13$
19. $-1/2$, $3/4$
21. 0.053233, 0.16606
23. -2.3404, -2.191

Section 3.5

1. 3/4
3. 1
5. −1/4
7. 1/3
9. −1
11. 17/31
13. 24/65
15. $5y - 2x = 1, \quad 2y + 5x = 12$
17. The first curve has slope −1 and the second has slope 1. The curves are orthogonal.
19. Let (x_0, y_0) be a point of intersection. Since both c and k are nonzero, neither x_0 nor y_0 is zero at a point of intersection. By implicit differentiation, the slope of $y^2 - x^2 = c$ at (x_0, y_0) is x_0/y_0, while the slope of $xy = k$ is $-y_0/x_0$. The curves are orthogonal.

Review Exercise Set 3.1

1. $(x^2 - 3x)(12x^2 - 2) + (4x^3 - 2x + 17)(2x - 3)$
3. $dy = (6x - 6)\,dx$
5. −144
7. $y' = 5(4x^3 - 7x)^4(12x^2 - 7)$

   $y'' = 120x(4x^3 - 7x)^4 + 20(12x^2 - 7)^2(4x^3 - 7x)^3$
9. $32y - x = 67$

Review Exercise Set 3.2

1. $\dfrac{x^2 + 3}{x} + 2x \ln x$
3. 0.98
5. $-\frac{1}{3}(x^3 - 3x + 2)^{-4/3}(3x^2 - 3)$
7. $\frac{81}{8}(3x + 4)^{-5/2}$
9. $y = -1$

More Challenging Exercises 3

1. $(fg)'(a) = f(a)g'(a) + f'(a)g(a) = 0 \cdot g'(a) + f'(a) \cdot 0 = 0$
3. Let $p(x) = (x - a)^2 q(x)$. Then $p(a) = 0 \cdot q(a) = 0$. Also

$$p'(x) = (x - a)^2 q'(x) + 2(x - a)q(x),$$

so

$$p'(a) = 0 \cdot q'(a) + 0 \cdot q(a) = 0.$$

5. $y = 2x + 2, \quad y = 6x - 14$
7. From $y = f(x)/g(x)$, we have $y \cdot g(x) = f(x)$. Then

$$y \cdot g'(x) + \frac{dy}{dx} g(x) = f'(x),$$

so

$$\frac{dy}{dx} = \frac{f'(x) - y \cdot g'(x)}{g(x)} = \frac{f'(x) - (f(x)/g(x))g'(x)}{g(x)}$$

$$= \frac{f'(x)g(x) - f(x)g'(x)}{g(x)}.$$

9. $\left.\dfrac{d(f(g(h(t))))}{dt}\right|_{t=t_1} = f'(g(h(t_1))) \cdot g'(h(t_1)) \cdot h'(t_1).$

CHAPTER 4

Section 4.1

1. $\sqrt{3}/2$

3. $-1/\sqrt{3}$

5. $-\sqrt{2}$

7. -2

9. 0

11. 1

13. 0

15. -1

17. Undefined

19. $\sqrt{2}$

21. $2\sqrt{2}/3$

23. $-1/\sqrt{24}$

25. $-\sqrt{15}$

27. $-4\sqrt{5}/9$

29. $-7/8$

31. a) $(u, -v)$ b) $\sin(-x) = -v = -\sin x; \quad \cos(-x) = u = \cos x$

33. a) $(v, -u)$ b) $\sin\left(x - \dfrac{\pi}{2}\right) = -u = -\cos x; \quad \cos\left(x - \dfrac{\pi}{2}\right) = v = \sin x$

35. $\sec(-x) = \dfrac{1}{\cos(-x)} = \dfrac{1}{\cos x} = \sec x$

37. $\sin\left(x - \dfrac{\pi}{2}\right) = \sin x \cos\left(-\dfrac{\pi}{2}\right) + \cos x \sin\left(-\dfrac{\pi}{2}\right) = (\sin x)(0) + (\cos x)(-1)$

$$= -\cos x$$

39. $\sec\left(x - \dfrac{\pi}{2}\right) = \dfrac{1}{\cos[x - (\pi/2)]} = \dfrac{1}{(\cos x)\cos(-\pi/2) - (\sin x)\sin(-\pi/2)}$

$$= \frac{1}{(\cos x)(0) - (\sin x)(-1)} = \frac{1}{\sin x} = \csc x$$

41. $\cos 2x = \cos^2 x - \sin^2 x = \cos^2 x - (1 - \cos^2 x) = 2\cos^2 x - 1,$

$\cos 2x = \cos^2 x - \sin^2 x = (1 - \sin^2 x) - \sin^2 x = 1 - 2\sin^2 x$

43. $\sqrt{39}$

Section 4.2

1. Amplitude 1, period 2π

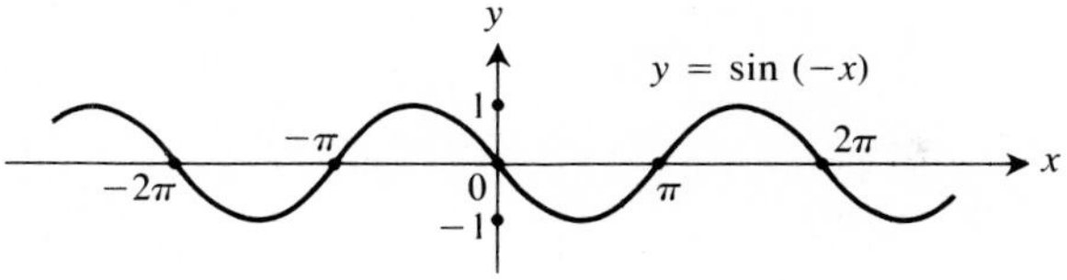

3. Amplitude 3, period $2\pi/3$

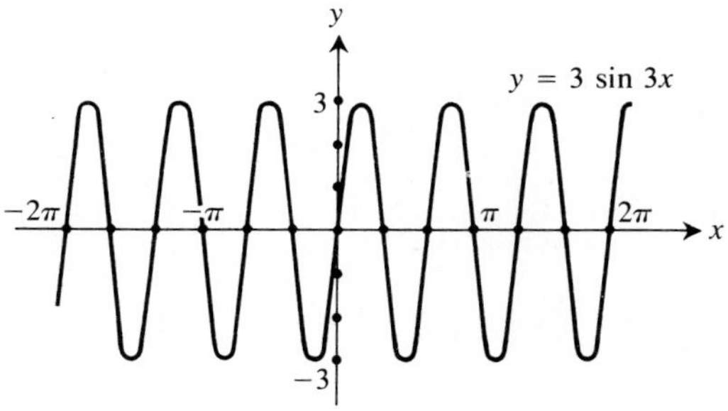

5. Amplitude 2, period 2π

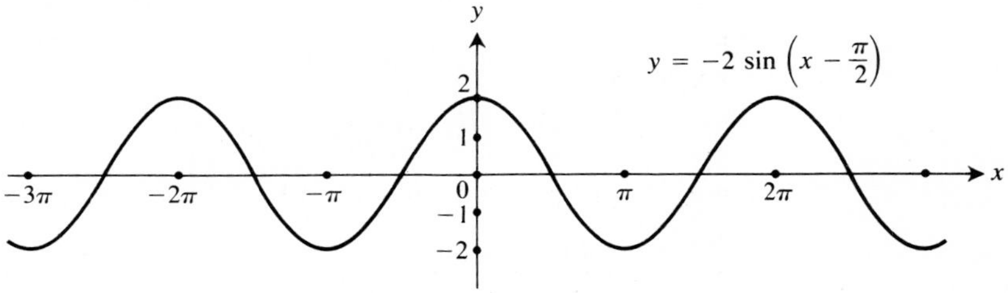

7. Amplitude 3, period $\pi/2$

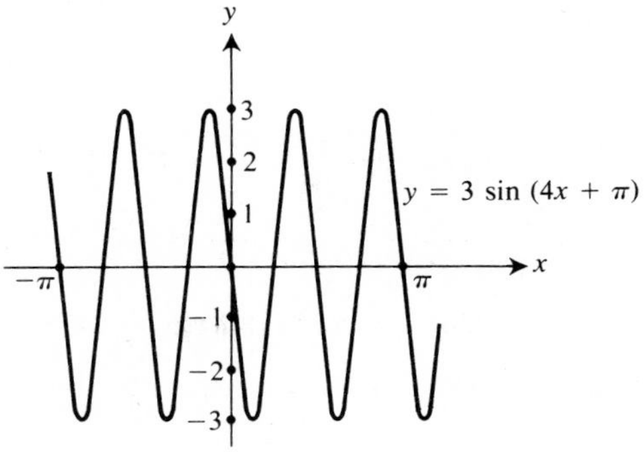

9. Amplitude 5, period 8π

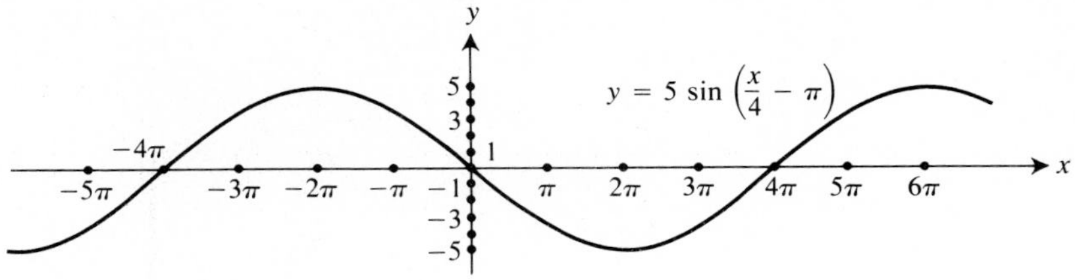

11. π;

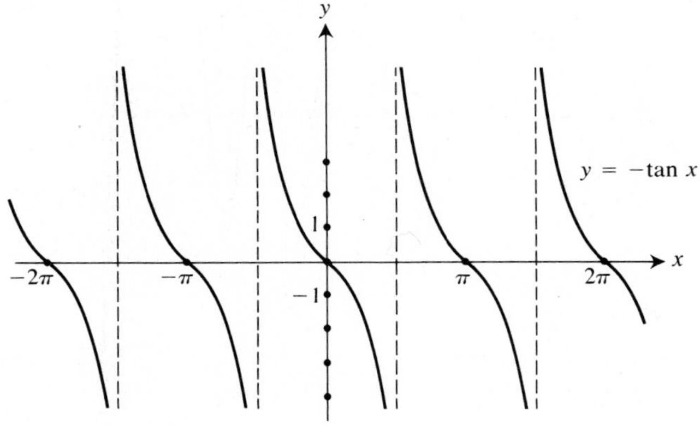

13. 2π;

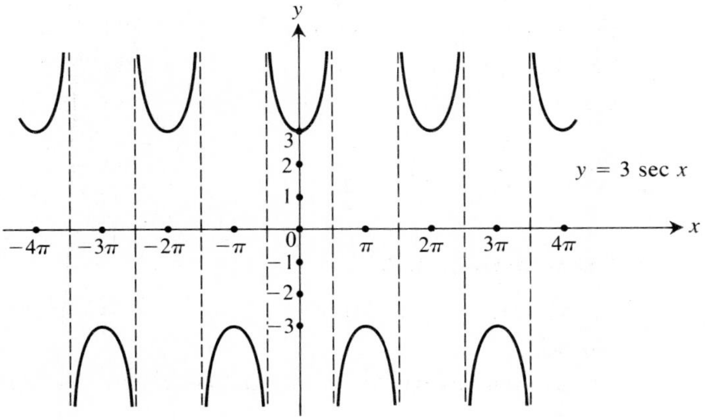

15. π;

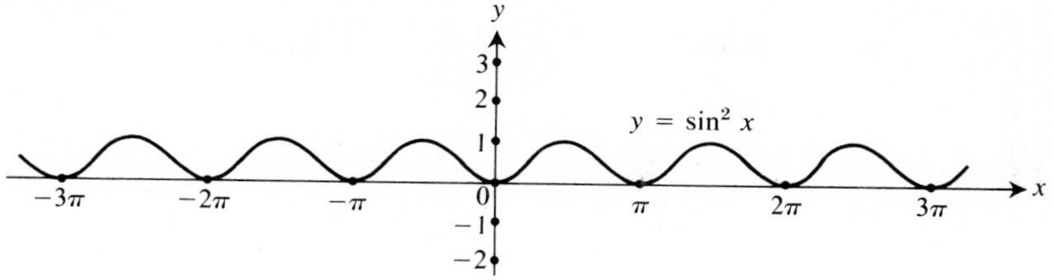

17. 2π;

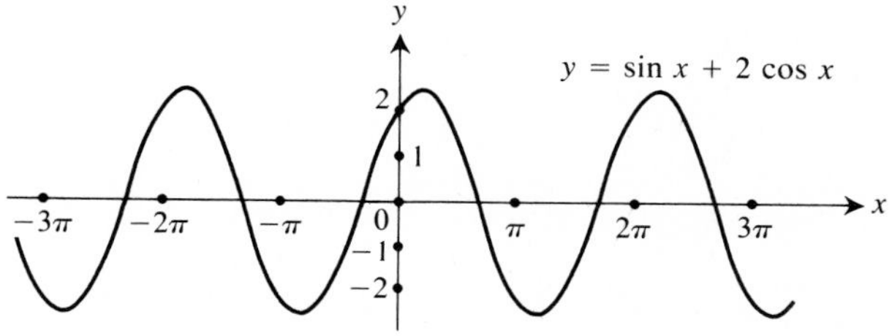

Section 4.3

1. Does not exist

3. 1

5. 1

7. 1

9. Does not exist

11. $-x \sin x + \cos x$

13. $(x^2 + 3x) \sec x \tan x + (2x + 3) \sec x$

15. $2 \sin x \cos x$

17. $2 \sec^2 x \tan x$

19. $[\cot x + x \csc^2 x]/\cot^2 x$

21. $2 \cos 2x$

23. $6 \cos (2x - 3) \sin (2x - 3)$

25. $-2 \sin^3 x \cos x + 2 \sin x \cos^3 x$

27. $\frac{1}{2}(1 + 2 \cos^2 x)^{-1/2}(-4 \cos x \csc^2 x)$

29. $3[\cos (\tan 3x)] \sec^2 3x$

31. 1

33. 0

35. 0

37. $y - \dfrac{1}{\sqrt{2}} = \dfrac{1}{\sqrt{2}}\left(x - \dfrac{\pi}{4}\right)$

39. $\dfrac{180 + \sqrt{3}\pi}{360}$

Review Exercise Set 4.1

1. a) $-1/\sqrt{3}$ b) $-1/\sqrt{2}$

3. $(\sqrt{3} - 1)/(2\sqrt{2})$

5. $\pi/3$

7. a) $6 \sin^2 2x \cos 2x$ b) $-3x^4 \csc x^3 \cot x^3 + 2x \csc x^3$

Review Exercise Set 4.2

1. a) $-1/2$ b) $-1/\sqrt{3}$

3. $-5/16$

5.

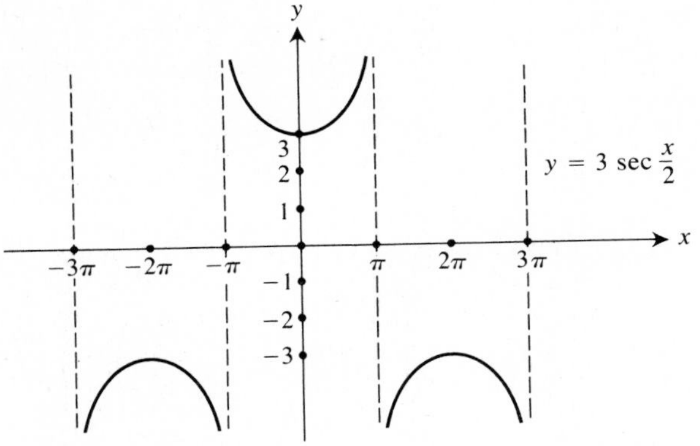

7. a) $2x \sec^2(x^2+1)$ b) $[(x-4)(2\sin x\cos x) - \sin^2 x]/[(x-4)^2]$

More Challenging Exercises 4

1. 1

3. 0

5. $\sqrt{3}/6$

CHAPTER 5

Section 5.1

1. $10\sqrt{3}$ in^2/min

3. $12/\sqrt{65}$ ft/sec

5. 15/7 ft/sec

7. $3/4\pi$ in./sec

9. 1 ft^2/min

11. a) $5\sqrt{3}/2$ ft/min
b) 25/2 ft^2/min

13. $-\pi l^3/60\sqrt{2}$ units^3/sec

Section 5.2

1. $\frac{97}{56} \approx 1.73214$

3. $\frac{59}{86} \approx 0.68605$

5. -2.652

7. 2.92402

9. 2.12937

11. Let $f(t)$ be Alice's height t years after she was born. Then f is a continuous function, $f(0) = 20$, and $f(t_1) = 69$ for some $t_1 > 0$. Hence, $f(t_0) = 48$ for some t_0 such that $0 < t_0 < t_1$ by Exercise 10.

13. We can write

$$f(x) = x^n\left(a_n + a_{n-1}\cdot\frac{1}{x} + \cdots + a_1\frac{1}{x^{n-1}} + \frac{a_0}{x^n}\right) \quad \text{for } x \neq 0.$$

If $a_n > 0$, then $\lim_{x\to\infty} f(x) = \infty$ and $\lim_{x\to-\infty} f(x) = -\infty$. Then there exists C such that $f(C) > 0$ while $f(-C) < 0$. By Theorem 5.1, $f(x_0) = 0$ for some x_0 such that $-C < x_0 < C$. If $a_n < 0$, the same arguments hold with some sign changes.

15. The number of hours of daylight is a continuous function $f(s)$ of the distance s from the North Pole along the 37° meridian. Let b be the distance from the North to the South Pole along this meridian. By elementary astronomy, $f(0) = 24$ while $f(b) = 0$. Thus there exists a point on the meridian some distance s_0 from the North Pole such that $f(s_0) = 19$.

Section 5.3

1. Maximum 16; minimum 0

3. Maximum 1; minimum 1/5

5. a) Maximum 2; minimum -6

b) Maximum -3; minimum -7

c) Maximum -3; minimum -7

7. a) Maximum 7/5; minimum 13/10

b) Maximum 3/2; minimum 1

c) Maximum 1; minimum 1/2

d) Maximum 3/2; minimum 1/2

9. a) Maximum $\sqrt{2}$; minimum 1

b) Maximum $\sqrt{2}$; minimum -1

c) Maximum $\sqrt{2}$; minimum -1

d) Maximum 1; minimum $-\sqrt{2}$

11. You may write

$$f(x) = x^n\left(a_n + \frac{a_{n-1}}{x} + \cdots + \frac{a_1}{x^{n-1}} + \frac{a_0}{x^n}\right), \quad x \neq 0.$$

Then $\lim_{x\to\infty} f(x) = \lim_{x\to-\infty} f(x) = \infty$. Consequently, there exists $x > 0$ such that $f(x) > f(1)$ if $|x| > c$. Then the minimum value assumed on $[-c, c]$, which exists by Theorem 5.2, is also the minimum value assumed on the whole x-axis.

13. -14

15. -27

17. Maximum -1.91134 at $x = 1 + (\sqrt{6}/3)$;

minimum -4.08866 at $x = 1 - (\sqrt{6})/3$

19. Maximum 0.342427 at $x = -0.523598$;

minimum -0.342427 at $x = 0.523598$

Section 5.4

1. $f(-3) = f(-2) = 10; \quad c = -1/2$

3. Let $f(x_1) = f(x_2) = \cdots = f(x_r) = 0$ where $a \le x_1 < x_2 < \cdots < x_r \le b$. By Rolle's Theorem, there exists c_i between x_i and x_{i+1} for $i = 1, \ldots, r - 1$ such that $f'(c_i) = 0$.

5. $f'(x) = 3x^2 - 3 = 3(x^2 - 1) > 0$ if $2 \le x \le 4$. Hence $f(x)$ has at most one zero in $[2, 4]$. Inspection shows that $f(3) = 0$, and there can be no other zero in $[2, 4]$.

7. 1

9. 49/4

11. a) It is the average rate of change of $f(x)$ over $[a, b]$.

b) It is the instantaneous rate of change of $f(x)$ at c.

c) If f is continuous on $[a, b]$ and differentiable for $a < x < b$, then there exists c where $a < c < b$ such that the instantaneous rate of change of $f(x)$ at c is the same as the average rate of change of $f(x)$ over $[a, b]$.

13. We easily compute that

$$\frac{f(x_2) - f(x_1)}{x_2 - x_1} = \frac{a(x_2{}^2 - x_1{}^2) + b(x_2 - x_1)}{x_2 - x_1} = a(x_2 + x_1) + b$$

and

$$f'\left(\frac{x_2 + x_1}{2}\right) = 2a\left(\frac{x_2 + x_1}{2}\right) + b = a(x_2 + x_1) + b$$

also. Example 1 illustrates this.

Section 5.5

1. -8

3. a) $\dfrac{b}{a} = -4$ b) $a = 3, \quad b = -12$ c) No

5.

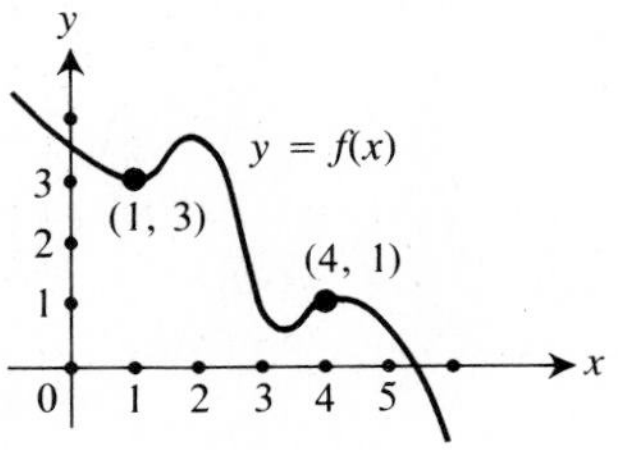

7.

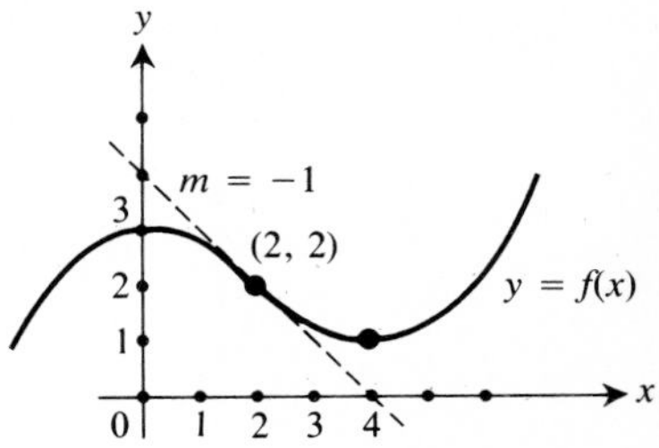

9. 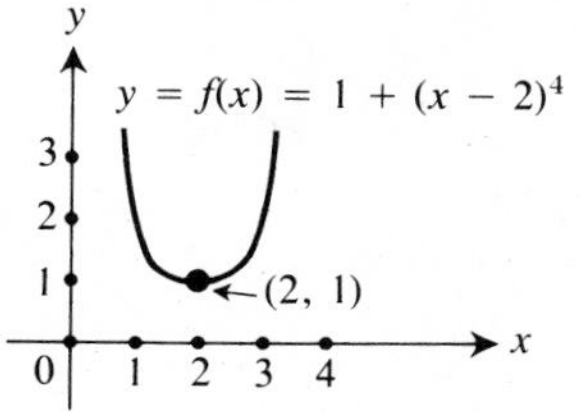

11. a) 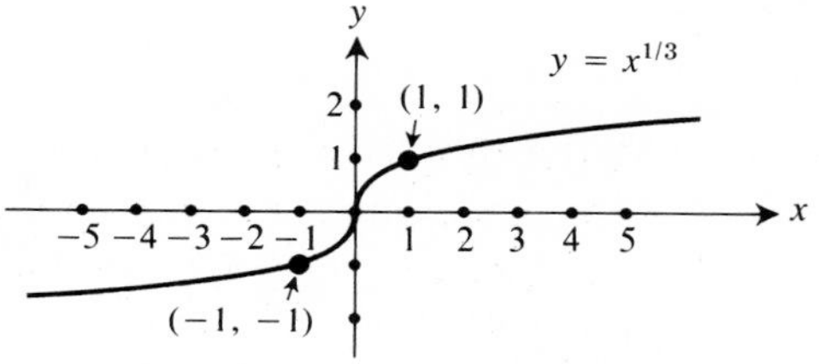

b) Yes c) No

13. a) $x \geq 3$

b) $x \leq 3$

c) None

d) -5 at $x = 3$

e) None

f) 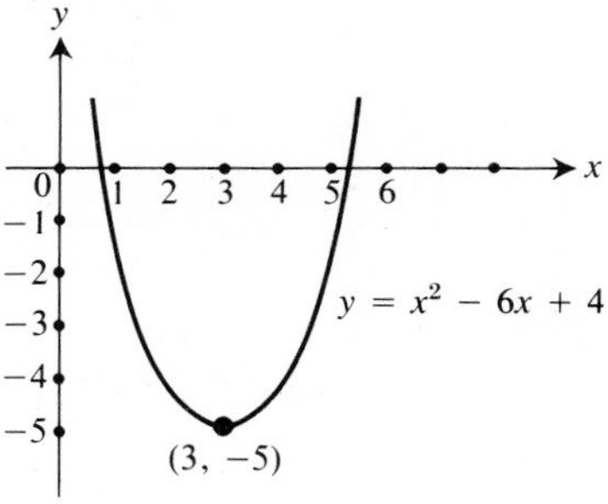

15. a) $x \leq -3$ or $x \geq 1$

b) $-3 \leq x \leq 1$

c) 5 at $x = -3$

d) $-17/3$ at $x = 1$

e) $(-1, -1/3)$

f) 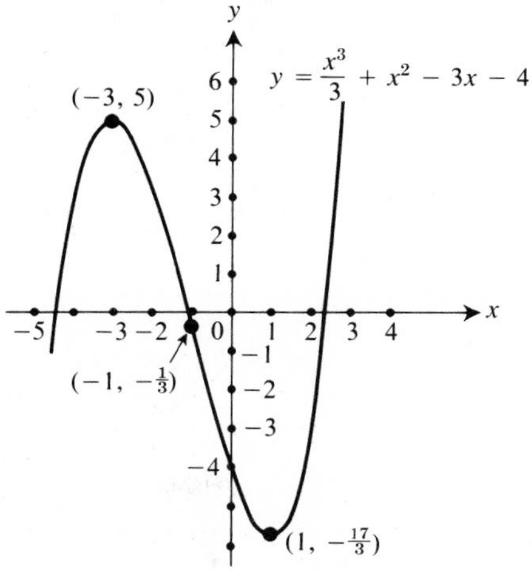

17. a) $x \leq 0$ or $x \geq 2$

b) $0 \leq x < 1$ or $1 < x \leq 2$

c) 0 at $x = 0$

d) 4 at $x = 2$

e) None

f)

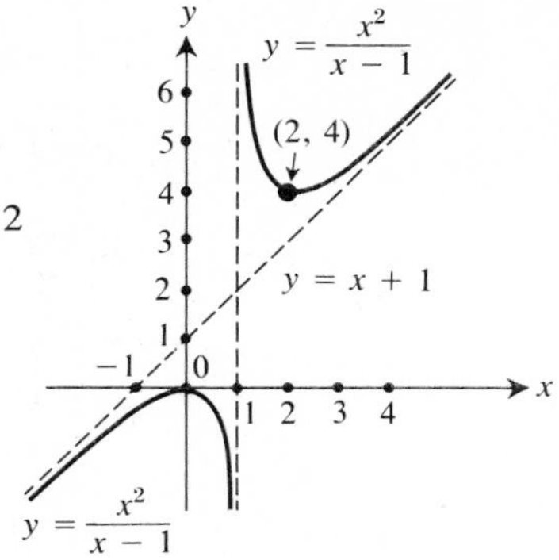

19. a) $x \leq -1$ or $x \geq 1$

b) $-1 \leq x \leq 1$

c) 5 at $x = -1$

d) -3 at $x = 1$

e) (0, 1)

f)

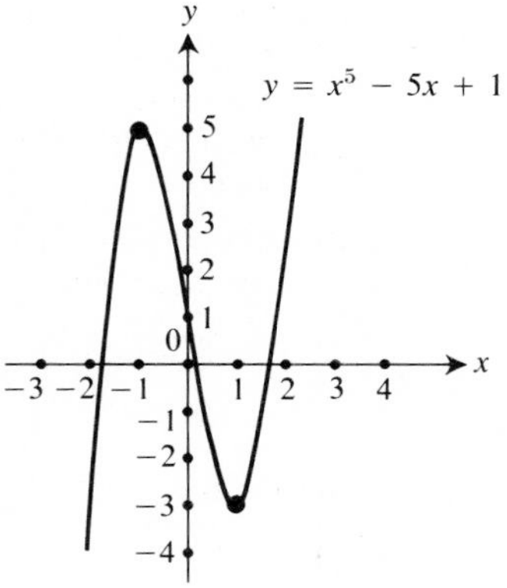

21. a) $-\dfrac{\pi}{6} + 2n\pi \leq x \leq \dfrac{7\pi}{6} + 2n\pi$ for each integer n

b) $\dfrac{7\pi}{6} + 2n\pi \leq x \leq \dfrac{11\pi}{6} + 2n\pi$ for each integer n

c) Where $x = (7\pi/6) + 2n\pi$

d) Where $x = (-\pi/6) + 2n\pi$

e) Where $x = (\pi/2) + n\pi$

f)

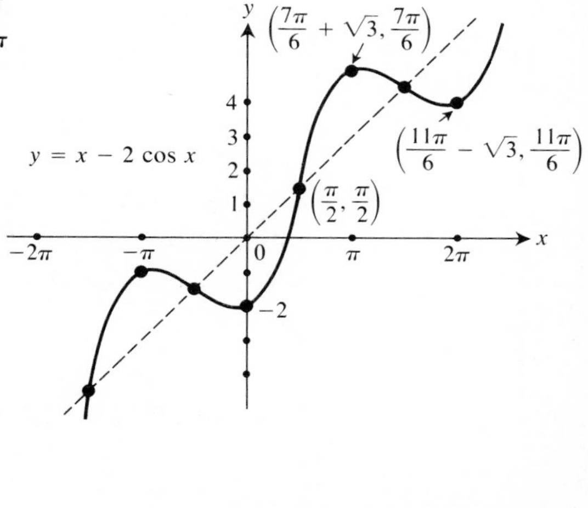

23.

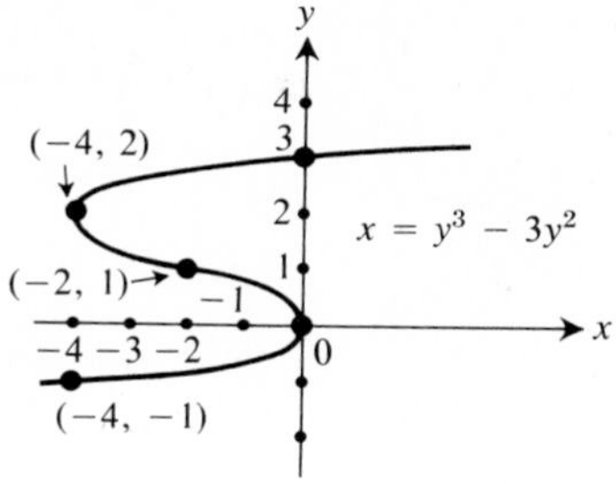

25.

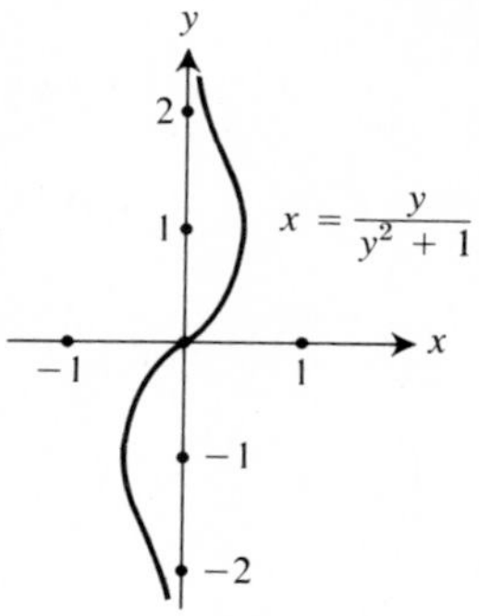

27. $-17 - (x - 2)^4$

29. If $g(a) > 0$, then $g(x) > 0$ near a since g is continuous. Since $f'(x) = g(x)(x - a)^2$, we see $f'(x) > 0$ for $x \neq a$ near a, so f is increasing at all points near a except possibly at a. Since f' exists, f is continuous near a, and it follows easily that f is increasing at a also. If $g(a) < 0$, all the inequalities are reversed, and f is decreasing near a.

31. Maximum (0.292893, 0.164214);
minimum (−0.5, −0.4375), (1.707107, −2.664214);
inflection points (−0.145497, −0.160058), and (1.145497, −1.45105)

Section 5.6

1. 25 ft^2

3. $x = 2$, $y = 4$

5. 6, 6

7. 108 in^2

9. 250 rods by 500 rods

11. 4 in. wide by 2 in. high

13. $\frac{4}{27}\pi a^2 b$

15. $3a^2\sqrt{3}/4$

17. $5\sqrt{3}$ miles

19. $6\sqrt{3}$ in. wide by $6\sqrt{6}$ in. deep

21. 2

23. a) $(A + B - b)x - a - (B + c)x^2$ b) $(A + B - b)/[2(B + c)]$

25. $(a^{2/3} + b^{2/3})^{3/2}$ ft

27. $24\sqrt{3}$ ft

Section 5.7

1. a) \$540/unit b) \$800/unit c) \$260/unit d) \$335/unit

3. a) \$90/stove b) 1000 stoves for \$40,000 profit

5. Average profit is $P(x)/x$. Differentiate and set equal to zero:

$$\frac{xP'(x) - P(x)}{x^2} = 0,$$

so $xP'(x) = P(x)$ and $P'(x) = P(x)/x$. That is, marginal profit equals average profit at an extremum of average profit.

7. The hypotheses give $S(I) + C(I) = I$. Then $S'(I) + C'(I) = 1$, so $C'(I) = 1 - S'(I)$.

9. 10 orders of 20 refrigerators each

Section 5.8

1. $2x + C$

3. $2x^4 - \frac{2}{3}x^3 + 4x + C$

5. $3x^{1/3} + C$

7. $\dfrac{x^5}{5} + \dfrac{2}{3}x^3 + x + C$

9. $2\sqrt{x+1} + C$

11. $\dfrac{1}{3}\sin 3x + C$

13. $8x - 19$

15. $\dfrac{x^2}{2} - \cos x + 4$

17. $\dfrac{1}{2}x^2 - \dfrac{2}{3}x^{3/2} + \dfrac{1}{6} + \pi$

Review Exercise Set 5.1

1. $3/10\pi$ ft/min

3. Maximum 19 at $x = 3$; minimum -1 at $x = 1$

5. $y = x^3 - 2x^2 + 5x - 6$

7. $r = 4\sqrt{6}$ units

Review Exercise Set 5.2

1. $3\sqrt{3}/2\sqrt{7}$ ft/min

3. Maximum 13 at $x = 3$; minimum -12 at $x = 2$

5. $s = t^3 - 4t^2 + 2t + 5$

7. $256/3\sqrt{3}$ units^2

More Challenging Exercises 5

1. Use 400 ft of fence for a square enclosure of 10,000 ft^2, and the rest for the circular enclosure. Total enclosed: 38,648 ft^2.

3. If f is n times differentiable for $a \leq x \leq b$ and $f(x)$ assumes the same value at $n + 1$ distinct points in $[a, b]$, then $f^{(n)}(c) = 0$ for some c where $a < c < b$.

5. Jog all the way.

7. 0.876726

9. $f(x) = x^{1/3}$, $a = 1$; then $a_{i+1} = -2a_i$

CHAPTER 6

Section 6.1

1. a) $a_0 + a_1 + a_2 + a_3$ b) $b_2^2 + b_3^2 + b_4^2 + b_5^2 + b_6^2$
c) $a_2 + a_4 + a_6 + a_8$ d) $a_8 + b_4^2 + a_{10} + b_5^2 + a_{12} + b_6^2$
e) $c + c^2 + c^3 + c^4 + c^5$ f) $2^{a_2} + 2^{a_3} + 2^{a_4}$

3. a) $\sum_{i=1}^{3} a_i b_i$ b) $\sum_{i=1}^{3} a_i b_{i+1}$ c) $\sum_{i=1}^{4} a_i^{\,i}$
d) $\sum_{i=1}^{3} a_i^{\,i+1}$ e) $\sum_{i=1}^{3} a_i b_{2i}^{\,2}$ f) $\sum_{i=1}^{3} a_i^{b_{3i}}$

5.
$$\begin{aligned}\sum_{i=1}^{n} (a_i + b_i)^2 &= (a_1 + b_1)^2 + \cdots + (a_n + b_n)^2\\ &= a_1^2 + 2a_1b_1 + b_1^2 + \cdots + a_n^2 + 2a_nb_n + b_n^2\\ &= a_1^2 + \cdots + a_n^2 + 2(a_1b_1 + \cdots + a_nb_n) + b_1^2 + \cdots + b_n^2\\ &= \sum_{i=1}^{n} a_i^2 + 2\sum_{i=1}^{n} a_ib_i + \sum_{i=1}^{n} b_i^2.\end{aligned}$$

7. 1.575

9. $S_2 = 5$, $s_2 = 1$

11. $S_4 \approx 0.76$, $s_4 \approx 0.63$

13. a)

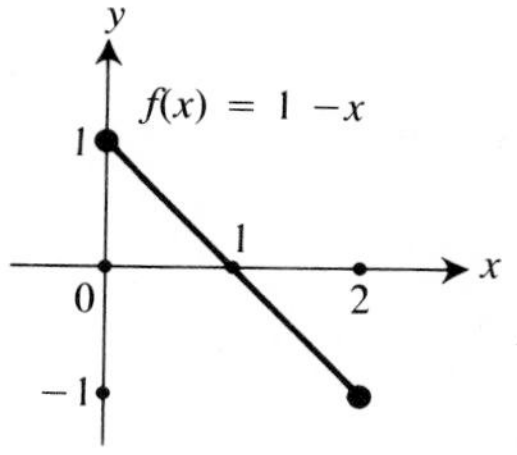

b) 1 c) −1 d) 0

15. $S_4 = (\pi/4)(2 + \sqrt{2})$, $s_4 = (\pi/4)(\sqrt{2})$

17.

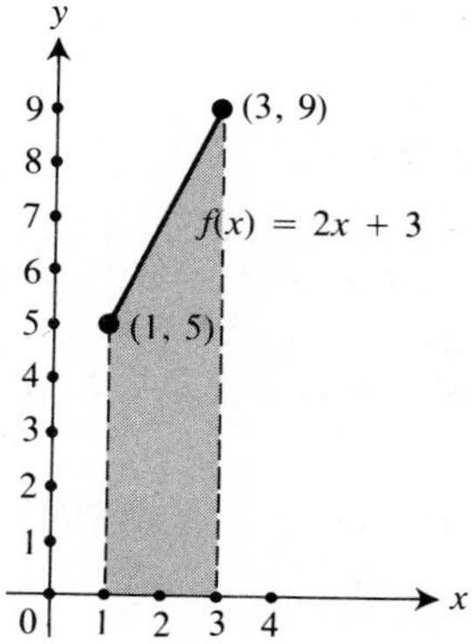

19.

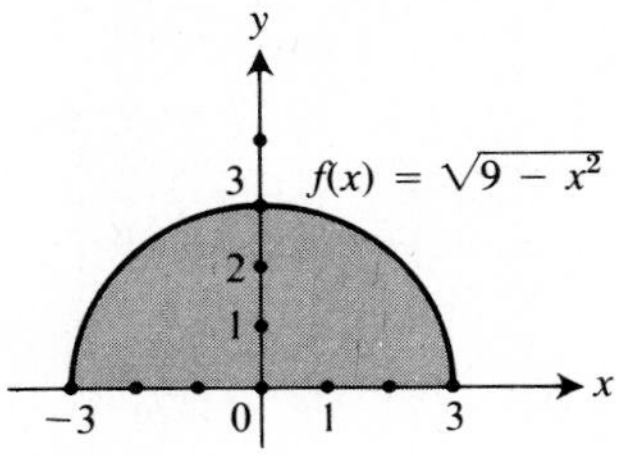

21.

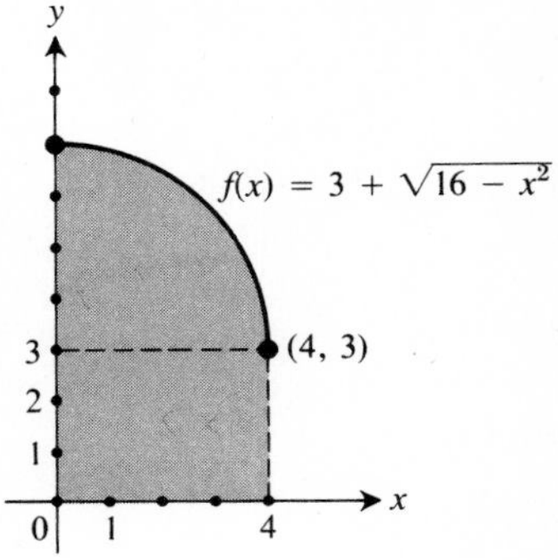

$12 + 4\pi$

23. 0

25. 2

27. 0

29. 4

31. 2

33.

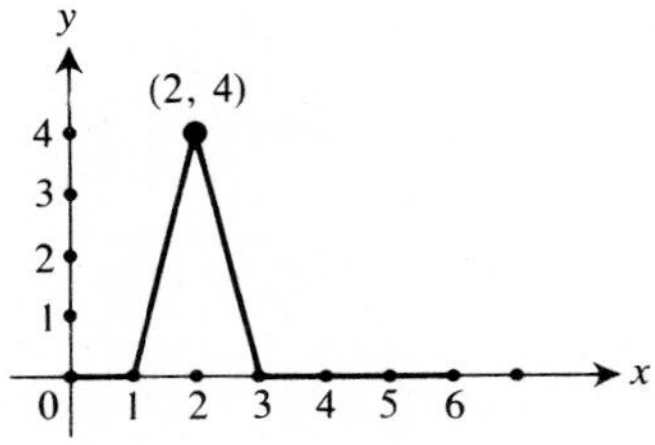

$S_3 = 4 \cdot 2 + 4 \cdot 2 + 0 \cdot 2 = 16$

$S_2 = 4 \cdot 3 + 0 \cdot 3 = 12$

35. 62/3

37. 3

39. 3.450386

Section 6.2

1. Refer to Theorem 6.4 to check your answer.

3. 1/4

5. 20/3

7. 2

9. 3

11. $3\sqrt{2}$

13. −20

15. 3/2

17. $-\pi/2$

19. $2 + (\pi/2)$

21. π

23. $\pi + (3\pi^3/4)$

25. $(3\pi^2/2) - 2\pi$

27. 8

29. $1/\sqrt{2}$

31. 88/15

33. 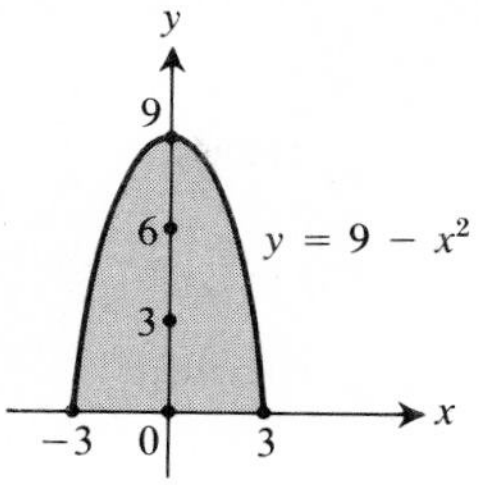

; 36

35. 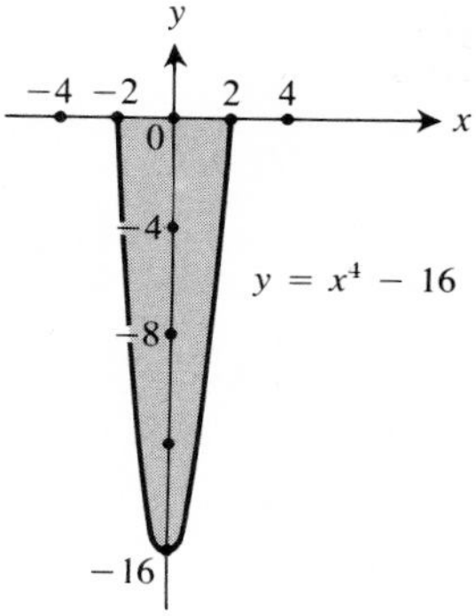

; 256/5

37. $\sqrt{t^2 + 1}$

39. $-\sqrt{t^2 + 1}$

41. $2\sqrt{t^2 + 1}$

43. $-\sqrt{t^2 + 1}$

45. $2t\sqrt{t^4 - 6t^2 + 10}$

Section 6.3

1. $\frac{1}{4}x^4 + \frac{4}{3}x^3 + C$

3. $\frac{1}{6}(x + 1)^6 + C$

5. $\frac{1}{8}(x^2 + 2)^4 + C$

7. $\sqrt{x^2 + 4} + C$

9. $\frac{1}{12}(x^3 + 4)^4 + C$

11. $\frac{1}{2}\sin^2 x + C$

13. $-\frac{1}{16}\cos^4 4x + C$

15. $-\frac{1}{2}\csc 2x + C$

17. $\frac{1}{6}\sec^3 2x + C$

19. $(1 + \cos x)^{-1} + C$

21. $\frac{1}{4}\sin 4x + C$

23. $2\sqrt{1 + \tan x} + C$

25. $-\frac{1}{10}\csc^5 2x + C$

27. Since $\cos 2\theta = 2\cos^2\theta - 1$, we have $\cos^2\theta = (\cos 2\theta + 1)/2$. Thus

$$\begin{aligned}\cos^4 x &= [(\cos 2x + 1)/2]^2\\ &= \tfrac{1}{4}(\cos^2 2x + 2\cos 2x + 1)\\ &= \tfrac{1}{4}[(\cos 4x + 1)/2] + \tfrac{1}{2}\cos 2x + \tfrac{1}{4}.\end{aligned}$$

This can be integrated without difficulty.

29. $2\tan y = x^2 + C$

31. $8\sqrt{u} = x^4 + C$

33. $3\cot y = \cos^3 x + C$

Section 6.4

1. $\frac{56}{3}$

3. $[-3/\sqrt{4 + x^2}] + C$

5. $-[\sqrt{(10 - x)}/x] + C$

7. $\frac{1}{8}(\pi - 2)$

9. $\frac{1}{12}\cos^5 2x \sin 2x + \frac{5}{48}\cos^3 2x \sin 2x + \frac{15}{192}\sin 4x + \frac{15}{48}x + C$

11. 0

13. $\frac{1}{14}\sin 7x + \frac{1}{6}\sin 3x + C$

15. $3 \tan \dfrac{x}{2} + C$

17. $\dfrac{1}{9}\sin 3x - \dfrac{x}{3}\cos 3x + C$

19. $\left(\dfrac{1}{4} - \dfrac{x^2}{2}\right)\cos 2x + \dfrac{x}{2}\sin 2x + C$

21. $\frac{1}{4}\tan 4x + C$

23. $-\frac{1}{2}(\cot x^2 + x^2) + C$

25. $\frac{1}{6}\sec^2 x^2 \tan x^2 + \frac{1}{3}\tan x^2 + C$

27. $\frac{1}{6}\sin^3 x^2 + C$

29. $-[\sqrt{4 - \sin^2 x}/(4 \sin x)] + C$

Section 6.5

1. a) 1.3524 b) 1.4583

3. a) 1.737 b) 1.7321

5. a) $\dfrac{49\pi}{120} = 1.2828$ b) $\dfrac{73\pi}{180} = 1.2741$

7. 33.0274

Review Exercise Set 6.1

1. $\dfrac{19}{20}$

3. $\dfrac{\sqrt{2t + t^2}}{2\sqrt{t}}$

5. -2

7. a) $\frac{1}{4}\sin^2 2x + C$ b) $\frac{1}{3}\sec^3 x + C$

9. $\dfrac{(12\sqrt{3} - 20)}{3}$

Review Exercise Set 6.2

1. $\dfrac{\pi}{2}\left[\sin\dfrac{\pi}{8} + \sin\dfrac{3\pi}{8} + \sin\dfrac{5\pi}{8} + \sin\dfrac{7\pi}{8}\right]$

3. $2\sin^2 2t + \sin^2 t$

5. -5

7. a) $-\frac{1}{9}\csc^3 3x + C$
   b) $-2\sqrt{\csc x} + C$

9. $(8 + 3\pi)/64$

More Challenging Exercises 6

1. 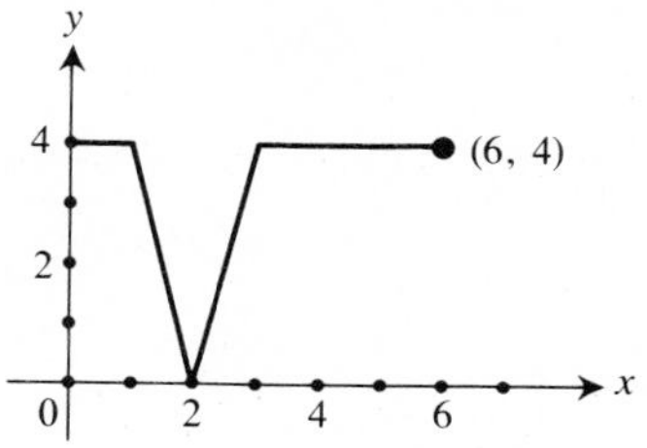

; $s_2 = 12$, $s_3 = 8$

3.
$$S_n = \frac{1}{n}\left[f\left(\frac{1}{n}\right) + f\left(\frac{2}{n}\right) + f\left(\frac{3}{n}\right) + \cdots + f\left(\frac{n-1}{n}\right) + f(1)\right],$$
$$s_n = \frac{1}{n}\left[f(0) + f\left(\frac{1}{n}\right) + f\left(\frac{2}{n}\right) + \cdots + f\left(\frac{n-1}{n}\right)\right]$$
(Subtract)
$$S_n - s_n = \frac{1}{n}[f(1) - f(0)] = f(1)/n$$

5. $\frac{b-a}{n}[f(a) - f(b)]$

7. $\frac{16}{3}$

9. $\frac{2}{5}$

11. 16/5

CHAPTER 7

Section 7.1

1. $\frac{32}{3}$

3. $\frac{1}{6}$

5. $\frac{4}{15}$

7. $\frac{44}{15}$

9. $\frac{9}{2}$

11. $\frac{64}{3}$

13. 4

15. $2\int_0^{\sqrt{3}} (-1 + \sqrt{4 - x^2})\, dx$

17. $\int_0^1 x^2\, dx + \int_1^{\sqrt{2}} \sqrt{2 - x^2}\, dx$

19. $-\frac{1}{3}$

21. $\frac{2}{\pi}$

23. 21.9919

Section 7.2

1. $16\pi/15$

3. $\int_0^h \pi\left(\frac{r}{h}x\right)^2 dx = \frac{\pi}{3}r^2h$

5. $\frac{\pi}{6}$

7. $3\pi/10$

9. $\pi^2/2$

11. $4a^3/\sqrt{3}$

13. $\frac{\pi}{3}(2a^3 + b^3 - 3a^2b)$

Section 7.3

1. $16\pi/15$

3. $\pi/6$

5. $3\pi/10$

7. $2\pi^2a^2b$

Section 7.4

1. $\frac{8}{27}(10^{3/2} - 1)$

3. $\frac{1}{6}(125 - 13^{3/2})$

5. $169/24$

7. 24

9. $\int_1^5 \sqrt{1 + 9x^4}\,dx$

11. $2\int_0^{\pi/2} \sqrt{\cos^2 t + 4\sin^2 2t}\,dt$

13. $\sqrt{17}/20$

15. 124.1327

17. 4.6472

Section 7.5

1. $\frac{\pi}{27}(10^{3/2} - 1)$

3. $\frac{253\pi}{20}$

5. $2\pi\frac{24\sqrt{3} + 4}{15}$

7. $\frac{\pi}{6}(37^{3/2} - 1)$

9. $\frac{\pi}{3}(29^{3/2} - 13^{3/2})$

11. $\int_0^2 2\pi(x + 3)\sqrt{1 + 4x^2}\,dx$

13. $\int_0^\pi 2\pi(2 - \sin y)\sqrt{1 + \cos^2 y}\,dy$

15. 37.7037

Section 7.6

1. a) i) -4 ii) 4 b) i) $5/2$ ii) $13/2$

3. a) i) $5/6$ ii) $5/6$ b) i) $2/3$ ii) 1
c) i) $3/2$ ii) $11/6$

5. a) i) $4/\pi$ ii) $4/\pi$ b) i) $4/\pi$ ii) $4\sqrt{2}/\pi$
c) i) 0 ii) $8\sqrt{2}/\pi$

7. a) $3t$ b) 6

9. a) $1 - \cos t$ b) $(3\pi + 2)/2$

Section 7.7

1. 64 ft-lbs

3. 12480π ft-lbs

5. $Gm_1m_2/2a$

7. 1,940,889.6 lbs

9. 998.4π lbs

Section 7.8

1. a) $kab^3/12$ b) $kab^4/20$

3. $\frac{3}{4}$

5. $\frac{6}{7}$

7. a) $ka^4/4$ b) $ka^5/5$

9. $3\int_1^4 y^2\sqrt{1 + (9y/4)}\, dy$

11. $4k\pi \int_b^a x\sqrt[3]{a^2 - x^2}\, dx$

Section 7.9

1. A plane annular region (a disk with a hole in it)

3. $\left(\dfrac{2a}{12 - 3\pi}, \dfrac{2a}{12 - 3\pi}\right)$

5. $\left(\dfrac{\pi}{2}, \dfrac{\pi}{8}\right)$

7. Let the maximum depth of water at the dam be h ft and let the width of the dam at depth x ft be $f(x)$ ft. Then the force on the dam is $\int_0^h (62.4)xf(x)\, dx = 62.4 \int_0^h xf(x)\, dx$ lb. Now $\int_0^h xf(x)\, dx$ is the moment of the plane region (with mass density 1) comprising the face of the dam about the axis formed by the top edge of the dam, and is therefore equal to sA. Thus the force is $(62.4)\ sA$ lb.

9. a) $1/\sqrt{3}$ b) $\dfrac{1}{\sqrt{3}}\sqrt{(a + 1)^3 - a^3}$

11. $\sqrt{2}\pi a^3$

Review Exercise Set 7.1

1. $\dfrac{4}{3}$

3. 288 ft-lbs

5. $\dfrac{7\pi}{9\sqrt{3}}$

7. $2\int_0^1 (y^3 + 3y^2)\sqrt{1 - y}\, dy$

Review Exercise Set 7.2

1. $\dfrac{125}{6}$

3. 208,000 lbs

5. $9\sqrt{2}\pi$

7. $\dfrac{20}{3\pi}m$

More Challenging Exercises 7

1. $(\pi/\sqrt{2})\displaystyle\int_{-2}^{2}(x^2 - 2x + 8)(4 - x^2)\,dx$ or $2\sqrt{2}\pi\displaystyle\int_{0}^{4}\sqrt{y}(y + 2)\,dy$; $\dfrac{704\sqrt{2}\pi}{15}$

3. $\dfrac{2\pi}{5}\displaystyle\int_{4}^{5}(2x^2 + 3x + 62)(25 - x^2)\,dx + \dfrac{18\pi}{5}\displaystyle\int_{0}^{4}(4x^2 + 3x + 30)\,dx$

5. $\dfrac{256\sqrt{2}}{5}$

7. $\dfrac{200}{3}$ hr

CHAPTER 8

Section 8.1

1. Yes; it is differentiable.

3. $y = x - 1$

5. 1.8

7. 2.5

9. 3.3

11. 0.47

13. $(1/2x)$, $x > 0$

15. $\sec^2 x/\tan x$, $\tan x > 0$

17. $\dfrac{2}{2x + 3} - \dfrac{2x}{x^2 + 4}$, $x > -\dfrac{3}{2}$

19. $(\ln x)(\cos x) + (\sin x)/x$

21. $\dfrac{1 - \ln x^2}{x^3}$, $x > 0$

23. $\dfrac{4x + 8}{x^2 + 4x} + \dfrac{9}{3x - 2}$, $x > \dfrac{2}{3}$

25. $\dfrac{\sec^2(\ln x)}{x}$

27. $\dfrac{1}{x} + \dfrac{1}{2x + 3}$, $x > 0$

29. $-2\tan x - 6\cot 2x$, $\sin 2x > 0$

31. $\frac{1}{2}\ln|2x + 3| + C$

33. $\frac{1}{3}\ln|\sin 3x| + C$

35. $\dfrac{-1}{x + 1} + C$

37. $\ln|\ln x| + C$

39. $\ln|\tan x| + C$

41. $\dfrac{-1}{9x + 3}$

43. $\sec x + C$

45. $\dfrac{\ln 3}{4} - \dfrac{1}{6}$

47. $\ln\frac{4}{3}$

49. $-2\sqrt{2} + 3\ln(3 + 2\sqrt{2})$

51. $2\ln\dfrac{2+\sqrt{3}}{\sqrt{2}+1}$

53. 1.6094868 with error 0.0000489

55. $x = 0.231286, 236.0637$

Section 8.2

1. 2

3. 0

5. -2

7. 6

9. 16

11. 9

13. $2e^{2x}$

15. $e^{2x}\cos x + 2e^{2x}\sin x$

17. $\dfrac{e^x}{x} + e^x(\ln 2x)$

19. $e^{\sec x}\sec x\tan x$

21. $\dfrac{-e^{1/x}}{x^2}$

23. 4

25. $-e^{\cos x} + C$

27. $\ln(1 + e^x) + C$

29.

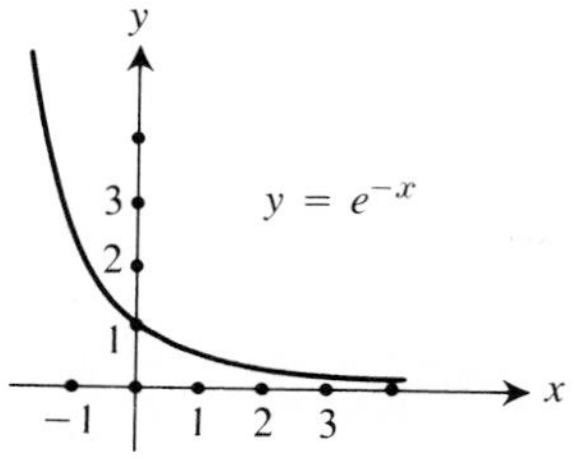

31. -0.90356

33. 0.4890435

Section 8.3

1. 64

3. 5

5. 9

7. $3(\ln 2)2^{3x}$

9. $3^{\sin x}[(x\cos x)(\ln 3) + 1]$

11. $(\sin x)^x[x\cot x + \ln(\sin x)]$

13. $-\sin(x^x)[x^x(\ln x + 1)]$

15. $1/((\ln 10)x)$

17. $2^x \cdot 3^x \cdot [(\ln 2) + 2(\ln 3)]$

19. $\dfrac{5^{x^2}}{7x}[2x(\ln 5) - (\ln 7)]$

21. $7^x \cdot 8^{-x^2} \cdot 100^x[(\ln 700) - 2x(\ln 8)]$

23. $-\dfrac{3^{-x}}{\ln 3} + C$

25. $\dfrac{2^{\sin x}}{\ln 2} + C$

27. Let $y = \log_a x$. Then $x = a^y$. Differentiating implicitly,

$$1 = a^y (\ln a) \cdot \frac{dy}{dx}, \quad \text{so} \quad \frac{dy}{dx} = \frac{1}{(\ln a)a^y} = \frac{1}{(\ln a)x}.$$

29. 0.687439

Section 8.4

1. $\dfrac{1600 \ln (3/2)}{\ln 2}$ yr

3. $\dfrac{2560}{9 \ln (4/3)}$ ft

5. a) At time t, the concentration of salt is $(f(t)/100)$ lb/gal. Since no salt is added and 4 gal/min of brine is being drawn off, we have

$$f'(t) = -4\frac{f(t)}{100} = -\frac{1}{25} f(t),$$

which is an equation of the form (1).

b) $\dfrac{200}{e}$ lb

7. $y = Ae^{kt} + B$, where A and B may be any constants

9. \$27,182.82

11. \$79,047

13. Year 4,029

Section 8.5

1. $\pi/2$

3. $-\pi/3$

5. $-5\pi/6$

7. $\pi/4$

9. $-\pi/2$

11. $-\pi$

13. $2/\sqrt{1 - 4x^2}$

15. $1/[2\sqrt{x}(1 + x)]$

17. $1/[x\sqrt{(1/x)^2 - 1}]$

19. $6(\tan^{-1} 2x)^2/(1 + 4x^2)$

21. $-1/[(1 + x^2)(\tan^{-1} x)^2]$

23. $2(x + \sin^{-1} 3x)[1 + (3/\sqrt{1 - 9x^2})]$

25. $\frac{1}{3} \sin^{-1} 3x + C$

27. $\pi/6$

29. $\pi/6$

31. $(16\pi/3) - 4\sqrt{3}$

33. $(\pi - 4\sqrt{3})/3$

35. $\pi/4$

37. Let $y = \cos^{-1} x$. Then $x = \cos y$ and

$$\frac{dy}{dx} = \frac{1}{dx/dy} = \frac{1}{-\sin y} = \frac{-1}{\sqrt{1 - \cos^2 y}} = \frac{-1}{\sqrt{1 - x^2}}.$$

(It is appropriate to substitute the *positive* quantity $\sqrt{1 - \cos^2 y}$ for $\sin y$ since $0 \le y \le \pi$, so $\sin y \ge 0$.)

39. Let $y = \cot^{-1} x$. Then $x = \cot y$ and

$$\frac{dy}{dx} = \frac{1}{dx/dy} = \frac{-1}{\csc^2 y} = \frac{-1}{1 + \cot^2 y} = \frac{-1}{1 + x^2}.$$

41. Let $y = \csc^{-1} x$. Then $x = \csc y$ and

$$\frac{dy}{dx} = \frac{1}{dx/dy} = \frac{-1}{\csc y \cot y} = \frac{-1}{\csc y\sqrt{\csc^2 y - 1}} = \frac{-1}{x\sqrt{x^2 - 1}}.$$

It is appropriate to substitute the *positive* quantity $\sqrt{\csc^2 y - 1}$ for $\cot y$ since

$$-\pi < y \le -\pi/2 \quad \text{or} \quad 0 < y \le \pi/2.$$

Section 8.6

1. Divide the relation $\cosh^2 x - \sinh^2 x = 1$ by $\cosh^2 x$.

3. $\sinh(-x) = \dfrac{e^{-x} - e^{-(-x)}}{2} = \dfrac{-e^x + e^{-x}}{2} = -\sinh x.$

5.
$$\sinh x \cosh y + \cosh x \sinh y = \frac{e^x - e^{-x}}{2} \cdot \frac{e^y + e^{-y}}{2} + \frac{e^x + e^{-x}}{2} \cdot \frac{e^y - e^{-y}}{2}$$
$$= \frac{2e^{x+y} - 2e^{-x-y}}{4} = \frac{e^{x+y} - e^{-x-y}}{2}$$
$$= \sinh(x + y).$$

7. $\sinh 2x = 2 \sinh x \cosh x$; $\cosh 2x = \cosh^2 x + \sinh^2 x$.

9. $D(\cosh x) = D\left(\dfrac{e^x + e^{-x}}{2}\right) = \dfrac{1}{2} D(e^x + e^{-x}) = \dfrac{1}{2}(e^x - e^{-x}) = \sinh x.$

11. $D(\operatorname{sech} x) = D\left(\dfrac{1}{\cosh x}\right) = \dfrac{-\sinh x}{\cosh^2 x} = -\tanh x \operatorname{sech} x.$

13. Let $y = \cosh^{-1} x$ so $x = \cosh y$. Then

$$\frac{dy}{dx} = \frac{1}{dx/dy} = \frac{1}{\sinh y} = \frac{1}{\sqrt{\cosh^2 y - 1}} = \frac{1}{\sqrt{x^2 - 1}}, \qquad x > 1.$$

(Since $y = \coth^{-1} x \ge 0$, $\sinh y \ge 0$, so the *positive* square root was appropriate.)

15. Let $y = \coth^{-1} x$ so $x = \coth y$. Then

$$\frac{dy}{dx} = \frac{1}{dx/dy} = \frac{-1}{\operatorname{csch}^2 y} = \frac{-1}{\coth^2 y - 1} = \frac{-1}{x^2 - 1} = \frac{1}{1 - x^2}, \qquad |x| > 1.$$

17. Let $y = \operatorname{csch}^{-1} x$ so $x = \operatorname{csch} y$. Then

$$\frac{dy}{dx} = \frac{1}{dx/dy} = \frac{-1}{\operatorname{csch} y \coth y} = \frac{-1}{(\operatorname{csch} y)(\pm\sqrt{1 + \operatorname{csch}^2 y})} = \frac{-1}{(x)(\pm\sqrt{1 + x^2})}.$$

If $x > 0$, then $y = \operatorname{csch}^{-1} x > 0$, so $\coth y > 0$ and the *plus sign* is appropriate. If $x < 0$, then $y = \operatorname{csch}^{-1} x < 0$ so $\coth y < 0$ and *minus sign* is appropriate. These two cases are both covered by the formula

$$D(\operatorname{csch}^{-1} x) = \frac{-1}{(|x|\sqrt{1 + x^2})}.$$

19. $2x \sinh (x^2)$

21. $-\dfrac{\operatorname{sech} \sqrt{x} \tanh \sqrt{x}}{2\sqrt{x}}$

23. $-\dfrac{\operatorname{csch} (\ln x) \coth (\ln x)}{x}$

25. $2 \sinh^3 x \cosh x + 2 \cosh^3 x \sinh x$

27. $\dfrac{2}{\sqrt{1 + 4x^2}}$

29. $2 \sec 2x$

31. $\dfrac{-2}{x\sqrt{1 - x^4}}$

33. $\dfrac{e^{2x}}{\sqrt{1 + e^{2x}}} + e^x \sinh^{-1} (e^x)$

35. $\dfrac{-3 \operatorname{csch}^2 3x}{1 - \cot^2 3x}, \quad |\cot 3x| < 1$

37. $\ln |\sinh x| + C$

39. $\frac{1}{3} \cosh (3x + 2) + C$

41. $\frac{1}{3} \tanh 3x + C$

43. $\ln |\sinh x| + C$

45. $\frac{1}{2} (\ln \frac{4}{3})$

47. $\frac{1}{4}\sqrt{1 + 4x^2} + C$

49. $-\operatorname{sech}^{-1} (e^x) + C$

51. $\dfrac{\sqrt{5}}{2} + 2 \sinh^{-1} \left(\dfrac{1}{2}\right) = \dfrac{\sqrt{5}}{2} + 2 \ln \left(\dfrac{1+\sqrt{5}}{2}\right)$

53. $\sqrt{16 + x^2} - 4 \sinh^{-1} \left|\dfrac{4}{x}\right| + C = \sqrt{16 + x^2} - 4 \ln \left(\dfrac{4 + \sqrt{16 + x^2}}{x}\right) + C$

55. $-\dfrac{9}{2} \sinh^{-1}\left(\dfrac{\sin x}{3}\right) + \dfrac{\sin x \sqrt{9 + \sin^2 x}}{2} + C$

$$= \frac{(\sin x)}{2}\sqrt{9 + \sin^2 x} - \frac{9}{2} \ln (\sin x + \sqrt{9 + \sin^2 x}) + C$$

57. $8 \cosh^{-1} \left(\dfrac{e^x}{4}\right) + \dfrac{e^x}{2}\sqrt{e^{2x} - 16} + C = \dfrac{e^x}{2}\sqrt{e^{2x} - 16} + 8 \ln |e^x + \sqrt{e^{2x} - 16}| + C$

59. $\dfrac{\cosh^2 4x \sinh 4x}{12} + \dfrac{1}{6} \sinh 4x + C$

61. $\cosh^{-1} x = \ln (x + \sqrt{x^2 - 1})$

63. $\coth^{-1} x = \dfrac{1}{2} \ln \left(\dfrac{x + 1}{x - 1}\right)$

65. $\operatorname{csch}^{-1} x = \dfrac{x}{|x|} \ln \left|\dfrac{1 + \sqrt{1 + x^2}}{x}\right|$

Review Exercise Set 8.1

1. a) $\ln x = \displaystyle\int_1^x \frac{1}{t}\, dt, \quad x > 0$

b) 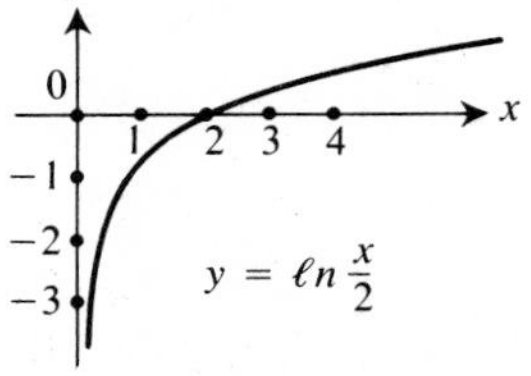

3. a) $e^{\tan x}\sec^2 x$ b) $\frac{7}{3}$

5. a) $-\dfrac{\ln 3}{\ln 24}$ b) 3

7. a) $-\pi/6$ b) $-\pi/3$

9. a) $\sinh x = \dfrac{e^x - e^{-x}}{2}$

b) $-6\,\text{sech}^3\,2x\tanh 2x$

Review Exercise Set 8.2

1. a) It is the unique number e satisfying $\ln e = 1$.

b)

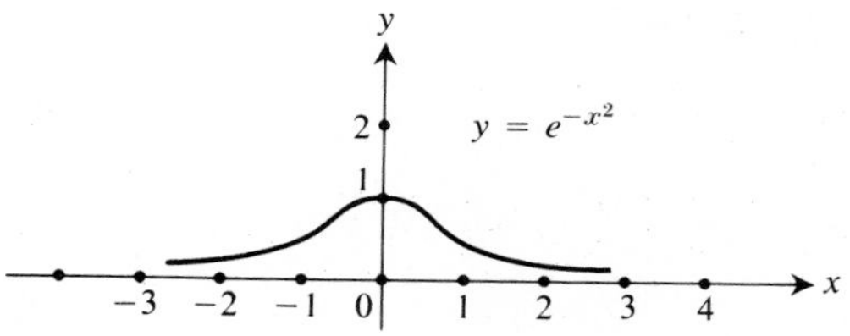

3. a) $\dfrac{e^{\sin^{-1} x}}{\sqrt{1 - x^2}}$ b) $\dfrac{1}{3}e^{\tan 3x} + C$

5. a) $10^{-1/5}$ b) 49

7. a) $\pi/3$ b) $-2\pi/3$

9. a)

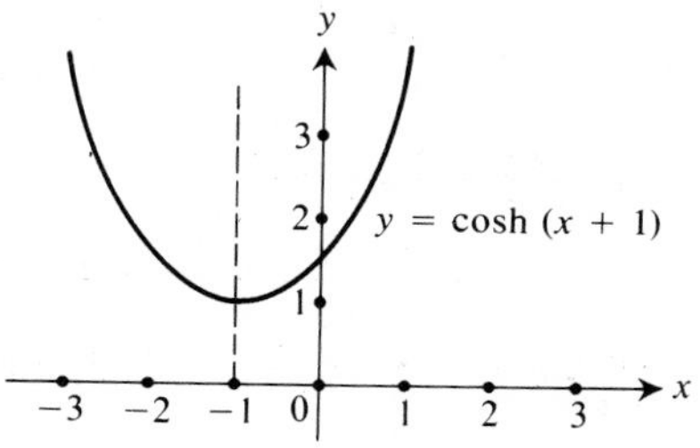

b) $-6\coth^2(2x + 1)\,\text{csch}^2(2x + 1)$

More Challenging Exercises 8

1. $x = 0, \ln 2$

3. $\ln n = \displaystyle\int_1^n \frac{1}{x}\,dx$. For this integral,

$$s_{n-1} = \frac{1}{2} + \frac{1}{3} + \cdots + \frac{1}{n} \quad \text{and} \quad S_{n-1} = 1 + \frac{1}{2} + \frac{1}{3} + \cdots + \frac{1}{n-1}.$$

5. $2\ln a$

7. If x is large enough, then e^x is greater than $f(x)$ for any particular polynomial function f.

CHAPTER 9

Section 9.1

1. $\cos x + x \sin x + C$

3. $-\frac{1}{3}e^{-x^3} + C$

5. $\frac{\pi}{4} - \frac{1}{2}$

7. $\frac{x^2}{2}(\ln x) - \frac{x^2}{4} + C$

9. $\frac{e^{ax}}{a^2 + b^2}(a \cos bx + b \sin bx) + C$

11. $x^3e^x - 3x^2e^x + 6xe^x - 6e^x + C$

13. $-\frac{1}{3}x^3 \cos x^3 + \frac{1}{3}\sin x^3 + C$

15. $x \ln(1 + x^2) - 2x + 2\tan^{-1} x + C$

17. $\frac{x^2}{2}\sec^{-1} x - \frac{1}{2}\sqrt{x^2 - 1} + C$

19. Take $u = x^n$ and $dv = e^{ax}\,dx$.

21. Take $u = \sin^{n-1} ax$ and $dv = \sin ax \cos^m ax\,dx$.

Section 9.2

1. $\frac{1}{2}\ln\left|\frac{x - 1}{x + 1}\right| + C$

3. $\frac{1}{2}x^2 + \ln|(x + 2)(x - 2)^3| + C$

5. $\frac{1}{x} + 3\ln\left|\frac{x - 1}{x}\right| + C$

7. $-\frac{4}{3}\ln|x| + \frac{13}{12}\ln|x - 3| + \frac{1}{4}\ln|x + 1| + C$

9. $\frac{-1}{3(x + 1)^2} - \frac{8}{9(x + 1)} + \frac{1}{27}\ln\left|\frac{x - 2}{x + 1}\right| + C$

11. $\ln\left|\frac{(x - 2)^2}{x + 2}\right| + \frac{1}{2}\ln(x^2 + 2) - \frac{1}{\sqrt{2}}\tan^{-1}(x/\sqrt{2}) + C$

13. $-\frac{3}{x - 1} - 2\ln|x - 1| + \frac{1}{2}\ln|x^2 + 3| + \frac{7}{\sqrt{3}}\tan^{-1}(x/\sqrt{3}) + C$

15. $-\frac{3}{x} + \frac{2 - x}{2(x^2 + 1)^2} - \frac{3x}{4x^2 + 4} - \frac{3}{4}\tan^{-1} x + C$

Section 9.3

1. a) $u = x \qquad dv = (x+1)^{1/2}\,dx$

$du = dx \qquad v = \frac{2}{3}(x+1)^{3/2}$

$$\int x\sqrt{x+1}\,dx = \tfrac{2}{3}x(x+1)^{3/2} - \int \tfrac{2}{3}(x+1)^{3/2}\,dx$$

$$= \tfrac{2}{3}x(x+1)^{3/2} - \tfrac{4}{15}(x+1)^{5/2} + C$$

b) $u = x^2 \qquad dv = x(4-x^2)^{1/2}\,dx$

$du = 2x\,dx \qquad v = -\frac{1}{3}(4-x^2)^{3/2}$

$$\int x^3\sqrt{4-x^2}\,dx = -\tfrac{1}{3}x^2(4-x^2)^{3/2} + \tfrac{1}{3}\int (4-x^2)^{3/2}\,2x\,dx$$

$$= -\tfrac{1}{3}x^2(4-x^2)^{3/2} - \tfrac{2}{15}(4-x^2)^{5/2} + C$$

3. $-\sqrt{4-x^2} - 3\sin^{-1}(x/2) + C$

5. $\sin^{-1}\left(\dfrac{\sin x}{2}\right) + C$

7. $\frac{1}{10}(3-2x)^{5/2} - \frac{1}{2}(3-2x)^{3/2} + C$

9. $2\sqrt{x-1} - 2\tan^{-1}\sqrt{x-1} + C$

11. $x - 2\sqrt{x} + C$

13. $x + 2 - 4\sqrt{x+2} + 4\ln|\sqrt{x+2} + 1| + C$

15. $\frac{1}{7}(x^2+1)^{7/2} - \frac{1}{5}(x^2+1)^{5/2} + C$

17. $\dfrac{6}{7}x^{7/6} - \dfrac{3}{2}x^{2/3} + 6x^{1/6} - 2\ln|1 + x^{1/6}| + \ln|x^{1/3} - x^{1/6} + 1|$

$$- 2\sqrt{3}\tan^{-1}\left(\frac{2x^{1/6} - 1}{\sqrt{3}}\right) + C$$

Section 9.4

1. $\ln\left|\dfrac{1 + \tan(x/2)}{1 - \tan(x/2)}\right| + C$

3. $-\ln|1 + \cos x| + C$

5. $-\dfrac{2}{\sqrt{3}}\tan^{-1}\left[\sqrt{3}\tan\left(\dfrac{\pi}{4} - \dfrac{x}{2}\right)\right] + C$

7. $-\cot x + \csc x + \ln|\sin x| + \ln|\csc x + \cot x| + C$

9. $\dfrac{x}{2} - \dfrac{1}{4}\ln|\cos x| - \dfrac{1}{4}\ln|\sec 2x + \tan 2x| + C$

11. $\dfrac{1}{4}\ln|\sec 2x + \tan 2x| + \dfrac{x}{2} + \dfrac{1}{4}\ln|\cos 2x| + C$

Section 9.5

1. Replacing n by $-n$ in (5) and (6) yields

$$\sin(mx - nx) = \sin mx \cos nx - \cos mx \sin nx \tag{5'}$$

and

$$\cos(mx - nx) = \cos mx \cos nx + \sin mx \sin nx. \tag{6'}$$

Subtracting (6) from (6′) gives (2), and adding (6) to (6′) gives (4). Adding (5) and (5′) gives (3).

3. $-\frac{1}{3}(\cos 2x)^{3/2} + \frac{1}{7}(\cos 2x)^{7/2} + C$

5. $2\sqrt{\sin x} - \frac{2}{5}(\sin x)^{2/5} + C$

7. $\sec x + \cos x + C$

9. $\dfrac{x}{8} - \dfrac{\sin 12x}{96} + C$

11. $\frac{1}{16}x - \frac{1}{48}\sin^3 2x - \frac{1}{64}\sin 4x + C$

13. $\tan x + \frac{1}{3}\tan^3 x + C$

15. $\frac{1}{3}\tan^3 x + \frac{1}{5}\tan^5 x + C$

17. $\frac{1}{6}\tan^2 3x + C$

19. $\tan x + \frac{2}{3}\tan^3 x + \frac{1}{5}\tan^5 x + C$

21. $\frac{1}{12}\tan^6 2x - \frac{1}{8}\tan^4 2x + \frac{1}{4}\tan^2 2x + \ln|\cos x| + C$

23. $-\frac{1}{4}\csc^4 x + C$

25.
$$\begin{aligned}\int \tan^n ax\,dx &= \int \tan^{n-2} ax\,(\sec^2 ax - 1)\,dx \\ &= \int \tan^{n-2} ax \sec^2 ax\,dx - \int \tan^{n-2} ax\,dx \\ &= \frac{\tan^{n-1} ax}{a(n-1)} - \int \tan^{n-2} ax\,dx\end{aligned}$$

27.
$$\begin{aligned}\int \sec^n ax\,dx &= \int (\sec^{n-2} ax)(1 + \tan^2 ax)\,dx \\ &= \int \sec^{n-2} ax\,dx + \int \sec^{n-2} ax \tan^2 ax\,dx.\end{aligned}$$

$$u = \tan ax \qquad dv = \sec^{n-2} ax \tan ax\,dx$$

$$du = a\sec^2 ax\,dx \qquad v = \frac{1}{a}\cdot\frac{\sec^{n-2} ax}{n-2}$$

$$\int \sec^{n-2} ax \tan^2 ax\,dx = \frac{1}{a(n-2)}\sec^{n-2} ax \tan ax - \frac{1}{n-2}\int \sec^n ax\,dx.$$

$$\int \sec^n ax\,dx = \int \sec^{n-2} ax\,dx + \frac{1}{a(n-2)}\sec^{n-2} ax \tan ax - \frac{1}{n-2}\int \sec^n ax\,dx$$

Solving this equation for $\int \sec^n ax\,dx$ gives the formula.

29.

$$u = \sin^{n-1} ax \qquad dv = \sin ax\,dx$$

$$du = a(n-1)\sin^{n-2} ax \cos ax \qquad v = -\frac{1}{a}\cos ax$$

$$\begin{aligned}\int \sin^n ax\,dx &= -\frac{1}{a}\sin^{n-1} ax\cos ax + (n-1)\int \sin^{n-2} ax\cos^2 ax\,dx \\ &= -\frac{1}{a}\sin^{n-1} ax\cos ax + (n-1)\int \sin^{n-2} ax(1-\sin^2 ax)\,dx \\ &= -\frac{1}{a}\sin^{n-1} ax\cos ax + (n-1)\int \sin^{n-2} ax\,dx \\ &\qquad - (n-1)\int \sin^n ax\,dx.\end{aligned}$$

Solving for $\int \sin^n ax\,dx$ yields the formula.

Section 9.6

1. $\sqrt{1+x^2}+C$

3. $\dfrac{(1+x^2)^{3/2}}{3} - \sqrt{1+x^2}+C$

5. $\frac{1}{3}(x^2-1)^{3/2}+\frac{1}{5}(x^2-1)^{5/2}+C$

7. $\sqrt{5+x^2}+C$

9. $\sqrt{x^2-16}-\ln|x+\sqrt{x^2-16}|+C$

11. $\dfrac{x}{2}\sqrt{1+4x^2}+\dfrac{1}{4}\ln|2x+\sqrt{1+4x^2}|+C$

13. $\dfrac{x-3}{2}\sqrt{x^2-6x+8}-\dfrac{1}{2}\ln|x-3+\sqrt{x^2-6x+8}|+C$

15. $\dfrac{1}{4\sqrt{2}}\tan^{-1}\dfrac{x+1}{\sqrt{2}}+C$

17. $\frac{3}{2}\ln|x^2+2x+2|-5\tan^{-1}(x+1)+C$

Section 9.7

1. 1/2

3. $\pi/2$

5. 1

7. $2\sqrt{2}$

9. 10/3

11. 1

13. Diverges

15. Diverges

17. Converges

19. Diverges

21. Converges

23. Converges

25. For $p \neq 1$,

$$\int_a^b (x-a)^{-p}\,dx = \lim_{h\to a+} \left.\frac{(x-a)^{1-p}}{1-p}\right]_h^b = \frac{(b-a)^{1-p}}{1-p} - \frac{1}{1-p}\lim_{h\to a+}(h-a)^{1-p}.$$

If $p < 1$, you have $1 - p > 0$, so $\lim_{h\to a+}(h-a)^{1-p} = 0$ and the integral converges. If $p > 1$, then $1 - p < 0$ and $\lim_{h\to a+}(h-a)^{1-p} = \infty$ and the integral diverges. If $p = 1$, then $\int_a^b [1/(x-a)]\,dx = \lim_{h\to a+} \ln|x-a|]_h^b = \ln|b-a| - \lim_{h\to a+}\ln|h-a| = \infty$, so the integral diverges for $p \geq 1$. The proof for $\int_a^b [1/(b-x)^p]\,dx$ is similar.

27. a) Yes b) $V = \int_0^1 2\pi x(1/\sqrt{x})\,dx = 4\pi/3$ c) Yes

Review Exercise Set 9.1

1. $\frac{x^2}{3}\sin 3x - \frac{2}{27}\sin 3x + \frac{2x}{9}\cos 3x + C$ **3.** $-\frac{5}{2}\ln|2x+3| + 5\ln|x-5| + C$

5. $6[\frac{1}{7}x^{7/6} - \frac{1}{6}x + \frac{1}{5}x^{5/6} - \frac{1}{4}x^{2/3} + \frac{1}{3}x^{1/2} - \frac{1}{2}x^{1/3} + x^{1/6} - \ln|x^{1/6}+1|] + C$

7. $-\frac{1}{2}\cos 2x + \frac{1}{2}\cos^2 2x - \frac{1}{10}\cos^5 2x + C$ **9.** $\frac{1}{9}\tan^3 3x + \frac{1}{15}\tan^5 3x + C$

11. $\frac{x}{2}\sqrt{16-9x^2} + \frac{8}{3}\sin^{-1}\left(\frac{3x}{4}\right) + C$

13. a) Diverges. $\int_0^1 \frac{1}{x^2}\,dx$ diverges by Theorem 9.1.

b) Diverges by Theorem 9.2. $(x+7)/(x^2-1) > 1/2x$ for large x, and $\frac{1}{2}\int_2^\infty (1/x)\,dx$ diverges by Theorem 9.1.

c)
$$\frac{\sqrt{2-x}}{x^2-4} = \frac{\sqrt{2-x}}{-(\sqrt{2-x})^2(x+2)} = -\frac{1}{\sqrt{2-x}(x+2)}.$$

Now

$$\frac{1}{\sqrt{2-x}(+2)} > \frac{1}{4}\cdot\frac{1}{\sqrt{2-x}} \quad \text{for } 1 \leq x < 2.$$

By Theorem 9.1, $\frac{1}{4}\int_1^2 \frac{1}{\sqrt{2-x}}\,dx$ diverges. By Theorem 9.2, the negative of the given integral diverges, so the given integral diverges.

Review Exercise Set 9.2

1. $\frac{x^2}{2}\ln x - \frac{x^2}{4} + C$ **3.** $2\ln|x-3| + \frac{3}{2}\tan^{-1} 2x + C$

5. $-\frac{4}{5}(x-1)^{5/4} - 3(x-1) - \frac{8}{3}(x-1)^{3/4} - 12(x-1)^{1/2} - 72(x-1)^{1/4}$
$- 216\ln|(x-1)^{1/4} - 3| + C$

7. $\frac{1}{3}\sin^3 x - \frac{1}{5}\sin^5 x + C$ **9.** $-\frac{1}{3}\cot^3 x - \frac{1}{5}\cot^5 x + C$

11. $2\sin^{-1}\left(\frac{x}{2}\right) - \frac{1}{2}x\sqrt{4 - x^2} + C$

13. a) Diverges since $\int_0^1 (1/x^3)\,dx$ diverges by Theorem 9.1.

b) Over $0 \le x < 1$, the integrand is less than $1/(x-1)^{2/3}$ which converges by Theorem 9.1. Over $1 < x \le 4$, it is less than $4/(x-1)^{2/3}$, which converges, by Theorem 9.1. Therefore it converges by Theorem 9.2.

c) $$\frac{1}{1+x^3} = \frac{1}{(x+1)(x^2-x+1)} > \frac{1}{6(x+1)} \quad \text{for } x \neq -1 \text{ but close to } -1.$$

Theorem 9.1 shows that

$$\int_{-1}^{-1+\epsilon} \frac{1}{6(x+1)}\,dx$$

diverges for each $\epsilon > 0$. Therefore the given integral diverges by Theorem 9.2.

More Challenging Exercises 9

1. $\int_0^x f'(t)\,dt = f(t)\Big|_0^x = f(x) - f(0)$. On the other hand, taking

$$u = f'(t) \qquad dv = 1 \cdot dt$$
$$du = f''(t)\,dt \qquad v = t - x,$$

you obtain

$$\int_0^x f'(t)\,dt = (t-x)f'(t)\Big|_0^x - \int_0^x f''(t)(t-x)\,dt$$
$$= 0 - (-x)f'(0) + \int_0^x f''(t)(x-t)\,dt.$$

Thus $f(x) - f(0) = f'(0)x + \int_0^x f''(t)(x-t)\,dt$, which gives the desired formula.

3. Suppose $\lim_{x\to\infty} f(x) = a > 0$. Then $f(x) > a/2$ for $x > b$ for some b. Thus $\lim_{h\to\infty}\int_b^h f(x)\,dx > \lim_{h\to\infty}\int_b^h (a/2)\,dx = \lim_{h\to\infty}(a/2)(h-b) = \infty$, so $\int_0^\infty f(x)\,dx$ diverges. A similar argument shows that $\int_0^\infty f(x)\,dx$ diverges if $\lim_{x\to\infty} f(x) = a < 0$. We have proved the equivalent contrapositive of the statement in the exercise.

5. a) If $f(x) = x$, then $\int_{-h}^h f(x)\,dx = \int_{-h}^h x\,dx = 0$, so the Cauchy principal value of $\int_{-\infty}^\infty f(x)\,dx$ is zero, although $\int_{-\infty}^\infty f(x)\,dx$ diverges.

b) If $\int_{-\infty}^\infty f(x)\,dx$ converges, then $\int_0^\infty f(x)\,dx$ and $\int_{-\infty}^0 f(x)\,dx$ converge, so $\lim_{h\to\infty}\int_0^h f(x)\,dx$ and $\lim_{h\to\infty}\int_{-h}^0 f(x)\,dx$ exist. Then $\int_{-\infty}^\infty f(x)\,dx = \lim_{h\to\infty}\int_0^h f(x)\,dx + \lim_{h\to\infty}\int_{-h}^0 f(x)\,dx = \lim_{h\to\infty}\int_0^h f(x)\,dx + \int_{-h}^0 f(x)\,dx = \lim_{h\to\infty}\int_{-h}^h f(x)\,dx$, which is the Cauchy principal value of $\int_{-\infty}^\infty f(x)\,dx$.

7. Let $f(x) = 0$ for x not in $[n - 1/10^{2n}, n + 1/10^{2n}]$ for each integer $n > 0$, and let the graph of f over $[n - 1/10^{2n}, n + 1/10^{2n}]$ consist of the line segment joining

$$(n - 1/10^{2n}, 0) \quad \text{and} \quad (n, 10^n)$$

and the line segment joining

$$(n, 10^n) \qquad \text{and} \qquad (n + 1/10^{2n}, 0).$$

The area inside this nth "spike" is $10^n(1/10^{2n}) = 1/10^n$. Thus

$$\int_0^{\infty} f(x)\,dx = \tfrac{1}{10} + \tfrac{1}{100} + \tfrac{1}{1000} + \cdots = 0.1 + 0.01 + 0.001 + \cdots = 0.1111\cdots$$

CHAPTER 10

Section 10.1

1. We have $\lim_{n\to\infty} a_n = \infty$ if for each real number γ there exists an integer N such that $a_n > \gamma$ provided that $n > N$.

3. Converges to 0

5. Converges to 1

7. Converges to 2

9. Diverges to ∞

11. Converges to 0

13. Diverges to ∞

15. Converges to 0

17. Converges to 0

19. Suppose that $\lim_{n\to\infty} a_n = c$, and let $d \neq c$. Let $\epsilon = |d - c|/2$. Then there exists an integer N_1 such that for all $n > N_1$, we have

$$|a_n - c| < \epsilon \qquad \text{or} \qquad |a_n - c| < |d - c|/2.$$

From this, one deduces

$$|a_n - d| > |d - c|/2 \quad \text{for all } n > N_1,$$

so for $\epsilon = |d - c|/2$, there can exist no N_2 such that $|a_n - d| < \epsilon$ for $n > N_2$. Hence $\lim_{n\to\infty} a_n \neq d$.

21. Let $\epsilon > 0$ be given, and find a positive integer N such that $(1/N) < \epsilon$. Then we have

$$|a_n - 0| = \left|\frac{(-1)^{n+1}}{n} - 0\right| = \left|\frac{1}{n}\right| < \epsilon$$

if $n > N$, so the sequence converges to 0.

23. Suppose the sequence has limit c. Then, taking $\epsilon = \frac{1}{2}$, there exists a positive integer N such that $|a_n - c| < \frac{1}{2}$ for $n > N$. In particular,

$$|a_{N+1} - c| < \tfrac{1}{2} \qquad \text{and} \qquad |a_{N+2} - c| < \tfrac{1}{2},$$

which implies $|a_{N+1} - a_{N+2}| < 1$. However, $|a_{N+1} - a_{N+2}| = 2$, so our assumption that the sequence has limit c must be false.

25. a) 13 b) 135 c) 1359 d) 13593

27. 1.64872

Section 10.2

1. a) 1 b) 1 c) 0 d) 0 e) 0 f) 0 g) 1

3. $a_1 = 1/2,\ a_2 = -1/6,\ a_3 = -1/12,\ a_4 = -1/20,\ a_5 = -1/30$

5. Converges to $\frac{3}{2}$

7. Converges to $\frac{1}{2}$

9. Diverges

11. Converges to $\frac{1}{3}$

13. Diverges to ∞

15. 60 ft

17. $1 + 1 + 1 + 1 + \cdots$ and $(-1) + (-1) + (-1) + (-1) + \cdots$

19. $\frac{7}{2}$

21. 6

23. 13

25. a) $1.333\ldots$ b) $\frac{4}{3}$ c) $\frac{4}{3}$

27. The series is

$$\left(\frac{2}{10}\right) + \left(\frac{2}{100}\right) + \left(\frac{2}{1000}\right) + \left(\frac{2}{10000}\right) + \cdots$$

and the sum of the series is $\frac{2}{9}$.

29. Let r be the real number represented by a certain unending decimal with a repeating pattern. By multiplying by a suitable power 10^m, we can assume that the decimal representing $10^m r$ is given by

$$10^m r = a_1 a_2 \cdots a_q \cdot b_1 b_2 \cdots b_s\, b_1 b_2 \cdots b_s\, b_1 b_2 \cdots b_s \cdots,$$

where each of a_i and b_j is an integer from 0 to 9, so that the repeating pattern of $10^m r$ starts immediately after the decimal point. Then

$$10^m r = (a_1 a_2 \cdots a_q) + \frac{b_1 b_2 \cdots b_s}{1 - [1/(10)^s]},$$

so

$$r = \frac{1}{(10)^m}\left((a_1 a_2 \cdots a_q) + \frac{b_1 b_2 \cdots b_s}{(10)^s - 1}\, 10^s\right),$$

which is a rational number. [*Note.* In this argument, $a_1 a_2 \cdots a_q$ and $b_1 b_2 \cdots b_s$ are decimal representations of integers, rather than products of q numbers or s numbers.]

Section 10.3

1. Converges to $\sqrt{5}$

3. Converges to $\frac{3}{4} + \sqrt{2} - \sqrt{3}$

5. Converges; comparison with $\sum_{n=1}^{\infty} (1/2^n)$

7. Converges; comparison with $\sum_{n=2}^{\infty} (1/2^n)$

9. Converges; comparison with $\sum_{n=1}^{\infty} (1/2^n)$

11. Diverges; $\lim_{n\to\infty} a_n = \frac{1}{3} \neq 0$

13. Converges; comparison with $\sum_{n=1}^{\infty} (1/5^n)$

15. Converges; sum of two convergent geometric series

17. Diverges; behaves like $\sum_{n=1}^{\infty} (1/n)$

19. Converges; behaves like geometric series with ratio $1/e^2$

21. Diverges; a_n does not approach 0 as $n \to \infty$

23. Diverges; $a_n \geq 1$ for all n

25. a) $s_1 = \frac{1}{2},\ s_2 = \frac{2}{3},\ s_3 = \frac{3}{4},\ s_4 = \frac{4}{5}$

b) $s_n = 1 - (1/(n+1)) = n/(n+1)$

c) $\lim_{n\to\infty} (n/(n+1)) = 1$, so the series converges to 1.

27. Let $a_n = -1$ and $b_n = 1/2^n$.

Section 10.4

1. Diverges

3. Diverges

5. Converges

7. $\displaystyle\int_2^{\infty} \frac{dx}{x\,(\ln x)} = \lim_{t\to\infty} \ln(\ln x)\Big]_2^t = \lim_{t\to\infty} \ln(\ln t) - \ln(\ln 2) = \infty$

9. Converges

11. Diverges

13. Converges

15. Converges

17. Converges

19. Diverges

21. Diverges

23. Diverges

25. Converges

27. Converges

29. Diverges

31. Diverges

33. Converges

35. Converges

37. Converges

39. Converges

41. The sum of the series lies in [2, 3].

43. Taking the limit of (7) as $s \to \infty$, we have

$$\int_r^{\infty} f(x)\,dx \leq \sum_{n=r}^{\infty} a_n \leq \left(\int_r^{\infty} f(x)\,dx\right) + a_r \leq \left(\int_r^{\infty} f(x)\,dx\right) + \epsilon.$$

From

$$\sum_{n=1}^{\infty} a_n = a_1 + \cdots + a_{r-1} + \sum_{n=r}^{\infty} a_n,$$

we at once have the desired result.

Section 10.5

1. $1 - 1 + 1 - 1 + 1 - 1 + \cdots$
3. $1 + \frac{1}{2} + \frac{1}{3} + \frac{1}{4} + \frac{1}{5} + \cdots$
5. Conditionally convergent
7. Absolutely convergent
9. Divergent
11. Conditionally convergent
13. Divergent
15. Divergent
17. Absolutely convergent
19. Absolutely convergent
21. a) Since $a_n = u_n + v_n$, this follows from Corollary 1 of Theorem 10.2 (Section 10.2).
    b) If say $\sum_{n=1}^{\infty} u_n$ converges, then convergence of $\sum_{n=1}^{\infty} a_n$ would imply convergence of $\sum_{n=1}^{\infty} (a_n - u_n) = \sum_{n=1}^{\infty} v_n$ by Corollary 1 of Theorem 10.2 (Section 10.2).
    c) Let $\sum_{n=1}^{\infty} a_n = 1 - \frac{1}{2} + \frac{1}{3} - \frac{1}{4} + \frac{1}{5} - \frac{1}{6} + \cdots$.
23. -0.9011
25. 1.6444

Review Exercise Set 10.1

1. See Definition 10.2 of Section 10.1.
3. 48
5. a) Diverges; $\lim_{n\to\infty} a_n = 1 \neq 0$ b) Diverges; behaves like $\sum_{n=2}^{\infty} (1/n)$ by Comparison Test 2.
7. The hypotheses of the theorem are satisfied.

$$\int_2^{\infty} \frac{1}{x^2 - x}\,dx = \int_2^{\infty} \left(-\frac{1}{x} + \frac{1}{x-1}\right) dx = \lim_{h\to\infty} (-\ln|x| + \ln|x - 1|\Big]_2^h$$

$$= \lim_{h\to\infty} \left(\ln\left|\frac{h-1}{h}\right| - \ln\frac{1}{2}\right) = \ln 1 - \ln\frac{1}{2} = \ln 2.$$

Converges.

9. a) Conditionally convergent. Satisfies alternating series test, but diverges absolutely. (Behaves like $\sum_{n=1}^{\infty} (1/\sqrt{n})$.)
   b) Converges absolutely by the integral test.

Review Exercise Set 10.2

1. Note that $(n - 1)/n = 1 - (1/n)$. Let $\epsilon > 0$ be given. Find N such that $1/N < \epsilon$. Then if $n > N$, we have $(1/n) < \epsilon$, so $|(1/n)| = |[1 - (1/n)] - 1| < \epsilon$. Thus $\{(n - 1)/n\}$ converges to 1.
3. 30 ft
5. a) Converges; Ratio test gives a ratio of $\frac{1}{3} < 1$.
   b) Converges; sum of convergent geometric series with ratios of 3/5 and 4/5.

7. The hypotheses are satisfied.

$$\int_2^{\infty} \frac{x+1}{x^2+1}\,dx = \lim_{h\to\infty}\left(\frac{1}{2}\ln|x^2+1| + \tan^{-1} x\right)\Bigg]_2^h$$

$$= \lim_{h\to\infty}\left[\frac{1}{2}\ln|h^2+1| + \tan^{-1} h - \frac{1}{2}\ln(5) - \tan^{-1} 2\right] = \infty.$$

Diverges.

9. a) Conditionally convergent. Satisfies the alternating series test, but diverges absolutely by the integral test.

b) Converges absolutely by Comparison Test 1, for the terms are at most $1/n^{3/2}$.

More Challenging Exercises 10

1. If $s_n = -t_n$, then $\{s_n\}$ is monotone increasing. By the fundamental property, either $\lim_{n\to\infty} s_n = \infty$, so that $\lim_{n\to\infty} - s_n = \lim_{n\to\infty} t_n = -\infty$, or $\lim_{n\to\infty} s_n = c$, so that $\lim_{n\to\infty} t_n = -c = d$.

3. From $m < b_n/a_n < M$ for all sufficiently large n, we have $ma_n < b_n < Ma_n$ for all $n > N$, for some integer N. Now $\sum_{n=1}^{\infty} ma_n$ and $\sum_{n=1}^{\infty} Ma_n$ converge if and only if $\sum_{n=1}^{\infty} a_n$ converges. It follows at once by the comparison test that $\sum_{n=1}^{\infty} a_n$ and $\sum_{n=1}^{\infty} b_n$ either both converge or both diverge.

5. Pick up *positive* ($\neq 0$) terms u_n of $\sum_{n=1}^{\infty} u_n$ in order, until a partial sum >17 is obtained. Then pick up *negative* ($\neq 0$) terms v_n in order until the partial sum becomes <17. Then pick up subsequent *positive* terms u_n in order until the partial sum becomes >17 again, then subsequent *negative* terms v_n until the partial sum becomes <17, etc.

7. Pick up *positive* ($\neq 0$) terms u_n in order until the partial sum becomes >1. Then pick up the first single *negative* ($\neq 0$) term v_n. Then pick up *positive* terms until the partial sum becomes >2, and then the next single *negative* term v_n. Then pick up *positive* terms u_n until the partial sum becomes >3, and then pick up the next single *negative* term, etc.

9. Pick up groups of successive *positive* ($\neq 0$) terms u_n and groups of successive *negative* ($\neq 0$) terms v_n alternately so that the partial sums, after picking up the groups, become successively ≥ 15, ≤ -6, ≥ 15, ≤ -6, etc.

CHAPTER 11

Section 11.1

1. $r = 1;\ -1 \leq x < 1$

3. $r = 1;\ -1 < x \leq 1$

5. $r = 1;\ -1 \leq x \leq 1$

7. $r = 5;\ -5 \leq x < 5$

9. $r = 3;\ -1 < x < 5$

11. $r = 1;\ 2 < x < 4$

13. $r = 3;\ -7 \le x \le -1$

15. $r = \frac{1}{2};\ -\frac{7}{2} < x \le -\frac{5}{2}$

17. $r = \frac{1}{3};\ \frac{11}{3} < x < \frac{13}{3}$

19. $\displaystyle\sum_{n=0}^{\infty} \frac{(x - \frac{5}{2})^n}{(\frac{3}{2})^n} = \sum_{n=0}^{\infty} \frac{(2x - 5)^n}{3^n}$

21. $\displaystyle\sum_{n=1}^{\infty} \frac{(x + 3)^n}{n \cdot 2^n}$

Section 11.2

1. $1 - \dfrac{x^2}{2!} + \dfrac{x^4}{4!} - \dfrac{x^6}{6!} + \dfrac{x^8}{8!} - \dfrac{x^{10}}{10!}$

3. $1 - \dfrac{(x - \pi/2)^2}{2!} + \dfrac{(x - \pi/2)^4}{4!} - \dfrac{(x - \pi/2)^6}{6!}$

5. $1 - (x - 1) + (x - 1)^2 - (x - 1)^3 + (x - 1)^4 - (x - 1)^5 + (x - 1)^6$

7. a) $2x - \dfrac{8x^3}{3!} + \dfrac{32x^5}{5!} - \dfrac{128x^7}{7!}$

b) The polynomial may be obtained by replacing x by $2x$ in the answer to Example 1.

9. a) $1 + x^2$

b) The polynomial is the portion of degree ≤ 2 of the polynomial obtained by replacing x by x^2 in the answer to Example 3.

c) $1 + x^2 + x^4 + x^6 + x^8$

11. a) $9 + 5\Delta x + (\Delta x)^2$ b) $9 + 5(x - 1) + (x - 1)^2$

c) $9 + 5(x - 1) + (x - 1)^2$

13. By the chain rule, $g^{(m)}(x_0) = c^m f^{(m)}(x_0)$ for $m \le n$, so the coefficient of x^m in the Taylor polynomial for g is c^m times the coefficient of x^m in the Taylor polynomial $T_n(x)$ for f. But this multiplication by c^m can also be achieved by forming $T_n(cx)$.

15. a) 26.46 b) 0.0036

17. a) 10π ft^3 b) $\dfrac{5\pi}{36}$ ft^3

19. a) $\dfrac{\pi}{90}$ b) $\dfrac{8\pi^3}{(9)(90)^3}$

Section 11.3

1. F T T T F T

3. $1 + x + \frac{3}{2}x^2 + \frac{x^3}{3!} + \cdots + \frac{x^n}{n!} + \cdots$ for $n \geq 3$

5. $x^2 - \frac{x^4}{3!} + \cdots + (-1)^n \frac{x^{2n+2}}{(2n+1)!} + \cdots$ for $n \geq 1$

7. $x + x^2 + x^3 + \cdots + x^{n+1} + \cdots$

9. $1 - \frac{x^6}{2!} + \cdots + (-1)^n \frac{x^{6n}}{(2n)!} + \cdots$

11. $1 + x - \frac{x^3}{3} - \frac{x^4}{6} - \frac{x^5}{30} - \cdots$

13. $1 - 2x + 3x^2 - \cdots + (-1)^n(n+1)x^n + \cdots$

15. $1 + \frac{x^2}{2!} + \frac{5x^4}{4!} + \cdots$

17. $1 + 2x + \frac{5}{2}x^2 + \frac{8}{3}x^3 + \cdots$

19. $x + \frac{x^3}{3!} + \frac{x^5}{5!} + \cdots + \frac{x^{2n+1}}{(2n+1)!} + \cdots$

21. $1 - x + \frac{x^2}{2!} - \frac{x^3}{3!} + \cdots + (-1)^n \frac{x^n}{n!} + \cdots$

23. $x + \frac{5x^3}{6} + \cdots$

25. a) $x - \frac{2}{3}x^3 + \frac{2}{15}x^5 + \cdots$

b) $x - \frac{2}{3}x^3 + \frac{2}{15}x^5 - \cdots + (-1)^n(2^{2n})\frac{x^{2n+1}}{(2n+1)!} + \cdots$

27. $1/x = -1/[1 - (x+1)]$; $-1 - (x+1) - (x+1)^2 - \cdots - (x+1)^n - \cdots$

29. a) Replace x by $-x$ in Eq. (4). b) The alternating series test is satisfied.

c) $|E_n(1)| = \left|\frac{n!}{(1+c)^{n+1}} \cdot \frac{1^{n+1}}{(n+1)!}\right| = \left|\frac{1}{(1+c)^{n+1}(n+1)}\right| \leq \left|\frac{1}{n+1}\right|,$

so $\lim_{n\to\infty} E_n(1) = 0$. It now follows from (a) that the alternating harmonic series converges to ln 2.

31. $C + x + \frac{x^3}{3} + \frac{x^5}{5 \cdot 2!} + \frac{x^7}{7 \cdot 3!} + \cdots + \frac{x^{2n+1}}{(2n+1)n!} + \cdots$

33. $\pi + x - \dfrac{x^5}{5 \cdot 2!} + \dfrac{x^9}{9 \cdot 4!} - \cdots + (-1)^n \dfrac{x^{4n+1}}{(4n+1)(2n)!} + \cdots$

35. a) $e^{ix}: 1 + ix - \dfrac{x^2}{2!} - i\dfrac{x^3}{3!} + \dfrac{x^4}{4!} + i\dfrac{x^5}{5!} - \dfrac{x^6}{6!} - \cdots + (i)^n \dfrac{x^n}{n!} + \cdots$

$e^{-ix}: 1 - ix - \dfrac{x^2}{2!} + i\dfrac{x^3}{3!} + \dfrac{x^4}{4!} - i\dfrac{x^5}{5!} - \dfrac{x^6}{6!} + \cdots + (-i)^n \dfrac{x^n}{n!} + \cdots$

b) and c) This is obvious from (a) and the series (2) and (3) for $\sin x$ and $\cos x$.

d) $\cos x = (e^{ix} + e^{-ix})/2$; $\sin x = (e^{ix} - e^{-ix})/2i$

e) They are the same except for the presence of i in certain places.

37. $2x + (1/(1+x))$

39. $3x + \cos x$

41. $\sin x + \cos x$

43. $(e^x - 1) + [1/(1-x)]$

45. $2/(1-x)^3$

47. $[1/(1+2x^2)] + 2x^2 - 1 = 4x^4/(1+2x^2)$

49. $n = 7$; error term gives $n = 7$ also.

Section 11.4

1. 1

3. 2

5. $-\infty$

7. 0

9. 1

11. 0

13. 0

15. 0

17. 0

19. -1

21. ∞

23. $\frac{1}{2}$

25. -1

27. ∞

29. 0

31. 0

33. ∞

35. ∞

37. e^3

39. ∞

41. 1.0000

43. 0.000

45. 0.5

Section 11.5

1. 6

3. 8

5. 1

7. $\frac{35}{16}$

9. 1

11. $\frac{5}{81}$

13. $1 + 0x + 0x^2 + \cdots + 0x^n + \cdots$; $\binom{0}{k} = 1$ for $k = 0$ and 0 for $k > 0$

15. $1 - 2x + 3x^2 - 4x^3 + 5x^4 - \cdots$; $r = 1$

17. $1 - \frac{1}{3}x^2 - \frac{1}{9}x^4 - \frac{5}{81}x^6 - \frac{10}{243}x^8 - \cdots; r = 1$

19. $1 + 5x + 5x^2 - \frac{5}{3}x^3 + \frac{5}{3}x^4 - \cdots; r = \frac{1}{3}$

21. Find a partial sum of $(1 - x)^{1/2}$ with $x = \frac{1}{3}$.

23. Find a partial sum of $(1 - x)^{-1/3}$ with $x = \frac{1}{2}$.

25. $\frac{1}{3} - \frac{1}{7 \cdot 3!} + \frac{1}{11 \cdot 5!} - \frac{1}{15 \cdot 7!}$ with error $< \frac{1}{19 \cdot 9!}$

27. $\frac{1}{4} - \frac{1}{8 \cdot 2!} + \frac{1}{12 \cdot 4!} - \frac{1}{16 \cdot 6!} + \frac{1}{20 \cdot 8!}$ with error $< \frac{1}{24 \cdot 10!}$

29. $\frac{1}{10} - \frac{1}{3 \cdot 10^3} + \frac{1}{5 \cdot 2! \cdot 10^5} - \frac{1}{7 \cdot 3! \cdot 10^7}$ with error $< \frac{1}{9 \cdot 4! \cdot 10^9}$

31. Checking coefficients of x^k, we need only show that

$$p\binom{p}{k} = \left[(k + 1)\binom{p}{k+1}\right] + \left[k\binom{p}{k}\right].$$

We have

$$(k + 1)\binom{p}{k+1} + k\binom{p}{k} = (k + 1)\frac{p(p - 1)\cdots(p - k)}{(k + 1)(k)\cdots(2)(1)} + k\frac{p(p - 1)\cdots(p - k + 1)}{k!}$$

$$= \frac{[p(p - 1)\cdots(p - k)] + k[p(p - 1)\cdots(p - k + 1)]}{k!}$$

$$= \frac{p(p - 1)\cdots(p - k + 1)}{k!}[p - k + k] = \binom{p}{k}p.$$

33. Series 1.462650; Simpson's rule 1.462654; difference 0.000004

35. 3.14159268

Review Exercise Set 11.1

1. $-8 \le x < -2$

3. $\frac{\sqrt{3}}{2} + \frac{1}{2}\left(x - \frac{\pi}{3}\right) - \frac{\sqrt{3}}{4}\left(x - \frac{\pi}{3}\right)^2 - \frac{1}{12}\left(x - \frac{\pi}{3}\right)^3 + \frac{\sqrt{3}}{48}\left(x - \frac{\pi}{3}\right)^4$

5. a) A function is analytic if it can be represented by a power series in a neighborhood of each point in its domain.

b) $x + x^3 + x^5 + x^7 + \cdots + x^{2n+1} + \cdots$

7. 6

9. $1 + \frac{1}{2}x^2 - \frac{1}{8}x^4 + \frac{1}{16}x^6 - \frac{5}{128}x^8$

Review Exercise Set 11.2

1. $\dfrac{-\sqrt{3}-1}{2} \le x \le \dfrac{\sqrt{3}-1}{2}$

3. $-2\left(x-\dfrac{\pi}{4}\right)+\dfrac{4}{3}\left(x-\dfrac{\pi}{4}\right)^3$

5. $x^5-\dfrac{x^7}{2}+\dfrac{x^9}{3}-\dfrac{x^{11}}{4}+\cdots+(-1)^n\dfrac{x^{2n+5}}{n+1}+\cdots$

7. $-\dfrac{1}{2}$

9. $1+\dfrac{x}{4}+\dfrac{x^2}{32}+\dfrac{5}{128}x^3$

More Challenging Exercises 11

1. $1/e$ **3.** ∞ **5.** 1/2 **7.** 0 **9.** 4/3

CHAPTER 12

Section 12.1

1.

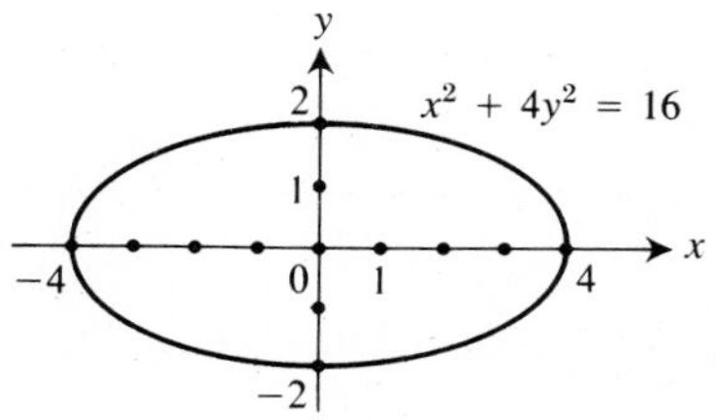

3.

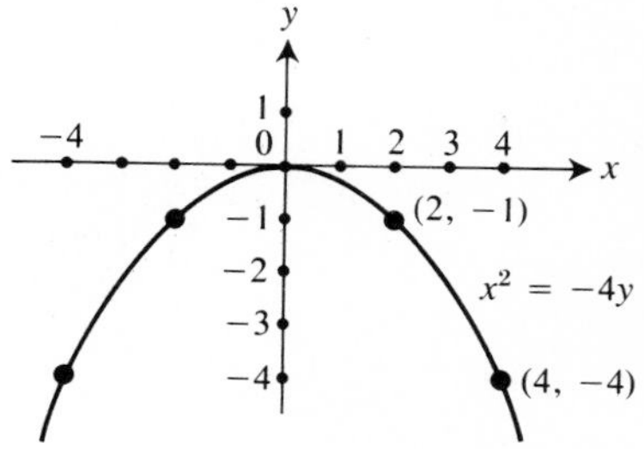

5.

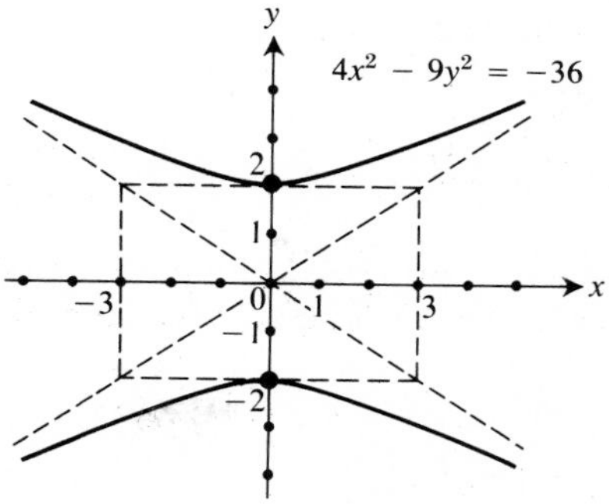

7.

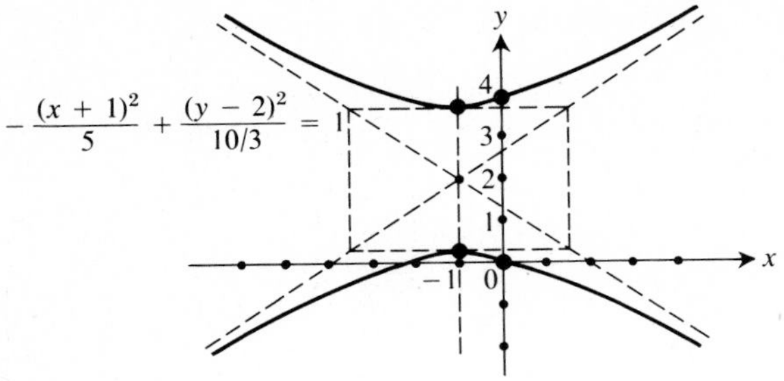

9.

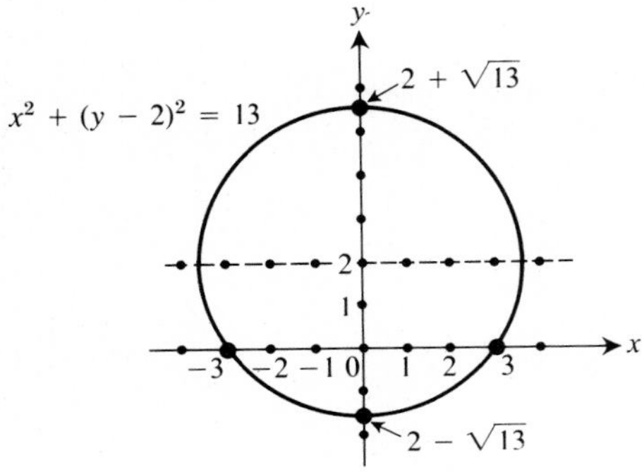

11.

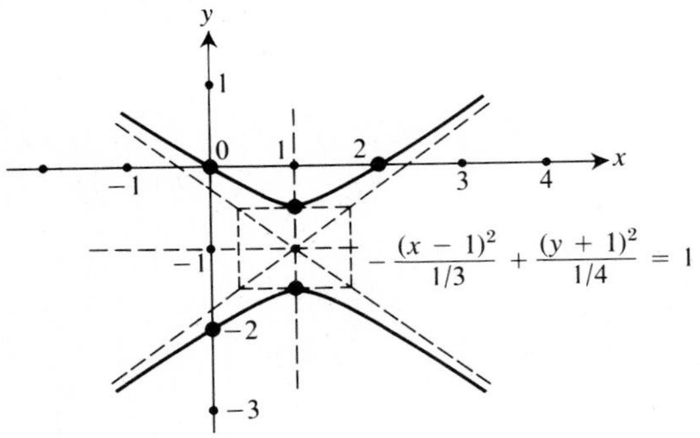

Section 12.2

1. Let P be a point on the hyperbola a distance d_1 from F_1 and a distance d_2 from F_2. Suppose $d_1 > d_2$. Then $d_1 - d_2 = 2a$. The triangle with vertices P, F_1, and F_2 has sides of lengths d_1, d_2, and $2c$. As suggested in the exercise, you then know that $d_1 < d_2 + 2c$. Therefore $d_1 - d_2 < 2c$, so $2a < 2c$ and $a < c$.

3. Let (x, y) be on the ellipse. Then

$$\left(\frac{\text{Distance to focus}}{\text{Distance to directrix}}\right)^2 = \frac{(x-c)^2+y^2}{[x-(a^2/c)]^2} = \frac{(x-c)^2+b^2[1-(x^2/a^2)]}{[x-(a^2/c)]^2}$$

$$= \frac{(x-c)^2+(a^2-c^2)[1-(x^2/a^2)]}{[x-(a^2/c)]^2}$$

$$= \frac{x^2-2cx+c^2+a^2-x^2-c^2+(c^2/a^2)x^2}{[x-(a^2/c)]^2}$$

$$= \frac{(c^2/a^2)x^2-2cx+a^2}{[x-(a^2/c)]^2} = \frac{(c^2/a^2)[x^2-2(a^2/c)x+(a^4/c^2)]}{[x-(a^2/c)]^2}$$

$$= \frac{c^2}{a^2}.$$

5. Foci $(\pm\sqrt{41}, 0)$
Directrices $x = \pm 25/\sqrt{41}$

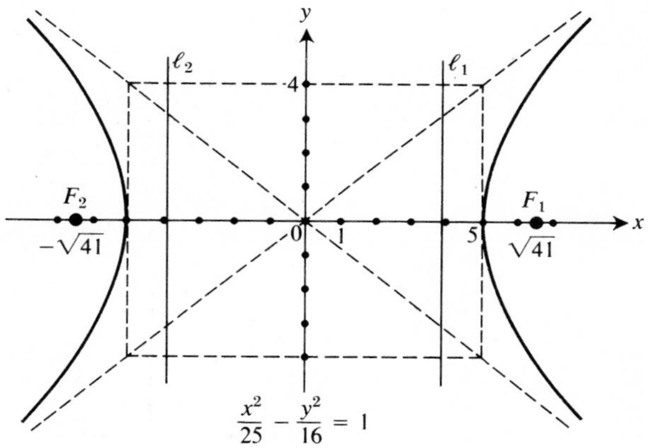

7. Foci $(-1 \pm \sqrt{2}, 2)$
Directrices $x = -1 \pm 2\sqrt{2}$

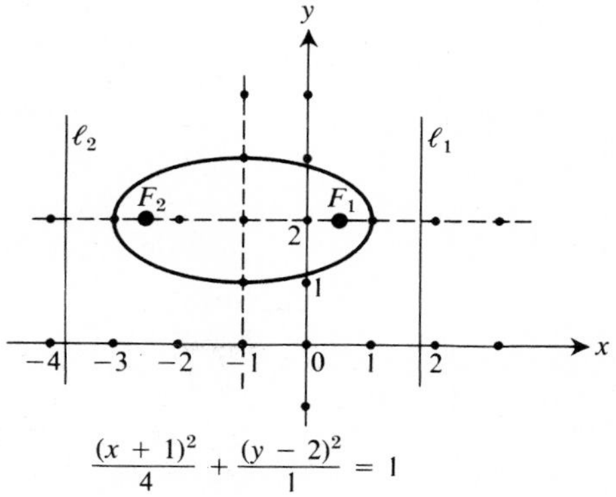

9. Focus $(3, -7)$
Directrix $x = 1$

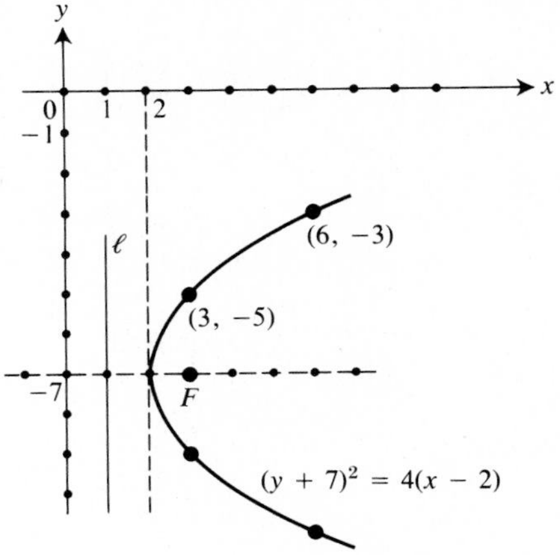

11. πab

13. $8x^2 - y^2 + 36x = -36$

15. $y^2 + 6x = 9$

17. $x^2 + 4x + 2y = 1$

19. $\dfrac{x^2}{36} + \dfrac{y^2}{27} = 1$

21. $\dfrac{x^2}{8} - \dfrac{y^2}{8} = 1$

23. $y^2 = 12x$

25. $\dfrac{(x+1)^2}{12} + \dfrac{(y-2)^2}{16} = 1$

27. $(x + 5)^2 = -8(y - 2)$

29. The line $x = p$ intersects $y^2 = 4px$ at $(0, 2p)$ and $(0, -2p)$. The distance between these points is $4p$.

Section 12.3

1. $x' = x \cos\theta + y \sin\theta$, $y' = -x \sin\theta + y \cos\theta$

3. Using Eq. (5) and trigonometric identities, we obtain $B'^2 - 4A'C' = (C - A)^2 \sin^2 2\theta + 2(C - A)B \sin 2\theta \cos 2\theta + B^2 \cos^2 2\theta - 4A^2 \cos^2\theta \sin^2\theta + 4AB \sin\theta \cos^3\theta - 4AC \cos^4\theta - 4AB \sin^3\theta \cos\theta + 4B^2 \sin^2\theta \cos^2\theta - 4BC \sin\theta \cos^3\theta - 4AC \sin^4\theta + 4CB \sin^3\theta \cos\theta - 4C^2 \sin^2\theta \cos^2\theta = B^2 - 4AC$.

5. $x'^2 - 5y'^2 = -10$

7. $5x'^2 + y'^2 - 4\sqrt{3}x' - 4y' = 36$

9. $11x'^2 + y'^2 + \dfrac{6}{\sqrt{10}}x' - \dfrac{2}{\sqrt{10}}y' = 24$

11. Parabola

13. Ellipse

15. Ellipse

17. Rotate through $\theta = 45°$. Then,

$$(x' + 1)^2 = -4(y' - 1)$$
$$\bar{x}^2 = -4\bar{y}$$

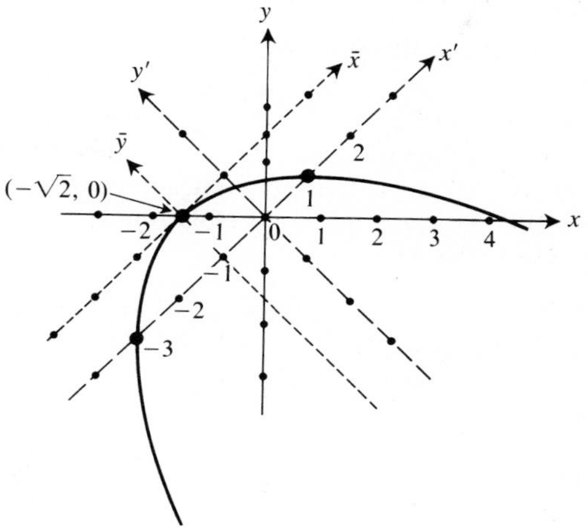

19. Rotate through $\theta = 45°$. Then,

$$\frac{(x' - 1)^2}{5} + \frac{(y' - 7)^2}{25} = 1$$
$$\frac{\bar{x}^2}{5} + \frac{\bar{y}^2}{25} = 1$$

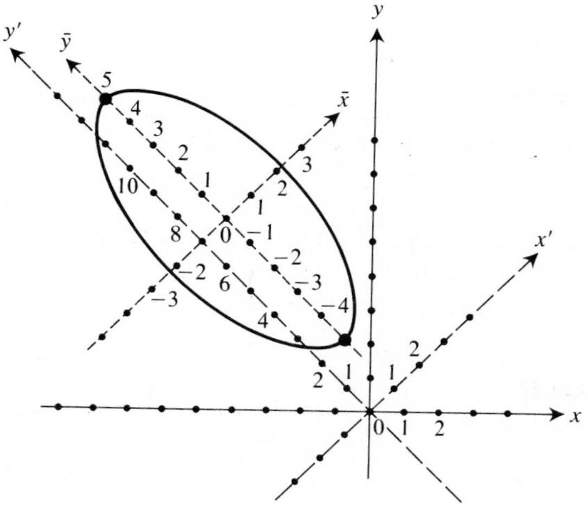

Section 12.4

1. For an ellipse, we have $c = \sqrt{a^2 - b^2} = a\sqrt{1 - (b/a)^2}$. It follows that if b/a is very near zero, we have $c/a \approx 1$.

3. As in Exercise 2, we could show that the point on an ellipse farthest from a given focus is the point farthest away on the major axis. The exercise now follows immediately from the fact that the center of the earth is at the focus of the ellipse.

5. Let the parabola be $y^2 = 4px$. The slope of the normal to the parabola at (x, y) is easily found to be $-y/2p$. The slope of the line from $(p, 0)$ to (x, y) is $y/(x - p)$. The tangent of an angle between lines whose slopes are m_1 and m_2 is $(m_1 - m_2)/(1 + m_1m_2)$ by the formula for $\tan(\theta_1 - \theta_2)$. Referring to the figure, you then see that

$$\tan\alpha = \frac{y/(x-p) + (y/2p)}{1 + (-y/2p)[y/(x-p)]}$$
$$= \frac{2py + xy - py}{2px - 2p^2 - y^2}$$
$$= \frac{py + xy}{2px - 2p^2 - 4px}$$
$$= \frac{y(p+x)}{-2p(p+x)} = \frac{-y}{2p}$$

and

$$\tan\beta = \frac{-(y/2p) - 0}{1 + (-y/2p)(0)} = \frac{-y}{2p}.$$

Thus $\alpha = \beta$.

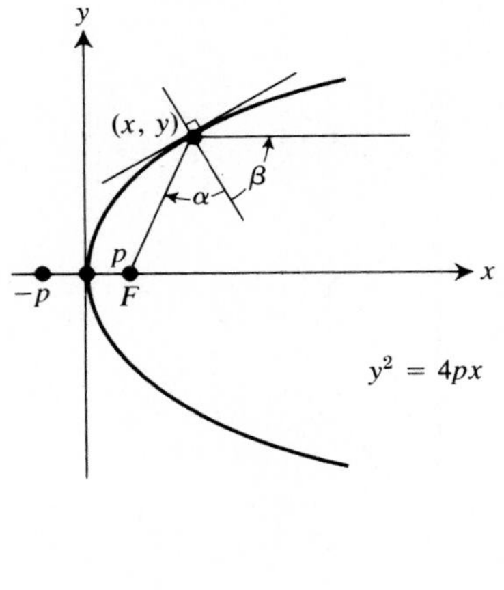

Section 12.5

1.$x = \sin t$
$y = \cos t$
$0 \le t \le \pi$

3.$x = \tan t$
$y = \sec t$
$-\dfrac{\pi}{2} < t < \dfrac{\pi}{2}$

5.$x = t^2$
$y = t^3$

$(1, 1)$
$(1, -1)$

7.$x = t - 3$
$y = t^2 + 1$

$(-3, 1)$

9.$x = t - 1$
$y = \ell n\, t$

11.$x = t^2$
$y = t^2 + 1$

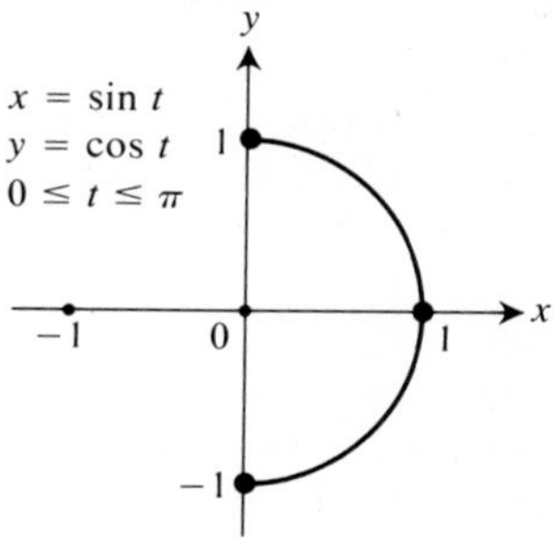

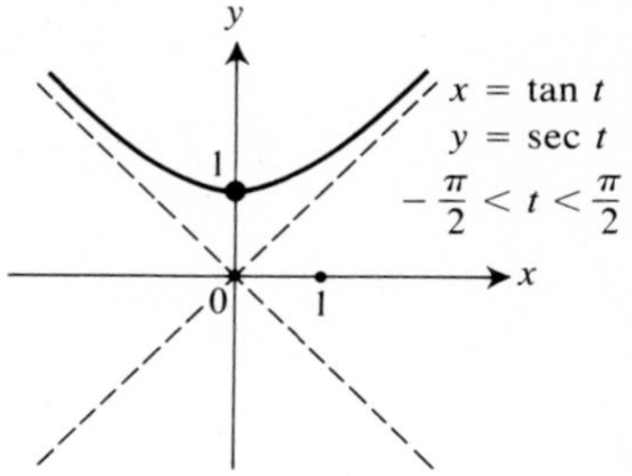

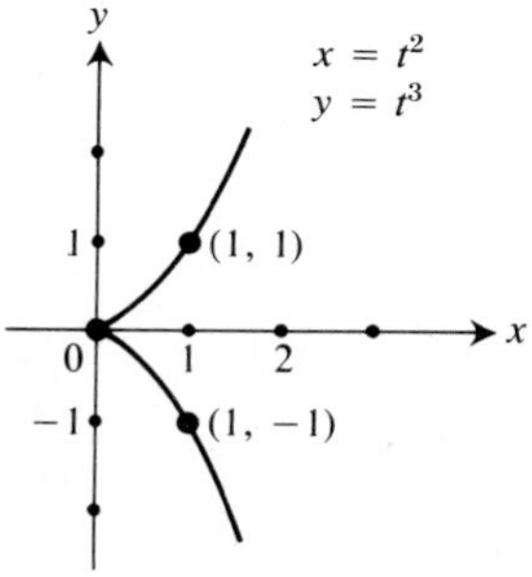

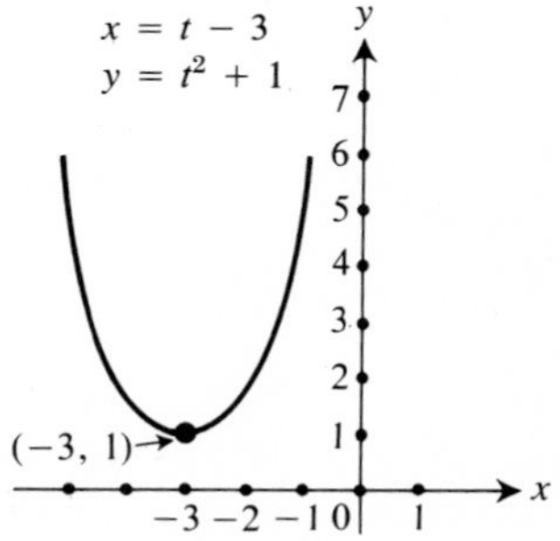

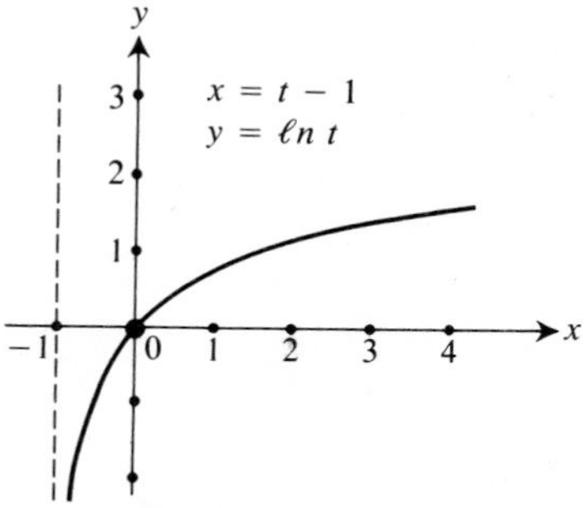

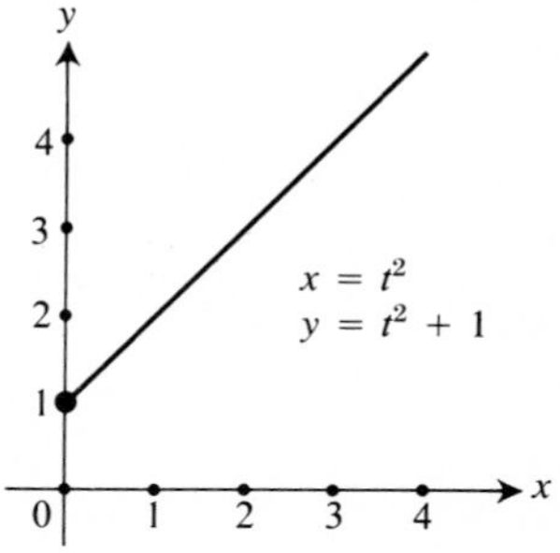

13. $x = \ln m,\ y = m \quad \text{for} \quad 0 < m < \infty$

15. $x = \dfrac{-1 + \sqrt{1 + 4d^2}}{2}, \qquad y = \left(\dfrac{-1 + \sqrt{1 + 4d^2}}{2}\right)^{1/2}, \qquad 0 \le d < \infty$

17. $x = a\theta - b \sin \theta,\ y = a - b \cos \theta$

19. $-\frac{2}{3}$

21. 1

23. $(1, 0)$

25. $\frac{1}{9}$

27. 24

29. $3a/2$

Section 12.6

1. 1

3. $1/2\sqrt{2}$

5. 4/25

7. $1/4a$

9. 1/5

11. Where $d\phi/ds$ is positive, ϕ increases as s increases, and increasing ϕ corresponds to counterclockwise rotation of the tangent line, so that the curve must bend to the left. Where $d\phi/ds$ is negative, ϕ decreases as s increases, and the curve bends to the right.

13. Zero

15. If the curvature is zero at each point, then by Formula (7) of the text, we have $d^2y/dx^2 = 0$ at each point. It follows from this differential equation that $y = Ax + B$ for some constants A and B.

17. $(x + \frac{13}{2})^2 + y^2 = \frac{81}{4}$

19. a) The center of the circle

b) The circles with center at the point

Review Exercise Set 12.1

1.1

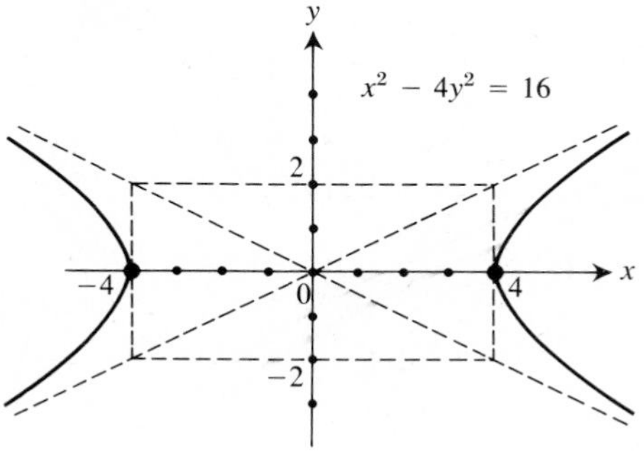

1.3

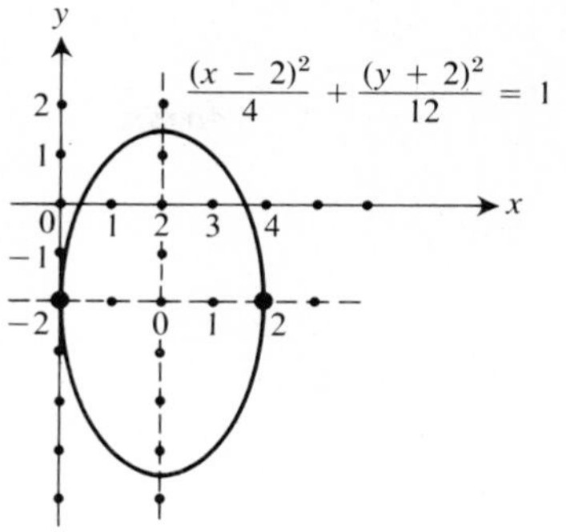

1. $3x^2 - y^2 + 14x + 4y = 9$

3. a) Hyperbola

b) Parabola

c) Ellipse

5. $\left.\frac{dy}{dx}\right|_{t=1} = \frac{1}{2}, \quad \left.\frac{d^2y}{dx^2}\right|_{t=1} = \frac{5}{4}$

7. $\int_0^3 \sqrt{4t^2 + \cos^2 t}\, dt$

9. $6\sqrt{3}/10^{3/2}$

Review Exercise Set 12.2

1.1

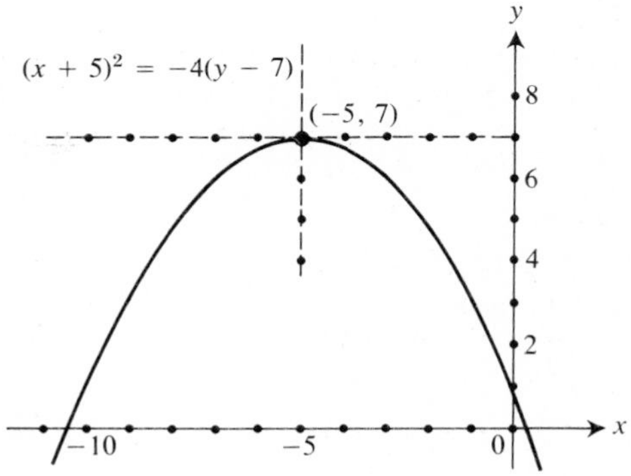

1.3

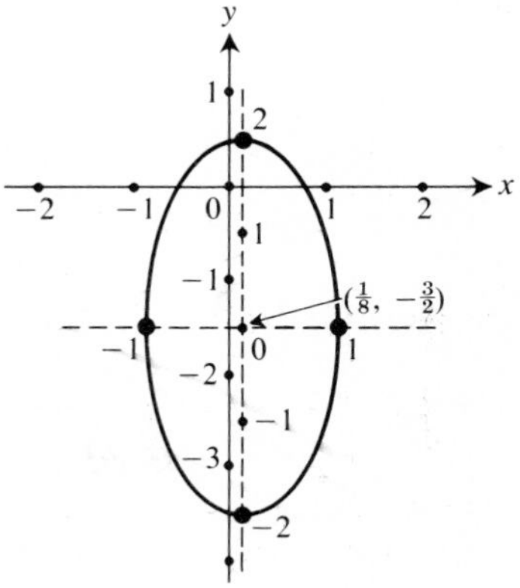

1. $9x^2 + 8y^2 + 18x - 14y = -14$

3. $\theta = \pi/4; \quad (y' - 1)^2 = x' + 3$

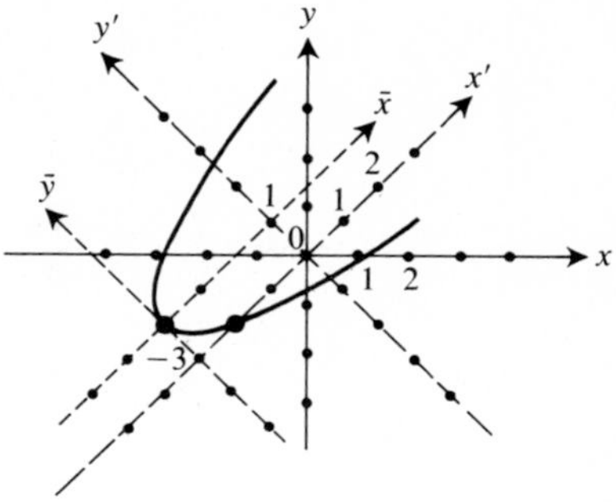

5. $\left.\dfrac{dy}{dx}\right|_{t=\pi/3} = -2\sqrt{3}, \quad \left.\dfrac{d^2y}{dx^2}\right|_{t=\pi/3} = -4$

7. $(15, 10\sqrt{15})$

9. $2/29^{3/2}$

More Challenging Exercises 12

1. a) A circle of radius a with center at $(0, 0)$

b) They become coincident at $(0, 0)$.

c) They recede to infinity.

d) 0

3. a) The lines $x = \pm a$

b) They recede to infinity.

c) The y-axis

d) ∞

5. a) The half lines on the x-axis where $|x| \geq a$

b) $(\pm a, 0)$

c) $x = \pm a$

d) 1

7. Since $1 + f'(x)^2 \geq 1$, we have

$$|\kappa(x)| = \left| -\frac{f''(x)}{(1 + f'(x)^2)^{3/2}} \right| \leq |f''(x)|.$$

9. $15a/8$

CHAPTER 13

Section 13.1

1. $(2\sqrt{2}, 2\sqrt{2})$

3. $(0, -6)$

5. $(-\sqrt{2}, -\sqrt{2})$

7. $(4, 0)$

9. $\left(2\sqrt{2}, \dfrac{\pi}{4} + 2n\pi\right)$ and $\left(-2\sqrt{2}, \dfrac{5\pi}{4} + 2n\pi\right)$

11. $\left(\sqrt{13}, \tan^{-1}\dfrac{3}{2} + \pi + 2n\pi\right)$ and $\left(-\sqrt{13}, \tan^{-1}\dfrac{3}{2} + 2n\pi\right)$

13. $d = \sqrt{(x_2 - x_1)^2 + (y_2 - y_1)^2} = \sqrt{(r_2 \cos \theta_2 - r_1 \cos \theta_1)^2 + (r_2 \sin \theta_2 - r_1 \sin \theta_1)^2}$
$= \sqrt{r_1^{\,2} + r_2^{\,2} - 2r_1r_2(\cos \theta_1 \cos \theta_2 + \sin \theta_1 \sin \theta_2)}$
$= \sqrt{r_1^{\,2} + r_2^{\,2} - 2r_1r_2 \cos (\theta_1 - \theta_2)}$

15. $2r \cos \theta + 3r \sin \theta = 5$

17. $(x - a)^2 + y^2 = a^2$

19. $e = 3/2;\ x = -5/3$

21. $32/\sqrt{3}$

Section 13.2

1.

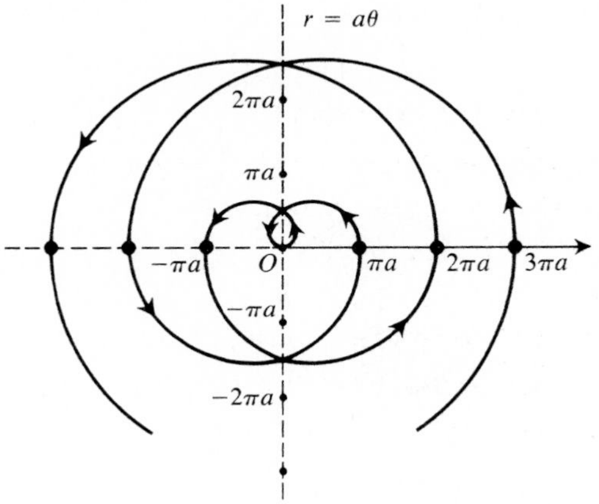

3.

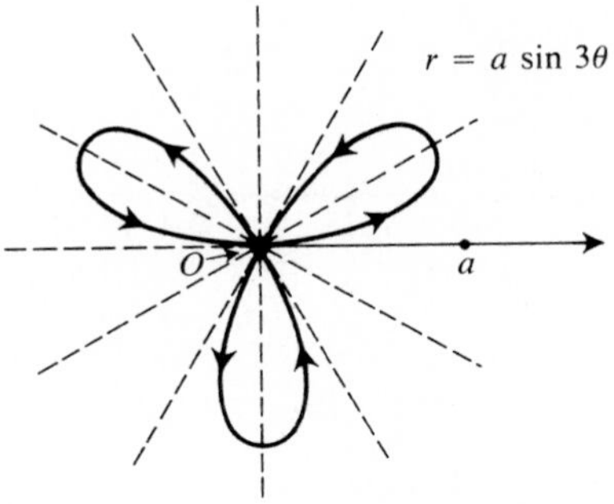

5.

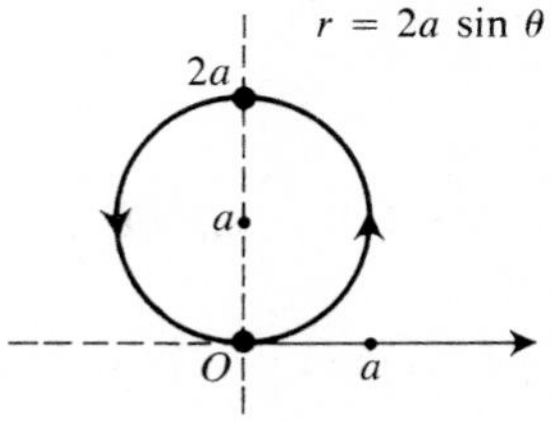

7.

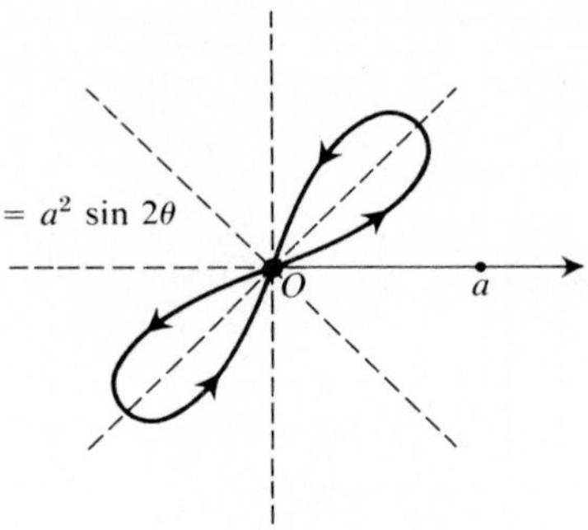
$r^2 = a^2 \sin 2\theta$
O
a

9.

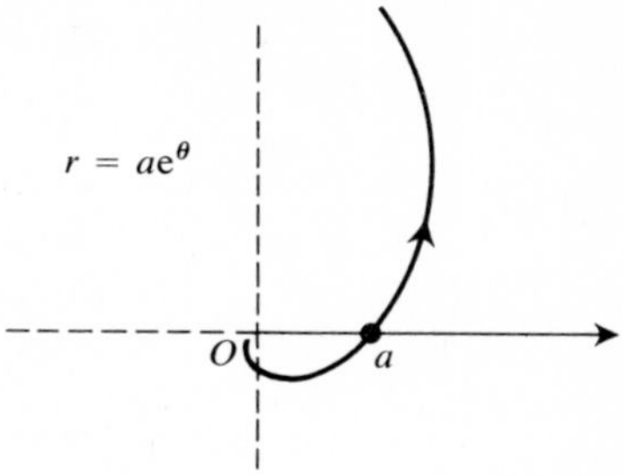
$r = ae^{\theta}$
O
a

11.

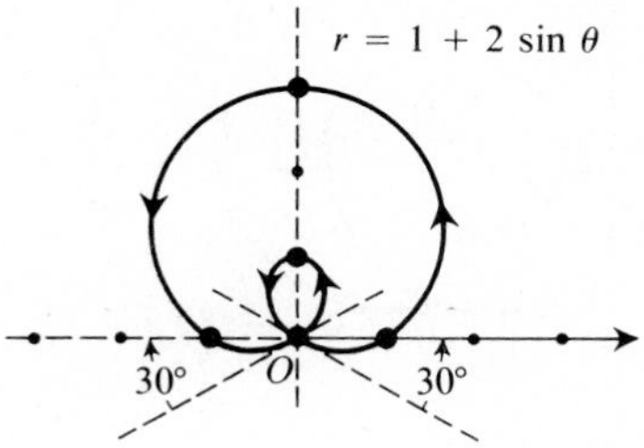
$r = 1 + 2 \sin \theta$
30°
O
30°

13.

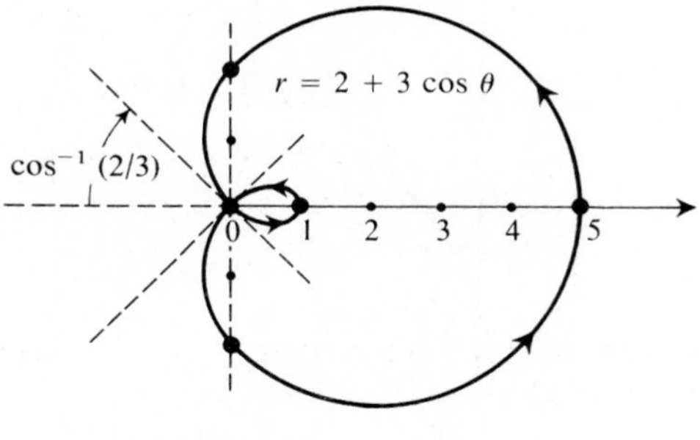
$r = 2 + 3 \cos \theta$
$\cos^{-1}(2/3)$
0
1
2
3
4
5

15.

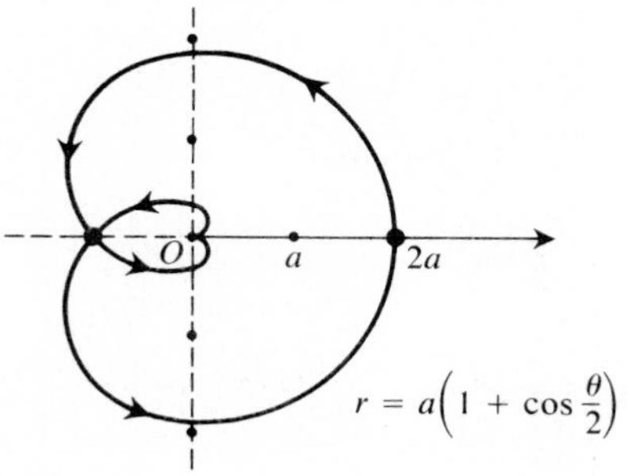
O
a
2a
$r = a\left(1 + \cos\frac{\theta}{2}\right)$

17. $(a, \pi/12)$, $(a, 5\pi/12)$, $(a, 13\pi/12)$, $(a, 17\pi/12)$

Section 13.3

1. πa^2

3. a^2

5. $a^2\left(\dfrac{2\pi}{3} + \sqrt{3}\right)$

7. $a^2\left(\dfrac{\pi}{2} - 1\right)$

Section 13.4

1. $-\tan^{-1}\left(-\dfrac{\pi}{2}\right)$

3. $(0, 0)$, $(3a/2, 2\pi/3)$, $(3a/2, 4\pi/3)$

5. $\pi/2$

7. $\dfrac{8}{3}a[(1 + \pi^2)^{3/2} - 1]$

9. $\dfrac{a}{2}\displaystyle\int_0^{\pi} \sqrt{1 + 3\cos^2(\theta/2)}\, d\theta$

11. $4\pi^2 a^2$

13. If a dot over a variable denotes differentiation with respect to a parameter, then

$$\kappa = \left|\frac{\dot{x}\ddot{y} - \dot{y}\ddot{x}}{(\dot{x}^2 + \dot{y}^2)^{3/2}}\right|.$$

Now $x = f(\theta)\cos\theta$ and $y = f(\theta)\sin\theta$. Thus, taking θ as parameter,

$$\begin{aligned}
\dot{x} &= -f(\theta)\sin\theta + f'(\theta)\cos\theta \\
\ddot{x} &= -f(\theta)\cos\theta - 2f'(\theta)\sin\theta + f''(\theta)\cos\theta \\
\dot{y} &= f(\theta)\cos\theta + f'(\theta)\sin\theta \\
\ddot{y} &= -f(\theta)\sin\theta + 2f'(\theta)\cos\theta + f''(\theta)\sin\theta.
\end{aligned}$$

Substituting these expressions in the formula for κ and simplifying yields the desired result.

Review Exercise Set 13.1

1. $[2, (5\pi/6) + 2n\pi]$,
$[-2, (11\pi/6) + 2n\pi]$

3. $x^2 + y^2 = x + y$

5. $(a/\sqrt{2}, \pi/6)$, $(a/\sqrt{2}, 5\pi/6)$, $(a/\sqrt{2}, 7\pi/6)$, $(a/\sqrt{2}, 11\pi/6)$

7. $(\pi/2) - \tan^{-1}(2/\sqrt{3})$

Review Exercise Set 13.2

1. $[-\sqrt{2}, (3\pi/4) + 2n\pi], [\sqrt{2}, (7\pi/4) + 2n\pi]$

3. $(x^2 + y^2)^2 = 2(x^2 + y^2) + 2xy, \quad (x, y) \neq (0, 0)$

5. $(a/\sqrt{2}, \pi/6), (a/\sqrt{2}, 5\pi/6), (a/\sqrt{2}, 7\pi/6), (a/\sqrt{2}, 11\pi/6)$

7. $\pi/2$

More Challenging Exercise 13

1. a) $(0, 0)$ b) 2 units time c) $r = \sqrt{2}e^{\pi/4}e^{-\theta}, \quad \theta \geq \pi/4$
d) 2 units e) It will get dizzy.

CHAPTER 14

Section 14.1

1. a)

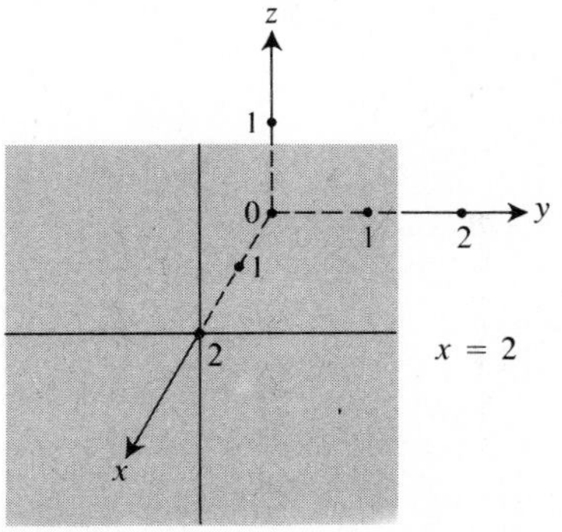

b)

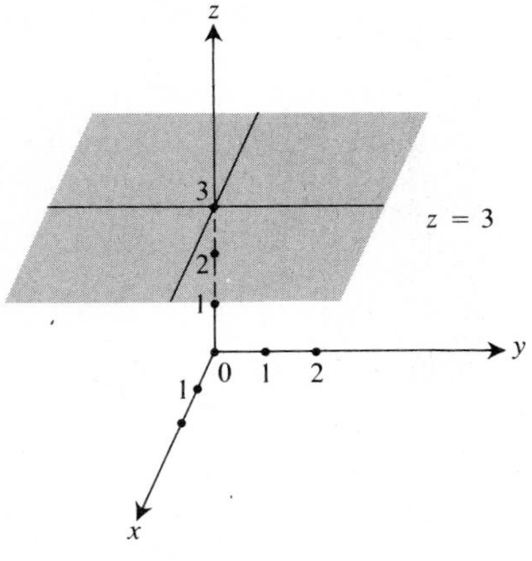

c) 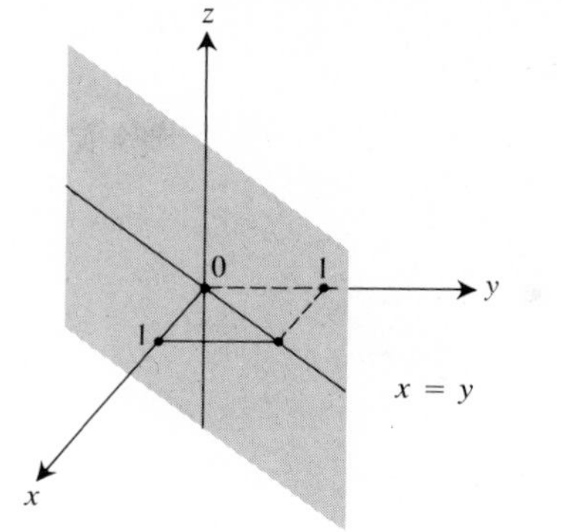

d) 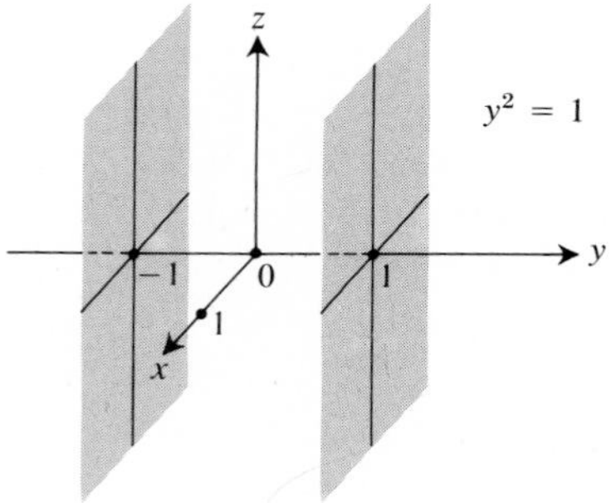

e) 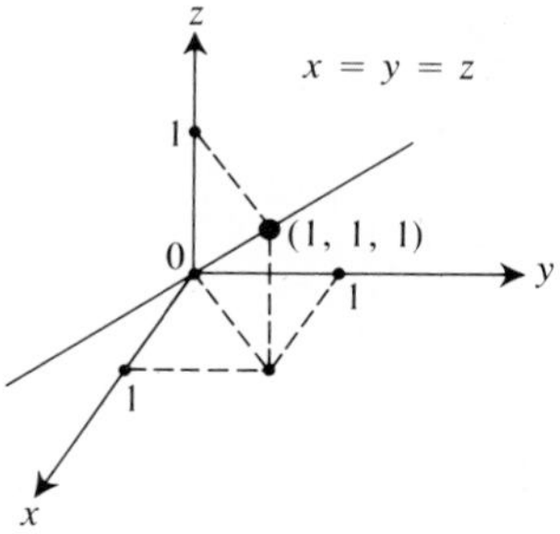

3. a) $(2, -4, 0)$
b) $(-2, 2, -1)$
c) $(6, -3, 1)$

5. $(x + 1)^2 + (y - 2)^2 + (z - 4)^2 = 19$

7. a) Center $(1, -1, 0)$, radius $\sqrt{2}$
b) Center $(3, 2, -4)$, radius 5

9. Same figure as answer 1(c).

11. 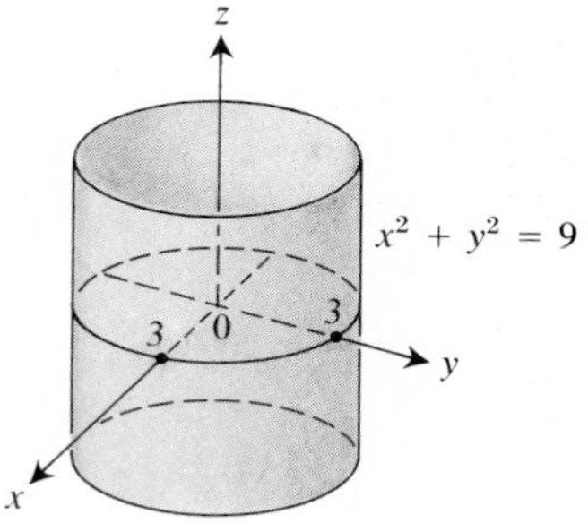

13. a) $(-\sqrt{2}, 0, \sqrt{2})$
b) $(0, 0, 0)$
c) $(2, 2\sqrt{3}, 0)$

15. $\rho = \sqrt{x^2 + y^2 + z^2}$

$$\phi = \cos^{-1} \frac{z}{\sqrt{x^2 + y^2 + z^2}}$$

$$\theta = \tan^{-1} \frac{y}{x}$$

17.

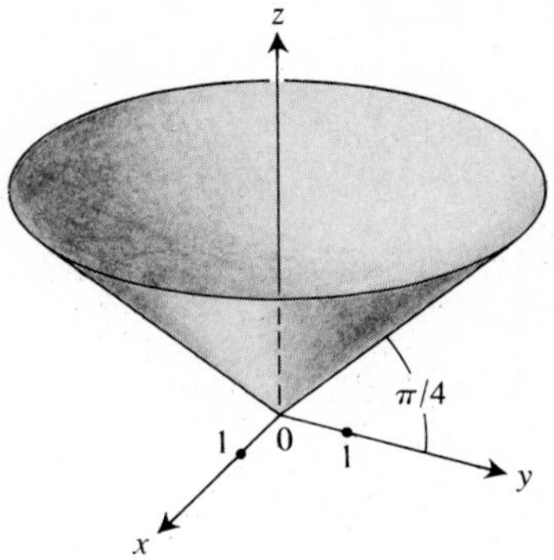

19. The plane $z = 0$

21. $\rho = 5$

Section 14.2

1. A plane $z = z_0$ intersects the surface in an ellipse if $z_0 > c$. Planes $x = x_0$ and $y = y_0$ intersect the surface in hyperbolas.

3.

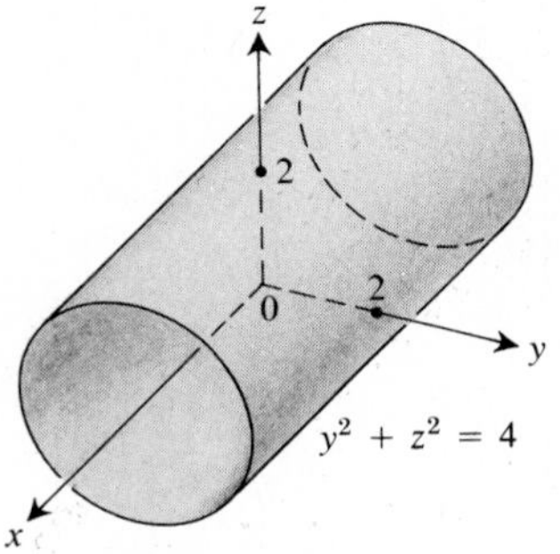

Right circular cylinder

5.

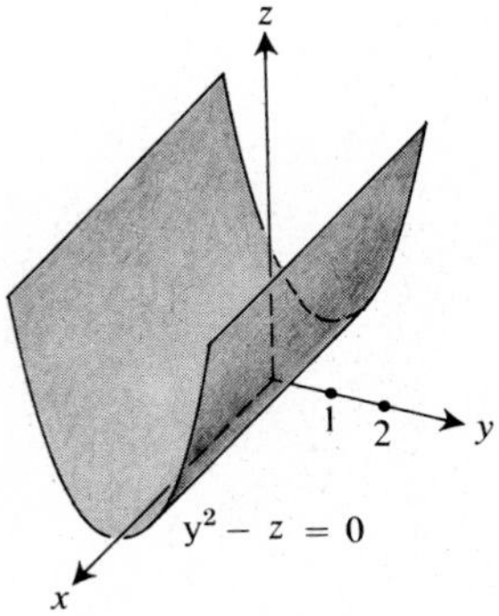

Parabolic cylinder

7.

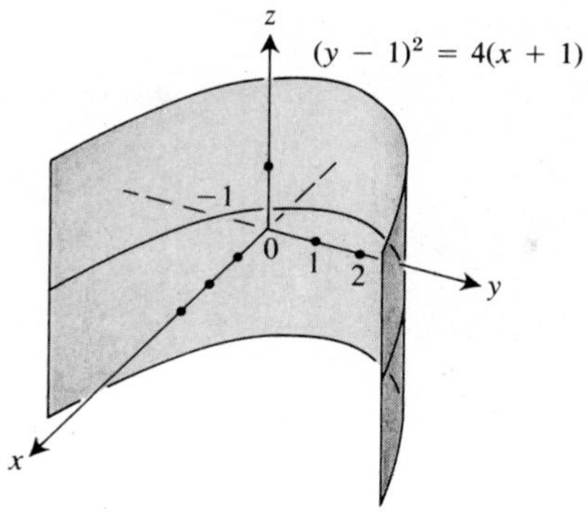

Parabolic cylinder

9.

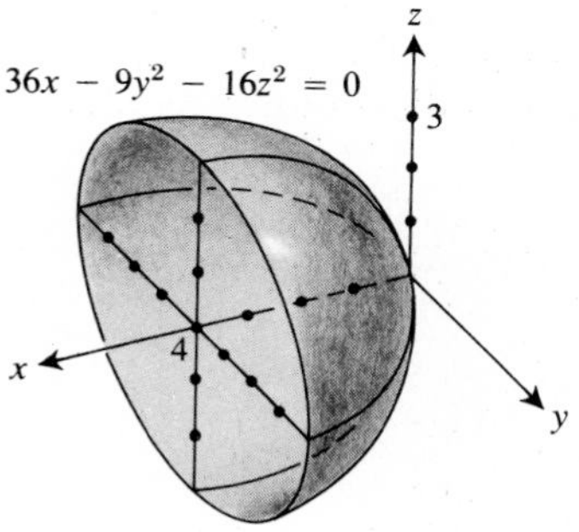

Elliptic paraboloid

11.

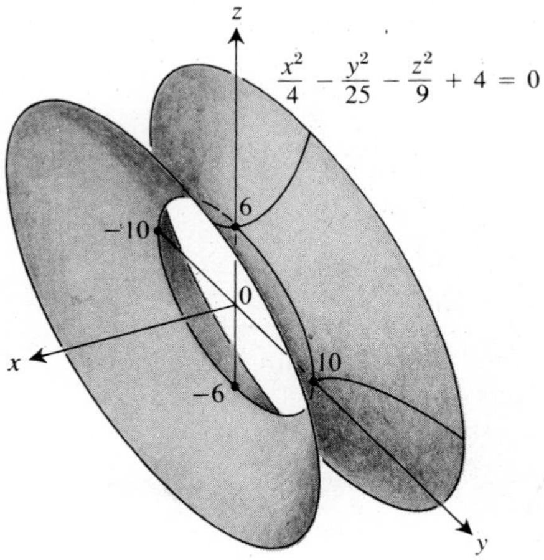

Hyperboloid of one sheet

13.$$\frac{x^2}{4} - \frac{y^2}{9} + z - 1 = 0$$

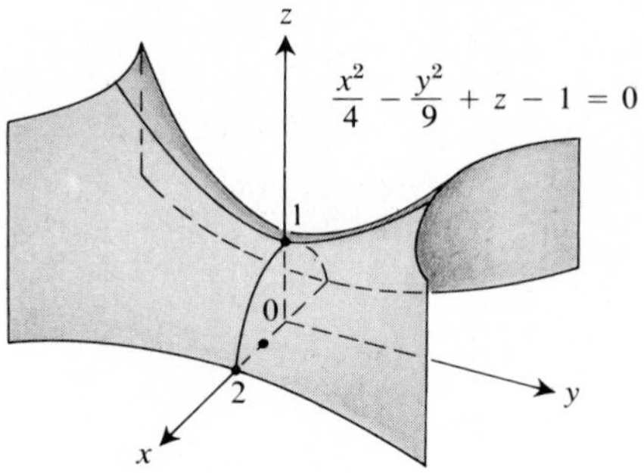

Hyperbolic paraboloid

15.$4y^2 = x^2 + 16z^2$

Elliptic cone

17.$x^2 + z^2 = 4(y + 2)$

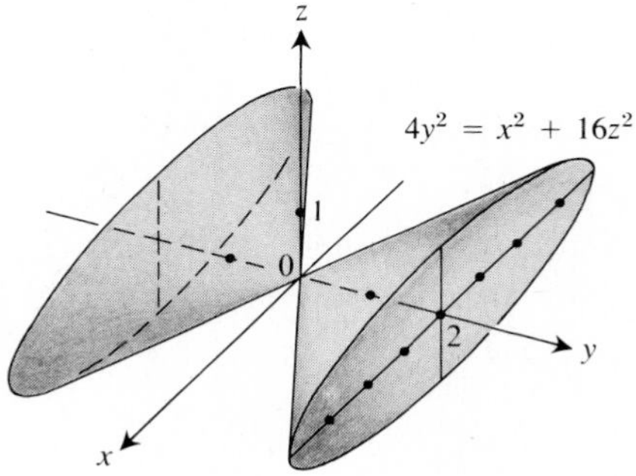

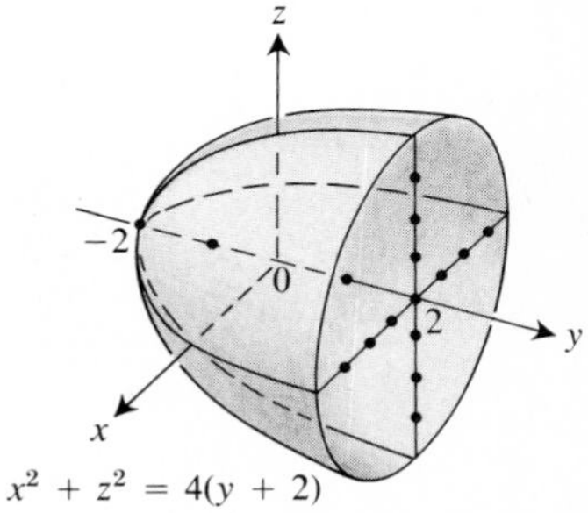

Circular paraboloid

Section 14.3

1.**3.** a) $\sqrt{17}$　　b) $\sqrt{17}$　　c) $2\sqrt{17}$　　d) 15

5. a) 3　　b) $-\frac{6}{5}$　　c) 0　　d) Impossible

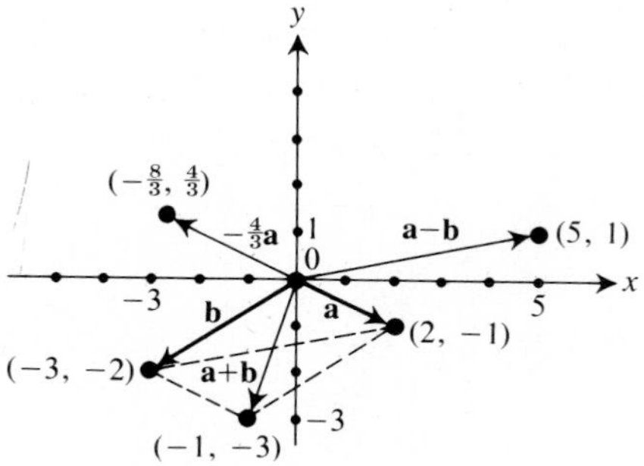

7. $\dfrac{1}{\sqrt{11}}(\boldsymbol{i} - \boldsymbol{j} + 3\boldsymbol{k})$

9. $(1/\sqrt{5})(\boldsymbol{i} + 2\boldsymbol{j})$ and $(1/\sqrt{2})(\boldsymbol{i} + \boldsymbol{k})$. (Many other answers are possible.)

11. $(3, -2, 4) - (1, -1, 4) = (2, -1, 0) = (-2, 1, 6) - (-4, 2, 6)$.
Also, $(3, -2, 4) - (-2, 1, 6) = (5, -3, -2) = (1, -1, 4) - (-4, 2, 6)$.

Section 14.4

1. $\pi/2$

3. $3\pi/4$

5. $\cos^{-1}\dfrac{-33}{15\sqrt{5}}$

7. $\cos^{-1}\dfrac{7}{\sqrt{65}}$

9. From Fig. 14.29, it suffices to show that if $|\boldsymbol{a}| = |\boldsymbol{b}|$, then $(\boldsymbol{a} + \boldsymbol{b}) \cdot (\boldsymbol{a} - \boldsymbol{b}) = 0$. But $(\boldsymbol{a} + \boldsymbol{b}) \cdot (\boldsymbol{a} - \boldsymbol{b}) = \boldsymbol{a} \cdot \boldsymbol{a} + \boldsymbol{b} \cdot \boldsymbol{a} - \boldsymbol{a} \cdot \boldsymbol{b} - \boldsymbol{b} \cdot \boldsymbol{b} = \boldsymbol{a} \cdot \boldsymbol{a} - \boldsymbol{b} \cdot \boldsymbol{b} = |\boldsymbol{a}|^2 - |\boldsymbol{b}|^2 = 0$.

11. $$[|\boldsymbol{a}|\boldsymbol{b} + |\boldsymbol{b}|\,\boldsymbol{a}] \cdot [|\boldsymbol{a}|\,\boldsymbol{b} - |\boldsymbol{b}|\,\boldsymbol{a}] = |\boldsymbol{a}|^2\,\boldsymbol{b} \cdot \boldsymbol{b} - |\boldsymbol{a}|\,|\boldsymbol{b}|\,\boldsymbol{b} \cdot \boldsymbol{a} + |\boldsymbol{b}|\,|\boldsymbol{a}|\,\boldsymbol{a} \cdot \boldsymbol{b} - |\boldsymbol{b}|^2\,\boldsymbol{a} \cdot \boldsymbol{a}$$
$$= |\boldsymbol{a}|^2\,\boldsymbol{b} \cdot \boldsymbol{b} - |\boldsymbol{b}|^2\,\boldsymbol{a} \cdot \boldsymbol{a} = |\boldsymbol{a}|^2\,|\boldsymbol{b}|^2 - |\boldsymbol{b}|^2\,|\boldsymbol{a}|^2 = 0.$$

13. Let $\boldsymbol{a} = \boldsymbol{i}$, $\boldsymbol{b} = 2\boldsymbol{i} - \boldsymbol{j}$, and $\boldsymbol{c} = 2\boldsymbol{i} + 7\boldsymbol{j}$. Then $\boldsymbol{a} \cdot \boldsymbol{b} = \boldsymbol{a} \cdot \boldsymbol{c} = 2$, but $\boldsymbol{b} \neq \boldsymbol{c}$.

15. Let $\boldsymbol{a} = a_1\boldsymbol{i} + a_2\boldsymbol{j} + a_3\boldsymbol{k}$, $\boldsymbol{b} = b_1\boldsymbol{i} + b_2\boldsymbol{j} + b_3\boldsymbol{k}$, and $\boldsymbol{c} = c_1\boldsymbol{i} + c_2\boldsymbol{j} + c_3\boldsymbol{k}$.

Then

$$\begin{aligned}\boldsymbol{a} \cdot (\boldsymbol{b} + \boldsymbol{c}) &= [a_1\boldsymbol{i} + a_2\boldsymbol{j} + a_3\boldsymbol{k}] \cdot [(b_1 + c_1)\boldsymbol{i} + (b_2 + c_2)\boldsymbol{j} + (b_3 + c_3)\boldsymbol{k}] \\ &= a_1(b_1 + c_1) + a_2(b_2 + c_2) + a_3(b_3 + c_3) \\ &= (a_1b_1 + a_2b_2 + a_3b_3) + (a_1c_1 + a_2c_2 + a_3c_3) \\ &= \boldsymbol{a} \cdot \boldsymbol{b} + \boldsymbol{a} \cdot \boldsymbol{c}.\end{aligned}$$

17. $$\begin{aligned}(\boldsymbol{a} + \boldsymbol{b}) \cdot (\boldsymbol{a} + \boldsymbol{b}) &= \boldsymbol{a} \cdot \boldsymbol{a} + \boldsymbol{a} \cdot \boldsymbol{b} + \boldsymbol{b} \cdot \boldsymbol{a} + \boldsymbol{b} \cdot \boldsymbol{b} && \text{[by (c)]} \\ &= \boldsymbol{a} \cdot \boldsymbol{a} + \boldsymbol{a} \cdot \boldsymbol{b} + \boldsymbol{a} \cdot \boldsymbol{b} + \boldsymbol{b} \cdot \boldsymbol{b} && \text{[by (b)]} \\ &= \boldsymbol{a} \cdot \boldsymbol{a} + 2(\boldsymbol{a} \cdot \boldsymbol{b}) + \boldsymbol{b} \cdot \boldsymbol{b}\end{aligned}$$

$$\begin{aligned}(\boldsymbol{a} - \boldsymbol{b}) \cdot (\boldsymbol{a} - \boldsymbol{b}) &= \boldsymbol{a} \cdot \boldsymbol{a} + \boldsymbol{a} \cdot (-\boldsymbol{b}) + (-\boldsymbol{b}) \cdot \boldsymbol{a} + (-\boldsymbol{b}) \cdot (-\boldsymbol{b}) && \text{[by (c)]} \\ &= \boldsymbol{a} \cdot \boldsymbol{a} - (\boldsymbol{a} \cdot \boldsymbol{b}) - (\boldsymbol{b} \cdot \boldsymbol{a}) + (\boldsymbol{b} \cdot \boldsymbol{b}) && \text{[by (d)]} \\ &= \boldsymbol{a} \cdot \boldsymbol{a} - (\boldsymbol{a} \cdot \boldsymbol{b}) - (\boldsymbol{a} \cdot \boldsymbol{b}) + (\boldsymbol{b} \cdot \boldsymbol{b}) && \text{[by (b)]} \\ &= \boldsymbol{a} \cdot \boldsymbol{a} - 2(\boldsymbol{a} \cdot \boldsymbol{b}) + (\boldsymbol{b} \cdot \boldsymbol{b})\end{aligned}$$

19. $\dfrac{3}{26}(\boldsymbol{i} + 3\boldsymbol{j} + 4\boldsymbol{k}), \dfrac{3}{\sqrt{26}}$

21. $\boldsymbol{0}, 0$

23. $$(\boldsymbol{a} - \boldsymbol{c}) \cdot \boldsymbol{b} = \left(\boldsymbol{a} - \frac{\boldsymbol{a} \cdot \boldsymbol{b}}{|\boldsymbol{b}|^2}\boldsymbol{b}\right) \cdot \boldsymbol{b} = \boldsymbol{a} \cdot \boldsymbol{b} - \frac{\boldsymbol{a} \cdot \boldsymbol{b}}{|\boldsymbol{b}|^2}(\boldsymbol{b} \cdot \boldsymbol{b}) = \boldsymbol{a} \cdot \boldsymbol{b} - \frac{\boldsymbol{a} \cdot \boldsymbol{b}}{|\boldsymbol{b}|^2}|\boldsymbol{b}|^2$$
$$= \boldsymbol{a} \cdot \boldsymbol{b} - \boldsymbol{a} \cdot \boldsymbol{b} = 0$$

Section 14.5

1. -15

3. 15

5. 61

7. a) $$\begin{vmatrix} a_1 & a_2 & a_3 \\ a_1 & a_2 & a_3 \\ c_1 & c_2 & c_3 \end{vmatrix} = a_1\begin{vmatrix} a_2 & a_3 \\ c_2 & c_3 \end{vmatrix} - a_2\begin{vmatrix} a_1 & a_3 \\ c_1 & c_3 \end{vmatrix} + a_3\begin{vmatrix} a_1 & a_2 \\ c_1 & c_2 \end{vmatrix}$$

$$= a_1a_2c_3 - a_1a_3c_2 - a_1a_2c_3 + a_2a_3c_1 + a_1a_3c_2 - a_2a_3c_1$$

$$= 0$$

b) $$\begin{vmatrix} a_1 & a_2 & a_3 \\ b_1 & b_2 & b_3 \\ a_1 & a_2 & a_3 \end{vmatrix} = a_1\begin{vmatrix} b_2 & b_3 \\ a_2 & a_3 \end{vmatrix} - a_2\begin{vmatrix} b_1 & b_3 \\ a_1 & a_3 \end{vmatrix} + a_3\begin{vmatrix} b_1 & b_2 \\ a_1 & a_2 \end{vmatrix}$$

$$= a_1a_3b_2 - a_1a_2b_3 - a_2a_3b_1 + a_1a_2b_3 + a_2a_3b_1 - a_1a_3b_2$$

$$= 0$$

9. $-6\boldsymbol{i} + 3\boldsymbol{j} + 5\boldsymbol{k}$

11. $\mathbf{0}$

13. 11

15. $\sqrt{374}$

17. 19/2

19. -6, $12\boldsymbol{i} + 4\boldsymbol{k}$

21. 20

23. 71/3

25. $(\boldsymbol{a} \times \boldsymbol{b}) \times \boldsymbol{c} = (\boldsymbol{a} \cdot \boldsymbol{c})\boldsymbol{b} - (\boldsymbol{b} \cdot \boldsymbol{c})\boldsymbol{a}$

27. Computation gives the determinant of a matrix with the first two rows the same, which is thus zero.

29. a) The vector $\boldsymbol{a} \times (\boldsymbol{b} \times \boldsymbol{c})$ is perpendicular to $\boldsymbol{b} \times \boldsymbol{c}$, which is in turn a vector perpendicular to the plane containing $\boldsymbol{b}$ and $\boldsymbol{c}$. Thus $\boldsymbol{a} \times (\boldsymbol{b} \times \boldsymbol{c})$ lies in this plane.

b) The argument is just like that in (a).

c) From (a) and (b), equal products $\boldsymbol{a} \times (\boldsymbol{b} \times \boldsymbol{c})$ and $(\boldsymbol{a} \times \boldsymbol{b}) \times \boldsymbol{c}$ would have to be parallel to $\boldsymbol{b}$. A quick sketch shows that $\boldsymbol{a} \times (\boldsymbol{b} \times \boldsymbol{c})$ is not, in general, parallel to $\boldsymbol{b}$.

Section 14.6

1. $x = 3 - 8t$,
$y = -2 + 4t$

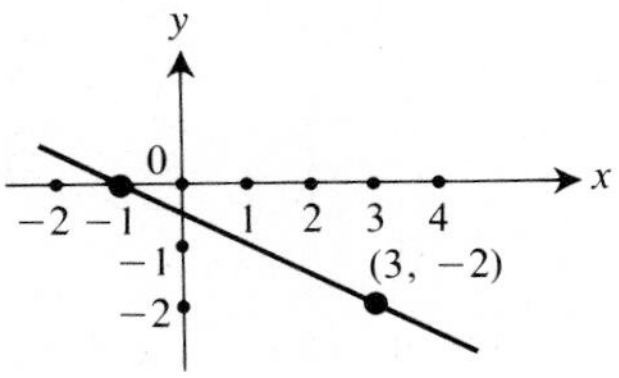

3. $x = 2 - t, \quad y = -1, \quad z = 4t$

5. $x = 5 + t, \quad y = -1 + 2t,$

7. a) $\cos^{-1}\dfrac{2}{\sqrt{5}}$ b) $\cos^{-1}\dfrac{2}{\sqrt{210}}$

9. Let the line have direction vector $\boldsymbol{d} = d_1\boldsymbol{i} + d_2\boldsymbol{j} + d_3\boldsymbol{k}$. Then $\cos\alpha = d_1/|\boldsymbol{d}|$, $\cos\beta = d_2/|\boldsymbol{d}|$, and $\cos\gamma = d_3/|\boldsymbol{d}|$. Thus

$$\cos^2\alpha + \cos^2\beta + \cos^2\gamma = \frac{d_1^2 + d_2^2 + d_3^2}{|\boldsymbol{d}|^2} = 1$$

11. a) $(\frac{1}{2}, \frac{3}{2})$ b) $(\frac{3}{2}, -2, \frac{5}{2})$

13. $(-\frac{3}{2}, -\frac{1}{2}, \frac{11}{4})$

15. $\cos\alpha = d_1/\sqrt{d_1^2 + d_2^2 + d_3^2}$, $\cos\beta = d_2/\sqrt{d_1^2 + d_2^2 + d_3^2}$,

$\cos\gamma = d_3/\sqrt{d_1^2 + d_2^2 + d_3^2}$

Section 14.7

1. $x - 2y + z = -7$

3. $3x - 2y + 7z = 39$

5. $3x - 7y + 3z = 0$

7. $x = -2 + t, \quad y = 1 - 2t, \quad z = 4t$

9. $x + y = 1$

11. $7x + 4y + 2z = 13$

13. $7x + 23y - z = 83$

15. $-7x + y + 11 = 0$

17. 0

19. $x = -9 + t, y = 16 - 3t, z = -t$

Review Exercise Set 14.1

1. a) $2\sqrt{22}$

b) $(x + 1)^2 + (y - 1)^2 + (z - 6)^2 = 21$

3.

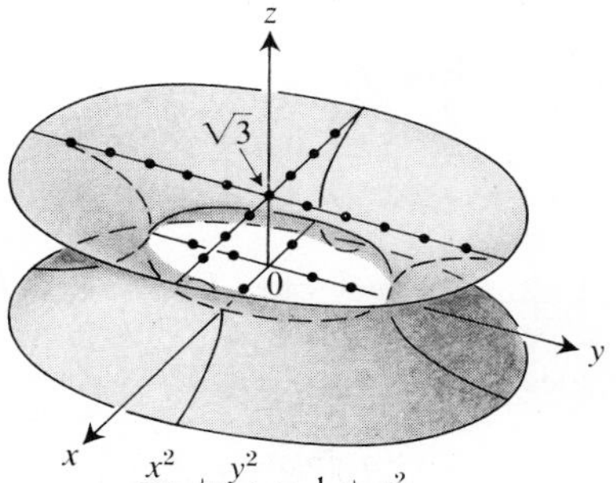

$\dfrac{x^2}{4} + \dfrac{y^2}{9} = 1 + z^2$

Hyperboloid of one sheet

5. $\cos^{-1}(-5/14)$

7. $-\boldsymbol{i} - 13\boldsymbol{j} - 9\boldsymbol{k}$

9. $-4\boldsymbol{i} - 2\boldsymbol{j} + 6\boldsymbol{k}$

11. $x = t, \quad y = -2t, \quad z = 3t$

13. $11/3$

15. $(\frac{37}{16}, \frac{59}{16}, \frac{31}{16})$

Review Exercise Set 14.2

1. a) $1 \pm 5\sqrt{5}$ b) Center $(1, 0, -2)$, radius 3

3. a)

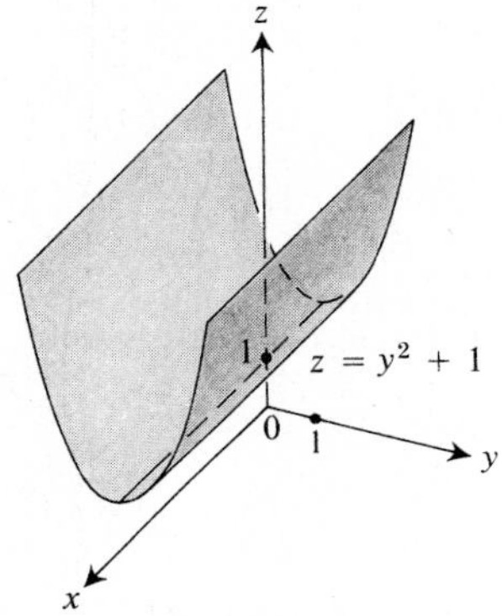

Parabolic cylinder

b)

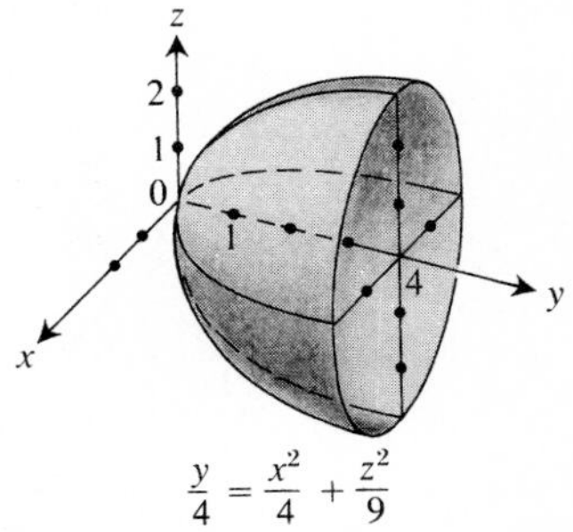

Elliptic paraboloid

5. $\cos^{-1}(7/3\sqrt{11})$

7. a) 0 b) $\boldsymbol{b}$ c) $\boldsymbol{0}$

9. 52

11. $x = -1 + 4t, \quad y = 5 - 6t, \quad z = 2 + 2t$

13. $19/\sqrt{10}$

15. $14x + 13y + 10z = 45$

More Challenging Exercises 14

1. $47/4\sqrt{14}$

3. $2x - 3y - z = 3$

5. 6

7. Theorem 14.2 applied to $(\boldsymbol{a} - x\boldsymbol{b}) \cdot (\boldsymbol{a} - x\boldsymbol{b}) \geq 0$ yields $(\boldsymbol{b} \cdot \boldsymbol{b})x^2 - 2(\boldsymbol{a} \cdot \boldsymbol{b})x + \boldsymbol{a} \cdot \boldsymbol{a} \geq 0$ for all x. Thus the equation $(\boldsymbol{b} \cdot \boldsymbol{b})x^2 - 2(\boldsymbol{a} \cdot \boldsymbol{b})x + (\boldsymbol{a} \cdot \boldsymbol{a}) = 0$ does *not* have two distinct real roots, so by the quadratic formula,

$$(-2(\boldsymbol{a} \cdot \boldsymbol{b}))^2 - 4(\boldsymbol{b} \cdot \boldsymbol{b})(\boldsymbol{a} \cdot \boldsymbol{a}) \leq 0,$$
$$4(\boldsymbol{a} \cdot \boldsymbol{b})^2 \leq 4(\boldsymbol{a} \cdot \boldsymbol{a})(\boldsymbol{b} \cdot \boldsymbol{b}),$$
$$(\boldsymbol{a} \cdot \boldsymbol{b})^2 \leq (\boldsymbol{a} \cdot \boldsymbol{a})(\boldsymbol{b} \cdot \boldsymbol{b}).$$

Since $\|\boldsymbol{a}\|^2 = \boldsymbol{a} \cdot \boldsymbol{a}$ and $\|\boldsymbol{b}\|^2 = \boldsymbol{b} \cdot \boldsymbol{b}$, this yields $|\boldsymbol{a} \cdot \boldsymbol{b}| \leq \|\boldsymbol{a}\|\,\|\boldsymbol{b}\|$.

CHAPTER 15

Section 15.1

1. By the chain rule and Eq. (8),

$$\frac{d\boldsymbol{r}}{ds} = \frac{dt}{ds}\frac{d\boldsymbol{r}}{dt} = \frac{dt}{ds}\boldsymbol{v} = \frac{dt}{ds}\left(\frac{ds}{dt}\boldsymbol{t}\right) = \boldsymbol{t}.$$

3. a) $2\boldsymbol{i} + 3\boldsymbol{j}$ b) $\sqrt{13}$ c) $0\boldsymbol{i} + 0\boldsymbol{j} = \boldsymbol{0}$

5. a) $-\boldsymbol{i}$ b) 1 c) $-4\boldsymbol{j}$

7. a) $\boldsymbol{i}$ b) 1 c) $-\boldsymbol{i} + \boldsymbol{j}$

9. a) $-\boldsymbol{i} - 2\boldsymbol{j}$ b) $\sqrt{5}$ c) $2\boldsymbol{i} + 6\boldsymbol{j}$

11. a) $4\boldsymbol{i} + 12\boldsymbol{j} - \boldsymbol{k}$ b) $\sqrt{161}$ c) $2\boldsymbol{i} + 12\boldsymbol{j}$

13. a) $-\boldsymbol{i} + \boldsymbol{j} + 2\boldsymbol{k}$ b) $\sqrt{6}$ c) $2\boldsymbol{i} - 2\boldsymbol{j} + 2\boldsymbol{k}$

15. 140

17. $\boldsymbol{a} \times \boldsymbol{b} = (-5t^3 + 24t^2 + 6t)\boldsymbol{i} + (t^4 + 16t^2)\boldsymbol{j} + (4t^3 + 16t^2 - 8t - 2)\boldsymbol{k},$

$$\frac{d\boldsymbol{a}}{dt} \times \boldsymbol{b} = (-10t^2 + 24t)\boldsymbol{i} + (2t^3 + 14t)\boldsymbol{j} + (4t^2 + 15t - 8)\boldsymbol{k},$$

$$\boldsymbol{a} \times \frac{d\boldsymbol{b}}{dt} = (-5t^2 + 24t + 6)\boldsymbol{i} + (2t^3 + 18t)\boldsymbol{j} + (8t^2 + 17t)\boldsymbol{k},$$

$$\frac{d(\boldsymbol{a} \times \boldsymbol{b})}{dt} = (-15t^2 + 48t + 6)\boldsymbol{i} + (4t^3 + 32t)\boldsymbol{j} + (12t^2 + 32t - 8)\boldsymbol{k}$$

$$= \frac{d\boldsymbol{a}}{dt} \times \boldsymbol{b} + \boldsymbol{a} \times \frac{d\boldsymbol{b}}{dt}$$

Section 15.2

1. By Newton's law that $\boldsymbol{F} = m\boldsymbol{a}$ and Eq. (10), we see that if $\dot{s} = 0$ when $t = t_0$, then $a_n = k\dot{s}^2 = 0$ when $t = t_0$, so $\boldsymbol{F} = ma_t\boldsymbol{t}$ has direction tangent to the curve at time $t = t_0$.

3. a) $a_t = 4/\sqrt{13}$, $a_n = 6/\sqrt{13}$ b) $6/(13)^{3/2}$

5. a) $a_t = 1$, $a_n = 2$ b) 2

7. a) $a_t = \sqrt{2}$, $a_n = \sqrt{2}$ b) $1/\sqrt{2}$

9. a) $a_t = 0$, $a_n = 2\sqrt{2}$ b) $\sqrt{2}$

Section 15.3

1. $\boldsymbol{v} = \boldsymbol{u}_r - \boldsymbol{u}_\theta$; $\boldsymbol{a} = 7\boldsymbol{u}_r + 2\boldsymbol{u}_\theta$

3. The ratio a^3/T^2 could be computed for a planet by astronomical observation. From Eq. (32), we have $M = (4\pi^2/G)(a^3/T^2)$, which could then be used to compute the mass M of the sun.

5. Using the notation of the hint, we have

$$2a = \text{apogee} + \text{perigee} + 2R,$$

so $2a$ is known if the apogee and perigee are known. From Eq. (32) we have

$$T = \frac{2\pi a^{3/2}}{\sqrt{GM}} = \left(\frac{\pi}{\sqrt{2GM}}\right)(2a)^{3/2},$$

where M is the mass of the earth.

7. If β is the area swept out in one unit of time, then by Kepler's second law, we must have $A(t) = \beta t$, so $dA/dt = \beta$.

9. Differentiating $r^2\dot{\theta} = 2\beta$ with respect to t, we have $r^2\ddot{\theta} + 2r\dot{r}\dot{\theta} = 0$, so $r(r\ddot{\theta} + 2\dot{r}\dot{\theta}) = 0$, which yields $r\ddot{\theta} + 2\dot{r}\dot{\theta} = 0$.

11. Differentiation of $r(1 - e\cos\theta) = B$ with respect to t yields $\dot{r}(1 - e\cos\theta) + r\dot{\theta}e\sin\theta = 0$.

13. Differentiation of the result in Exercise 12 with respect to t yields

$$\ddot{r}B + (2\beta e\cos\theta)\ddot{\theta} = 0,$$

so by Exercise 8, we have

$$\ddot{r} = \frac{-(2\beta e\cos\theta)\dot{\theta}}{B} = \frac{-(2\beta e\cos\theta)(r^2\dot{\theta})}{r^2 B}$$
$$= \frac{-(2\beta e\cos\theta)(2\beta)}{r^2 B} = -\frac{4\beta^2}{r^2}\frac{e\cos\theta}{B}.$$

Since $r = B/(1 - e\cos\theta)$, we have

$$\frac{1}{r} = \frac{1}{B} - \frac{e\cos\theta}{B},$$

so

$$-\frac{e\cos\theta}{B} = \frac{1}{r} - \frac{1}{B}.$$

We then have

$$\ddot{r} = \frac{4\beta^2}{r^2}\left(\frac{1}{r} - \frac{1}{B}\right).$$

15. By Exercise 10, we have

$$\boldsymbol{F} = m\boldsymbol{a} = m(\ddot{r} - r\dot{\theta}^2)\boldsymbol{u}_r.$$

Since $\boldsymbol{u}_r$ is directed away from the sun, Exercise 14 (with the negative sign) yields

$$|\boldsymbol{F}| = m\left(\frac{4\beta^2}{B}\cdot\frac{1}{r^2}\right) = \frac{4m\beta^2/B}{r^2},$$

and the force is directed toward the sun.

17. From $dA/dt = \beta$, we have $A = \beta t + C$, so β units of area are swept out per unit time. In T units of time, therefore, βT units of area are swept out; the area of the ellipse is thus βT.

19. We have

$$|\boldsymbol{F}| = \frac{4m\beta^2}{Br^2} = \frac{4m\pi^2a^2b^2}{T^2r^2B}.$$

Now by Exercise 16, we have

$$B = a(1 - e^2) = a\left(\frac{b^2}{a^2}\right) = \frac{b^2}{a}.$$

Then

$$|\boldsymbol{F}| = \frac{4m\pi^2a^2b^2}{T^2t^2(b^2/a)} = \frac{4m\pi^2a^3}{T^2r^2}.$$

21. If $a^3/T^2 = KM$ where M is the mass at the center of the central force field and K is a constant, then Exercise 19 yields

$$|\boldsymbol{F}| = \frac{4\pi^2mMK}{r^2} = \frac{mM(4\pi^2K)}{r^2}.$$

This yields Newton's universal law of gravitation, where $G = 4\pi^2K$.

Section 15.4

1. $\kappa = 0,\ \tau = 0$

3. 6

5. $\boldsymbol{t} = \dfrac{1}{\sqrt{1 + 2t^2}}(\boldsymbol{i} + t\boldsymbol{j} - t\boldsymbol{k}),\quad \boldsymbol{n} = \dfrac{1}{\sqrt{2 + 4t^2}}(-2t\boldsymbol{i} + \boldsymbol{j} - \boldsymbol{k}),\quad \boldsymbol{b} = \dfrac{1}{\sqrt{2}}(\boldsymbol{j} + \boldsymbol{k})$

7. $y + z = 0$

9. From $y = t^2$ and $z = -t^2$, we obtain at once $y + z = 0$, which is the equation of a plane.

11. 27

13. $\boldsymbol{t} = \frac{1}{9}(4\boldsymbol{i} + 8\boldsymbol{j} + \boldsymbol{k})$,
$\boldsymbol{n} = \frac{1}{9}(-7\boldsymbol{i} + 4\boldsymbol{j} - 4\boldsymbol{k})$

15. $\frac{2}{243}$

17. $\boldsymbol{\delta} \times \boldsymbol{t} = (\tau\boldsymbol{t} + \kappa\boldsymbol{b}) \times \boldsymbol{t} = \tau(\boldsymbol{t} \times \boldsymbol{t}) + \kappa(\boldsymbol{b} \times \boldsymbol{t}) = \tau\boldsymbol{0} + \kappa\boldsymbol{n} = \kappa\boldsymbol{n}$

$\boldsymbol{\delta} \times \boldsymbol{n} = (\tau\boldsymbol{t} + \kappa\boldsymbol{b}) \times \boldsymbol{n} = \tau(\boldsymbol{t} \times \boldsymbol{n}) + \kappa(\boldsymbol{b} \times \boldsymbol{n}) = \tau\boldsymbol{b} + \kappa(-\boldsymbol{t}) = -\kappa\boldsymbol{t} + \tau\boldsymbol{b}$

$\boldsymbol{\delta} \times \boldsymbol{b} = (\tau\boldsymbol{t} + \kappa\boldsymbol{b}) \times \boldsymbol{b} = \tau(\boldsymbol{t} \times \boldsymbol{b}) + \kappa(\boldsymbol{b} \times \boldsymbol{b}) = \tau(-\boldsymbol{n}) + \kappa\boldsymbol{0} = -\tau\boldsymbol{n}$

19. From $\ddot{\boldsymbol{r}} = \ddot{s}\boldsymbol{t} + \dot{s}^2\kappa\boldsymbol{n}$, we obtain

$$\begin{aligned}
\dddot{\boldsymbol{r}} &= \ddot{s}\dot{\boldsymbol{t}} + \dddot{s}\boldsymbol{t} + \dot{s}^2[\kappa\dot{\boldsymbol{n}} + \dot{\kappa}\boldsymbol{n}] + 2\dot{s}\ddot{s}\kappa\boldsymbol{n} \\
&= \ddot{s}\left(\frac{ds}{dt}\frac{d\boldsymbol{t}}{ds}\right) + \dddot{s}\boldsymbol{t} + \dot{s}^2\left[\kappa\frac{ds}{dt}\frac{d\boldsymbol{n}}{ds} + \dot{\kappa}\boldsymbol{n}\right] + 2\dot{s}\ddot{s}\kappa\boldsymbol{n} \\
&= \dot{s}\ddot{s}\kappa\boldsymbol{n} + \dddot{s}\boldsymbol{t} + \dot{s}^2[\kappa\dot{s}(-\kappa\boldsymbol{t} + \tau\boldsymbol{b}) + \dot{\kappa}\boldsymbol{n}] + 2\dot{s}\ddot{s}\kappa\boldsymbol{n} \\
&= (\dddot{s} - \dot{s}^3\kappa^2)\boldsymbol{t} + (3\dot{s}\ddot{s}\kappa + \dot{s}^2\dot{\kappa})\boldsymbol{n} + \dot{s}^3\kappa\tau\boldsymbol{b}.
\end{aligned}$$

21. a) $\dfrac{|\dot{\boldsymbol{r}} \times \ddot{\boldsymbol{r}}|}{|\dot{\boldsymbol{r}}|^3} = \dfrac{|\dot{s}^3\kappa|}{|\dot{s}|^3} = |\kappa| = \kappa.$ b) $\dfrac{(\dot{\boldsymbol{r}} \times \ddot{\boldsymbol{r}}) \cdot \dddot{\boldsymbol{r}}}{|\dot{\boldsymbol{r}} \times \ddot{\boldsymbol{r}}|^2} = \dfrac{\dot{s}^6\kappa^2\tau}{|\dot{s}^3\kappa|^2} = \tau.$

23. The Frenet formula $d\boldsymbol{t}/ds = \kappa\boldsymbol{n}$ yields $d\boldsymbol{t}/ds = \boldsymbol{0}$ if $\kappa = 0$. But then $\boldsymbol{t}$ is constant, so $\boldsymbol{t} = c_1\boldsymbol{i} + c_2\boldsymbol{j} + c_3\boldsymbol{k}$. Thus $d\boldsymbol{r}/ds = c_1\boldsymbol{i} + c_2\boldsymbol{j} + c_3\boldsymbol{k}$, so in terms of the parameter s, we obtain $\boldsymbol{r} = (c_1 s + d_1)\boldsymbol{i} + (c_2 s + d_2)\boldsymbol{j} + (c_3 s + d_3)\boldsymbol{k}$, and the locus has parametric equations $x = c_1 s + d_1$, $y = c_2 s + d_2$, $z = c_3 s + d_3$, and is a line.

25. Following the hint, we have

$$\begin{aligned}\frac{dw}{ds} &= \boldsymbol{t}\cdot\frac{d\bar{\boldsymbol{t}}}{ds} + \frac{d\boldsymbol{t}}{ds}\cdot\bar{\boldsymbol{t}} + \boldsymbol{n}\cdot\frac{d\bar{\boldsymbol{n}}}{ds} + \frac{d\boldsymbol{n}}{ds}\cdot\bar{\boldsymbol{n}} + \boldsymbol{b}\cdot\frac{d\bar{\boldsymbol{b}}}{ds} + \frac{d\boldsymbol{b}}{ds}\cdot\bar{\boldsymbol{b}}\\ &= \boldsymbol{t}\cdot(\kappa\bar{\boldsymbol{n}}) + \kappa\boldsymbol{n}\cdot\bar{\boldsymbol{t}} + \boldsymbol{n}\cdot(-\kappa\bar{\boldsymbol{t}} + \tau\bar{\boldsymbol{b}}) + (-\kappa\boldsymbol{t} + \tau\boldsymbol{b})\cdot\bar{\boldsymbol{n}} + \boldsymbol{b}\cdot(-\tau\bar{\boldsymbol{n}}) + (-\tau\boldsymbol{n})\cdot\bar{\boldsymbol{b}}\\ &= 0.\end{aligned}$$

Thus w is a constant, and since $\boldsymbol{t}$, $\boldsymbol{n}$, and $\boldsymbol{b}$ are equal to their respective barred counterparts at 0, we have $w(0) = 1 + 1 + 1 = 3$. Since $\boldsymbol{t}, \bar{\boldsymbol{t}}, \boldsymbol{n}, \bar{\boldsymbol{n}}, \boldsymbol{b}, \bar{\boldsymbol{b}}$ are unit vectors, we have $\boldsymbol{t}\cdot\bar{\boldsymbol{t}} \le 1$ and $\boldsymbol{t}\cdot\bar{\boldsymbol{t}} = 1$ if and only if $\boldsymbol{t} = \bar{\boldsymbol{t}}$, with similar results holding for $\boldsymbol{n}$ and $\bar{\boldsymbol{n}}$, and for $\boldsymbol{b}$ and $\bar{\boldsymbol{b}}$. Thus $w = 3$ implies $\boldsymbol{t} = \bar{\boldsymbol{t}}$, $\boldsymbol{n} = \bar{\boldsymbol{n}}$, and $\boldsymbol{b} = \bar{\boldsymbol{b}}$ for all values of the parameter s. From $\boldsymbol{t} = \bar{\boldsymbol{t}}$, we have $d\boldsymbol{r}/ds = d\bar{\boldsymbol{r}}/ds$, so $\boldsymbol{r}(s) = \bar{\boldsymbol{r}}(s) + \boldsymbol{c}$, and from $\boldsymbol{r}(0) = \bar{\boldsymbol{r}}(0)$, we have $\boldsymbol{c} = \boldsymbol{0}$. Thus $\boldsymbol{r}(s) = \bar{\boldsymbol{r}}(s)$, so our curves are the same in terms of the arc length parameter.

Review Exercise Set 15.1

1. a) $\boldsymbol{r} = (\sin 2t)\boldsymbol{i} + (\cos t)\boldsymbol{j}$ b) $\boldsymbol{v} = (-1/\sqrt{2})\boldsymbol{j}$ c) $1/\sqrt{2}$ d) $-4\boldsymbol{i} - (1/\sqrt{2})\boldsymbol{j}$

3. $\boldsymbol{v} = \frac{1}{4}\boldsymbol{u}_r + \frac{3}{4}\boldsymbol{u}_\theta$; $\boldsymbol{a} = -\frac{10}{32}\boldsymbol{u}_r + \frac{15}{128}\boldsymbol{u}_\theta$

5. A (possibly degenerate) ellipse, parabola, or hyperbola with focus at the center of the force field; i.e., a second-degree plane curve

7. a) $\boldsymbol{i} + (3/\sqrt{2})\boldsymbol{j} + (3/\sqrt{2})\boldsymbol{k}$ b) $\sqrt{10}$ c) $(-3/\sqrt{2})\boldsymbol{j} + (3/\sqrt{2})\boldsymbol{k}$

9. a) 3/10 b) 1/10

Review Exercise Set 15.2

1. a) $\frac{1}{3}\boldsymbol{i} + \frac{1}{4}\boldsymbol{j}$ b) 5/12 c) $-\frac{1}{27}\boldsymbol{i} - \frac{1}{32}\boldsymbol{j}$

3. 60 ft/sec

5. See the start of Section 15.3.2.

7. The torsion τ is the unique scalar such that $d\boldsymbol{b}/ds = -\tau\boldsymbol{n}$.

9. $\kappa = 14/17\sqrt{17}$; $\tau = 6/49$

CHAPTER 16

Section 16.1

1. $f_x = 3$, $f_y = 4$

3. $f_x = 2x$, $f_y = 2y$

5. $f_x = \dfrac{e^{x/y}}{y}$, $f_y = \dfrac{-xe^{x/y}}{y^2}$

7. $f_x = y^2 + \dfrac{6x}{y^3}$, $f_y = 2xy - \dfrac{9x^2}{y^4}$

9. $f_x = (x^2 + 2xy)(2x) + (y^3 + x^2)(2x + 2y)$,

$f_y = (x^2 + 2xy)(3y^2) + (y^3 + x^2)(2x)$

11. $f_x = \dfrac{2xy - 2xy^2}{(x^2 + y)^2}$, $f_y = \dfrac{2y(x^2 + y) - (x^2 + y^2)}{(x^2 + y)^2}$

13. $f_x = 2x \sec^2(x^2 + y^2)$, $f_y = 2y \sec^2(x^2 + y^2)$

15. $f_x = 2xye^{xy^2} \sec(x^2y) \tan(x^2y) + y^2e^{xy^2} \sec(x^2y)$

$f_y = x^2e^{xy^2} \sec(x^2y) \tan(x^2y) + 2xye^{xy^2} \sec(x^2y)$

17. $f_x = \dfrac{2 \cot y^2}{2x + y}$, $f_y = -2y \ln(2x + y) \csc^2 y^2 + \dfrac{\cot y^2}{2x + y}$

19. $f_x = 3y \sec^3 x \tan x + y^2$, $f_y = \sec^3 x + 2xy$

21. $f_x = \dfrac{y^2}{1 + x^2y^4}$, $f_y = \dfrac{2xy}{1 + x^2y^4}$

23. $2xy$

25. ye^{yz}

27. $24 \sec^2(2x + y - 3z) \tan(2x + y - 3z)$

29. a) $f_{xy} = f_{yx} = 2x$ b) $f_{xy} = f_{yx} = 6x^2y - \dfrac{2x}{y^2}$ c) $f_{xy} = f_{yx} = 0$

Section 16.2

1. $3x + 16y - z = 21$

3. $x = 1 + 2t$, $y = -1 + t$, $z = t$

5. $8x_1 - 28x_2 + x_3 = -32$; $x_1 = -2 + 8t$, $x_2 = 1 - 28t$, $x_3 = 12 + t$

7. 23.6

9. 3.01

11. Increased by 126π ft^3

Section 16.3

1. $(8, 4, -2)$; $8dx + 4dy - 2dz$

3. $(\frac{1}{2} + \pi, \frac{1}{4} + \pi e^{4\pi}, 4e^{4\pi} + 2)$; $(\frac{1}{2} + \pi)\, dx + (\frac{1}{4} + \pi e^{4\pi})\, dy + (4e^{4\pi} + 2)\, dz$

5. $6\pi + \dfrac{1}{\pi}$; $\left(6\pi + \dfrac{1}{\pi}\right) dx$

7. 2.04

9. $\epsilon_1 = \Delta x - 2 \cdot \Delta y$; $\epsilon_2 = -\Delta y$

11. 264.25

Section 16.4

1. a) $x = 1, \quad y = 2, \quad z = 5/4$ b) $15/4$
   c) $z = t^4 + (t+1)^{-2}$ d) $15/4$
3. 2
5. 0
7. 4800π in.^{3}/min
9. $-300m$ units/unit time
11. $1/5$ lb/ft^2/min
13. $-17.008G$
15. $bc(\partial w/\partial x) = ac(\partial w/\partial y) = ab(\partial w/\partial z)$
17. We have

$$\frac{\partial w}{\partial x} = f'(u)\cdot y^2 \quad \text{and} \quad \frac{\partial w}{\partial y} = f'(u)\cdot 2xy,$$

from which the desired result follows at once.

19. We have

$$\frac{\partial w}{\partial x} = f'(u)\left(\frac{1}{y}\right) \quad \text{and} \quad \frac{\partial w}{\partial y} = f'(u)\left(\frac{-x}{y^2}\right),$$

from which we at once obtain the desired result.

21. If f is a differentiable function of two variables and $f(tx, ty) = t^k f(x, y)$ for all t, then

$$x \cdot f_x(x, y) + y \cdot f_y(x, y) = k \cdot f(x, y).$$

23. If $f(\boldsymbol{x}) = f(x_1, x_2, x_3)$ is differentiable and $f(t\boldsymbol{x}) = t^k f(\boldsymbol{x})$, then

$$x_1 f_{x_1}(\boldsymbol{x}) + x_2 f_{x_2}(\boldsymbol{x}) + x_3 f_{x_3}(\boldsymbol{x}) = kf(\boldsymbol{x}).$$

Section 16.5

1. $-32/\sqrt{5}$
3. -9
5. Maximum rate of $2\sqrt{5}$ in the direction of $2\boldsymbol{i} + \boldsymbol{j}$.
   Minimum rate of $-2\sqrt{5}$ in the direction of $-2\boldsymbol{i} - \boldsymbol{j}$.
7. Let $f(x, y) = x^2 - y^2$. Then $\boldsymbol{f}'(0, 0) = (0, 0)$ and all directional derivatives $\boldsymbol{f}'(0, 0)\cdot \boldsymbol{u}$ at $(0, 0)$ are therefore zero.
9. $\boldsymbol{i}$
11. $\boldsymbol{i}$ and $-\boldsymbol{i}$
13. $\boldsymbol{j}$ and $-\boldsymbol{j}$

Section 16.6

1. $2/(\pi - 2)$
3. 0
5. 0
7. $-1/7$

9. 2 **11.** 0

13. Tangent line: $3x + 5y = -2$; normal line: $x = 1 + 3t$, $y = 1 - 5t$

15. Tangent plane: $14x + y + 10z = -4$; normal line: $x = -1 + 14t$, $y = t$, $z = 1 + 10t$

17. $\frac{1}{8}$

19. At a point (x, y, z) of intersection, a vector normal to the first surface is $2x\mathbf{i} - 4y\mathbf{j} + 2z\mathbf{k}$, while a vector perpendicular to the second surface is $yz\mathbf{i} + xz\mathbf{j} + xy\mathbf{k}$. These vectors are perpendicular.

Review Exercise Set 16.1

1. a) $z^2 + y^2ze^{xz}$ b) $y^2z^2e^{xz}$ c) $2yze^{xz}$

3. $(-10, 17, -12)$ **5.** 12

7. $-\frac{2784}{5}$ **9.** $\frac{7}{13}$

Review Exercise Set 16.2

1. a) $x_1x_3 \cos x_2x_3 - 2x_2x_1^3$ b) $x_2 \cos x_2x_3$

3. $(1, 0)$ **5.** 10

7. $\frac{12}{5}$ **9.** $\frac{3}{8}$

More Challenging Exercises 16

1.
$$\begin{aligned} f(x + \Delta x, y + \Delta y, z + \Delta z) - f(x, y, z) &= [f(x + \Delta x, y + \Delta y, z + \Delta z) - f(x, y + \Delta y, z + \Delta z)] \\ &\quad + [f(x, y + \Delta y, z + \Delta z) - f(x, y, z + \Delta z)] \\ &\quad + [f(x, y, z + \Delta z) - f(x, y, z)] \end{aligned}$$

3. $(x + y) - [(x + y)^3/3!]$ **5.** $1 + xy + \frac{1}{2}x^2y^2$

7. -0.03 with error ≤ 0.0037

9. The binomial theorem of algebra states that

$$(a + b)^n = \sum_{i=1}^{n} \binom{n}{i} a^i b^{n-i},$$

where

$$\binom{n}{i} = \frac{n!}{i!(n - i)!}.$$

Thus the coefficient of $(x - x_0)^h(y - y_0)^k$ in the given expression for $T_n(x, y)$ is

$$\frac{1}{(h+k)!}\binom{h+k}{h}\frac{\partial^{h+k}f}{\partial x^h\partial_y k}\bigg|_{(x_0,y_0)}$$

Since

$$\frac{1}{(h+k)!}\binom{h+k}{h} = \frac{1}{(h+k)!}\cdot\frac{(h+k)!}{h!k!} = \frac{1}{h!k!},$$

the result follows immediately.

CHAPTER 17

Section 17.1

1. Local maximum of 1 wherever $xy = (\pi/2) + 2n\pi$; local minimum of -1 wherever $xy = (3\pi/2) + 2n\pi$
3. No local maximum; local minimum of 1 at $(0, 0)$.
5. No local maximum; no local minimum
7. No local maximum; local minimum of -4 at $(0, 2)$
9. Local maximum of 48 at $(0, -2)$; local minimum of -20 at $(2, 2)$
11. No local maximum; local minimum of -12 at $(1, -1, 2)$ and $(-1, -1, 2)$
13. a) $-x^2 - y^4$ b) $x^2 + y^4$ c) $x^2 + y^3$
    d) If $AC - B^2 = 0$, further examination of f is needed to determine its behavior at (x_0, y_0).
15. Maximum assumed at $(1, 1)$ and $(-1, -1)$; minimum assumed at $(1, -1)$ and $(-1, 1)$
17. Maximum assumed at $(1, -1)$ and $(-1, 1)$; minimum assumed at $(0, 0)$. [Seen from $x^2 + y^2 - xy = (\frac{1}{2}x - y)^2 + \frac{3}{4}x^2$.]

Section 17.2

1. $2/\sqrt{2}$
3. $\left(\frac{10}{49}, \frac{-15}{49}, \frac{30}{49}\right)$
5. $\frac{32}{81}\pi a^3$
7. $\left(\frac{6}{116}, \frac{-3}{116}, \frac{15}{116}\right)$
9. $\left(\frac{58}{19}, \frac{-18}{19}, \frac{44}{19}\right)$

Section 17.3

1. Exact; $\frac{x^3}{3} - \frac{y^2}{2} + C$
3. Exact; $x^2y + C$
5. Not exact
7. Exact; $x^2y + e^{-y} + C$
9. Exact; $\tan xy + y + C$
11. Exact; $x^2y^3 - 3x + 2y^2 + C$
13. Exact; $-\tan^{-1}\left(\frac{y}{z}\right) + C, \quad z > 0$
15. Exact; $\sin xyz - 3xz^2 + y^3 + C$

Section 17.4

1. $\frac{1}{12}(5^{3/2} - 1)$
3. $\frac{7}{2}$
5. $-\frac{5}{3}$
7. a) $\frac{\pi}{8} + \frac{5}{6} - \frac{2\sqrt{2}}{3}$ b) $\frac{3\pi^2 + 12}{256}$
9. 28
11. $\frac{2\pi + 21}{2}$
13. 8
15. $\tan^{-1}4$

Section 17.5

1. $\frac{2}{3}$
3. $\frac{3}{4}$
5. 9
7. 162
9. 18
11. $\frac{315}{2}$
13. -6π

Review Exercise Set 17.1

1. There are none
3. Closest $(-1, \sqrt{3}, 0)$; farthest $(-1, -\sqrt{3}, 0)$
5. $y^2e^x + x^3y - \cos y + C$
7. $\frac{976}{27}$
9. When $M(x, y)\,dx + N(x, y)\,dy$ is an exact differential. If M and N have continuous partial derivatives and their common domain has no holes in it, then it is true if and only if $\partial M/\partial y = \partial N/\partial x$.

Review Exercise Set 17.2

1. Relative maximum of -2 at $(1, -1)$
3. $\left(\frac{136}{35}, \frac{-12}{35}, \frac{-20}{35}\right)$
5. $x^2 \sin y - x^3y + \frac{4}{3}y^3 + C$

7. $\ln(\sqrt{2}+1) + \dfrac{2-2\sqrt{2}}{3}$ **9.** 0

More Challenging Exercises 17

1. a) $E_1(x, y)$
$= \frac{1}{2}[f_{xx}(c_1, c_2)\cdot(x-x_0)^2 + 2f_{xy}(c_1, c_2)\cdot(x-x_0)(y-y_0) + f_{yy}(c_1, c_2)\cdot(y-y_0)^2]$

b) We have then $f(x, y) = f(x_0, y_0) + E_1(x, y)$, from which the result is immediate.

3. Consider

$$[A(\Delta x) + B(\Delta y)]^2 + (AC - B^2)(\Delta y)^2. \tag{1}$$

Suppose $A \neq 0$. If $AC - B^2 > 0$, then (1) is >0 for all $(\Delta x, \Delta y) \neq (0, 0)$. If $AC - B^2 < 0$, then for $\Delta y \neq 0$ and $\Delta x = (-B/A)\,\Delta y$, you see that (1) <0, while for $\Delta y = 0$ and $\Delta x \neq 0$, (1) is >0. If $AC - B^2 = 0$, then (1) assumes the value zero at points where $A(\Delta x) + B(\Delta y) = 0$. With the hint, this shows that

$$A(\Delta x)^2 + 2B(\Delta x)(\Delta y) + C(\Delta y)^2 \tag{2}$$

is >0 for all $(\Delta x, \Delta y) \neq (0, 0)$ if $A > 0$ and $AC - B^2 > 0$. Similarly, if $A < 0$ and $AC - B^2 > 0$, then (2) is <0 for $(\Delta x, \Delta y) \neq (0, 0)$. Clearly if $A \neq 0$ and $AC - B^2 < 0$, then (1) can assume both positive and negative values, and thus (2) can also. If $A = 0$ and $C \neq 0$, then a similar argument can be made using

$$C[A(\Delta x)^2 + 2B(\Delta x)(\Delta y) + C(\Delta y)^2] = (AC - B^2)(\Delta x)^2 + [B(\Delta x) + C(\Delta y)]^2.$$

If both A and C are zero, then (2) reduces to $2B(\Delta x)(\Delta y)$, which can assume both positive and negative values if $B \neq 0$.

5. $\partial P/\partial x = \partial Q/\partial w$, $\partial P/\partial y = \partial R/\partial w$, $\partial P/\partial z = \partial S/\partial w$, $\partial Q/\partial y = \partial R/\partial x$, $\partial Q/\partial z = \partial S/\partial x$, $\partial R/\partial z = \partial S/\partial y$

7. a) $\dfrac{\partial M}{\partial y} = \dfrac{\partial N}{\partial x} = \dfrac{x^2 - y^2}{(x^2 + y^2)^2}$

b) $\displaystyle\int \frac{y}{x^2 + y^2}\,dx = \tan^{-1}\left(\frac{x}{y}\right) + A$ and $\displaystyle\int \frac{-x}{x^2 + y^2}\,dy = \tan^{-1}\left(\frac{x}{y}\right) + B$

show that $F(x, y)$ has the desired form for $y \neq 0$. Since $\lim_{y\to 0+} \tan^{-1}(1/y) + A = (\pi/2) + A$ and $\lim_{y\to 0-} \tan^{-1}(1/y) + B = -(\pi/2) + B$, you see you must have $(\pi/2) + A = -(\pi/2) + B$, so $B = A + \pi$. You then see that you must define $F(x, y) = A + (\pi/2)$ for $y = 0$, $x > 0$, in order to have $F(x, y)$ continuous at $(x, 0)$ for $x > 0$. But then

$$\lim_{y\to 0+} F(-1, y) = A - \frac{\pi}{2} \quad \text{while} \quad \lim_{y\to 0-} F(-1, y) = A + \frac{3\pi}{2},$$

so it is impossible to define $F(x, 0)$ for $x < 0$ to make $F(x, y)$ continuous there.

CHAPTER 18

Section 18.1

1. The integral $\int_R f(x, y)\,dx\,dy$ is equal to the volume of the region over R and under the surface $z = f(x, y)$, where $f(x, y) \geq 0$, minus the volume of the region under R and over the surface where $f(x, y) \leq 0$.

3. a) $S_n = d(b - a)(d - c)$; $s_n = c(b - a)(d - c)$ b) No

c) The maximum span of the cells in a partition of R must be small to ensure that a Riemann sum for f is close to $\iint_R f(x, y)\,dx\,dy$.

5.

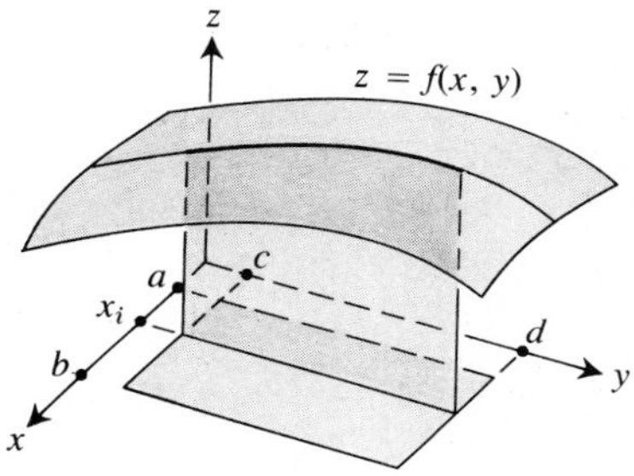

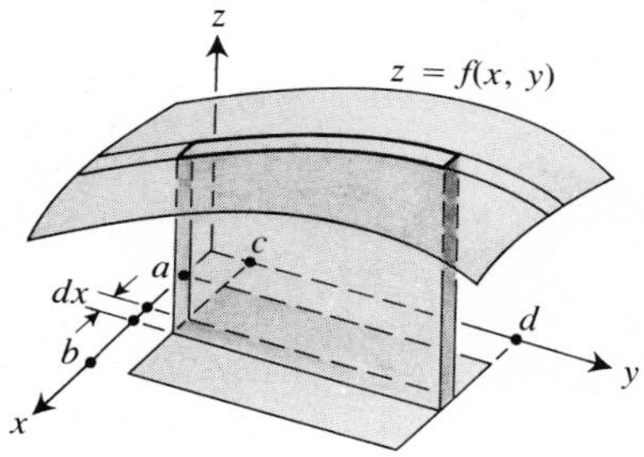

7. a) The first integration can be performed with respect to any one of three variables, the next with respect to any of the two remaining variables, and the last with respect to the remaining variable to give $3 \cdot 2 \cdot 1 = 3! = 6$ orders in all.

b) $2^3 3! = 48$

c) Half of them have the value $\iint_R f(x, y)\,dx\,dy$ and the other half have the value $-\iint_R f(x, y)\,dx\,dy$.

9. 32

11. 10,240

13. 128

15. 48

17. $\pi^2/4$

19. π

21. $\frac{95}{4}$

23. 32

Section 18.2

1. $\frac{1}{2}$

3. $\frac{1}{6}$

5. $\frac{1}{15}$

7. 4π

9. $\displaystyle\int_{-1}^{1}\int_{0}^{\sqrt{1-x^2}} 4xy\,dy\,dx$

11. $\displaystyle\int_{0}^{1}\int_{\sin^{-1} y}^{\pi-\sin^{-1} y} x^2y\,dx\,dy$

13. $\int_0^1 \int_y^{\sqrt{x}} y \cos x \, dx \, dy$

15. $\int_{-2}^{0} \int_0^{(x+2)/2} x \cos^2 y \, dy \, dx + \int_0^1 \int_0^{1-x} x \cos^2 y \, dy \, dx$

17. $\int_0^1 \int_{-\sqrt{1-z}}^{\sqrt{1-z}} \int_{-\sqrt{1-y^2-z}}^{\sqrt{1-y^2-z}} xyz^2 \, dx \, dy \, dz$

19. $\frac{7}{6}$

21. $e - \frac{3}{2}$

23. 16π

Section 18.3

1. $2a^2 + \frac{a^2\pi}{4}$

3. $a^2\left(\frac{\pi}{6} - \frac{\sqrt{3}}{8}\right)$

5. $\frac{\pi a^3}{3}$

7. a) A plane containing the z-axis
b) A plane perpendicular to the z-axis

9. $\frac{3\pi}{2}$

11. $\frac{\pi}{3}(64 - 24\sqrt{3})$

13. $\frac{320\pi}{3}$

Section 18.4

1. $\frac{\pi}{3}[2a^3 - 3a^2b + b^3]$

3. $\frac{4\pi}{9}$

5. $\frac{4}{5}\pi a^5$

Section 18.5

1. a) $\frac{1}{6}$ b) $(\frac{2}{3}, \frac{3}{4})$

3. a) $\frac{2}{3}k\pi a^3$ b) $\frac{2}{3}k\pi a^4$ c) $\frac{2\sqrt{2}}{3}k\pi a^4$ d) $\frac{2}{5}k\pi a^5$

5. a) $\frac{k\pi a^4}{6}$ b) $\left(0, 0, \frac{4}{5}a\right)$ c) $\frac{k\pi a^5}{6}$ d) $\frac{3k\pi a^5}{10}$ e) $\frac{k\pi a^6}{15}$

7. $\frac{28k\pi a^5}{15}$

9. $\left(0, 0, \frac{31}{18}\right)$

11. $\left(0, 0, \dfrac{3a}{8(2-\sqrt{2})}\right)$

13. The first moment of a body in space about the plane $z = -a$ is $M_{xy} + ma$.

15. a) $1/\sqrt{3}$ b) $\dfrac{1}{\sqrt{3}}\sqrt{(a+1)^3 - a^3}$

Section 18.6

1. $\dfrac{4}{15}(33 - 9\sqrt{3} - 4\sqrt{2})$

3. $\pi a^2(2 - \sqrt{2})$

5. $\dfrac{\pi a^2}{6}(5\sqrt{5} - 1)$

Review Exercise Set 18.1

1. $S_2 = 40$, $s_2 = -24$

3. $\displaystyle\int_0^4 \int_{-\sqrt{y}}^{\sqrt{y}} (x^2 - 3xy)\, dx\, dy$

5. $$\int_{-2}^{2} \int_{-\sqrt{4-x^2}}^{\sqrt{4-x^2}} \int_{x^2+y^2}^{4+\sqrt{4-x^2-y^2}} 1 \cdot dz\, dy\, dx = 4\int_0^2 \int_0^{\sqrt{4-x^2}} \int_{x^2+y^2}^{4+\sqrt{4-x^2-y^2}} 1 \cdot dz\, dy\, dx$$
$$= \int_0^{2\pi} \int_0^2 \int_{r^2}^{4+\sqrt{4-r^2}} r\, dz\, dr\, d\theta$$

7. 16π

9. $\pi a^6/6$

Review Exercise Set 18.2

1. -8

3. $\displaystyle\int_0^4 \int_0^{\sqrt{z}} \int_0^{\sqrt{z-y^2}} x^2 z\, dx\, dy\, dz$

5. $\frac{8}{3}$

7. $2^8 \cdot 23\pi/105$

9. $(0, 0, \frac{1}{3})$

More Challenging Exercises 18

1. $\pi^2 a^4/2$

3. $2\pi^2 a^3$

5. $\displaystyle\sum_{i,j=1}^{n} \left[\left(5 + \frac{5i}{n}\right)^2 + 3\left(5 + \frac{5i}{n}\right)\left(-2 + \frac{4j}{n}\right)\right] \cdot \frac{20}{n^2}$

7. $\frac{2}{3}$

9. a) $dx = \cos\theta\, dr - r\sin\theta\, d\theta,$
$dy = \sin\theta\, dr + r\cos\theta\, d\theta$

b) $r\, dr \wedge d\theta$

11. 2541/25

CHAPTER 19

Section 19.1

1. 4π units mass/unit time

3. Both integrals equal $-\frac{8}{3}$

5. Both integrals equal $\frac{1}{12}$

7. -96π units mass/unit time (The plates were *three* units apart.)

9. $2\pi ab$

11. By Green's Theorem,

$$\frac{1}{2}\oint_{\partial G}(x\,dy - y\,dx) = \frac{1}{2}\iint_G (1+1)\,dx\,dy = \iint_G 1\cdot dx\,dy = \text{area of } G.$$

13. $\oint_{\partial G}\left(\frac{\partial f}{\partial x}dy - \frac{\partial f}{\partial y}dx\right)$

Section 19.2

1. Both integrals equal -120.

3. a) By Green's Theorem (as in Eq. (9) of Section 19.1),

$$\iint_G (\nabla\cdot \boldsymbol{E})\,dx\,dy = \oint_{\partial G}(\boldsymbol{E}\cdot\boldsymbol{n})\,ds = \oint_{\partial G}(\boldsymbol{F}\cdot\boldsymbol{n})\,ds = \iint_G (\nabla\cdot\boldsymbol{F})\,dx\,dy.$$

b) By Green's Theorem (as in Eq. (10) of Section 19.2),

$$\iint_G (\text{curl } \boldsymbol{E})\,dx\,dy = \oint_{\partial G}(\boldsymbol{E}\cdot\boldsymbol{t})\,ds = \oint_{\partial G}(\boldsymbol{F}\cdot\boldsymbol{t})\,ds = \iint_G (\text{curl } \boldsymbol{F})\,dx\,dy.$$

5. Let the left border of B be given by $x = h(y)$ and let the right border be $x = k(y)$ for $r \le y \le s$. Then

$$\begin{aligned}\oint_{\partial G} Q(x,y)\,dy &= \int_r^s Q(k(y),y)\,dy + \int_s^r Q(h(y),y)\,dy\\ &= \int_r^s [Q(k(y),y) - Q(h(y),y)]\,dy = \int_r^s Q(x,y)\Big]_{x=h(y)}^{x=k(y)}\,dy\\ &= \int_r^s\int_{x=h(y)}^{x=k(y)}\frac{\partial Q}{\partial x}\,dx\,dy = \iint_G \frac{\partial Q}{\partial x}\,dx\,dy.\end{aligned}$$

7. Let H be the dark-shaded region in Fig. 19.10 between γ_1 and γ_2. We may regard H as a region with border $\gamma_2 + \gamma_3 - \gamma_1 - \gamma_3$, which can be decomposed into a finite number of simple subregions. By Green's Theorem and the fact that curl $\boldsymbol{F} = 0$, we have

$$0 = \iint_H (\operatorname{curl} \boldsymbol{F})\,dx\,dy = \int_{\gamma_2+\gamma_3-\gamma_1-\gamma_3} (\boldsymbol{F}\cdot\boldsymbol{t})\,ds$$
$$= \oint_{\gamma_2} (\boldsymbol{F}\cdot\boldsymbol{t})\,ds + \int_{\gamma_3} (\boldsymbol{F}\cdot\boldsymbol{t})\,ds - \oint_{\gamma_1} (\boldsymbol{F}\cdot\boldsymbol{t})\,ds - \int_{\gamma_3} (\boldsymbol{F}\cdot\boldsymbol{t})\,ds$$
$$= \oint_{\gamma_2} (\boldsymbol{F}\cdot\boldsymbol{t})\,ds - \oint_{\gamma_1} (\boldsymbol{F}\cdot\boldsymbol{t})\,ds.$$

Thus $\oint_{\gamma_1} (\boldsymbol{F}\cdot\boldsymbol{t})\,ds = \oint_{\gamma_2} (\boldsymbol{F}\cdot\boldsymbol{t})\,ds$.

9. $\operatorname{curl} \nabla f = \operatorname{curl}\left(\frac{\partial f}{\partial x}\boldsymbol{i} + \frac{\partial f}{\partial y}\boldsymbol{j}\right) = \frac{\partial(\partial f/\partial y)}{\partial x} - \frac{\partial(\partial f/\partial x)}{\partial y} = \frac{\partial^2 f}{\partial x\,\partial y} - \frac{\partial^2 f}{\partial y\,\partial x} = 0$

if second partial derivatives are continuous.

11. -2

13. a) $4x$ b) 2

15. If the force field $\boldsymbol{F}(x, y) = P(x, y)\boldsymbol{i} + Q(x, y)\boldsymbol{j} = a\boldsymbol{i} + b\boldsymbol{j}$, then $\partial Q/\partial x = \partial P/\partial y = 0$, so curl $\boldsymbol{F} = \partial Q/\partial x - \partial P/\partial y = 0$. The work is independent of the path joining A and B; if $A = (x_1, y_1)$ and $B = (x_2, y_2)$, then the work is $a(x_2 - x_1) + b(y_2 - y_1)$.

17. a) Let $\boldsymbol{E} = E_1\boldsymbol{i} + E_2\boldsymbol{j}$ and $\boldsymbol{F} = F_1\boldsymbol{i} + F_2\boldsymbol{j}$.
Then $a\boldsymbol{E} + b\boldsymbol{F} = (aE_1 + bF_1)\boldsymbol{i} + (aE_2 + bF_2)\boldsymbol{j}$.

$$\frac{\partial(aE_2 + bF_2)}{\partial x} - \frac{(aE_1 + bF_1)}{\partial y} = a\frac{\partial E_2}{\partial x} + b\frac{\partial F_2}{\partial x} - a\frac{\partial E_1}{\partial y} - b\frac{\partial F_1}{\partial y}$$
$$= a\left(\frac{\partial E_1}{\partial x} - \frac{\partial E_2}{\partial y}\right) + b\left(\frac{\partial F_1}{\partial x} - \frac{\partial F_2}{\partial y}\right)$$
$$= a\cdot 0 + b\cdot 0 = 0.$$

b) $au + bv + C$ for any constant C

19. $W(A, B) = \int_a^b (\boldsymbol{F}\cdot d\boldsymbol{r}) = \int_a^b m\boldsymbol{v}'(t)\cdot\left(\frac{d\boldsymbol{r}}{dt}\right)dt = m\int_a^b [\boldsymbol{v}'(t)\cdot\boldsymbol{v}]\,dt.$

Now

$$\frac{d(|\boldsymbol{v}(t)|^2)}{dt} = \frac{d(\boldsymbol{v}(t)\cdot\boldsymbol{v}(t))}{dt} = 2\boldsymbol{v}'(t)\cdot(t).$$

Thus

$$W(A, B) = \frac{m}{2}\int_a^b \frac{d(|\boldsymbol{v}(t)|^2)}{dt}\,dt = \frac{1}{2}m|\boldsymbol{v}(b)|^2 - \frac{1}{2}m|\boldsymbol{v}(a)|^2.$$

21. a) $-\dfrac{kx}{(x^2 + y^2)^{3/2}}\boldsymbol{i} - \dfrac{ky}{(x^2 + y^2)^{3/2}}\boldsymbol{j}$ b) $\dfrac{-k}{\sqrt{x^2 + y^2}}$ c) $\dfrac{k}{\sqrt{13}} - \dfrac{k}{2}$

Section 19.3

1. $\mathbf{0}$

3. $\frac{1}{y}\boldsymbol{i} + (e^y - x\cos xy)\boldsymbol{k}$

5. Both integrals equal 4π.

7. 0

9. -48π

11. a) $\boldsymbol{F}\cdot\boldsymbol{t} = 0$ so $\oint_{\partial G}(\boldsymbol{F}\cdot\boldsymbol{t})\,ds = 0$, so $\iint_G [(\mathbf{curl}\,\boldsymbol{F})\cdot\boldsymbol{n}]\,dS = 0$ by Stokes' Theorem.

b) $(\mathbf{curl}\,\boldsymbol{F})\cdot\boldsymbol{n} = 0$, so $\iint_G [(\mathbf{curl}\,\boldsymbol{F})\cdot\boldsymbol{n}]\,dS = 0$, so $\oint_{\partial G}(\boldsymbol{F}\cdot\boldsymbol{t})\,ds = 0$.

13. $\mathbf{curl}\,\boldsymbol{F} = \mathbf{0}$, so $\iint_G [(\mathbf{curl}\,\boldsymbol{F})\cdot\boldsymbol{n}]\,dS = 0$, so $\oint_{\partial G}(\boldsymbol{F}\cdot\boldsymbol{t})\,ds = 0$.

15. a) If $\boldsymbol{F} = F_1\boldsymbol{i} + F_2\boldsymbol{j} + F_3\boldsymbol{k}$, then

$$\mathbf{curl}\,\boldsymbol{F} = \begin{vmatrix} \boldsymbol{i} & \boldsymbol{j} & \boldsymbol{k} \\ \frac{\partial}{\partial x} & \frac{\partial}{\partial y} & \frac{\partial}{\partial z} \\ F_1 & F_2 & F_3 \end{vmatrix} = \left(\frac{\partial F_3}{\partial y} - \frac{\partial F_2}{\partial z}\right)\boldsymbol{i} + \left(\frac{\partial F_1}{\partial z} - \frac{\partial F_3}{\partial x}\right)\boldsymbol{j} + \left(\frac{\partial F_2}{\partial x} - \frac{\partial F_1}{\partial y}\right)\boldsymbol{k}.$$

Then

$$\begin{aligned}\nabla\cdot(\mathbf{curl}\,\boldsymbol{F}) &= \left(\frac{\partial^2 F_3}{\partial x\,\partial y} - \frac{\partial^2 F_2}{\partial x\,\partial z}\right) + \left(\frac{\partial^2 F_1}{\partial y\,\partial z} - \frac{\partial^2 F_3}{\partial y\,\partial x}\right) + \left(\frac{\partial^2 F_2}{\partial z\,\partial x} - \frac{\partial^2 F_1}{\partial z\,\partial y}\right) \\ &= \left(\frac{\partial^2 F_3}{\partial x\,\partial y} - \frac{\partial^2 F_3}{\partial y\,\partial x}\right) + \left(\frac{\partial^2 F_2}{\partial z\,\partial x} - \frac{\partial^2 F_2}{\partial x\,\partial z}\right) + \left(\frac{\partial^2 F_1}{\partial y\,\partial z} - \frac{\partial^2 F_1}{\partial z\,\partial y}\right) \\ &= 0\end{aligned}$$

if second partial derivatives are continuous.

b) $$\begin{vmatrix} \boldsymbol{i} & \boldsymbol{j} & \boldsymbol{k} \\ \frac{\partial}{\partial x} & \frac{\partial}{\partial y} & \frac{\partial}{\partial z} \\ \frac{\partial f}{\partial x} & \frac{\partial f}{\partial y} & \frac{\partial f}{\partial z} \end{vmatrix} = \left(\frac{\partial^2 f}{\partial y\,\partial z} - \frac{\partial^2 f}{\partial z\,\partial y}\right)\boldsymbol{i} + \left(\frac{\partial^2 f}{\partial z\,\partial x} - \frac{\partial^2 f}{\partial x\,\partial z}\right)\boldsymbol{j} + \left(\frac{\partial^2 f}{\partial x\,\partial y} - \frac{\partial^2 f}{\partial y\,\partial x}\right)\boldsymbol{k}$$

$$= 0\boldsymbol{i} + 0\boldsymbol{j} + 0\boldsymbol{k} = \mathbf{0}$$

if second partial derivatives are continuous.

17. Remove a very small "disk" from G and let the remaining surface be H. Then

$$\iint_G [(\nabla\times\boldsymbol{F})\cdot\boldsymbol{n}]\,dS \approx \iint_H [(\nabla\times\boldsymbol{F})\cdot\boldsymbol{n}]\,dS,$$

since only a very small part of G was removed. In fact, the integrals can be made as nearly equal as you please by removing a sufficiently small part of G. But

$$\iint_H [(\nabla \times \boldsymbol{F}) \cdot \boldsymbol{n}]\, dS = \oint_{\partial H} (\boldsymbol{F} \cdot \boldsymbol{t})\, ds \approx 0,$$

since ∂H is a very short curve. By removing a sufficiently small part of G, ∂H can be made as short as you please, and $\oint_{\partial H} (\boldsymbol{F} \cdot \boldsymbol{t})\, ds$ as close to 0 as you please. Thus, taking the limit as smaller and smaller parts of G are removed, you see that $\iint_G [(\nabla \times \boldsymbol{F}) \cdot \boldsymbol{n}]\, dS$ must be zero.

19. $\iiint_G [\nabla \cdot (\boldsymbol{curl\, F})]\, dx\, dy\, dz = \iint_{\partial G} [(\boldsymbol{curl\, F}) \cdot \boldsymbol{n}]\, dS$ by Exercise 18. Now ∂G is a surface that is the entire boundary of the three-dimensional region G. By Exercise 17, $\iint_{\partial G} [(\boldsymbol{curl\, F}) \cdot \boldsymbol{n}]\, dS = \iint_{\partial G} [(\nabla \times \boldsymbol{F}) \cdot \boldsymbol{n}]\, dS = 0$.

Review Exercise Set 19.1

1. See Theorem 19.4 of Section 19.2.

3. Let $\boldsymbol{F}$ be a continuously differentiable vector field, and let G be a suitable region in space. Then $\iiint_G (\nabla \cdot \boldsymbol{F})\, dx\, dy\, dz = \iint_{\partial G} (\boldsymbol{F} \cdot \boldsymbol{n})\, dS$ where $\boldsymbol{n}$ is the unit outward normal vector. For a flux vector $\boldsymbol{F}$, this states that the integral of the divergence of the flux vector is equal to the integral of the normal component of the flux vector over the surface. Roughly speaking, integrating the rate at which mass is leaving little volume elements throughout G gives the same result as finding the rate that mass leaves G through the surface ∂G.

5. a) $\partial Q/\partial x = \partial P/\partial y$ b) $\partial P/\partial x + \partial Q/\partial y = 0$

7. -162π

Review Exercise Set 19.2

1. Let $\boldsymbol{F}$ be a continuously differentiable vector field in the plane and let G be a plane region that can be decomposed into a finite number of simple subregions. Then $\iint_G (\nabla \cdot \boldsymbol{F})\, dx\, dy = \oint_{\partial G} (\boldsymbol{F} \cdot \boldsymbol{n})\, ds$. Also $\iint_G (\text{curl } \boldsymbol{F})\, dx\, dy = \oint_{\partial G} (\boldsymbol{F} \cdot \boldsymbol{t})\, ds$.

3. By the divergence theorem,

$$\iint_{\partial G} \left[\left(\frac{x^3}{3}\boldsymbol{i} + \frac{y^3}{3}\boldsymbol{j} + 0\boldsymbol{k}\right) \cdot \boldsymbol{n}\right] dS = \iiint_G (x^2 + y^2)\, dx\, dy\, dz$$

and the second integral gives the moment of inertia about the z-axis if mass density is 1. [The fact that G is a ball with center at the origin is irrelevant.]

5. $(xz - y)\boldsymbol{i} - yz\boldsymbol{j}$ **7.** $\frac{1}{2}$

More Challenging Exercises 19

1. A computation shows that

$$\nabla \times \boldsymbol{F} = \left(\frac{\partial R}{\partial y} - \frac{\partial Q}{\partial z}\right)\boldsymbol{i} + \left(\frac{\partial P}{\partial z} - \frac{\partial R}{\partial x}\right)\boldsymbol{j} + \left(\frac{\partial Q}{\partial x} - \frac{\partial P}{\partial y}\right)\boldsymbol{k}.$$

The normal unit vector $\boldsymbol{n}$ at (x, y, z) on G can be written as $\boldsymbol{n} = (\cos\alpha)\boldsymbol{i} + (\cos\beta)\boldsymbol{j} + (\cos\gamma)\boldsymbol{k}$ where α, β, and γ are the angles which $\boldsymbol{n}$ makes with the x, y, and z-axes. Consider the contribution $\iint_G [(\partial R/\partial y) - (\partial Q/\partial z)] \times (\cos\alpha)\, dS$ to $\iint_G [(\nabla \times \boldsymbol{F}) \cdot \boldsymbol{n}]\, dS$. Since α is the angle between $\boldsymbol{n}$ and $\boldsymbol{i}$, it is also the angle between the tangent plane at (x, y, z) and the yz-plane. Thus $(\cos\alpha)\, dS$ is the area of the projection of a surface element of area dS onto the y,z-plane. But this is the meaning of $dy\, dz$ in $\iint_G [(\partial R/\partial y) - (\partial Q/\partial z)]\, dy\, dz$, which thus equals $\iint_G[(\partial R/\partial y) - (\partial Q/\partial z)](\cos\alpha)\, dS$. A similar analysis with other components of $\nabla \times \boldsymbol{F}$ shows that

$$\iint_G [(\nabla \times \boldsymbol{F}) \cdot \boldsymbol{n}]\, dS = \iint_G \left[\left(\frac{\partial R}{\partial y} - \frac{\partial Q}{\partial z}\right) dy\, dz + \left(\frac{\partial P}{\partial z} - \frac{\partial R}{\partial x}\right) dz\, dx + \left(\frac{\partial Q}{\partial x} - \frac{\partial P}{\partial y}\right) dx\, dy\right].$$

Since $\boldsymbol{t} = (dx/ds)\boldsymbol{i} + (dy/ds)\boldsymbol{j} + (dz/ds)\boldsymbol{k}$, it is clear that

$$\oint_{\partial G} (\boldsymbol{F} \cdot \boldsymbol{t})\, ds = \oint_{\partial G} (P\, dx + Q\, dy + R\, dz).$$

3. a) $-3x\, dx \wedge dy$ b) $(x + z^3)\, dx \wedge dy \wedge dz$

5. $$f(B) - f(A) = \int_\gamma \left(\frac{\partial f}{\partial x} dx + \frac{\partial f}{\partial y} dy + \frac{\partial f}{\partial z} dz\right)$$

7. $$\oint_{\partial G} (P\, dx + Q\, dy + R\, dz) = \iint_G \left[\left(\frac{\partial R}{\partial y} - \frac{\partial Q}{\partial z}\right) dy\, dz + \left(\frac{\partial P}{\partial z} - \frac{\partial R}{\partial x}\right) dz\, dx + \left(\frac{\partial Q}{\partial x} - \frac{\partial P}{\partial y}\right) dx\, dy\right]$$

which is the form of Stokes' Theorem given in Exercise 1.

CHAPTER 20

Section 20.1

1.

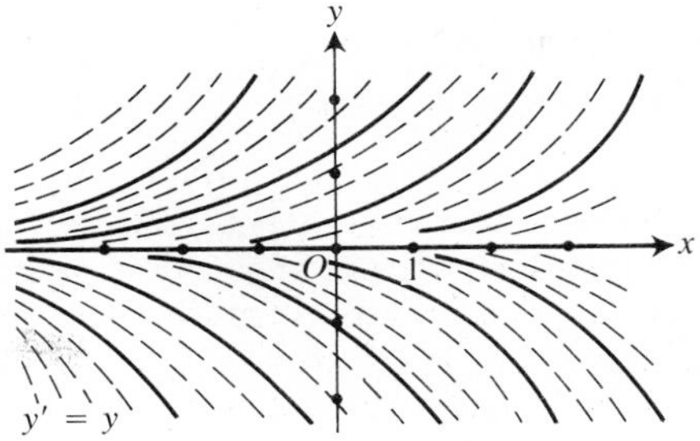

3.

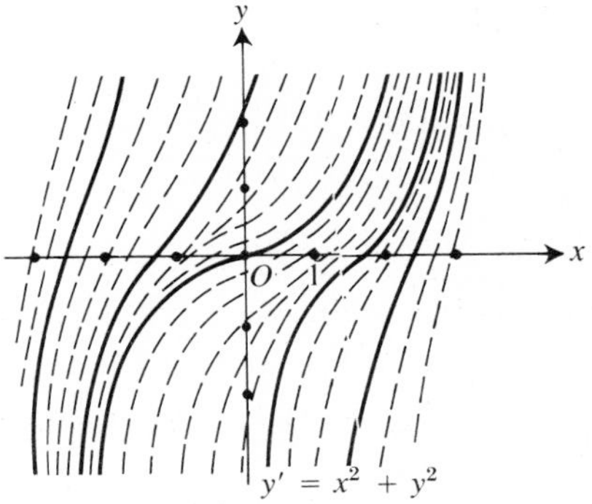

5.

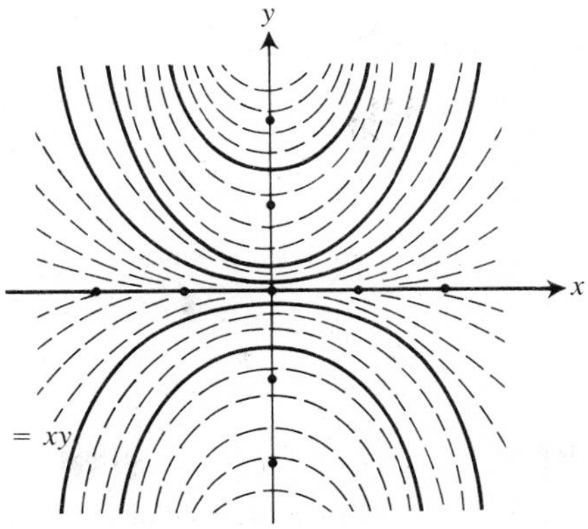

7. 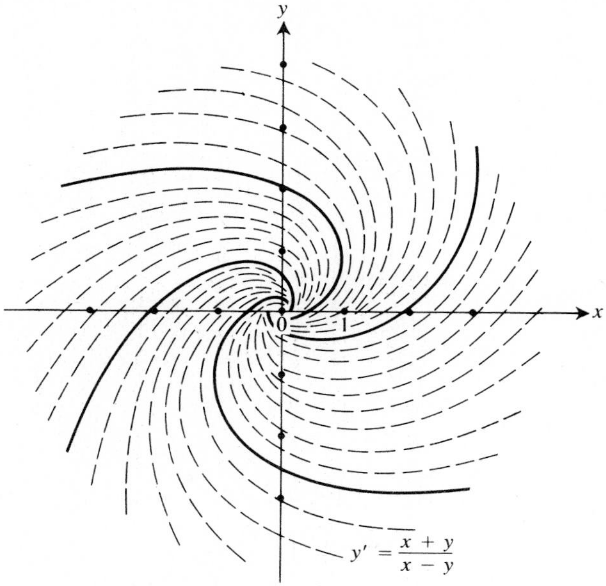

9. a) All $r > 0$ b) $\frac{\pi}{2}$

Section 20.2

1. $-\frac{1}{y} = \frac{1}{2}x^2 + C$

3. $\tan^{-1} y = \frac{1}{3}x^3 + C$

5. $\sin^{-1} y = \frac{1}{2}x^2 + C$

7. $Cx\left(1 - \frac{y^2}{x^2}\right) = 1$

9. $\frac{y}{x} = \ln|x| + C$

11. $y = x + \frac{x^2}{2} - \frac{5}{2}$

13. $y^2 = x^2 + 5$

15. $26x^2 - y^2 = 1$

17. $y + x = C$

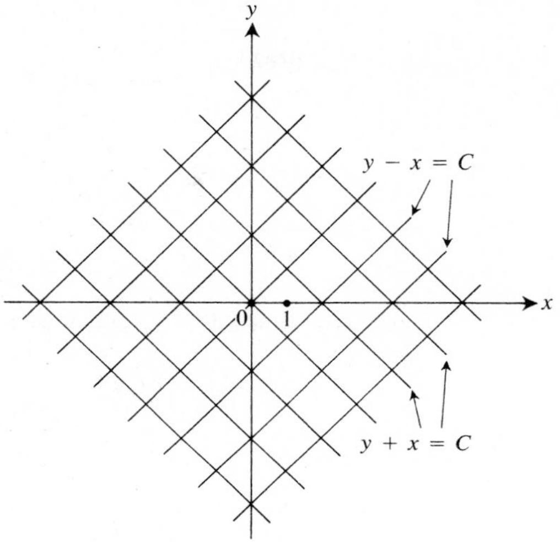

19. $y = Ce^{2x}$

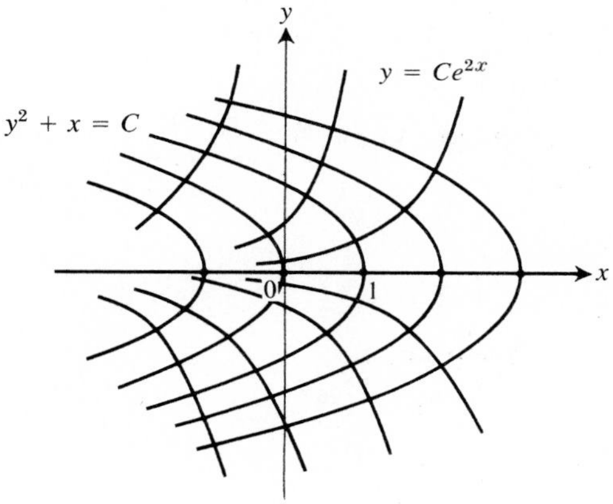

Section 20.3

1. $x \cos y + y = C$

3. $xy^2 + \ln |y| = C$

5. $xe^y - \sin xy = C$

7. $\ln |x| - \dfrac{1}{xy} + \dfrac{3}{y} = C$

9. $\dfrac{x^2}{2} + \dfrac{x}{y} - \ln |y| = C$

11. $\dfrac{4}{x} + \dfrac{y}{x} + 3y = C$

13. $\ln|x| + x^2y^3 + \dfrac{y^3}{3} = C$

15. Let $F' = f$. Now

$$\begin{aligned} d(F(G(x, y))) &= \frac{\partial(F(G(x, y)))}{\partial x}\,dx + \frac{\partial(F(G(x, y)))}{\partial y}\,dy \\ &= F'(G(x, y))\frac{\partial G}{\partial x}\,dx + F'(G(x, y))\frac{\partial G}{\partial y}\,dy \\ &= f(G(x, y))\mu M\,dx + f(G(x, y))\mu N\,dy. \end{aligned}$$

This shows that $f(G(x, y))\mu$ is an integrating factor.

17. From

$$N\frac{\partial \mu}{\partial x} - M\frac{\partial \mu}{\partial y} = \mu\left(\frac{\partial M}{\partial y} - \frac{\partial N}{\partial x}\right),$$

we obtain

$$N\left(\frac{1}{\mu}\frac{\partial \mu}{\partial x}\right) - M\left(\frac{1}{\mu}\frac{\partial \mu}{\partial y}\right) = \frac{\partial M}{\partial y} - \frac{\partial N}{\partial x} \quad \text{or} \quad N\frac{\partial(\ln|\mu|)}{\partial x} - M\frac{\partial(\ln|\mu|)}{\partial y} = \frac{\partial M}{\partial y} - \frac{\partial N}{\partial x}.$$

19. a) $xy^2 + x - y^3 = C$ b) $xy + \sin y = C$

Section 20.4

1. $y = Ce^{x^2/2}$

3. $y = Ce^{-x} + \dfrac{3}{2}e^x$

5. $y = Ce^{-2x} + \dfrac{x^2}{2}e^{-2x} + \dfrac{3}{2}$

7. $y = C\csc x - 3x\cot x - \cot x + 3$, $0 < x < \pi$

9. $y = -\dfrac{2}{9}e^{3x} - \dfrac{x}{3} - \dfrac{7}{9}$

11. $y = -e^{-\tan^{-1}x} + 3$

13. $y = e^{x^2/2}\left(C + \dfrac{x^3}{3} - \dfrac{x^5}{5\cdot 2!} + \dfrac{x^7}{7\cdot 2^2\cdot 2!} - \dfrac{x^9}{9\cdot 2^3\cdot 3!} + \dfrac{x^{11}}{11\cdot 2^4\cdot 4!} - \cdots\right)$

15. a) $i = e^{-Rt/L}\left(i_0 + \dfrac{1}{L}\displaystyle\int_0^t e^{Ru/L}E(u)\,du\right)$

b) From (a), we obtain

$$i = e^{-Rt/L}\left(i_0 + \frac{E}{R}e^{Rt/L} - \frac{E}{R}\right) = \left(i_0 - \frac{E}{R}\right)e^{-Rt/L} + \frac{E}{R}.$$

Thus $\lim_{t\to\infty} i = E/R$.

c) i approaches zero.

d) It decays exponentially from its value at $t = t_0$.

17. a) Differentiation of $v = y^{1-n}$ yields $v' = (1-n)y'/y^n$. Multiplying the given differential equation by $(1-n)/y^n$, we obtain $v' + (1-n)p(x)v = (1-n)g(x)$, which is linear.

b) $y^2\left(Ce^{-2x^2} - \frac{5}{2}\right) = 1$

Section 20.5

1. (This is a slightly tedious but routine exercise in differentiation.)

3. $y = Ce^{-2x}$

5. $y = C_1e^{-5x/2} + C_2e^{-x/2}$

7. $y = C_1 + e^{-x/2}(C_2 + C_3x)$

9. $y = C_1 \cos \sqrt{3}x + C_2 \sin \sqrt{3}x$

11. $y = C_1e^x + e^{-x/2}\left(C_2 \cos \frac{\sqrt{3}}{2}x + C_3 \sin \frac{\sqrt{3}}{2}x\right)$

13. $y = C_1 + C_2x + C_3e^{-2x} + C_4 \cos \sqrt{2}x + C_5 \sin \sqrt{2}x$

15. $y = e^{-x}(C_1 + C_2x) + e^{-x/2}\left(C_3 \cos \frac{\sqrt{7}}{2}x + C_4 \sin \frac{\sqrt{7}}{2}x\right)$

17. $y = -3e^{3x} + 4e^{2x}$

19. $y = e^{2x} + e^{-x}\left(\cos \sqrt{3}x - \frac{1}{\sqrt{3}} \sin \sqrt{3}x\right)$

Section 20.6

1. $y = C_1e^{-x} + C_2e^{3x} - \frac{4}{3}x + \frac{8}{9}$

3. $y = e^{-x}(C_1 + C_2x) + \frac{1}{2}x^2e^{-x}$

5. $y = C_1e^x + C_2e^{2x} + \frac{x}{2} - \frac{3}{4}$

7. $y = C_1 \cos 2x + C_2 \sin 2x + \frac{1}{3} \sin x$

9. $y = C_1 + C_2x + C_3e^{-3x} + \frac{1}{18}x^3 - \frac{1}{18}x^2 + \frac{1}{54}e^{3x}$

11. $y = e^{-2x}(C_1 + C_2x) + \frac{15}{289}e^{2x} \cos x + \frac{8}{289}e^{2x} \sin x$

13. $y = C_1 + C_2x + C_3x^2 + C_4e^{2x} - \frac{1}{240}x^6 - \frac{3}{80}x^5 - \frac{3}{32}x^4 - \frac{3}{16}x^3$

15. Height: $s = 3512 - 512e^{-t/4} - 128t$; terminal velocity: -128 ft/sec

17. The solutions of the characteristic equation $mr^2 + cr + k = 0$ are

$$r_1 = \frac{-c + \sqrt{c^2 - 4km}}{2m}$$

and

$$r_2 = \frac{-c - \sqrt{c^2 - 4km}}{2m}$$

Since $c^2 > 4km$, we have $r_2 < r_1 < 0$. The general solution of $m\ddot{s} + c\dot{s} + ks = 0$ is then $s = C_1e^{r_1t} + C_2e^{r_2t}$. From the initial conditions $s(0) = h$ and $\dot{s}(0) = 0$, one obtains as solution for the given problem

$$s = \frac{-r_2h}{r_1 - r_2}e^{r_1t} + \frac{r_1h}{r_1 - r_2}e^{r_2t}.$$

Since $r_1 < 0$ and $r_2 < 0$, we have $\lim_{t\to\infty}s(t) = 0$, so the body approaches its position of rest as $t \to \infty$. For s to be zero, our solution shows we would need to have $e^{(r_1-r_2)t} = r_1/r_2$. Since $r_1 - r_2 > 0$ and $0 < r_1/r_2 < 1$, we see that there is no $t > 0$ satisfying this condition, so the body never crosses its position of rest.

19. The solutions of the characteristic equation $mr^2 + cr + k = 0$ are complex, and if $a = \sqrt{4km - c^2}/2m$, the solutions of the equation have the form

$$s = e^{(-c/2m)r}[C_1 \cos at + C_2 \sin at]$$

for constants C_1 and C_2. The desired conclusion now follows at once.

21. If $k = 0$, our equation may be written as

$$\ddot{s} + \frac{c}{a}s = \frac{1}{a}\sin kt = 0.$$

We see that any solution of this equation is of the form

$$s = C_1 \cos\left(\sqrt{\frac{c}{a}}\,t\right) + C_2 \sin\left(\sqrt{\frac{c}{a}}\,t\right)$$

for suitable constants C_1 and C_2. The period of this oscillatory motion is

$$2\pi\sqrt{a}/\sqrt{c}$$

If $k = \sqrt{c/a}$, then any solution of our equation is of the form

$$s = (C_1 + At)\cos\left(\sqrt{\frac{c}{a}}\,t\right) + C_2 \sin\left(\sqrt{\frac{c}{a}}\,t\right)$$

for suitable constants C_1, C_2, and $A \neq 0$. The amplitude $\sqrt{(C_1 + At)^2 + C_2^2}$ then increases without bound as $t \to \infty$.

Section 20.7

1. 0

3. $$y = a_0\left(1 - \frac{x^2}{2!} + \frac{3x^4}{4!} - \frac{3 \cdot 5x^6}{6!} + \cdots + (-1)^n\frac{(3)(5)\cdots(2n-1)}{(2n)!}x^{2n} + \cdots\right) = a_0e^{-x^2/2}$$

5. $$y = a_0\left(1 - \frac{2}{2!}(x+1)^2 - \frac{6}{3!}(x+1)^3 - \cdots - \frac{2^n - 2}{n!}(x+1)^n + \cdots\right)$$
$$+ a_1\left((x+1) + \frac{3}{2!}(x+1)^2 + \frac{7}{3!}(x+1)^3 + \cdots + \frac{2^n - 1}{n!}(x+1)^n + \cdots\right)$$
$$= a_0(2e^{x+1} - e^{2x+2}) + a_1(e^{2x+2} - e^{x+1})$$

7. $$y = a_0\left(1 + \frac{x^3}{3!} - \frac{2x^6}{6!} + \cdots + (-1)^{n+1}\frac{(2)(5)\cdots(3n-4)}{(3n)!}x^{3n} + \cdots\right)$$
$$+ a_1x + a_2\left(x^2 - \frac{x^5}{5!} + \frac{4x^8}{8!} - \cdots + (-1)^{n+1}\frac{(4)(7)\cdots(3n-5)}{(3n-1)!}x^{3n-1} + \cdots\right)$$
for $n \geq 2$

9. $y = a_0(x+1)$

Section 20.8

1. $$y = a_0\left(1 - x^2 + \frac{x^4}{3} - \frac{x^6}{3 \cdot 5} + \cdots + (-1)^n\frac{x^{2n}}{(3)(5)\cdots(2n-1)} + \cdots\right)$$
$$+ a_1\left(x - \frac{x^3}{2} + \frac{x^5}{2 \cdot 4} - \frac{x^7}{2 \cdot 4 \cdot 6} + \cdots + (-)^n\frac{x^{2n+1}}{(2)(4)\cdots(2n)} + \cdots\right)$$

3. $y = a_0(1 + \frac{1}{6}x^3 - \frac{1}{20}x^5 + \frac{1}{180}x^6 + \cdots) + a_1(x - \frac{1}{3}x^3 + \frac{1}{12}x^4 + \frac{1}{10}x^5 - \frac{1}{30}x^6 + \cdots)$

5. $$y = a_0 + a_1x + \left(\frac{x^4}{12} - \frac{x^8}{56 \cdot 3!} + \frac{x^{12}}{132 \cdot 5!} - \cdots + (-1)^n\frac{x^{4n+4}}{(4n+4)(4n+3)(2n+1)!} + \cdots\right)$$

7. $$y = a_0 + a_1x + \left(\frac{x^5}{5 \cdot 4} - \frac{x^7}{7 \cdot 6 \cdot 3} + \frac{x^9}{9 \cdot 8 \cdot 5} - \cdots + (-1)^n\frac{x^{2n+5}}{(2n+5)(2n+4)(2n+1)} + \cdots\right)$$

9. $y = a_0 + a_1(x-1) + a_2[(x-1)^2 + \frac{1}{3}(x-1)^3]$

11. $$y = a_0\left(1 - \frac{x^2}{2!} + \frac{x^4}{4!} - \frac{x^6}{6!} + \cdots + (-1)^n\frac{x^{2n}}{(2n)!} + \cdots\right)$$
$$+ a_1\left(x - \frac{x^3}{3!} + \frac{x^5}{5!} - \frac{x^7}{7!} + \cdots + (-1)^n\frac{x^{2n+1}}{(2n+1)!} + \cdots\right)$$
$$+ \left(\frac{x^3}{3!} - \frac{2x^5}{5!} + \frac{3x^7}{7!} - \frac{4x^9}{9!} + \cdots\right)$$
$$= a_0\cos x + a_1\sin x + (\tfrac{1}{2}\sin x - \tfrac{1}{2}x\cos x)$$

Review Exercise Set 20.1

1.

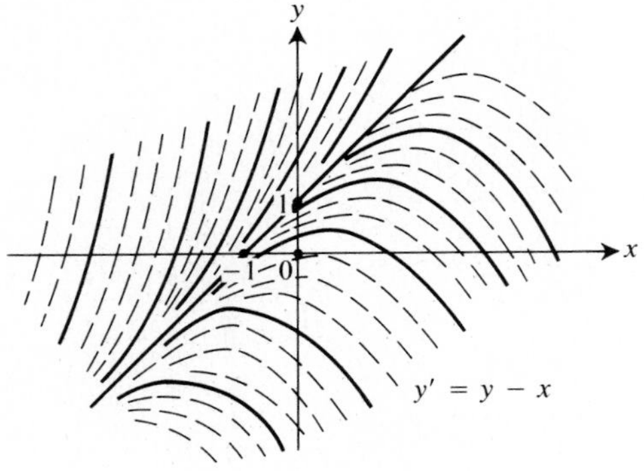

3. $\ln(2\sin^2 y) = x^2$

5. $x^3 + 3xy^2 = C$

7. $y = 1 + Ce^{\cos x}$

9. $y = e^x(C_1 \cos 2x + C_2 \sin 2x) + \frac{1}{8}e^{3x}$

Review Exercise Set 20.2

1. (See Theorem 20.1, Section 20.1.)

3. $\frac{1}{2}\ln\left(\left(\frac{y}{x}\right)^2 + 1\right) + \tan^{-1}\left(\frac{y}{x}\right) = -\ln|x| + C$

5. $xy + \sin y - \cos x = C$

7. $y = C_1 + C_2x + C_3e^{2x}$

9. $y = C_1e^x + C_2e^{2x} + \frac{1}{10}\sin x + \frac{3}{10}\cos x$

index

68. (a) $\int \sin ax \cos bx\, dx = -\frac{\cos(a+b)x}{2(a+b)} - \frac{\cos(a-b)x}{2(a-b)} + C, \qquad a^2 \neq b^2$

(b) $\int \sin ax \sin bx\, dx = \frac{\sin(a-b)x}{2(a-b)} - \frac{\sin(a+b)x}{2(a+b)}, \qquad a^2 \neq b^2$

(c) $\int \cos ax \cos bx\, dx = \frac{\sin(a-b)x}{2(a-b)} + \frac{\sin(a+b)x}{2(a+b)}, \qquad a^2 \neq b^2$

69. $\int \sin ax \cos ax\, dx = -\frac{\cos 2ax}{4a} + C$

70. $\int \sin^n ax \cos ax\, dx = \frac{\sin^{n+1} ax}{(n+1)a} + C, \qquad n \neq -1$

71. $\int \frac{\cos ax}{\sin ax}\, dx = \frac{1}{a} \ln|\sin ax| + C$

72. $\int \cos^n ax \sin ax\, dx = -\frac{\cos^{n+1} ax}{(n+1)a} + C, \qquad n \neq -1$

73. $\int \frac{\sin ax}{\cos ax}\, dx = -\frac{1}{a} \ln|\cos ax| + C$

74. $\int \sin^n ax \cos^m ax\, dx = -\frac{\sin^{n-1} ax \cos^{m+1} ax}{a(m+n)} + \frac{n-1}{m+n} \int \sin^{n-2} ax \cos^m ax\, dx, \qquad n \neq -m$ (If $n = -m$, use No. 92.)

75. $\int \sin^n ax \cos^m ax\, dx = \frac{\sin^{n+1} ax \cos^{m-1} ax}{a(m+n)} + \frac{m-1}{m+n} \int \sin^n ax \cos^{m-2} ax\, dx, \qquad m \neq -n$ (If $m = -n$, use No. 93.)

76. $\int \frac{dx}{b + c \sin ax} = \frac{-2}{a\sqrt{b^2 - c^2}} \tan^{-1}\left[\sqrt{\frac{b-c}{b+c}} \tan\left(\frac{\pi}{4} - \frac{ax}{2}\right)\right] + C, \qquad b^2 > c^2$

77. $\int \frac{dx}{b + c \sin ax} = \frac{-1}{a\sqrt{c^2 - b^2}} \ln\left|\frac{c + b \sin ax + \sqrt{c^2 - b^2} \cos ax}{b + c \sin ax}\right| + C, \qquad b^2 < c^2$

78. $\int \frac{dx}{1 + \sin ax} = -\frac{1}{a} \tan\left(\frac{\pi}{4} - \frac{ax}{2}\right) + C$

79. $\int \frac{dx}{1 - \sin ax} = \frac{1}{a} \tan\left(\frac{\pi}{4} + \frac{ax}{2}\right) + C$

80. $\int \frac{dx}{b + c \cos ax} = \frac{2}{a\sqrt{b^2 - c^2}} \tan^{-1}\left[\sqrt{\frac{b-c}{b+c}} \tan \frac{ax}{2}\right] + C, \qquad b^2 > c^2$

81. $\int \frac{dx}{b + c \cos ax} = \frac{1}{a\sqrt{c^2 - b^2}} \ln\left|\frac{c + b \cos ax + \sqrt{c^2 - b^2} \sin ax}{b + c \cos ax}\right| + C, \qquad b^2 < c^2$

82. $\int \frac{dx}{1 + \cos ax} = \frac{1}{a} \tan \frac{ax}{2} + C$

83. $\int \frac{dx}{1 - \cos ax} = -\frac{1}{a} \cot \frac{ax}{2} + C$

84. $\int x \sin ax\, dx = \frac{1}{a^2} \sin ax - \frac{x}{a} \cos ax + C$

85. $\int x \cos ax\, dx = \frac{1}{a^2} \cos ax + \frac{x}{a} \sin ax + C$

86. $\int x^n \sin ax\, dx = -\frac{x^n}{a} \cos ax + \frac{n}{a} \int x^{n-1} \cos ax\, dx$

87. $\int x^n \cos ax\, dx = \frac{x^n}{a} \sin ax - \frac{n}{a} \int x^{n-1} \sin ax\, dx$

88. $\int \tan ax\, dx = -\frac{1}{a} \ln|\cos ax| + C$

89. $\int \cot ax\, dx = \frac{1}{a} \ln|\sin ax| + C$

90. $\int \tan^2 ax\,dx = \frac{1}{a}\tan ax - x + C$

91. $\int \cot^2 ax\,dx = -\frac{1}{a}\cot ax - x + C$

92. $\int \tan^n ax\,dx = \frac{\tan^{n-1} ax}{a(n-1)} - \int \tan^{n-2} ax\,dx, \qquad n \neq 1$

93. $\int \cot^n ax\,dx = -\frac{\cot^{n-1} ax}{a(n-1)} - \int \cot^{n-2} ax\,dx, \qquad n \neq 1$

94. $\int \sec ax\,dx = \frac{1}{a}\ln|\sec ax + \tan ax| + C$

95. $\int \csc ax\,dx = -\frac{1}{a}\ln|\csc ax + \cot ax| + C$

96. $\int \sec^2 ax\,dx = \frac{1}{a}\tan ax + C$

97. $\int \csc^2 ax\,dx = -\frac{1}{a}\cot ax + C$

98. $\int \sec^n ax\,dx = \frac{\sec^{n-2} ax \tan ax}{a(n-1)} + \frac{n-2}{n-1}\int \sec^{n-2} ax\,dx, \qquad n \neq 1$

99. $\int \csc^n ax\,dx = -\frac{\csc^{n-2} ax \cot ax}{a(n-1)} + \frac{n-2}{n-1}\int \csc^{n-2} ax\,dx, \qquad n \neq 1$

100. $\int \sec^n ax \tan ax\,dx = \frac{\sec^n ax}{na} + C, \qquad n \neq 0$

101. $\int \csc^n ax \cot ax\,dx = -\frac{\csc^n ax}{na} + C, \qquad n \neq 0$

102. $\int \sin^{-1} ax\,dx = x\sin^{-1} ax + \frac{1}{a}\sqrt{1 - a^2x^2} + C$

103. $\int \cos^{-1} ax\,dx = x\cos^{-1} ax - \frac{1}{a}\sqrt{1 - a^2x^2} + C$

104. $\int \tan^{-1} ax\,dx = x\tan^{-1} ax - \frac{1}{2a}\ln(1 + a^2x^2) + C$

105. $\int x^n \sin^{-1} ax\,dx = \frac{x^{n+1}}{n+1}\sin^{-1} ax - \frac{a}{n+1}\int \frac{x^{n+1}\,dx}{\sqrt{1 - a^2x^2}}, \qquad n \neq -1$

106. $\int x^n \cos^{-1} ax\,dx = \frac{x^{n+1}}{n+1}\cos^{-1} ax + \frac{a}{n+1}\int \frac{x^{n+1}\,dx}{\sqrt{1 - a^2x^2}}, \qquad n \neq -1$

107. $\int x^n \tan^{-1} ax\,dx = \frac{x^{n+1}}{n+1}\tan^{-1} ax - \frac{a}{n+1}\int \frac{x^{n+1}\,dx}{1 + a^2x^2}, \qquad n \neq -1$

Integrals involving exponential and logarithmic functions

108. $\int e^{ax}\,dx = \frac{1}{a}e^{ax} + C$

109. $\int b^{ax}\,dx = \frac{1}{a}\frac{b^{ax}}{\ln b} + C, \qquad b > 0,\quad b \neq 1$

110. $\int xe^{ax}\,dx = \frac{e^{ax}}{a^2}(ax - 1) + C$

111. $\int x^n e^{ax}\,dx = \frac{1}{a}x^n e^{ax} - \frac{n}{a}\int x^{n-1}e^{ax}\,dx$

112. $\int x^n b^{ax}\,dx = \frac{x^n b^{ax}}{a\ln b} - \frac{n}{a\ln b}\int x^{n-1}b^{ax}\,dx, \qquad b > 0,\quad b \neq 1$

113. $\int e^{ax}\sin bx\,dx = \frac{e^{ax}}{a^2 + b^2}(a\sin bx - b\cos bx) + C$

114. $\int e^{ax}\cos bx\,dx = \frac{e^{ax}}{a^2 + b^2}(a\cos bx + b\sin bx) + C$